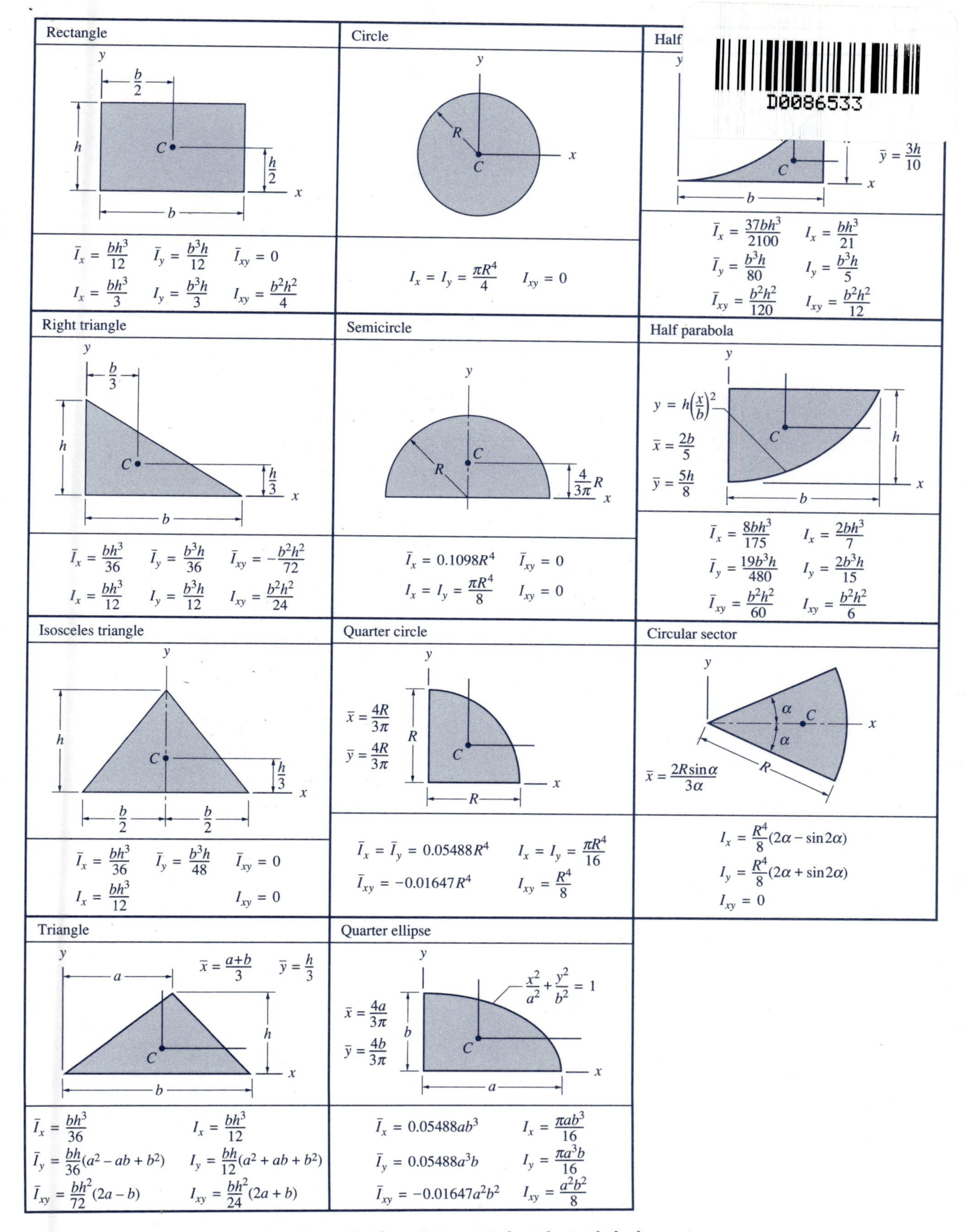

Area Moments of Inertia (continued on the inside back cover)

ENGINEERING MECHANICS

Dynamics

ENGINEERING MECHANICS

Dynamics

Andrew Pytel

The Pennsylvania State University

Jaan Kiusalaas

The Pennsylvania State University

HarperCollins*CollegePublishers*

Sponsoring Editor: John Lenchek
Project Editor: Carol Zombo
Art Administrator: Jess Schaal
Text Design and Project Management: Publication Services
Cover Design: Kay Fulton
Production Administrator: Randee Wire
Printer and Binder: R.R. Donnelley & Sons Company

Engineering Mechanics: Dynamics

Library of Congress Cataloging-in-Publication Data

Pytel, Andrew.
Engineering mechanics: dynamics / Andrew Pytel, Jaan Kiusalaas.
p. cm.
Includes index.
ISBN 0-06-045276-5
1. Dynamics. 2. Mechanics, Applied. I. Kiusalaas, Jaan. II. Title.
TA352.P97 1993
620.1'04—dc20 93-21044
CIP

95 96 9 8 7 6 5 4

To: Jean, Leslie, Lori, John, Nicholas

and

To: Judy, Nicholas, Jennifer, Timothy

Contents

*Indicates optional articles.

Preface

This textbook and its companion volume, *Statics,* were written to help students master the principles and applications of engineering mechanics. We wrote these books for two reasons. Firstly, we believed that the pedagogy that we had developed over many years of teaching might be beneficial to others. Secondly, engineering mechanics textbooks must provide a framework to help faculty meet the challenge of incorporating numerical methods *throughout* engineering curricula.

In preparing the texts, we have, of course, drawn upon the experiences of the many scientists and engineers who have contributed to the theory of mechanics and its many applications. Furthermore, we have benefited greatly from the labors of the outstanding authors who have preceded us.

This book contains a number of optional articles, which are indicated by an asterisk (*). Any of these topics can be omitted without interfering with the presentation of dynamics.

Discussed below are what we believe to be the salient features of *Dynamics*.

Expanded Discussion of Sample Problems

To help students master the intricacies of engineering problem analysis, our discussion of each sample problem is comprehensive. When appropriate, we compare the number of unknowns with the number of available equations, thereby reinforcing the importance of a unified method of analysis, a concept that was used throughout *Statics*.

Carefully Selected Homework Problems

One of the primary strengths of any dynamics book must be the homework problems. Therefore, we have selected our problems very carefully to ensure a balanced presentation. We have included many problems for which the solutions are relatively straightforward in order to help a student master problem-solving techniques. In addition, many problems have direct engineering relevance, which we hope that students will find both interesting and challenging.

An asterisk (*) preceding a problem number indicates that the solution requires advanced knowledge that may not yet have been attained by the student.

Integrated Presentation of Particle Kinematics and Kinetics

The second chapter of this text, Chapter 12, integrates the coverage of particle kinematics and kinetics. Article 12.2 introduces particle kinematics in rectangular coordinates. The remainder of the chapter is devoted primarily to deriving (with the aid of free-body diagrams) and solving the equations of motion. We find this arrangement of topics superior to the traditional introductory chapters devoted exclusively to particle kinematics. By applying Newton's laws of motion at the beginning, the student is immediately introduced to practical problems, where the equations of motion are derived, and not "given." Furthermore, the student appreciates early the role played by free-body diagrams in the study of dynamics.

Numerical Methods (Optional Coverage)

One of the main tasks of dynamic analysis is the derivation and solution of equations of motion. Unfortunately, these equations are usually nonlinear, requiring (except for a few special cases) numerical solutions. In the past, students were not familiar with the required numerical techniques; hence dynamics textbooks focused primarily on the computation of instantaneous values, e.g., the forces and accelerations *at a particular instant.* With the ever-increasing availability of personal computers and mathematical software, the situation has been rapidly changing. Many of today's students of dynamics have already been introduced to numerical integration in other courses. In addition, educators are being called upon to use computer methods throughout engineering curricula, wherever they can enhance the education of their students.

In this text, the following three articles are devoted to applying numerical methods to dynamics:

12.8 Numerical Integration of an Equation of Motion

12.9 Numerical Solution of Coupled Equations of Motion

13.5 Numerical Solution: Generalized Coordinates

Each of these articles is written assuming that the reader has little or no knowledge of numerical integration. In addition, each article is followed by sample problems and homework problems that require a student to derive the equations of motion, and solve them numerically. Homework problems requiring numerical solutions appear, where appropriate, throughout the text. A number of the homework problems, drawn from engineering practice, lend themselves to assignment as group projects.

The emphasis is on computer *applications*, not computer *programming*. Only one homework problem requires that a computer program be written, and the student can then use this program throughout the remainder of the course, if necessary. However, students should be encouraged to use any available software that can perform numerical integration, preferably a program that has plotting capability. (One such program is contained on the floppy disk that is attached inside the back cover of this book.)

The instructor should consider the discussion and application of numerical integration to be optional, choosing to cover as much, or as little, of this material as deemed appropriate. All references to numerical methods are marked with a disk icon (💾). In most instances, the problems requiring numerical integration have two parts: (a) Show that the equation of motion is . . . ; and (b) integrate the equation of motion. An instructor who elects to not introduce numerical analysis may find it beneficial to assign part (a) as an exercise in problem formulation, and omit part (b). This technique would alert students to the fact that dynamics problems, even those appearing to be comparatively simple, frequently lead to equations of motion that are difficult to solve.

Acknowledgments

During the more than several years that we have been writing these textbooks, we have had the support and encouragement of many fine people. We would especially like to recognize our indebtedness to Mr. Mu-Jen Yang and Dr. Nagesh Sonti for checking the solutions of the problems, our colleagues in the Department of Engineering Science and Mechanics, and the staff at HarperCollins. We are also grateful to the following reviewers for their many valuable suggestions:

Beckry Abdel-Magid, Winona State University
Anil K. Bajaj, Purdue University
W. E. Baker, University of New Mexico
Thomas Burton, Washington State University
Mohammad Dehghani, Ohio University
George Flowers, Auburn University
Robert A. Freeman, University of Texas—Austin
S. C. Gambrell, Jr., University of Alabama
Norman W. Garrick, University of Connecticut
Vernal H. Kenner, Ohio State University
E. Harry Law, Clemson University
Mohammad Mahinfalah, North Dakota State University
Robert Merrill, Rochester Institute of Technology
Satish Nair, University of Missouri—Columbia
Jamal F. Nayfeh, University of Central Florida
Saeed B. Niku, California Poly Tech
William W. Predebon, Michigan Tech

(*continued*)

Edward W. Price, North Dakota State University
P. K. Raju, Auburn University
John M. Vance, Texas A&M
Arthur N. Willoughby, Morgan State University
D. W. Yannitell, Louisiana State University
Shen Yi Luo, University of Nevada—Reno
Tong Zhou, California State University—Sacramento

ENGINEERING MECHANICS

Dynamics

11 Introduction to Dynamics

11.1 Introduction

In this chapter we begin our study of classical dynamics, that is, the motion of bodies governed by Newton-Euler laws.* The general subdivisions of classical dynamics are shown in Fig. 11.1.

Dynamics is concerned with the motion of particles and rigid bodies. When the motion of a body is such that rotation is of no consequence, the body is said to be analyzed as a *particle*. Of course, whether or not it is meaningful to employ the particle idealization depends on the application. For example, the rotation of the earth about its axis may be neglected, with insignificant loss of accuracy, when studying the earth's orbit around the sun. However, this axial rotation is obviously an important factor when investigating the effect of the moon on the earth's oceans. In the first case, the earth can be treated as a particle, but in the second case it cannot.

All bodies deform (i.e., change shape) when they are subjected to forces. If the deformations are so small that they have no appreciable effect on the motion of the body as a whole, the body is said to be *rigid*. A *rigid body* is, therefore, a body for which the deformations can be neglected with no loss of accuracy in the analysis of motion.

As indicated in Fig. 11.1, the two main branches of dynamics are kinematics and kinetics. *Kinematics* is the study of the geometry of motion, without concern for the forces that produce the motion. *Kinetics* uses the laws of Newton and Euler to relate the forces and the motion.

Kinematics is an important subject in its own right. For example, the kinematics of gears and linkages is a major topic in mechanical engineering. To understand kinetics requires a prior knowledge of kinematics, since it would be

* Sir Isaac Newton is credited with laying the foundation of classical mechanics with the publication of *Principia* in 1687. However, the laws of motion as we use them today were developed by Leonhard Euler and his contemporaries more than sixty years later. In particular, the laws governing the motion of finite bodies are attributable to Euler.

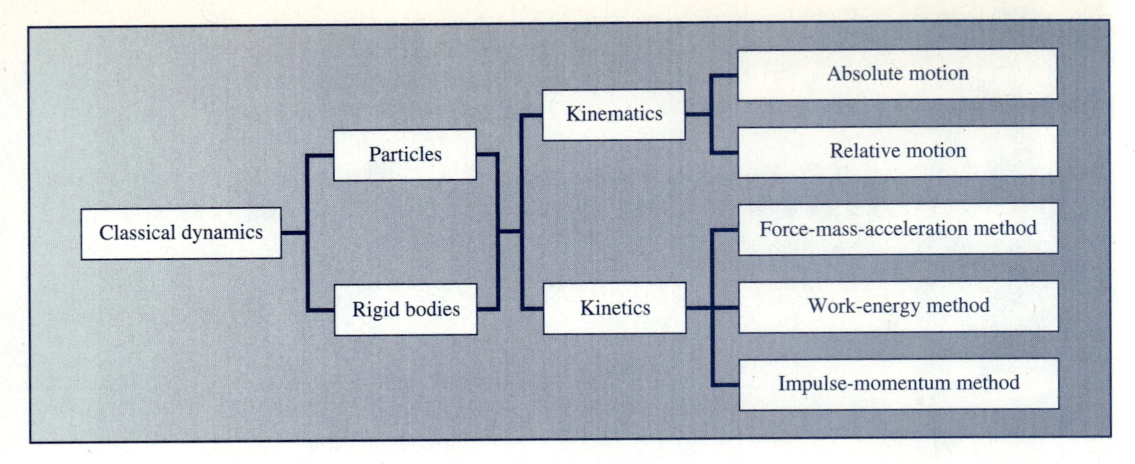

Fig. 11.1

impossible to relate the forces to the motion if the motion could not be described in the first place. Therefore, a study of dynamics always begins with the fundamentals of kinematics.

Referring again to Fig. 11.1, we see that kinematics is subdivided into two categories: absolute motion and relative motion. *Absolute motion* describes the motion as observed from a fixed reference frame. *Relative motion*, as the name suggests, is motion that is observed from a reference frame that is itself moving.

Figure 11.1 also lists the three methods of kinetic analysis: (1) the force-mass-acceleration (FMA) method, (2) the work-energy method, and (3) the impulse-momentum method. As we will see, these methods are not independent of each other, because each is derived directly from Newton-Euler laws of motion.

The *FMA method* uses Newton-Euler laws to relate the forces that act on a body to its accelerations. The accelerations can then be integrated to determine the velocities and positions of the body as functions of time.

The *work-energy method* relates the work done by the applied forces to the change in the kinetic energy of the body. When the change in speed of the body is of primary interest, this method can be particularly efficient, because only those forces that do work need be considered in the analysis.

The *impulse-momentum method* relates the impulses of the applied forces to the change in momentum of the body. This method is useful when the time dependence of the forces is unknown, as is often the case in problems involving colliding bodies.

In the next article, we present a review of vector differentiation, followed by definitions of the kinematic quantities for a particle: displacement, velocity, and acceleration. This chapter concludes with a discussion of the fundamentals of Newtonian mechanics, which form the basis for kinetic analysis.

Chapter 12 is devoted to the kinematics and kinetics (FMA method) of particle motion utilizing rectangular coordinates. The kinematics and kinetics (FMA method) of particle motion employing curvilinear coordinates are discussed in Chapter 13. Chapter 14 applies the work-energy and impulse-momentum methods for particle motion. The remainder of this text extends the principles of kinematics and kinetics to systems of particles and rigid bodies.

11.2 Derivatives of Vector Functions

A knowledge of vector calculus is a prerequisite for the study of dynamics. Here we discuss the derivatives of vectors; integration is introduced later as needed.

The vector **A** is said to be a vector function of a scalar parameter u if the magnitude and direction of **A** depend on u (in dynamics, time is frequently chosen to be the scalar parameter). This functional relationship is denoted by $\mathbf{A}(u)$. If the scalar variable changes from the value u to $(u + \Delta u)$, the vector **A** will change from $\mathbf{A}(u)$ to $\mathbf{A}(u + \Delta u)$. Therefore, the change in the vector **A** can be written as

$$\Delta\mathbf{A} = \mathbf{A}(u + \Delta u) - \mathbf{A}(u) \tag{11.1}$$

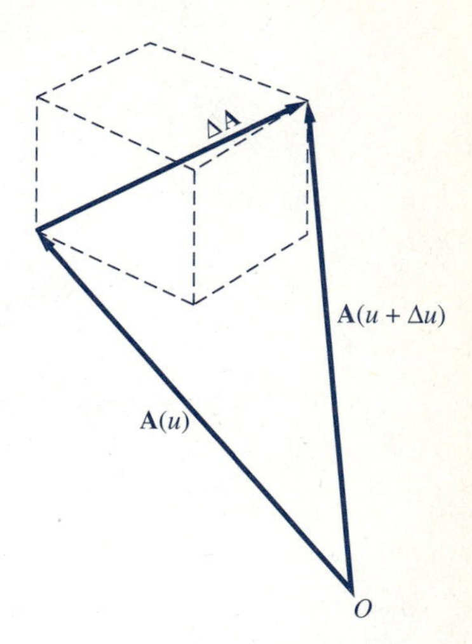

Fig. 11.2

As seen in Fig. 11.2, $\Delta\mathbf{A}$ is due to a change in both the magnitude and the direction of the vector **A**.

The derivative of **A** with respect to the scalar u is defined as

$$\frac{d\mathbf{A}}{du} = \lim_{\Delta u \to 0} \frac{\Delta\mathbf{A}}{\Delta u} = \lim_{\Delta u \to 0} \frac{\mathbf{A}(u + \Delta u) - \mathbf{A}(u)}{\Delta u} \tag{11.2}$$

assuming that the limit exists.

The above definition resembles the derivative of the scalar function $y(u)$, which is defined as

$$\frac{dy}{du} = \lim_{\Delta u \to 0} \frac{\Delta y}{\Delta u} = \lim_{\Delta u \to 0} \frac{y(u + \Delta u) - y(u)}{\Delta u} \tag{11.3}$$

When dealing with a vector function, the magnitude of the derivative $|d\mathbf{A}/du|$ must not be confused with the derivative of the magnitude $d|\mathbf{A}|/du$. In general, these two derivatives will not be equal. For example, if the magnitude of a vector **A** is constant, then $d|\mathbf{A}|/du = 0$. However, $|d\mathbf{A}/du|$ will not equal zero unless the direction of **A** is also constant.

The following useful identities can be derived from the definitions of derivatives (**A** and **B** are assumed to be vector functions of the scalar u, and m is also a scalar):

$$\frac{d(m\mathbf{A})}{du} = m\frac{d\mathbf{A}}{du} + \frac{dm}{du}\mathbf{A} \tag{11.4}$$

$$\frac{d(\mathbf{A} + \mathbf{B})}{du} = \frac{d\mathbf{A}}{du} + \frac{d\mathbf{B}}{du} \tag{11.5}$$

$$\frac{d(\mathbf{A} \cdot \mathbf{B})}{du} = \mathbf{A} \cdot \frac{d\mathbf{B}}{du} + \frac{d\mathbf{A}}{du} \cdot \mathbf{B} \tag{11.6}$$

$$\frac{d(\mathbf{A} \times \mathbf{B})}{du} = \mathbf{A} \times \frac{d\mathbf{B}}{du} + \frac{d\mathbf{A}}{du} \times \mathbf{B} \tag{11.7}$$

11.3 Position, Velocity, and Acceleration of a Point

In this article we study the *kinematics*—that is, the geometry of motion—of a point (in kinematics, the terms *point* and *particle* are used interchangeably). *Kinetics*, the study of the relationship between the motion and the forces that produce it, is introduced in the next article.

Figure 11.3 shows the path of a point that moves along a smooth curve that is fixed in space.* A convenient method for locating the moving point is to introduce its *position vector* $\mathbf{r}(t)$, a time-dependent vector drawn from the fixed reference point O to the moving point. Let the location of the moving point be A at time t, and B at time $(t + \Delta t)$, where Δt is a finite time interval. The corresponding position vectors are $\mathbf{r}(t)$ and $\mathbf{r}(t + \Delta t)$. The *displacement* $\Delta\mathbf{r}$ of the point during the time interval Δt is defined to be the vector drawn from A to B. Referring to Fig. 11.3, we see that the displacement is equal to the change in the position vector, i.e.,

$$\Delta\mathbf{r} = \mathbf{r}(t + \Delta t) - \mathbf{r}(t) \tag{11.8}$$

Note that the displacement of a point is a vector, and it is associated with a finite time interval.

A second method of locating a moving point is to specify its *path coordinate*, defined as the distance measured along the path from a fixed reference point to the moving point. This method is also shown in Fig. 11.3, where s is the path coordinate, measured from the fixed point E. The path coordinate of the moving point is a scalar function of time, denoted by $s(t)$. As the point moves from A to B during the time interval Δt, its path coordinate changes from $s(t)$ to $s(t + \Delta t)$. The length of the path between A and B is equal to the change in s, given by

$$\Delta s = s(t + \Delta t) - s(t) \tag{11.9}$$

Inspection of Fig. 11.3 reveals that Δs is greater than $|\Delta\mathbf{r}|$, the magnitude of the displacement vector. The only exception occurs when the path between A and B is a straight line, in which case $\Delta s = |\Delta\mathbf{r}|$.

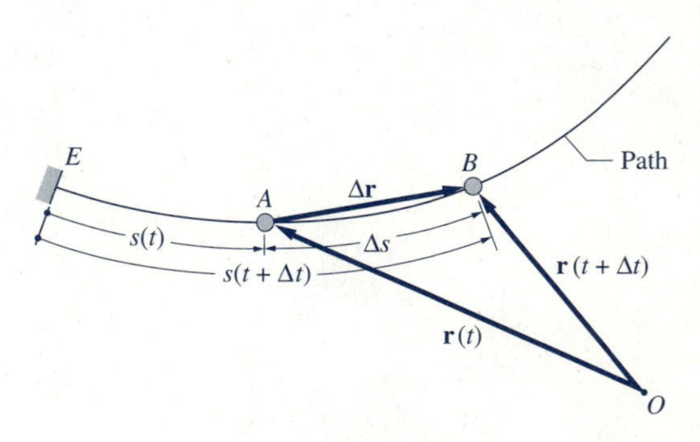

Fig. 11.3

* By considering the path to be fixed in space, we are restricting our attention to the absolute motion of a point. Relative motion is discussed in Chapter 15.

Be careful not to confuse the path length Δs with the distance traveled by the point during the time interval Δt. The distance traveled is equal to Δs only if the point moves from A to B without reversing its direction of motion. If the direction of motion changes during Δt, the distance traveled will obviously be greater than Δs.

The velocity $\mathbf{v}$ of a moving point at time t is defined to be the time derivative of its position vector, that is,

$$\mathbf{v}(t) = \lim_{\Delta t \to 0} \frac{\Delta \mathbf{r}}{\Delta t} = \frac{d\mathbf{r}}{dt} = \dot{\mathbf{r}} \tag{11.10}$$

where a dot over a variable indicates differentiation with respect to time. Because velocity is the derivative of a vector, it is also a vector. Referring to Fig. 11.3 we see that B approaches A as Δt approaches zero, which means that the direction of $\Delta \mathbf{r}$ approaches the direction of the tangent to the path at A. Therefore, *the velocity is always tangent* to the path.

Further examination of Fig. 11.3 reveals that $|\Delta \mathbf{r}| \to \Delta s$ as $\Delta t \to 0$, from which we conclude that the magnitude of the velocity is given by

$$v = \frac{ds}{dt} = \dot{s} \tag{11.11}$$

The magnitude of the velocity is also referred to as the *speed.* The unit of velocity is distance per unit time—for example, meters per second (m/s) or feet per second (ft/s).

Figure 11.4(a) shows the velocities $\mathbf{v}(t)$ and $\mathbf{v}(t + \Delta t)$ of a point at positions A and B, respectively. Note that $\mathbf{v}(t)$ is tangent to the path at A and $\mathbf{v}(t + \Delta t)$ is tangent to the path at B. The difference $\Delta \mathbf{v}$ between these two velocities represents the change in the velocity during the time interval Δt, that is,

$$\Delta \mathbf{v} = \mathbf{v}(t + \Delta t) - \mathbf{v}(t) \tag{11.12}$$

As seen in Fig. 11.4(b), $\Delta \mathbf{v}$ is caused by changes in both the magnitude and the direction of the velocity vector.

The acceleration $\mathbf{a}$ of the point at time t is defined to be

$$\mathbf{a}(t) = \lim_{\Delta t \to 0} \frac{\Delta \mathbf{v}}{\Delta t} = \frac{d\mathbf{v}}{dt} = \dot{\mathbf{v}} \tag{11.13a}$$

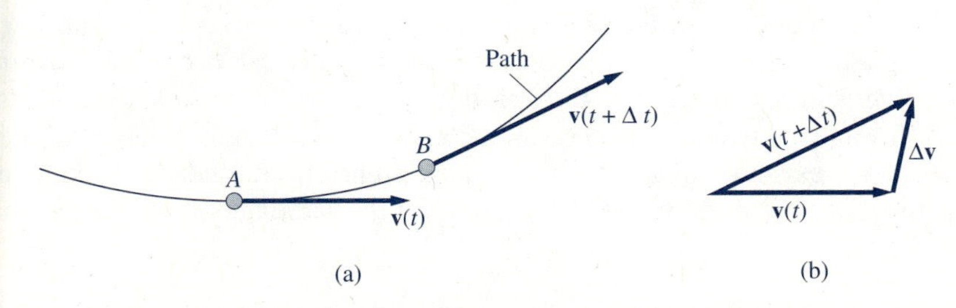

Fig. 11.4

Acceleration is obviously also a vector quantity. Using $\mathbf{v} = \dot{\mathbf{r}}$, the acceleration can also be expressed as

$$\mathbf{a}(t) = \frac{d^2\mathbf{r}}{dt^2} = \ddot{\mathbf{r}} \tag{11.13b}$$

In other words, the acceleration of a point is the time rate of change of its velocity vector, or equivalently, the second time derivative of its position vector. The unit of acceleration is distance per unit time squared (e.g., m/s^2 or ft/s^2).

The direction of the acceleration coincides with the direction of $\Delta\mathbf{v}$ as Δt approaches zero. From Fig. 11.4(b) we see that the direction of $\Delta\mathbf{v}$, and therefore the direction of **a**, will in general *not be tangent* to the path.

The kinematic variables that we have defined are sometimes referred to as linear displacement, linear velocity, and linear acceleration to distinguish them from angular displacement, angular velocity, and angular acceleration of rigid bodies, which are defined in Chapter 16. However, the term *linear* is usually omitted when referring to the motion of a point.

You must remember that velocity and acceleration are vectors, each possessing magnitude and direction. However, when the meanings are clear from the context, "velocity" and "acceleration" are often used in place of "magnitude of velocity" and "magnitude of acceleration," respectively. For example, the acceleration due to gravity is often given as 32.2 ft/s^2. Actually, it is the magnitude of this acceleration that is being stated.

11.4 Newtonian Mechanics*

a. Scope of Newtonian mechanics

In 1687, Sir Isaac Newton (1642–1727) published his celebrated laws of motion in *Principia (Mathematical Principles of Natural Philosophy)*. Without a doubt, this work ranks among the most influential scientific books ever published. We should not think, however, that its publication immediately established classical mechanics. Newton's work on mechanics dealt primarily with celestial mechanics and was thus limited to particle motion. Another two hundred or so years elapsed before rigid-body dynamics, fluid mechanics, and the mechanics of deformable bodies were developed. Each of these areas required new axioms before it could assume a usable form.

Nevertheless, Newton's work is the foundation of classical, or Newtonian, mechanics. His efforts have even influenced two other branches of mechanics born at the beginning of the twentieth century: relativistic and quantum mechanics. *Relativistic mechanics* addresses phenomena that occur on a cosmic scale (velocities approaching the speed of light, strong gravitational fields, etc.). It removes two of the most objectionable postulates of Newtonian mechanics: the existence of a fixed or inertial reference frame and the assumption that time is an absolute variable, "running" at the same rate in all parts of the

* This article, which is the same as Art. 1.2 in *Statics*, is repeated here because of its relevance to our study of dynamics.

universe. (There is evidence that Newton himself was bothered by these two postulates.) *Quantum mechanics* is concerned with particles on the atomic or subatomic scale. It removes two cherished concepts of classical mechanics: determinism and continuity. Quantum mechanics is essentially a probabilistic theory; instead of predicting an event, it determines the likelihood that an event will occur. Moreover, according to this theory, the events occur in discrete steps (called *quanta*) rather than in a continuous manner.

Relativistic and quantum mechanics, however, have by no means invalidated the principles of Newtonian mechanics. In the analysis of the motion of bodies encountered in our everyday experience, both theories converge on the equations of Newtonian mechanics. Thus the more esoteric theories actually reinforce the validity of Newton's laws of motion.

b. Newton's laws for particle motion

Here we limit our discussion to the motion of a single particle, an idealization for which the mass of the body is assumed to be concentrated at a point. Using modern terminology, Newton's laws of particle motion may be stated as follows.

1. If a particle is at rest (or moving with constant velocity in a straight line), it will remain at rest (or continue to move with constant velocity in a straight line), unless acted on by a force.
2. A particle acted on by a force will accelerate in the direction of the force. The magnitude of the acceleration is proportional to the magnitude of the force and inversely proportional to the mass of the particle.
3. For every action, there is an equal and opposite reaction; that is, the forces of interaction between two particles are equal in magnitude and opposite in direction.

Although the first law is simply a special case of the second law, it is customary to state the first law separately because of its importance to the subject of statics.

c. Inertial reference frames

When applying Newton's second law, attention must be paid to the coordinate system in which the accelerations are measured. An *inertial reference* frame (also known as a Newtonian or Galilean reference frame) is defined to be any rigid coordinate system in which Newton's laws of particle motion relative to that frame are valid with an acceptable degree of accuracy. In most design applications used on the surface of the earth, an inertial frame can be approximated with sufficient accuracy by attaching the coordinate system to the earth. In the study of earth satellites, a coordinate system attached to the sun usually suffices. For interplanetary travel, it is necessary to use coordinate systems attached to the so-called fixed stars.

It can be shown that any frame that is translating with constant velocity relative to an inertial frame is itself an inertial frame. It is a common practice to omit the word *inertial* when referring to frames for which Newton's laws obviously apply.

d. Systems of units

If a particle of mass m is acted on by a force $\mathbf{F}$, Newton's second law states that

$$\mathbf{F} = km\mathbf{a}$$

where k is a constant of proportionality and $\mathbf{a}$ is the acceleration vector of the particle. It is conventional to choose k to be dimensionless and equal to unity, resulting in the familiar form

$$\boxed{\mathbf{F} = m\mathbf{a}} \tag{11.14}$$

Note that the units of force, mass, and acceleration are not arbitrary in Eq. (11.14). If the units for any two of the quantities are specified, the equation will determine the units of the third quantity. Choosing the dimensions to be length $[L]$, time $[T]$, force $[F]$, and mass $[M]$, dimensional homogeneity of Eq. (11.14) can be satisfied in one of the following two ways.

$$[M] = \left[\frac{FT^2}{L}\right] \quad \text{or} \quad [F] = \left[\frac{ML}{T^2}\right] \tag{11.15}$$

A system of units that satisfies the first of these dimensional relationships—where the units of force, length, and time are taken to be base units and the units of mass are derived—is called a *gravitational* system. An *absolute* system is one that satisfies the second of the dimensional relationships shown in Eqs. (11.15)—where the units of mass, length, and time are chosen to be base units, and the units of force are derived.

Throughout this text, both U.S. Customary units and SI units (from *Système international d'unités*) are used. The following table lists the units in these two systems.

	U.S. Customary Units		SI Units	
Quantity	**Unit**	**Symbol**	**Unit**	**Symbol**
Force	pound*	lb	newton†	N
Length	foot*	ft	meter*	m
Time	second*	s	second*	s
Mass	slug†	—	kilogram*	kg

*Base unit †Derived unit

Note that U.S. Customary units constitute a gravitational system, whereas SI units are absolute. In both systems, the base units are defined with respect to standard bodies or with respect to reproducible physical phenomena. For example, the standard pound is defined to be the weight of a certain platinum bar that is kept in Washington, D.C., and the standard second is defined to be the duration of a specified number of radiation cycles of a certain isotope.

In U.S. Customary units, the derived unit of mass (the mass that will be accelerated at 1 ft/s^2 by a force of 1 lb) is called a *slug*. From Newton's second law, $m = F/a$, we find

$$1 \text{ slug} = \frac{1 \text{ lb}}{1 \text{ ft/s}^2} = 1 \text{ lb} \cdot \text{s}^2/\text{ft}$$

In SI units, the force that will accelerate 1 kg at the rate of 1 m/s^2 is called a newton (N). From $F = ma$, we have

$$1 \text{ N} = (1 \text{ kg})(1 \text{ m/s}^2) = 1 \text{ kg} \cdot \text{m/s}^2$$

Strict guidelines have been adopted for the use of SI units, the more important of which are listed on the inside front cover of this book.

When using Newton's laws of motion, it is important to distinguish between weight and mass. The weight W of a body is the force that attracts the body to the earth. According to Newton's second law, the weight is given by

$$\boxed{W = mg} \tag{11.16}$$

where g is the acceleration due to gravity at the location of the body. Near the surface of the earth, the accepted values of g are 32.2 ft/s^2 and 9.81 m/s^2. In Newtonian mechanics, mass is treated as a constant, but the weight of the body will depend on the "local" acceleration due to gravity. Since weight is a force, it must be expressed in pounds for U.S. Customary units, and in newtons when using SI units.

e. Law of gravitation

In addition to his many other accomplishments, Newton also proposed the law of universal gravitation. Consider two particles of mass m_A and m_B that are separated by a distance R, as shown in Fig. 11.5. The law of gravitation states that the two particles are attracted to each other by forces of magnitude F that act along the line connecting the particles, where

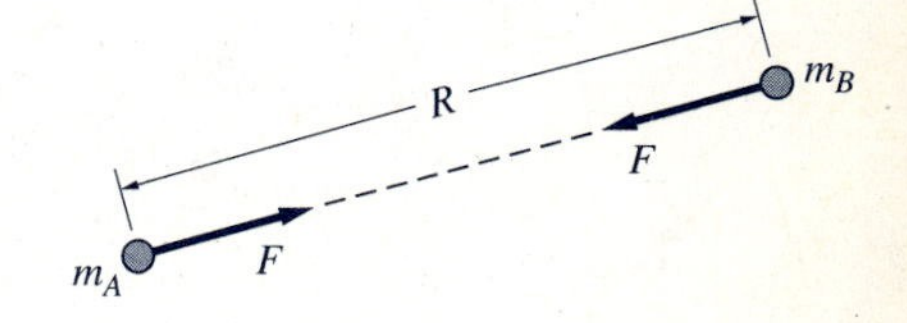

Fig. 11.5

$$\boxed{F = K\frac{m_A m_B}{R^2}} \tag{11.17}$$

The universal gravitational constant K is equal to 3.44×10^{-8} $ft^4/(lb \cdot s^4)$ or 6.67×10^{-11} $m^3/(kg \cdot s^2)$. Although this law is valid for particles, Newton showed that it is also applicable to spherical bodies provided that their masses are distributed uniformly. (When attempting to derive this result, Newton was forced to develop calculus.)

If we let $m_A = M_e$ (the mass of the earth), $m_B = m$ (the mass of a body), and $R = R_e$ (the mean radius of the earth), then F in Eq. (11.17) will be the weight W of the body. Comparing $W = KM_e m/R_e^2$ with $W = mg$, we find that $g = KM_e/R_e^2$. Of course, adjustments may be necessary in the value of g for some applications in order to account for local variation of the gravitational attraction.

12 Dynamics of a Particle: Rectangular Coordinate System

12.1 Introduction

The preceding chapter has introduced the basic kinematic variables for particle motion: position, velocity, and acceleration. In this chapter we consider the kinematics and kinetics (force-mass-acceleration method) of particle motion that is described relative to a fixed rectangular (xyz) coordinate system. (Recall that kinematics refers to the geometry of motion, whereas kinetics relates the forces to the motion.) The next chapter discusses particle dynamics using curvilinear coordinate systems.

In Art. 12.2 the kinematics of particle motion are expressed in terms of their rectangular components. The general form of the equations of motion in rectangular coordinates is then described in Art. 12.3. Article 12.4 introduces the force-mass-acceleration (FMA) method of kinetic analysis. The FMA technique determines the equations of motion for a particle by applying Newton's second law to the free-body diagram of the particle. This leads us to a discussion of the methods of solving (integrating) the equations of motion. The kinetic applications in this chapter are limited to the special cases of rectilinear motion (straight path), and curvilinear motion (curved path) that can be described as the superposition of rectilinear motions.

The last two articles of this chapter discuss numerical methods for solving equations of particle motion. Although this topic is optional, it is of great practical importance. The equations of motion for most dynamics problems cannot be solved by analytical means, but solutions can always be computed numerically.

12.2 Kinematics—Rectangular Coordinates

a. General motion

The position (**r**), velocity (**v**), and acceleration (**a**) of a point (particle) are defined in the preceding chapter without reference to a coordinate system.

Therefore the definitions are applicable to all frames of reference. However, the mathematical expression for **r**, **v**, and **a** depend on the specific coordinate system used. The simplest expressions are obtained for the rectangular (xyz) coordinate system, which is used throughout the remainder of this chapter. Although in principle rectangular components can be employed in the analysis of all problems, you will later discover that other coordinate systems may be more convenient.

Figure 12.1(a) shows the path of particle A, which moves in a fixed rectangular reference frame. Letting **i**, **j**, and **k** be the base vectors (unit vectors), the position vector of the particle can be written as

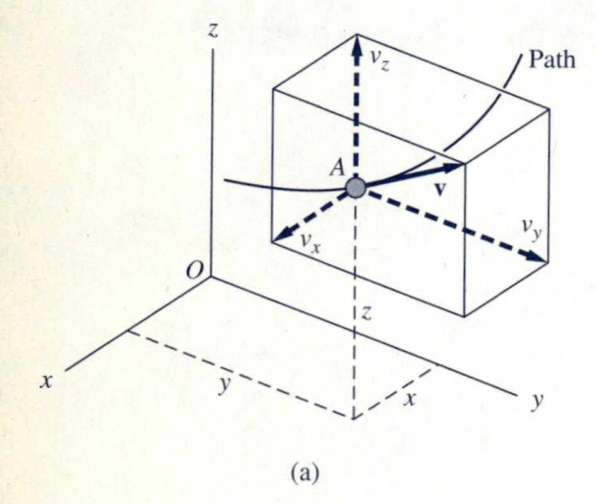

$$\mathbf{r}(t) = x\mathbf{i} + y\mathbf{j} + z\mathbf{k} \tag{12.1}$$

where x, y, and z are the time-dependent rectangular coordinates of the particle.

Applying the definition of velocity, Eq. (11.10), and the chain rule of differentiation, Eq. (11.4), we obtain

$$\mathbf{v} = \frac{d\mathbf{r}}{dt} = \frac{d}{dt}(x\mathbf{i} + y\mathbf{j} + z\mathbf{k})$$

$$= x\frac{d\mathbf{i}}{dt} + \dot{x}\mathbf{i} + y\frac{d\mathbf{j}}{dt} + \dot{y}\mathbf{j} + z\frac{d\mathbf{k}}{dt} + \dot{z}\mathbf{k}$$

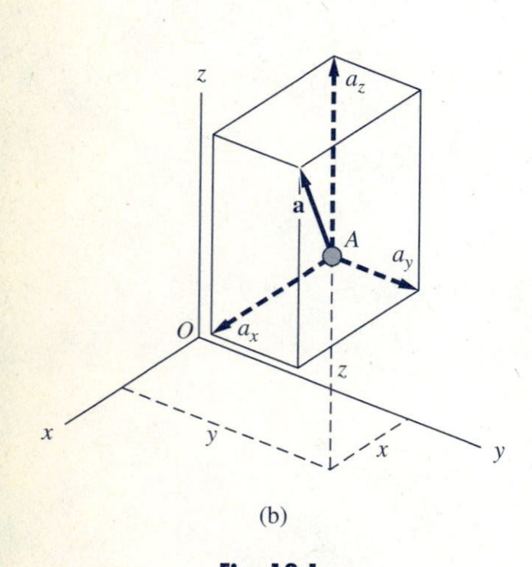

Fig. 12.1

Because the xyz-frame is assumed to be fixed,* the base vectors do not change, giving $d\mathbf{i}/dt = d\mathbf{j}/dt = d\mathbf{k}/dt = \mathbf{0}$. Therefore, the rectangular representation of the velocity is

$$\mathbf{v} = v_x\mathbf{i} + v_y\mathbf{j} + v_z\mathbf{k} \tag{12.2}$$

where the components are

$$v_x = \dot{x} \qquad v_y = \dot{y} \qquad v_z = \dot{z} \tag{12.3}$$

These components are shown in Fig. 12.1(a).

Similarly, the definition of acceleration, Eq. (11.13a), yields

$$\mathbf{a} = \frac{d\mathbf{v}}{dt} = \frac{d}{dt}(v_x\mathbf{i} + v_y\mathbf{j} + v_z\mathbf{k})$$

$$= \dot{v}_x\mathbf{i} + \dot{v}_y\mathbf{j} + \dot{v}_z\mathbf{k}$$

Therefore, in a rectangular coordinate system the acceleration is

$$\mathbf{a} = a_x\mathbf{i} + a_y\mathbf{j} + a_z\mathbf{k} \tag{12.4}$$

where the components, shown in Fig. 12.1(b), are

$$a_x = \dot{v}_x = \ddot{x} \qquad a_y = \dot{v}_y = \ddot{y} \qquad a_z = \dot{v}_z = \ddot{z} \tag{12.5}$$

* This assumption is actually overly restrictive. As we show in Art. 16.7, the results remain valid if the coordinate system moves without rotating.

b. Plane motion

Plane motion occurs often enough in engineering applications to warrant special attention. Figure 12.2(a) shows the path of a particle A that moves in the xy-plane. To obtain the two-dimensional rectangular components of **r**, **v**, and **a**, we set $z = 0$ in Eqs. (12.1)–(12.5). The results are

$$\begin{aligned} \mathbf{r} &= x\mathbf{i} + y\mathbf{j} \\ \mathbf{v} &= v_x\mathbf{i} + v_y\mathbf{j} \quad \text{where } v_x = \dot{x}, \quad v_y = \dot{y} \\ \mathbf{a} &= a_x\mathbf{i} + a_y\mathbf{j} \quad \text{where } a_x = \dot{v}_x = \ddot{x}, \quad a_y = \dot{v}_y = \ddot{y} \end{aligned} \tag{12.6}$$

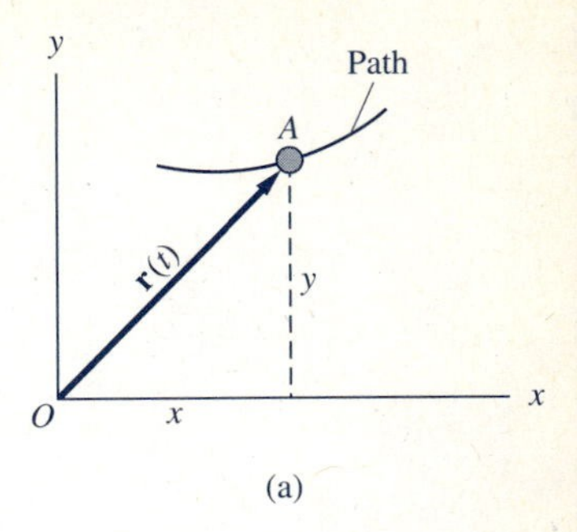

(a)

Figure 12.2(b) shows the rectangular components of the velocity. The angle θ, which defines the direction of **v**, can be obtained from

$$\tan\theta = \frac{v_y}{v_x} = \frac{dy/dt}{dx/dt} = \frac{dy}{dx} \tag{12.7}$$

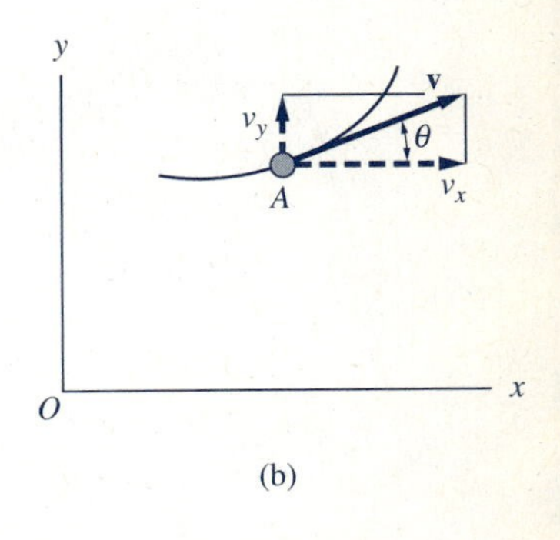

(b)

Since the slope of the path is also equal to dy/dx, we see from Eq. (12.7) that **v** is tangent to the path, a result that was pointed out in the preceding chapter.

The rectangular components of **a** are shown in Fig. 12.2(c). The angle β that defines the direction of **a** can be computed from

$$\tan\beta = \frac{a_y}{a_x} = \frac{d^2y/dt^2}{d^2x/dt^2} \tag{12.8}$$

Since β is generally not equal to θ, the acceleration is not necessarily tangent to the path.

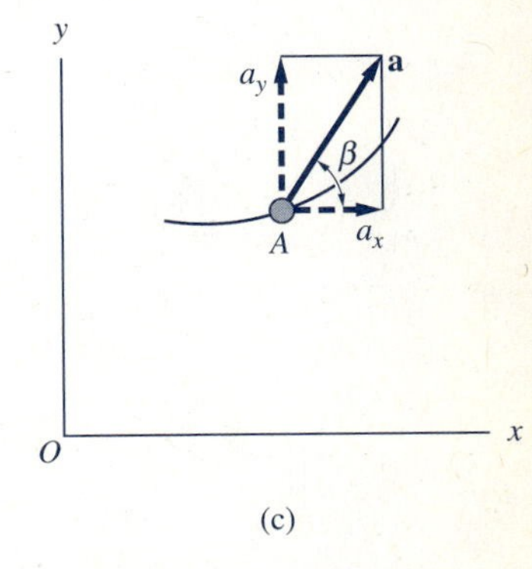

(c)

Fig. 12.2

c. Rectilinear motion

If the path of a particle is a straight line, the motion is called *rectilinear*. An example of rectilinear motion, in which the particle A moves along the x-axis, is depicted in Fig. 12.3. In this case, we set $y = 0$ in Eq. (12.6), obtaining $\mathbf{r} = x\mathbf{i}$, $\mathbf{v} = v_x\mathbf{i}$, and $\mathbf{a} = a_x\mathbf{i}$. Each of these vectors is directed along the path (i.e., the motion is one-dimensional). Since the subscripts are no longer needed, the equations for rectilinear motion along the x-axis are usually written as

$$\begin{aligned} \mathbf{r} &= x\mathbf{i} \\ \mathbf{v} &= v\mathbf{i} \quad \text{where } v = \dot{x} \\ \mathbf{a} &= a\mathbf{i} \quad \text{where } a = \dot{v} = \ddot{x} \end{aligned} \tag{12.9}$$

In some problems, it is more convenient to express the acceleration in terms of velocity and position, rather than velocity and time. This change of variable can be accomplished by the chain rule of differentiation: $a = dv/dt = (dv/dx)(dx/dt)$. Noting that $dx/dt = v$, we obtain

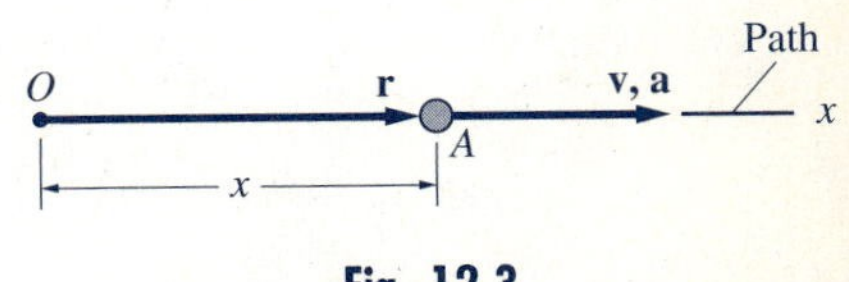

Fig. 12.3

$$a = v\frac{dv}{dx} \tag{12.10}$$

Sample Problem 12.1

The position of a particle that moves along the x-axis is defined by $x = -3t^2 + 12t - 6$ ft, where t is in seconds. For the time interval $t = 0$ to $t = 3$ seconds, (1) plot the position, velocity, and acceleration as functions of time; (2) calculate the distance traveled; and (3) determine the displacement of the particle.

Solution

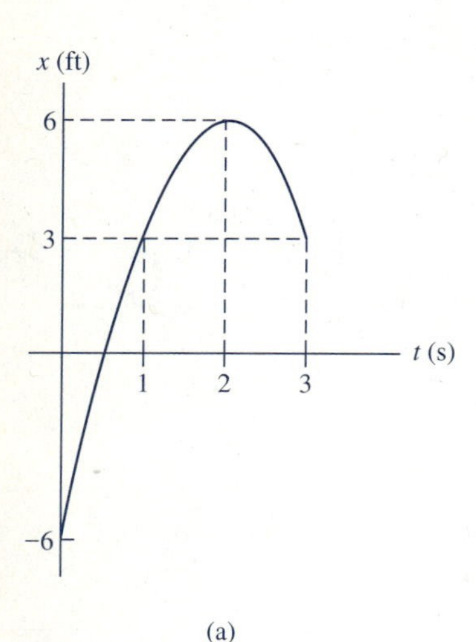

(a)

Part 1

Since the motion is rectilinear, the velocity and acceleration may be calculated as follows.

$$x = -3t^2 + 12t - 6 \text{ ft} \qquad \text{(a)}$$

$$v = \frac{dx}{dt} = -6t + 12 \text{ ft/s} \qquad \text{(b)}$$

$$a = \frac{dv}{dt} = \frac{d^2x}{dt^2} = -6 \text{ ft/s}^2 \qquad \text{(c)}$$

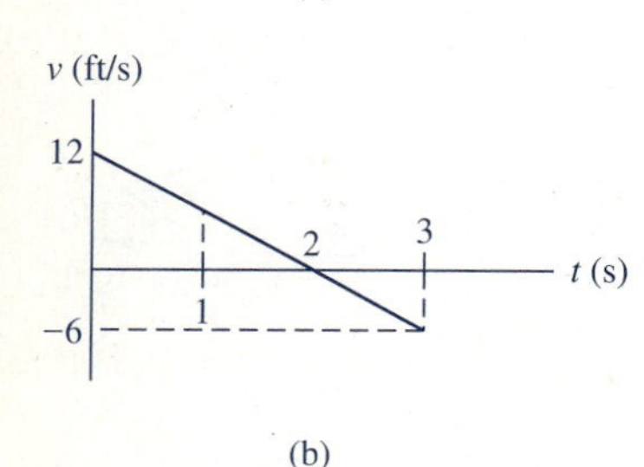

(b)

These functions are plotted in Figs. (a)–(c) for the prescribed time interval $t = 0$ to $t = 3$ s. Note that the plot of x is parabolic, so that successive differentiations yield a linear function for the velocity and a constant value for the acceleration. The time corresponding to the maximum (or minimum) value of x can be found by setting $dx/dt = 0$, or utilizing Eq. (b), $v = -6t + 12 = 0$, which gives $t = 2$ s. Substituting $t = 2$ s into Eq. (a) we find

$$x_{\text{max}} = -3(2)^2 + 12(2) - 6 = 6 \text{ ft}$$

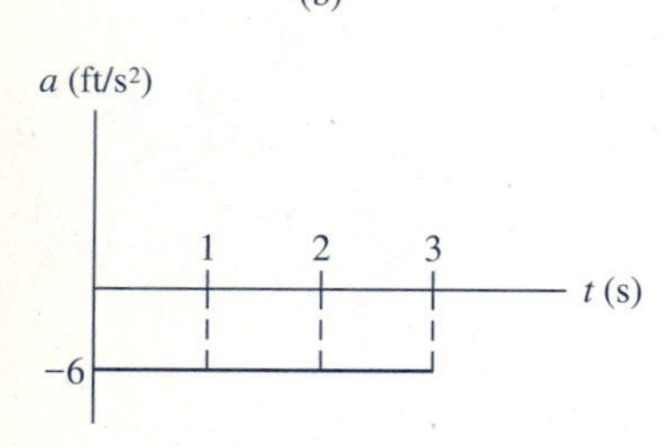

(c)

Part 2

Figure (d) shows how the particle moves during the time interval $t = 0$ to $t = 3$ s. When $t = 0$, the particle leaves $A(x = -6 \text{ ft})$ moving to the right. When $t = 2$ s, the particle comes to a stop at $B(x = 6 \text{ ft})$. Then it moves to the left, arriving at $C(x = 3 \text{ ft})$ when $t = 3$ s. Therefore, the distance traveled is equal to the distance that the point moves to the right ($\overline{AB}$), plus the distance it moves to the left ($\overline{BC}$), which gives

$$d = \overline{AB} + \overline{BC} = 12 + 3 = 15 \text{ ft} \qquad \textit{Answer}$$

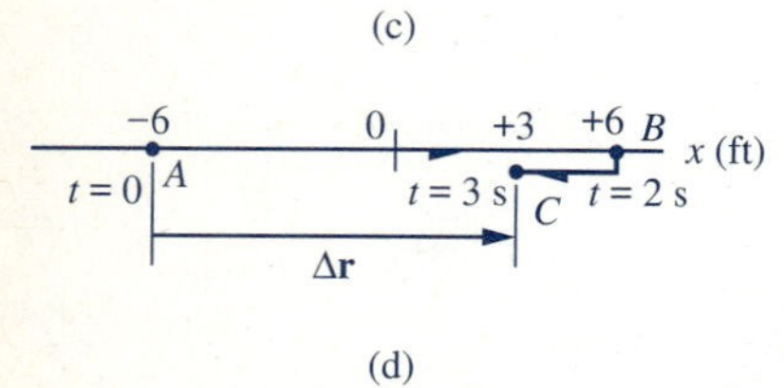

(d)

Part 3

The displacement during the time interval $t = 0$ to $t = 3$ s is the vector drawn from the initial position of the point to its final position. This vector, indicated as $\Delta\mathbf{r}$ in Fig. (d), is

$$\Delta\mathbf{r} = 9\mathbf{i} \text{ ft} \qquad \textit{Answer}$$

Observe that the total distance traveled (15 ft) is greater than the magnitude of the displacement vector (9 ft) because the direction of the motion changes during the time interval.

Sample Problem 12.2

Pin P at the end of the telescoping rod in Fig. (a) slides along the fixed parabolic path $y^2 = 40x$, where x and y are measured in millimeters. The y coordinate of P varies with time t (measured in seconds) according to $y = 4t^2 + 6t$ mm. When $y = 30$ mm, compute (1) the velocity vector of P; and (2) the acceleration vector of P.

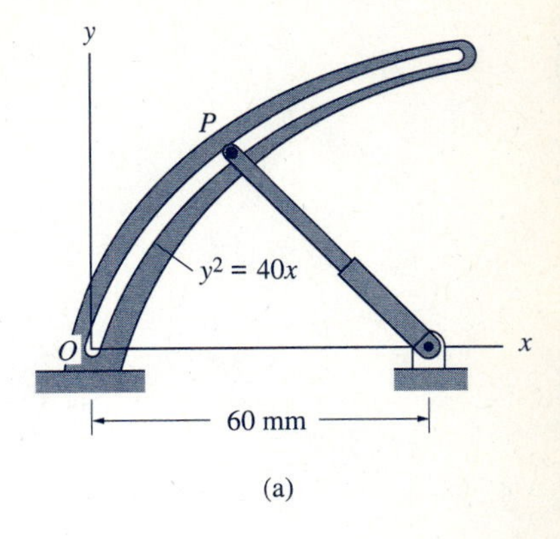

(a)

Solution

Part 1

Substituting

$$y = 4t^2 + 6t \text{ mm} \tag{a}$$

into the equation of the path and solving for x, we obtain

$$x = \frac{(4t^2 + 6t)^2}{40} = 0.40t^4 + 1.20t^3 + 0.90t^2 \text{ mm} \tag{b}$$

The rectangular components of the velocity vector thus are

$$v_x = \dot{x} = 1.60t^3 + 3.60t^2 + 1.80t \text{ mm/s} \tag{c}$$

$$v_y = \dot{y} = 8t + 6 \text{ mm/s} \tag{d}$$

Setting $y = 30$ mm in Eq. (a) and solving for t gives $t = 2.090$ s. Substituting this value of time into Eqs. (c) and (d), we obtain

$$v_x = 34.1 \text{ mm/s} \quad \text{and} \quad v_y = 22.7 \text{ mm/s}$$

Consequently, the velocity vector at $y = 30$ mm is

$$\mathbf{v} = 34.1\mathbf{i} + 22.7\mathbf{j} \text{ mm/s} \qquad \textit{Answer}$$

The pictorial representation of this result is shown below and also in Fig. (b).

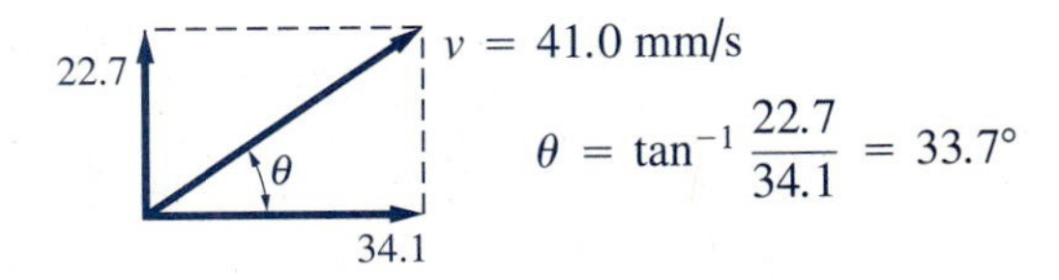

By evaluating the slope of the path, dy/dx, at $y = 30$ mm, it is easy to verify that the velocity vector determined above is indeed tangent to the path.

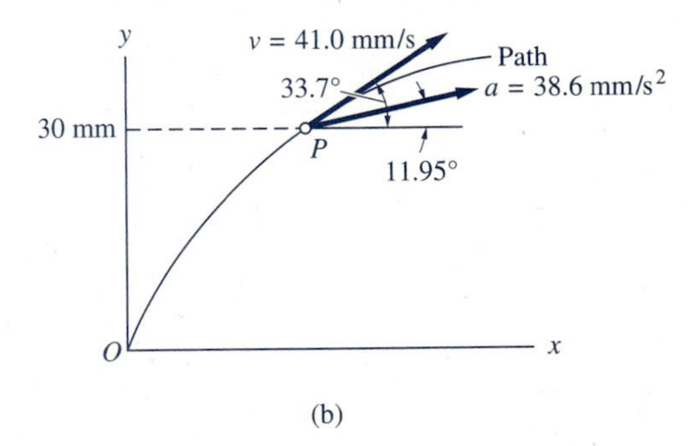

(b)

Part 2

From Eqs. (c) and (d), we can determine the components of the acceleration vector by differentiation:

$$a_x = \dot{v}_x = 4.80t^2 + 7.20t + 1.80 \text{ mm/s}^2$$
$$a_y = \dot{v}_y = 8 \text{ mm/s}^2$$

Substituting $t = 2.090$ s, we obtain

$$a_x = 37.8 \text{ mm/s}^2 \quad \text{and} \quad a_y = 8 \text{ mm/s}^2$$

Therefore, the acceleration vector at $y = 30$ mm is

$$\mathbf{a} = 37.8\mathbf{i} + 8\mathbf{j} \text{ mm/s}^2 \qquad \textit{Answer}$$

The pictorial representation of **a** is

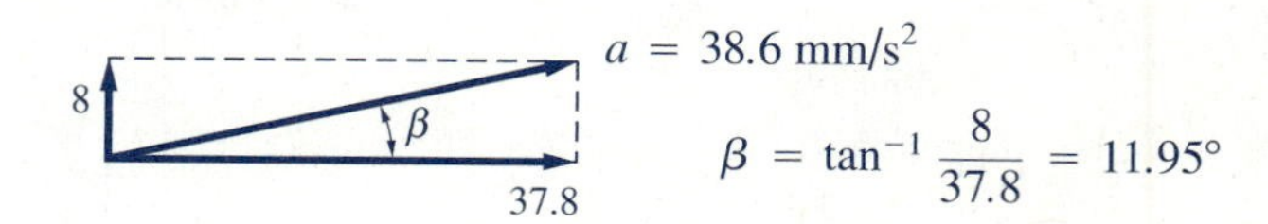

From the drawing of the acceleration vector in Fig. (b) we see that the direction of **a** is not tangent to the path.

Sample Problem 12.3

The rectangular coordinates describing the spatial motion of a point are

$$x = R\cos\omega t \qquad y = \frac{R}{2}\sin 2\omega t \qquad z = \frac{R}{2}(1 - \cos 2\omega t)$$

where R and ω are constants. (1) By calculating the magnitude of the position vector **r**, show that the path lies on a sphere of radius R, centered at the origin of the coordinate system. (2) Determine the rectangular components and the magnitudes of the velocity and acceleration vectors.

Solution

Part 1

The magnitude of the position vector can be calculated using $r^2 = x^2 + y^2 + z^2$. Substituting the given expressions for the rectangular coordinates, we obtain

$$r^2 = R^2\left[\cos^2\omega t + \frac{1}{4}\sin^2 2\omega t + \frac{1}{4}(1 - 2\cos 2\omega t + \cos^2 2\omega t)\right]$$

$$= R^2\left[\cos^2\omega t + \frac{1}{2}(1 - \cos 2\omega t)\right]$$

where we used the trigonometric identity $\sin^2 2\omega t + \cos^2 2\omega t = 1$. Utilizing the trigonometric formula $(1/2)(1 - \cos 2\omega t) = \sin^2 \omega t$, we obtain $r^2 = R^2(\cos^2 \omega t + \sin^2 \omega t) = R^2$. Because the magnitude of the position vector is equal to the constant R, the path must lie on a sphere of that radius, centered at the origin of the coordinate system.

Part 2

The components of **v** can be found by differentiating the given expressions for the rectangular coordinates.

$$v_x = \dot{x} = -R\omega \sin \omega t$$
$$v_y = \dot{y} = R\omega \cos 2\omega t \qquad \textit{Answer}$$
$$v_z = \dot{z} = R\omega \sin 2\omega t$$

Therefore,

$$\begin{aligned} v^2 &= v_x^2 + v_y^2 + v_z^2 = R^2\omega^2(\sin^2 \omega t + \cos^2 2\omega t + \sin^2 2\omega t) \\ &= R^2\omega^2(\sin^2 \omega t + 1) \end{aligned}$$

and the magnitude of **v** is

$$v = R\omega\sqrt{\sin^2 \omega t + 1} \qquad \textit{Answer}$$

The components of the acceleration vector are

$$a_x = \dot{v}_x = -R\omega^2 \cos \omega t$$
$$a_y = \dot{v}_y = -2R\omega^2 \sin 2\omega t \qquad \textit{Answer}$$
$$a_z = \dot{v}_z = 2R\omega^2 \cos 2\omega t$$

from which

$$\begin{aligned} a^2 &= a_x^2 + a_y^2 + a_z^2 \\ &= R^2\omega^4(\cos^2 \omega t + 4\sin^2 2\omega t + 4\cos^2 2\omega t) \\ &= R^2\omega^4(\cos^2 \omega t + 4) \end{aligned}$$

Therefore, the magnitude of **a** is

$$a = R\omega^2\sqrt{\cos^2 \omega t + 4} \qquad \textit{Answer}$$

Sample Problem 12.4

A circular cam of radius $R = 16$ mm is pivoted at O, thus producing an eccentricity of $R/2$. Using geometry, it can be shown that the relationship between x, the position coordinate of the follower A, and the angle θ is

$$x(\theta) = \frac{R}{2}\left(\cos\theta + \sqrt{\cos^2\theta + 3}\right)$$

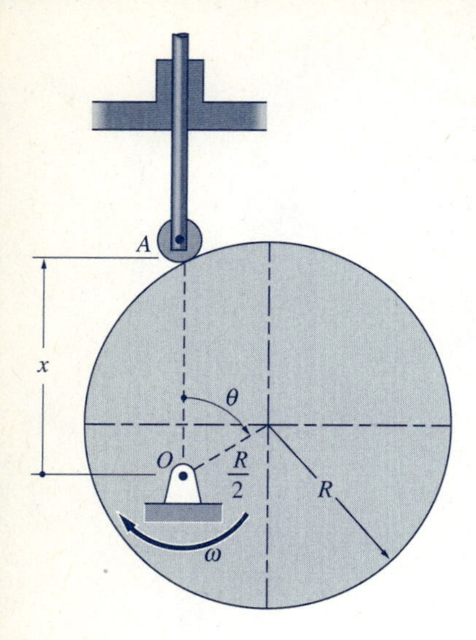

The cam is rotating about O with constant clockwise angular speed $\omega = d\theta/dt = 2000$ rev/min. (1) Use numerical differentiation with $\Delta\theta = 1.0°$ to compute the speed of the follower when $\theta = 45°$. (2) Compare your answer from Part 1 with the result obtained analytically. (3) Find the acceleration of the follower when $\theta = 0$, using numerical differentiation with $\Delta\theta = 1.0°$. (Numerical differentiation with finite differences is summarized in Appendix E.)

Solution

Substituting $R = 0.016$ m into the given expression for $x(\theta)$, we have

$$x(\theta) = 0.008\left(\cos\theta + \sqrt{\cos^2\theta + 3}\right) \text{ m} \tag{a}$$

The given angular velocity is

$$\omega = 2000 \text{ rev/min} = 12\,000 \text{ deg/s} = 209.4 \text{ rad/s} \tag{b}$$

Part 1

Applying the chain rule for differentiation, we can write the speed of the follower as

$$v = \frac{dx}{dt} = \frac{dx}{d\theta}\frac{d\theta}{dt} = \frac{dx}{d\theta}\omega \tag{c}$$

which eliminates time as an explicit variable. Using the central difference formula, Eq. (E.1) in Appendix E for $dx/d\theta$, we obtain

$$v_i \approx \frac{x_{i+1} - x_{i-1}}{2\Delta\theta}\omega \tag{d}$$

When $\theta = 45°$ and $\Delta\theta = 1.0°$, we have $x_{i+1} = x(\theta + \Delta\theta) = x(46°)$ and $x_{i-1} = x(\theta - \Delta\theta) = x(44°)$. Consequently, Eq. (d) becomes

$$v(45°) \approx \frac{x(46°) - x(44°)}{2(1.0)}\omega$$

where v is in m/s if x is in meters and ω is in deg/s. Evaluating $x(46°)$ and $x(44°)$ from Eq. (a) and substituting $\omega = 12\,000$ deg/s, we have

$$v(45°) \approx \frac{0.020\,486\,54 - 0.020\,758\,61}{2}(12\,000)$$

$$= -1.6324 \text{ m/s} \qquad \textit{Answer} \tag{e}$$

Part 2

The analytic expression for the speed of the follower is derived by substituting the derivative $dx/d\theta$, obtained by differentiating Eq. (a), into Eq. (c). The result is

$$v = \frac{dx}{d\theta}\omega = 0.008\left(-\sin\theta + \frac{(1/2)(-2\sin\theta\cos\theta)}{\sqrt{\cos^2\theta + 3}}\right)\omega \tag{f}$$

Substituting $\theta = 45°$ and $\omega = 209.4$ rad/s gives

$$v = -1.6326 \text{ m/s} \qquad \textit{Answer} \quad \text{(g)}$$

Comparing Eqs. (e) and (g), we see that the speed obtained in Part 1 using $\Delta\theta = 1.0°$ is very close to the analytical solution.

Part 3

To evaluate the acceleration of the follower, we again employ the chain rule for differentiation to eliminate time as an explicit variable:

$$a = \frac{dv}{dt} = \frac{dv}{d\theta}\frac{d\theta}{dt} = \frac{dv}{d\theta}\omega$$

Substituting $v = (dx/d\theta)\omega$ from Eq. (c) gives

$$a = \frac{d}{d\theta}\left(\frac{dx}{d\theta}\omega\right)\omega = \frac{d^2x}{d\theta^2}\omega^2 + 2\frac{dx}{d\theta}\frac{d\omega}{d\theta}\omega$$

Since ω is constant, the last term equals zero, and we are left with

$$a = \frac{d^2x}{d\theta^2}\omega^2 \qquad \text{(h)}$$

Utilizing the central difference formula, Eq. (E.2) in Appendix E for $d^2x/d\theta^2$, Eq. (h) becomes

$$a_i \approx \frac{x_{i+1} - 2x_i + x_{i+1}}{(\Delta\theta)^2}\omega^2 \qquad \text{(i)}$$

When $\theta = 0$ and $\Delta\theta = 1°$, we have $x_{i-1} = x(-1°)$, $x_i = x(0)$ and $x_{i+1} = x(+1°)$. Therefore, Eq. (i) becomes

$$a(0) \approx \frac{x(-1°) - 2x(0) + x(+1°)}{(1.0)^2}\omega^2$$

which gives the acceleration a in m/s^2 if x is in meters and ω is in deg/s. Evaluating $x(-1°)$, $x(0)$, and $x(+1°)$ from Eq. (a), and substituting $\omega = 12\,000$ deg/s, the preceding equation yields

$$\begin{aligned} a(0) &\approx [0.023\,998\,17 - 2(0.024\,000\,0) + 0.023\,998\,17](12\,000)^2 \\ &= -526 \text{ m/s}^2 \qquad \textit{Answer} \end{aligned}$$

In this case the analytical solution for the acceleration would be tedious to compute, because the expression for $a = dv/dt$, obtainable by differentiating Eq. (f), would be complicated.

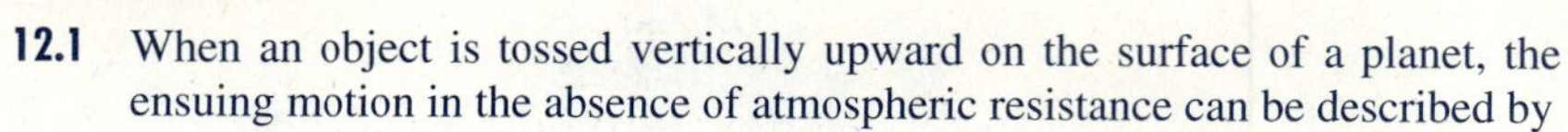

PROBLEMS

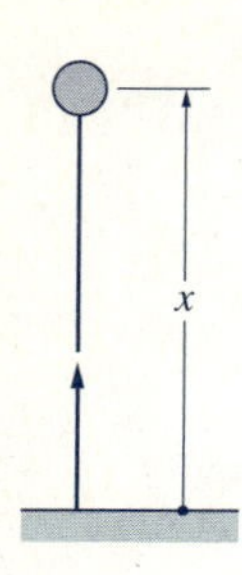

Fig. P12.1

12.1 When an object is tossed vertically upward on the surface of a planet, the ensuing motion in the absence of atmospheric resistance can be described by

$$x = -(1/2)gt^2 + v_0 t$$

where g and v_0 are constants. (a) Derive the expressions for the velocity and acceleration of the object. Use the results to show that v_0 is the initial speed of the body and that g represents the gravitational acceleration. (b) Derive the maximum height reached by the object and the total time of flight. (c) Evaluate the results of Part (b) for $v_0 = 60$ mi/h and $g = 32.2$ ft/s^2 (surface of the earth).

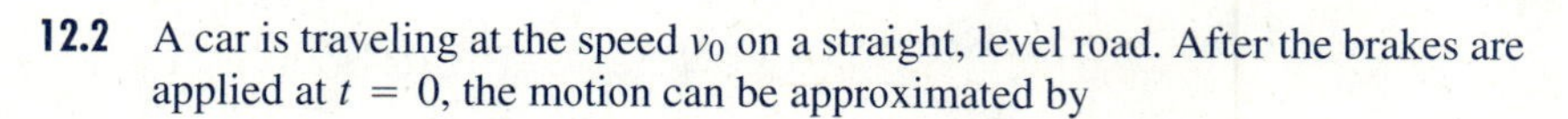

12.2 A car is traveling at the speed v_0 on a straight, level road. After the brakes are applied at $t = 0$, the motion can be approximated by

$$x = \frac{t^3}{100} - t^2 + 17t$$

where x is the distance traveled in meters and t is the time in seconds. (a) Determine v_0. (b) Find the time and distance required for the car to stop. (c) Compute the maximum deceleration during braking.

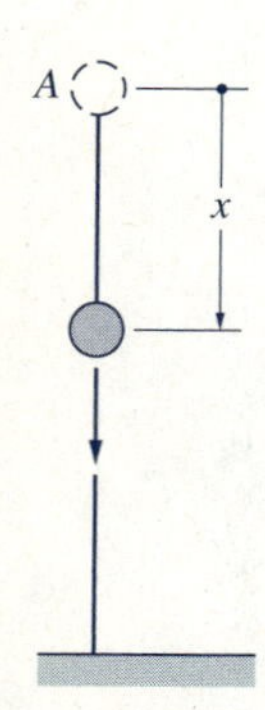

Fig. P12.3

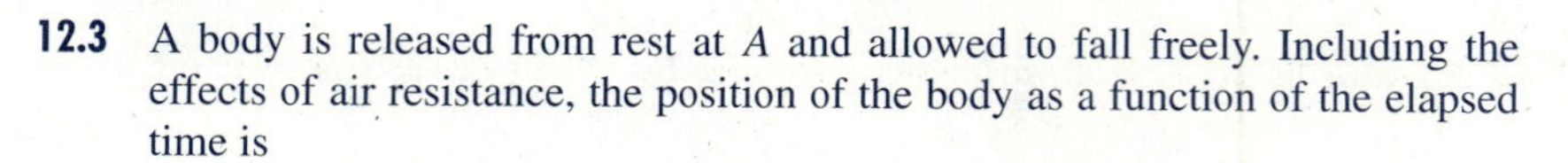

12.3 A body is released from rest at A and allowed to fall freely. Including the effects of air resistance, the position of the body as a function of the elapsed time is

$$x = v_0(t - t_0 + t_0 e^{-t/t_0})$$

where v_0 and t_0 are constants. (a) Derive the expression for the speed of the body. Use the result to explain why v_0 is called the *terminal velocity.* (b) Derive the expressions for the acceleration of the body as a function of t and as a function of v. Use the results to show that $t_0 = v_0/g$ where g is the gravitational acceleration. (Hint: When $v = 0$, $a = g$.)

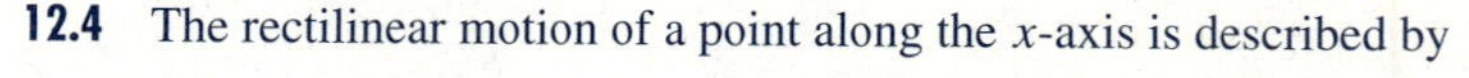

12.4 The rectilinear motion of a point along the x-axis is described by

$$x = x_0\left(\frac{t}{t_0} - 1\right)\left(\frac{t}{t_0}\right)^2$$

where $x_0 = 9$ in. and $t_0 = 0.3$ s. For the time interval $t = 0$ to $t = t_0$: (a) plot the position, velocity, and acceleration as functions of time; (b) determine the distance traveled.

12.5 The position of a point moving along the x-axis is described by

$$x = x_0\left(e^{-t/t_0} + \frac{t}{2t_0}\right)$$

where x_0 and t_0 are constants. For the time interval $t = 0$ to t_0: (a) plot the position, velocity, and acceleration versus time; and (b) compute the distance traveled.

12.6 The circular cam of radius R and eccentricity $R/2$ rotates clockwise with a constant angular speed ω. The resulting vertical motion of the flat follower A can be shown to be

$$x = R[1 + (1/2)\cos \omega t]$$

(a) Obtain the velocity and acceleration of the follower as function of t. (b) If ω were doubled, how would the maximum velocity and maximum acceleration of the follower be changed? (c) What is the distance traveled by the follower during one revolution of the cam?

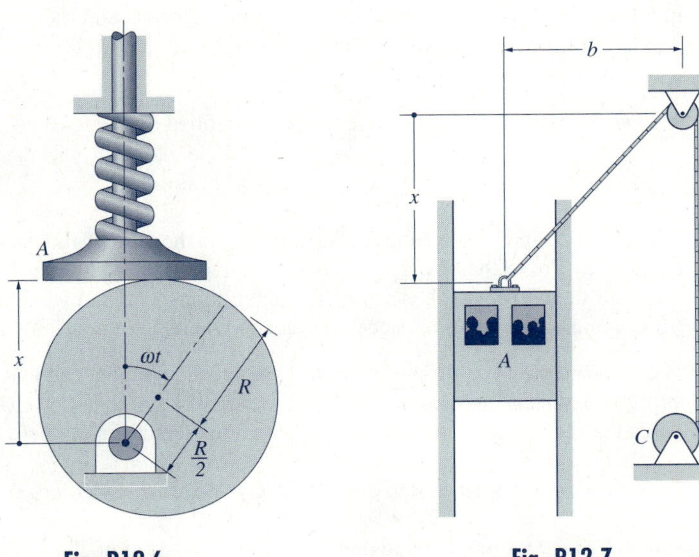

Fig. P12.6 **Fig. P12.7**

12.7 The elevator A is lowered by a cable that runs over pulley B. If the cable unwinds from the winch C at the constant speed v_0, the motion of the elevator is

$$x = \sqrt{(v_0 t - b)^2 - b^2}$$

Determine the velocity and acceleration of the elevator in terms of (a) the time t; and (b) the distance x.

12.8 A missile is launched from the surface of a planet with the speed v_0 at $t = 0$. According to the theory of universal gravitation, the speed v of the missile after launch is given by

$$v^2 = 2gr_0[(r_0/r) - 1] + v_0^2$$

where g is the gravitational acceleration on the surface of the planet and r_0 is the mean radius of the planet. (a) Determine the acceleration of the missile in terms of r. (b) Find the *escape velocity*, i.e., the minimum value of v_0 for which the missile will not return to the planet. (c) Using the result of Part (b), calculate the escape velocity for earth, where $g = 32.2\ \text{ft/s}^2$ and $r_0 = 3960$ mi.

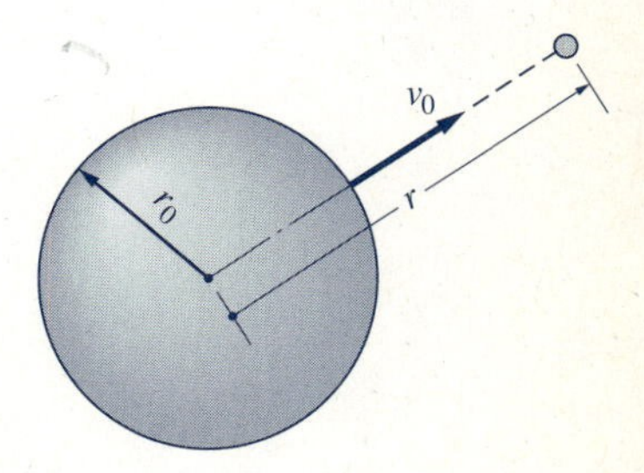

Fig. P12.8

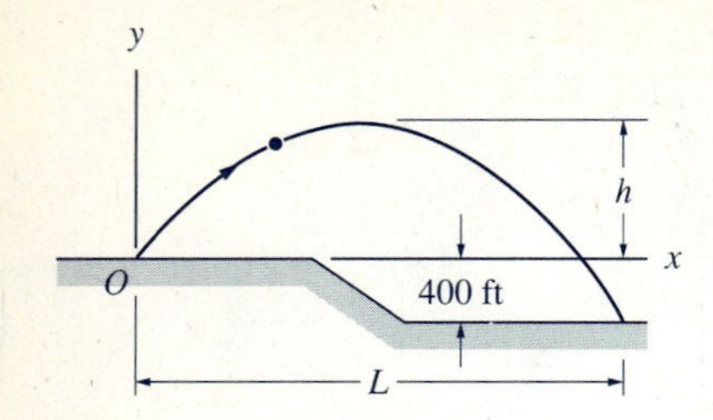

Fig. P12.9

12.9 A projectile fired at O follows a parabolic trajectory, given in parametric form by

$$x = 283t \qquad y = 316t - 16.1t^2$$

where x and y are measured in feet and t in seconds. Calculate (a) the acceleration vector throughout the flight; (b) the velocity vector at O; (c) the maximum height h; and (d) the range L.

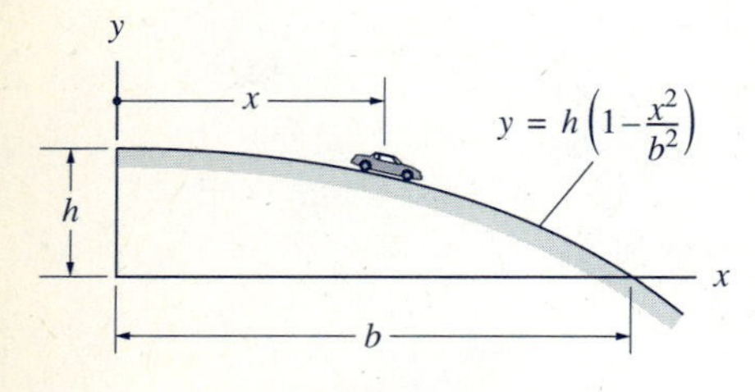

Fig. P12.10

12.10 An automobile goes down a hill that has the parabolic cross section shown. Assuming that the horizontal component of the velocity vector has a constant magnitude v_0, determine (a) the expression for the speed of the automobile in terms of x; (b) the magnitude and direction of the acceleration; and (c) the maximum speed and maximum acceleration in $0 \leq x \leq b$ if $v_0 = 80$ km/h, $h = 200$ m, and $b = 1600$ m.

12.11 The position of a particle in plane motion is defined by

$$x = a \cos \omega t \qquad y = b \sin \omega t$$

where $a > b$, and ω is a constant. (a) Show that the path of the particle is an ellipse. (b) Prove that the acceleration vector is always directed toward the center of the ellipse. (c) Determine the maximum speed and where it occurs. (d) Calculate the maximum acceleration and where it occurs.

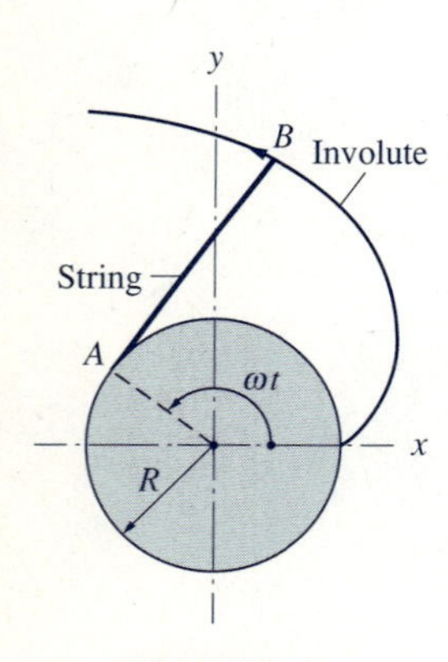

Fig. P12.12

12.12 When a taut string is unwound from a stationary cylinder, the end B of the string generates a curve known as the *involute* of a circle. If the string is unwound at the constant angular speed ω, the equation of the involute is

$$x = R \cos \omega t + R\omega t \sin \omega t \qquad y = R \sin \omega t - R\omega t \cos \omega t$$

where R is the radius of the cylinder. (a) Find the speed of B as a function of time, and show that the velocity vector is always perpendicular to the string. (b) Calculate the angle between the velocity and acceleration vectors of B when $\omega t = \pi/2$.

12.13 When a wheel of radius R rolls with a constant angular velocity ω, the point B on the circumference of the wheel traces out a curve known as a *cycloid*, the equation of which is

$$x = R(\omega t - \sin \omega t) \qquad y = R(1 - \cos \omega t)$$

(a) Show that the velocity vector of B is always perpendicular to $\overline{BC}$, and show that its magnitude is proportional to $\overline{BC}$. (b) Show that the acceleration vector of B is directed along $\overline{BG}$, and show that its magnitude is constant.

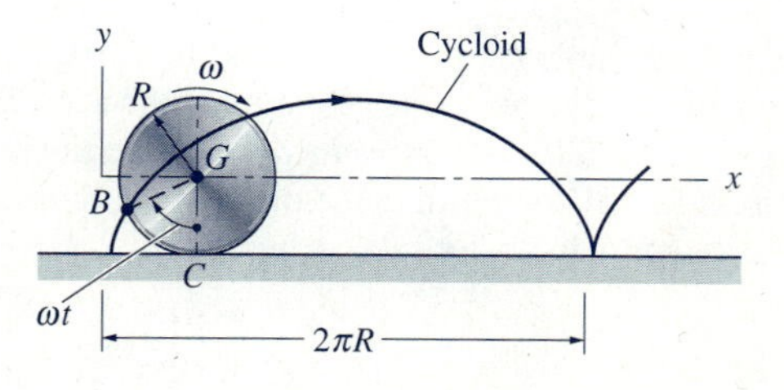

Fig. P12.13

12.14 When a particle moves along the helix shown, the components of its position vector are

$$x = R\cos\omega t \qquad y = R\sin\omega t \qquad z = -(h/2\pi)\omega t$$

where ω is constant. Show that the velocity and acceleration have constant magnitudes, and compute their values if $R = 1.2$ m, $h = 0.75$ m, and $\omega = 4\pi$ rad/s.

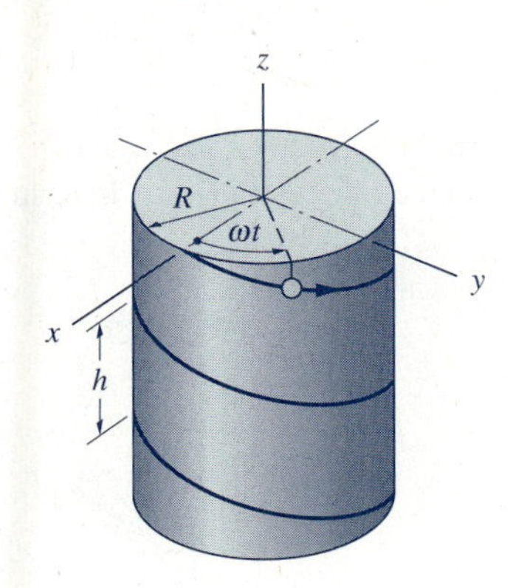

Fig. P12.14

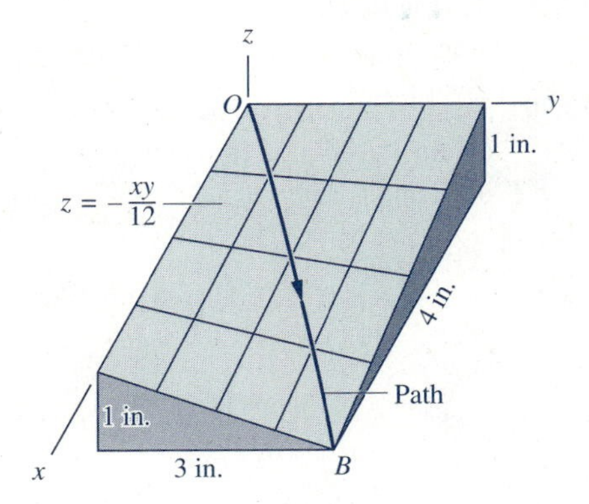

Fig. P12.15

12.15 Path OB of a particle lies on the hyperbolic paraboloid shown. The description of motion is

$$x = \frac{4}{5}v_0 t \qquad y = \frac{3}{5}v_0 t \qquad z = -\frac{1}{25}v_0^2 t^2$$

where the coordinates are measured in inches, and v_0 is a constant. Determine (a) the velocity and acceleration when the particle is at O; (b) the velocity and acceleration when the particle is at B; and (c) the angle between the path and the xy-plane at B.

12.16 The spatial motion of a particle is described by

$$x = 3t^2 + 4t \qquad y = -4t^2 + 3t \qquad z = -6t + 9$$

where the coordinates are measured in feet and the time in seconds. (a) Determine the velocity and acceleration vectors of the particle as functions of time. (b) Verify that the particle is undergoing plane motion (the motion is not in a coordinate plane) by showing that the unit vector perpendicular to the plane formed by **v** and **a** is constant.

12.17 The three-dimensional motion of a point is described by

$$x = R\cos\omega t \qquad y = R\sin\omega t \qquad z = \frac{R}{2}\sin 2\omega t$$

where R and ω are constants. (a) Sketch the path of the point. (b) Calculate the maximum speed and maximum acceleration of the point. (c) Find the angle between **a** and the tangent to the path when the acceleration has its maximum value.

12.18 A rocket engine takes 8 seconds after firing to reach its full thrust. Assuming that the rocket was fired at $t = 0$, use the following time-elevation data to estimate the velocity and acceleration (a) at $t = 0$; and (b) at $t = 8$ s.

time (s)	0	1	2	3	4	5	6	7	8
elevation (ft)	0	4	31	102	233	439	731	1116	1600

12.19 For the mechanism shown, determine (a) the velocity of slider C in terms of θ and $\dot{\theta}$; and (b) the acceleration of C in terms of θ, $\dot{\theta}$, and $\ddot{\theta}$.

12.20 The pin attached to the sliding collar A engages the slot in bar OB. Determine (a) the speed of A in terms of θ and $\dot{\theta}$; and (b) the acceleration of A in terms of θ, and $\dot{\theta}$, and $\ddot{\theta}$.

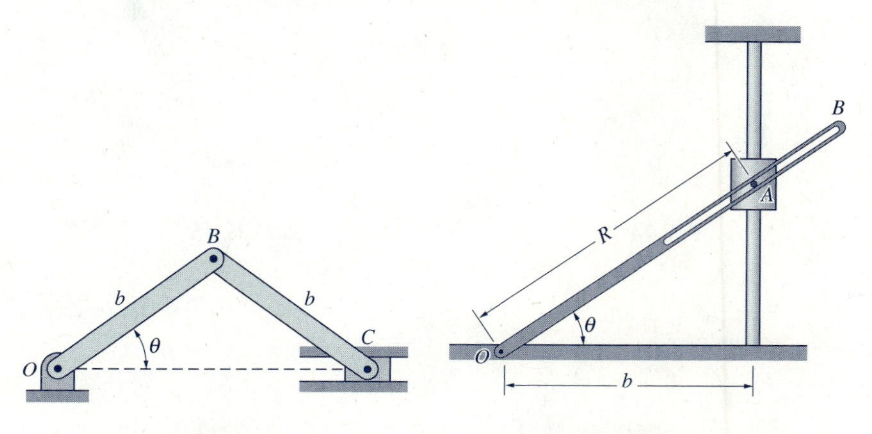

Fig. P12.19 **Fig. P12.20, P12.21**

12.21 Rod OB oscillates about O, its angular position being $\theta = (\pi/4)\sin 4\pi t$, where θ is in radians and t in seconds. Find the velocity and acceleration of pin A at $t = 0.05$ s by numerical differentiation, using (a) $\Delta t = 0.005$ s; and (b) $\Delta t = 0.001$ s. Compare your answers with the analytical solutions $v = 9.96b$ and $a = -11.81b$, obtainable from the results of Prob. 12.20.

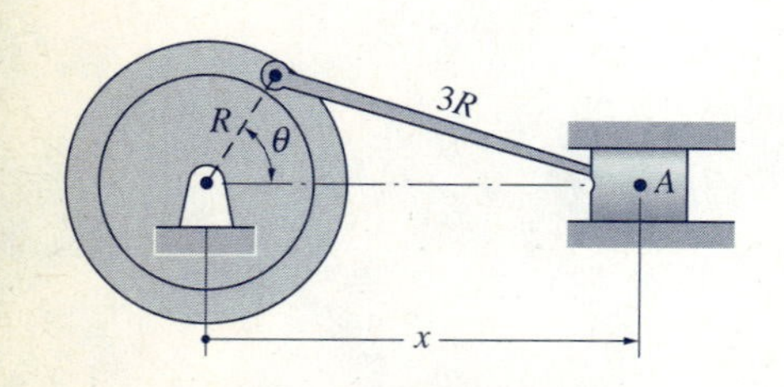

Fig. P12.22, P12.23

12.22 The position coordinate of piston A can be shown to be related to the crank angle θ of the flywheel by

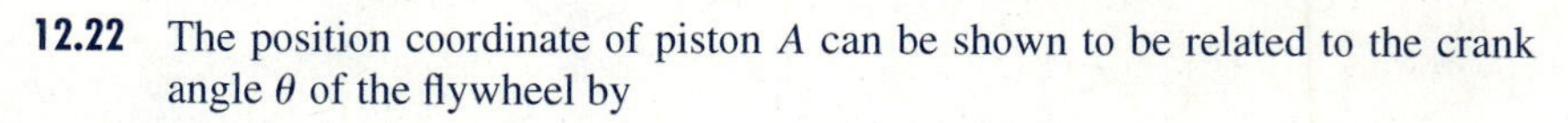

$$x = R\left(\cos\theta + \sqrt{9 - \sin^2\theta}\right)$$

The flywheel rotates at the constant angular speed ω, which gives $\theta = \omega t$. (a) Derive the expression for the velocity of the piston as a function of θ. (b) Use a plot of the results of Part (a) to estimate the maximum speed and maximum acceleration of the piston.

12.23 The piston of the mechanism described in Prob. 12.22 reaches its maximum speed when $\theta = 1.27$ rad and its maximum acceleration when $\theta = 0$. Use numerical differentiation to compute both of these maxima to at least 3 significant digits.

12.24 The cam rotates with a constant angular speed $\omega = 1200$ rev/min. Using the cam profile given below, (a) plot the speed of the follower A during a 30° rotation of the cam from the position shown; (b) estimate the acceleration of the follower when the cam is in the position shown.

θ (deg)	0	5	10	15	20	25	30
r (mm)	55.0	54.8	54.2	53.1	51.7	49.9	47.6

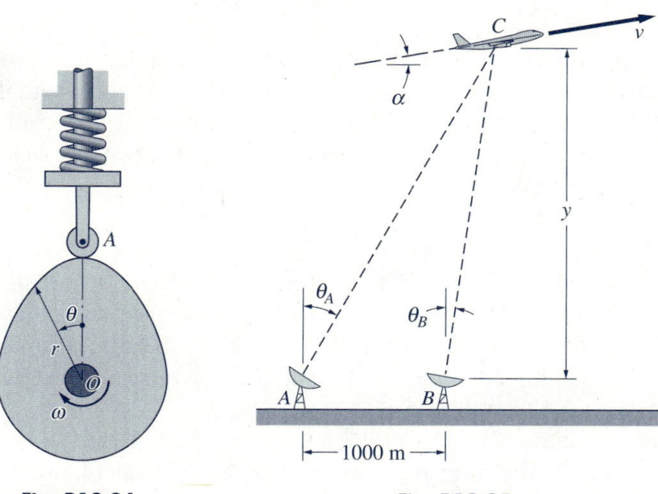

Fig. P12.24 **Fig. P12.25**

***12.25** The plane C is being tracked by radar stations A and B. At the instant shown, the triangle ABC lies in the vertical plane, and the radar readings are $\theta_A = 30°$, $\theta_B = 22°$, $\dot{\theta}_A = 0.026$ rad/s, and $\dot{\theta}_B = 0.032$ rad/s. Determine (a) the altitude y; (b) the speed v; and (c) the climb angle α of the plane at this instant.

12.3 Equations of Motion

When rectangular coordinates are used, the scalar form of Newton's second law $\mathbf{F} = m\mathbf{a}$ becomes

$$F_x = m\ddot{x} \qquad F_y = m\ddot{y} \qquad F_z = m\ddot{z} \tag{12.11}$$

where F_x, F_y, and F_z are the rectangular components of the resultant force $\mathbf{F}$ that acts on the particle. Equations (12.11) are known as the *equations of motion* of the particle.

If the motion of the particle is known, Eqs. (12.11) can be used to compute the forces that cause the motion. In other words, if $x(t)$, $y(t)$, and $z(t)$ are known functions of time, the second time derivatives of these functions will determine the force components F_x, F_y, and F_z, as indicated by Eqs. (12.11).

A more common problem in dynamics is to determine the motion that results from given forces. Since in general the forces are functions of time, position, and velocity of the particle, determination of the motion is equivalent to solving the following differential equations.

$$\begin{aligned} F_x(x, y, z, \dot{x}, \dot{y}, \dot{z}; t) &= m\ddot{x} \\ F_y(x, y, z, \dot{x}, \dot{y}, \dot{z}; t) &= m\ddot{y} \\ F_z(x, y, z, \dot{x}, \dot{y}, \dot{z}; t) &= m\ddot{z} \end{aligned} \tag{12.12}$$

To say that the forces are "given" means that F_x, F_y, and F_z are known expressions of the arguments listed in Eqs. (12.12). For example, $F_x = 20x - 15t$ lb is a given force component. However, even if the forces are given, it may not be possible to evaluate them at a given value of time unless x, y, and z are also known at that time. In other words, although the forces may be given, they can in general be evaluated only after Eqs. (12.12) have been solved, that is, after the motion has been determined.

Because the force components may be nonlinear functions of their arguments, Eqs. (12.12) will generally constitute a coupled set of nonlinear ordinary differential equations. Since only a small number of such problems can be analyzed by analytical methods, numerical methods have become the standard means of solving equations of motion. Several numerical techniques are described in Arts. 12.8 and 12.9.

Thus far we have seen that problems may be divided into two categories: (1) determining the forces from known motion and (2) calculating the motion from given forces. However, many problems of practical importance are of the mixed type, where partial information is known about both the motion and the forces. For example, in a particular problem the applied forces may be given, but the reactions may depend on the kinematic constraints imposed on the motion.

12.4 Force-Mass-Acceleration Method; Mass-Acceleration Diagrams

The *force-mass-acceleration* (FMA) method is a standard tool for obtaining the equations of motion. The method employs two diagrams that graphically represent Newton's second law: $\mathbf{F} = m\mathbf{a}$. The first of these is the *free-body diagram* (FBD), which shows all the forces acting on the particle; the resultant force vector represents the left side of the equation. The second diagram, which we call the *mass-acceleration diagram* (MAD) shows the *inertia vector* $m\mathbf{a}$ of the particle, i.e., the right side of the equation. According to Newton's second law, these two diagrams are statically equivalent; that is, the resultant $\mathbf{F}$ of the forces on the FBD is equal to $m\mathbf{a}$.

The FMA method is illustrated Fig. 12.4(a). The FBD displays the forces $\mathbf{F}_1, \mathbf{F}_2, \mathbf{F}_3, \ldots$ that act on the particle of mass m, whereas the MAD shows the inertia vector. The equal sign between the two diagrams is used to indicate that the two systems are equivalent. It is now relatively easy to write the equations of motion from the conditions of equivalence. If the rectangular components of the inertia vector are employed, the FMA method takes the form shown in

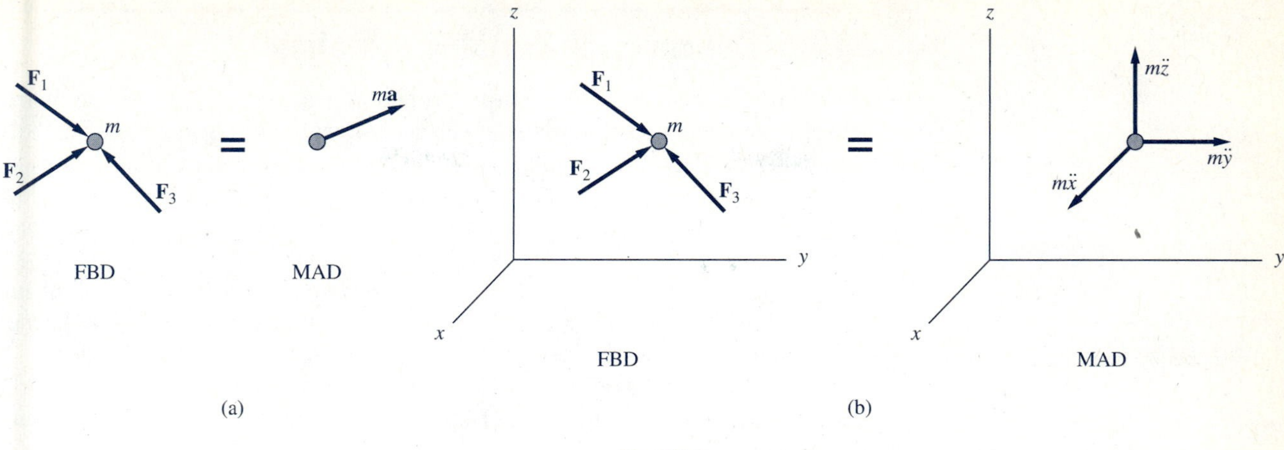

Fig. 12.4

Fig. 12.4(b); applications employing curvilinear coordinates are discussed in the next chapter.

The free-body diagram is as important in dynamics as it is in statics. It identifies all the forces that act on the particle in a clear and concise manner, it defines the notation used for unknown quantities, and it displays the known quantities. The mass-acceleration diagram serves a similar purpose. It also defines the notation for the unknowns, and it shows the known magnitudes and directions. But perhaps the greatest benefit of the MAD is that it focuses our attention on the kinematics required to describe the inertia vector. After all, it is kinematics that enables us to decide which components of the acceleration vector are known beforehand and which components are unknown.

In summary, the FMA method consists of the following steps.

Step 1: Draw the free-body diagram (FBD) of the particle that shows all forces acting on the particle.
Step 2: Use kinematics to analyze the acceleration of the particle.
Step 3: Sketch the mass-acceleration diagram (MAD) for the particle that displays the components of the inertia vector $m\mathbf{a}$, utilizing the results of Step 2.
Step 4: Referring to the FBD and MAD, relate the forces to the acceleration using Newton's second law: $\mathbf{F} = m\mathbf{a}$.

It should be mentioned that the FBD-MAD method is a variation of what is referred to as d'Alembert's principle (formulated by d'Alembert in 1743). Following this principle, the inertia vector is reversed ($-m\mathbf{a}$ is often called the *inertia force*) and placed on the free-body diagram. This modified FBD is interpreted as the graphical equivalent of $\mathbf{F} - m\mathbf{a} = \mathbf{0}$, and it can be analyzed using the principles of statics (in fact, the expression "dynamic equlibrium" is used to describe the equation $\mathbf{F} - m\mathbf{a} = \mathbf{0}$). Although d'Alembert's principle is a powerful technique that is useful in more advanced applications (see *Methods of Analytical Dynamics,* L. Meirovitch, McGraw-Hill, 1970, p. 65), we do not use it here because it offers no advantage at the level of this text.

12.5 Dynamics of Rectilinear Motion

To illustrate the analysis of rectilinear motion, we consider a block that is free to slide along a smooth horizontal surface, as shown in Fig. 12.5(a). The horizontal force P is assumed to be a known function of time t, position coordinate x, and velocity v of the block. If several horizontal forces act on the block simultaneously (one of the forces would be friction if the contact surfaces are rough), P should be interpreted as the resultant of these forces.

Note that the motion of the block is rectilinear; its path is a straight line parallel to the horizontal surface. Consequently, the velocity $v = \dot{x}$ and acceleration $a = \dot{v} = \ddot{x}$ are also parallel to the horizontal surface at all times.

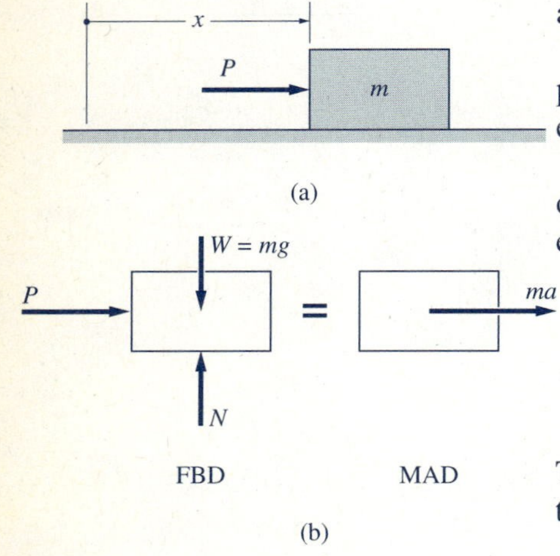

Fig. 12.5

The free-body diagram (FBD) and the mass-acceleration diagram (MAD) of the block are shown in Fig. 12.5(b). These diagrams yield the following equations of motion:

$$\Sigma F_x = ma_x \qquad \xrightarrow{+} \qquad P = ma$$

$$\Sigma F_y = ma_y \qquad +\uparrow \qquad N - mg = 0$$

The second equation simply establishes that $N = mg$, whereas the first equation describes the motion of the block.

We now focus our attention on solving the following equation for the velocity v and position x.

$$\boxed{P = ma} \tag{12.13}$$

In the general case, in which P is a function of t, x, and v, it is usually impossible to obtain an analytical solution. Therefore, the problem must be analyzed numerically by one of the methods described in Arts. 12.8 and 12.9. However, if P depends on only a single variable (t, x, or v), the problem can be solved by straightforward integration, as explained in the following.

Case 1: P is a function of time only: $P = P(t)$

Substituting $a = dv/dt$ in Eq. (12.13), we get

$$P(t) = m\frac{dv}{dt}$$

or

$$\boxed{dv = \frac{1}{m}P(t)\,dt} \tag{12.14}$$

Integration with respect to t yields the velocity as a function of time:

$$v = \frac{1}{m}\int P(t)\,dt + C_1 \tag{12.15}$$

where C_1 is a constant of integration. Replacing v by dx/dt and integrating again, we get the position as a function of time:

$$x = \frac{1}{m}\int\int [P(t)\,dt]\,dt + C_1 t + C_2 \qquad (12.16)$$

To evaluate the constants of integration C_1 and C_2, we must know two conditions on the motion, e.g., the initial conditions $x(0)$ and $v(0)$.

Case 2: *P* is a function of position only: $P = P(x)$

In this case, we replace a in Eq. (12.13) by Eq. (12.10): $a = v\,dv/dx$, which gives

$$P(x) = mv\frac{dv}{dx}$$

This equation can be rearranged so that the variables are separated; i.e., x and v appear on different sides of the equation:

$$\boxed{v\,dv = \frac{1}{m}P(x)\,dx} \qquad (12.17)$$

Both sides of Eq. (12.17) can be integrated, yielding

$$\frac{v^2}{2} = \frac{1}{m}\int P(x)\,dx + C_3 \qquad (12.18)$$

where C_3 is a constant of integration. Therefore, the velocity is

$$v(x) = \sqrt{\frac{2}{m}\int P(x)\,dx + 2C_3} \qquad (12.19)$$

We could now replace v by dx/dt in Eq. (12.19), separate the variables x and t, and integrate again to obtain $x(t)$. Unfortunately, the integration may be difficult owing to the presence of the square root.

Case 3: *P* is a function of velocity only: $P = P(v)$

One approach is to start with Eq. (12.17).

$$v\,dv = \frac{1}{m}P(v)\,dx$$

and separate the variables x and t, obtaining

$$\boxed{dx = m\frac{v\,dv}{P(v)}} \qquad (12.20)$$

Integration now yields position as a function of velocity:

$$x(v) = m \int \frac{v\,dv}{P(v)} + C_4 \tag{12.21}$$

A second approach begins with Eq. (12.14):

$$dv = \frac{1}{m} P(v)\,dt$$

After separating the variables, we get

$$\boxed{dt = \frac{m\,dv}{P(v)}} \tag{12.22}$$

which, upon integration, leads to the following relationship between time and velocity

$$t(v) = m \int \frac{dv}{P(v)} + C_5 \tag{12.23}$$

Sample Problem 12.5

The 300-N block A in Fig. (a) is at rest on the rough horizontal plane before the force P is applied at $t = 0$. Find the velocity and position of the block when $t = 5$ s. The magnitude of P is $80t$ N, where t is the time in seconds, and its direction is constant. The coefficients of static and kinetic friction are $\mu_s = 0.4$ and $\mu_k = 0.2$, respectively.

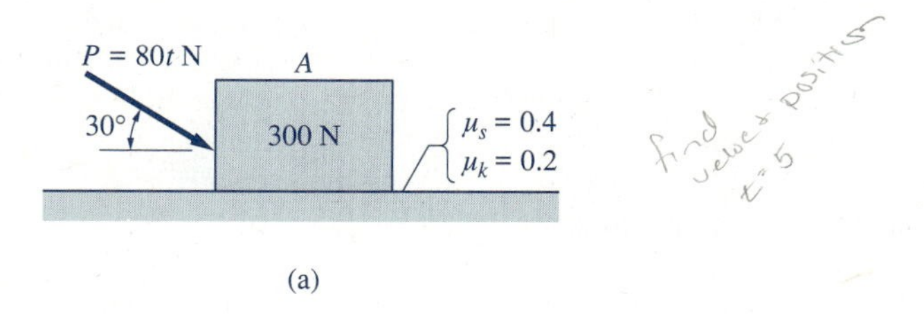

(a)

Solution

The FBD of the block is shown in Fig. (b), where N_A and F_A are the normal and friction forces exerted on the block by the rough plane. Figure (b) also shows the MAD. Because the motion is rectilinear, $a_y = 0$, and the magnitude of the inertia vector is $ma_x = ma$.

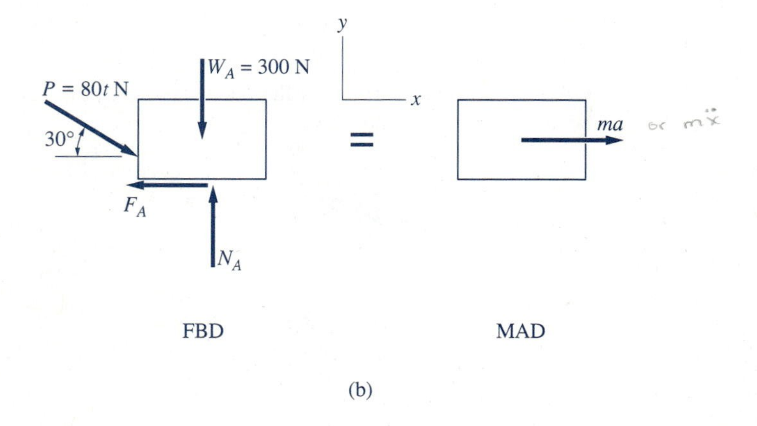

(b)

The friction force F_A is zero at $t = 0$, because $P = 0$ at that time. As t increases, the block will remain at rest until the state of impending motion is reached, that is, until the static friction force reaches its limiting value $F_A = \mu_s N_A = 0.4N_A$. Our first task is, therefore, to find the time when motion begins. From the FBD in Fig. (b) we obtain the following equations of static equilibrium of the block.

$$\Sigma F_y = 0 \quad +\uparrow \quad -80t \sin 30^\circ + N_A - 300 = 0$$

$$N_A = 300 + 40t \text{ N} \tag{a}$$

$$\Sigma F_x = 0 \quad \xrightarrow{+} \quad 80t \cos 30^\circ - F_A = 0$$

$$F_A = 69.28t \text{ N} \tag{b}$$

On substituting $F_A = \mu_s N_A = 0.4N_A$ in Eq. (b) and using Eq. (a) to eliminate N_A, we obtain

$$0.4(300 + 40t) = 69.28t$$

$$t = 2.252 \text{ s} \tag{c}$$

This is the time when the block starts to move.

When $t > 2.252$ s, the acceleration and the velocity are both directed to the right. The kinetic friction force $F_A = \mu_k N_A = 0.2N_A$ is directed to the left, that is, opposite to the velocity. Observe that Eq. (a), which is a static equilibrium equation, remains valid during the motion. Since there is no acceleration in the y direction, $\Sigma F_y = 0$ is true whether the block moves or remains at rest. Using Eq. (a), the kinetic friction force is

$$F_A = 0.2N_A = 0.2(300 + 40t) = 60 + 8.0t \text{ N}$$

However, we must replace Eq. (b) by Newton's second law of motion. Referring to the FBD and MAD, we find that

$$\Sigma F_x = ma_x \qquad \xrightarrow{+} \qquad 80t \cos 30° - (60 + 8.0t) = \frac{300}{9.81}a$$

from which the acceleration is

$$a = 2.004t - 1.962 \text{ m/s}^2 \tag{d}$$

The velocity and position coordinate of the block can now be found by direct integration of the acceleration, as follows.

$$v = \int a\, dt = \int (2.004t - 1.962)\, dt$$

which gives

$$v = 1.002t^2 - 1.962t + C_1 \tag{e}$$

and

$$x = \int v\, dt = \int (1.002t^2 - 1.962t + C_1)\, dt$$

from which we obtain

$$x = 0.3340t^3 - 0.9810t^2 + C_1 t + C_2 \tag{f}$$

In Eqs. (e) and (f), C_1 and C_2 are constants of integration to be found from the initial conditions. The initial velocity was given to be zero. However, we are still free to choose the origin of the coordinate axis, i.e., the value of the initial position coordinate x. The most convenient choice is $x = 0$. Therefore, the initial conditions are

$$v = 0 \quad \text{and} \quad x = 0 \quad \text{when } t = 2.252 \text{ s} \tag{g}$$

Substituting the initial condition for v into Eq. (e) yields

$$0 = 1.002(2.252)^2 - 1.962(2.252) + C_1$$
$$C_1 = -0.6632 \text{ m/s}$$

Substituting this result and the initial condition for x into Eq. (f), we find

$$0 = 0.3340(2.252)^3 - 0.9810(2.252)^2 - 0.6632(2.252) + C_2$$
$$C_2 = 2.654 \text{ m}$$

When the values for C_1 and C_2 are substituted into Eqs. (e) and (f), the velocity and position coordinate become

$$v = 1.002t^2 - 1.962t - 0.6632 \text{ m/s} \quad \text{(h)}$$
$$x = 0.3340t^3 - 0.9810t^2 - 0.6632t + 2.654 \text{ m} \quad \text{(i)}$$

When $t = 5$ s, Eqs. (h) and (i) yield

$$v = 14.58 \text{ m/s} \quad \text{and} \quad x = 16.56 \text{ m} \qquad \textit{Answer}$$

We could easily show that 16.56 m is also the distance traveled by the block during the time interval $t = 2.252$ s to $t = 5$ s. If we were to set $v = 0$ in Eq. (h) and solve for t, we would find that there is no root, other than $t = 2.252$ s, that lies within the time interval. Since the block does not come to rest once motion has begun, the direction of motion cannot change. Therefore, the block travels directly from $x = 0$ to $x = 16.56$ m without reversing the direction of its motion.

Sample Problem 12.6

Figure (a) shows a block that slides along a smooth, horizontal plane. The position coordinate x is measured from the unstretched position of the linear spring, which has a stiffness (force that causes unit deformation) equal to k. (1) Derive the equation of motion for the block. (2) Determine the velocity of the block as a function of x. (3) If the velocity of the block is v_0 when $x = 0$, find the position of the block when it first comes to rest.

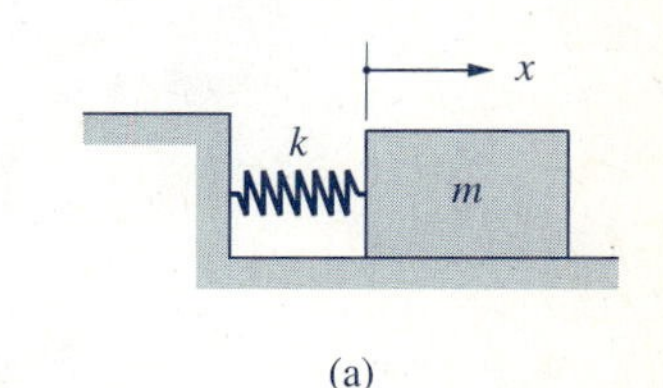

(a)

Solution

Part 1

The FBD of the block for an arbitrary value of x is shown in Fig. (b), where N is the normal force exerted by the smooth plane and $P_s = kx$ is the force caused by the spring. Figure (b) also shows the MAD. Because the motion occurs only in the x direction, we have $a_y = 0$, and the magnitude of the inertia vector is $ma_x = ma$. Referring to the FBD and MAD, the equation of motion is

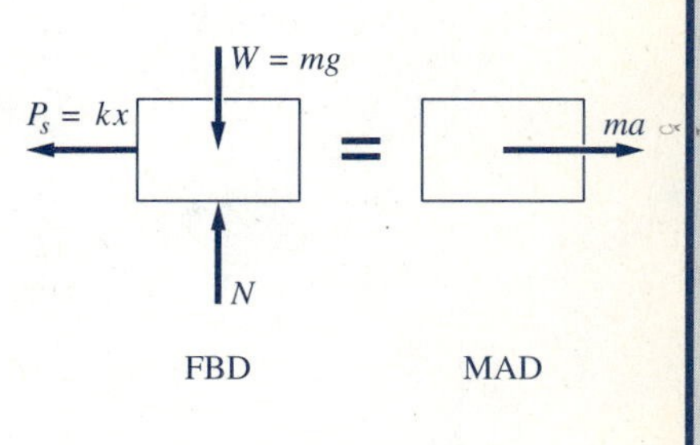

(b)

$$\Sigma F_x = ma_x \qquad \xrightarrow{+} \qquad -kx = ma$$

Therefore the equation of motion may be written as

$$a = -\frac{k}{m}x \qquad \textit{Answer} \text{ (a)}$$

Part 2

We can determine the velocity as a function of position by letting x be the independent variable. Using $a = v\,dv/dx$ from Eq. (12.10), Eq. (a) becomes

$$v\frac{dv}{dx} = -\frac{k}{m}x$$

Separating the variables x and v by rearranging the terms, we find

$$v\,dv = -\frac{k}{m}x\,dx \qquad \text{(b)}$$

Integration of Eq. (b) yields

$$\frac{v^2}{2} = -\frac{kx^2}{2m} + C \qquad \text{(c)}$$

The constant of integration C can be evaluated from the initial condition: $v = v_0$ when $x = 0$, yielding $C = v_0^2/2$. Therefore, the velocity can be expressed as

$$v = \sqrt{(-k/m)x^2 + v_0^2} \qquad \textit{Answer} \quad \text{(d)}$$

Part 3

The position of the block when it comes to rest is found by setting $v = 0$ in Eq. (d), the result being

$$x = v_0\sqrt{m/k} \qquad \textit{Answer}$$

Sample Problem 12.7

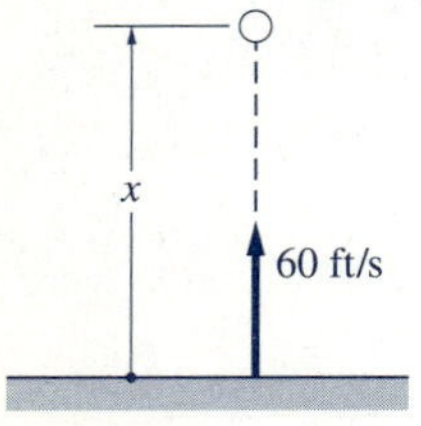

(a)

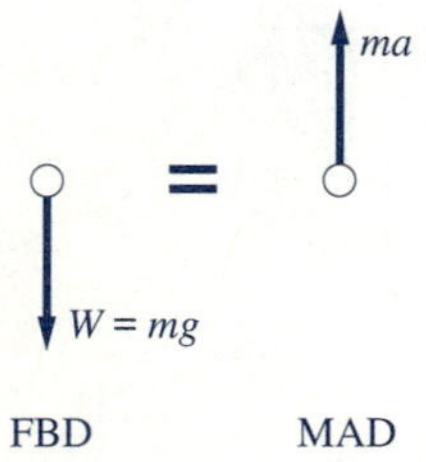

(b)

The ball shown in Fig. (a) weighs 4.8 oz and is thrown upward with an initial velocity of 60 ft/s. Calculate the maximum height reached by the ball if (1) air resistance is negligible; and (2) the air gives rise to a resisting force F_D, known as *aerodynamic drag*, that opposes the velocity. Assume that $F_D = cv^2$, where $c = 4 \times 10^{-5}$ lb · s²/ft².

Solution

Part 1

When air resistance is neglected, the only force acting on the ball during flight is its weight W, shown in the FBD in Fig. (b). Because the motion is rectilinear, the magnitude of the inertia vector is $ma_x = ma$, as shown in the MAD in Fig. (b). Applying Newton's second law, we have

$$\Sigma F_x = ma_x \qquad +\uparrow \quad -mg = ma$$

from which we find that the acceleration is

$$a = -g = -32.2 \text{ ft/s}^2 \qquad \text{(a)}$$

The acceleration, which is a constant because the FBD does not vary with time, may be integrated with respect to time to obtain the position and velocity as follows:

$$v = \int a\,dt = \int (-32.2)\,dt = -32.2t + C_1 \text{ ft/s} \tag{b}$$

$$x = \int v\,dt = \int (-32.2t + C_1)\,dt = -16.1t^2 + C_1 t + C_2 \text{ ft} \tag{c}$$

The constants of integration, C_1 and C_2, are evaluated by applying the initial conditions $x = 0$ and $v = 60$ ft/s when $t = 0$, the results being $C_1 = 60$ ft/s and $C_2 = 0$. Therefore, the velocity and position are given by

$$v = -32.2t + 60 \text{ ft/s} \tag{d}$$

$$x = -16.1t^2 + 60t \text{ ft} \tag{e}$$

The ball reaches its maximum height when $dx/dt = 0$, that is, when $v = 0$. Letting $v = 0$ in Eq. (d), we obtain

$$0 = -32.2t + 60 \qquad \text{or} \qquad t = 1.863 \text{ s}$$

Substituting this value of time into Eq. (e), the maximum height reached by the ball in the absence of air resistance is

$$x_{max} = -16.1(1.863)^2 + 60(1.863) = 55.9 \text{ ft} \qquad \textit{Answer}$$

In this case the acceleration, and thus the velocity and position, are independent of the weight of the ball.

Part 2

When aerodynamic drag is considered, the FBD and MAD of the ball during its upward flight are as shown in Fig. (c). Observe that the drag force F_D, which always opposes the velocity, acts downward because the positive direction for v is upward (the same as the positive direction for x). From Newton's second law, we obtain the following equation of motion.

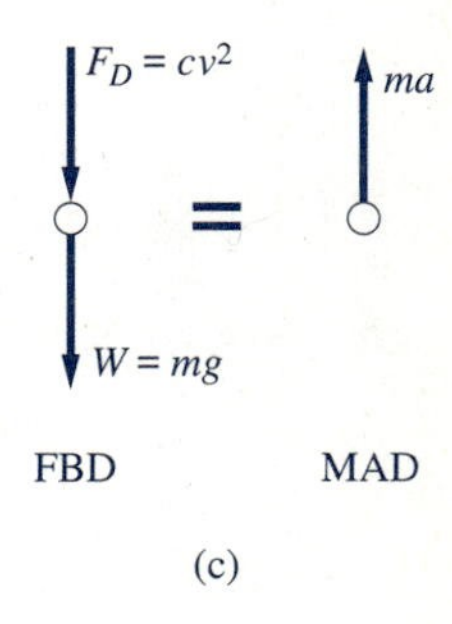

(c)

$$\Sigma F_x = ma_x \qquad +\uparrow \qquad -mg - cv^2 = ma \tag{f}$$

The complete solution (x and v as functions of t) of Eq. (f) is best computed by one of the numerical methods described in Art. 12.8. However, it is possible to derive the velocity as a function of position by direct integration if the independent variable is changed from t to x. Substituting $a = v\,dv/dx$ from Eq. (12.10), Eq. (f) becomes

$$-mg - cv^2 = m\frac{dv}{dx}v$$

in which the variables x and v may be separated as follows.

$$dx = -\frac{mv\,dv}{mg + cv^2} \tag{g}$$

Integrating both sides of this equation (using a table of integrals if necessary), we obtain

$$x = -\frac{m}{2c}\ln(mg + cv^2) + C_3 \tag{h}$$

where C_3 is a constant of integration. Substituting the numerical values $mg = W = 4.8/16 = 0.3$ lb, $m = 0.3/32.2 = 932 \times 10^{-5}$ slugs, and $c = 4 \times 10^{-5}$ lb $\cdot$ s^2/ft^2, Eq. (h) becomes

$$x = -116.5\ \ln[0.3 + (4 \times 10^{-5})v^2] + C_3 \text{ m} \tag{i}$$

Applying the initial condition, $v = 60$ m/s when $x = 0$, we find that $C_3 = -94.6$ m. Therefore, the solution for x is

$$x = -116.5\ \ln[0.3 + (4 \times 10^{-5})v^2] - 94.6 \text{ m} \tag{j}$$

Since the maximum height of the ball occurs when $v = 0$, we have

$$x_{max} = -116.5\ \ln 0.3 - 94.6 = 45.7 \text{ m} \qquad \textit{Answer} \tag{k}$$

Of course, this value is smaller than the maximum height obtained in Part 1, where air resistance was neglected.

To summarize, we have used the FMA approach to determine the equations of motion for both parts of this sample problem. When air resistance was neglected in Part 1, the acceleration was simply $-g$, independent of the weight of the ball. The velocity and position could be determined in terms of t by integration. The inclusion of aerodynamic drag in Part 2 introduced the additional term $-cv^2$ into the equation of motion with the result that the acceleration depended on c, v, and W. For this case, the complete solution $x(t)$ and $v(t)$ was not determined (it would be very tedious to do so). However, $x(v)$ was readily obtained, from which we computed the maximum height of the ball.

PROBLEMS

12.26 The pendulum AB is suspended from a cart that has a constant acceleration a to the right. Determine the constant angle θ of the pendulum.

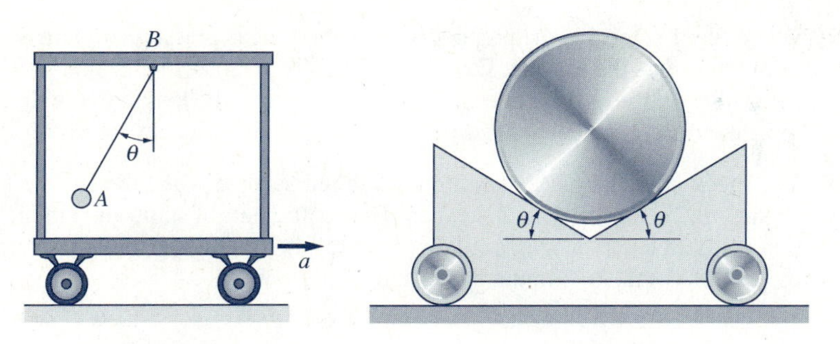

Fig. P12.26 **Fig. P12.27**

12.27 A uniform cylinder is placed in a V-notched cradle. What is the largest horizontal acceleration that the cradle may have without causing the cylinder to climb out of the cradle?

12.28 The cam described in Prob. 12.6 is rotating at the constant angular speed of $\omega = 3000$ rev/min. The force in the spring keeps the cam in contact with the follower A at all times. The radius R of the cam is 3 in., and the follower A weighs 0.5 lb. Calculate the smallest spring force that will maintain the contact when the follower is in its highest position.

12.29 The 0.2 kg collar A is free to slide along the smooth rod OB. A pin attached to the collar slides in the smooth guide slot, confining the motion of the collar to the horizontal direction. In the position $\theta = 30°$, the acceleration of the collar is 12 m/s^2 to the left. For this position compute (a) the force exerted on the collar by the rod OB; and (b) the contact force between the pin and the guide slot.

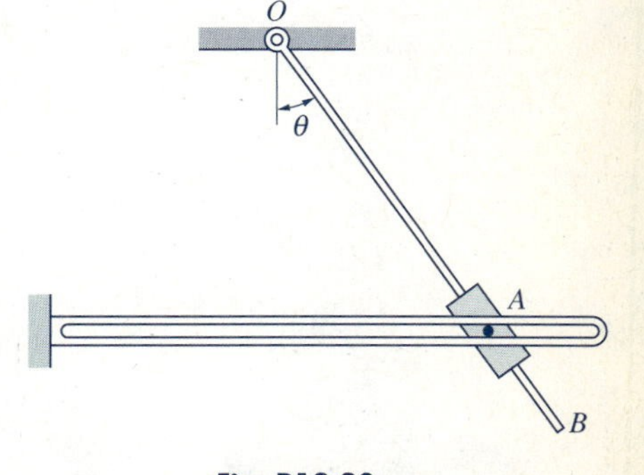

Fig. P12.29

12.30 A car is traveling at the speed v_0 on a level road. (a) Derive an expression for the minimum stopping distance if the coefficient of static friction between the road and the tires is μ. (b) Calculate this distance for $v_0 = 60$ mi/h and $\mu = 0.6$.

12.31 A 2000-kg rocket is launched vertically from the surface of the earth. The engine shuts off after providing a constant thrust of 60 kN for the first 20 seconds. Neglect the reduction in the mass of the rocket due to the burning of the fuel and the variation of the gravitational acceleration with altitude. Calculate (a) the speed and altitude of the rocket at the end of the powered portion of the flight; (b) the time required (measured from the time of launch) for the rocket to reach its maximum altitude; and (c) the maximum altitude.

12.32 A 3000-lb rocket sled is propelled along a straight test track. The rocket engine fires for 4 seconds producing a thrust of $F = 1200$ lb, and then shuts down. Assuming that the sled starts from rest and that the coefficient of kinetic friction is 0.05, determine (a) the maximum speed reached by the sled; and (b) the distance traveled by the sled before stopping.

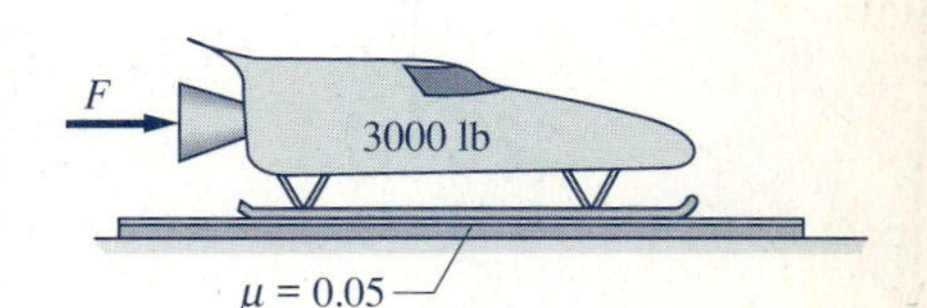

Fig. P12.32, P12.33

12.33 Solve Prob. 12.32 if the rocket thrust decays exponentially according to $F(t) = 1000\, e^{-0.2t}$, where $F(t)$ is in pounds and t is the time in seconds measured from the instant the rocket engine was fired.

12.34 A 0.05-kg bullet is fired from a rifle at $t = 0$. During the time that the bullet is inside the barrel, it is propelled by a force that can be approximated by

$$F = F_0\left(\sqrt{\frac{t_0}{t}} - \mu\right)$$

where the first term is the propulsive force of the gas pressure, and the second term represents the friction. For $F_0 = 75$ kN, $t_0 = 80 \times 10^{-6}$ s, and $\mu = 0.2$, determine (a) the optimum length of the rifle barrel (the length that will maximize the muzzle velocity); and (b) the corresponding muzzle velocity.

12.35 The block of weight W is resting on a rough surface when the force $F(t) = W \sin(\pi t/2)$ is applied for 2 seconds. The coefficients of static and kinetic friction are 0.25 and 0.20, respectively. Find (a) the time when the block starts to move; and (b) the maximum speed and the time when it occurs.

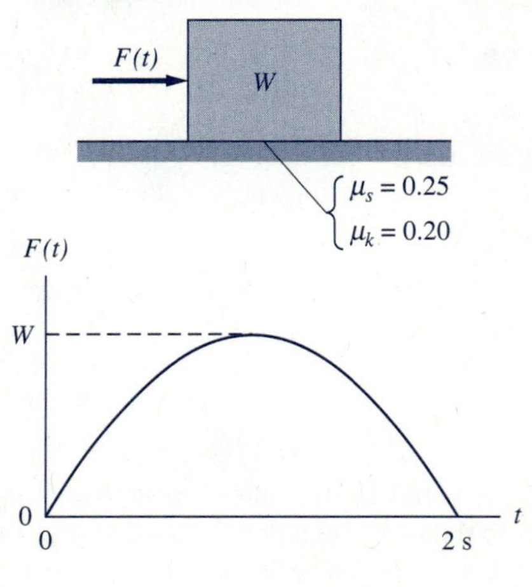

Fig. P12.35

12.36 A ship of mass m is sailing with speed v_0 on a straight course when the engines are stopped. Assume that the drag force is $F_D = kv^2$, where k is a constant and v is the speed of the ship. Determine the distance the ship will coast before the speed is reduced to $0.5v_0$.

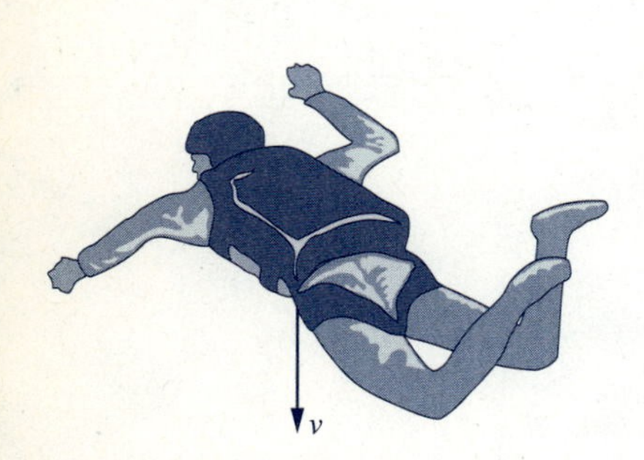

Fig. P12.37, P12.38

12.37 The drag force acting on a 135-lb skydiver in the "spread" position shown can be approximated by $F_D = 0.00436v^2$ where F_D is in pounds and v is in feet per second (from *Scientific and Engineering Problem-Solving with the Computer*, W. R. Bennett, Jr., Prentice-Hall, New York, 1976). Assuming that the skydiver follows a vertical path, determine (a) the terminal velocity; (b) the velocity after a 1000-ft fall.

12.38 Find the time for the skydiver described in Prob. 12.37 to reach a speed of 100 mi/h.

12.39 According to Stoke's Law, the drag force exerted on a sphere moving through a fluid is $F_D = 6\pi\eta Rv$, where R is the radius of the sphere and v is its speed. The constant η is a property of the fluid called the viscosity. A glass sphere (weight density $\gamma_g = 162$ lb/ft^3) with radius $R = 0.10$ in. is released from rest in a container of SAE 10 oil (weight density $\gamma_f = 56$ lb/ft^3 and $\eta = 0.001$ lb · s/ft^2). Calculate (a) the terminal velocity of the sphere; and (b) the distance that the sphere must fall before it reaches 99% of its terminal velocity.

12.40 Determine the time required for the glass sphere in Prob. 12.39 to reach 99% of its terminal velocity.

12.41 A linear spring of stiffness k is to be designed to stop a 20-Mg railroad car traveling at 8 km/h within 300 mm after impact. (a) Find the smallest value of k that will produce the desired result. (b) How long will it take the spring found in Part (a) to stop the car?

Fig. P12.41

12.42 Solve Part (a) of Prob. 12.41 if two springs with identical stiffness k are nested as shown. Note that one spring is 200 mm shorter than the other.

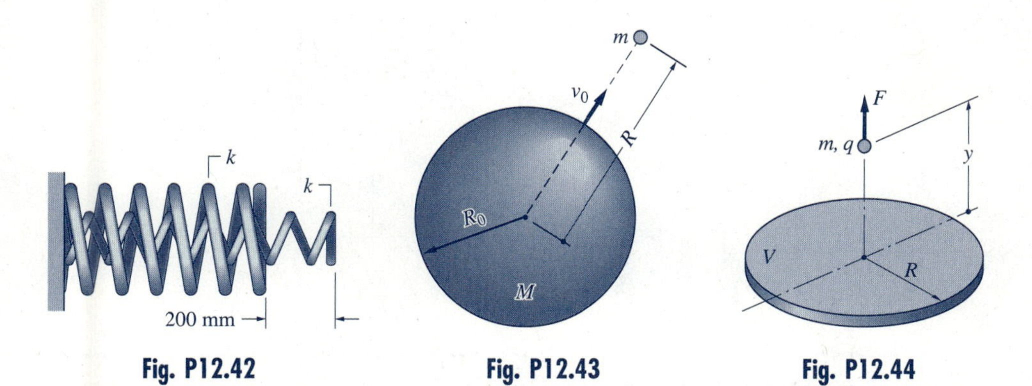

Fig. P12.42 Fig. P12.43 Fig. P12.44

12.43 According to the law of gravitation, Eq. (11.17), the force acting on a particle of mass m located at the distance R from the center of a planet of mass M is $F = KmM/R^2$. (a) Show that this force is equivalent to $F = mg(R_0/R)^2$, where R_0 is the radius of the planet and g is the gravitational acceleration at its surface. (b) If the mass m is launched vertically from the surface of the earth ($R_0 =$ 3960 mi and $g = 32.2$ ft/s^2) with the initial velocity $v_0 = 5000$ ft/s, how high above the surface of the earth will it rise if air resistance is neglected?

12.44 The disk of radius R meters carries a charge with the electrostatic potential V volts. A particle of charge q coulombs lies on the axis of the disk at a distance y meters from the disk. It can be shown that the repulsive force F (in newtons) acting on the particle is

$$F = \frac{Vq}{R}\left(1 - \frac{y}{\sqrt{R^2 + y^2}}\right)$$

If a particle starts from the center of the plate with zero velocity, determine its speed at $y = R$. Neglect the effect of gravity.

12.45 The elevator of weight W rides in a vertical shaft with negligible friction, operated by a cable that passes over the pulley C. The tension in the cable is kept constant at $F = 1.2W$. If the elevator starts from rest in position A, calculate its maximum displacement x_{max}.

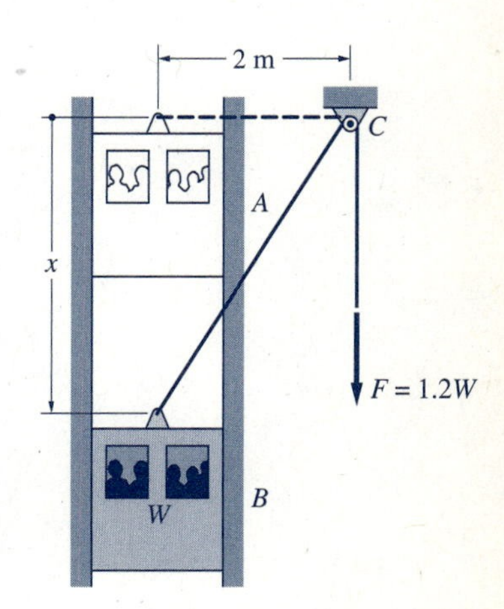

Fig. P12.45

12.46 A catapult is made of two elastic bands, each 5 in. long when unstretched. Each band behaves like a linear spring of stiffness 36 lb/ft. If a 2.5-oz rock is fired from the position shown, determine the speed of the rock when it leaves the catapult at D.

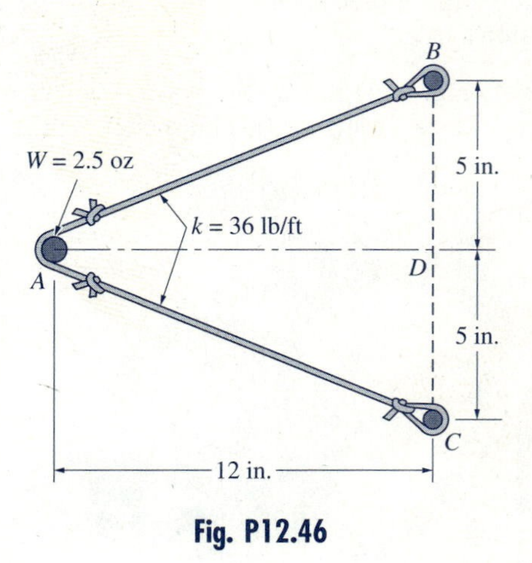

Fig. P12.46

12.6 Superposition of Rectilinear Motions

In the preceding article we applied the force-mass-acceleration method to the rectilinear motion of a particle (path is a straight line). Here we extend the analysis to a special case of curvilinear motion (path is a curve), where the rectangular components of the motion are independent of each other.

Consider the case of curvilinear motion where the resultant force **F** acting on the particle is such that its x component is a function of x, $\dot{x}$, and t only; its y component is a function of y, $\dot{y}$, and t only; and its z component is a function of z, $\dot{z}$, and t only. For this situation, the equations of motion, Eqs. (12.12), have the following form:

$$\begin{aligned} F_x(x,\ \dot{x};\ t) &= m\,\ddot{x} \\ F_y(y,\ \dot{y};\ t) &= m\,\ddot{y} \\ F_z(z,\ \dot{z};\ t) &= m\,\ddot{z} \end{aligned} \qquad (12.24)$$

These differential equations are said to be *uncoupled*, because the motion in any one of the coordinate directions does not affect the motion in the other two directions. The motion can thus be considered to be the vector superposition of three simultaneous rectilinear motions, one along each of the coordinate directions. Therefore, the analysis consists of an extension of the tools that were employed in the preceding article for rectilinear motion: free-body and mass-acceleration diagrams, equations of motion, integration techniques, etc.

If the particle moves in a coordinate plane, say the xy-plane, the motion described in Eqs. (12.24) can be treated as the vector superposition of two rectilinear motions, one in the x-direction, the other in the y-direction ($\ddot{z} = 0$). As illustrated in the following sample problems, this technique is particularly useful when analyzing the motion of projectiles.

Sample Problem 12.8

As shown in Fig. (a), a projectile of weight W is fired from point O at a 30° angle to the horizontal. It lands at A, a distance of 600 ft from O, as measured along the inclined plane. Determine the initial velocity $\mathbf{v}_0$ and the maximum height h reached by the projectile.

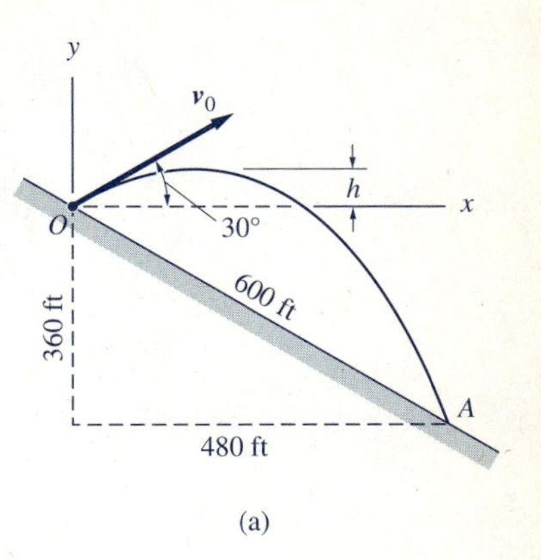

(a)

Solution

The FBD of the projectile contains only its weight W, as shown in Fig. (b). The MAD in Fig. (b) shows the rectangular components of the inertia vector. According to Newton's second law, we have

$$\Sigma F_x = ma_x \qquad a_x = 0$$

$$\Sigma F_y = ma_y \qquad +\uparrow \quad -W = \frac{W}{g}a_y$$

$$a_y = -g$$

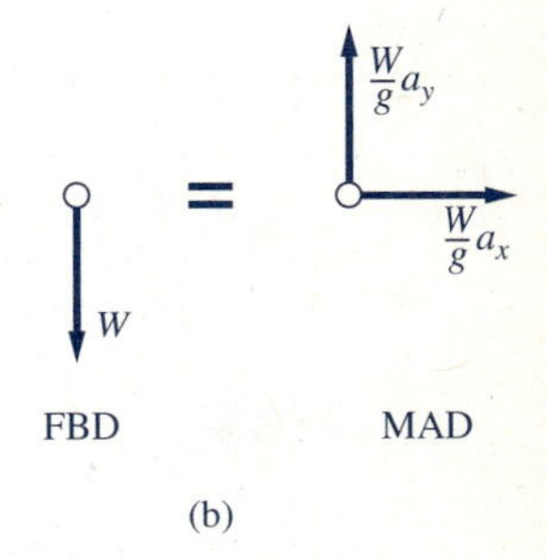

(b)

Since a_x and a_y are constants, the solution to these equations is obtained by integration, as shown in the following.

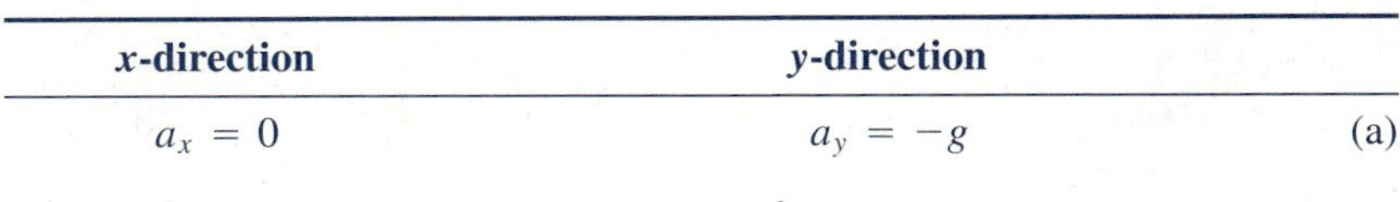

x-direction	*y*-direction	
$a_x = 0$	$a_y = -g$	(a)
$v_x = \int a_x\,dt = C_1$	$v_y = \int a_y\,dt = -gt + C_3$	(b)
$x = \int v_x\,dt = C_1 t + C_2$	$y = \int v_y\,dt = -\frac{gt^2}{2} + C_3 t + C_4$	(c)

The constants of integration C_1, C_2, C_3, and C_4 are to be determined from the boundary conditions specified in the problem statement. Equations (c) are parametric equations (t is the parameter) of a parabola lying in the xy-plane. Therefore, a projectile follows a parabolic path if air resistance is neglected.

To evaluate the four constants C_1 to C_4, we must identify four boundary conditions imposed on the motion. Letting $t = 0$ be the instant of firing, examination of the problem statement and Fig. (a) reveals that the four conditions are as follows:

1. $x = 0$ when $t = 0$, which gives $C_2 = 0$ from Eqs. (c).
2. $y = 0$ when $t = 0$, which gives $C_4 = 0$ from Eqs. (c).
3. $(v_0)_y/(v_0)_x = \tan 30°$ (note that $\mathbf{v}_0$ equals $\mathbf{v}$ at $t = 0$), which gives $C_3 = C_1 \tan 30°$ using Eqs. (b).
4. $x = 480$ ft when $y = -360$ ft.

Substituting condition **4** and $g = 32.2\ \text{ft/s}^2$ into Eqs. (c), and letting t_A represent the time when the projectile reaches A, we get

$$480 = C_1 t_A \quad \text{and} \quad -360 = -16.1t_A^2 + C_3 t_A$$

Combining these two equations with $C_3 = C_1 \tan 30°$ from condition **3** gives us three equations that can be solved for the three unknowns: C_1, C_3, and t_A. Omitting the algebraic details, the results are

$$C_1 = 76.30 \text{ ft/s} \qquad C_3 = 44.05 \text{ ft/s} \qquad t_A = 6.291 \text{ s}$$

Substituting C_1 to C_4 into the expressions in Eqs. (a)–(c), we obtain the following description of the the motion.

$$a_x = 0 \qquad a_y = -32.2 \text{ ft/s}^2 \tag{d}$$

$$v_x = 76.30 \text{ ft/s} \qquad v_y = -32.2t + 44.05 \text{ ft/s} \tag{e}$$

$$x = 76.30t \text{ ft} \qquad y = -16.1t^2 + 44.05t \text{ ft} \tag{f}$$

All other characteristics of the motion can now be determined from Eqs. (d)–(f).

The initial velocity vector $\mathbf{v}_0$ is equal to $\mathbf{v}$ at $t = 0$. From Eqs. (e) we thus obtain $(v_0)_x = 76.30$ ft/s and $(v_0)_y = 44.05$ ft/s. Therefore the initial velocity vector is

y
44.05
θ
x
76.30

$$v_0 = 88.1 \text{ ft/s}$$ *Answer*

Note that $\theta = \tan^{-1}(44.05/76.30) = 30°$, as expected.

The maximum height h is equal to the value of y when $v_y = 0$. Letting t_1 be the time when this occurs, the second of Eqs. (e) gives us

$$0 = -32.2t_1 + 44.05 \quad \text{or} \quad t_1 = 1.368 \text{ s}$$

Substituting this value for t_1 into the second of Eqs. (f), we find that

$$h = -16.1(1.368)^2 + 44.05(1.368) = 30.1 \text{ ft}$$ *Answer*

In summary, the following steps may be used to analyze the motion of a projectile.

Analysis of Projectiles (Air Resistance Neglected)

1. Using the FMA approach, determine the rectangular components of the acceleration of the projectile (these components will be constant).
2. Integrate the acceleration components with respect to time to obtain expressions for $v_x(t)$, $v_y(t)$, $x(t)$, and $y(t)$. Treating the integrals as indefinite integrals, these expressions will contain four constants of integration.
3. From the problem statement, determine the four boundary conditions on the motion.

4. Use the boundary conditions from Step 3 to obtain four equations that will contain the four unknown constants of integration. Solve these equations for the constants and substitute them into the expressions for $x(t)$, $y(t)$, $v_x(t)$, and $v_y(t)$.
5. Compute whatever characteristics of the motion are of interest—initial velocity, final velocity, maximum height, etc.

Sample Problem 12.9

A projectile of mass m is fired from point O with the initial velocity v_0 as shown in Fig. (a). The aerodynamic drag F_D is proportional to the speed of the projectile: $F_D = cv$, where c is a constant. (1) Derive the equations of motion. (2) Verify that the solution of the equations of motion is

$$x = C_1 e^{-ct/m} + C_2 \quad \text{and} \quad y = C_3 e^{-ct/m} - \frac{mgt}{c} + C_4$$

where C_1 to C_4 are constants. (3) Find the maximum height h and the range R. Plot the path of the projectile given that $W = 2$ lb, $c = 0.04$ lb · s/ft, $v_0 = 80$ ft/s, and $\theta = 30°$.

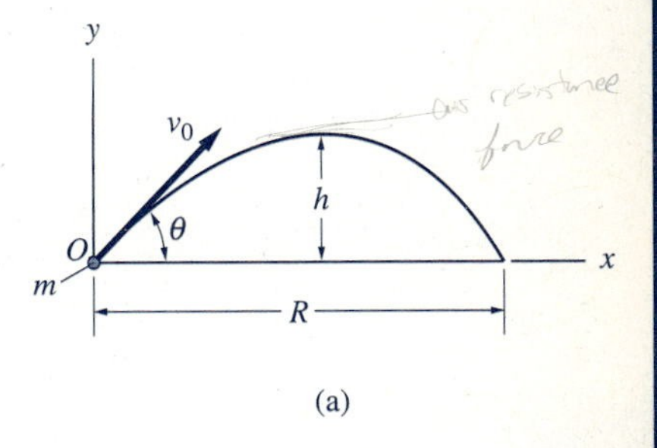

(a)

Solution

Part 1

Figure (b) shows the FBD and MAD of the particle. The direction of the drag F_D in the FBD is opposite to the direction of the velocity vector (tangent to the path), and its rectangular components are cv_x and cv_y. The rectangular components of the inertia vector, ma_x and ma_y, are shown in the MAD. Newton's second law gives the two equations of motion

mg Path cv_x cv_y $F_D = cv$ ma_y ma_x

FBD MAD

(b)

$$\Sigma F_x = ma_x \qquad \xrightarrow{+} \qquad -cv_x = ma_x \qquad \textit{Answer} \quad \text{(a)}$$

$$\Sigma F_y = ma_y \qquad +\uparrow \qquad -mg - cv_y = ma_y \qquad \textit{Answer} \quad \text{(b)}$$

Part 2

To verify that the expressions for $x(t)$ and $y(t)$ given in the problem statement satisfy the equations of motion, we must first evaluate their derivatives.

$$x = C_1 e^{-ct/m} + C_2 \qquad y = C_3 e^{-ct/m} - \frac{mgt}{c} + C_4 \qquad \text{(c)}$$

$$v_x = \dot{x} = -C_1 \frac{c}{m} e^{-ct/m} \qquad v_y = \dot{y} = -C_3 \frac{c}{m} e^{-ct/m} - \frac{mg}{c} \qquad \text{(d)}$$

$$a_x = \ddot{x} = C_1 \left(\frac{c}{m}\right)^2 e^{-ct/m} \qquad a_y = \ddot{y} = C_3 \left(\frac{c}{m}\right)^2 e^{-ct/m} \qquad \text{(e)}$$

Substitution of the above results into Eqs. (a) and (b) verifies that Eqs. (c) are indeed the solution, since they satisfy the equations of motion.

Part 3

Using the given numerical values for c and W, we have $c/m = 0.04(32.2)/2 = 0.6440\ \text{s}^{-1}$, and $mg/c = 2/0.04 = 50$ ft/s. Substituting these values into Eqs. (c) and (d), and assuming that time t is measured in seconds, we obtain

$$x = C_1 e^{-0.6440t} + C_2 \text{ ft} \qquad y = C_3 e^{-0.6440t} - 50t + C_4 \text{ ft} \tag{f}$$

$$v_x = -0.6440 C_1 e^{-0.6440t} \text{ ft/s} \qquad v_y = -0.6440 C_3 e^{-0.6440t} - 50 \text{ ft/s} \tag{g}$$

From the problem statement, we deduce that the motion must satisfy the following initial conditions.

1. $t = 0,\ x = 0$
2. $t = 0,\ y = 0$
3. $t = 0,\ v_x = 80 \cos 30° = 69.28$ ft/s
4. $t = 0,\ v_y = 80 \sin 30° = 40.00$ ft/s

The equations obtained by substituting these four conditions into Eqs. (f) and (g) can be solved for the constants of integration. Omitting the algebraic details, the results are $C_1 = -107.58$ ft, $C_2 = 107.58$ ft, $C_3 = -139.75$ ft, and $C_4 = 139.75$ ft. Substituting these values into Eqs. (f) and (g), we find that

$$x = 107.58(1 - e^{-0.6440t}) \text{ ft} \tag{h}$$

$$y = 139.75(1 - e^{-0.6440t}) - 50t \text{ ft} \tag{i}$$

and

$$v_x = 69.28 e^{-0.6440t} \text{ ft/s} \tag{j}$$

$$v_y = 90 e^{-0.6440t} - 50 \text{ ft/s} \tag{k}$$

The maximum value of y occurs when $v_y = 0$. Letting this time be t_1, Eq. (k) yields

$$0 = 90 e^{-0.6440t_1} - 50$$

from which we find

$$t_1 = -\frac{\ln(50/90)}{0.6440} = 0.9127 \text{ s}$$

Substituting $t = t_1 = 0.9127$ s into Eq. (i), we get for the maximum value of y

$$y_{\max} = h = 139.75\left[1 - e^{-0.6440(0.9127)}\right] - 50(0.9127)$$

$$= 16.48 \text{ ft} \qquad \textit{Answer}$$

To evaluate the range R, we must first find the time (other than zero) when $y = 0$. Letting t_2 be this time, Eq. (i) yields

$$0 = 139.75\left(1 - e^{-0.6440t_2}\right) - 50t_2$$

This equation must be solved numerically (e.g., by Newton's method given in Appendix B) or by trial and error. The result, which may be verified by substitution, is $t_2 = 2.047$ s. The corresponding value of x is the range R. From Eq. (h), we find that

$$R = 107.58\left[1 - e^{-0.6440(2.047)}\right] = 78.8 \text{ ft} \qquad \textit{Answer}$$

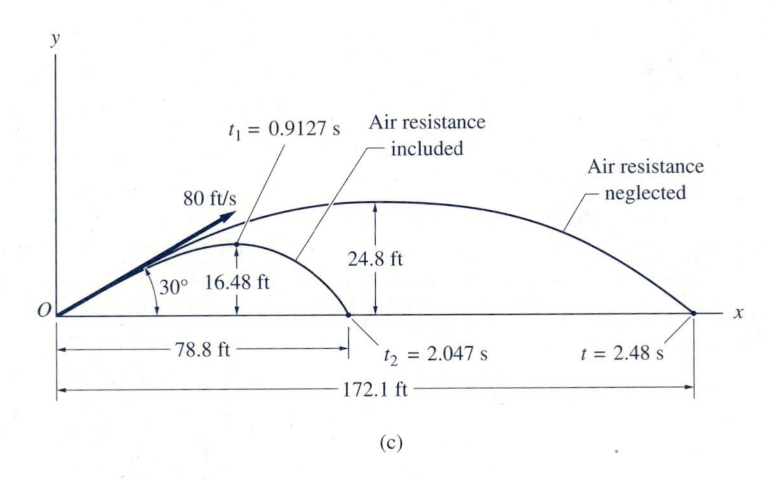

(c)

Plotting x and y from Eqs. (h) and (i), respectively, gives the path shown in Fig. (c). For purposes of comparison, the parabolic path that would occur if air resistance were neglected is also shown. As would be expected, both the maximum height and range are reduced if air resistance is included, and the path is no longer parabolic.

PROBLEMS

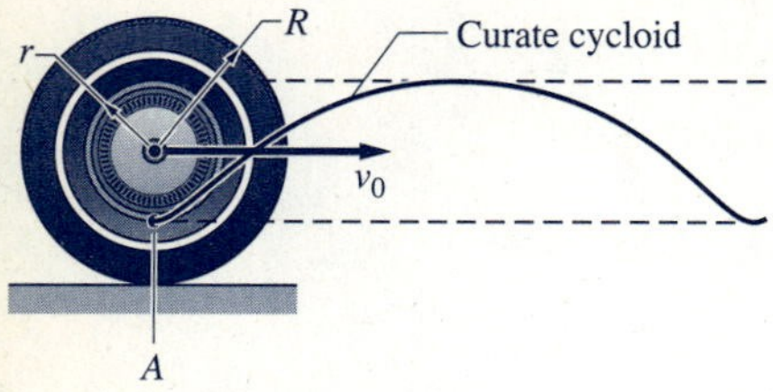

Fig. P12.47

12.47 A 2-oz balancing weight A is attached to the rim of a car wheel. When the car travels with the constant speed v_0, the path of the weight is the curate cycloid described by

$$x = v_0 t - r\sin(v_0 t/R)$$
$$y = R - r\cos(v_0 t/R)$$

Using $v_0 = 60$ mi/h, $R = 1.25$ ft, and $r = 0.8$ ft, compute the force required to hold the weight onto the wheel.

12.48 The slider of mass m moves along the smooth guide rod ABC, propelled by the horizontal force $F(t)$. The position of the slider is given by $x = b\sin(2\pi t/t_0)$, $y = (b/4)[1 + \cos(4\pi t/t_0)]$, where t_0 is a constant (the period of oscillation). (a) Show that the guide rod is a parabola. (b) Assuming that ABC lies in a vertical plane, determine the expression for F when the mass is at point C.

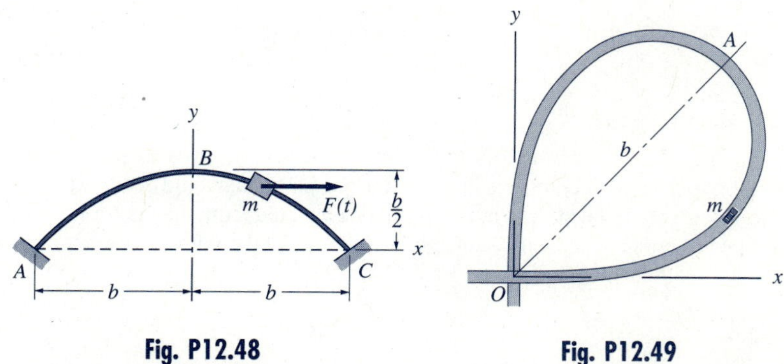

Fig. P12.48

Fig. P12.49

12.49 A car of mass m travels along the cloverleaf interchange. The position of the car is described by

$$x = \frac{b}{2}\left(\sin\frac{t}{t_0} + \sin\frac{3t}{t_0}\right)$$
$$y = \frac{b}{2}\left(\cos\frac{t}{t_0} - \cos\frac{3t}{t_0}\right)$$

where $b = 240$ m and t_0 is constant (it can be shown that $\pi t_0/4$ is the time of travel between O and A). If the static coefficient of friction between the tires and the road is $\mu = 0.6$, determine the smallest possible value of t_0 and the corresponding maximum speed reached by the car on the interchange.

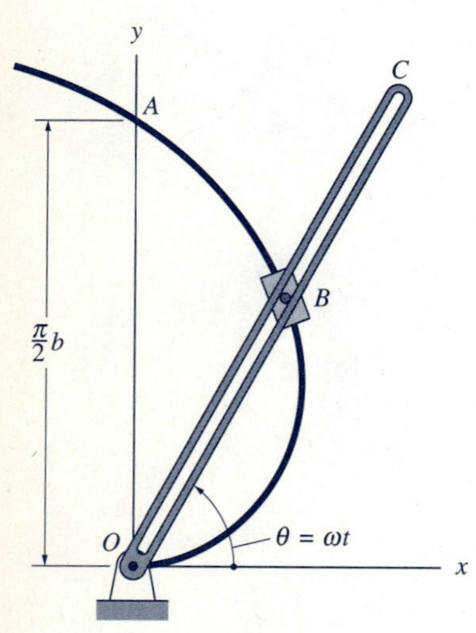

Fig. P12.50

12.50 The collar B of mass m moves along a guide rod that has been bent into the shape of a spiral. A pin on the collar slides in the slotted arm OC, which is rotating counterclockwise at the angular speed ω. The resulting motion of the collar is described by

$$x = b\omega t\cos\omega t \qquad y = b\omega t\sin\omega t$$

When the collar is at A, determine the contact force applied to the collar by (a) the guide rod; and (b) arm OC. Neglect friction and gravity.

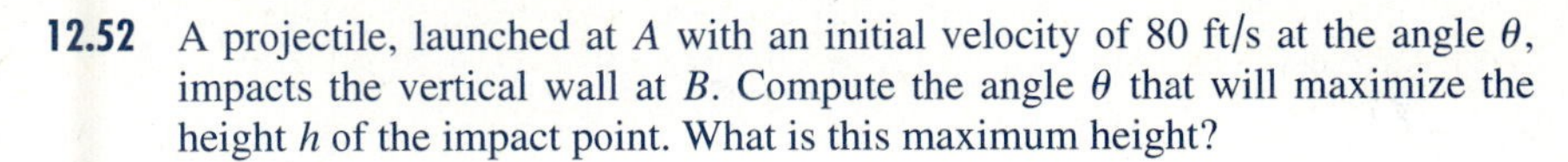

12.51 A projectile, launched at A with an initial velocity of 80 ft/s at the angle $\theta = 60°$, impacts the vertical wall at B. Neglecting air resistance, calculate (a) the height h; and (b) the velocity vector at B.

12.52 A projectile, launched at A with an initial velocity of 80 ft/s at the angle θ, impacts the vertical wall at B. Compute the angle θ that will maximize the height h of the impact point. What is this maximum height?

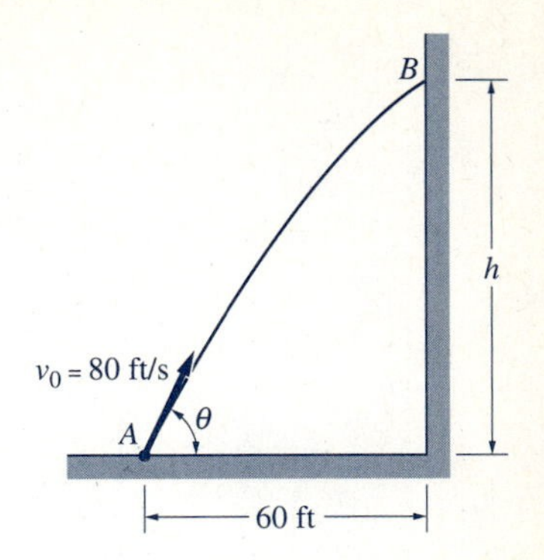

Fig. P12.51, P12.52

12.53 A projectile launched at A is to hit a target at B with air resistance neglected. (a) Derive the expression that determines the required angle of elevation θ. (b) Find the two solutions θ_1 and θ_2 of the expression derived in Part (a) if $v_0 = 300$ m/s and $L = 5$ km.

12.54 The volleyball player serves the ball from point A with an initial velocity v_0 at the angle θ to the horizontal. If the ball is to just clear the net C and land on the baseline B, determine the values of v_0 and θ.

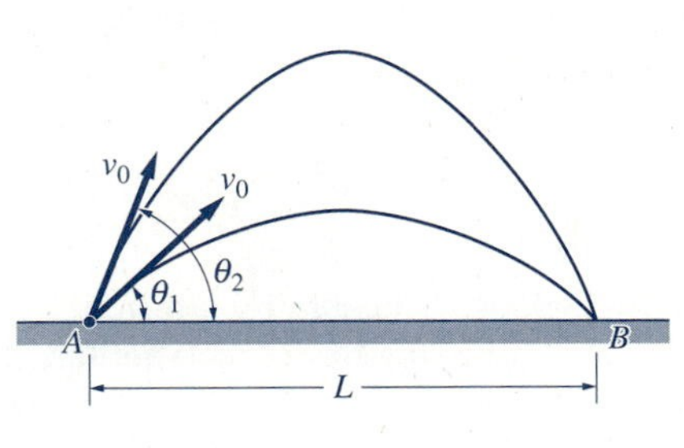

Fig. P12.53

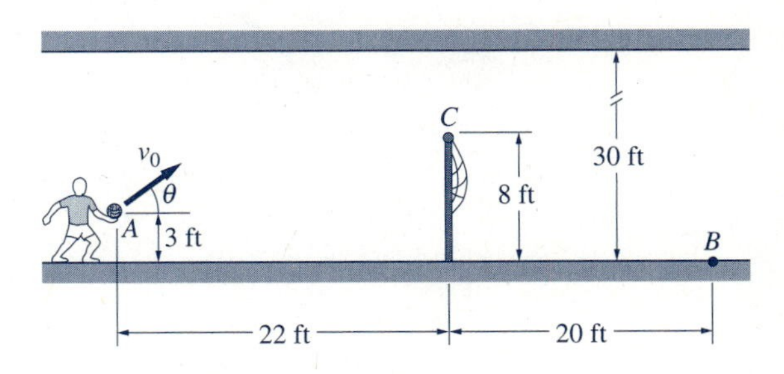

Fig. P12.54, P12.55

12.55 The volleyball player serves the ball from point A with an initial velocity v_0 at the angle θ to the horizontal. If the ball is to land as close as possible behind the net C, determine the values of v_0 and θ. Note that the trajectory of the ball is limited by the height of the ceiling.

12.56 A projectile is fired up the inclined plane with the initial velocity shown. Compute the maximum height h, measured perpendicular to the plane, that is reached by the projectile. Neglect air resistance. (Hint: Use the xy-coordinate system shown.)

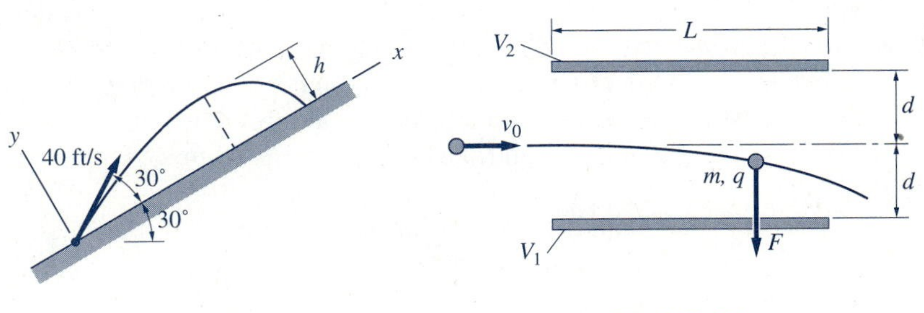

Fig. P12.56

Fig. P12.57

12.57 A particle of mass m (kg) carrying a charge q (coulombs) enters the space between two charged plates with the horizontal velocity v_0 (m/s) as shown. Neglecting gravitational acceleration, the force acting on the particle while it is between the plates is $F = q\Delta V/2d$, where $\Delta V = V_2 - V_1$ is the electrostatic potential difference (volts) between the plates. Derive the expression for the largest ΔV that may be applied if the particle is not to hit one of the plates.

12.58 A projectile of mass m is launched at O with the initial speed v_0 at the angle α to the horizontal. The aerodynamic drag force acting on the projectile during its flight is $\mathbf{F}_D = -c\mathbf{v}$, where c is a constant. If point A is the peak of the trajectory, derive the expressions for (a) the time required to reach A; (b) the speed of the projectile at A.

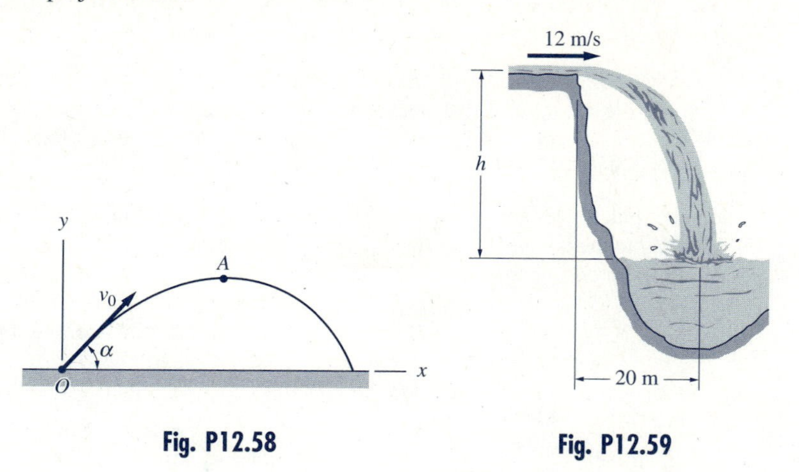

Fig. P12.58 **Fig. P12.59**

12.59 The velocity of water at the top of the waterfall is 12 m/s, directed horizontally. The horizontal distance between the top and the base of the waterfall is 20 m. It takes 2.75 s for the water to complete the fall. Assuming that the aerodynamic drag force acting on the water is proportional to its speed, find the height h of the waterfall. Hint: start with the equation of the trajectory given in Sample Problem 12.9:

$$x = C_1 e^{-ct/m} + C_2 \quad \text{and} \quad y = C_3 e^{-ct/m} - \frac{mgt}{c} + C_4$$

and compute m/c before evaluating h.

*12.7 Motion Diagrams by the Area Method

Figure 12.6 shows the time plots of acceleration, velocity, and position, called *motion diagrams,* for a particle in rectilinear motion along the x-axis. This article develops the geometric relationships among the motion diagrams that form the basis for the *area method* for analyzing motion.* The area method enables us to construct the velocity and position diagrams from a given acceleration diagram.

Since $a = dv/dt$ and $v = dx/dt$ for rectilinear motion, we deduce that the slopes of the velocity and position diagrams can be determined in the following manner:

1. The slope of the velocity diagram at time t_i is equal to the acceleration at that time; that is, $(dv/dt)_i = a_i$ as shown in Fig. 12.6(b).
2. The slope of the position diagram at time t_i is equal to the velocity at that time, that is, $(dx/dt)_i = v_i$ as shown in Fig. 12.6(c).

* The area method for constructing motion diagrams is similar to the area method for shear and moment diagrams for beams (see Art. 6.4).

Consider next the time interval that begins at time t_0 and ends at time t_n, as shown in Fig. 12.6. The initial and final values of acceleration, velocity, and position are labeled a_0, v_0, x_0 and a_n, v_n, and x_n, respectively. Rewriting $a = dv/dt$ as $dv = a\,dt$ and integrating between t_0 and t_n yields

$$\int_{v_0}^{v_n} dv = \int_{t_0}^{t_n} a(t)\,dt$$

Recognizing the right-hand side of this equation as the area of the acceleration diagram between t_0 and t_n, we have

$$v_n - v_0 = \text{area of the } a\text{-}t \text{ diagram}\Big]_{t_0}^{t_n} \tag{12.25}$$

Similarly, rewriting $v = dx/dt$ as $dx = v\,dt$, and integrating between t_0 and t_n, we obtain

$$\int_{x_0}^{x_n} dx = \int_{t_0}^{t_n} v(t)\,dt$$

Since the right-hand side of this equation is the area of the velocity diagram between t_0 and t_n, we arrive at

$$x_n - x_0 = \text{area of the } v\text{-}t \text{ diagram}\Big]_{t_0}^{t_n} \tag{12.26}$$

Equations (12.25) and (12.26) can be restated in the following manner.

3. The increase in velocity during a given time interval is equal to the area of the *a-t* diagram for that time interval [the shaded area in Fig. 12.6(a)].
4. The increase in position coordinate during a given time interval is equal to the area of the *v-t* diagram for that time interval [the shaded area in Fig. 12.6(b)].

The geometric relationships **1–4** are useful tools for analyzing rectilinear motion numerically if one of the motion diagrams has been obtained experimentally. The area method is also convenient if the acceleration diagram is made up of straight lines. In that case, the *v-t* diagram will be made up of parabolas, whereas the *x-t* diagrams will consist of cubic polynomials.

The relationships **1–4** have been stated for rectilinear motion. However, they also apply to the special case of curvilinear motion that can be described as the superposition of rectilinear motions, one along each of the coordinate axes (Art. 12.6). In the two-dimensional case, for example, the motions can be represented by two sets of motion diagrams: a_x-t, v_x-t, x-t, and a_y-t, v_y-t, y-t.

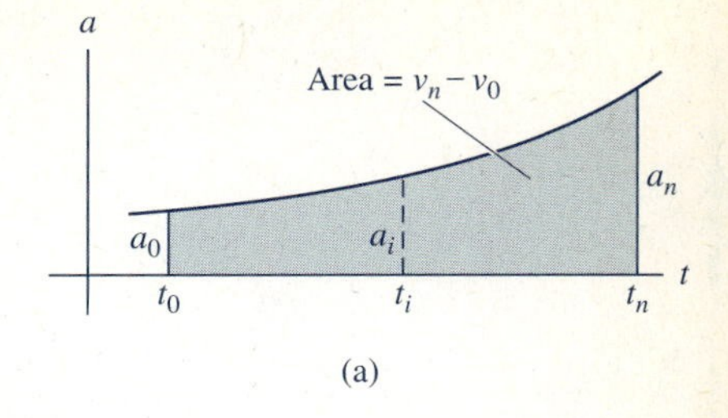

(a)

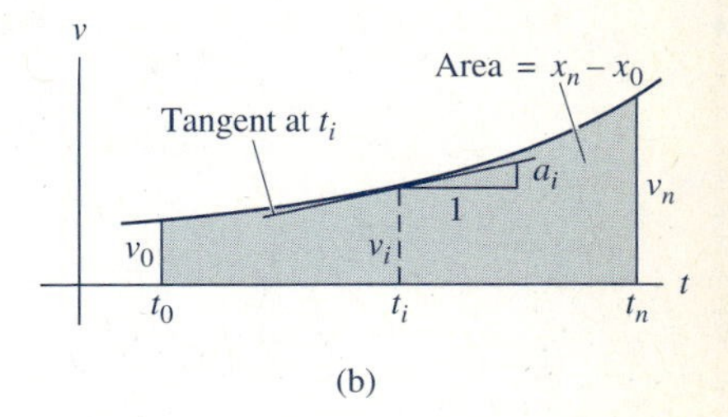

(b)

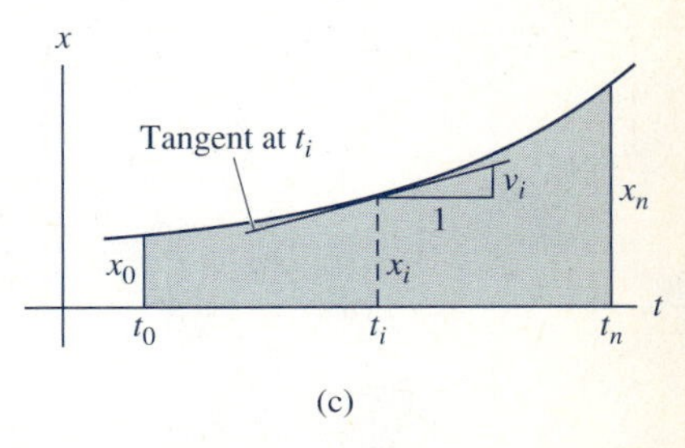

(c)

Fig. 12.6

Sample Problem 12.10

The 5-kg block in Fig. (a) is at rest at $x = 0$ and $t = 0$ when the force $P(t)$ is applied. The variation of $P(t)$ with time is shown in Fig. (b). Friction between the block and the horizontal plane can be neglected. (1) Use the area method to construct the motion diagrams. (2) Determine the velocity and position of the block at $t = 5$ s.

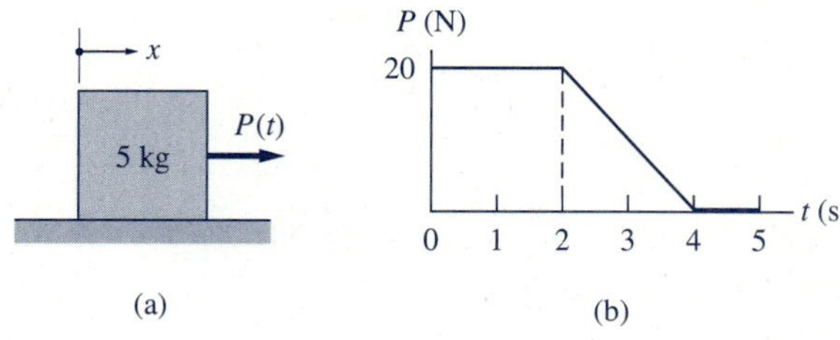

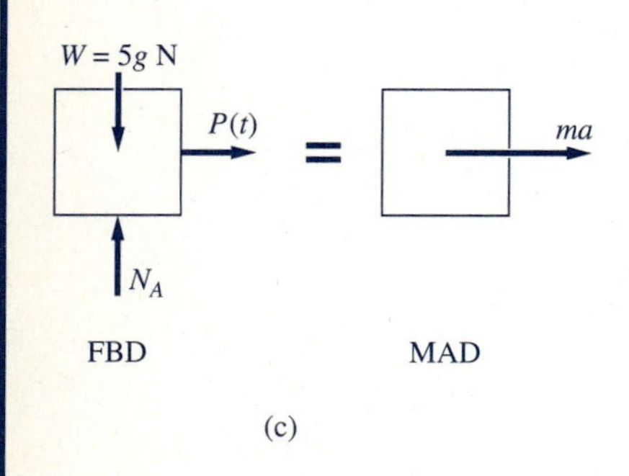

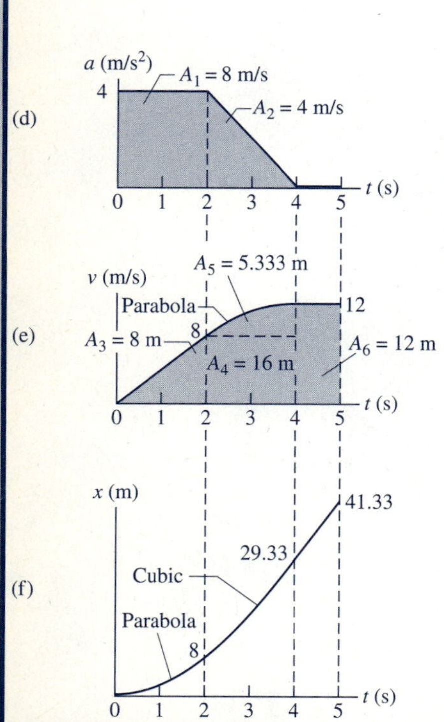

Solution

Part 1

a-t Diagram

The free-body diagram (FBD) of the block is shown in Fig. (c). The weight of the block is $W = 5g$ N, and N_A is the normal force exerted by the horizontal plane. Figure (c) also shows the mass-acceleration diagram (MAD) of the block; the inertia vector is horizontal because the motion is rectilinear. Applying Newton's second law

$$\Sigma F_x = ma_x \qquad \xrightarrow{+} \quad P = 5a$$

gives

$$a = \frac{P}{5}$$

The resulting *a-t* diagram is shown in Fig. (d).

In the remainder of the solution, we will use subscripts on a, v, and x to indicate the values of these variables at various times. For example, $v_0, v_1, v_2, \ldots$ will refer to the velocities at $t = 0, 1\text{ s}, 2\text{ s}, \ldots$, respectively.

v-t Diagram

Before constructing the *v-t* diagram, we compute the areas under the *a-t* diagram in Fig. (d): $A_1 = 2(4) = 8$ m/s and $A_2 = (1/2)(2)(4) = 4$ m/s. The velocities v_2, v_4, and v_5 are found by applying Eq. (12.25) (recall that $v_0 = 0$):

$$v_2 = v_0 + \text{area of the } a\text{-}t \text{ diagram}\Big]_{t=0}^{t=2\text{ s}}$$

$$= v_0 + A_1 = 0 + 8 = 8 \text{ m/s}$$

$$v_4 = v_2 + \text{area of the } a\text{-}t \text{ diagram}\Big]_{t=2\text{ s}}^{t=4\text{ s}}$$

$$= v_2 + A_2 = 8 + 4 = 12 \text{ m/s}$$

$$v_5 = v_4 + \text{area of the } a\text{-}t \text{ diagram}\Big]_{t=4\text{ s}}^{t=5\text{ s}}$$

$$= v_4 + 0 = 12 \text{ m/s}$$

The values v_0, v_2, v_4, and v_5 are then plotted in Fig. (e). The shape of the v-t diagram connecting these points is deduced from $a = dv/dt$, that is, the acceleration is equal to the slope of the v-t diagram.

1. *Time interval from* $t = 0$ *to* $t = 2$ s. From Fig. (d) we see that the acceleration is equal to $+4$ m/s^2 throughout the interval. Therefore, the v-t diagram is a straight line with a constant slope equal to $+4$ m/s^2—see Fig. (e).
2. *Time interval from* $t = 2$ s *to* $t = 4$ s. From Fig. (d), we observe that the acceleration is positive and is decreasing in a linear manner (a-t curve is a sloping straight line). Therefore, the slope of the v-t diagram is also positive and decreasing, that is, the v-t diagram is a parabola as shown in Fig. (e). When sketching this parabola, observe that (1) a_2 is positive, indicating that the slope of the v-t diagram at $t = 2$ is positive; (2) a_4 is zero, indicating that the slope of the v-t diagram at $t = 4$ s is zero. Since the acceleration diagram is continuous everywhere, the velocity diagram has a continuous slope (i.e., the diagram is smooth with no discontinuities).
3. *Time interval from* $t = 4$ s *to* $t = 5$ s. The acceleration is equal to zero throughout the time interval. Therefore, the slope of the v-t diagram is also equal to zero, which means that the v-t diagram is a straight horizontal line—see Fig. (e).

x-t Diagram

We begin by computing the component areas under the v-t diagram in Fig. (e): $A_3 = (1/2)(2)(8) = 8$ m; $A_4 = (2)(8) = 16$ m; $A_5 = (2/3)(2)(4) = 5.333$ m; $A_6 = (1)(12) = 12$ m. The positions x_2, x_4, and x_5 are then computed from Eq. (12.26), starting with the known value $x_0 = 0$:

$$x_2 = x_0 + \text{area of the } a\text{-}t \text{ diagram}\Big]_{t=0}^{t=2\text{ s}}$$

$$= 0 + A_3 = 0 + 8 = 8 \text{ m}$$

$$x_4 = x_2 + \text{area of the } a\text{-}t \text{ diagram}\Big]_{t=2\text{ s}}^{t=4\text{ s}}$$

$$= 8 + (A_4 + A_5) = 8 + 16 + 5.333 = 29.33 \text{ m}$$

$$x_5 = x_4 + \text{area of the } a\text{-}t \text{ diagram}\Big]_{t=4\text{ s}}^{t=5\text{ s}}$$

$$= 29.33 + A_6 = 29.33 + 12 = 41.33 \text{ m}$$

After plotting the points x_0, x_2, x_4, and x_5 in Fig. (f), the shape of the connecting curve can be determined from $v = dx/dt$; i.e., the slope of the x-t diagram is equal to the velocity.

1. *Time interval from* $t = 0$ *to* $t = 2$ s. From Fig. (e), we see that the velocity is positive and increasing in a linear manner. Consequently, the slope of the x-t diagram is also positive and increasing, as shown in Fig. (f). Since the v-t diagram is a straight line, the x-t diagram is a parabola. As a check on the concavity of this parabola, note that $v_0 = 0$ and v_2 is positive, which tells us that the slope of the x-t diagram is zero at $t = 0$ and positive at $t = 2$ s.
2. *Time interval from* $t = 2$ s *to* $t = 4$ s. Because the velocity in Fig. (e) is a parabola with positive increasing values, the x-t diagram is a cubic (or third-degree) curve with positive increasing slope, as shown in Fig. (f). The concavity of this cubic curve can be verified by noting that $v_2 < v_4$, which indicates that the slope of the x-t diagram at $t = 2$ s is less than the slope at $t = 4$ s.
3. *Time interval from* $t = 4$ s *to* $t = 5$ s. From Fig. (e) we see that the velocity equals the constant value of $+12$ m/s. Therefore, the slope of the x-t diagram is also constant and equal to $+12$ m/s. As shown in Fig. (f), the x-t diagram is, therefore, a straight line. Since the v-t diagram has no discontinuities, the x-t diagram is also smooth.

Part 2

After the motion diagrams in Figs. (d)–(f) have been constructed, it is a simple matter to determine a, v, or x at a given value of time. In particular, from Figs. (e) and (f) we see that

$$v_5 = 12 \text{ m/s} \quad \text{and} \quad x_5 = 41.3 \text{ m} \qquad \textit{Answer}$$

Sample Problem 12.11

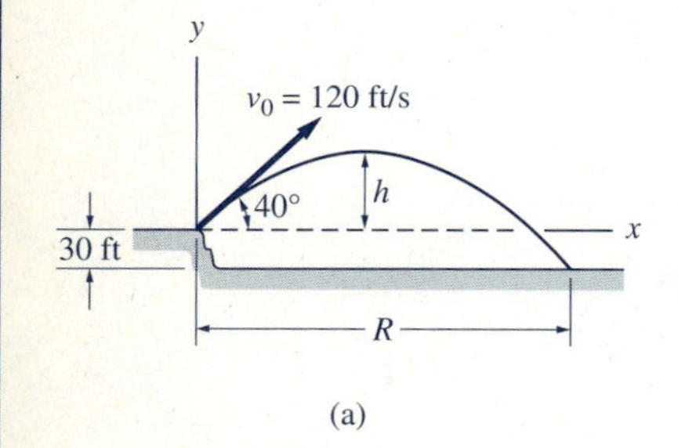

(a)

A golf ball is driven off a tee that is elevated 30 ft above the fairway. The initial velocity vector $\mathbf{v}_0$ of the ball is shown in Fig. (a). (1) Construct the motion diagrams using the area method. (2) Determine the maximum height h of the ball above the tee, the range R, and the velocity vector of the ball when it hits the fairway. Neglect air resistance.

Solution

Introductory Comments

Because air resistance is neglected, the only force acting on the ball is its weight. Therefore, Newton's second law gives $a_y = -g = -32.2$ ft/s^2 and $a_x = 0$. Since the motions in the x and y directions are uncoupled, it is convenient to consider the curvilinear motion of the ball as a vector superposition of the motions in the x and y directions (see Art 12.6). Therefore, there will be two sets of motion diagrams—one set for the vertical direction and another for the horizontal direction.

For future reference, we introduce the following times: $t_0 = 0$, when the ball is hit; t_1, when the ball reaches its maximum height; and t_2, when the ball hits the fairway. In addition, subscripts 0, 1, and 2 are used to indicate the values of x, y, v_x, and v_y at these times.

From Fig. (a), we have the following five boundary conditions imposed on the motion.

1. $x_0 = 0$
2. $y_0 = 0$
3. $y_2 = -30$ ft
4. $(v_y)_0 = 120 \sin 40° = 77.13$ ft/s
5. $(v_x)_0 = 120 \cos 40° = 91.93$ ft/s

Part 1

Motion Curves for y-Direction

The a_y-t diagram is shown in Fig. (b), where a_y is equal to the constant value of -32.2 ft/s^2.

From $a_y = dv_y/dt$, we conclude that the v_y-t diagram is a straight line with a slope equal to -32.2 ft/s^2, as shown in Fig. (c). The initial value of this diagram is $(v_y)_0 = 77.13$ ft/s. Note that $(v_y)_1 = 0$ because t_1 represents the time when the ball reaches its maximum height.

Since $v_y = dy/dt$, we deduce that the y-t diagram is a parabola as shown in Fig. (d). From the boundary conditions we know that $y_0 = 0$ and $y_2 = -30$ ft. Note that this curve is smooth, because there are no discontinuities in the v_y-t diagram.

Applying the area method to the motion diagrams in Figs. (b)–(d), we arrive at the following equations.

$(v_y)_1 - (v_y)_0 = A_1$	$0 - 77.13 = -32.2t_1$	(a)
$(v_y)_2 - (v_y)_1 = A_2$	$(v_y)_2 - 0 = -32.2(t_2 - t_1)$	(b)
$y_1 - y_0 = A_3$	$y_1 - 0 = \frac{1}{2}(77.13)t_1$	(c)
$y_2 - y_1 = A_4$	$-30 - y_1 = -\frac{1}{2}(32.2)(t_2 - t_1)^2$	(d)

Solving Eqs. (a)–(d) for the four unknowns, we find $t_1 = 2.395$ s, $t_2 = 5.152$ s, $(v_y)_2 = -88.78$ ft/s and $y_1 = 92.37$ ft.

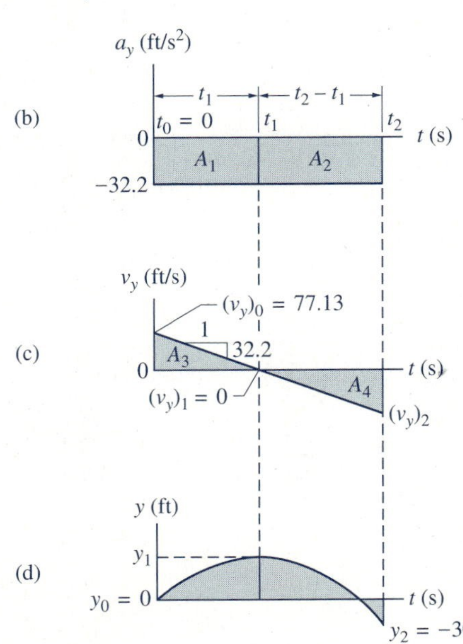

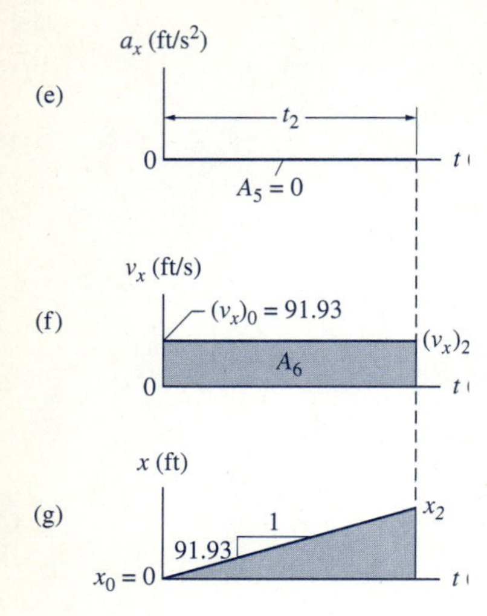

Motion Curves for x-Direction

The a_x-t diagram, with $a_x = 0$, is shown in Fig. (e). From $a_x = dv_x/dt$, we conclude that the v_x-t diagram in Fig. (f) is a horizontal straight line with the initial value $(v_x)_0 = 91.93$ ft/s. Using $v_x = dx/dt$, we find that the x-t diagram shown in Fig. (g) is an inclined straight line with the slope 91.93 ft/s.

The following equations result from applying the area method to the motion diagrams in Figs. (e)–(f).

$(v_x)_2 - (v_x)_0 = A_5$	$(v_x)_2 - 91.93 = 0$ (e)
$x_2 - x_0 = A_6$	$x_2 - 0 = 91.93t_2$ (f)

Solving Eq. (e) gives $(v_x)_2 = 91.93$ ft/s. Substituting $t_2 = 5.152$ s into Eq. (f) yields $x_2 = 474$ ft.

The final motion curves are shown in Figs. (h)–(m).

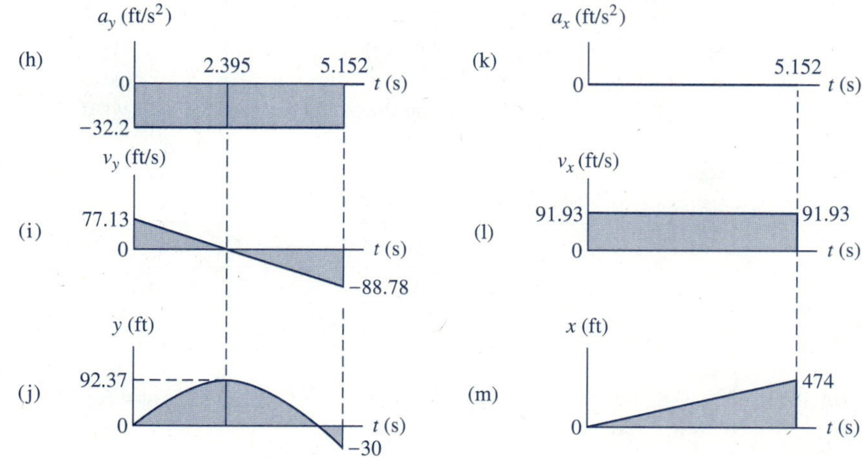

Part 2

The maximum height of the ball above the tee is obtained from Fig. (j).

$$h = y_1 = 92.37 \text{ ft} \qquad \textit{Answer}$$

From Fig. (m) the range is

$$R = x_2 = 474 \text{ ft} \qquad \textit{Answer}$$

From Figs. (i) and (l), we see that $(v_y)_2 = -88.78$ ft/s and $(v_x)_2 = 91.93$ ft/s. Therefore, the velocity vector of the ball when it hits the fairway is

$$v_2 = 127.8 \text{ ft/s} \qquad \textit{Answer}$$

PROBLEMS

Solve the following problems using the area method. Sketch the motion diagrams for each problem.

12.60 Solve Prob. 12.31.

12.61 Solve Prob. 12.32.

12.62 Solve Prob. 12.51.

12.63 A missile is launched horizontally at A with the speed $v_0 = 188$ m/s. Knowing that the horizontal distance traveled by the missile is $d = 1200$ m, calculate the launch height h and the time of flight.

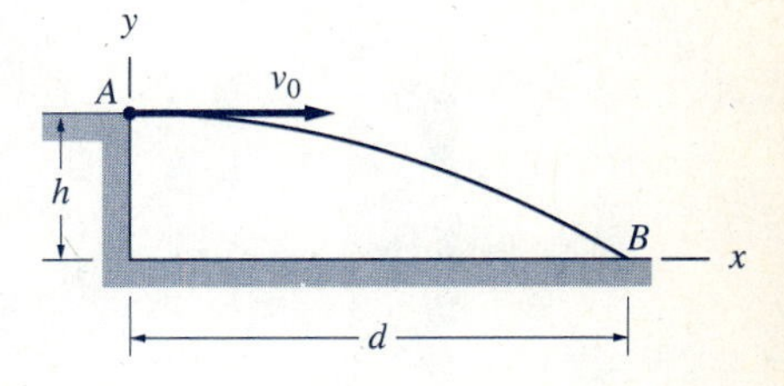

Fig. P12.63, P12.64

12.64 A projectile is launched horizontally at A with the speed v_0. The time of flight is 8 s, and the path of the projectile at B is inclined at 15° with the horizontal. Determine v_0, the range d, and the launch height h. Use U.S. Customary units.

12.65 A projectile is launched with the initial speed $v_0 = 200$ m/s in the direction shown. Find the time of flight, the range d, and the maximum height h of the projectile above the inclined plane.

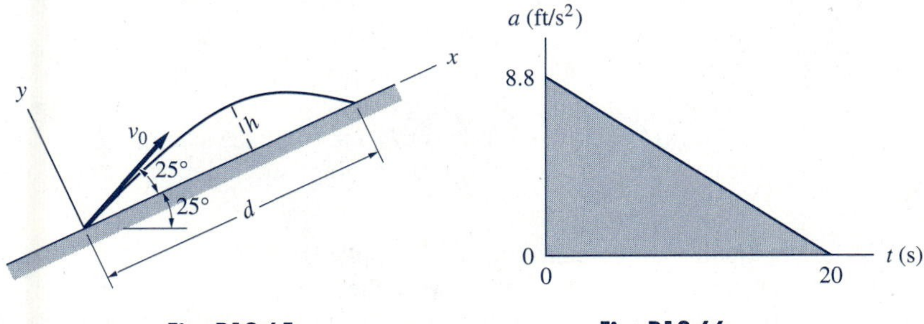

Fig. P12.65 Fig. P12.66

12.66 A car that is initially at rest accelerates along a straight, level road according to the diagram shown. Determine (a) the maximum speed; and (b) the distance traveled by the car when the maximum speed is reached.

12.67 A subway train stops at two stations that are 1.2 miles apart. The maximum acceleration and deceleration of the train are 6.6 ft/s² and 5.5 ft/s², respectively, and the maximum allowable speed is 45 mi/h. Find the shortest possible time of travel between the two stations.

12.68 A train is brought to an emergency stop in 16 seconds, the deceleration being as shown in the diagram. Compute (a) the speed of the train before the brakes were applied; and (b) the stopping distance.

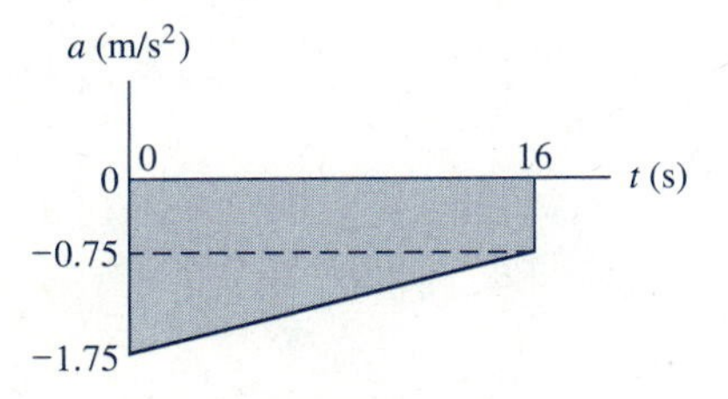

Fig. P12.68

12.69 The 800-kg rocket is launched vertically. The thrust F varies parabolically with time as shown in the diagram. Compute the maximum speed of the rocket.

12.70 A particle, at rest when $t = 0$, undergoes the periodic acceleration shown. Determine the velocity and distance traveled when (a) $t = 3t_0$; and (b) $t = 3.5t_0$.

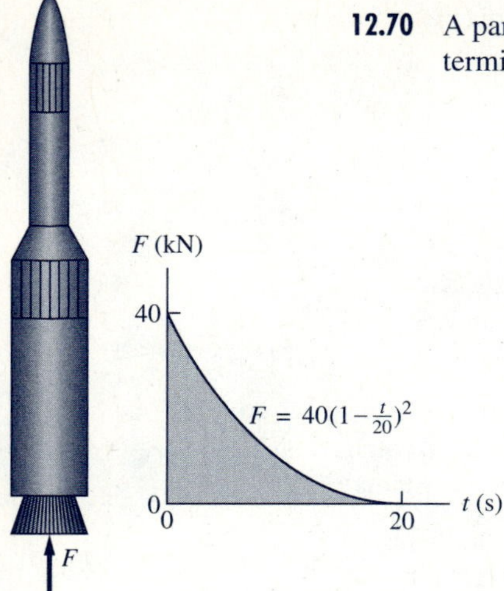

Fig. P12.69

Fig. P12.70

12.71 The 8-lb block is at rest on a rough surface when $t = 0$. For $t > 0$, the periodic horizontal force $P(t)$ of amplitude P_0 is applied to the block. Note that the period of $P(t)$ is 0.5 s. (a) Calculate the value of P_0 for which the average acceleration during each period is zero. (b) What is the average speed during each period in part (a)?

12.72 The amplitude of the periodic force that is applied to the 8-lb block is $P_0 = 6$ lb. The coefficient of kinetic friction between the block and the horizontal surface is 0.2. If the velocity of the block at $t = 0$ was 2 ft/s to the right, determine (a) the velocity of the block at $t = 0.7$ s; and (b) the displacement of the block from $t = 0$ to 0.7 s.

12.73 An accelerometer is used to measure the vertical acceleration of a car as it drives along a test track. For the data shown, determine the change in elevation of the car during the 1-s interval. Approximate the area under the diagram as the sum of trapezoidal areas.

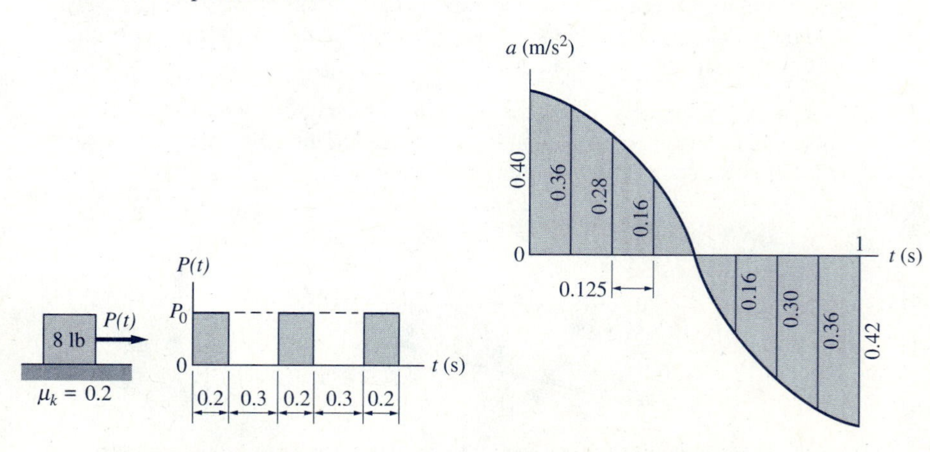

Fig. P12.71, P12.72

Fig. P12.73

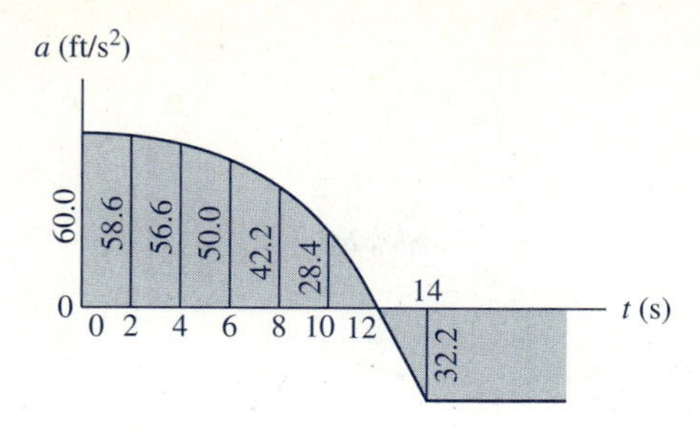

Fig. P12.74

12.74 A rocket is fired vertically from the surface of the earth. The engine burns for 14 s, resulting in the acceleration shown in the diagram. Compute (a) the maximum speed; (b) the maximum height; and (c) the time when the maximum height occurs. Approximate the area under the diagram as the sum of trapezoidal areas.

12.8 Numerical Integration of an Equation of Motion

a. Rationale

The differential equations of motion that are encountered in the study of dynamics can be solved analytically in relatively few cases. Even when analytic solutions are possible, the work required may be tedious because of the complexity of the equations.

The primary mathematical difficulty that arises in the solution of dynamics problems is the *nonlinearity* of the equations of motion. That is, the displacements and velocities appear in the expressions for the accelerations in nonlinear form, such as v^2, xv, $\sin x$, etc. Unfortunately, the topic of nonlinear differential equations has not been nearly as productive as the study of linear equations. For single-degree-of-freedom problems, i.e., problems that are governed by only one equation of motion, nonlinearity does not necessarily preclude analytical solutions. For example, several cases of nonlinear rectilinear motion were solved in Art. 12.4. However, if the number of degrees of freedom—and therefore, the number of equations of motion—is two or more, numerical methods usually provide the only practical means of solution.

This article introduces numerical integration for one-degree-of-freedom particle motion.* (The terms "integration" and "solution" are used interchangeably for differential equations.) We begin with a discussion of the special case where the acceleration is a function of time only; i.e., $a = f(t)$. Then we consider the general case where the acceleration can be a function of velocity, position, and time; i.e., $a = f(v, x, t)$.

* It must be pointed out that the numerical integration procedures discussed here are not limited to the analysis of rectilinear motion. These methods are equally applicable to the solution of any ordinary second-order differential equation, or, as explained in the next article, a coupled system of such equations.

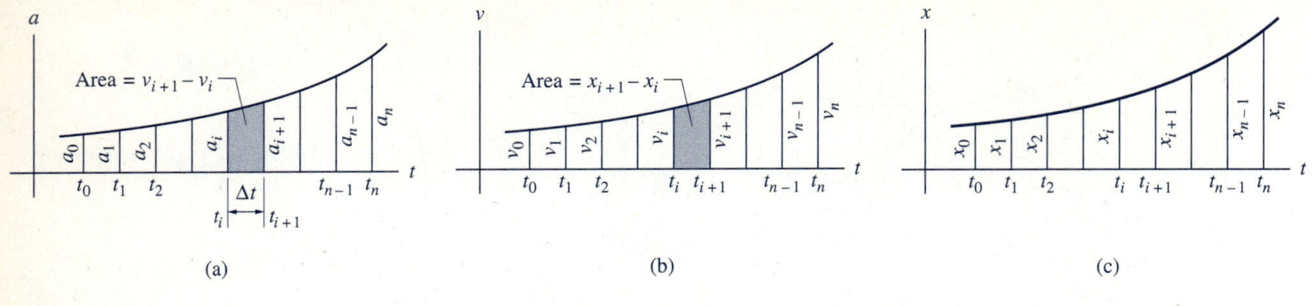

Fig. 12.7

b. Special case: $a = f(t)$

The acceleration of a particle that moves along the x-axis will be a function of time only if the forces do not depend explicitly on velocity or position. Then we can plot the a-t curve as shown in Fig. 12.7(a). The v-t and x-t curves, to be computed from the a-t curve by the area method, are shown in Fig. 12.7 parts (b) and (c), respectively.

If $f(t)$ cannot be integrated analytically, or if values for $f(t)$ are known only at a discrete number of points, it is necessary to compute the areas under the a-t and v-t diagrams by numerical methods. [Of course, one may use numerical integration even if $f(t)$ can be integrated analytically.] To this end, we divide the areas under the diagrams into n strips, called *panels*, each of width Δt, as shown in Fig. 12.7. The values of time that separate the panels are denoted by $t_0, t_1, \ldots, t_i, \ldots, t_n$, and the corresponding values of accelerations are written as $a_0, a_1, \ldots, a_i, \ldots, a_n$. Similar notation is used in the v-t and x-t diagrams.

According to Eq. (12.25), the area of the shaded panel in Fig. 12.7(a) is equal to the change in the velocity during the time interval from t_i to t_{i+1}, i.e.,

$$v_{i+1} = v_i + \text{area of } a\text{-}t \text{ panel}\Big]_{t_i}^{t_{i+1}} \tag{12.27}$$

Similarly, from Eq. (12.26) we see that the area of the shaded panel in Fig. 12.7(b) is equal to the change in the position from time t_i to t_{i+1}. Therefore,

$$x_{i+1} = x_i + \text{area of } v\text{-}t \text{ panel}\Big]_{t_i}^{t_{i+1}} \tag{12.28}$$

The task of numerical integration is as follows. Given the initial values v_0 and x_0, compute $v_1, v_2, \ldots, v_n$ and $x_1, x_2, \ldots, x_n$ using Eqs. (12.27) and (12.28). To implement the numerical method, we need a technique for estimating the area of each panel.

The simplest estimate for the area of a panel is to approximate it by a rectangle, as shown in Fig. 12.8(a). Using this approximation, Eqs. (12.27) and (12.28) yield

$$\boxed{v_{i+1} = v_i + a_i \Delta t} \tag{12.29}$$

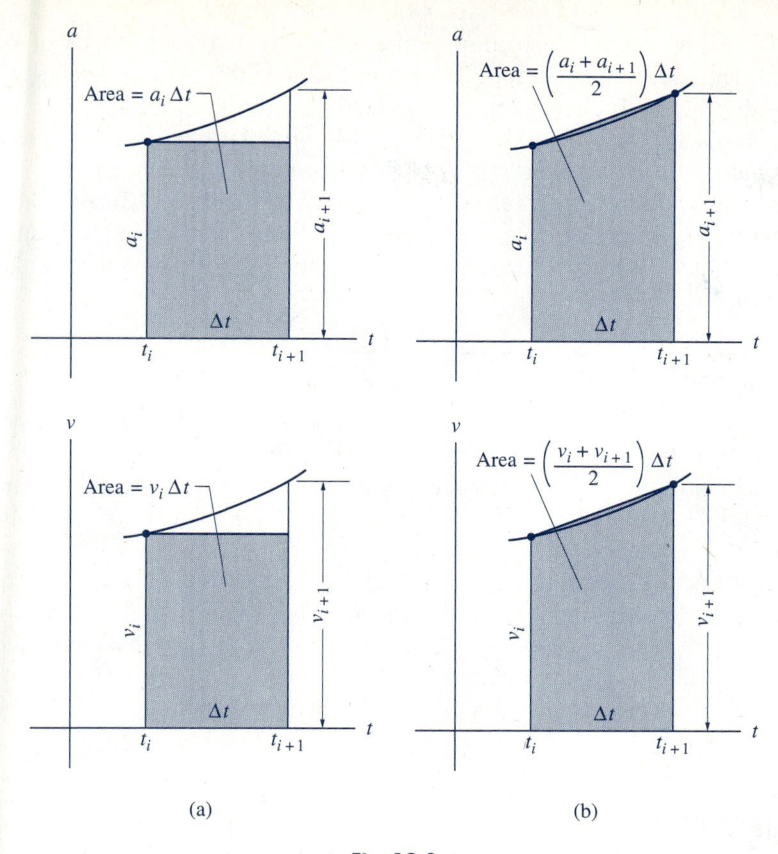

Fig. 12.8

and

$$x_{i+1} = x_i + v_i \Delta t \tag{12.30}$$

Increased accuracy is achieved if the panels are approximated by trapezoids as shown in Fig. 12.8(b), in which case Eqs. (12.27) and (12.28) become

$$v_{i+1} = v_i + \frac{a_i + a_{i+1}}{2} \Delta t \tag{12.31}$$

and

$$x_{i+1} = x_i + \frac{v_i + v_{i+1}}{2} \Delta t \tag{12.32}$$

Equations (12.31) and (12.32) are equivalent to the Trapezoidal Rule discussed in Appendix A.

Accuracy could be further increased by employing Simpson's rule (see Appendix A), where the curve bounding two adjacent panels is approximated by a parabola.

The computational procedure consists of repeated applications of Eqs. (12.29) and (12.30), or Eqs. (12.31) and (12.32), starting with $i = 0$ and ending with $i = n - 1$. Observe that the numerical procedure can be started only if the initial values v_0 and x_0 are known.* In addition, we see that there are two choices for implementing the numerical integration technique: (1) first determine the entire *v-t* curve, then compute the *x-t* curve, or (2) compute the corresponding values of v and x concurrently, thereby generating both curves simultaneously. Which of these techniques is employed is a matter of personal preference.

c. The general case: $a = f(v, x, t)$

In general, the forces that act on a particle can be functions of velocity, position, and time. If the forces are known (i.e., known functions of v, x, and t), the acceleration found by Newton's second law will also be a known function of these variables. Therefore, we will be able to evaluate $a = f(v, x, t)$ for given values of v, x, and t.

Numerical integration of the general equation of motion $a = f(v, x, t)$, where f is a known function, is almost identical to the procedure outlined previously for the special case $a = f(t)$. The numerical method that corresponds to the rectangular approximation is called the *Euler method.* The *modified Euler method* corresponds to the trapezoidal approximation. Both of these methods belong to a family of integration techniques known as the *Runge-Kutta methods.* These integration techniques are described below.

1. Euler Method In the Euler method, each panel under the *a-t* and *v-t* curves is approximated by a rectangle. Therefore, the following equations that were derived for the special case $a = f(t)$ are still valid.

$$\boxed{v_{i+1} = v_i + a_i \Delta t} \qquad \text{(12.29, repeated)}$$

$$\boxed{x_{i+1} = x_i + v_i \Delta t} \qquad \text{(12.30, repeated)}$$

Consequently, the computational procedure used for the special case $a = f(t)$ is also valid, with one exception. The *v-t* and *x-t* curves must be computed simultaneously because the evaluation of $a_i = f(v_i, x_i, t_i)$ that appears in Eq. (12.29) requires that all three of the arguments, including x_i, must be known. (Recall that for $a = f(t)$, one could compute the entire *v-t* diagram without considering the *x-t* diagram since only the value of t_i is required.)

Figure 12.9 illustrates the graphical interpretation of Euler's method. As shown in Fig. 12.9(a), all three curves are initially unknown. However, a single point on each curve, representing the initial values v_0 and x_0 and the computed value of $a_0 = f(v_0, x_0, t_0)$ is known.

The area of the first panel under the *a-t* diagram, $a_0 \Delta t$, is then used to compute $v_1 = v_0 + a_0 \Delta t$. Similarly, the area under the *v-t* diagram, $v_0 \Delta t$, yields $x_1 = x_0 + v_0 \Delta t$. Finally, the acceleration is calculated from the equation of motion $a_1 = f(v_1, x_1, t_1)$, which completes the first integration step. The re-

* Problems of this type are called "initial value problems."

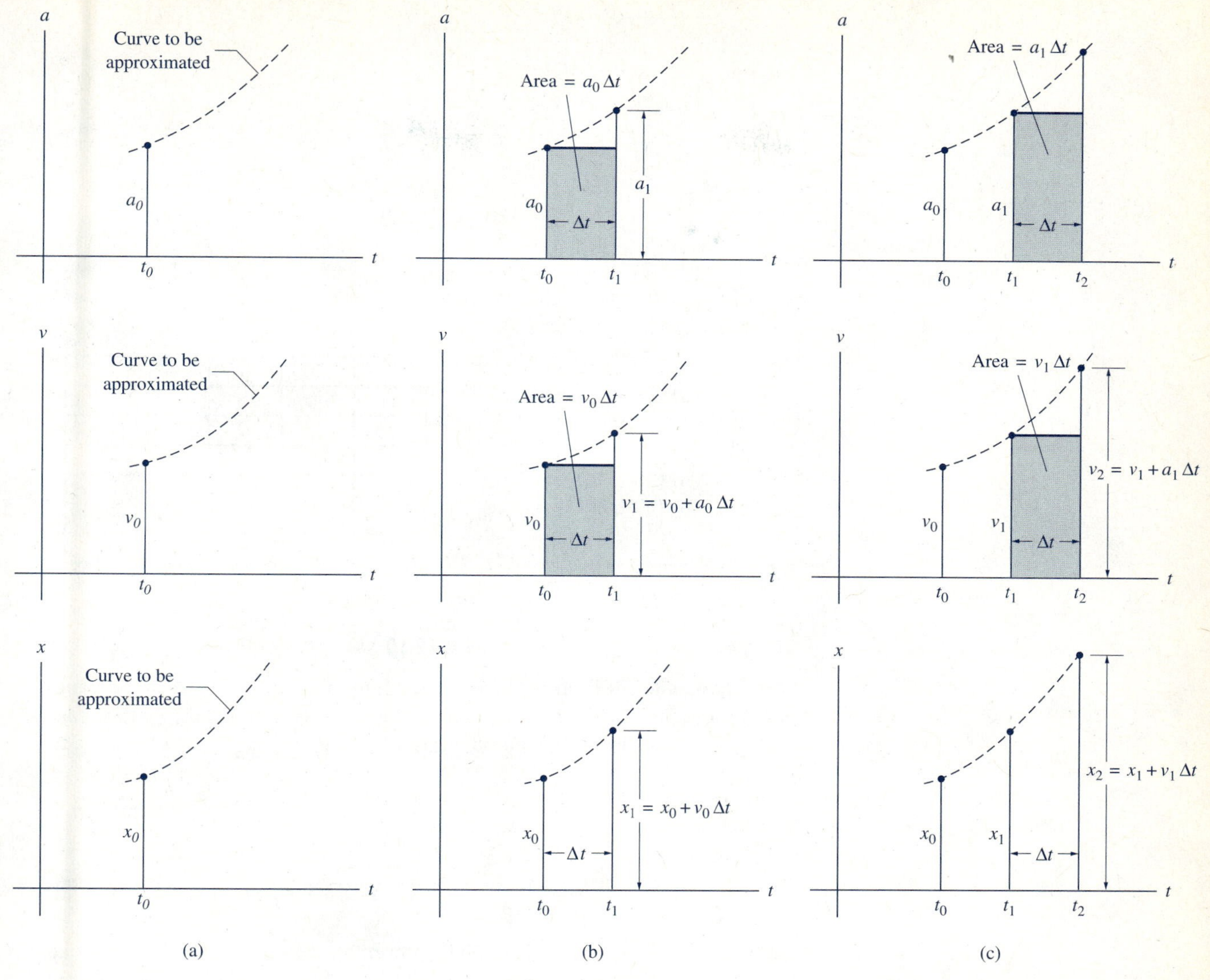

Fig. 12.9

sults are shown in Fig. 12.9(b). Note that there are now two known points on each of the three curves. The second integration step, illustrated in Fig. 12.9(c), repeats all of the operations of the first step. The rectangular approximation for the area under the *a*-*t* diagram gives $v_2 = v_1 + a_1\Delta t$, the approximating rectangle under the *v*-*t* diagram yields $x_2 = x_1 + v_1\Delta t$, and the equation of motion gives $a_2 = f(v_2, x_2, t_2)$. This scheme is repeated until t_n, the end of the time interval of interest is reached. In this manner, approximate values for a, v, and x are obtained at the times $t_1, t_2, \ldots, t_n$.

The algorithm that will compute the values of a_i, v_i, and x_i by Euler's method is presented in Fig. 12.10(a). The table in Fig. 12.10(b) is convenient for recording the numerical values if the algorithm is implemented with a calculator. If a computer program is used, its output should also be able to reproduce this table.

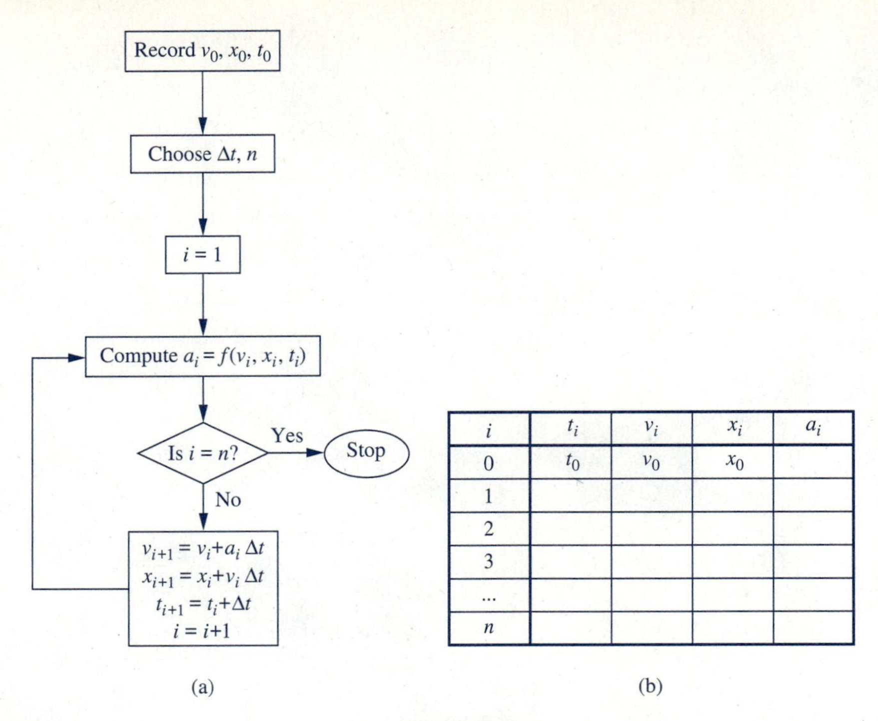

i	t_i	v_i	x_i	a_i
0	t_0	v_0	x_0	
1				
2				
3				
...				
n				

Fig. 12.10

2. Modified Euler Method The modified Euler method is based on the trapezoidal approximation. When this approximation was used for the special case $a = f(t)$, the equations for one integration step were

$$v_{i+1} = v_i + \frac{a_i + a_{i+1}}{2}\Delta t \qquad \text{(12.31, repeated)}$$

$$x_{i+1} = x_i + \frac{v_i + v_{i+1}}{2}\Delta t \qquad \text{(12.32, repeated)}$$

The presence of a_{i+1} in Eq. (12.31) presented no difficulty in the special case $a = f(t)$, since it could be evaluated for any value of time from $a_{i+1} = f(t_{i+1})$. Then v_{i+1} and x_{i+1} could be evaluated from Eqs. (12.31) and (12.32), respectively.

A major problem arises when we attempt to apply Eqs. (12.31) and (12.32) to the general case. From $a_{i+1} = f(v_{i+1}, x_{i+1}, t_{i+1})$ we see that the evaluation of a_{i+1} requires that v_{i+1} and x_{i+1} must be known beforehand. However, a_{i+1} is itself required to compute v_{i+1} from Eq. (12.31). Similarly, v_{i+1} is needed to calculate x_{i+1} from Eq. (12.32).

This problem is overcome by using the Euler method to predict, or estimate, the values of v_{i+1} and x_{i+1} that are required in Eqs. (12.31) and (12.32). Denoting these predictions by a caret (the symbol ^), we have from Eqs. (12.29) and (12.30)

$$\hat{v}_{i+1} = v_i + a_i \Delta t \qquad \text{(12.33a)}$$

$$\hat{x}_{i+1} = x_i + v_i \Delta t \qquad \text{(12.33b)}$$

Substituting these predicted values into Eqs. (12.31) and (12.32), we obtain

$$v_{i+1} = v_i + \frac{a_i + \hat{a}_{i+1}}{2}\Delta t \tag{12.34a}$$

$$x_{i+1} = x_i + \frac{v_i + \hat{v}_{i+1}}{2}\Delta t \tag{12.34b}$$

The accelerations in Eqs. (12.33a) and (12.34a) are evaluated, as in the Euler method, from the equation of motions as follows.

$$a_i = f(v_i, x_i, t_i) \tag{12.35a}$$

$$\hat{a}_{i+1} = f(\hat{v}_{i+i}, \hat{x}_{i+1}, t_{i+1}) \tag{12.35b}$$

Each integration step in the modified Euler method can be seen to consist of two parts. First, the predicted values are computed from Eqs. (12.33); next, the "improved" values are calculated using Eqs. (12.34). Apart from this modification, the computational details are identical to the Euler method, as can be seen from the flow diagram of Fig. 12.11(a). If the procedure is carried out by hand computation with the aid of a calculator, the results should be organized into a table, such as shown in Fig. 12.11(b).

3. Higher-order Runge-Kutta Methods As mentioned previously, Euler's method and the modified Euler method belong to a family of integration techniques known as the Runge-Kutta method. The members of this family are ranked, or ordered, according to their inherent accuracy. For example, the Euler method can be considered to be the first-order (the least accurate) member of the family. The modified Euler method is equivalent to the second-order Runge-Kutta method.

The integration technique most commonly used by engineers is the fourth-order Runge-Kutta method in which the areas of the panels are approximated by Simpson's rule (see Appendix A).* The derivation of the formulas for the fourth-order method will not be presented here because it requires a knowledge of advanced numerical methods.† These higher-order numerical methods are readily available, because a large volume of computer software is devoted to the solution of initial value problems. Many of the computer programs use the fourth-order Runge-Kutta method, but other equally accurate techniques are also commonly employed. The main differences among the various integration schemes lie in the manner by which the areas under the curves are approximated.

* Unless otherwise noted, the numerical integrations in this text were performed using the fourth-order Runge-Kutta method. If a different method is used, the results should still agree with those given in the text.

† A complete derivation of the fourth-order Runge-Kutta method can be found in most textbooks on numerical methods. For example, see *Applied Numerical Methods for Digital Computation,* 4th ed., by M. L. James, G. M. Smith, and J. C. Wolford. HarperCollins, New York, 1993.

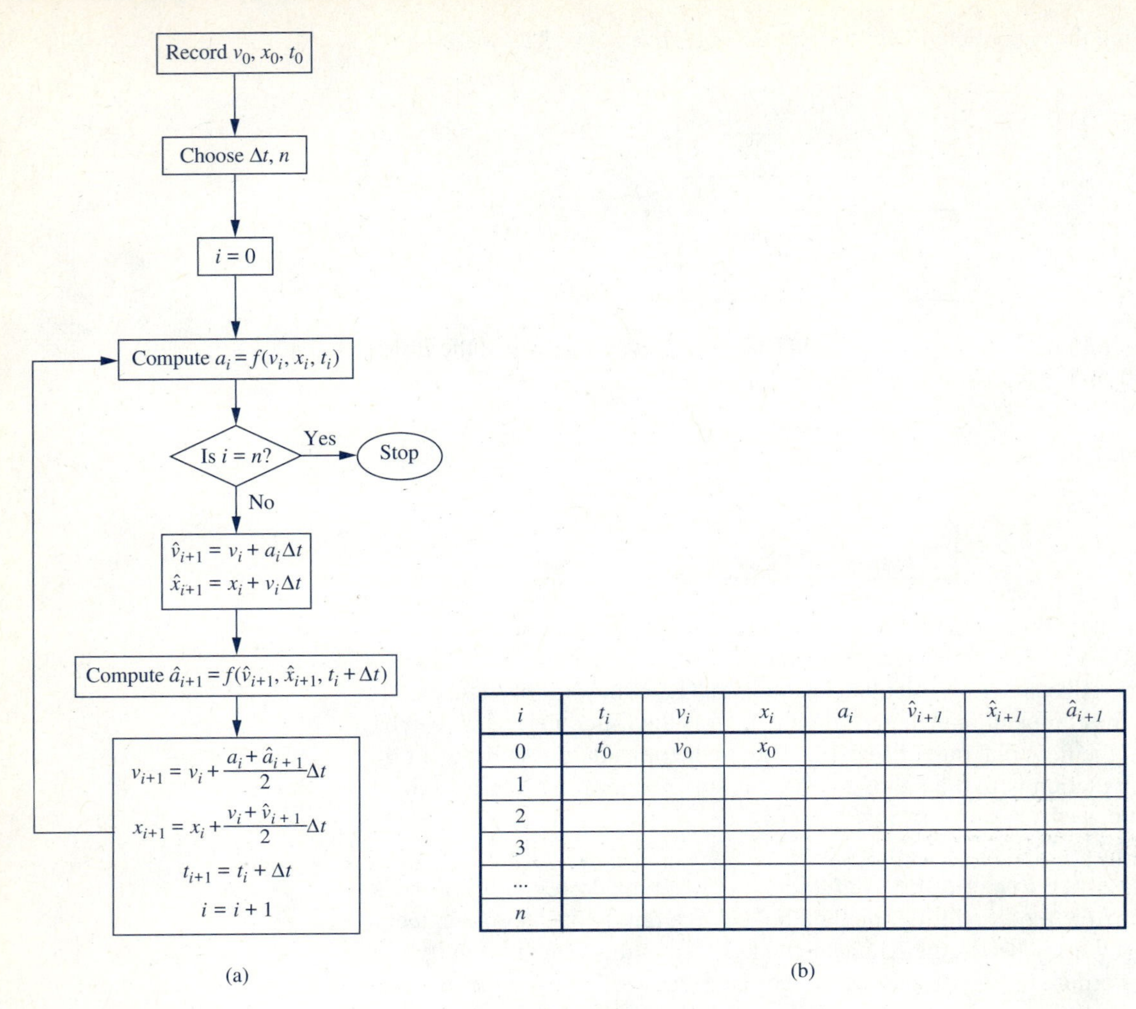

i	t_i	v_i	x_i	a_i	$\hat{v}_{i+1}$	$\hat{x}_{i+1}$	$\hat{a}_{i+1}$
0	t_0	v_0	x_0				
1							
2							
3							
...							
n							

Fig. 12.11

d. Sources of error

The accuracy of a numerical integration is determined by errors from two possible sources—truncation and round-off. Truncation errors arise because the areas used to represent the integrals are only approximations. For the first-order formulas, for which the integrals are approximated by rectangles, it can be shown that the truncation error is proportional to $(\Delta t)^2$. Therefore, the truncation error can be minimized by using a very small Δt.

Round-off errors are caused by the rounding off of numbers that occurs in all computers and calculators. Unlike truncation errors, round-off errors increase as Δt is made smaller.

To illustrate the nature of round-off errors, let us re-examine the first-order formula given in Eq. (12.30). For example, let $v_0 = 1$ and $a_0 = f(x_0, v_0, t_0) = 1$. Moreover, assume that the computer rounds off all numbers to seven significant digits. Suppose that we had $\Delta t = 5.555\,556 \times 10^{-3}$. The computation for v_1 would then yield $v_1 = 1.000\,000 + 5.555\,556 \times 10^{-3} = 1.005\,556$. Note that only four of the original seven figures in Δt contributed to v_1, i.e., three significant digits were lost. If Δt were made smaller, say $\Delta t = 5.555\,556 \times 10^{-6}$, it is not hard to see that six significant digits would disappear.

Round-off errors can be reduced by using double-precision arithmetic, but not all programming languages provide that option. Another solution is to avoid very small values of Δt, but in that case we have to accept a truncation error that is too large. The best compromise is to use higher-order integration formulas, such as the fourth-order Runge-Kutta method, that have small intrinsic truncation error and thus do not require that Δt be very small.

If accuracy is not of paramount importance, the modified Euler method is adequate. This technique is capable of yielding results accurate to within three significant digits with most programming languages. Moreover, the algorithm is simple enough to be implemented on a programmable calculator. Euler's method should be avoided because of its excessive truncation error in most problems.

e. Choice of time increment

When using the various numerical integration methods, particular attention must be paid to the selection of the integration step size Δt. Actually, the optimum value of Δt depends on the solution. If $x(t)$ or $v(t)$ vary rapidly, a small value of Δt is called for. On the other hand, a slowly varying solution may be obtained with sufficient accuracy using a larger Δt. In other words, the informed choice of Δt requires some knowledge of the solution. If this information is not available at the outset, it must be acquired by trial and error, that is, by experimenting with different values of Δt.

In some problems, excessively large Δt may result in numerical instability of the solution. Instability is the result of truncation errors being magnified in each successive integration step, thereby causing the cumulative error to increase exponentially. Instability of the solution will eventually result in overflow error (the numbers generated become larger than the biggest number that can be stored in the computer's memory).

Sample Problem 12.12

The air bubble shown in Fig. (a) breaks loose from the bottom of a shallow dish of water and rises to the surface. The equation of motion of the bubble, determined by its buoyancy and the viscous drag of the water, is $a = 80 - 16v$, where a is the acceleration in ft/s^2 and v is the velocity in ft/s. The position coordinate x is measured upward from the bottom of the dish. By hand computation, determine the velocity of the bubble when it reaches the surface of the water and compare your answer with the analytical solution $v = 3.988$ ft/s. Use (1) the Euler method with $\Delta t = 0.01$ s, and (2) the modified Euler method with $\Delta t = 0.02$ s. Use the tables in Figs. 12.10 and 12.11, respectively, to report your results.

3 in.

x

(a)

Solution

Part 1

The equation of motion is given as $a = f(v) = 80 - 16v$ ft/s^2, and we deduce that the initial conditions are $v_0 = 0$ and $x_0 = 0$ when $t = t_0 = 0$. From Fig. 12.10(a) we see that the equations required to carry out the numerical integration using Euler's method are

$$v_{i+1} = v_i + a_i \Delta t$$

$$x_{i+1} = x_i + v_i \Delta t$$

where a_i can be computed from the equation of motion

$$a_i = f(v_i) = 80 - 16v_i$$

Note that the time interval $\Delta t = 0.1$ s has been specified.

The table in Fig. (b) shows the results of the computations for the velocity, position, and acceleration of the bubble as it moves up through the water. The computations begin by entering the initial conditions, shown in Fig. (b). For illustrative

n	t_i (s)	v_i (ft/s)	x_i (ft)	a_i (ft/s^2)
0	$t_0 = 0.0000$	$v_0 = 0.0000$	$x_0 = 0.0000$	$a_0 = 80 - 16v_0$ $= 80 - 16(0)$ $= 80.000$
1	$t_1 = t_0 + \Delta t$ $= 0 + 0.01$ $= 0.01000$	$v_1 = v_0 + a_0 \Delta t$ $= 0 + 80.000(0.01)$ $= 0.8000$	$x_1 = x_0 + v_0 \Delta t$ $= 0 + 0(0.01)$ $= 0.0000$	$a_1 = 80 - 16v_1$ $= 80 - 16(0.8000)$ $= 67.200$
2	$t_2 = t_1 + \Delta t$ $= 0.01 + 0.01$ $= 0.0200$	$v_2 = v_1 + a_1 \Delta t$ $= 0.8000 + 67.200(0.01)$ $= 1.4720$	$x_2 = x_1 + v_1 \Delta t$ $= 0 + (0.8000)(0.01)$ $= 0.0080$	$a_2 = 80 - 16v_2$ $= 80 - 16(1.4720)$ $= 56.448$
3	0.0300	2.0365	0.0227	47.416
4	0.0400	2.5106	0.0431	39.830
5	0.0500	2.9089	0.0682	33.457
6	0.0600	3.2345	0.0973	28.104
7	0.0700	3.5245	0.1297	23.607
8	0.0800	3.7606	0.1650	19.830
9	0.0900	3.9589	0.2026	16.657
10	0.1000	4.1255	0.2422	13.992
11	0.1100	4.2654	0.2834	11.753

(b)

purposes, all of the computations are shown for $n = 0$, 1, and 2. The remaining entries in the table show only the results of the calculations. (If a programmable calculator is used, a relatively short program can be written that will eliminate much of the work required to complete the table.)

Inspecting the entries for $n = 10$ and $n = 11$ in Fig. (b), we see that $x = 0.25$ ft, which corresponds to the bubble reaching the surface of the water, does not occur exactly at the end of one of our time intervals. Therefore, linear interpolation should be used to estimate the velocity when $x = 0.25$ ft. Letting v be the velocity when $x = 0.25$ ft, the linear interpolation shown in Fig. (c) yields

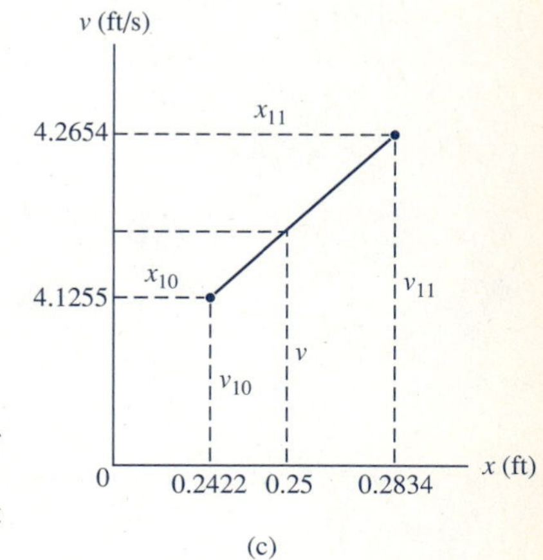

(c)

$$\frac{v_{11} - v_{10}}{x_{11} - x_{10}} = \frac{v - v_{10}}{0.25 - x_{10}}$$

which yields

$$\frac{4.2654 - 4.1255}{0.2834 - 0.2422} = \frac{v - 4.1255}{0.25 - 0.2422}$$

from which we obtain

$$v = 4.152 \text{ ft/s} \qquad \textit{Answer}$$

Comparing this result with the value obtained analytically, 3.988 ft/s, we find that the percent error is

$$\% \text{ error} = \frac{4.152 - 3.988}{3.988} \times 100 = 4.11\%$$

Part 2

According to Fig. 12.11, the equations required for numerical integration by the modified Euler method are

$$\hat{v}_{i+1} = v_i + a_i \Delta t$$

$$\hat{x}_{i+1} = x_i + v_i \Delta t$$

$$v_{i+1} = v_i + \frac{a_i + \hat{a}_{i+1}}{2} \Delta t$$

$$x_{i+1} = x_i + \frac{v_i + \hat{v}_{i+1}}{2} \Delta t$$

The equation of motion is used to compute a_i and $\hat{a}_{i+1}$, as follows.

$$a_i = f(v_i) = 80 - 16 v_i$$

$$\hat{a}_{i+1} = f(v_{i+1}) = 80 - 16 v_{i+1}$$

The specified time interval $\Delta t = 0.02$ s is twice as large as that used in Part 1.

Figure (d) shows the numerical values that are computed by the modified Euler method. The computations begin by entering the initial conditions, t_0, v_0, and x_0. All details of the computations are shown for $n = 0$ and 1. The remainder of the table shows only the results.

n	t_i (s)	v_i (ft/s)	x_i (ft)	a_i (ft/s^2)	$\hat{v}_{i+1}$ (ft/s)	$\hat{x}_{i+1}$ (ft)	$\hat{a}_{i+1}$ (ft/s^2)
0	$t_0 = 0.0000$	$v_0 = 0.0000$	$x_0 = 0.0000$	$a_0 = 80 - 16v_0 = 80 - 16(0) = 80.000$	$\hat{v}_1 = v_0 + a_0\Delta t = 0 + 80(0.02) = 1.6000$	$\hat{x}_1 = x_0 + v_0\Delta t = 0 + 0 = 0.0000$	$\hat{a}_1 = 80 - 16\hat{v}_1 = 80 - 16(1.6000) = 54.400$
1	$t_1 = t_0 + \Delta t = 0 + 0.02 = 0.0200$	$v_1 = v_0 + \frac{a_0 + \hat{a}_1}{2}\Delta t = 0 + \frac{80 + 54.400}{2}(0.02) = 1.3440$	$x_1 = x_0 + \frac{v_0 + \hat{v}_1}{2}\Delta t = 0 + \frac{0 + 1.600}{2}(0.02) = 0.0160$	$a_1 = 80 - 16v_1 = 80 - 16(1.3440) = 58.496$	$\hat{v}_2 = v_1 + a_1\Delta t = 1.3440 + 58.496(0.02) = 2.5139$	$\hat{x}_2 = x_1 + v_1\Delta t = 0.0160 + 1.3440(0.02) = 0.0429$	$\hat{a}_2 = 80 - 16\hat{v}_2 = 80 - 16(2.5139) = 39.778$
2	0.0400	2.3267	0.0546	42.772	3.1822	0.1011	29.085
3	0.0600	3.0453	0.1097	31.275	3.6708	0.1706	21.267
4	0.0800	3.5707	0.1768	22.868	4.0281	0.2482	15.551
5	0.1000	3.9549	0.2528	16.721	4.2893	0.3319	11.371

(d)

From the table we see that the bubble reaches the surface ($x = 0.25$ ft) between $t_4 = 0.08$ s and $t_5 = 0.10$ s. Once again, linear interpolation can be used to find the velocity v when $x = 0.25$ ft. Following the same procedure as used in Part (1), the result of the interpolation is

$$v = 3.941 \text{ ft/s} \qquad \textit{Answer}$$

Comparing this value with the analytic solution 3.988 ft/s, we see that the percentage error is

$$\% \text{ error} = \frac{3.941 - 3.988}{3.988} \times 100 = -1.18\%$$

The amount of computation involved in each of the two methods used in this sample problem was approximately the same, as judged by the total number of entries in the tables. (The modified Euler method required about twice the arithmetic for each integration step, as compared to the Euler method. But the step size Δt used in the modified Euler method was twice as large as that used in the Euler method.) However, we observe that the result obtained with the modified Euler method is considerably more accurate. It is generally true that a higher-order method will produce better accuracy than a lower-order method for a given amount of computational effort.

Sample Problem 12.13

The coefficients of static and kinetic friction between the block in Fig (a) and the horizontal surface are the same, $\mu = 0.2$. The spring attached to the block has the stiffness $k = 30$ N/m and it is unstretched when $x = 0$. When $t = 0$, the block is at $x = 0$ and is moving to the right with a speed of $v = 6$ m/s. (1) Derive an equation of motion for the block that is valid for both positive and negative values of v. (2) Determine the times and the corresponding positions when the block comes to rest during the time period $t = 0$ to $t = 1.2$ s. (3) Sketch the plot of acceleration vs. position for $t = 0$ to $t = 1.2$ s.

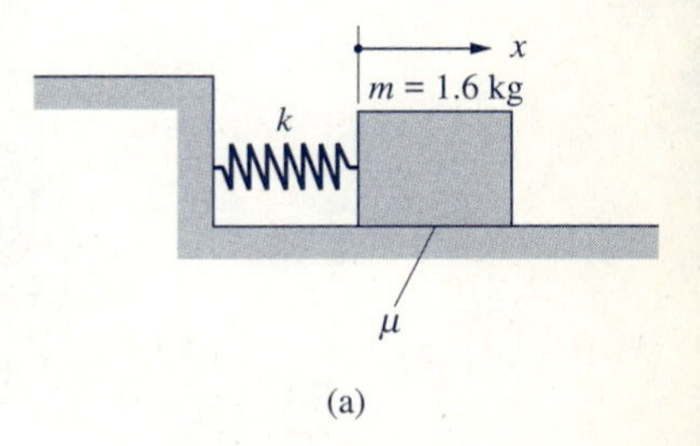

(a)

Solution

Part 1

The free-body diagrams and the mass-acceleration diagrams of the block moving to the right and to the left are shown in Figs. (b) and (c), respectively. Although the MADs are identical (the positive direction of the acceleration is to the right in both cases), the FBDs differ because the kinetic friction force always opposes the motion (velocity). Summation of forces in the y-direction yields $\Sigma F_y = ma_y = 0$, which gives $N_1 = mg$. Summing forces in the x-direction gives

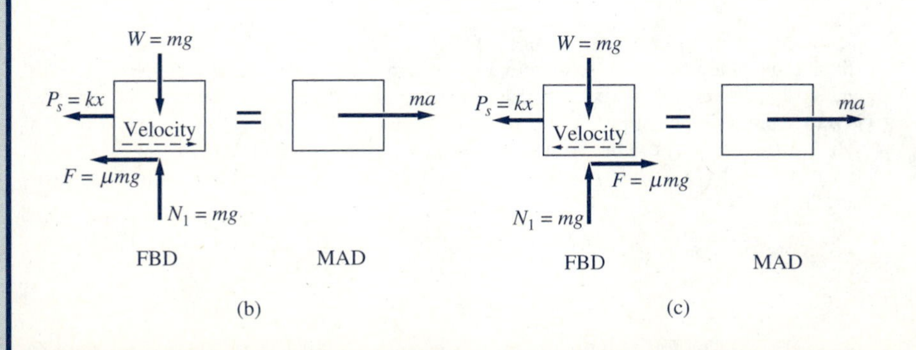

(b) (c)

$$\Sigma F_x = ma_x \quad \xrightarrow{+} \quad -kx \pm \mu mg = ma \tag{a}$$

which can be written as

$$a = -\frac{k}{m}x \pm \mu g \tag{b}$$

The plus sign on the friction term in Eqs. (a) and (b) applies when the block is moving to the left, and the minus sign is to be used when it is moving to the right.*

Substituting the given values for k, m, and μ, and assuming that x is measured in meters, Eq. (b) becomes

$$a = -\frac{30}{1.6}x \pm (0.2)(9.81) = -18.75x \pm 1.962 \text{ m/s}^2 \tag{c}$$

The easiest method of programming equations that possess a dual sign, such as Eq. (c), is to employ the "sign" function that is built into most programming languages.† Referring to this function as "sgn," the properties of sgn v are defined to be

$$\text{sgn } v = \begin{cases} 1 & \text{if } v > 0 \\ 0 & \text{if } v = 0 \\ -1 & \text{if } v < 0 \end{cases} \tag{d}$$

Using sgn v, the equation of motion in Eq. (c) can be written as

$$a = \ddot{x} = -18.75x - 1.962 \text{ sgn } v \text{ m/s}^2 \qquad \textit{Answer} \tag{e}$$

which is valid for both positive and negative values of v.

Part 2

Observe that the equation of motion in Eq. (e) gives the acceleration as a function of the position x, i.e., $a = a(x)$. Because $a(x)$ in Eq. (e) is simply a linear function of x, we could integrate $a\,dx = v\,dv$ analytically to obtain v as a function of x. The positions where the block stops could then be obtained by setting $v = 0$ and solving for x. However, this analysis would not determine the times when the block stops. If time t is to be treated as the independent variable, the differential equation of motion in Eq. (e) must be integrated numerically. [An analytical solution is not possible because the variables x and t in Eq. (e) are not separable.]

The numerical solution of Eq. (e) was accomplished with the fourth-order Runge-Kutta method using the time increment $\Delta t = 0.02$ s and the initial conditions $t_0 = 0$, $x_0 = 0$, and $v_0 = 6$ m/s. The resulting velocity-position curve is shown in Fig. (d). Observe that in the time interval $t = 0$ to 1.2 s, the block stops twice, that is, there are two points on the curve where $v = 0$. The values

* By using $F = \pm\mu N_1 = \pm\mu mg$ for the friction force, we assume that the block is always slipping, i.e., the velocity is never zero, or it is zero for only an instant. This presumes, of course, that the spring force is large enough to overcome the friction force when the velocity is zero, thereby perpetuating the motion.

† In BASIC, SGN(v) has the form described in Eq. (d). In FORTRAN, the corresponding function is SIGN(z,v), which returns the value of the first argument with the sign of the second argument.

of t and x corresponding to $v = 0$, which can be obtained from the numerical output (not shown here), are

$$t_1 = 0.346 \text{ s} \qquad x_1 = 1.28 \text{ m}$$
$$t_2 = 1.072 \text{ s} \qquad x_2 = -1.08 \text{ m} \qquad \textit{Answer}$$

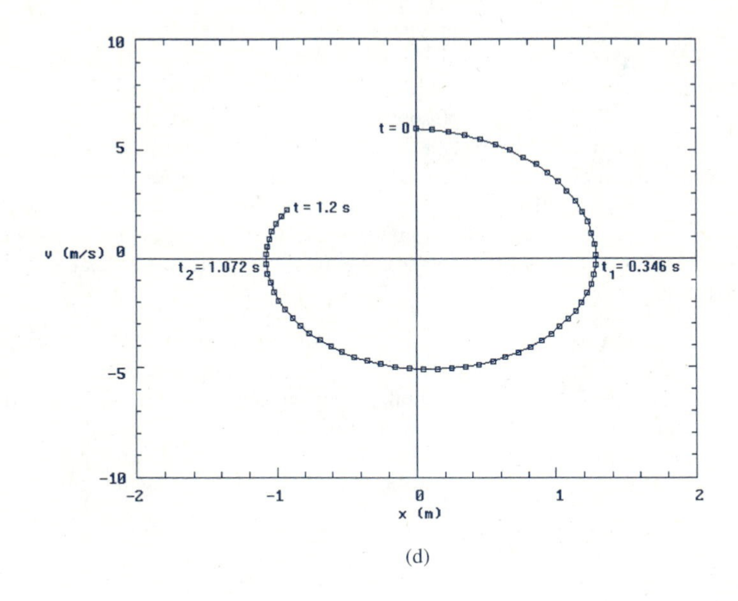

(d)

Part 3

According to Eq. (c), the plot of acceleration vs. position consists of the two straight lines

$$a = -18.75x - 1.962 \text{ m/s}^2 \quad \text{for} \quad v > 0$$
$$a = -18.75x + 1.962 \text{ m/s}^2 \quad \text{for} \quad v < 0$$

as shown in Fig. (e). These lines have the same slope (-18.75 s^{-2}), but different intercepts with the a-axis (-1.962 and $+1.962 \text{ m/s}^2$). The acceleration "jumps" between these lines whenever the velocity reverses sign at the positions x_1 and x_2 found in Part 2.

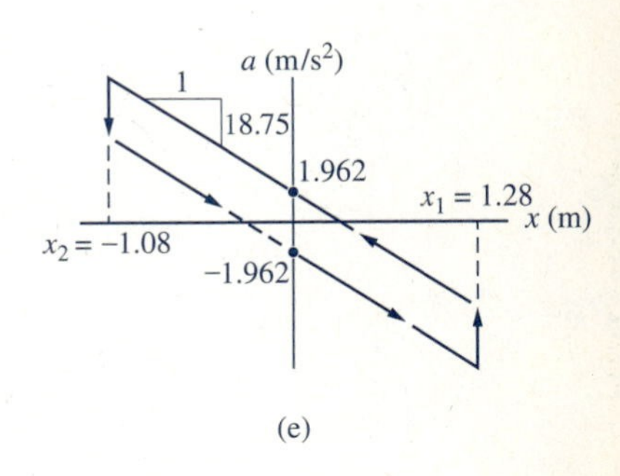

(e)

PROBLEMS

General Instructions

Problems 12.75–80 are to be solved by hand computation using either the Euler method or the modified Euler method, as directed by the problem statement. The results should be reported using the tables shown in Figs. 12.10(b) or 12.11(b).

Problem 12.81 requires that a computer program be written which performs numerical integration by the modified Euler method. (This program can be used to solve subsequent problems, wherever applicable.)

Problems 12.82–85 are to be solved by the modified Euler method on a computer. (The program written in Prob. 12.81 may be used for this purpose.)

Problems 12.86–90 are to be solved by the numerical method that is recommended by your instructor. (Although not absolutely necessary, a computer program with plotting capability would be helpful for performing the plots requested in each problem.) The time increment Δt should be small enough to yield results that are accurate to within three significant digits.

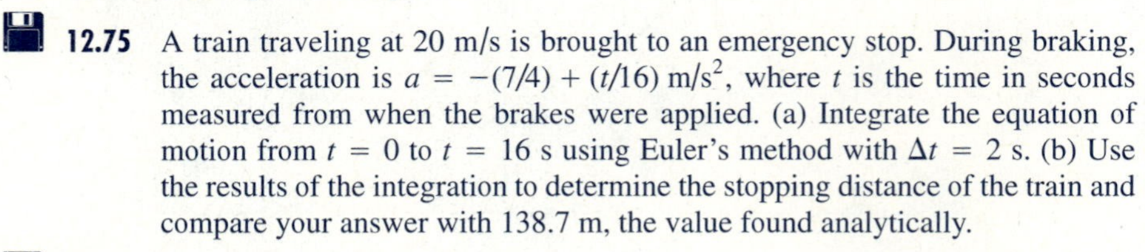

12.75 A train traveling at 20 m/s is brought to an emergency stop. During braking, the acceleration is $a = -(7/4) + (t/16)$ m/s^2, where t is the time in seconds measured from when the brakes were applied. (a) Integrate the equation of motion from $t = 0$ to $t = 16$ s using Euler's method with $\Delta t = 2$ s. (b) Use the results of the integration to determine the stopping distance of the train and compare your answer with 138.7 m, the value found analytically.

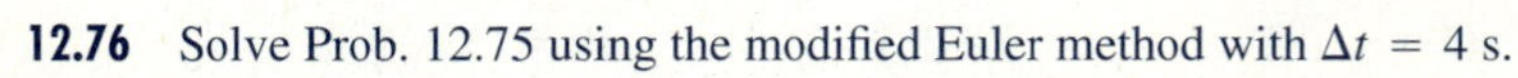

12.76 Solve Prob. 12.75 using the modified Euler method with $\Delta t = 4$ s.

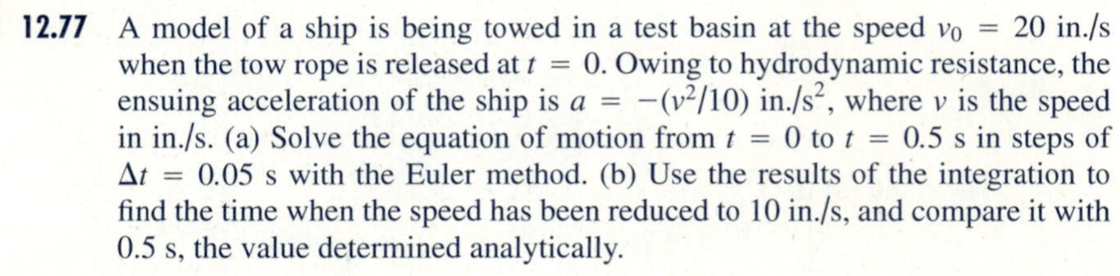

12.77 A model of a ship is being towed in a test basin at the speed $v_0 = 20$ in./s when the tow rope is released at $t = 0$. Owing to hydrodynamic resistance, the ensuing acceleration of the ship is $a = -(v^2/10)$ in./s^2, where v is the speed in in./s. (a) Solve the equation of motion from $t = 0$ to $t = 0.5$ s in steps of $\Delta t = 0.05$ s with the Euler method. (b) Use the results of the integration to find the time when the speed has been reduced to 10 in./s, and compare it with 0.5 s, the value determined analytically.

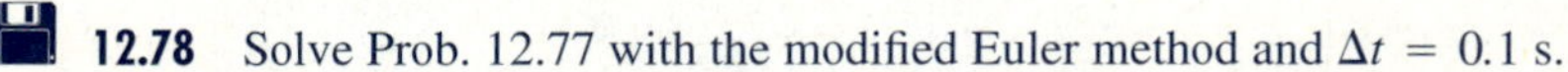

12.78 Solve Prob. 12.77 with the modified Euler method and $\Delta t = 0.1$ s.

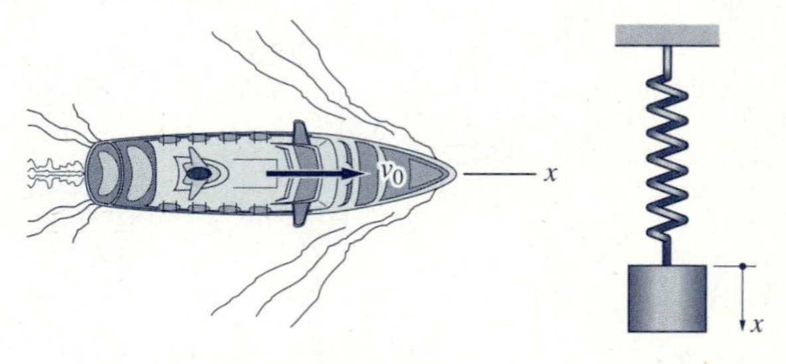

Fig. P12.77, P12.78 **Fig. P12.79, P12.80**

12.79 The mass suspended from the spring is released from rest at $x = 0.1$ m. The acceleration of the mass is $a = -36x$ m/s^2, where x (measured in meters) is the displacement of the mass from its equilibrium position $x = 0$. (a) Solve the equation of motion from $t = 0$ to $t = 0.3$ s in steps of $\Delta t = 0.05$ s using the Euler method. (b) From the results of the integration determine the time when the mass passes through the equilibrium position $x = 0$ for the first time. Compare your answer with the value found analytically: $(\pi/12)$ s.

12.80 Solve Prob. 12.79 with the modified Euler method using $\Delta t = 0.05$ s.

12.81 Write a computer program that performs numerical integration with the modified Euler method. The output should appear as a table similar to that shown below. Test the program by integrating the equation of motion $a = -36x$ from $t = 0$ to $t = 0.25$ s in steps of $\Delta t = 0.025$ s, starting with the initial conditions $x_0 = 0.1$, $v_0 = 0$. Your results should be identical to those presented in the following table.

i	$t(i)$	$x(i)$	$v(i)$	$a(i)$
0	0.000	0.100 00	0.000 00	−3.600 00
1	0.025	0.098 88	−0.090 00	−3.559 50
2	0.050	0.095 51	−0.177 98	−3.438 46
3	0.075	0.089 99	−0.261 93	−3.239 60
4	0.100	0.082 43	−0.339 98	−2.967 41
5	0.125	0.073 00	−0.410 34	−2.628 05
6	0.150	0.061 92	−0.471 42	−2.229 18
7	0.175	0.049 44	−0.521 85	−1.779 82
8	0.200	0.035 84	−0.560 47	−1.290 13
9	0.225	0.021 42	−0.586 42	−0.771 19
10	0.250	0.006 52	−0.599 10	−0.234 74

12.82 The acceleration of the rocket described in Prob. 12.31 is given by

$$a = \begin{cases} 20.19 \text{ m/s}^2 & \text{when } t \leq 20 \text{ s} \\ -9.81 \text{ m/s}^2 & \text{when } t > 20 \text{ s} \end{cases}$$

Using the sgn function, this acceleration may be written as the single equation

$$a = 15 \operatorname{sgn}(20 - t) + 5.19 \text{ m/s}^2$$

Determine (a) the time required for the rocket to reach its maximum altitude after being launched from rest on the surface of the earth; and (b) the maximum altitude. Use $\Delta t = 5$ s and compare your answers with the values obtained analytically: 61.2 s and 12 350 m.

12.83 The acceleration of the rocket sled in Prob. 12.33 is $a = 32.2[(1/3)e^{-0.2t} - 0.05]$ ft/s^2 where t is the time in seconds measured from when the sled starts from rest. Compute the time and the distance traveled when the sled comes to a stop, using $\Delta t = 2$ s. Compare your answer with 33.3 s and 627 ft, the values obtained analytically.

12.84 The free vertical fall of the skydiver described in Prob. 12.37 is governed by the equation

$$a = 32.2(1 - 32.2 \times 10^{-6} v^2) \text{ ft/s}^2$$

where v is the speed measured in feet per second. The positive directions of v and a are directed downward. Compute the time that it takes the skydiver to reach 100 mph after jumping. Use $\Delta t = 0.5$ s and compare your answer with 6.55 s, the value obtained analytically.

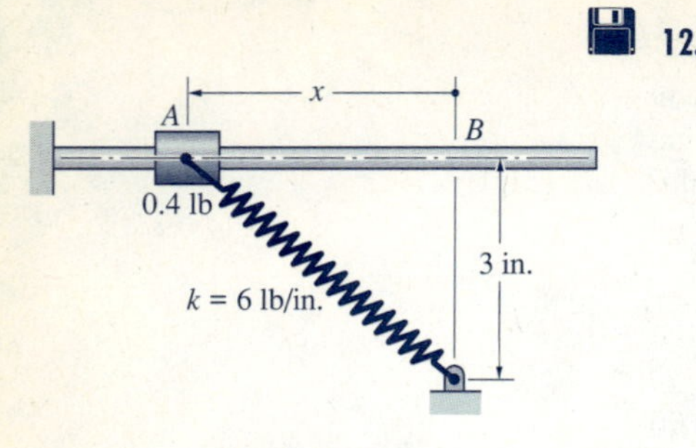

Fig. P12.85

12.85 The free length of the spring that is attached to the 0.4-lb slider A is 5 in. The slider is released from rest when $x = 8$ in. Neglecting friction, the ensuing motion is governed by the equation

$$a = -5796\left(1 - \frac{5}{\sqrt{x^2 + 9}}\right)x \text{ in./s}^2$$

where x is measured in inches. Using $\Delta t = 0.025$ s, determine the speed of the slider when it reaches point B. Compare your answer with 223 in./s, the value found analytically.

12.86 Solve Prob. 12.85 if the slider A is released from rest at $x = 6.5$ in. Plot the velocity vs. x for $0 \le x \le 6.5$ in.

12.87 A 250-lb object is released from rest at 30 000 ft above the surface of the earth. The aerodynamic drag force F_D acting on the object can be approximated by

$$F_D = 0.156\, v^2 e^{-(3.211\times10^{-5})x} \text{ lb}$$

where v is the speed (ft/s) and x is the elevation (ft). The exponential term accounts for the variation of air density with elevation. (W. R. Bennett, Jr., *Scientific and Engineering Problem-Solving with the Computer,* Prentice-Hall, 1976.) As a result, the acceleration of the object can be shown to be

$$a = -32.2\left[1 - \left(6.24 \times 10^{-4}\right)v^2 e^{-(3.211\times10^{-5})x}\right] \text{ft/s}^2$$

(a) Determine the maximum speed of the object and the elevation at which this speed occurs. (b) Plot the velocity vs. elevation from the time of release until the maximum speed is attained.

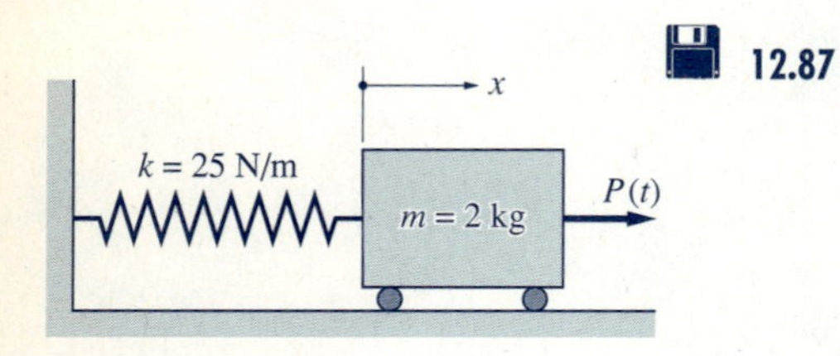

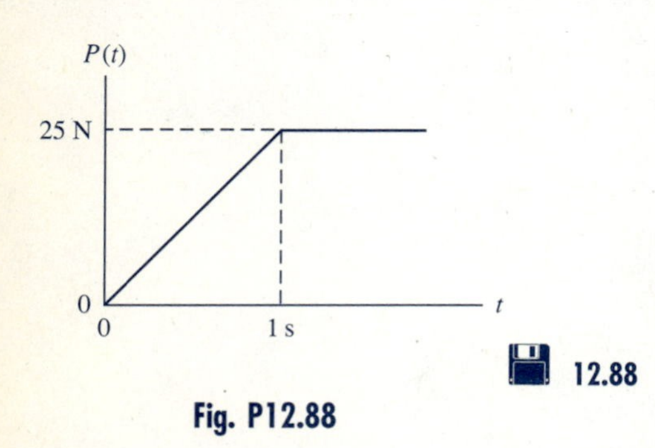

Fig. P12.88

12.88 The 2-kg block is at rest in the position shown with the spring unstretched when the force $P(t)$ is applied. The resulting acceleration (m/s^2) of the block is

$$a = \begin{cases} 12.5(t - x) & \text{if } t \le 1.0 \text{ s} \\ 12.5(1 - x) & \text{if } t > 1.0 \text{ s} \end{cases}$$

where x is measured in meters. (a) Determine the maximum displacement and maximum velocity of the block. (b) Plot the velocity vs. displacement for the time interval $0 \le t \le 5$ s.

12.89 An object is fired vertically upward with the initial speed $v_0 = 220$ ft/s. It is known that the aerodynamic drag force is proportional to the speed squared and that the terminal speed is $v_\infty = 440$ ft/s. (a) Derive the single equation of motion that is valid for both upward and downward flight of the object. (b) Compute the maximum elevation. (c) Plot the speed versus elevation.

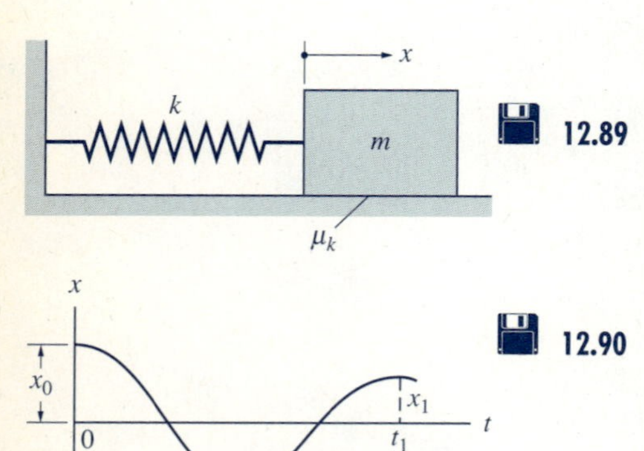

Fig. P12.90

12.90 The 5-kg block is attached to a spring of stiffness $k = 2.4$ kN/m. The coefficient of kinetic friction between the block and the horizontal surface is $\mu_k = 0.4$. The block is released from rest at $x = 0.10$ m, where x is measured from the position where the spring is unstretched. The resulting plot of x vs. t is shown in the figure. (a) Derive the equation of motion of the mass. (b) Determine t_1 and x_1 and verify that the result agrees with the analytical relationship $x_0 - x_1 = 4\mu_k gm/k$. (c) Plot x vs. t for $0 \le t \le t_1$.

12.9 Numerical Solution of Coupled Equations of Motion

a. Multiple degrees of freedom

The equation of motion for a system with one degree of freedom (DOF), such as a particle undergoing rectilinear motion, has the general form $a = f(v, x, t)$, or $\ddot{x} = f(\dot{x}, x, t)$. The equation states that, in general, the acceleration a is a function of the velocity v, the position coordinate x, and the time t. The numerical methods for integrating second-order differential equations of this type are explained in Art. 12.8. Recall that the initial values x_0 and v_0 had to be given before the integration could be started.

Consider next a system that possesses two DOFs, e.g., the motion of a projectile or two coupled particles undergoing rectilinear motions (see Sample Problem 12.14). For this case, two coordinates, which we label as x_1 and x_2, will be required to specify the position of the system. There are also two equations of motion of the form

$$\begin{aligned} a_1 &= f_1(v_1, v_2, x_1, x_2, t) \quad \text{or} \quad \ddot{x}_1 = f_1(\dot{x}_1, \dot{x}_2, x_1, x_2, t) \\ a_2 &= f_2(v_1, v_2, x_1, x_2, t) \quad \text{or} \quad \ddot{x}_2 = f_2(\dot{x}_1, \dot{x}_2, x_1, x_2, t) \end{aligned} \tag{12.36}$$

Equations of this type are said to be *coupled* because the acceleration associated with one position coordinate depends on one or more variables associated with the other position coordinate, e.g., a_1 depends on v_2. Therefore, it is not possible to solve one equation independently of the other. For a two-DOF system, four initial conditions are required (two for each DOF); namely, $(x_1)_0$, $(v_1)_0$, $(x_2)_0$, and $(v_2)_0$ must be known at some value of time (usually $t = 0$).

The analysis of two DOF systems can easily be extended to a system with N DOFs. Such a system is governed by N second-order equations of motion, one equation for each DOF. Letting $x_1, x_2, \ldots, x_N$ be the independent position coordinates, the N equations have the form

$$\begin{aligned} a_1 &= f_1(v_1, v_2, \ldots, v_N, x_1, x_2, \ldots, x_N, t) \quad \text{or} \\ \ddot{x}_1 &= f_1(\dot{x}_1, \dot{x}_2, \ldots, \dot{x}_N, x_1, x_2, \ldots, x_N, t) \\ a_2 &= f_2(v_1, v_2, \ldots, v_N, x_1, x_2, \ldots, x_N, t) \quad \text{or} \\ \ddot{x}_2 &= f_2(\dot{x}_1, \dot{x}_2, \ldots, \dot{x}_N, x_1, x_2, \ldots, x_N, t) \\ &\vdots \\ a_N &= f_N(v_1, v_2, \ldots, v_N, x_1, x_2, \ldots, x_N, t) \quad \text{or} \\ \ddot{x}_N &= f_N(\dot{x}_1, \dot{x}_2, \ldots, \dot{x}_N, x_1, x_2, \ldots, x_N, t) \end{aligned} \tag{12.37}$$

The integration of these equations requires $2N$ initial values: $(x_1)_0$ and $(v_1)_0$, $(x_2)_0$ and $(v_2)_0, \ldots, (x_N)_0$ and $(v_N)_0$.

Coupled differential equations can be solved analytically only if the equations are linear, and even then the solution may be tedious. If the equations are nonlinear, it is usually impossible to determine the solutions analytically, leaving numerical methods as the only practical means of solution.

b. Methods of solution

The techniques discussed in Art. 12.8 (Euler's method, the modified Euler method, and the higher-order Runge-Kutta methods) will work for any number of equations with only minor modifications. For the sake of brevity, our discussion will be limited to the solution of two equations. The extension to more than two equations will be obvious. The discussions of sources of error and choice of time increment in Art. 12.8 remain valid regardless of the number of equations being solved.

1. Euler's method According to Eq. (12.36), the accelerations at any time t_i can be obtained from

$$\begin{aligned}(a_1)_i &= f_1[(v_1)_i, (v_2)_i, (x_1)_i, (x_2)_i, t_i] \\ (a_2)_i &= f_2[(v_1)_i, (v_2)_i, (x_1)_i, (x_2)_i, t_i]\end{aligned} \tag{12.38}$$

provided that all positions and velocities are known at time t_i. Clearly, this is the case at the initial time t_0, where the initial values are known. After the accelerations have been computed, the velocities and positions at time $t_{i+1} = t_i + \Delta t$ can be evaluated from Eqs. (12.29) and (12.30);

$$\begin{aligned}(v_1)_{i+1} &= (v_1)_i + (a_1)_i\Delta t \\ (v_2)_{i+1} &= (v_2)_i + (a_2)_i\Delta t \\ (x_1)_{i+1} &= (x_1)_i + (v_1)_i\Delta t \\ (x_2)_{i+1} &= (x_2)_i + (v_2)_i\Delta t\end{aligned} \tag{12.39}$$

Numerical integration of two equations of motion by Euler's method is thus essentially the same procedure that was detailed in the flow diagram in Fig. 12.10(a). The only difference is that there are now two accelerations, velocities, and positions to be computed at each step.

2. Modified Euler method To extend the modified Euler method described in Art. 12.8 to two coupled DOFs, we must again make provision for two equations at each step of the integration. All other details outlined in Fig. 12.11 remain unchanged. Each step of the integration procedure thus contains the following stages. First, compute the accelerations from Eq. (12.38), and then evaluate the predicted velocities and positions from Eq. (12.33);

$$\begin{aligned}(\hat{v}_1)_{i+1} &= (v_1)_i + (a_1)_i\Delta t \\ (\hat{v}_2)_{i+1} &= (v_2)_i + (a_2)_i\Delta t \\ (\hat{x}_1)_{i+1} &= (x_1)_i + (v_1)_i\Delta t \\ (\hat{x}_2)_{i+1} &= (x_2)_i + (v_2)_i\Delta t\end{aligned} \tag{12.40}$$

Next, compute the corresponding accelerations from Eq. (12.35b),

$$\begin{aligned}(\hat{a}_1)_{i+1} &= f_1[(\hat{v}_1)_{i+1}, (\hat{v}_2)_{i+1}, (\hat{x}_1)_{i+1}, (\hat{x}_2)_{i+1}, t_{i+1}] \\ (\hat{a}_2)_{i+1} &= f_2[(\hat{v}_1)_{i+1}, (\hat{v}_2)_{i+1}, (\hat{x}_1)_{i+1}, (\hat{x}_2)_{i+1}, t_{i+1}]\end{aligned} \tag{12.41}$$

Finally, the corrected values of velocities and positions are given by Eq. (12.34):

$$\begin{aligned}
(v_1)_{i+1} &= (v_1)_i + \tfrac{1}{2}[(a_1)_i + (\hat{a}_1)_{i+1}]\Delta t \\
(v_2)_{i+1} &= (v_2)_i + \tfrac{1}{2}[(a_2)_i + (\hat{a}_2)_{i+1}]\Delta t \\
(x_1)_{i+1} &= (x_1)_i + \tfrac{1}{2}[(v_1)_i + (\hat{v}_1)_{i+1}]\Delta t \\
(x_2)_{i+1} &= (x_2)_i + \tfrac{1}{2}[(v_2)_i + (\hat{v}_2)_{i+1}]\Delta t
\end{aligned} \tag{12.42}$$

Due to the large number of computations, it is not practical to implement the modified Euler method by hand calculation.

3. Higher-order Runge-Kutta methods The fourth-order Runge-Kutta method mentioned in Art. 12.8 is also applicable to more than one equation of motion. As noted previously, it is the preferred numerical integration technique if appropriate computer software is available, because of its superior accuracy compared to the lower-order methods.

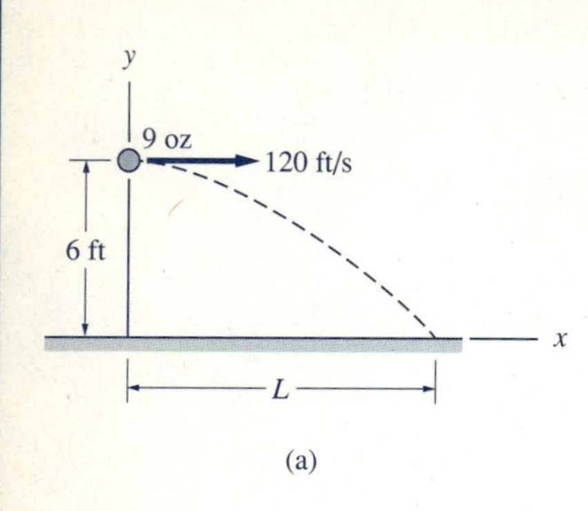

(a)

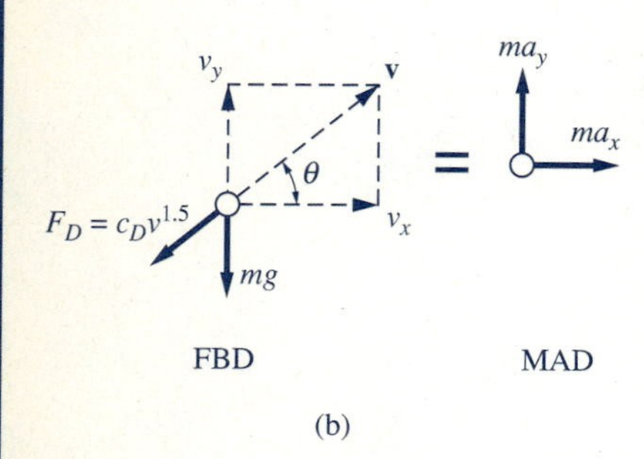

(b)

Sample Problem 12.14

The 9-oz ball shown in Fig. (a) is thrown horizontally with the velocity 120 ft/s from a height of 6 ft. The aerodynamic drag force acting on the ball is $F_D = c_D v^{1.5}$, where F_D is in lb, v is the velocity in ft/s, and $c_D = 0.0012$ lb · (s/ft)$^{1.5}$. (1) Using the xy-coordinate system shown, derive the equations of motion and state the initial conditions. (2) Using hand computation and Euler's method with $\Delta t = 0.10$ s, determine the time of flight and the range L. Note: The analytic solution is $t = 0.655$ s and $L = 63.0$ ft.

Solution

Part 1

The free-body diagram (FBD) of the ball in flight is shown in Fig. (b). Note that the rectangular components of the velocity vector **v** are shown acting in the positive coordinate directions and that the direction of $\mathbf{F}_D$ opposes the velocity **v**. Figure (b) also shows the mass-acceleration diagram (MAD) that contains the rectangular components of the inertia vector. Writing Newton's second law for the x-direction yields

$$\Sigma F_x = ma_x \qquad \xrightarrow{+} \qquad -c_D v^{1.5}\cos\theta = ma_x$$

Noting that $v_x = v\cos\theta$, this becomes

$$a_x = -(c_D/m)v_x v^{0.5} \tag{a}$$

Newton's second law for the y-direction yields

$$\Sigma F_y = ma_y \qquad +\uparrow \qquad -c_D v^{1.5}\sin\theta - mg = ma_y$$

Introducing $v_y = v\sin\theta$, we get

$$a_y = -(c_D/m)v_y v^{0.5} - g \tag{b}$$

Substituting $v = (v_x^2 + v_y^2)^{1/2}$ and

$$\frac{c_D}{m} = \frac{0.0012}{(9/16)/32.2} = 0.068\,693\ (\text{ft} \cdot \text{s})^{-0.5}$$

into Eqs. (a) and (b), the equations of motion become

$$a_x = f_x(v_x, v_y) = -0.068\,693 v_x(v_x^2 + v_y^2)^{1/4}\ \text{ft/s}^2 \tag{c}$$

$$a_y = f_y(v_x, v_y) = -0.068\,693 v_y(v_x^2 + v_y^2)^{1/4} - 32.2\ \text{ft/s}^2 \tag{d}$$

Answer

Note that these equations are coupled since a_x and a_y depend on both v_y and v_x.

Inspection of Fig. (a) reveals that the four initial conditions are

$$\text{at } t = 0: \qquad x = 0 \qquad y = 6\ \text{ft} \qquad v_x = 120\ \text{ft/s} \qquad v_y = 0$$

Answer

Observe that two of these conditions apply to motion in the x-direction, and two to motion in the y-direction.

Part 2

Letting $x_1 = x$, $x_2 = y$, $v_1 = v_x$, and $v_2 = v_y$ in Eq. (12.39) yields the following formulas for Euler's method:

$$(v_x)_{i+1} = (v_x)_i + (a_x)_i \Delta t \quad \text{(e)}$$

$$(v_y)_{i+1} = (v_y)_i + (a_y)_i \Delta t \quad \text{(f)}$$

$$x_{i+1} = x_i + (v_x)_i \Delta t \quad \text{(g)}$$

$$y_{i+1} = y_i + (v_y)_i \Delta t \quad \text{(h)}$$

The following table lists the values computed by Euler's method using Eqs. (c)–(h).

i	t (s)	x (ft)	v_x (ft/s)	a_x (ft/s^2)	y (ft)	v_y (ft/s)	a_y (ft/s^2)
0	0.000	0.000[1]	120.0000[2]	−90.2993[5]	6.0000[3]	0.0000[4]	−32.2000[6]
1	0.1000	12.0000[7]	110.9701[8]	−80.3180[11]	6.0000[9]	−3.2200[10]	−29.8694[12]
2	0.2000	23.0978	102.9383	−71.8078	5.6780	−6.2069	−27.8702
3	0.3000	33.3908	95.7575	−64.5097	5.0573	−8.9940	−26.1410
4	0.4000	42.9666	89.3065	−58.2179	4.1579	−11.6081	−24.6328
5	0.5000	51.8972	83.4847	−52.7673	2.9971	−14.0713	−23.3061
6	0.6000	60.2457	78.2080	−48.0245	1.5900	−16.4019	−22.1282
7	0.7000	68.0665	73.4056	−43.8806	−0.0502	−18.6148	−21.0724

Explanatory Notes

Entries* (1)–(4) are the initial conditions.

Entry (5)—from Eq. (c):

$$(a_x)_0 = -0.068\,693(v_x)_0[(v_x)_0^2 + (v_y)_0^2]^{1/4}$$
$$= -0.068\,693(120)(120^2 + 0)^{1/4} = -90.2993 \text{ ft/s}^2$$

Entry (6)—from Eq. (b):

$$(a_y)_0 = -0.068\,693(v_y)_0[(v_x)_0^2 + (v_y)_0^2]^{1/4} - 32.2$$
$$= -0.068\,693(0) - 32.2 = -32.2000 \text{ ft/s}^2$$

Entry (7)—from Eq. (g):

$$x_1 = x_0 + (v_x)_0 \Delta t$$
$$= 0 + (120.0000)(0.10) = 12.0000 \text{ ft}$$

Entry (8)—from Eq. (e):

$$(v_x)_1 = (v_x)_0 + (a_x)_0 \Delta t$$
$$= 120.0000 + (-90.2993)(0.10) = 110.9701 \text{ ft/s}$$

Entry (9)—from Eq. (h):

$$y_1 = v_0 + (v_y)_0 \Delta t$$
$$= 6.0000 + 0 = 6.0000 \text{ ft}$$

* Refer to superscripts in the table.

Entry (10)—from Eq. (f):

$$(v_y)_1 = (v_y)_0 + (a_y)_0\Delta t$$
$$= 0 + (-32.2000)(0.10) = -3.2200 \text{ ft/s}$$

Entry (11)—from Eq. (c):

$$(a_x)_1 = -0.068\,693(v_x)_1[(v_x)_1^2 + (v_y)_1^2]^{1/4}$$
$$= -0.068\,693(110.9701)[(110.9701)^2 + (-3.2200)^2]^{1/4}$$
$$= -80.3180 \text{ ft/s}^2$$

Entry (12)—from Eq. (d):

$$(a_y)_1 = -0.068\,693(v_y)_1[(v_x)_1^2 + (v_y)_1^2]^{1/4} - 32.2$$
$$= -0.068\,693(-3.2200)[(110.9701)^2 + (-3.2200)^2]^{1/4} - 32.2$$
$$= -29.8694 \text{ ft/s}^2$$

The remaining entries in the table can be computed by repeating the foregoing steps.

The computations are terminated at $i = 7$ since y_7 is negative (-0.0502 ft), which means that the ball hits the ground before $t = 0.70$ s.

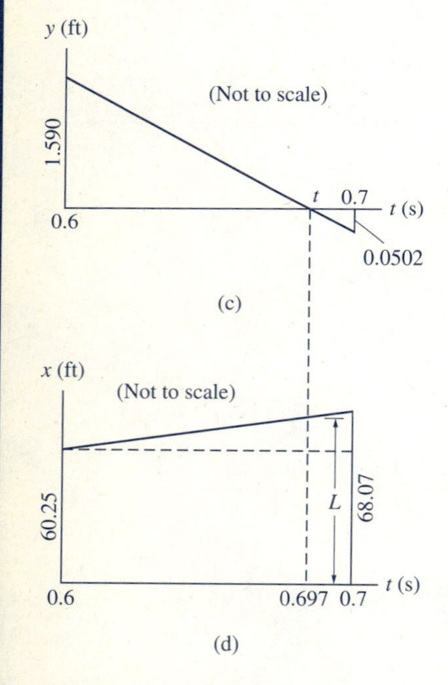

(c)

(d)

The linear interpolation shown in Fig. (c) can be used to calculate the time t corresponding to $y = 0$. From similar triangles, we have

$$\frac{t - 0.6}{1.590} = \frac{0.7 - 0.6}{1.590 + 0.0502}$$

from which we find that the time of flight is

$$t = 0.697 \text{ s}$$ *Answer*

Linear interpolation can also be used to compute the range L, i.e., the value of x when $t = 0.697$ s. From Fig. (d), similar triangles yield

$$\frac{L - 60.25}{0.697 - 0.600} = \frac{68.07 - 60.25}{0.7 - 0.6}$$

which gives

$$L = 67.8 \text{ ft}$$ *Answer*

The above answers for t and L are in error by approximately 6.4% and 7.6%, respectively, when compared with the analytic values. Accuracy would be improved by increasing the number of integration steps, i.e., selecting a value for Δt that is smaller than 0.10 s.

Sample Problem 12.15

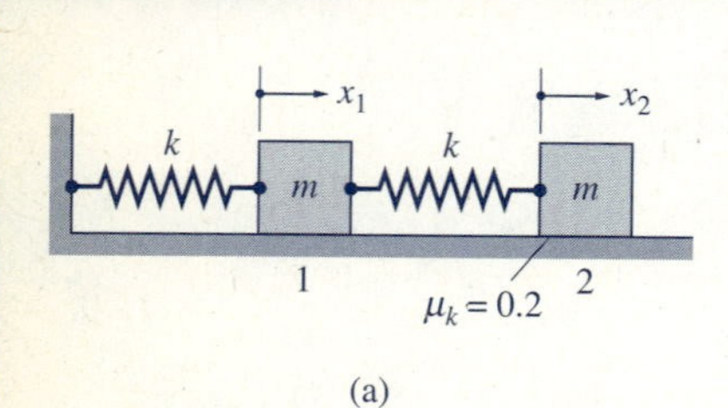

(a)

The two blocks of mass $m = 0.4$ kg shown in Fig. (a) are attached to springs of stiffness $k = 36$ N/m. The blocks are held at rest on the rough surface with each spring initially compressed by 0.10 m, and then released when time $t = 0$.

(1) Derive the equation of motion for each block, and state the initial conditions. (2) Solve the equations of motion for the time interval $t = 0$ and $t = 0.5$ s. (3) Plot the velocity-time diagram for each block.

Solution

Part 1

The free-body diagram (FBD) and mass-acceleration diagram (MAD) of each block are shown in Figs. (b) and (c). The coordinates x_1 and x_2 are measured from the positions where the springs are unstretched. Since there is no vertical acceleration, the normal force under each block equals the weight of the block, i.e., $N_1 = N_2 = mg$. The BASIC function sgn v is used to ensure that the direction of the friction force on each block is opposite to the direction of its velocity.* Writing Newton's second law for each block, we obtain

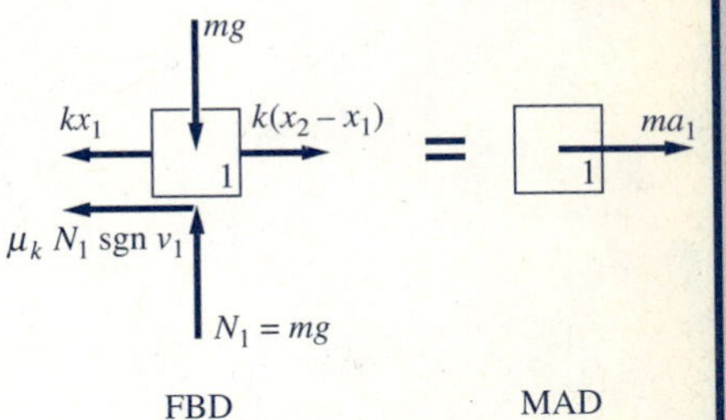

(b)

Block 1

$$\Sigma F_x = ma_1 \qquad \xrightarrow{+} \qquad k(x_2 - x_1) - kx_1 - \mu_k mg \text{ sgn } v_1 = ma_1$$

which yields

$$a_1 = \frac{k}{m}(x_2 - 2x_1) - \mu_k g \text{ sgn } v_1 \tag{a}$$

mg
$k(x_2 - x_1)$
2
=
ma_2
2
$F_2 = \mu_k N_2$ sgn v_2
$N_2 = mg$
FBD
MAD

(c)

Block 2

$$\Sigma F_x = ma_2 \qquad \xrightarrow{+} \qquad -k(x_2 - x_1) - \mu_k mg \text{ sgn } v_2 = ma_2$$

from which

$$a_2 = \frac{k}{m}(x_1 - x_2) - \mu_k g \text{ sgn } v_2 \tag{b}$$

Substituting $m = 0.4$ kg, $\mu_k = 0.2$, $k = 36$ N/m, and $g = 9.81$ m/s^2 into Eqs. (a) and (b), and simplifying, the equations of motion become

$$a_1 = 90(x_2 - 2x_1) - 1.962 \text{ sgn } v_1 \text{ m/s}^2 \tag{c}$$

$$a_2 = 90(x_1 - x_2) - 1.962 \text{ sgn } v_2 \text{ m/s}^2 \tag{d}$$

Answer

The initial conditions (at $t = 0$) are

$x_1 = -0.10$ m, $x_2 = -0.20$ m (since each spring is initially compressed 0.10 m) *Answer*

$v_1 = 0$, $v_2 = 0$ (since each block is released from rest)

Note: Although the initial velocity of each block is identically zero, it is necessary to enter small values for $(v_1)_0$ and $(v_2)_0$ so that the function sgn v will not be zero at $t = 0$.

* The properties of sgn v, which are discussed in Sample Problem 12.13, are sgn $v = 1$ if $v > 0$; sgn $v = 0$ if $v = 0$; sgn $v = -1$ if $v < 0$. The equivalent function in FORTRAN is SIGN(z,v), which returns the value of the first argument with the sign of the second argument.

Part 2

The solutions to the equations of motion, Eqs. (c) and (d), were computed using the fourth-order Runge-Kutta method with the time increment $\Delta t = 0.005$ s. This gives 100 steps in the time interval $t = 0$ to $t = 0.5$ s, but only the results of every second step were saved to reduce the volume of output.

The maximum velocity of each block occurs when its acceleration is zero. It is necessary to use linear interpolation to find the time when the acceleration is zero. Inspection of the computer output (the entire output is not shown here) reveals that the data immediately before and after a_1 changes its sign are as follows:

t (s)	v_1 (m/s)	a_1 (m/s^2)
0.31	0.6932	0.0256
0.32	0.6896	−0.7500

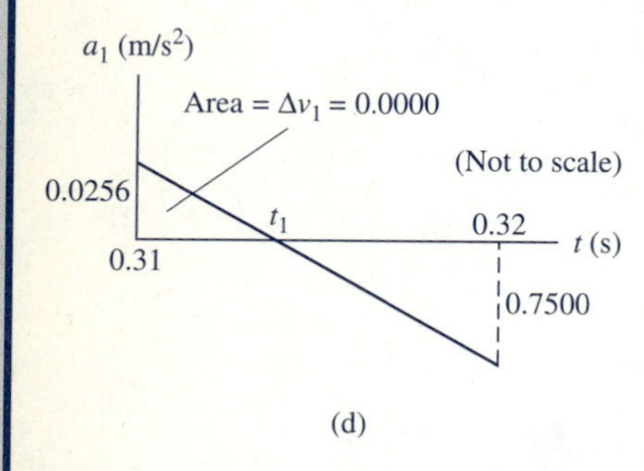

(d)

The data are also shown in Fig. (d), where t_1 denotes the time when $a_1 = 0$. Utilizing similar triangles, we get

$$\frac{t_1 - 0.31}{0.0256} = \frac{0.32 - 0.31}{0.0256 + 0.7500}$$

the solution of which is $t_1 = 0.3103$ s.

From Fig. (d), we also obtain the change in v_1 between $t = 0.31$ s and $t = 0.3103$ s.

$$\Delta v_1 = \frac{1}{2}(0.0256)(0.3103 - 0.31) = 0.0000$$

Therefore, the maximum speed of block 1 and the time that it occurs are

$$(v_1)_{\max} = 0.693 \text{ m/s} \quad \text{at } t = 0.310 \text{ s} \qquad \textit{Answer}$$

Block 2 can be treated in a similar manner. The recorded data immediately before and after a_2 changes its sign are

t (s)	v_2 (m/s)	a_2 (m/s^2)
0.17	0.7554	0.2666
0.18	0.7556	−0.2106

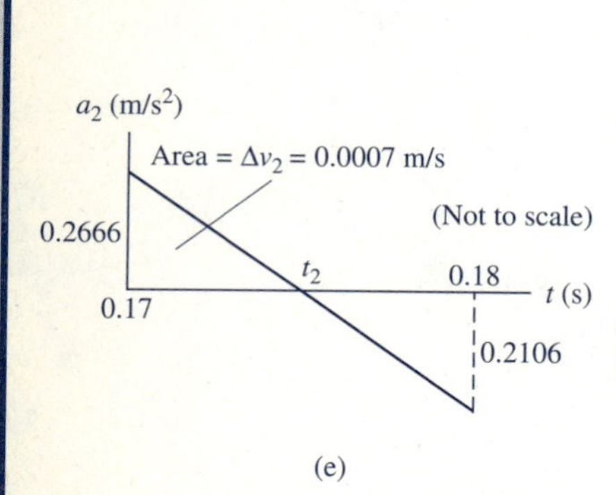

(e)

From the similar triangles in Fig. (e), we obtain

$$\frac{t_2 - 0.17}{0.2666} = \frac{0.18 - 0.17}{0.2666 + 0.2106}$$

from which $t_2 = 0.1756$ s. The change in v_2 during the time interval $t = 0.17$ to $t = 0.1756$ s is

$$\Delta v_2 = \frac{1}{2}(0.1756 - 0.17)(0.2666) = 0.0007 \text{ m/s}$$

Therefore the maximum speed of block 2 and the corresponding time are

$$(v_2)_{\max} = (v_2)_{0.17} + \Delta v_2 = 0.7554 + 0.0007$$
$$= 0.756 \text{ m/s} \quad \text{at } t = 0.176 \text{ s} \qquad \textit{Answer}$$

Part 3

The program that was used to solve the equations of motion had plotting capability, which was utilized to produce the velocity-time diagrams shown in Fig. (f). The points in each plot represent 0.01-s time intervals.

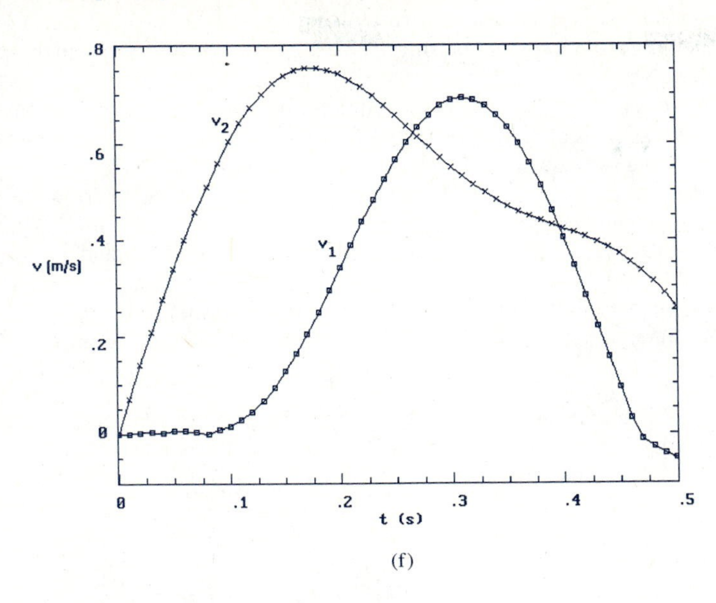

(f)

PROBLEMS

General Instructions

Problems 12.91–93 are to be solved by hand computation using Euler's method. The results should be reported using a table similar to the one shown in the solution to Sample Problem 12.14.

Problem 12.94 requires that a computer program be written that solves two coupled equations of motion by the modified Euler method. (This program can be used to solve subsequent problems, wherever applicable.)

Problems 12.95 and 12.96 are to be solved using the modified Euler method on a computer. (The program written in Prob. 12.94 may be used for this purpose.)

Problems 12.97–101 are to be solved by the numerical method that is recommended by your instructor. (Although not absolutely necessary, a computer program with plotting capability would be helpful.) The time increment Δt should be small enough to yield results that are accurate to within three significant digits.

12.91 A 0.1-kg rock is thrown at a wall from a distance of 30 m at an elevation of 2 m with the initial velocity shown. The aerodynamic drag acting on the rock is $F_D = 0.0005v^2$, where F_D is in newtons and the velocity v is in m/s. (a) Show that the equations of motion are

$$a_x = -0.005v_x\sqrt{v_x^2 + v_y^2}\ \text{m/s}^2$$

$$a_y = -0.005v_y\sqrt{v_x^2 + v_y^2} - 9.81\ \text{m/s}^2$$

(b) Use Euler's method with $\Delta t = 0.25$ s to find the height h where the rock hits the wall and the speed of impact. Note: the analytical solution is $h = 24.0$ m, $v = 16.3$ m/s.

12.92 The 0.01-kg particle travels freely on a smooth, horizontal surface lying in the xy-plane. The force **F** acting on the particle is always directed away from the origin O, its magnitude being $F = 0.005/d^2$ N, where d is the distance in meters of the particle from O. At time $t = 0$, the position of the particle is $x = 0.3$ m, $y = 0.4$ m, and its velocity is $\mathbf{v} = -2\mathbf{j}$ m/s. (a) Derive the equations of motion of the particle, and state the initial conditions. (b) Use Euler's method with $\Delta t = 0.04$ s to determine the x-coordinate and the speed of the particle when it crosses the x-axis.

12.93 Solve Prob. 12.92 if the force **F** is directed toward the origin O.

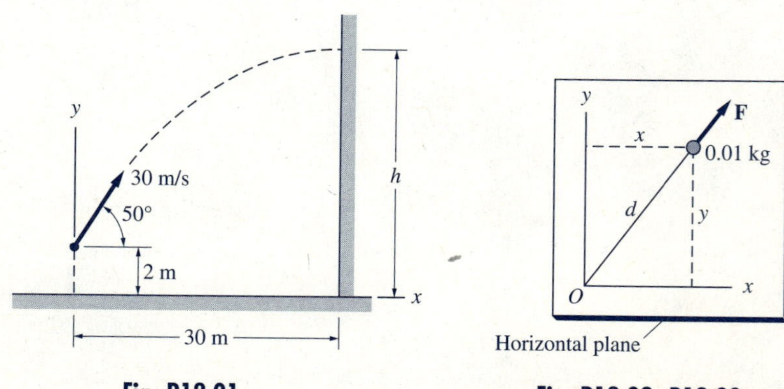

Fig. P12.91

Fig. P12.92, P12.93

12.94 Write a computer program that solves two coupled equations of motion by the modified Euler method. Test your program by integrating the following equations from $t = 0$ to $t = 1$ using $\Delta t = 0.1$.

$$a_1 = 12(x_2 - 2x_1) \qquad a_2 = 12(x_1 - x_2)$$

The initial conditions at $t = 0$ are $x_1 = 1$, $x_2 = 0$, $v_2 = v_1 = 0$. Your output should agree with the values given in the following table.

i	$t(i)$	$x_1(i)$	$x_2(i)$	$v_1(i)$	$v_2(i)$	$a_1(i)$	$a_2(i)$
0	0.0000	1.0000	0.0000	0.0000	0.0000	−24.0000	12.0000
1	0.1000	0.8800	0.0600	−2.4000	1.2000	−20.4000	9.8400
2	0.2000	0.5380	0.2292	−4.0800	1.9680	−10.1616	3.7056
3	0.3000	0.0792	0.4445	−4.4885	1.9757	3.4337	−4.3840
4	0.4000	−0.3525	0.6202	−3.4879	1.1494	15.9018	−11.6720
5	0.5000	−0.6218	0.6768	−1.4102	−0.2960	23.0437	−15.5824
6	0.6000	−0.6476	0.5692	1.0456	−1.9211	22.3729	−14.6019
7	0.7000	−0.4312	0.3041	3.0421	−3.2033	13.9973	−8.8234
8	0.8000	−0.0570	−0.0603	3.8846	−3.7109	0.6431	0.0404
9	0.9000	0.3347	−0.4312	3.2601	−3.2511	−13.2078	9.1911
10	1.0000	0.5947	−0.7104	1.3531	−1.9413	−22.7970	15.6607

12.95 Solve Prob. 12.91 by the modified Euler method using $\Delta t = 0.15$ s.

12.96 Solve Prob. 12.92 by the modified Euler method using $\Delta t = 0.025$ s.

12.97 A 1.0-kg ball is kicked with an initial velocity of 30 m/s into a 20-m/s headwind. The aerodynamic drag force acting on the ball is $\mathbf{F}_D = -0.5\mathbf{v}$ N. The resulting acceleration of the ball is

$$\mathbf{a} = -(10 + 0.5v_x)\mathbf{i} - (9.81 + 0.5v_y)\mathbf{j} \text{ m/s}^2$$

where the components of the velocity are in m/s. (a) Determine the horizontal distance of travel b and the time of flight. (b) Plot the trajectory of the ball (y vs. x).

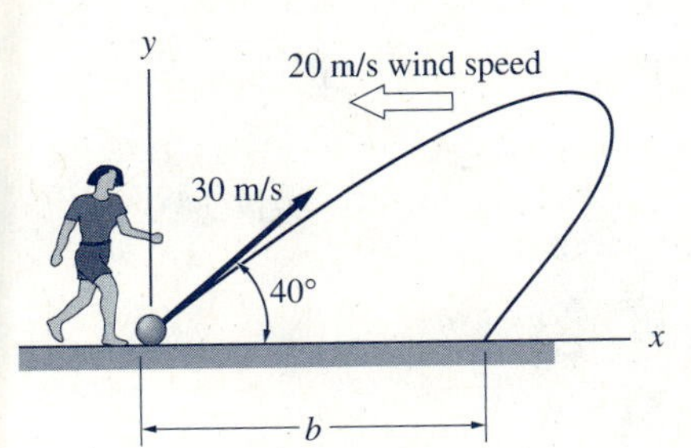

Fig. P12.97

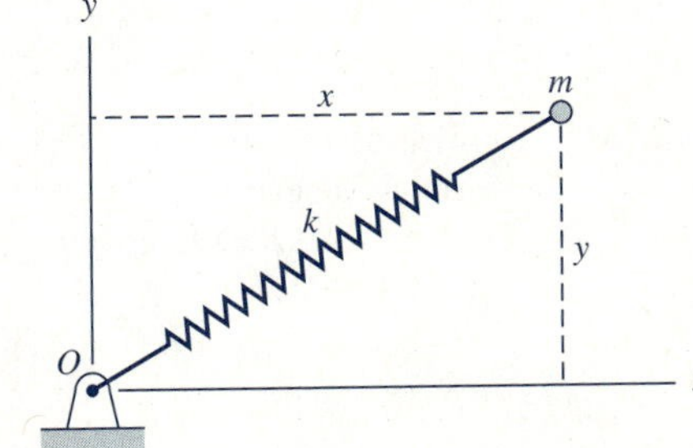

Fig. P12.98–P12.99

12.98 The mass $m = 0.25$ kg, attached to a linear spring (stiffness $k = 10$ N/m, free length $L_0 = 0.5$ m), moves in the vertical plane. The spring can resist both tension and compression. The mass is released from rest at $x = 0.5$ m, $y = -0.5$ m. (a) Derive the equations of motion and state the initial conditions. (b) Solve the equations numerically for $0 \le t \le 2$ s, and plot the trajectory of the mass. What is the range of x and y during this time period?

12.99 Solve Prob. 12.98 if the mass is released from rest at $x = y = 0.5$ m and the spring is unable to resist compression.

12.100 The plane motion of a table-tennis ball in Fig. (a) is governed by the three components of acceleration given in Fig. (b). In addition to the gravitational acceleration g, there are the effects of the aerodynamic drag a_D and aerodynamic lift a_L (the lift is caused by the Bernoulli effect, i.e., the difference in air pressure caused by the spin of the ball). Realistic approximations for these accelerations are $a_D = 0.05v^2$ and $a_L = 0.16\omega v$, where v is in ft/s and the spin ω is in rev/s. The ball leaves the paddle at table height with the initial velocity $v_0 = 60$ ft/s inclined at $\theta_0 = 60°$ with the horizontal, and a topspin $\omega = +10$ rev/s (the spin may be assumed to be constant). (a) Derive the equations of motion for the x- and y-directions. (b) Assuming that the ball lands on the table, solve the equations of motion for the duration of the flight. Determine the time of flight and the horizontal distance traveled. (c) Plot the trajectory of the ball.

12.101 Solve Prob. 12.100 if $v_0 = 60$ ft/s, $\theta_0 = 0$, and $\omega = -10$ rev/s (backspin).

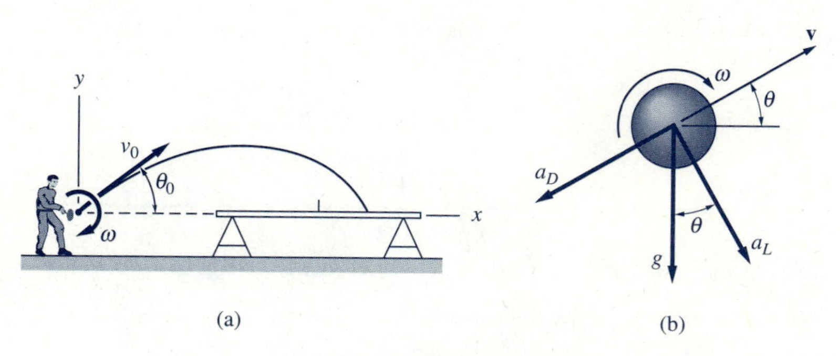

Fig. P12.100

REVIEW PROBLEMS

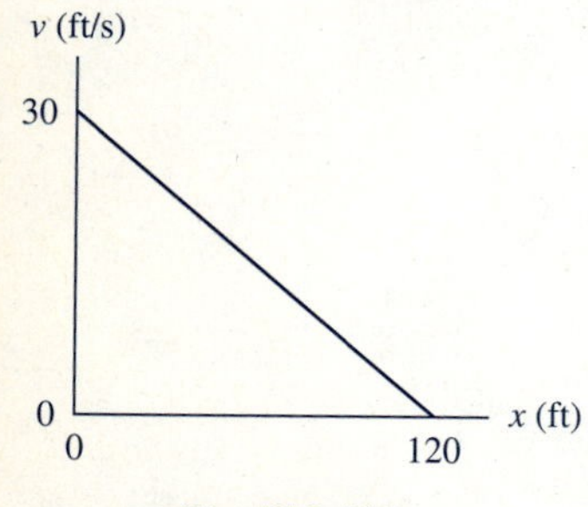

Fig. P12.102

12.102 A car traveling at 30 ft/s applies its brakes at $t = 0$ and $x = 0$. During braking, the plot of speed vs. distance traveled is a straight line, as shown. Determine the acceleration of the car as a function of x.

12.103 For the car described in Prob. 12.102, determine x as a function of t.

12.104 A projectile is launched at O with the velocity v_0 inclined at 60° to the horizontal. Determine the smallest value of v_0 for which the projectile will clear the wall AB. Neglect air resistance.

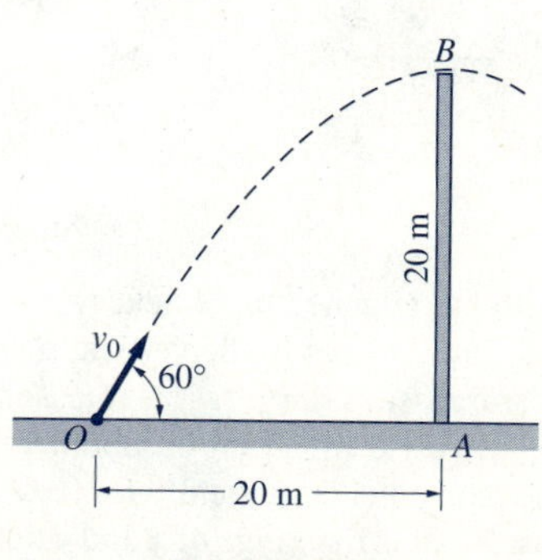

Fig. P12.104

12.105 The bar AB rotates at the constant angular velocity $\dot{\theta}$ while pushing the parcel of mass m along the horizontal surface. Determine the largest value of $\dot{\theta}$ for which the parcel will maintain contact with the surface. Assume that θ varies from 0° to 60°, and neglect friction.

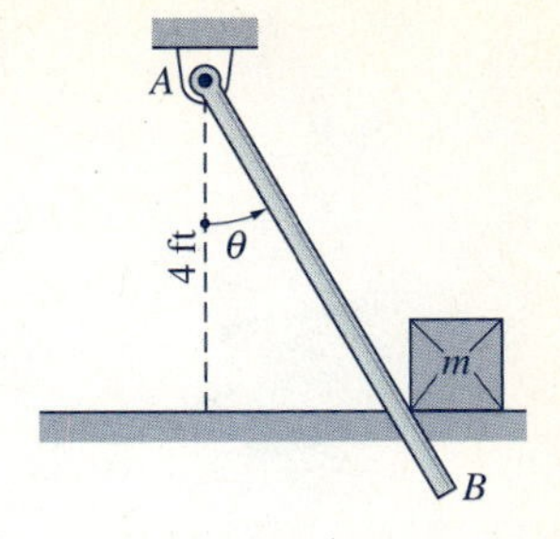

Fig. P12.105

12.106 A 1200-kg rocket is launched vertically from the surface of the earth. During the 20-second burn the thrust F of the engine varies with time, as shown in the plot. Determine the maximum height reached by the rocket. Assume the gravitational acceleration and the mass of the rocket to be constant.

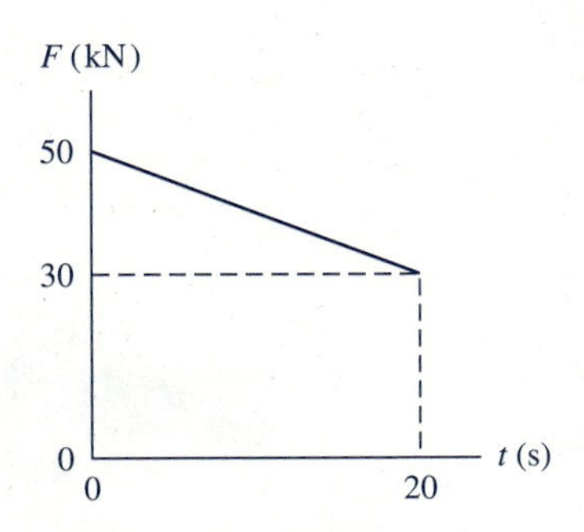

Fig. P12.106

12.107 A particle is in rectilinear motion. Its acceleration a and position coordinate x are related by $a = -a_0 \exp(x/x_0)$, where $a_0 = 8$ m/s^2 and $x_0 = 0.2$ m. If the initial conditions at $t = 0$ are $v = 6$ m/s, $x = 0$, determine the maximum value of x.

12.108 A 10-lb parcel is dropped from a height with no initial velocity. During the fall, the drag force F_D acting on the parcel is proportional to its velocity v: $F_D = kv$, where k is a constant. If the terminal velocity of the parcel is 200 ft/s, determine (a) the value of the constant k; and (b) the time when the velocity of the parcel reaches 190 ft/s.

12.109 A golf ball is hit down a 20° incline, as shown. Determine the angle θ that maximizes the range R for a given initial velocity v_0. Neglect air resistance.

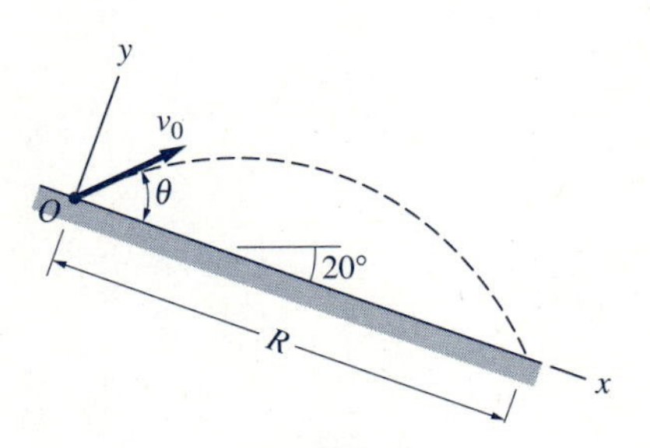

Fig. P12.109

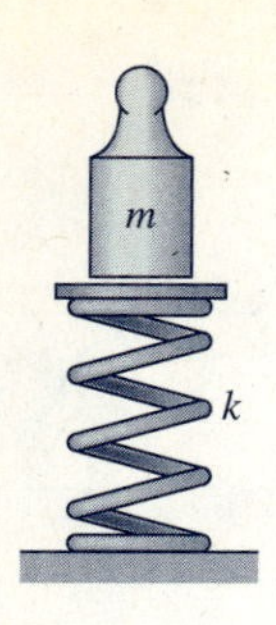

Fig. P12.110

12.110 The mass m is released with zero velocity on top of a spring of stiffness k. Derive the expressions for (a) the maximum force in the spring; and (b) the maximum velocity of the mass.

12.111 The figure shows the path of a particle, the motion of which is described by $x = b \sin \omega t$ and $y = b \exp(-\omega t/2)$, where b and ω are constants. Determine the magnitude and direction of the acceleration at points (a) A; and (b) B.

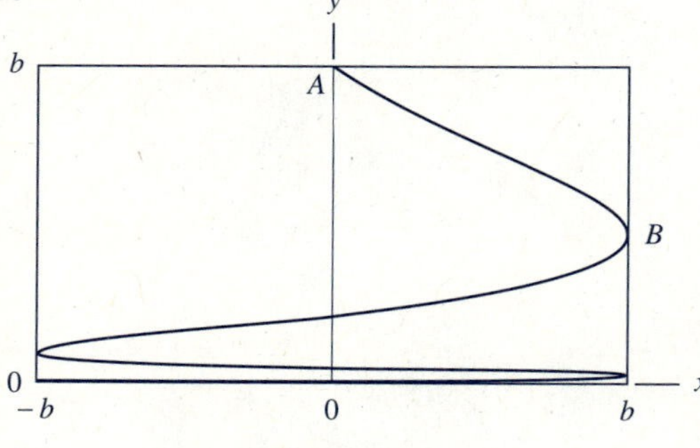

Fig. P12.111

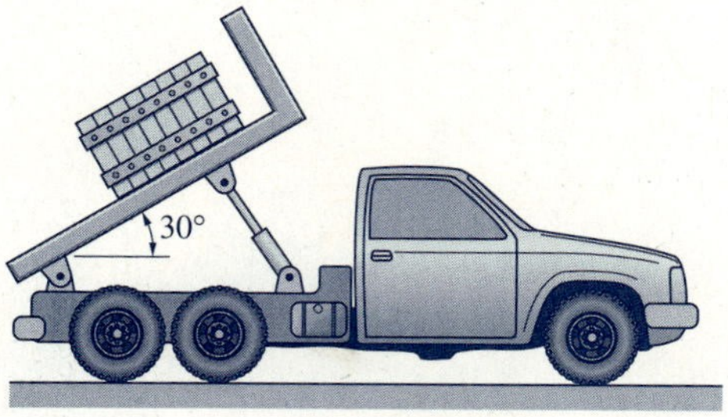

Fig. P12.112

12.112 The coefficient of static friction between the bed of the dump truck and the crate is 0.64. In order to slide the crate off the truck when the bed is in the position shown, the truck must accelerate to the right. Determine the smallest acceleration that would cause the crate to slide.

13 Particle Motion: Curvilinear Coordinates

13.1 Introduction

In the preceding chapter we have described the dynamics of particles using a rectangular reference frame. Although all problems could, in principle, be handled with rectangular coordinates, there are many cases that are more efficiently analyzed by employing a curvilinear coordinate system.

This chapter is devoted to two systems of curvilinear coordinates—path (normal and tangential) coordinates and cylindrical coordinates (including the special case of polar coordinates). Articles 13.2 and 13.3 describe particle kinematics (geometry of motion) relative to these coordinate systems. In Art. 13.4, we present Newton's second law in terms of curvilinear coordinates. Since the laws of kinetics are unaffected by the choice of coordinates, no new principles are contained in Art. 13.4.

This chapter concludes with the extension of the numerical methods previously discussed (Arts. 12.8 and 12.9) to generalized coordinates.

13.2 Kinematics: Path (Normal-Tangential) Coordinates

a. Plane motion

1. Geometric Preliminaries Figure 13.1 illustrates the curvilinear motion of a particle in the xy-plane. The following list explains the contents of the figure and defines some of the terminology used in this article.

- A: position of the particle at time t
- B: position of the particle at time $t + dt$
- $d\mathbf{r}$: differential displacement of the particle during the time interval dt [the position vector $\mathbf{r}(t)$ from O to A is not shown in the figure]
- AC and BC: lines that are perpendicular to the path at A and B, respectively

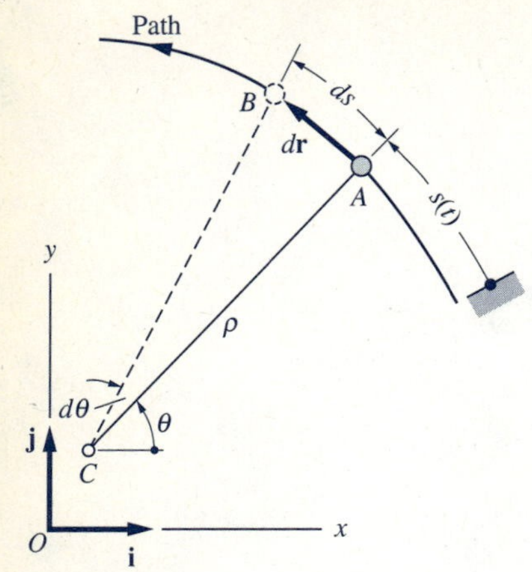

$s(t)$: *path coordinate* (distance measured along the path from a fixed reference point)
ds: differential path length between A and B
θ: angle (in radians) measured from the x-axis to the line AC
$d\theta$: differential angle between lines AC and BC
C: *center of curvature* of the arc ds (located at the intersection of AC and BC)
ρ: *radius of curvature* of the path at A (length of AC)*
$1/\rho$: *curvature* of the path at A

Fig. 13.1

From Fig. 13.1 we see that the radius of curvature is related to the differential arc length ds by

$$ds = \rho\, d\theta \tag{13.1}$$

The base vectors $\mathbf{e}_n$ and $\mathbf{e}_t$ associated with the path coordinate system are shown at position A in Fig. 13.2. The unit vector $\mathbf{e}_t$ is tangent to the path, and it points in the direction of motion. The unit vector $\mathbf{e}_n$ is perpendicular to $\mathbf{e}_t$, i.e., normal to the path, and it is directed toward the center of curvature C. Note that when the particle moves to position B the base vectors have rotated in order to line up with the tangent and normal to the path at B. Thus we can consider the base vectors to be attached to the particle, their directions determined by the lines that are tangent and normal to the path at the location of the particle.

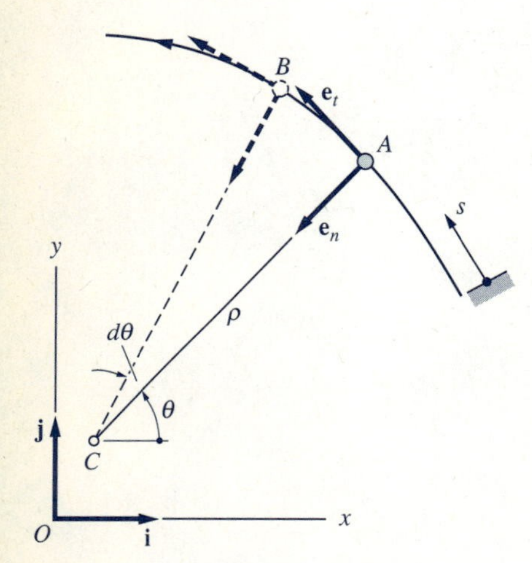

Fig. 13.2

2. Derivatives of the Unit Base Vectors Although their magnitudes are constant (equal to one), the base vectors possess nonzero time derivatives because their directions vary, as illustrated in Fig. 13.2.

The time derivatives of the base vectors can be computed by first relating the vectors to the fixed xy-frame. Referring to Fig. 13.3, we find that

$$\begin{aligned}\mathbf{e}_t &= -\sin\theta\mathbf{i} + \cos\theta\mathbf{j}\\ \mathbf{e}_n &= -\cos\theta\mathbf{i} - \sin\theta\mathbf{j}\end{aligned} \tag{13.2}$$

Differentiating each side with respect to time, and noting that $d\mathbf{i}/dt = d\mathbf{j}/dt = \mathbf{0}$, we obtain

$$\frac{d\mathbf{e}_t}{dt} = (-\cos\theta\mathbf{i} - \sin\theta\mathbf{j})\dot{\theta} \qquad \frac{d\mathbf{e}_n}{dt} = (\sin\theta\mathbf{i} - \cos\theta\mathbf{j})\dot{\theta}$$

from which we conclude that

$$\dot{\mathbf{e}}_t = \dot{\theta}\mathbf{e}_n \qquad \dot{\mathbf{e}}_n = -\dot{\theta}\mathbf{e}_t \tag{13.3}$$

The time derivative of θ, i.e., $\dot{\theta}$, is called the *angular velocity* of the line AC in Fig. 13.3.

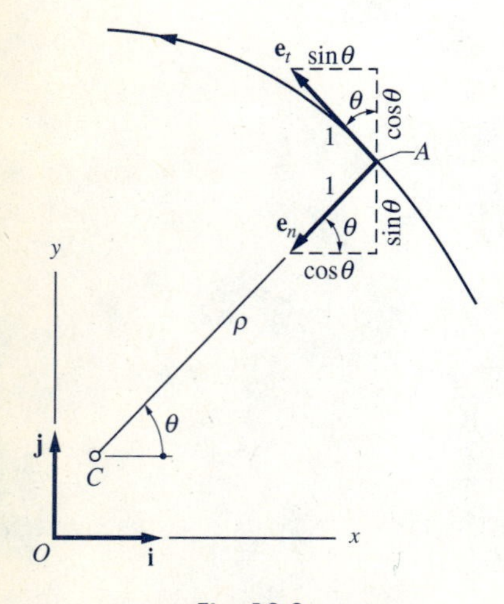

Fig. 13.3

* If the equation of the path is known, the radius of curvature at a given point can be computed from $\rho = [1 + (dy/dx)^2]^{3/2}/|d^2y/dx^2| = [1 + (dx/dy)^2]^{3/2}/|d^2x/dy^2|$.

The base vectors and their derivatives are shown in Fig. 13.4 parts (a) and (b). The derivatives are perpendicular to the respective vectors, which reflects the fact that the direction of each base vector changes, but not the magnitude. A change in the magnitude of a vector, which is equivalent to the change in length of the vector, would occur in the direction of the vector.

Another useful equation is found by inspection of Fig. 13.1. Since $|d\mathbf{r}| = ds$, and because $d\mathbf{r}$ is tangent to the path at A, we have $d\mathbf{r} = ds\,\mathbf{e}_t$, from which we find that

$$\frac{d\mathbf{r}}{ds} = \mathbf{e}_t \tag{13.4}$$

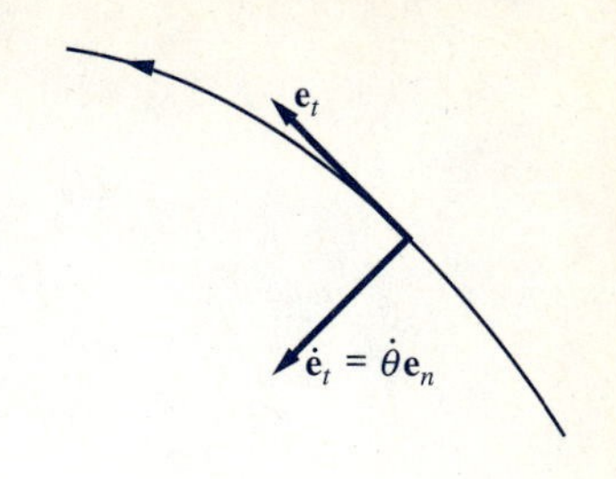

(a)

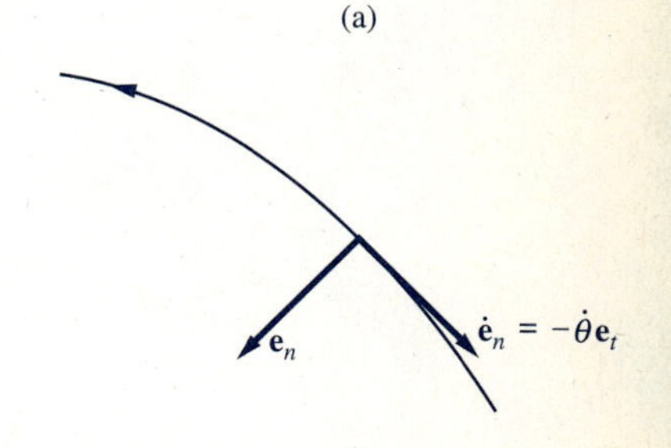

(b)

Fig. 13.4

3. Velocity and Acceleration Vectors In Eq. (11.10) the particle velocity was defined to be $\mathbf{v} = d\mathbf{r}/dt$. Using the chain rule of differentiation, $\mathbf{v}$ becomes

$$\mathbf{v} = \frac{d\mathbf{r}}{dt} = \frac{d\mathbf{r}}{ds}\frac{ds}{dt} = v\mathbf{e}_t \tag{13.5}$$

where $v = ds/dt$ is the magnitude of the velocity vector (or the speed) of the particle. Equation (13.5) shows that the velocity vector is tangent to the path, an important property that we have mentioned frequently. Since $ds = \rho\, d\theta$ from Eq. (13.1), the speed can also be written as $v = \rho\, d\theta/dt$, or

$$v = \rho\dot{\theta} \tag{13.6}$$

In Eq. (11.13a) the particle acceleration was defined as $\mathbf{a} = d\mathbf{v}/dt$. Substituting $\mathbf{v} = v\mathbf{e}_t$ from Eq. (13.5), we obtain

$$\mathbf{a} = \frac{d\mathbf{v}}{dt} = \frac{d}{dt}(v\mathbf{e}_t) = \dot{v}\mathbf{e}_t + v\dot{\mathbf{e}}_t = \dot{v}\mathbf{e}_t + v\dot{\theta}\mathbf{e}_n$$

where we substituted $\dot{\mathbf{e}}_t = \dot{\theta}\mathbf{e}_n$ from Eq. (13.3). The acceleration vector can, therefore, be written as

$$\mathbf{a} = a_t\mathbf{e}_t + a_n\mathbf{e}_n \tag{13.7}$$

1.95

where $a_t = \dot{v}$ and $a_n = v\dot{\theta}$ are the tangential and normal components of the acceleration, respectively. Utilizing Eq. (13.6), we obtain other useful expressions for a_t and a_n:

$$\begin{aligned} a_t &= \dot{v} = \rho\ddot{\theta} + \dot{\rho}\dot{\theta} \\ a_n &= v\dot{\theta} = \rho\dot{\theta}^2 = \frac{v^2}{\rho} \end{aligned} \tag{13.8}$$

The term $\ddot{\theta}$ is called the *angular acceleration* of the line AC in Fig. 13.3.

Time can be eliminated as an explicit variable in the expression for a_t using the chain rule of differentiation*

$$a_t = \frac{dv}{dt} = \frac{dv}{ds}\frac{ds}{dt} = v\frac{dv}{ds} \tag{13.9}$$

The velocity and acceleration vectors are shown in Fig. 13.5 parts (a) and (b), respectively. From Eq. (13.8) we see that the tangential component of acceleration, $\mathbf{a}_t$, is due to the *change in magnitude* of $\mathbf{v}$. If the speed is increasing, $\mathbf{a}_t$ and $\mathbf{v}$ will be in the same direction; if the speed is decreasing, the direction of $\mathbf{a}_t$ will be opposite to that of $\mathbf{v}$. If the speed is constant, then $a_t = 0$. From Eq. (13.8) we also see that the normal component of acceleration, $\mathbf{a}_n$, is caused by the rate of *change in the direction* of $\mathbf{v}$ as measured by $\dot{\theta}$. The normal component of acceleration is always directed toward the center of curvature of the path. (If the path is a straight line, then $a_n = 0$.)

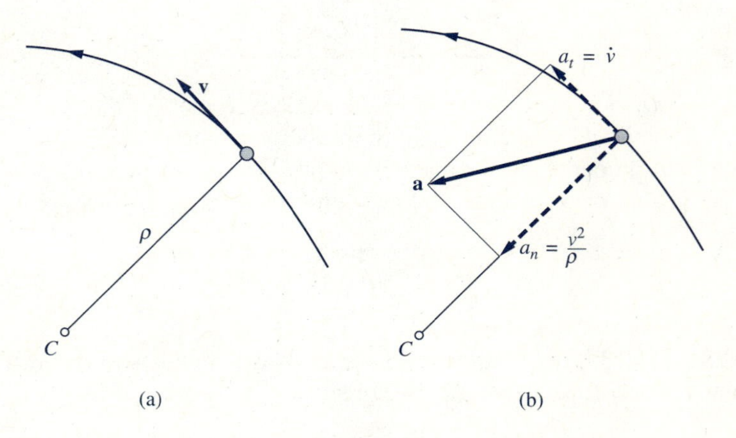

Fig. 13.5

We must point out the possibility of confusing the following three equations: (1) $\mathbf{a} = d\mathbf{v}/dt$—the fundamental definition of the acceleration vector; (2) $a = dv/dt$—the acceleration for rectilinear motion; (3) $a_t = dv/dt$—the tangential component of the acceleration vector in plane curvilinear motion. Confusion can be avoided by paying meticulous attention to the notation (do not omit subscripts or confuse vectors with scalars).

4. Motion Along a Circular Path Particle motion along a circular path occurs frequently enough to warrant special attention. As we show in Chapter 16, the equations for circular motion play an important role in the kinematics of rigid bodies.

* Equation (13.9) is very similar to the equation for rectilinear motion: $a = v\,dv/dx$, Eq. (12.10). However, remember that Eq. (13.9) applies only to the tangential component of the acceleration, not the total acceleration.

Sample Problem 13.1

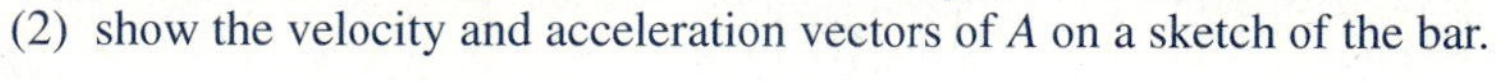

The angle between the 2-m bar shown in Fig. (a) and the x-axis varies according to $\theta(t) = 0.3t^3 - 1.6t + 3$ rad, where t is the time in seconds. When $t = 2$ s, (1) determine the magnitudes of the velocity and acceleration of end A; and (2) show the velocity and acceleration vectors of A on a sketch of the bar.

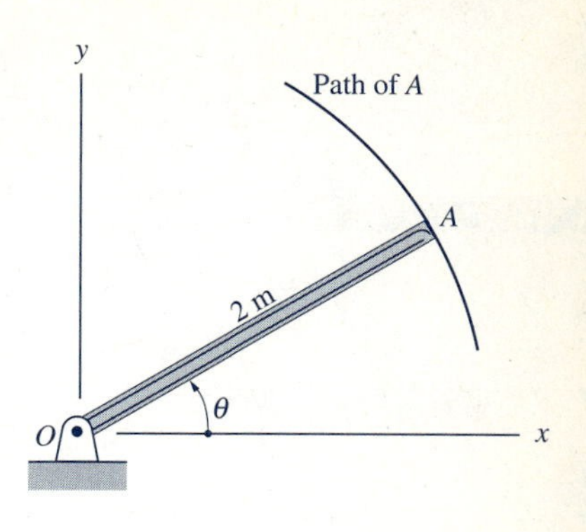

(a)

Solution

Part 1

It is convenient to use normal and tangential components because the path of A is a circle (centered at point O, of radius $R = 2$ ft).

The angular velocity and acceleration of the bar are the derivatives of θ: $\dot{\theta} = 0.9t^2 - 1.6$ rad/s and $\ddot{\theta} = 1.8t$ rad/s². At $t = 2$ s we find that

$$\dot{\theta}|_{t=2\text{ s}} = 0.9(2)^2 - 1.6 = 2.00 \text{ rad/s}$$

$$\ddot{\theta}|_{t=2\text{ s}} = 1.8(2) = 3.60 \text{ rad/s}^2$$

Because $\dot{\theta}$ and $\ddot{\theta}$ are positive, their directions are the same as the positive direction for θ, i.e., counterclockwise.

From Eqs. (13.10), the magnitude of the velocity of A is

$$v = R\dot{\theta} = 2(2.00) = 4.00 \text{ m/s} \qquad \textit{Answer}$$

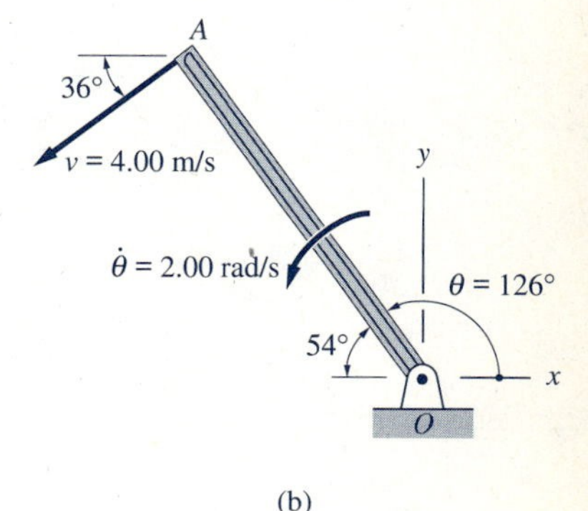

(b)

and the normal and tangential components of the acceleration of A are

$$a_n = R\dot{\theta}^2 = 2(2.00)^2 = 8.00 \text{ m/s}^2$$

$$a_t = R\ddot{\theta} = 2(3.60) = 7.20 \text{ m/s}^2$$

Therefore, the magnitude of the acceleration of A is

$$a = \sqrt{a_n^2 + a_t^2} = \sqrt{(8.00)^2 + (7.20)^2} = 10.76 \text{ m/s}^2 \qquad \textit{Answer}$$

Part 2

On substituting $t = 2$ s into the expression for $\theta(t)$, we find that the angular position of the bar at $t = 2$ s is

$$\theta|_{t=2\text{ s}} = (0.3)(2)^3 - 1.6(2) + 3 = 2.20 \text{ rad} = 126°$$

The velocity vector of end A is shown in Fig. (b). The magnitude of $\mathbf{v}$ is 4.00 m/s, as computed in Part 1, and the vector is tangent to the circular path, its direction being consistent with the direction of $\dot{\theta}$.

Figure (c) shows the directions of the normal and tangential components of the acceleration vector as determined in Part 1. Note that a_n is normal to the path and directed toward the point O, the center of the path. The direction of a_t is tangent to the path, consistent with the direction for $\ddot{\theta}$. The acceleration vector of magnitude 10.76 m/s² is also shown in Fig. (c), where the angle between $\mathbf{a}$ and $\mathbf{a}_t$ was found to be

$$\alpha = \tan^{-1}(a_n/a_t) = \tan^{-1}(8.00/7.20) = 48.0°$$

(c)

Observe that if $\dot{\theta}$ had been negative, the direction of $\mathbf{v}$ would be opposite to that shown in Fig. (b). If $\ddot{\theta}$ had been negative, the direction of $\mathbf{a}_t$ would be opposite to that shown in Fig. (c). However, $\mathbf{a}_n$ is directed toward point O regardless of the sign of $\dot{\theta}$ or $\ddot{\theta}$.

Sample Problem 13.2

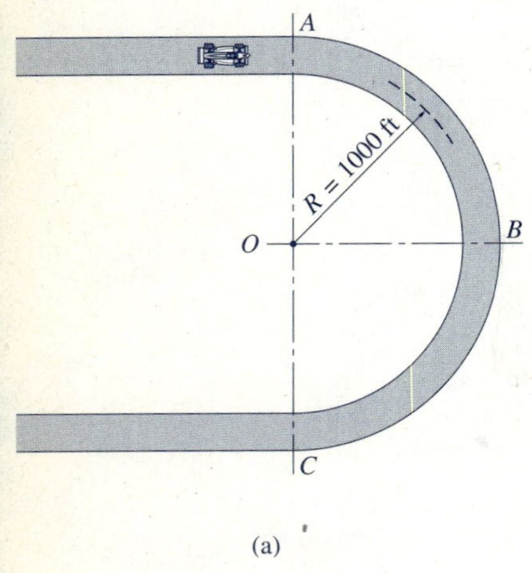

(a)

The racing car shown in Fig. (a) is traveling at 90 mi/h when it enters the semicircular curve at A. The driver increases the speed at a uniform rate, emerging from the curve at C at 120 mi/h. Determine the magnitude of the acceleration vector when the car is at B.

Solution

Because the racing car follows a circular path, it is convenient to describe its motion using path coordinates. As shown in Fig. (b), we let s be the distance measured along the path from A toward C.

The magnitude of the tangential component of acceleration is *constant* between A and C, since the speed increases at a uniform rate. Therefore, the integration of $a_t\, ds = v\, dv$ yields

$$\frac{v^2}{2} = a_t s + C \qquad \text{(a)}$$

where C is a constant of integration. The two constants a_t and C can be evaluated using the following two conditions on the motion:

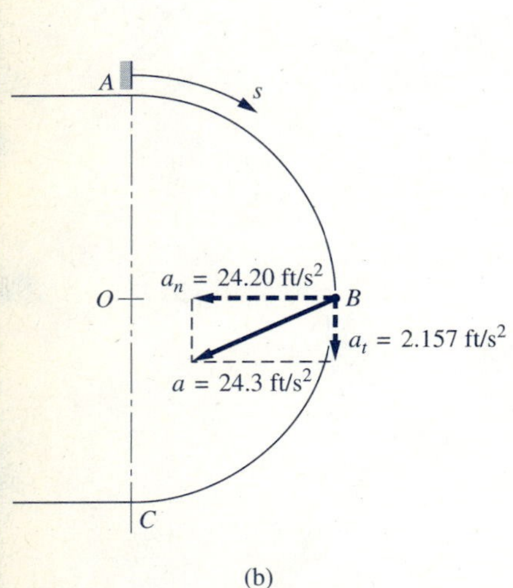

(b)

1. At A: $s = 0$, $v = 132.0$ ft/s (90 mi/h)
2. At C: $s = \pi R = 1000\pi$ ft, $v = 176.0$ ft/s (120 mi/h)

Substituting condition **1** into Eq. (a), we find

$$\frac{(132.0)^2}{2} = 0 + C$$

from which the constant of integration is

$$C = 8712\ (\text{ft/s})^2 \qquad \text{(b)}$$

Substituting condition **2** and the value of C into Eq. (a) gives

$$\frac{(176.0)^2}{2} = a_t(1000\pi) + 8712$$

Solving for a_t yields

$$a_t = 2.157\ \text{ft/s}^2 \qquad \text{(c)}$$

As shown in Fig. (b), the direction of a_t is downward at B, i.e., in the direction of increasing speed.

On substituting the values of C and a_t into Eq. (a), the relationship between the speed v and the distance s is found to be

$$\frac{v^2}{2} = 2.157s + 8712 \tag{d}$$

To compute the speed of the car at B, we substitute $s = \pi R/2 = 500\pi$ ft into Eq. (d), the result being

$$\frac{v^2}{2} = 2.157(500\pi) + 8712$$

$$v = 155.56 \text{ ft/s} \tag{e}$$

From Eq. (13.10), the normal component of the acceleration at B is

$$a_n = \frac{v^2}{R} = \frac{(155.56)^2}{1000} = 24.20 \text{ ft/s}^2 \tag{f}$$

directed toward the center of curvature of the path, namely point O, as indicated in Fig. (b).

The magnitude of the acceleration vector is

$$a = \sqrt{(24.20)^2 + (2.157)^2} = 24.3 \text{ ft/s}^2 \qquad \textit{Answer}$$

with the direction shown in Fig. (b).

Sample Problem 13.3

A particle travels along the branch of the hyperbola $(x/a)^2 - (y/b)^2 = 1$ shown in Fig. (a). If the constant speed of the particle is v_0, determine the magnitude of the acceleration of the particle when it is at A.

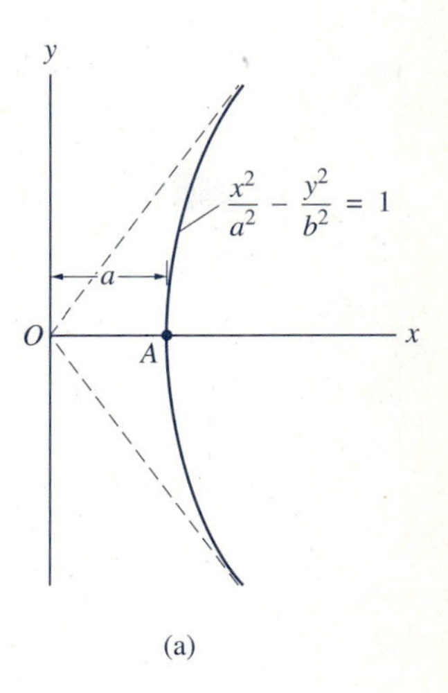

(a)

Solution

Because the speed of the particle is constant, the tangential component of the acceleration is zero at all points along the path; i.e., the acceleration has only a normal component, given by Eq. (13.8):

$$|\mathbf{a}| = a_n = \frac{v_0^2}{\rho} \tag{a}$$

where ρ is the radius of curvature of the path at the point of interest. The radius of curvature at point A can be computed from

$$\rho_A = \frac{\left[1 + \left(\frac{dx}{dy}\right)_A^2\right]^{3/2}}{\left|\left(\frac{d^2x}{dy^2}\right)_A\right|} \tag{b}$$

Differentiating the equation of the hyperbola with respect to y yields

$$\frac{2x}{a^2}\left(\frac{dx}{dy}\right) - \frac{2y}{b^2} = 0$$

from which we find that

$$\frac{dx}{dy} = \frac{a^2 y}{b^2 x} \tag{c}$$

At point A, where $x = a$ and $y = 0$, we have

$$\left(\frac{dx}{dy}\right)_A = 0 \tag{d}$$

as expected. Taking the derivative of Eq. (c) with respect to y gives

$$\frac{d^2x}{dy^2} = \frac{a^2}{b^2}\frac{x - y\left(\dfrac{dx}{dy}\right)}{x^2} = \frac{a^2}{b^2x^2}\left[x - y\left(\frac{dx}{dy}\right)\right] \tag{e}$$

Substituting $x = a$, $y = 0$, and $dx/dy = 0$, we find that

$$\left(\frac{d^2x}{dy^2}\right)_A = \frac{a^2}{b^2a^2}(a - 0) = \frac{a}{b^2} \tag{f}$$

Substituting Eqs. (d) and (f) into Eq. (b), the radius of curvature of the path at A is

$$\rho_A = \frac{(1 + 0)^{3/2}}{a/b^2} = \frac{b^2}{a} \tag{g}$$

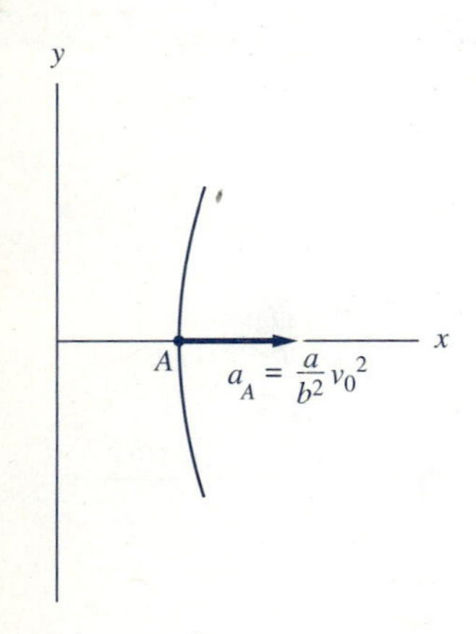

(b)

Equation (a) yields the magnitude of the acceleration at A.

$$a_A = \frac{v_0^2}{\rho_A} = \frac{av_0^2}{b^2} \qquad \textit{Answer}$$

Because the acceleration is due only to a_n, the acceleration vector at A is directed toward the center of curvature of the hyperbola at A, as shown in Fig. (b).

Since the radius of curvature of the hyperbola has its minimum value at A, the acceleration at A is also maximum acceleration of the particle. [The fact that the minimum value of ρ occurs at A may be established by examining the general expression for ρ that is obtained by substituting Eqs. (c) and (e) into Eq. (b).]

PROBLEMS

13.1 Pulley A is attached to the crankshaft of an automobile engine. If the crankshaft rotates at the constant angular speed of 2000 rev/min, determine the maximum acceleration of any point of the V-belt that runs around the three pulleys.

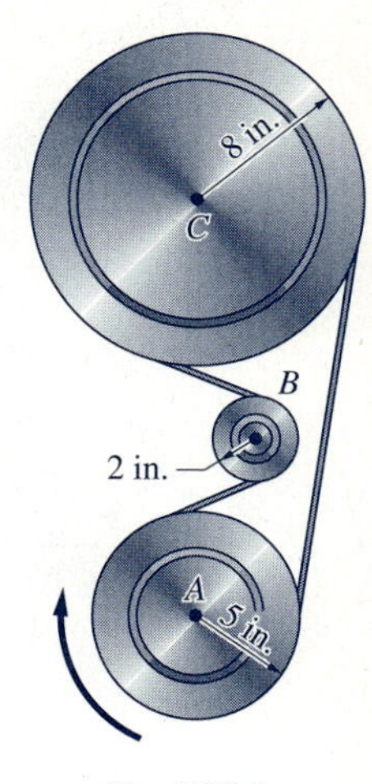

Fig. P13.1

13.2 At the instant shown, the angular speed and acceleration of rod OB are $\dot{\theta} =$ 8 rad/s and $\ddot{\theta} =$ 24 rad/s^2, respectively, both counterclockwise. Calculate (a) the velocity vectors of points A and B on the rod; and (b) the acceleration vectors of A and B.

13.3 The angular velocity of rod OB varies as $\dot{\theta} = 8 - 12t^2$ rad/s where t is in seconds. Compute the magnitude of the acceleration of point B at (a) $t = 0$; and (b) $t = 1.0$ s.

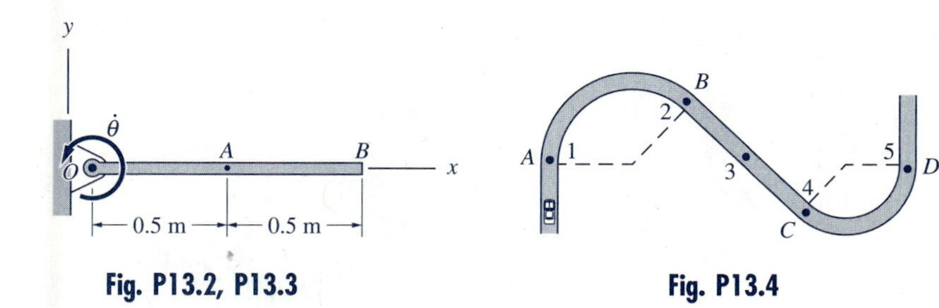

Fig. P13.2, P13.3

Fig. P13.4

13.4 A car drives through portion AB of the S-curve at constant speed, decelerates in BC, and accelerates in CD. Show the approximate direction of the acceleration vector at each of the five points indicated.

13.5 The car of a rollercoaster starts from rest at A and moves with negligible friction. Show the approximate direction of the acceleration vector at points 1 to 5.

13.6 The speed of the belt is changed at a uniform rate from 0 to 2 m/s during a time interval of 0.2 second. Calculate (a) the distance traveled by the belt during the 0.2-second interval; and (b) the maximum acceleration of any point on the belt.

13.7 The rate of change of speed of the belt is given by $0.06(10 - t)$ m/s^2, where t is in seconds. The speed of the belt is 0.8 m/s at $t = 0$. When the normal acceleration of a point in contact with the pulley is 40 m/s^2, determine (a) the speed of the belt; (b) the time required to reach that speed; and (c) the distance traveled by the belt.

Fig. P13.5

Fig. P13.6, P13.7

13.8 A motorist entering the exit ramp of a highway at 40 km/h immediately applies the brake so that the magnitude of the acceleration of the car at A is 1.5 m/s^2. If the tangential acceleration is maintained, how far will the car travel before coming to a stop?

13.9 The speed of a car that starts from rest at A varies according to $(90 + s)/450$ m/s^2, where s is the distance in meters measured along the curve from A. Compute the acceleration vector of the car when $s = 180$ m.

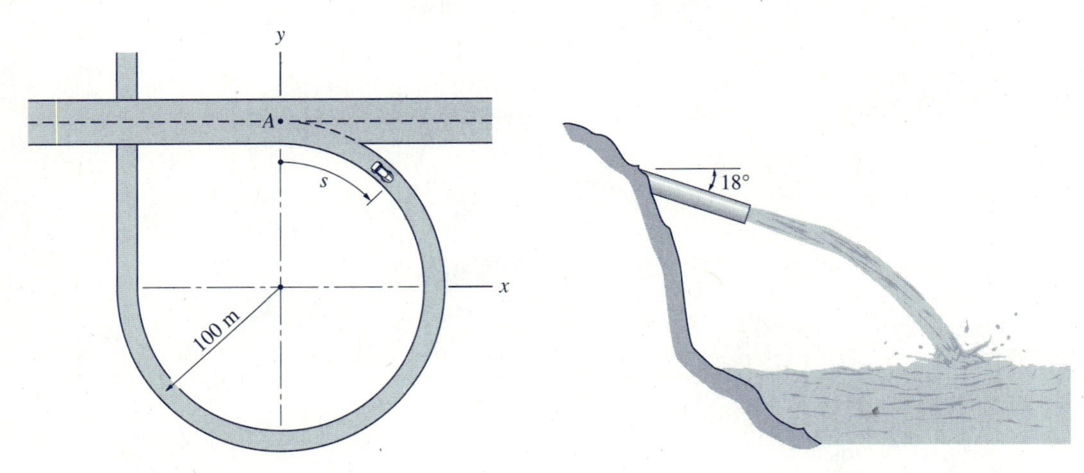

Fig. P13.8, P13.9

Fig. P13.10

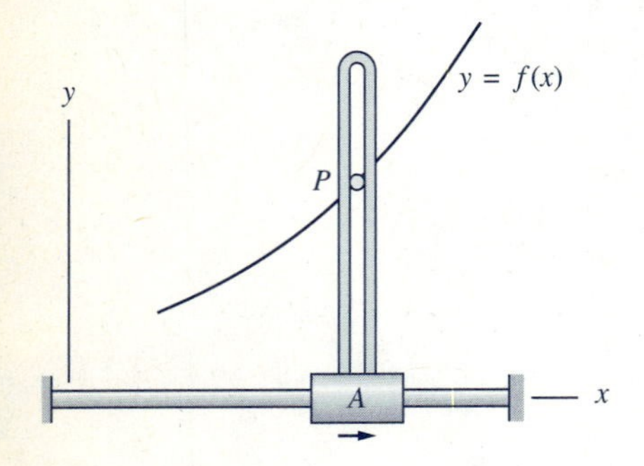

Fig. P13.11, P13.12

13.10 The inclined discharge pipe emits a stream of liquid waste. The radius of curvature of the stream as it leaves the pipe is measured to be 8.5 ft. Determine the velocity and the tangential acceleration of the liquid at the point of exit.

13.11 Pen P of the flatbed plotter traces the curve $y = x^3/128$, where x and y are measured in inches. When $x = 8$ in., the speed of slider A is 2.4 in./s. For this position, calculate (a) the speed of P; and (b) the normal component of the acceleration of P.

13.12 Pen P of the flatbed plotter traces the curve $y = f(x)$ as the slider A moves with the constant speed 0.2 m/s. When P is in the position shown, its acceleration is $\mathbf{a} = 0.3\mathbf{e}_t + 0.4\mathbf{e}_n$ m/s^2. Determine (a) the slope; and (b) the radius of curvature of the curve $f(x)$ at this position.

13.13 A particle moves with the constant speed v_0 along the parabola $y = Ax^2 + Bx + C$. Find the maximum acceleration and the corresponding x-coordinate.

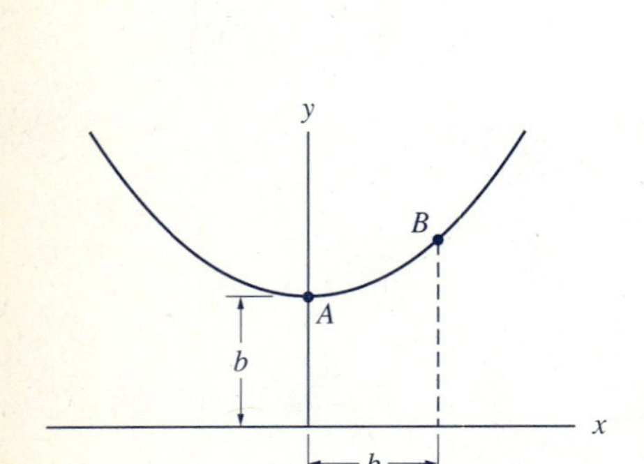

Fig. P13.14

13.14 A particle travels with constant speed along the curve $y = (b/2)(e^{x/b} + e^{-x/b})$. Compute the magnitude of its acceleration at (a) point A; and (b) point B.

13.15 A particle moves with constant speed v_0 along the ellipse $(x/a)^2 + (y/b)^2 = 1$, where $a > b$. Determine the maximum acceleration of the particle.

***13.16** A transition curve, such as AB, is used by railroads to provide a smooth transition between straight track and a circular curve. The curvature of AB is increased linearly with distance traveled: $1/\rho = ks$, where $k = 1.0 \times 10^{-6}\ \text{ft}^{-2}$. A locomotive enters the curve at A with the speed of 60 ft/s and decelerates at a constant rate to a stop at B. Calculate the magnitude of the maximum acceleration of the locomotive and the value of s where it occurs.

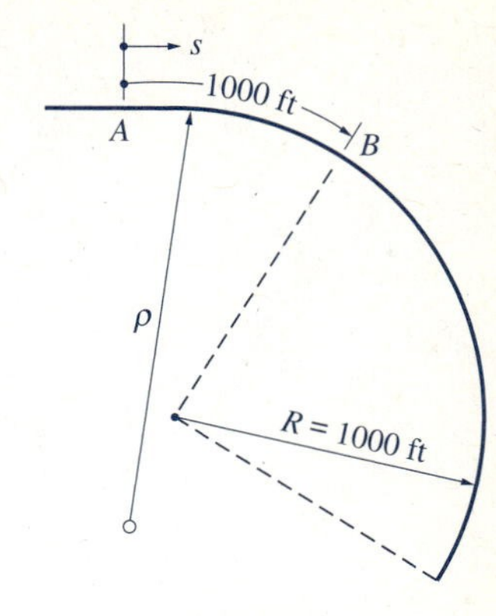

Fig. P13.16

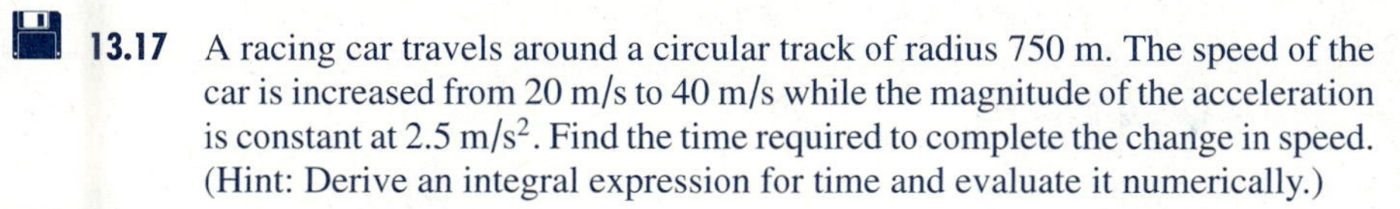

13.17 A racing car travels around a circular track of radius 750 m. The speed of the car is increased from 20 m/s to 40 m/s while the magnitude of the acceleration is constant at 2.5 m/s^2. Find the time required to complete the change in speed. (Hint: Derive an integral expression for time and evaluate it numerically.)

13.18 For the racing car described in Prob. 13.17, compute the distance traveled during the time that the speed changed. (Hint: Derive an integral expression for the distance and evaluate it numerically.)

13.3 Kinematics: Polar and Cylindrical Coordinates

a. Plane motion (polar coordinates)

1. Geometric Preliminaries Figure 13.9 shows the polar coordinates R and θ that specify the position of particle A that is moving in the xy-plane. The radial coordinate R is the length of the *radial line OA*, and θ is the angle between the x-axis and the radial line. (In plane motion the polar coordinate R is equal to the magnitude of the position vector $\mathbf{r}$ of the particle.)

The base vectors $\mathbf{e}_R$ and $\mathbf{e}_\theta$ of the polar coordinate system are also shown in Fig. 13.9. The vector $\mathbf{e}_R$ is directed along the radial line, pointing away from O; $\mathbf{e}_\theta$ is perpendicular to $\mathbf{e}_R$, in the direction of increasing θ.

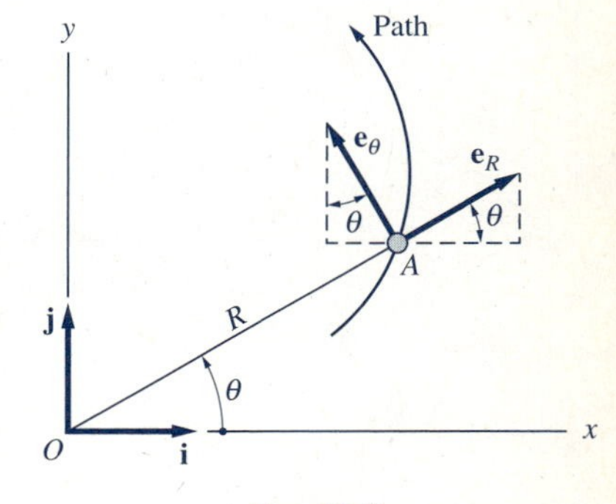

Fig. 13.9

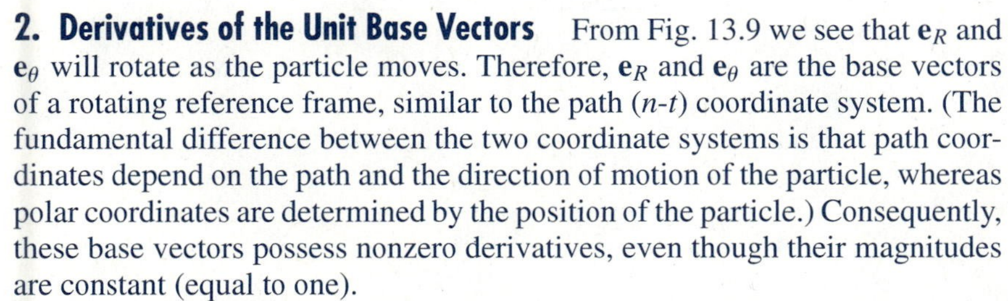

2. Derivatives of the Unit Base Vectors From Fig. 13.9 we see that $\mathbf{e}_R$ and $\mathbf{e}_\theta$ will rotate as the particle moves. Therefore, $\mathbf{e}_R$ and $\mathbf{e}_\theta$ are the base vectors of a rotating reference frame, similar to the path (n-t) coordinate system. (The fundamental difference between the two coordinate systems is that path coordinates depend on the path and the direction of motion of the particle, whereas polar coordinates are determined by the position of the particle.) Consequently, these base vectors possess nonzero derivatives, even though their magnitudes are constant (equal to one).

As in the preceding article, the time derivatives of the unit base vectors can be determined by first relating the vectors to the xy-coordinate system. From Fig. 13.9 we find that

$$
\begin{aligned}
\mathbf{e}_R &= \cos\theta\mathbf{i} + \sin\theta\mathbf{j} \\
\mathbf{e}_\theta &= -\sin\theta\mathbf{i} + \cos\theta\mathbf{j}
\end{aligned}
\tag{13.11}
$$

Differentiating with respect to time while noting that $d\mathbf{i}/dt = d\mathbf{j}/dt = \mathbf{0}$ (the xy-frame is fixed) yields

$$
\frac{d\mathbf{e}_R}{dt} = (-\sin\theta\mathbf{i} + \cos\theta\mathbf{j})\dot{\theta} \qquad \frac{d\mathbf{e}_\theta}{dt} = (-\cos\theta\mathbf{i} - \sin\theta\mathbf{j})\dot{\theta}
$$

Comparing these results with Eq. (13.11), we find that

$$\dot{\mathbf{e}}_R = \dot{\theta}\mathbf{e}_\theta \qquad \dot{\mathbf{e}}_\theta = -\dot{\theta}\mathbf{e}_R \tag{13.12}$$

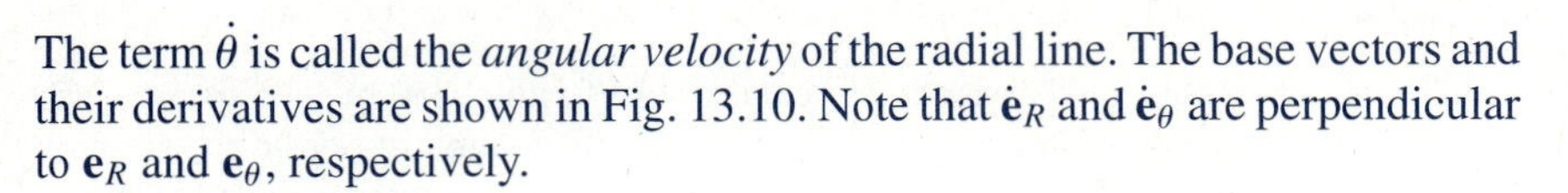
The term $\dot{\theta}$ is called the *angular velocity* of the radial line. The base vectors and their derivatives are shown in Fig. 13.10. Note that $\dot{\mathbf{e}}_R$ and $\dot{\mathbf{e}}_\theta$ are perpendicular to $\mathbf{e}_R$ and $\mathbf{e}_\theta$, respectively.

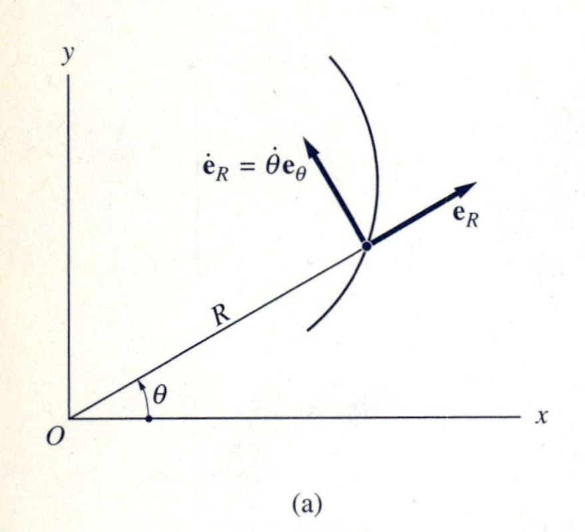

(a)

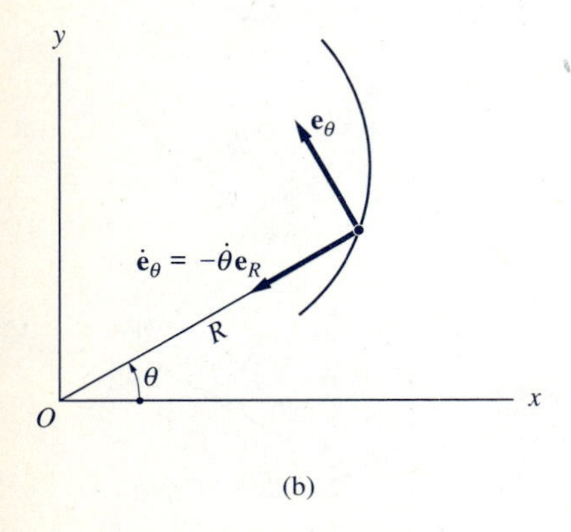

(b)

Fig. 13.10

3. Velocity and Acceleration Vectors The position vector $\mathbf{r}$ of the particle can be written in polar coordinates as

$$\mathbf{r} = R\mathbf{e}_R \tag{13.13}$$

Since the velocity vector is, by definition, $\mathbf{v} = d\mathbf{r}/dt$, we have

$$\mathbf{v} = \frac{d\mathbf{r}}{dt} = \frac{d}{dt}(R\mathbf{e}_R) = \dot{R}\mathbf{e}_R + R\dot{\mathbf{e}}_R$$

Substituting for $\dot{\mathbf{e}}_R$ from Eqs. (13.12) gives

$$\mathbf{v} = v_R\mathbf{e}_R + v_\theta\mathbf{e}_\theta \tag{13.14}$$

where

$$v_R = \dot{R} \qquad v_\theta = R\dot{\theta} \tag{13.15}$$

The components v_R and v_θ are called the *radial* and *transverse* components of the velocity, respectively.

The acceleration vector is computed as follows:

$$\begin{aligned}\mathbf{a} = \frac{d\mathbf{v}}{dt} &= \frac{d}{dt}(\dot{R}\mathbf{e}_R + R\dot{\theta}\mathbf{e}_\theta) \\ &= (\ddot{R}\mathbf{e}_R + \dot{R}\dot{\mathbf{e}}_R) + (\dot{R}\dot{\theta}\mathbf{e}_\theta + R\ddot{\theta}\mathbf{e}_\theta + R\dot{\theta}\dot{\mathbf{e}}_\theta)\end{aligned}$$

The term $\ddot{\theta}$ is called the *angular acceleration* of the radial line. Substituting for $\dot{\mathbf{e}}_R$ and $\dot{\mathbf{e}}_\theta$ from Eqs. (13.12) and rearranging terms, this equation becomes

$$\mathbf{a} = a_R\mathbf{e}_R + a_\theta\mathbf{e}_\theta \tag{13.16}$$

where the radial and transverse components of acceleration are given by

$$a_R = \ddot{R} - R\dot{\theta}^2 \qquad a_\theta = R\ddot{\theta} + 2\dot{R}\dot{\theta} \tag{13.17}$$

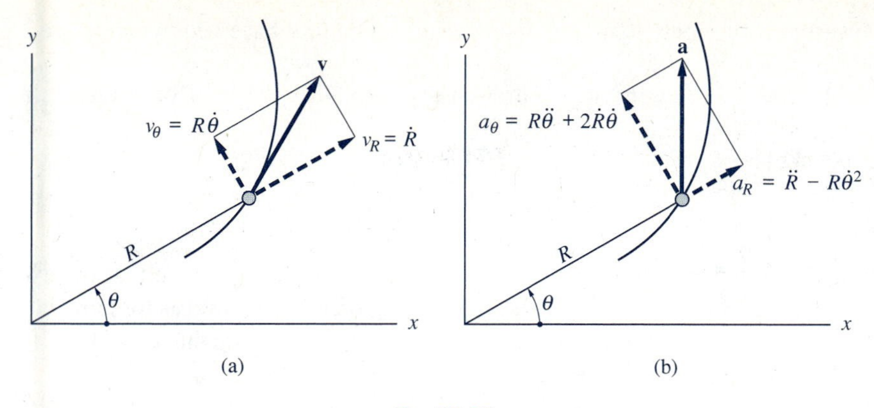

Fig. 13.11

The polar components of the velocity and acceleration vectors are shown in Fig. 13.11. Note again that the positive components point in the directions of the base vectors.

For the special case where the path is a circle, the polar coordinate R equals the radius of the circle (a constant). Therefore, from Eqs. (13.14)–(13.17), the velocity and acceleration vectors are

$$\begin{aligned} \mathbf{v} &= R\dot{\theta}\mathbf{e}_\theta \\ \mathbf{a} &= -R\dot{\theta}^2\mathbf{e}_R + R\ddot{\theta}\mathbf{e}_\theta \end{aligned} \tag{13.18}$$

These expressions are in agreement with Eqs. (13.10), where path coordinates were used. (Note that $\mathbf{e}_R = -\mathbf{e}_n$ and $\mathbf{e}_\theta = \mathbf{e}_t$ for motion on a circular path.)

b. Space motion (cylindrical coordinates)

The cylindrical coordinates shown in Fig. 13.12 may be used to specify the location of a particle A that is moving in space. The three cylindrical coordinates consist of the polar coordinates R and θ, and the axial coordinate z (which

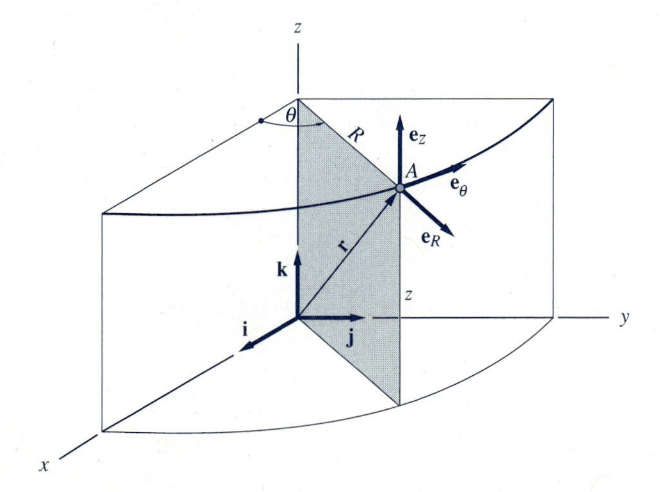

Fig. 13.12

corresponds to the rectangular coordinate z). The unit base vectors are $\mathbf{e}_R$, $\mathbf{e}_\theta$, and $\mathbf{e}_z$, where $\mathbf{e}_z = \mathbf{k}$.

Expressed in cylindrical coordinates, the position vector $\mathbf{r}$ of the particle in Fig. 13.12 is

$$\boxed{\mathbf{r} = R\mathbf{e}_R + z\mathbf{e}_z} \tag{13.19}$$

Comparing this with $\mathbf{r} = R\mathbf{e}_R$ for polar coordinates, we conclude that the expressions for $\mathbf{v}$ and $\mathbf{a}$ in cylindrical coordinates will be the same as for polar coordinates, except for additional terms due to $z\mathbf{e}_z$. Recognizing that $\dot{\mathbf{e}}_z = \dot{\mathbf{k}} = \mathbf{0}$, Eqs. (13.14)–(13.17) can be easily modified to yield

$$\boxed{\begin{aligned} \mathbf{v} &= \dot{R}\mathbf{e}_R + R\dot{\theta}\mathbf{e}_\theta + \dot{z}\mathbf{e}_z \\ \mathbf{a} &= (\ddot{R} - R\dot{\theta}^2)\mathbf{e}_R + (R\ddot{\theta} + 2\dot{R}\dot{\theta})\mathbf{e}_\theta + \ddot{z}\mathbf{e}_z \end{aligned}} \tag{13.20}$$

Sample Problem 13.4

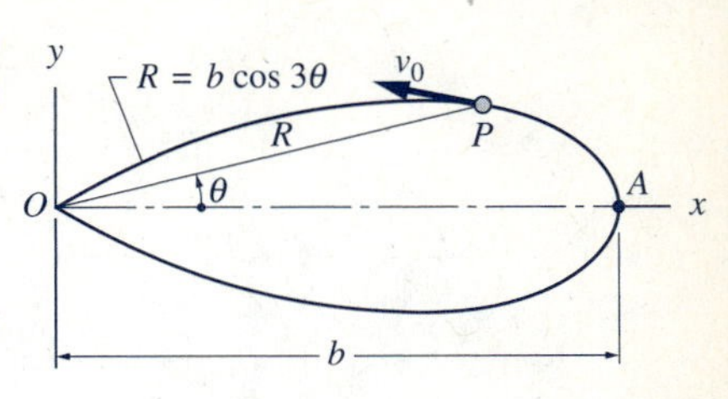

The particle P is traveling with constant speed v_0 along a wire, the polar description of which is $R = b\cos 3\theta$. Determine the acceleration vector of the particle when it is at point A.

Solution

The polar components of acceleration depend on R, $\dot{R}$, $\ddot{R}$, $\dot{\theta}$, and $\ddot{\theta}$. The expressions for R and its derivatives at A are obtained as follows:

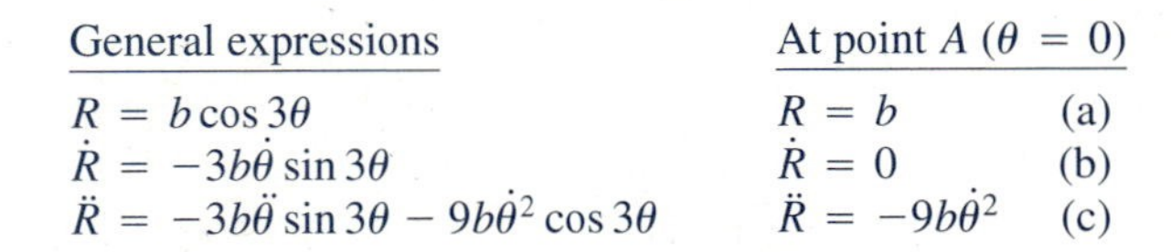

General expressions	At point A ($\theta = 0$)	
$R = b\cos 3\theta$	$R = b$	(a)
$\dot{R} = -3b\dot{\theta}\sin 3\theta$	$\dot{R} = 0$	(b)
$\ddot{R} = -3b\ddot{\theta}\sin 3\theta - 9b\dot{\theta}^2\cos 3\theta$	$\ddot{R} = -9b\dot{\theta}^2$	(c)

At point A we have $v_R = \dot{R} = 0$, which means that $v_\theta = v_0$. Since $v_\theta = R\dot{\theta} = b\dot{\theta}$ at A, we find that the angular velocity of the radial line is

$$\dot{\theta} = \frac{v_0}{b} \quad \text{at } A \tag{d}$$

The fact that the speed is constant also provides information about the angular acceleration $\ddot{\theta}$ of the radial line. Taking the time derivative of $v_0^2 = v_R^2 + v_\theta^2 = \dot{R}^2 + R^2\dot{\theta}^2$ we obtain

$$2\dot{R}\ddot{R} + 2R\dot{R}\dot{\theta}^2 + 2R^2\dot{\theta}\ddot{\theta} = 0$$

Evaluating this equation at A with the aid of Eqs. (a)–(d), we get

$$0 + 0 + 2b^2\frac{v_0}{b}\ddot{\theta} = 0$$

which gives

$$\ddot{\theta} = 0 \quad \text{at } A \tag{e}$$

Using Eqs. (a)–(e), the polar components of the acceleration of the particle at point A become

$$\begin{aligned} a_R &= \ddot{R} - R\dot{\theta}^2 \\ &= -9b\dot{\theta}^2 - b\dot{\theta}^2 = -10b\dot{\theta}^2 = -10\frac{v_0^2}{b} \\ a_\theta &= R\ddot{\theta} + 2\dot{R}\dot{\theta} = 0 + 0 = 0 \end{aligned}$$

Noting that $\mathbf{e}_R = \mathbf{i}$ when the particle is at A, the acceleration vector at that point is

$$\mathbf{a} = a_R\mathbf{e}_R = -\frac{10v_0^2}{b}\mathbf{i} \qquad \textit{Answer}$$

Sample Problem 13.5

Pin P is attached to the rim of a rotating disk and slides in the slotted bar OA as shown in Fig. (a). When $\beta = 45°$, the angular velocity of OA is $\dot{\theta} = 0.5$ rad/s. For this position, determine $\dot{\beta}$ and $\dot{R}$ (the angular velocity of the disk and the velocity of P relative to the arm OA).

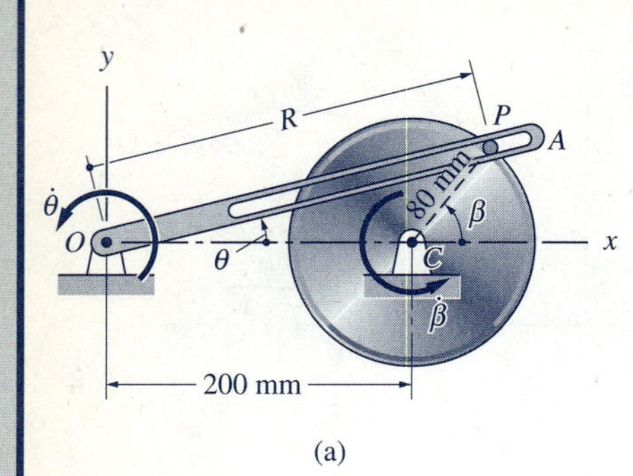

(a)

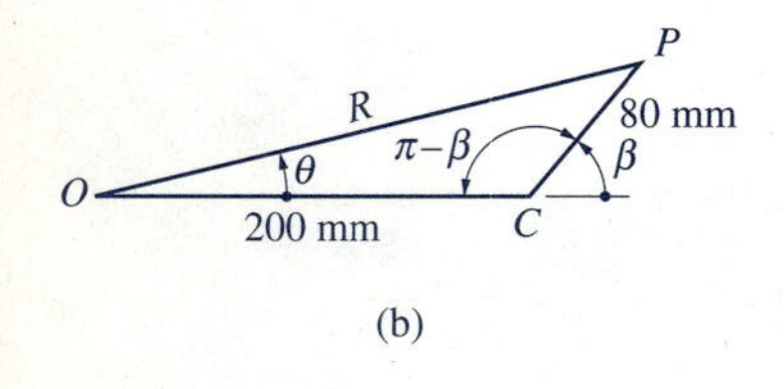

(b)

Solution

Applying the law of cosines and the law of sines to the triangle in Fig. (b), we obtain the following geometric relationships among R, θ, and β:

$$\text{Law of cosines: } R^2 = 200^2 + 80^2 + 2(200)(80)\cos\beta \qquad \text{(a)}$$

$$\text{Law of sines: } \sin\beta = \frac{R\sin\theta}{80} \qquad \text{(b)}$$

In arriving at Eqs. (a) and (b) we used the trigonometric identities $\cos(\pi - \beta) = -\cos\beta$ and $\sin(\pi - \beta) = \sin\beta$. The time derivatives of Eqs. (a) and (b) are, respectively,

$$2R\dot{R} = -32\,000(\sin\beta)\dot{\beta} \qquad \text{(c)}$$

$$(\cos\beta)\dot{\beta} = \frac{\dot{R}\sin\theta + R(\cos\theta)\dot{\theta}}{80} \qquad \text{(d)}$$

The variables R, $\dot{R}$, θ, and $\dot{\beta}$ can be evaluated at $\beta = 45°$ from Eqs. (a)–(d) using the following steps.

Substituting $\beta = 45°$ into Eq. (a) yields

$$R = 262.7 \text{ mm} \qquad \text{(e)}$$

With this value of R and $\beta = 45°$, Eq. (b) gives

$$\theta = 12.44° \qquad \text{(f)}$$

Substituting $\beta = 45°$, $R = 262.7$ mm, $\theta = 12.44°$, and $\dot{\theta} = 0.5$ rad/s into Eqs. (c) and (d), the equations become

$$\dot{R} = -43.07\dot{\beta} \qquad \text{(g)}$$

and

$$\dot{\beta} = (3.808 \times 10^{-3})\dot{R} + 2.267 \qquad \text{(h)}$$

Solving Eqs. (g) and (h) simultaneously, we find that

$$\dot{R} = -83.9 \text{ mm/s} \quad \text{and} \quad \dot{\beta} = 1.948 \text{ rad/s} \qquad \textit{Answer}$$

Since $\dot{R}$ is negative, the velocity of P relative to the arm OA is directed toward O. The positive sign for $\dot{\beta}$ means that the disk is rotating counterclockwise.

Sample Problem 13.6

The passenger car of an amusement park ride is connected by AB to the vertical mast OC. During a certain time interval, the mast is rotating at the constant rate $\dot{\theta}$ while the arm AB is being elevated at the constant rate $\dot{\phi}$. Determine the cylindrical components of the velocity and acceleration of the car during this time interval.

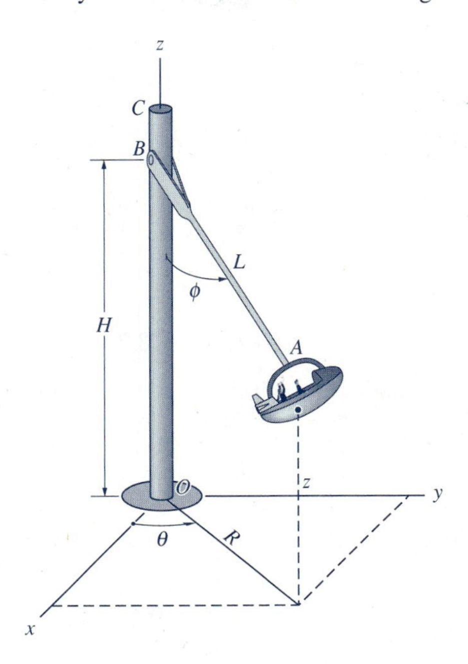

Solution

Referring to the figure, we see that the R- and z-coordinates of the car are $R = L\sin\phi$ and $z = H - L\cos\phi$. Using the fact that $\dot{\phi}$ is constant, differentiation with respect to time yields

$$\dot{R} = L\cos\phi\,\dot{\phi} \qquad \dot{z} = L\sin\phi\,\dot{\phi}$$
$$\ddot{R} = -L\sin\phi\,\dot{\phi}^2 \qquad \ddot{z} = L\cos\phi\,\dot{\phi}^2$$

Using Eqs. (13.20), the cylindrical components of the velocity are

$$\begin{aligned} v_R &= \dot{R} = L\cos\phi\,\dot{\phi} \\ v_\theta &= R\dot{\theta} = L\sin\phi\,\dot{\theta} \\ v_z &= \dot{z} = L\sin\phi\,\dot{\phi} \end{aligned}$$

Answer

Recognizing that $\dot{\theta}$ is constant, the components of the acceleration in Eqs. (13.20) become

$$\begin{aligned} a_R &= \ddot{R} - R\dot{\theta}^2 = -L\sin\phi\,\dot{\phi}^2 - L\sin\phi\,\dot{\theta}^2 \\ &= -L\sin\phi\,(\dot{\phi}^2 + \dot{\theta}^2) \\ a_\theta &= R\ddot{\theta} + 2\dot{R}\dot{\theta} = 0 + 2L\cos\phi\,\dot{\phi}\dot{\theta} \\ a_z &= \ddot{z} = L\cos\phi\,\dot{\phi}^2 \end{aligned}$$

Answer

PROBLEMS

13.19 The rod OB rotates counterclockwise about O at the constant angular speed of 45 rev/min while the collar A slides toward B with the constant speed 2 ft/s, measured relative to the rod. When collar A is in the position $R = 0.8$ ft, $\theta = 0$, calculate (a) the velocity vector; and (b) the acceleration vector.

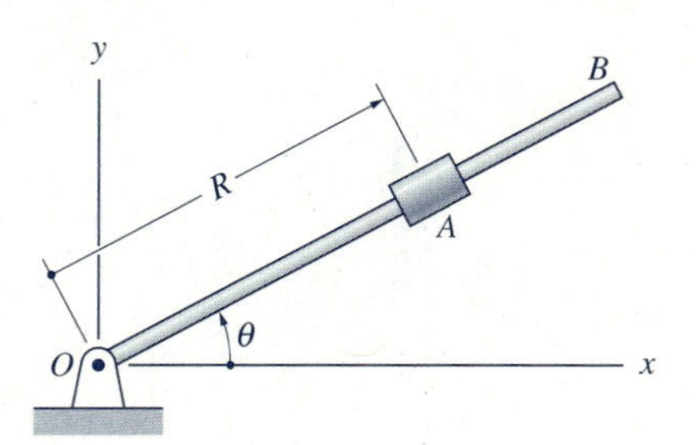

Fig. P13.19–P13.21

13.20 The rod OB rotates counterclockwise with the constant angular speed $\dot{\theta} = \omega$. The position of the collar A on the rod varies as $R = R_0 e^{\omega t}$, where R_0 is a constant. Determine the velocity and acceleration vectors of the collar as functions of time.

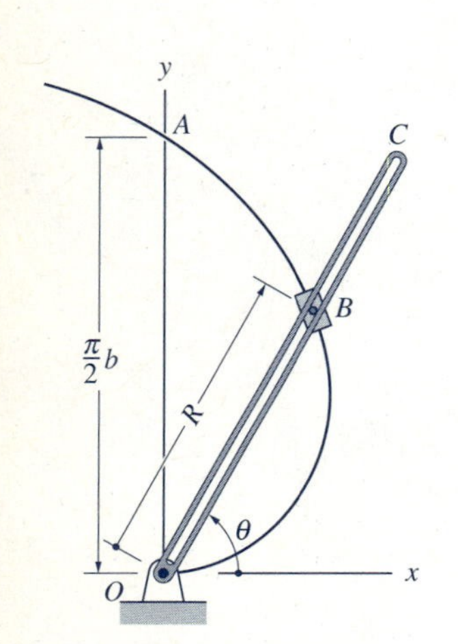

Fig. P13.23, P13.24

13.21 The motion of rod OB is described by $\dot{\theta} = \alpha t$, where $\alpha = 1.2$ rad/s^2 is the constant angular acceleration of the rod. The position of the collar A on the rod is $R = v_0 t$, where $v_0 = 0.8$ m/s is the constant outward speed of the collar relative to the rod. Calculate the velocity and acceleration vectors of the collar as functions of time.

13.22 The plane motion of a particle described in polar coordinates is $\theta = \omega t$, $R = b\sqrt{\omega t}$, where ω and b are constants. When $\theta = \pi$, determine (a) the velocity vector of the particle; (b) the acceleration vector of the particle; and (c) the radius of curvature of the path.

13.23 The collar B slides along a guide rod that has been bent in the shape of the spiral: $R = b\theta$. A pin on the collar slides in the slotted arm OC, which is rotating at the constant angular speed $\dot{\theta} = \omega$. Determine the magnitude of the acceleration of the collar when it is in position A.

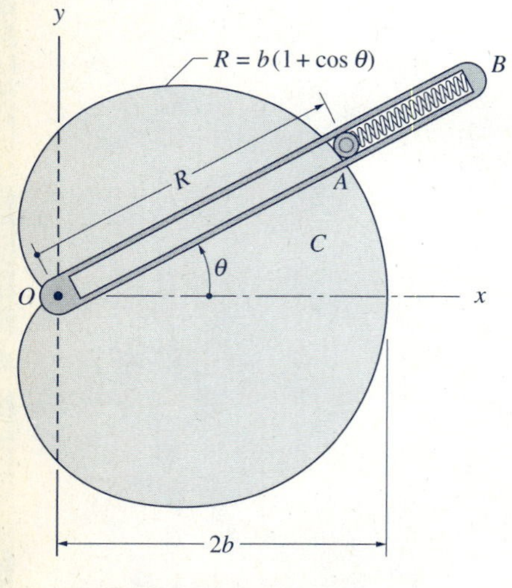

Fig. P13.25–P13.27

13.24 If the speed of the collar B in Prob. 13.23 is constant at v_0, determine the angular speed $\dot{\theta}$ of the rod OC in terms of v_0, b, and θ.

13.25 The slotted arm OB rotates about the pin at O. The ball A in the slot is pressed against the stationary cam C by the spring. If the angular speed of OB is $\dot{\theta} = \omega$, where ω is a constant, calculate the maximum magnitudes of (a) the velocity of A; (b) the acceleration of A; (c) $\dot{R}$ (the velocity of A relative to OB); and (d) $\ddot{R}$ (the acceleration of A relative to OB).

13.26 The angular position of the slotted arm in Prob. 13.25 is given by $\theta = \pi \sin \omega t$, where ω is a constant. Determine the velocity and acceleration vectors of ball A when $\theta = \pi/2$.

13.27 If the ball A in Prob. 13.25 travels around the cam with constant speed v_0, determine (a) $\dot{\theta}$ (the angular speed of OB); and (b) $\dot{R}$ (the speed of ball A relative to OB). Express your results in terms of v_0, b, and θ.

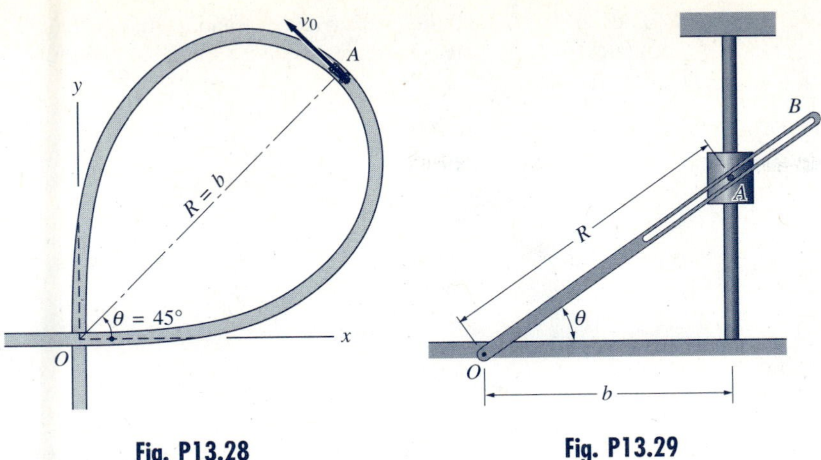

Fig. P13.28 Fig. P13.29

***13.28** The curved portion of the cloverleaf highway interchange is defined by $R^2 = b^2 \sin 2\theta$, $0 \le \theta \le 90°$. If a car travels along the curve at the constant speed v_0, determine its acceleration at position A.

13.29 The pin attached to the sliding collar A engages the slot in bar OB. Using polar coordinates, determine the speed of A in terms of θ and $\dot{\theta}$. (Note: The solution using rectangular coordinates was requested in Prob. 12.20.)

13.30 The helicopter is tracked by radar, which records R, θ, and $\dot{\theta}$ at regular time intervals. The readings at a certain instant are $R = 8050$ ft, $\theta = 38.4°$, and $\dot{\theta} = 0.0367$ rad/s. If the helicopter is in level flight, calculate the elevation h and speed of the helicopter at this instant.

13.31 The telescopic arm of the robot slides in the mount A, which rotates about a horizontal axis at O. End B of the arm traces the vertical line shown with the constant speed v_B. In terms of v_B, b, and θ, determine expressions for (a) $\dot{\theta}$ and $\dot{R}$; and (b) $\ddot{\theta}$ and $\ddot{R}$.

13.32 The telescopic arm of the robot in Prob. 13.31 is being extended at the constant rate $\dot{R}$ as end B traces the vertical line shown. Derive the expressions for the magnitudes of the velocity and acceleration of end B in terms of $\dot{R}$, b, and θ.

13.33 The collar A slides along the rod OB, which is rotating counterclockwise. (a) Find $\dot{R}$ in terms of R, θ, and $\dot{\theta}$. (b) Determine the velocity vector of A when OB is vertical, given that the speed of end B is 0.4 m/s in that position.

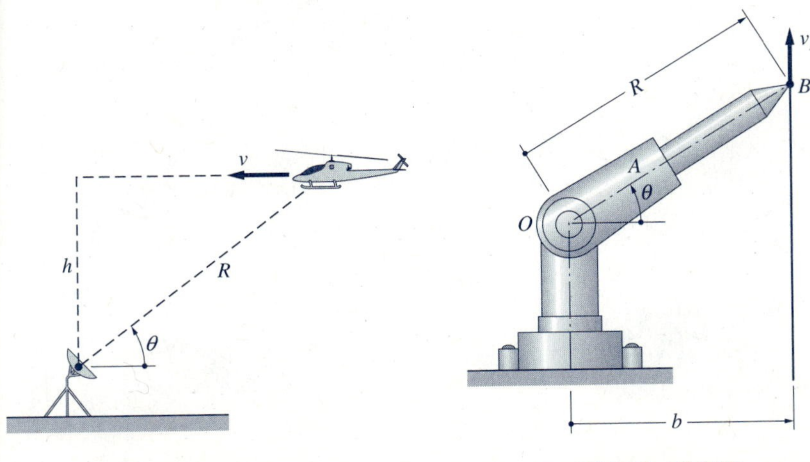

Fig. P13.30 Fig. P13.31, P13.32

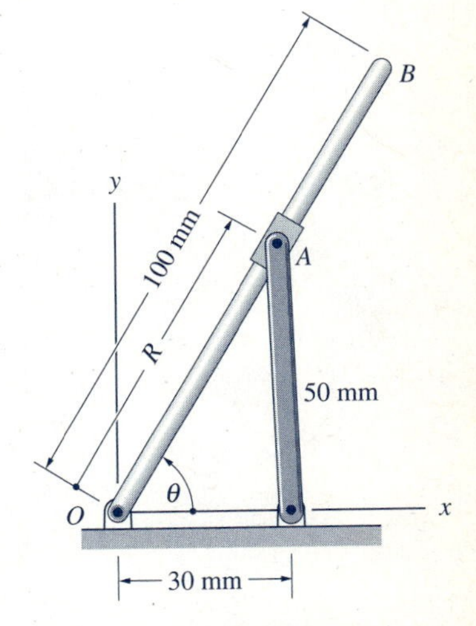

Fig. P13.33

13.34 The pin P slides in slots in both the rotating arm OA and the fixed circular bar BC. If OA rotates with the constant angular speed $\dot{\theta} = 2$ rad/s, find the speed of P when $\theta = 60°$.

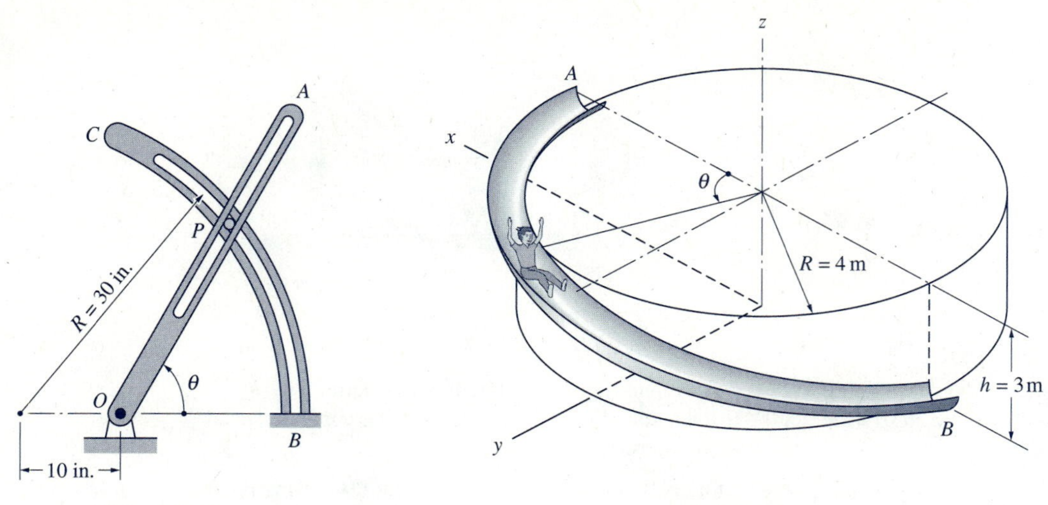

Fig. P13.34

Fig. P13.35

13.35 A child slides down the helical water slide AB. The description of motion in cylindrical coordinates is $R = 4$ m; $\theta = \omega^2 t^2$ and $z = h[1 - (\omega^2 t^2/\pi)]$, where $h = 3$ m and $\omega = 0.75$ rad/s. Compute the magnitudes of the velocity vector and acceleration vector when the child is at B.

13.36 The rod OB rotates about the z-axis with the constant angular speed $\dot{\theta} = 4$ rad/s while the slider A moves up the rod at the constant speed $\dot{s} = 6$ ft/s. Determine the magnitudes of the velocity and acceleration vectors of A when $s = 2$ ft.

13.37 The rotating water sprinkler has a constant angular speed of $\dot{\theta} = 6$ rad/s about the z-axis. The speed of the water relative to the curved tube OA is 7.5 ft/s. Compute the magnitudes of the water velocity and acceleration vectors just below the nozzle at A.

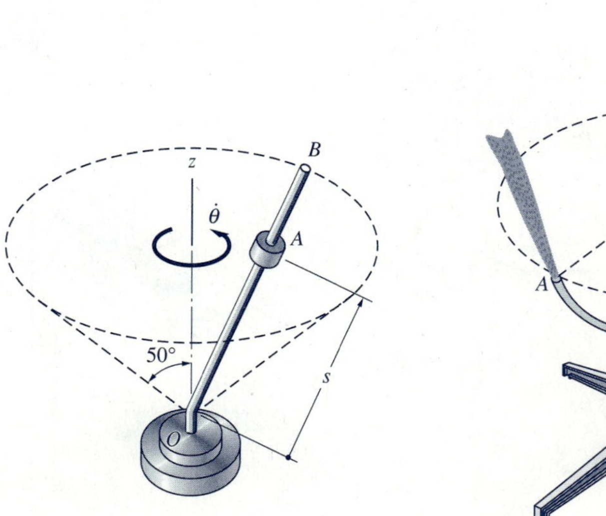

Fig. P13.36

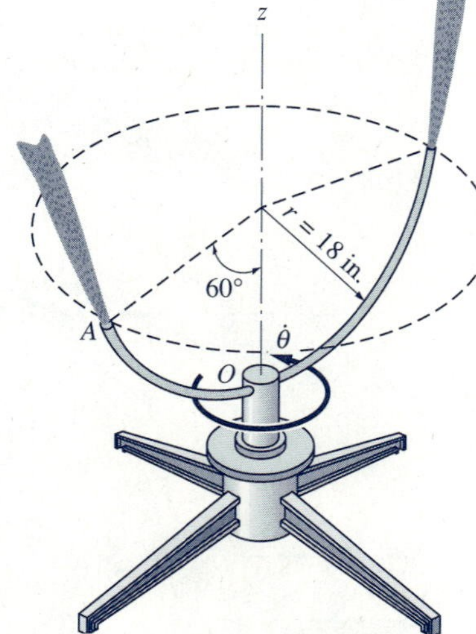

Fig. P13.37

13.38 The path of the particle that is moving on the surface of a cone is defined by

$$R = \frac{h}{2\pi}\theta \tan \beta \qquad z = \frac{h}{2\pi}\theta$$

where R, θ, and z are the cylindrical coordinates. If the motion of the particle is such that $\dot{\theta} = \omega$ (constant), determine the following as functions of θ: (a) the speed of the particle; and (b) the cylindrical components of the acceleration vector.

***13.39** Solve Prob. 13.38 if the motion of the particle is such that $R\dot{\theta} = v_0$ (constant).

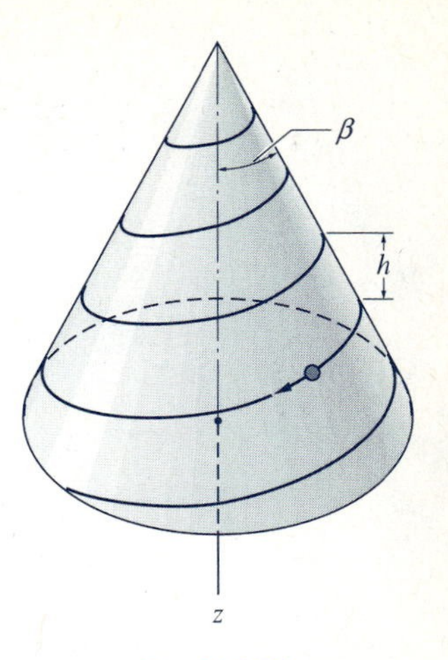

Fig. P13.38

13.4 Force-Mass-Acceleration Method: Curvilinear Coordinates

Recall from Art. 12.4 that the force-mass-acceleration (FMA) method of kinetic analysis consists of the following three steps: (1) construct the free-body diagram for the particle; (2) use kinematics to describe the acceleration vector; and (3) combine the results of the first two steps into Newton's second law: $\mathbf{F} = m\mathbf{a}$. Here we examine problems that are more easily solved by employing curvilinear components of acceleration, rather than the rectangular components used in Chapter 12. Otherwise, the material covered in this article differs little from that presented in Art. 12.4.

a. Path (*n*-*t*) coordinates

Let F_t and F_n be the tangential and normal components, respectively, of the resultant force acting on a particle that is undergoing plane motion. Using the *n*-*t* components of acceleration given in Eqs. (13.8) and (13.9), the equations of motion for the particle become

$$\boxed{\begin{aligned} F_t &= ma_t \\ F_n &= ma_n \end{aligned}} \qquad (13.21)$$

where $a_t = \dot{v} = \rho\ddot{\theta} + \dot{\rho}\dot{\theta} = v\, dv/ds$ and $a_n = v\dot{\theta} = \rho\dot{\theta}^2 = v^2/\rho$. As explained in Art. 13.2, normal-tangential coordinates are of limited use unless the motion occurs in a plane.

b. Cylindrical (polar) coordinates

Letting F_R, F_θ, and F_z be the cylindrical components of the resultant force acting on a particle, and using the cylindrical components of acceleration given in Eqs. (13.20), the equations of motion become

$$\boxed{\begin{aligned} F_R &= ma_R \\ F_\theta &= ma_\theta \\ F_z &= ma_z \end{aligned}} \qquad (13.22)$$

where $a_R = \ddot{R} - \dot{R}\dot{\theta}^2$, $a_\theta = R\ddot{\theta} + 2\dot{R}\dot{\theta}$, and $a_z = \ddot{z}$. If motion occurs only in the xy-plane, then $\ddot{z} = 0$ (polar coordinates).

As was the case with rectangular components in Chapter 12, it is advantageous to employ the concept of a mass-acceleration diagram (MAD), which is a sketch of the particle showing its inertia vector $m\mathbf{a}$. Figure 13.13 summarizes the FBD-MAD method for particle motion using normal-tangential and cylindrical coordinates. As we have indicated previously, when applying Newton's second law $\mathbf{F} = m\mathbf{a}$, the free-body diagram (FBD) displays the terms of the left side of the equation, and the mass-acceleration diagram (MAD) enables us to properly identify the terms on the right side.

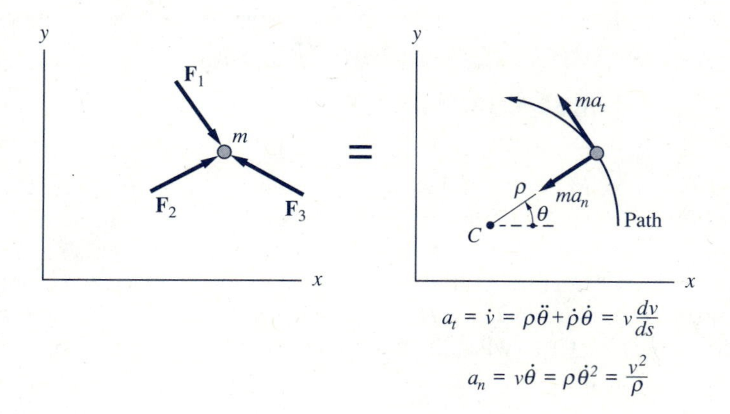

(a) Path (n-t) coordinates

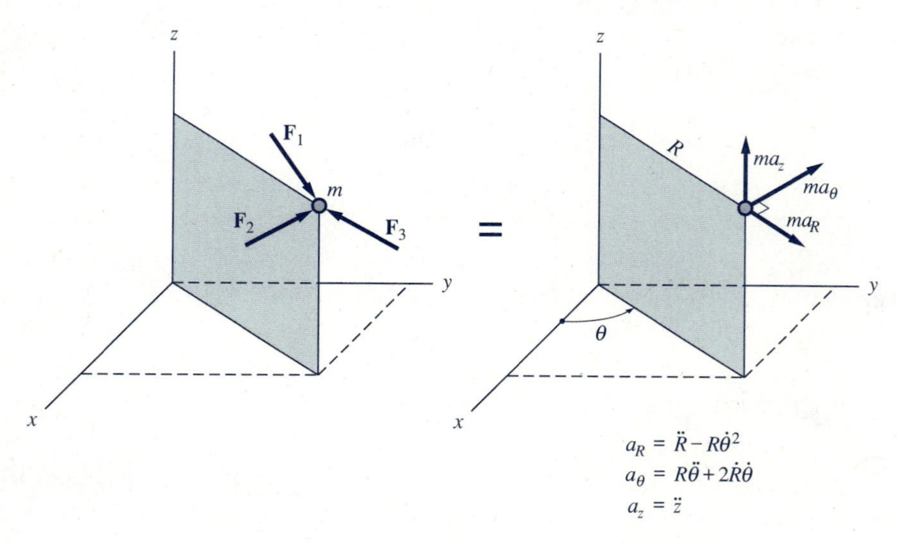

(b) Cylindrical coordinates

Fig. 13.13

Sample Problem 13.7

Two 16-in. cables AB and AC restrain the 6-lb ball A shown in Fig. (a). End B is attached directly to the vertical shaft, but end C is connected to the 10-lb slider, which is free to move on the shaft. When the angular speed $\dot{\theta}$ of the shaft is constant, A travels on a horizontal circle with the slider being stationary. Find the constant angular speed of the shaft for which $\alpha = 30°$.

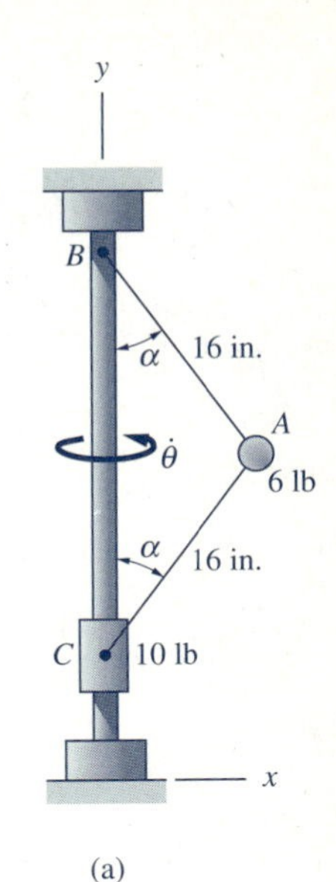

(a)

Solution

Figure (b) shows the free-body diagram of the slider, where T_{AC} is the tension in cable AC and N is the contact force exerted by the shaft. Since the slider is in equilibrium when $\dot{\theta}$ is constant, summing forces in the y-direction yields

$$\Sigma F_y = 0 \qquad +\uparrow \quad T_{AC}\cos 30° - 10 = 0$$

which gives

$$T_{AC} = 11.547 \text{ lb}$$

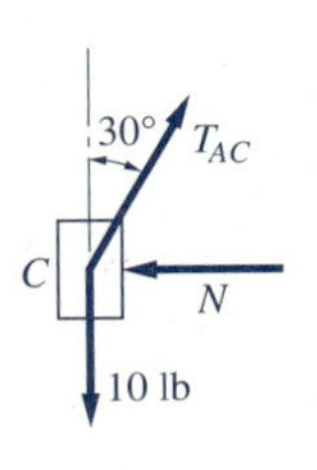

(b)

The free-body diagram (FBD) and mass-acceleration diagram (MAD) for the ball A are shown in Fig. (c). The FBD contains the weight of the ball and the two cable tensions. Because the ball moves with constant speed on a horizontal circle, the inertia vector in the MAD has only a normal component. Summing forces in the y-direction gives

$$\Sigma F_y = 0 \qquad +\uparrow \quad T_{AB}\cos 30° - T_{AC}\cos 30° - 6 = 0$$

Substituting the value for T_{AC} and solving for T_{AB}, we find

$$T_{AB} = \frac{11.547\cos 30° + 6}{\cos 30°} = 18.475 \text{ lb}$$

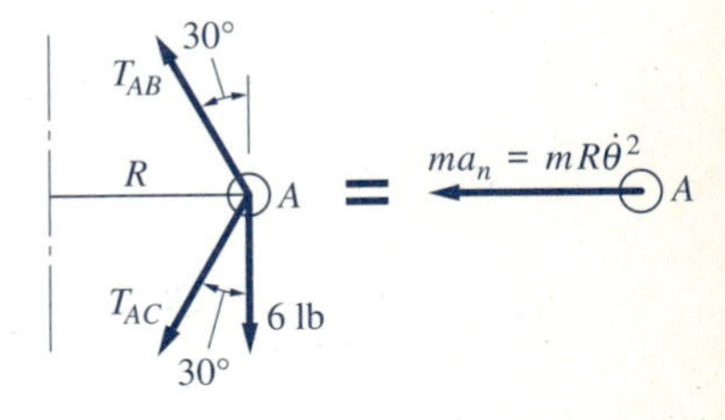

(c)

Recognizing that the radius of the circular path of A is $R = 16\sin\alpha = 16\sin 30° = 8$ in. $= 8/12$ ft, we obtain the equation of motion

$$\Sigma F_n = ma_n = mR\dot{\theta}^2$$

$$\xleftarrow{+} \quad T_{AB}\sin 30° + T_{AC}\sin 30° = \frac{6}{32.2}\frac{8}{12}\dot{\theta}^2$$

$$(18.475 + 11.547)\sin 30° = \frac{48}{(32.2)(12)}\dot{\theta}^2$$

which gives

$$\dot{\theta} = 10.99 \text{ rad/s} \qquad \textit{Answer}$$

Sample Problem 13.8

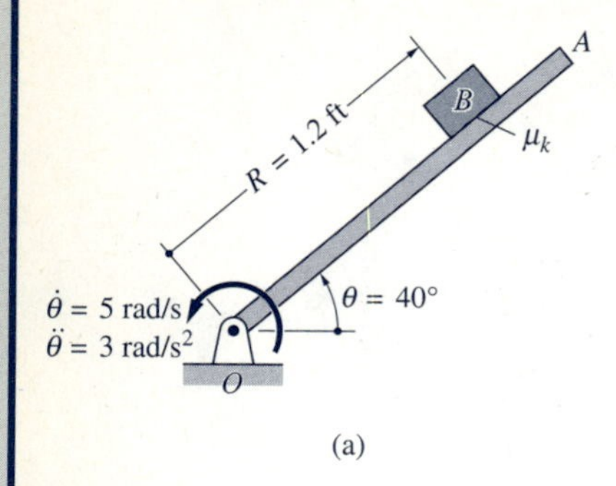

(a)

The 0.5-lb block B shown in Fig. (a) slides along the rotating bar OA. The coefficient of kinetic friction between B and OA is $\mu_k = 0.2$. In the position shown, $\dot{R} = 2$ ft/s, $\dot{\theta} = 5$ rad/s, and $\ddot{\theta} = 3$ rad/s^2. For this position, determine $\ddot{R}$, the acceleration of B relative to bar OA.

Solution

Referring to the free-body diagram (FBD) in Fig. (b), we see that there are three forces acting on B: its weight of 0.5 lb, the normal force N exerted by the bar OA, and the kinetic friction force $F = \mu_k N$. The direction of F is opposite to $\dot{R}$, the velocity of B relative to OA. The mass-acceleration diagram (MAD) of B is also shown in Fig. (b), where the inertia vector $m\mathbf{a}$ is described in terms of its polar components.

40° 0.5 lb

$ma_\theta = m(R\ddot{\theta} + 2\dot{R}\dot{\theta})$

$ma_R = m(\ddot{R} - R\dot{\theta}^2)$

$F = 0.2N$ N

FBD = MAD

(b)

Inspection of Fig. (b) reveals that there are only two unknowns in the FBD-MAD diagrams—N and $\ddot{R}$, the other variables having been specified. Therefore, the two unknowns can be computed from the two equations of motion. For the direction perpendicular to the bar, we have

$$\Sigma F_\theta = m(R\ddot{\theta} + 2\dot{R}\dot{\theta})$$

$$+\nwarrow \quad N - 0.5\cos 40° = \frac{0.5}{32.2}[(1.2)(3) + 2(2)(5)]$$

which gives

$$N = 0.7495 \text{ lb}$$

For the radial direction, we obtain

$$\Sigma F_R = m(\ddot{R} - R\dot{\theta}^2)$$

$$+\nearrow \quad -0.5\sin 40° - 0.2N = \frac{0.5}{32.2}[\ddot{R} - (1.2)(5)^2]$$

Substituting $N = 0.7495$ lb and solving for $\ddot{R}$ yields

$$\ddot{R} = -0.351 \text{ ft/s}^2 \qquad \textit{Answer}$$

The minus sign means that the acceleration of B relative to bar OA is directed toward point O.

Sample Problem 13.9

The vertical shaft AB in Fig. (a) rotates in bearings at A and B. The 0.6-kg slider P can move freely along the smooth bar OD, which is rigidly joined to AB at a 30° angle. At a certain instant when $r = 1.2$ m, it is known that $\dot{\theta} = 4$ rad/s, $\ddot{\theta} = 0$, and the velocity of P relative to OD is $\dot{r} = 4$ m/s. At this instant, determine (1) the magnitude of the contact force exerted on P by OD; and (2) $\ddot{r}$, the acceleration of P relative to OD.

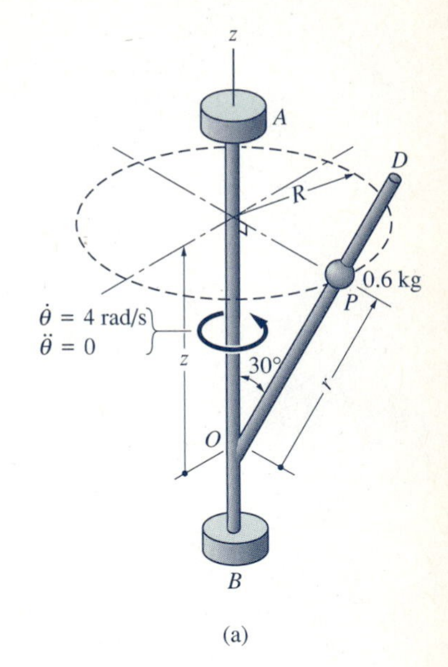

(a)

Solution

Part 1

The free-body diagram (FBD) of the slider P at the instant of interest is shown in Fig. (b), where its weight is $mg = 0.6(9.81) = 5.886$ N. It is convenient to decompose the contact force exerted by OD (which is normal to OD) into two components: N_1, which is perpendicular to OD and passes through the z-axis; and N_2, which is perpendicular to both OD and N_1. The mass-acceleration diagram (MAD) of the slider P is shown in Fig. (c), where the inertia vector $m\mathbf{a}$ is expressed in terms of its cylindrical components.

From Fig. (a) we obtain $R = r \sin 30° = 1.2 \sin 30° = 0.60$ m and $z = r \cos 30°$. Differentiating with respect to time, and substituting $\dot{r} = 4$ m/s, we have

$$\dot{R} = \dot{r} \sin 30° = 4 \sin 30° = 2.00 \text{ m/s}$$

$$\ddot{R} = \ddot{r} \sin 30°$$

$$\dot{z} = \dot{r} \cos 30°$$

$$\ddot{z} = \ddot{r} \cos 30°$$

Since the values of $\dot{\theta}$ and $\ddot{\theta}$ are given, we see that $\ddot{r}$ is the only kinematical variable that is unknown. The FBD in Fig. (b) contains two unknowns (N_1 and N_2), so that we have a problem involving three unknowns, which can be determined from the three available equations of motion for the slider.

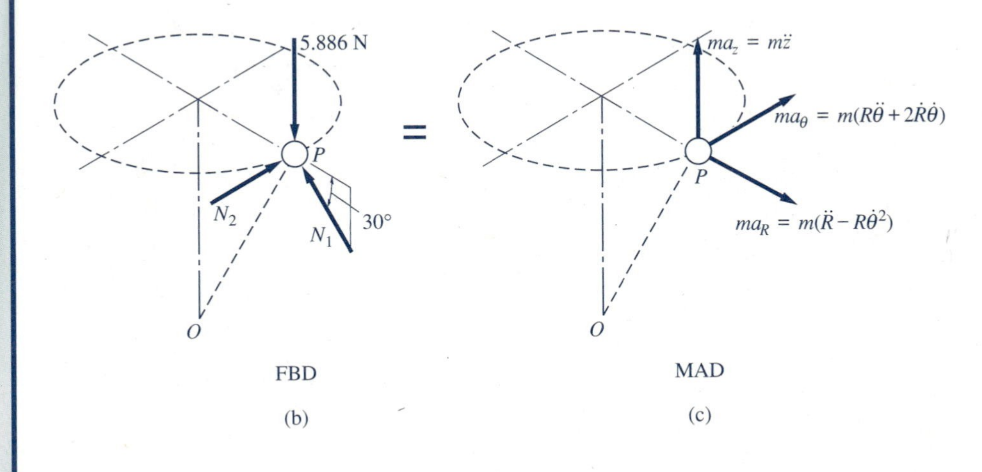

(b) (c)

Referring to Figs. (b) and (c), the equations of motion are

$$\Sigma F_R = ma_R = m(\ddot{R} - R\dot{\theta}^2)$$

$$\overset{+}{\searrow} \quad -N_1 \cos 30^\circ = 0.6[\ddot{r} \sin 30^\circ - 0.60(4)^2] \tag{a}$$

$$\Sigma F_\theta = ma_\theta = m(R\ddot{\theta} + 2\dot{R}\dot{\theta})$$

$$+\nearrow \quad N_2 = 0.6[0 + 2(2.00)(4)] = 9.600 \text{ N} \tag{b}$$

$$\Sigma F_z = ma_z = m\ddot{z}$$

$$+\uparrow \quad N_1 \sin 30^\circ - 5.886 = 0.6(\ddot{r} \cos 30^\circ) \tag{c}$$

Solving Eqs. (a) and (c) simultaneously yields

$$\ddot{r} = -3.70 \text{ m/s}^2$$ *Answer*

and $N_1 = 7.931$ N. Therefore the magnitude of the contact force exerted by OD is

$$N = \sqrt{N_1^2 + N_2^2} = \sqrt{(7.931)^2 + (9.600)^2} = 12.45 \text{ N}$$ *Answer*

PROBLEMS

13.40 A car travels over the crest A of a hill where the radius of curvature is 200 ft. Find the maximum speed for which the wheels will stay in contact with the road.

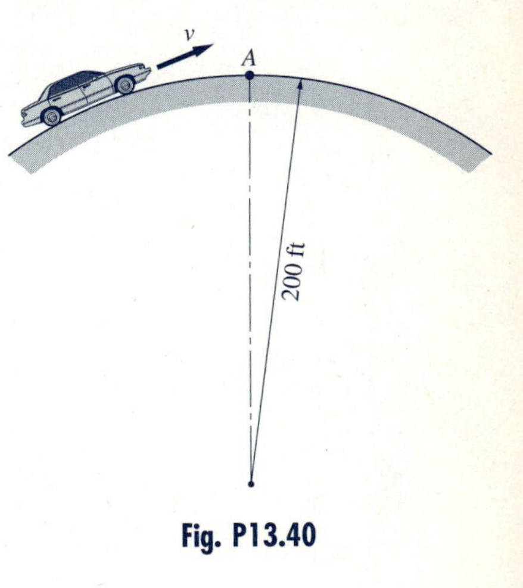

Fig. P13.40

13.41 The 7.5-kg box that is sliding down a circular chute reaches point A with a speed of 3 m/s. The kinetic coefficient of friction between the box and the chute is 0.25. When the box is at A, calculate (a) the normal force acting between it and the chute; and (b) its rate of change of speed.

13.42 The path of a ball that rolls around a smooth, circular track is a circle of radius R. The track is banked at the angle β. Determine the speed of the ball.

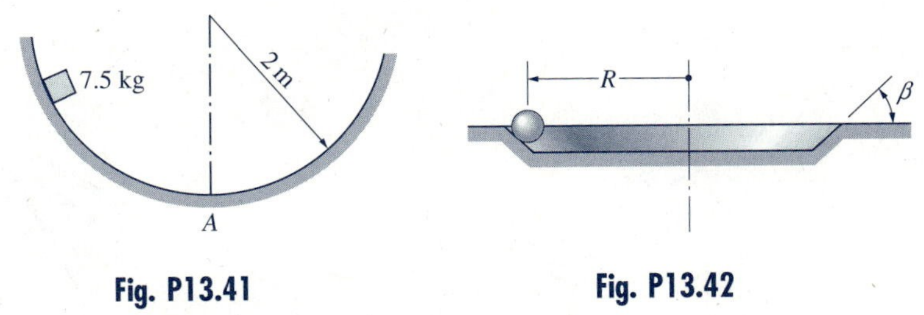

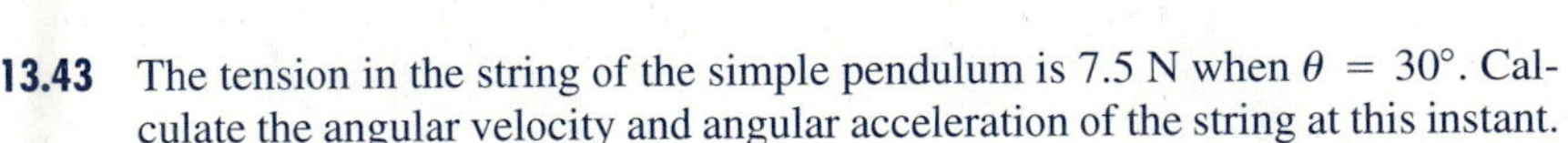

Fig. P13.41

Fig. P13.42

13.43 The tension in the string of the simple pendulum is 7.5 N when $\theta = 30°$. Calculate the angular velocity and angular acceleration of the string at this instant.

13.44 The pendulum is released from rest with $\theta = 30°$. (a) Derive the equation of motion using θ as the independent variable. (b) Determine the speed of the bob as a function of θ.

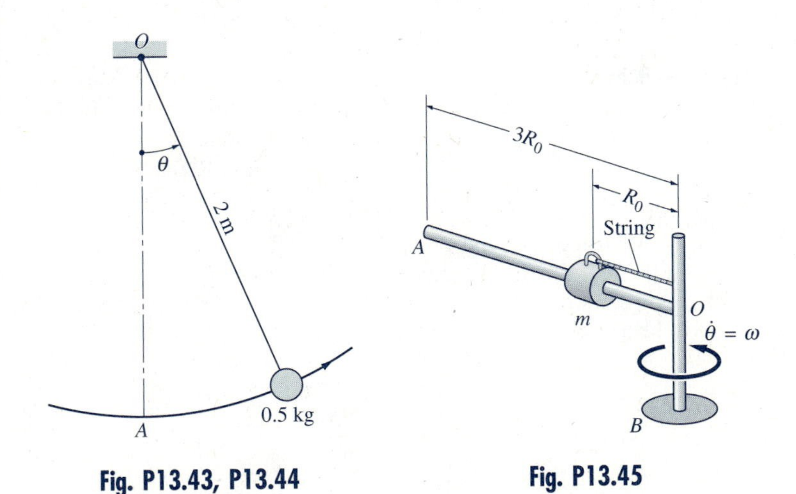

Fig. P13.43, P13.44

Fig. P13.45

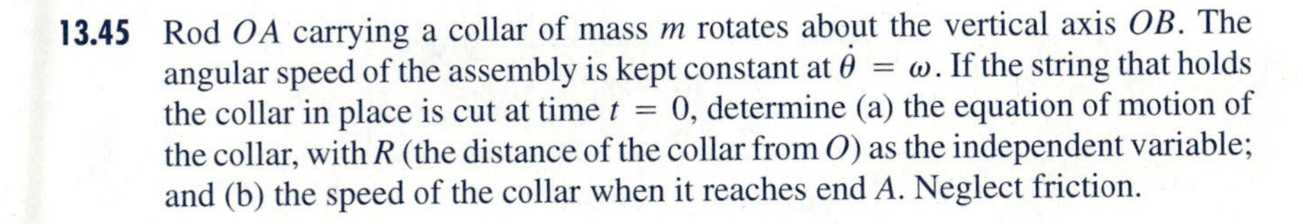

13.45 Rod OA carrying a collar of mass m rotates about the vertical axis OB. The angular speed of the assembly is kept constant at $\dot{\theta} = \omega$. If the string that holds the collar in place is cut at time $t = 0$, determine (a) the equation of motion of the collar, with R (the distance of the collar from O) as the independent variable; and (b) the speed of the collar when it reaches end A. Neglect friction.

13.46 The pendulum is connected to the vertical shaft by a pin at A. The mass of the bob is 1.2 kg, and the mass of the arm AB is negligible. The shaft rotates with a constant angular speed, causing the bob to travel in a horizontal circle. If $\theta = 85°$, determine (a) the tensile force in AB; and (b) the speed v of the bob.

13.47 If the speed of the bob of the pendulum described in Prob. 13.46 is $v = 2$ m/s, compute (a) the angle θ; and (b) the tensile force in arm AB.

13.48 The 360-lb wrecking ball is initially held at rest by the two cables AB and BC. Calculate the force in cable AB (a) before cable BC is released; and (b) just after cable BC is released.

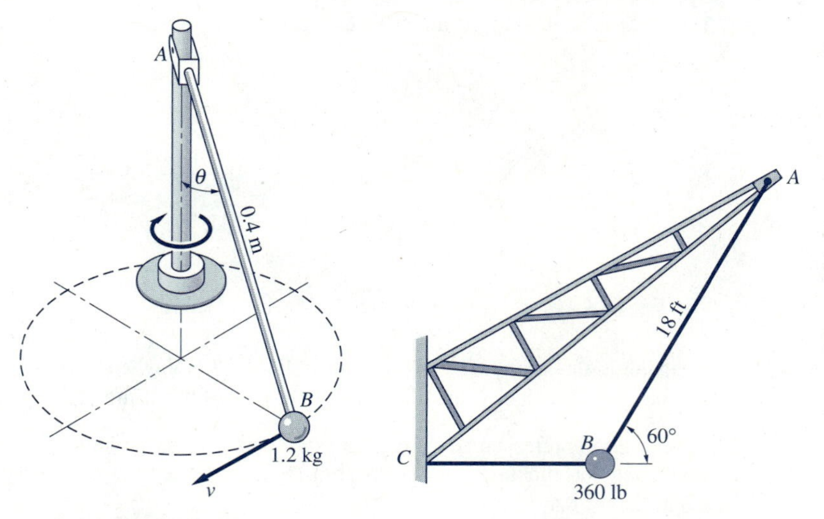

Fig. P13.46, P13.47 **Fig. P13.48**

13.49 The geosynchronous orbit of an earth satellite is a circle with a period of 23 h 56 min. Determine (a) the radius of the orbit; and (b) the speed of the satellite (SI units). Use $KM_e = 3.98 \times 10^{14}$ m^3/s^2, where K is the universal gravitational constant and M_e is the mass of the earth. [Hint: Refer to Eq. (11.17).]

13.50 The 0.5-kg disk slides on a smooth horizontal surface. The elastic cord connected between the disk and point O has a stiffness of 100 N/m and a free length of 0.75 m. The disk is given the initial velocity v_1, directed as shown, when it is 1.25 m from O. Determine the value of v_1 for which the path of the disk will be a circle.

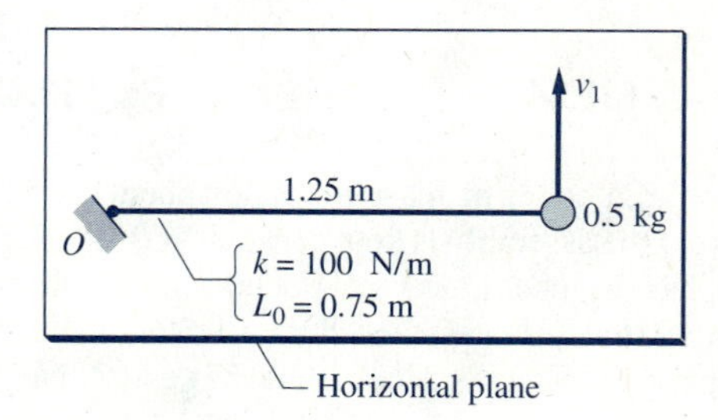

Fig. P13.50

13.51 The 1.0-in. diameter marble rolls in a groove that has the shape of a parabola in the horizontal xy-plane. The cross section of the groove is a rectangle of width 0.8 in. Neglecting friction, find the largest speed for which the marble will stay in the groove.

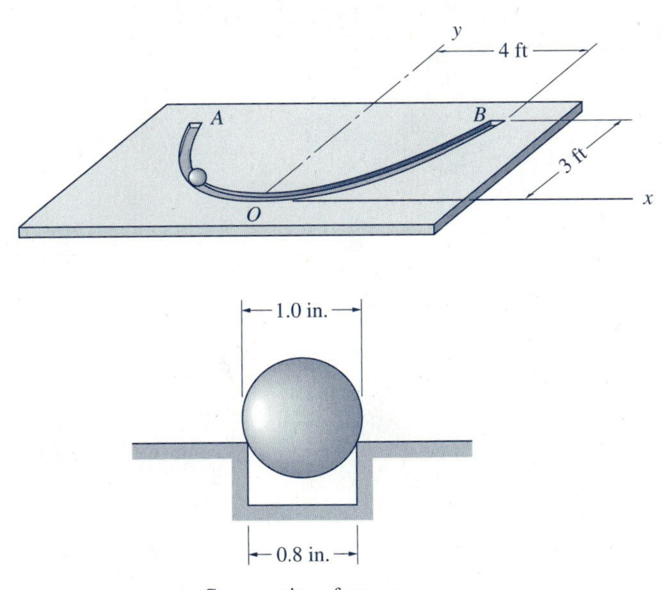

Cross section of groove

Fig. P13.51

13.52 The block A of mass m and the rotating arm OB lie on a smooth, horizontal table. The static coefficient of friction between the mass and the arm is μ. The system starts from rest at $\theta = 0$ under the action of the external couple C_0, which causes the angular acceleration $\ddot{\theta}$ of the arm to be constant. Determine the angle θ at the instant when A begins to slide relative to OB.

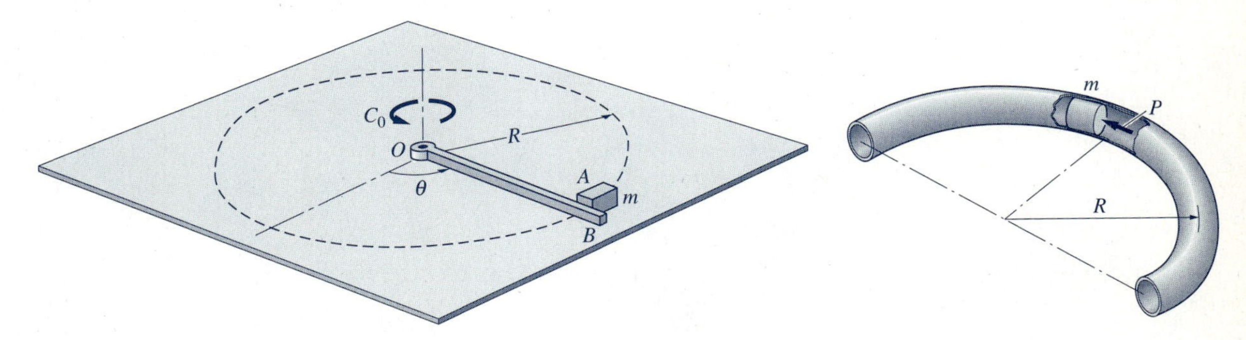

Fig. P13.52 **Fig. P13.53**

13.53 In the pneumatic delivery system shown, air pressure forces the cylindrical container of mass m through the fixed curved tube of radius R. The tube lies in the horizontal plane and the kinetic coefficient of friction between the cylinder and tube is μ. Determine the propulsive force P required to maintain a constant speed of v_0 for the cylinder.

13.54 Friction between the shoes A and the housing B enables the centrifugal clutch to transmit torque from shaft 1 to shaft 2. Centrifugal force holds the shoes against the housing, where the coefficient of static friction between the shoes and the housing is 0.8. Each shoe weighs 1.0 lb, and the weights of the other parts of the shoe assemblies may be neglected. (a) Calculate the initial tension in each clutch spring C so that contact between the shoes and the housing occurs only when $\dot{\theta} \geq 500$ rev/min. (b) Determine the maximum torque that can be transmitted when $\dot{\theta} = 2000$ rev/min.

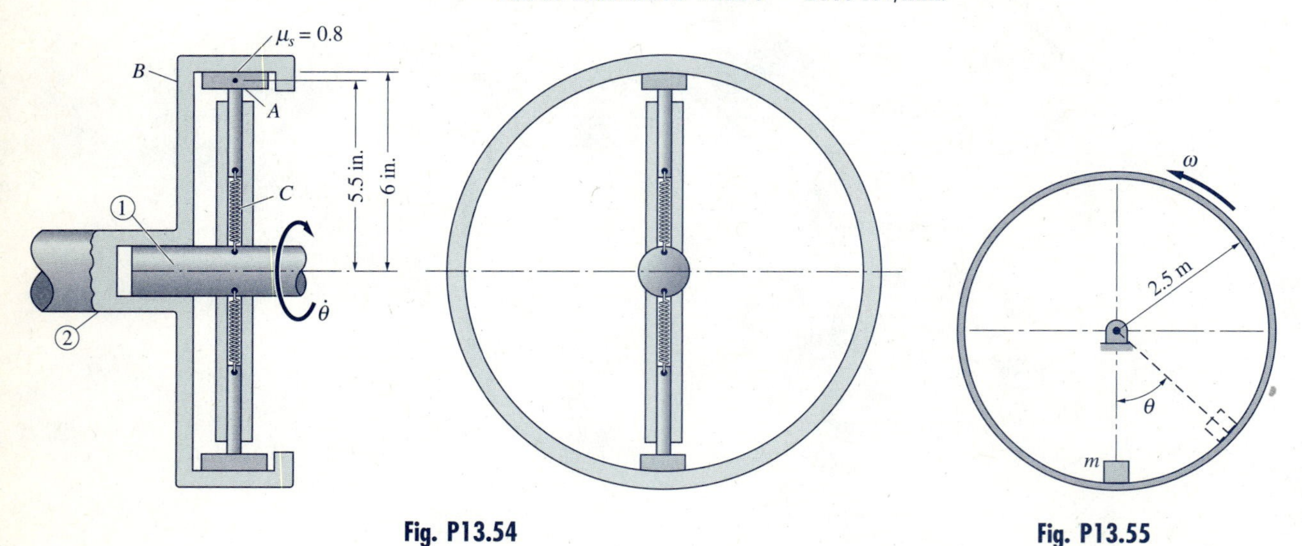

Fig. P13.54 **Fig. P13.55**

13.55 A package of mass m is placed inside a drum that rotates in the vertical plane at the constant angular speed $\omega = 1.36$ rad/s. If the package reaches the position $\theta = 45°$ before slipping, determine the static coefficient of friction between the package and the drum.

13.56 The conical vessel is spinning about the vertical axis at 120 rev/min. Determine the range of R for which a particle of mass m will not slide relative to the vessel. The coefficient of static friction between the particle and the vessel is 0.25.

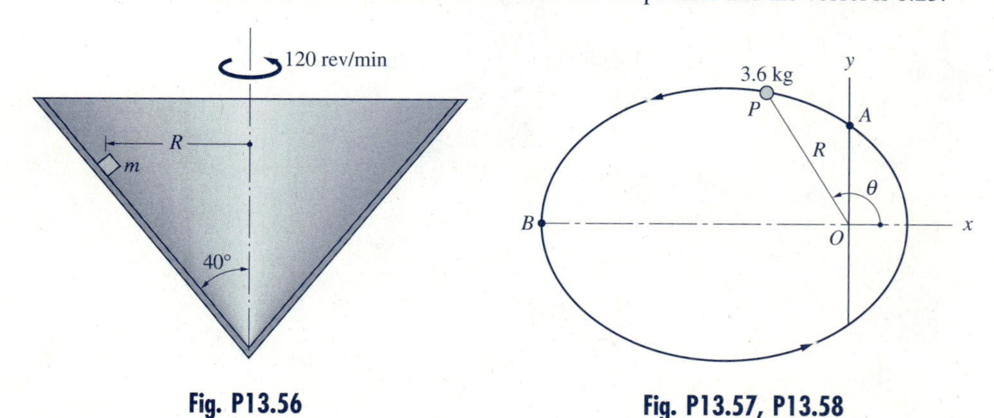

Fig. P13.56 **Fig. P13.57, P13.58**

13.57 The path of particle P is an ellipse given by $R = R_0/(1 + e\cos\theta)$, where $R_0 = 0.5$ m and $e = 2/3$. Assuming that the angular speed of line OP is constant at 20 rad/s, calculate the polar components of the force that acts on the particle when it is at (a) position A; and (b) position B.

13.58 Solve Prob. 13.57 if particle P moves so that $v_\theta = R\dot{\theta} = 10$ m/s (constant).

13.59 The 2.4-lb follower is attached to the end of a light telescopic rod that is pivoted at O. The follower is pressed against a smooth spiral surface by a spring of stiffness $k = 8$ lb/ft and free length $L_0 = 3$ ft. The equation of the spiral, which lies in the horizontal plane, is $R = b\theta/2\pi$, where $b = 1.2$ ft and θ is in radians. Immediately after the rod is released from rest in position OA, determine (a) the angular acceleration of the rod; and (b) the contact force between the follower and the spiral surface.

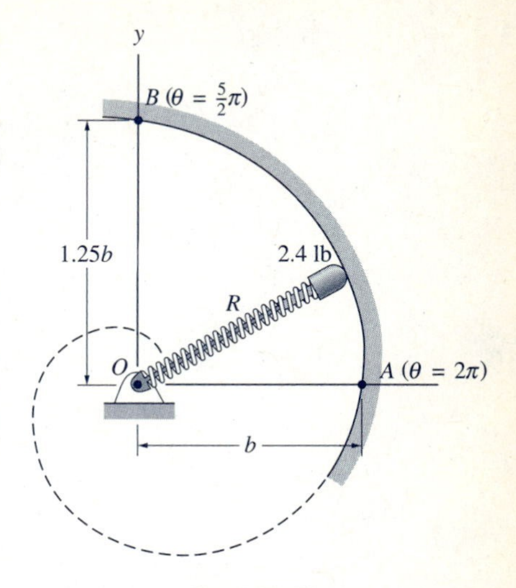

Fig. P13.59

13.60 The 0.10-kg ball A, which slides in a slot in the rotating arm OB, is kept in contact with the stationary cam C by a compression spring of stiffness k. The spring exerts a force of 2 N on the ball when the arm is stationary in position OP. If the arm rotates with the constant angular speed $\dot{\theta} = 20$ rad/s, calculate the minimum spring stiffness k that will maintain contact between the ball and the cam when the arm is in position OQ. Neglect friction and assume that the assembly lies in the horizontal plane.

***13.61** The arm OB of the system described in Prob. 13.60 rotates with the constant speed $\dot{\theta} = 20$ rad/s. When $\theta = 60°$, the force exerted by the spring on the ball A is 12.5 N. For this position, determine the contact force between (a) the ball and the cam; (b) the ball and the slot.

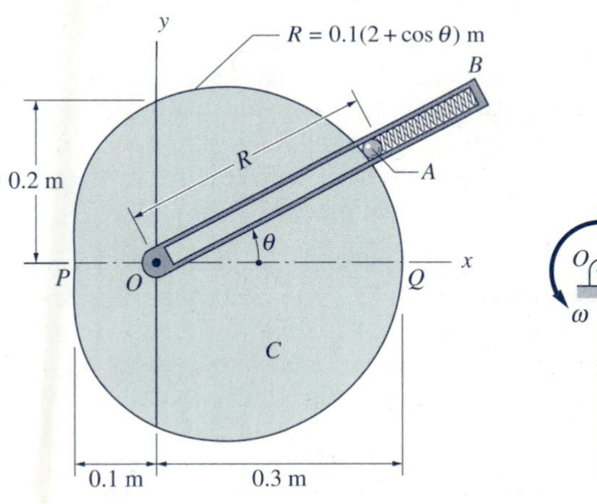

Fig. P13.60, P13.61

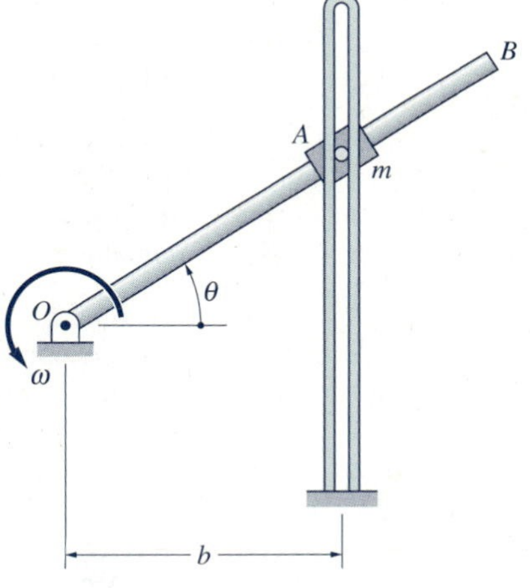

Fig. P13.62

13.62 The collar A of mass m slides on the weightless rod OB, which rotates with a constant angular velocity $\dot{\theta} = \omega$. A pin attached to the collar engages the fixed vertical slot. Neglecting friction, determine (a) the force exerted on the pin by the slot; (b) the force exerted on the collar by the rod. Express you answers in terms of θ, ω, m, b, and g.

13.63 The mass C is connected by two wires to the vertical shaft AB. Rotation of the shaft causes the mass to travel in the horizontal circle shown. Calculate the speed v_0 of the mass that would result in equal tensions in the wires.

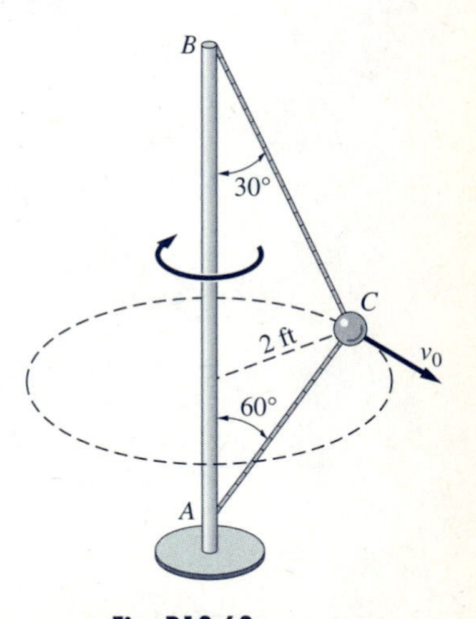

Fig. P13.63

13.64 A 0.6-kg particle slides down a rough helical wire at the constant speed $v_0 = 2$ m/s. (a) Determine the cylindrical components of the force exerted by the wire on the particle. (b) Find the coefficient of kinetic friction between the particle and the wire.

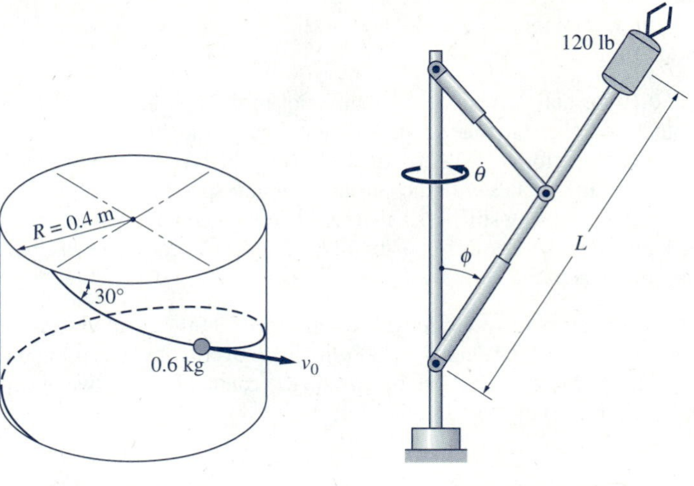

Fig. P13.64 **Fig. P13.65, P13.66**

13.65 The telescopic arm of the mechanical manipulator rotates about the vertical axis with the constant angular speed $\dot{\theta} = 8$ rad/s. The angle ϕ is kept constant at 45°, but the length of the arm varies as $L = 6 + 2\sin(2\dot{\theta}t)$ ft, where t is in seconds. Compute the cylindrical components of the force exerted by the arm on the 120-lb manipulator head as functions of time.

13.66 The telescopic arm of the mechanical manipulator rotates about the vertical axis with the constant angular speed $\dot{\theta} = 8$ rad/s. At the same time, the arm is extended and lowered at the constant rates $\dot{L} = 4$ ft/s and $\dot{\phi} = 2$ rad/s, respectively. Determine the cylindrical components of the force that the arm exerts on the 120-lb manipulator head when $L = 6$ ft and $\phi = 45°$.

13.5 Numerical Solution: Generalized Coordinates

a. Generalized velocity and acceleration

The numerical integration techniques in Arts. 12.8 and 12.9 were discussed in the context of rectangular coordinates. For example, the general form of the equations of motion for a two-degree-of-freedom system in Eqs. (12.36) was

$$\begin{aligned} a_1 &= f_1(v_1, v_2, x_1, x_2, t) \quad \text{or} \quad \ddot{x}_1 = f_1(\dot{x}_1, \dot{x}_2, x_1, x_2, t) \\ a_2 &= f_2(v_1, v_2, x_1, x_2, t) \quad \text{or} \quad \ddot{x}_2 = f_2(\dot{x}_1, \dot{x}_2, x_1, x_2, t) \end{aligned} \qquad \text{(12.36, repeated)}$$

In these equations, the x_i's were interpreted as rectangular coordinates that specified the configuration of the system.

Review of Arts. 12.8 and 12.9 reveals that the coordinates need not be rectangular. The form of the equations of motion as well as the integration formulas are, in fact, valid for any independent set of coordinates. Therefore, x_i in Eqs. (12.36) may represent any set of *generalized coordinates.** In that case, v_i and a_i are called *generalized velocities* and *generalized accelerations*, respectively.

If x_i represent rectangular coordinates, then v_i and a_i are components of the velocity and acceleration vectors, respectively. However, this is not true for all generalized coordinates. For example, let the generalized coordinates of a particle be the polar coordinates $x_1 = R$ and $x_2 = \theta$. We now have the following situation:

Components of velocity vector **v**	Generalized velocities
$v_R = \dot{R}$	$v_1 = \dot{x}_1 = \dot{R}$
$v_\theta = R\dot{\theta}$	$v_2 = \dot{x}_2 = \dot{\theta}$

Components of acceleration vector **a**	Generalized accelerations
$a_R = \ddot{R} - R\dot{\theta}^2$	$a_1 = \dot{v}_1 = \ddot{x}_1 = \ddot{R}$
$a_\theta = R\ddot{\theta} + 2\dot{R}\dot{\theta}$	$a_2 = \dot{v}_2 = \ddot{x}_2 = \ddot{\theta}$

In this case, only the generalized velocity v_1 is equal to its vector counterpart v_R. The differences between the components of vectors **v** and **a** and the generalized velocities and accelerations must be kept in mind when interpreting the results of the integration formulas in Arts. 12.8 and 12.9.

b. Choice of generalized coordinates

In principle, any problem can be formulated and solved in any coordinate system. However, for many problems there exists a coordinate system that is intrinsically more convenient to use than the alternatives. For example, rectangular coordinates are well-suited for the description of projectile motion, but polar coordinates are more convenient for describing orbital motion. In these "preferred" coordinate systems, the equations of motion are easier to derive and are less complex. These advantages are, however, not of paramount importance if the equations are to be solved numerically—numerical methods work just as well on complicated equations as they do on simple equations.

Therefore, when numerical techniques are used, considerations other than algebraic simplicity may determine the most suitable coordinate system. For example, the equations of motion for an earth satellite are formulated most easily in polar coordinates. The output of the numerical integration procedure would thus be R, $\dot{R}$, $\ddot{R}$, θ, $\dot{\theta}$, and $\ddot{\theta}$ at different values of time. If one wishes to plot the orbit of the satellite, it is necessary to convert R and θ into rectangular coordinates (unless, of course, the software includes polar plotting capability). Therefore, it might be advantageous to formulate and solve the problem in rectangular coordinates, even though the corresponding equations of motion would be somewhat more complicated.

* In Art 10.2, generalized coordinates were defined to be any set of kinematically independent parameters (not necessarily restricted to position coordinates) that completely describes the configuration of a system of particles.

Path (normal-tangential) coordinates are not useful for numerical solutions owing to geometrical difficulties. It is simpler to formulate the problem using rectangular or cylindrical coordinates and then convert the results to path coordinates if necessary. For example, if one wishes to calculate the normal component of acceleration of a projectile at a certain time, the easiest method is to compute the velocity and radius of curvature using rectangular coordinates and then substitute the results into the equation $a_n = v^2/\rho$.

c. Methods of solution

All the integration techniques described in Arts. 12.8 and 12.9 are also valid for generalized coordinates; therefore, there is no need to repeat them here. However, as mentioned, be careful not to confuse the generalized velocities and accelerations with the components of the velocity and acceleration vectors.

Sample Problem 13.10

The particle of mass $m = 0.3$ kg shown in Fig. (a) is attached to an ideal spring and moves on a smooth horizontal plane. The stiffness of the spring is $k = 28.1$ N/m, and it is unstretched when the particle is at A. The particle is launched at A with the velocity $\mathbf{v}_0$ as shown. (1) Derive the equations of motion, and state the initial conditions. (2) Integrate the equations of motion from $t = 0$ to 1.5 s, and plot R vs. θ. (3) Verify that $R_{max} = 400$ mm, and find the angle θ_1 where R_{max} occurs for the first time.

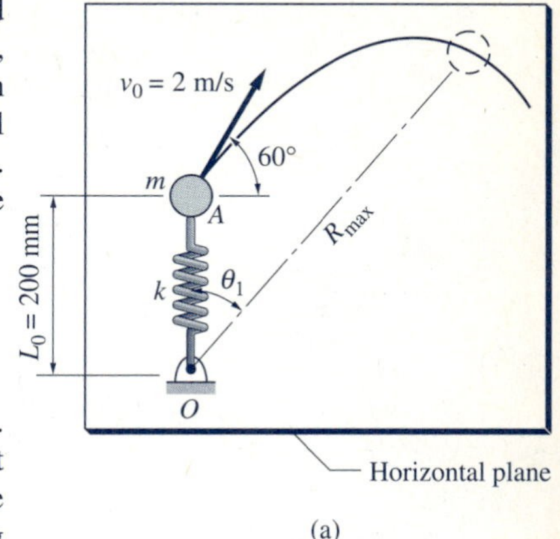

(a)

Solution

Part 1

The free-body diagram for an arbitrary position of the particle is shown in Fig. (b); the only force acting in the plane of motion is the spring force F_s. The weight of the particle and the force exerted by the smooth plane are perpendicular to the plane of the motion and are not shown on the FBD. Since the particle is undergoing central-force motion (F_s is always directed toward O), it is convenient to formulate the equations of motion using the polar coordinates R and θ shown on the FBD. Because L_0 is the unstretched length of the spring, its elongation is $\delta = R - L_0$, which gives $F_s = k\delta = k(R - L_0)$.

e_R

e_θ

$F_s = k(R - L_0)$

R

θ

O

FBD (forces perpendicular to plane of motion not shown)

=

$ma_R = m(\ddot{R} - R\dot{\theta}^2)$

$ma_\theta = m(R\ddot{\theta} + 2\dot{R}\dot{\theta})$

R

θ

O

MAD

(b)

The mass-acceleration diagram for the arbitrary position of the particle is also shown in Fig. (b).

Referring to the diagrams in Fig. (b), the equations of motion are

$$\Sigma F_R = ma_R \qquad +\nearrow \qquad -k(R - L_0) = m(\ddot{R} - R\dot{\theta}^2) \tag{a}$$

$$\Sigma F_\theta = ma_\theta \qquad \searrow+ \qquad 0 = m(R\ddot{\theta} + 2\dot{R}\dot{\theta}) \tag{b}$$

Substituting $m = 0.3$ kg, $k = 28.1$ N/m and $L_0 = 0.2$ m, and solving Eqs. (a) and (b) for $\ddot{R}$ and $\ddot{\theta}$, the equations of motion become

$$\ddot{R} = R\dot{\theta}^2 - 93.667R + 18.733 \text{ m/s}^2 \tag{c}$$

Answer

$$\ddot{\theta} = -\frac{2\dot{R}\dot{\theta}}{R} \text{ rad/s}^2 \tag{d}$$

Note that Eqs. (c) and (d) are now in the same form as Eqs. (12.36).

The given initial conditions are at $t = 0$:

$$R = L_0 = 0.2 \text{ m}$$

$$\theta = 0$$

$$\dot{R} = v_R = 2\sin 60° = 1.732 \text{ m}$$

$$\dot{\theta} = \frac{v_\theta}{L_0} = \frac{2\cos 60°}{0.2} = 5.000 \text{ rad/s}$$

Answer

Observe that there are four initial conditions—two for the radial coordinate and two for the transverse coordinate.

Part 2

The equations of motion given in Eqs. (c) and (d) were integrated with the fourth-order Runge-Kutta method from $t = 0$ to 1.5 s using the time increment $\Delta t = 0.01$ s. Therefore, 150 integration steps were completed, and the results after each step were recorded.

The plot of R versus θ for the time interval $t = 0$ to 1.5 s is shown in Fig. (c). It can be seen that R_{max}, the maximum value of R, occurs three times in the first revolution of the spring, i.e., in the interval $\theta = 0$ to $\theta = 2\pi = 6.28$ rad. We also note that at $\theta = 6.28$ rad, when the spring has returned to its initial angular position, the corresponding value of R is not equal to its original value of 0.2 m.

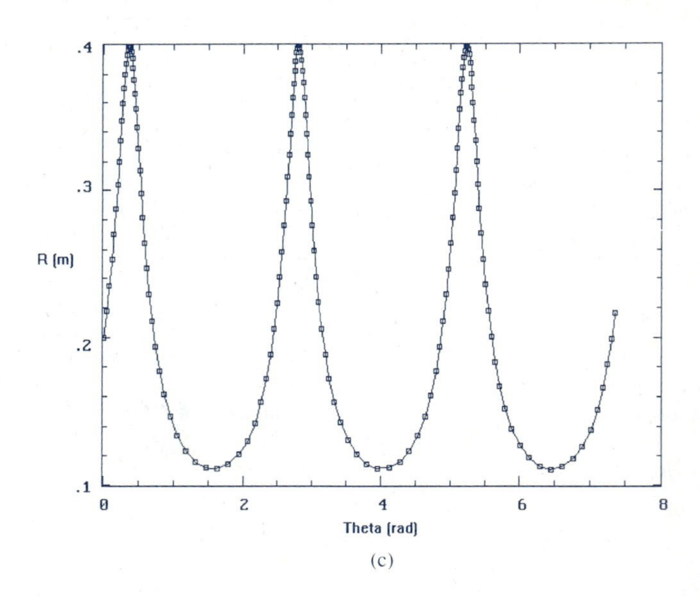

(c)

Part 3

To determine the values of R_{max} and the corresponding θ with precision, we must inspect the numerical output of the computer program to find the polar coordinates of the positions where $\dot{R} = 0$. This is found to occur for the first time between $t = 0.16$ s and $t = 0.17$ s, as indicated by the sign change of $\dot{R}$:

t (s)	R (m)	$\dot{R}$ (m/s)	θ (rad)
$0.1600E + 00$	$0.3995E + 00$	$0.1491E + 00$	$0.3505E + 00$
$0.1700E + 00$	$0.4001E + 00$	$-0.3186E - 01$	$0.3630E + 00$

From the data, it can be seen that $R_{max} = 400$ mm with a three significant digit accuracy.

To find the angle θ where R_{max} occurs, we use linear interpolation of the data for $\dot{R}$ and θ, as shown in Fig. (d). The angle θ_1 in the figure corresponds to the value of θ when $\dot{R} = 0$. By similar triangles, we find

$$\frac{\theta_1 - 0.3505}{0.1491} = \frac{0.3630 - 0.3505}{0.1491 + 0.0319}$$

which yields

$$\theta_1 = 0.3608 \text{ rad} = 20.7^\circ \qquad \textit{Answer}$$

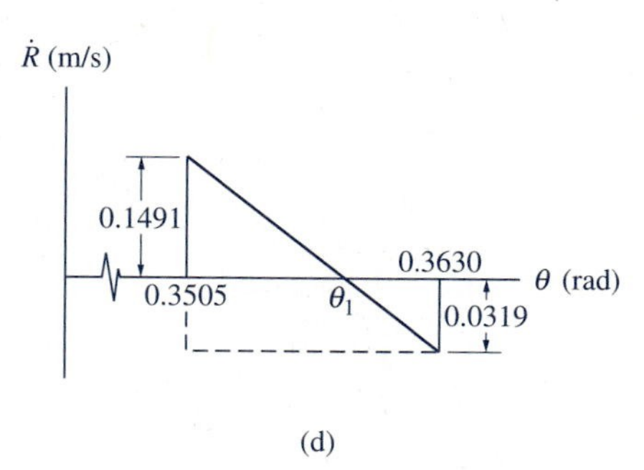

(d)

PROBLEMS

Fig. P13.67, P13.68

13.67 The equation of motion for the simple pendulum can be shown to be $\ddot{\theta} = -(g/L)\sin\theta$. Given that $L = 9.81$ m and that the pendulum is released from rest at $\theta = 60°$, determine the time required for the pendulum to reach the position $\theta = 0$. Use $\Delta t = 0.10$ s and compare your answer with the analytical solution of 1.686 s.

13.68 Solve Prob. 13.67 if the pendulum is released from rest at $\theta = 90°$, and plot θ vs. t for $90° \geq \theta \geq 0$.

13.69 A 75-kg ski jumper starts his run at the top of a parabolic track. The coefficient of kinetic friction between the skis and the track is $\mu = 0.12$. (a) Show that the equation of motion for the ski jumper is

$$\ddot{x} = (g + 2k\dot{x}^2)(\mu - 2kx)/(1 + 4k^2x^2)$$

where $k = (12/225)\ \mathrm{m}^{-1}$, and state the initial conditions. (b) Solve the equation numerically from $t = 0$ until the skier reaches point O; plot x vs. t. (c) Use the numerical solution to determine the speed of the skier and the normal force between the skis and the track when the skier is at O.

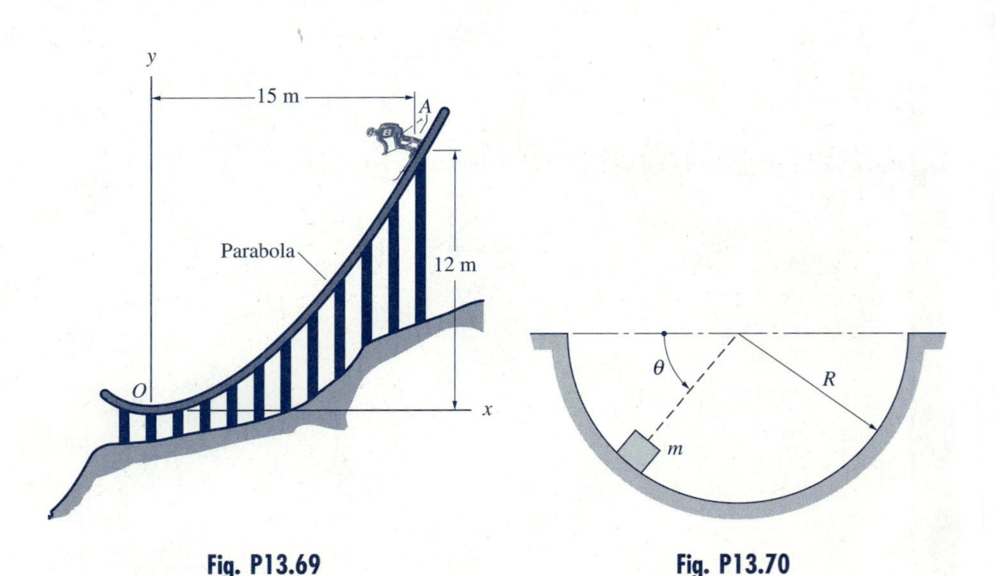

Fig. P13.69

Fig. P13.70

13.70 The block of mass m is released from rest at $\theta = 0$ and allowed to slide on the circular surface. The kinetic coefficient between the block and the surface is μ. (a) Show that the equation of motion of the block is

$$\ddot{\theta} = (g/R)(\cos\theta - \mu\sin\theta) - \mu\dot{\theta}^2 \qquad (\dot{\theta} \geq 0)$$

(b) Using $R = 2$ m and $\mu = 0.3$, determine by numerical integration the value of θ where the block comes to rest for the first time.

13.71 Refer to the system described in Prob. 13.59. (a) Show that the equation of motion is

$$\ddot{\theta} = \frac{(k/m)[2\pi(L_0/b) - \theta] - \theta\dot{\theta}^2}{1 + \theta^2}$$

where m is the mass of the follower, and state the initial conditions (let $t = 0$ be the time of release). (b) Solve the equation numerically from the time of release until the rod reaches position OB; plot θ vs. t. (c) Use the numerical solution to determine the angular speed of the rod in position OB and the time required to reach that position.

13.72 A peg attached to the rotating disk causes the slope angle of the table OB to vary as $\theta = \theta_0 \cos \omega t$, where $\theta_0 = 15°$ and $\omega = \pi$ rad/s. At $t = 0$, the particle A is placed on the table at $R = 2$ ft with no velocity relative to the table. (a) Derive the equation of motion for the particle, and state the initial conditions. (b) Integrate the equations numerically from the time of release until the particle moves off the table; plot R vs. t. (c) Determine the time when A slides off the table and the velocity of A relative to the table at that instant. Neglect friction.

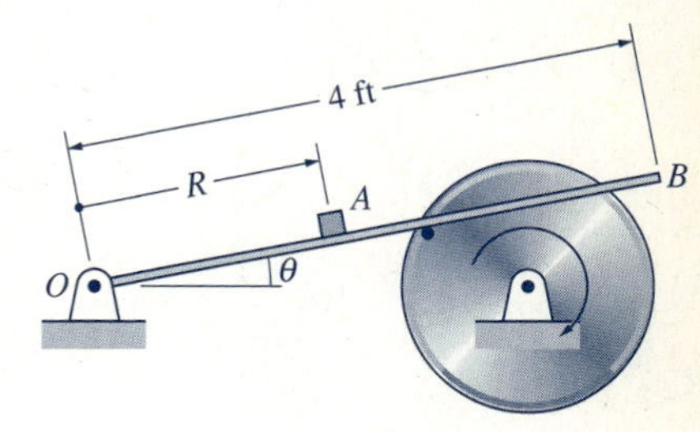

Fig. P13.72

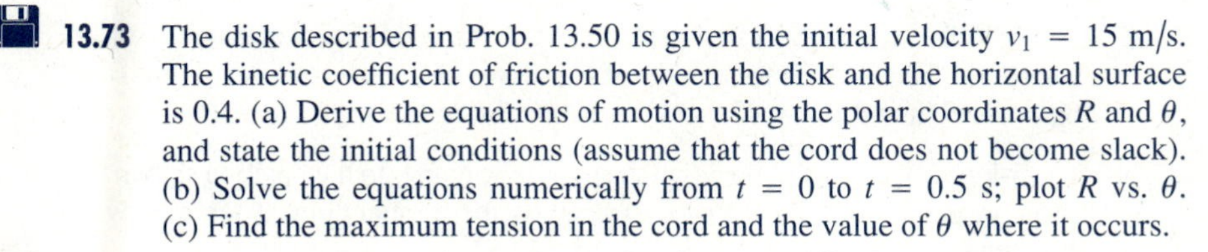

13.73 The disk described in Prob. 13.50 is given the initial velocity $v_1 = 15$ m/s. The kinetic coefficient of friction between the disk and the horizontal surface is 0.4. (a) Derive the equations of motion using the polar coordinates R and θ, and state the initial conditions (assume that the cord does not become slack). (b) Solve the equations numerically from $t = 0$ to $t = 0.5$ s; plot R vs. θ. (c) Find the maximum tension in the cord and the value of θ where it occurs.

13.74 The particle of mass m slides inside the smooth, conical vessel. The particle is launched at $t = 0$ with the velocity $v_1 = 10$ ft/s, tangent to the rim of the vessel. (a) Show that the equations of motion, with θ and z as the independent coordinates, are

$$\ddot{\theta} = -\frac{2\dot{z}\dot{\theta}}{z}$$

$$\ddot{z} = \frac{z\dot{\theta}^2 \tan^2\beta - g}{1 + \tan^2\beta}$$

and state the initial conditions. (b) Using $\beta = 20°$, solve the equations of motion numerically from $t = 0$ to $t = 2$ s; plot z vs. θ. (c) From the numerical solution, find the vertical distance h below which the particle will not travel.

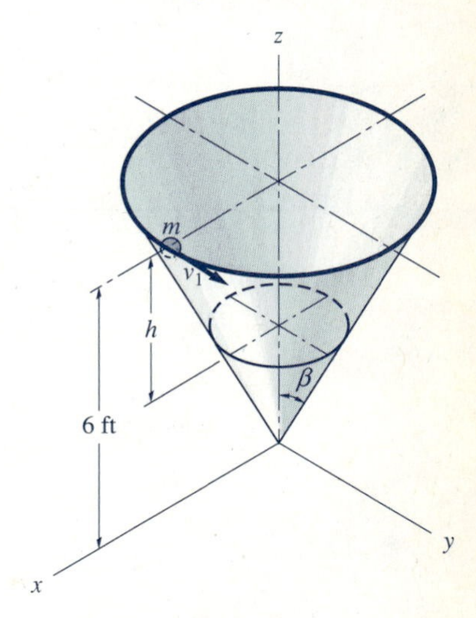

Fig. P13.74

13.75 The 0.25-kg mass, which is attached to an elastic cord of stiffness 10 N/m and free length 0.5 m, is free to move in the vertical plane. The mass is released from rest at $\theta = 0$ with the cord undeformed. (a) Derive the equations of motion, and state the initial conditions. (b) Solve the equations numerically for the time of release until the cord becomes vertical for the first time; plot R and $\dot{\theta}$ vs. θ. (c) Find the maximum speed of the mass during the period of integration.

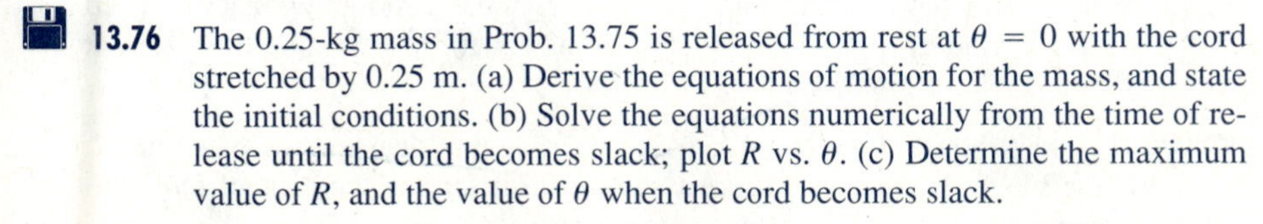

13.76 The 0.25-kg mass in Prob. 13.75 is released from rest at $\theta = 0$ with the cord stretched by 0.25 m. (a) Derive the equations of motion for the mass, and state the initial conditions. (b) Solve the equations numerically from the time of release until the cord becomes slack; plot R vs. θ. (c) Determine the maximum value of R, and the value of θ when the cord becomes slack.

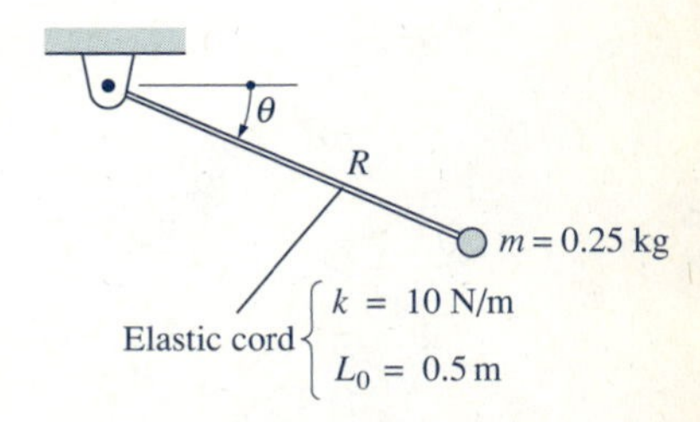

Fig. P13.75, P13.76

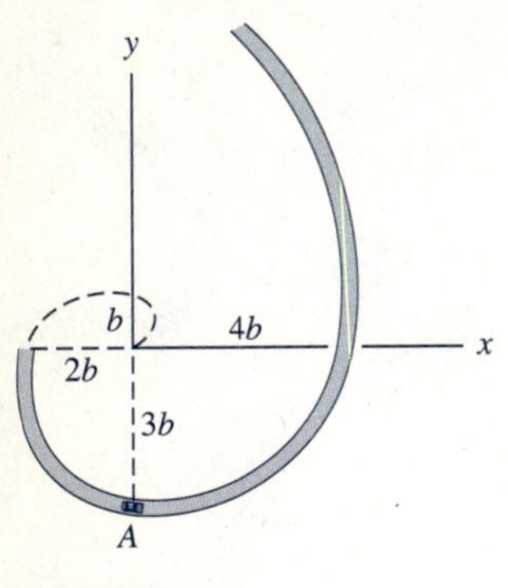

Fig. P13.77

REVIEW PROBLEMS

13.77 The car travels on a curve that has the shape of the spiral $R = (2b/\pi)\theta$, where $b = 10$ m. If $\dot{\theta} = 0.5$ rad/s (constant), determine the speed of the car and the magnitude of the acceleration when $\theta = 3\pi/2$ rad (point A).

13.78 Bar AB starts from rest at $\theta = 0$ with the constant angular acceleration $\ddot{\theta} = 6$ rad/s^2. The block of mass m begins sliding on the bar when $\theta = 45°$. Determine the coefficient of static friction between the block and the bar.

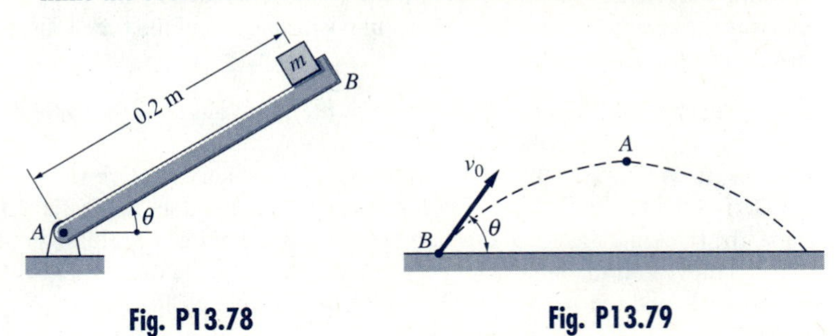

Fig. P13.78 Fig. P13.79

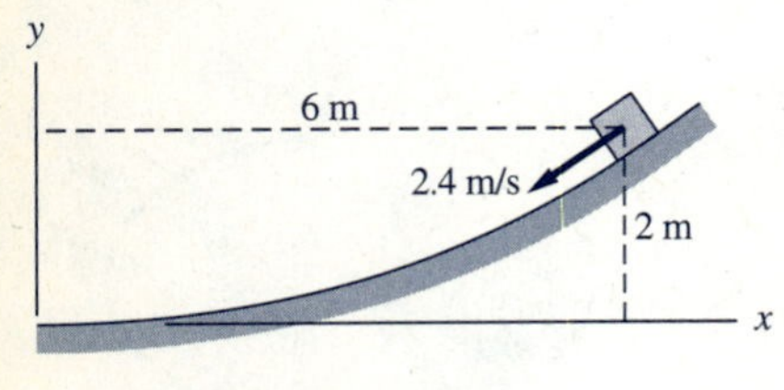

Fig. P13.80

13.79 A projectile is launched at B with the velocity v_0 inclined at angle θ to the horizontal. Knowing that the radii of curvature of the trajectory at A and B are $\rho_A = 40.8$ m and $\rho_B = 63.1$ m, determine θ and v_0. Neglect air resistance.

13.80 The 5-kg package is sliding down the parabolic chute. In the position shown, the speed of the package is 2.4 m/s. Determine the normal contact force between the chute and the package in this position.

13.81 The 9-in. radius spool is started from rest in the position shown. The angular speed of the spool varies with time, as specified on the diagram. Determine the time required to wind up the 150 ft of cable.

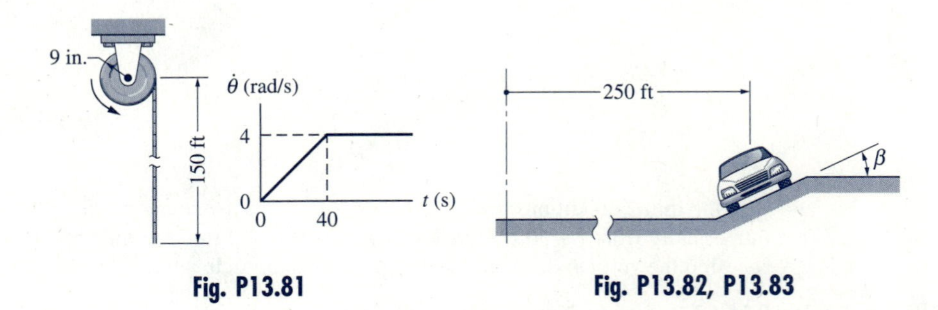

Fig. P13.81 Fig. P13.82, P13.83

13.82 The car travels at a constant speed on a circular, banked track. If the bank angle is $\beta = 10°$ and the coefficient of static friction between the track and the tires is 0.85, determine the maximum possible speed of the car.

13.83 The car travels on a circular, unbanked ($\beta = 0$) track. The coefficient of static friction between the track and the tires is 0.85. Assuming that the speed of the car increases at the constant rate of 7.2 ft/s^2, determine the speed at which the car begins to slide.

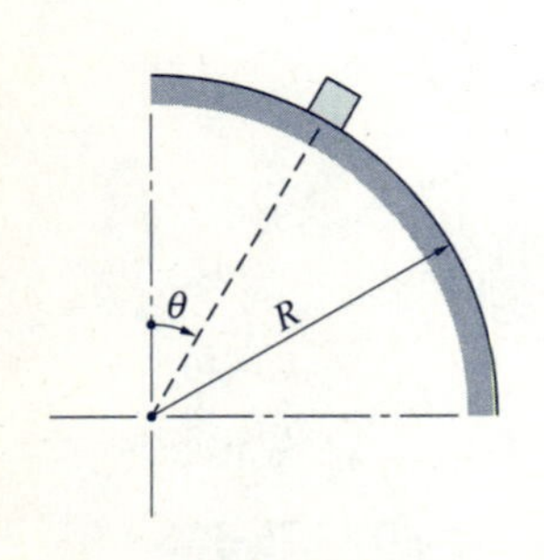

Fig. P13.84

13.84 The small block slides without friction on a cylindrical surface of radius R. If the block is released from rest at $\theta = 0$, determine the contact force between the block and the surface in terms of θ.

13.85 The mass m is suspended from two wires, as shown, when wire AB is cut. If $\beta = 30°$, determine the force in wire BC (a) before AB is cut; and (b) just after AB is cut. (c) For what value of β would the results of parts (a) and (b) be the same?

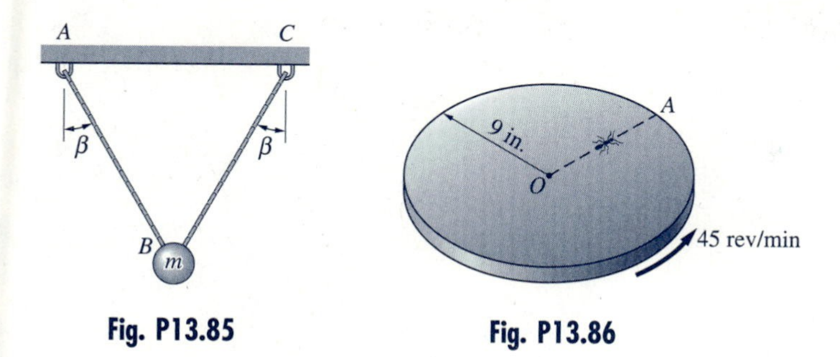

Fig. P13.85 **Fig. P13.86**

13.86 An insect runs on a disk that rotates at the constant speed of 45 rev/min. Relative to the disk, the speed of the insect is constant at 0.5 ft/s, and the track is the radial line OA. Determine the smallest coefficient of friction that would allow the insect to reach point A of the rim of the disk.

13.87 The motion of the particle on the surface of a cylinder is described by $\theta = \pi \sin \omega t$, $z = R \cos \omega t$, where R is the radius of the cylinder and ω is a constant. Note that if the cylinder is unrolled, the path of the particle is an oval, as shown in the z vs. $R\theta$ plot. Determine the acceleration of the particle when (a) $\omega t = 0$ (point A); and (b) $\omega t = \pi/2$ (point B). Show the acceleration vectors on a sketch of the cylinder.

13.88 The mass at C is attached to the vertical pole AB by two wires. The assembly is rotating about AB at the constant angular speed $\dot{\theta}$. If the force in wire BC is twice the force in AC, determine the value of $\dot{\theta}$.

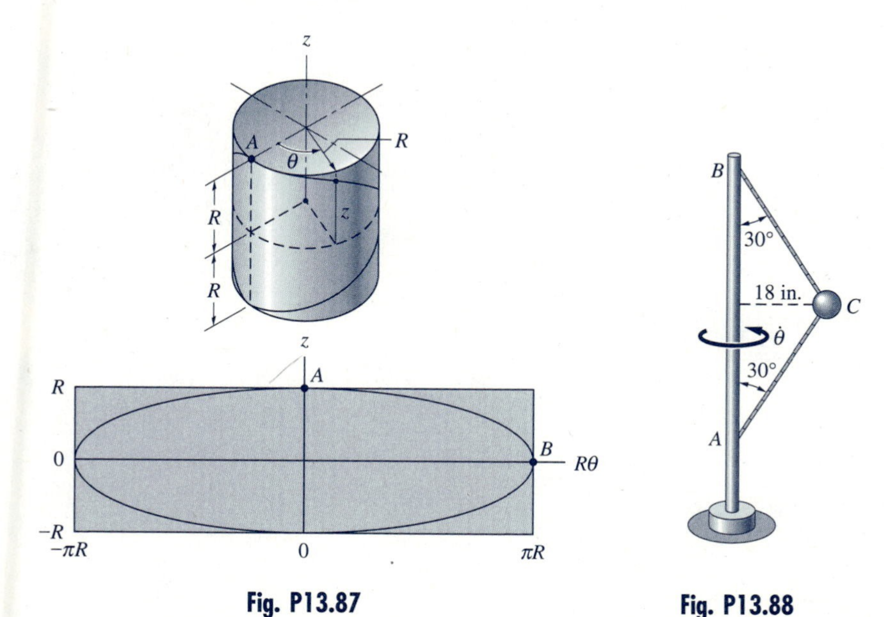

Fig. P13.87 **Fig. P13.88**

14 Work-Energy and Impulse-Momentum Principles for a Particle

14.1 Introduction

In the force-mass-acceleration method of kinetic analysis, which we employed in Chapters 12 and 13, the equations of motion of a particle were determined by applying Newton's laws. The computation of the velocity and/or position of the particle required the integration of the equations of motion with respect to time or position. In this chapter, we discuss the two additional methods of kinetic analysis for particle motion: the work-energy method and the impulse-momentum method, each of which is based on a principle of the same name.

The work-energy principle relates the work done by applied forces to the change in kinetic energy of the particle. The work-energy method of analysis is particularly convenient if one is interested in only the change in velocity and if the work done by the forces can be easily computed. The conservation of mechanical energy is a special case of the work-energy principle that can be applied to motions produced by conservative force systems.

The impulse-momentum principle relates the impulse of the forces (the integral of the forces over a time interval) to the change in the momentum of the particle. The impulse-momentum method of analysis can be advantageous if the force-time variation is known in graphical, rather than analytical, form. The impulse—i.e., the area under the force-time curve—can frequently be obtained without the need to determine an analytic expression for the force.

As in the previous two chapters, our attention here is focused on the motion of a single particle. Systems of particles are discussed in the next chapter.

14.2 Work of a Force

a. Definition of work

We begin by defining the work of a force, a concept that plays a fundamental role in the work-energy principle that is derived in the next article. Then we consider the work of three special forces: constant force, force exerted by an ideal spring, and gravitational force.

Let point A, the point of application of a force $\mathbf{F}$, follow the path $\mathcal{L}$ shown in Fig. 14.1. The position vector of A (measured from a fixed point O) is denoted by $\mathbf{r}$ at time t and $\mathbf{r} + d\mathbf{r}$ at time $t + dt$. Note that $d\mathbf{r}$, the displacement of the point during the infinitesimal time interval dt, is tangent to the path at A. The differential work dU done by the force $\mathbf{F}$ as its point of application undergoes the displacement $d\mathbf{r}$ is defined to be*

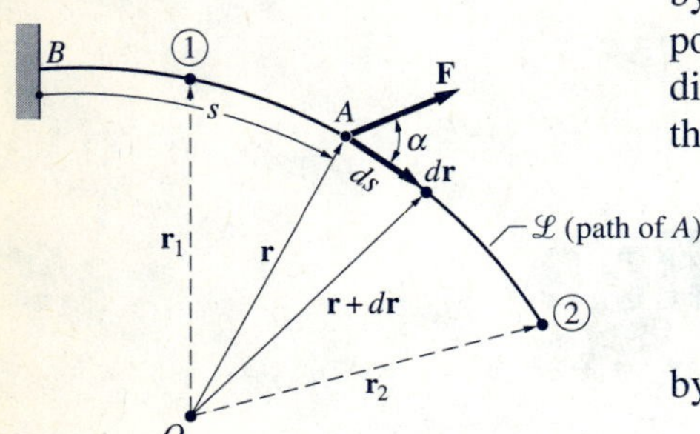

Fig. 14.1

$$dU = \mathbf{F} \cdot d\mathbf{r} \tag{14.1}$$

The work done by $\mathbf{F}$ as point A moves from position ① to ② is obtained by integrating Eq. (14.1) along the path $\mathcal{L}$:

$$U_{1\text{-}2} = \int_{\mathcal{L}} dU = \int_{\mathcal{L}} \mathbf{F} \cdot d\mathbf{r} \tag{14.2}$$

Using the notations $ds = |d\mathbf{r}|$ and $F = |\mathbf{F}|$, the work may be written as

$$U_{1\text{-}2} = \int_{s_1}^{s_2} F \cos\alpha \, ds \tag{14.3}$$

where α is the angle between $\mathbf{F}$ and $d\mathbf{r}$ as shown in Fig. 14.1, and s_1 and s_2 are the path lengths measured from a fixed point, such as B, to positions ① and ②, respectively. The unit of work is (unit of force) × (unit of distance): N · m, lb · ft, etc. Observe that work is a *scalar* quantity that can be positive, negative, or zero, depending on the angle α.

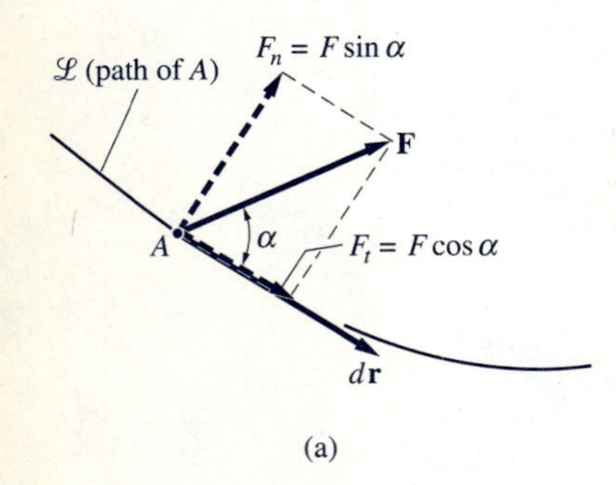

(a)

If $\mathbf{F}$ is decomposed into components that are normal and tangential to the path at A—$F_n = F \sin\alpha$ and $F_t = F \cos\alpha$, as shown in Fig. 14.2(a)—Eq. (14.3) becomes

$$U_{1\text{-}2} = \int_{s_1}^{s_2} F_t \, ds \tag{14.4}$$

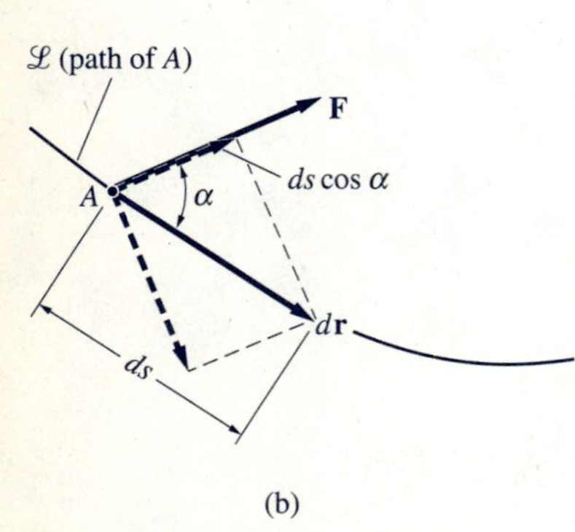

(b)

Fig. 14.2

Note that the normal component of $\mathbf{F}$ does not do work; i.e., $\mathbf{F}_n \cdot d\mathbf{r} = 0$ (the vectors $\mathbf{F}_n$ and $d\mathbf{r}$ are mutually perpendicular). The tangential component F_t is called the *working component* of $\mathbf{F}$. Therefore, the work may be interpreted as the integral of (working component of force) × (magnitude of differential displacement).

From Fig. 14.2(b) we see that $ds \cos\alpha$ is the component of $d\mathbf{r}$ in the direction of $\mathbf{F}$. This component is called the *work-absorbing component* of the differential displacement. Therefore, the work may also be interpreted as the integral of (magnitude of force) × (work-absorbing component of the differential displacement).

* This definition is similar to the definition for virtual work given in Chapter 10. However, here we are dealing with real displacements, as opposed to the virtual displacements that are involved in Chapter 10.

Another useful expression for the work done by a force is obtained by using the representation of the dot product $\mathbf{F}\cdot d\mathbf{r}$ in rectangular coordinates:

$$U_{1\text{-}2} = \int_{\mathcal{L}} (F_x\,dx + F_y\,dy + F_z\,dz) = \int_{x_1}^{x_2} F_x\,dx + \int_{y_1}^{y_2} F_y\,dy + \int_{z_1}^{z_2} F_z\,dz \tag{14.5}$$

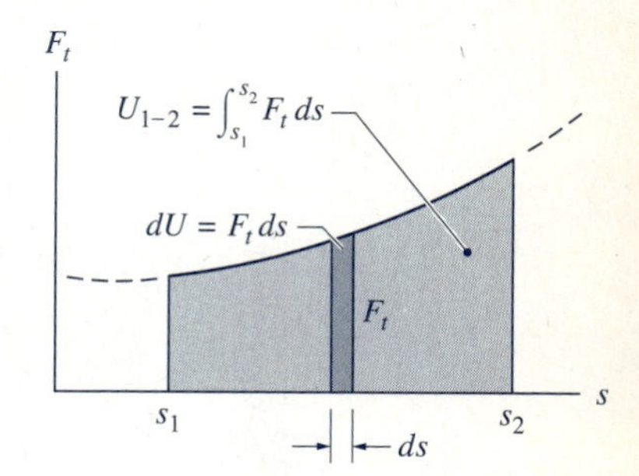

Fig. 14.3

where dx, dy, and dz are the components of $d\mathbf{r}$, and (x_1, y_1, z_1) and (x_2, y_2, z_2) are the coordinates of points ① and ②, respectively.

If the integrals for work are difficult to evaluate, or if the relationship between $\mathbf{F}$ and $\mathbf{r}$ is known only at a discrete number of points, as is often the case with experimental data, the integrals may be evaluated numerically. In that case, it is useful to know that the integral in Eq. (14.4) represents the area under the F_t-s diagram, as shown in Fig. 14.3. However, the work can be computed analytically with relative ease for the following three cases, which are of considerable practical importance.

b. Work of a constant force

Figure 14.4 shows a force $\mathbf{F}$ that is constant in both magnitude and direction. From Eq. (14.5), the work done by $\mathbf{F}$ as its point of application travels from position ① to position ② is

$$U_{1\text{-}2} = F_x \int_{x_1}^{x_2} dx + F_y \int_{y_1}^{y_2} dy + F_z \int_{z_1}^{z_2} dz$$

where the components of $\mathbf{F}$ have been taken outside the integral signs because they are constant. On integrating, the work becomes

$$\begin{aligned} U_{1\text{-}2} &= F_x(x_2 - x_1) + F_y(y_2 - y_1) + F_z(z_2 - z_1) \\ &= \mathbf{F}\cdot\Delta\mathbf{r} \end{aligned}$$

where $\Delta\mathbf{r}$ is the displacement vector from position ① to position ②, as shown in Fig. 14.4.

Let β be the angle between $\mathbf{F}$ and $\Delta\mathbf{r}$, and let $\Delta d = d_2 - d_1$ be the displacement in the direction of $\mathbf{F}$, measured from an arbitrary reference line

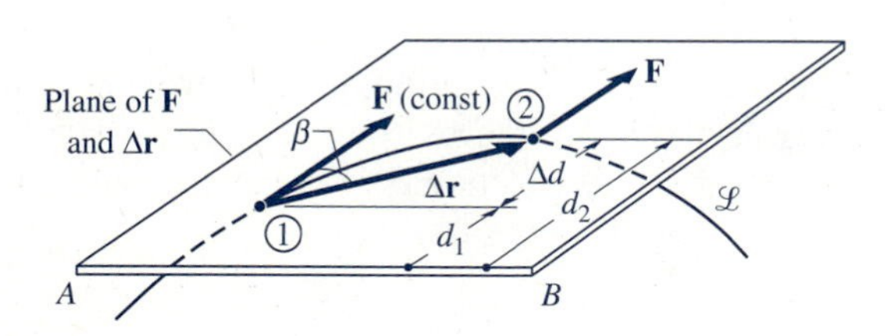

Fig. 14.4

AB. Using the definition of the dot product, the work done by **F** becomes $U_{1\text{-}2} = \mathbf{F}\cdot\Delta\mathbf{r} = F|\Delta\mathbf{r}|\cos\beta$, which may be written as*

$$\boxed{U_{1\text{-}2} = F(d_2 - d_1) = F\Delta d} \tag{14.6}$$

Note that Δd is *not* the displacement of the point of application of **F**; it is the work-absorbing component of the displacement. The work U_{1-2} will be positive or negative, depending on whether Δd is positive or negative.

From Eq. (14.6) we see that the work done by a constant force depends only on the initial and final positions of its point of application; that is, the work is independent of the path $\mathcal{L}$. (Forces for which the work is independent of the path are called *conservative forces*. See Art. 14.4.)

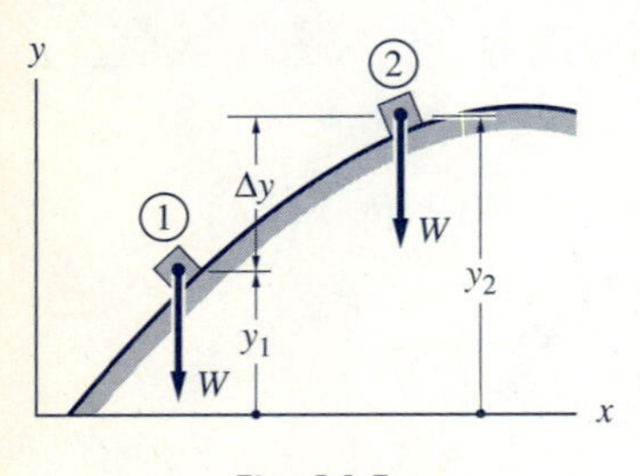

Fig. 14.5

If an object stays in close proximity to the surface of the earth, its weight may be considered to be a constant force, and Eq. (14.6) can be used to calculate its work. Figure 14.5 shows such an object of weight W, which moves up an incline from position ① to ②. Observe that the change in elevation, $\Delta y = y_2 - y_1$, is also the work-absorbing component of the displacement. Therefore, from Eq. (14.6), the work done by W as it moves from ① to ② is

$$\boxed{U_{1\text{-}2} = -W(y_2 - y_1) = -W\Delta y} \tag{14.7}$$

The minus sign in this equation is due to the fact that W and the positive y-coordinate have opposite directions.

As is the case for all constant forces, the work done by W is independent of the path. Therefore, the contour of the incline between positions ① and ② in Fig. 14.4 does not affect the calculation of the work. Note that the location of the x-axis, i.e., the reference line from which the elevation y is measured, is irrelevant because Δy represents the *difference* in two elevations.

c. Work of an ideal spring

This discussion of springs will be limited to ideal springs that have the following properties: (1) the weight of the spring is negligible, and (2) the deformation (elongation or contraction) of the spring is proportional to the force that causes it. Therefore, the force F that is required to elongate the spring by an amount equal to δ is

$$F = k\delta \tag{14.8}$$

where k is called the *stiffness* of the spring, or the *spring constant*. The stiffness k is measured in units of force per unit length (lb/ft, N/m, etc.). Note that Eq. (14.8) is valid for both positive δ (elongation) and negative δ (compression).

* Equation (14.6) is the basis for the common "definition" of work as "force times distance." It is important to remember that this equation is valid only for a constant force; the general definition of work is given by Eq. (14.2).

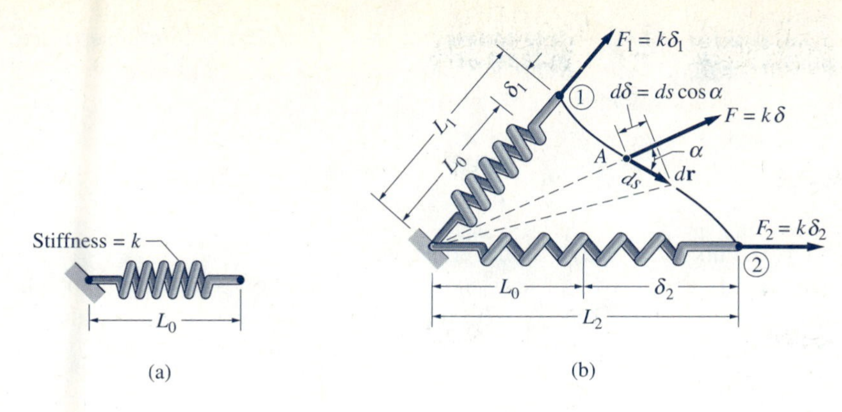

Fig. 14.6

The undeformed, or free length L_0 of an ideal spring is shown in Fig. 14.6(a). Figure 14.6(b) shows the spring in two deformed positions, indicated as ① and ②. The corresponding forces acting on the spring are F_1 and F_2, and the lengths of the deformed spring are L_1 and L_2. Since the spring is ideal, we have $F_1 = k\delta_1$ and $F_2 = k\delta_2$, where $\delta_1 = L_1 - L_0$ and $\delta_2 = L_2 - L_0$. (Avoid the common mistake of confusing the length of a spring with its deformation.)

The work done on the spring as it is deformed from position ① to ② can be calculated from Eq. (14.3).

$$U_{1\text{-}2} = \int_{s_1}^{s_2} F \cos\alpha \, ds$$

Noting from Fig. 14.6(b) that $ds \cos\alpha = d\delta$ (the differential deformation of the spring) and substituting $F = k\delta$, we obtain

$$U_{1\text{-}2} = \int_{\delta_1}^{\delta_2} k\delta \, d\delta = \frac{1}{2}k(\delta_2^2 - \delta_1^2) \tag{14.9}$$

Equation (14.9) represents the work done *on* the spring. The force exerted *by* the spring is equal and opposite to the force acting on the spring. This means that the work done by the spring has an opposite sign to the work computed from Eq. (14.9). Therefore, the work done *by* the spring is

$$\boxed{U_{1\text{-}2} = -\frac{1}{2}k(\delta_2^2 - \delta_1^2)} \tag{14.10}$$

Observe that the work done on (or by) the spring depends only on the spring constant and the initial and final deformations δ_1 and δ_2. Consequently, the work is independent of the path followed by the end of the spring.

d. Work of gravitational forces

If motion of an object is restricted to the vicinity of the surface of the earth, its weight may be considered to be a constant gravitational force. Therefore, the work may be calculated by using Eq. (14.7). Otherwise, Newton's law of universal gravitation, Eq. (11.17), must be used to calculate the work done by gravitational forces. According to this law, two masses m_A and m_B, separated by a distance R, are attracted toward each other by a force of magnitude F that acts along the line between them, where

$$F = K\frac{m_A\, m_B}{R^2} \qquad \text{(11.17, repeated)}$$

and K is the universal gravitational constant.

In Fig. 14.7, the two masses m_A and m_B are shown moving along the indicated paths from ① to ②. The path lengths are denoted by s_A and s_B,

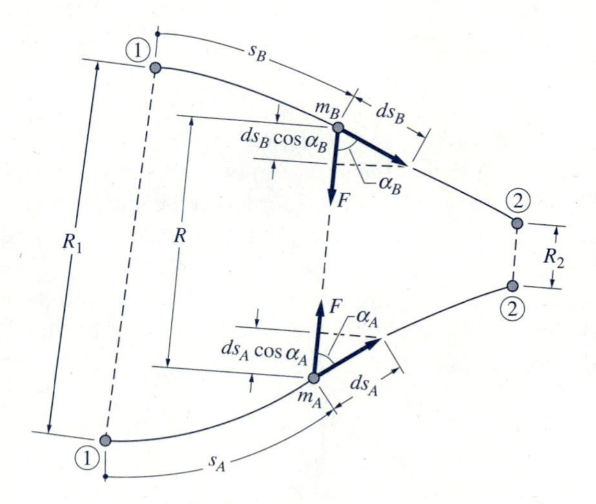

Fig. 14.7

respectively. The differential work done by F is obtained by summing the work done on m_A and m_B.

$$dU = F(ds_A \cos \alpha_A + ds_B \cos \alpha_B)$$

where α_A and α_B are the angles between $\mathbf{F}$ and the paths of m_A and m_B, respectively. By inspection of Fig. 14.7, we find that the term in parentheses represents a differential change in R. Note that this change is negative (the distance is reduced) in the figure. Hence, the differential work can be written as

$$dU = F(-dR) = -K\frac{m_A m_B}{R^2}\, dR$$

Integrating dU between positions ① and ②, where the distances between the masses are R_1 and R_2, respectively, we have

$$U_{1\text{-}2} = -\int_{R_1}^{R_2} K\frac{m_A m_B}{R^2}\, dR = -K m_A m_B \int_{R_1}^{R_2} \frac{dR}{R^2}$$

from which we find that

$$\boxed{U_{1\text{-}2} = K m_A m_B \left(\frac{1}{R_2} - \frac{1}{R_1}\right)} \qquad (14.11)$$

Equation (14.11) can be used to compute the work done by the earth on a satellite that is moving above its surface. In this application, $m_A = M_e$ (the mass of the earth) and R_1 and R_2 are the initial and final altitudes of the satellite, as measured from the center of the earth.

From Eq. (14.11) we see that the work done by gravitational forces is independent of the path followed by the satellite. The work is determined only by the initial and final distances between m_A and m_B.

14.3 Principle of Work and Kinetic Energy

The principle of work and kinetic energy (also known as the work-energy principle) states that the work done by all forces acting on a particle equals the change in the kinetic energy of the particle. This principle, which forms the basis for the work-energy method of kinetic analysis, is derived by integrating Newton's second law along the path of the particle.

We begin by substituting Newton's second law, $\mathbf{F} = m\mathbf{a} = m(d\mathbf{v}/dt)$, into the expression for the differential work, Eq. (14.1), which yields

$$dU = \mathbf{F}\cdot d\mathbf{r} = m\frac{d\mathbf{v}}{dt}\cdot d\mathbf{r}$$

Substituting $d\mathbf{r} = \mathbf{v}\,dt$ yields

$$dU = m\frac{d\mathbf{v}}{dt}\cdot\mathbf{v}\,dt = \frac{1}{2}m\frac{d}{dt}(\mathbf{v}\cdot\mathbf{v})\,dt = \frac{1}{2}m\,d(\mathbf{v}\cdot\mathbf{v})$$

Since $\mathbf{v}\cdot\mathbf{v} = v^2$, this equation becomes

$$dU = \frac{1}{2}m\,d(v^2) = d\left(\frac{1}{2}mv^2\right) \tag{14.12}$$

The *kinetic energy* of the particle is defined to be

$$\boxed{T = \frac{1}{2}mv^2} \tag{14.13}$$

which means that Eq. (14.12) may be written as

$$dU = dT \tag{14.14}$$

When the particle moves along its path between positions ① and ②, the integral of dU is the work done by the resultant force acting on the particle. Therefore, taking the integral of Eq. (14.14) between ① and ②, we get

$$\boxed{U_{1-2} = \int_{T_1}^{T_2} dT = T_2 - T_1 = \Delta T} \tag{14.15}$$

where $T_1 = \frac{1}{2}mv_1^2$ and $T_2 = \frac{1}{2}mv_2^2$ are the kinetic energies of the particle in positions ① and ②, respectively.

Equation (14.15) is the *work-energy principle* (or balance of work and energy) for particle motion: The work done by the resultant force acting on a particle as it moves from position ① to position ② equals the change in kinetic energy of the particle between these two positions. When this principle is used in kinetic analysis, the method is referred to as the *work-energy method.*

Since the work-energy principle results from integrating Newton's second law, it is not an independent principle. Therefore, any problem that can be solved by the work-energy method can, in theory, be solved by the force-mass-acceleration (FMA) method.

The main advantages of the work-energy method when compared with the FMA method are as follows:

1. In problems where the work can be calculated without integration (as in the special cases discussed in the preceding article), the change in speed of the particle may be easily obtained with a minimum of computation. (If integration must be used to compute the work, the work-energy method will, in general, have no advantage over integrating the acceleration determined by the FMA method.)
2. Only forces that do work need be considered; nonworking forces do not appear in the analysis.
3. If the final position, i.e., position ②, is chosen to be an arbitrary position, the work-energy method will determine the speed as a function of position of the particle.

The following points should be kept in mind when applying the work-energy principle to the motion of a particle:

1. Work of a force is a *scalar* quantity (positive, negative, or zero) that is associated with a *change in the position* of the point of application of the force. (The phrase "work at a given position" is meaningless.)
2. Kinetic energy is a *scalar* quantity (always positive) associated with the speed of a particle at a given instant of time. The units of kinetic energy are the same as the units of work: $\text{N} \cdot \text{m}$, $\text{lb} \cdot \text{ft}$, etc.
3. The work energy principle, $U = \Delta T$, is a *scalar* equation. Although this is an obvious point, a common error is to apply the work-energy principle separately in the x-, y-, or z-directions, which is, of course, incorrect.
4. An *active force diagram,* which shows only the forces that do work, may be used in place of the conventional free-body diagram. A convenient method for determining the active force diagram is to first draw the FBD of the particle in an arbitrary position and then delete any forces that do not do work.

Sample Problem 14.1

The collar A of mass $m = 1.8$ kg shown in Fig. (a) slides in the vertical plane along a smooth rod. A cable runs over a pulley at B and is pulled with a constant horizontal force P. The collar starts from rest in position ①. (1) Calculate the speed of the collar when it reaches position ② if $P = 20$ N. (2) What is the smallest value of P for which the collar will reach position ②?

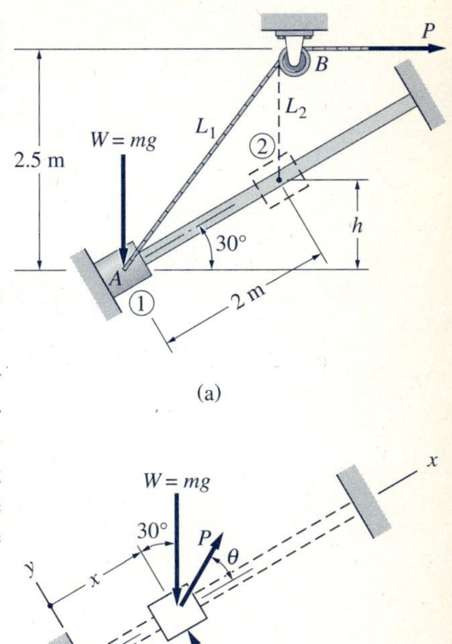

(a)

(b)

Solution

Preliminary Discussion

Since we are interested in the change in speed of the collar between two given positions (and because it is relatively easy to compute the work done on the collar), the work-energy method is the most convenient means of analyzing this problem.

Figure (a) is also the active-force diagram, because only the forces P and W do work as the collar slides along the rod. To further illustrate this point, consider the FBD of the collar in Fig. (b) that is drawn when the collar is at an arbitrary distance x from the support. There are three forces acting on the collar: its weight $W = mg$, the cable tension P, and the normal force N_A. It is important to observe that θ, the angle between P and the rod, varies with the position of the collar. We now consider the work done by each of these forces.

Work Done by W

The work done by the weight as the collar moves from ① and ② is easy to compute, because W is constant in magnitude and direction.

$$U_{1-2} = -mgh \tag{a}$$

where h is the vertical distance shown in Fig. (a).

Work Done by P

The force P shown in Fig. (b) also does work, since its point of application moves with the collar. During a differential movement dx of the collar, the differential work done by P is $dU = P\cos\theta\, dx$. The evaluation of the resulting integral between positions ① and ② would be complicated, because θ must first be expressed in terms of x. However, if we look at Fig. (a), we see that the work done by P is simply

$$U_{1-2} = P(L_1 - L_2) \tag{b}$$

because the force P acting at the end of the cable has constant magnitude and direction, and it moves through the distance $(L_1 - L_2)$ as the collar moves from ① to ②.

Work Done by N_A

Using the FBD in Fig. (b), $\Sigma F_y = ma_y = 0$ gives $N_A = W\cos 30° - P\sin\theta$. Observe that N_A is also a function of θ; i.e., it varies with the position of the collar. Fortunately, however, N_A does not do work, because it is perpendicular to the path of the collar throughout the motion.

A Note Concerning Friction

If friction between the rod and the collar were not negligible, we would have to include the kinetic friction force $F_k = \mu_k N_A$, directed opposite to the velocity, on the FBD in Fig. (b). Since N_A varies with the position coordinate x of the collar, F_k would also vary with x. Therefore, integration would be required to compute the work done by the friction force. In this case, the work-energy method would have no advantage over integrating the equation of motion (i.e., the acceleration) that could be found by the FMA method. However, if N_A were constant, the kinetic friction force would also be constant, and the work done by friction could be easily computed. (See Sample Problem 14.2.)

Part 1

The speed of the collar in position ② can be found by using the work-energy principle, Eq. (14.15),

$$U_{1-2} = T_2 - T_1 = \frac{1}{2}(1.8)v_2^2 - 0 \tag{c}$$

where we have substituted $T_2 = \frac{1}{2}mv_2^2$ and $T_1 = 0$ (the collar is at rest in position ①). The total work done is obtained by summing the contributions of P and W given in Eqs. (a) and (b), respectively:

$$U_{1-2} = -mgh + P(L_1 - L_2) \tag{d}$$

From Fig. (a) we see that $h = 2\sin 30° = 1.0$ m, $L_1 = \sqrt{(2\cos 30°)^2 + (2.5)^2} = 3.041$ m, and $L_2 = 2.5 - 1.0 = 1.5$ m. Substituting these values together with $P = 20$ N into Eq. (d), we obtain

$$U_{1-2} = -(1.8)(9.81)(1.0) + 20(3.041 - 1.5) = 13.162 \text{ N} \cdot \text{m} \tag{e}$$

Therefore, the work-energy principle, Eq. (c), becomes

$$13.162 = 0.9v_2^2$$

from which we obtain

$$v_2 = 3.82 \text{ m/s} \qquad \textit{Answer}$$

Part 2

In this part of the problem, ② is also a rest position, so that $T_2 = T_1 = 0$. The problem is thus reduced to finding the value of P for which the total work done on the collar between positions ① and ② is zero. Referring to Eq. (d), we have

$$-mgh + P(L_1 - L_2) = 0$$

or

$$P = \frac{mgh}{L_1 - L_2} = \frac{(1.8)(9.81)(1.0)}{3.041 - 1.5} = 11.46 \text{ N} \qquad \textit{Answer}$$

Observe that if P were less than 11.46 N, U_{1-2} would be negative, thereby giving an imaginary speed (square root of a negative number) for position ②. This result indicates that it is not possible for the collar to reach position ② for values of P less than 11.46 N.

Sample Problem 14.2

The block of mass $m = 1.6$ kg slides along the horizontal rough plane as shown in Fig. (a). The position coordinate x is measured from the undeformed position of the ideal spring, which has a stiffness of $k = 30$ N/m. At $x = 0$ the block is initially moving to the right with speed $v_1 = 6$ m/s. (1) Find the value of x when the block first comes to rest. (2) Show that the block does not remain at rest in the position found in Part 1. (3) Calculate the speed of the block when it reaches $x = 0$ for the second time.

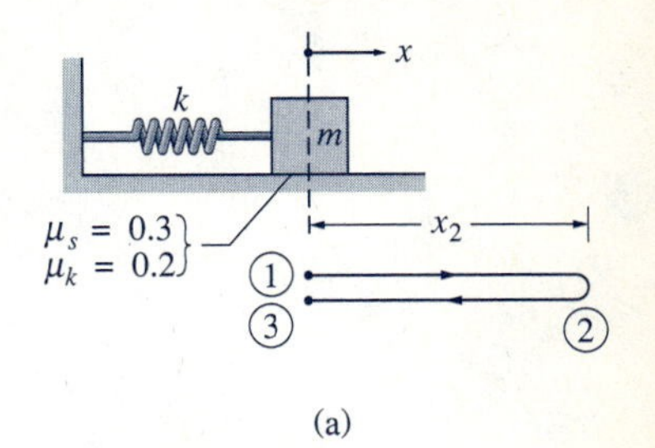

(a)

Solution

Part 1

Figure (b) shows the FBD of the block when it is moving to the right on the rough plane. The forces acting on the block are its weight $W = mg = 1.6(9.81) = 15.696$ N, the spring force $P_s = kx$, the normal force N_A, and the kinetic friction force $F_k = \mu_k N_A$. Observe that the direction of F_k opposes the direction of the motion. Since the sum of the forces in the vertical direction must be zero, we find that $N_A = W = 15.696$ N, which gives $F_k = 0.2(15.696) = 3.139$ N. Note that N_A, and therefore F_k, are constant throughout the motion. (The simplification arising from a constant normal force was discussed in the solution of Sample Problem 14.1.) Consequently, the work done by F_k will be $-(F_k) \times$ (distance moved by the block).

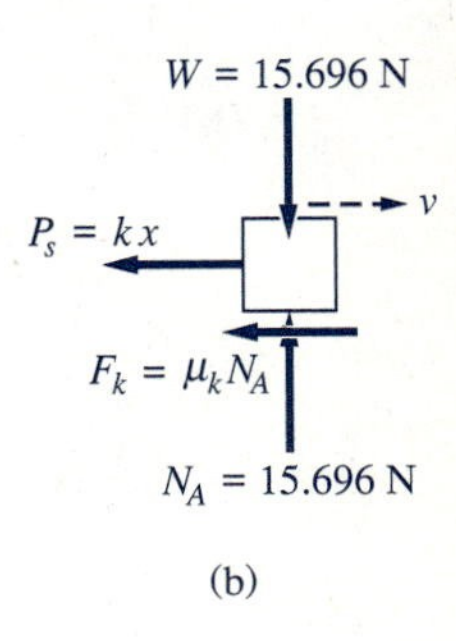

(b)

Applying the work-energy principle between the initial position ① and the position ② where the block comes to rest, we obtain

$$U_{1\text{-}2} = T_2 - T_1$$

$$-\frac{1}{2}k(x_2^2 - x_1^2) - F_k x_2 = \frac{1}{2}mv_2^2 - \frac{1}{2}mv_1^2$$

Substituting the given numerical values for k, F, m, and v_1, and recognizing that $x_1 = 0$ and $v_2 = 0$, yields

$$-\frac{1}{2}(30)(x_2^2 - 0) - 3.139x_2 = 0 - \frac{1}{2}(1.6)(6)^2$$

The two roots of this equation are $x_2 = 1.2850$ m and -1.4942 m. Since the motion was assumed to be directed to the right, only the positive root is meaningful; i.e.,

$$x_2 = 1.285 \text{ m} \qquad \textit{Answer}$$

Part 2

The FBD of the block at rest in position ② ($x_2 = 1.285$ m) is shown in Fig. (c). The spring force $P_s = kx_2 = 30(1.285) = 38.55$ N, which tends to pull the block to the left, is resisted by the static friction force F_s. The block will remain at rest if F_s is less than or equal to its maximum possible value, that is, if $F_s \leq \mu_s N_A = 0.3(15.696) = 4.709$ N. Assuming that the block is in equilibrium, $\Sigma F_x = 0$ gives $F_s = P_s = 38.55$ N. Since this value is greater than 4.709 N, we conclude that the block will not remain at rest in position ②, but will begin moving to the left.

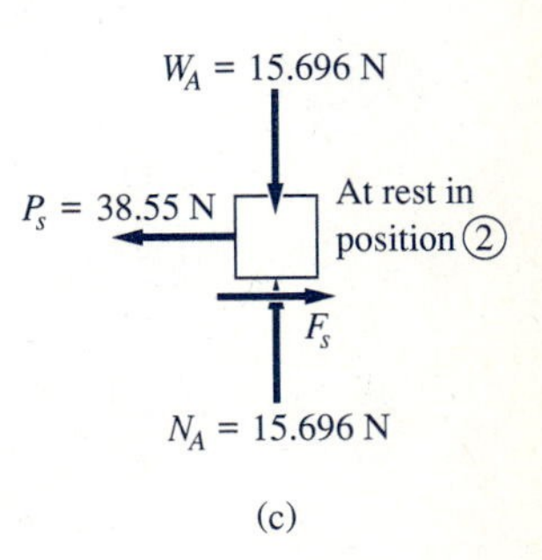

(c)

Part 3

To calculate the speed of the block as it passes the position $x = 0$ for the second time—position ③ in Fig. (a)—we apply the work-energy principle between positions ① and ③ (applying the principle between ② and ③ would yield the same result). The net work done by the spring is zero between ① and ③, because the spring is undeformed in both positions. (The negative work done by the spring from ① to ② cancels its positive work done from ② to ③.) Therefore, only the kinetic friction force does work between positions ① and ③.

Observe that since F_k always opposes motion, its work is negative between positions ① and ②, as well as between positions ② and ③. Thus the total work done by F_k is $-2F_k x_2$. Applying the work-energy principle between ① and ③, we have

$$U_{1\text{-}3} = T_3 - T_1$$

$$-2F_k x_2 = \frac{1}{2}m(v_3^2 - v_1^2)$$

$$-2(3.139)(1.285) = \frac{1}{2}(1.6)(v_3^2 - 6^2)$$

from which we find that

$$v_3 = 5.09 \text{ m/s} \qquad \textit{Answer}$$

If v_3 had turned out to be an imaginary number (square root of a negative number), we would conclude that it was not possible for the block to reach position ③. We would then know that the block had come to rest somewhere between ② and ③. If this were the case, another straightforward application of the work-energy principle would determine this rest position.

Sample Problem 14.3

The 2-lb collar slides along the smooth rod shown in Fig. (a). The free length of the spring that is attached to the collar is 6 ft and its stiffness is 8 lb/ft. If the collar is moving down the rod with speed $v_A = 12$ ft/s when it is at A, calculate the speed of the collar when it reaches B.

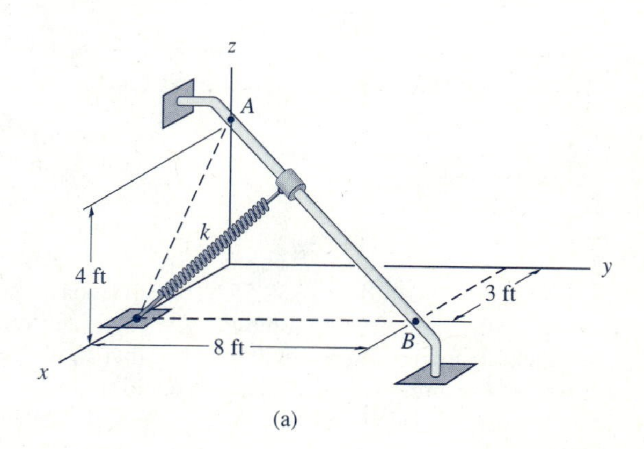

(a)

Solution

The free-body diagram (FBD) of the collar when it is at an arbitrary distance r from A is shown in Fig. (b). In addition to its weight W, the collar is acted on by the spring force F_s and the normal force N. Of the three forces in the FBD, only W and F_s do work on the collar during its motion along the rod. The normal force N does not do work because it is perpendicular to AB, the path of the collar.

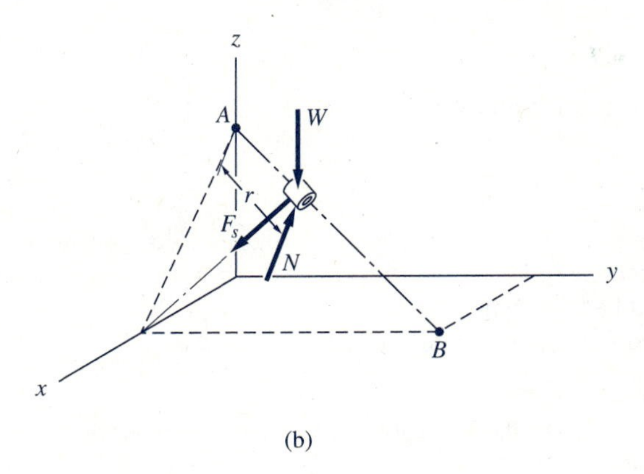

(b)

The work done on the collar as it moves from A to B thus consists of the following two parts.

Work done by W: from Eq. (14.7)

$$U_{A-B} = -W(z_B - z_A) = -(2)(0 - 4) = 8 \text{ lb} \cdot \text{ft} \tag{a}$$

Work done by spring: from Eq. (14.10)

$$U_{A-B} = -\frac{1}{2}k(\delta_B^2 - \delta_A^2) \tag{b}$$

The undeformed length of spring was given as $L_0 = 6$ ft. Letting L_A and L_B be the lengths of the spring when the collar is at A and B, respectively, the corresponding deformations are found to be

$$\delta_A = L_A - L_0 = \sqrt{3^2 + 4^2} - 6 = -1.0 \text{ ft}$$
$$\delta_B = L_B - L_0 = 8 - 6 = 2.0 \text{ ft}$$

(The spring is compressed 1.0 ft when the collar is at A and stretched 2.0 ft when it is at B.) Substituting the values for δ_A and δ_B into Eq. (b), the work performed by the spring is

$$U_{A-B} = -\frac{1}{2}(8)[(2.0)^2 - (-1.0)^2] = -12 \text{ lb} \cdot \text{ft} \tag{c}$$

The total work done on the collar is, therefore,

$$U_{A-B} = 8 - 12 = -4 \text{ lb} \cdot \text{ft} \tag{d}$$

Substituting Eq. (d) into the work-energy principle

$$U_{A\text{-}B} = T_B - T_A = \frac{1}{2}\frac{W}{g}(v_B^2 - v_A^2)$$

yields

$$-4 = \frac{1}{2}\frac{2}{32.2}(v_B^2 - 12^2)$$

Solving for v_B, we find that

$$v_B = 3.90 \text{ ft/s} \qquad \textit{Answer}$$

If the analysis had predicted that v_B was imaginary, i.e., the square root of a negative number, we would conclude that the collar came to rest before reaching B. This position could be then found by another application of the work-energy principle, with the speed in the new position set equal to zero.

Concluding Remarks

The ease with which v_B has been obtained indicates how powerful the work-energy method can be in the solution of some problems. Inspection of the FBD in Fig. (b) reveals that the spring force F_s is a rather complicated function of r. Therefore, the acceleration of the collar is also a complicated function of r. However, by using the work-energy method, we were able to compute the change in speed between A and B without having to derive and integrate the equation of motion. The reason is that the corresponding integration was already performed in the derivation of the work-energy principle.

PROBLEMS

14.1 (a) Compute the work done by each force given in the following list as its point of application moves from ① to ③ along the straight line connecting ① and ③. (b) Repeat part (a) if the path consists of the straight line segments ①–② and ②–③. (x and y are in feet.)

1. $\mathbf{F} = 30\mathbf{i} - 10\mathbf{j}$ lb
2. $\mathbf{F} = 3x\mathbf{i} - y\mathbf{j}$ lb
3. $\mathbf{F} = 3y\mathbf{i} - x\mathbf{j}$ lb

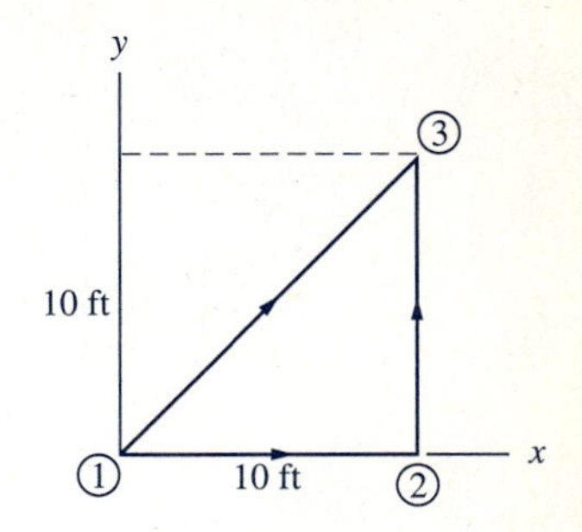

Fig. P14.1

14.2 Compute the work of the force $\mathbf{F} = (F_0/b^3)(xy^2\mathbf{i} + x^2y\mathbf{j})$ as its point of application moves from ① to ② along (a) the line $y = x$; and (b) the parabola $y = x^2/b$.

14.3 Repeat Prob. 14.2 for the force $\mathbf{F} = (F_0/b^3)(x^2y\mathbf{i} + xy^2\mathbf{j})$.

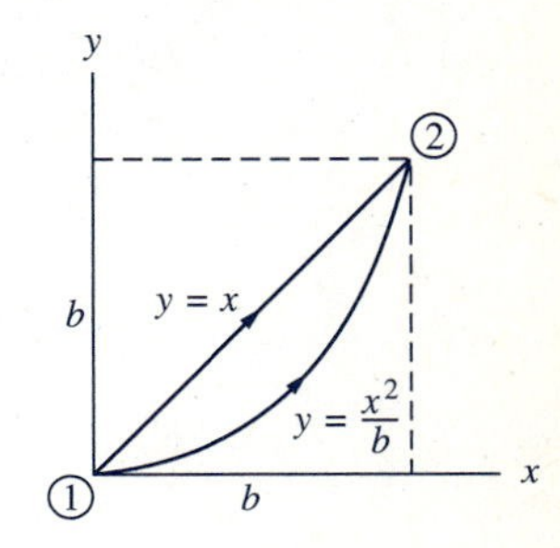

Fig. P14.2, P14.3

14.4 The collar of weight W slides on a smooth circular arc of radius R. The ideal spring attached to the collar has the free length $L_0 = R$ and stiffness k. When the slider moves from A to B, compute (a) the work done by the spring; and (b) the work done by the weight.

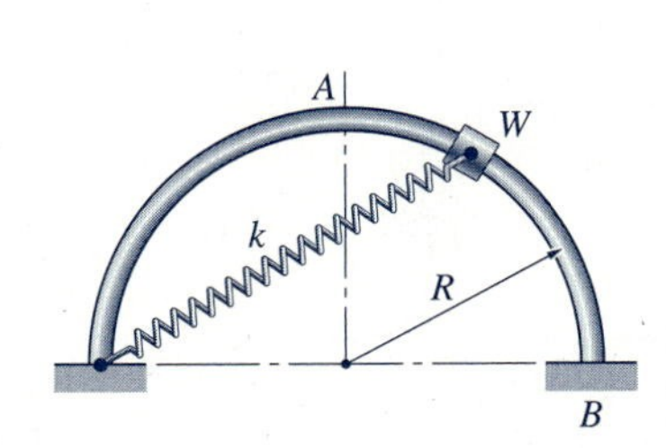

Fig. P14.4

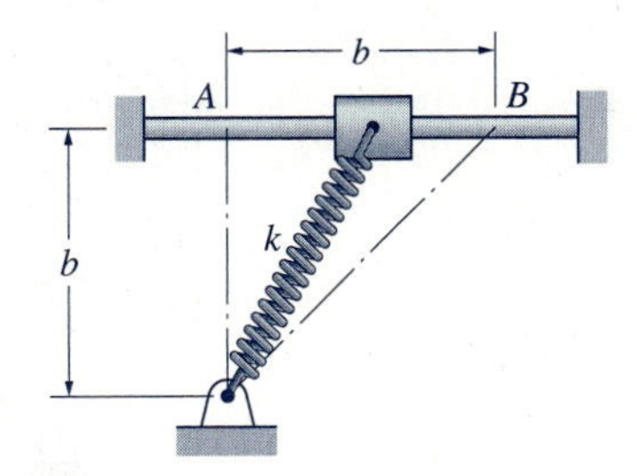

Fig. P14.5, P14.6

14.5 Derive the expression for the work done by the ideal spring when the slider moves from A to B. Assume that the free length of the spring is (a) $L_0 = b$; and (b) $L_0 = 0.75b$.

14.6 The coefficient of kinetic friction between the slider and the rod is μ, and the free length of the spring is $L_0 = b$. Derive the expression for the work done by the friction force on the slider as it moves from A to B. Neglect the weight of the slider.

14.7 A crate of weight W is dragged across a rough floor from A to B by the constant vertical force P acting at the end of the rope. Calculate the work done on the crate by (a) the force P; and (b) the friction force between the crate and the floor. Assume that the crate does not lift off the floor.

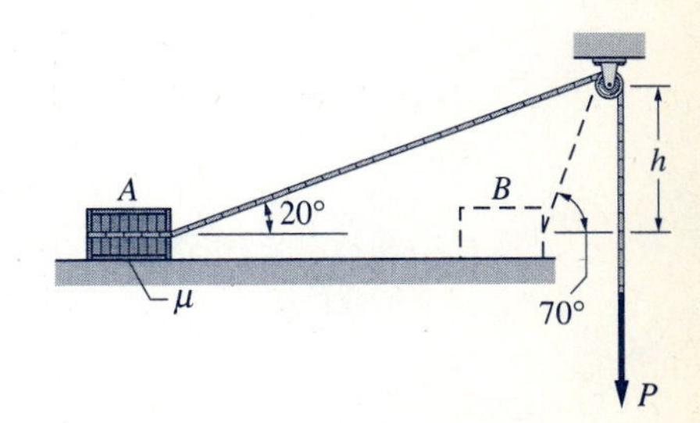

Fig. P14.7

14.8 Solve Prob. 12.30 by the work-energy method.

14.9 Solve Prob. 12.43, part (b) by the work-energy method.

14.10 Solve Prob. 12.45 by the work-energy method.

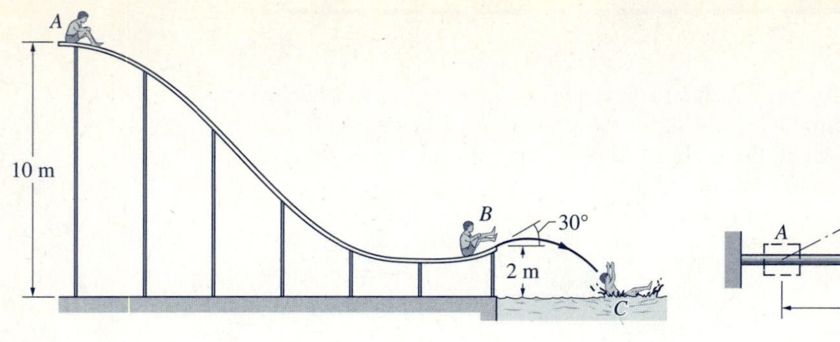

Fig. P14.11

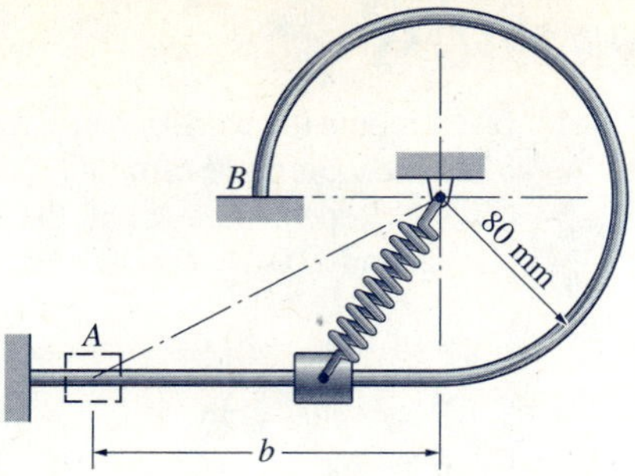

Fig. P14.12

14.11 A boy slides down a water chute, starting from rest at A. Neglecting friction, determine his velocity vector (a) at the end B of the chute; and (b) on entering the water at C.

14.12 The 0.31-kg mass slides on a smooth wire that lies in the vertical plane. The ideal spring attached to the mass has a free length of 80 mm and its stiffness is 120 N/m. Calculate the smallest value of the distance b if the mass is to reach the end of the wire at B after being released from rest at A.

14.13 The 3-lb collar moves from A to B along a smooth rod. The stiffness of the spring is k and its free length is 8 in. Compute the value of k so that the slider arrives at B with a speed of 2 ft/s after being released from rest at A.

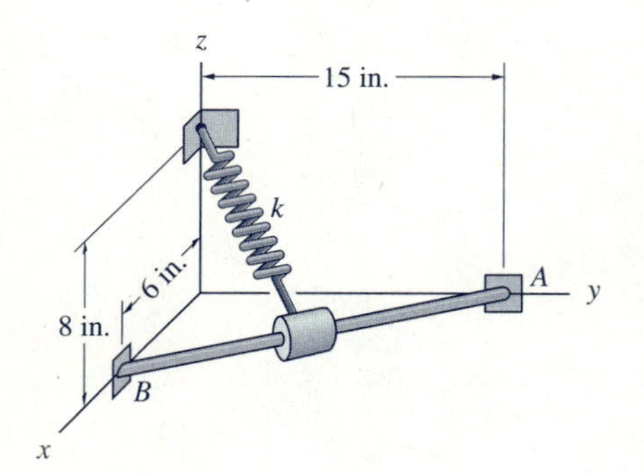

Fig. P14.13

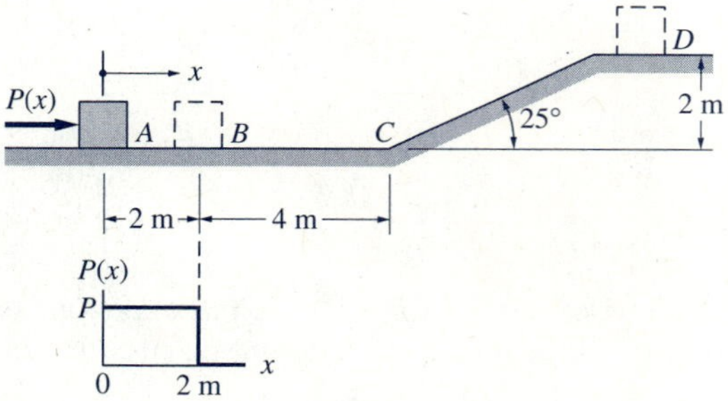

Fig. P14.14, P14.15

14.14 An 8-kg package, initially at rest at A, is propelled between A and B by a constant force P. Neglecting friction, find the smallest value of P for which the package will reach D.

14.15 Solve Prob. 14.14 assuming that the coefficient of kinetic friction between the package and the contact surfaces is 0.1.

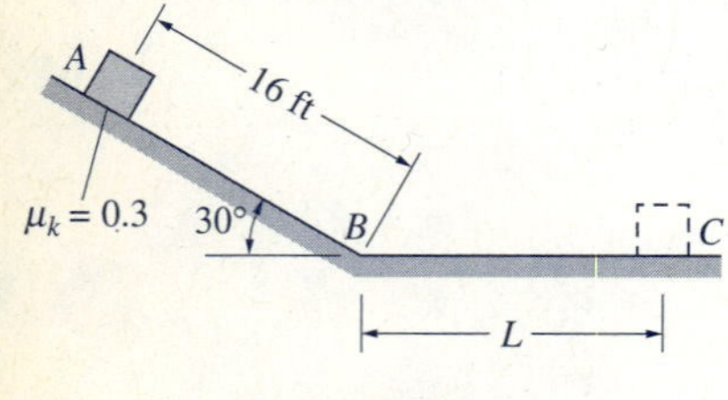

Fig. P14.16

14.16 The coefficient of kinetic friction between the package and the surface is 0.3. If the package is released from rest at A, compute (a) the speed of the package at B; and (b) the distance L that the package travels on the horizontal surface before coming to rest at C.

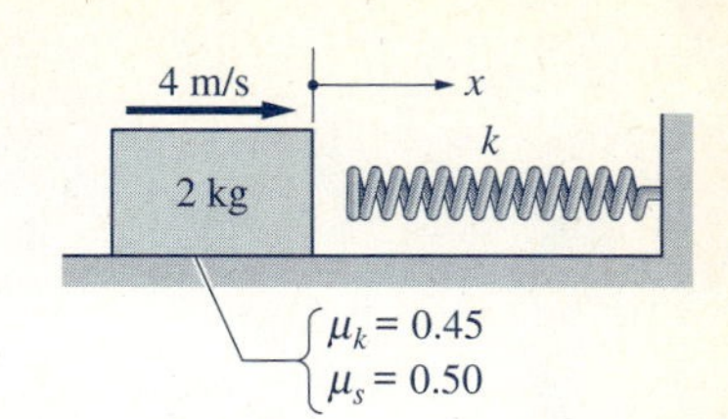

Fig. P14.17–P14.19

14.17 The 2-kg block hits the spring with a speed of 4 m/s. (a) Determine the maximum compressive deformation of the spring. (b) What is the total distance traveled by the block before it comes to a permanent stop? Use $k = 8$ N/m and the coefficients of friction shown.

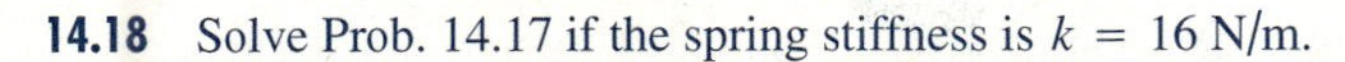

14.18 Solve Prob. 14.17 if the spring stiffness is $k = 16$ N/m.

14.19 Solve Prob. 14.17 if the spring stiffness is $k = 32$ N/m.

14.20 The block of mass m is released from rest in the position shown. The free length of the spring is L_0 and its stiffness is k. The coefficients of static and kinetic friction are both equal to μ, where $\mu < \tan\theta$. (a) Compute the maximum displacement of the block. (b) Find the maximum speed of the block and the position where it occurs.

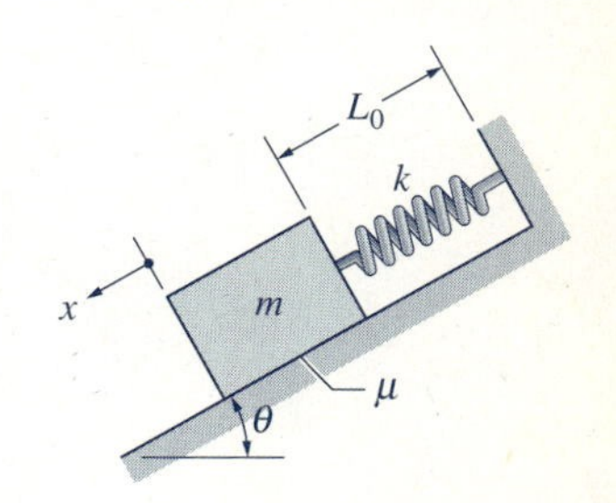

Fig. P14.20

14.21 The diagram shows the relationship between the force F and deformation x for an energy-absorbing car bumper. Determine the maximum deformation of the bumper if a 2400-lb car hits a rigid wall at a speed of (a) 3 mi/h; and (b) 5 mi/h.

14.22 The diagram shows how the force F required to push an arrow slowly through a bale of hay varies with the distance of penetration x. Assuming that the F-x plot is independent of the speed of penetration, calculate the exit speed of a 3-oz arrow if its speed at entry is 200 ft/s.

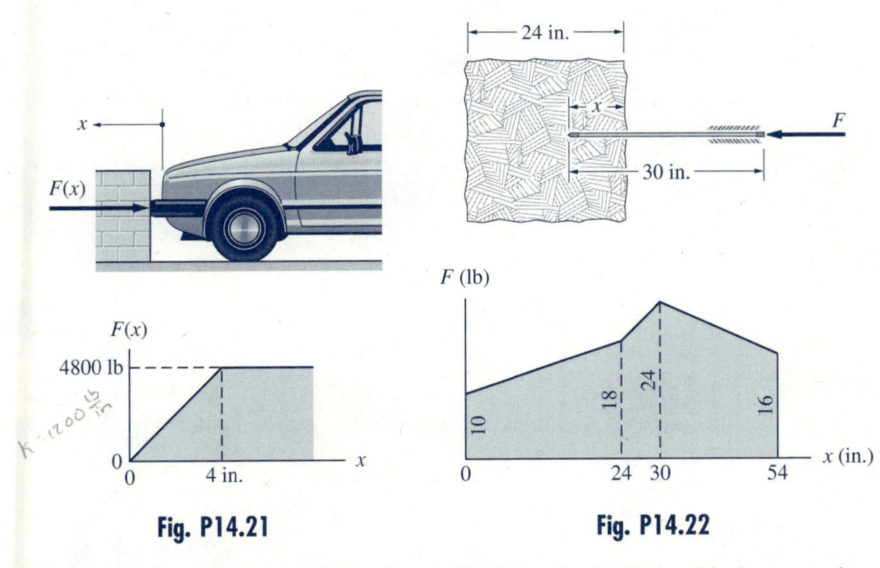

Fig. P14.21

Fig. P14.22

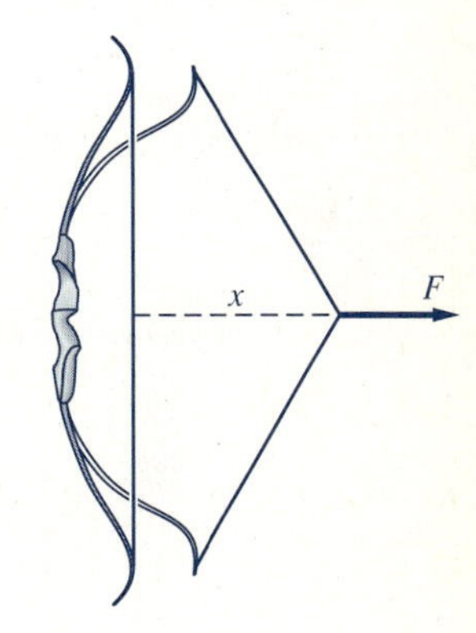

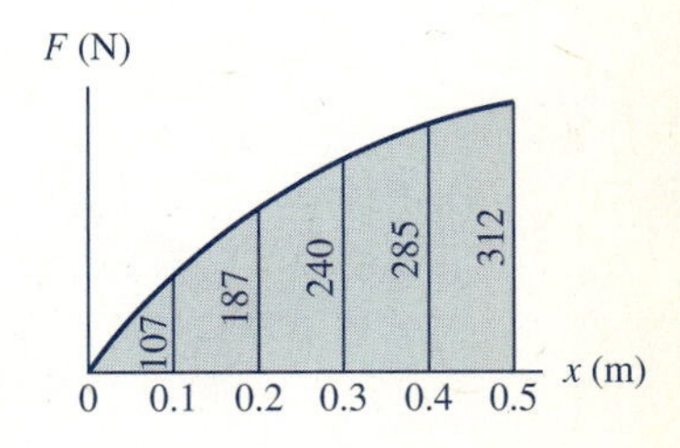

Fig. P14.23

14.23 The diagram shows the experimentally determined relationship between the pull F and draw x of a bow. Find the speed at which a 0.085-kg arrow leaves the bow, assuming that the draw was 0.5 m.

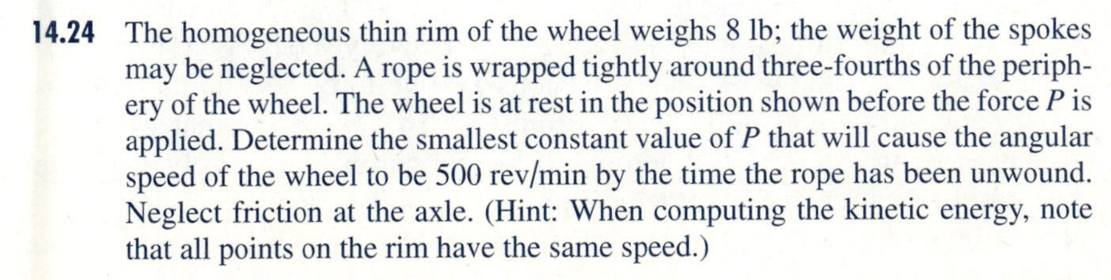

14.24 The homogeneous thin rim of the wheel weighs 8 lb; the weight of the spokes may be neglected. A rope is wrapped tightly around three-fourths of the periphery of the wheel. The wheel is at rest in the position shown before the force P is applied. Determine the smallest constant value of P that will cause the angular speed of the wheel to be 500 rev/min by the time the rope has been unwound. Neglect friction at the axle. (Hint: When computing the kinetic energy, note that all points on the rim have the same speed.)

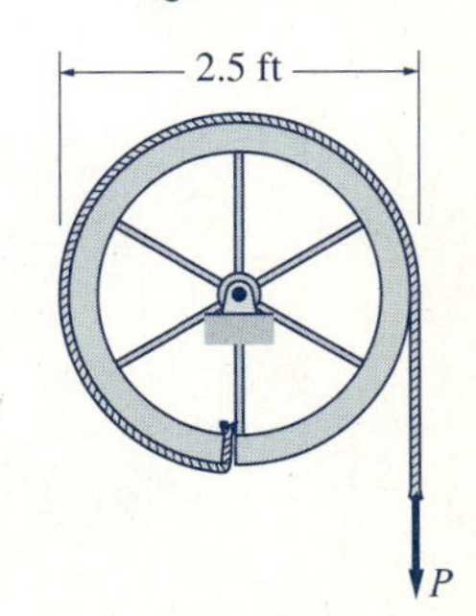

Fig. P14.24

14.4 Conservative Forces and the Conservation of Mechanical Energy

It is often convenient to describe the effects of conservative forces in terms of their potential energies. Roughly speaking, potential energy is the capacity of a conservative force to do work, a property shared with the potentials of other forms of energy, such as chemical, electrical, and thermal (heat). The principle of conservation of energy, one of the fundamental postulates in mechanics, states that the total energy (the sum of all forms of energy) remains constant for a closed system. The form of the energy may change—e.g., electrical energy may be converted to mechanical energy—but the total energy can neither be created nor destroyed.

In this text, we restrict our attention to mechanical energy, defined to be the sum of the potential and kinetic energies. If all forces acting on a particle, body, or closed system of bodies are conservative, mechanical energy is conserved, a concept known as the *principle of conservation of mechanical energy.*

This article discusses the application of the principle of conservation of mechanical energy, which is simply a restatement of the work-energy principle, $U_{1-2} = \Delta T$, for conservative force systems. Although the energy principle may be easier to apply than the work-energy method in some problems, its use is limited, because it is not valid for nonconservative forces, such as kinetic friction.

a. Conservative forces and potential energy

A force $\mathbf{F}$ is said to be *conservative* if its work—given by Eq. (14.2), $U_{1-2} = \int_{\mathcal{L}} \mathbf{F}\cdot d\mathbf{r}$—is a function of the initial and final positions of its point of application. Therefore, the work done by a conservative force is independent of the path followed by its point of application. The line integral $\int_{\mathcal{L}} \mathbf{F}\cdot d\mathbf{r}$ can be a function of the endpoints of the path only if the integrand is an exact differential of some function $-V(\mathbf{r})$, i.e., if the integrand can be written as

$$\mathbf{F}\cdot d\mathbf{r} = -dV \tag{14.16}$$

(The reason for the minus sign will be explained shortly.) Integrating Eq. (14.16) along the path $\mathcal{L}$ connecting positions ① and ② (see Fig. 14.1), we obtain

$$U_{1-2} = \int_{\mathcal{L}} \mathbf{F}\cdot d\mathbf{r} = -\int_{V_1}^{V_2} dV = -(V_2 - V_1) = -\Delta V \tag{14.17}$$

The function V, which is a function of position only, is called the *potential function* or *potential energy* of the force. In Eq. (14.17), V_1 and V_2 are the potential energies at positions ① and ②, respectively; i.e., $V_1 = V(\mathbf{r}_1)$ and $V_2 = V(\mathbf{r}_2)$. Thus the work done by a conservative force is equal to the negative change in its potential energy. From Eq. (14.17) we see that when the force does positive work, its potential energy decreases. Conversely, negative work results in an increase in the potential energy. Therefore, the introduction of the minus sign in Eq. (14.16) permits us to interpret potential energy as the capacity of the force to do work.

Let $\mathscr{L}_1$ and $\mathscr{L}_2$ be two different paths between positions ① and ②, as shown in Fig. 14.8. The work $U_{1\text{-}2}$ of a conservative force will be the same for each path; i.e., $\int_{\mathscr{L}_1} \mathbf{F}\cdot d\mathbf{r} = \int_{\mathscr{L}_2} \mathbf{F}\cdot d\mathbf{r}$, since the integrals depend only on the endpoints. Note that a closed path would be formed if the force were to move from ① to ② along $\mathscr{L}_1$ and then move back to ① along the path $\mathscr{L}_2$. The total work done by **F** would be $\int_{\mathscr{L}_1} \mathbf{F}\cdot d\mathbf{r} - \int_{\mathscr{L}_2} \mathbf{F}\cdot d\mathbf{r} = 0$, which shows that the work done by a conservative force is zero when its point of application moves around a closed path. This property is written as

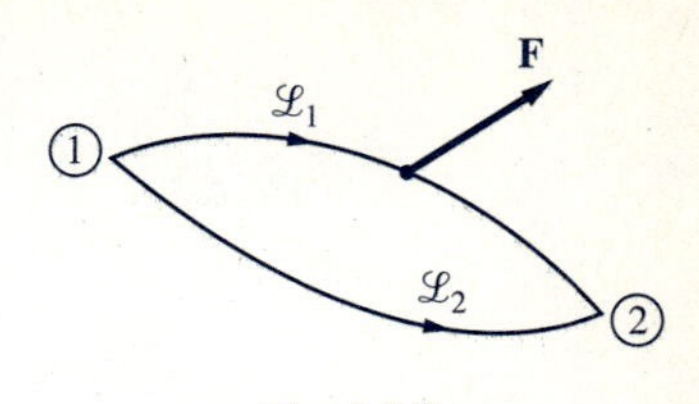

Fig. 14.8

$$\oint \mathbf{F}\cdot d\mathbf{r} = 0 \tag{14.18}$$

where the circle on the integral sign indicates that the integral is to be taken around a closed path.

An important property of a conservative force is that its components can be derived from its potential energy. Consider a conservative force **F** that acts at a point with rectangular coordinates (x, y, z). Using $\mathbf{F} = F_x\mathbf{i} + F_y\mathbf{j} + F_z\mathbf{k}$ and $d\mathbf{r} = dx\,\mathbf{i} + dy\,\mathbf{j} + dz\,\mathbf{k}$, we obtain

$$dV = -\mathbf{F}\cdot d\mathbf{r} = -(F_x\,dx + F_y\,dy + F_z\,dz) \tag{14.19}$$

Since dV is an exact differential of potential energy V, it may be written as

$$dV = \frac{\partial V}{\partial x}dx + \frac{\partial V}{\partial y}dy + \frac{\partial V}{\partial z}dz \tag{14.20}$$

Comparing Eqs. (14.19) and (14.20), the rectangular components of **F** become

$$F_x = -\frac{\partial V}{\partial x} \qquad F_y = -\frac{\partial V}{\partial y} \qquad F_z = -\frac{\partial V}{\partial z} \tag{14.21}$$

Using the gradient operator of vector calculus, Eq. (14.21) can be abbreviated as $\mathbf{F} = -\mathbf{grad}\,V$. Note that only conservative forces are derivable from a potential function in the manner indicated in Eq. (14.21). Nonconservative forces do not possess a potential.

b. Conservation of mechanical energy

If all the forces acting on a particle are conservative, then the work done on the particle as it moves from position ① to position ② is, according to Eq. (14.17), $U_{1\text{-}2} = -(V_2 - V_1)$. Substituting this into the work-energy principle $U_{1\text{-}2} = \Delta T$, we obtain $-(V_2 - V_1) = T_2 - T_1$, or

$$V_1 + T_1 = V_2 + T_2 \tag{14.22}$$

Letting the total mechanical energy* E be the sum of the kinetic and potential energies; i.e.,

$$E = T + V \tag{14.23}$$

* The emphasis here is on mechanical energy. Other forms of energy, such as heat, are excluded from our discussion.

Eq. (14.22) becomes

$$E_1 = E_2 \quad \text{or} \quad \Delta E = 0 \tag{14.24}$$

This equation is called the *principle of conservation of mechanical energy.*

Since the work done by a kinetic friction force is not independent of the path, kinetic friction is a nonconservative force. Therefore, when kinetic friction is present, the total mechanical energy is not conserved, but is reduced by the negative work done by the friction force. This energy is not lost; it is transformed into thermal energy in the form of heat. In other words, the total energy is still conserved; it is the form of the energy that has changed.

Because the total mechanical energy E is constant if all of the forces are conservative, we conclude that

$$\boxed{\frac{dE}{dt} = 0} \tag{14.25}$$

This equation is sometimes useful for deriving the equation of motion.

c. Computation of potential energy

The potential energies of conservative forces can be computed by comparing their work, derived by the methods of Art. 14.2, with the definition of potential energy in Eq. (14.17).

1. Potential Energy of a Constant Force The work done by a constant force **F** (constant in magnitude and direction) was previously shown to be

$$U_{1\text{-}2} = F(d_2 - d_1) = F\Delta d \qquad \text{(14.6, repeated)}$$

where F is the magnitude of the force and Δd represents the work-absorbing displacement of its point of application, as shown in Fig. 14.4. Equating the expressions for $U_{1\text{-}2}$ in Eqs. (14.6) and (14.17), we have $F(d_2 - d_1) = -(V_2 - V_1)$. Consequently, we conclude that the *potential energy of a constant force* is

$$V_f = -Fd + C$$

where C is an arbitrary constant.* Choosing $V_f = 0$ when $d = 0$, then $C = 0$ and the potential energy becomes

$$\boxed{V_f = -Fd} \tag{14.26}$$

If we assume that motion is restricted to small distances from the surface of the earth, the weight of an object may be treated as a constant force. The work done by the weight W in Fig. 14.5 was given by

$$U_{1\text{-}2} = -W(y_2 - y_1) \qquad \text{(14.7, repeated)}$$

* Expressions for potential energy invariably contain an arbitrary constant. However, since only the change in the potential energy is of interest, the arbitrary constant always cancels out in the analysis. Therefore the constant may be chosen to be any value that is mathematically convenient.

Comparing this equation with $U_{1-2} = -(V_2 - V_1)$, we conclude that the potential energy of W, called the *gravitational potential energy,* is equal to

$$V_g = Wy + C$$

where C is an arbitrary constant. The value of C is determined when a reference, or *datum,* is chosen for the potential energy. If the potential energy is chosen to be zero at $y = 0$, then $C = 0$ and the gravitational potential energy is

$$\boxed{V_g = Wy \quad \text{(for constant weight } W)} \tag{14.27}$$

When applying Eq. (14.27), the positive direction of y must be vertically upward.

2. Elastic Potential Energy According to Eq. (14.10), the work done by an ideal spring* when its free end moves from position ① to ② is (see Fig. 14.6)

$$U_{1-2} = -\frac{1}{2}k(\delta_2^2 - \delta_1^2) \qquad \text{(14.10, repeated)}$$

where δ is the elongation of the spring, measured from its free length L_0. The potential energy of the spring, also called *elastic potential energy,*† is obtained by comparing this equation with $U_{1-2} = -(V_2 - V_1)$. Then we see that the elastic potential energy is

$$V_e = \frac{1}{2}k\delta^2 + C$$

where C is an arbitrary constant. Letting the undeformed ($\delta = 0$) position be the datum for elastic potential energy, we have

$$\boxed{V_e = \frac{1}{2}k\delta^2} \tag{14.28}$$

Observe that the elastic potential energy is always positive.

3. Gravitational Potential Energy Equation (14.27) can be used to calculate the gravitational potential energy of a weight only if the variation of g is negligible. For motions that violate this restriction, the constant weight must be replaced by the force obtained from Newton's law of gravitation $F = Km_Am_B/R^2$. The work done by F was found to be

$$U_{1-2} = Km_Am_B\left(\frac{1}{R_2} - \frac{1}{R_1}\right) \qquad \text{(14.11, repeated)}$$

where m_A and m_B are the masses being attracted toward each other (see Fig. 14.7). Comparing this equation with $U_{1-2} = -(V_2 - V_1)$, we see that

* Observe that we consider the work done *by* the spring, not the work done *on* the spring. Potential energy of a spring refers to the ability of a spring to do work on the body to which it is attached, not to the effect that the body has on the spring.

† *Elasticity* refers to the ability of a deformed body to return to its undeformed shape when loads are removed.

the gravitational potential energy of the system consisting of masses m_A and m_B is

$$V_g = -\frac{K m_A m_B}{R} + C$$

It is customary to assume that the arbitrary constant C is zero, so that the gravitational potential energy of the system becomes

$$V_g = -\frac{K m_A m_B}{R} \tag{14.29}$$

Observe that the gravitational potential energy is always a negative quantity that approaches zero as R approaches infinity.

Sample Problem 14.4

The figure shows a 10-lb collar that slides along the smooth vertical rod under the actions of gravity and an ideal spring. The spring has a stiffness of 60 lb/ft and its free length is 2.5 ft. The collar is released from rest in position ①. Use the principle of conservation of mechanical energy to determine (1) the speed of the collar when it is in position ②; and (2) the distance h that locates position ③, the highest position reached by the collar.

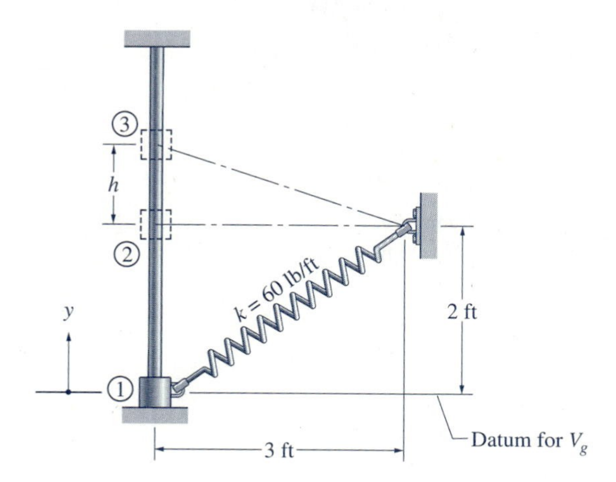

Solution

Preliminary Calculations

Without drawing the free-body diagram of the collar, we recognize that only the weight and the spring do work on the collar as it moves along the rod. (The normal force between the collar and the rod is a nonworking force.) Since both the weight and the spring force are conservative, we are justified in using the principle of conservation of mechanical energy. The total potential energy is equal to the sum of the gravitational (V_g) and elastic (V_e) potential energies.

As shown in the figure, we choose the datum position for V_g (where $V_g = 0$) to be ①, with the positive y-direction being upward. Consequently, Eq. (14.27) yields $V_g = Wy$. Next, we choose the datum for zero elastic potential energy of the spring to be its undeformed position. We then obtain from Eq. (14.28) $V_e = \frac{1}{2}k\,\delta^2$, where δ is the deformation of the spring. The kinetic energy of the collar v is $T = \frac{1}{2}mv^2$, as given by Eq. (14.13).

The following table shows the computation of potential and kinetic energies at positions ①, ②, and ③.

	Position ①	**Position ②**	**Position ③**
$V_g = Wy$	$10(0) = 0$	$10(2) = 20$ lb · ft	$10(h+2)$ lb · ft
$\delta = L - L_0$	$\sqrt{(2)^2 + (3)^2} - 2.5 = 1.1056$ ft	$3 - 2.5 = 0.5$ ft	$\sqrt{h^2 + (3)^2} - 2.5$ ft
$V_e = \frac{1}{2}k\delta^2$	$\frac{1}{2}(60)(1.1056)^2 = 36.67$ lb · ft	$\frac{1}{2}(60)(0.5)^2 = 7.50$ lb · ft	$\frac{1}{2}(60)\left(\sqrt{h^2+9} - 2.5\right)^2 =$ $30h^2 - 150\sqrt{h^2+9} + 457.50$
$T = \frac{1}{2}mv^2$	$0\ (v_1 = 0)$	$\frac{1}{2}\frac{10}{32.2}v_2^2 = 0.155\,28v_2^2$ lb · ft	$0\ (v_3 = 0)$

Part 1

Applying the principle of conservation of mechanical energy between positions ① and ②, we have

$$(V_g)_1 + (V_e)_1 + T_1 = (V_g)_2 + (V_e)_2 + T_2$$

Substituting the appropriate values from the above table yields

$$0 + 36.67 + 0 = 20 + 7.50 + 0.155\,28v_2^2$$

from which we obtain

$$v_2 = 7.68 \text{ ft/s} \qquad \textit{Answer}$$

Part 2

Applying the principle of conservation of mechanical energy between positions ① and ③ (② and ③ could also be used) gives

$$(V_g)_1 + (V_e)_1 + T_1 = (V_g)_3 + (V_e)_3 + T_3$$

On substitution for the appropriate values from the table, we obtain

$$0 + 36.67 + 0 = 10(h + 2) + \left(30h^2 - 150\sqrt{h^2 + 9} + 457.50\right) + 0$$

which, on simplification, becomes

$$30h^2 + 10h - 150\sqrt{h^2 + 9} + 440.83 = 0$$

Solving this equation by a numerical method (e.g., secant method), or by trial and error yields

$$h = 0.675 \text{ ft} \qquad \textit{Answer}$$

Since all the forces are conservative, the collar will theoretically continue to move between ① and ③ forever. In practice, friction (such as the friction force between the collar and the rod, or internal friction in the spring) will gradually dissipate mechanical energy in the form of heat, causing the collar to come to rest eventually.

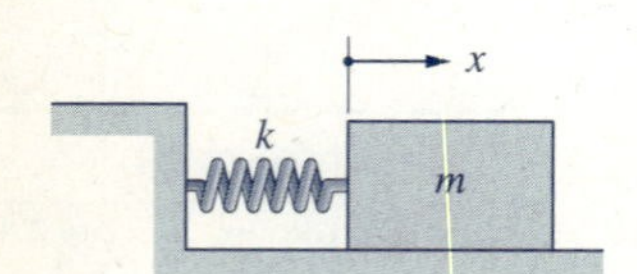

Sample Problem 14.5

The figure shows a block of mass m that slides along a smooth horizontal plane. The position coordinate x is measured from the undeformed position of the linear spring of stiffness k. Derive the equation of motion, $a = f(x)$, using the principle of conservation of mechanical energy. (This problem was solved by the force-mass-acceleration method in Sample Problem 12.6.)

Solution

Since only the conservative spring force does work as the block moves, we are justified in using the principle of conservation of mechanical energy.

The kinetic energy T of the block in terms of its speed v is

$$T = \frac{1}{2}mv^2 = \frac{1}{2}m\dot{x}^2 \tag{a}$$

Since the position coordinate x is measured from the undeformed position of the spring, it also corresponds to δ, the deformation of the spring. Therefore, using the undeformed position of the spring as the datum, the elastic potential energy is

$$V_e = \frac{1}{2}k\delta^2 = \frac{1}{2}kx^2 \tag{b}$$

Combining Eqs. (a) and (b), the total mechanical energy E is

$$E = T + V_e = \frac{1}{2}m\dot{x}^2 + \frac{1}{2}kx^2 \tag{c}$$

Since the total mechanical energy is conserved, we have from Eq. (14.25)

$$\frac{dE}{dt} = m\dot{x}\ddot{x} + kx\dot{x} = 0$$

or

$$\dot{x}(m\ddot{x} + kx) = 0 \tag{d}$$

Ignoring the static solution $\dot{x} = 0$, and recognizing that the acceleration is $a = \ddot{x}$, the equation of motion becomes

$$a = -\frac{k}{m}x \qquad \textit{Answer}$$

This agrees with the result obtained in Sample Problem 12.6 by the force-mass-acceleration method.

PROBLEMS

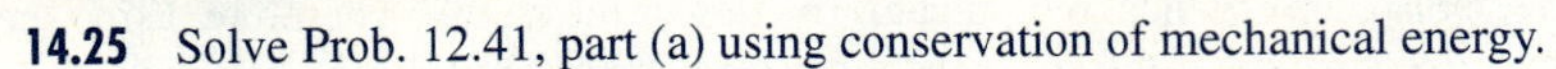

14.25 Solve Prob. 12.41, part (a) using conservation of mechanical energy.

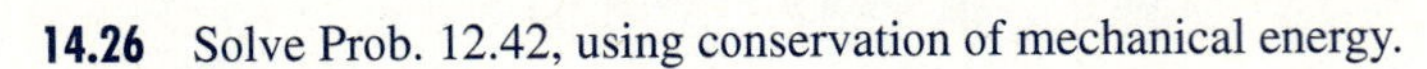

14.26 Solve Prob. 12.42, using conservation of mechanical energy.

14.27 Solve Prob. 12.46, using conservation of mechanical energy.

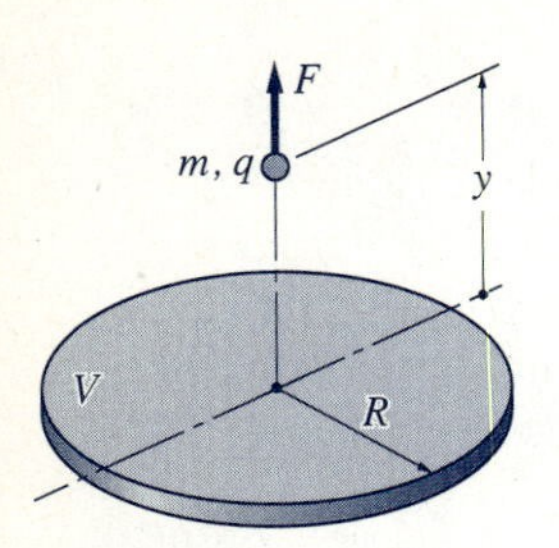

Fig. P14.28

14.28 The disk of radius R meters carries a charge with the electrostatic potential V volts. A particle of mass m kg and charge q coulombs lies on the axis of the disk at a distance y meters from the disk. The electrostatic force F acting on the particle is given in Prob. 12.44. (a) Show that the potential energy of the particle is

$$-\frac{Vq}{R}\left(y - \sqrt{R^2 + y^2}\right)$$

(b) Use the principle of conservation of mechanical energy to find the speed of the particle at $y = R$ after the particle is released from rest at $y = 0$. [Note: Part (b) was solved by the force-mass-acceleration method in Prob. 12.44.]

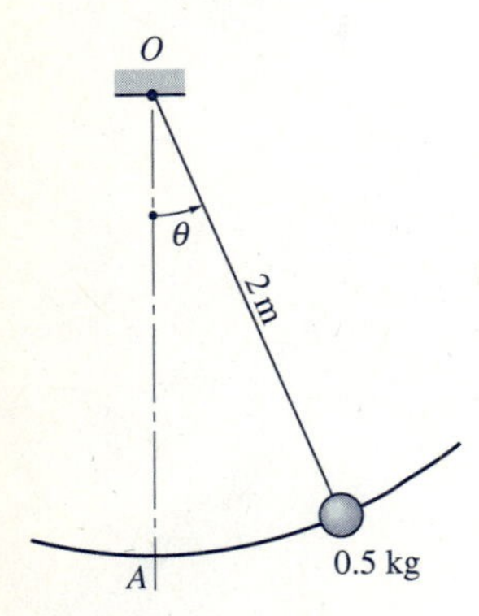

Fig. P14.29

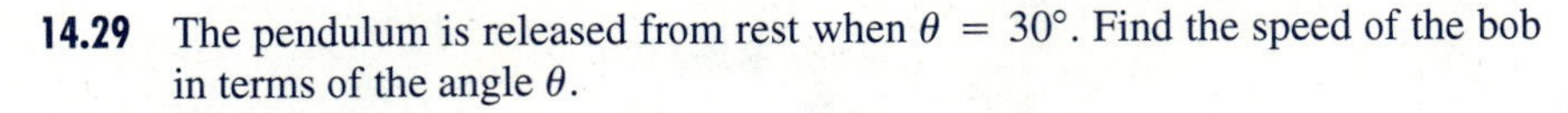

14.29 The pendulum is released from rest when $\theta = 30°$. Find the speed of the bob in terms of the angle θ.

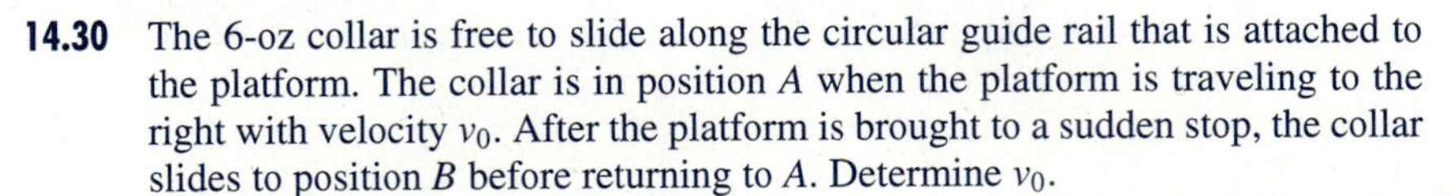

14.30 The 6-oz collar is free to slide along the circular guide rail that is attached to the platform. The collar is in position A when the platform is traveling to the right with velocity v_0. After the platform is brought to a sudden stop, the collar slides to position B before returning to A. Determine v_0.

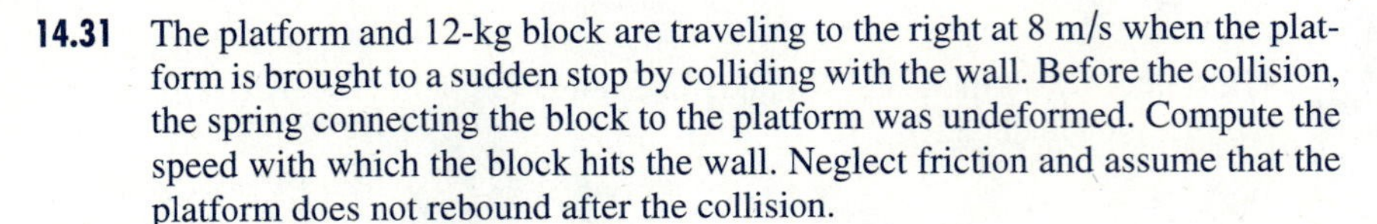

14.31 The platform and 12-kg block are traveling to the right at 8 m/s when the platform is brought to a sudden stop by colliding with the wall. Before the collision, the spring connecting the block to the platform was undeformed. Compute the speed with which the block hits the wall. Neglect friction and assume that the platform does not rebound after the collision.

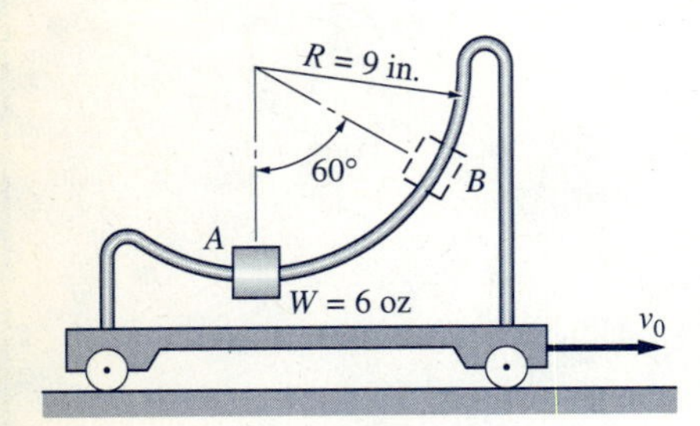

Fig. P14.30

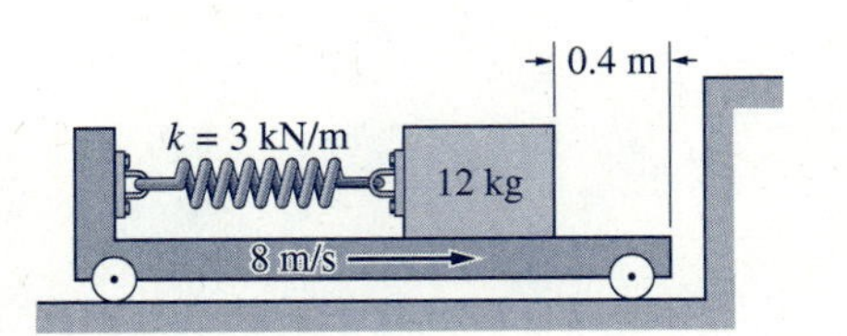

Fig. P14.31

Fig. P14.32

14.32 The 2400-lb car is descending the 10° incline at 60 mi/h when the driver applies the brakes. What constant braking force F will bring the car to a stop in 180 ft?

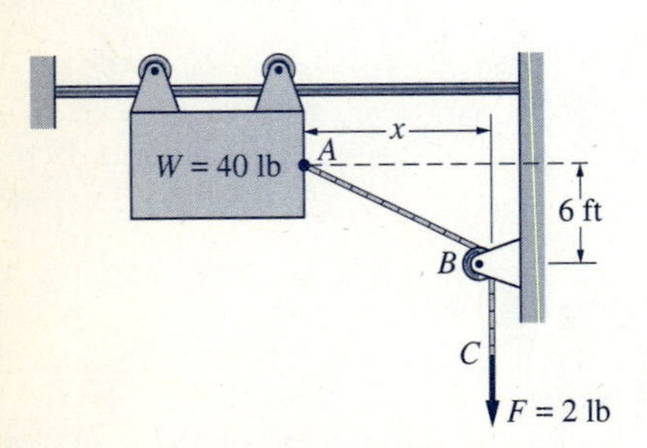

Fig. P14.33, P14.34

14.33 The 40-lb sliding door is suspended from frictionless rollers that run on a horizontal rail. The door is closed by the 2-lb constant force applied to the end of rope ABC. If the door starts from rest at $x = 8$ ft, calculate its speed when $x = 0$.

14.34 Determine the speed of the door in Prob. 14.33 as a function of the distance x.

14.35 The rope ABC passes over a smooth peg B that is attached to the 80-lb block. A constant vertical force of 20 lb is applied at the end of the rope. If the velocity of the block is 5 ft/s to the right when $x = 6$ ft, calculate (a) the maximum value of x; and (b) the speed of the block when it hits the wall at A.

14.36 Derive the equation of motion $a = f(x)$ for the block described in Prob. 14.35.

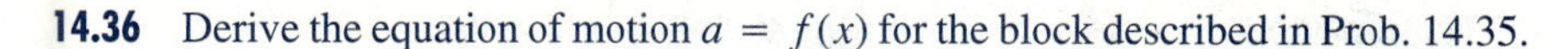

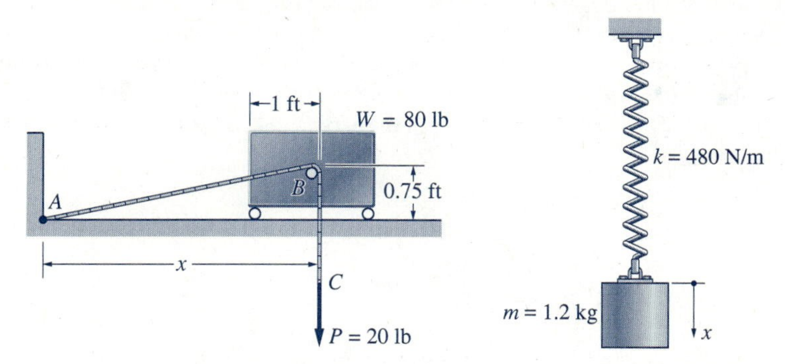

Fig. P14.35, P14.36 **Fig. P14.37, P14.38**

14.37 The 1.2-kg mass is released from rest when the spring is undeformed. Determine (a) the maximum displacement of the mass; and (b) the maximum speed of the mass and the corresponding displacement.

14.38 Derive the equation of motion $a = f(x)$ for the mass-spring system described in Prob. 14.37.

14.39 The spring has a free length of 150 mm and a stiffness of 220 N/m. The 0.2-kg collar attached to the spring slides on a smooth horizontal rod. If the mass is released from rest when $\theta = 45°$, find its speed when (a) $\theta = 0$; and (b) $\theta = 30°$.

14.40 Derive the equation of motion $a = f(x)$ for the collar described in Prob. 14.39.

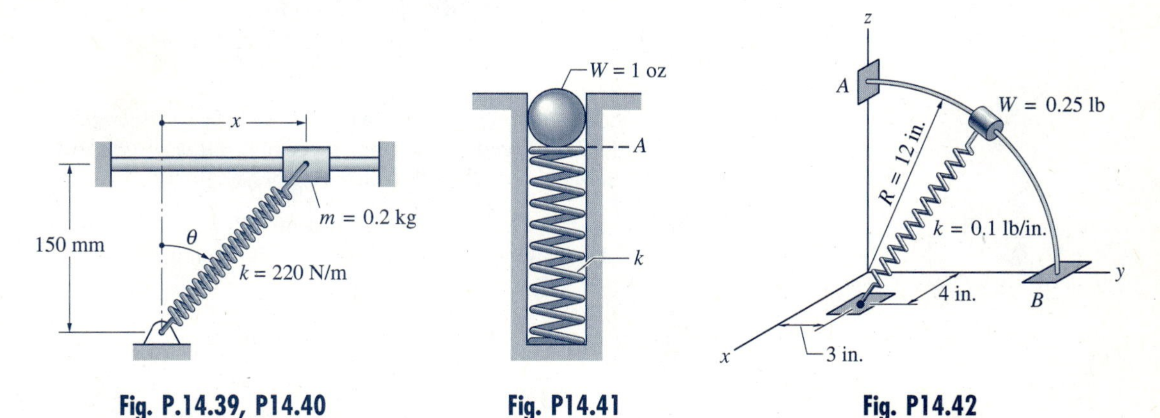

Fig. P.14.39, P14.40 **Fig. P14.41** **Fig. P14.42**

14.41 The 1.0-oz ball is fired by compressing the spring 6 in. from its undeformed position A and then releasing it. If the ball reaches a maximum height of 47.5 ft above A, determine the spring stiffness k.

14.42 The 0.25-lb collar slides on the smooth circular rod AB, which lies in a vertical plane. The spring attached to the collar has a free length of 9 in., and its stiffness is 0.1 lb/in. If the collar is released from rest at A, determine the speed with which it will arrive at B.

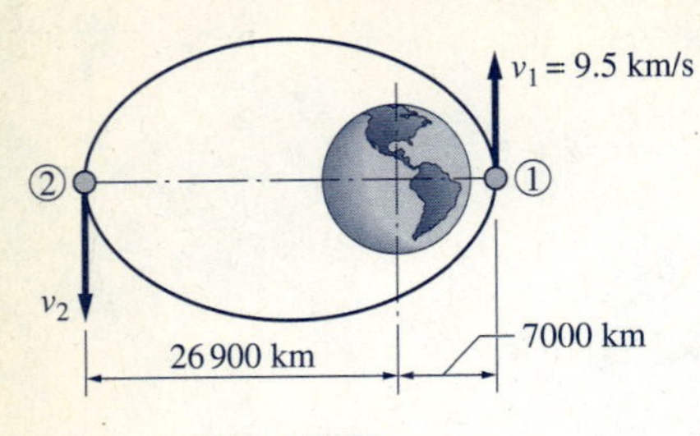

Fig. P14.43

14.43 A satellite is in an elliptical orbit around the earth. If the speed of the satellite at the perigee (position ①) is 9.5 km/s, calculate its speed at the apogee (position ②). Use $KM_e = 3.98 \times 10^{14}$ m^3/s^2, where M_e is the mass of the earth and K is the universal gravitational constant.

14.44 The firing of retrorockets causes the earth satellite to slow down to 3 km/s when it reaches its perigee A. As a result, the satellite leaves its orbit and descends toward the earth. The satellite burns up when it enters the earth's atmosphere at B, 100 km above the earth's surface. If the speed of the satellite was observed to be 7.70 km/s at B, determine the height h of the perigee. Use $KM_e = 3.98 \times 10^{14}$ m^3/s^2, where M_e is the mass of the earth and K is the universal gravitational constant.

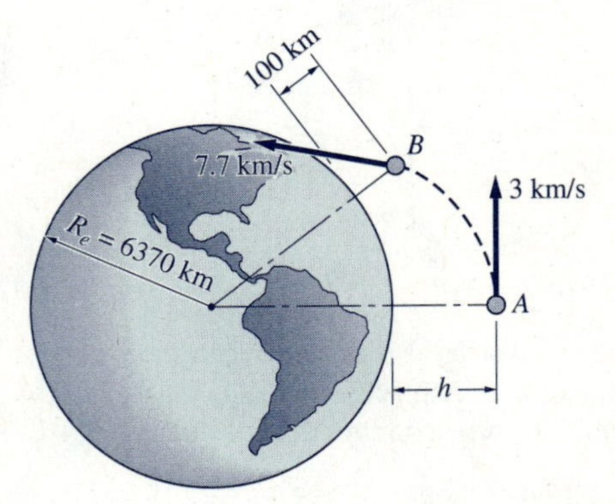

Fig. P14.44

14.45 Compute the energy required to launch a 500-kg communications satellite into geosynchronous orbit around the earth (a circular, equatorial orbit with a period of 23 hr 56 min). The satellite is to be launched in the direction of the earth's rotation from the equator. The geosynchronous orbit requires an altitude of 35 800 km from the surface of the earth and a velocity of 3070 m/s. Use $KM_e = 3.98 \times 10^{14}$ m^3/s^2 and $R_e = 6370$ km, where M_e and R_e are the mass and radius of the earth, respectively, and K is the universal gravitational constant.

14.46 The bob of the pendulum is to travel a complete circle about the pivot point O. What is the smallest speed of the bob in position A if it is suspended from (a) a rigid, weightless rod; and (b) a flexible string?

14.47 The restraining cable BC is suddenly released, allowing the 360-lb wrecking ball to swing freely. Find the tension in cable AB when it is vertical.

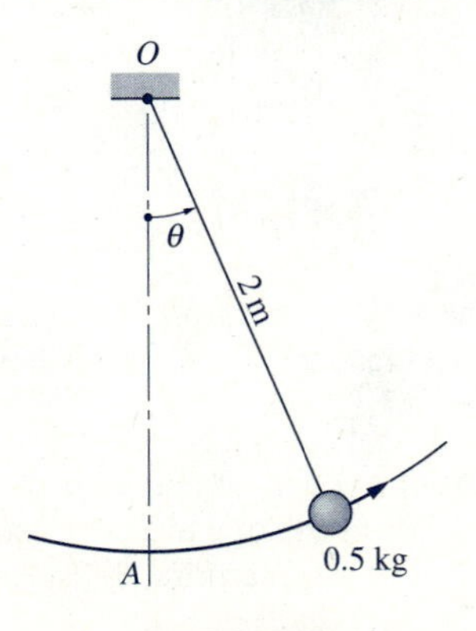

Fig. P14.46

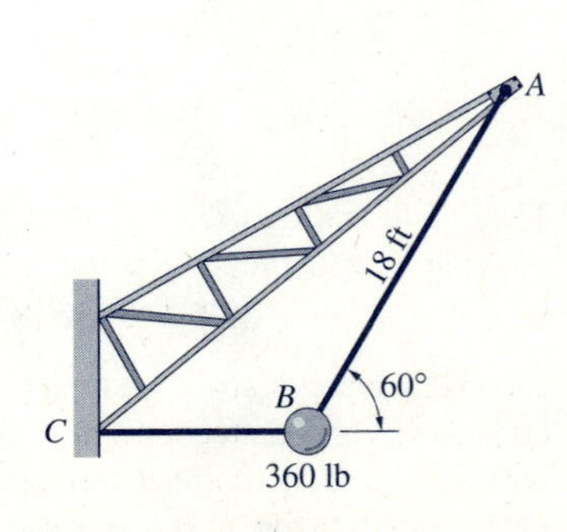

Fig. P14.47

14.48 A 75-kg ski jumper starts his run at the top of a parabolic track. Assuming negligible friction, calculate the maximum contact force between the track and the skis during the downhill run.

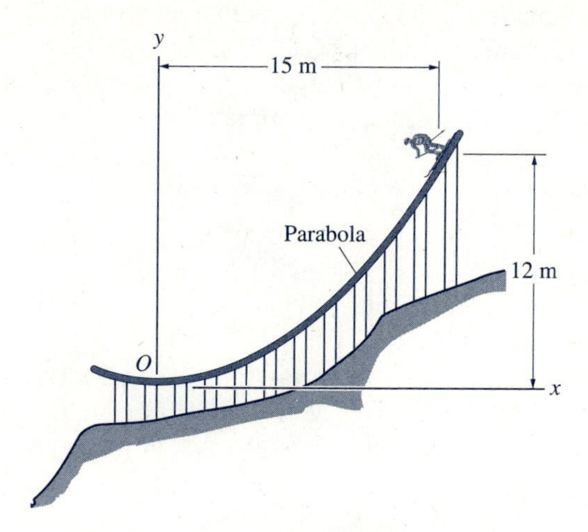

Fig. P14.48

14.49 The 1.0-kg particle is released from rest at $\theta = 15°$ and slides down the smooth, cylindrical surface of radius 2 m. Find the angle θ at which the particle leaves the surface.

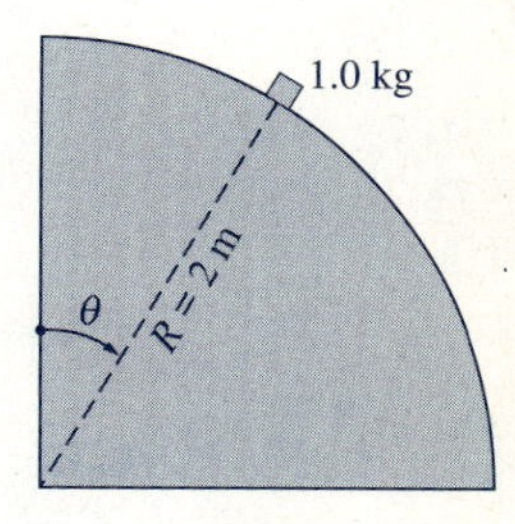

Fig. P14.49

14.50 The 2.4-lb follower is attached to the end of a light telescopic rod that is pivoted at O. The follower is pressed against a smooth spiral surface by a spring of stiffness $k = 8$ lb/ft and free length $L_0 = 3$ ft. The equation of the spiral, which lies in the horizontal plane, is $R = b\theta/(2\pi)$, where $b = 1.2$ ft and θ is in radians. If the rod is released from rest in position OA, find its angular speed in position OB.

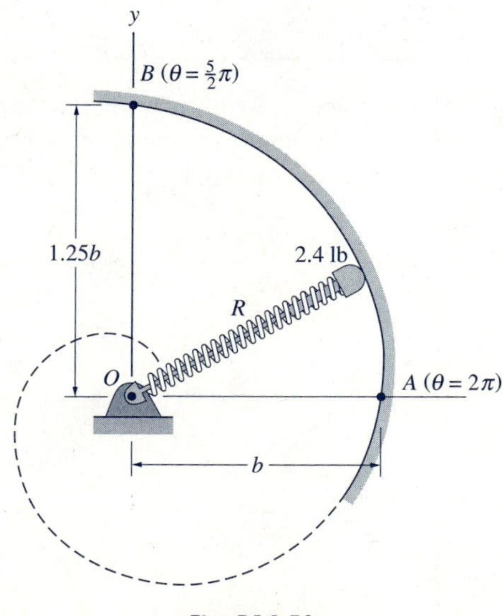

Fig. P14.50

14.5 Power and Efficiency

From a dynamics viewpoint, a machine is a device that does work. It is not meaningful to discuss the capacity of a machine in terms of the work that can be done by the machine. Both large and small machines are capable of doing the same amount of work; the difference is that a large machine will do the work in less time. Therefore, machines are rated in terms of power, defined as the *rate at which work is done.*

Letting P be the power and U the work, we have

$$P = \frac{dU}{dt} \tag{14.30}$$

Power is a scalar quantity, its units being (units of work)/(units of time). In SI units, power is measured in watts (1 W = 1 J/s = 1 N · m/s). In U.S. Customary units, power is measured in lb · in./s, lb · ft/s, or horsepower (1 hp = 550 lb · ft/s).

Using $dU = \mathbf{F} \cdot d\mathbf{r}$ from Eq. (14.1), we obtain

$$P = \frac{\mathbf{F} \cdot d\mathbf{r}}{dt} = \mathbf{F} \cdot \mathbf{v} \tag{14.31}$$

Therefore the power of a force is equal to the dot product of the force and the velocity of its point of application.

The output power of a machine is the rate at which the machine is performing work; the input power is the rate at which energy is being supplied to the machine. The ratio of these two quantities (usually multiplied by 100 to express it as a percentage) is called the *efficiency* of the machine; that is,

$$\eta = \frac{\text{Output power}}{\text{Input power}} \times 100\% \tag{14.32}$$

In a real machine the output power is always less than the input power because of unavoidable loss of mechanical energy caused by friction, vibration, etc. Therefore the efficiency of a machine is always less than 100%.

Sample Problem 14.6

A 3000-lb automobile accelerates under constant power from 60 mi/h to 90 mi/h on a $\frac{1}{4}$-mi straight, level test track. (1) Determine the horsepower that is delivered to the drive wheels. (2) What is the power output of the engine if the efficiency of the drivetrain is 82%?

Solution

Part 1

Let F be the driving force, i.e., the force between the driving wheels and the pavement that produces the acceleration, as shown in the figure. Since F and the velocity v of the automobile are in the same direction, the power that must be supplied to the drive wheels becomes, according to Eq. (14.31), $P = Fv$. From Newton's second law, we have $F = ma = m(dv/dt) = m(dv/ds)(ds/dt) = mv\,dv/ds$. Therefore, the power becomes

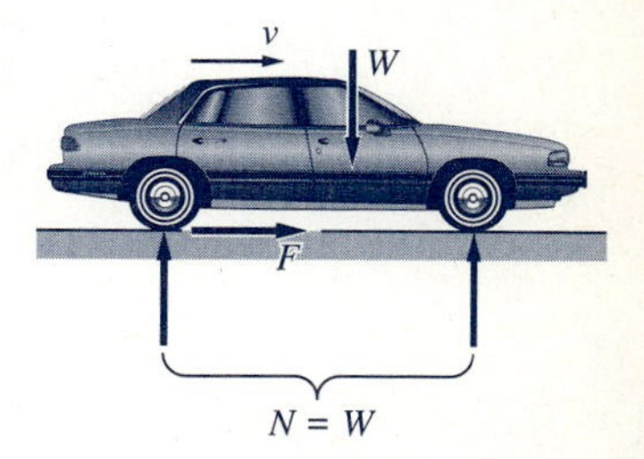

$$P = Fv = mv^2\frac{dv}{ds}$$

where m is the mass of the automobile and s is the distance measured along the track. Multiplying both sides of this equation by ds and integrating (note that P and m are constants) yields

$$P\int ds = m\int v^2\,dv$$

$$Ps = \frac{mv^3}{3} + C$$

where C is the constant of integration. Substituting $m = W/g = 3000/32.2$ slugs, we have

$$Ps = 31.06v^3 + C \tag{a}$$

Letting $s = 0$ correspond to the position at which $v = 88$ ft/s (60 mi/h), Eq. (a) yields

$$0 = 31.06(88)^3 + C$$

from which we obtain

$$C = -21.17 \times 10^6 \text{ lb}\cdot\text{ft}^2/\text{s} \tag{b}$$

Substituting $s = 1320$ ft (1/4 mi) and $v = 132$ ft/s (90 mi/h) together with the value of C from Eq. (b) into Eq. (a) gives

$$P(1320) = 31.06(132)^3 - (21.17 \times 10^6)$$

$$P = 38\,080 \text{ lb}\cdot\text{ft/s}$$

Therefore, the horsepower that is supplied to the drive wheels is

$$P = 38\,080 \text{ lb}\cdot\text{ft/s} \times \frac{1 \text{ hp}}{550 \text{ lb}\cdot\text{ft/s}} = 69.2 \text{ hp} \qquad \textit{Answer}$$

Part 2

Since the efficiency of the drivetrain is 82%, the input power to the drivetrain (i.e., the output power of the engine) is

$$P_{in} = \frac{P}{\eta} = \frac{69.2}{0.82} = 84.4 \text{ hp} \qquad \textit{Answer}$$

Sample Problem 14.7

A 1800-kg delivery van is traveling at the speed $v_0 = 26$ m/s when the driver applies the brakes, causing all four wheels to lock. The friction force between the tires and the road causes the van to skid to a stop. Calculate the power of the friction force as a function of time if the coefficient of kinetic friction is 0.6.

Solution

As shown in the figure, the free-body diagram of the skidding van contains the following forces: the weight mg of the van, the resultant normal reaction N between the road and the four wheels, and the kinetic friction force $F = 0.6N$. Summing forces in the y-direction yields

$$\Sigma F_y = 0 \qquad +\uparrow \qquad N - mg = 0$$

$$N = mg = 1800(9.81) = 17.658 \text{ kN} \tag{a}$$

For the x-direction, we find

$$\Sigma F_x = ma_x \qquad \xrightarrow{+} \qquad F = ma$$

$$0.6N = 0.6(mg) = ma$$

from which the acceleration of the van is found to be

$$a = 0.6g = 0.6(9.81) = 5.886 \text{ m/s}^2 \rightarrow \tag{b}$$

Note that the acceleration does not depend on time.

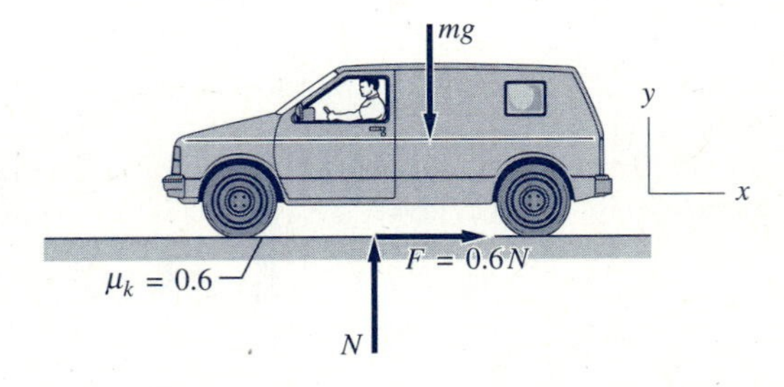

The velocity of the van is found from $a = dv/dt$ as follows.

$$v = \int a\,dt = \int 5.886\,dt = 5.886t + C \text{ m/s}$$

where C is the constant of integration. For the units to be consistent, t must be in seconds. Applying the condition $v = -26$ m/s when $t = 0$ gives $C = -26$ m/s. Therefore, the velocity is

$$v = 5.886t - 26 \text{ m/s} \rightarrow \tag{c}$$

Note that the van comes to a stop when $v = 0$, i.e., when $t = 26/5.886 = 4.42$ s.

The power P of the friction force can be found from Eq. (14.31):

$$P = Fv \tag{d}$$

Substituting $F = 0.6N = 0.6(17.658) = 10.595$ kN and v from Eq. (c) into Eq. (d), we find that the power of the friction force is

$$P = Fv = 10.595(5.886t - 26) = 62.4t - 275 \text{ kN} \cdot \text{m/s}$$

or

$$P = 62.4t - 275 \text{ kW} \qquad (0 \le t \le 4.42 \text{ s}) \qquad \textit{Answer}$$

Note that P is the power supplied to the car by the friction force. Since the velocity of the car and the friction force have opposite directions, the value of P is negative during the period of braking.

PROBLEMS

14.51 The block of mass m is at rest at time $t = 0$ when the force F is applied. (a) If F is constant, determine its power as a function of t. (b) If the power of F is constant, determine F as a function of t.

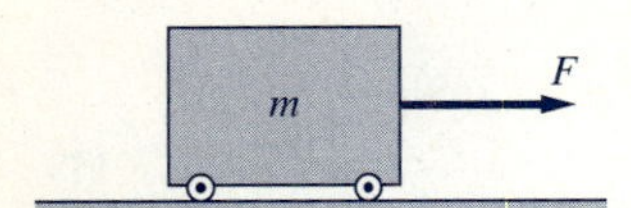

Fig. P14.51

14.52 The mass of the block in Prob. 14.51 is $m = 12$ kg. The block is initially at rest when the force $F(t) = 60 \cos \pi t$ is applied for 0.5 second, where F is in newtons and t is in seconds. Sketch the plot of the power supplied to the block versus time. What is the maximum power, and at what time does it occur?

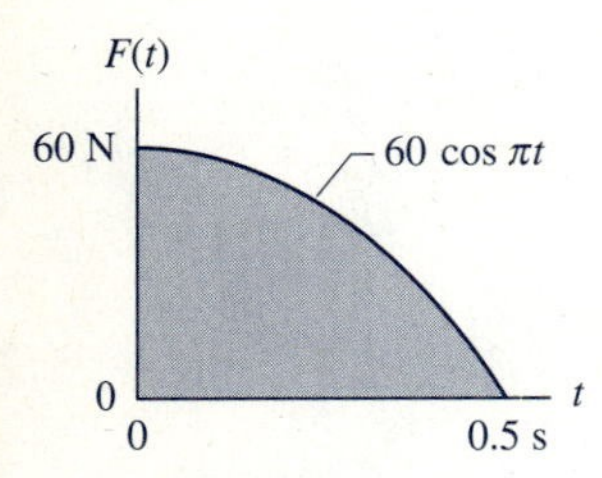

Fig. P14.52

14.53 An electric hoist lifts a 500-kg mass at a constant speed of 0.76 m/s while consuming 4.5 kW of power. (a) Compute the efficiency of the hoist. (b) At what constant speed can the hoist lift a 750-kg mass with the same power consumption as in part (a)?

14.54 An electric hoist consumes 5 kW of power at a constant rate to lift the 500-kg mass. Assuming that the hoist is 80% efficient, determine (a) the maximum speed reached by the mass; and (b) the acceleration of the mass when the speed has reached one-half of its maximum value.

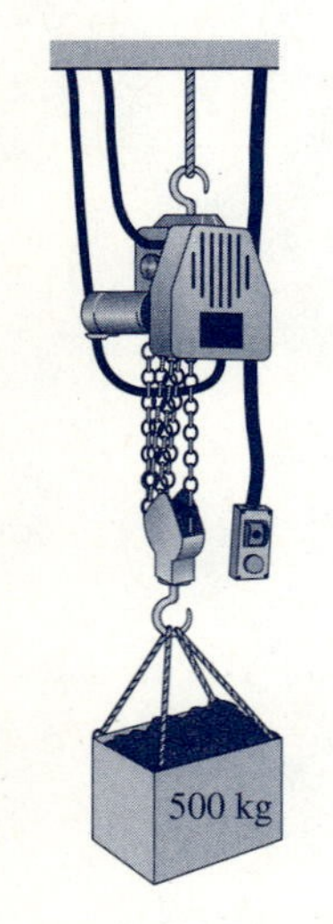

Fig. P14.53, P14.54

14.55 A 165-lb man pedals a 35-lb bicycle up a 5° incline at 6 mi/h. If the bicycle is 95% efficient, calculate the man's horsepower.

14.56 The hydrodynamic force that opposes the motion of a boat is proportional to the square of the speed of the boat. If 18 hp are required to move the boat at a constant speed of 15 ft/s, what is the power requirement for a constant speed of 20 ft/s?

14.57 A compact car travels 32 miles per 1 U.S. gallon of gasoline when cruising at a constant speed of 55 mi/h on a level road. Assuming that the combined efficiency of the engine and drivetrain is 48%, determine the resisting force (resulting from aerodynamic and rolling resistance) that opposes the motion. The heat value of a U.S. gallon of gasoline is 140×10^3 BTU, where 1 BTU $= 778$ lb · ft.

14.58 The aerodynamic force that opposes the motion of a car is proportional to $v^{2.5}$, where v is the car's speed. The force caused by rolling resistance is a constant. If 19.3 hp is required to move the car at a constant speed of 50 mi/h and 32.3 hp at 60 mi/h, what is the horsepower required for the car to travel at 70 mi/h?

14.59 The 4-kg block accelerates from rest as a result of the action of the force $F(t) = 12[1 - (t/2)]$, where F is in newtons and t is in seconds. The duration of the force is 2 seconds. Calculate (a) the power of F as a function of t; and (b) the maximum power and the time when it occurs.

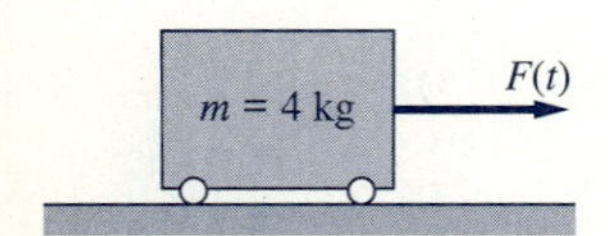

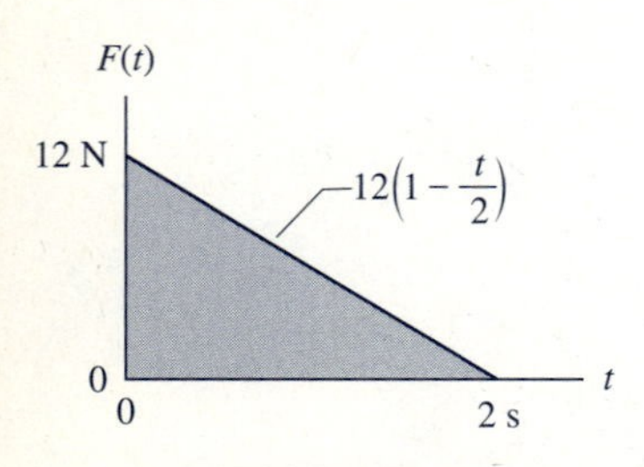

Fig. P14.59

14.60 The 6-lb cart is given an initial displacement of $x = 6$ in. and then released from rest. The position coordinate x is measured from the undeformed position of the spring. Determine (a) the power of the spring force as a function of x; and (b) the maximum power of the spring force and the value of x where it occurs.

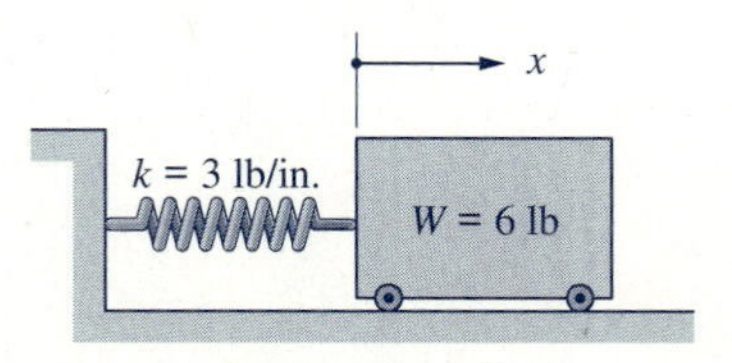

Fig. P14.60

14.61 The diagram shows a typical P-v relationship for a gasoline-powered car that is accelerating in first gear, where P is the power of the drive force F and v is the speed of the car. Determine the maximum value of F and the speed where it occurs. (Hint: Note that the line from the origin to any point on the curve has the slope $F = P/v$.)

14.62 If the mass of the car in Prob. 14.61 is 1500 kg, compute the time that it takes for the car to accelerate from 2 m/s to 5 m/s.

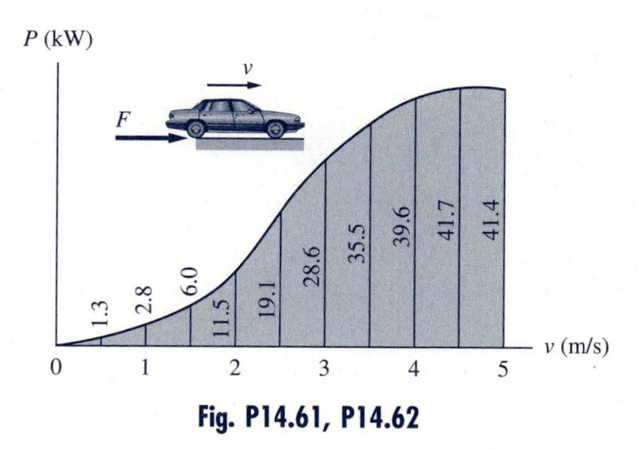

Fig. P14.61, P14.62

14.6 Principle of Impulse and Momentum

The method of analysis introduced in this article is based on the principle of impulse and momentum. Like the work-energy principle, the impulse-momentum principle is derivable from Newton's second law of motion. We begin with the definitions of the impulse of a force and the momentum of a particle.*

a. Impulse of a force

The *impulse* $\mathbf{L}_{1\text{-}2}$ of a force $\mathbf{F}$ for the time interval t_1 to t_2 is defined as

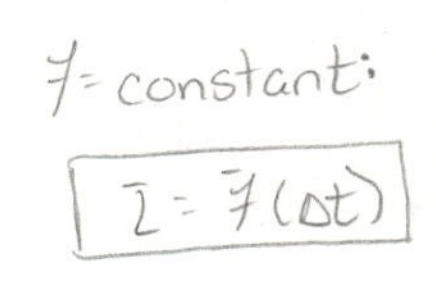

$$\mathbf{L}_{1\text{-}2} = \int_{t_1}^{t_2} \mathbf{F}\, dt \tag{14.33}$$

Impulse is a vector quantity, its rectangular components being

$$(L_{1\text{-}2})_x = \int_{t_1}^{t_2} F_x\, dt \qquad (L_{1\text{-}2})_y = \int_{t_1}^{t_2} F_y\, dt \qquad (L_{1\text{-}2})_z = \int_{t_1}^{t_2} F_z\, dt \tag{14.34}$$

where F_x, F_y, and F_z are the components of $\mathbf{F}$. As shown in Fig. 14.9, $(L_{1\text{-}2})_x$ is equal to the area under the F_x-t diagram between t_1 and t_2. Similarly, $(L_{1\text{-}2})_y$ and $(L_{1\text{-}2})_z$ are the areas under the F_y-t and F_z-t diagrams, respectively. This

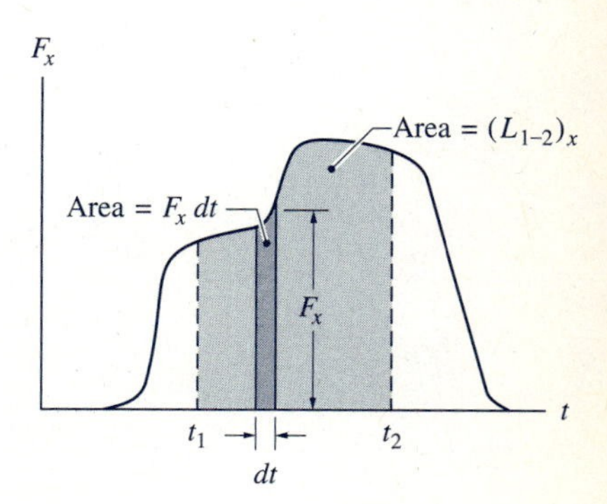

Fig. 14.9

* They are more accurately called *linear* impulse and *linear* momentum, to distinguish them from *angular* impulse and *angular* momentum, which are discussed in the next article. However, it is common to omit "linear" when the meaning is clear from the context.

knowledge is very useful when computing the impulse of a force when its time dependence is given in graphical or numerical form. The units of impulse are (units of force) × (units of time), e.g., N · s, lb · s.

In the work-energy methods described in the preceding articles, only forces that did work entered the analysis. Consequently, the free-body diagram could be replaced by an active-force diagram. However, free-body diagrams must always be used when calculating impulses, because a force has an impulse even if it does no work.

An important special case arises if the force **F** is *constant* in magnitude and direction. The impulse of the force then reduces to

$$\mathbf{L}_{1\text{-}2} = \mathbf{F}\int_{t_1}^{t_2} dt = \mathbf{F}(t_2 - t_1) = \mathbf{F}\,\Delta t \qquad (\mathbf{F}\text{ constant}) \tag{14.35}$$

The impulse of a constant force is thus equal to the product of the force and the time interval, the impulse being in the same direction as the force.

b. Momentum of a particle and momentum diagrams

The momentum **p** of a particle of mass m at an instant of time is defined as

$$\mathbf{p} = m\mathbf{v} \tag{14.36}$$

where **v** is the velocity vector of the particle at that instant. The momentum of a particle is a vector quantity that acts in the same direction as the velocity vector. The unit of momentum is (unit of mass) × (unit of velocity), e.g., kg · m/s, slug · ft/s (or equivalently, N · s, lb · s, etc.). Therefore, the unit of momentum is the same as the unit of impulse.

The *momentum diagram* for a particle is a sketch of the particle showing its momentum vector $m\mathbf{v}$. Momentum diagrams are useful tools in the analysis of problems using the principle of impulse and momentum.

c. Force-momentum relationship

If the mass of the particle is constant, Newton's second law states that $\mathbf{F} = m\mathbf{a}$. If the mass varies with time, this law takes the form

$$\mathbf{F} = \frac{d}{dt}(m\mathbf{v})$$

Substituting $\mathbf{p} = m\mathbf{v}$ yields

$$\mathbf{F} = \frac{d\mathbf{p}}{dt} \tag{14.37}$$

Equation (14.37) is, in fact, the general form of Newton's second law: The resultant force is equal to the rate of change of the momentum. Therefore, $\mathbf{F} = m\mathbf{a}$ should be considered as a special case that is valid for constant mass only. Equation (14.37) is particularly useful when analyzing the motion of rockets, the mass of which varies because fuel is being burned.

d. Impulse-momentum principle

Multiplying both sides of Eq. (14.37) by dt and integrating from time t_1 to t_2, we obtain

$$\int_{t_1}^{t_2} \mathbf{F}\,dt = \int_{\mathbf{p}_1}^{\mathbf{p}_2} d\mathbf{p} = \mathbf{p}_2 - \mathbf{p}_1 \tag{14.38}$$

where $\mathbf{p}_1$ and $\mathbf{p}_2$ represent the momenta at t_1 and t_2, respectively. Since the left side of this equation is the impulse of $\mathbf{F}$ over the time interval t_1 to t_2, we have

$$\boxed{\mathbf{L}_{1\text{-}2} = \mathbf{p}_2 - \mathbf{p}_1 = \Delta\mathbf{p}} \tag{14.39}$$

Equation (14.39) is called the *impulse-momentum principle,* or the *balance of impulse and momentum.* When this principle is applied to the analysis of motion, the technique is called the *impulse-momentum method.*

If a rectangular coordinate system is used, Eq. (14.39) is equivalent to the three following scalar equations:

$$\begin{aligned} (L_{1\text{-}2})_x &= \Delta p_x = (mv_x)_2 - (mv_x)_1 \\ (L_{1\text{-}2})_y &= \Delta p_y = (mv_y)_2 - (mv_y)_1 \\ (L_{1\text{-}2})_z &= \Delta p_z = (mv_z)_2 - (mv_z)_1 \end{aligned} \tag{14.40}$$

You must be careful not to confuse the work-energy principle, $U_{1\text{-}2} = T_2 - T_1$, with the impulse-momentum principle, $\mathbf{L}_{1\text{-}2} = \mathbf{p}_2 - \mathbf{p}_1$. Some of the important differences between the two methods are:

Work ($U_{1\text{-}2}$) is a scalar quantity associated with a force and a change in position.

Impulse ($\mathbf{L}_{1\text{-}2}$) is a vector quantity associated with a force and a time interval.

Kinetic energy (T) is a scalar quantity associated with a mass and its speed at an instant of time.

Momentum ($\mathbf{p}$) is a vector quantity associated with a mass and its velocity vector at an instant of time.

Most importantly, note that the work-energy principle, $U_{1\text{-}2} = \Delta T$, is a *scalar* relationship, whereas the impulse-momentum principle, $\mathbf{L}_{1\text{-}2} = \Delta\mathbf{p}$, is a *vector* relationship.

e. Conservation of momentum

From the impulse-momentum principle, Eq. (14.39), it can be seen that if the impulse acting on a particle is zero during a given time interval, the momentum of the particle will be conserved during that interval. In other words, if $\mathbf{L}_{1\text{-}2} = \mathbf{0}$, then

$$\boxed{\mathbf{p}_1 = \mathbf{p}_2 \quad \text{or} \quad \Delta\mathbf{p} = \mathbf{0}} \tag{14.41}$$

Equation (14.41) is called the *principle of conservation of momentum.* Observe that this principle is valid only if the impulse of the resultant force acting on the particle is zero. If there is no resultant force acting on a particle, the resultant impulse will obviously be zero, and momentum will be conserved. However, it is possible for the impulse of a force, i.e., its time integral, to be zero even if the force is not zero.

Since momentum is a vector quantity, it is possible for one or two of its components to be conserved, even though the total momentum itself is not conserved. For example, we see from Eq. (14.40) that the momentum in the x-direction will be conserved if $(L_{1-2})_x = 0$, regardless of the values of $(L_{1-2})_y$ and $(L_{1-2})_z$.

The conservation of momentum principle is very useful in the analysis of impact and other interactions between particles.

Sample Problem 14.8

At time $t = 0$, the velocity of the 0.5-kg particle in Fig. (a) is 10 m/s to the right. In addition to its weight (the xy-plane is vertical), the particle is acted on by the force $\mathbf{P}(t)$. The direction of $\mathbf{P}(t)$ is constant throughout the motion, but its magnitude varies with time in the manner shown in Fig. (b). Calculate the velocity of the particle when $t = 4$ s.

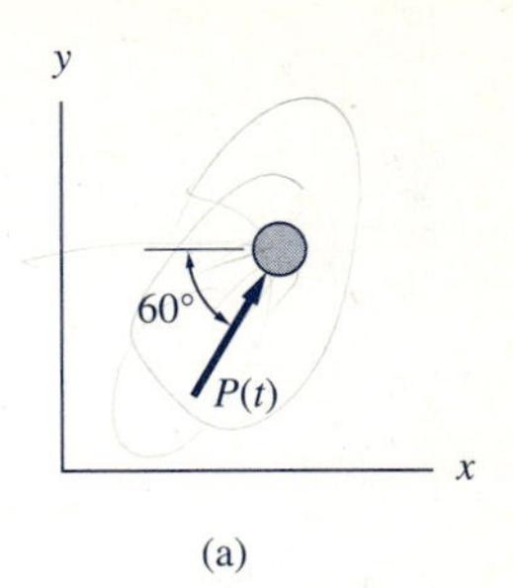

(a)

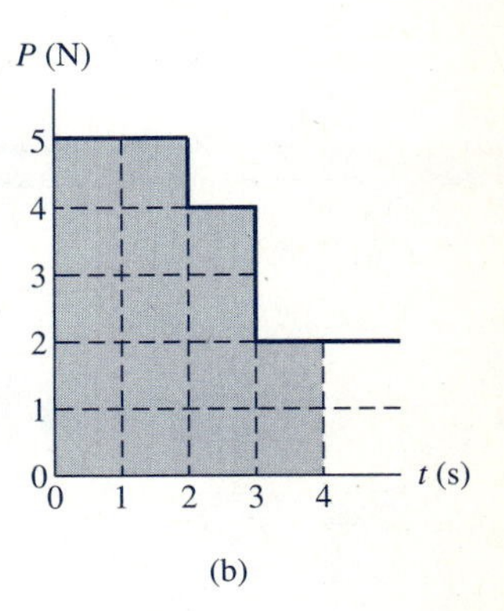

(b)

Solution

The impulse-momentum method is the most direct means of solving this problem because (1) the impulse-momentum method deals directly with the change in momentum (velocity) during a time interval, and (2) the impulses of the forces can be easily computed. Although the force-mass-acceleration method could be used, the solution would be much more cumbersome. Due to the discontinuous nature of $P(t)$, the acceleration would be different for the time intervals 0–2 s, 2–3 s, and 3–4 s, leading to three separate equations of motion that would have to be integrated.

The diagrams required to solve the problem by the impulse-momentum method are shown in Figs. (c)–(e), where the subscripts 1 and 2 correspond to $t = 0$ and 4 s, respectively. The FBD of the particle at an arbitrary time t is shown in Fig. (c). Figure (d) shows the momentum diagram when $t = 0$, where $\mathbf{p}_1 = m\mathbf{v}_1 = 0.5(10)\mathbf{i} = 5\mathbf{i}$ N · s. The momentum diagram when $t = 4$ s is shown in Fig. (e), where both components of $m\mathbf{v}_2$ were drawn in the positive coordinate directions.

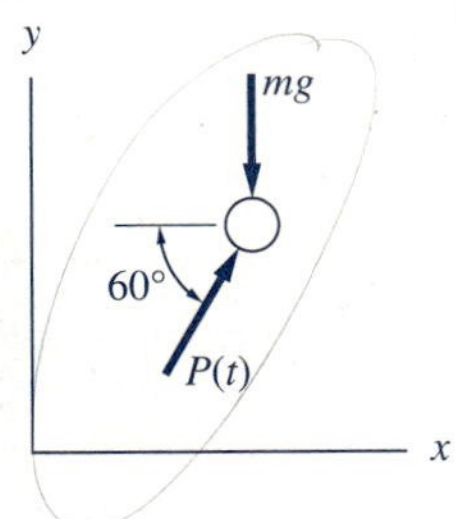

(c) FBD at time t

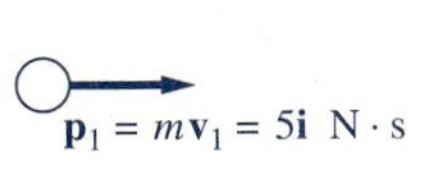

(d) Momentum diagram ($t = 0$)

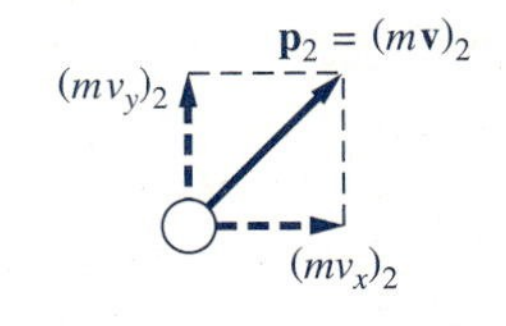

(e) Momentum diagram ($t = 4$ s)

Since the weight is a constant force, the components of its impulse are easily computed from Eq. (14.35):

$$(L_{1\text{-}2})_x = 0$$
$$(L_{1\text{-}2})_y = -mg\Delta t = -0.5(9.81)(4) = -19.620 \text{ N} \cdot \text{s} \qquad \text{(a)}$$

The area under the P-t diagram in Fig. (b) between 0 and 4 s is $5(2)+4(1)+2(1) = 16$ N · s. Because the direction of $\mathbf{P}(t)$ is constant, the components of its impulse are

$$(L_{1\text{-}2})_x = 16(\cos 60°) = 8.0 \text{ N} \cdot \text{s}$$
$$(L_{1\text{-}2})_y = 16(\sin 60°) = 13.856 \text{ N} \cdot \text{s} \qquad \text{(b)}$$

Substituting Eqs. (a) and (b) in the impulse-momentum principle, Eq. (14.40), yields for the x-direction

$$(L_{1\text{-}2})_x = (mv_x)_2 - (mv_x)_1 \qquad \xrightarrow{+} \quad 8.0 = 0.5(v_x)_2 - 5$$

from which we obtain

$$(v_x)_2 = 26.00 \text{ m/s} \qquad \text{(c)}$$

For the y-direction, we get

$$(L_{1\text{-}2})_y = (mv_y)_2 - (mv_y)_1 \qquad +\uparrow \quad 13.856 - 19.620 = 0.5(v_y)_2 - 0$$

which gives us

$$(v_y)_2 = -11.53 \text{ m/s} \tag{d}$$

The corresponding velocity vector at time $t = 4$ s is shown below.

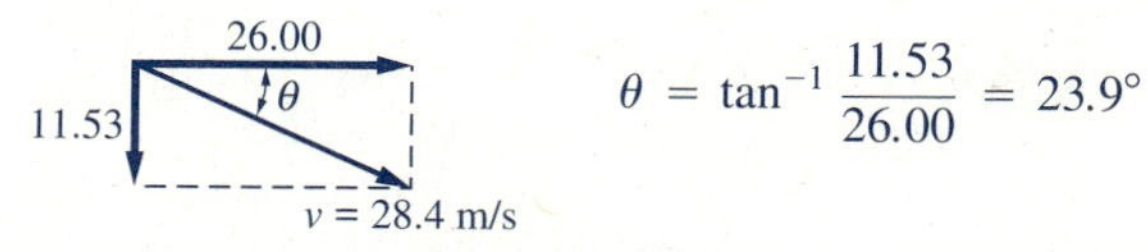

$$\theta = \tan^{-1}\frac{11.53}{26.00} = 23.9° \qquad \textit{Answer}$$

Sample Problem 14.9

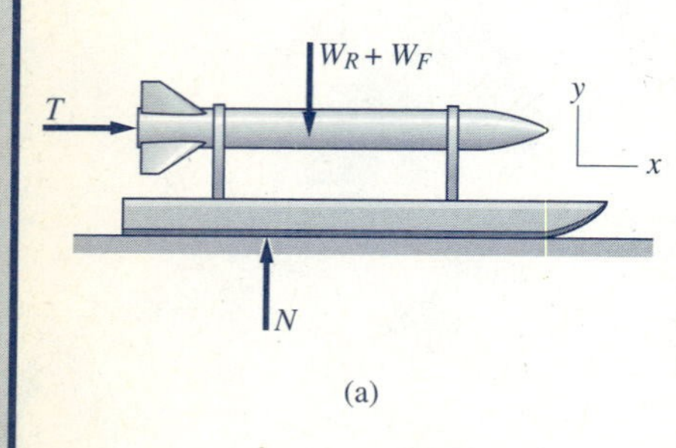

(a)

(b)

Figure (a) shows the free-body diagram of a rocket that is being test-fired on a horizontal track (friction and air resistance are neglected). The combined weight of the rocket and the test sled is $W_R = 200$ lb, W_F is the weight of the fuel inside the rocket (it varies with time), N is the reaction with the track, and T is the thrust. Figure (b) shows the thrust-time variation during the final 10 s of powered motion. At $t = 0$ the weight of the remaining fuel is 600 lb, and the velocity of the assembly is 60 ft/s to the right. Calculate the velocity of the assembly 10 s later, when the fuel has been completely consumed.

Solution

We apply the impulse-momentum method to the assembly consisting of the rocket, the test sled, and the fuel inside the rocket. From Eqs. (14.40), the impulse-momentum equations for the x- and y-directions are

$$(L_{1\text{-}2})_x = (mv_x)_2 - (mv_x)_1 \tag{a}$$

$$(L_{1\text{-}2})_y = (mv_y)_2 - (mv_y)_1 \tag{b}$$

The subscripts 1 and 2 will be used to indicate the values of the variables at $t = 0$ and 10 s, respectively.

Note that in Eq. (b) we have $(L_{1\text{-}2})_y = 0$, because the y-component of the velocity vector of the assembly is zero. Therefore, it yields $N = W_R + W_F$ (a fact that could also have been deduced directly from $\Sigma F_y = 0$).

When applying the impulse-momentum equation for the x-direction, we must be careful to note that the weight of the fuel drops from 600 lb to zero during the time interval of interest. Also observe that the thrust is the only force on the FBD that has an impulse in the x-direction. The magnitude of this impulse is equal to the area under the parabola in Fig. (b). Therefore, for the x-direction, we have

$$(L_{1\text{-}2})_x = (mv_x)_2 - (mv_x)_1$$

$$\overset{+}{\rightarrow} \quad \frac{1}{3}(600)(10) = \frac{200}{32.2}(v_x)_2 - \frac{(200 + 600)}{32.2}(60)$$

which yields

$$(v_x)_2 = 562 \text{ ft/s} \qquad \textit{Answer}$$

PROBLEMS

14.63 A constant horizontal force **P** (not shown) acts on the 4-lb body as it slides along a smooth, horizontal table. During the time interval $t = 5$ s to $t = 7.5$ s, the velocity changes as indicated in the figure. Determine the magnitude and direction of **P**.

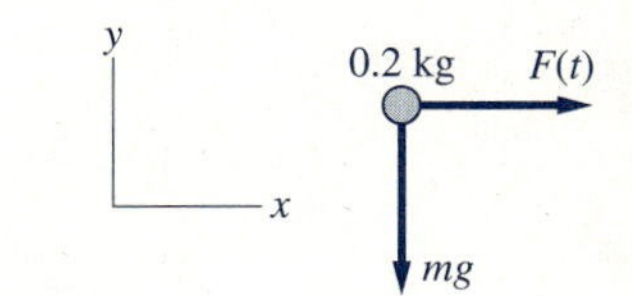

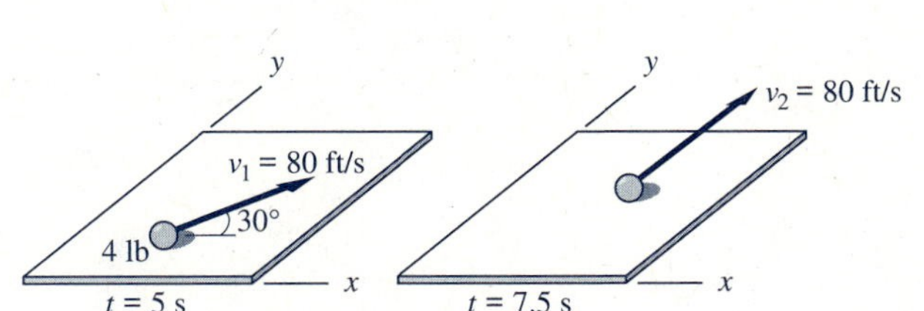

Fig. P14.63

F(t)
1.2 N
0
0 1 3 4 t (s)

Fig. P14.64

14.64 The 0.2-kg mass moves in the vertical xy-plane. At time $t = 0$, the velocity of the mass is $8\mathbf{j}$ m/s. In addition to its weight, the mass is acted on by the force $\mathbf{F}(t) = F(t)\mathbf{i}$, where the magnitude of the force varies with time as shown in the figure. Determine the velocity vector of the mass at $t = 4$ s.

14.65 Find the smallest time in which a 2400-lb automobile traveling at 60 mi/h can be stopped on a straight road if the static coefficient of friction between the tires and the road is 0.6. Assume that the road is level.

14.66 Solve Prob. 14.65 if the car is descending a hill that is inclined at 5° with the horizontal.

14.67 A 0.09-kg tennis ball traveling at 15 m/s rebounds in the opposite direction at 20 m/s after being hit by a racket. During the 0.032-s period of contact, the magnitude of the force $F(t)$ exerted by the racket has the parabolic time dependence shown. Calculate F_{max}, the maximum value of $F(t)$. Because of the short time of contact, the impulse of the weight of the ball can be neglected.

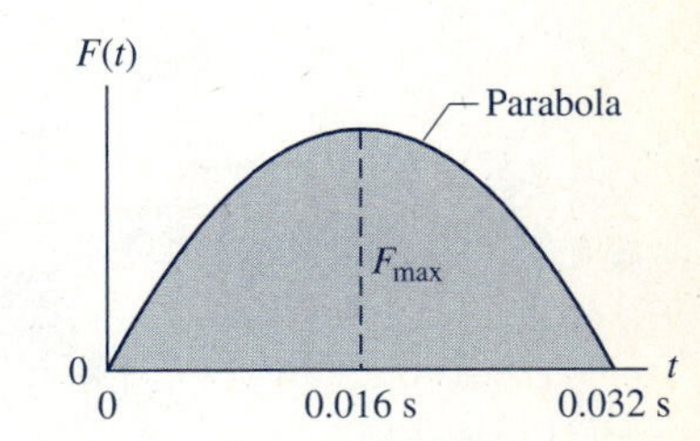

Fig. P14.67

14.68 A 2.5-oz ball hits a smooth, rigid, horizontal surface with the speed $v_1 = 30$ ft/s at the angle $\theta_1 = 70°$. The angle of rebound is $\theta_2 = 62°$. Compute (a) the speed of the ball immediately after the rebound; and (b) the resultant impulse acting on the ball during its time of contact with the surface.

14.69 A ball with a mass of 0.06 kg hits a smooth, rigid, horizontal surface with the speed $v_1 = 20$ m/s at the angle $\theta_1 = 45°$. The impulse acting on the ball during its contact with the surface is 1.5 N · s. Find the rebound speed v_2 and the corresponding angle θ_2.

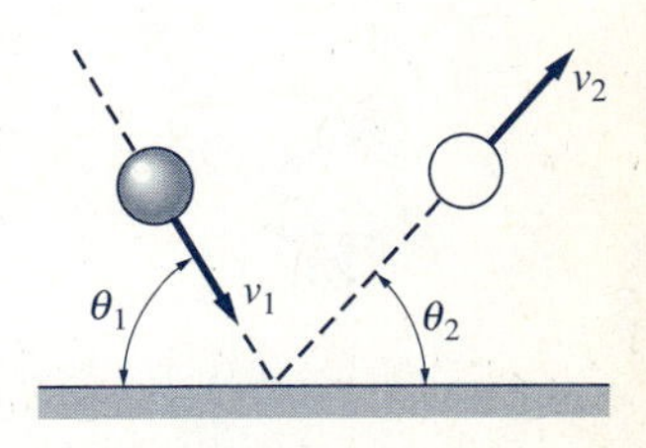

Fig. P14.68, P14.69

14.70 The 100-kg block is at rest on a smooth surface when the force $P(t)$ is applied. Determine (a) the maximum velocity of the block and the corresponding time; and (b) the velocity of the block at $t = 6$ s.

14.71 The 100-kg block is at rest on a rough surface when the force $P(t)$ is applied. The coefficients of static and kinetic friction are both equal to 0.3. Compute (a) the maximum velocity of the block and the corresponding time; and (b) the time when the block comes to rest.

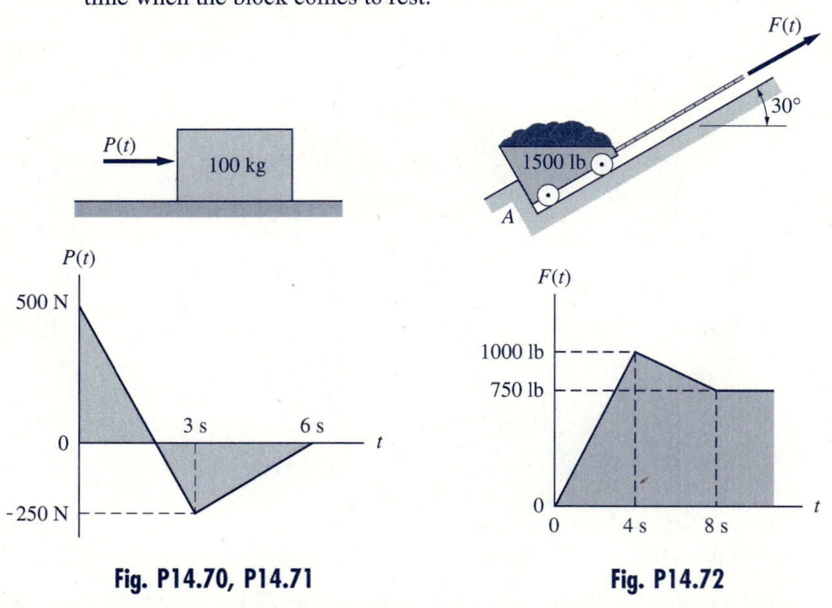

Fig. P14.70, P14.71 **Fig. P14.72**

14.72 The 1500-lb cable car is resting against the stop at A when the force $F(t)$ is applied to the cable. Neglecting friction, compute (a) the time when the car starts to move; and (b) the speed of the car at $t = 8$ s.

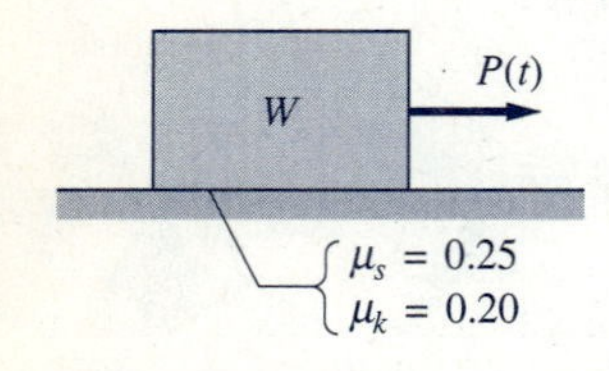

Fig. P14.73

14.73 The block of weight W is at rest on a rough surface at time $t = 0$. During the period from $t = 0$ to 2 s, the block is acted on by the sinusoidally varying force $P(t) = W \sin(\pi t/2)$. Using the coefficients of friction shown, find (a) the time when the block starts to move; and (b) the velocity of the block, measured in ft/s, at $t = 2$ s.

14.74 A parcel is lowered onto a conveyor belt that is moving at 4 m/s. If the coefficient of kinetic friction between the parcel and the belt is 0.25, calculate the time that it takes for the parcel to reach the speed of the belt.

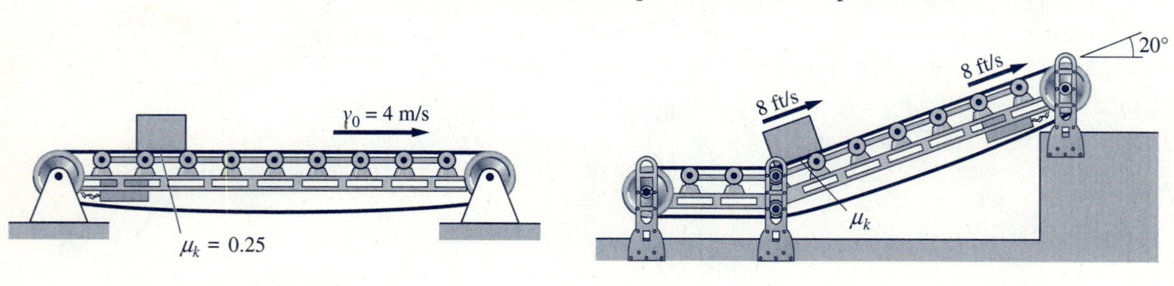

Fig. P14.74 **Fig. P14.75**

14.75 A parcel starts up the inclined portion of the conveyor with the same velocity as the conveyor belt, namely, 8 ft/s. However, the parcel slides on the belt due to insufficient friction, reaching its maximum height in 4 s. Determine the coefficient of kinetic friction between the parcel and the conveyor belt.

14.76 The mass of the rocket sled is 1200 kg without fuel. The engine produces a constant thrust of 10 kN for 12 s, during which time it burns up its 240 kg of fuel at a constant rate. Assuming that the sled started from rest, determine its speed when (a) half the fuel has been used; and (b) all the fuel has been used. Neglect friction.

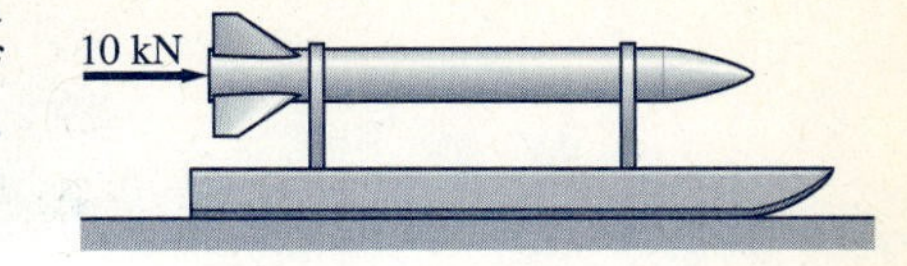

Fig. P14.76, P14.77

14.77 For the rocket sled in Prob. 14.76, determine (a) the time (measured from the instant of ignition) when the velocity reaches 50 m/s; and (b) the acceleration at the time found in part (a).

14.78 The total mass of a rocket on its launch pad is 200 kg, of which 125 kg is fuel. During powered flight the engine produces a constant thrust of 4 kN while burning fuel at the rate of 6.25 kg/s. If the rocket is launched vertically from rest, find its maximum upward velocity. Assume that the gravitational acceleration g is constant throughout the flight.

14.79 A rocket generates a constant 625-lb thrust during its vertical flight. Because of the consumption of fuel, the weight of the rocket varies during a 12-s period, as shown in the following table. If the velocity of the rocket at the beginning of the 12-s period is 250 ft/s, what is its velocity at the end of this period? Assume that the gravitational acceleration remains constant at $g = 32.2$ ft/s^2.

t (s)	0	2	4	6	8	10	12
W (lb)	300	292	274	255	227	196	150

14.80 The elevator of weight W starts from rest at time $t = 0$ and reaches its operating speed in 3 s. The acceleration is accomplished by varying the tension in cable AB (by means of a control circuit) in the manner shown on the diagram. Determine the operating speed of the elevator in m/s.

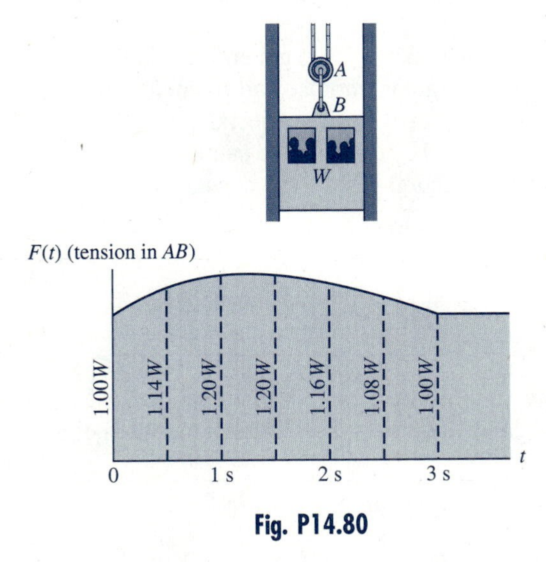

Fig. P14.80

14.7 Principle of Angular Impulse and Momentum

a. Angular impulse of a force

In Eq. (14.35) the linear impulse of a force **F** during a time interval from t_1 to t_2 is defined to be

$$\mathbf{L}_{1-2} = \int_{t_1}^{t_2} \mathbf{F}\, dt$$

Here we define the angular impulse (or the moment of the impulse) of a force about a point. Let **r** be the vector drawn from an arbitrary point A to the point of application of the force **F**. The *angular impulse* of **F** about A for the time interval from t_1 to t_2 is defined as

$$(\mathbf{A}_A)_{1-2} = \int_{t_1}^{t_2} \mathbf{r} \times \mathbf{F}\, dt = \int_{t_1}^{t_2} \mathbf{M}_A\, dt \tag{14.42}$$

where $\mathbf{M}_A = \mathbf{r} \times \mathbf{F}$ is the moment of **F** about A. Angular impulse is obviously a vector quantity, and its units are lb · ft · s, N · m · s, etc.

Since $\mathbf{M}_A = M_x\mathbf{i} + M_y\mathbf{j} + M_z\mathbf{k}$, where M_x, M_y, and M_z are the moments of **F** about rectangular axes passing through A, the angular impulse of a force may be written as

$$(\mathbf{A}_A)_{1-2} = \mathbf{i}\int_{t_1}^{t_2} M_x\, dt + \mathbf{j}\int_{t_1}^{t_2} M_y\, dt + \mathbf{k}\int_{t_1}^{t_2} M_z\, dt \tag{14.43}$$

The coefficients of **i**, **j**, and **k** in this equation are called the angular impulses about the x-, y-, and z-axes, respectively.

Since the direction of $\mathbf{M}_A$ will, in general, continuously change throughout the time interval, the angular impulse and moment will not necessarily act in the same direction. However, for the special case where the direction of $\mathbf{M}_A$ is constant, $(\mathbf{A}_A)_{1-2}$ and $\mathbf{M}_A$ do have the same direction. Furthermore, if both the magnitude and direction of $\mathbf{M}_A$ are constant, the angular impulse in Eq. (14.42) becomes

$$(\mathbf{A}_A)_{1-2} = \mathbf{M}_A \int_{t_1}^{t_2} dt = \mathbf{M}_A(t_2 - t_1) = \mathbf{M}_A \Delta t \quad (\mathbf{M}_A \text{ constant}) \tag{14.44}$$

The angular impulse of a system of forces about a point is obtained by summing the angular impulse of each force about the point.

b. Angular momentum of a particle

In Eq. (14.36) the linear momentum of a particle of mass m was defined as $\mathbf{p} = m\mathbf{v}$, where **v** is the velocity vector of the particle. Figure 14.10 illustrates the angular momentum of a particle P. Point A is an arbitrary point moving with the velocity $\mathbf{v}_A$, and point O is fixed in the inertial reference frame. The position vector of P is $\mathbf{r}_P$, and $\mathbf{r}_{P/A}$ denotes its position vector relative to A. The *angular momentum* of a particle about a point is defined to be the moment of its linear momentum about that point (for this reason, angular momentum is frequently referred to as the moment of momentum). Therefore, the angular momenta of the particle P about points O and A are

$$\boxed{\mathbf{h}_O = \mathbf{r}_P \times m\mathbf{v}} \tag{14.45}$$

and

$$\boxed{\mathbf{h}_A = \mathbf{r}_{P/A} \times m\mathbf{v}} \tag{14.46}$$

Fig. 14.10

respectively, where $\mathbf{v} = \dot{\mathbf{r}}_P$ is the velocity of the particle. Angular momentum is a vector quantity, its units being that of distance times linear momentum, i.e., the same as the units for angular impulse (lb · ft · s, N · m · s, etc.). The vector representing the angular momentum about a point is perpendicular to the plane formed by the velocity vector and the point, as indicated for $\mathbf{h}_A$ in Fig. 14.10.

Using the rectangular components shown in Fig. 14.11(a), we have $\mathbf{r}_{P/A} = x\mathbf{i} + y\mathbf{j} + z\mathbf{k}$, $\mathbf{v} = v_x\mathbf{i} + v_y\mathbf{j} + v_z\mathbf{k}$, and the angular momentum of P about A becomes

$$\mathbf{h}_A = \mathbf{r}_{P/A} \times m\mathbf{v} = \begin{vmatrix} \mathbf{i} & \mathbf{j} & \mathbf{k} \\ x & y & z \\ mv_x & mv_y & mv_z \end{vmatrix}$$

Expanding the determinant gives

$$\boxed{\mathbf{h}_A = h_x\mathbf{i} + h_y\mathbf{j} + h_z\mathbf{k}} \tag{14.47a}$$

where

$$\boxed{\begin{aligned} h_x &= m(yv_z - zv_y) \\ h_y &= m(zv_x - xv_z) \\ h_z &= m(xv_y - yv_x) \end{aligned}} \tag{14.47b}$$

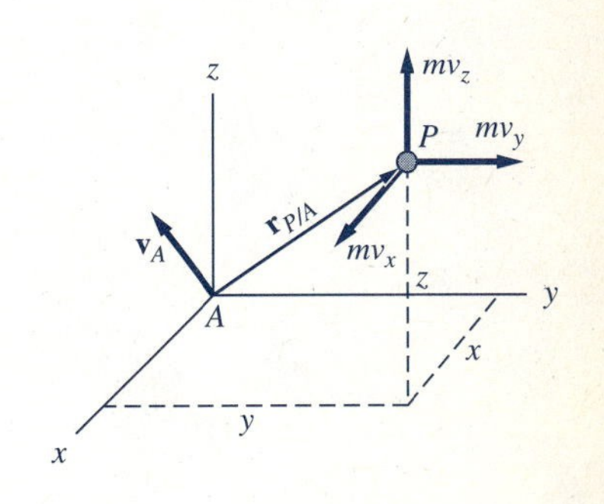

(a)

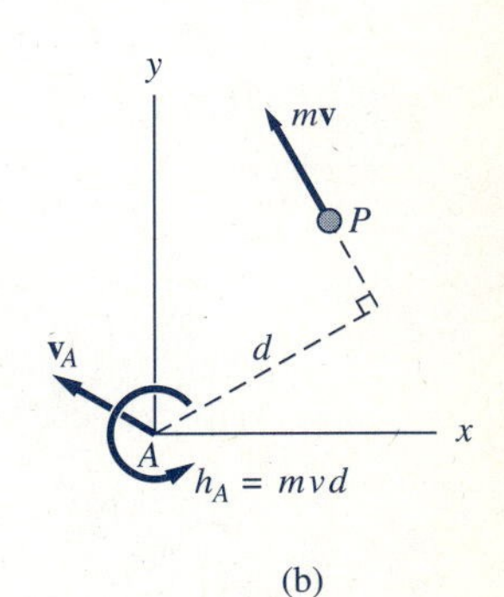

(b)

Fig. 14.11

are called the angular momenta of P about the x-, y-, and z-axes, respectively.

If the motion of the particle is confined to the xy-plane, as shown in Fig. 14.11(b), it is frequently convenient to calculate the magnitude of the angular momentum of P about A from the scalar equation

$$\boxed{h_A = mvd} \tag{14.48}$$

where d is the perpendicular distance from A to the momentum (or velocity) vector of P. When using Eq. (14.48), it is important that the direction (clockwise or counterclockwise) of the angular momentum also be recorded.

c. Moment-angular momentum relationship

The angular momentum of a particle about an arbitrary point A was defined in Eq. (14.46) to be $\mathbf{h}_A = \mathbf{r}_{P/A} \times m\mathbf{v}$. Differentiating this expression with respect to time gives

$$\dot{\mathbf{h}}_A = \mathbf{r}_{P/A} \times m\mathbf{a} + \dot{\mathbf{r}}_{P/A} \times m\mathbf{v}$$

where $\mathbf{a} = \dot{\mathbf{v}}$ is the acceleration of the particle. Utilizing Newton's second law, $\mathbf{F} = m\mathbf{a}$, where $\mathbf{F}$ is the resultant force acting on the particle, we have $\mathbf{r}_{P/A} \times m\mathbf{a} = \mathbf{r}_{P/A} \times \mathbf{F}$, which represents the resultant moment of $\mathbf{F}$ about point A, denoted by $\mathbf{M}_A$. Therefore, the preceding equation may be written as

$$\mathbf{M}_A = \dot{\mathbf{h}}_A - \dot{\mathbf{r}}_{P/A} \times m\mathbf{v} \tag{14.49}$$

which we call the *moment-angular momentum relationship.*

The important special case

$$\mathbf{M}_A = \dot{\mathbf{h}}_A \tag{14.50}$$

is obtained when the last term in Eq. (14.49) vanishes. This occurs if one of the following conditions is satisfied.

1. A is a fixed point; i.e., $\mathbf{v}_A = \mathbf{0}$.
   Proof—From Fig. 14.10 we see that $\mathbf{r}_P = \mathbf{r}_A + \mathbf{r}_{P/A}$, the time derivative of which gives $\mathbf{v} = \dot{\mathbf{r}}_P = \mathbf{v}_A + \dot{\mathbf{r}}_{P/A}$. Therefore, if $\mathbf{v}_A = \mathbf{0}$, we have $\dot{\mathbf{r}}_{P/A} \times m\mathbf{v} = \mathbf{v} \times m\mathbf{v} = \mathbf{0}$.
2. The vectors $\mathbf{r}_{P/A}$ and $\mathbf{v}$ are parallel.
   Proof—Since the cross product of two parallel vectors vanishes, we get $\dot{\mathbf{r}}_{P/A} \times m\mathbf{v} = \mathbf{0}$.

Because the second of these cases is of limited value in the solution of problems, we omit reference to it in the following discussions.

d. Angular impulse-momentum principle

In Eq. (14.50) we saw that the equation $\mathbf{M}_A = \dot{\mathbf{h}}_A$ is valid for a particle provided A is a fixed point. Multiplying each side of the equation by dt and integrating over the time interval t_1 to t_2 yields

$$\int_{t_1}^{t_2} \mathbf{M}_A \, dt = \int_{(\mathbf{h}_A)_1}^{(\mathbf{h}_A)_2} d\mathbf{h}_A$$

where $(\mathbf{h}_A)_1$ and $(\mathbf{h}_A)_2$ are the angular momenta about A at times t_1 and t_2, respectively. Recognizing that the left side of this equation is, by definition, the resultant angular impulse about A [see Eq. (14.42)], the equation may be written as

$$(\mathbf{A}_A)_{1\text{-}2} = (\mathbf{h}_A)_2 - (\mathbf{h}_A)_1 = \Delta\mathbf{h}_A \qquad (A\text{: fixed point}) \tag{14.51}$$

Equation (14.51) is called the *angular impulse-momentum principle,* or the *balance of angular impulse and momentum.*

Note the similarities between Eq. (14.51) and the principle of linear impulse-momentum: $\mathbf{L}_{1-2} = \mathbf{p}_2 - \mathbf{p}_1$. Both are vector equations that relate an impulse occuring over a time period to the change in momentum that occurs during that time period.

e. Conservation of angular momentum

If the angular impulse about A is zero, it follows from Eq. (14.51) that the angular momentum of a particle is conserved about A. In other words,

$$\text{if } (\mathbf{A}_A)_{1-2} = 0, \quad \text{then } (\mathbf{h}_A)_1 = (\mathbf{h}_A)_2 \qquad (A\text{: fixed point}) \tag{14.52}$$

which is known as the *principle of conservation of angular momentum.* Observe that angular momentum about a fixed point is conserved during a given time interval if and only if the angular impulse about that point is zero during that time interval. Since Eq. (14.52) is a vector equation, it is possible for the angular momentum about an axis passing through A to be conserved, even though the total angular momentum about A may not be conserved.

Sample Problem 14.10

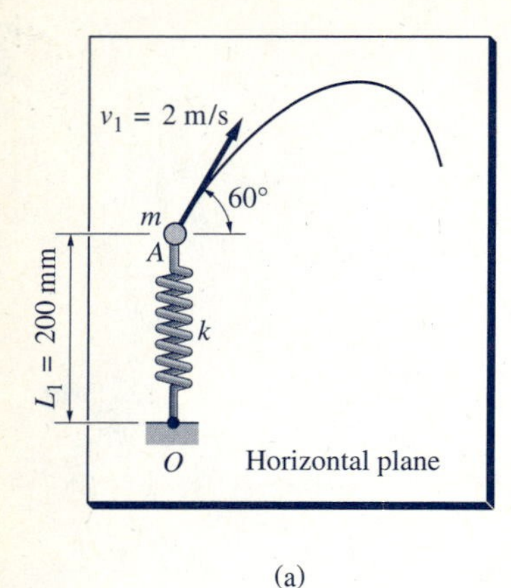

(a)

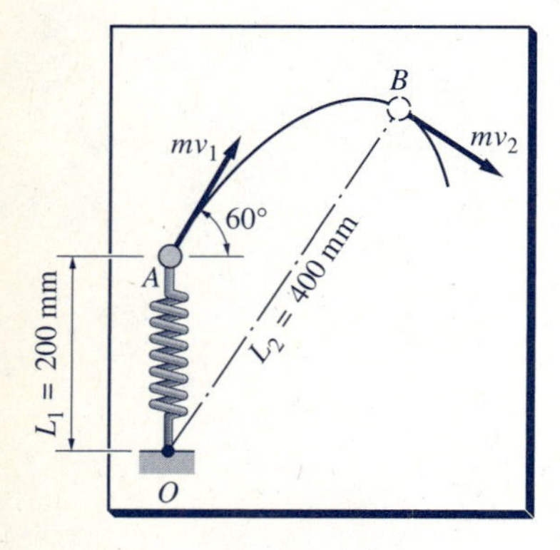

(b) Momentum diagram

The particle of mass $m = 0.3$ kg shown in Fig. (a) moves on a smooth horizontal plane. One end of the linear spring is attached to the particle, and the other end is attached to the fixed point O. If the particle is launched from position A with the velocity $\mathbf{v}_1$ as shown, determine the spring stiffness k if the maximum distance between the path of the particle and point O is 400 mm. The spring is undeformed when the particle is at A.

Solution

The diagram in Fig. (b) shows the momentum vectors $m\mathbf{v}_1$ and $m\mathbf{v}_2$ when the particle is at points A and B, respectively. Point B is the position of the particle when the length of the spring equals its maximum value $L_2 = 400$ mm. Note that the direction of $\mathbf{v}_2$ is tangent to the path of the particle; in other words, the velocity vector is perpendicular to the spring.

This problem is solved most easily by the application of the principles of conservation of angular momentum and conservation of mechanical energy. Because only points A and B are of interest, it is not necessary to use the force-mass-acceleration method, thus avoiding the complexities that would arise as a result of the variable spring force.

Noting that the spring force is always directed toward O, we conclude that angular momentum about O is conserved. Referring to the momentum vectors in Fig. (b), and recalling that the angular momentum equals the moment of the linear momentum ($h_O = mvd$), we obtain

$$(h_O)_1 = (h_O)_2 \qquad \circlearrowleft{+} \qquad (mv_1 \cos 60°)L_1 = mv_2 L_2$$

which yields

$$v_2 = \frac{v_1 \cos 60° L_1}{L_2} = \frac{2\cos 60°(200)}{400} = 0.500 \text{ m/s}$$

Since the system consisting of the particle and the spring is conservative, the conservation of mechanical energy (the work-energy principle could also be used) yields

$$T_1 + V_1 = T_2 + V_2$$

$$\frac{1}{2}mv_1^2 + \frac{1}{2}k\delta_1^2 = \frac{1}{2}mv_2^2 + \frac{1}{2}k\delta_2^2$$

where δ_1 and δ_2 are the deformations of the spring when the particle is at A and B, respectively. Substituting $\delta_1 = 0$, $\delta_2 = L_2 - L_1 = 200$ mm $= 0.200$ m, and the values for m, v_1, and v_2, we obtain

$$\frac{1}{2}(0.3)(2)^2 + 0 = \frac{1}{2}(0.3)(0.500)^2 + \frac{1}{2}k(0.200)^2$$

from which the spring stiffness is found to be

$$k = 28.1 \text{ N/m} \qquad \textit{Answer}$$

PROBLEMS

14.81 The ball A weighing 24 lb is attached to the rod OA, which has negligible weight. Determine the couple C applied to OA that will cause the angular acceleration $\ddot{\theta} = 4 \text{ rad/s}^2$.

14.82 The rod OA of negligible weight carries the 24-lb ball and is acted on by a couple that varies with time according to $C = 20 \sin(\pi t/4)$ lb · ft, where t is in seconds. If $\dot{\theta} = 20$ rad/s when $t = 0$, determine (a) the angular speed $\dot{\theta}$ as a function of time; and (b) the maximum angular speed.

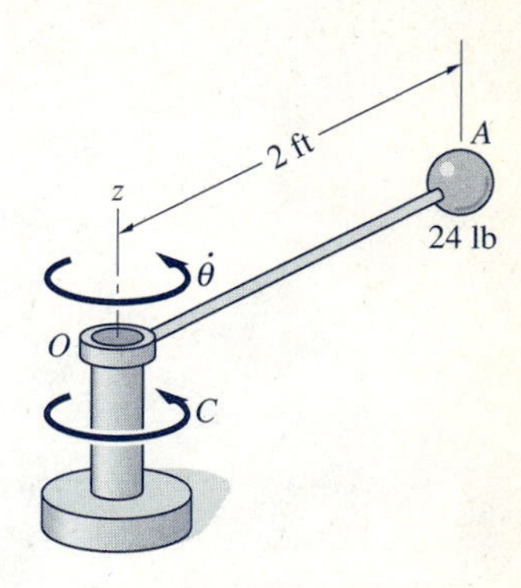

Fig. P14.81, P14.82

14.83 The rim of the wheel has a mass of 16 kg; the mass of the spokes is negligible. A rope wound around the rim is pulled by the force F, which varies with time as shown. The wheel is at rest when $t = 0$. Calculate (a) the angular velocity of the wheel when $t = 0.5$ s; and (b) the maximum angular acceleration of the wheel. Neglect friction.

14.84 As the rod OB rotates about the z-axis, the force F pulls end C of the string AOC downward at the constant speed shown, thereby moving slider A toward O. If the angular speed of OB is $\dot{\theta} = 8$ rad/s when $R = 2$ ft, calculate $\dot{\theta}$ and $\ddot{\theta}$ when $R = 1.0$ ft. Assume that the mass of OB is negligible and neglect friction.

14.85 As the rod OB rotates about the z-axis, the force F pulls end C of the string AOC downward, thereby moving the 2-lb slider A toward O. The variable force F maintains the speed of end C constant at $v_C = 0.25$ ft/s. If $\dot{\theta} = 8$ rad/s, when $R = 2$ ft, determine the magnitudes of (a) the force F; and (b) the force exerted on the slider by the rod at this instant. Neglect the mass of rod OB and friction.

14.86 A couple (not shown) is applied to the rod OB of the assembly described in Prob. 14.85, causing it to rotate with constant angular speed $\dot{\theta} = 8$ rad/s. The slider A moves toward O, because end C of the string AOC is pulled downward at the constant speed shown. Determine the magnitude and direction of the couple when $R = 2$ ft.

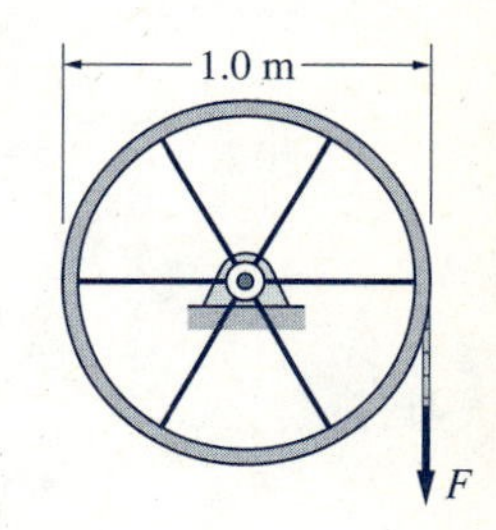

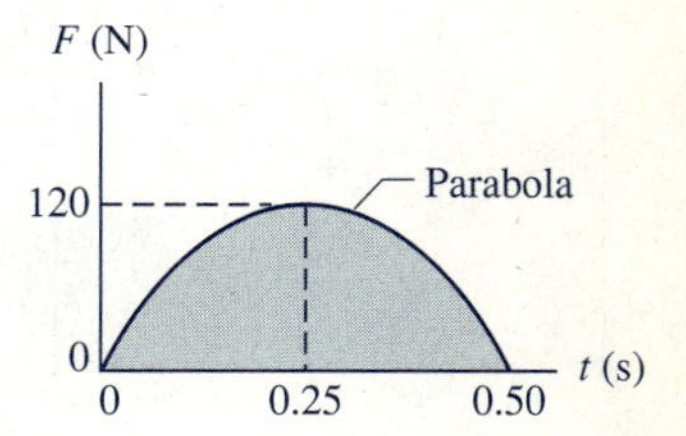

Fig. P14.83

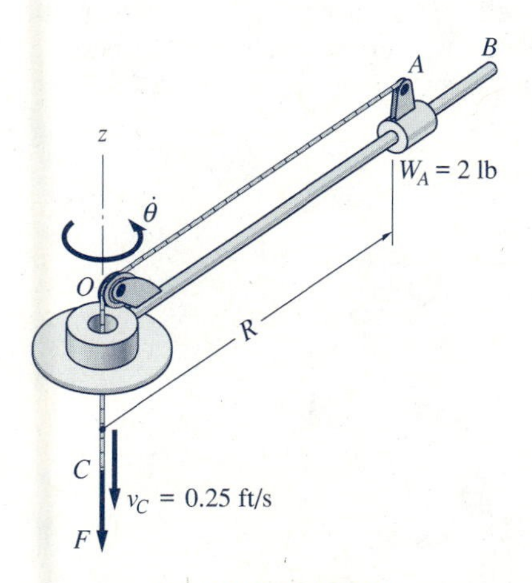

Fig. P14.84–P14.86

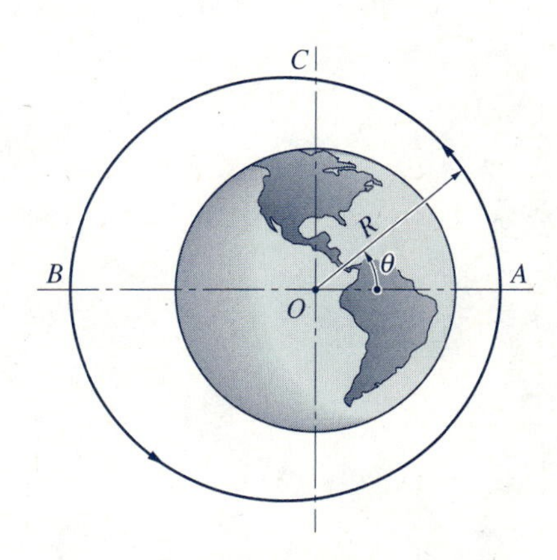

Fig. P14.87, P14.88

14.87 The path of the earth satellite is the ellipse $(1/R) = (119.3 \times 10^{-9}) \times (1 + 0.161 \cos\theta)$, where R is in meters. If the speed of the satellite at position A is 8000 m/s, compute its speed at position B.

14.88 Calculate the speed when the earth satellite described in Prob. 14.87 is in position C.

14.89 The particle is attached to an ideal spring that is pinned to the smooth horizontal table at O. The particle is launched from A in the y-direction with the speed v_A. If the velocity of the particle at B is $\mathbf{v}_B = 3.66\mathbf{i} - 5.72\mathbf{j}$ m/s, determine v_A.

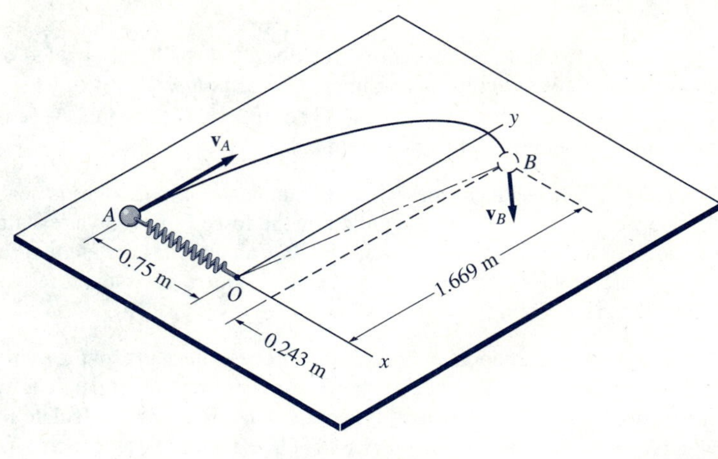

Fig. P14.89

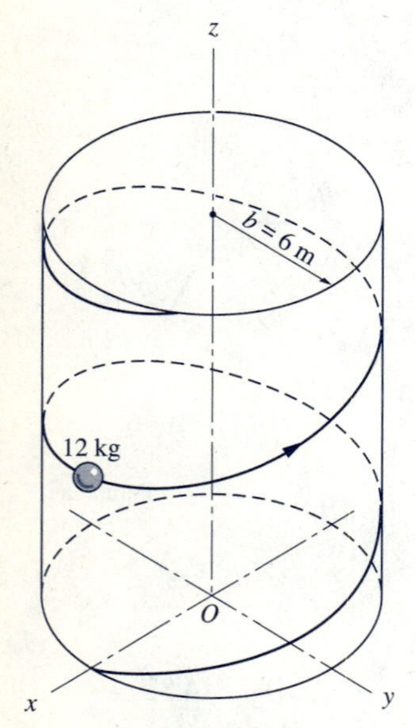

Fig. P14.90, P14.91

14.90 The 12-kg mass travels along a helix of radius b according to $x = b\cos\omega t$, $y = b\sin\omega t$, $z = (b/4)\omega t$, where $b = 6$ m and $\omega = 0.25$ rad/s. Calculate the rectangular components of the angular momentum vector about the origin O when $t = 40$ s.

14.91 The motion of the 12-kg mass in Prob. 14.90 can be described in cylindrical coordinates as $R = b$, $\theta = \omega t$, $z = (b/4)\omega t$. Determine the cylindrical components of the angular momentum vector about the origin O when $t = 40$ s.

14.92 The eccentricity e of an elliptical orbit is defined to be $e = (R_2 - R_1)/(R_2 + R_1)$. If $e = 0.6$ and the speed of an earth satellite at the perigee (point closest to the earth) is $v_1 = 8$ km/s, determine the elevation of the satellite at the perigee. Use $KM_e = 3.98 \times 10^{14}$ m^3/s^2, where K is the universal gravitational constant and M_e is the mass of the earth.

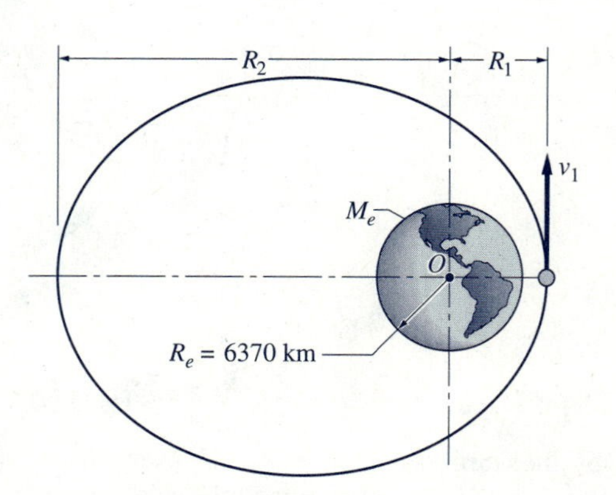

Fig. P14.92

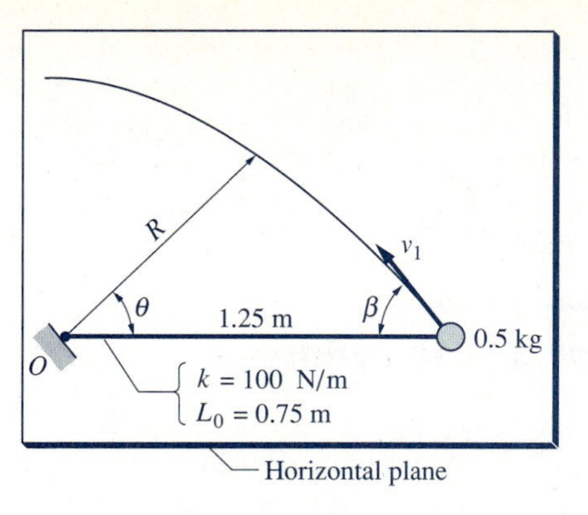

Fig. P14.93, P14.94

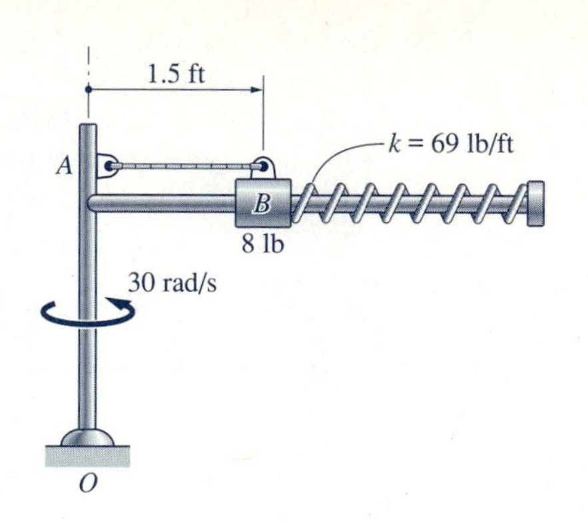

Fig. P14.95

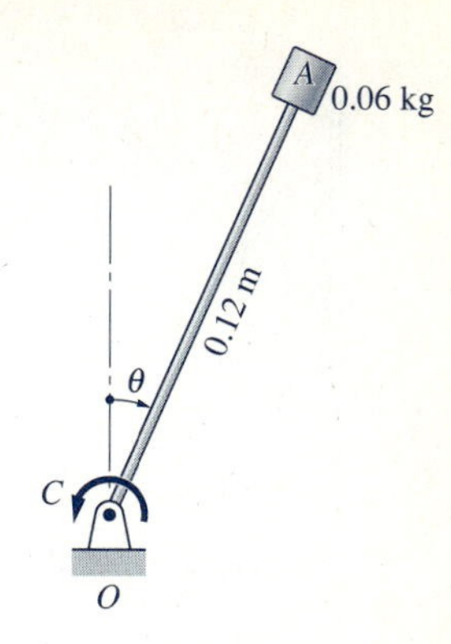

Fig. P14.96, P14.97

14.93 The 0.5-kg disk slides on a smooth horizontal surface. The elastic cord connected between the disk and point O has a stiffness of 100 N/m and a free length of 0.75 m. The disk is given the initial velocity v_1, directed as shown, when it is 1.25 m from O. If $\beta = 70°$, determine the smallest v_1 for which the cord will always remain taut.

14.94 If the disk described in Prob. 14.93 is launched with $v_1 = 15$ m/s and $\beta = 70°$, calculate the maximum and minimum distances from point O to the disk during the ensuing motion. (Make the verifiable assumption that the cord does not become slack.)

14.95 The system is rotating about axis OA with an angular speed of 30 rad/s. Initially, the 8-lb collar B is restrained by a cord, and the spring is undeformed. If the cord is cut, determine the maximum displacement of the collar relative to the rod and the corresponding angular velocity of the system. Neglect friction.

14.96 The mass A is attached to the end of a light rod that rotates in a vertical plane about O. A torsion spring (not shown) applies the couple $C = 0.5\theta^2$ N·m, where θ is in radians. The rod is released from rest when $\theta = \pi/2$. When $\theta = 0$, determine (a) the angular speed of the rod; and (b) the axial force in the rod.

14.97 For the system described in Prob. 14.96, (a) derive the equation of motion and state the initial conditions; (b) solve the equation numerically from the time of release until the rod reaches the $\theta = 0$ position, and plot θ vs. t; and (c) use the numerical solution to find the time required to reach the $\theta = 0$ position and the angular velocity of the bar OA in that position.

14.98 The telescopic arm of the mechanical manipulator rotates about the vertical axis with the constant angular speed $\dot{\theta} = 8$ rad/s. At the same time, the arm is extended and lowered at the constant rates $\dot{L} = 4$ ft/s and $\dot{\phi} = 2$ rad/s, respectively. When $L = 6$ ft and $\phi = 45°$, calculate the couple C that must be acting on the vertical shaft. Assume that the weight of the supporting mechanism is negligible compared to the 120-lb manipulator head.

14.99 The particle of mass m slides inside the smooth, conical vessel. If the initial velocity of the particle $v_1 = 10$ ft/s is tangent to the rim of the vessel, the particle will not travel below the circle that lies a vertical distance h beneath the rim. Calculate the distance h.

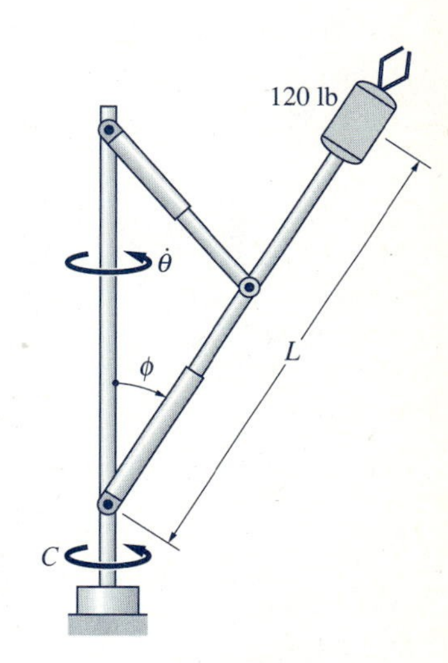

Fig. P14.98

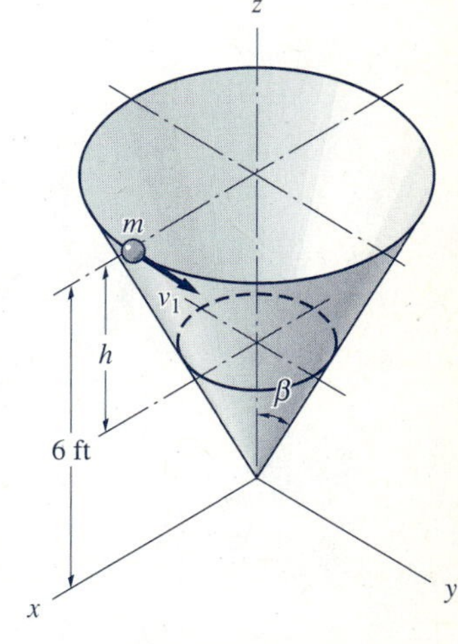

Fig. P14.99

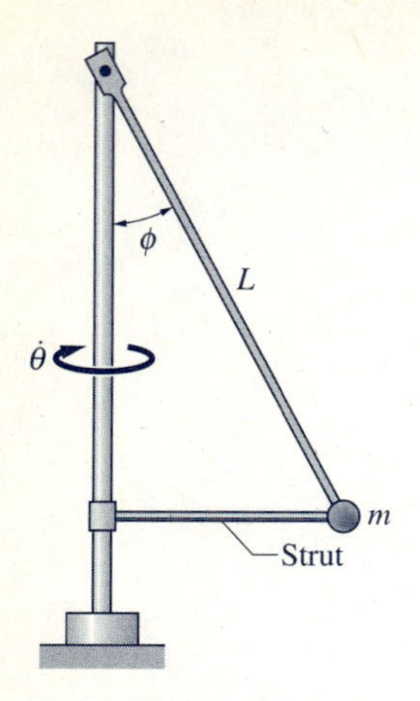

Fig. P14.100

14.100 The pendulum consists of the particle of mass m and the light arm of length L pinned to the vertical shaft. The light horizontal strut maintains the angle ϕ at a fixed value. The system is initially rotating freely about the vertical axis with the angular speed $\dot{\theta}$ when the strut is removed. Calculate (a) $\ddot{\theta}$; and (b) $\ddot{\phi}$ immediately after removal of the strut.

*14.8 Space Motion Under a Gravitational Force

Space motion under the action of a gravitational force comes under the category of central-force motion. The equations governing central-force motion can be readily derived from the principles of conservation of angular momentum and conservation of mechanical energy.

a. Central-force motion

Consider a moving particle that is acted on only by a force $\mathbf{F}$ that is always directed toward a fixed point O. For this case, called *central-force motion,* we have

$$\mathbf{M}_O = \mathbf{r} \times \mathbf{F} = \dot{\mathbf{h}}_O = \mathbf{0}$$

where $\mathbf{r}$ is the position vector of the particle, drawn from O. From Eq. (14.45), with $\mathbf{r}_P$ replaced by $\mathbf{r}$, we have

$$\boxed{\mathbf{h}_O = \mathbf{r} \times m\mathbf{v} = \text{constant}} \tag{14.53}$$

Note that this equation can be valid only if $\mathbf{r}$ and $\mathbf{v}$ always lie in the same plane. Therefore, central-force motion is plane motion with constant angular momentum about point O.

b. Motion under gravitational attraction

The remainder of this article analyzes the motion of bodies (e.g., satellites) that move under the gravitational attraction of a planet (or the sun). We confine our attention to trajectories for which the gravitational attraction of the planet is the only force that needs to be considered. Letting m be the mass of the body and M the mass of the planet, we assume that $m \ll M$, which means that the planet may be considered to be fixed in the analysis.

Since the only force acting on the body is its weight, which is always directed toward the center of the planet, the body undergoes central-force motion, as described in Eq. (14.53). It is convenient to describe this motion, which is confined to a plane, in terms of polar coordinates R and θ, as shown in Fig. 14.12. The nonrotating xy reference frame has its origin at F, the center of the planet, also called the *focus* of the trajectory.

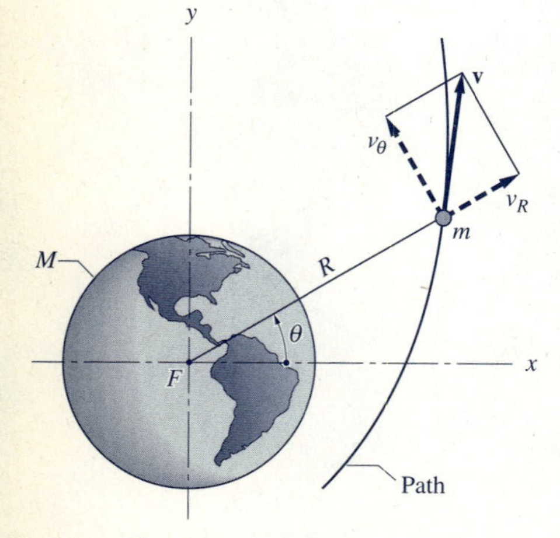

Fig. 14.12

From Eq. (14.53) we know that the angular momentum of the body about F is conserved, which means that $R(mv_\theta) = R(mR\dot{\theta})$ is constant. Letting h_0 be the angular momentum about F per unit mass of the body,* we have

* The subscript 0 (the number zero) on h, and later on E, is used to denote constants that are associated with unit mass. It must not be confused with the letter O used to refer to a point O.

$$h_0 = Rv_\theta = R^2\dot{\theta} \qquad \text{(a constant)} \tag{14.54}$$

The kinetic energy of the body is $T = \frac{1}{2}mv^2$, and its potential energy, according to Eq. (14.29), is $V_g = -KMm/R$ (where K is the universal gravitational constant). Because the gravitational attraction is a conservative force, the total energy of the body is conserved. Letting E_0 be the total energy per unit mass—i.e., $E_0 = (T + V_g)/m$—we have

$$E_0 = \frac{1}{2}v^2 - \frac{KM}{R} \qquad \text{(a constant)} \tag{14.55}$$

Note that the total energy per unit mass may be positive, negative, or zero.

c. Equation of the trajectory

We next determine the equation of the path (trajectory) of the body in the form $R = R(\theta)$. We begin by substituting $v^2 = v_R^2 + v_\theta^2$ into Eq. (14.55) to obtain

$$E_0 = \frac{1}{2}(v_R^2 + v_\theta^2) - \frac{KM}{R} \tag{a}$$

With the help of Eq. (14.54), v_R and R can be eliminated from Eq. (a). Applying the chain rule for differentiation to Eq. (14.54), we get

$$\dot{R}v_\theta + R\dot{v}_\theta = 0 \tag{b}$$

Substituting

$$\dot{R} = v_R \quad \text{and} \quad \dot{v}_\theta = \frac{dv_\theta}{d\theta}\dot{\theta} = \frac{dv_\theta}{d\theta}\frac{v_\theta}{R}$$

Eq. (b) becomes

$$v_R v_\theta = -R\left(\frac{dv_\theta}{d\theta}\frac{v_\theta}{R}\right)$$

which yields

$$v_R = -\frac{dv_\theta}{d\theta} \tag{c}$$

Substituting $R = h_0/v_\theta$ [see Eq. (14.54)] and Eq. (c) into Eq. (a), we find

$$E_0 = \frac{1}{2}\left[\left(\frac{dv_\theta}{d\theta}\right)^2 + v_\theta^2\right] - \frac{KMv_\theta}{h_0} \tag{d}$$

The solution of Eq. (d) for $d\theta$ is

$$d\theta = \pm\frac{dv_\theta}{\left(2E_0 + \dfrac{2KMv_\theta}{h_0} - v_\theta^2\right)^{1/2}} \tag{e}$$

Noting that E_0 is constant, Eq. (e) can readily be integrated (see a table of integrals) to yield

$$\theta = \pm \sin^{-1}\left(\frac{v_\theta - \dfrac{KM}{h_0}}{\dfrac{KM}{h_0}\left[1 + 2E_0\left(\dfrac{h_0}{KM}\right)^2\right]^{1/2}} \right) - \alpha_0 \tag{f}$$

where α_0 is a constant of integration. The dual sign means that for every point on the path identified by v_θ and θ, there is a second point given by v_θ and $(\theta + \pi)$. Therefore, the minus sign will be omitted from here on without loss of generality.

Inverting Eq. (f), we obtain

$$\sin(\theta + \alpha_0) = \frac{v_\theta - \dfrac{KM}{h_0}}{\dfrac{KM}{h_0} e} \tag{14.56}$$

where

$$e = \sqrt{1 + 2E_0\left(\frac{h_0}{KM}\right)^2} \tag{14.57}$$

can be shown to represent the eccentricity of the trajectory. It can be proven that e is always a real number; i.e., the term under the radical in Eq. (14.57) cannot be negative.

Solving Eq. (14.56) for v_θ, we obtain

$$v_\theta = \frac{KM}{h_0}[1 + e\sin(\theta + \alpha_0)] \tag{14.58}$$

Since $R = h_0/v_\theta$, according to Eq. (14.54), we find that

$$R = \frac{h_0^2}{KM[1 + e\sin(\theta + \alpha_0)]} \tag{14.59}$$

We see from Eq. (14.59) that the smallest value of R occurs when $\sin(\theta + \alpha_0) = 1$; i.e., when $\theta + \alpha_0 = \pi/2$. Letting this position correspond to $\theta = 0$ (the x-axis), as shown in Fig. 14.13, the constant of integration is $\alpha_0 = \pi/2$. Since $\sin(\theta + \alpha_0) = \sin[\theta + (\pi/2)] = \cos\theta$, Eqs. (14.58) and (14.59) become

$$v_\theta = \frac{KM}{h_0}(1 + e\cos\theta) \tag{14.60}$$

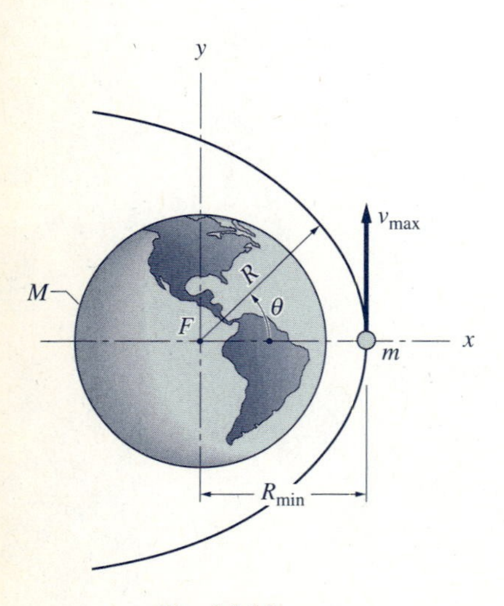

Fig. 14.13

and

$$R = \frac{h_0^2}{KM(1 + e\cos\theta)} \tag{14.61}$$

Equation (14.61), which is the equation of the trajectory, represents a conic section (circle, ellipse, parabola, or hyperbola) in polar coordinates, where e is the eccentricity of the curve and F (the center of the planet) is the focus.

Note that Eqs. (14.60) and (14.61) are valid only if the x-axis is chosen so that R is minimized when $\theta = 0$. If this is not the case, Eqs. (14.58) and (14.59) must be used.

From Eq. (14.61) we see that the minimum value of R is

$$R_{\text{min}} = R\big|_{\theta=0} = \frac{h_0^2}{KM(1 + e)} \tag{14.62}$$

It follows that $\dot{R} = v_R = 0$ when $\theta = 0$, which means that $v = v_\theta$ at this position. Furthermore, inspection of Eq. (14.55) reveals that v is largest when R is smallest, from which we conclude that the maximum velocity is

$$v_{\text{max}} = v_\theta\big|_{\theta=0} = \frac{KM(1 + e)}{h_0} \tag{14.63}$$

directed as shown in Fig. 14.13. Substituting $h_0 = R_{\text{min}}v_{\text{max}}$ into Eq. (14.63) and solving for v_{max} yields

$$v_{\text{max}} = \sqrt{\frac{KM(1 + e)}{R_{\text{min}}}} \tag{14.64}$$

Numerical data for selected members of our planetary system are presented in Table 14.1. These data are to be used when solving the homework problems at the end of this article.

Table 14.1 Selected Solar System Constants

Universal gravitational constant:
$K = 6.672 \times 10^{-11}\ \text{m}^3 \cdot \text{kg}^{-1} \cdot \text{s}^{-2}\ (1.0688 \times 10^{-9}\ \text{ft}^3 \cdot \text{lb}^{-1} \cdot \text{s}^{-2})$

	Mean equatorial radius		Mass/weight	
Body	**km**	**mi**	**kg**	**lb**
Sun	696 000	432 000	$1.989\,1 \times 10^{30}$	4.385×10^{30}
Moon	1 738	1 080	$0.073\,483 \times 10^{24}$	$0.162\,002 \times 10^{24}$
Mercury	2 439	1 516	$0.330\,22 \times 10^{24}$	$0.728\,01 \times 10^{24}$
Venus	6 052	3 761	$4.869\,0 \times 10^{24}$	$10.734\,3 \times 10^{24}$
Earth	6 378.140	3 963.192	$5.974\,2 \times 10^{24}$	$13.170\,9 \times 10^{24}$
Mars	3 393.4	2 108.6	$0.641\,91 \times 10^{24}$	$1.415\,17 \times 10^{24}$

The SI values are taken from *The Astronomical Almanac for the Year 1987,* Nautical Almanac Office. The values in U.S. Customary units are conversions from the SI values, using the factors 1.0 lbm = 0.453 592 37 kg, and 1.0 mi (U.S. statute) = 1.609 344 km.

d. Classification of trajectories

It has been pointed out that Eq. (14.61) represents a conic section, which means that the trajectory must be one of the following curves, depending on the value of e.

Case I:	$e = 0$	circle
Case II:	$0 < e < 1$	ellipse
Case III:	$e = 1$	parabola
Case IV:	$e > 1$	hyperbola

If the trajectory is circular or elliptical, the body is said to be captured by the planet. The body is then known as a satellite, and its trajectory is called an orbit. For the other two cases, the gravitational pull of the planet is not strong enough to capture the body.

Details of the four types of trajectories are presented in Table 14.2. The following discussion summarizes the entries in this table and presents several additional useful formulas.

1. Energy per Unit Mass E_0 The tabulated ranges of E_0 were determined by inspection of Eq. (14.57). It should be noted that for a given value of h_0, E_0 is smallest for a circular orbit ($e = 0$), and E_0 increases as e increases.

Table 14.2 Classification of Trajectories

Case	e	Path	E_0	Velocity	Distance from focus	Drawing of path
I	$e = 0$	Circle	$E_0 < 0$	$v_{circ} = \sqrt{\dfrac{KM}{R_{circ}}}$	$R_{circ} = \dfrac{h_0^2}{KM}$	
II	$0 < e < 1$	Ellipse	$E_0 < 0$	$v_{max} = \sqrt{\dfrac{KM(1+e)}{R_{min}}}$ $v_{min} = \sqrt{\dfrac{KM(1-e)}{R_{max}}}$ $= \sqrt{\dfrac{(1-e)^2}{1+e} \cdot \dfrac{KM}{R_{min}}}$	$R_{min} = \dfrac{h_0^2}{KM(1+e)}$ $R_{max} = \dfrac{h_0^2}{KM(1-e)}$	
III	$e = 1$	Parabola	$E_0 = 0$	$v_{max} = v_{esc}$ $= \sqrt{\dfrac{2KM}{R_{min}}}$	$R_{min} = \dfrac{h_0^2}{2KM}$ $R_{max} = \infty$ at $\theta = \pi$	
IV	$e > 1$	Hyperbola	$E_0 > 0$	$v_{max} = \sqrt{\dfrac{KM(1+e)}{R_{min}}}$	$R_{min} = \dfrac{h_0^2}{KM(1+e)}$ $R_{max} = \infty$ at $\cos\theta_1 = -\dfrac{1}{e}$	

2. Radial Coordinate Substituting $e = 0$ into Eq. (14.61) yields a constant value for R, which means that the corresponding conic section is a circle (Case I). Letting R_{circ} be the radius of the circular orbit, Eq. (14.61) becomes

$$R_{circ} = \frac{h_0^2}{KM} \tag{14.65}$$

The tabulated expressions for R_{min} for the other cases were found directly from Eq. (14.62).

The maximum value of R occurs when the denominator in Eq. (14.61) is minimized. For $0 < e < 1$, this occurs when $\theta = \pi$, which gives the tabulated expression for R_{max}. Since R_{max} is finite, we conclude that the corresponding trajectory is an ellipse (Case II).

For $e = 1$, the minimum value of the denominator in Eq. (14.61) is zero, leading to the conclusion that $R_{max} = \infty$, occurring at $\theta = \pi$. Therefore, the conic is a parabola (Case III).

When $e > 1$, the minimum value of the denominator in Eq. (14.61) is again zero, which gives $R_{max} = \infty$ when $\theta = \theta_1$, where $\cos\theta_1 = -1/e$. Thus the trajectory is a hyperbola (Case IV).

3. Velocity For a circular orbit (Case I) the velocity v_{circ} is constant throughout the motion. Substituting $e = 0$ and $R = R_{circ}$ into Eq. (14.60) yields

$$v_{circ} = \sqrt{\frac{KM}{R_{circ}}} \tag{14.66}$$

The expressions listed for v_{max} for Cases II and IV are simply restatements of Eq. (14.64). For Case III (parabola), substitution of $e = 1$ into Eq. (14.64) gives

$$v_{max} = v_{esc} = \sqrt{\frac{2KM}{R_{min}}} \tag{14.67}$$

This velocity is known as the *escape velocity,* because it is the minimum velocity for which the body will not be captured by the planet, given the value of R_{min}.

For a given value of R_{min}, the following relationships between the velocities follow from Table 14.2.

Circle ($e = 0$): $v_{circ} = \sqrt{\dfrac{KM}{R_{circ}}}$

Ellipse ($0 < e < 1$): $v_{esc} > v_{max} > v_{circ}$
$v_{min} < v_{circ}$

Parabola ($e = 1$): $v_{esc} = \sqrt{2}\, v_{circ}$

Hyperbola ($e > 1$): $v_{max} > v_{esc}$

These equations provide a convenient means of determining the type of trajectory when the initial conditions for the motion are known.

e. Properties of elliptical orbits

An elliptical orbit that is centered at point O is shown in Fig. 14.14. Note that *perigee* and *apogee* are the names given to the locations of R_{max} and R_{min},

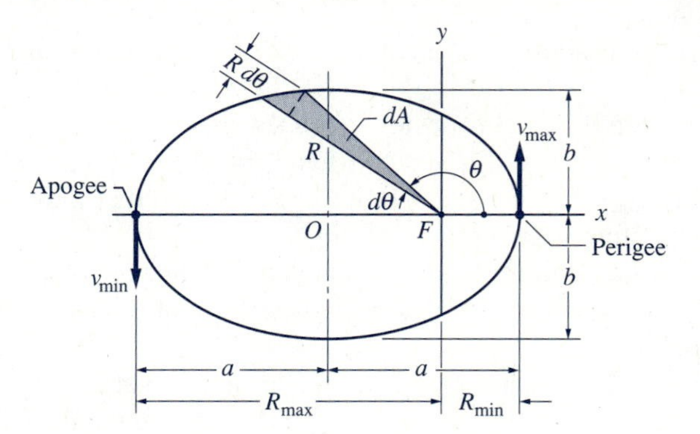

Fig. 14.14

respectively.* It can be shown that the geometric interpretation of the eccentricity is

$$e = \frac{R_{max} - R_{min}}{R_{max} + R_{min}} \tag{14.68}$$

The length of the major semiaxis in Fig. 14.14 is given by

$$a = \frac{R_{max} + R_{min}}{2} = \frac{h_0^2}{KM(1 - e^2)} \tag{14.69}$$

From analytic geometry, the length of the minor semiaxis may be written as

$$b = \sqrt{R_{max}R_{min}} = \frac{h_0^2}{KM\sqrt{1 - e^2}} \tag{14.70}$$

Using Eqs. (14.69) and (14.70), the area A of the ellipse can be written as

$$A = \pi ab = \frac{\pi}{2}(R_{max} + R_{min})\sqrt{R_{max}R_{min}} = \frac{\pi h_0^4}{(KM)^2(1 - e^2)^{3/2}} \tag{14.71}$$

The period τ of an elliptical (or circular) orbit is the time required to complete one revolution around the path. The period may be related to the area of the ellipse by noting that the differential area shown in Fig. 14.14 may be expressed as $dA = (1/2)R(R\,d\theta)$. The rate at which the area is swept over by

* The terms *perigee* and *apogee* are used only for orbital motion with the earth as the focus. When the attracting body is not the earth, the corresponding terms are *periapsis* and *apoapsis*, respectively.

the line between the focus and the satellite, namely dA/dt, is called the *areal velocity.* Utilizing Eq. (14.54), we have

$$\frac{dA}{dt} = \frac{1}{2}R^2\dot{\theta} = \frac{h_0}{2} \tag{14.72}$$

Note that the areal velocity is constant. This is one of the celebrated laws published by Johannes Kepler (1571–1630), based on astronomical observations. Integrating with respect to time from $t = 0$ to $t = \tau$ gives $A = h_0\tau/2$. Therefore, using Eq. (14.71), the period is

$$\boxed{\tau = \frac{2A}{h_0} = \frac{2\pi h_0^3}{(KM)^2(1 - e^2)^{3/2}}} \tag{14.73}$$

A special case of an elliptical path is the ballistic trajectory, where the ellipse intersects the surface of the planet. Such a trajectory is followed by all projectiles if air resistance is neglected. If the elevation of the projectile is small enough so that the variation of gravitational force with height may be neglected, then the trajectory assumes the familiar parabolic form. Hence, the parabolic trajectory is an approximation of the true elliptical path, valid for small changes of elevation.

Sample Problem 14.11

The eccentricity of the earth's orbit around the sun is 0.017 and the period of the orbit is 365.26 days. Calculate the maximum and minimum values of (1) the earth's distance from the center of the sun; and (2) the velocity of the earth around the sun.

Solution

Preliminary Calculations

Using Table 14.1, the constant KM_s, where M_s is the mass of the sun, is found to be

$$KM_s = (6.672 \times 10^{-11})(1.9891 \times 10^{30}) = 1.3271 \times 10^{20} \text{ m}^3/\text{s}^2$$

The given period of the orbit is

$$\tau = 365.26 \text{ days} \times \frac{24 \text{ h}}{1 \text{ day}} \times \frac{3600 \text{ s}}{1 \text{ h}} = 31.56 \times 10^6 \text{ s}$$

Solving Eq. (14.73) for h_0 (the constant angular momentum per unit mass of the earth about the center of the sun) yields

$$\begin{aligned} h_0 &= \left[\frac{(KM_s)^2\tau(1-e^2)^{3/2}}{2\pi}\right]^{1/3} \\ &= \left[\frac{(1.3271 \times 10^{20})^2(31.56 \times 10^6)[1-(0.017)^2]^{3/2}}{2\pi}\right]^{1/3} \\ &= 4.455 \times 10^{15} \text{ m}^2/\text{s} \end{aligned}$$

Part 1

Using the formulas for Case II in Table 14.2 and the constants determined above, the maximum and minimum distances of the earth from the center of the sun are

$$\left.\begin{matrix} R_{\text{min}} \\ R_{\text{max}} \end{matrix}\right\} = \frac{h_0^2}{KM_s(1 \pm e)} = \frac{(4.455 \times 10^{15})^2}{(1.3271 \times 10^{20})(1 \pm 0.017)}$$

from which we find that

$$R_{\text{max}} = 1.522 \times 10^{11} \text{ m} \qquad R_{\text{min}} = 1.471 \times 10^{11} \text{ m} \qquad \textit{Answer}$$

Part 2

Referring to Case II of Table 14.2 again, the maximum velocity of the earth is given by

$$\begin{aligned} v_{\text{max}} &= \sqrt{\frac{KM_s(1+e)}{R_{\text{min}}}} = \sqrt{\frac{(1.3271 \times 10^{20})(1+0.017)}{1.471 \times 10^{11}}} \\ &= 30.29 \times 10^3 \text{ m/s} = 30.29 \text{ km/s} \qquad \textit{Answer} \end{aligned}$$

The minimum velocity of the earth, again from the formulas in Table 14.1, is

$$v_{\min} = \sqrt{\frac{KM_s(1-e)}{R_{\max}}} = \sqrt{\frac{(1.3271 \times 10^{20})(1-0.017)}{1.522 \times 10^{11}}}$$

$$= 29.28 \times 10^3 \text{ m/s} = 29.28 \text{ km/s} \qquad \textit{Answer}$$

Note that the mean velocity of the earth is $(1/2)(v_{\max} + v_{\min}) = (1/2) \times (30.29 + 29.28) = 29.79$ km/s, which agrees within four significant digits with the commonly accepted value.

An alternative method for computing $v_{\max}$ and $v_{\min}$ is to use the fact that h_0 is constant, that is, to solve the equations $h_0 = v_{\max} R_{\min} = v_{\min} R_{\max}$.

PROBLEMS

14.101 Assuming that the orbit of the moon around the earth is a circle (its eccentricity is actually 0.055), and knowing that the period of the orbit is 27.3 days, compute the distance in miles between the centers of the earth and moon.

14.102 Calculate the maximum and minimum distances in miles between the centers of the earth and moon, taking the eccentricity of the moon's orbit into account. Use the data given in Prob. 14.101.

14.103 The orbit of Phobos, a Martian moon, has an eccentricity of 0.018 and a major semiaxis of length 9380 km. Determine the orbital period of Phobos.

14.104 A comet at its closest approach is 125×10^6 km from the center of the sun (between the orbits of earth and Venus), where its speed is 45.5 km/s. (a) Show that the comet has an elliptical orbit around the sun. (b) Determine the period of the orbit in years (1 year = 365.26 days). (c) Compute the maximum distance between the comet and the sun, and compare it with the radius of the solar system (the orbital radii of Neptune and Pluto, the two outermost planets, are 4510 and 5890×10^6 km, respectively).

14.105 A 16-Mg spacecraft is in orbit around the moon. The spacecraft's maximum and minimum altitudes above the lunar surface are 120 and 360 km, respectively. Neglecting the gravitational effect of the earth, determine the minimum energy required for the spacecraft to escape lunar gravity in order to return to earth.

14.106 The engines of the spacecraft in the orbit described in Prob. 14.105 develop a constant thrust of 120 kN. When the spacecraft is at its lowest altitude of 120 km, the engines are fired briefly with the thrust opposing the motion. Calculate the length of the burn required to place the craft in a circular orbit with an altitude of 120 km.

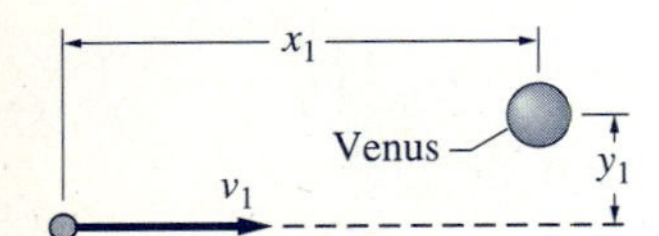

Fig. P14.107, P14.108

14.107 As the spacecraft approaches the planet Venus its speed is $v_1 = 8000$ ft/s when $x_1 = 5.0 \times 10^8$ ft and $y_1 = 1.0 \times 10^8$ ft. Determine (a) the type of trajectory that the spacecraft will travel; (b) the minimum distance between the surface of Venus and the trajectory; and (c) the maximum speed of the craft.

14.108 For the spacecraft described in Prob. 14.107 determine (a) the largest speed v_1 that would produce an elliptical orbit, assuming that an elliptical orbit were possible; and (b) whether an elliptical orbit is possible without the craft hitting the surface of Venus.

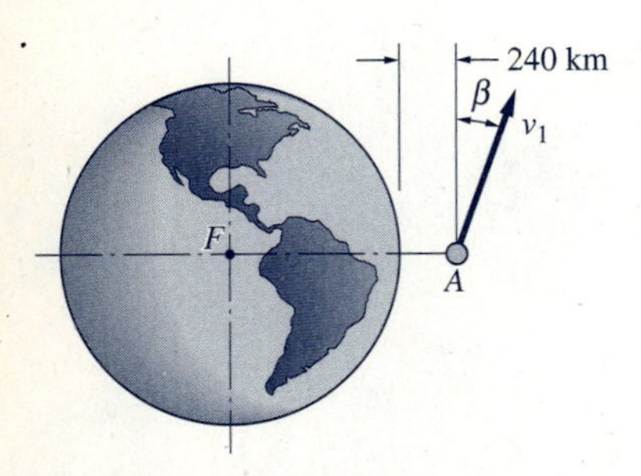

Fig. P14.109, P14.110

14.109 An earth satellite is inserted into its orbit at A with the speed $v_1 = 9200$ m/s, in the direction $\beta = 5°$. (a) Show that the trajectory is an ellipse. (b) Calculate the angle from the line FA to the major axis of the orbit. (c) Calculate the smallest distance between the orbit and the surface of the earth.

14.110 The speed of the spacecraft at A is $v_1 = 11.2$ km/s in the direction $\beta = 12°$. (a) Show that the trajectory is hyperbolic. (b) Find the terminal speed of the craft, ignoring the gravitational attraction of the sun.

14.111 A 2400-lb communications satellite is traveling in a circular "parking" orbit of radius 4160 miles (orbit 1). The satellite must traverse the elliptical orbit AB in order to reach the desired geosynchronous circular orbit of radius 26 200 miles (orbit 2). Determine the impulses that the satellite must receive in positions A and B during its transfer between the orbits.

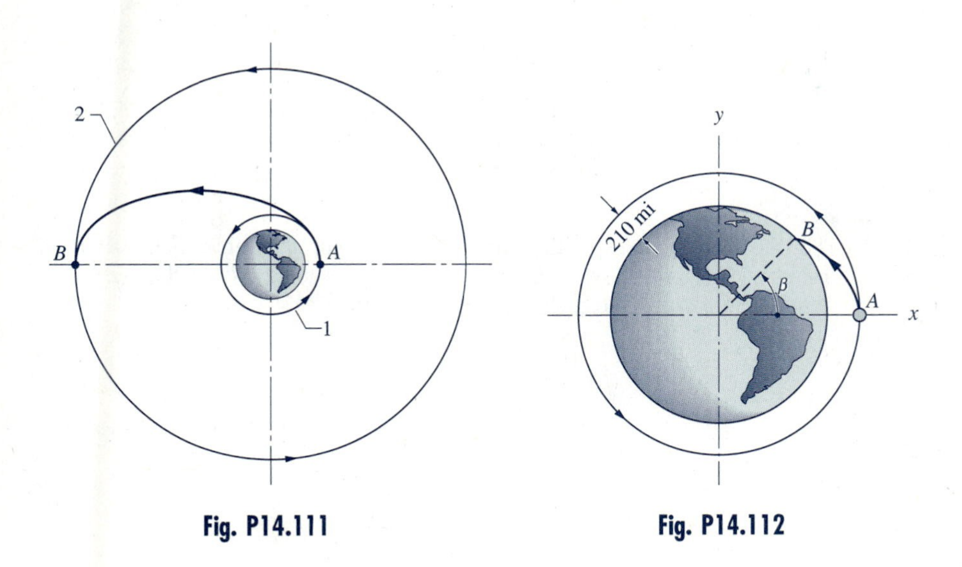

Fig. P14.111 **Fig. P14.112**

14.112 A spacecraft is traveling in a circular orbit around the earth at an altitude of 210 miles. When the craft reaches point A, its engines are fired for a short period, reducing its speed by 7.5%. The resulting path is the crash trajectory AB. Determine the angle β, measured from the nonrotating x-axis, which locates the landing site.

14.113 A ballistic missile fired from the North Pole lands at the equator, after reaching a maximum altitude of 360 km above the surface of the earth. Neglecting air resistance, find the firing angle (measured from the vertical) and the initial speed of the missile.

14.114 A satellite is launched into orbit around earth at an altitude $H_0 = 480$ mi with the initial speed $v_0 = 17\,500$ mi/h in the direction shown. (a) Derive the equations of motion for the satellite, and state the initial conditions. (b) Solve the equations numerically from the time of launch to the time when the satellite returns to the launch position; plot R vs. t. (c) Determine the orbital period (time to execute one orbit). (d) Find the highest and lowest altitudes reached by the satellite.

14.115 A spacecraft has the velocity $v_0 = 15\,000$ mi/h in the direction shown when its altitude is $H_0 = 480$ mi. (a) Derive the equations of motion for the spacecraft, and state the initial conditions. (b) Solve the equations numerically until the time when the satellite hits the earth; plot R vs. θ. (c) Find the angle θ at the impact point.

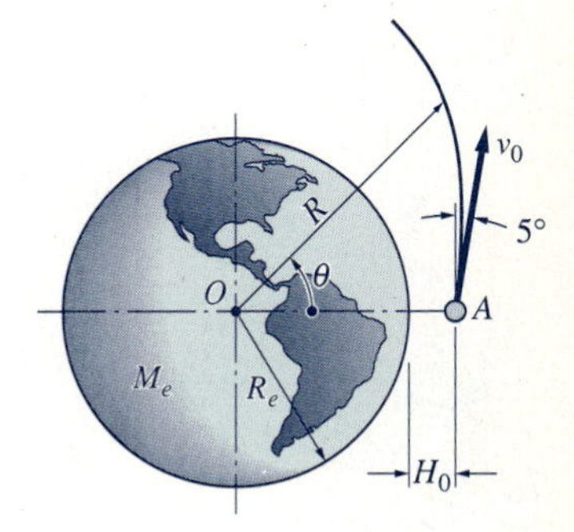

Fig. P14.114, P14.115

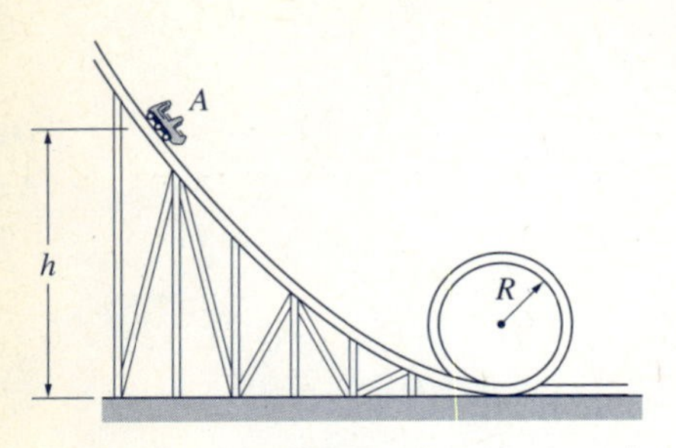

Fig. P14.116

REVIEW PROBLEMS

14.116 The car of a rollercoaster travels with negligible friction on the track shown. If the car starts from rest at A, determine the smallest ratio h/R that is necessary for the car to stay in contact with the track.

14.117 The 1200-lb box A is hoisted by the motor-driven winch at C. In the position shown, the tension in the cable is 1300 lb and the box is moving up at 3 ft/s. Determine the power output of the motor driving the winch at this instant.

14.118 The coefficients of static and kinetic friction between the 18-kg crate and the horizontal surface are $\mu_s = 0.3$ and $\mu_k = 0.25$, respectively. The crate is at rest when the horizontal, time-dependent force P is applied. Determine the speed of the crate at $t = 10$ s.

14.119 The 8-lb mass A slides on the vertical rod OA with negligible friction. The spring AB has a free length of 3 ft, and its stiffness is 4.5 lb/ft. If the mass is released from rest in the position shown, determine its speed at O.

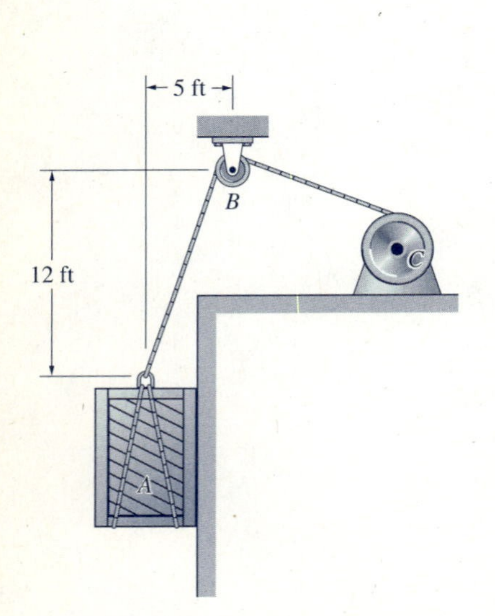

Fig. P14.117

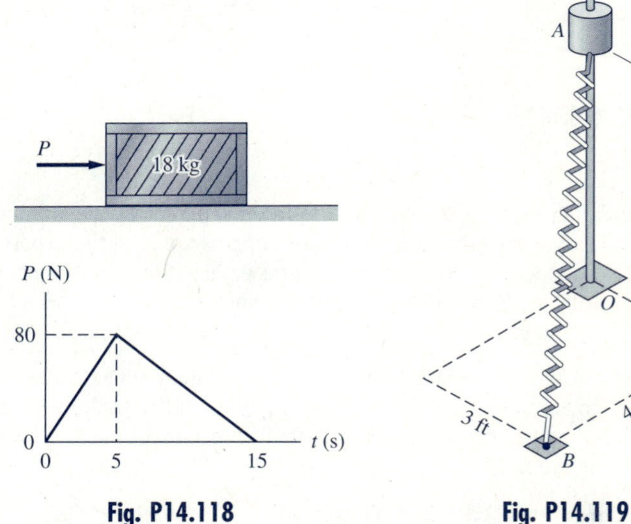

Fig. P14.118

Fig. P14.119

14.120 A package of mass M is placed with zero velocity on a conveyor belt that is moving at 2 m/s. The coefficient of kinetic friction between the package and the belt is 0.4. Determine the distance that the package travels before it reaches the speed of the belt.

14.121 Determine the time required for the package in Prob. 14.120 to reach the speed of the conveyor belt.

14.122 The weight of a rocket at launch is 150 lb, half of which is the weight of fuel. During the powered portion of the flight, the rate of fuel consumption is 6.5 lb/s and the thrust of the engine is 450 lb. Determine the speed of the rocket as a function of time during the powered flight. Assume the flight path to be vertical, and neglect air resistance and changes in gravitational acceleration.

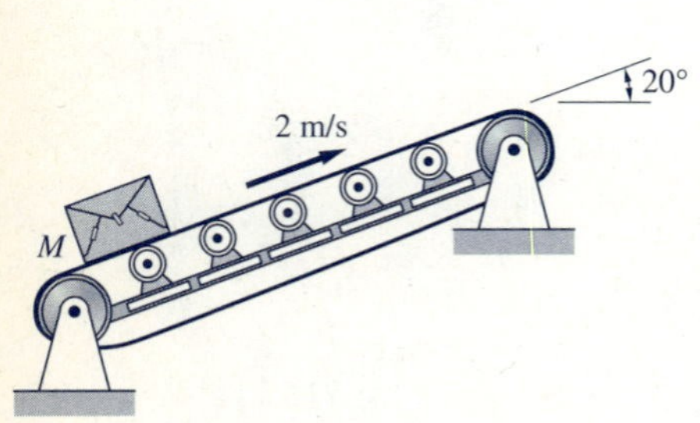

Fig. P14.120

14.123 An unpowered space vehicle of mass m travels past the planet Venus along the trajectory AB. If the coordinates of A and B are $(-5.0, -1.0) \times 10^8$ m and $(3.960, 3.382) \times 10^8$ m, respectively, determine the speed of the vehicle at B.

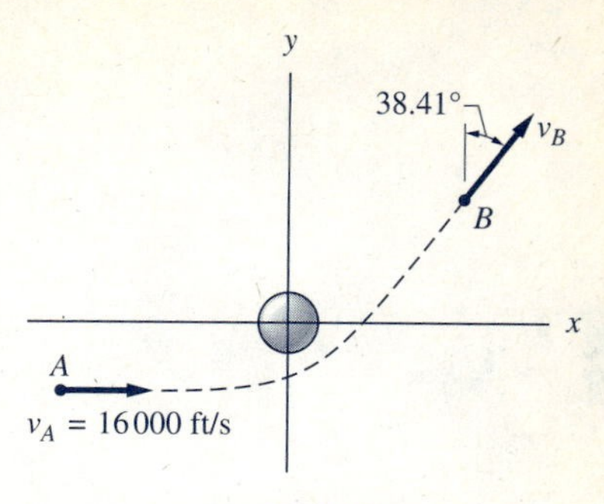

Fig. P14.123

14.124 The relationship between the axial force F and the resulting deformation x of the nonlinear spring is $F = kx(1 + x^2/b^2)$, where $k = 12$ kN/m and $b = 0.025$ m. After the 6-kg mass hits the free end of the spring with the speed v_0, the maximum deformation of the spring is observed to be 0.032 m. Determine the value of v_0.

14.125 The mass m is attached to a string and swings in a horizontal circle of radius $R = 0.25$ m when $L = 0.5$ m. The length L is then shortened by pulling the string slowly through the hole A in a table until the speed of the mass has doubled. Determine the corresponding value of L.

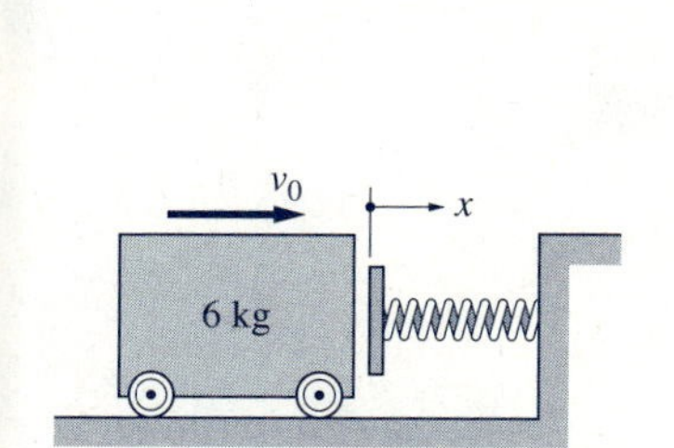

Fig. P14.124

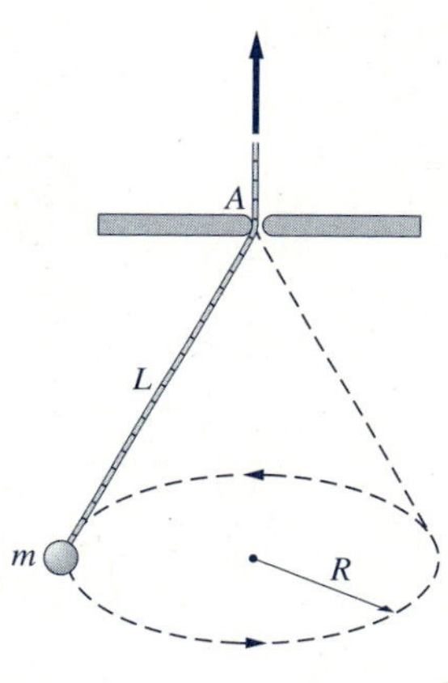

Fig. P14.125

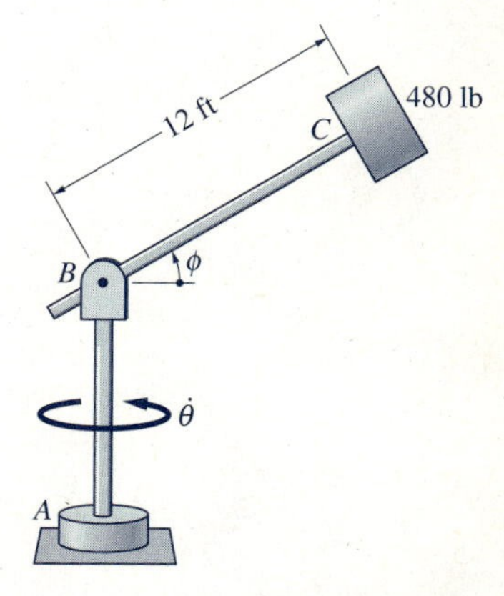

Fig. P14.126

14.126 The assembly shown rotates about the axis AB with the constant angular velocity $\dot{\theta} = 1.5$ rad/s while an internal mechanism raises the arm BC at the constant rate $\dot{\phi} = 0.4$ rad/s. A brake installed at A is responsible for keeping $\dot{\theta}$ constant. Determine (a) the couple that the brake must apply about AB as a function of ϕ; and (b) the energy that the brake must absorb as ϕ varies from 0° to 90°.

14.127 The 0.15-kg mass is attached to the elastic cord OA, which has a free length of 250 mm and a stiffness of 40 N/m. The mass is launched on a frictionless, horizontal table from position A with the velocity of 2 m/s perpendicular to the cord. When the mass reaches position B, where the cord becomes slack for the first time, determine (a) the speed of the mass; and (b) the angle β between the cord and the velocity vector.

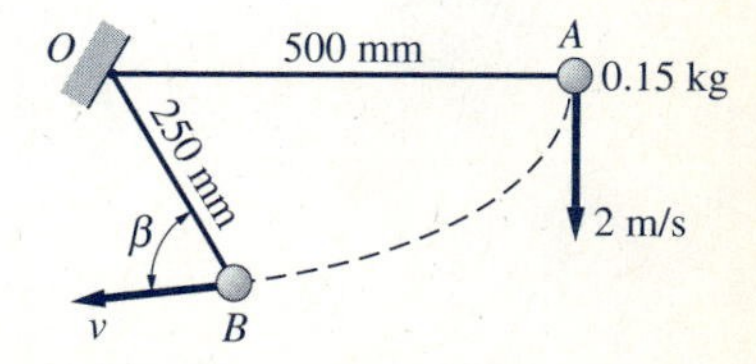

Fig. P14.127

15 Dynamics of Particle Systems

15.1 Introduction

Up to this point, our discussion of dynamics has focused on the three methods of analysis of particle motion: force-mass-acceleration, work-energy, and impulse-momentum. In this chapter we extend these methods to analysis of systems containing two or more particles.

Before the kinetics of systems of particles can be discussed, it is necessary to consider the kinematics of relative motion between particles. Relative motion provides a convenient description of the geometric constraints that are usually present in systems of particles. The kinematics of relative motion, combined with one or more of the kinetic methods, is sufficient for the analysis of most particle systems.

This chapter introduces two new topics—the impact of particles and mass flow. Impact refers to collisions of particles and is characterized by a very short time of contact during which the contact forces reach very large values. Mass flow applies to problems where mass enters or leaves the system, such as fluid flow through pipes and rocket propulsion.

15.2 Kinematics of Relative Motion

Absolute motion denotes motion that is described in a fixed, or inertial, reference frame. *Relative motion*, on the other hand, is measured with respect to a reference point, or reference frame, that is itself in motion. The laws of kinetics, such as Newton's second law, are invariably restricted to absolute motion. However, in kinematics it is frequently convenient to employ relative motion.

Figure 15.1(a) shows the paths of two particles A and B. At time t, the velocities of the particles are denoted by $\mathbf{v}_A$ and $\mathbf{v}_B$, and their position vectors measured from the origin O of the fixed XYZ axes are $\mathbf{r}_A$ and $\mathbf{r}_B$. The vector $\mathbf{r}_{B/A}$, drawn from A to B, is called the *relative position vector* (position of B

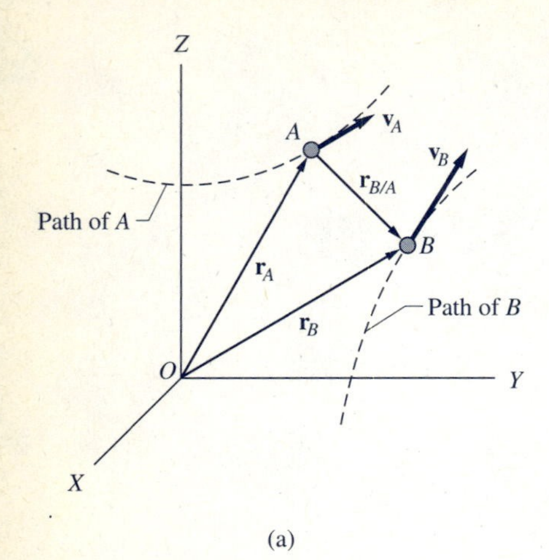

(a)

(b)

Fig. 15.1

relative to A). Note that particle A, from which $\mathbf{r}_{B/A}$ is drawn, constitutes a moving reference point.

From Fig. 15.1(a) we see that the relative and absolute position vectors are related by

$$\mathbf{r}_B = \mathbf{r}_A + \mathbf{r}_{B/A} \tag{15.1}$$

Taking the time derivative of this equation and introducing the notation

$$\mathbf{v}_{B/A} = \frac{d\mathbf{r}_{B/A}}{dt} \tag{15.2}$$

yields the relative velocity equation,

$$\mathbf{v}_B = \mathbf{v}_A + \mathbf{v}_{B/A} \tag{15.3}$$

The vector $\mathbf{v}_{B/A}$, shown in Fig. 15.1(b), is called the *relative velocity vector* (velocity of B relative to A).

Differentiating each side of Eq. (15.3) with respect to time gives the relative acceleration equation:

$$\mathbf{a}_B = \mathbf{a}_A + \mathbf{a}_{B/A} \tag{15.4}$$

where

$$\mathbf{a}_{B/A} = \frac{d\mathbf{v}_{B/A}}{dt} = \frac{d^2\mathbf{r}_{B/A}}{dt^2} \tag{15.5}$$

is called the *relative acceleration vector* (acceleration of B relative to A).

From Fig. 15.1(a) it is obvious that the vector from A to B is the negative of the vector from B to A, an observation that leads us to the following identities.

$$\begin{aligned} \mathbf{r}_{B/A} &= -\mathbf{r}_{A/B} \\ \mathbf{v}_{B/A} &= -\mathbf{v}_{A/B} \\ \mathbf{a}_{B/A} &= -\mathbf{a}_{A/B} \end{aligned} \tag{15.6}$$

Therefore, the kinematic variables that relate the motion of B to A are equal in magnitude but opposite in direction to the corresponding variables that relate the motion of A to B.

It is often convenient to describe the relative motion utilizing a coordinate system that moves with the reference particle. In Fig. 15.2 the reference frame XYZ is fixed in space, and the xyz-axes are attached to (and move with) the reference particle A. Letting x, y, and z be the rectangular coordinates of particle B measured relative to the xyz-frame, the expression for the relative position vector is

$$\mathbf{r}_{B/A} = x\mathbf{i} + y\mathbf{j} + z\mathbf{k} \tag{15.7}$$

where **i**, **j**, and **k** are the base vectors of the xyz-frame. Therefore, the relative velocity and acceleration vectors are found to be

$$\mathbf{v}_{B/A} = \dot{\mathbf{r}}_{B/A} = (\dot{x}\mathbf{i} + \dot{y}\mathbf{j} + \dot{z}\mathbf{k}) + (x\dot{\mathbf{i}} + y\dot{\mathbf{j}} + z\dot{\mathbf{k}}) \tag{15.8}$$

and

$$\begin{aligned}\mathbf{a}_{B/A} = \dot{\mathbf{v}}_{B/A} = \ddot{\mathbf{r}}_{B/A} &= (\ddot{x}\mathbf{i} + \ddot{y}\mathbf{j} + \ddot{z}\mathbf{k}) \\ &+ 2(\dot{x}\dot{\mathbf{i}} + \dot{y}\dot{\mathbf{j}} + \dot{z}\dot{\mathbf{k}}) \\ &+ (x\ddot{\mathbf{i}} + y\ddot{\mathbf{j}} + z\ddot{\mathbf{k}})\end{aligned} \tag{15.9}$$

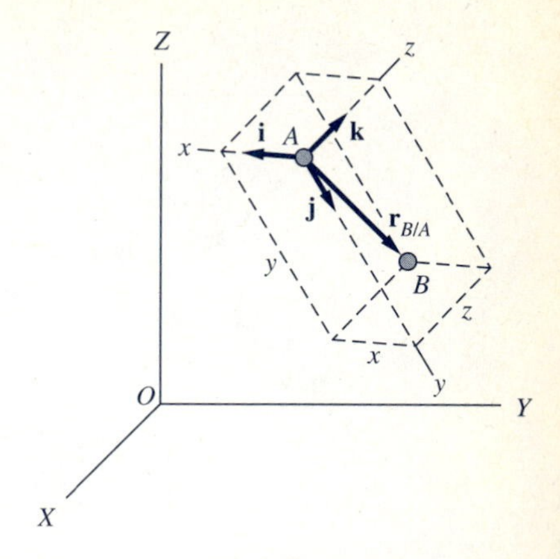

Fig. 15.2

Note that $\dot{\mathbf{i}}, \dot{\mathbf{j}}, \dot{\mathbf{k}}$ and $\ddot{\mathbf{i}}, \ddot{\mathbf{j}}, \ddot{\mathbf{k}}$, the time derivatives of the base vectors, are not necessarily zero. Although **i**, **j**, and **k** have constant magnitudes (equal to unity), their directions may vary with time, in which case their time derivatives will not vanish. Derivatives of base vectors are discussed in Art. 16.7.

For the special case where the xyz reference frame is not rotating, the derivatives of **i**, **j**, and **k** are identically zero, because both the magnitude and direction of each base vector are constant. Therefore, Eqs. (15.8) and (15.9) yield

$$\mathbf{v}_{B/A} = (\dot{x}\mathbf{i} + \dot{y}\mathbf{j} + \dot{z}\mathbf{k}) \tag{15.10}$$

and

$$\mathbf{a}_{B/A} = (\ddot{x}\mathbf{i} + \ddot{y}\mathbf{j} + \ddot{z}\mathbf{k}) \tag{15.11}$$

From Eqs. (15.10) and (15.11) we conclude that $\mathbf{v}_{B/A}$ and $\mathbf{a}_{B/A}$ can be interpreted as the velocity and acceleration of particle B that would be seen by a *nonrotating* observer who moves with particle A.

15.3 Kinematics of Constrained Motion

Before discussing the kinematics of constrained motion, it is necessary to introduce the following terminology:*

Kinematic constraints: Geometric constraints imposed on the motion of particles are called kinematic constraints.

Equations of constraint: Equations of constraint are the mathematical expressions that describe the kinematic constraints on particles in terms of their position coordinates (angles or distances) or their derivatives.

Kinematically independent coordinates: Position coordinates that are not subject to constraints are called kinematically independent coordinates, or simply, independent coordinates.

Degrees of freedom: The number of degrees of freedom (DOF) of a system of particles is the number of kinematically independent position coordinates that are required to completely describe its configuration.

* These terms were introduced in Art 10.3 as a preliminary to the discussion of virtual work. The definitions are paraphrased here because of their importance in dynamics.

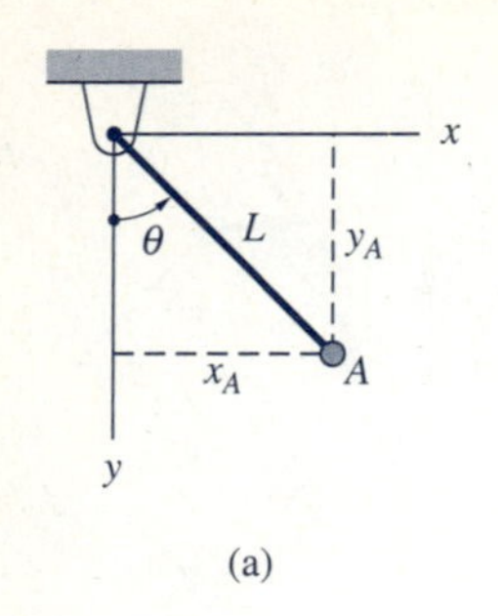

(a)

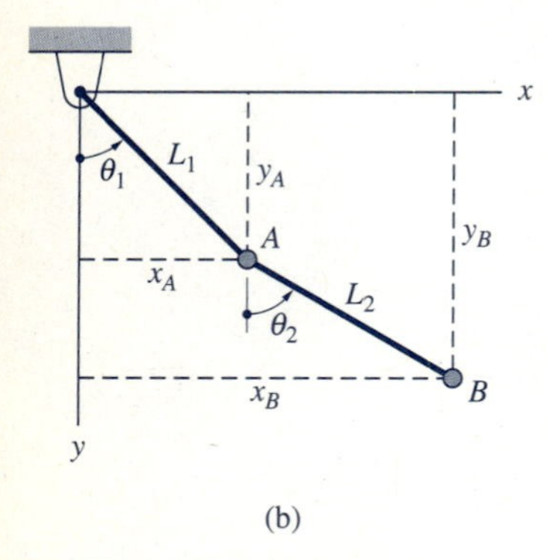

(b)

Fig. 15.3

As an illustration, consider the simple pendulum shown in Fig. 15.3(a). It consists of a mass A on an inextensible string of length L. The number of DOF is one, because the angular coordinate θ is sufficient to locate the mass. The rectangular coordinates x_A and y_A could also be used to locate A, but note that they must satisfy the equation of constraint $x_A^2 + y_A^2 = L^2$.

The double pendulum shown in Fig. 15.3(b) is an example of a two-DOF system, because a minimum of two coordinates, such as θ_1 and θ_2, are required to locate the two masses. If the four coordinates x_A, y_A, x_B, and y_B are used, they will be subject to two equations of constraint:

$$x_A^2 + y_A^2 = L_1^2 \quad \text{and} \quad (x_B - x_A)^2 + (y_B - y_A)^2 = L_2^2$$

In each of the preceding examples, we observe that (number of coordinates selected) − (number of equations of constraint) = (number of DOF). This is a characteristic of all but a special class of mechanical systems.*

In previous chapters, we have considered many cases of constrained motion of a single object (a car that moves along a straight road, a collar that slides on a rod, etc.). For systems of particles, there can also be kinematic constraints on the relative motion between the particles in addition to constraints on the absolute motion of individual particles. For example, Fig. 15.4 shows a system consisting of two particles A and B connected by a rigid bar of length L, that slides on a horizontal table. The table imposes a constraint on the absolute motion of the particles: The position vectors $\mathbf{r}_A$ and $\mathbf{r}_B$ must lie in the xy-plane. In addition, the rigid bar imposes a constraint on the relative motion of the particles: $|\mathbf{r}_{B/A}| = L$.

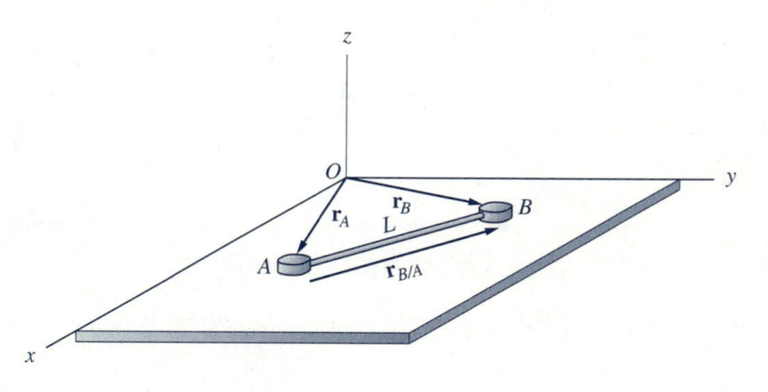

Fig. 15.4

* Systems for which the number of required coordinates is always larger than the number of degrees of freedom are called *non-holonomic* systems. See D. T. Greenwood, *Classical Dynamics*, Prentice-Hall, 1977, p. 10.

Sample Problem 15.1

Two airplanes A and B are flying at constant speeds at the same altitude. The positions of the planes at $t = 0$ are as shown in the figure. Assuming that the xy-coordinate system is fixed in space, determine (1) the velocity vector of A relative to B; (2) the position vector of A relative to B as a function of time; and (3) the smallest distance between the planes and the time when this occurs.

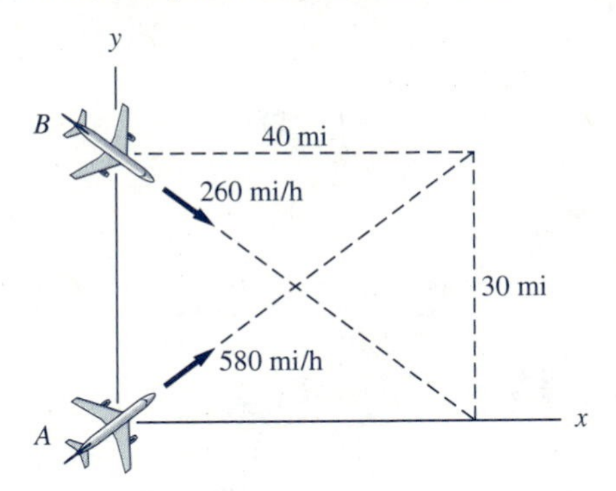

Solution

Part 1

The velocity vectors of the planes are

$$\mathbf{v}_A = 580\left(\frac{40\mathbf{i} + 30\mathbf{j}}{50}\right) = 464\mathbf{i} + 348\mathbf{j} \text{ mi/h}$$

$$\mathbf{v}_B = 260\left(\frac{40\mathbf{i} - 30\mathbf{j}}{50}\right) = 208\mathbf{i} - 156\mathbf{j} \text{ mi/h}$$

The velocity vector of A relative to B is

$$\mathbf{v}_{A/B} = \mathbf{v}_A - \mathbf{v}_B = (464\mathbf{i} + 348\mathbf{j}) - (208\mathbf{i} - 156\mathbf{j})$$
$$= 256\mathbf{i} + 504\mathbf{j} \text{ mi/h} \qquad \textit{Answer}$$

Part 2

The position vector of A relative to B can be found by integrating the relative velocity vector (note that $\mathbf{i}$ and $\mathbf{j}$ are constant, since the xy-coordinate system is fixed):

$$\mathbf{r}_{A/B} = \int \mathbf{v}_{A/B}\, dt = \int (256\mathbf{i} + 504\mathbf{j})\, dt = 256t\mathbf{i} + 504t\mathbf{j} + \mathbf{r}_0$$

where $\mathbf{r}_0$ is the relative position vector $\mathbf{r}_{A/B}$ evaluated at $t = 0$ (a constant of integration). By inspection of the figure, we see that $\mathbf{r}_0 = -30\mathbf{j}$ mi. Therefore, the position vector of A relative to B becomes

$$\mathbf{r}_{A/B} = 256t\mathbf{i} + (504t - 30)\mathbf{j} \text{ mi} \qquad \textit{Answer}$$

Part 3

The distance between the airplanes is

$$|\mathbf{r}_{A/B}| = \sqrt{(256t)^2 + (504t - 30)^2} \text{ mi} \qquad \text{(a)}$$

The minimum distance occurs at the time t for which $|d\mathbf{r}_{A/B}|/dt = 0$, i.e., when

$$\frac{1}{2}\frac{2(256)^2 t + 2(504t - 30)(504)}{\sqrt{(256t)^2 + (504t - 30)^2}} = 0$$

which yields

$$t = 0.047\,32 \text{ h} = 170.4 \text{ s} \qquad \textit{Answer}$$

Substituting this value of t into Eq. (a) gives

$$|\mathbf{r}_{A/B}|_{\min} = 13.59 \text{ mi} \qquad \textit{Answer}$$

Sample Problem 15.2

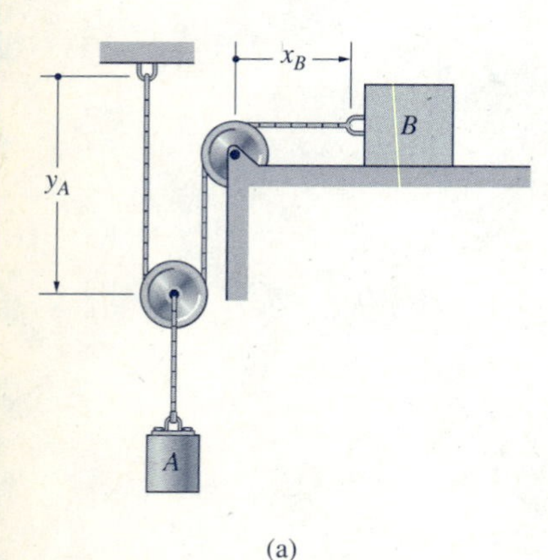

(a)

Figure (a) shows a system consisting of two blocks A and B connected by an inextensible cable that runs around two pulleys. Determine the kinematic relationships between the velocities and the accelerations of the blocks.

Solution

The system shown in Fig. (a) has one degree of freedom because one coordinate, e.g., y_A or x_B, determines its configuration. We will explain two methods for relating the velocities and accelerations of the blocks; both methods follow from the fact that the length of the cable is constant.

Method I

It is convenient to number the pulleys and to indicate the fixed distance h, as shown in Fig. (b). Letting L be the length of the cable, we have

$$L = y_A + \begin{pmatrix}\text{length of cable wrapped} \\ \text{around pulley 1}\end{pmatrix} + (y_A - h)$$

$$+ \begin{pmatrix}\text{length of cable wrapped} \\ \text{around pulley 2}\end{pmatrix} + x_B$$

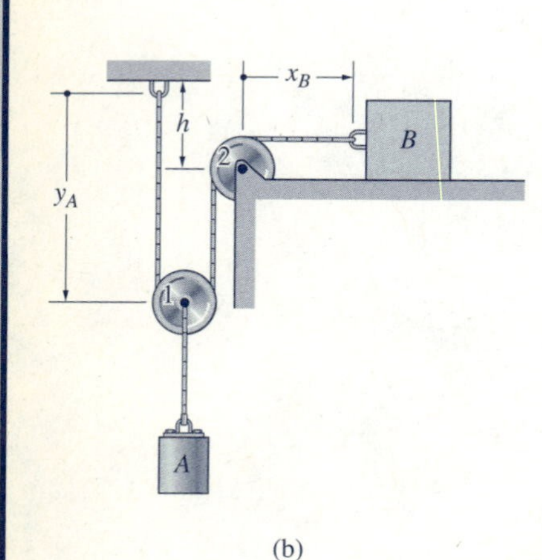

(b)

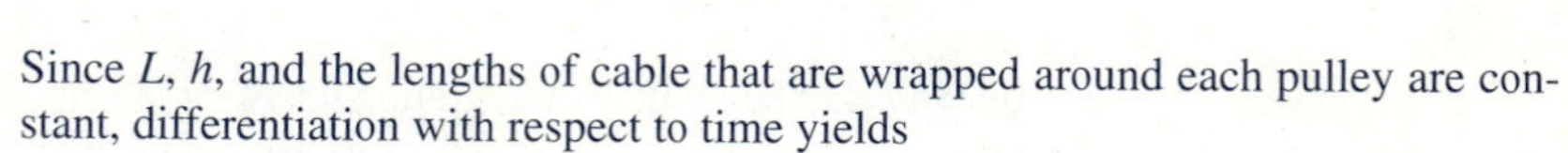

Since L, h, and the lengths of cable that are wrapped around each pulley are constant, differentiation with respect to time yields

$$\frac{dL}{dt} = v_A + 0 + v_A + 0 + v_B = 0$$

which gives

$$v_B = -2v_A \qquad \textit{Answer}$$

Differentiation of this equation with respect to time yields

$$a_B = -2a_A \qquad \textit{Answer}$$

Method II

The kinematic relationships between the motions of the blocks can also be found by giving the coordinates y_A and x_B the differential increments dy_A and dx_B, respectively, as shown in Fig. (c). It can be seen that the corresponding increment in the length of the cable is

$$dL = 2dy_A + dx_B$$

However, since the length L is constant, it imposes the constraint $dL = 0$, which means that

$$2dy_A + dx_B = 0$$

Division by dt yields $2v_A + v_B = 0$, which leads to the same kinematic relationships for the velocities and accelerations that are obtained by Method I.

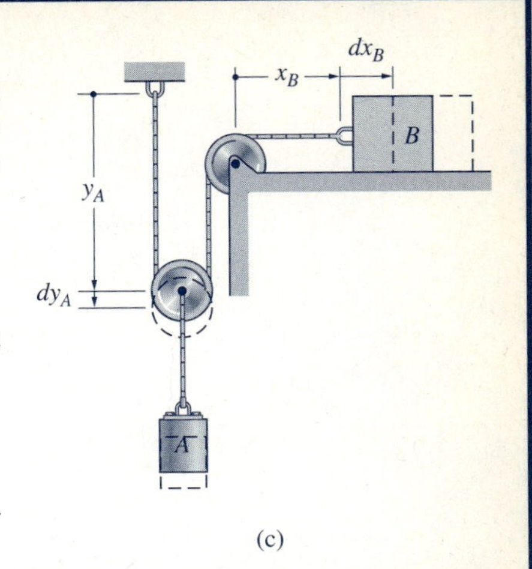

(c)

Sample Problem 15.3

The two collars A and B are pinned to a rigid rod of length L. Collar A is moving to the right with constant velocity v_A. Determine the velocity and acceleration of collar B as functions of v_A and the angle θ.

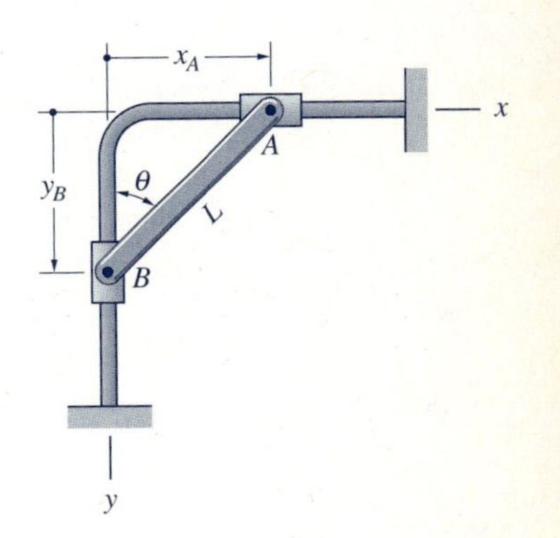

Solution

The system has one degree of freedom because one coordinate (e.g., the angle θ) determines its configuration. This problem will be solved by the following two methods: (I) differentiating the equation of constraint that relates x_A and y_B; and (II) relating the motions of A and B using θ as the independent coordinate.

Method I

The equation of constraint that relates the coordinates x_A and y_B is

$$x_A^2 + y_B^2 = L^2 \tag{a}$$

Differentiating this equation with respect to time—noting that $\dot{x}_A = v_A$ and $\dot{y}_B = v_B$—we obtain $2x_Av_A + 2y_Bv_B = 0$, which reduces to

$$x_Av_A + y_Bv_B = 0 \tag{b}$$

Taking the time derivative of Eq. (b), we get

$$(x_Aa_A + v_A^2) + (y_Ba_B + v_B^2) = 0 \tag{c}$$

where we used $\dot{v}_A = a_A$ and $\dot{v}_B = a_B$.

Solving Eq. (b) for the velocity of B yields

$$v_B = -v_A\frac{x_A}{y_B} = -v_A\frac{L\sin\theta}{L\cos\theta}$$

or

$$v_B = -v_A \tan\theta \qquad \textit{Answer} \quad \text{(d)}$$

Since $a_A = 0$, the acceleration of B from Eq. (c) is

$$a_B = -\frac{v_A^2 + v_B^2}{y_B}$$

Substituting for v_B from Eq. (d) and $y_B = L\cos\theta$, we obtain

$$a_B = -\frac{v_A^2 + (-v_A \tan\theta)^2}{L\cos\theta} = -\frac{v_A^2(1 + \tan^2\theta)}{L\cos\theta}$$

Using the identity $(1 + \tan^2\theta) = 1/\cos^2\theta$, this equation reduces to

$$a_B = -\frac{v_A^2}{L\cos^3\theta} \qquad \textit{Answer} \quad \text{(e)}$$

Method II

The position coordinates of the collars are related to the independent coordinate θ as follows:

$$x_A = L\sin\theta \quad \text{and} \quad y_B = L\cos\theta \qquad \text{(f)}$$

Differentiation with respect to time gives

$$v_A = \dot{x}_A = L\dot{\theta}\cos\theta \quad \text{and} \quad v_B = \dot{x}_B = -L\dot{\theta}\sin\theta \qquad \text{(g)}$$

Using Eqs. (g), we obtain $v_B/v_A = -L\dot{\theta}\sin\theta/L\dot{\theta}\cos\theta$, from which the velocity of B is

$$v_B = -v_A \tan\theta \qquad \textit{Answer} \quad \text{(h)}$$

Using this equation, the acceleration of B is given by

$$a_B = \dot{v}_B = -v_A \frac{d}{dt}\tan\theta - a_A \tan\theta$$

Substituting $a_A = 0$ and $d(\tan\theta)/dt = \dot{\theta}\sec^2\theta$ yields

$$a_B = -v_A\dot{\theta}\sec^2\theta$$

From the first of Eqs. (g), we find $\dot{\theta} = v_A/L\cos\theta$. Therefore the expression for the acceleration of B becomes

$$a_B = -v_A \sec^2\theta \frac{v_A}{L\cos\theta} = -\frac{v_A^2}{L\cos^3\theta} \qquad \textit{Answer} \quad \text{(i)}$$

The answers in Eqs. (h) and (i) agree, of course, with those found previously in Eqs. (d) and (e).

PROBLEMS

15.1 The airspeed of a plane is 560 mi/h, directed north. If the wind speed is 60 mi/h in the direction shown, determine the ground speed and the course (angle θ) of the plane.

15.2 A boat with a cruising speed (speed of boat relative to the water) of 26 km/h is crossing a river that has a current of 10 km/h. (a) Find the course, determined by the angle θ, that the boat must steer in order to follow a straight line from A to C. (b) Find the time required for the boat to complete the crossing.

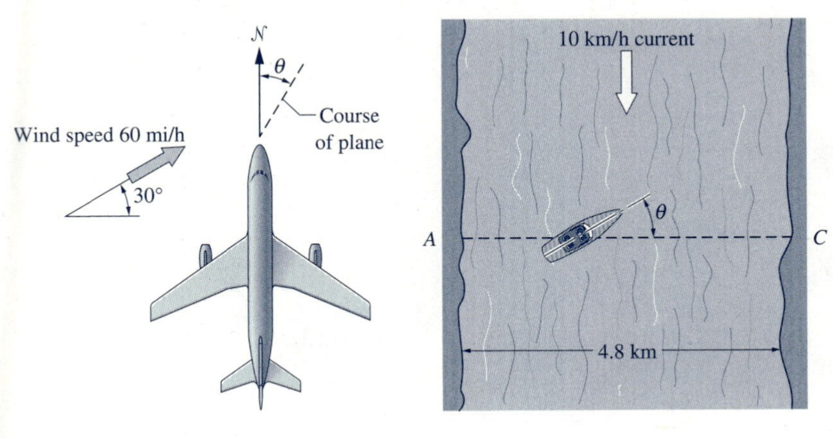

Fig. P15.1 **Fig. P15.2**

15.3 Two billiard balls A and B, initially at rest, are hit at the same instant and roll along the paths AC and BC. If the velocities of the balls are as shown in the figure, determine the angle θ if the balls are to collide.

15.4 When a stationary car is pointing into the wind, the streaks made by raindrops on the side windows are inclined at $\theta = 15°$ with the vertical. When the car is driven at 30 mi/h into the wind, the angle θ increases to 75°. Find the speed of the raindrops.

15.5 Two cars A and B traveling at constant speeds are in the positions shown at time $t = 0$. Determine (a) the velocity of A relative to B; (b) the position vector of A relative to B as a function of time; and (c) the smallest distance between the cars and the time when this occurs.

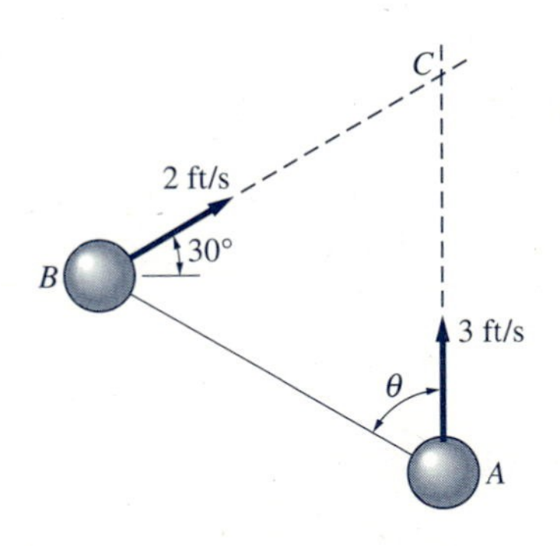

Fig. P15.3

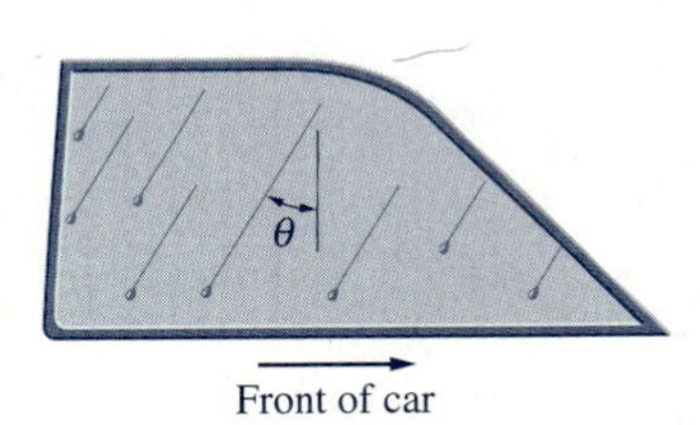

Fig. P15.4

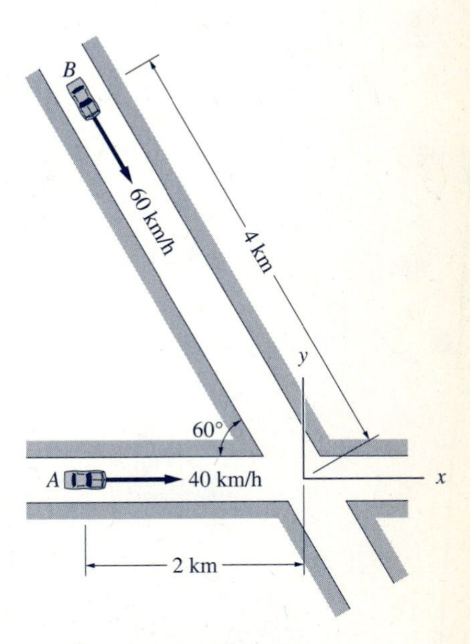

Fig. P15.5

15.6 Each of the airplanes maintains a constant course and constant ground speed after being launched simultaneously from an aircraft carrier. Three minutes after launch, the planes are flying at the same altitude and are in the relative positions shown, where $\theta = 30°$ and $r = 6$ mi. If the heading of plane B is $\alpha = 70°$, determine the ground speed of each plane.

15.7 The airplane A is flying due west at a constant altitude with a constant speed of 1120 km/h. The radar in plane A detects a second plane B, flying at the same altitude, that is located at $\theta = 44.1°$ and $r = 28.3$ km. Five minutes after the initial sighting, the radar readings are $\theta = 11.4°$ and $r = 68.9$ km. Calculate the speed and the heading angle α of plane B.

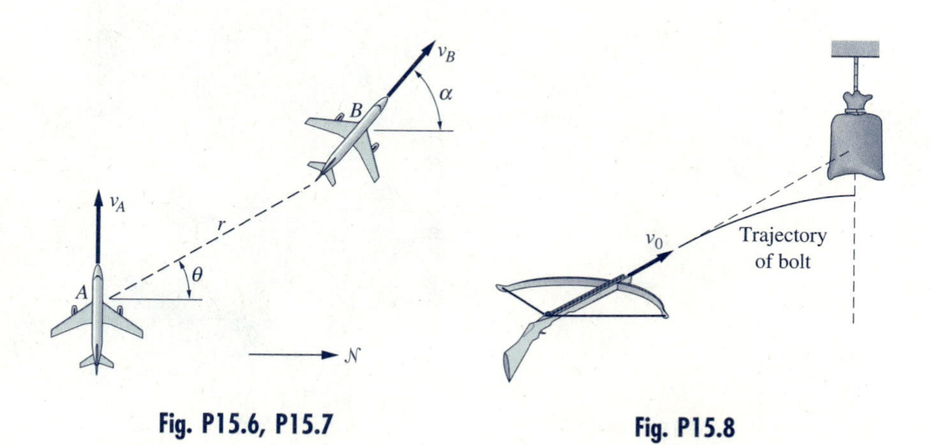

Fig. P15.6, P15.7 **Fig. P15.8**

15.8 The crossbow is aimed at the sandbag, which is suspended from a cord. At the instant the cord is cut, the crossbow is fired. Show that the bolt will always hit the sandbag, regardless of the initial velocity v_0 of the bolt.

15.9 Two projectiles A and B are launched simultaneously in the same vertical plane with the initial positions and velocities shown in the figure. If the projectiles collide 8 s after launch, determine (a) their relative velocity at collision; and (b) the initial velocity vector of A.

15.10 In Figs. (a) and (b), block A is moving to the right with the velocity 6 in./s. Compute the velocity of block B in each case.

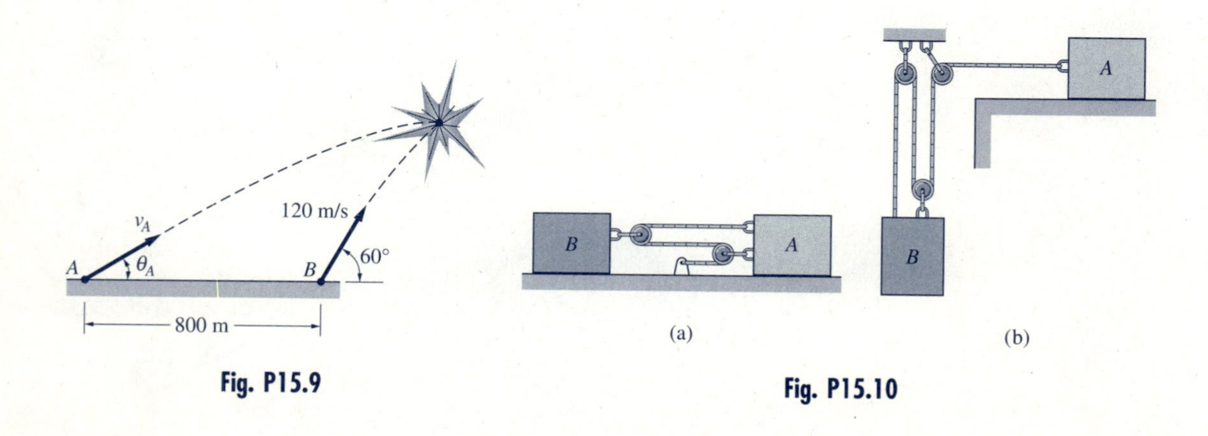

Fig. P15.9 **Fig. P15.10**

15.11 In Figs. (a) and (b), the velocity of block A relative to block B is 360 mm/s, directed downward. Calculate the velocities of both blocks in each case.

15.12 Determine the velocity of block B if block A is moving upward with the velocity 16 in./s.

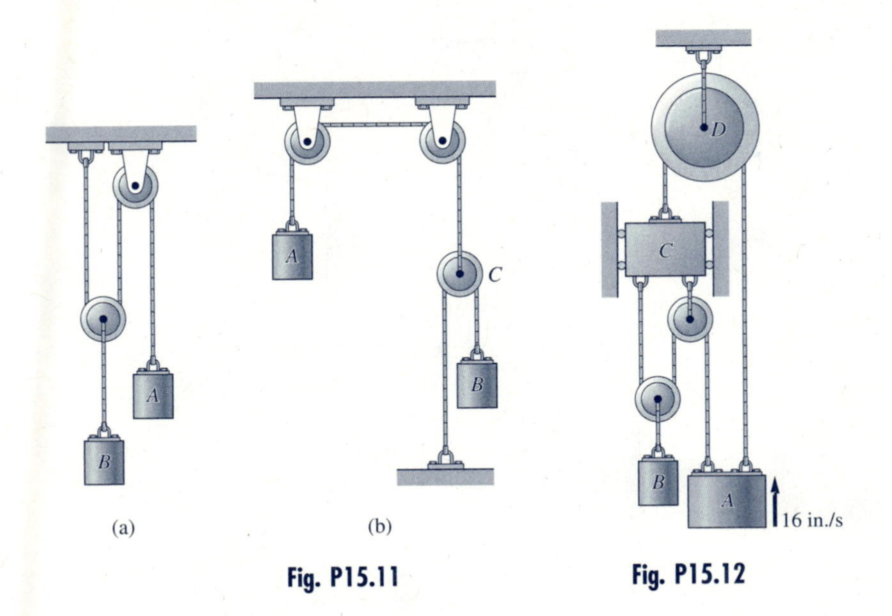

Fig. P15.11 **Fig. P15.12**

15.13 The two tapered blocks A and B are at rest in the position shown. If the acceleration of A is 2 m/s^2 to the right, calculate the acceleration of B.

15.14 The pin P at the end of the plunger A engages the semicircular slot in the sliding panel B. If A is moving to the right with the constant velocity 0.4 ft/s, determine the velocity and acceleration of B when $x = 1.0$ ft.

15.15 The bucket B is being raised by a mule moving at the constant speed v_A. Compute the velocity and acceleration of the bucket as functions of x.

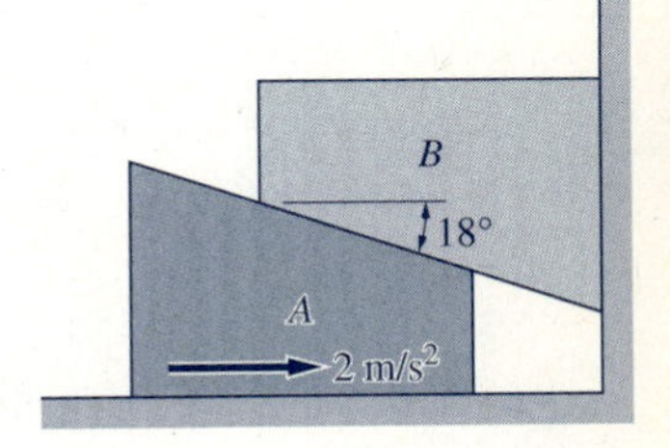

Fig. P15.13

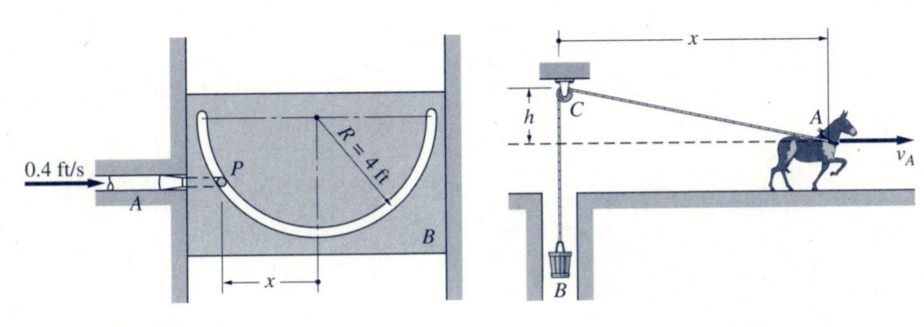

Fig. P15.14 **Fig. P15.15**

15.16 The velocity of the slider A as a function of its position is $v_A = 4 - (x/4)$ m/s, where x is in meters. Determine the velocity and acceleration of block B when $x = 4$ m.

15.17 Sliders A and B are connected by the 8-in. bar. Given that $v_A = 8$ in./s and $a_A = -16$ in./s^2 when $x = 3$ in., find the velocity and acceleration of B at this instant.

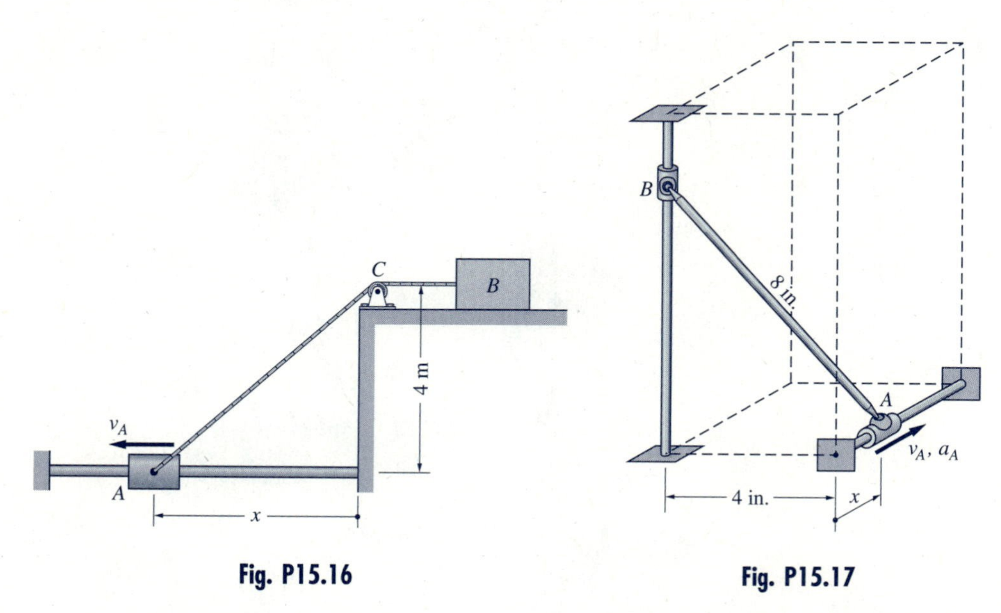

Fig. P15.16 **Fig. P15.17**

***15.18** In order to gain speed, the waterskier B follows a straight-line path that differs by 20° from the course being followed by the towing boat A. Determine the angle θ for which the speed of the skier is twice the speed of the boat.

15.19 Two cars travel at constant speeds $v_A = 36$ ft/s and $v_B = 45$ ft/s around a circular track. When the cars are in the positions shown, determine the magnitudes of $\mathbf{v}_{B/A}$ and $\mathbf{a}_{B/A}$.

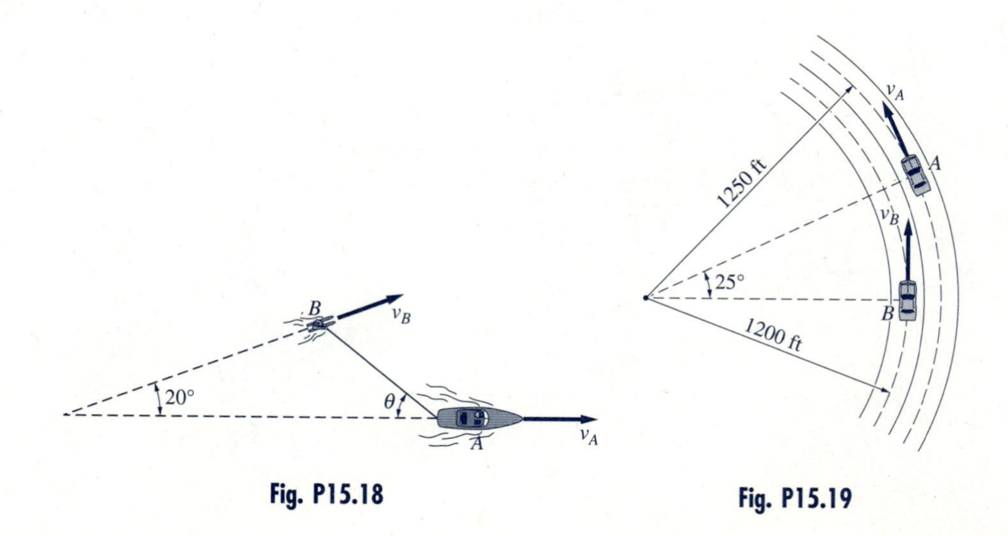

Fig. P15.18 **Fig. P15.19**

15.20 The pin A at the end of crank OA engages a slot in slider B. If OA rotates with the constant angular speed $\dot{\theta} = \omega_0$, calculate the velocity and acceleration of pin A relative to slider B in terms of R, ω_0 and θ.

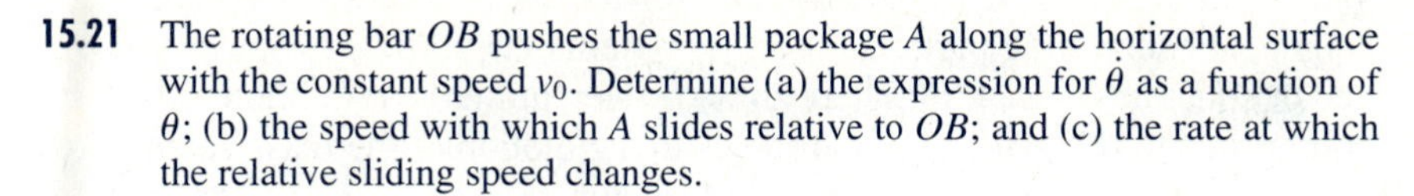

15.21 The rotating bar OB pushes the small package A along the horizontal surface with the constant speed v_0. Determine (a) the expression for $\dot{\theta}$ as a function of θ; (b) the speed with which A slides relative to OB; and (c) the rate at which the relative sliding speed changes.

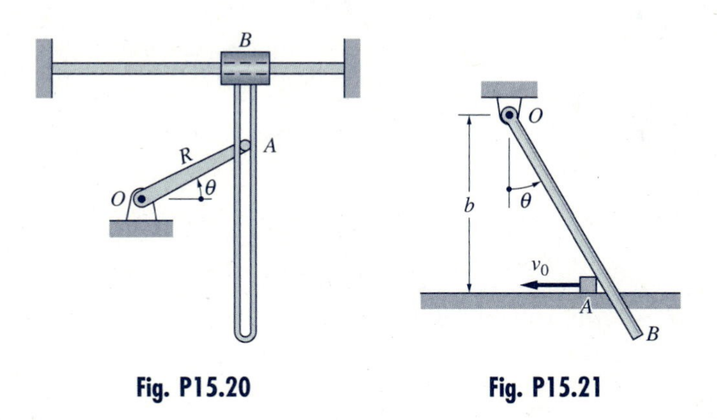

Fig. P15.20 **Fig. P15.21**

15.22 The T-shaped rod rotates about point O. When $\theta = 0$, the angular velocity and acceleration of the rod are $\dot{\theta} = 2$ rad/s and $\ddot{\theta} = -10$ rad/s^2, respectively. Determine the velocity and acceleration of end B relative to end A at this instant. Express your answers in terms of rectangular components.

15.23 When $\theta = 30°$, end A of bar AB is sliding to the left with the speed 1.2 m/s. Calculate the speed of end B at this instant.

***15.24** Collar A slides along bar OB, which is free to rotate about O. If the collar travels along the circular path of radius b with the constant speed v_0, determine the speed and acceleration with which the collar slides relative to the rod at (a) point C; and (b) point D.

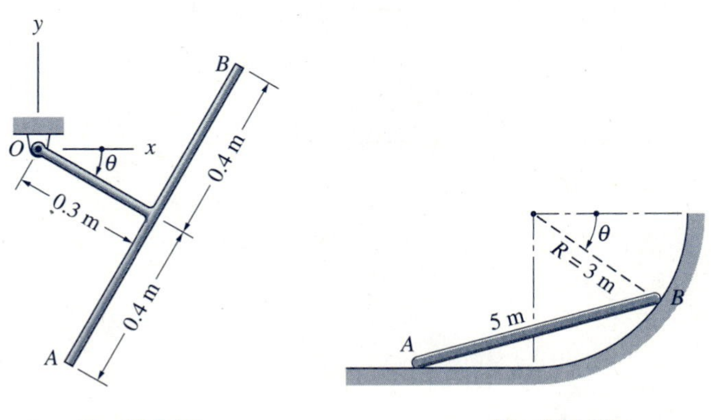

Fig. P15.22 **Fig. P15.23**

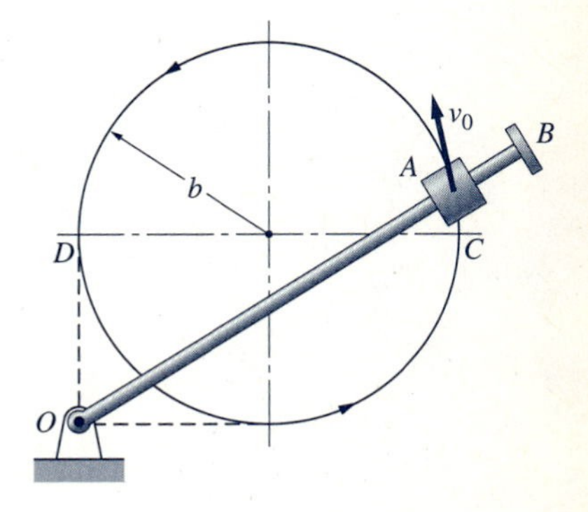

Fig. P15.24

15.4 Equations of Motion of the Mass Center

When analyzing a system of particles, Newton's second law can be applied, of course, to each of the particles. Thus a system of n particles is governed by n vector equations of motion. There are, however, numerous situations where it is appropriate to relate the external forces to the motion of the mass center of the system. In these cases, the equations of motion can be simplified considerably. But before we can proceed further, we must define the mass center of a system of particles.*

Figure 15.5 shows a system of n particles. The mass of the ith particle is denoted by m_i, and its position vector is $\mathbf{r}_i = x_i\mathbf{i} + y_i\mathbf{j} + z_i\mathbf{k}$. The center of mass G is defined to be the point whose position vector is given by

$$\mathbf{r}_G = \frac{1}{m}\sum_{i=1}^{n} m_i\mathbf{r}_i \tag{15.12}$$

where $m = \Sigma_{i=1}^{n} m_i$ is the total mass of the system. The rectangular components of $\mathbf{r}_G$ are†

$$x_G = \frac{1}{m}\sum_{i=1}^{n} m_i x_i \qquad y_G = \frac{1}{m}\sum_{i=1}^{n} m_i y_i \qquad z_G = \frac{1}{m}\sum_{i=1}^{n} m_i z_i \tag{15.13}$$

If the particles of the system are moving, the velocity and acceleration vectors of the mass center will be $\mathbf{v}_G = d\mathbf{r}_G/dt$ and $\mathbf{a}_G = d\mathbf{v}_G/dt = d^2\mathbf{r}_G/dt^2$, respectively (we assume that the xyz-coordinate system is a fixed reference system).

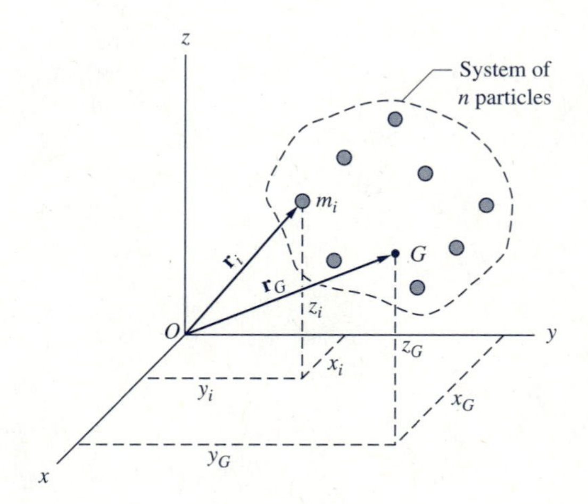

Fig. 15.5

* As mentioned in Art. 8.5, engineers often use *center of gravity* and *center of mass* interchangeably because the two points coincide for most cases of engineering importance.

† The defining equations for the mass center of a system of particles, Eqs. (15.12) and (15.13), are similar to the corresponding equations for the mass center of a body given in Eqs. (8.16). The main difference is that for systems of particles the integrations are replaced by summations.

Figure 15.6(a) shows the free-body diagram of a closed system of n particles labeled 1, 2, ..., n.* The vectors $\mathbf{F}_1, \mathbf{F}_2, \ldots, \mathbf{F}_n$ represent the *external* forces that act on each of the particles. The external forces are caused by interaction of the particles with the external world (i.e., their sources are external to the system). Examples of external forces that may act on a particle are its weight, its interactions with other particles that are not included in the system, and support reactions.

In addition to external forces, particles of the system may also be subjected to forces that are *internal* to the system. For example, two particles could be connected together by a spring, collide with each other, or carry electrical charges that cause them to repel or attract each other. It is not necessary to show internal forces on the FBD of the system in Fig. 15.6(a), because interactions between particles always occur as pairs of forces that are equal in magnitude, opposite in direction, and have collinear lines of action (Newton's third law). Therefore, the internal forces cancel. However, the motion of an individual particle is determined by both the external and the internal forces.

Figure 15.6(b) shows the pair of internal forces that act between the ith and jth particles. The force $\mathbf{f}_{ij}$ represents the internal force acting on the ith particle that is caused by the jth particle. Similarly, $\mathbf{f}_{ji}$ is the internal force acting on the jth particle caused by the ith particle. [Each particle can have an interaction with every other particle in the system, but only one such interaction is shown in Fig. 15.6(b).] According to Newton's third law,

$$\mathbf{f}_{ij} = -\mathbf{f}_{ji} \qquad (i \neq j) \tag{15.14}$$

and the two forces are collinear.†

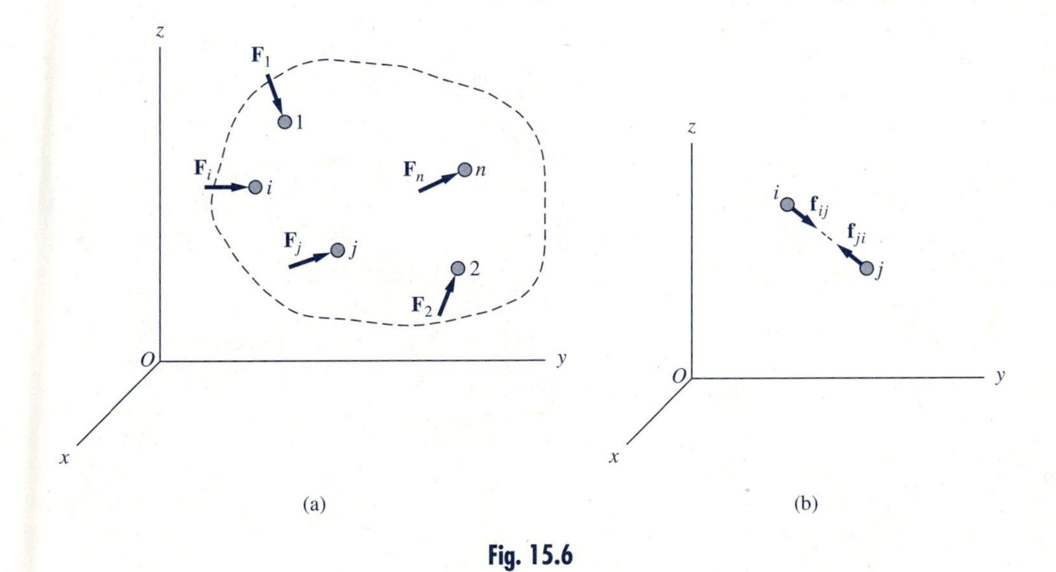

Fig. 15.6

* A system is "closed" if no particles enter or leave the system. The reason for this restriction will become apparent later. As the particles move, the boundary of the system can be imagined to be a flexible "pouch" that always encloses the same particles.

† Note that $i = j$ is excluded in the Eq. (15.14) because $\mathbf{f}_{ii}$, the force exerted on the ith particle by the ith particle, would be meaningless.

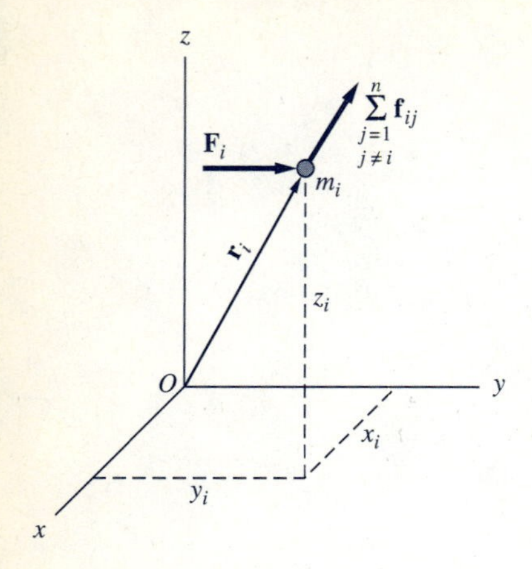

Fig. 15.7

The FBD of a typical (ith) particle is shown in Fig. 15.7. The force $\mathbf{F}_i$ represents the resultant external force acting on the particle (including its weight). The force

$$\sum_{\substack{j=1\\ j\neq i}}^{n} \mathbf{f}_{ij}$$

is the sum of all the internal forces acting on the particle that are caused by interactions with all the other particles in the system. From the FBD we deduce that the equation of motion for the ith particle is

$$\mathbf{F}_i + \sum_{\substack{j=1\\ j\neq i}}^{n} \mathbf{f}_{ij} = m_i\ddot{\mathbf{r}}_i \qquad (i = 1, 2, \ldots, n) \tag{15.15}$$

Because there are n particles in the system, Eq. (15.15) represents n vector equations. Summing all n of these equations yields

$$\sum_{i=1}^{n} \mathbf{F}_i + \sum_{i=1}^{n}\sum_{\substack{j=1\\ j\neq i}}^{n} \mathbf{f}_{ij} = \sum_{i=1}^{n} m_i\ddot{\mathbf{r}}_i \tag{15.16}$$

This rather formidable vector equation can be simplified by noting the following.

1. $\Sigma_{i=1}^{n}\mathbf{F}_i = \mathbf{R}$, where $\mathbf{R}$ is resultant external force acting on the system (including the total weight of the particles).
2. From Eq. (15.14), we see that

$$\sum_{i=1}^{n}\sum_{\substack{j=1\\ j\neq i}}^{n} \mathbf{f}_{ij} = \mathbf{0} \tag{15.17}$$

   This summation vanishes because, as mentioned previously, the internal forces occur in pairs that are equal in magnitude and oppositely directed. (This conclusion would not be valid if the system were not closed, i.e., if particles entered or left the system.)
3. The right side of Eq. (15.16) can be simplified by using Eq. (15.12).

$$\sum_{i=1}^{n} m_i\ddot{\mathbf{r}}_i = \frac{d^2}{dt^2}\sum_{i=1}^{n} m_i\mathbf{r}_i = \frac{d^2}{dt^2}m\mathbf{r}_G = m\mathbf{a}_G \tag{15.18}$$

   where we have used the fact that the total mass m of the system of particles is constant.

   Applying these results, Eq. (15.16) can be written as

$$\boxed{\mathbf{R} = m\mathbf{a}_G} \tag{15.19}$$

Comparing Eq. (15.19) and Newton's second law for particle motion, $\mathbf{F} = m\mathbf{a}$, we see that the mass center G of the system accelerates as if it were a particle of mass equal to the total mass of the system, acted on by the resultant external force.

Observe that Eq. (15.19) is valid whether or not **R** passes through the center of mass G. In general, there is no advantage in relating the *moment* of **R** to the motion of the system. However, as discussed in Chapter 17, moment equations play an important role in the analysis of rigid-body motion.

Sample Problem 15.4

The man shown in Fig. (a) walks from the left end to the right end of the uniform plank, which is initially at rest on a smooth sheet of ice. Determine the distance D that the man will have moved when he reaches the right end. The masses of the man and the plank are m_M and m_P, respectively, and friction between the plank and the ice may be neglected.

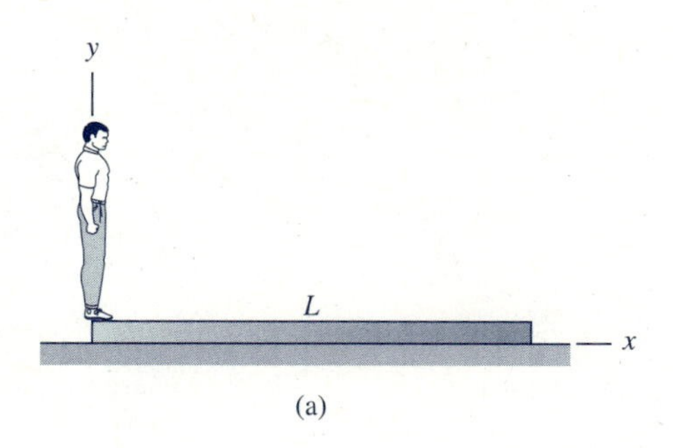

(a)

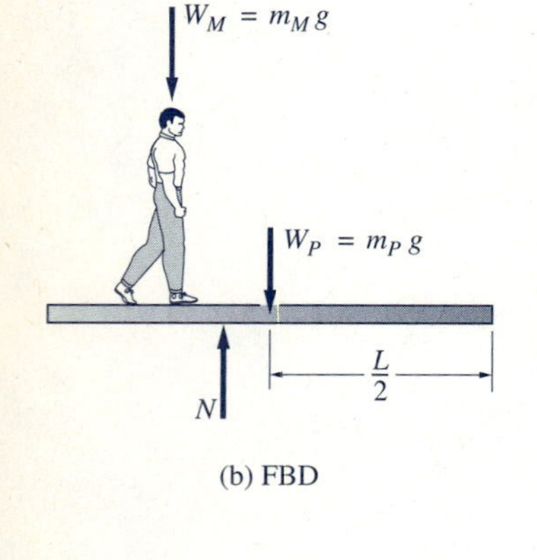

(b) FBD

Solution

The free-body diagram of the system containing the walking man and the plank is shown in Fig. (b). The only forces that appear on this FBD are the weights of the man and the plank, W_M and W_P, respectively, and the normal force N. The normal and the friction forces that act between the man and the plank do not appear on the FBD, because they are internal to the system.

From the FBD in Fig. (b) we see that there are no forces acting on the system in the x-direction. Therefore, according to $\mathbf{R} = m\mathbf{a}_G$ the center of mass G of the system remains stationary, as indicated in Figs. (c) and (d).

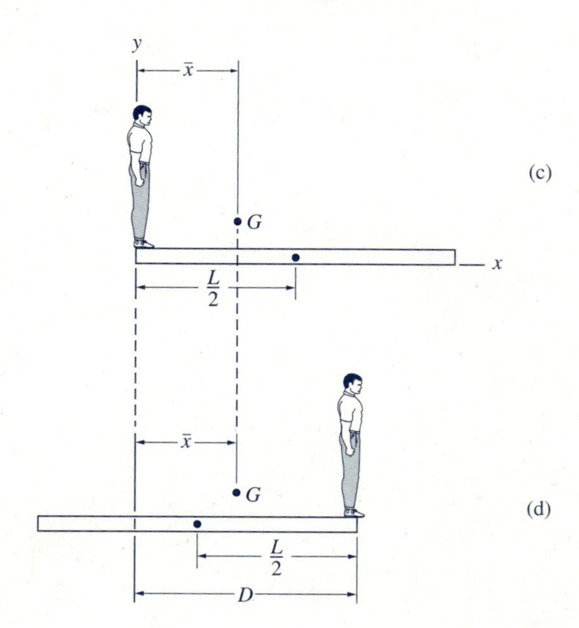

We first compute the x-coordinate of G, denoted by $\bar{x}$, when the man is at the left end of the plank. Referring to Fig. (c), we have

$$m\bar{x} = \sum_{i=1}^{n} m_i x_i \qquad (m_M + m_P)\bar{x} = m_M(0) + m_P \frac{L}{2}$$

which gives

$$\bar{x} = \frac{m_P L}{2(m_M + m_P)} \tag{a}$$

Repeating the procedure when the man has reached the right end of the plank, as shown in Fig. (d), we get

$$(m_M + m_P)\bar{x} = m_M D + m_P \left(D - \frac{L}{2}\right)$$

from which

$$\bar{x} = D - \frac{m_P L}{2(m_M + m_P)} \tag{b}$$

Equating the right-hand sides of Eqs. (a) and (b), and solving for D, yields

$$D = \frac{m_P L}{m_M + m_P} \qquad \textit{Answer}$$

Observe that every step taken by the man results in his moving to the right and the plank moving to the left. The magnitudes of these movements are in the proper ratio to ensure that the mass center of the system does not move horizontally.

We can also solve the problem by noting that the distance between the man and the mass center G must be the same in Figs. (c) and (d), because of the symmetry of the two configurations. In other words, $\bar{x} = D - \bar{x}$, or $D = 2\bar{x}$. Substitution for $\bar{x}$ from Eq. (a) would result in the same value of D as before.

Sample Problem 15.5

Figure (a) shows a 120-lb woman who is standing on a scale as she rides in an elevator that weighs 2000 lb. Determine the scale reading and the corresponding acceleration of the elevator if the tension in the cable is (1) $T = 1100$ lb; and (2) $T = 900$ lb. Neglect the weights of the scale and the support pulley.

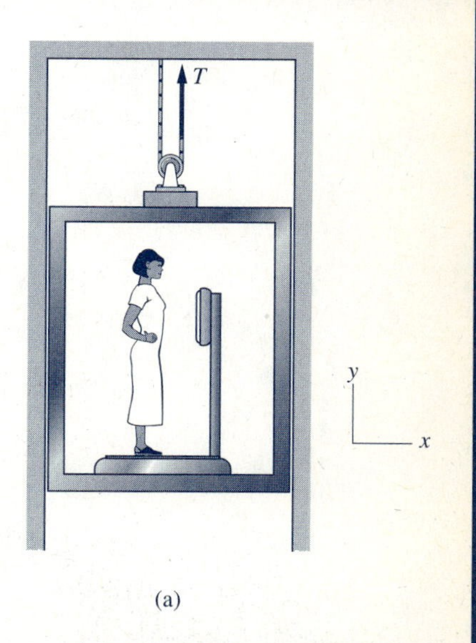

(a)

Solution

Preliminary Calculations

Figure (b) shows the free-body diagram (FBD) of the system consisting of the woman and the elevator. The only external forces are the weights and cable tensions. The force acting between the woman and the scale does not appear because it is an internal force. The mass-acceleration diagram (MAD) for the system is also shown in Fig. (b). Since the woman and the elevator have the same acceleration a, assumed to be upward, the inertia vector equals the total mass of the system multiplied by a. Newton's second law yields

$$\Sigma F_y = ma_y \qquad +\uparrow \quad 2T - 120 - 2000 = \frac{120 + 2000}{g}a$$

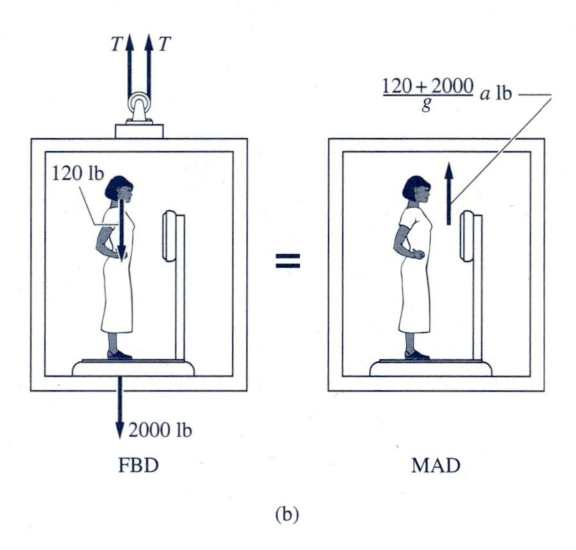

(b)

Therefore, the acceleration is

$$a = \frac{(2T - 2120)g}{2120} \text{ ft/s}^2 \tag{a}$$

where T is measured in pounds and $g = 32.2$ ft/s^2.

To determine the force that acts between the woman and the scale, we must isolate the woman from the scale. The FBD of the woman is shown in Fig. (c), where N is the force exerted on her by the scale. This force is, of course, equal and opposite to the force exerted by the woman on the scale. The MAD of the woman is also shown in Fig. (c), where a is again directed upward, which is consistent with our previous assumption. From Newton's second law we obtain

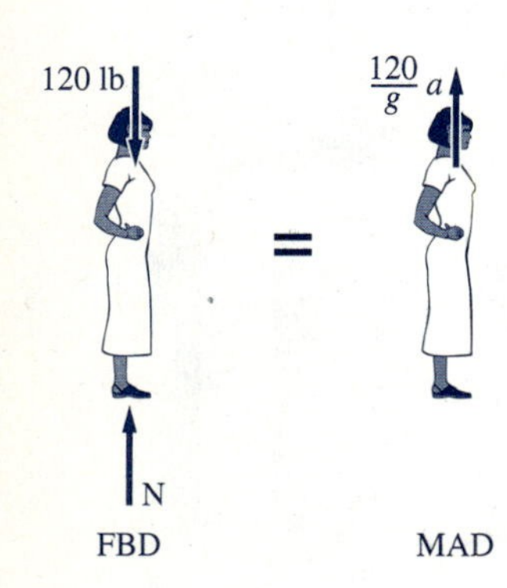

(c)

$$\Sigma F_y = ma_y \qquad +\uparrow \quad N - 120 = \frac{120}{g}a$$

from which the relation between N and a (ft/s^2) is

$$N = \frac{120a}{g} + 120 \text{ lb} \tag{b}$$

Substituting Eq. (a) into Eq. (b) and simplifying, we find the relationship between N and T to be

$$N = 0.113\,21T \tag{c}$$

Part 1

If $T = 1100$ lb, the scale reading from Eq. (c) is

$$N = 0.113\,21T = 0.113\,21(1100) = 124.5 \text{ lb} \qquad \textit{Answer}$$

and the corresponding acceleration from Eq. (a) is

$$a = \frac{(2T - 2120)g}{2120} = \frac{[2(1100) - 2120](32.2)}{2120} = 1.215 \text{ ft/s}^2 \qquad \textit{Answer}$$

Since a is positive, it is in the assumed direction, i.e., upward.

Part 2

Using $T = 900$ lb in Eqs. (a) and (c), we get

$$N = 0.113\,21T = 0.113\,21(900) = 101.9 \text{ lb} \qquad \textit{Answer}$$

and

$$a = \frac{(2T - 2120)g}{2120} = \frac{[2(900) - 2120](32.2)}{2120} = -4.86 \text{ ft/s}^2 \qquad \textit{Answer}$$

Since a is negative, it is directed opposite to the assumed direction, i.e., downward.

Observe that the woman exerts a force that is greater than her weight when the acceleration is upward, and less than her weight when the acceleration is downward.

Sample Problem 15.6

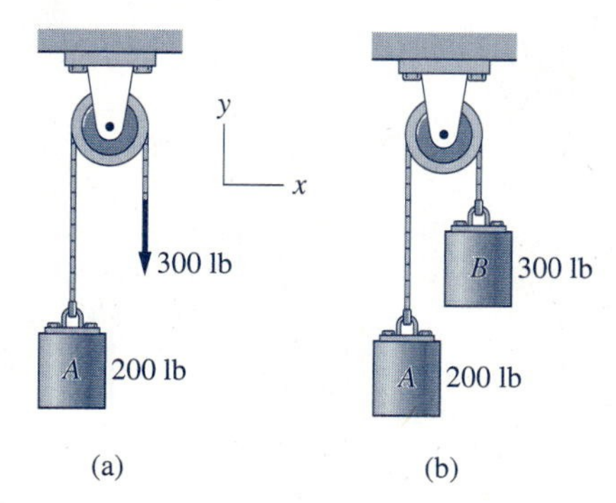

The 300-lb force in Fig. (a) is applied to the cable that is attached to the 200-lb block A. In Fig. (b), this force is replaced by a 300-lb block B. Neglecting the weight of the pulley, determine the acceleration of A and the tension in the cable for both cases.

Solution

System in Fig. (a)

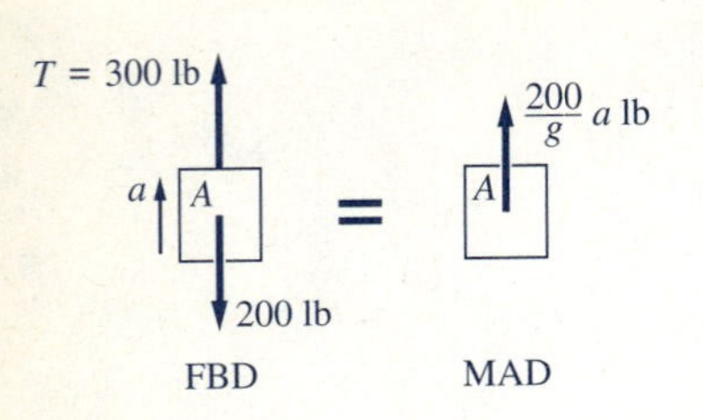

(c)

Figure (c) shows the free-body diagram (FBD) of block *A*. Since the mass of the pulley is to be neglected, the tension is the same throughout the cable, which gives

$$T = 300 \text{ lb} \qquad \textit{Answer}$$

Figure (c) also shows the mass-acceleration diagram (MAD) for block *A*, where its acceleration *a* is assumed to be upward. Newton's second law gives

$$\Sigma F_y = ma_y \qquad +\uparrow \quad 300 - 200 = \frac{200}{g}a$$

from which the acceleration is

$$a = \frac{g}{2} = 16.1 \text{ ft/s}^2 \qquad \textit{Answer}$$

System in Fig. (b)

The FBDs and MADs of the blocks are shown in Fig. (d). The kinematics of the inextensible cable determine that the acceleration of *A* is equal in magnitude to the acceleration of *B*, but in the opposite direction. We assumed the acceleration of *A* to be upward. Therefore the equation of motion of block A is

$$\Sigma F_y = ma_y \qquad +\uparrow \quad T - 200 = \frac{200}{g}a$$

For block B, we have

$$\Sigma F_y = ma_y \qquad +\downarrow \quad 300 - T = \frac{300}{g}a$$

Solving these two equations simultaneously, we obtain

$$T = 240 \text{ lb} \quad \text{and} \quad a = \frac{g}{5} = 6.44 \text{ ft/s}^2 \qquad \textit{Answer}$$

Note that applying a 300-lb force to the end of the cable is not equivalent to attaching a 300-lb weight.

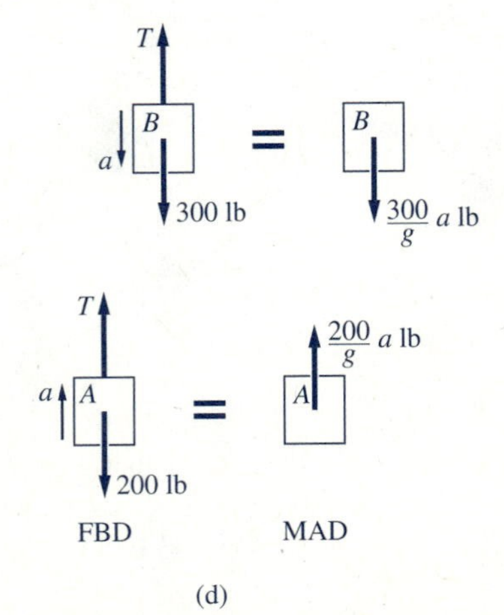

(d)

Sample Problem 15.7

For the system shown in Fig. (a), find the acceleration of block A in terms of the gravitational acceleration g. Assume that $m_A = 2m_B$ and that the masses of the pulleys are negligible.

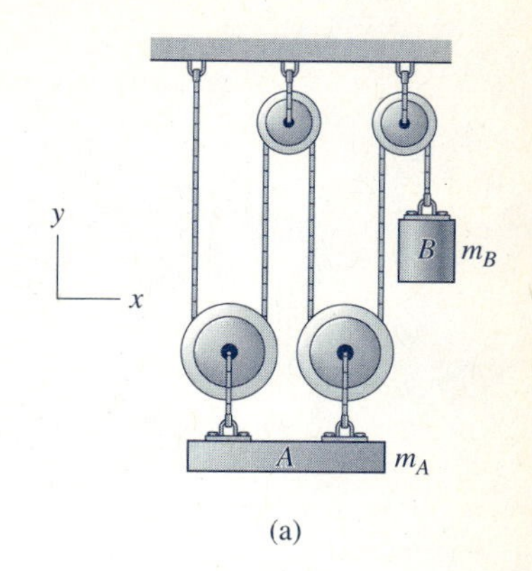

(a)

Solution

The free-body diagrams of block A (with two pulleys attached) and block B are shown in Fig. (b). Since the masses of the pulleys are neglected, the tension T is constant throughout the cable. Figure (b) also shows the mass-acceleration diagram for each block. Note that we have assumed that a_A and a_B are directed upward and downward, respectively. Applying Newton's second law, we obtain for block A

$$\Sigma F_y = ma_y \qquad +\uparrow \quad 4T - m_A g = m_A a_A \tag{a}$$

For block B,

$$\Sigma F_y = ma_y \qquad +\downarrow \quad m_B g - T = m_B a_B \tag{b}$$

Using either of the two methods explained in Sample Problem 15.2, it can be shown that the kinematic relationship between the accelerations is

$$a_A = \frac{a_B}{4} \tag{c}$$

Substituting $m_A = 2m_B$ into Eq. (a), Eqs. (a)–(c) yield

$$a_A = \frac{g}{9} \qquad \textit{Answer}$$

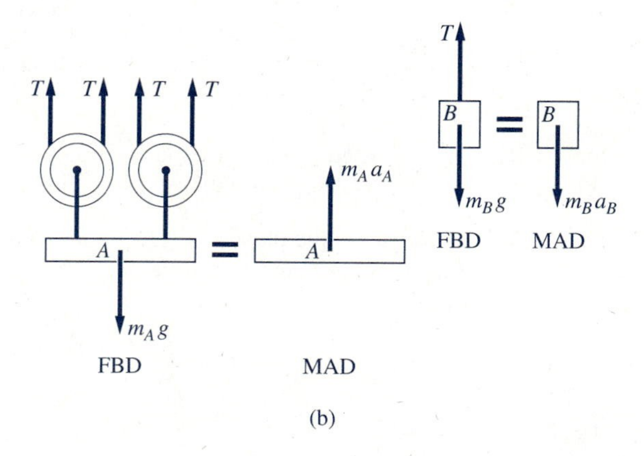

(b)

PROBLEMS

15.25 A 0.08-lb bullet is fired from a rifle that is clamped to a trolley. The combined weight of the rifle and trolley is 20 lb. After firing, the velocity of the trolley is known to be $v_A = 4.8$ ft/s to the left. Calculate (a) v_B, the velocity of the bullet after firing; and (b) the muzzle velocity (the velocity of the bullet relative to the barrel of the rifle).

15.26 A 160-lb man A jumps off an 80-lb stationary cart B. Immediately after takeoff, the velocity of the man relative to the cart is as shown in the figure. Determine the velocity vectors of the man and the cart.

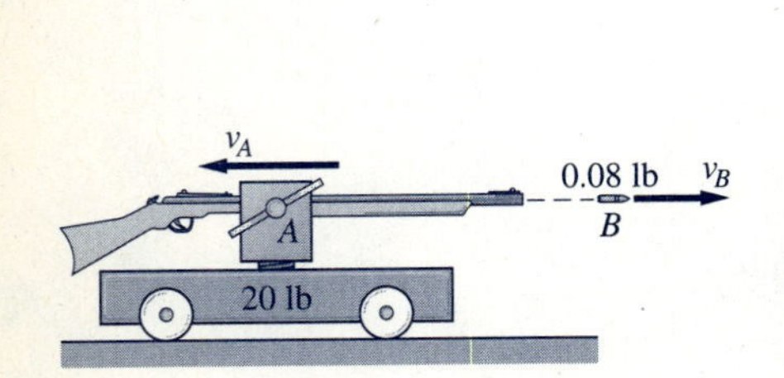

Fig. P15.25

Fig. P15.26

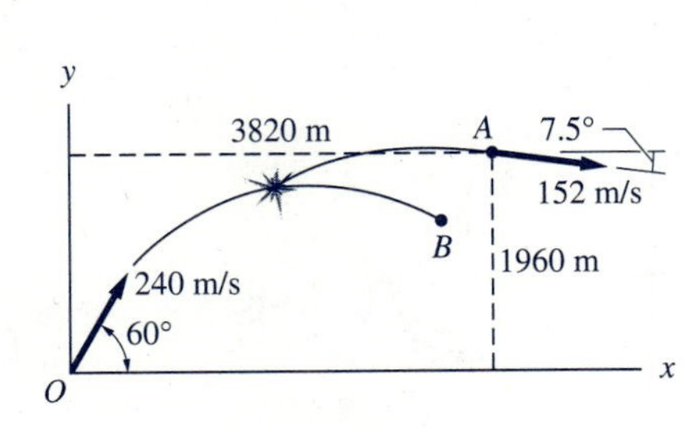

Fig. P15.27

***15.27** An 80-kg projectile is launched from point O at $t = 0$ with the velocity shown. During flight, the projectile explodes into two parts, A and B, of masses 24 kg and 56 kg, respectively. The parts remain in the xy-plane. If the position and velocity of A at $t = 30$ s are as shown, find the position and velocity of B at this time.

15.28 Trailer B is hitched to the four-wheel-drive truck A. The coefficient of static friction between the truck tires and the road is 0.9. Determine (a) the maximum possible acceleration; and (b) the corresponding tensile force in the trailer hitch.

15.29 The two packages are released from rest in the position shown. Find the relative velocity between the packages just before they collide.

15.30 The two packages slide down the plane in contact with each other. Calculate the acceleration of the packages and the normal force between them. Friction between the packages is negligible.

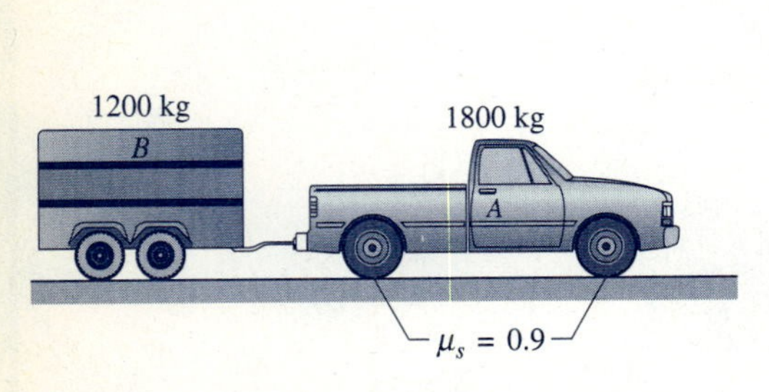

Fig. P15.28

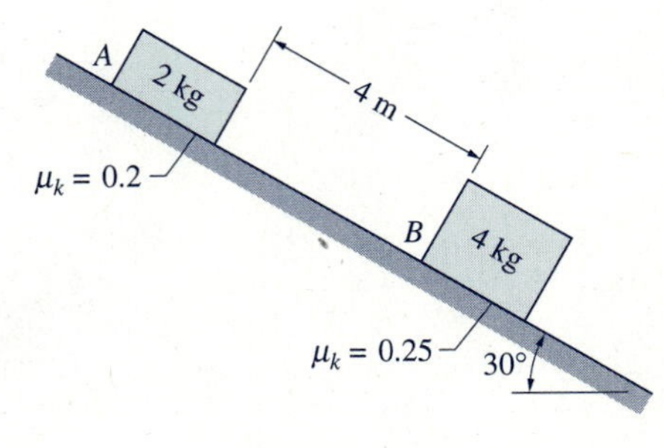

Fig. P15.29

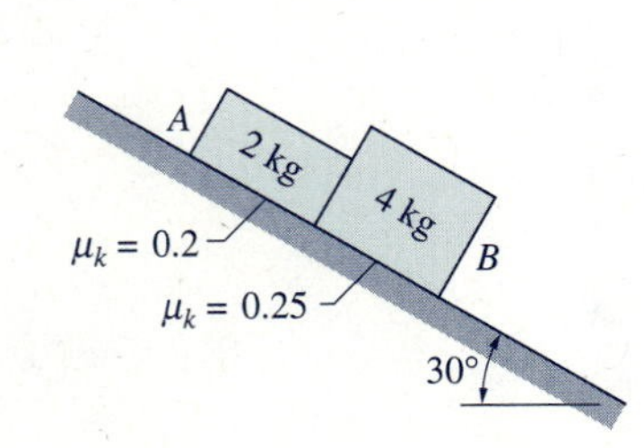

Fig. P15.30

***15.31** The two blocks are released from rest simultaneously in the positions shown. The free length of the spring connecting the blocks is 6 in. Determine which block is the first to hit the stop at C and the speed of that block just before it hits C. Neglect friction.

15.32 The static coefficient of friction between the 100-lb crate and the 300-lb cart is 0.24. Find the maximum force P that may be applied to the crate without causing it to slip on the cart.

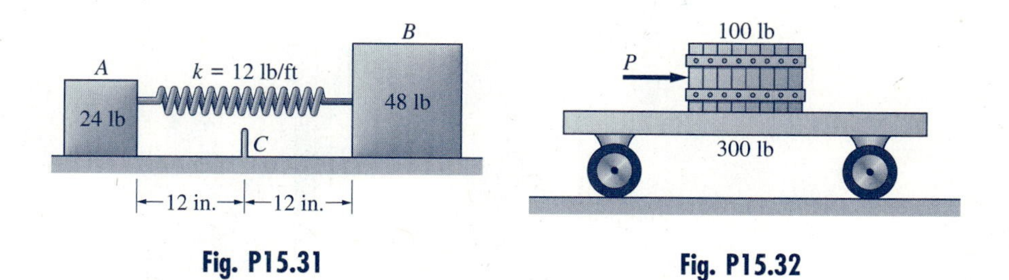

Fig. P15.31

Fig. P15.32

15.33 Determine the tension in the cable connecting blocks A and B after the 40-lb force is applied. The blocks were at rest before the force was applied.

15.34 The blocks A and B slide on a rough horizontal table, the coefficient of kinetic friction being 0.3 for both blocks. Neglecting the weight of the pulley C, compute the accelerations of A, B, and C after the 24-kN horizontal force is applied.

15.35 The system consisting of blocks A and B, and the weightless pulley C, is pulled upward by the constant 2.35-kN force. Determine the force in the cable joining A and B.

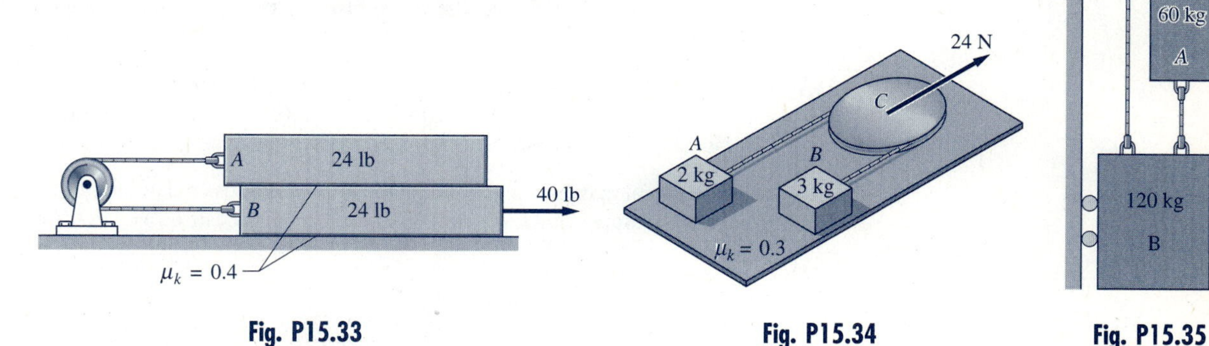

Fig. P15.33

Fig. P15.34

Fig. P15.35

15.36 Plank A rests on top of B as shown in Fig. (a) when the 48-lb force is applied to B. Calculate the displacement of each plank when their relative displacement has reached 4 ft, as shown in Fig. (b).

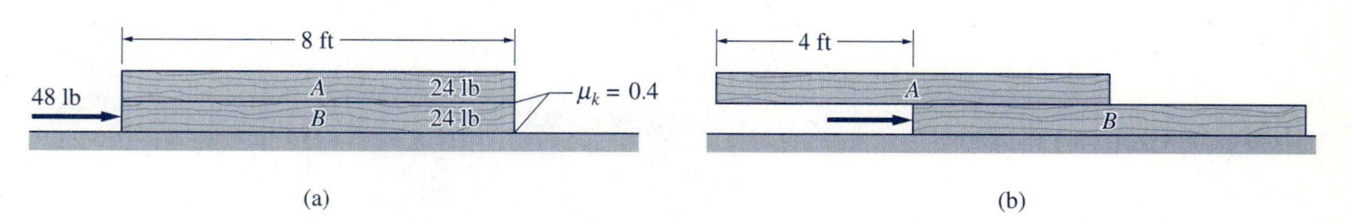

Fig. P15.36

15.37 Find the acceleration of block A in terms of the gravitational acceleration g, given that $m_A = 2m_B$. Neglect the masses of the pulleys.

15.38 The acceleration of collar C is $g/4$, upward. Find the accelerations of blocks A and B, which weigh 7 lb and 2 lb, respectively.

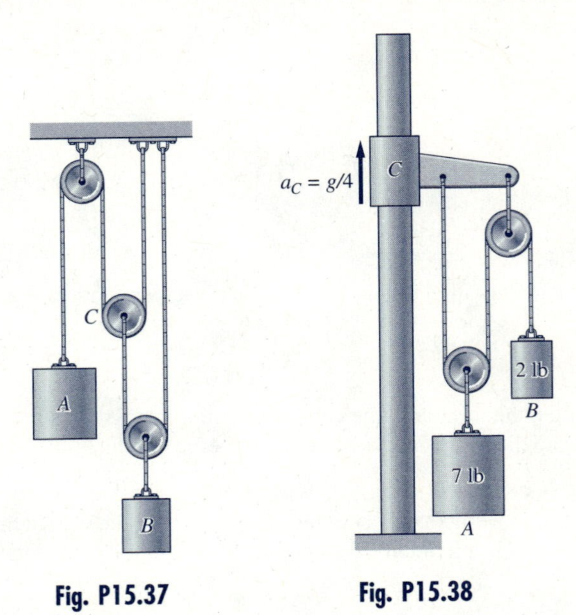

Fig. P15.37 **Fig. P15.38**

15.39 The slider B is free to move in the slotted block A. The system is initially at rest in the position shown, when A is given the constant acceleration $a_A = 10$ m/s^2 to the right. Determine the speed of B relative to A when it reaches the top of the slot.

15.40 Two identical masses A and B are connected by a rigid rod of negligible mass. The length of the rod is $L = 2$ m. Compute the acceleration of each mass immediately after the system is released from rest at $\theta = 30°$. Neglect friction.

15.41 Two identical masses A and B are joined by a weightless rod of length L. Show that the equation of motion of the system is $\ddot{\theta} = (g/L)\sin\theta$. Neglect friction.

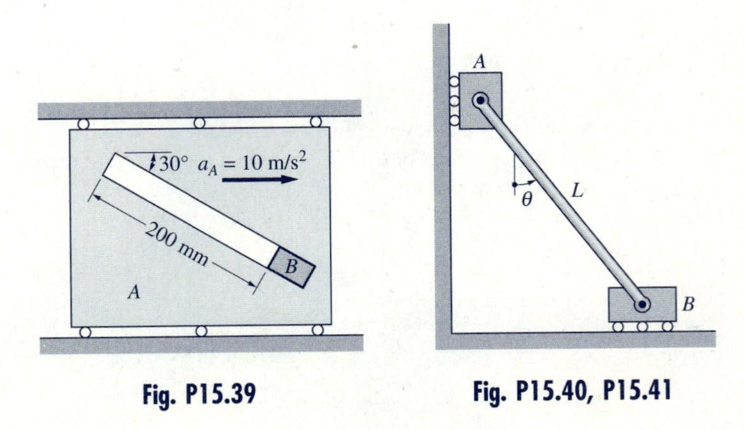

Fig. P15.39 **Fig. P15.40, P15.41**

15.42 The sliding panel A is connected by cables to the two counterweights B. Find the acceleration of A immediately after the system is released from rest in the position shown. Neglect friction.

15.43 The truck A is about to tow the trailer B from a standing start. The truck has four-wheel drive, and the static coefficient of friction between its tires and the road is 0.8. Rolling resistance of the trailer is negligible. Determine the maximum possible initial acceleration of the truck if $x = 10$ m.

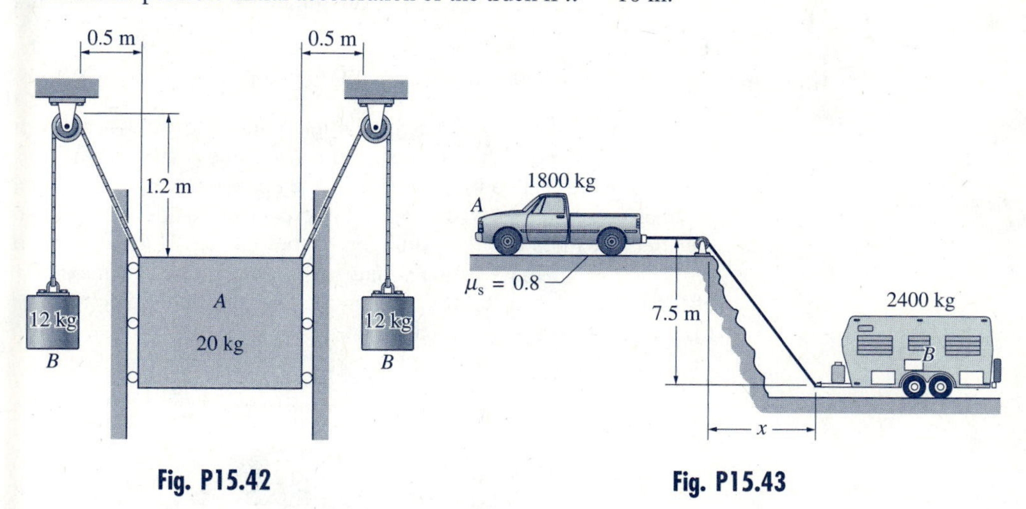

Fig. P15.42 **Fig. P15.43**

15.44 Two identical blocks A and B are released from rest in the position shown. Calculate a_B and $a_{A/B}$ in terms of the gravitational acceleration g. Neglect friction.

15.45 The stiffness of the spring that is attached to the two 5-lb blocks is 2 lb/in. The system is initially at rest on the smooth surface when the constant 1.2-lb force is applied at $t = 0$. (a) Derive the equation of motion for each block, and state the initial conditions. (b) Using Euler's method with $\Delta t = 0.02$ s, determine the speed of each block and the force P in the spring when $t = 0.1$ s. (Note: The analytic solution is $v_1 = 7.23$ in./s, $v_2 = 2.05$ in./s, $P = 0.712$ lb.)

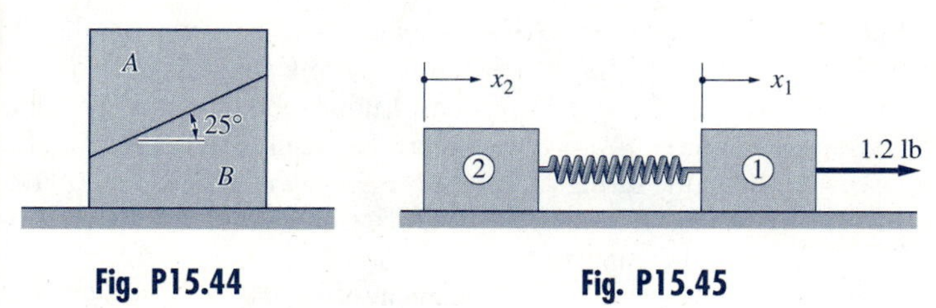

Fig. P15.44 **Fig. P15.45**

15.46 Solve Prob. 15.45 with the modified Euler method using $\Delta t = 0.01$ s.

15.47 The figure shows the top view of the two particles A and B that slide on a smooth horizontal table. The particles carry identical electric charges, which give rise to the repulsive force $F = c/d^2$, where $c = 0.005\ \text{N} \cdot \text{m}^2$ and d is the distance between the particles in meters. At time $t = 0$, it is known that $d = 0.5$ m, A is at rest, and B is traveling toward A at the speed of 2 m/s. (a) Derive the equation of motion, and state the initial conditions. (b) Use Euler's method with $\Delta t = 0.05$ s to calculate the minimum value of d and the speed of each particle at that instant. (Note: The analytic solution is $d_{\text{min}} = 0.227$ m, $v_A = v_B = 800$ mm/s.)

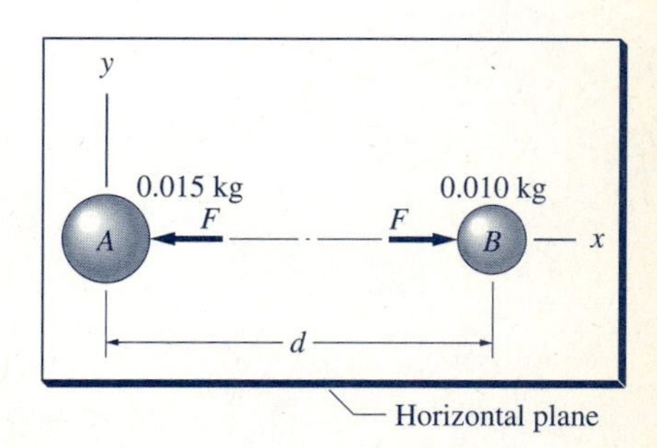

Fig. P15.47

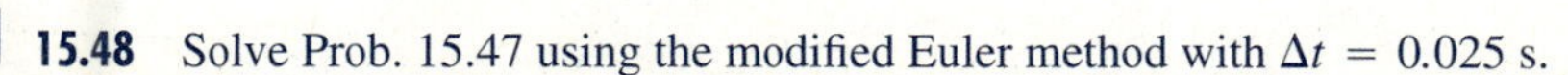

15.48 Solve Prob. 15.47 using the modified Euler method with $\Delta t = 0.025$ s.

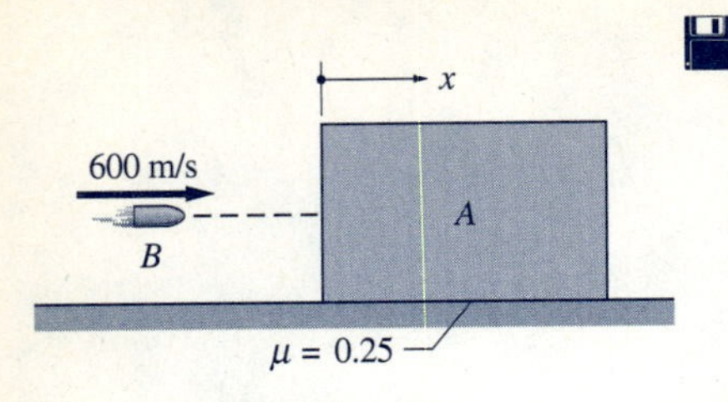

Fig. P15.49

15.49 The 0.025-kg bullet B traveling at 600 m/s hits and becomes embedded in the 15-kg block A, which was initially at rest on the smooth surface. The force between A and B during the embedding phase is $F = 50v_{B/A}$, where F is in newtons and the relative velocity is in meters per second. (a) Determine the equations of motion for A and B during the embedding phase, and state the initial conditions. (b) Use Euler's method with $\Delta t = 0.2$ ms to compute the velocity of B and the distance moved by A during 1.0 ms following the initial contact. (Note: The analytical solution is $v_B = 81.8$ m/s and $x_A = 0.567$ mm.)

15.50 Solve Prob. 15.49 with the modified Euler method using $\Delta t = 0.1$ ms.

15.51 The two railroad cars are coasting with the velocities shown when they collide. The bumpers of the cars are ideal springs, the combined stiffness of two bumpers being 20 000 lb/ft. (a) Derive the equation of motion for each of the two railroad cars. (b) Assuming that the contact begins at $t = 0$, solve the equations of motion for the duration of the impact, which is approximately 0.4 s. (c) From the solution found in (b), determine the maximum value of the contact force, the time of contact, and the final velocity of each car. (d) Plot the velocity of each car versus time.

Fig. P15.51

15.52 The two blocks are released from rest at $t = 0$ with the spring stretched by 20 mm. (a) Derive the equations of motion for each block assuming that A slips relative to B. (b) Solve the equations of motion for the time interval $t = 0$ to $t = 0.2$ s. (c) From the solution found in (b), find the maximum speed of each block and the corresponding times. (d) Plot the velocity of each block versus time.

15.53 The 3-lb block A and table B are at rest when a shaker attached to the table is activated at time $t = 0$. The shaker causes the following oscillation of the table: $x_B = \sin 10t$ ft, where t is in seconds. Lubrication on top of the table causes a viscous drag force $F_D = c_D v_{B/A}$ to act between A and B. The coefficient of viscosity c_D and the spring stiffness k are as shown in the figure. (a) Derive the equation of motion for block A. (b) Solve the equation of motion from $t = 0$ to $t = 1.0$ s. (c) From the solution found in (b), determine the maximum acceleration and maximum displacement of A, and the corresponding times. (d) Plot the acceleration of A versus time.

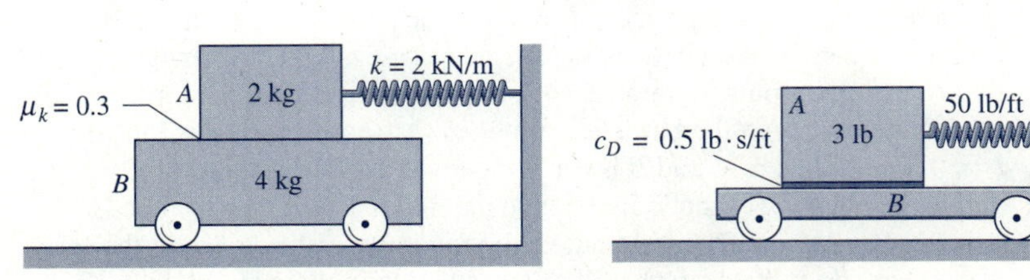

Fig. P15.52

Fig. P15.53

15.54 The rough, horizontal disk of radius 2.5 ft rotates at the constant angular speed $\omega = 45$ rev/min. The particle A is placed on the disk with no initial velocity at $t = 0$, $R = 1.0$ ft, and $\theta = 0$. (a) Show that the equations of motion for the disk are

$$\ddot{R} = R\dot{\theta}^2 - \frac{\mu_k g \dot{R}}{v} \qquad \ddot{\theta} = -\frac{2\dot{R}\dot{\theta}}{R} - \frac{\mu_k g(\dot{\theta} - \omega)}{v}$$

where

$$v = \left[\dot{R}^2 + R^2\left(\dot{\theta} - \omega\right)^2\right]^{1/2}$$

and state the initial conditions. (b) Solve the equations numerically for the period of time that the particle stays on the disk; plot R vs. θ. (c) Use the numerical solution to find the speed of the particle and the corresponding value of θ when it is about to slide off the disk.

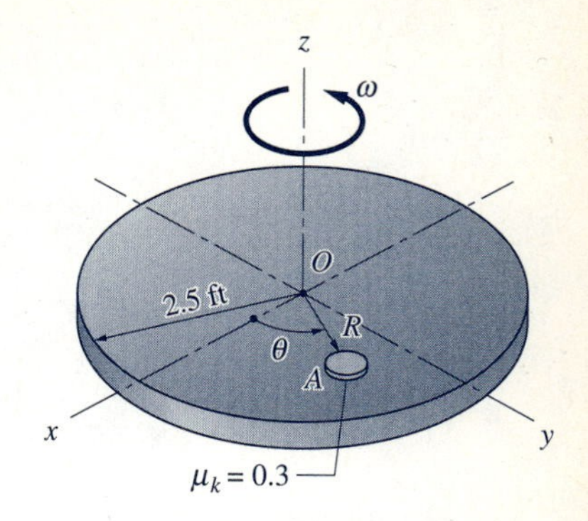

Fig. P15.54

15.5 Work-Energy Principles

This article extends the work-energy methods for a single particle, presented in Arts. 14.2–14.4, to a system of particles.

a. Work done on a system of particles

Consider the system of n particles shown in Fig. 15.8, where $\mathbf{F}_1, \mathbf{F}_2, \ldots, \mathbf{F}_n$ are external forces as described in Art. 15.4. The paths followed by the particles are shown as dotted lines, with ① and ② indicating the initial and final positions of each particle, respectively.

As shown in Fig. 15.8, the forces acting on the ith particle are the external force $\mathbf{F}_i$ (which includes the weight of the particle), and the resultant internal force $\sum_{\substack{j=1 \\ j \neq i}}^{n} \mathbf{f}_{ij}$. Let $\mathbf{r}_i(t)$ and $\mathscr{L}_i$ be the position vector and path of the ith

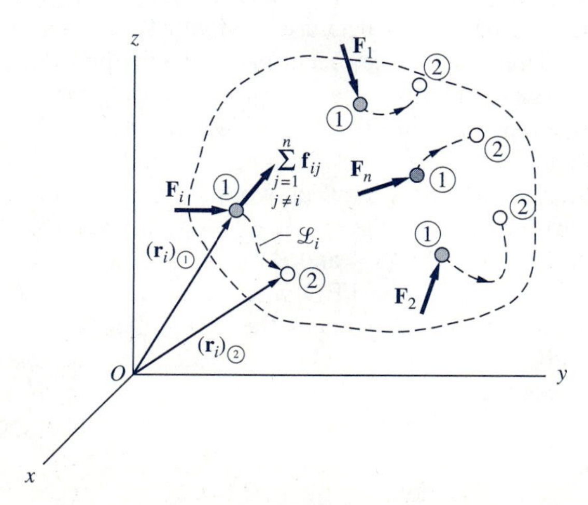

Fig. 15.8

particle, respectively. According to Eq. (14.2), the work done on this particle is

$$(U_{1\text{-}2})_i = \int_{\mathcal{L}_i} \left(\mathbf{F}_i + \sum_{\substack{j=1 \\ j\neq i}}^{n} \mathbf{f}_{ij} \right) \cdot d\mathbf{r}_i \tag{15.20}$$

The total work done on the system of n particles (i.e., the sum of the work done on each of the particles) can be expressed as

$$U_{1\text{-}2} = \sum_{i=1}^{n} (U_{1\text{-}2})_i = \sum_{i=1}^{n} \int_{\mathcal{L}_i} \mathbf{F}_i \cdot d\mathbf{r}_i + \sum_{i=1}^{n} \sum_{\substack{j=1 \\ j\neq i}}^{n} \int_{\mathcal{L}_i} \mathbf{f}_{ij} \cdot d\mathbf{r}_i \tag{15.21}$$

The first term on the right side of this equation equals $(U_{1\text{-}2})_{\text{ext}}$, the work done by all of the external forces acting on the system. The second term represents $(U_{1\text{-}2})_{\text{int}}$, the total work done by the internal forces. Therefore, the total work done on the system of particles becomes

$$\boxed{U_{1\text{-}2} = (U_{1\text{-}2})_{\text{ext}} + (U_{1\text{-}2})_{\text{int}}} \tag{15.22}$$

The work of the external forces, $(U_{1\text{-}2})_{\text{ext}}$, can be computed by the methods explained in Art. 14.2.

It should be noted that the total work done by the weights of the particles can be calculated by assuming that the total weight acts at the center of gravity of the system. Choosing the z-axis to be vertical, we find from the definition of center of gravity that $W\bar{z} = W_1 z_1 + W_2 z_2 + \cdots + W_n z_n$, where the total weight is $W = W_1 + W_2 + \cdots + W_n$, and z_i is the z-coordinate of the particle of weight W_i. Differentiation yields $W\,d\bar{z} = W_1\,dz_1 + W_2\,dz_2 + \cdots + W_n\,dz_n$. The left side of this equation is the differential work done by the total weight acting at the center of gravity, and the right side is the total differential work done by the weights of the particles.

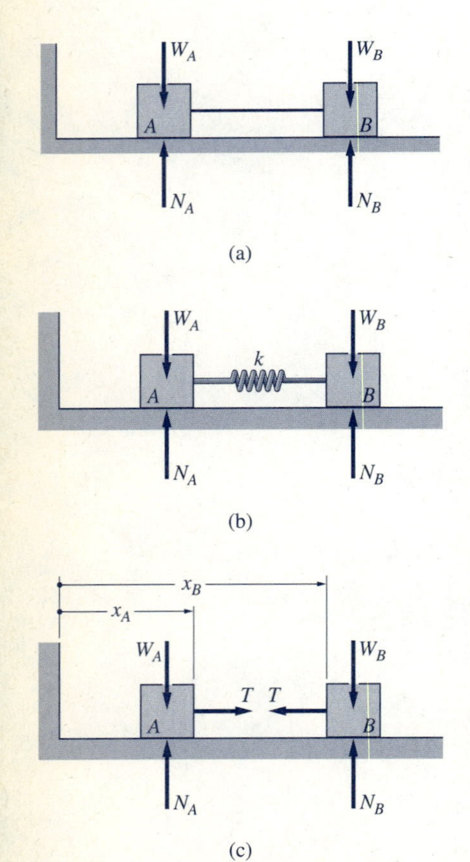

Fig 15.9

b. Work of internal forces

The calculation of the work done by internal forces, $(U_{1\text{-}2})_{\text{int}}$, can often be simplified as a result of the special nature of internal forces. As we have already seen in Eq. (15.14), internal forces occur in equal and opposite collinear pairs; i.e., $\mathbf{f}_{ij} = -\mathbf{f}_{ji}$. The total work done by a pair of internal forces will be zero if $d\mathbf{r}_i = d\mathbf{r}_j$, since then the work done by $\mathbf{f}_{ij}$ will cancel the work done by $\mathbf{f}_{ji}$. In this case, the internal forces are said to be *workless*.

To illustrate the difference between workless internal forces and internal forces that do work, consider the systems shown in Fig. 15.9 parts (a) and (b), which consist of two blocks A and B that slide along a smooth horizontal plane. Figure 15.9(a) shows the FBD when the blocks are connected by an inextensible string; Fig. 15.9(b) shows the FBD when the connection is an ideal spring of stiffness k. These two FBDs are identical because both systems are subject to the same external forces: W_A, W_B, N_A, and N_B. Note that in each case $U_{\text{ext}} = 0$, since the external forces are perpendicular to the direction of motion.

To compute the work done by internal forces, it is necessary to analyze the forces acting on each block separately, as shown in the FBD of Fig. 15.9(c).

The force T represents either the tension in the string for the system in Fig. 15.9(a) or the tension in the spring for the system in Fig. 15.9(b). During a differential movement of the system, the work done by T on blocks A and B is $T\,dx_A$ and $-T\,dx_B$, respectively. Therefore, the total work done by T on the system is $T(dx_A - dx_B)$. If the connection between the blocks is an inextensible string, then $dx_A = dx_B$, and the total work done by the string on the system is zero. However, if the connection is a linear spring, dx_A will generally not be equal to dx_B, because the spring can deform. We conclude that the spring force is capable of doing work on the system even though it is an internal force.

In summary, *rigid* internal connections, such as inextensible strings and pinned joints, perform equal and opposite work on the bodies they connect, which results in zero net work being performed on the system. On the other hand, *deformable* internal connections, which include springs and friction surfaces that slide, are capable of doing work on a system. The conclusion is that the total work done on a system of particles can include the work of both external and internal forces.

c. Principle of work and kinetic energy

Applying the work-energy principle, Eq. (14.15), to the arbitrary ith particle of the system, we have

$$(U_{1\text{-}2})_i = (\Delta T)_i \tag{15.23}$$

where $(U_{1\text{-}2})_i$ is the work done on the particle and $(\Delta T)_i$ is the change in its kinetic energy. If the system contains n particles, there will be n equations similar to Eq. (15.23). Adding all of these scalar equations, and using Eq. (15.22), we find that

$$\boxed{(U_{1\text{-}2})_{\text{ext}} + (U_{1\text{-}2})_{\text{int}} = \Delta T} \tag{15.24}$$

where T, the kinetic energy of the system, is defined to be the sum of the kinetic energies of all of the particles, i.e.,

$$\boxed{T = \sum_{i=1}^{n} T_i = \sum_{i=1}^{n} \frac{1}{2} m_i v_i^2} \tag{15.25}$$

d. Conservation of mechanical energy

If all the forces, internal as well as external, are conservative (see Art. 14.4), the mechanical energy of the system is conserved; i.e.,

$$\boxed{V_1 + T_1 = V_2 + T_2} \tag{15.26}$$

where V_1 and V_2 are the initial and final potential energies of the system (the sum of the potential energies of all the forces, both external and internal, that are capable of doing work on the system), and T_1 and T_2 are the initial and final kinetic energies of the system (the sum of the kinetic energies of all particles).

15.6 Principle of Impulse and Momentum

In this article we extend the impulse-momentum principles for a particle, discussed in Art. 14.6, to systems of particles.

a. Linear momentum

A system of n particles is shown in Fig. 15.10. The position vectors of the arbitrary ith particle and the mass center G of the system, both measured from the fixed point O, are $\mathbf{r}_i$ and $\mathbf{r}_G$, respectively. According to Eq. (14.36), the linear momentum of the ith particle, also shown in Fig. 15.10, is $\mathbf{p}_i = m_i\mathbf{v}_i$, where m_i is the mass of the particle and $\mathbf{v}_i = \dot{\mathbf{r}}_i$ is its velocity.

The linear momentum (or simply the momentum) $\mathbf{p}$ of a system of n particles is defined to be the sum of the linear momenta of all the particles, which can be written as

$$\mathbf{p} = \sum_{i=1}^{n} \mathbf{p}_i = \sum_{i=1}^{n} m_i\mathbf{v}_i \qquad (15.27)$$

For a closed system—i.e., a system that always contains the same particles—we can relate the linear momentum of the system to the motion of its mass center G. According to Eq. (15.12), the location of G is given by $m\mathbf{r}_G = \Sigma_{i=1}^{n} m_i\mathbf{r}_i$, where m is the total mass of the system. Since m and m_i are constants, taking the time derivative of both sides of this equation yields $m\mathbf{v}_G = \Sigma_{i=1}^{n} m_i\mathbf{v}_i$. Therefore, the linear momentum in Eq. (15.27) becomes

$$\mathbf{p} = m\mathbf{v}_G \qquad (15.28)$$

We see that the linear momentum of a system of particles can be computed by assuming that the system consists of a single particle, with mass equal to the mass of the entire system, moving with the velocity of the mass center G.

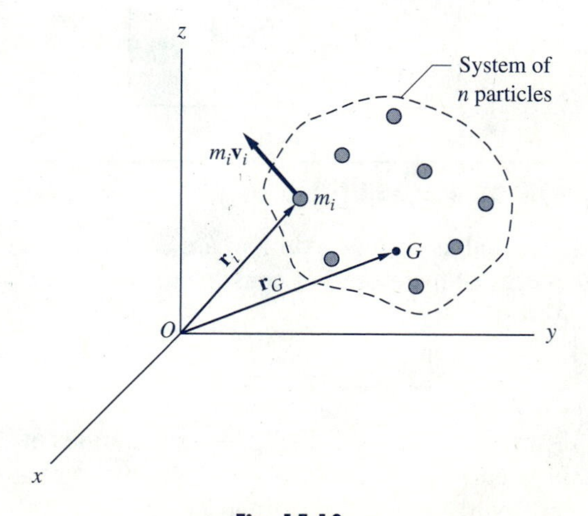

Fig. 15.10

b. Force-momentum relationship

Equation (15.19) stated that the motion of a closed system of particles is governed by $\mathbf{R} = m\mathbf{a}_G$, where $\mathbf{R}$ is the resultant external force acting on the system, m is the total mass of the system, and $\mathbf{a}_G$ is the acceleration of the mass center. Because the mass within a closed system is constant, this equation may be rewritten as

$$\mathbf{R} = m\mathbf{a}_G = m\frac{d\mathbf{v}_G}{dt} = \frac{d}{dt}(m\mathbf{v}_G)$$

Utilizing Eq. (15.28), the above force-momentum relationship for a closed system of particles becomes

$$\boxed{\mathbf{R} = \frac{d\mathbf{p}}{dt}} \tag{15.29}$$

This equation is similar to Eq. (14.37), the force-momentum equation for a single particle, namely, $\mathbf{F} = d\mathbf{p}/dt$.

c. Impulse-momentum principle

Multiplying both sides of Eq. (15.29) by dt and integrating between times t_1 and t_2, we find that

$$\int_{t_1}^{t_2} \mathbf{R}\,dt = \int_{\mathbf{p}_1}^{\mathbf{p}_2} d\mathbf{p} = \mathbf{p}_2 - \mathbf{p}_1 \tag{15.30}$$

where $\mathbf{p}_1$ and $\mathbf{p}_2$ denote the linear momenta of the system at $t = t_1$ and t_2, respectively. Since the left side of Eq. (15.30) is the linear impulse of the external forces, we can write

$$\boxed{(\mathbf{L}_{1\text{-}2})_{\text{ext}} = \mathbf{p}_2 - \mathbf{p}_1 = \Delta\mathbf{p}} \tag{15.31}$$

which is the *linear impulse-momentum principle* for a system of particles. Note the similarities between this principle and Eq. (14.39), $\mathbf{L}_{1\text{-}2} = \Delta\mathbf{p}$, the linear impulse-momentum principle for a single particle. Equation (15.31) is, of course, a vector equation that is equivalent to three scalar equations.

d. Conservation of momentum

From Eq. (15.31), we see that if the linear impulse of the external forces is zero, linear momentum of the system is conserved. In other words, if $(\mathbf{L}_{1\text{-}2})_{\text{ext}} = \mathbf{0}$, we obtain

$$\boxed{\mathbf{p}_1 = \mathbf{p}_2 \quad \text{or} \quad \Delta\mathbf{p} = \mathbf{0}} \tag{15.32}$$

which is the *principle of conservation of linear momentum* for a system of particles. Equation (15.32) is similar to the principle of conservation of linear momentum for a particle, Eq. (14.41).

It should be noted that if the linear momentum of each particle of a system is conserved, then the linear momentum of the entire system is also conserved. However, the converse of this statement is not necessarily true: If the linear momentum of a system is conserved, it does not imply that the linear momentum of each particle is conserved.

Because Eq. (15.32) is a vector relationship, it is possible for a component of the linear momentum of a system to be conserved, even though the resultant linear momentum vector is not conserved.

15.7 Principle of Angular Impulse and Momentum

The angular impulse-momentum principles for a particle are discussed in Art. 14.7. Here we extend these principles to a system of particles.

a. Angular momentum

The angular momentum of a system of particles is equal to the sum of the angular momenta of all of the particles in the system.

Consider again the system of n particles shown in Fig. 15.10. Referring to the arbitrary ith particle of the system, we let m_i be its mass, $\mathbf{r}_i$ its position vector measured from a fixed point O, and $\mathbf{r}_{i/A}$ its position vector relative to an arbitrary point A, as illustrated in Fig. 15.11. The angular momentum of the ith particle about A equals the moment of its linear momentum $\mathbf{p}_i = m_i\mathbf{v}_i$ about A. The angular momentum of the system about A thus becomes

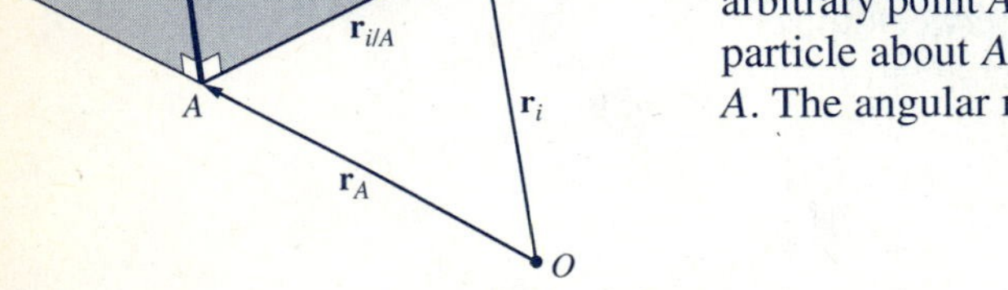

Fig. 15.11

$$\mathbf{h}_A = \sum_{i=1}^{n} (\mathbf{h}_A)_i = \sum_{i=1}^{n} \mathbf{r}_{i/A} \times m_i\mathbf{v}_i \tag{15.33}$$

b. Moment-angular momentum relationship

Differentiating the expression for $\mathbf{h}_A$ given in Eq. (15.33) with respect to time and assuming that the system of particles is closed—i.e., it always contains the same particles—we obtain

$$\dot{\mathbf{h}}_A = \sum_{i=1}^{n} \mathbf{r}_{i/A} \times m_i\mathbf{a}_i + \sum_{i=1}^{n} \dot{\mathbf{r}}_{i/A} \times m_i\mathbf{v}_i \tag{a}$$

where $\mathbf{a}_i = \dot{\mathbf{v}}_i$ is the acceleration of the ith particle.

According to Newton's second law, $m_i\mathbf{a}_i$ is equal to the resultant force acting on the ith particle. Therefore, the first sum on the right side of Eq. (a) represents the moment about point A of the forces that act on all the particles in the system, i.e., the resultant moment acting on the system about A. As explained in Art. 15.4, the forces acting on a particle can be classified as being external or internal to the system (see Fig. 15.7). The internal forces contribute nothing to the resultant moment, since they occur as equal and opposite collinear pairs. Consequently, we have

$$\sum_{i=1}^{n} \mathbf{r}_{i/A} \times m_i\mathbf{a}_i = (\Sigma\mathbf{M}_A)_{\text{ext}} \tag{b}$$

where $(\Sigma \mathbf{M}_A)_{ext}$ is the resultant moment about A of the forces that are *external* to the system.

Substituting $\mathbf{r}_{i/A} = \mathbf{r}_i - \mathbf{r}_A$ (refer to Fig. 15.11), the second term on the right-hand side of Eq. (a) becomes

$$\sum_{i=1}^{n} \dot{\mathbf{r}}_{i/A} \times m_i \mathbf{v}_i = \sum_{i=1}^{n} (\dot{\mathbf{r}}_i - \dot{\mathbf{r}}_A) \times m_i \dot{\mathbf{r}}_i$$

$$= \sum_{i=1}^{n} \dot{\mathbf{r}}_i \times m_i \dot{\mathbf{r}}_i - \dot{\mathbf{r}}_A \times \sum_{i=1}^{n} m_i \dot{\mathbf{r}}_i \qquad \text{(c)}$$

Since $\dot{\mathbf{r}}_i \times m_i \dot{\mathbf{r}}_i = \mathbf{0}$ and $\Sigma_{i=1}^{n} m_i \dot{\mathbf{r}}_i = \Sigma_{i=1}^{n} m_i \mathbf{v}_i = m\mathbf{v}_G$, where m is the total mass of the system and $\mathbf{v}_G$ is the velocity of its mass center G, Eq. (c) becomes

$$\sum_{i=1}^{n} \dot{\mathbf{r}}_{i/A} \times m_i \mathbf{v}_i = -\mathbf{v}_A \times m\mathbf{v}_G \qquad \text{(d)}$$

Substituting Eqs. (b) and (d) into Eq. (a), the *moment-angular momentum relationship* for a closed system of particles becomes

$$\boxed{(\Sigma \mathbf{M}_A)_{ext} = \dot{\mathbf{h}}_A + \mathbf{v}_A \times m\mathbf{v}_G} \qquad (15.34)$$

An important special case of this equation is

$$\boxed{(\Sigma \mathbf{M}_A)_{ext} = \dot{\mathbf{h}}_A} \qquad (15.35)$$

which is valid for each of the following cases.

1. $\mathbf{v}_A = \mathbf{0}$
2. A is the mass center G
3. $\mathbf{v}_G = \mathbf{0}$
4. $\mathbf{v}_A$ is parallel to $\mathbf{v}_G$

It is easy to verify that the second term on the right side of Eq. (15.34) vanishes for each of these cases.

c. Angular impulse-momentum principle

Comparing Eqs. (14.50) and (15.35), we see that the equation $\Sigma \mathbf{M}_A = \dot{\mathbf{h}}_A$ applies to both a single particle and a system of particles, provided $\Sigma \mathbf{M}_A$ and $\dot{\mathbf{h}}_A$ are interpreted correctly. As we have noted, there are several choices for the point A for which this equation applies. However, the only ones of practical interest are (1) A is a fixed point, or (2) A is the mass center. Assuming that one of these cases applies, we may integrate $\Sigma \mathbf{M}_A \, dt = d\mathbf{h}_A$ over the time interval t_1 to t_2 as follows.

$$\int_{t_1}^{t_2} \sum \mathbf{M}_A \, dt = \int_{(\mathbf{h}_A)_1}^{(\mathbf{h}_A)_2} d\mathbf{h}_A$$

where $(\mathbf{h}_A)_1$ and $(\mathbf{h}_A)_2$ are the angular momenta about A at times t_1 and t_2, respectively. Recognizing that the left side of this equation is by definition the angular impulse about A, the equation may be written as

$$(\mathbf{A}_A)_{1\text{-}2} = (\mathbf{h}_A)_2 - (\mathbf{h}_A)_1 = \Delta\mathbf{h}_A \qquad (A\text{: fixed point or mass center}) \tag{15.36}$$

Equation (15.36) is called the *angular impulse-momentum principle.* Note that, practically speaking, this principle is valid only if A is a fixed point or the mass center of a closed system of particles. When applying this principle to a system of particles, $(\mathbf{A}_A)_{1\text{-}2}$ is to be interpreted as the angular impulse of forces that are *external* to the system.

d. Conservation of angular momentum

If the angular impulse about A is zero, it follows from Eq. (15.36) that the angular momentum of the system of particles is conserved about A. In other words,

$$\text{if } (\mathbf{A}_A)_{1\text{-}2} = \mathbf{0}, \quad \text{then } (\mathbf{h}_A)_1 = (\mathbf{h}_A)_2 \qquad (A\text{: fixed point or mass center}) \tag{15.37}$$

which is known as the *principle of conservation of angular momentum.* Observe that angular momentum about a fixed point, or about the mass center, is conserved during a given time interval if and only if the angular impulse about that point is zero throughout that time interval. When applying Eq. (15.37) to a system of particles, $(\mathbf{A}_A)_{1\text{-}2}$ refers, of course, to the angular impulse of the external forces. Since Eq. (15.37) is a vector equation, it is possible for the angular momentum about an axis passing through A to be conserved, even though the total angular momentum about point A may not be conserved.

Sample Problem 15.8

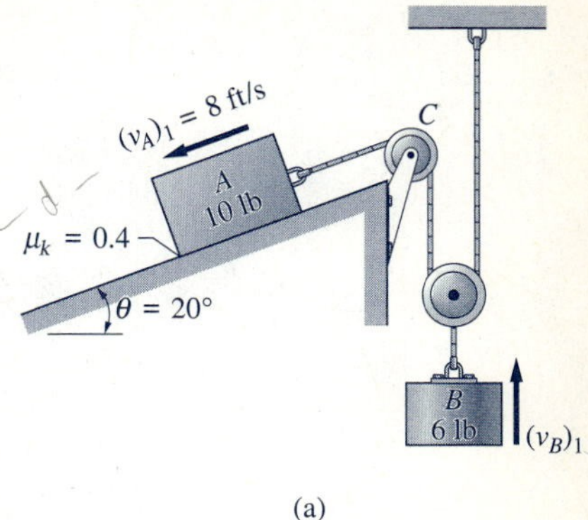

(a)

As shown in Fig. (a), the 10-lb block A and the 6-lb block B are connected by a cable that runs around the two pulleys of negligible weight. The kinetic coefficient of friction between the inclined plane and block A is $\mu_k = 0.4$. If block A is given the initial velocity $(v_A)_1 = 8$ ft/s down the plane, determine the distance d (measured from the initial position) that it will move down the incline before the system comes to rest.

Solution

The work-energy method is convenient for analyzing this problem because a change in speed is to be related to the distance moved. We will apply this method to the system consisting of both blocks, both pulleys, and the cable. The only forces that do work on this system are the weights of the blocks, W_A and W_B, and the friction force F_A that acts on block A. The pin reaction on the pulley at C is workless because the pin does not move. The cable does work on each block; however, the net work performed on the system is zero since the cable is assumed to be inextensible (the work done by the cable on A is canceled by the work it does on B).

The friction force F_A, directed opposite to the motion, is found by analyzing the free-body diagram of block A, which is shown in Fig. (b). Setting the sum of the forces in the y direction equal to zero gives $N_A = W_A \cos 20°$, from which $F_A = \mu_k N_A = 0.4\,(W_A \cos 20°) = 0.4(10 \cos 20°) = 3.759$ lb.

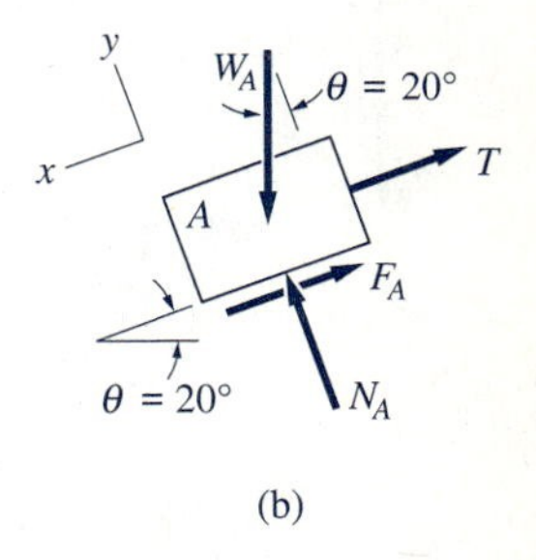

(b)

From kinematic analysis we see that the distance that A moves down the plane is twice the distance that B moves upward. Therefore, it follows that $v_A = 2v_B$, which gives the initial velocity of B to be 4 ft/s upward.

Letting ① and ② refer to the initial and final positions of the system, respectively, the work-energy principle of Eq. (15.24) yields for the system

$$(U_{1\text{-}2})_{\text{ext}} + (U_{1\text{-}2})_{\text{int}} = \Delta T = T_2 - T_1$$

$$\left[(W_A \sin\theta)d - F_A d - W_B\left(\frac{d}{2}\right)\right] + 0 = 0 - \left[\frac{1}{2}m_A(v_A)_1^2 + \frac{1}{2}m_B(v_B)_1^2\right]$$

Observe that both the work done by internal forces, $(U_{1\text{-}2})_{\text{int}}$, and the final kinetic energy, T_2, are zero. Note also that the distance moved by B has been set equal to $d/2$, i.e., one-half the distance moved by A. Substituting the numerical values, the above equation becomes

$$(10 \sin 20°)d - 3.759d - 6\left(\frac{d}{2}\right) = -\frac{1}{2}\frac{10}{32.2}(8)^2 - \frac{1}{2}\frac{6}{32.2}(4)^2$$

from which we find that the distance moved down the incline by A is

$$d = 3.42 \text{ ft} \qquad \textit{Answer}$$

This problem could also have been solved by using two work-energy equations, one for each block. In that case, the work done by the cable force would appear in each of these equations. However, when the two equations are added together, the work of the cable force would cancel out.

Sample Problem 15.9

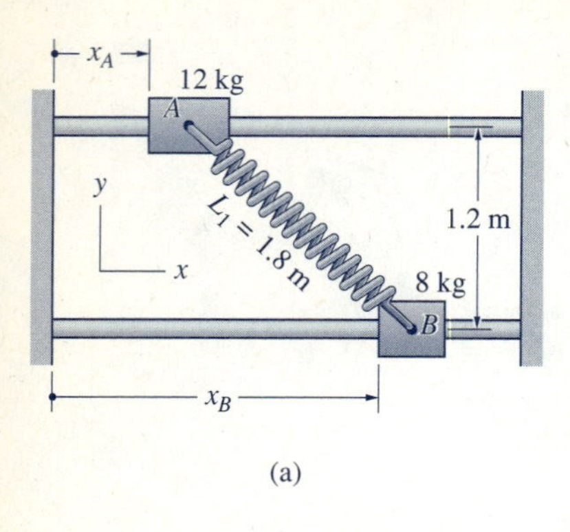

(a)

The two collars A and B shown in Fig. (a) slide along smooth bars that lie in the same vertical plane and are 1.2 m apart. The masses of A and B are 12 kg and 8 kg, respectively. The stiffness of the spring is $k = 100$ N/m and its free length is $L_0 = 1.2$ m. If the system is released from rest in the position shown in Fig. (a), where the spring has been stretched to the length $L_1 = 1.8$ m, calculate the maximum speed reached by each of the collars.

Solution

When the system is released from rest in the position shown in Fig. (a), which we will refer to as ①, the tension in the spring pulls the collars toward each other. Since the free length of the spring is identical to the distance between the rails, the spring will be unstretched when A is directly above B. This position, which we shall denote as ②, is thus the position where the speeds of the collars are maximized. After passing through position ②, the tension in the spring reduces the speeds of the collars, eventually bringing them to a temporary stop. The motion then reverses itself, with the system returning to position ①, since the system is conservative.

We will analyze the system consisting of the two collars and the spring. Figure (b) shows the free-body diagram of this system for an arbitrary position. The only external forces that act on this system are the weights of the collars, W_A and W_B, and the normal forces, N_A and N_B, that are provided by the smooth rails. The force in the spring does not appear on the FBD because it is an internal force.

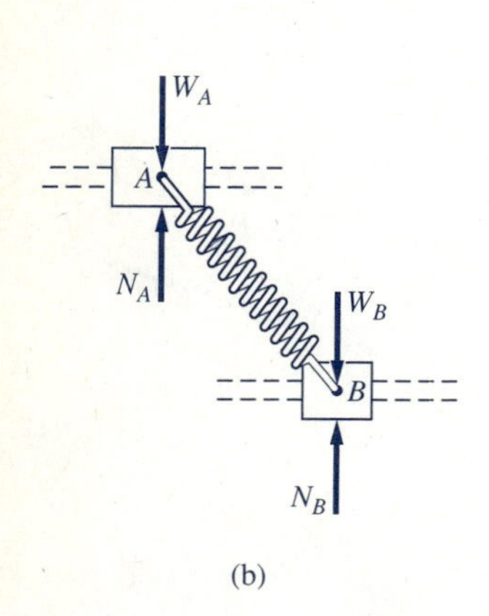

(b)

From the FBD in Fig. (b) we see that $(U_{1-2})_{\text{ext}} = 0$ because each external force is perpendicular to the path of the collar. Therefore, the work-energy principle,* Eq. (15.24), yields

$$(U_{1-2})_{\text{ext}} + (U_{1-2})_{\text{int}} = \Delta T = T_2 - T_1$$

$$0 + (U_{1-2})_{\text{int}} = T_2 - 0$$

where we have substituted $T_1 = 0$ for the initial kinetic energy. Using Eq. (14.10) to compute $(U_{1-2})_{\text{int}}$, the work done by the spring force, and recognizing that the kinetic energy of the system is the sum of the kinetic energy for each collar, this equation becomes

$$-\frac{1}{2}k\left(\delta_2^2 - \delta_1^2\right) = \frac{1}{2}m_A\,(v_A)_2^2 + \frac{1}{2}m_B\,(v_B)_2^2$$

Substituting numerical values—noting that the spring deformations are $\delta_1 = L_1 - L_0 = 1.8 - 1.2 = 0.6$ m and $\delta_2 = 0$—we obtain

$$-\frac{1}{2}(100)[0 - (0.6)^2] = \frac{1}{2}(12)(v_A)_2^2 + \frac{1}{2}(8)(v_B)_2^2$$

which may be simplified to

$$6(v_A)_2^2 + 4(v_B)_2^2 = 18 \qquad \text{(a)}$$

* The principle of conservation of mechanical energy, Eq. (15.26), could also have been used, because the system is conservative.

A second equation relating the final velocities of A and B is obtained by applying the impulse-momentum principle to the system. From the FBD in Fig. (b), we see that there are no external forces, i.e., no external impulses acting on the system in the x-direction. Therefore, the momentum of the system is conserved in the x-direction. (The change in momentum in the vertical direction for the system is also zero, but this is of no interest here.) Since the momentum in the x-direction in position ① is zero, it must also be zero in position ②. Computing the momentum of the system by adding the momenta of both collars, we have

$$\xrightarrow{+} \quad (p_x)_2 = m_A(v_A)_2 + m_B(v_B)_2 = 0$$

where both velocities were assumed to be directed to the right. Substituting values for the masses, this equation becomes

$$12(v_A)_2 + 8(v_B)_2 = 0 \qquad \text{(b)}$$

Solving Eqs. (a) and (b) simultaneously, the maximum speeds of the collars are found to be

$$(v_A)_2 = \pm 1.095 \text{ m/s} \quad \text{and} \quad (v_B)_2 = \mp 1.643 \text{ m/s} \qquad \textit{Answer}$$

The duality of signs in these answers indicates that if collar A is moving to the right as it passes through position ②, collar B is moving to the left, and vice versa.

Sample Problem 15.10

The 12-kg block A in Fig. (a) is released from rest at the top of the 2-kg wedge B (position ①). Determine the velocities of A and B when the block has reached the bottom of the inclined face of B as shown in Fig. (b) (position ②). Neglect friction.

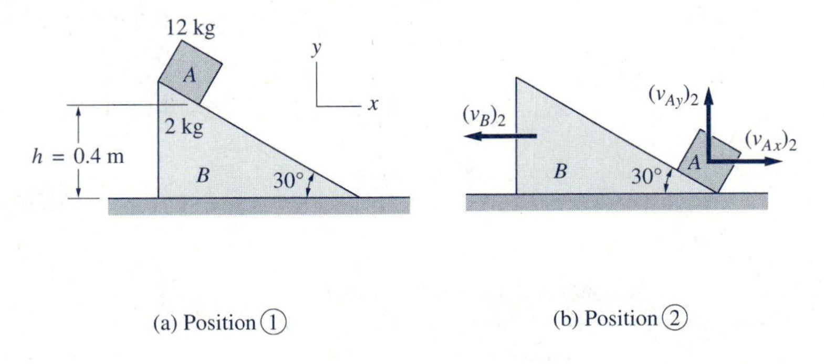

(a) Position ①   (b) Position ②

Solution

Our solution consists of the following steps applied to the system consisting of the block A and the wedge B:

Step 1: Apply the principle of conservation of mechanical energy.
Step 2: Apply the principle of conservation of linear momentum in the x-direction.
Step 3: Relate the velocities of A and B using kinematics.
Step 4: Solve the equations that result from Steps 1–3.

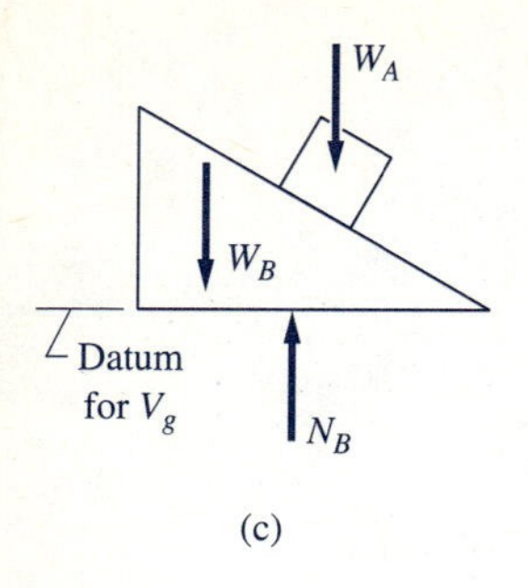

(c)

The order in which Steps 1–3 are performed is immaterial. The assumed directions for the velocities of A and B in position ② are shown in Fig. (b). The free-body diagram of the system in an arbitrary position is shown in Fig. (c).

Step 1: Conservation of Mechanical Energy

Since friction is neglected, the only force that does work on the system between positions ① and ② is the weight of A. This means that the system is conservative, the potential energy consisting of the gravitational potential energy V_g of block A only. Choosing the horizontal plane as the datum for V_g, as indicated in Fig. (c), the principle of conservation of mechanical energy,* Eq. (15.26), gives

$$V_1 + T_1 = V_2 + T_2$$

$$m_A g h + 0 = 0 + \frac{1}{2} m_A (v_A)_2^2 + \frac{1}{2} m_B (v_B)_2^2$$

Note that $T_1 = 0$ and $V_2 = 0$. Substituting numerical values, this equation becomes

$$12(9.81)(0.4) = \frac{1}{2}(12)(v_A)_2^2 + \frac{1}{2}(2)(v_B)_2^2$$

which after simplification can be written as

$$6(v_A)_2^2 + (v_B)_2^2 = 47.09 \qquad \text{(a)}$$

Step 2: Conservation of Linear Momentum in the *x*-direction

From the free-body diagram in Fig. (c) we see that there are no forces, and thus no impulses, that act on the system in the x-direction. Therefore the x-component of the momentum of the system is conserved. (The normal force that acts between A and B has an x-component, but this force is internal to the system.) Since the x-component of the momentum of the system in position ① is zero, the x-component of the momentum of the system in position ② is also zero. Using the velocities shown in Fig. (b), we obtain

$$\xrightarrow{+} \quad (p_x)_2 = m_A (v_{Ax})_2 - m_B (v_B)_2 = 0$$

Substituting the values for m_A and m_B, we find that

$$12(v_{Ax})_2 - 2(v_B)_2 = 0 \qquad \text{(b)}$$

Step 3: Relate the Velocities of *A* and *B* Using Kinematics

The velocities of A and B must satisfy the relative velocity equation $\mathbf{v}_A = \mathbf{v}_{A/B} + \mathbf{v}_B$. The kinematic constraint is that $\mathbf{v}_{A/B}$ is directed along the inclined face of B. Assuming that $\mathbf{v}_{A/B}$ is directed down the inclined face, we have

$$\mathbf{v}_{A/B} = \overset{30^\circ}{\searrow} \; v_{A/B}$$

Using the velocities shown in Fig. (b), the relative velocity equation for position ② becomes

$$(\mathbf{v}_A)_2 = (\mathbf{v}_{A/B})_2 + (\mathbf{v}_B)_2$$

$$(v_{Ax})_2\mathbf{i} + (v_{Ay})_2\mathbf{j} = (v_{A/B})_2 \cos 30^\circ\mathbf{i} - (v_{A/B})_2 \sin 30^\circ\mathbf{j} - (v_B)_2\mathbf{i}$$

* The work-energy principle, Eq. (15.24), could also have been used.

Equating the vector components gives the following two scalar equations:

$$(v_{Ax})_2 = (v_{A/B})_2 \cos 30^\circ - (v_B)_2 \tag{c}$$

$$(v_{Ay})_2 = -(v_{A/B})_2 \sin 30^\circ \tag{d}$$

Step 4: Solution of the Equations That Result from Steps 1–3

Careful inspection reveals that Eqs. (a)–(d) represent four equations that contain the four unknowns: $(v_{Ax})_2$, $(v_{Ay})_2$, $(v_B)_2$, and $(v_{A/B})_2$. Omitting the algebraic details, the solution yields

$$(v_{Ax})_2 = 0.580 \text{ m/s} \qquad (v_{Ay})_2 = -2.34 \text{ m/s}$$
$$(v_B)_2 = 3.48 \text{ m/s} \qquad (v_{A/B})_2 = 4.69 \text{ m/s}$$

Answer

Since the sign of $(v_{Ay})_2$ is negative, its direction is opposite to what is shown in Fig. (b).

The speed of A in position ② is, therefore,

$$(v_A)_2 = \sqrt{(v_{Ax})_2^2 + (v_{Ay})_2^2} = \sqrt{(0.580)^2 + (2.34)^2}$$
$$= 2.41 \text{ m/s}$$

Answer

Sample Problem 15.11

The assembly shown in Fig. (a) consists of two small balls, each of mass m, that slide on a smooth, rigid frame AOB of negligible mass. The support at O permits free rotation of the frame about the z-axis. The frame is initially rotating with the angular velocity ω_1 while strings hold the balls at the radial distance R_1. The strings are subsequently cut simultaneously, permitting the balls to slide toward the stops at A and B, which are located at the radial distance R_2. Determine ω_2, the final angular velocity of the assembly, assuming that the balls do not rebound after striking the stops.

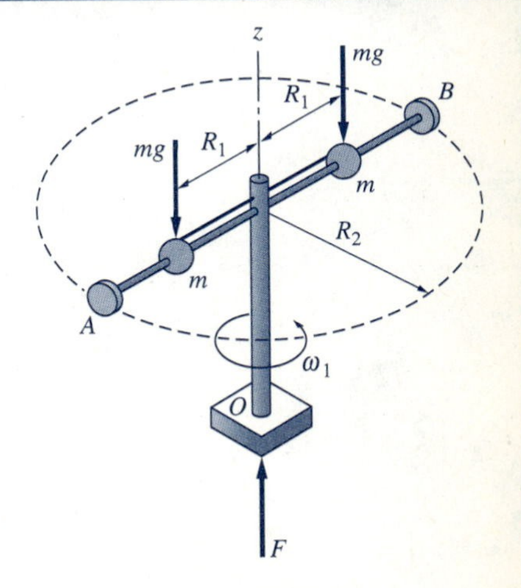

(a) Free-body diagram

Solution

The free-body diagram (FBD) of the assembly before the strings are cut is shown in Fig. (a). The only external forces acting on the assembly are the weights mg of the balls and the vertical support force F at O. (Symmetry precludes other forces or moments exerted by the support at O.) The forces acting between the balls and the rod, and the forces in the strings, are all internal forces that do not appear on the FBD of the system.

From Fig. (a) we see that the moment of the external forces about the z-axis is zero since these forces are parallel to the axis. After the strings have been cut, this FBD will change with time because the balls move away from the z-axis. However, the weight vectors always remain parallel to the z-axis, which means that the angular impulse about the z-axis continues to be zero as the balls move along the rod. When the balls hit the stops, the resulting impact forces are internal to the FBD of the assembly. Consequently, there will never be an angular impulse acting on the assembly about the z-axis, which means that angular momentum about that axis is always conserved.

The momentum diagrams of the assembly at times t_1 and t_2 are shown in Figs. (b) and (c), respectively, where t_1 is a time before the strings were cut, and t_2 is a time after the balls have come to rest relative to the rod. Only the linear momentum of each ball is shown, since the mass of the frame is negligible. The velocities of the balls are related to the angular velocities by $v_1 = R_1\omega_1$ and $v_2 = R_2\omega_2$.

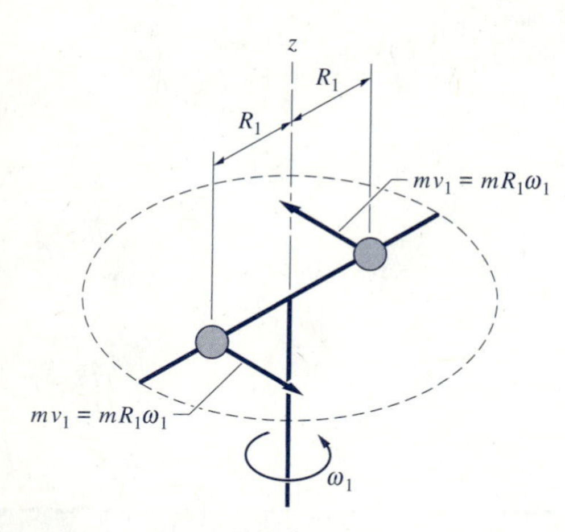

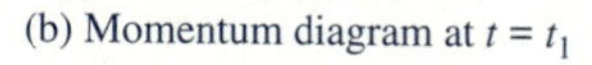
(b) Momentum diagram at $t = t_1$

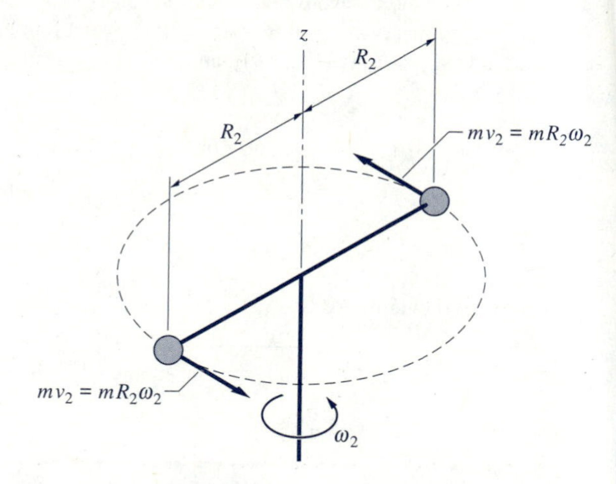

(c) Momentum diagram at $t = t_2$

Since angular momentum equals the moment of the linear momentum, conservation of angular momentum about the z-axis yields

$$(h_z)_1 = (h_z)_2$$

$$2(mR_1\omega_1)R_1 = 2(mR_2\omega_2)R_2$$

from which we find

$$\omega_2 = (R_1/R_2)^2\omega_1 \qquad \textit{Answer}$$

PROBLEMS

15.55 Solve Prob. 15.26 using the principle of conservation of linear momentum.

15.56 Solve Prob. 15.31 using the conservation of linear momentum and the conservation of mechanical energy.

15.57 Solve Prob. 15.41 using the conservation of mechanical energy.

15.58 The compressive force in the spring equals 2 lb when the system is at rest in the position shown. If the cord is cut, find the velocities of masses A and B when the spring force becomes zero. Neglect friction.

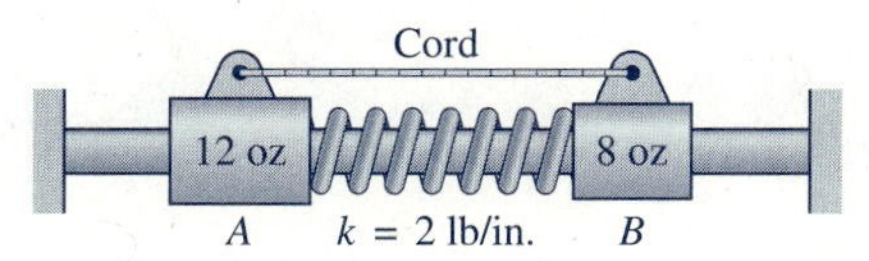

Fig. P15.58

15.59 The package A lands on the stationary cart B with the horizontal velocity $v_0 = 2.5$ m/s. The coefficient of kinetic friction between A and B is 0.25, and the rolling resistance of B may be neglected. (a) Determine the velocity of B after the sliding of A relative to B has stopped. (b) Find the total distance that A slides relative to B.

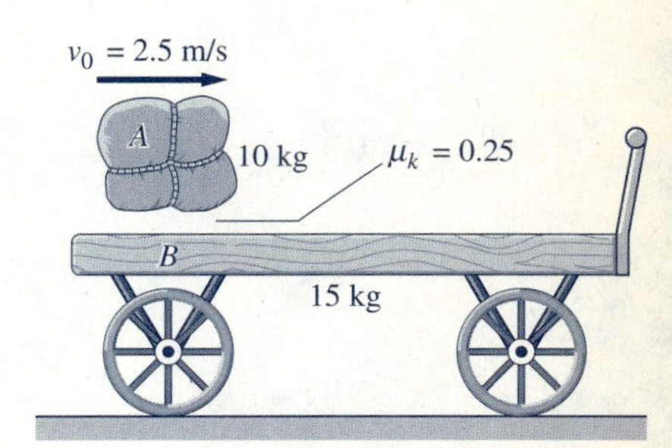

Fig. P15.59

15.60 The two blocks A and B are connected by a cable that runs around two pulleys of negligible weight. If the system is released from rest in the position shown, calculate the velocity of A when it hits the floor. The kinetic coefficient of friction between B and the horizontal surface is 0.24.

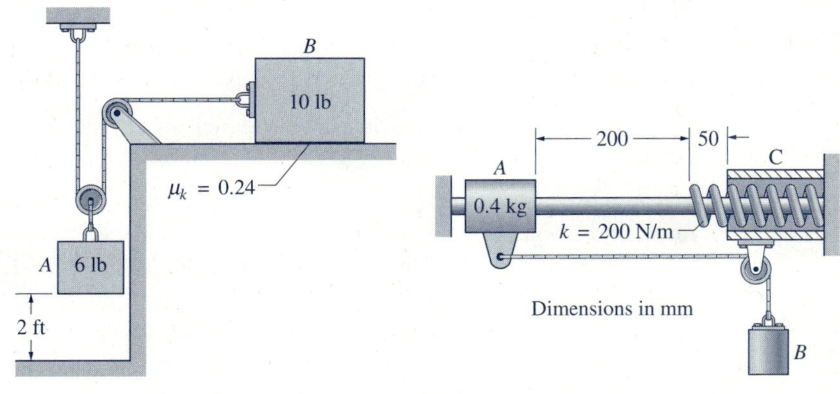

Fig. P15.60 **Fig. P15.61**

15.61 The system is released from rest in the position shown. Neglecting friction, find the mass of B that will cause A to reach the left end of the retaining tube C with zero velocity. Note that the stiffness of the spring in the retaining tube is 200 N/m.

15.62 Masses A and B are connected by a rope that passes over a pulley of negligible mass. The collar C is too large to fit through the opening at D, but it is free to slide relative to B. Note that the surface under A is rough. If the system is released from rest in the position shown, determine (a) the velocity of A just before C hits the constriction at D; and (b) the total distance that A travels before coming to rest.

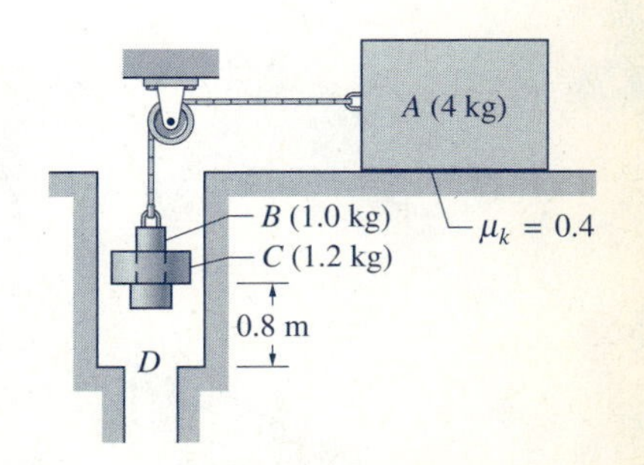

Fig. P15.62

15.63 The block A and cart B are stationary when the constant force $P = 1.0$ lb is applied. Find the speed of the cart when the block has moved 2 ft relative to the cart. Neglect rolling resistance of the cart, and note that the surface between the block and cart is rough.

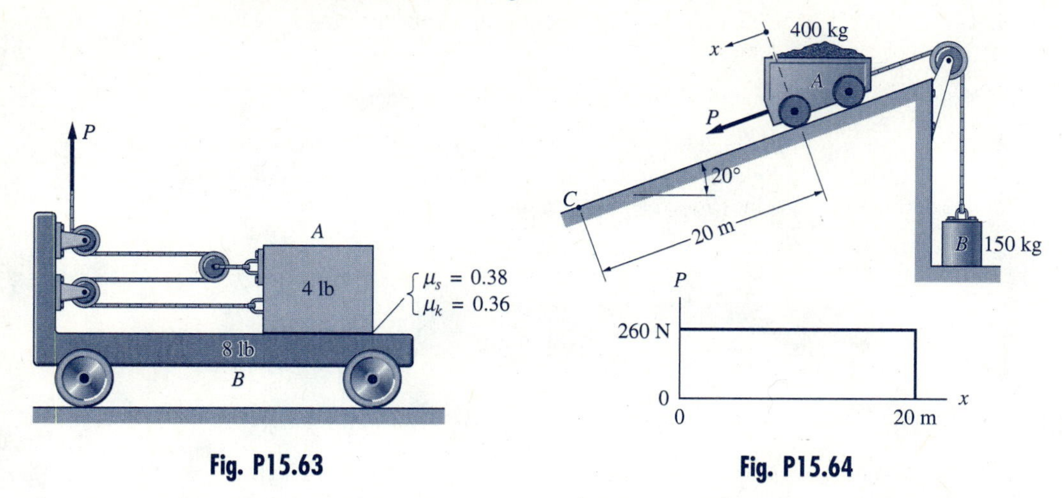

Fig. P15.63 **Fig. P15.64**

15.64 The cable car A and counterweight B are at rest in the position shown when the force P is applied to the car. As shown in the graph, the magnitude of P has the constant value of 260 N until the car passes point C, at which time it drops to zero. Determine (a) the maximum distance traveled by the car before coming to rest again; and (b) the maximum velocity of the car. Neglect friction.

15.65 The system is released from rest when $\theta = 0$. Determine the ratio m_A/m_B of the two masses for which the system will come to rest again when $\theta = 60°$. Neglect friction.

15.66 The 5-ft chain AB weighs 15 lb. If the chain is released from rest in the position shown, calculate the speed with which end B hits the floor. Neglect friction.

15.67 Re-solve Prob. 15.66 if the coefficient of kinetic friction between the chain and the tabletop is 0.4.

15.68 The system is released from rest in the position shown. Determine the distance L for which A would come to rest just before it hits B. (Hint: You must first compute the speed of the system when B hits the floor. Why?)

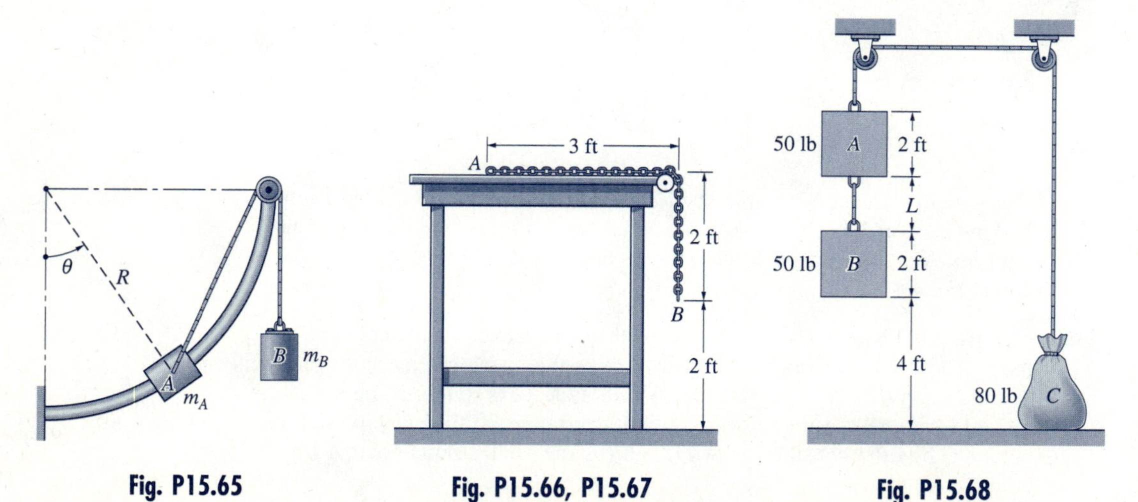

Fig. P15.65 **Fig. P15.66, P15.67** **Fig. P15.68**

15.69 The spring between the cart A and slider B is held in compression by the restraining cable. The cable is cut when the system is at rest. Knowing that slider B leaves its guide-rod with the relative velocity $v_{B/A} = 15$ m/s, determine the velocities of A and B at this same instant.

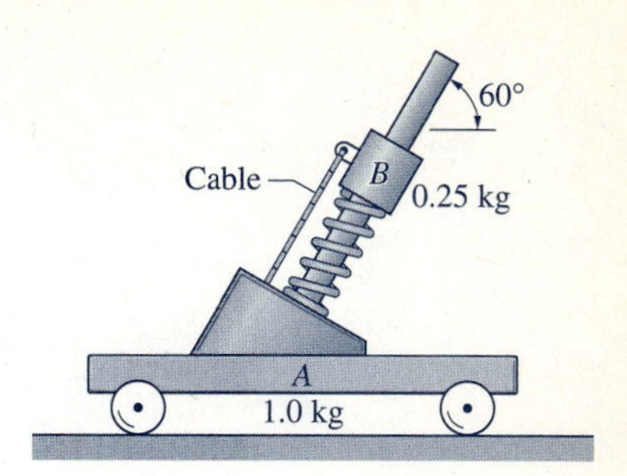

Fig. P15.69

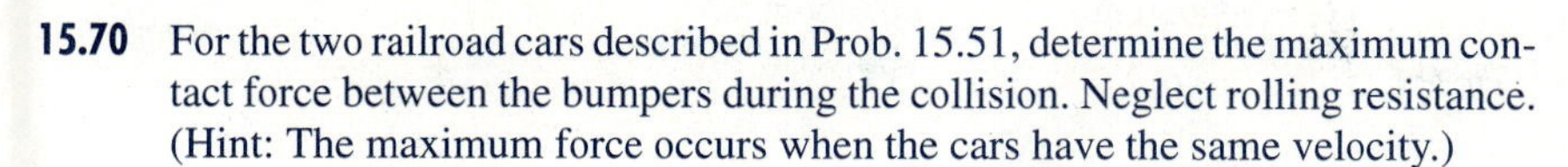

15.70 For the two railroad cars described in Prob. 15.51, determine the maximum contact force between the bumpers during the collision. Neglect rolling resistance. (Hint: The maximum force occurs when the cars have the same velocity.)

15.71 For the railroad cars described in Prob. 15.51, compute (a) the velocity of each car after the collision; and (b) the impulse that each car received during the collision.

15.72 The system is at rest when the simple pendulum is released from the position shown. When the pendulum reaches the vertical position for the first time, compute the absolute speed of (a) the pendulum's bob A; and (b) the carriage B. Neglect rolling resistance of the carriage.

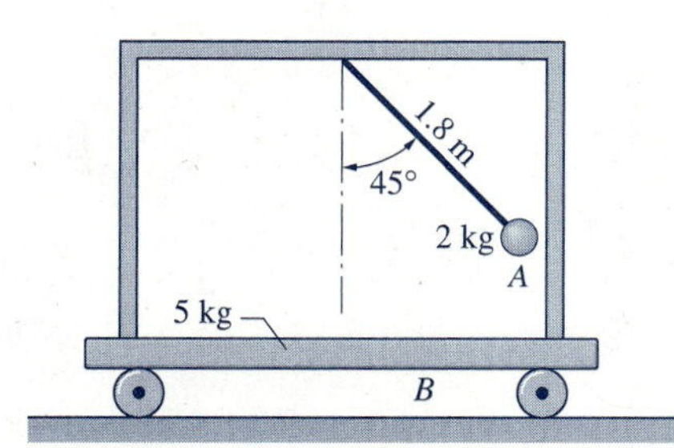

Fig. P15.72

15.73 A 120-lb projectile explodes during its flight and breaks into three pieces with the weights $W_A = 60$ lb, $W_B = 40$ lb, and $W_C = 20$ lb. The velocities of the pieces 0.5 s after the explosion are $\mathbf{v}_A = -36\mathbf{i} + 12\mathbf{j} + 48\mathbf{k}$ ft/s, $\mathbf{v}_B = -30\mathbf{j} + 21\mathbf{k}$ ft/s, and $\mathbf{v}_C = 56\mathbf{i} - 8\mathbf{j} - 12\mathbf{k}$ ft/s. Determine the velocity of the projectile just before the explosion. Be sure to include the effect of gravity, which acts in the negative z-direction.

15.74 A system consists of three particles that have the velocities shown at a certain instant. Compute the following for the system: (a) the coordinates of the mass center G; (b) the linear momentum; (c) the angular momentum about the origin; and (d) the angular momentum about G.

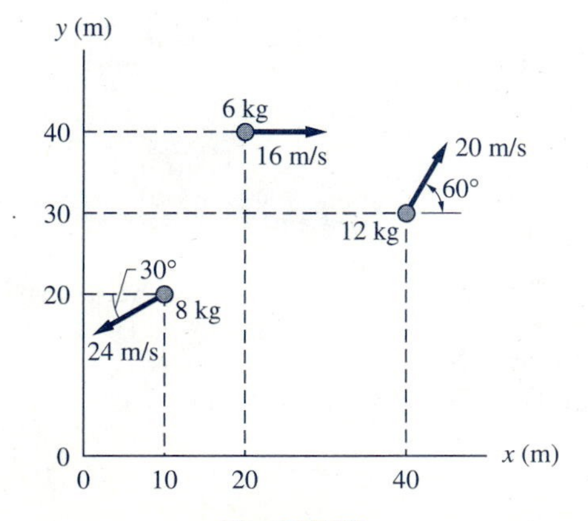

Fig. P15.74

15.75 At the instant shown, the velocities of the three particles are $\mathbf{v}_A = -12\mathbf{i} + 3\mathbf{j} + 16\mathbf{k}$ ft/s, $\mathbf{v}_B = -10\mathbf{j} + 8\mathbf{k}$ ft/s, and $\mathbf{v}_C = 18\mathbf{i} - 4\mathbf{k}$ ft/s. If each particle weighs 1.5 lb, calculate the following for the system: (a) the coordinates of the mass center G; (b) the linear momentum; (c) the angular momentum about the origin; and (d) the angular momentum about B.

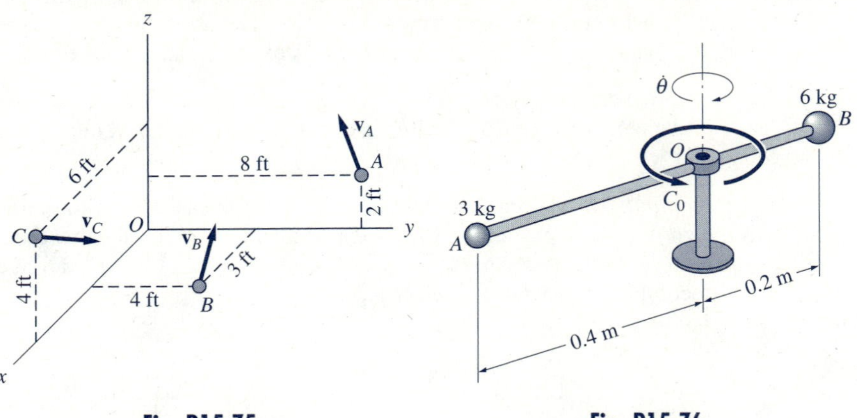

Fig. P15.75

Fig. P15.76

15.76 The rigid assembly, consisting of the two masses attached to a massless rod, rotates about the vertical axis at O. The assembly is initially rotating freely at the angular speed $\dot{\theta}_0 = 120$ rad/s, when the constant couple $C_0 = 8$ N · m that opposes the motion is applied. Find (a) the time required to stop the assembly; and (b) the number of revolutions made by the assembly before coming to rest.

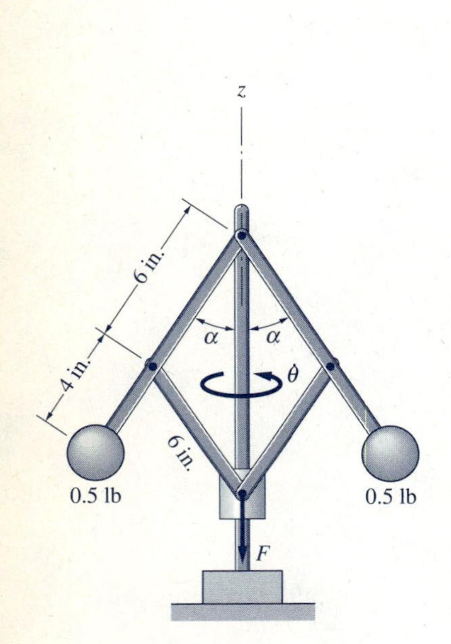

Fig. P15.77

15.77 A flyball speed governor consists of the two 0.5-lb weights and a supporting linkage of negligible weight. The position of the weights can be changed by adjusting the magnitude of the force F acting on the sliding collar. The entire assembly is initially rotating about the z-axis at 500 rev/min with the supporting arms inclined at $\alpha = 60°$. (a) Compute the initial angular momentum of the system about the z-axis. (b) Find the angular speed of the assembly when the angle α is changed to 30°. Neglect friction.

15.78 The two masses are connected by a rigid rod of negligible mass. The assembly is falling vertically without spinning when it strikes the obstruction C with the speed v_0. Assuming no rebound, determine the angular speed of the assembly immediately after the impact.

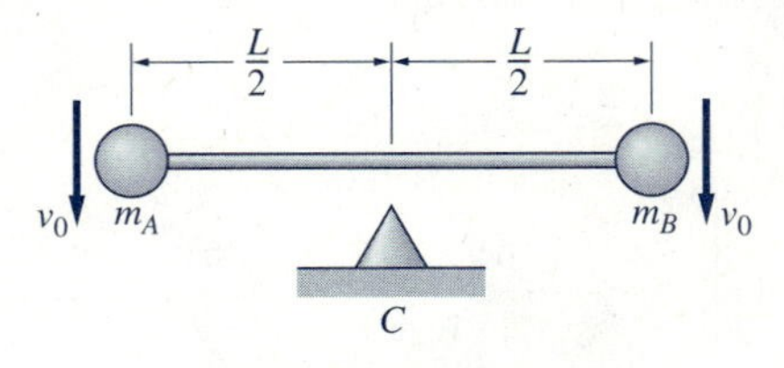

Fig. P15.78

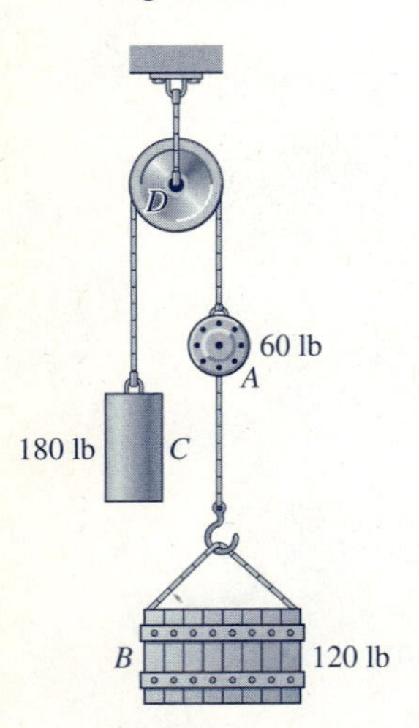

Fig. P15.79

15.79 The system consists of the electric hoist A, the crate B, and the counterweight C, all of which are suspended from the pulley D. The system is stationary when the hoist A is turned on, causing it to start rewinding the rope connecting A and B at the rate of 2 ft/s. Determine the resulting velocities of A, B, and C.

15.80 The two particles A and B, connected by a rigid rod of negligible mass, are initially at rest on a smooth horizontal surface. If A is given the initial velocity v_0 as shown, determine the velocities of A and B when the assembly has rotated through 90°.

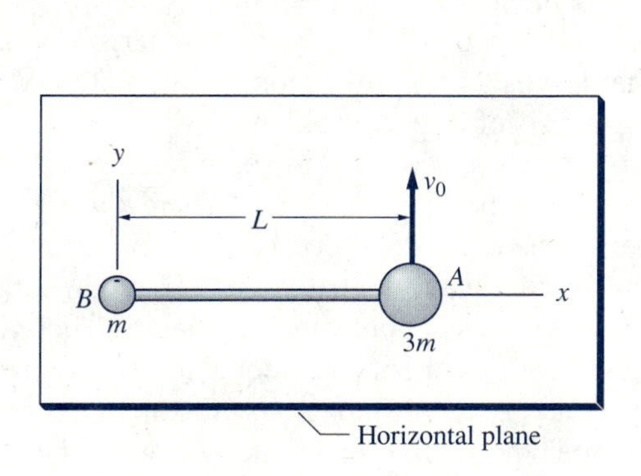

Fig. P15.80

(a)

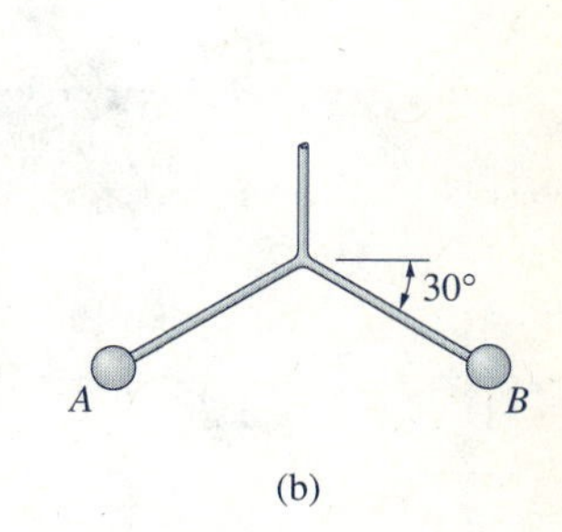

(b)

Fig. 15.81

15.81 The assembly consists of three particles A, B, and C (each of mass m) that are connected to a light, rigid frame. The assembly is initially rotating counterclockwise about its mass center D with the angular velocity ω_0. If rod DC suddenly breaks when the assembly is in the position shown in (a), determine the velocities of masses A and B when they reach the position shown in (b).

15.82 The assembly consisting of two identical masses and a supporting frame of negligible mass rotates freely about the z-axis. Mass A is attached to the frame, but mass B is free to slide on the smooth horizontal bar. An ideal spring of stiffness k and free length R_0 is connected between A and B. Determine all possible combinations of R and $\dot{\theta}$ for which B remains at rest relative to the frame.

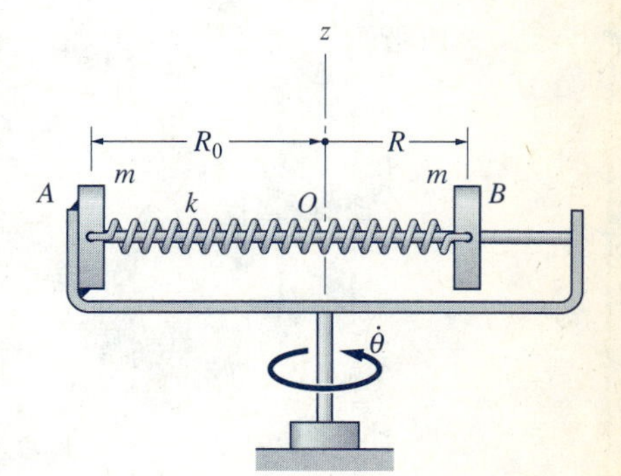

Fig. P15.82–P15.84

15.83 Mass B of the assembly described in Prob. 15.82 is released from the position $R = R_0$ when $\dot{\theta} = \dot{\theta}_0$. (a) Find the speed of B relative to the frame when it passes point O. (b) Determine the range of $\dot{\theta}_0$ for which B will not reach O.

15.84 The mass B of the assembly described in Prob. 15.82 is released at $t = 0$ when $R = R_0$ and $\dot{\theta} = \dot{\theta}_0$. (a) Show that the equations of motion are

$$\ddot{R} = R\left(\dot{\theta}^2 - \frac{k}{m}\right) \qquad \ddot{\theta} = -\frac{2R\dot{R}\dot{\theta}}{R_0^2 + R^2}$$

and state the initial conditions. (b) Solve the equations numerically from $t = 0$ to 0.2 s, using the following data: $m = 0.25$ kg, $k = 300$ N/m, $R_0 = 0.4$ m, and $\dot{\theta}_0 = 30$ rad/s; plot $\dot{\theta}$ vs. R. (c) Use the numerical solution to determine the range of R and $\dot{\theta}$.

15.8 Plastic Impact

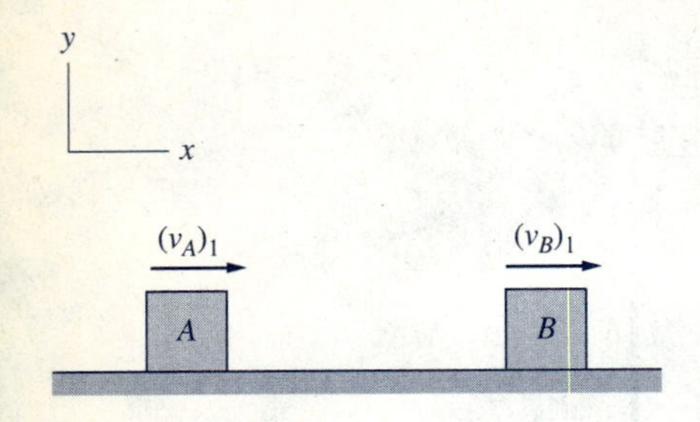

(a) Velocities before impact: $(v_A)_1 > (v_B)_1$

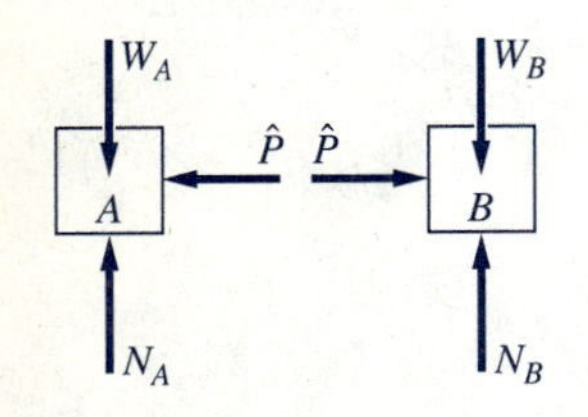

(b) FBD during impact

(c) Velocities after impact

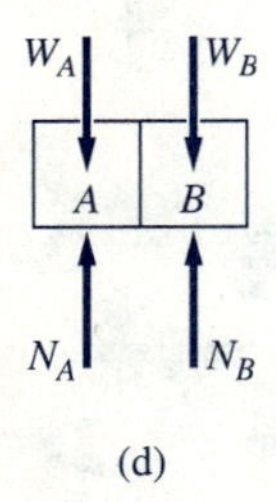

(d)

Fig. 15.12

One of the more complicated problems in dynamics is the impact, or collision, between objects. Since the magnitudes of the contact forces during an impact are usually unknown, the force-mass-acceleration method of analysis cannot be used. Furthermore, because mechanical energy is generally lost during the impact (energy may be converted into sound or heat), the work-energy approach cannot be applied directly. That leaves the impulse-momentum method as the only usable technique for the analysis of impact problems.

To explain the nature of impact, it is convenient to begin with the situation depicted in Fig. 15.12(a), where two smooth blocks A and B are sliding to the right on a horizontal plane. Before the blocks collide, their velocities are $(v_A)_1$ and $(v_B)_1$, where $(v_A)_1 > (v_B)_1$. During the time that the blocks are in contact, equal and opposite time-dependent forces act between the blocks. These contact or *impact forces*, being caused by the collision, are equal to zero before and after the impact. One of the primary goals of impact analysis is to determine the velocities $(v_A)_2$ and $(v_B)_2$ of the blocks after the impact, knowing the initial velocities $(v_A)_1$ and $(v_B)_1$.

The free-body diagram of each block during the impact is shown in Fig. 15.12(b). In addition to the weights W_A and W_B and the normal forces N_A and N_B, the blocks are subjected to the equal and opposite impact forces $\hat{P}$. (We will use a caret ^ above a letter to indicate an impact force.) Drawing the FBDs during the impact and identifying the impact forces are very important steps in the analysis of impact problems.

Summing forces in the y-direction for either block gives $N_A = W_A$ and $N_B = W_B$ throughout the motion. Applying the impulse-momentum principle, $(L_{1\text{-}2})_x = \Delta p_x$, to block A, and using the velocities shown in Fig. 15.12 parts (a) and (c), we obtain

$$\xrightarrow{+} \quad -\int_{t_1}^{t_2} \hat{P}\, dt = m_A(v_A)_2 - m_A(v_A)_1 \tag{15.38}$$

where $t = t_1$ to t_2 is the duration of the impact. Similarly, for block B we obtain

$$\xrightarrow{+} \quad \int_{t_1}^{t_2} \hat{P}\, dt = m_B(v_B)_2 - m_B(v_B)_1 \tag{15.39}$$

Note that the integrals in Eqs. (15.38) and (15.39) represent the impulses of the impact force $\hat{P}$.

Alternatively, we could analyze the system containing both blocks, the FBD of which is shown in Fig. 15.12(d). Since the external impulse acting on the system during the impact is zero (the impact force $\hat{P}$ is internal to the system, and the impulses of the normal forces cancel the impulses of the weights), the momentum vector of the system is conserved. The momentum balance in the x-direction for the system yields $(p_1)_x = (p_2)_x$; i.e.,

$$\xrightarrow{+} \quad m_A(v_A)_1 + m_B(v_B)_1 = m_A(v_A)_2 + m_B(v_B)_2 \tag{15.40}$$

Observe that Eq. (15.40) can also be obtained by adding Eqs. (15.38) and (15.39).

Assuming that the initial velocities are known, we see that Eq. (15.40) contains two unknown velocities: $(v_A)_2$ and $(v_B)_2$. To complete the analysis, we need another equation that takes into account the deformation characteristics of the impacting bodies* (the collision of steel blocks will obviously differ from the impact of rubber blocks). For the present, we will consider only the special case of *plastic impact*, where the velocities of the two blocks are the same immediately after the impact.

For plastic impact, we thus have the additional equation

$$(v_A)_2 = (v_B)_2 \tag{15.41}$$

which, combined with Eq. (15.40), will yield the solution for $(v_A)_2$ and $(v_B)_2$. Once the final velocities have been determined, either Eq. (15.38) or (15.39) can be used to calculate the impulse of the impact force $\hat{P}$ that acts between the blocks.

In order to illustrate the analysis of plastic impact, consider the graphs of $\hat{P}$, v_A, and v_B shown in Fig. 15.13. The magnitude of the impact force $\hat{P}$ is zero except for the impact interval $\Delta t = t_2 - t_1$. The initial velocities, $(v_A)_1$ and $(v_B)_1$, are assumed to be given, and we know that $(v_A)_2 = (v_B)_2$ for plastic impact. The graphs of $\hat{P}$ and the velocities during the period of contact are shown with dotted lines in Fig. 15.13 to emphasize that these functions are unknown. However, the area under the $\hat{P}$-t diagram can be computed, since it represents the impulse between the blocks.

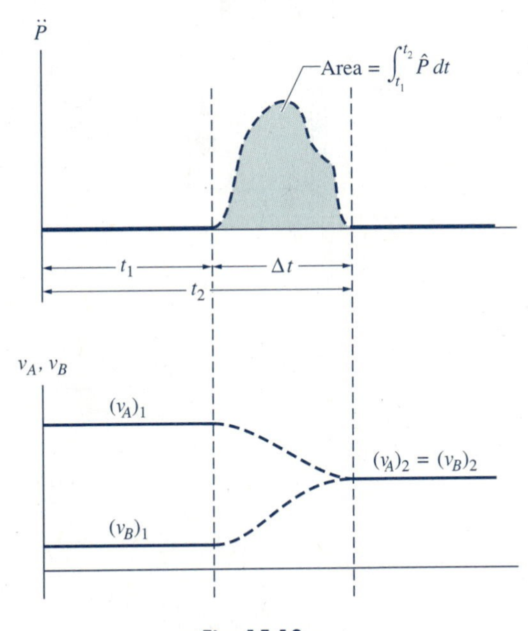

Fig. 15.13

Observe that the analysis presented here is concerned only with velocities immediately before and immediately after impact, and the impulse of the impact forces. It cannot deal with the variations of velocities and impact forces that occur during the impact.

* Note that this is one of the few situations that we have encountered for which the rigid-body model must be abandoned. There is no analysis that is valid for the impact of two "rigid" objects.

Also note that the duration of the impact, Δt in Fig. 15.13, does not have to be known, because it does not enter into the analysis. However, there is a class of impact problems that requires the assumption that the duration of impact is very small. These problems are discussed in the next article.

15.9 Impulsive Motion

The analysis presented in the preceding article is independent of the duration of the impact Δt. However, there are many impact problems that can be solved only if Δt is so small that the displacement of the bodies during the impact period can be neglected. Any motion that satisfies this condition is called *impulsive motion*. Of course, the results obtained by assuming impulsive motion are only approximations. The analysis is exact only in the idealized case where $\Delta t \to 0$.

Reconsider the plastic impact of the two blocks depicted in Fig. 15.12 with the idealization $\Delta t \to 0$. Figure 15.13, which shows a finite duration of impact, is replaced by the diagram shown in Fig. 15.14. Note that the impact interval Δt in Fig. 15.13 has been replaced by the infinitesimal time period dt in Fig. 15.14.

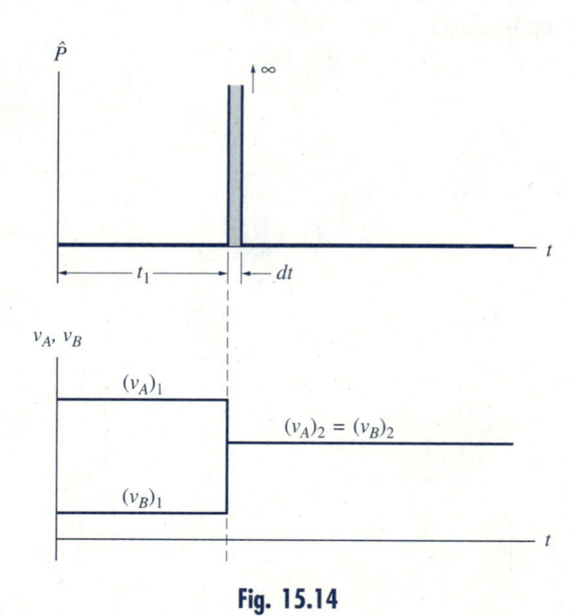

Fig. 15.14

From inspection of Fig. 15.14, we conclude the following.

1. **The magnitudes of impact forces are infinite:** The impulse of the impact force $\hat{P}$ is finite, because it is determined by the impulse-momentum equations, Eqs. (15.38) and (15.39). But since $\hat{P}$ acts over the infinitesimal time interval dt, the magnitude of $\hat{P}$ must become infinite for its impulse to remain finite. A force of infinite magnitude that acts over an infinitesimal time

interval and exerts a finite impulse (area under the force-time curve) is called an *impulsive force*.*

2. **The impulses of finite forces are negligible:** If the magnitude of a force is finite, its impulse during the impact period is negligible. For example, the weight W of a particle is an example of a finite, or nonimpulsive, force. During the time interval Δt, the impulse of W—i.e., $W\Delta t$—approaches zero as $\Delta t \to 0$.
3. **The accelerations of the blocks are infinite during impact:** Because the changes in velocities are assumed to occur within an infinitesimal time period, the v-t diagrams exhibit jump discontinuities at the time of impact, as shown in Fig. 15.14. Since $a = dv/dt$, it follows that the "jumps" correspond to infinite accelerations.
4. **The blocks are in the same location before and after the impact:** Since $\Delta t \to 0$, the distances moved by the blocks during the impact are infinitesimal (the velocities are finite).

The four steps in the analysis of the impact problems are listed below. They apply to impulsive as well as nonimpulsive motion.

Step 1: Draw the FBDs of the impacting particles. Particular attention must be paid to identifying the impact forces—use a special symbol, such as a caret, to label each of the impact forces.

Step 2: Draw the momentum diagrams of the particles at the instant immediately before the impact. (Recall that the momentum diagram of a particle is a sketch of the particle showing its momentum vector.)

Step 3: Draw the momentum diagrams for the particles at the instant immediately after the impact.

Step 4: Using the diagrams drawn in Steps 1–3, derive and solve the appropriate impulse-momentum equations.

* An impulsive force is an example of a Dirac delta function, or a *spike*. By way of illustration, consider the rectangle of width Δx and height $1/\Delta x$, shown in (a). The area of this rectangle is 1, independent of the value of Δx. If the width Δx becomes the infinitesimal dx, the height approaches infinity, but the area is still 1. The result of this limiting procedure is the delta function shown in (b).

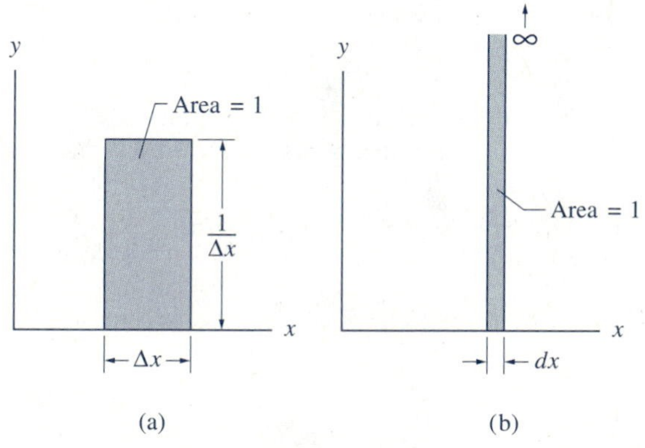

Sample Problem 15.12

As shown in Fig. (a), the 2-lb slider A is moving to the right at the speed $(v_A)_1 = 6$ ft/s when it collides with the 8-lb slider B, which is moving to the left with $(v_B)_1 = 4$ ft/s. The length of the spring attached to B is $L_1 = 20$ in. when the impact occurs, and its free length and stiffness are $L_0 = 14$ in. and $k = 0.5$ lb/in., respectively. The impact is plastic and the duration of the impact is negligible. Compute (1) the velocity v_2 of the sliders immediately after the impact; (2) the percentage of kinetic energy that is lost during impact; (3) the impulse of the impact force; and (4) the velocity v_3 of the slider B when the spring is vertical. Neglect friction.

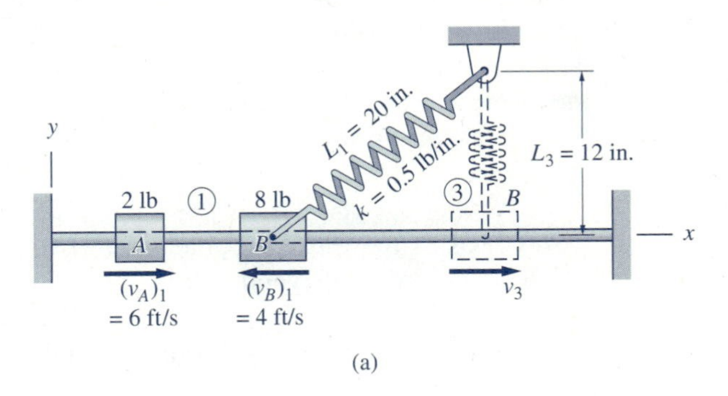

(a)

Solution

The three positions of interest are ①—immediately before impact; ②—immediately after impact; and ③—the position where the spring is vertical. Positions ① and ③ are indicated in Fig. (a). Because the duration of the impact is neglected, position ② coincides with position ①; i.e., the sliders are considered to be in the same locations immediately before and immediately after the impact.

Part 1

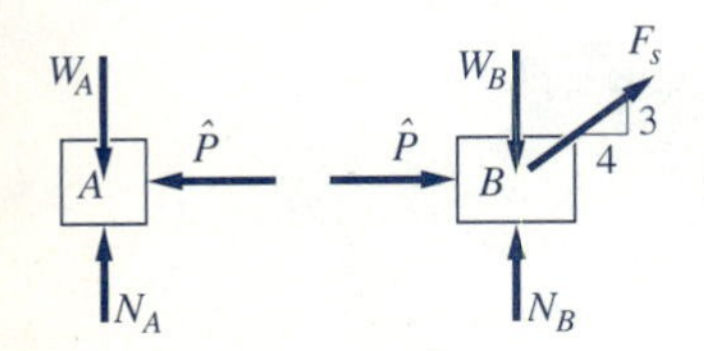

(b) Free-body diagrams during impact

The forces that act on each slider during the impact are shown on the free-body diagrams in Fig. (b): the weights W_A and W_B, the normal forces N_A and N_B, the spring force F_s, and the impact force $\hat{P}$. Since the duration of the impact is neglected, the motion during the impact is impulsive. Therefore, the impulses of all finite (bounded) forces are negligible. Careful consideration of each force to determine whether it is finite or impulsive leads to the following conclusions.

1. The weights W_A and W_B are finite forces because they are constants.
2. The spring force F_s is a finite force. Since the spring force is proportional to the deformation of the spring, it can be impulsive only if the deformation is infinite.
3. The normal forces N_A and N_B are finite. Applying the equation $\Sigma F_y = 0$ to each FBD in Fig. (b), we find that N_A and N_B are finite because W_A, W_B, and F_s are finite.
4. Since there are no constraints on the magnitude of the impact force $\hat{P}$, it is an impulsive force.

Because the analysis of impulsive motion is concerned only with impulsive forces, it is convenient to redraw the FBDs showing only these forces [see Fig. (c)]. The momentum diagrams that display the momentum vectors immediately before and immediately after impact are shown in Figs. (d) and (e), respectively. Note that v_2, the common velocity of A and B after impact, is assumed to be directed to the right in Fig. (e).

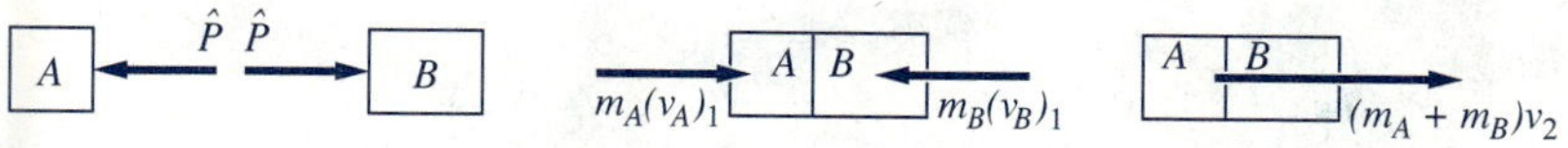

(c) FBDs (impulsive forces only)

(d) Momentum diagram before impact

(e) Momentum diagram after impact

In Fig. (c) we see that there are no external impulses acting on the system (the force $\hat{P}$ is an internal force). From Eq. (15.31), $(\mathbf{L}_{1\text{-}2})_{ext} = \Delta\mathbf{p}$, we conclude that the momentum vector for the system is conserved. Referring to the momentum diagrams in Figs. (d) and (e), the balance of momentum in the x-direction for the system gives

$$(p_x)_1 = (p_x)_2$$

$$\xrightarrow{+} \quad m_A(v_A)_1 - m_B(v_B)_1 = (m_A + m_B)v_2$$

Substituting the given numerical data yields

$$\frac{2}{g}(6) - \frac{8}{g}(4) = \left(\frac{2}{g} + \frac{8}{g}\right)v_2$$

from which the velocity after impact is

$$v_2 = -2.00 \text{ ft/s} \qquad \textit{Answer}$$

The minus sign indicates that the sliders are moving to the left immediately after impact.

Part 2

The kinetic energy lost during impact is the difference between the kinetic energies of the system before and after the impact, which gives

$$\begin{aligned}
\Delta T &= T_2 - T_1 \\
&= \frac{1}{2}(m_A + m_B)v_2^2 - \left[\frac{1}{2}m_A(v_A)_1^2 + \frac{1}{2}m_B(v_B)_1^2\right] \\
&= \frac{1}{2}\left[\frac{2}{32.2} + \frac{8}{32.2}\right](-2)^2 - \left[\frac{1}{2}\frac{2}{32.2}(6)^2 + \frac{1}{2}\frac{8}{32.2}(4)^2\right] \\
&= 0.6211 - 3.1056 = -2.4845 \text{ lb} \cdot \text{ft}
\end{aligned}$$

Therefore, the percentage of kinetic energy lost during impact is

$$\%\text{ loss} = \frac{|\Delta T|}{T_1} \times 100\% = \frac{2.4845}{3.1056} \times 100\% = 80.0\% \qquad \textit{Answer}$$

Part 3

The impulse of $\hat{P}$ is found by applying the impulse-momentum equation to either slider during impact. Referring to Figs. (c)–(e), we find for slider A

$$(L_{1\text{-}2})_x = (p_x)_2 - (p_x)_1 = m_A\Big[v_2 - (v_A)_1\Big]$$

$$\xrightarrow{+} \qquad -\int \hat{P}\,dt = \frac{2}{32.2}\Big[(-2) - 6\Big]$$

from which we find that

$$\int \hat{P}\,dt = 0.497 \text{ lb} \cdot \text{s} \qquad \textit{Answer}$$

Applying the impulse-momentum equation to slider B would, of course, yield an identical result. Note that the analysis determines only the impulse of $\hat{P}$; the magnitude of $\hat{P}$ remains indeterminate.

Part 4

After impact the sliders move to the left, but B is eventually brought to rest by the spring force. Slider B then moves to the right, reaching position ③ with the velocity v_3. With v_2 having been computed, either the work-energy principle or the principle of conservation of mechanical energy can be used to calculate v_3. Applying the work-energy principle between positions ② and ③ for slider B gives $U_{2\text{-}3} = \Delta T = T_3 - T_2$. Note that the spring force is the only force that does work on slider B. Therefore, from Eq. (14.10) we have $U_{2\text{-}3} = -(1/2)k(\delta_3^2 - \delta_2^2)$, where δ_2 and δ_3 are the deformations of the spring in positions ② and ③, respectively. These deformations are $\delta_2 = L_2 - L_0 = L_1 - L_0 = 20 - 14 = 6$ in. and $\delta_3 = L_3 - L_0 = 12 - 14 = -2$ in.

Substituting the numerical values into the work-energy equation, we get

$$U_{2\text{-}3} = T_3 - T_2$$

$$-\frac{1}{2}k\left(\delta_3^2 - \delta_2^2\right) + 0 = \frac{1}{2}m_B\left(v_3^2 - v_2^2\right)$$

$$-\frac{1}{2}(0.5)\left[(-2)^2 - 6^2\right]\frac{1}{12} = \frac{1}{2}\frac{8}{32.2}\left[v_3^2 - (-2)^2\right]$$

The factor (1/12) on the left side converts the units for work from lb · in. to lb · ft, the units for kinetic energy on the right side. Solving for the velocity of B in position ③, we obtain

$$v_3 = 3.06 \text{ ft/s} \qquad \textit{Answer}$$

A common error in the analysis of problems of this type is attempting to apply the work-energy principle (or equivalently, the principle of conservation of mechanical energy) to block B between positions ① and ③ directly. In this problem, the equation $U_{1\text{-}3} = T_3 - T_1$ is invalid since it does not account for the kinetic energy of B lost during the impact.

Sample Problem 15.13

The 10-kg wedge B shown in Fig. (a) is being held at rest by the stop at C when it is struck by the 50-g bullet A, which is traveling horizontally with the speed $(v_A)_1 = 900$ m/s. The duration of the impact is negligible. Assuming that all impacts are plastic and neglecting friction, calculate (1) the velocity with which the wedge starts up the incline; and (2) the impulse of each impulsive force.

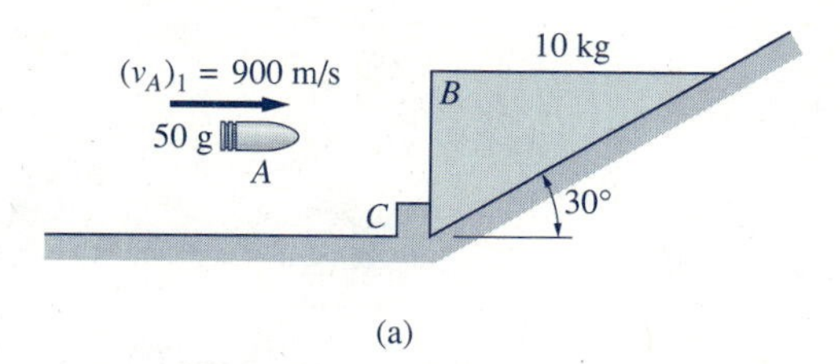

(a)

Solution

Preliminary Discussion

The two positions of interest are ①—immediately before impact, and ②—immediately after impact. Because the duration of the impact is negligible, motion is impulsive, with the bullet and wedge each occupying the same position before and after impact. Observe that there are actually two impacts in this problem—the impact between A and B and the impact between B and the smooth incline. Assuming plastic impacts is equivalent to stating that (a) the bullet becomes embedded in the block (the depth of penetration is neglected); and (b) the wedge stays in contact with the incline.

Figure (b) shows the free-body diagrams of A and B during impact. Also shown are the xy- and ab-axes that will be used in our analysis. The forces appearing on these FBDs are W_A and W_B—the weights of A and B; F—the force at C that prevents B from sliding down the incline; $\hat{P}_x$ and $\hat{P}_y$—the components of the impact force $\hat{\mathbf{P}}$ that acts between A and B (from Newton's third law, these components occur in equal and opposite pairs); and $\hat{N}$—the normal impact force exerted on B by the smooth incline.

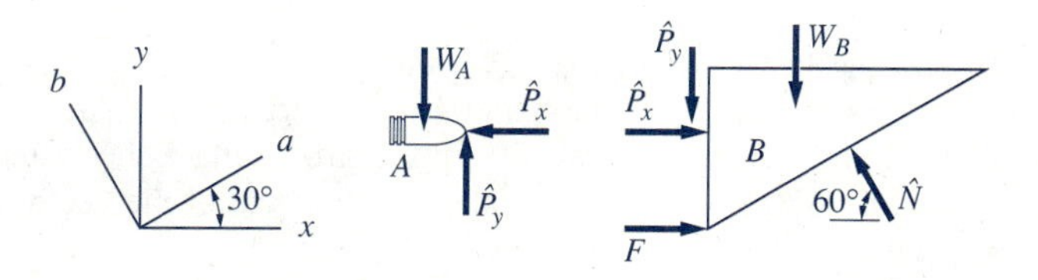

(b) Free-body diagrams during impact

The impact forces $\hat{\mathbf{P}}$ and $\hat{N}$ are the only impulsive forces. The weights W_A and W_B are finite forces, since they are constant. The force F is also a finite force, because its magnitude is bounded: Before impact, F is equal to the static value required to maintain equilibrium of the wedge; after the impact, $F = 0$ since contact with the stop is lost.

The FBDs showing only the impulsive forces are shown in Fig. (c). The momentum diagrams displaying the momenta of A and B immediately before and immediately after impact are shown in Figs. (d) and (e). In Fig. (e), observe that the direction of v_2, the common velocity of A and B after the impact, is directed up the incline.

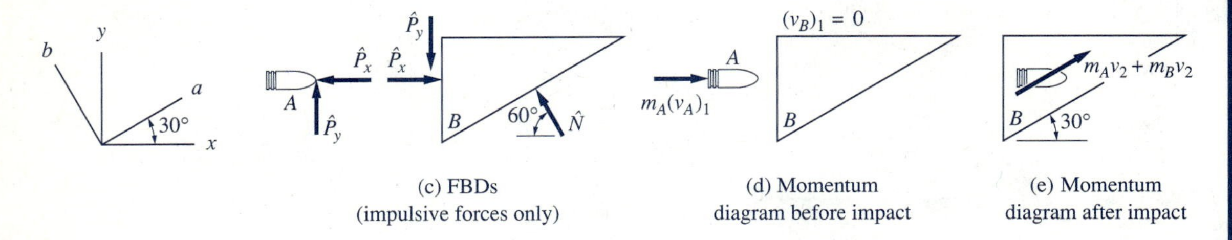

(c) FBDs (impulsive forces only)

(d) Momentum diagram before impact

(e) Momentum diagram after impact

Examination of Figs. (c)–(e) reveals that there are a total of four unknowns: the impulse of $\hat{P}_x$, the impulse of $\hat{P}_y$, the impulse of $\hat{N}$, and the final velocity v_2. Equating the impulse vectors to the change in the momentum vectors for A and B individually will yield four scalar equations that can be solved for the four unknowns. However, a more efficient solution is obtained by initially considering the system consisting of both A and B.

Part 1

From the FBDs in Fig. (c), we note that $\hat{N}$ is the only impulsive force that exerts an external impulse on the system consisting of A and B ($\hat{\mathbf{P}}$ is an internal force). Since $\hat{N}$ acts in the b-direction, momentum of the system is conserved in the a-direction:

$$
\begin{aligned}
(p_a)_1 &= (p_a)_2 \\
\overset{+}{\nearrow} \quad m_A(v_A)_1 \cos 30^\circ &= (m_A + m_B)v_2 \\
0.050(900) \cos 30^\circ &= (0.050 + 10)v_2
\end{aligned}
$$

from which the common velocity of A and B immediately after the impact is

$$v_2 = 3.88 \text{ m/s} \qquad \textit{Answer}$$

Part 2

The impulse of $\hat{N}$ can be found by considering the change in the b-component of the momentum of the system. The momentum of the system in the b-direction before impact is due only to the b component of $m_A(\mathbf{v}_A)_1$; after impact, the system has no momentum in the b-direction. Therefore, the b-component of the vector equation $(\mathbf{L}_{1\text{-}2})_{\text{ext}} = \Delta\mathbf{p}$ is

$$
\begin{aligned}
\overset{+}{\nwarrow} \quad \int \hat{N}\, dt &= (p_b)_2 - (p_b)_1 \\
&= 0 - [-m_A(v_A)_1 \sin 30^\circ] = 0.050(900) \sin 30^\circ \\
&= 22.5 \text{ N} \cdot \text{s} \qquad \textit{Answer}
\end{aligned}
$$

The impulses of $\hat{P}_x$ and $\hat{P}_y$ can be computed by solving the two scalar impulse-momentum equations for bullet A only. Referring to Figs. (c)–(e), the x-component of the impulse-momentum equation is

$$(L_{1\text{-}2})_x = (p_x)_2 - (p_x)_1$$

$$\overset{+}{\rightarrow} \quad -\int \hat{P}_x \, dt = m_A v_2 \cos 30^\circ - m_A (v_A)_1$$

$$= 0.050(3.88)\cos 30^\circ - 0.050(900)$$

from which

$$\int \hat{P}_x \, dt = 44.8 \text{ N} \cdot \text{s} \qquad \textit{Answer}$$

Similarly, the y-component of the impulse-momentum equation for bullet A gives

$$(L_{1\text{-}2})_y = (p_y)_2 - (p_y)_1$$

$$+\uparrow \quad \int \hat{P}_y \, dt = m_A v_2 \sin 30^\circ - 0$$

$$= 0.050(3.88)\sin 30^\circ$$

$$= 0.0970 \text{ N} \cdot \text{s} \qquad \textit{Answer}$$

We may check our solutions by considering the impulse-momentum equation for wedge B only. For example, for Figs. (d) and (e) we note that the b-component of the momentum of B is zero, both before and after impact. Therefore, using Fig. (c) to obtain the b-component of the impulse acting on B, we obtain

$$(L_{1\text{-}2})_b = (p_b)_2 - (p_b)_1 = 0 - 0$$

$$+\nwarrow \quad \int \hat{N} \, dt - \int \hat{P}_x \, dt \sin 30^\circ - \int \hat{P}_y \, dt \cos 30^\circ = 0$$

The fact that our answers for the impulses satisfy this equation provides a check on our solution.

Sample Problem 15.14

Two masses A and B shown in Fig. (a) are attached to a light, rigid bar OB, which is initially hanging at rest in the vertical position. A bullet C, of mass 0.02 kg, is fired into B with the velocity $\mathbf{v}_1$. The bullet passes through B, exiting with the velocity $\mathbf{v}_2$. Neglecting the duration of the impact, calculate ω_2, the angular velocity of the bar immediately after the impact.

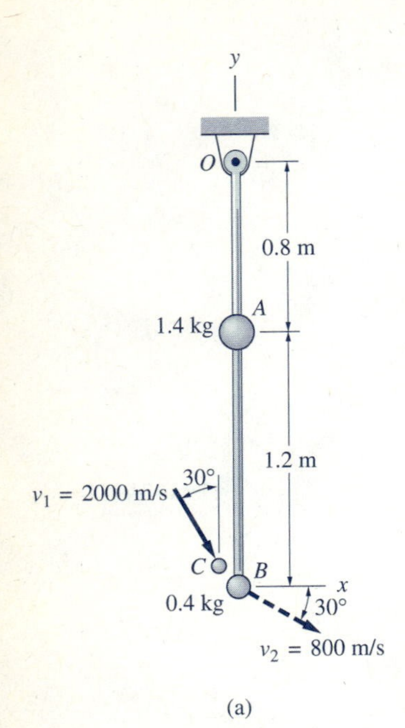

(a)

Solution

Figure (b) shows the free-body diagram (FBD) of the system consisting of the bar and the bullet during impact. The FBD shows only the external impulsive forces, which are the components $\hat{O}_x$ and $\hat{O}_y$ of the impulsive pin reaction at O. The weights of A, B, and C do not appear on the FBD because, being finite forces, their impulses can be neglected. The impact force between bullet C and mass B does not appear on the FBD because it is internal to the system.

Figures (c) and (d) display the momentum diagrams of the system immediately before and after the impact. The system occupies the same position in Figs. (b)–(d), because the duration of the impact is being neglected.

The momentum diagram in Fig. (c) displays only the initial linear momentum of C, since the bar was at rest before the impact. After the impact, the bar is rotating about O with the counterclockwise angular velocity ω_2, as shown in Fig. (d). In addition to the linear momentum of the bullet, this diagram contains the linear momenta of masses A and B. Because the masses move on circles centered at O, their velocities are computed from $v = R\omega_2$, where R is the distance from O to the respective mass.

Inspection of Figs. (b)–(d) shows that the total number of unknowns is three: ω_2 and the linear impulses of $\hat{O}_x$ and $\hat{O}_y$. There are also a total of three independent impulse-momentum equations: two linear and one angular.

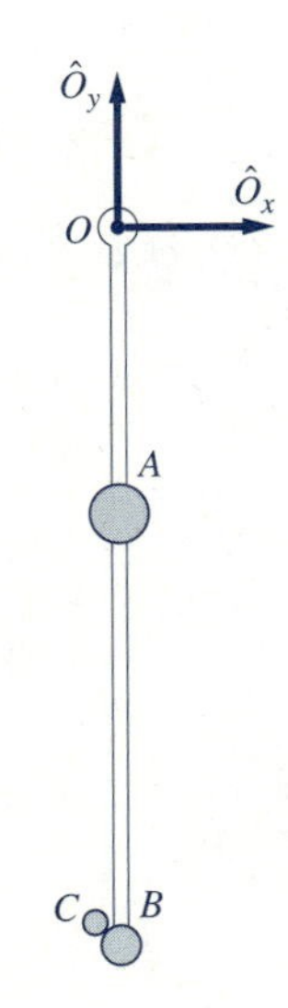

(b) FBD during impact
(external impulsive forces only)

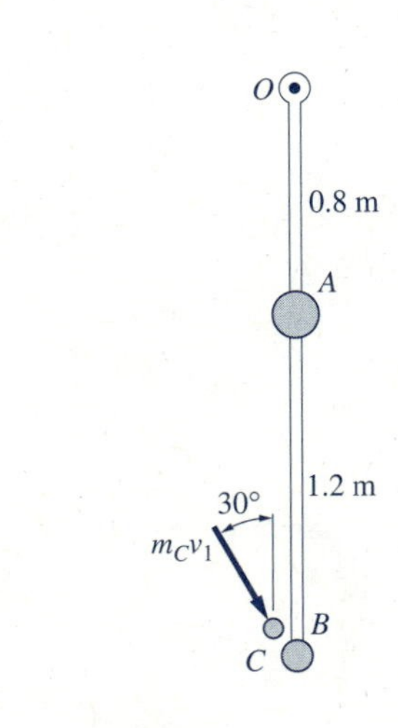

(c) Momentum before impact

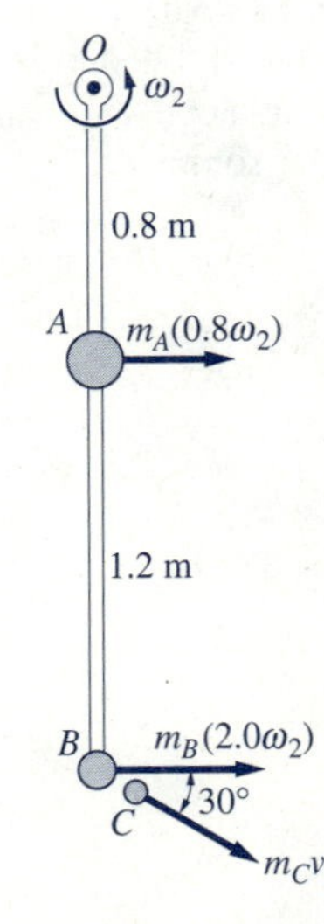

(d) Momenta after impact

Since we are required to find only ω_2, it is not necessary to derive and solve all three of the equations. The easiest method for computing ω_2 is to observe that the angular impulse about the fixed point O is zero for the system ($\hat{O}_x$ and $\hat{O}_y$ have no moment about O). Therefore, h_O, the angular momentum of the system about that point, is conserved during the impact. Since angular momentum equals the moment of linear momentum, the following equation can be written directly from Figs. (c) and (d).

$$(h_O)_1 = (h_O)_2$$

$$\circlearrowleft + \quad 2.0(m_C v_1 \sin 30^\circ) = 2.0(m_C v_2 \cos 30^\circ) + 2.0[m_B(2.0\omega_2)] + 0.8[m_A(0.8\omega_2)]$$

which becomes

$$2.0[(0.02)(2000 \sin 30^\circ)] = 2.0[(0.02)(800) \cos 30^\circ] + 2.0[(0.4)(2.0\omega_2)] + 0.8[(1.4)(0.8\omega_2)]$$

Solving this last equation, we find

$$\omega_2 = 4.92 \text{ rad/s} \qquad \textit{Answer}$$

Although not required in this problem, the impulses of $\hat{O}_x$ and $\hat{O}_y$ could be found by applying the linear impulse-momentum equations to the system in Figs. (b)–(d). However, to compute the impulse that acts between the bullet and mass B during the impact, it would be necessary to analyze either the bullet or the bar separately.

PROBLEMS

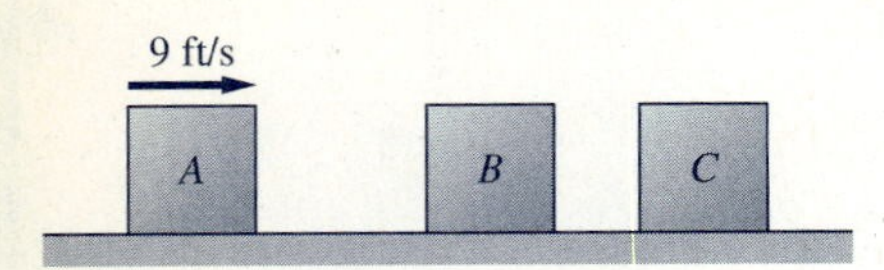

Fig. P15.85

15.85 The three identical blocks weigh 2.5 lb each. Initially, A is sliding to the right with a speed of 9 ft/s, and B and C are at rest. Assuming that all collisions are plastic, determine the impulse given to block A by block B during (a) the collision of A with B; and (b) the collision of A and B with C. Neglect friction.

15.86 The 12-kg block A is hanging in the stationary position shown when it is hit by the 20-g bullet B, which is traveling horizontally ($\alpha = 0$). After the impact, the block and the embedded bullet swing through the angle $\theta = 36°$ before coming to rest. Calculate the initial speed of the bullet.

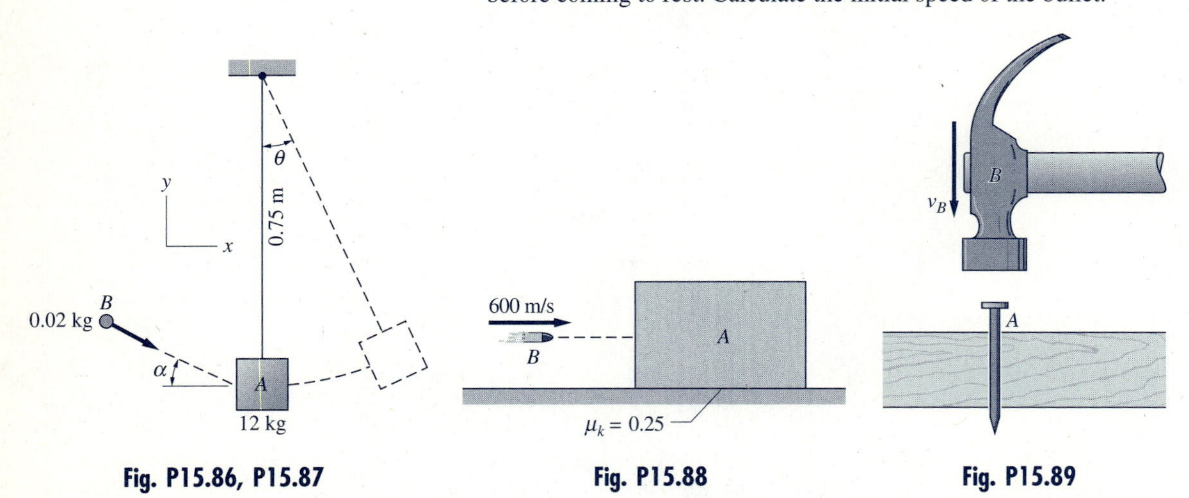

Fig. P15.86, P15.87

Fig. P15.88

Fig. P15.89

15.87 The 20-g bullet B is traveling at 980 m/s at an angle $\alpha = 20°$ when it hits and becomes embedded in the 12-kg block A. Before impact, the block was stationary in the postion $\theta = 0$. Find the velocity vector of the block immediately after the impact, assuming that the block is suspended from (a) a rigid rod; (b) an elastic (i.e., deformable) rope.

15.88 The 25-g bullet B hits the 15-kg stationary block A with a horizontal velocity of 600 m/s. The kinetic coefficient of friction between the block and the horizontal surface is 0.25. Determine (a) the total distance moved by the block after the impact; and (b) the percentage of mechanical energy lost during the impact. Assume that the bullet becomes embedded in the block.

15.89 The 80-g nail A is being driven through a board with the 400-g hammer B. The motion of the nail is resisted by a 2400-N friction force between the board and the nail. Calculate (a) the speed of the hammerhead required to advance the nail 10 mm with a single blow; and (b) the corresponding impulse between the hammerhead and the nail. Assume that the hammer does not rebound after striking the nail.

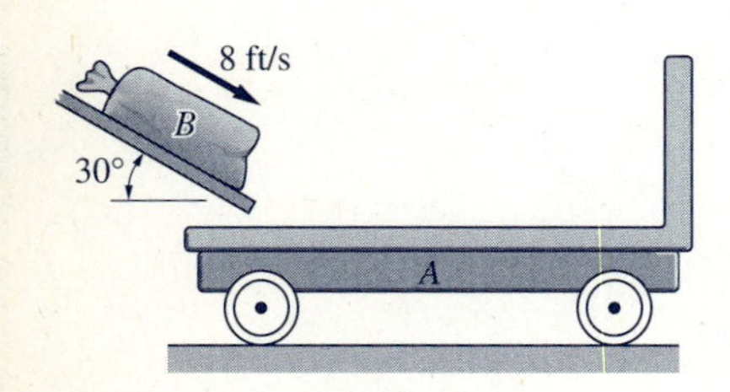

Fig. P15.90

15.90 The 80-lb bag B slides down a chute and lands on the 240-lb stationary cart A with a velocity of 8 ft/s directed as shown. Neglecting rolling resistance of the cart, determine (a) the speed of the cart after the bag comes to rest on it; and (b) the percentage of mechanical energy lost during the impact.

15.91 Two cars A and B enter an intersection with initial velocities directed as shown. After the collision, the cars become hooked together and skid 24 ft in the direction shown before coming to a stop. Knowing that the coefficient of kinetic friction between the road and the tires is 0.65 for each car, find the speed of each car before the accident.

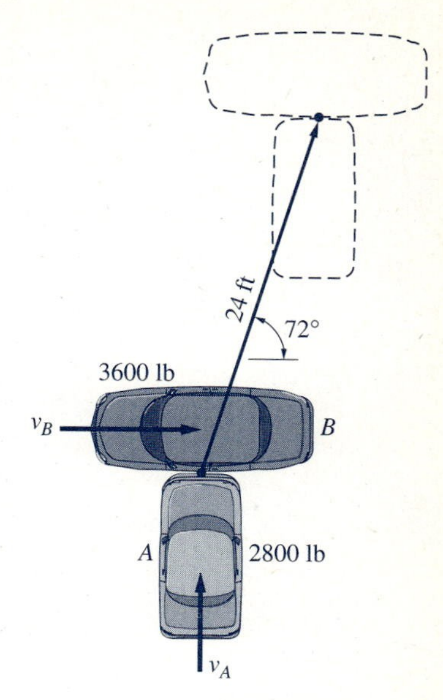

Fig. P15.91

15.92 The 12-lb sandbag B is falling vertically when it is hit by the 3-oz arrow A traveling at 120 ft/s in the direction shown. The speed and elevation of the sandbag at the time of impact are 12 ft/s and 10 ft, respectively. Determine the distance d that locates the spot where the bag hits the floor.

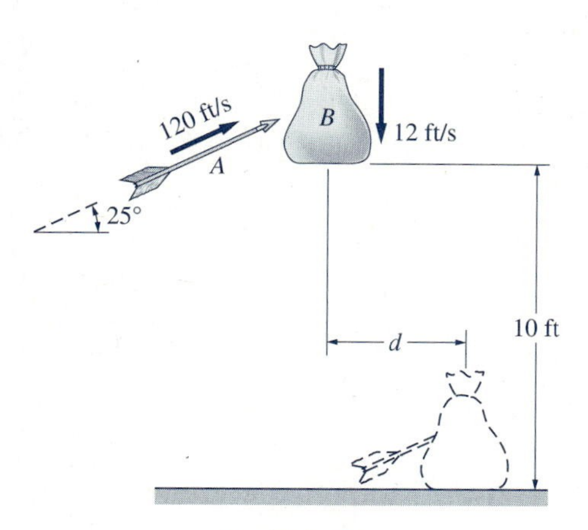

Fig. P15.92

15.93 The muzzle speed of the 15-kg shell A is 640 m/s and the mass of gunbarrel B is 320 kg. The spring is designed to limit the recoil of the barrel, which slides in its mount C, to 0.75 m. Neglecting friction in the mount, calculate the smallest acceptable stiffness of the spring.

15.94 The 60-lb sack of grain A is dropped on a spring scale from the height $h = 0$. The platform B of the scale weighs 12 lb and the combined spring stiffness of all three springs is 1200 lb/ft. Compute the maximum reading of the scale.

15.95 Solve Prob. 15.94 if sack A is dropped from the height $h = 6$ in.

15.96 The speed of the sled is v just before it reaches the sharp corner at the bottom of the hill. When the sled hits the corner, it receives a vertical impulse from the ground that changes its speed to u, directed along the horizontal surface. In terms of v, the mass m of the sled, and the slope angle α, calculate (a) u; and (b) the energy lost due to the impact.

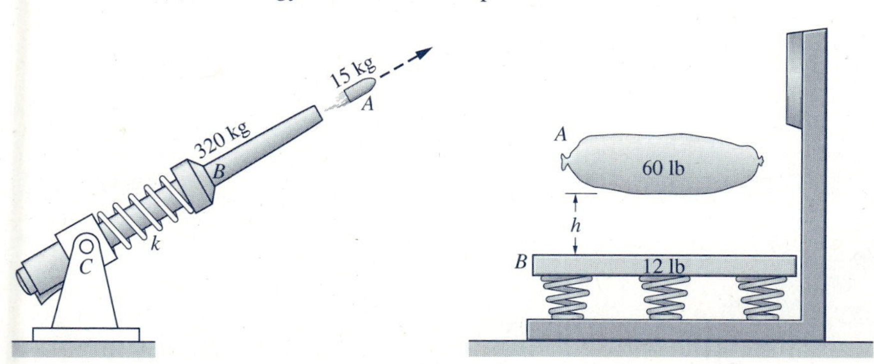

Fig. P15.93

Fig. P15.94, P15.95

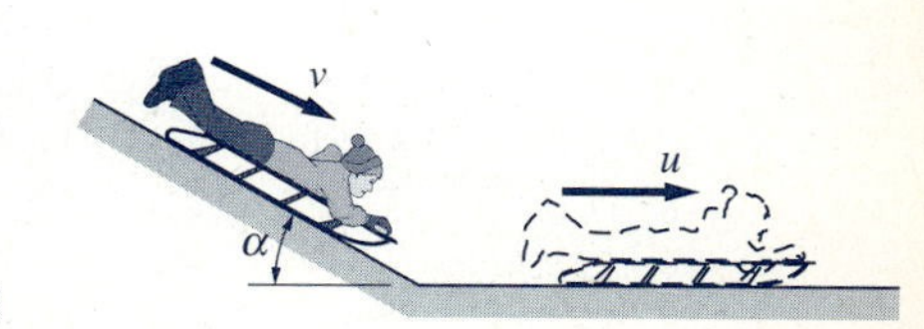

Fig. P15.96

15.97 The 60-kg notched cradle A is carrying the 12-kg pipe B at a speed of 8 m/s when it hits the rigid wall. Assuming that all impacts are plastic, determine (a) the speed of the pipe immediately after the impact; (b) the impulse between the cradle and the wall; and (c) the impulse between the cradle and the ground.

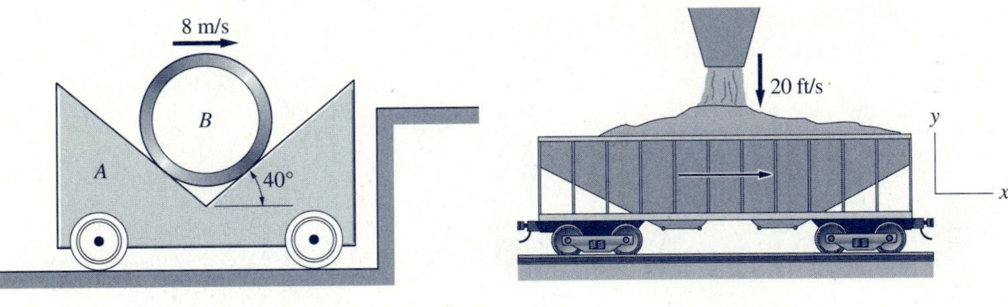

Fig. P15.97 **Fig. P15.98**

15.98 A 10 000-lb empty railroad car is coasting at 0.5 mph when 7500 lb of grain is dumped into it from a grain elevator. The speed of the falling grain is 20 ft/s. Compute (a) the speed of the loaded car; and (b) the impulse vector exerted on the car by the grain. Neglect rolling resistance of the wheels.

***15.99** Blocks B and C are at rest when B is hit by the 1.5-oz bullet A traveling at 1200 ft/s. Assuming that all impacts are plastic, calculate the velocity of each block immediately after the bullet has become embedded in block B. Neglect friction.

15.100 The ballistic pendulum consists of three equally spaced blocks attached to a light rod. Each of the upper two blocks has a mass m_1, whereas the mass of the lowest block is m_2. A bullet of mass m_0 is fired at the middle block and becomes embedded in it. Determine the ratio m_2/m_1 for which there will be no impulsive reaction at O due to the impact. Assume that m_0 is much smaller than the other masses.

15.101 The acrobat A jumps onto the end of the seesaw, thereby propelling acrobat B vertically upward. Neglecting the weight of the seesaw, find the maximum height reached by B.

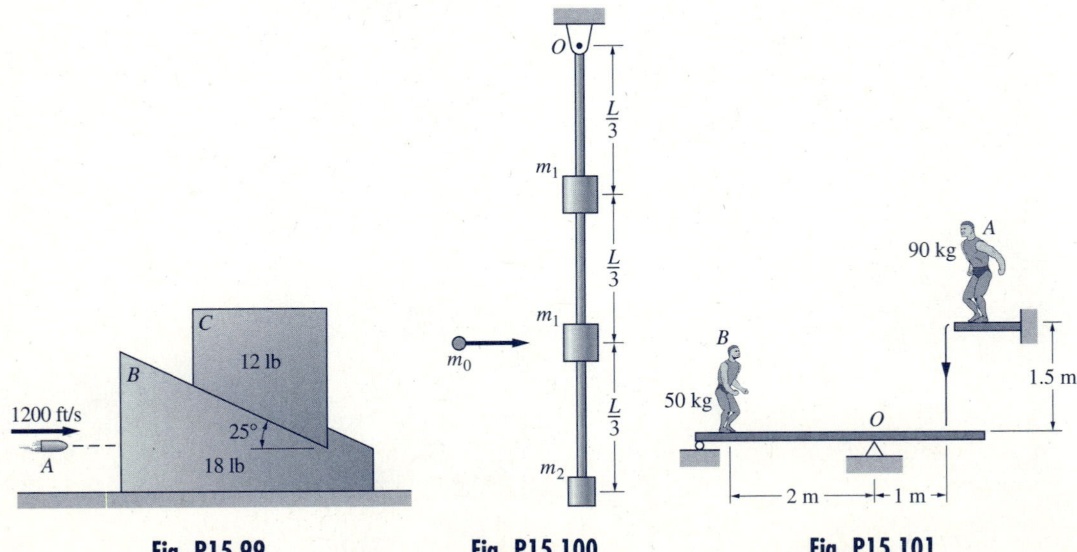

Fig. P15.99 **Fig. P15.100** **Fig. P15.101**

15.10 Elastic Impact

Most bodies possess the ability to return either totally or partially to their original shape when released from a deformed position, a property known as *elasticity.* Therefore, there are two stages in the impact of elastic particles. Initially there is the deformation stage wherein the particle is compressed by the impact force. This stage is followed by a recovery stage during which the particle returns totally or partially to its undeformed shape. When two elastic particles collide, the recovery stage causes the particles to rebound, or move apart, after the impact (the recovery stage is absent during plastic impact). If the particles are perfectly elastic, they will return to their original shape with no energy lost during the impact. A more common occurrence is for the impacting particles to be left partially deformed by the relatively large impact forces, in which case a fraction of the initial kinetic energy is lost due to the permanent deformation. Kinetic energy can also be lost in the generation of heat and sound during the impact. Before characterizing the impact of two elastic particles, it is necessary to distinguish between direct and oblique impact.

Figure 15.15 shows the impact between two circular disks A and B that are sliding on a smooth horizontal plane. The line that is perpendicular to the contact surface (the x-axis) is called the *line of impact.* The velocities of the

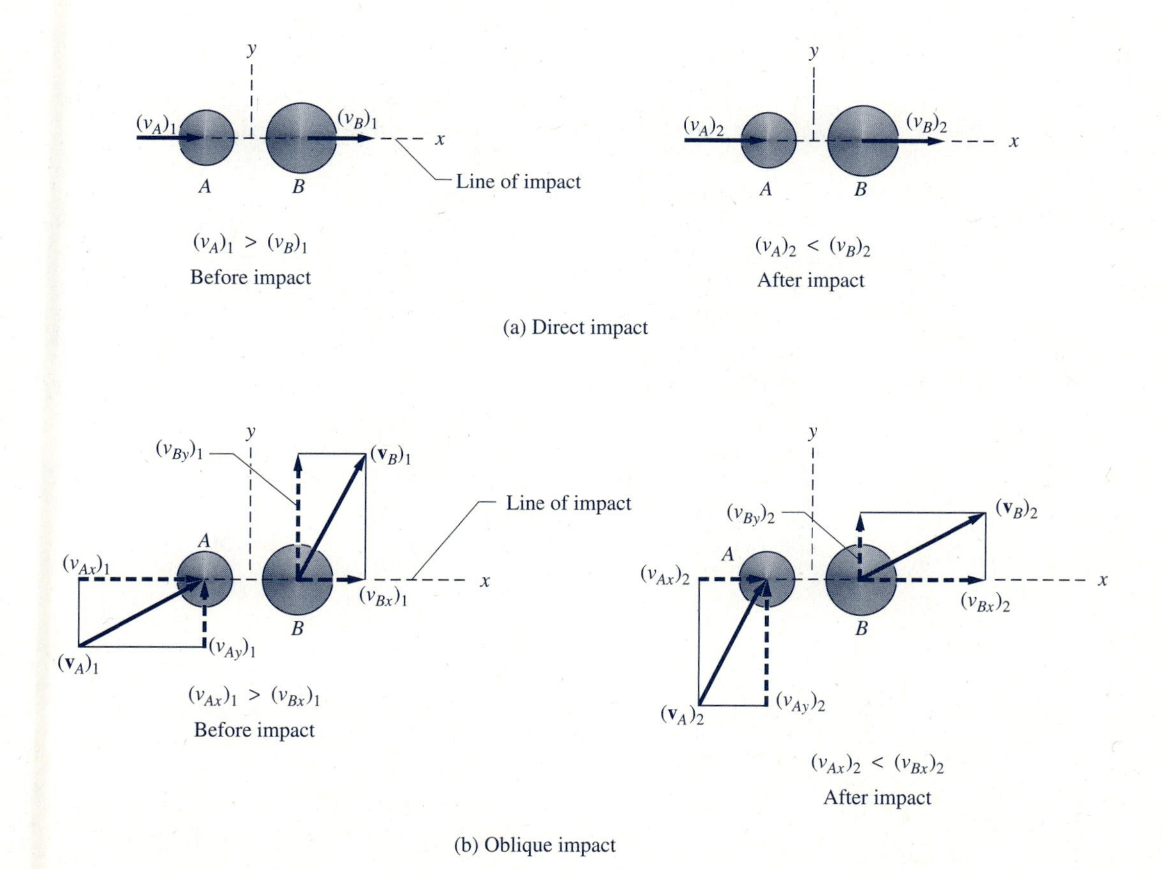

Fig. 15.15

particles before the impact are denoted by $(v_A)_1$ and $(v_B)_1$. When both initial velocities are directed along the line of impact, as shown in Fig. 15.15(a), the impact is called *direct impact.* Otherwise, the impact is referred to as *oblique,* as depicted in Fig. 15.15(b). Thus direct impact is equivalent to a head-on collision, and oblique impact refers to a glancing blow. Note that the linear impulse acting on the systems is zero for both cases shown in Fig. 15.15, since no external forces act on either system.

For the direct impact in Fig. 15.15(a), the momentum balance in the x-direction gives

$$\xrightarrow{+} \quad m_A(v_A)_1 + m_B(v_B)_1 = m_A(v_A)_2 + m_B(v_B)_2 \tag{15.42}$$

where $(v_A)_2$ and $(v_B)_2$ are the velocities after impact.

The *coefficient of restitution e* is an experimental constant that characterizes the "elasticity" of the impacting bodies. It is defined as

$$e = \frac{v_{\text{sep}}}{v_{\text{app}}} \tag{15.43}$$

where for direct impact

$v_{\text{sep}} = (v_B)_2 - (v_A)_2$ is the *velocity of separation* (the rate at which the distance between the particles increases after the impact)

$v_{\text{app}} = (v_A)_1 - (v_B)_1$ is the *velocity of approach* (the rate at which the distance between the particles decreases before the impact)

For an impact to occur in Fig. 15.15(a), we must have $v_{\text{app}} > 0$; if the particles are to separate after the impact (omitting the possibility that A could pass through B), we have $v_{\text{sep}} > 0$. Thus e will generally be a nonnegative number.

From Eq. (15.43) we see that $e = 0$ corresponds to plastic impact. If $e = 1$, the impact is called *perfectly elastic,* a situation for which no energy is lost during the impact (see Prob. 15.102). For most impacts the values of e lie between 0 and 1. (A negative coefficient of restitution indicates that one body has passed through the other, such as a bullet passing through a plate.)

Oblique impact is analyzed by assuming that e has the same value as for direct impact; however, in the defining equation for e, the velocities are replaced by their components that are directed along the line of impact. Therefore, if the line of impact coincides with the x-axis as shown in Fig. 15.15(b), we have

$$\begin{aligned} v_{\text{sep}} &= (v_{Bx})_2 - (v_{Ax})_2 \\ v_{\text{app}} &= (v_{Ax})_1 - (v_{Bx})_1 \end{aligned} \tag{15.44}$$

If the particles are smooth, the impact force acting on either particle will be directed along the x-axis, with no y-component. For this case, the y-components of the velocities of both A and B will not change during the impact.

The value of e is usually considered to be constant, given the material of the colliding particles. But this is only an approximation to reality. Experimental evidence indicates that the coefficient of restitution actually depends on many factors (the magnitude of the relative velocity of approach, the condition of the impacting surfaces, etc.).

Sample Problem 15.15

The 1.8-kg block A shown in Fig. (a) is sliding toward the right with the velocity $(v_A)_1 = 1.2$ m/s when it hits block B, which is moving to the left with $(v_B)_1 = 2.0$ m/s. The impact causes block A to stop. If the coefficient of restitution for the impact is 0.5, determine (1) the velocity of B after the impact; and (2) the mass of B. Neglect friction.

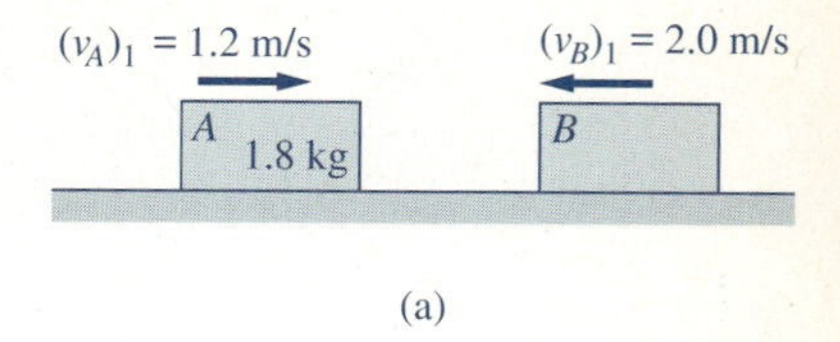

(a)

Solution

Part 1

The velocity of B after impact, $(v_B)_2$, can be calculated using the defining equation for the coefficient of restitution: $e = v_{sep}/v_{app}$. Referring to Fig. (a), we see that the velocity of approach is $v_{app} = (v_A)_1 + (v_B)_1 = 1.2 + 2.0 = 3.2$ m/s. Noting that block A is at rest after the impact, the velocity of separation is $v_{sep} = (v_B)_2$. Therefore, we have

$$0.5 = \frac{(v_B)_2}{3.2}$$

from which we find

$$(v_B)_2 = 1.6 \text{ m/s} \qquad \textit{Answer}$$

Part 2

The free-body diagrams of blocks A and B during impact, shown in Fig. (b), contain the following forces: the weights W_A and W_B, the contact forces N_A and N_B, and the impact force $\hat{P}$. Since $\hat{P}$ is internal to the system of both blocks, we conclude that the net impulse acting on the system in the x-direction is zero.

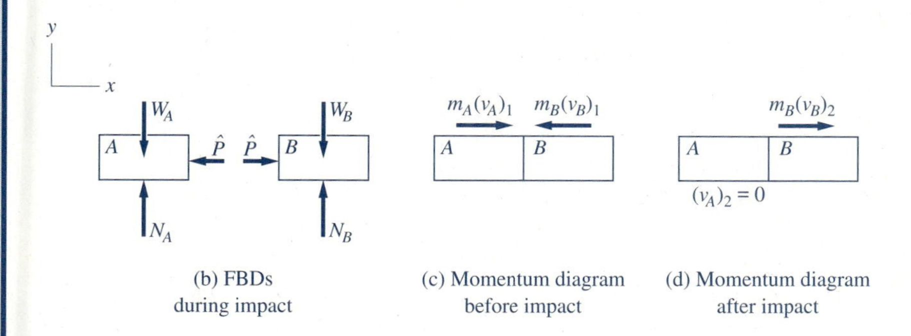

(b) FBDs during impact

(c) Momentum diagram before impact

(d) Momentum diagram after impact

Referring to Figs. (c) and (d), conservation of linear momentum in the x-direction for the system yields

$$(p_x)_1 = (p_x)_2$$

$$\overset{+}{\rightarrow} \quad m_A(v_A)_1 - m_B(v_B)_1 = m_B(v_B)_2$$

Substitution of the numerical values gives

$$1.8(1.2) - m_B(2.0) = m_B(1.6)$$

from which we find

$$m_B = 0.6 \text{ kg} \qquad \textit{Answer}$$

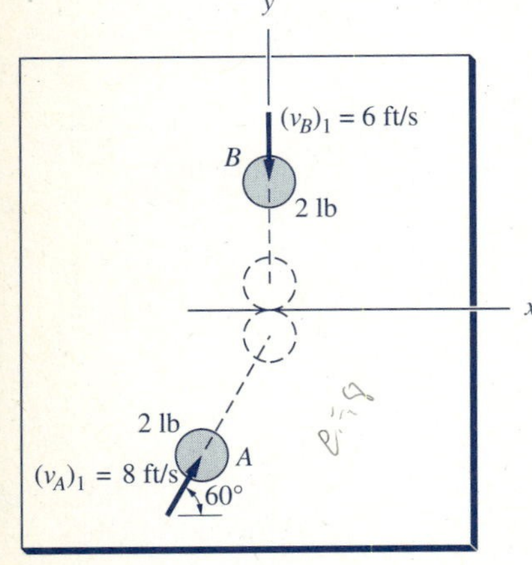

(a)

Sample Problem 15.16

Two identical, smooth disks A and B weighing 2 lb each are sliding across a horizontal tabletop when they collide with the initial velocities $(v_A)_1 = 8$ ft/s and $(v_B)_1 = 6$ ft/s, directed as shown in Fig. (a). If the coefficient of restitution for the impact is $e = 0.8$, calculate $(\mathbf{v}_A)_2$ and $(\mathbf{v}_B)_2$, the velocity vectors of the disks immediately after the impact.

Solution

As shown in the free-body diagrams in Fig. (b), the only forces acting in the xy-plane during the impact are the impact forces $\hat{P}$, which are oppositely directed on A and B. (The weights of the disks and the normal forces exerted by the tabletop are perpendicular to the horizontal xy-plane.) Because the disks are smooth, $\hat{P}$ will be directed along the y-axis, which is the line of impact. Assuming that the duration of the impact is negligible, the motion is impulsive and the disks will be located at the origin of the coordinate system before and after the impact.

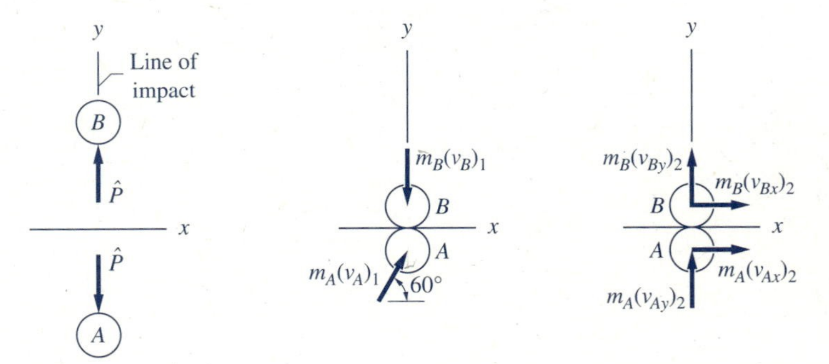

(b) FBDs during impact (forces acting in xy-plane only)

(c) Momentum diagram before impact

(d) Momentum diagram after impact

The momentum diagrams that display the momentum vectors for A and B immediately before and after the impact are shown in Figs. (c) and (d), respectively. Note that in Fig. (d) all components of the final velocities have been assumed to act in positive coordinate directions.

From Figs. (b)–(d) we see that there are five unknowns: $\int \hat{P}\,dt$ (the impulse of the impact force), $(v_{Ax})_2$, $(v_{Ay})_2$, $(v_{Bx})_2$, and $(v_{By})_2$. Furthermore, we see that there are five independent equations: two components of the vector impulse-momentum equation, $\mathbf{L}_{1\text{-}2} = \Delta\mathbf{p}$, for each of the two disks (a total of four equations), plus the coefficient of restitution equation, Eq. (15.43). These five equations could, of course, be used to solve for the five unknowns. However, since we are not required to find $\int \hat{P}\,dt$, a more efficient solution is obtained by also analyzing the system that contains both disks. Thus our solution consists of the following four parts.

1. Apply $(\mathbf{L}_{1-2})_x = \Delta p_x$ to Disk A

From Fig. (b) we see that there is no impulse on A in the x-direction. Therefore, the x-component of the momentum of A is conserved. Referring to the momentum diagrams for disk A in Figs. (c) and (d), we obtain

$$(p_x)_1 = (p_x)_2$$

$$\xrightarrow{+} \quad m_A(v_A)_1 \cos 60° = m_A(v_{Ax})_2$$

Canceling m_A and substituting $(v_A)_1 = 8$ ft/s, this equation yields

$$(v_{Ax})_2 = (v_A)_1 \cos 60° = 8 \cos 60° = 4.00 \text{ ft/s} \tag{a}$$

2. Apply $(\mathbf{L}_{1-2})_x = \Delta p_x$ to Disk B

Using the same argument as given above for disk A, we find that

$$(v_{Bx})_2 = (v_{Bx})_1 = 0 \tag{b}$$

(Observe that the x-components of the velocities of both A and B are unchanged by the impact because $\hat{P}$ is directed along the y-axis.)

3. Apply $(\mathbf{L}_{1-2})_{ext} = \Delta \mathbf{p}$ to the System Containing Both Disks

From Fig. (b) we see that there are no external forces, and thus no external impulses that act on the system during impact ($\hat{P}$ is an internal force). Therefore, the momentum of the system is conserved. (The fact that the x-component of momentum is conserved is already clear from Eqs. (a) and (b): The x-component of momentum is conserved for each disk, hence it is obviously conserved for the system.) Referring to Figs. (c) and (d), conservation of the y-component of momentum for the system becomes

$$(p_y)_1 = (p_y)_2$$

$$+\uparrow \quad m_A(v_A)_1 \sin 60° - m_B(v_B)_1 = m_A(v_{Ay})_2 + m_B(v_{By})_2$$

$$\frac{2}{g}(8) \sin 60° - \frac{2}{g}(6) = \frac{2}{g}(v_{Ay})_2 + \frac{2}{g}(v_{By})_2$$

which can be reduced to

$$(v_{Ay})_2 + (v_{By})_2 = 0.9282 \tag{c}$$

4. Use the Coefficient of Restitution Equation

Noting that the line of impact is the y-axis, the velocity of approach [see Fig. (a)] is $v_{app} = (v_A)_1 \sin 60° + (v_B)_1 = 8 \sin 60° + 6 = 12.928$ ft/s. Referring to Fig. (d), we deduce that the velocity of separation is $v_{sep} = (v_{By})_2 - (v_{Ay})_2$. Therefore, $e = v_{sep}/v_{app}$ becomes

$$0.8 = \frac{(v_{By})_2 - (v_{Ay})_2}{12.928}$$

which after simplification may be written as

$$(v_{By})_2 - (v_{Ay})_2 = 10.342 \tag{d}$$

Solving Eqs. (c) and (d) simultaneously, we obtain

$$(v_{Ay})_2 = -4.71 \text{ ft/s} \quad \text{and} \quad (v_{By})_2 = 5.63 \text{ ft/s} \tag{e}$$

From the values given in Eqs. (a), (b), and (e), the velocities of disks A and B after impact are as follows.

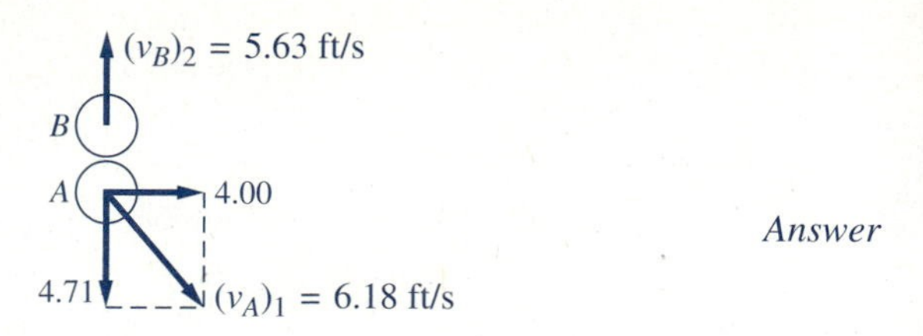

Answer

If needed, the impulse of the impact force, $\int \hat{P}\,dt$, could be computed by applying the equation $(L_{1\text{-}2})_y = \Delta p_y$ to either disk A or disk B.

Sample Problem 15.17

As shown in Fig. (a), a ball A of mass m is dropped with the velocity v_1 onto a smooth, rigid plane that is inclined at the angle θ with the horizontal. Determine the coefficient of restitution for which the rebound velocity v_2 will be horizontal.

(a)

Solution

As shown in the free-body diagram in Fig. (b), the forces acting on the ball during impact are its weight W and the impact force $\hat{N}$, which is normal to the smooth plane. The y-axis, which is perpendicular to the plane, is the line of impact. We will neglect the duration of the impact, which means that the motion is impulsive. It follows that the impulse of W, which is a finite force, can be neglected, and that the ball is in the same location before and after the impact.

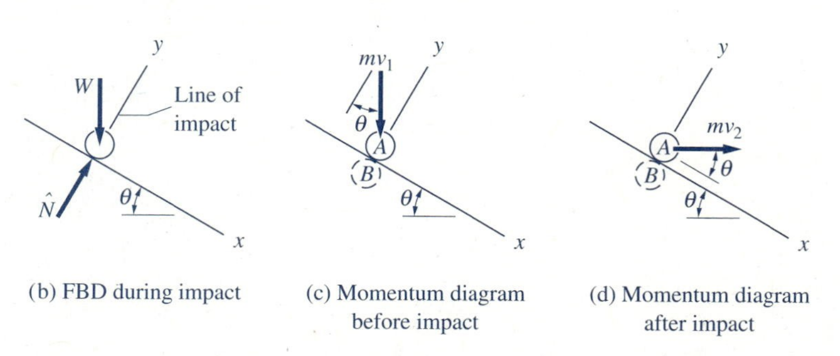

(b) FBD during impact (c) Momentum diagram before impact (d) Momentum diagram after impact

Since the description of impact requires two particles, we imagine that a second particle B is embedded in the rigid plane, as shown in the momentum diagrams in Figs. (c) and (d). Because the velocity of B is always zero, the momentum diagrams before and after impact contain only the momentum of ball A.

From Fig. (b) we see that there will be no impulse acting on the ball in the x-direction during the impact (remember that the impulse of W is being neglected). Therefore, the x-component of the momentum of A is conserved. Referring to the momentum diagrams in Figs. (c) and (d), we obtain

$$(p_x)_1 = (p_x)_2$$

$$mv_1 \sin\theta = mv_2 \cos\theta$$

which yields

$$v_2 = v_1 \tan\theta \qquad \text{(a)}$$

Recognizing that the velocity of B is always zero, and referring to Figs. (c) and (d), we see that the velocity of approach is $v_{app} = v_1 \cos\theta$ and the velocity of separation is $v_{sep} = v_2 \sin\theta$. Therefore, we obtain

$$e = \frac{v_{sep}}{v_{app}} = \frac{v_2 \sin\theta}{v_1 \cos\theta} = \frac{v_2}{v_1} \tan\theta \qquad \text{(b)}$$

Substituting $v_2 = v_1 \tan\theta$ from Eq. (a) into Eq. (b), we find the required value of e to be

$$e = \tan^2\theta \qquad \textit{Answer}$$

PROBLEMS

15.102 Prove that no energy is lost during direct impact of two particles if $e = 1$.

15.103 Solve Prob. 15.89, assuming that the impact between the hammer and the nail is elastic with $e = 0.5$. Also determine the speed of the hammerhead immediately after the impact.

15.104 The two identical blocks are sliding on a smooth horizontal surface with the initial velocities $(v_A)_1$ and $(v_B)_1$, where $(v_A)_1 > (v_B)_1$. Show that the speeds of the blocks after impact are $(v_A)_2 = (1/2)(1 - e)(v_A)_1 + (1/2)(1 + e)(v_B)_1$ and $(v_B)_2 = (1/2)(1 + e)(v_A)_1 + (1/2)(1 - e)(v_B)_1$.

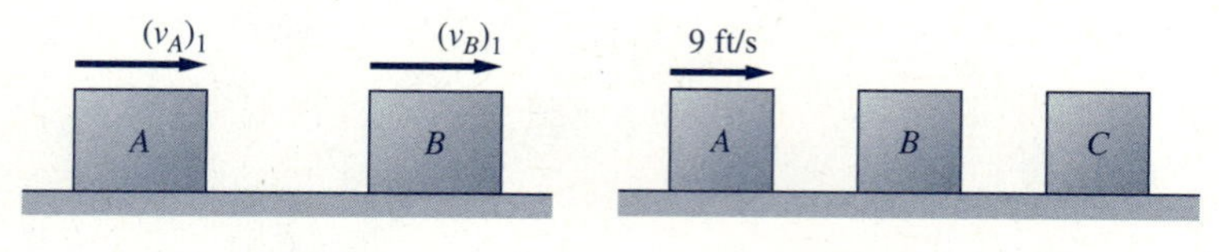

Fig. P15.104 **Fig. P15.105, P15.106**

15.105 The three identical blocks slide on a smooth horizontal plane. The initial velocity of A is 9 ft/s, whereas B and C are at rest. Use the formulas given in Prob. 15.104 to compute the speed of each block after all collisions have occurred. Assume elastic impacts with $e = 0.6$.

15.106 Solve Prob. 15.105 if the initial speed of block C is 4 ft/s, directed to the right.

15.107 Two identical coins are placed on a rough, horizontal surface as shown in (a). After coin A is propelled into the stationary coin B with the initial velocity $(v_A)_1$, the coins come to rest in the positions shown in (b). Determine the coefficient of restitution for the impact between the coins.

15.108 The elastic ball is bounced off a smooth, rigid surface. Show that the relationship between the angle of incidence and the rebound angle is $\tan\theta_2 = e\tan\theta_1$, where e is the coefficient of restitution.

15.109 The elastic ball bounces down a flight of rigid, smooth stairs. Determine the height h of the steps if all the trajectories between successive steps are identical. Use $e = 0.85$ and the initial velocity shown.

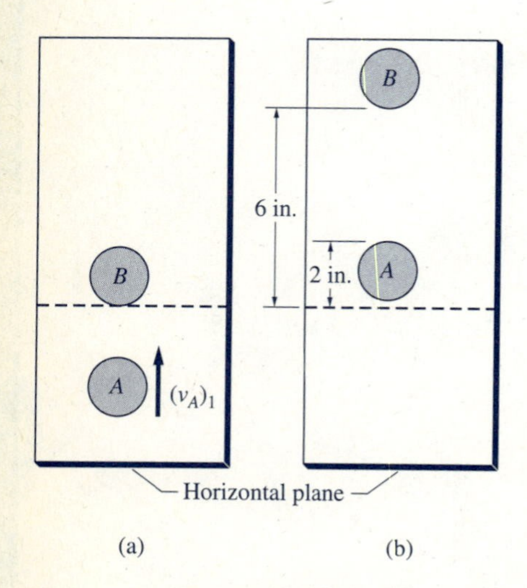

Fig. P15.107

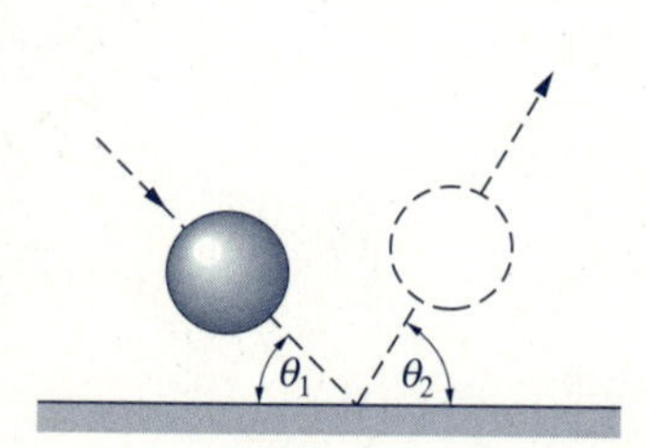

Fig. P15.108

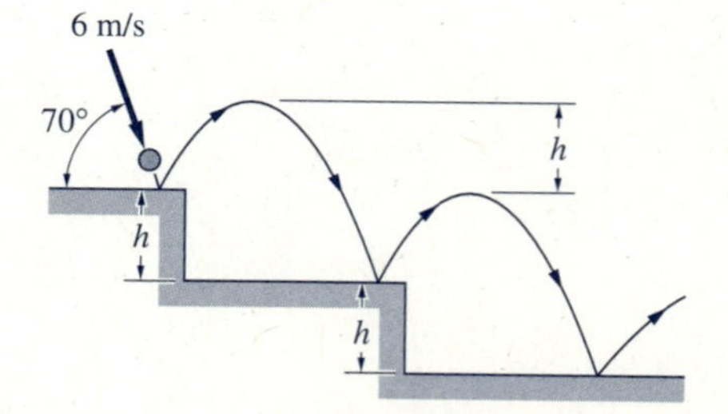

Fig. P15.109

15.110 Two cars traveling with the velocities shown collide at an intersection. The coefficient of restitution is 0.25 for the impact, and the contacting surfaces are smooth. Calculate the velocity of each car after the impact.

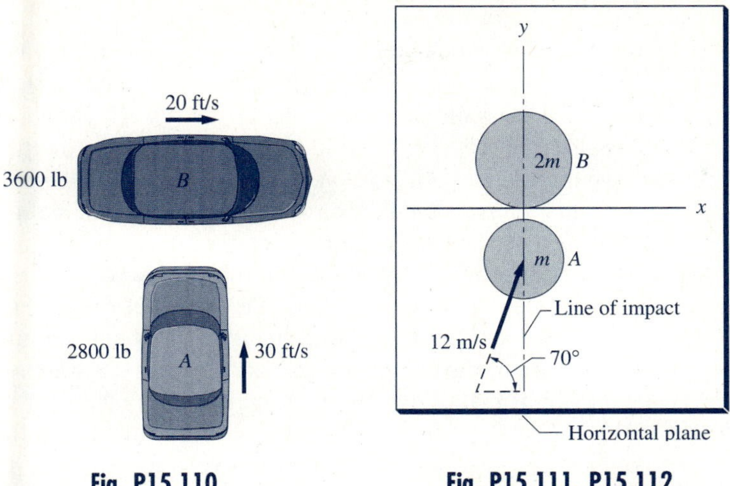

Fig. P15.110

Fig. P15.111, P15.112

15.111 The two disks A and B lie on a horizontal surface. Disk A is propelled into B, which is initially stationary, with the velocity shown. If $e = 0.75$, calculate the velocity of each disk after the impact.

15.112 The two disks A and B lie on a horizontal surface. Disk A is propelled into B, which is initially stationary, with the velocity shown. After the impact, the motion of A is directed along the x-axis. Determine (a) the coefficient of restitution for the impact; and (b) the velocity of each disk immediately after the collision.

15.113 A Super Ball is dropped from the height h_0 onto a rigid floor. If the coefficient of restitution is 0.985, find the number of bounces made by the ball before its height of rebound is reduced to $h_0/2$.

15.114 A 0.01-lb airgun pellet is fired at a stationary target consisting of two 0.5-lb weights connected by a light rod. The target can spin about a smooth pin at O. The pellet hits the lower weight with the velocity $v_0 = 300$ ft/s. If the coefficient of restitution is 0.5, determine the angular velocity of the target immediately after impact.

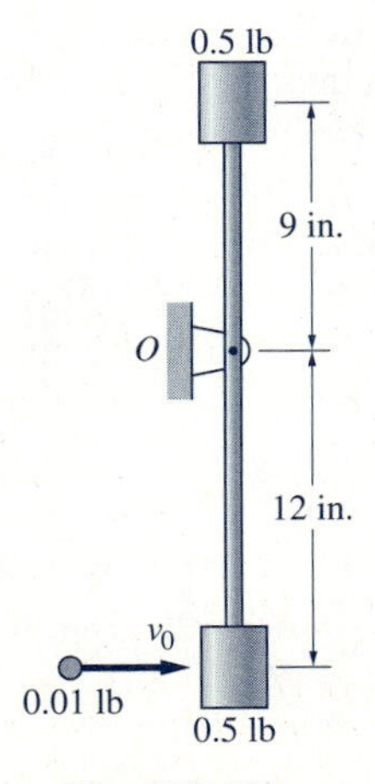

Fig. P15.114

*15.11 Mass Flow

a. Control volume

Up to this point, we have restricted our attention to systems that always contained the same particles. In other words, we assumed that the system was closed in the sense that mass did not enter or leave the system. In this article, we apply the impulse-momentum principle to *mass flow,* where the particles continuously move through a spatial region, called the *control volume.*

An example of mass flow is the flow of water through a section of pipe, as shown in Fig. 15.16(a). Here a convenient choice for the control volume V is the interior of the pipe. The water enters the control volume with the velocity $\mathbf{v}_{in}$ and exits with the velocity $\mathbf{v}_{out}$. Since the momentum of the water in the pipe is changed, a force equal to the rate of change of this momentum must exist between the surface of the control volume and the flowing water.

A second example of mass flow is the jet engine of an aircraft shown in Fig. 15.16(b). The engine takes in air ($\mathbf{v}_{in} \approx \mathbf{0}$), which is mixed with fuel and ignited. The combustion gases are expelled at the velocity $\mathbf{v}_{out}$. The control volume V is the interior of the engine, which is itself moving with the velocity $\mathbf{v}$. In this case two changes occur within the control volume: The velocity of the air is increased from zero to $\mathbf{v}_{out}$, and the velocity of the combusted fuel is changed from $\mathbf{v}$ to $\mathbf{v}_{out}$. The corresponding rate of change of momentum gives rise to a force between the engine and the flowing mass called the *thrust.*

Strictly speaking, the analysis of mass flow belongs in the realm of *fluid mechanics,* which is beyond the scope of this text. We will restrict our discussion to problems that do not require the specialized knowledge and techniques of fluid mechanics. In particular, we assume throughout that the inflow velocity ($\mathbf{v}_{in}$) and outflow velocity ($\mathbf{v}_{out}$) are constant across the inlet and outlet areas of the control volume, respectively.

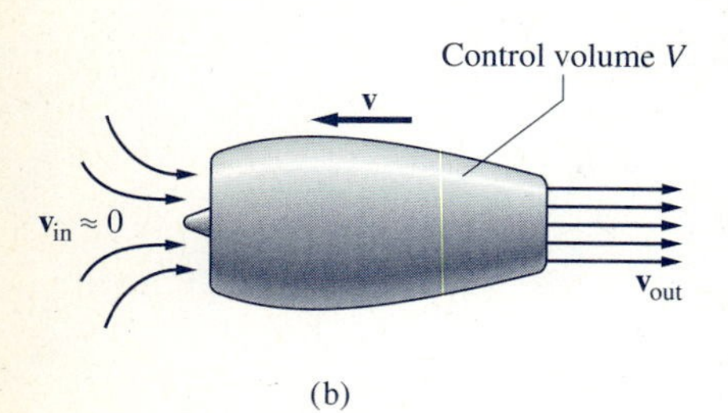

Fig. 15.16

b. Impulse-momentum principle

We will now formulate the impulse-momentum principle, based on the concept of a control volume.

Figure 15.17(a) shows a control volume V consisting of the region inside a vessel, a convenient choice of V for our present discussion. In general, any region of space may be selected as the control volume, with the best choice determined by the problem being considered. However, you must clearly identify the control volume at the beginning of the analysis, since it determines the equations that will enter into the solution.

We now consider the momentum of the system consisting of all the particles that are inside the control volume V at time t, as indicated in Fig. 15.17(a). [When the shape of the control volume is maintained by a vessel, as in Fig. 15.17(a), the vessel may also be included as part of the system, if convenient.] Note that unlike the control volume, the system always contains the same particles: i.e., the mass of the system is constant. Let the momentum of the system at time t be denoted by $\mathbf{p}$, and let the momentum of the mass within the control volume at time t be $\mathbf{p}_V$. Because the system is inside V at time t, we have $\mathbf{p} = \mathbf{p}_V$.

Consider next the momentum of the system at time $t + \Delta t$, where Δt is a small time interval. As shown in Fig. 15.17(b), some of the particles of the

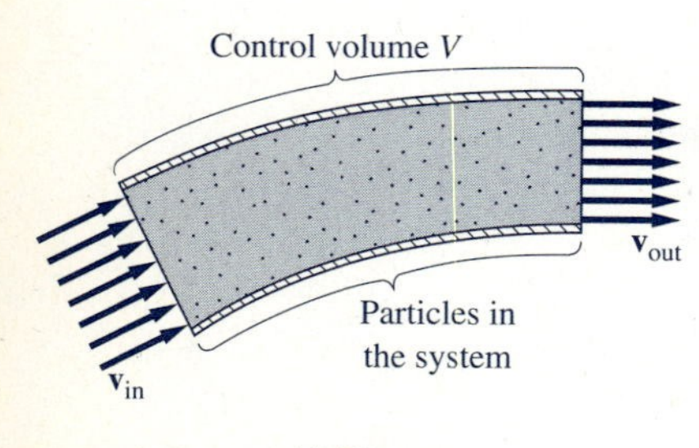

(a) Time t

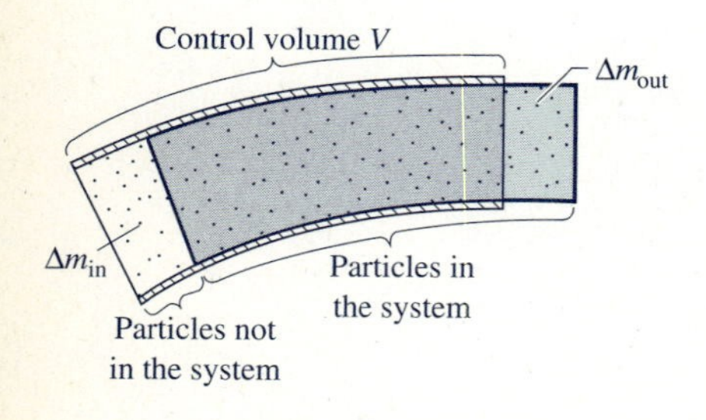

(b) Time $t + \Delta t$

Fig. 15.17

system have left the control volume V, and other particles, not part of the system, have entered V. The masses of these particles are labeled as Δm_{out} and Δm_{in}, respectively. Since the outlet and inlet velocities are assumed to be spatially constant, the momenta of the particles leaving and entering the control volume are, respectively,

$$\Delta \mathbf{p}_{out} = \Delta m_{out} \mathbf{v}_{out} \tag{a}$$

$$\Delta \mathbf{p}_{in} = \Delta m_{in} \mathbf{v}_{in} \tag{b}$$

The momentum of the system has changed from $\mathbf{p}$ to $\mathbf{p} + \Delta\mathbf{p}$, and the momentum of the particles that are now in V is $\mathbf{p}_V + \Delta\mathbf{p}_V$. Since the momentum of the particles leaving V may not be the same as the momentum of the particles entering V, $\Delta\mathbf{p}$ is not necessarily equal to $\Delta\mathbf{p}_V$. From Fig. 15.17(b), we see that the momentum of the system equals the momentum of the particles within V plus the momentum of the particles leaving V (these are part of the system) minus the momentum of the particles entering V (these are not part of the system); i.e.,

$$\mathbf{p} + \Delta\mathbf{p} = \mathbf{p}_V + \Delta\mathbf{p}_V + \Delta\mathbf{p}_{out} - \Delta\mathbf{p}_{in} \tag{c}$$

Substituting from Eqs. (a) and (b), and recalling that $\mathbf{p} = \mathbf{p}_V$, we obtain

$$\Delta\mathbf{p} = \Delta\mathbf{p}_V + \Delta m_{out}\mathbf{v}_{out} - \Delta m_{in}\mathbf{v}_{in} \tag{d}$$

The rate at which the momentum of the system changes at time t is given by

$$\dot{\mathbf{p}} = \lim_{\Delta t \to 0} \frac{\Delta\mathbf{p}}{\Delta t}$$

which yields

$$\boxed{\dot{\mathbf{p}} = \dot{\mathbf{p}}_V + \dot{m}_{out}\mathbf{v}_{out} - \dot{m}_{in}\mathbf{v}_{in}} \tag{15.45}$$

where

$$\dot{m}_{out} = \lim_{\Delta t \to 0} \frac{\Delta m_{out}}{\Delta t} \quad \text{and} \quad \dot{m}_{in} = \lim_{\Delta t \to 0} \frac{\Delta m_{in}}{\Delta t}$$

are the mass flow rates out of and into the control volume, respectively. (The units for mass flow rate are mass per unit time, e.g., slugs/s or kg/s.)

Equation (15.45) is a form of *Reynold's transport theorem:* The rate at which the momentum of a system changes equals the rate at which the momentum inside the control volume changes plus the net rate at which the momentum is flowing out of the control volume.

If $\mathbf{R}$ is the resultant force acting on the system at time t (when the entire system is contained within the control volume), then the impulse-momentum principle states that $\mathbf{R} = \dot{\mathbf{p}}$, which, using Eq. (15.45), becomes

$$\boxed{\mathbf{R} = \dot{\mathbf{p}}_V + \dot{m}_{out}\mathbf{v}_{out} - \dot{m}_{in}\mathbf{v}_{in}} \tag{15.46}$$

The following two examples of mass flow represent applications of Eq. (15.46) that deserve special attention because of their practical importance.

c. Deflection of a steady fluid stream

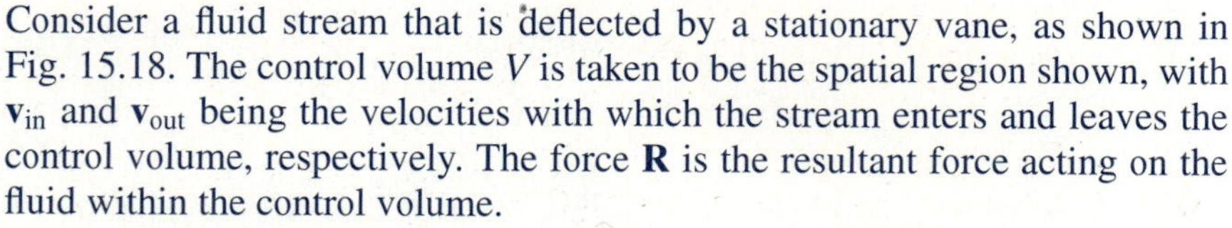

Consider a fluid stream that is deflected by a stationary vane, as shown in Fig. 15.18. The control volume V is taken to be the spatial region shown, with $\mathbf{v}_{in}$ and $\mathbf{v}_{out}$ being the velocities with which the stream enters and leaves the control volume, respectively. The force $\mathbf{R}$ is the resultant force acting on the fluid within the control volume.

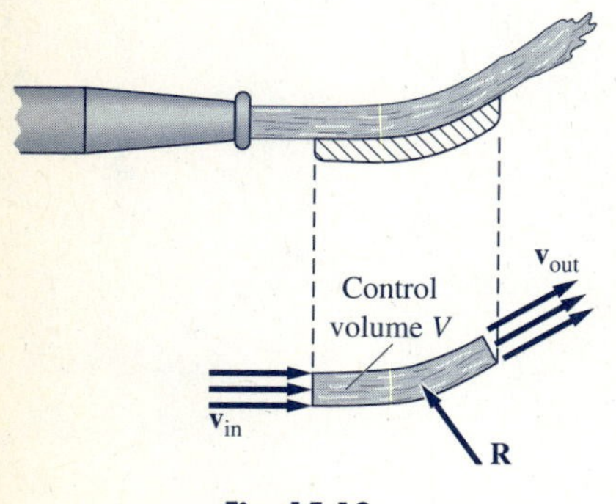

Fig. 15.18

For steady flow, $\dot{m}_{in} = \dot{m}_{out} = \dot{m}$ (a constant), which means that there is no accumulation of mass in V. Furthermore, the momentum of the fluid in the control volume is also constant; i.e., $\dot{\mathbf{p}}_V = \mathbf{0}$. Making these substitutions into Eq. (15.46) gives

$$\mathbf{R} = \dot{m}(\mathbf{v}_{out} - \mathbf{v}_{in}) \tag{15.47}$$

Observe that the resultant force $\mathbf{R}$ acting on the fluid is constant in the case of steady flow.

d. Rocket propulsion

Figure 15.19(a) shows a rocket in vertical flight. The mass of the rocket, including its contents, is denoted by $M(t)$, and its velocity is $\mathbf{v}(t) = v(t)\mathbf{j}$, where t is time. The rocket is expelling gases at the rate $\dot{m}$ with the constant nozzle velocity $\mathbf{u}$ *relative to the rocket.*

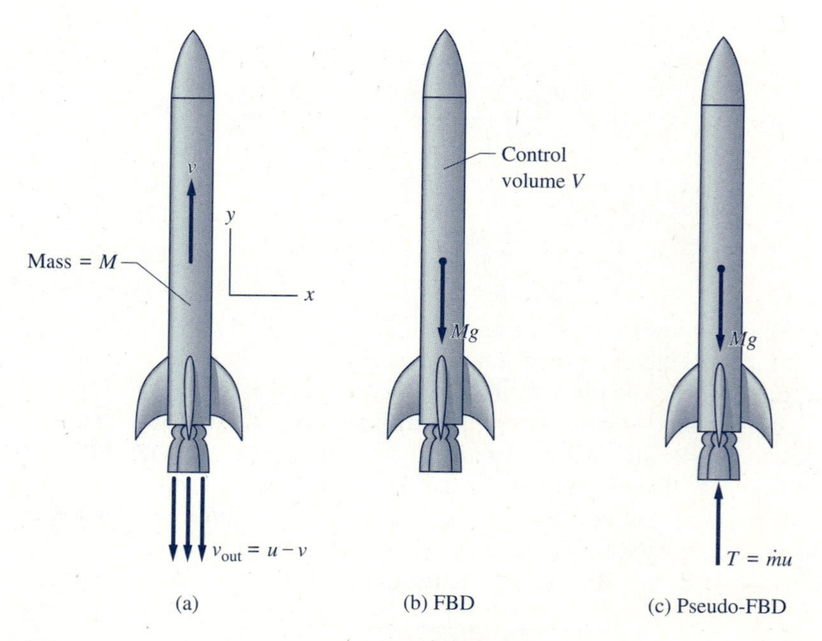

Fig. 15.19

We choose the control volume V to be the rocket and its interior. Since the rocket consumes only the fuel that it carries (there is no air intake), the rates of mass flow are

$$\dot{m}_{out} = \dot{m} \quad \text{and} \quad \dot{m}_{in} = 0 \tag{h}$$

The momentum of the mass within the control volume is $\mathbf{p}_V = Mv\mathbf{j}$, which on differentiation with respect to time yields

$$\dot{\mathbf{p}}_V = \dot{M}v\mathbf{j} + M\dot{v}\mathbf{j} = -\dot{m}v\mathbf{j} + M\dot{v}\mathbf{j} \tag{i}$$

where we substituted $\dot{M} = -\dot{m}$ (this is the rate at which fuel is consumed). The velocity of the expelled gas can be written as

$$\mathbf{v}_{out} = (v - u)\mathbf{j} \tag{j}$$

The free-body diagram (FBD) of the rocket and its contents is shown in Fig. 15.19(b). The only force acting on the rocket is its total weight (we neglect air resistance and the pressure of the gases at the exit from the nozzle).

Substituting $\mathbf{R} = -Mg\mathbf{j}$ and Eqs. (h)–(j) into Eq. (15.46), we obtain

$$-Mg\mathbf{j} = -\dot{m}v\mathbf{j} + M\dot{v}\mathbf{j} + \dot{m}(v - u)\mathbf{j}$$

which simplifies to

$$\boxed{\dot{m}u - Mg = M\dot{v}} \tag{15.48}$$

Letting $T = \dot{m}u$, called the *thrust* of the rocket engine, the equation of motion, Eq. (15.48), becomes

$$\boxed{T - Mg = M\dot{v}} \tag{15.49}$$

It is interesting to consider the "pseudo"-FBD shown in Fig. 15.19(c), where the thrust is incorrectly shown as an external force. This diagram yields the correct equation of motion, provided you use "force equals mass times acceleration," which is actually an invalid equation, because the mass M of the rocket is not constant.

Sample Problem 15.18

Figure (a) shows water entering a 60° horizontal bend in a pipe with the velocity $v_{in} = 20$ ft/s. As the water passes through the bend, its pressure drops from $p_{in} = 6$ lb/in.^{2} to $p_{out} = 4$ lb/in.^{2}, and the pipe diameter increases from $d_{in} = 8$ in. to $d_{out} = 10$ in. Determine the force exerted on the bend by the water. (Water weighs 62.4 lb/ft^3.)

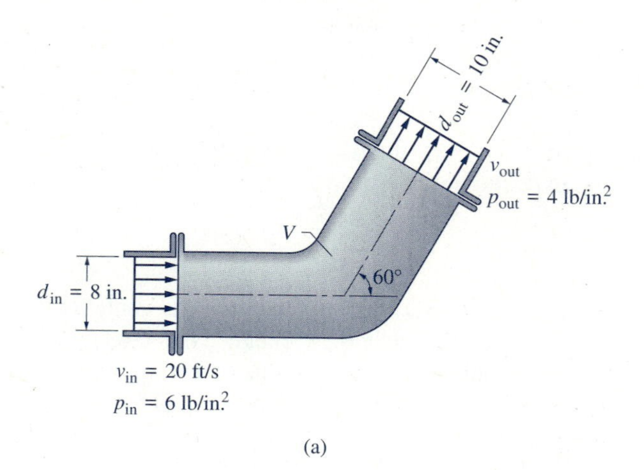

(a)

Solution

As shown in Fig. (a), we let the control volume V be the interior region of the pipe that is contained between the inlet and outlet cross-sectional areas of the pipe section.

Assuming steady flow, we have $A_{in}v_{in} = A_{out}v_{out}$ (A refers to the cross-sectional area of the pipe). Therefore, the velocity of the water on leaving the control volume is

$$v_{out} = v_{in}\left(\frac{d_{in}}{d_{out}}\right)^2 = 20\left(\frac{8}{10}\right)^2 = 12.80 \text{ ft/s}$$

The rate of mass flow through a cross section of the pipe is $\dot{m} = A\rho v$, where ρ is the mass density of water and v represents the velocity of water. Using the values of A and v at the inlet (the outlet parameters could also be used), we obtain

$$\dot{m} = \frac{\pi d_{in}^2}{4}\rho v_{in} = \frac{\pi}{4}\left(\frac{8}{12}\right)^2\left(\frac{62.4}{32.2}\right)(20) = 13.529 \text{ slug/s}$$

Using Eq. (15.47), the resultant force acting on the water in the control volume is

$$\begin{aligned}\mathbf{R} &= \dot{m}(\mathbf{v}_{out} - \mathbf{v}_{in}) \\ &= 13.529[12.80(\cos 60°\mathbf{i} + \sin 60°\mathbf{j}) - 20.0\mathbf{i}] \\ &= -184.0\mathbf{i} + 150.0\mathbf{j} \text{ lb}\end{aligned}$$

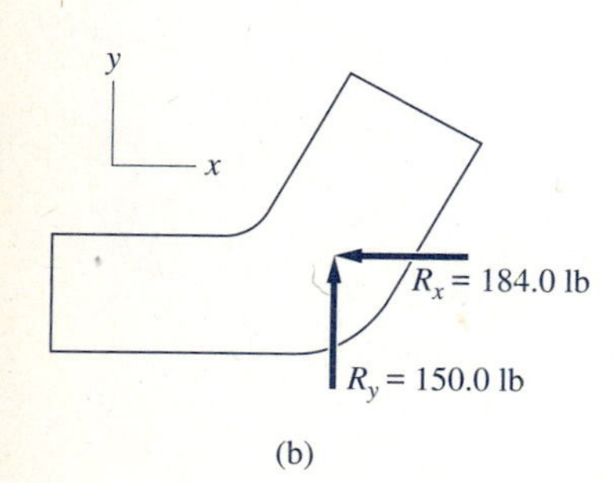

(b)

The rectangular components of **R** are shown in Fig. (b).

The free-body diagram (FBD) of the water in the control volume, shown in Fig. (c), contains Q_x and Q_y, the components of the force exerted on the water by the pipe, and P_{in} and P_{out}, the forces due to the inlet and outlet pressures (the weight of the water would also be included if the bend were in any plane other than the horizontal). Because the pressure distribution is constant for steady flow, the forces due to the pressure are

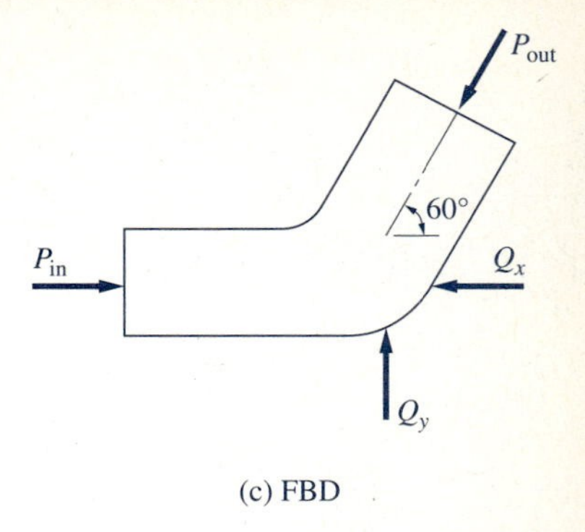

(c) FBD

$$P_{\text{in}} = p_{\text{in}}A_{\text{in}} = 6\frac{\pi(8)^2}{4} = 301.6 \text{ lb}$$

$$P_{\text{out}} = p_{\text{out}}A_{\text{out}} = 4\frac{\pi(10)^2}{4} = 314.2 \text{ lb}$$

Using these values, and comparing Figs. (b) and (c), we obtain

$$R_x = Q_x + P_{\text{out}}\cos 60^\circ - P_{\text{in}}$$
$$R_y = Q_y - P_{\text{out}}\sin 60^\circ$$

which yield

$$Q_x = R_x + P_{\text{in}} - P_{\text{out}}\cos 60^\circ$$
$$= -184.0 + 301.6 - 314.2\cos 60^\circ = -39.5 \text{ lb}$$
$$Q_y = R_y + P_{\text{out}}\sin 60^\circ = 150.0 + 314.2\sin 60^\circ = 422 \text{ lb}$$

Therefore, the force exerted by the bend on the water is $\mathbf{Q} = -39.5\mathbf{i} + 422\mathbf{j}$ lb. Letting $\mathbf{F}$ be the force exerted on the pipe by the water, we note that $\mathbf{F} = -\mathbf{Q}$, which gives

$$\mathbf{F} = 39.5\mathbf{i} - 422\mathbf{j} \text{ lb} \qquad \textit{Answer}$$

PROBLEMS

15.115 A rocket in vertical flight has a total mass M_0 at liftoff and consumes fuel (including oxidizer) at the constant rate $\dot{m}$. If the velocity of the exhaust gases relative to the rocket is u, show that the velocity of the rocket at time t after the liftoff is

$$v = u \ln\left(\frac{M_0}{M_0 - \dot{m}t}\right) - gt$$

Neglect air resistance and the change of gravity with altitude.

15.116 A Saturn V rocket weighs 6.2×10^6 lb at liftoff. Its first stage develops 7.5×10^6 lb of thrust while consuming 4.4×10^6 lb of fuel during its 2.5-min burn. Determine (a) the velocity of the exhaust gas relative to the rocket; and (b) the velocity of the rocket at the end of the burn. Assume vertical flight, and neglect air resistance and the change of gravity with altitude. (Hint: See Prob. 15.115.)

15.117 The V-2 rocket of World War II fame weighed 14 tons at liftoff, including 9.5 tons of fuel (alcohol and oxygen). The powered flight lasted 65 s, and the rocket reached a maximum velocity of 3600 mi/h in vertical flight. Calculate the thrust of the rocket engine. (Hint: See Prob. 15.115.)

15.118 If the V-2 rocket described in Prob. 15.117 is launched vertically, determine (a) its altitude at the end of the powered flight; and (b) its maximum altitude. Neglect air resistance and the change in gravity with altitude. (Hint: See Prob. 15.115.)

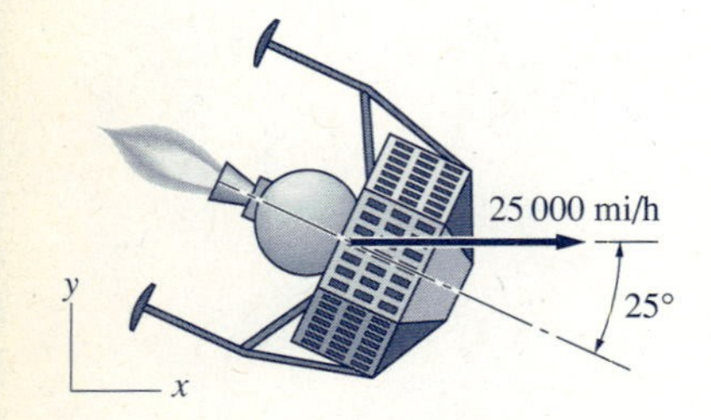

Fig. P15.119

15.119 The space probe weighing 1200 lb is traveling at a constant speed of 25 000 mi/h when its thruster is fired for 100 s. During the firing, the fuel consumption is 5 lb/s, and the gases are expelled at 4000 mi/h relative to the vehicle. If the line of thrust is inclined at 25° to the initial direction of travel, determine the final velocity of the probe. Neglect the effect of gravity.

15.120 The firehose is discharging a jet of water against the flat plate. The diameter of the jet is $d = 50$ mm, and the speed of the water is 60 m/s. Calculate the force exerted by the jet on the plate if the plate is (a) stationary; (b) moving to the right at 5 m/s. (The density of water is 1000 kg/m^3.)

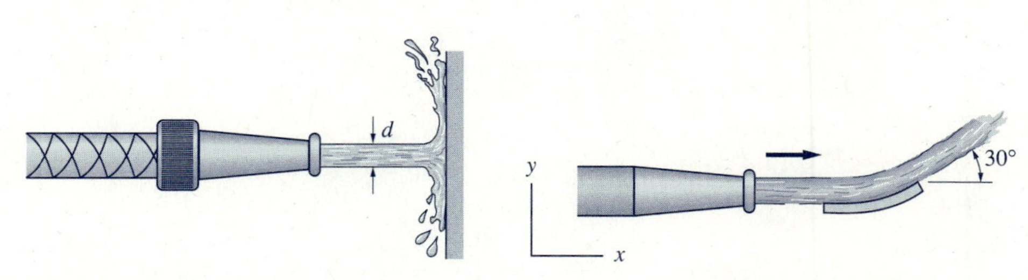

Fig. P15.120 **Fig. P15.121, P15.122**

15.121 The jet of water is deflected by a stationary vane through 30° as shown. The flow rate is 240 gal/min, and the speed of the jet is 40 ft/s. Determine the force exerted by the jet on the vane. (Water weighs 8.34 lb/gal.)

15.122 Solve Prob. 15.121 if the vane is moving to the right at a constant velocity of 12 ft/s.

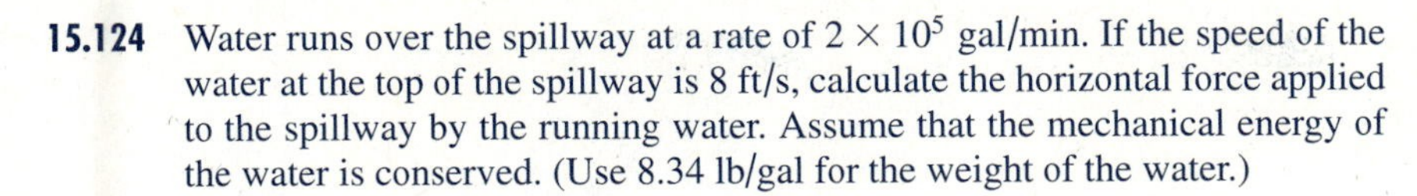

15.123 The 0.15-kg ball is held stationary by a jet of water as shown. At the nozzle, the speed of the jet is 6 m/s, and its diameter is $d = 12$ mm. Find the height h of the ball. (The density of water is 1000 kg/m^3.) (Hint: Use the energy method to determine the speed of the jet when it hits the ball.)

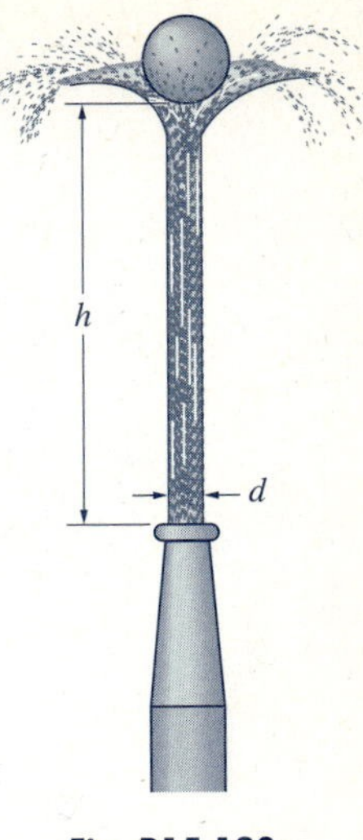

Fig. P15.123

15.124 Water runs over the spillway at a rate of 2×10^5 gal/min. If the speed of the water at the top of the spillway is 8 ft/s, calculate the horizontal force applied to the spillway by the running water. Assume that the mechanical energy of the water is conserved. (Use 8.34 lb/gal for the weight of the water.)

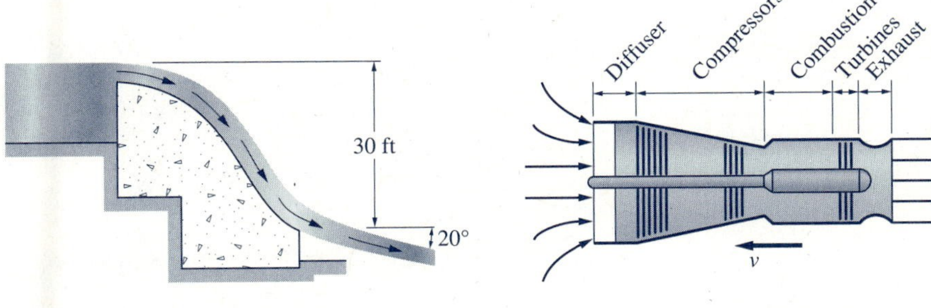

Fig. P15.124

Fig. P15.125

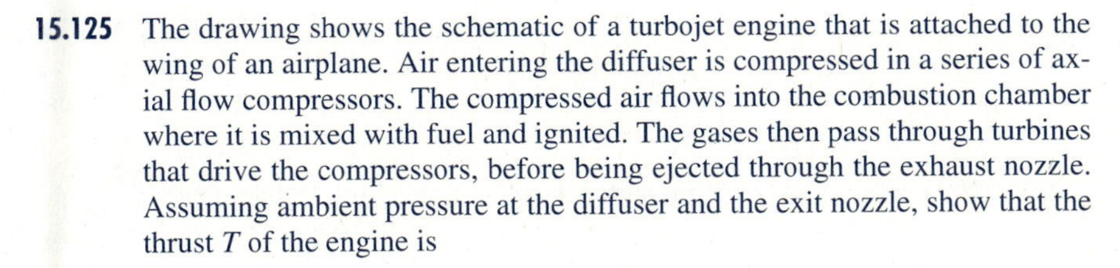

15.125 The drawing shows the schematic of a turbojet engine that is attached to the wing of an airplane. Air entering the diffuser is compressed in a series of axial flow compressors. The compressed air flows into the combustion chamber where it is mixed with fuel and ignited. The gases then pass through turbines that drive the compressors, before being ejected through the exhaust nozzle. Assuming ambient pressure at the diffuser and the exit nozzle, show that the thrust T of the engine is

$$T = \dot{m}(u - v) + \dot{m}_f u$$

where u is the velocity of the exhaust gas relative to the engine, v is the velocity of the airplane, $\dot{m}$ is the rate of air flow through the engine, and $\dot{m}_f$ is the rate of fuel consumption.

15.126 An airplane is powered by a single turbojet engine. At a constant speed of 880 km/h in level flight, the rates of air intake and fuel consumption are 100 kg/s and 1.5 kg/s, respectively. The burnt mixture is exhausted at 700 m/s relative to the airplane. Determine the aerodynamic drag force acting on the airplane. (Hint: See Prob. 15.125.)

15.127 When operating at its maximum thrust, a turbojet engine mounted on an airplane has an airflow rate of 260 lb/s and consumes fuel at the rate of 4.2 lb/s. If the engine produces a maximum thrust of 17 500 lb during a static test, calculate the maximum thrust when the airspeed is 550 mi/h. Assume that the rates of airflow and fuel consumption are not affected by the airspeed. (Hint: See Prob. 15.125.)

15.128 Water enters the reducer section of the pipe at the speed of 10 ft/s and a gage pressure of 5 lb/in.^{2}. If the gage pressure on exit is 2.27 lb/in.^{2}, determine the horizontal force applied by the water to the reducer. (Water weighs 62.4 lb/ft^3.)

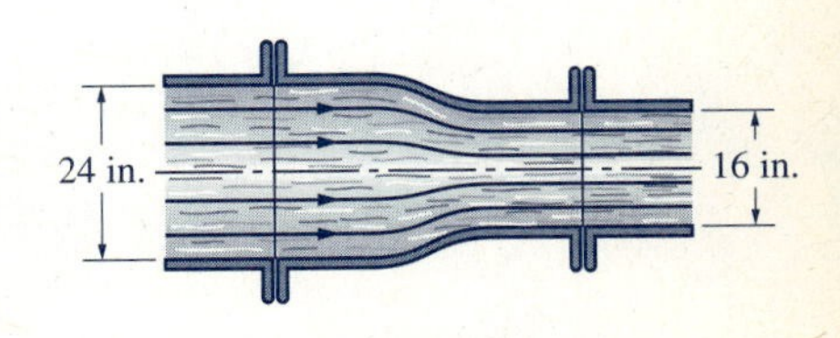

Fig. P15.128

15.129 Water enters the horizontal bend in the pipe at a velocity of 4.8 m/s. The entrance and exit gage pressures are 23 kPa and 32 kPa, respectively. Find the horizontal force applied by the water to the bend in the pipe. (The mass density of water is 1000 kg/m^3.)

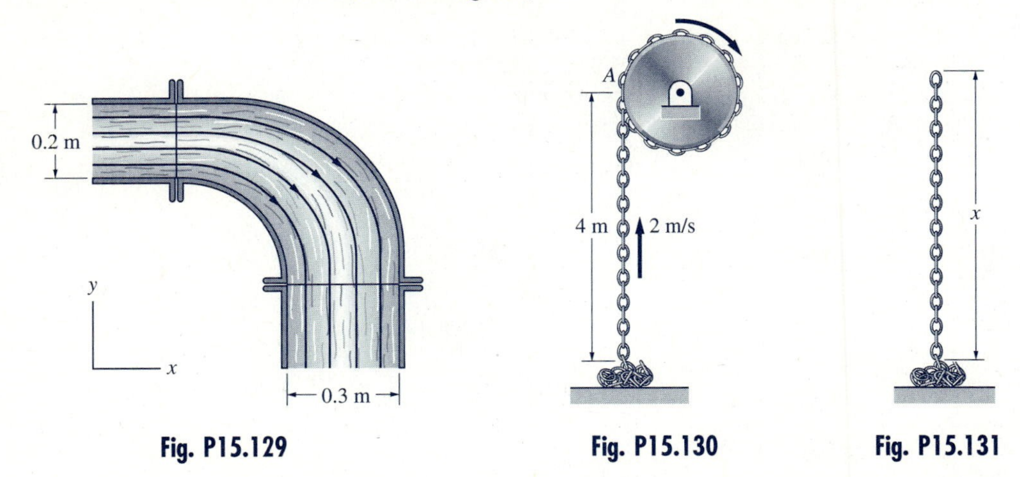

Fig. P15.129 Fig. P15.130 Fig. P15.131

15.130 The chain of mass 3.6 kg/m is raised at the constant speed of 2 m/s. Determine the tension in the chain at A.

15.131 The chain of length L and mass per unit length ρ is released from rest with $x = L$ (lower end of the chain just touching the ground). Determine the reactive force exerted by the ground on the chain as a function of x. (Hint: The vertical portion of the chain is in free fall.)

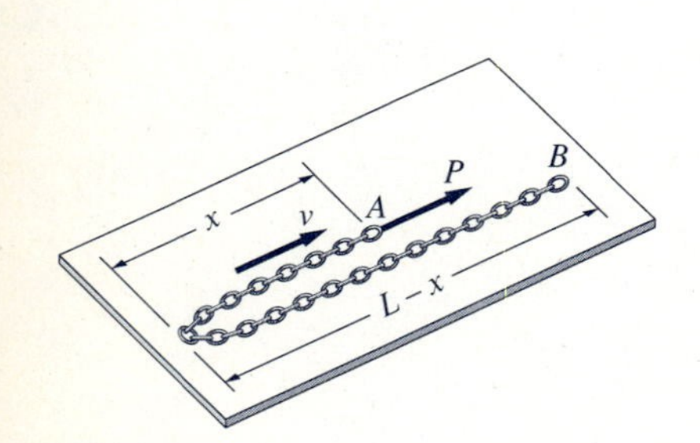

Fig. P15.132

15.132 The chain AB, of length L and mass per unit length ρ, is laid out in a straight line on a smooth horizontal surface. A constant force P is then applied to end A at time $t = 0$, causing A to move to the right with velocity v as shown. (a) Determine v as a function of x. (b) Show that only half of the work done by P goes into the kinetic energy of the chain, and explain what happens to the other half.

15.133 A vertical chute discharges coal onto a conveyor at the rate of 60 kg/s. If the speed of the conveyor belt is constant at 3 m/s, find the power required to drive the conveyor. Neglect friction.

15.134 Each nozzle of the turbine discharges water at the rate of 0.8 lb/s with the velocity $u = 18$ ft/s relative to the nozzle. Find (a) the power generated by the turbine in terms of the constant nozzle velocity v; and (b) the maximum power and the corresponding value of v. Neglect the velocity with which the water enters the nozzles.

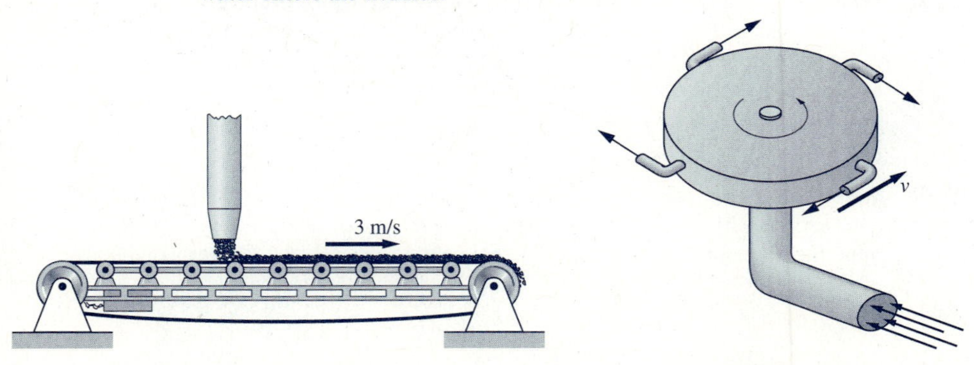

Fig. P15.133 Fig. P15.134

REVIEW PROBLEMS

15.135 Neglecting friction, determine the acceleration of each block when the 80-lb horizontal force is applied to block A.

15.136 The projectiles A and B are launched simultaneously at $t = 0$ with the velocities shown. Assuming both trajectories lie in the same vertical plane, determine (a) the relative position vector $\mathbf{r}_{B/A}$ as a function of t; and (b) the smallest distance between the projectiles. Neglect air resistance.

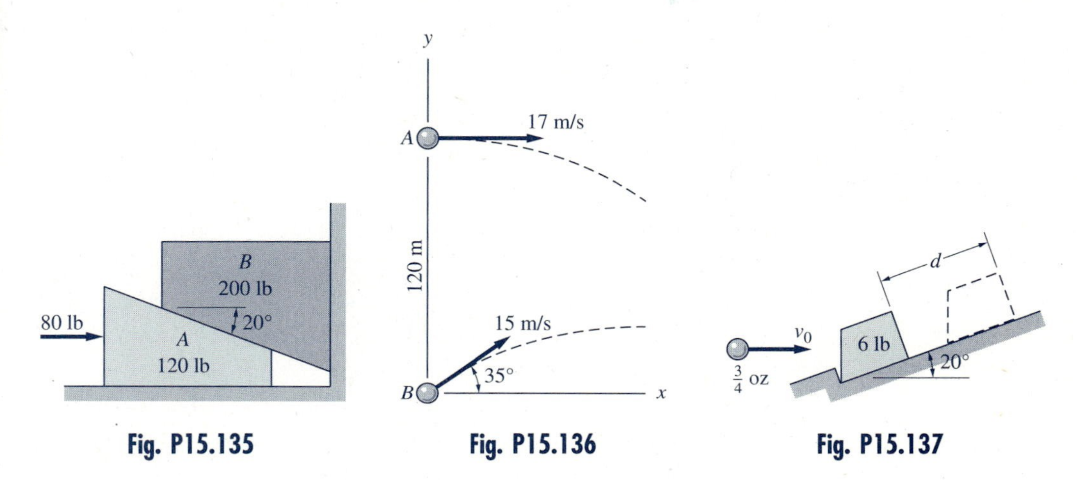

Fig. P15.135 **Fig. P15.136** **Fig. P15.137**

15.137 The $\frac{3}{4}$-oz bullet strikes the stationary 6-lb block with the horizontal velocity v_0 and becomes embedded. If the maximum displacement of the block after the impact is $d = 1.2$ ft, determine v_0. Neglect friction.

15.138 The airspeed of the helicopter in level flight is 125 mi/h, directed east. If the ground speed is 160 mi/h in the direction 25° north of east, determine the magnitude and direction of the wind velocity.

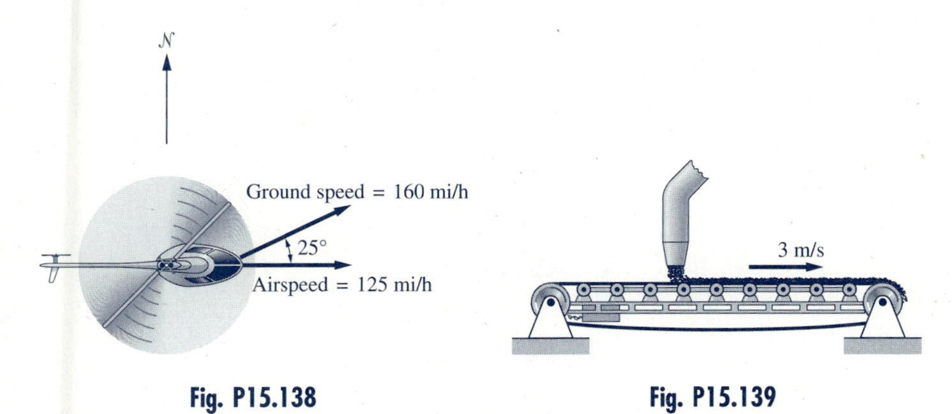

Fig. P15.138 **Fig. P15.139**

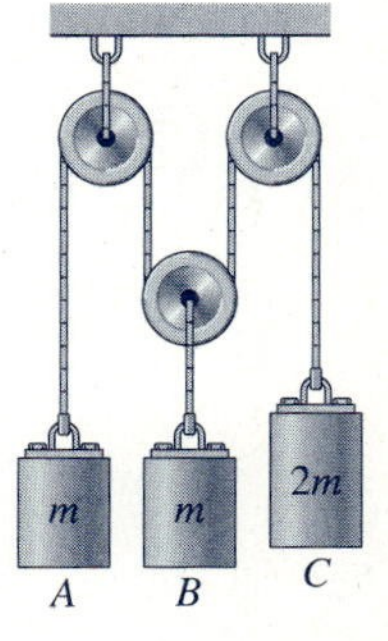

Fig. P15.140

15.139 The vertical chute discharges coal at the constant rate of 50 kg/s onto a conveyor belt that moves at the constant speed of 3 m/s. Neglecting friction, determine the power needed to operate the conveyor.

15.140 The three masses are suspended from the cable-and-pulley system shown. Determine the acceleration of each mass in terms of the gravitational acceleration g. Neglect the masses of the pulleys.

15.141 The 620-lb rocket A is in unpowered flight near the surface of the earth when it breaks into two pieces—the 185-lb nose cone B and the 435-lb thruster C. The figure shows the velocities of the two parts 60 s after the breakup (note that A, B, and C are in the same vertical plane). Determine the velocity (magnitude and direction) of the rocket just before the breakup.

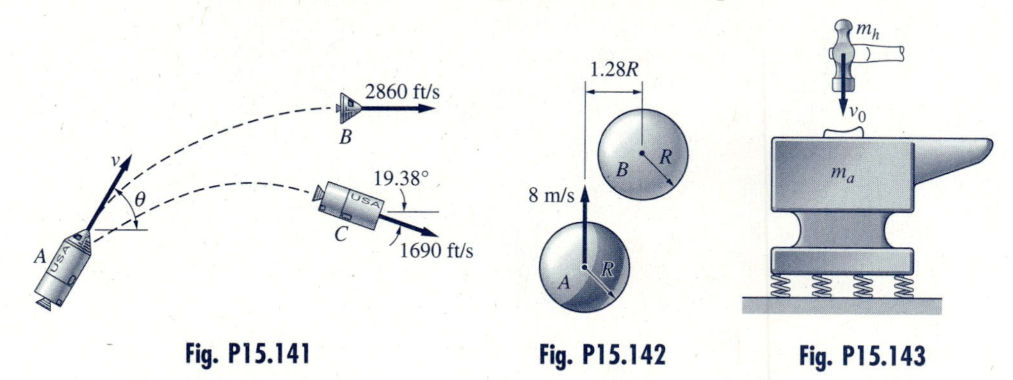

Fig. P15.141 Fig. P15.142 Fig. P15.143

15.142 Two smooth, identical billiard balls of radius R are on a horizontal table. After ball A hits ball B with the velocity of 8 m/s, the speed of B is 5.52 m/s. Determine (a) the coefficient of restitution between the balls; and (b) the speed of ball A after the impact.

15.143 A small piece of red-hot iron is placed on the anvil and struck with the hammer moving at the velocity v_0. The masses of the hammer and the anvil are m_h and m_a, respectively, and the coefficient of restitution between the hammer and hot iron is e. If the anvil sits on an elastic base (as indicated by the springs in the figure), derive the expression for the impulse delivered by the hammer to the hot metal.

15.144 Bar AB of the mechanism shown rotates counterclockwise with the constant angular velocity $\dot{\theta} = \omega_0$. Determine the velocity of collar C relative to bar AB in the position where $R = 0.6b$.

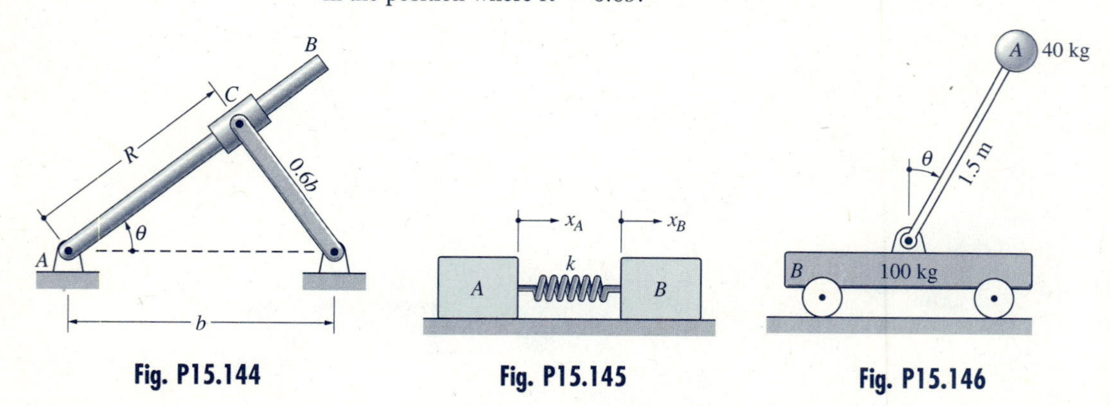

Fig. P15.144 Fig. P15.145 Fig. P15.146

15.145 The identical blocks A and B of mass m each are at rest on a smooth, horizontal surface. The spring of stiffness k connecting the blocks is undeformed. If block B is suddenly given an initial velocity v_0 to the right, determine (a) the relationship between the relative velocity $v_{B/A}$ and the relative displacement $x_{B/A}$; and (b) the maximum value of $x_{B/A}$.

15.146 The 100-kg cart B is free to roll on the horizontal surface. The cart is at rest when the 40-kg inverted pendulum A is released with negligible initial velocity from the vertical ($\theta = 0$) position. Determine the speed of the cart when the pendulum has reached the position $\theta = 45°$.

16 Planar Kinematics of Rigid Bodies

16.1 Introduction to Rigid-Body Motion

The kinematics of relative motion of particles (points) was discussed in Art. 15.2. We introduced $\mathbf{r}_{B/A}$, the relative position vector between points A and B, and the relative velocity and acceleration relations

$$\mathbf{v}_B = \mathbf{v}_A + \mathbf{v}_{B/A} \qquad \text{(15.3, repeated)}$$

$$\mathbf{a}_B = \mathbf{a}_A + \mathbf{a}_{B/A} \qquad \text{(15.4, repeated)}$$

The motions of A and B are kinematically independent if there are no constraints imposed on the relative motion terms $\mathbf{v}_{B/A}$ and $\mathbf{a}_{B/A}$. Here we consider the restrictions that exist on the relative motion if points A and B belong to the same rigid body.

Rigidity imposes the constraints that the distances between points in the body and the angles between lines inscribed in the body are constant. Using these conditions, it can be shown (see Appendix D) that the relative velocity between two points A and B in a rigid body must be of the form

$$\boxed{\mathbf{v}_{B/A} = \dot{\mathbf{r}}_{B/A} = \boldsymbol{\omega} \times \mathbf{r}_{B/A}} \qquad (16.1)$$

where $\boldsymbol{\omega}$ is called the *angular velocity* of the body. The angular velocity is a free vector that does not depend on the choice of points A and B. The units for angular velocity are radians per unit time, e.g., radians per second (rad/s).

Taking the time derivative of both sides of Eq. (16.1), we get

$$\mathbf{a}_{B/A} = \dot{\mathbf{v}}_{B/A} = (\boldsymbol{\omega} \times \dot{\mathbf{r}}_{B/A}) + (\dot{\boldsymbol{\omega}} \times \mathbf{r}_{B/A}) \qquad (16.2a)$$

which yields

$$\mathbf{a}_{B/A} = (\boldsymbol{\omega} \times \mathbf{v}_{B/A}) + (\boldsymbol{\alpha} \times \mathbf{r}_{B/A}) \qquad (16.2b)$$

$$= \boldsymbol{\omega} \times (\boldsymbol{\omega} \times \mathbf{r}_{B/A}) + (\boldsymbol{\alpha} \times \mathbf{r}_{B/A}) \qquad (16.2c)$$

where $\boldsymbol{\alpha} = \dot{\boldsymbol{\omega}}$ is called the *angular acceleration* of the body. The units for angular acceleration are radians per unit time squared, e.g., radians per second squared (rad/s^2). Since the angular velocity and angular acceleration are independent of the choice of the points A and B, they are properties of the body as a whole. The directions of $\boldsymbol{\omega}$ and $\boldsymbol{\alpha}$ will, in general, be different, as illustrated in Fig. 16.1(a).

The relative velocity vector in Eq. (16.1) is perpendicular to both $\boldsymbol{\omega}$ and $\mathbf{r}_{B/A}$, as shown in Fig. 16.1(b). Figure 16.1 parts (c) and (d) show the components of $\mathbf{a}_{B/A}$ that are given in Eq. (16.2c). Observe that the first component, shown in Fig. 16.1(c), is perpendicular to $\boldsymbol{\omega}$ and $\mathbf{v}_{B/A}$, and directed toward the line OA (the line that passes through A and is parallel to $\boldsymbol{\omega}$); the second component is perpendicular to both $\boldsymbol{\alpha}$ and $\mathbf{r}_{B/A}$, as illustrated in Fig. 16.1(d). In general, the two components of $\mathbf{a}_{B/A}$ in Eq. (16.2c) will not be perpendicular to each other.

From the above discussion, we see that motion of any point of a rigid body, such as B, can be written as the motion of a *reference point* (point A in the foregoing equations) plus the relative motion terms that arise from the angular motion of the body.

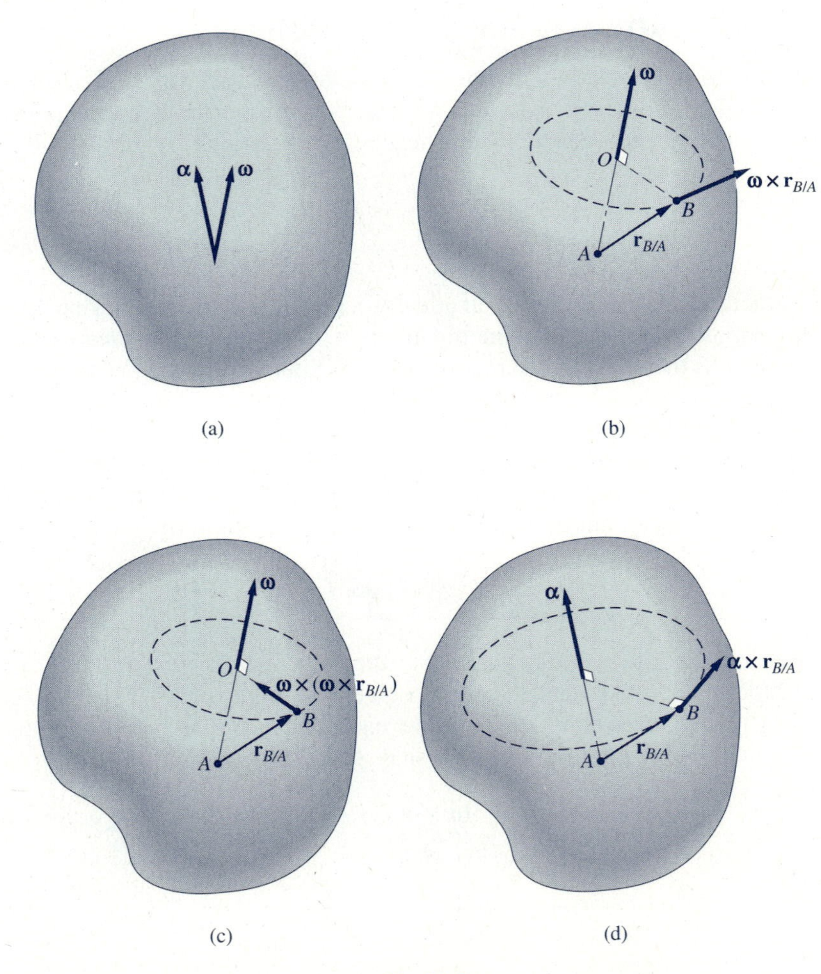

Fig. 16.1

Two important special cases of rigid-body motion are translation and plane motion.

Translation: A rigid body is said to be translating if all lines in the body remain parallel to their original positions. Therefore, translation implies the absence of rotation ($\boldsymbol{\omega} = \boldsymbol{\alpha} = \mathbf{0}$). Consequently, the velocities and accelerations of all points of the body are equal, so the motion of any one point of the body determines the motion of the entire body.

Plane Motion: A rigid body is undergoing plane motion if all points in the body remain a constant distance from a fixed reference plane, called the *plane of the motion.* In plane motion the direction of $\boldsymbol{\omega}$ is always perpendicular to the plane of the motion. This ensures that the relative velocity vector $\mathbf{v}_{B/A}$ of any point B in the body is parallel to the plane of the motion.

This chapter is devoted to the kinematics of plane motion of a rigid body. The kinetics of plane motion is the subject of the next two chapters. Three-dimensional motion of a rigid body is treated in Chapter 19.

16.2 Rotation About a Fixed Axis

Rotation about a fixed axis is a special case where one line in the body, called the *axis of rotation,* is fixed in space. Since the body is rigid, the path of each point (except points on the axis of rotation) is a circle with its center on the axis of rotation. Therefore, rotation about a fixed axis comes under the category of plane motion—all points remain a fixed distance from a reference plane that is perpendicular to the axis of rotation.

Figure 16.2(a) shows a rigid body that is rotating about the fixed axis OA. The circular path of point B of the body lies in a plane perpendicular to the axis of rotation. Point O is the center of the circle, and R is its radius. The reference point A is a point on the axis of rotation, and $\mathbf{r}$ is the position vector of B, measured from A. Because the body is rigid, the angle β between $\mathbf{r}$ and the axis of rotation is constant. The angular position coordinate θ of point B is

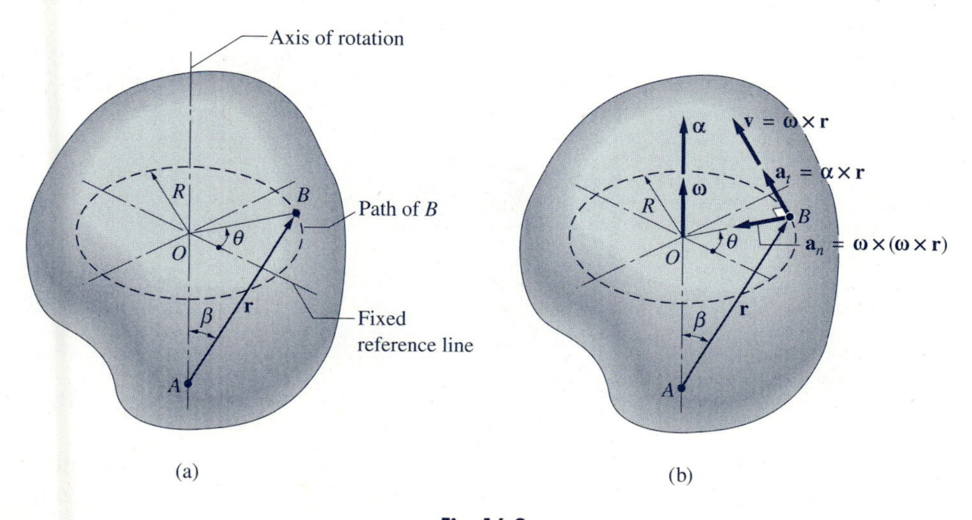

Fig. 16.2

measured from an arbitrary fixed reference line that lies in the plane of motion of B.

Since we have plane motion, the angular velocity of the body must be normal to the plane of motion, i.e., parallel to the axis of rotation, as shown in Fig. 16.2(b). As the direction of $\boldsymbol{\omega}$ is fixed, the angular acceleration $\boldsymbol{\alpha}$ of the body is also parallel to the axis of rotation, its magnitude being $\alpha = \dot{\omega}$.

Recalling that A is a fixed point, the velocity and acceleration vectors of B are given by Eqs. (16.1) and (16.2) (the subscripts having been omitted)

$$\mathbf{v} = \boldsymbol{\omega} \times \mathbf{r} \tag{16.3}$$

and

$$\mathbf{a} = \boldsymbol{\omega} \times (\boldsymbol{\omega} \times \mathbf{r}) + (\boldsymbol{\alpha} \times \mathbf{r}) \tag{16.4}$$

These vectors are shown in Fig. 16.2(b). Their directions can be deduced from the properties of the cross product: (1) $\mathbf{v}$ is tangent to the path of B; (2) the acceleration component $\boldsymbol{\omega} \times (\boldsymbol{\omega} \times \mathbf{r})$ is directed toward O, i.e., normal to the path; and (3) the acceleration component $\boldsymbol{\alpha} \times \mathbf{r}$ is tangent to the path. In Fig. 16.2(b) we refer to the acceleration components as $\mathbf{a}_n$ and $\mathbf{a}_t$, because they turn out to be identical to the normal and tangential acceleration components of a particle discussed in Art. 13.2.

The magnitude of the velocity vector is $v = |\boldsymbol{\omega} \times \mathbf{r}| = \omega r \sin\beta$. The magnitudes of the acceleration components are $a_n = |\boldsymbol{\omega} \times (\boldsymbol{\omega} \times \mathbf{r})| = \omega|\boldsymbol{\omega} \times \mathbf{r}| = \omega^2 r \sin\beta$ and $a_t = |\boldsymbol{\alpha} \times \mathbf{r}| = \alpha r \sin\beta$. Recognizing from Fig. 16.2(a) that $r \sin\beta = R$, we obtain

$$\boxed{\begin{aligned} v &= R\omega \\ a_n &= R\omega^2 = \frac{v^2}{R} = v\omega \\ a_t &= R\alpha \end{aligned}} \tag{16.5}$$

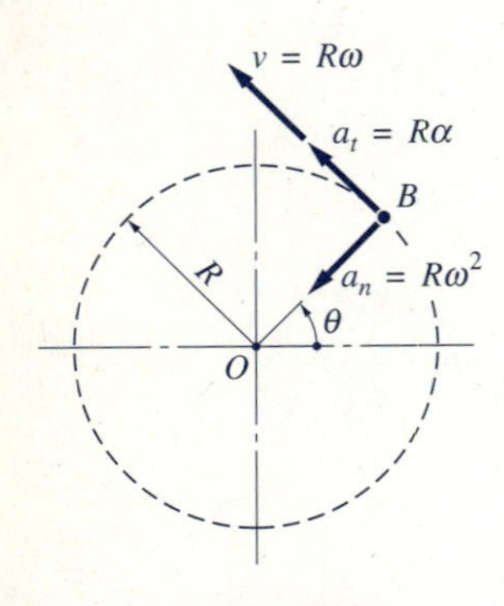

Fig. 16.3

Figure 16.3 shows the directions of the above components, viewed perpendicular to the plane of motion.

The equations of a particle moving along a circular path—Eqs. (13.10)—were $v = R\dot{\theta}$, $a_t = R\ddot{\theta}$, and $a_n = R\dot{\theta}^2$. Comparing with Eqs. (16.5) we conclude that

$$\boxed{\omega = \dot{\theta} \quad \text{and} \quad \alpha = \dot{\omega} = \ddot{\theta}} \tag{16.6}$$

The angular position of a rigid body that is rotating about a fixed axis is specified by the scalar function $\theta(t)$. The angular displacement of the body during the finite time interval Δt is defined as

$$\Delta\theta = \theta(t + \Delta t) - \theta(t) \tag{16.7}$$

The angular displacement can also be represented as a vector, since it has magnitude ($\Delta\theta$), direction (along the axis of rotation with its sense determined by the right-hand rule), and obeys the parallelogram law for addition (the sum of collinear vectors is a special case of the parallelogram law).

However, it can be demonstrated that angular displacements are not vectors for the general case. Figure 16.4(a) shows the initial position of a book that is given 90° angular rotations $\Delta\theta_x$ and $\Delta\theta_y$ about the x- and y-axes, respectively. Figure 16.4 parts (b) and (c) show the results if the rotations are performed in the order $\Delta\theta_x$ followed by $\Delta\theta_y$, and $\Delta\theta_y$ followed by $\Delta\theta_x$, respectively. As can be seen, the final orientation of the book depends on the order in which the rotations are performed; i.e., finite rotations in general do not obey the commutative property of vector addition. Consequently, they are not vectors. However, it is possible to show that *differential* angular displacements are vectors, even if the motion is three-dimensional.

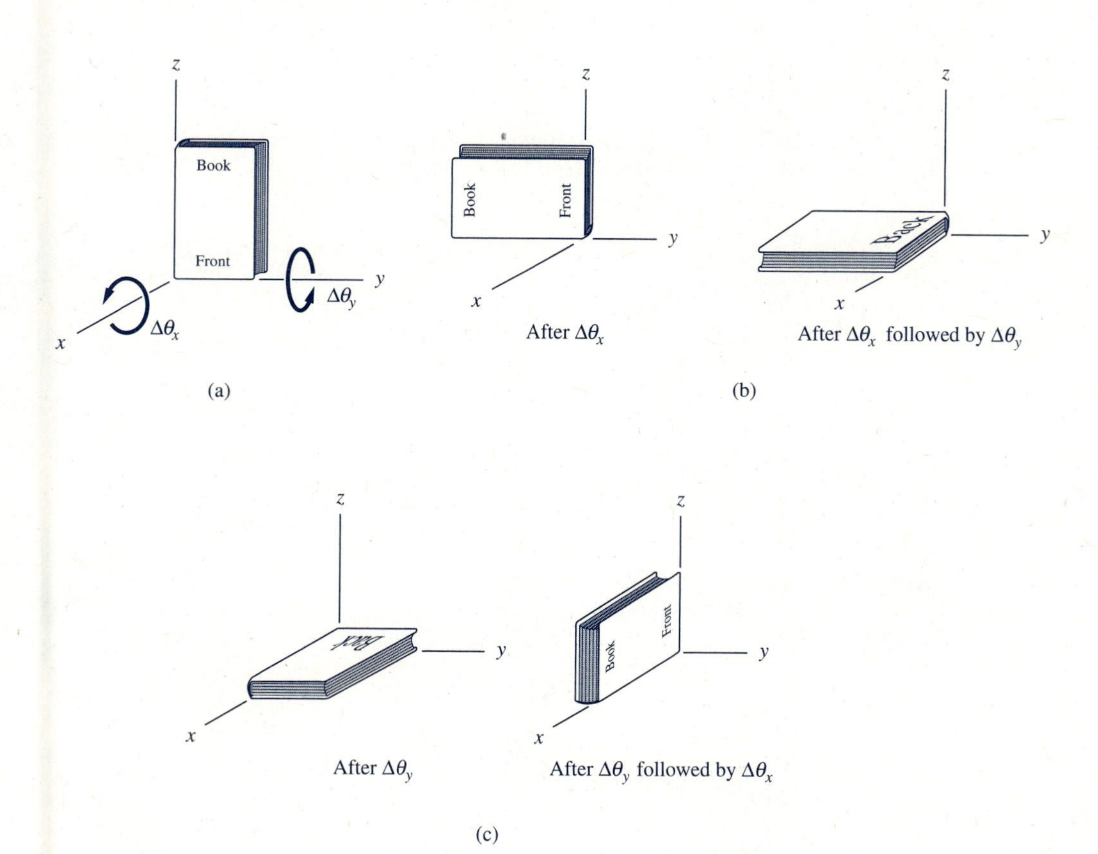

Fig. 16.4

Sample Problem 16.1

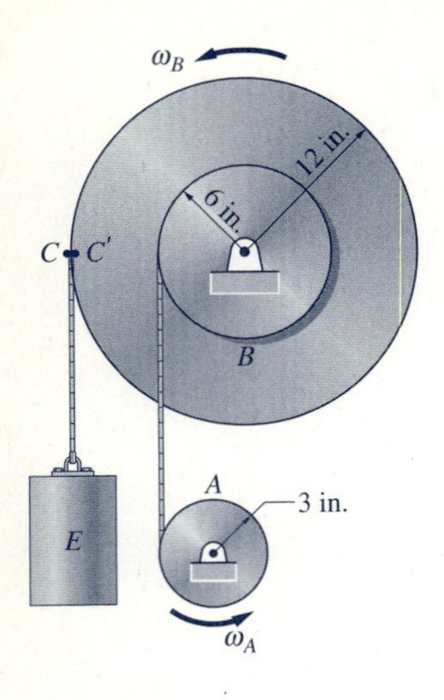

The cable-driven hoist, consisting of the drive pulley A connected to the compound pulley B, is used to lower block E. The angular acceleration of pulley A is given by $\alpha_A = 3t^2$ rad/s^2 counterclockwise, where t is measured from the instant when the system begins to move. Assuming that the cables do not slip on the pulleys, determine (1) the angular velocity and position of pulley B as functions of time; (2) the angular displacement of B during the time interval $t = 0$ to $t = 2$ s; and (3) the velocity and acceleration of block E when $t = 2$ s.

Solution

If a cable does not slip, every point on the cable that is in contact with a pulley has the same velocity and acceleration as the corresponding contact point on the pulley. An exception is the location where a cable leaves a pulley, such as the points C (on the cable) and C' (on the pulley), as indicated in the figure. Since the points are moving on different paths (the path of C becomes a vertical line; the path of C' is a circle), their acceleration vectors cannot be equal. However, the acceleration of C is equal to the *tangential component* of the acceleration of C', since this component of acceleration equals the rate of change of speed.

Part 1

Since there is no slipping, the velocity of any point on the cable connecting A and B is found from Eq. (16.5): $v = R\omega$. Therefore, $v = 3\omega_A = 6\omega_B$, which gives $\omega_B = 0.5\omega_A$. On differentiation with respect to time, we get $\alpha_B = 0.5\alpha_A = 0.5(3t^2) = 1.5t^2$ rad/s^2.

Utilizing $\alpha = \dot{\omega}$ and $\omega = \dot{\theta}$, we obtain

$$\omega_B = \int \alpha_B\, dt = \int 1.5t^2\, dt = 0.5t^3 + C_1 \qquad \text{(a)}$$

$$\theta_B = \int \omega_B\, dt = \int (0.5t^3 + C_1)\, dt = 0.125t^4 + C_1 t + C_2 \qquad \text{(b)}$$

Applying the initial conditions $t = 0$, $\theta_B = 0$ (an arbitrary choice) and $t = 0$, $\omega_B = 0$, the constants of integration are found to be $C_1 = C_2 = 0$. Therefore, the angular velocity and position of pulley B are

$$\omega_B = 0.5t^3 \text{ rad/s} \qquad \textit{Answer} \quad \text{(c)}$$

$$\theta_B = 0.125t^4 \text{ rad} \qquad \textit{Answer} \quad \text{(d)}$$

Part 2

From Eq. (d) we find that the angular position of pulley B at $t = 2$ s is $\theta_B = 0.125(2)^4 = 2.00$ rad. Since the initial value of θ_B was chosen to be zero, θ_B represents the angular displacement of the pulley, i.e.,

$$\Delta\theta_B = 2.00 \text{ rad} = 114.6^\circ \circlearrowleft \qquad \textit{Answer}$$

Part 3

Using $\omega_B = 0.5t^3$ and $\alpha_B = 1.5t^2$, the velocity and acceleration of E at $t = 2$ s are found to be

$$v_E = v_C = R_B\omega_B = 12(0.5)(2)^3 = 48.0 \text{ in./s} \qquad \mathbf{v}_E = 48.0 \text{ in./s} \downarrow \quad \textit{Answer}$$

$$a_E = a_C = R_B\alpha_B = 12(1.5)(2)^2 = 72.0 \text{ in./s}^2 \qquad \mathbf{a}_E = 72.0 \text{ in./s}^2 \downarrow \quad \textit{Answer}$$

The directions of the velocity and acceleration vectors for E are downward, as determined by the counterclockwise directions for ω_B and α_B, respectively.

Sample Problem 16.2

The rigid body, consisting of the arm OA attached to the shaft BC, is rotating in bearings at B and C. When the body is in the position shown in Fig. (a), its angular velocity and angular acceleration are 3 rad/s and 2 rad/s², respectively, directed as shown. For this position, determine the velocity and acceleration vectors of point A using (1) vector equations; and (2) scalar equations.

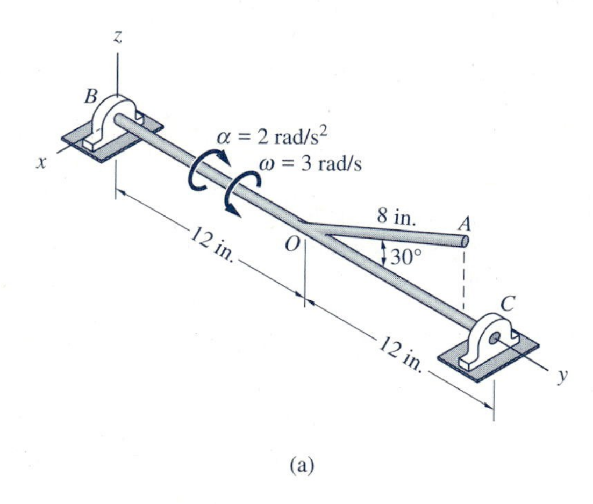

(a)

Solution

Part 1

Applying the right-hand rule, the angular velocity and angular acceleration vectors of the body are

$$\boldsymbol{\omega} = 3\mathbf{j} \text{ rad/s} \quad \text{and} \quad \boldsymbol{\alpha} = -2\mathbf{j} \text{ rad/s}^2$$

The position vector of A can be measured from the fixed point O (any point on BC could be used as a reference, since all points on the axis of the shaft are fixed), which gives

$$\mathbf{r}_A = \mathbf{r}_{A/O} = 8\cos 30°\mathbf{j} + 8\sin 30°\mathbf{k} = 6.928\mathbf{j} + 4.000\mathbf{k} \text{ in.}$$

The velocity vector of point A is found from Eq. (16.3) to be

$$\mathbf{v}_A = \boldsymbol{\omega} \times \mathbf{r}_A = 3\mathbf{j} \times (6.928\mathbf{j} + 4.000\mathbf{k})$$
$$= 12.00\mathbf{i} \text{ in./s} \qquad \textit{Answer}$$

From Eq. (16.4) the acceleration vector of A is

$$\mathbf{a}_A = \boldsymbol{\omega} \times (\boldsymbol{\omega} \times \mathbf{r}_A) + (\boldsymbol{\alpha} \times \mathbf{r}_A)$$
$$= (3\mathbf{j}) \times (12.00\mathbf{i}) + (-2\mathbf{j}) \times (6.928\mathbf{j} + 4.000\mathbf{k})$$
$$= -36.0\mathbf{k} - 8.00\mathbf{i} \text{ in./s}^2 \qquad \textit{Answer}$$

The absence of $\mathbf{j}$ components in the solutions for $\mathbf{v}_A$ and $\mathbf{a}_A$ is due to the fact that the plane of the motion is perpendicular to the y-axis.

Part 2

Figure (b) shows the plane of motion of point A (looking from C toward B). Note that the radius of the path of A is $R = 8 \sin 30° = 4.0$ in. The magnitudes of the velocity and acceleration components of A can be found from Eq. (16.5).

$$v_A = R\omega = 4(3) = 12.00 \text{ in./s}$$
$$(a_A)_n = R\omega^2 = 4(3)^2 = 36.0 \text{ in./s}^2$$
$$(a_A)_t = R\alpha = 4(2) = 8.00 \text{ in./s}^2$$

The directions of the velocity and acceleration components of A shown in Fig. (b) are determined as follows: $\mathbf{v}_A$ is tangent to the path, its direction found from the counterclockwise direction of ω; $(\mathbf{a}_A)_n$ is directed toward the center of curvature of the path; and $(\mathbf{a}_A)_t$ is tangent to the path, its direction found from the clockwise direction of α. Therefore, using vector notation, we obtain

$$\mathbf{v}_A = 12.00\mathbf{i} \text{ in./s} \qquad \textit{Answer}$$

and

$$\mathbf{a}_A = -36.0\mathbf{k} - 8.00\mathbf{i} \text{ in./s}^2 \qquad \textit{Answer}$$

which, of course, agree with the results found in Part 1.

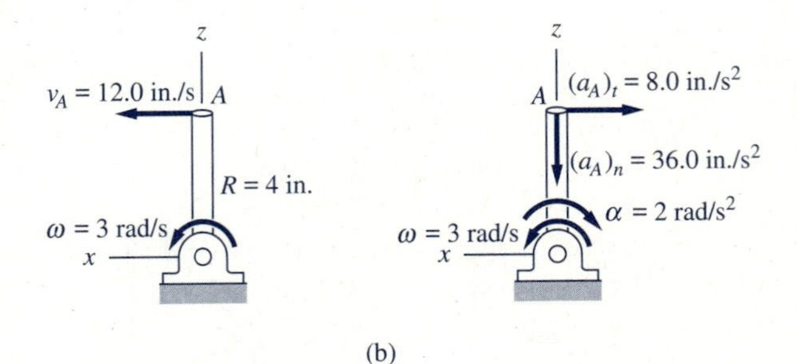

(b)

PROBLEMS

16.1 An electric motor is running at 5000 rev/min when the power is cut off. Determine the number of revolutions made by the motor as it coasts to a stop in 90 s with constant deceleration.

16.2 A carousel accelerates uniformly from 2 rev/min to 8 rev/min while turning through four revolutions. Find the time required for this change in angular speed.

16.3 A gyroscope that is mounted in air bearings is rotating at 20 000 rev/min. After the power source is disconnected, 8.32 min are required for the speed to be reduced by 50%. Assuming that the angular acceleration of the gyroscope is proportional to the negative of its angular speed, calculate the angular speed one hour after the power is cut off.

16.4 The motor of an industrial fan is programmed to produce the angular acceleration shown in the diagram when the fan is turned on at $t = 0$. Assuming that the fan is at rest when $t = 0$, determine (a) the operating speed of the fan; and (b) the number of revolutions required to reach the operating speed.

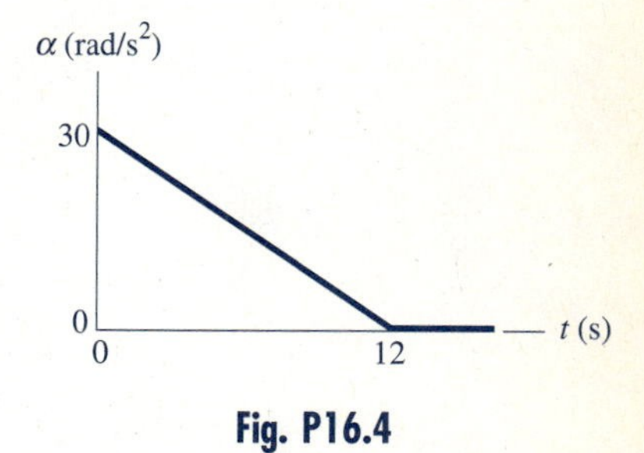

Fig. P16.4

16.5 The belt-driven hoist lifts the load A at the constant speed $v_A = 3$ ft/s. Find the angular speeds of pulley B and the electric motor C.

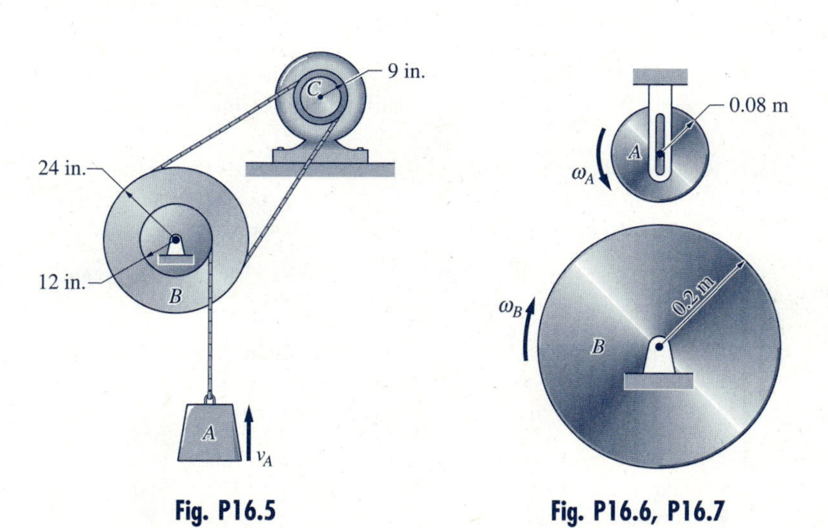

Fig. P16.5 **Fig. P16.6, P16.7**

16.6 Just before the two friction wheels are brought into contact, B is rotating clockwise at 18 rad/s and A is stationary. Slipping between the wheels takes place during the first 6 seconds of contact, during which the angular speed of each wheel changes uniformly. If the final angular speed of B is 12 rad/s, determine (a) the angular acceleration of A during the period of slipping; and (b) the number of revolutions made by A before it reaches its final speed.

16.7 The two friction wheels are rotating at 18 rad/s in the directions shown when they are brought together. Following the initial contact, there is a period of slipping during which A turns through six revolutions before reaching its final angular speed of 30 rad/s. Determine the constant angular acceleration of each wheel during the slipping period.

16.8 The crank ABC rotates about the fixed axis AB. At the instant shown, the angular velocity of the crank is $\omega = 9$ rad/s and is decreasing at the rate of 48 rad/s^2. For this instant, calculate the magnitudes of the velocity and acceleration of end C using (a) vector equations; and (b) scalar equations.

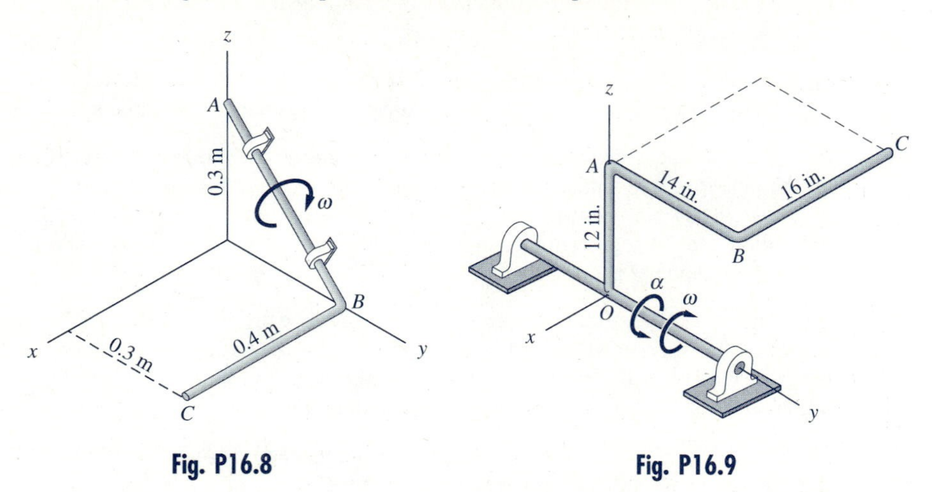

Fig. P16.8 **Fig. P16.9**

16.9 In the position shown, rod $OABC$ is rotating about the y-axis with the angular velocity $\omega = 2.4$ rad/s and angular acceleration $\alpha = 7.2$ rad/s^2 in the directions shown. For this position, compute the velocity and acceleration vectors of point C using (a) vector equations; and (b) scalar equations.

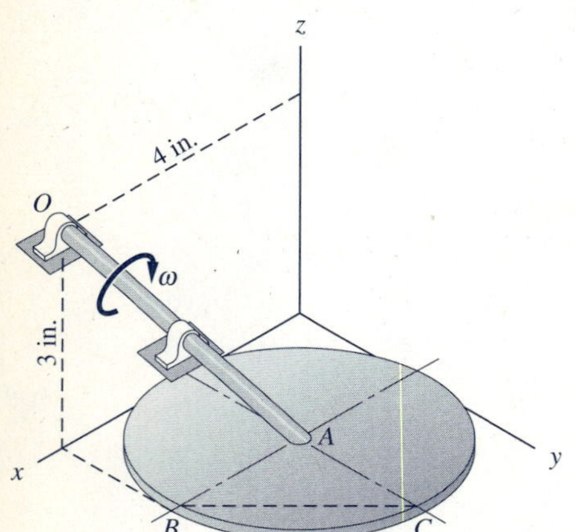

Fig. P16.10, P16.11

16.10 The circular plate is rigidly connected to the shaft OA. If OA is rotating with the constant angular speed $\omega = 16$ rad/s, determine the magnitudes of the velocity and acceleration of point B in the position shown.

16.11 For the circular plate rotating as described in Prob. 16.10, calculate (a) $\mathbf{v}_{C/B}$; (b) $\mathbf{a}_{C/B}$; and (c) component of $\mathbf{a}_{C/B}$ in the direction of the line BC.

16.12 The bent rod is rotating about the axis AC. In the position shown, the angular speed of the rod is $\omega = 4$ rad/s and is increasing at the rate of 9 rad/s^2. For this position, determine the velocity and acceleration vectors at point B.

16.13 The rigid body is rotating about a fixed axis that is perpendicular to the xy-plane. At a certain instant, the velocities of points A and B in the body are as shown in the figure, where v_B is the unknown speed of point B. For this instant, compute (a) v_B; and (b) the angular speed of the body.

16.14 Find the angular velocity vector of the triangular plate at the instant when its corners have the velocities shown.

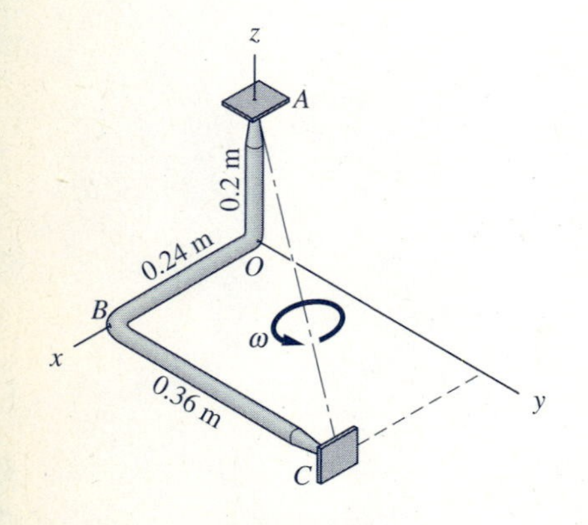

Fig. P16.12

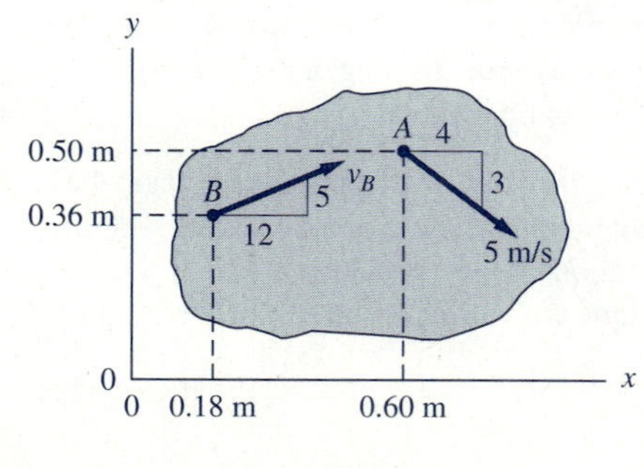

Fig. P16.13

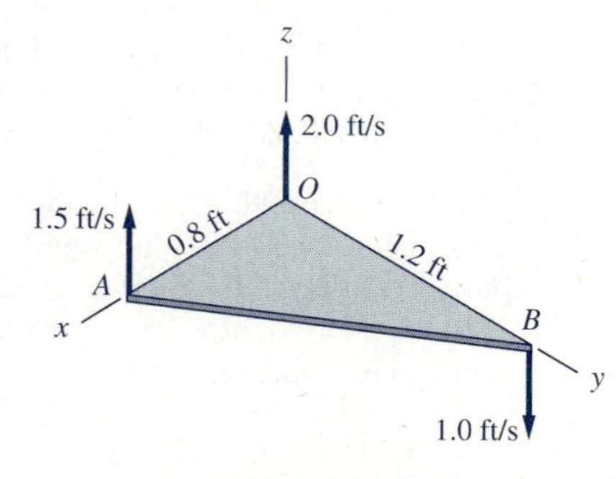

Fig. P16.14

16.15 The rigid body is rotating about a fixed axis. At the instant shown, the velocity vectors of points A and B in the body are $\mathbf{v}_A = 6\mathbf{i} + 10\mathbf{k}$ ft/s and $\mathbf{v}_B = -3\mathbf{i} + 4\mathbf{j} - w\mathbf{k}$ ft/s, respectively. What is the velocity component w at this instant?

16.16 The take-up reel of a tape recorder rotates at the constant angular speed $\omega_0 = 2000$ rev/min. If the thickness of the tape is $h = 1/2000$ in., determine the acceleration a_T of the straight portion of the tape.

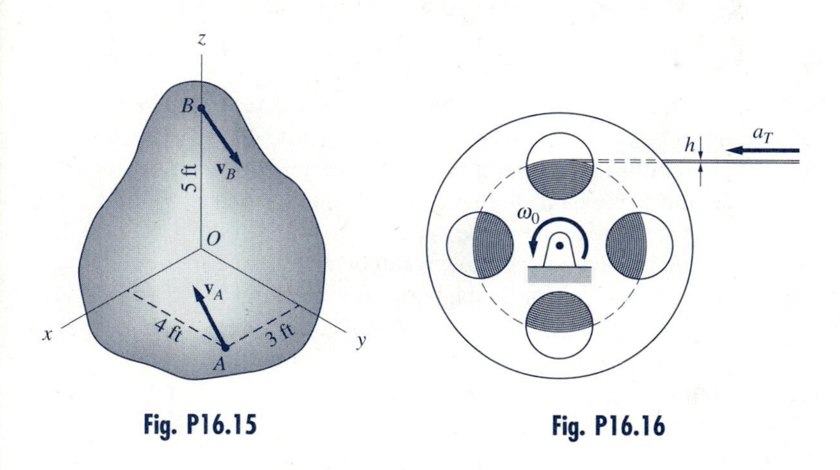

Fig. P16.15 **Fig. P16.16**

16.3 Kinematics of Plane Motion

Plane motion of a rigid body refers to the special case in which all points in the body remain a constant distance from a fixed reference plane. In kinematics, the reference plane, or any parallel plane, is called the plane of the motion. (In kinetics, the plane of the motion is the plane in which the center of gravity moves.) The various types of plane motion are described below.

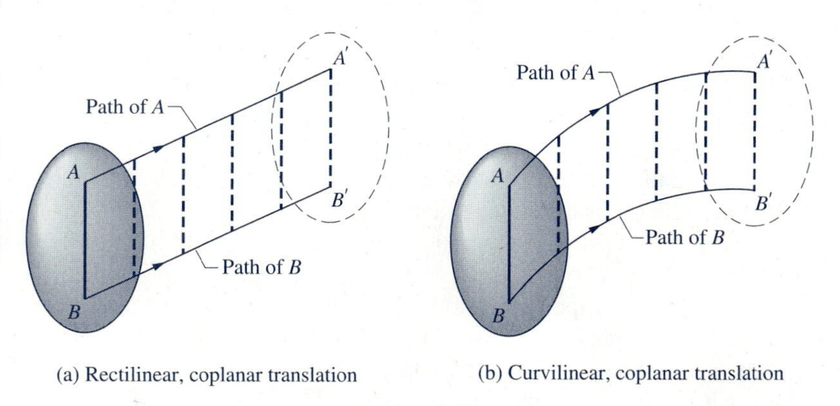

(a) Rectilinear, coplanar translation (b) Curvilinear, coplanar translation

Fig. 16.5

Coplanar translation Coplanar translation refers to the special case of plane motion in which the body does not rotate. Therefore, all lines in the body remain parallel to their initial positions. Figure 16.5 shows two rigid bodies undergoing rectilinear and curvilinear coplanar translation, respectively.* In rectilinear

* When illustrating plane motion, it is common practice to show a representative cross section of the body that is parallel to the plane of motion. However, do not lose sight of the fact that the body has three dimensions.

translation, Fig. 16.5(a), all points of the body follow parallel, straight-line paths. In curvilinear translation, Fig. 16.5(b), the points of the body follow curved paths. In both cases, however, every line in the body (such as $A'B'$) remains parallel to its initial position (such as AB).

Note that during translation the motion of any one point of the body determines the motion of all points. Therefore, the kinematics of coplanar translation reduces to the kinematics of a single point, a topic that has been discussed at length in previous chapters on particle motion.

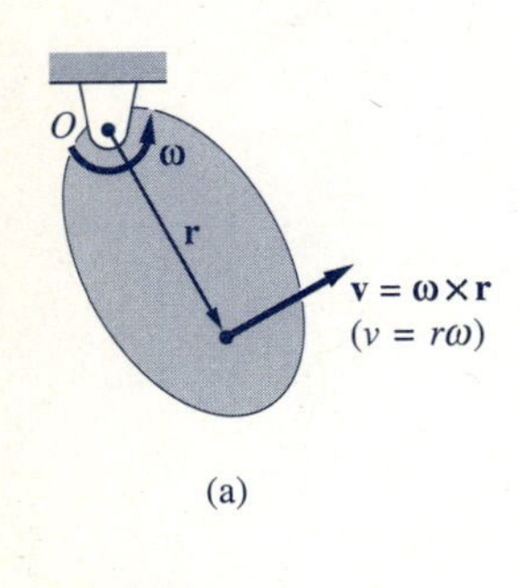

(a)

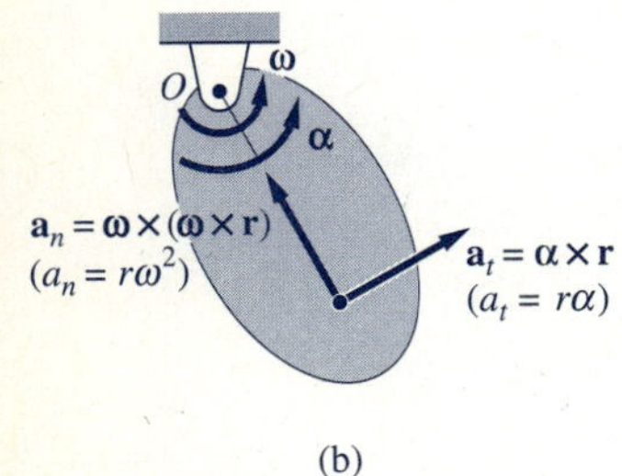

(b)

Fig. 16.6

Rotation about a fixed axis The special case of plane motion in which a rigid body is rotating about a fixed axis is depicted in Fig. 16.6. Although the body appears to be rotating about point O, it must be remembered that the rotation takes place about a fixed axis that passes through O and is perpendicular to the plane of the motion (the plane of the paper is the plane of the motion). Utilizing the results of the preceding article, we can determine the motion of any point in terms of the angular motion of the body, as shown in Fig. 16.6. Observe that the converse is also true: If the velocity and acceleration vectors of any one point not on the axis of rotation are known, the angular velocity and angular acceleration of the body can be computed. As was the case in translation, the motion of one point thus determines the motion of all points of the body.

General plane motion General plane motion refers to the plane motion of a rigid body that is neither pure translation nor pure rotation. As an illustration of general plane motion, Fig. 16.7(a) shows a disk traveling in the plane of the paper, where $\mathbf{v}_A$ is the velocity of its center A and ω is the angular velocity of the disk. Here the motion of one point does *not* necessarily determine the motion of the body. This is seen from Eq. (16.1): $\mathbf{v}_{B/A} = \mathbf{v}_B - \mathbf{v}_A = \boldsymbol{\omega} \times \mathbf{r}_{B/A}$. Clearly, both $\mathbf{v}_A$ and $\boldsymbol{\omega}$ must be known in order to compute $\mathbf{v}_B$.

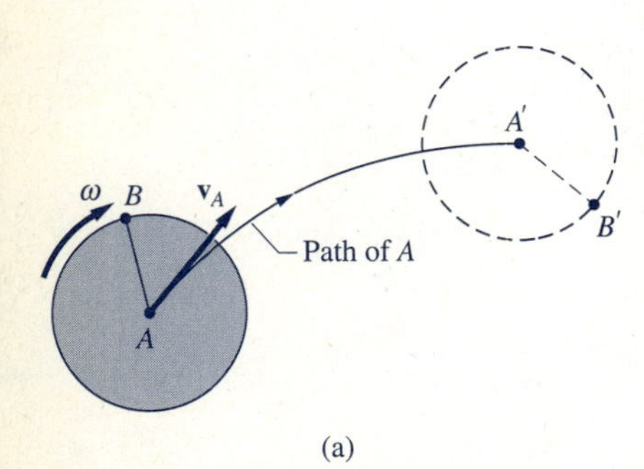

(a)

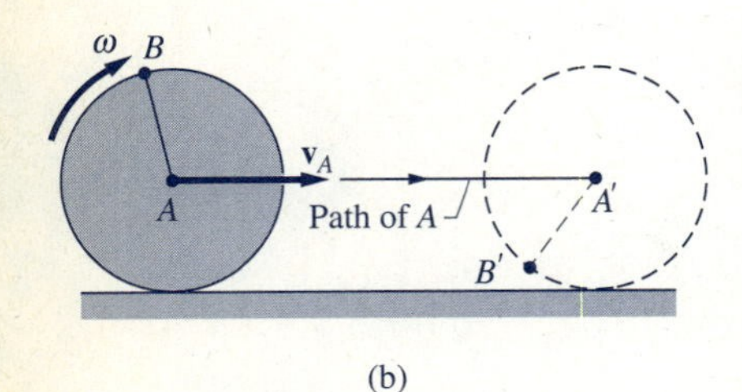

(b)

Fig. 16.7

Constrained plane motion In many instances, the plane motion of a rigid body—or a system of connected rigid bodies—is subject to kinematic constraints that restrict its motion. A disk that rolls on a horizontal surface without slipping, as illustrated in Fig. 16.7(b), is an example of constrained plane motion. As explained in the next article, the no-slip condition imposes a relationship (kinematic constraint) between v_A and ω, so that these two variables are no longer kinematically independent.

Our earlier discussion of kinematics of constrained motion for systems of particles is also applicable to rigid bodies (or systems of rigid bodies). This discussion will not be repeated here. It is recommended that the reader review the following terms presented in Art. 15.3: kinematic constraints, equations of constraint, independent coordinates, and degrees of freedom.

The two methods of analyzing constrained plane motion of rigid bodies are (1) differentiation of the equations of constraint, and (2) relative motion analysis. The first method is useful for determining velocities and accelerations for an arbitrary position of the body or system; the second method is best suited for analyzing the motion at a specific position. The following sample problems illustrate the method of differentiation of the equations of constraint. Relative motion analysis is the subject of the remaining articles of this chapter.

Sample Problem 16.3

The position of the bar AB is controlled by the horizontal rod CD. If the velocity of CD is v_0 (a constant) directed to the left, derive the equations for the angular velocity and angular acceleration of AB in terms of the angle θ. Solve by differentiating the appropriate equation of constraint.

Solution

This system has a single degree of freedom. A convenient choice for the generalized coordinate is the angular position θ of bar AB. From the figure we see that the equation of constraint that relates the position of CD to θ is

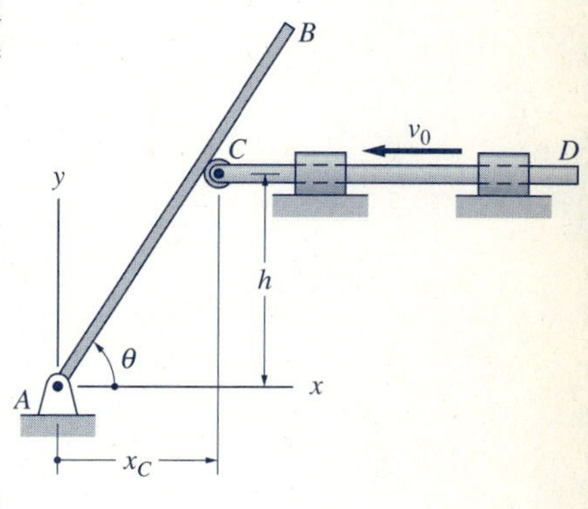

$$x_C = h \cot \theta$$

Differentiation with respect to time yields

$$\dot{x}_C = -h\dot{\theta} \operatorname{cosec}^2 \theta$$

Recognizing that $\dot{x}_C = -v_0$, the angular velocity of bar AB becomes

$$\omega = \dot{\theta} = \frac{v_0}{h} \sin^2 \theta \qquad \textit{Answer} \quad \text{(a)}$$

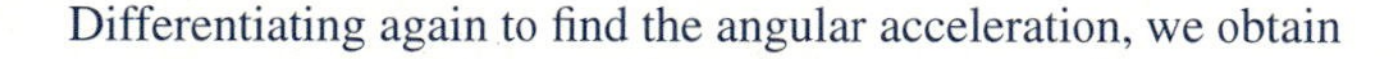

Differentiating again to find the angular acceleration, we obtain

$$\alpha = \dot{\omega} = \frac{v_0}{h} 2 \sin\theta \cos\theta \cdot \dot{\theta}$$

Substituting $\dot{\theta}$ from Eq. (a) yields

$$\alpha = \frac{v_0}{h} 2 \sin\theta \cos\theta \left(\frac{v_0}{h} \sin^2 \theta\right)$$

which becomes

$$\alpha = \frac{2v_0^2}{h^2} \sin^3 \theta \cos \theta \qquad \textit{Answer} \quad \text{(b)}$$

Sample Problem 16.4

When $\theta = 30°$, the angular velocity of bar AB is 3 rad/s clockwise, and its angular acceleration is zero. By differentiating the equation of constraint, compute the following when $\theta = 30°$: (1) the angular velocity and angular acceleration of bar BC; and (2) the velocity and acceleration of the slider C.

Solution

Preliminary Calculations

We note that the mechanism has a single degree of freedom, with θ being a convenient choice for the generalized coordinate. Referring to the figure, we see that

$$\omega_{AB} = \dot{\theta} \qquad \alpha_{AB} = \ddot{\theta} \quad \text{(positive counterclockwise)} \qquad \text{(a)}$$

and

$$\omega_{BC} = \dot{\beta} \qquad \alpha_{BC} = \ddot{\beta} \quad \text{(positive clockwise)} \tag{b}$$

Projecting the lengths of AB and BC onto the y-axis, we see that the equation of constraint that relates the angles θ and β is

$$8 \sin \beta = 5 \sin \theta + 4 \tag{c}$$

Differentiation of both sides of this equation with respect to time gives

$$8 \cos \beta \cdot \dot{\beta} = 5 \cos \theta \cdot \dot{\theta} \tag{d}$$

or

$$\dot{\beta} = \frac{5 \cos \theta \cdot \dot{\theta}}{8 \cos \beta} \tag{e}$$

Differentiating Eq. (d) with respect to time yields

$$8(\cos \beta \cdot \ddot{\beta} - \sin \beta \cdot \dot{\beta}^2) = 5(\cos \theta \cdot \ddot{\theta} - \sin \theta \cdot \dot{\theta}^2)$$

or

$$\ddot{\beta} = \frac{5(\cos \theta \cdot \ddot{\theta} - \sin \theta \cdot \dot{\theta}^2) + 8 \sin \beta \cdot \dot{\beta}^2}{8 \cos \beta} \tag{f}$$

According to the figure, the x-coordinate of C is given by

$$x_C = 5 \cos \theta + 8 \cos \beta \tag{g}$$

Since the path of C is rectilinear, the velocity of C is

$$v_C = \dot{x}_C = -5 \sin \theta \cdot \dot{\theta} - 8 \sin \beta \cdot \dot{\beta} \tag{h}$$

and its acceleration is

$$a_C = \dot{v}_C = -5(\sin \theta \cdot \ddot{\theta} + \cos \theta \cdot \dot{\theta}^2) - 8(\sin \beta \cdot \ddot{\beta} + \cos \beta \cdot \dot{\beta}^2) \tag{i}$$

Part 1

When $\theta = 30°$, then $\omega_{AB} = \dot{\theta} = -3$ rad/s, $\alpha_{AB} = \ddot{\theta} = 0$, and the angle β from Eq. (c) is

$$\beta = \sin^{-1} \frac{5 \sin \theta + 4}{8} = \sin^{-1} \frac{5 \sin 30° + 4}{8} = 54.34° \tag{j}$$

Substituting the known values into Eq. (e), the angular velocity of BC becomes

$$\omega_{BC} = \dot{\beta} = \frac{5 \cos 30° \cdot (-3)}{8 \cos 54.34°} = -2.785 \text{ rad/s} \qquad \textit{Answer} \tag{k}$$

The negative sign means that the sense of ω_{BC} is counterclockwise.

Similarly, the angular acceleration of BC is found from Eq. (f) to be

$$\alpha_{BC} = \ddot{\beta} = \frac{5[(0 - \sin 30° \cdot (-3)^2] + 8 \sin 54.34° \cdot (-2.785)^2}{8 \cos 54.34°}$$

$$= 5.985 \text{ rad/s}^2 \qquad \textit{Answer} \quad (1)$$

Since α_{BC} is positive, its direction is clockwise.

Part 2

Substituting the values computed above into Eq. (h), the velocity of C is

$$v_C = -5 \sin 30° \cdot (-3) - 8 \sin 54.34° \cdot (-2.785)$$

$$= 25.60 \text{ in./s} \qquad \textit{Answer}$$

From Eq. (i) the acceleration of C is

$$a_C = -5[0 + \cos 30° \cdot (-3)^2] - 8[\sin 54.34° \cdot (5.985) + \cos 54.34° \cdot (-2.785)^2]$$

$$= -114.0 \text{ in./s}^2 \qquad \textit{Answer}$$

The signs indicate that the velocity of C is directed to the right and that its acceleration is directed to the left.

PROBLEMS

16.17 Characterize the motion of bodies A and B of each mechanism shown as (1) translation, (2) rotation about a fixed axis, or (3) general plane motion.

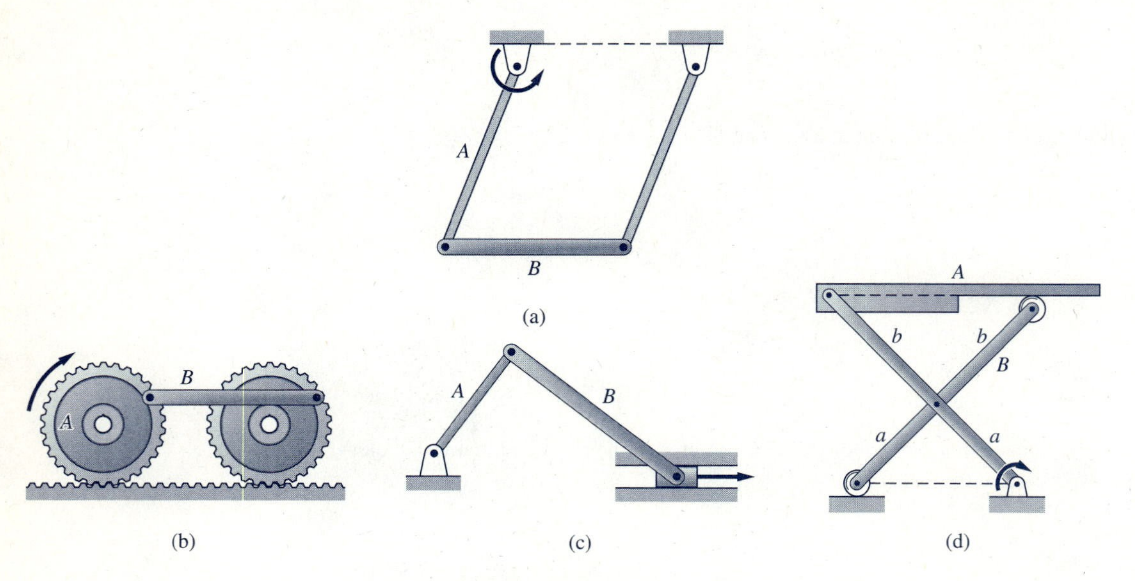

Fig. P16.17

Note: Problems 16.18–16.32 are to be solved by differentiating the equation of constraint.

16.18 When the crank AB of the scotch yoke is in the position $\theta = 50°$, its angular velocity is 12 rad/s, and its angular acceleration is 180 rad/s^2, both clockwise. Calculate the velocity and acceleration of the sliding rod D in this position.

16.19 The arm of the robot consists of two links OA and AB. When $\theta_1 = 30°$ and $\theta_2 = 45°$, the angular velocities and accelerations of the links are $\dot{\theta}_1 = 0$, $\dot{\theta}_2 = -1.2$ rad/s, $\ddot{\theta}_1 = 0.8$ rad/s^2, and $\ddot{\theta}_2 = 0$. Determine the velocity and acceleration vectors of point B in that position.

16.20 The link OA of the robot's arm is rotating clockwise with the constant angular velocity $\dot{\theta}_1 = 0.8$ rad/s. At the same time, end B of arm AB is tracing the vertical line $x = 0.8$ m. Determine the angular velocity and acceleration of link AB when $\theta_1 = 30°$. Assume that $\theta_2 < 90°$.

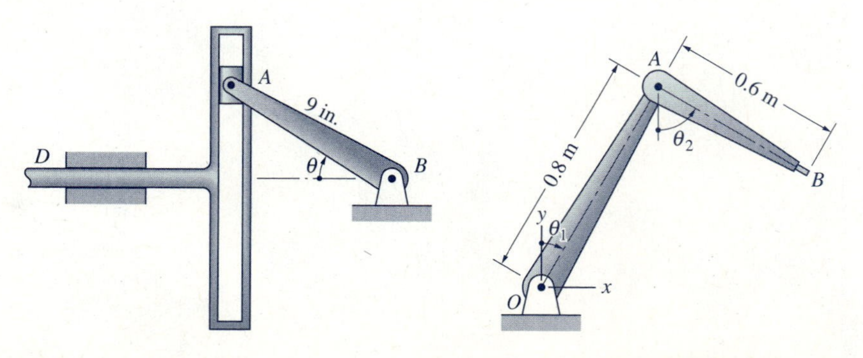

Fig. P16.18 **Fig. P16.19, P16.20**

16.21 The rod AD is sliding in the fixed collar E at the constant speed v_0. Find (a) the angular velocity; and (b) the angular acceleration of bar AB as functions of the angle θ.

16.22 The rod AD is sliding in the fixed collar E at the constant speed v_0. Determine (a) the angular velocity; and (b) the angular acceleration of bar AB as functions of the angle θ. Note that the axis of the disk is attached to the collar E.

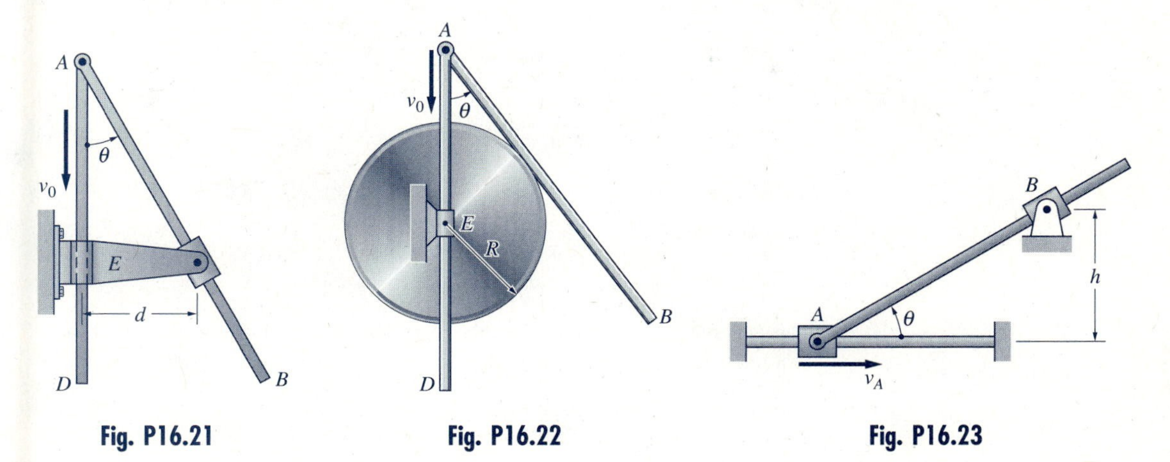

Fig. P16.21 **Fig. P16.22** **Fig. P16.23**

16.23 The slider A is moving to the right with the constant speed v_A. Determine (a) the angular velocity; and (b) the angular acceleration of bar AB in terms of the angle θ.

16.24 The circular cam has a radius of R and an eccentricity $e = R$. The follower A is kept in contact with the surface of the cam by a compression spring. Assuming that the cam starts from rest at $\theta = 0$ and accelerates at the constant rate $\ddot{\theta} = \alpha_0$, find the acceleration of the follower as a function of θ.

16.25 The radius of the circular cam is $R = 100$ mm and its eccentricity is $e = 60$ mm. If the angular speed of the cam is 1000 rev/min, calculate the velocity of the follower A when $\theta = 60°$.

16.26 For the linkage shown, find the ratio $\dot{\theta}_1/\dot{\theta}_2$ in the position $\theta_1 = \theta_2 = 0$.

16.27 The two ropes supporting the bar ABC are rotating at the constant angular speed $\dot{\theta}$. Determine the magnitude and direction of (a) the velocity of point B; and (b) the acceleration of point B.

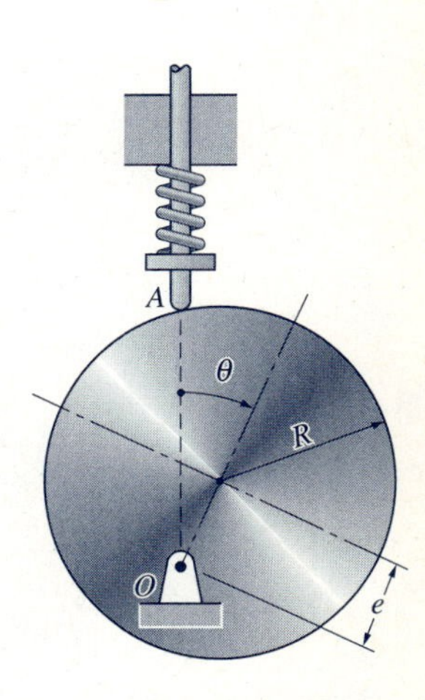

Fig. P16.24, P16.25

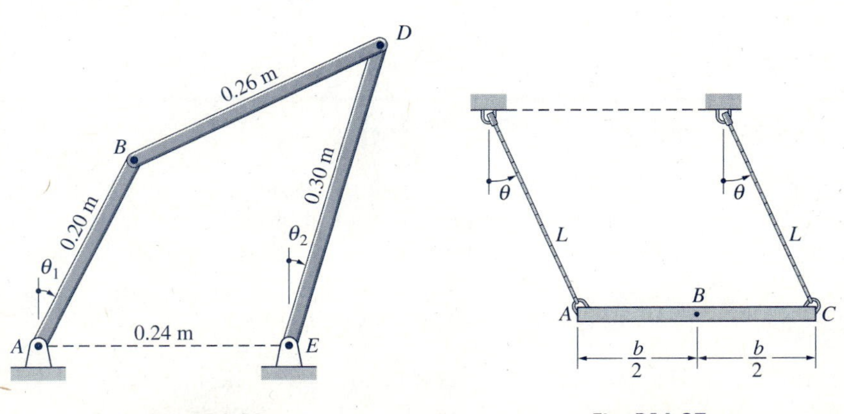

Fig. P16.26 **Fig. P16.27**

16.28 The crank OB rotates at the constant clockwise angular speed ω_0. (a) Derive an expression for the velocity v_A of the slider as a function of the crank angle θ. (b) From a plot of the expression found in part (a), estimate the largest value of v_A and the corresponding value of θ.

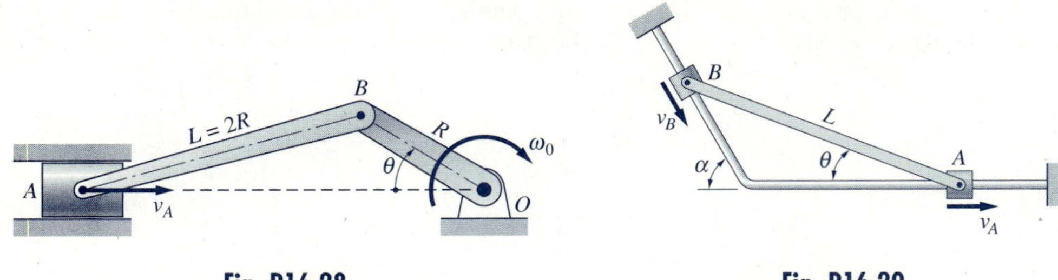

Fig. P16.28

Fig. P16.29

16.29 If the velocity of slider A is constant, derive the expressions for (a) the velocity; and (b) the acceleration of slider B in terms of the angle θ.

16.30 The eccentrically mounted circular cam has a constant angular speed $\dot{\beta} = 8$ rad/s. Calculate the angular speed of the follower rod AB when $\theta = 30°$.

16.31 The pendulum AB is suspended from the sliding collar A. Determine the expression for the speed of the bob B in terms of v_A (the velocity of the collar), θ, and $\dot{\theta}$.

16.32 For the double pendulum shown, derive the expression for the speed of bob B in terms of L, θ_1, θ_2, $\dot{\theta}_1$, and $\dot{\theta}_2$.

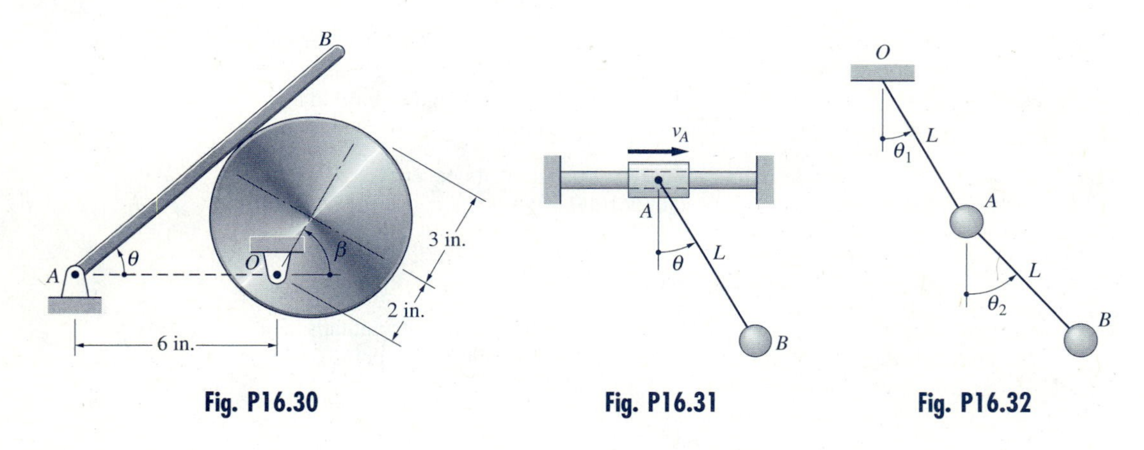

Fig. P16.30

Fig. P16.31

Fig. P16.32

16.4 Method of Relative Velocity

Figure 16.8(a) shows a rigid body undergoing general plane motion. If A and B are two points on the body, then according to Eq. (16.1) the velocity of B relative to A is

$$\mathbf{v}_{B/A} = \boldsymbol{\omega} \times \mathbf{r}_{B/A} \qquad (v_{B/A} = r_{B/A}\omega) \tag{16.8}$$

where $\boldsymbol{\omega}$ is the angular velocity vector of the body. The magnitude of $\mathbf{v}_{B/A}$ given in parentheses follows from the fact that $\boldsymbol{\omega}$ is perpendicular to $\mathbf{r}_{B/A}$.

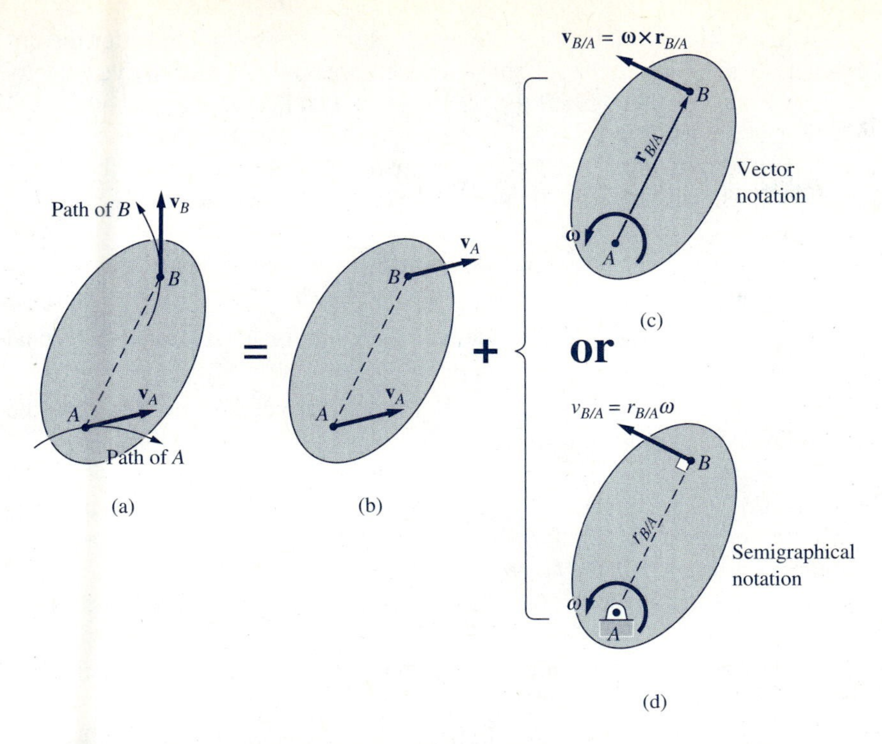

Plane motion = Translation + Rotation about A

Fig. 16.8

Since by definition $\mathbf{v}_{B/A} = \mathbf{v}_B - \mathbf{v}_A$, we can write

$$\boxed{\mathbf{v}_B = \mathbf{v}_A + \boldsymbol{\omega} \times \mathbf{r}_{B/A}} \tag{16.9}$$

Equation (16.9) is illustrated in Fig. 16.8. Note that plane motion is equivalent to a rigid-body translation with the velocity of all points equal to the velocity of reference point A, as seen in Fig. 16.8(b), plus a rigid-body rotation about reference point A, shown in Fig. 16.8(c). (The rotation is actually about an axis that passes through A and is perpendicular to the plane of the motion.) Remember that angular velocity is a property of the body that does not depend on the choice of reference point.

The relative velocity vector, i.e., the term due to the rotation, may be computed using either the vector notation of Fig. 16.8(c) or the semigraphical notation shown in Fig. 16.8(d). The choice of the notation is a matter of personal preference. In vector notation, the vector $\mathbf{v}_{B/A}$ is computed from the cross product $\boldsymbol{\omega} \times \mathbf{r}_{B/A}$. When semigraphical notation is used, the magnitude of $\mathbf{v}_{B/A}$ is calculated using $r_{B/A}\omega$, and its direction is found by considering the reference point A to be fixed at this instant. The sense of the vector $\mathbf{v}_{B/A}$ is then found from the direction of the angular velocity.

The relative velocity $\mathbf{v}_{B/A}$ can be interpreted as the velocity of B as it would appear to an observer who is translating with A, but not rotating. (Rotating reference frames are discussed in Art. 16.8.) The path of B relative to this nonrotating observer is a circle of radius $r_{B/A}$, with $\mathbf{v}_{B/A}$ being tangent to the circle.

The method of relative velocity consists of solving Eq. (16.9) for the unknowns. In applying Eq. (16.9), it is important to recall that the relative velocity term $\boldsymbol{\omega} \times \mathbf{r}_{B/A}$ is valid only if the two points are on the same rigid body.

For plane motion, Eq. (16.9) is equivalent to two scalar equations (e.g., equating the horizontal and vertical components of both sides of the vector equation). The number of variables is five (assuming that $\mathbf{r}_{B/A}$ is given):

$\mathbf{v}_B$: two variables (magnitude and direction, or horizontal and vertical components).

$\mathbf{v}_A$: two variables (magnitude and direction, or horizontal and vertical components).

$\mathbf{v}_{B/A}$: one variable (magnitude, or the angular velocity ω). Note that the direction of $\mathbf{v}_{B/A}$is known, since it is always perpendicular to $\mathbf{r}_{B/A}$.

Clearly, Eq. (16.9) cannot be solved unless three of the above five variables are known beforehand. A point for which the magnitude or direction of the velocity vector is known is called a *kinematically important point* (for velocity).

The steps in the application of the relative velocity method are as follows.

Step 1: Identify two kinematically important points on the same rigid body.
Step 2: Write the relative velocity equation for the chosen points, identifying the unknown variables (either vector or semigraphical notation can be used).
Step 3: If the number of unknowns is two, solve the relative velocity equation.

If the number of unknowns is greater than two, it may still be possible to solve the problem by considering the motion of other kinematically important points.

Rolling without slipping

Figure 16.9(a) shows a circular disk of radius R that is rolling on a horizontal surface with angular velocity ω and angular acceleration α, both clockwise. Observe that the path of the center O is a straight line parallel to the surface. Rolling without slipping occurs if the contact point C on the disk has no veloc-

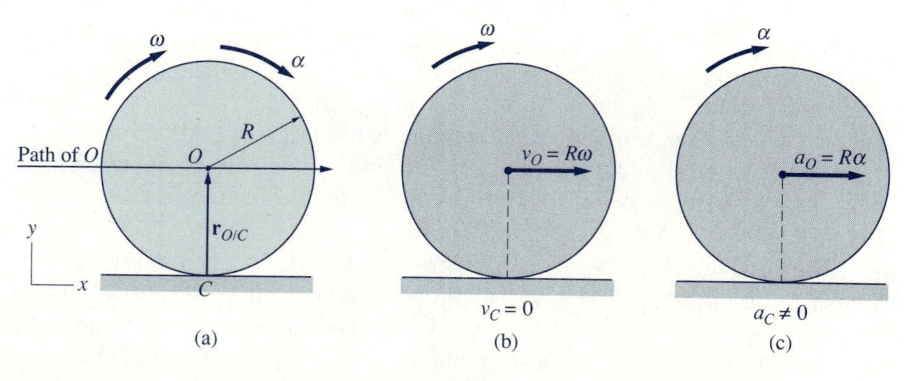

Fig. 16.9

ity; i.e., the disk does not slide along the surface. This case deserves special attention because it occurs in many engineering applications.

Applying the relative velocity equation, Eq. (16.9), to points C and O, where O is the center of the disk, we have

$$\mathbf{v}_O = \mathbf{v}_C + \boldsymbol{\omega} \times \mathbf{r}_{O/C}$$

Substituting $\mathbf{v}_C = \mathbf{0}$, $\boldsymbol{\omega} = -\omega\mathbf{k}$, and $\mathbf{r}_{O/C} = R\mathbf{j}$, we get

$$\mathbf{v}_O = (-\omega\mathbf{k}) \times (R\mathbf{j}) = R\omega\mathbf{i} \qquad (16.10a)$$

As expected, this result shows that the velocity of the center O is parallel to the surface on which the disk rolls, its magnitude being

$$\boxed{v_O = R\omega} \qquad (16.10b)$$

as shown in Fig. 16.9(b).

It is convenient here to derive the acceleration of O, although this information is not used until Art. 16.6. The acceleration of O can be obtained by differentiation of Eq. (16.10a). Noting that R and $\mathbf{i}$ are constants, we get

$$\mathbf{a}_O = \dot{\mathbf{v}}_O = R\alpha\mathbf{i} \qquad (16.11a)$$

Thus the acceleration of O is also parallel to the surface of rolling, and its magnitude is

$$\boxed{a_O = R\alpha} \qquad (16.11b)$$

as shown in Fig. 16.9(c). It should be noted that although the velocity of C is zero, its acceleration is *not zero.*

Sample Problem 16.5

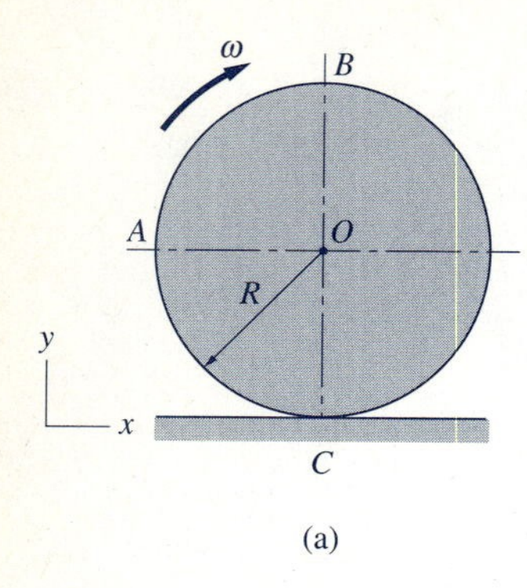

(a)

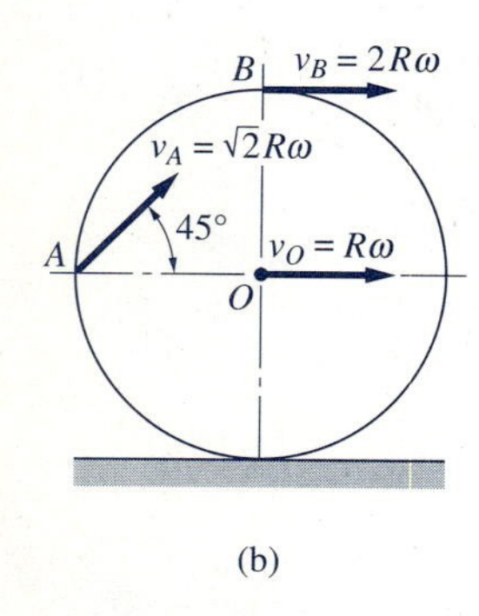

(b)

Figure (a) shows a wheel of radius R that is rolling without slipping with the clockwise angular velocity ω. For the position shown, determine the velocity vectors for (1) point A; and (2) point B.

Solution

Introductory Comments

This problem will be solved using both semigraphical and vector notations. (The results are, of course, identical regardless of which notation is used.) We choose point O (the center of the wheel) as the reference point, since its velocity is known from Eq. (16.10) to be $\mathbf{v}_O = R\omega\mathbf{i}$, as shown in Fig. (b). The reader may find it instructive to repeat the solution using the point of contact C ($\mathbf{v}_C = \mathbf{0}$) as the reference point.

Solution I (using semigraphical notation)

Part 1

When semigraphical notation is used to write the relative velocity equation between A and O, $\mathbf{v}_{A/O}$ is computed by assuming that point O is fixed. Therefore, the equation becomes

$$\mathbf{v}_A = \mathbf{v}_O + \mathbf{v}_{A/O}$$

$R\omega$ (→) $R\omega$ (↑ at A, with R and ω about fixed O)

from which the velocity of A is found to be

$$\mathbf{v}_A = \sqrt{2}R\omega \ \angle 45^\circ \qquad \textit{Answer}$$

The velocity of A is shown in Fig. (b).

Part 2

Using semigraphical notation, the relative velocity equation between points B and O is

$$\mathbf{v}_B = \mathbf{v}_O + \mathbf{v}_{B/O}$$

$R\omega$ (→) $R\omega$ (→ at B, with R and ω about fixed O)

which yields

$$\mathbf{v}_B = 2R\omega \rightarrow \qquad \textit{Answer}$$

The velocity of B is also shown in Fig. (b).

Solution II (using vector notation)

Part 1

When vector notation is used, the relative velocity equation between A and O is

$$\mathbf{v}_A = \mathbf{v}_O + \boldsymbol{\omega} \times \mathbf{r}_{A/O}$$

Substituting $\mathbf{v}_O = R\omega\mathbf{i}$, $\boldsymbol{\omega} = -\omega\mathbf{k}$, and $\mathbf{r}_{A/O} = -R\mathbf{i}$ —Fig. (c)—we get

$$\begin{aligned}\mathbf{v}_A &= R\omega\mathbf{i} + (-\omega\mathbf{k}) \times (-R\mathbf{i}) \\ &= R\omega\mathbf{i} + R\omega\mathbf{j} \qquad \textit{Answer}\end{aligned}$$

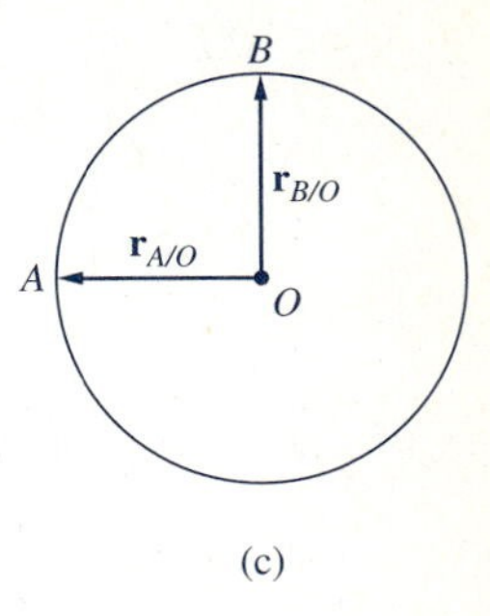

(c)

Part 2

The relative velocity equation between B and O is

$$\mathbf{v}_B = \mathbf{v}_O + \boldsymbol{\omega} \times \mathbf{r}_{B/O}$$

Substituting $\mathbf{v}_O = R\omega\mathbf{i}$, $\boldsymbol{\omega} = -\omega\mathbf{k}$, and $\mathbf{r}_{B/O} = R\mathbf{j}$ —Fig. (c)—we obtain

$$\begin{aligned}\mathbf{v}_B &= R\omega\mathbf{i} + (-\omega\mathbf{k}) \times (R\mathbf{j}) \\ &= 2R\omega\mathbf{i} \qquad \textit{Answer}\end{aligned}$$

Sample Problem 16.6

The angular velocity of bar AB in Fig. (a) is 3 rad/s clockwise in the position shown. Determine the angular velocity of bar BC and the velocity of the slider C in the same position. (Note: This problem was solved in Sample Problem 16.4 by differentiating the equation of constraint.)

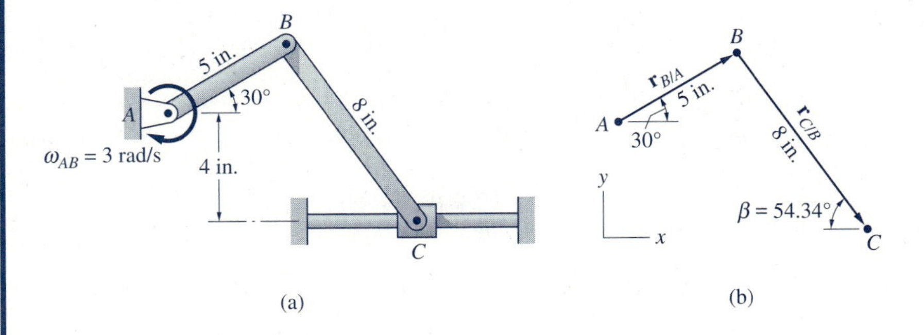

(a) (b)

Solution

Introductory Comments

We note from the geometry in Fig. (b) that $5 \sin 30° + 4 = 8 \sin \beta$, from which we find that $\beta = 54.34°$ for the position shown.

Following are the kinematically important points.

A: It is a fixed point.

C: Its path is a horizontal straight line.

B: Its path is a circle with center at A, and it connects bars AB and BC.

Because B and C are kinematically important points of the same rigid bar BC, it seems reasonable to investigate the relative velocity equation $\mathbf{v}_C = \mathbf{v}_B + \mathbf{v}_{C/B}$ (the equivalent equation $\mathbf{v}_B = \mathbf{v}_C + \mathbf{v}_{B/C}$ could also be used).

Two solutions are presented below—one using semigraphical notation, the other employing vector notation. The relative position vectors used in the solutions are shown in Fig. (b).

Solution I (using semigraphical notation)

Using semigraphical notation, the relative velocity equation for B and C is

$$\mathbf{v}_C = \mathbf{v}_B + \mathbf{v}_{C/B} \qquad \text{(a)}$$

Comments on Eq. (a):

1. $\mathbf{v}_C$ is assumed to be directed to the right.
2. $\mathbf{v}_B$ is found by recognizing that the path of B is a circle centered at A. Its magnitude is thus $v_B = r_{B/A}\omega_{AB} = 5(3) = 15$ in./s, directed at right angles to AB. The sense of $\mathbf{v}_B$ is determined by the given direction of ω_{AB}.
3. $\mathbf{v}_{C/B}$ is obtained by considering B to be a fixed point at this instant, which gives $v_{C/B} = r_{C/B}\omega_{BC} = 8\omega_{BC}$. The direction of the vector is perpendicular to BC, with its sense determined by the (assumed) counterclockwise direction of ω_{BC}.

Since there are a total of two unknowns (v_C and ω_{BC}), Eq. (a) can be solved. Equating x- and y-components, we obtain

$$\xrightarrow{+} \quad v_C = 15\sin 30^\circ + 8\omega_{BC}\sin 54.34^\circ \qquad \text{(b)}$$

$$+\uparrow \quad 0 = -15\cos 30^\circ + 8\omega_{BC}\cos 54.34^\circ \qquad \text{(c)}$$

The solution is

$$v_C = 25.60 \text{ in./s} \quad \text{and} \quad \omega_{BC} = 2.785 \text{ rad/s} \qquad \textit{Answer}$$

The positive signs mean that the assumed directions of v_C and ω_{BC} are correct.

Solution II (using vector notation)

The relative velocity equation for points B and C in vector notation is

$$\begin{aligned}\mathbf{v}_C &= \mathbf{v}_B + \mathbf{v}_{C/B} \\ &= \boldsymbol{\omega}_{AB} \times \mathbf{r}_{B/A} + \boldsymbol{\omega}_{BC} \times \mathbf{r}_{C/B} \qquad \text{(d)}\end{aligned}$$

From the given information and inspection of Fig. (b), we obtain

$$\begin{aligned}\mathbf{v}_C &= v_C\mathbf{i} \quad \text{(assuming the velocity of } C \text{ to be directed to the right)} \\ \boldsymbol{\omega}_{AB} &= -3\mathbf{k} \text{ rad/s} \\ \mathbf{r}_{B/A} &= 5\cos 30°\mathbf{i} + 5\sin 30°\mathbf{j} \\ &= 4.330\mathbf{i} + 2.500\mathbf{j} \text{ in.}\end{aligned}$$

$$\begin{aligned}\boldsymbol{\omega}_{BC} &= \omega_{BC}\mathbf{k} \quad \text{(assuming the direction of } \omega_{BC} \text{ to be counterclockwise)} \\ \mathbf{r}_{C/B} &= 8\cos 54.34°\mathbf{i} - 8\sin 54.34°\mathbf{j} \\ &= 4.664\mathbf{i} - 6.500\mathbf{j} \text{ in.}\end{aligned}$$

Inspection of these expressions reveals that there are only two unknowns: v_C and ω_{BC}. Therefore, Eq. (d) can be solved, since it is equivalent to two scalar equations.

Substituting the above expressions into Eq. (d) yields

$$\begin{aligned}v_C\mathbf{i} &= (-3\mathbf{k}) \times (4.330\mathbf{i} + 2.500\mathbf{j}) \\ &\quad + (\omega_{BC}\mathbf{k}) \times (4.664\mathbf{i} - 6.500\mathbf{j}) \\ &= -12.990\mathbf{j} + 7.500\mathbf{i} + 4.664\omega_{BC}\mathbf{j} + 6.500\omega_{BC}\mathbf{i}\end{aligned}$$

Equating the coefficients of **i** and **j** yields

$$\begin{aligned}v_C &= 7.500 + 6.500\omega_{BC} \\ 0 &= -12.990 + 4.664\omega_{BC}\end{aligned}$$

Solving the two equations, we find that

$$v_C = 25.60 \text{ in./s} \quad \text{and} \quad \omega_{BC} = 2.785 \text{ rad/s} \qquad \textit{Answer}$$

The positive signs indicate that the assumed directions of $\mathbf{v}_C$ and $\boldsymbol{\omega}_{BC}$ are correct.

Sample Problem 16.7

In the position shown in Fig. (a), the angular velocity of bar AB is 2 rad/s clockwise. Calculate the angular velocities of bars BC and CD for this position.

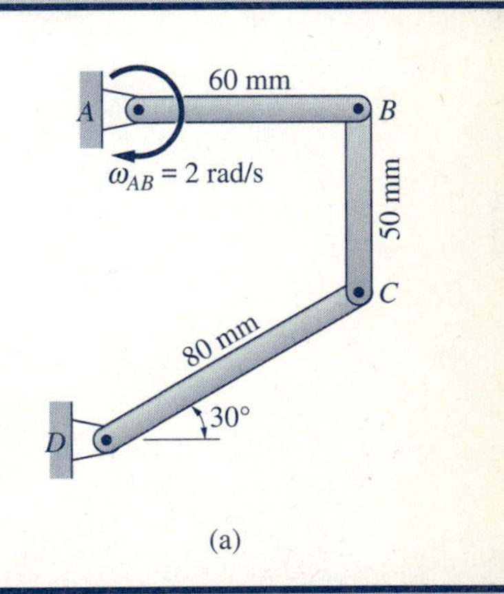

(a)

Solution

Introductory Comments

A mechanism of the type shown in Fig. (a) is called a *four-bar linkage.* (The ground joining the supports at A and D is considered to be the fourth bar.)

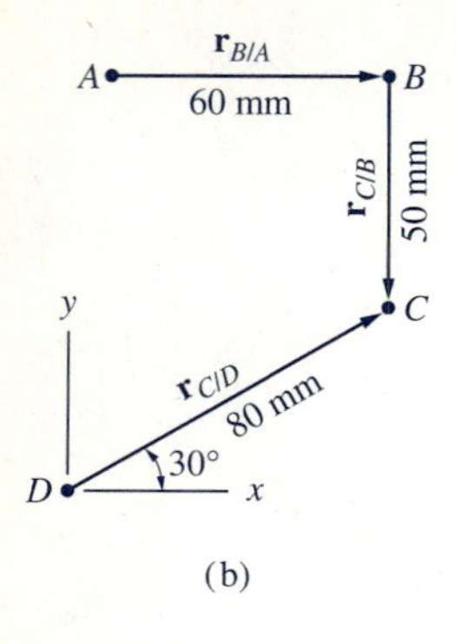

(b)

From Fig. (a) we observe that the following are the kinematically important points.

A: It is a fixed point.

B: Its path is a circle with center at A; it also connects bars AB and BC.

D: It is a fixed point.

C: Its path is a circle centered at D; it also connects bars BC and CD.

Because B and C are kinematically important points belonging to the same rigid body, we are led to consider the relative velocity equation $\mathbf{v}_C = \mathbf{v}_B + \mathbf{v}_{C/B}$ (the equivalent equation $\mathbf{v}_B = \mathbf{v}_C + \mathbf{v}_{B/C}$ could also be used). Two solutions are presented—one using semigraphical notation, the other using vector notation. The relative position vectors used in the solution are shown in Fig. (b).

Solution I (using semigraphical notation)

The relative velocity equation for B and C is

$$\mathbf{v}_C \quad = \quad \mathbf{v}_B \quad + \quad \mathbf{v}_{C/B} \tag{a}$$

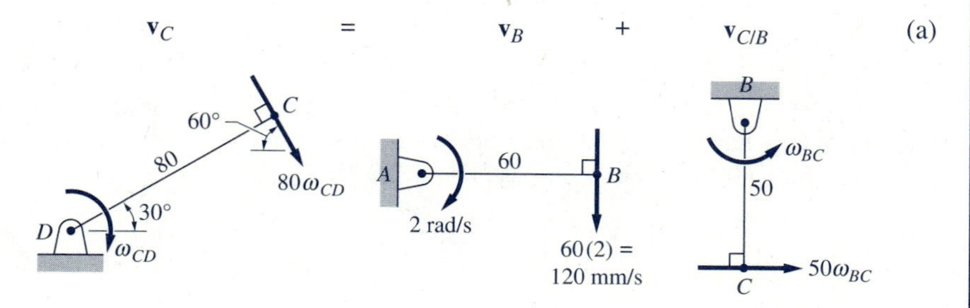

Comments on Eq. (a):

1. $\mathbf{v}_C$ is found by recognizing that the path of C is a circle centered at D. Therefore, its magnitude is $v_C = r_{C/D}\omega_{CD} = 80\omega_{CD}$, its direction being perpendicular to CD. The sense of the $\mathbf{v}_C$ is found from the assumed (clockwise) direction of ω_{CD}.
2. $\mathbf{v}_B$ is found by noting that the path of B is a circle centered at A. Its magnitude is thus $v_B = r_{B/A}\omega_{AB} = 60(2) = 120$ mm/s. The direction of $\mathbf{v}_B$ is perpendicular to AB, and its sense is found from the given clockwise direction of ω_{AB}.
3. $\mathbf{v}_{C/B}$ is constructed by considering B to be a fixed point at this instant, which gives $v_{C/B} = r_{C/B}\omega_{BC} = 50\omega_{BC}$. The direction of $\mathbf{v}_{C/B}$ is perpendicular to BC, and its sense is found from the assumed (counterclockwise) direction of ω_{BC}.

Equation (a) contains a total of two unknowns, ω_{CD} and ω_{BC}, which can be found from the two equivalent scalar equations. Equating x- and y-components of Eq. (a), we obtain

$$\overset{+}{\rightarrow} \quad 80\omega_{CD}\cos 60° = 0 + 50\omega_{BC} \tag{b}$$

$$+\uparrow \quad -80\omega_{CD}\sin 60° = -120 + 0 \tag{c}$$

Solving these equations, we get

$$\omega_{CD} = 1.732 \text{ rad/s} \quad \text{and} \quad \omega_{BC} = 1.386 \text{ rad/s} \qquad \textit{Answer}$$

The positive signs indicate that the assumed directions of ω_{CD} and ω_{BC} are correct.

Solution II (using vector notation)

The relative velocity equation for points B and C is

$$\mathbf{v}_C = \mathbf{v}_B + \mathbf{v}_{C/B}$$

$$\boldsymbol{\omega}_{CD} \times \mathbf{r}_{C/D} = \boldsymbol{\omega}_{AB} \times \mathbf{r}_{B/A} + \boldsymbol{\omega}_{BC} \times \mathbf{r}_{C/B} \qquad \text{(d)}$$

Utilizing the given information and the vectors shown in Fig. (b), we obtain

$$\boldsymbol{\omega}_{CD} = -\omega_{CD}\mathbf{k} \quad \text{(the direction of } \omega_{CD} \text{ is assumed to be clockwise)}$$

$$\mathbf{r}_{C/D} = 80\cos 30°\mathbf{i} + 80\sin 30°\mathbf{j}$$

$$= 69.28\mathbf{i} + 40.00\mathbf{j} \text{ mm}$$

$$\boldsymbol{\omega}_{AB} = -2\mathbf{k} \text{ rad/s}$$

$$\mathbf{r}_{B/A} = 60\mathbf{i} \text{ mm}$$

$$\boldsymbol{\omega}_{BC} = \omega_{BC}\mathbf{k} \quad \text{(the direction of } \omega_{BC} \text{ is assumed to be counterclockwise)}$$

$$\mathbf{r}_{C/B} = -50\mathbf{j} \text{ mm}$$

We see that these expressions contain only two unknowns: ω_{CD} and ω_{BC}. Therefore, Eq. (d), being equivalent to two scalar equations, can be solved for these unknowns.

Substituting the above expressions into Eq. (d) yields

$$(-\omega_{CD}\mathbf{k}) \times (69.28\mathbf{i} + 40.00\mathbf{j}) = (-2\mathbf{k}) \times (60\mathbf{i}) + (\omega_{BC}\mathbf{k}) \times (-50\mathbf{j})$$

which becomes

$$-69.28\omega_{CD}\mathbf{j} + 40.00\omega_{CD}\mathbf{i} = -120\mathbf{j} + 50\omega_{BC}\mathbf{i}$$

Equating the coefficients of **i** and **j** yields

$$40.00\omega_{CD} = 50\omega_{BC}$$

$$-69.28\omega_{CD} = -120$$

The solution of these two equations is

$$\omega_{CD} = 1.732 \text{ rad/s} \quad \text{and} \quad \omega_{BC} = 1.386 \text{ rad/s} \qquad \textit{Answer}$$

The positive answers indicate that the assumed directions of ω_{CD} and ω_{BC} are correct.

PROBLEMS

16.33 The velocity vectors of points A and B on a rigid body are parallel. Show that the angular velocity of the body must be zero, unless $\theta = 90°$.

16.34 The wheel rolls on its 12-in. radius hub without slipping. If the angular velocity of the wheel in the position shown is 3 rad/s clockwise, determine the velocities of points B and D on the rim of the wheel.

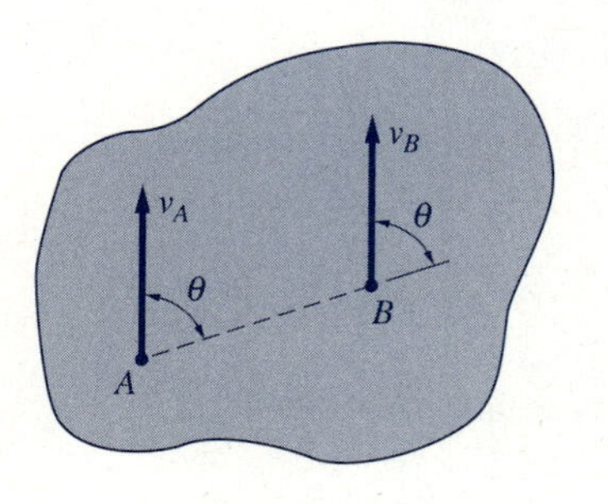

Fig. P16.33

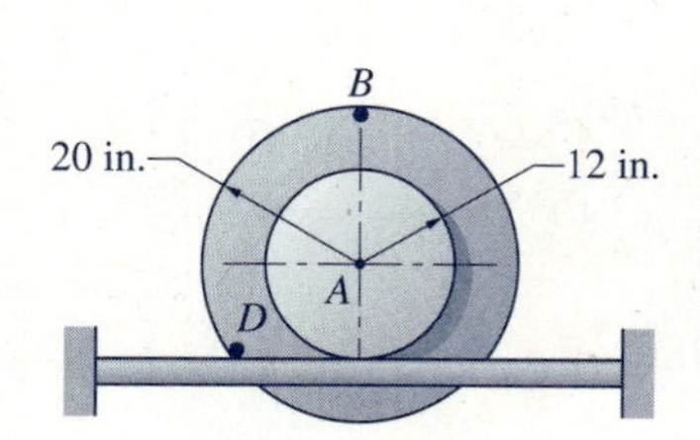

Fig. P16.34

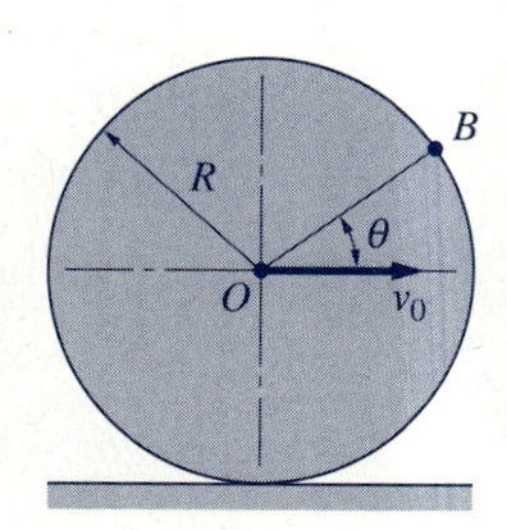

Fig. P16.35

16.35 The wheel rolls without slipping to the right with constant angular velocity. The velocity of the center of the wheel is v_0. Determine the speed of point B on the rim as a function of its angular position θ.

16.36 The arm AB joining the two friction wheels rotates with the constant angular velocity ω_0. Assuming that wheel A is stationary and that there is no slipping between the wheels, determine the angular velocity of wheel B.

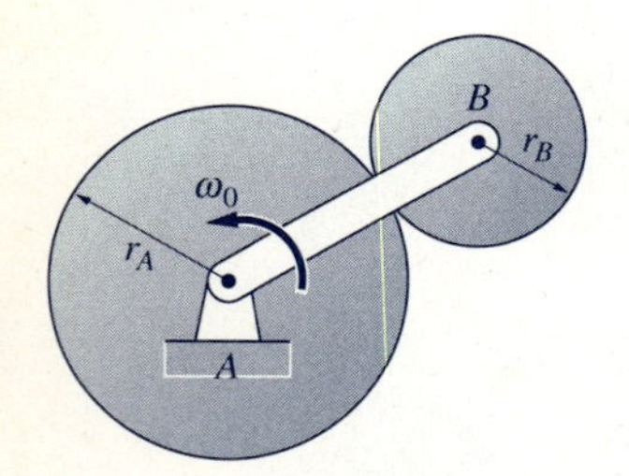

Fig. P16.36, P16.37

16.37 Solve Prob. 16.36 if wheel A is rotating clockwise with the angular velocity $\omega_A = 2\omega_0$.

16.38 Gear A of the planetary gear train is rotating clockwise at $\omega_A = 12$ rad/s. Calculate the angular velocities of gear B and the arm AB. Note that the outermost gear C is stationary.

16.39 The bar AB is rotating counterclockwise with the constant angular speed ω_0. (a) Find the velocities of ends A and B as functions of θ. (b) Differentiate the results of part (a) to determine the accelerations of A and B in terms of θ.

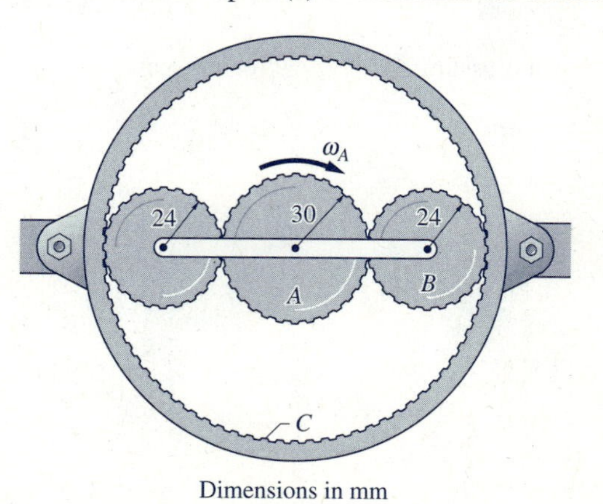

Fig. P16.38

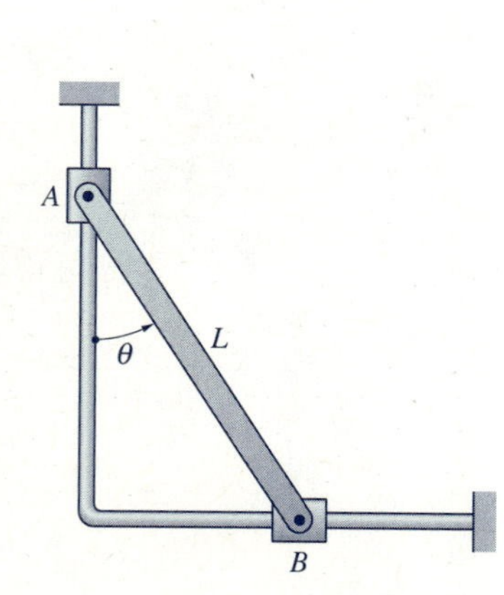

Fig. P16.39

16.40 End A of bar AD is pushed to the right with the constant velocity $v_A = 1.2$ ft/s. (a) Determine the angular velocity of AD as a function of θ. (b) By differentiating the result of part (a), find the angular acceleration of AD.

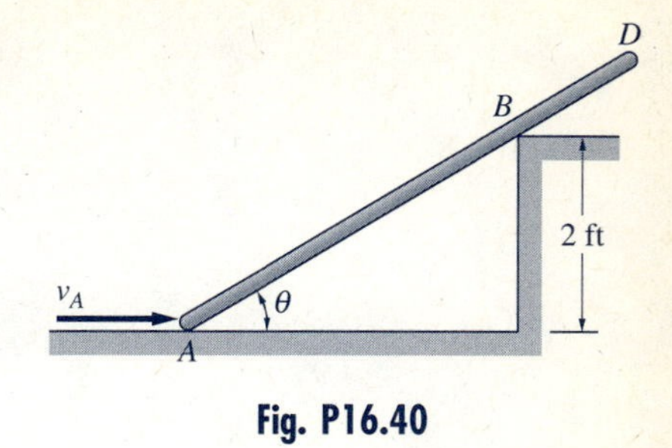

Fig. P16.40

16.41 The angular speed of link AB in the position shown is 3.2 rad/s clockwise. Compute the angular speeds of links BC and CD in this position.

16.42 The link AB of the mechanism rotates with the constant angular speed of 4 rad/s counterclockwise. Calculate the angular velocities of links BD and DE in the position shown.

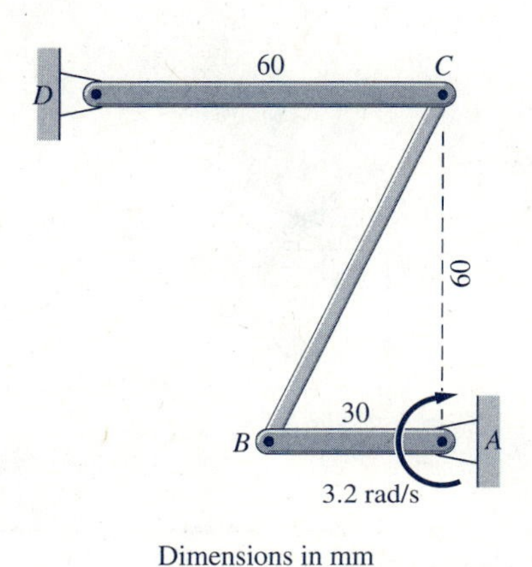

Fig. P16.41

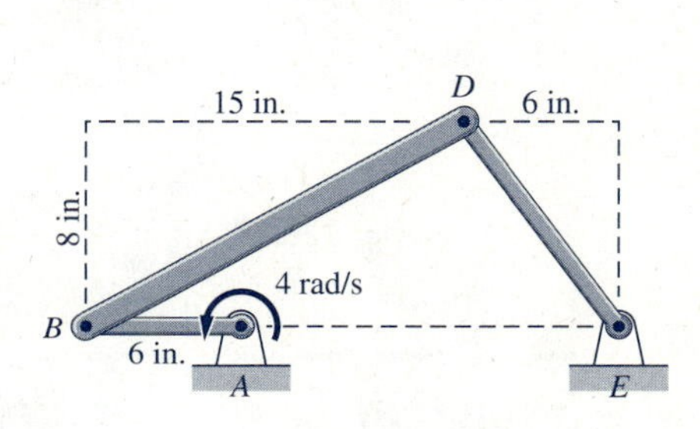

Fig. P16.42

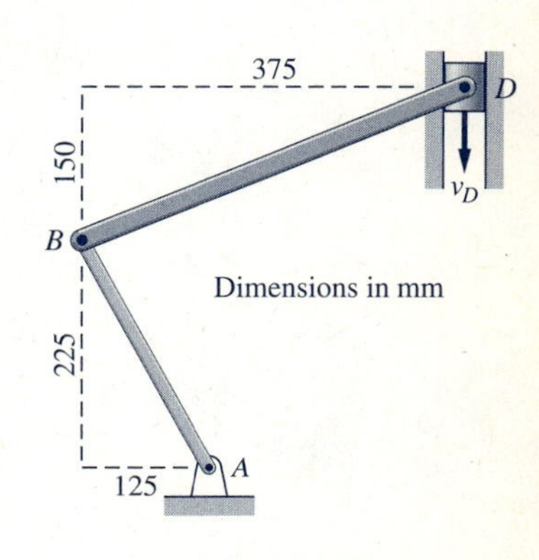

Fig. P16.43

16.43 When the mechanism is in the position shown, the velocity of slider D is $v_D = 1.25$ m/s. Determine the angular velocities of bars AB and BD at this instant.

16.44 The flywheel rotates counterclockwise about O with the constant angular velocity ω_0. When $\theta = 120°$, the speed of the piston A is 4 m/s. Determine ω_0.

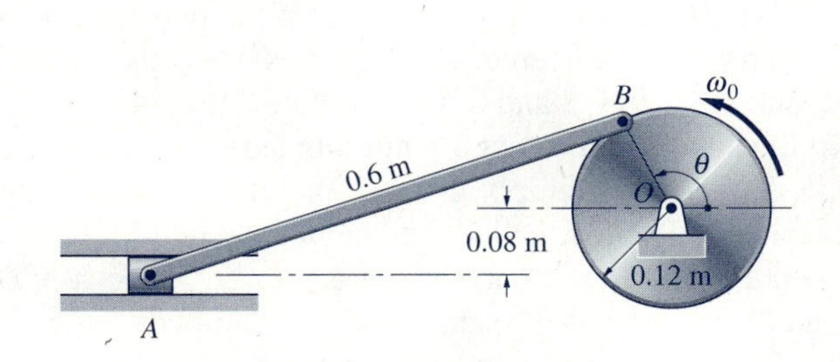

Fig. P16.44

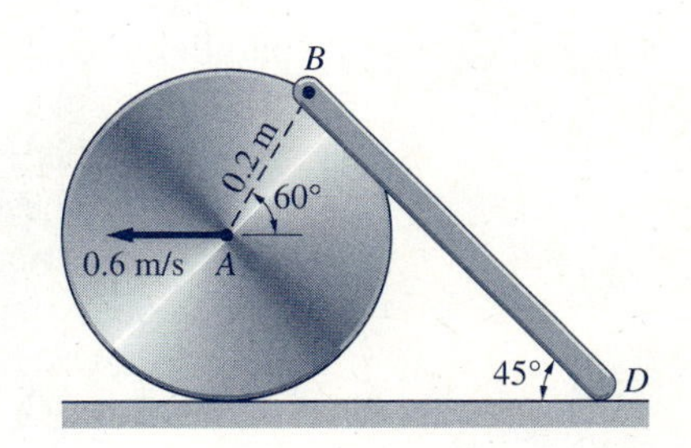

Fig. P16.45

16.45 In the position shown, the velocity of the center of the wheel, which is rolling without slipping, is 0.6 m/s to the left. For this position, compute the angular velocity of bar BD and the velocity of end D.

16.46 Crank AB rotates with a constant counterclockwise angular velocity of 16 rad/s. Calculate the angular velocity of bar BE when $\theta = 60°$.

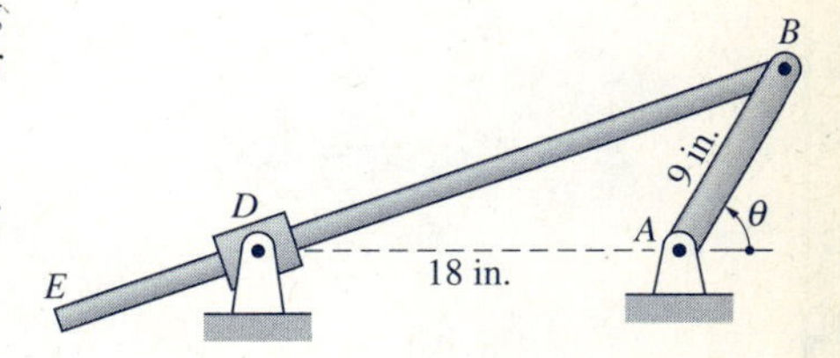

Fig. P16.46

16.47 The hydraulic cylinder E raises point B at the rate of 40 mm/s. Determine the velocity vectors of points A and D in the position shown.

16.48 In the position shown, the speeds of corners A and B of the right triangular plate are $v_A = 3$ m/s and $v_B = 2.4$ m/s. Find (a) the angle α; and (b) the speed of corner D.

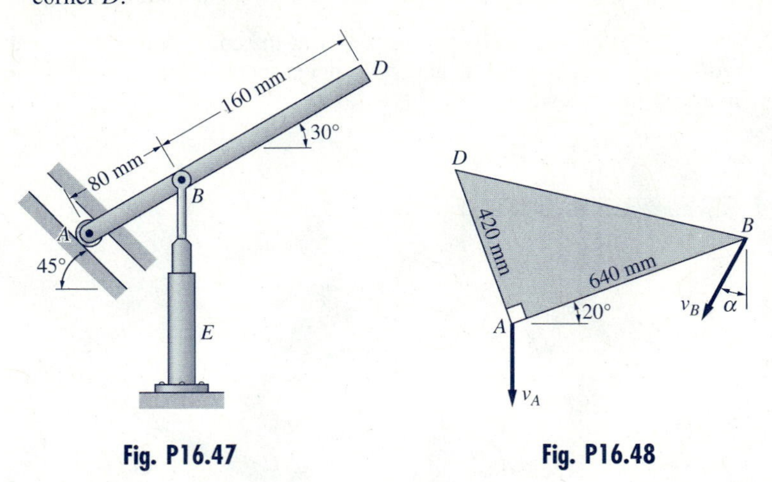

Fig. P16.47 **Fig. P16.48**

16.5 Instant Center for Velocities

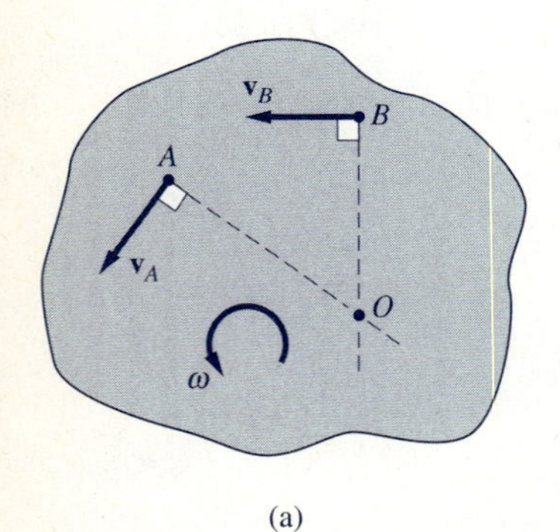

(a)

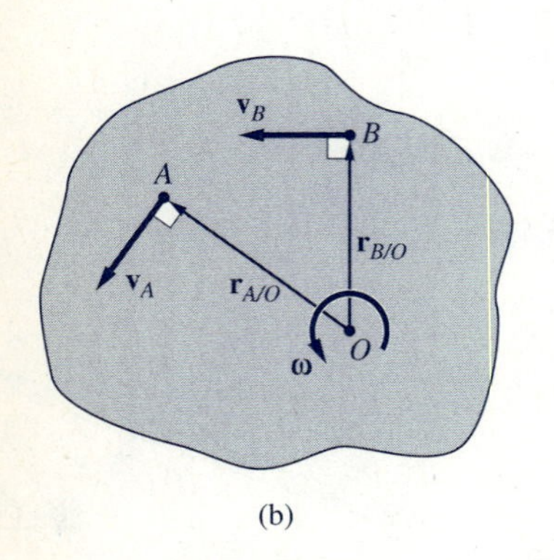

(b)

Fig. 16.10

The instant center for velocities of a body undergoing plane motion is defined to be the point that has zero velocity at the instant under consideration.* This point may be either in a body or outside the body (in the "body extended"). It is often convenient to use the instant center of the body in computing the velocities of points in the body.

Figure 16.10(a) shows a rigid body that is undergoing plane motion. It is assumed that the velocity vectors for points A and B, respectively, are not parallel to each other, and that the angular velocity ω of the body is counterclockwise. In order to find the instant center for velocities, we construct a line through A that is perpendicular to $\mathbf{v}_A$, and a line through B that is perpendicular to $\mathbf{v}_B$. These two lines will intersect at a point labeled O in the figure. If point O does not lie in the body, we simply imagine that the body is enlarged to include it, the expanded body being called the "body extended."

We now show that point O is the instant center, i.e., that the velocity of O is zero. Because A, B, and O are points in the same rigid body (or body extended), we can write the following relative velocity equations.

$$\mathbf{v}_O = \mathbf{v}_A - \boldsymbol{\omega} \times \mathbf{r}_{A/O} \tag{a}$$

$$\mathbf{v}_O = \mathbf{v}_B - \boldsymbol{\omega} \times \mathbf{r}_{B/O} \tag{b}$$

where $\mathbf{r}_{A/O}$ and $\mathbf{r}_{B/O}$ are the relative position vectors shown in Fig. 16.10(b). Since both the angular velocity vector $\boldsymbol{\omega}$ and $\mathbf{r}_{A/O}$ are perpendicular to $\mathbf{v}_A$, it

* Three "centers" are sometimes used in the kinematic analysis of plane motion: the instant center of rotation for virtual motion (see Art. 10.6), the instant center for velocities, and the instant center for accelerations. Each of these points is called simply the "instant center" when it is clear from the context which center is being used. The discussion of instant center for velocities presented here parallels the discussion of instant center of rotation for virtual motion in Art. 10.6.

follows that $\boldsymbol{\omega} \times \mathbf{r}_{A/O}$ in Eq. (a) is parallel to $\mathbf{v}_A$. This means that $\mathbf{v}_O$ is parallel to $\mathbf{v}_A$. Similarly, it can be shown from Eq. (b) that $\mathbf{v}_O$ is parallel to $\mathbf{v}_B$. Because a nonzero vector cannot be parallel to two different directions simultaneously, we conclude that $\mathbf{v}_O = \mathbf{0}$. Therefore, point O is indeed the instant center for velocities.

In general, the instant center for velocities is not a fixed point. Therefore, the acceleration of the instant center for velocities is not necessarily zero. However, if a point of a body is fixed, that point is obviously the instant center for both velocities and accelerations.

If the instant center O is chosen as the reference point, the relative velocity equation for any other point A of the body, $\mathbf{v}_A = \mathbf{v}_O + \boldsymbol{\omega} \times \mathbf{r}_{A/O}$, reduces to

$$\boxed{\mathbf{v}_A = \boldsymbol{\omega} \times \mathbf{r}_{A/O} \qquad (v_A = r_{A/O}\omega)} \tag{16.12}$$

This equation has the same form as Eq. (16.3) for rotation about a fixed axis. Consequently, the relative velocity equations, with O as the reference point, are the same as those for a body rotating about an axis through O. We must reiterate, however, that for general plane motion the instant center is not a fixed point. Therefore, the analogy with rotation about a fixed axis is valid only for velocities at a *given instant* of time (hence the term "instant" center). The analogy does not apply to accelerations.

The following two rules, which follow directly from Eq. (16.12), apply when the instant center is used.

1. The magnitude of the velocity of any point of the body is proportional to the distance of the point from the instant center.
2. The direction of the velocity vector of any point must be consistent with the direction of the angular velocity of the body.

The construction shown in Fig. 16.10(a) for locating the instant center is obviously valid only if $\mathbf{v}_A$ and $\mathbf{v}_B$ are not parallel. Figure 16.11 illustrates the methods for locating the instant center when $\mathbf{v}_A$ and $\mathbf{v}_B$ are parallel.

In Fig. 16.11(a) the perpendicular lines to the velocity vectors are parallel, which means that the instant center is an infinite distance from A and B. This leads to the conclusion that $\omega = 0$; i.e., the body is translating with $\mathbf{v}_A = \mathbf{v}_B$.

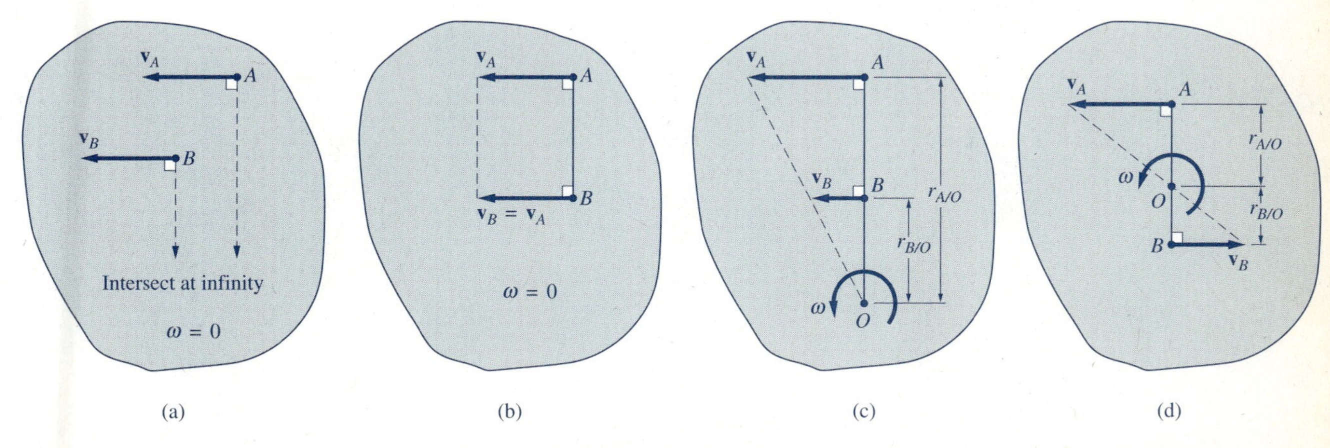

Fig. 16.11

In Fig. 16.11 parts (b)–(d) the lines joining A and B are perpendicular to both velocity vectors, and the usual construction for locating the instant center does not work. In the case where $\mathbf{v}_A = \mathbf{v}_B$, as in Fig. 16.11(b), the instant center is at infinity, so that $\omega = 0$ (i.e., the body is translating). If the magnitudes of the velocities are not equal, the instant center is located using the constructions in Fig. 16.11 parts (c) and (d), which are simply applications of rules 1 and 2 just stated.

The question arises whether the simplification of velocity analysis based on the instant center is worth the labor it takes to find the instant center in the first place. A general rule of thumb is to begin by locating the instant center on a sketch of the body. If the distances between the kinematically important points on the body and the instant center are fairly easy to calculate, the analysis is well suited for utilizing the instant center. However, if a prohibitive amount of trigonometry is required to compute the location of the instant center, the relative velocity method of the preceding article would be the preferred method of solution. Of course, there are many problems that require the same amount of labor using either method, in which case personal preference would dictate which method to use.

Sample Problem 16.8

When the mechanism in Fig. (a) is in the position shown, the velocity of bar AB is $\omega_{AB} = 3$ rad/s clockwise. Utilizing instant centers for velocities, calculate the angular velocity of bar BC and the velocity of slider C for this position. (This problem was solved previously as Sample Problem 16.4 by differentiating the equation of constraint, and as Sample Problem 16.6 by using relative velocity.)

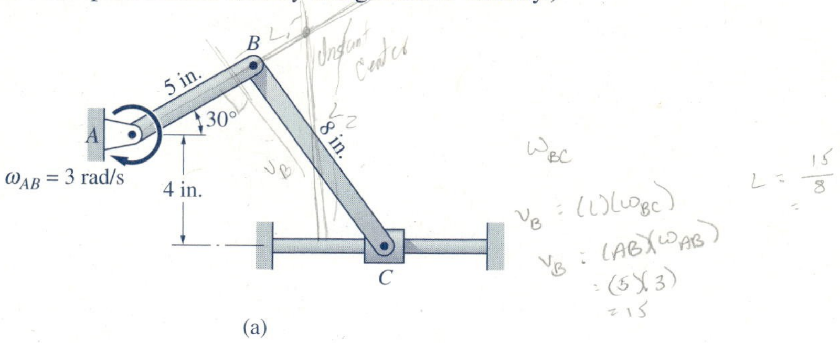

(a)

Solution

We must first locate the instant centers of the two rigid bodies AB and BC. Since A is a fixed point, it is obviously the instant center of bar AB. The instant center of BC, labeled O in Fig. (b), is located at the point of intersection of the lines that are perpendicular to the velocity vectors of B and C. Note that $\mathbf{v}_B$ is perpendicular to AB, and $\mathbf{v}_C$ is horizontal. Therefore, O is located at the intersection of line AB and the vertical line that passes through C.

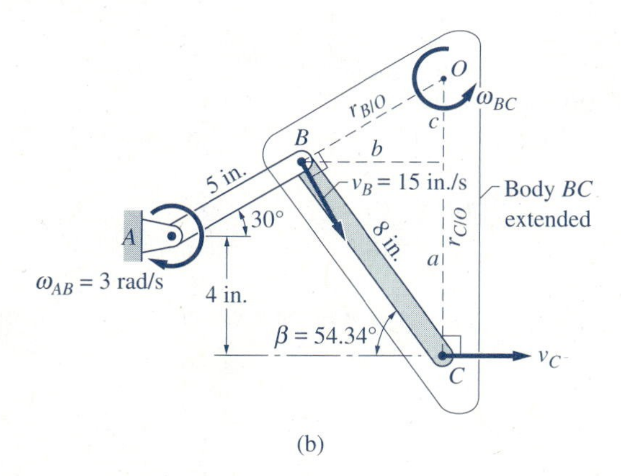

(b)

From the geometry in Fig. (b), we note that $5 \sin 30° + 4 = 8 \sin \beta$, from which we find that $\beta = 54.34°$. Therefore, the distances a, b, and c are

$$a = 8 \sin 54.34° = 6.500 \text{ in.}$$

$$b = 8 \cos 54.34° = 4.664 \text{ in.}$$

$$c = b \tan 30° = 4.664 \tan 30° = 2.693 \text{ in.}$$

The distances to B and C from O are

$$r_{B/O} = c/\sin 30° = 2.693/\sin 30° = 5.386 \text{ in.}$$

$$r_{C/O} = a + c = 6.500 + 2.693 = 9.193 \text{ in.}$$

Now that the instant center for each bar has been found, we can easily compute the required velocities.

Considering the motion of AB (the instant center is at A), we find that $v_B = r_{B/A}\omega_{AB} = 5(3) = 15$ in./s, directed as shown in Fig. (b). Analyzing the motion of BC (the instant center is at O) yields

$$\omega_{BC} = \frac{v_B}{r_{B/O}} = \frac{15}{5.386} = 2.785 \text{ rad/s}$$

$$\boldsymbol{\omega}_{BC} = 2.79 \text{ rad/s} \circlearrowleft \qquad \textit{Answer}$$

and

$$v_C = r_{C/O}\omega_{BC} = 9.193(2.785) = 25.60 \text{ in./s}$$

$$\mathbf{v}_C = 25.6 \text{ in./s} \rightarrow \qquad \textit{Answer}$$

Sample Problem 16.9

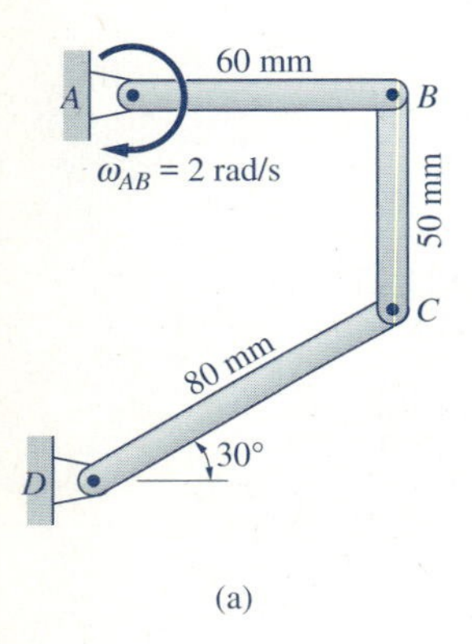

(a)

When the linkage in Fig. (a) is in the position shown, the angular velocity of bar AB is $\omega_{AB} = 2$ rad/s clockwise. For this position, determine the angular velocities of bars BC and CD and the velocity of C using the instant centers for velocities. (This problem was solved previously as Sample Problem 16.7 using relative velocity.)

Solution

Because A and D are fixed points, they are the instant centers for bars AB and CD, respectively. The instant center for bar BC, labeled O in Fig. (b), is located at the point of intersection of the lines that are perpendicular to the velocity vectors of B and C. Since $\mathbf{v}_B$ and $\mathbf{v}_C$ are perpendicular to AB and CD, respectively, one of the lines coincides with AB and the other coincides with CD.

The distances to B and C from O, found from the triangle OBC, are

$$r_{B/O} = 50/\tan 30° = 86.60 \text{ mm}$$

$$r_{C/O} = 50/\sin 30° = 100 \text{ mm}$$

The instant centers A, O, and D can now be used to compute the required angular velocities directly from Fig. (b).

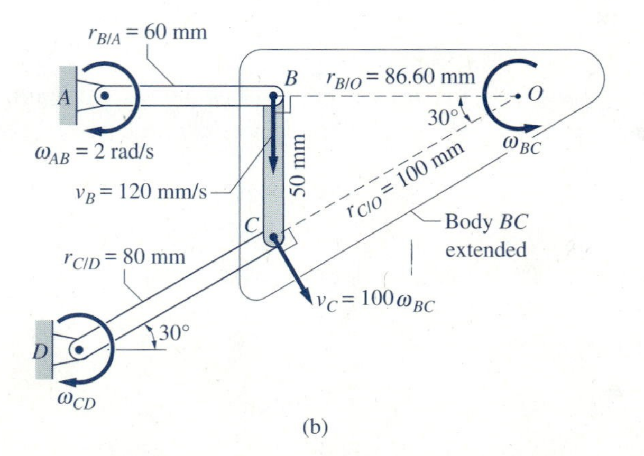

(b)

Considering the motion of AB (the instant center is at A), we find that $v_B = r_{B/A}\omega_{AB} = 60(2) = 120$ mm/s, directed as shown in Fig. (b). Analyzing the motion of BC (the instant center is at O) yields

$$\omega_{BC} = \frac{v_B}{r_{B/O}} = \frac{120}{86.60} = 1.386 \text{ rad/s}$$

$$\boldsymbol{\omega}_{BC} = 1.386 \text{ rad/s} \circlearrowleft \qquad \textit{Answer}$$

and

$$v_C = r_{C/O}\omega_{BC} = 100(1.386) = 138.6 \text{ mm/s}$$

$$\mathbf{v}_C = 138.6 \text{ mm/s} \ \angle 30° \qquad \textit{Answer}$$

Because C is also a point on bar CD (the instant center is at D), the angular velocity of bar CD is

$$\omega_{CD} = \frac{v_C}{r_{C/D}} = \frac{138.6}{80} = 1.733 \text{ rad/s}$$

$$\boldsymbol{\omega}_{CD} = 1.733 \text{ rad/s} \circlearrowright \qquad \textit{Answer}$$

Sample Problem 16.10

The wheel in Fig. (a) rolls without slipping with the constant clockwise angular velocity $\omega_0 = 1.6$ rad/s. Calculate the angular velocity of bar AB and the velocity of the slider B when the mechanism is in the position shown. Use the instant centers for velocities.

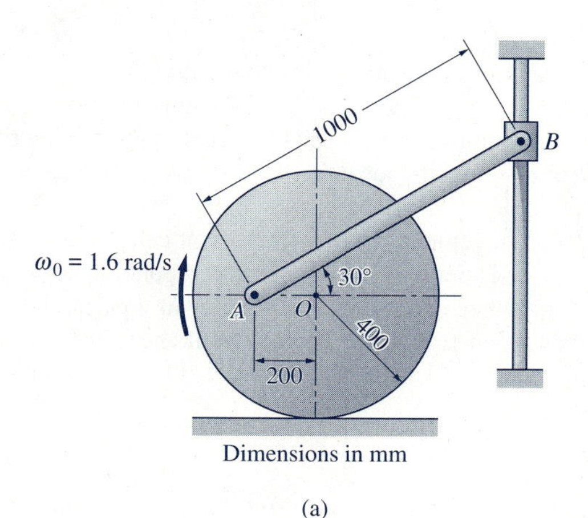

(a)

Solution

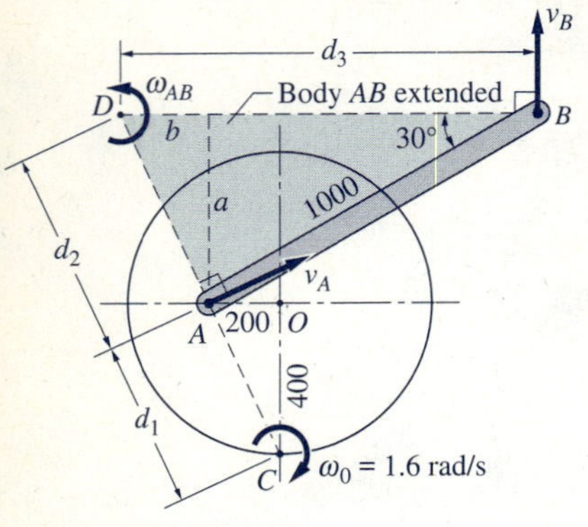

The velocity vectors, distances, and points required to solve this problem are shown in Fig. (b).

Because the wheel rolls without slipping, its instant center is at the point of contact C. Therefore, $\mathbf{v}_A$ is perpendicular to the line AC. Note that the slope of AC equals $\overline{OC}/\overline{AO} = 400/200 = 2/1$.

The instant center of bar AB, labeled D in Fig. (b), is located at the intersection of the lines that are perpendicular to $\mathbf{v}_A$ and $\mathbf{v}_B$. Because $\mathbf{v}_A$ is perpendicular to AC, D also lies on AC. The line BD, which is drawn perpendicular to $\mathbf{v}_B$, is horizontal because $\mathbf{v}_B$ is vertical (it is not necessary to know here that the sense of $\mathbf{v}_B$ is upward).

Referring to Fig. (b), the distances of interest are computed as follows.

(b)

$$d_1 = \sqrt{\overline{AO}^2 + \overline{OC}^2} = \sqrt{200^2 + 400^2} = 447.2 \text{ mm}$$

$$a = 1000 \sin 30^\circ = 500 \text{ mm}$$

$$b = a/2 = 250 \text{ mm} \quad \text{(from the slope of the line } DAC\text{)}$$

$$d_2 = \sqrt{a^2 + b^2} = \sqrt{500^2 + 250^2} = 559.0 \text{ mm}$$

$$d_3 = 1000 \cos 30^\circ + b = 1000 \cos 30^\circ + 250 = 1116 \text{ mm}$$

Using the distances d_1, d_2, and d_3 and $\omega_0 = 1.6$ rad/s, we find that

$$v_A = d_1\omega_0 = 447.2(1.6) = 715.5 \text{ mm/s}$$

$$\mathbf{v}_A = 716 \text{ mm/s} \quad \text{(slope 1 up, 2 across)} \qquad \textit{Answer}$$

$$\omega_{AB} = \frac{v_A}{d_2} = \frac{715.5}{559.0} = 1.280 \text{ rad/s}$$

$$\boldsymbol{\omega}_{AB} = 1.280 \text{ rad/s} \circlearrowleft \qquad \textit{Answer}$$

$$v_B = d_3\omega_{AB} = 1116(1.280) = 1428 \text{ mm/s}$$

$$\mathbf{v}_B = 1428 \text{ mm/s} \uparrow \qquad \textit{Answer}$$

The sense of $\mathbf{v}_A$ was found by considering that the angular velocities of the wheel is clockwise and its instant center is at C. The counterclockwise direction of ω_{AB} was deduced by inspection of the sense of $\mathbf{v}_A$ and the location of the instant center D. The direction of ω_{AB} and the location of B relative to D determine that the sense of $\mathbf{v}_B$ is upward.

It is frequently convenient to show the instant centers of more than one body on the same sketch, as is done in Fig. (b). However, one must be careful to use the proper instant center when discussing the velocity of a particular point. For example, a common error when referring to Fig. (b) would be to write $v_C = (d_1 + d_2)\omega_{AB}$, which is incorrect, because D is the instant center of bar AB, which does not include point C.

PROBLEMS

Note: The following problems are to be solved using the instant centers for velocities.

16.49 The end of the cord that is wrapped around the hub of the wheel is pulled to the right with the velocity $v_0 = 28$ in./s. Find the angular velocity of the wheel, assuming no slipping.

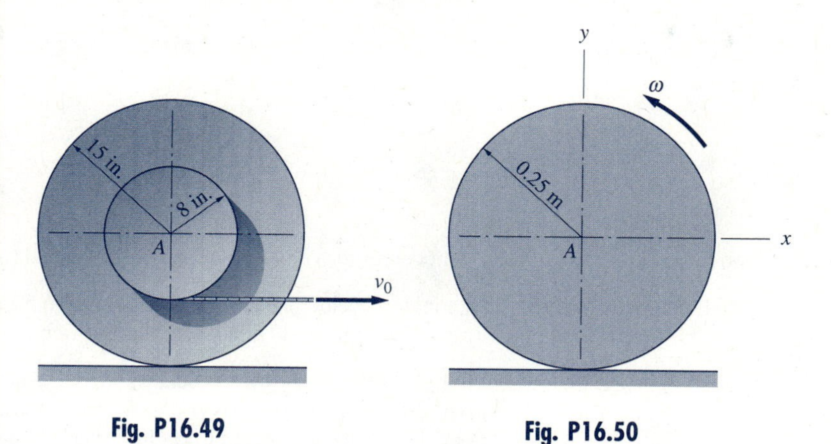

Fig. P16.49 **Fig. P16.50**

16.50 The wheel rolls without slipping with the angular velocity $\omega = 12$ rad/s. Determine the coordinates of the point B on the wheel for which the velocity vector is $\mathbf{v}_B = -2.4\mathbf{i} + 0.7\mathbf{j}$ m/s.

16.51 A 500-mm diameter wheel rolls and slips on a horizontal plane. The angular velocity of the wheel is $\omega = 12$ rad/s (counterclockwise), and the velocity of the center of the wheel is 1.8 m/s to the left. (a) Find the instant center for the velocities of the wheel. (b) Calculate the velocity of the point on the wheel that is in contact with the plane.

16.52 Determine the coordinates of the instant center for velocities of the bar AB in (a) and (b).

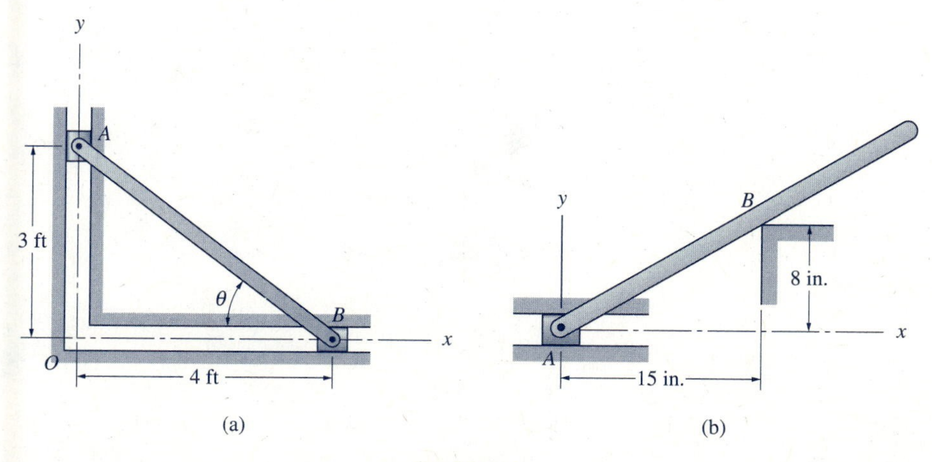

Fig. P16.52

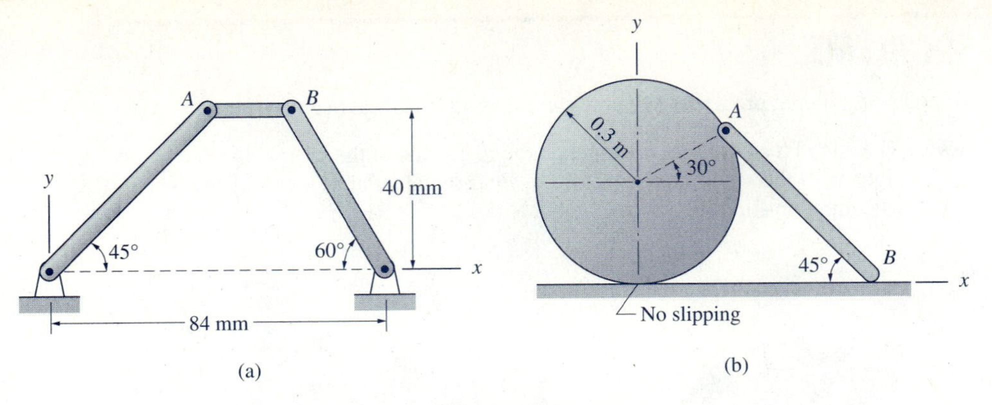

Fig. P16.53

16.53 Find the coordinates of the instant center for velocities of bar AB in (a) and (b).

16.54 Sketch the locus of the instant center of velocities of bar AB in Fig. P16.52(a) as θ varies from 0° to 90°. (This type of curve is called a *space centrode.*)

16.55 The 8-ft wooden plank is tumbling as it falls in the vertical plane. When the plank is in a horizontal position, the velocities of ends A and B are as shown in the figure. For this position, determine the location of the instant center for velocities, the angular velocity of the plank, and the velocity of the mid-point G.

16.56 For the triangular plate undergoing plane motion, $\mathbf{v}_A$ and the direction of $\mathbf{v}_B$ are known. Calculate the angular speed of the plate and the speeds of corners B and C.

16.57 Crank AB is rotating at the constant angular speed $\omega_0 = 25$ rad/s. Find the angular velocity of bar DE and the velocity of end E in the position shown.

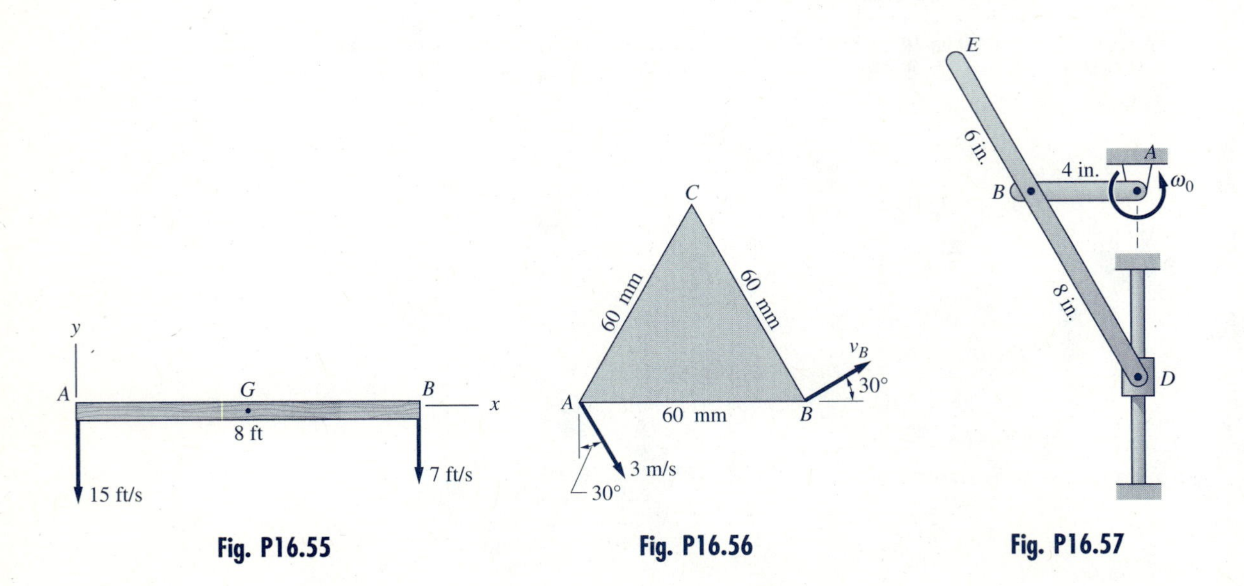

Fig. P16.55 Fig. P16.56 Fig. P16.57

16.58 When bar AB is in the position shown, end B is sliding to the right with a velocity of 0.6 m/s. Determine the velocity of end A in this position.

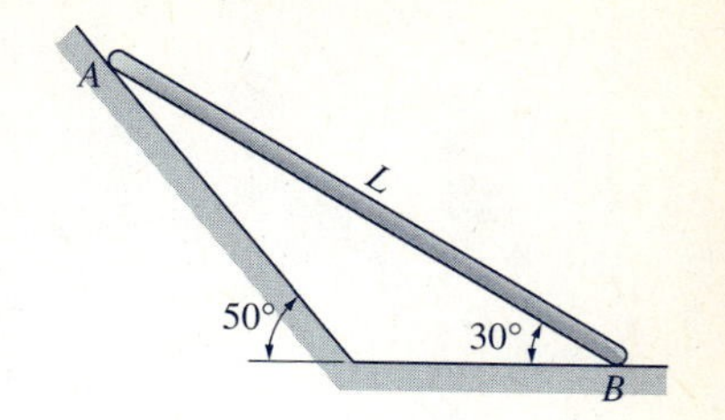

Fig. P16.58

16.59 Bar AB of the mechanism rotates clockwise with the angular velocity ω_0. Compute the angular velocities of bars BD and DE for the position shown.

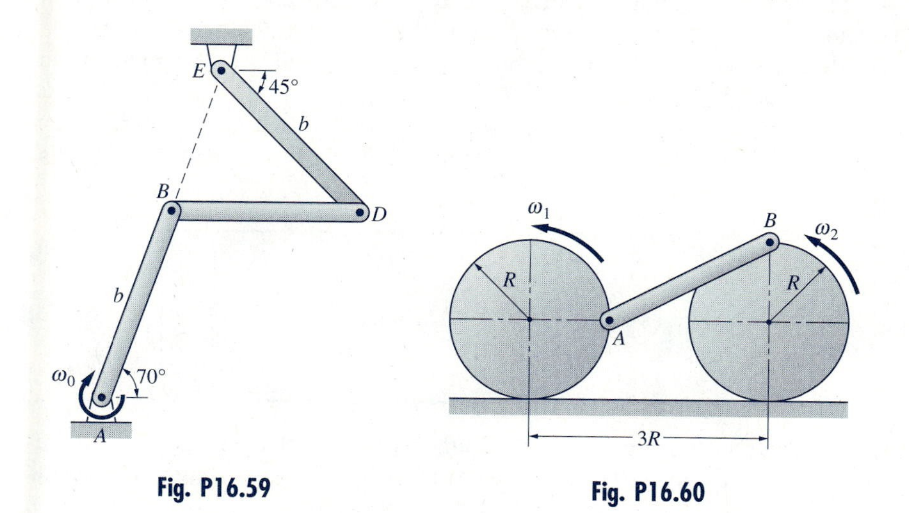

Fig. P16.59

Fig. P16.60

16.60 The connecting rod AB joins the two identical wheels, which are rolling without slipping. Determine ω_2/ω_1, the ratio of the angular velocities of the wheels in the position shown.

16.61 In the position shown, the horizontal component of velocity of point E is 1.2 m/s to the right. Find the angular velocity of each of the three bars for this position.

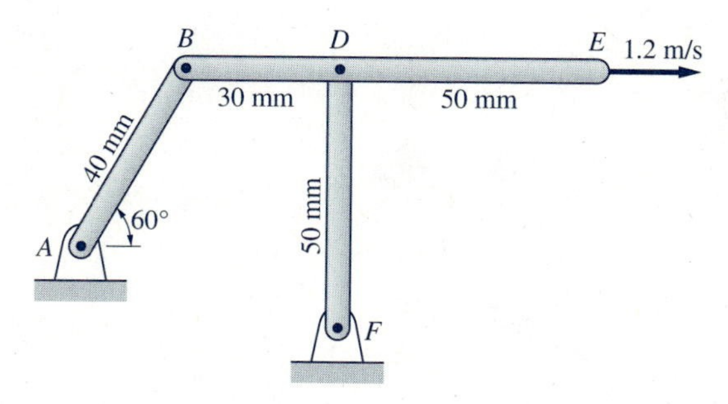

Fig. P16.61

16.62 For the mechanism described in Prob. 16.46, locate the instant center of velocities of bar BE graphically from a scale drawing of the mechanism when $\theta = 60°$. Using the distances measured on the scale drawing, calculate the angular velocity of bar BE and the velocity of BE relative to the collar.

16.63 Using a scale drawing of bar AD of the mechanism described in Prob. 16.47, locate the instant center of velocities of AD graphically. Compute the velocity vectors of points A and D by measuring the appropriate distances and angles on the scale drawing.

16.6 Method of Relative Acceleration

Figure 16.12(a) shows a rigid body that is undergoing plane motion, where $\mathbf{a}_A$ and $\mathbf{a}_B$ are the accelerations of points A and B on the body. The angular velocity $\boldsymbol{\omega}$ and the angular acceleration $\boldsymbol{\alpha}$ of the body are assumed to be counterclockwise. Note that for plane motion, the angular velocity and the angular acceleration vectors are parallel, both being perpendicular to the plane of motion.

According to Eq. (16.2c), the acceleration of B relative to A can be represented as the vector sum of its normal component $(\mathbf{a}_{B/A})_n$ and its tangential component $(\mathbf{a}_{B/A})_t$:

$$\mathbf{a}_{B/A} = (\mathbf{a}_{B/A})_n + (\mathbf{a}_{B/A})_t \tag{16.13a}$$

$$(\mathbf{a}_{B/A})_n = \boldsymbol{\omega} \times (\boldsymbol{\omega} \times \mathbf{r}_{B/A}) \qquad [(a_{B/A})_n = r_{B/A}\omega^2] \tag{16.13b}$$

$$(\mathbf{a}_{B/A})_t = \boldsymbol{\alpha} \times \mathbf{r}_{B/A} \qquad [(a_{B/A})_t = r_{B/A}\alpha] \tag{16.13c}$$

The reason for referring to these components as normal and tangential will become apparent shortly. The magnitudes of the components shown in brackets

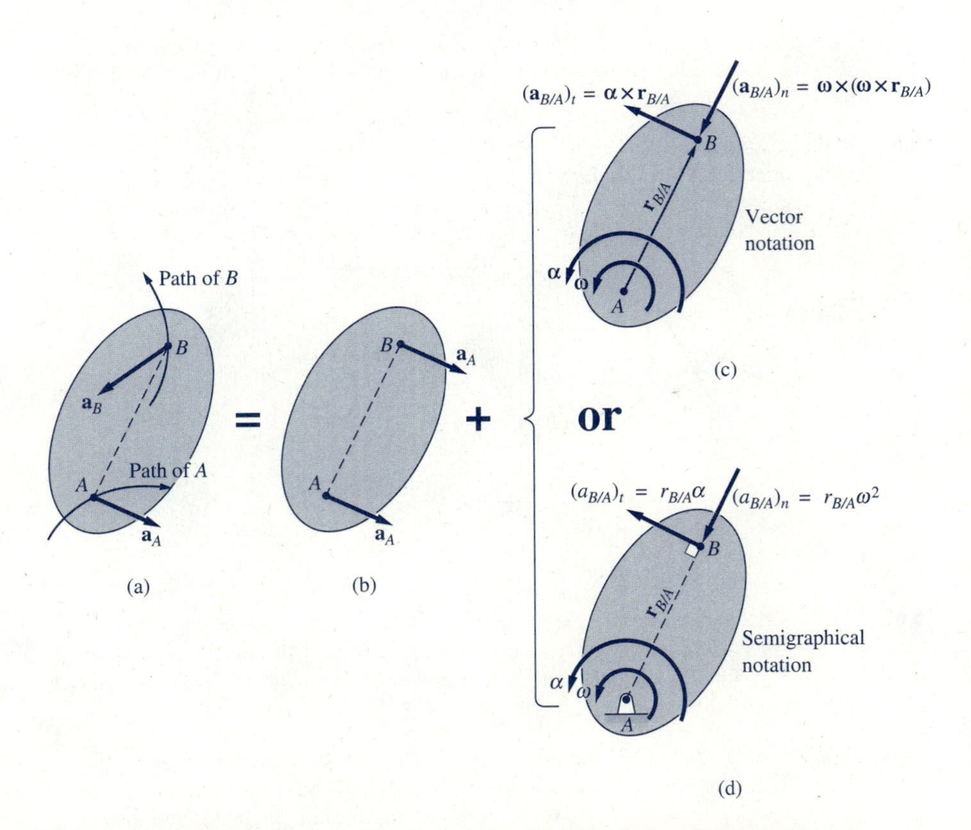

Fig. 16.12

in Eqs. (16.13) may be verified from the properties of the cross product. Substituting Eqs. (16.13) into $\mathbf{a}_B = \mathbf{a}_A + \mathbf{a}_{B/A}$, the relative acceleration equation becomes

$$\mathbf{a}_B = \mathbf{a}_A + \boldsymbol{\omega} \times (\boldsymbol{\omega} \times \mathbf{r}_{B/A}) + \boldsymbol{\alpha} \times \mathbf{r}_{B/A} \qquad (16.14)$$

Figure 16.12 illustrates each of the three terms on the right-hand side of Eq. (16.14). We see once again that plane motion can be represented as the superposition of a rigid body translation, where the accelerations of all points on the body are equal to the acceleration of the reference point A, and a rigid-body rotation about A.

As shown in Fig. 16.12 parts (c) and (d), the relative acceleration terms can be evaluated using either vector or semigraphical notation. When vector notation is used, the vectors $(\mathbf{a}_{B/A})_n$ and $(\mathbf{a}_{B/A})_t$ are computed formally from the vector products in Eqs. (16.13). In semigraphical notation, the magnitudes $(a_{B/A})_n$ and $(a_{B/A})_t$ are calculated from the expressions in brackets in Eqs. (16.13). The directions are found by considering the reference point A to be fixed at this instant, as shown in Fig. 16.12(d). The normal component is always directed toward the reference point, whereas the sense of the tangential component is determined by the direction of the angular acceleration.

The relative acceleration $\mathbf{a}_{B/A}$ can be interpreted as the acceleration of B as seen by an observer who is translating with A. Since the path of B relative to this nonrotating observer is a circle of radius $r_{B/A}$, we see that the directions of $(\mathbf{a}_{B/A})_n$ and $(\mathbf{a}_{B/A})_t$ are normal and tangential to the relative path.

The method of relative acceleration consists of solving the relative acceleration equation, Eq. (16.14), for the unknowns. For plane motion, Eq. (16.14) is equivalent to two scalar equations containing the following six variables (assuming that $\mathbf{r}_{B/A}$ is known).

$\mathbf{a}_B$: two variables (e.g., magnitude and direction, horizontal and vertical components).

$\mathbf{a}_A$: two variables (e.g., magnitude and direction, horizontal and vertical components).

$(\mathbf{a}_{B/A})_n$: one variable (magnitude, or the angular velocity ω). The direction of $(\mathbf{a}_{B/A})_n$ is always toward A.

$(\mathbf{a}_{B/A})_t$: one variable (magnitude, or the angular acceleration α). The direction of $(\mathbf{a}_{B/A})_t$ is always perpendicular to $\mathbf{r}_{B/A}$.

If the angular velocity ω is not known, it must be computed beforehand using either relative velocity or the instant center for velocities. This reduces the potential number of unknowns in Eq. (16.14) from six to five. Consequently, three more variables must be known before the two scalar equations can be solved. This requirement restricts the choice of A and B to kinematically important points. A kinematically important point (for acceleration) is a point at which the magnitude or direction of the acceleration is known.

The relative acceleration method parallels the relative velocity method, except that here it may be necessary to compute the angular velocity first. The steps in the analysis are as follows.

Step 1: If the angular velocity of the body is not given, compute it by relative velocity analysis or the method of instant centers.

Step 2: Identify two kinematically important points (for acceleration) on the same rigid body.

Step 3: Write the relative acceleration equation for the chosen points, identifying the unknown variables (either vector or semigraphical notation can be used).

Step 4: If the number of unknowns is two, solve the relative acceleration equation.

If the number of unknowns is greater than two, it may still be possible to solve the problem by considering the motion of other kinematically important points.

It should be mentioned that the point that has zero acceleration is called the instant center for accelerations. In general, the instant center for velocities and the instant center for accelerations are not the same point. It can be shown that these two instant centers coincide only if the angular velocity of the body is zero or if the body rotates about a fixed axis. In principle, the instant center for accelerations can be found for any body undergoing plane motion. However, the difficulty in locating this point usually outweighs the advantages gained by its use.

Sample Problem 16.11

The wheel of radius R shown in Fig. (a) is rolling without slipping. At the instant shown, its angular velocity and angular acceleration are ω and α, both clockwise. Determine the acceleration vectors of (1) point C, the point of contact on the wheel; and (2) point A.

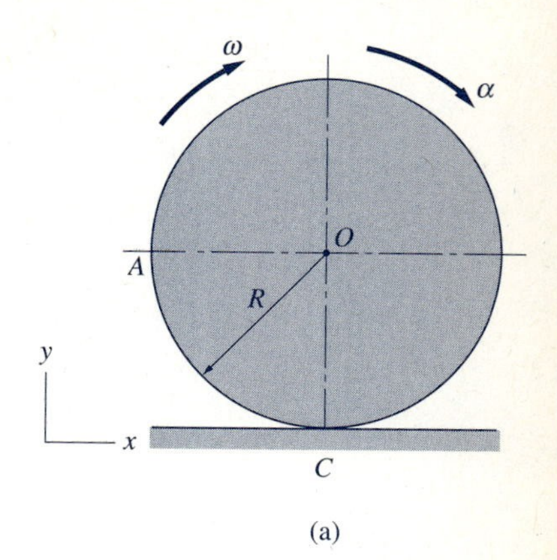

(a)

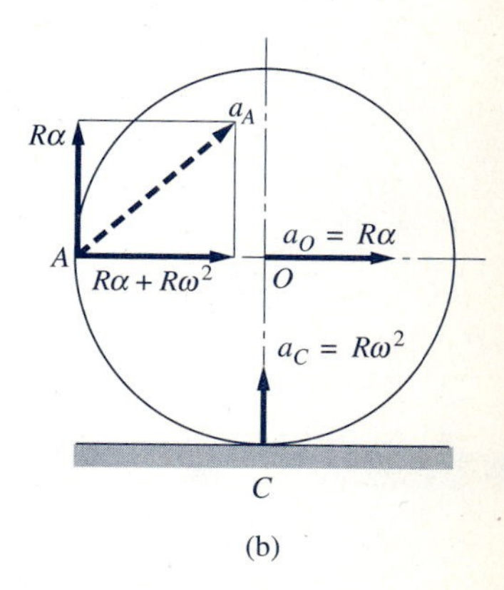

(b)

Solution

Introductory Comments

This problem will be solved using both semigraphical and vector notations. We choose point O (the center of the wheel) as the reference point, because its acceleration is known from Eq. (16.11) to be $\mathbf{a}_O = R\alpha\mathbf{i}$, as shown in Fig. (b).

Solution I (using semigraphical notation)

Part 1

When semigraphical notation is used to write the relative acceleration equation between C and O, $\mathbf{a}_{C/O}$ is found by assuming that point O is fixed. Therefore, we have

$$\mathbf{a}_C = \mathbf{a}_O + \mathbf{a}_{C/O}$$

$R\alpha$ (→) $\qquad$ O, ω, α, R, C, $R\alpha$ (←), $R\omega^2$ (↑)

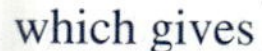

which gives

$$\mathbf{a}_C = R\omega^2 \uparrow \qquad \textit{Answer}$$

This result is shown in Fig. (b). It should be noted that although C is the instant center for velocities, its acceleration is not zero.

Part 2

The relative acceleration equation between A and O is

$$\mathbf{a}_A = \mathbf{a}_O + \mathbf{a}_{A/O}$$

$R\alpha$ (→) $\qquad$ $R\alpha$ (↑), $R\omega^2$ (→), A, R, α, ω, O

which yields

$$\mathbf{a}_A = R\alpha\ (\uparrow) + R(\alpha + \omega^2)\ (\rightarrow) \qquad \textit{Answer}$$

The acceleration vector of A is also shown in Fig. (b).

Solution II (using vector notation)

Part 1

Using vector notation, the relative acceleration equation between C and O is

$$\mathbf{a}_C = \mathbf{a}_O + \boldsymbol{\alpha} \times \mathbf{r}_{C/O} + \boldsymbol{\omega} \times (\boldsymbol{\omega} \times \mathbf{r}_{C/O})$$

$\mathbf{r}_{A/O}$ O A $\mathbf{r}_{C/O}$ C

(c)

Substituting $\mathbf{r}_{C/O} = -R\mathbf{j}$ [see Fig. (c)], $\boldsymbol{\omega} = -\omega\mathbf{k}$, and $\boldsymbol{\alpha} = -\alpha\mathbf{k}$, this equation becomes

$$\begin{aligned}\mathbf{a}_C &= R\alpha\mathbf{i} + (-\alpha\mathbf{k}) \times (-R\mathbf{j}) + (-\omega\mathbf{k}) \times [(-\omega\mathbf{k}) \times (-R\mathbf{j})] \\ &= R\alpha\mathbf{i} - R\alpha\mathbf{i} + (-\omega\mathbf{k}) \times (-R\omega\mathbf{i}) \\ &= R\omega^2\mathbf{j}\end{aligned}$$

Answer

Part 2

The relative acceleration equation between A and O becomes, on substituting $\mathbf{r}_{A/O} = -R\mathbf{i}$ [see Fig. (c)],

$$\begin{aligned}\mathbf{a}_A &= \mathbf{a}_O + \boldsymbol{\alpha} \times \mathbf{r}_{A/O} + \boldsymbol{\omega} \times (\boldsymbol{\omega} \times \mathbf{r}_{A/O}) \\ &= R\alpha\mathbf{i} + (-\alpha\mathbf{k}) \times (-R\mathbf{i}) + (-\omega\mathbf{k}) \times [(-\omega\mathbf{k}) \times (-R\mathbf{i})] \\ &= R\alpha\mathbf{i} + R\alpha\mathbf{j} + (-\omega\mathbf{k}) \times (R\omega\mathbf{j}) \\ &= R\alpha\mathbf{i} + R\alpha\mathbf{j} + R\omega^2\mathbf{i} \\ &= R(\alpha + \omega^2)\mathbf{i} + R\alpha\mathbf{j}\end{aligned}$$

Answer

Sample Problem 16.12

Bar AB of the mechanism shown in Fig. (a) is rotating clockwise with a constant angular velocity of 3 rad/s. Determine the angular acceleration of bar BC and the acceleration of the slider C at the instant when bar AB makes an angle of 30° with the horizontal as shown. (This problem was solved in Sample Problem 16.4 by differentiating the equation of constraint.)

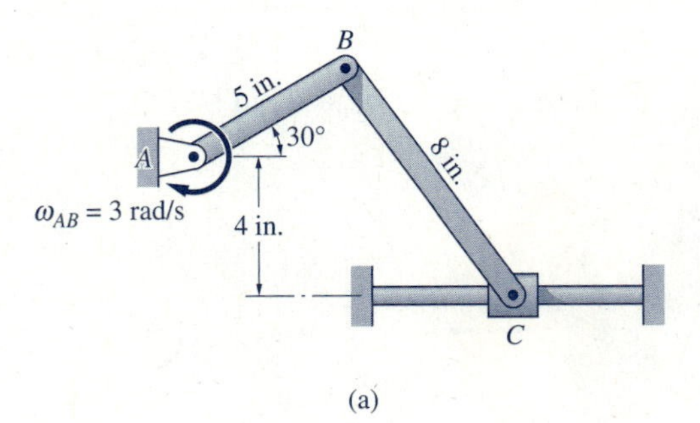

(a)

Solution

Introductory Comments

We will solve this problem using semigraphical notation (Solution I) and vector notation (Solution II). Since the angular velocity of bar BC is not given, it must be calculated before the accelerations can be found. We assume that this has already been done, the result being $\omega_{BC} = 2.785$ rad/s counterclockwise.* Furthermore, we assume that the value of angle β shown in Fig. (b) has been computed by trigonometry.

Clearly, B and C are the kinematically important points on bar BC: The acceleration of B can be computed from the prescribed motion of bar AB, and the path of point C is known. Therefore, the problem can be solved from the relative acceleration equation between points B and C.

We assume that the acceleration of C is directed to the right and that the angular acceleration of BC is counterclockwise. The angular acceleration of AB is zero.

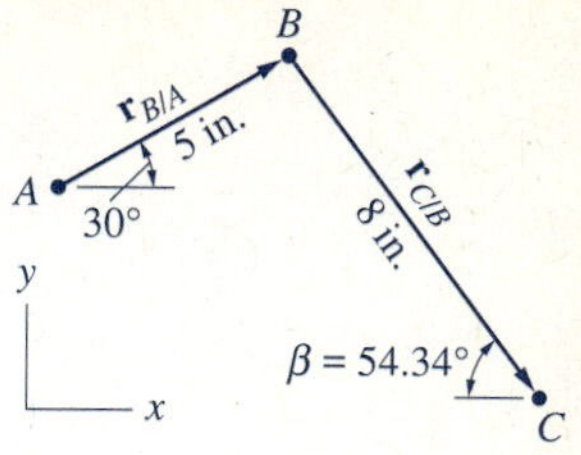

(b)

Solution I (using semigraphical notation)

The acceleration equation of C relative to B is

$$\mathbf{a}_C = \mathbf{a}_B + \mathbf{a}_{C/B} \tag{a}$$

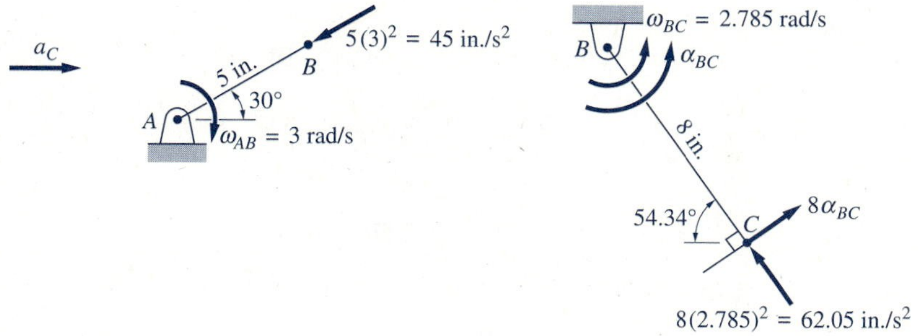

Inspection of Eq. (a) reveals that it contains two unknowns: a_C and α_{BC}. Equating the horizontal and vertical components, we obtain the following two scalar equations.

$$\xrightarrow{+} \quad a_C = -45\cos 30° + 8\alpha_{BC}\sin 54.34° - 62.05\cos 54.34° \tag{b}$$

$$+\uparrow \quad 0 = -45\sin 30° + 8\alpha_{BC}\cos 54.34° + 62.05\sin 54.34° \tag{c}$$

the solution of which is $a_C = -114.0$ in./s^2, and $\alpha_{BC} = -5.99$ rad/s^2. Therefore,

$$\mathbf{a}_C = 114.0 \text{ in./s}^2 \leftarrow \qquad \boldsymbol{\alpha}_{BC} = 5.99 \text{ rad/s}^2 \circlearrowright \qquad \textit{Answer}$$

Note that if ω_{BC} had not been determined previously, Eqs. (b) and (c) would contain ω_{BC} as a third unknown, making the equations unsolvable.

Solution II (using vector notation)

The relative position vectors shown in Fig. (b) are

$$\begin{aligned}\mathbf{r}_{B/A} &= 5\cos 30°\mathbf{i} + 5\sin 30°\mathbf{j} \\ &= 4.330\mathbf{i} + 2.500\mathbf{j} \text{ in.}\end{aligned} \tag{d}$$

* See the solution of either Sample Problem 16.4 (method of relative velocity), or Sample Problem 16.6 (instant centers).

and

$$\mathbf{r}_{C/B} = 8\cos 54.34°\mathbf{i} - 8\sin 54.34°\mathbf{j}$$
$$= 4.664\mathbf{i} \text{ in.} - 6.500\mathbf{j} \text{ in.} \tag{e}$$

The relative acceleration equation that solves the problem is

$$\mathbf{a}_C = \mathbf{a}_B + \mathbf{a}_{C/B} \tag{f}$$

Since the direction of $\mathbf{a}_C$ is horizontal, we have

$$\mathbf{a}_C = a_C\mathbf{i} \tag{g}$$

The acceleration of B in Eq. (f) can be determined by noting that B moves on a circular path centered at A. Therefore,

$$\mathbf{a}_B = \boldsymbol{\alpha}_{AB} \times \mathbf{r}_{B/A} + \boldsymbol{\omega}_{AB} \times (\boldsymbol{\omega}_{AB} \times \mathbf{r}_{B/A})$$

Substituting $\boldsymbol{\alpha}_{AB} = \mathbf{0}$, $\boldsymbol{\omega}_{AB} = -3\mathbf{k}$ rad/s, and $\mathbf{r}_{B/A}$ from (d), we get

$$\mathbf{a}_B = \mathbf{0} + (-3\mathbf{k}) \times [(-3\mathbf{k}) \times (4.330\mathbf{i} + 2.500\mathbf{j})]$$
$$= (-3\mathbf{k}) \times (-12.99\mathbf{j} + 7.500\mathbf{i})$$
$$= -38.97\mathbf{i} - 22.50\mathbf{j} \text{ in./s}^2 \tag{h}$$

According to Eqs. (16.12) and (16.13), the acceleration of C relative to B is

$$\mathbf{a}_{C/B} = \boldsymbol{\alpha}_{BC} \times \mathbf{r}_{C/B} + \boldsymbol{\omega}_{BC} \times (\boldsymbol{\omega}_{BC} \times \mathbf{r}_{C/B})$$

Substituting $\boldsymbol{\alpha}_{BC} = \alpha_{BC}\mathbf{k}$, $\boldsymbol{\omega}_{BC} = 2.785\mathbf{k}$ rad/s, and $\mathbf{r}_{C/B}$ from Eq. (e), this equation becomes

$$\mathbf{a}_{C/B} = (\alpha_{BC}\mathbf{k}) \times (4.664\mathbf{i} - 6.500\mathbf{j})$$
$$+ (2.785\mathbf{k}) \times [(2.785\mathbf{k}) \times (4.664\mathbf{i} - 6.500\mathbf{j})]$$
$$= 4.664\alpha_{BC}\mathbf{j} + 6.500\alpha_{BC}\mathbf{i}$$
$$+ (2.785\mathbf{k}) \times (12.99\mathbf{j} + 18.10\mathbf{i})$$
$$= 4.664\alpha_{BC}\mathbf{j} + 6.500\alpha_{BC}\mathbf{i} - 36.18\mathbf{i} + 50.41\mathbf{j} \text{ in./s}^2 \tag{i}$$

Substituting Eqs. (g), (h), and (i) into Eq. (f) and equating coefficients of $\mathbf{i}$ and $\mathbf{j}$, we obtain the following scalar equations.

$$a_C = -38.97 + 6.500\alpha_{BC} - 36.18 \tag{j}$$
$$0 = -22.50 + 4.664\alpha_{BC} + 50.41 \tag{k}$$

Solving Eqs. (j) and (k) simultaneously gives $a_C = -114.0$ in./s^2, and $\alpha_{BC} = -5.99$ rad/s^2. Therefore,

$$\mathbf{a}_C = -114.0\mathbf{i} \text{ in./s}^2 \quad \text{and} \quad \boldsymbol{\alpha}_{BC} = -5.99\mathbf{k} \text{ rad/s}^2$$ *Answer*

Sample Problem 16.13

When the linkage in Fig. (a) is in the position shown, bar AB is rotating with angular velocity $\omega_{AB} = 2.4$ rad/s and angular acceleration $\alpha_{AB} = 1.5$ rad/s^2, both counterclockwise. Determine the angular accelerations of bars BC and CD for this position.

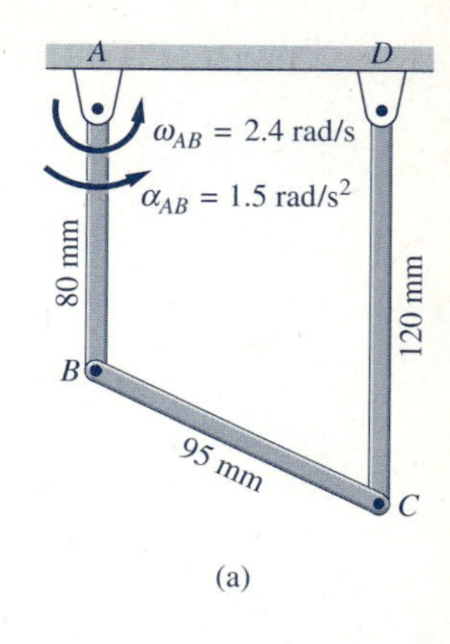

(a)

Solution

Preliminary Calculations

This problem will be solved using semigraphical notation (Solution I) and vector notation (Solution II).

Inspection of the linkage in Fig. (a) reveals that A, B, C, and D are the kinematically important points: A and D are fixed, and the paths of B and C (which are points on the same rigid bar BC) are known to be circles centered at A and D, respectively. Figure (c) shows the relative position vectors between the kinematically important points. The angle between BC and the horizontal was found to be $\theta = \sin^{-1}(40/95) = 24.90°$.

Before the angular accelerations can be found, the angular velocities of bars BC and CD must be known. The velocities can be determined using either the relative velocity method or the instant centers for velocities, with the latter being more convenient for this problem.

Because both $\mathbf{v}_B$ and $\mathbf{v}_C$ are horizontal, as shown in Fig. (b), the instant center for bar BC is at infinity. Therefore, BC is translating at this instant; i.e., $\omega_{BC} = 0$.

The magnitude of $\mathbf{v}_B$ is found to be $v_B = r_{B/A}\omega_{AB} = 80(2.4) = 192.0$ mm/s, the sense being to the right since ω_{AB} is directed counterclockwise. As the bar BC is translating, it follows that $v_C = v_B = 192.0$ mm/s (all points of a translating body possess the same velocities), also directed to the right. Therefore $\omega_{CD} = v_C/r_{C/D} = 192.0/120 = 1.6$ rad/s with a counterclockwise direction, as shown in Fig. (b). Summarizing these results in vector notation, we have

$$\boldsymbol{\omega}_{AB} = 2.4\mathbf{k} \text{ rad/s} \qquad \boldsymbol{\omega}_{BC} = \mathbf{0} \qquad \boldsymbol{\omega}_{CD} = 1.6\mathbf{k} \text{ rad/s} \tag{a}$$

Assuming that α_{BC} and α_{CD} are both counterclockwise, the angular accelerations are

$$\boldsymbol{\alpha}_{AB} = 1.5\mathbf{k} \text{ rad/s}^2 \qquad \boldsymbol{\alpha}_{BC} = \alpha_{BC}\mathbf{k} \text{ rad/s}^2 \qquad \boldsymbol{\alpha}_{CD} = \alpha_{CD}\mathbf{k} \text{ rad/s}^2 \tag{b}$$

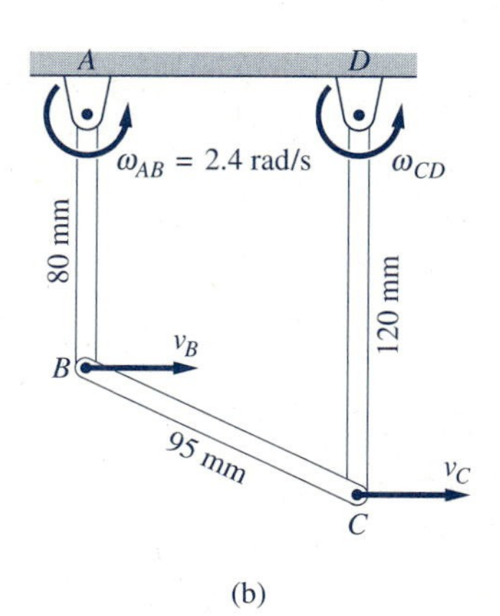

(b)

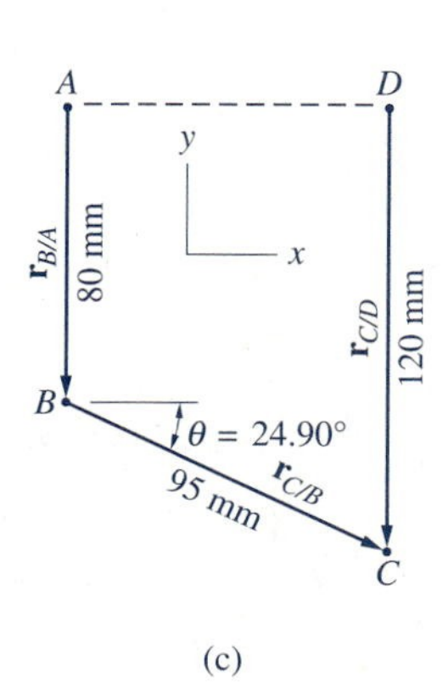

(c)

Solution I (using semigraphical notation)

The relative acceleration equation between points B and C is $\mathbf{a}_C = \mathbf{a}_B + \mathbf{a}_{C/B}$. The expression for the relative acceleration $\mathbf{a}_{C/B}$ is obtained by imagining that point B is fixed at the instant of concern. The accelerations $\mathbf{a}_B$ and $\mathbf{a}_C$ are derived from the fact that bars AB and CD rotate about the fixed points A and D, respectively. Therefore, the relative acceleration equation becomes

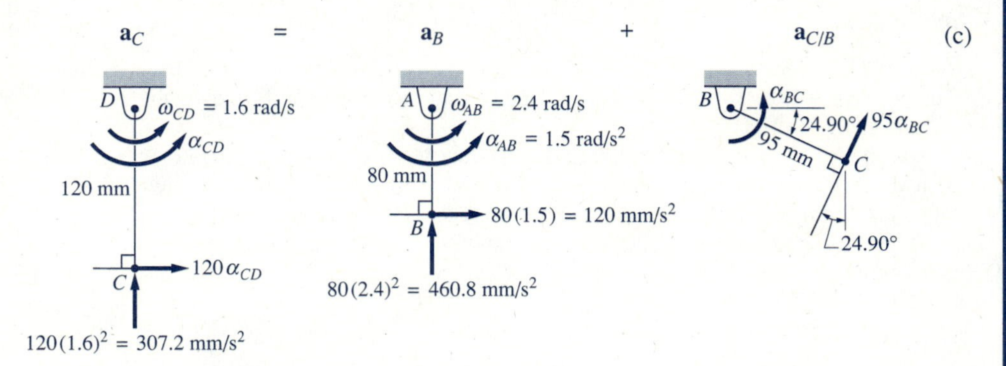

Inspection of this vector equation reveals that there are two unknowns: α_{BC} and α_{CD}, which can be found by solving the two equivalent scalar equations. Equating horizontal and vertical components of Eq. (c):

$$\xrightarrow{+} \quad 120\alpha_{CD} = 120 + 95\alpha_{BC} \sin 24.90^\circ \tag{d}$$

$$+\uparrow \quad 307.2 = 460.8 + 95\alpha_{BC} \cos 24.90^\circ \tag{e}$$

Solving Eqs. (d) and (e) gives $\alpha_{BC} = -1.783 \text{ rad/s}^2$ and $\alpha_{CD} = 0.406 \text{ rad/s}^2$, i.e.,

$$\boldsymbol{\alpha}_{BC} = 1.783 \text{ rad/s}^2 \circlearrowright \qquad \boldsymbol{\alpha}_{CD} = 0.406 \text{ rad/s}^2 \circlearrowleft \qquad \textit{Answer}$$

Solution II (using vector notation)

The relative position vectors shown in Fig. (c) can be written in vector form as

$$\left.\begin{aligned} \mathbf{r}_{B/A} &= -80\mathbf{j} \text{ mm} \\ \mathbf{r}_{C/B} &= 95 \cos 24.90^\circ \mathbf{i} - 95 \sin 24.90^\circ \mathbf{j} \\ &= 86.17\mathbf{i} - 40.00\mathbf{j} \text{ mm} \\ \mathbf{r}_{C/D} &= -120\mathbf{j} \text{ mm} \end{aligned}\right\} \tag{f}$$

The relative acceleration equation between C and B is

$$\mathbf{a}_C = \mathbf{a}_B + \mathbf{a}_{C/B} \tag{g}$$

Because the path of C is a circle centered at D, we have

$$\mathbf{a}_C = \boldsymbol{\alpha}_{CD} \times \mathbf{r}_{C/D} + \boldsymbol{\omega}_{CD} \times (\boldsymbol{\omega}_{CD} \times \mathbf{r}_{C/D}) \tag{h}$$

Substituting the vectors in Eqs. (a), (b), and (f), the result is

$$\begin{aligned} \mathbf{a}_C &= (\alpha_{CD}\mathbf{k}) \times (-120\mathbf{j}) + (1.6\mathbf{k}) \times [(1.6\mathbf{k}) \times (-120\mathbf{j})] \\ &= 120\alpha_{CD}\mathbf{i} + (1.6\mathbf{k}) \times (192.0\mathbf{i}) \\ &= 120\alpha_{CD}\mathbf{i} + 307.2\mathbf{j} \text{ mm/s}^2 \end{aligned} \tag{i}$$

Noting that B moves on a circular path centered at A, we conclude that

$$\begin{aligned}
\mathbf{a}_B &= \boldsymbol{\alpha}_{AB} \times \mathbf{r}_{B/A} + \boldsymbol{\omega}_{AB} \times (\boldsymbol{\omega}_{AB} \times \mathbf{r}_{B/A}) \\
&= (1.5\mathbf{k}) \times (-80\mathbf{j}) + (2.4\mathbf{k}) \times [(2.4\mathbf{k}) \times (-80\mathbf{j})] \\
&= 120\mathbf{i} + (2.4\mathbf{k}) \times (192.0\mathbf{i}) \\
&= 120\mathbf{i} + 460.8\mathbf{j} \text{ mm/s}^2 \qquad \text{(j)}
\end{aligned}$$

From Eqs. (16.13), we obtain

$$\begin{aligned}
\mathbf{a}_{C/B} &= \boldsymbol{\alpha}_{BC} \times \mathbf{r}_{C/B} + \boldsymbol{\omega}_{BC} \times (\boldsymbol{\omega}_{BC} \times \mathbf{r}_{C/B}) \\
&= (\alpha_{BC}\mathbf{k}) \times (86.17\mathbf{i} - 40.00\mathbf{j}) + \mathbf{0} \\
&= 86.17\alpha_{BC}\mathbf{j} + 40.00\alpha_{BC}\mathbf{i} \qquad \text{(k)}
\end{aligned}$$

Substituting Eqs. (i)–(k) into Eq. (g), and equating the coefficients of **i** and **j**, yields the following scalar equations.

$$120\alpha_{CD} = 120 + 40.00\alpha_{BC} \qquad \text{(l)}$$

$$307.2 = 460.8 + 86.17\alpha_{BC} \qquad \text{(m)}$$

Solving Eqs. (l) and (m) gives $\alpha_{BC} = -1.783 \text{ rad/s}^2$ and $\alpha_{CD} = 0.406 \text{ rad/s}^2$, or

$$\boldsymbol{\alpha}_{BC} = -1.783\mathbf{k} \text{ rad/s}^2 \quad \text{and} \quad \boldsymbol{\alpha}_{CD} = 0.406\mathbf{k} \text{ rad/s}^2 \qquad \textit{Answer}$$

PROBLEMS

16.64 The rectangular block, which is suspended from two identical rods AB and DE, is released from rest in the position shown. If the initial angular acceleration of each rod is 2 rad/s^2 counterclockwise, calculate the initial acceleration of G, the center of the block.

16.65 In the position shown, the collar A is moving to the left, and its speed is decreasing at the rate of 36 in./s^2. Knowing that the acceleration of G, the midpoint of bar AB, is zero, determine (a) the angular velocity and angular acceleration of bar AB; and (b) the acceleration of end B.

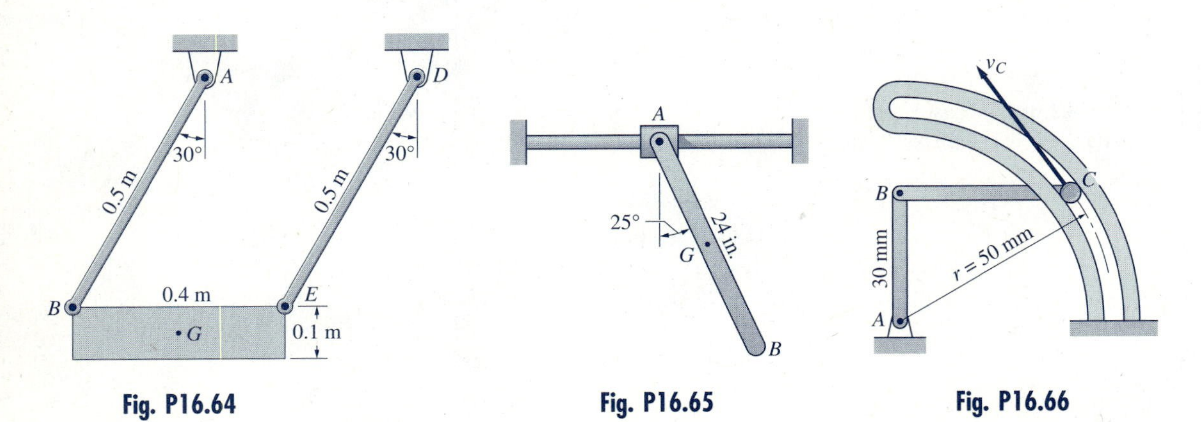

Fig. P16.64 **Fig. P16.65** **Fig. P16.66**

16.66 The pin C at the end of bar BC moves in a circular slot. The velocity of C is v_C = 0.15 m/s in the position shown, and it is increasing at the rate of 0.25 m/s^2. For this position, compute (a) the angular accelerations of bars AB and BC; and (b) the magnitude of the acceleration of pin C.

16.67 The wheel rolls on its 12-in. radius hub without slipping. The angular velocity of the wheel is 3 rad/s. Determine the acceleration of point D on the rim of the wheel if the angular acceleration of the wheel is α = 6.75 rad/s^2 (a) clockwise; and (b) counterclockwise.

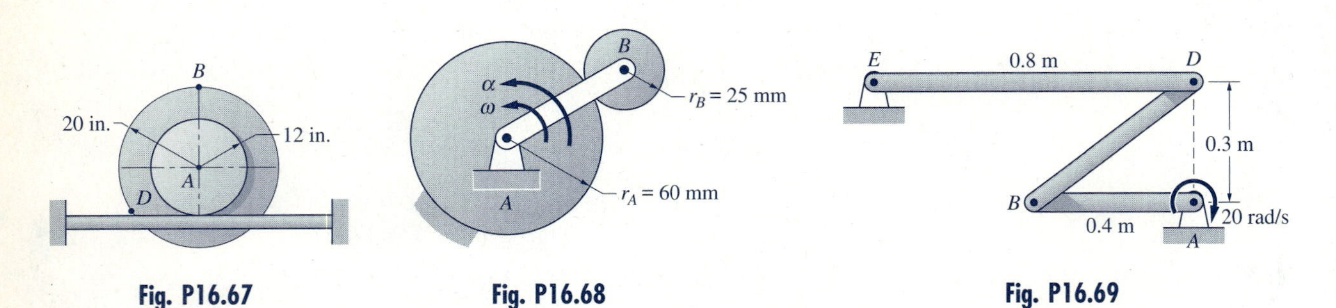

Fig. P16.67 **Fig. P16.68** **Fig. P16.69**

16.68 The arm joining the two friction wheels A and B has the angular velocity ω = 5 rad/s and the angular acceleration α = 12.5 rad/s^2, both counterclockwise. Assuming that A is stationary and that there is no slipping, calculate the acceleration of the point on B that is in contact with A.

16.69 Bar AB is rotating clockwise with a constant angular velocity of 20 rad/s. For the position shown, determine the angular accelerations of bars BD and DE.

16.70 Bar AB is released from rest in the position shown. If the initial angular acceleration of the bar is α (clockwise), determine the acceleration of the midpoint G immediately after release. Express your answer in terms of α, θ, and L.

16.71 Slider D is moving to the right with the constant velocity $v_D = 2.4$ ft/s. For the position shown, find the angular acceleration of bar BD and the acceleration of G, the midpoint of BD.

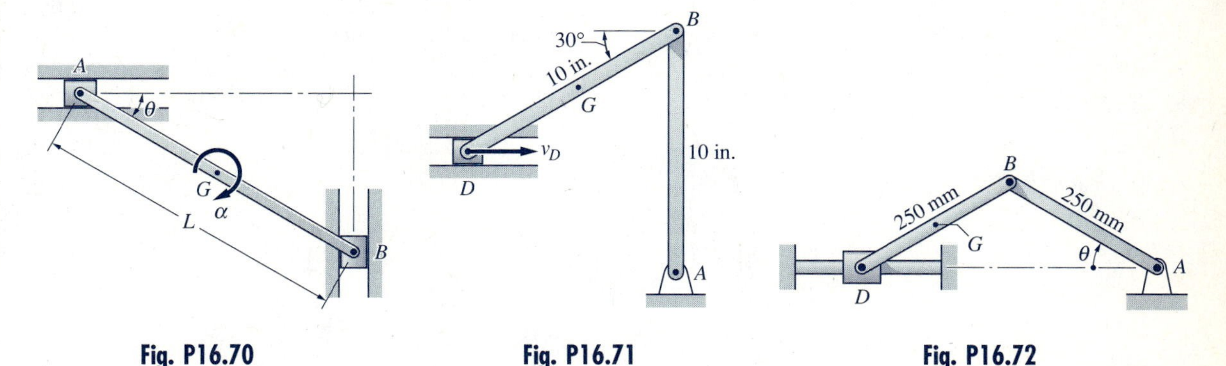

Fig. P16.70

Fig. P16.71

Fig. P16.72

16.72 The angular position of bar AB varies with time according to $\theta = (\pi/3)\sin 2\pi t$ radians, where t is in seconds. Calculate the acceleration of G, the midpoint of BD, when $t = 1.2$ s.

16.73 The wheel rolls without slipping. In the position shown, the angular velocity of the wheel is 4 rad/s counterclockwise, and its angular acceleration is 5 rad/s^2 clockwise. Find the angular acceleration of rod AB and the acceleration of the slider at B.

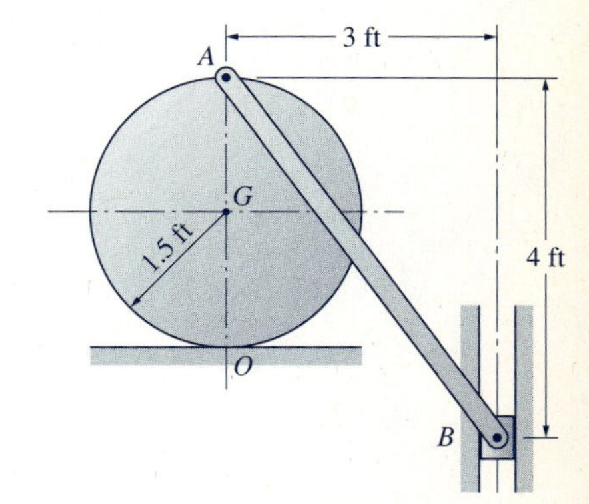

Fig. P16.73

16.74 The disk is rotating counterclockwise with the constant angular speed 2 rad/s. For the position shown, find the angular accelerations of bars AB and BD.

16.75 The weight A is suspended from a cord that is wrapped around the hub of wheel B. Given the velocity and acceleration of A in the figure, determine the angular accelerations of bars DE and EF in the position shown.

16.76 Link AB of the mechanism shown rotates with the constant angular speed 4 rad/s counterclockwise. For the position shown, calculate the angular accelerations of BD and DE.

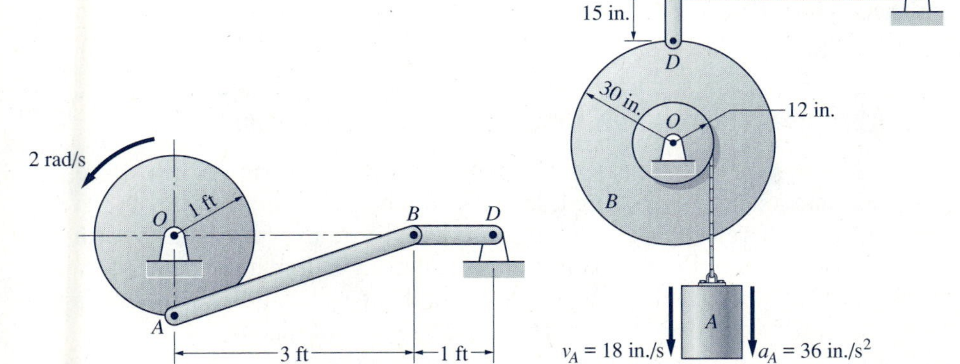

Fig. P16.74

Fig. P16.75

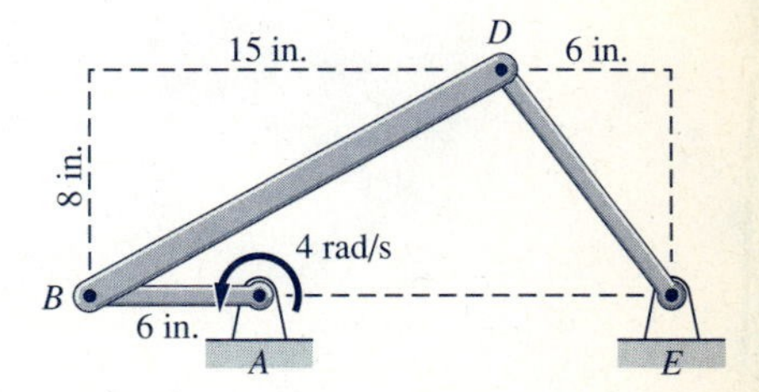

Fig. P16.76

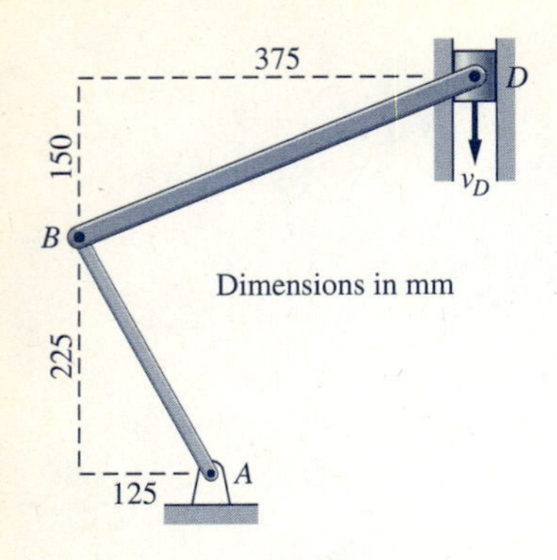

Fig. P16.77

16.77 When the mechanism is in the position shown, the velocity of slider D is v_D = 1.25 m/s, and its acceleration is zero. (a) Verify that the angular velocities of bars AB and BD are ω_{AB} = 1.818 rad/s counterclockwise and ω_{BD} = 2.727 rad/s clockwise. (b) Determine the acceleration vector of point B.

16.78 The wheel rolls to the left without slipping, with the velocity of its center A being constant at 0.6 m/s. For the position shown, (a) verify that the angular velocity of bar BD is 0.8039 rad/s clockwise; and (b) compute the angular acceleration of BD and the acceleration of point D.

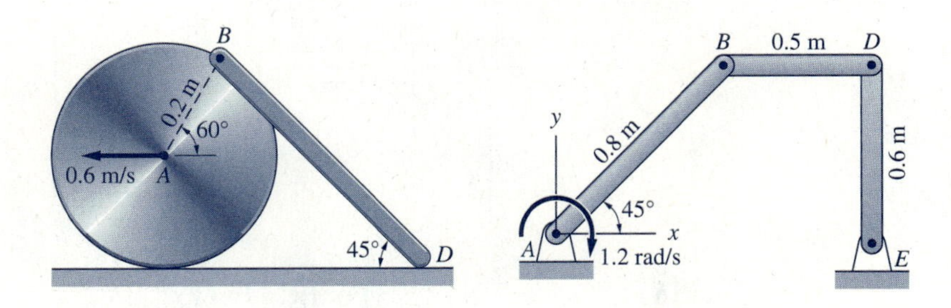

Fig. P16.78

Fig. P16.79

16.79 Bar AB of the mechanism rotates with the constant angular velocity 1.2 rad/s clockwise. For the position shown, (a) verify that the angular velocities of the other two bars are ω_{BD} = 1.358 rad/s counterclockwise and ω_{DE} = 1.131 rad/s clockwise; and (b) determine the acceleration vector of point D.

16.7 Absolute and Relative Derivatives of Vectors

a. Introductory comments

The rest of this chapter discusses problems that require the analysis of the relative motion between two points that are not on the same rigid body. For problems of this type, it is convenient to describe the motion relative to a reference frame that is embedded in one of the bodies, i.e., a frame that moves with the body. This article discusses the derivatives of vectors that are described in such a moving coordinate system. The next article applies these concepts to plane motion of rigid bodies that are joined by nonrigid connections, such as a pin sliding in a slot.

b. Absolute and relative time derivatives of vectors

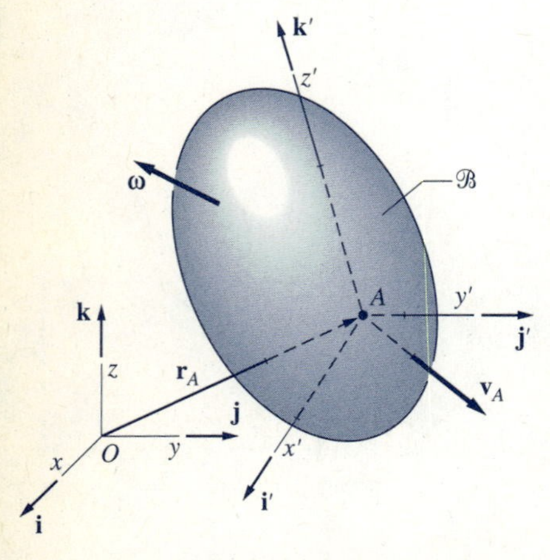

Fig. 16.13

Figure 16.13 shows two rectangular reference frames. The xyz-frame is fixed in space and its base vectors are $\mathbf{i}$, $\mathbf{j}$, and $\mathbf{k}$. The origin A of the $x'y'z'$-system is moving with velocity $\mathbf{v}_A = \dot{\mathbf{r}}_A$, and the coordinate axes are rotating with the angular velocity $\boldsymbol{\omega}$. The base vectors of the $x'y'z'$-frame are denoted by $\mathbf{i}'$, $\mathbf{j}'$, and $\mathbf{k}'$. It is convenient to imagine that the $x'y'z'$-coordinate system is embedded in a rigid body $\mathcal{B}$ that is both rotating and translating.

We distinguish between two types of derivatives of a time-dependent vector $\mathbf{V}$. The *absolute* derivative of $\mathbf{V}$ (or simply, the derivative of $\mathbf{V}$) measures the rate of change of $\mathbf{V}$ relative to the fixed reference frame. The *relative* derivative of $\mathbf{V}$ refers to the rate of change of $\mathbf{V}$ with respect to the $x'y'z'$-frame, i.e., the change in $\mathbf{V}$ as seen by an observer who translates and rotates with the body $\mathcal{B}$. The absolute derivative, which is the only derivative we have been using up to this point, and the relative derivative will be denoted as follows:

$$\text{absolute derivative} = \frac{d\mathbf{V}}{dt} = \dot{\mathbf{V}} \tag{16.15}$$

$$\text{relative derivative} = \left(\frac{d\mathbf{V}}{dt}\right)_{/\mathcal{B}} \tag{16.16}$$

where the subscript "$/\mathcal{B}$" denotes that the derivative is measured relative to the body $\mathcal{B}$.

c. Derivatives of the unit base vectors

Since the xyz-coordinate system is fixed in space, the *absolute* derivatives of its unit base vectors are zero; i.e.,

$$\frac{d\mathbf{i}}{dt} = \mathbf{0} \qquad \frac{d\mathbf{j}}{dt} = \mathbf{0} \qquad \frac{d\mathbf{k}}{dt} = \mathbf{0} \tag{16.17}$$

Furthermore, since the unit vectors $\mathbf{i}'$, $\mathbf{j}'$, and $\mathbf{k}'$ are attached to the body $\mathcal{B}$, it follows that their *relative* derivatives vanish; i.e.,

$$\left(\frac{d\mathbf{i}'}{dt}\right)_{/\mathcal{B}} = \mathbf{0} \qquad \left(\frac{d\mathbf{j}'}{dt}\right)_{/\mathcal{B}} = \mathbf{0} \qquad \left(\frac{d\mathbf{k}'}{dt}\right)_{/\mathcal{B}} = \mathbf{0} \tag{16.18}$$

Of particular interest to us are the absolute derivatives of $\mathbf{i}'$, $\mathbf{j}'$, and $\mathbf{k}'$. Setting $\mathbf{r}_{B/A}$ in Eq. (16.1) equal to $\mathbf{i}'$, $\mathbf{j}'$, and $\mathbf{k}'$ in turn, we find that

$$\frac{d\mathbf{i}'}{dt} = \boldsymbol{\omega} \times \mathbf{i}' \qquad \frac{d\mathbf{j}'}{dt} = \boldsymbol{\omega} \times \mathbf{j}' \qquad \frac{d\mathbf{k}'}{dt} = \boldsymbol{\omega} \times \mathbf{k}' \tag{16.19}$$

Observe that these derivatives depend only on the rotation of the $x'y'z'$ reference frame and not on its translation.

d. Relationships between absolute and relative derivatives

An arbitrary vector $\mathbf{V}$ may be expressed in terms of its components in the fixed (xyz) reference frame as

$$\mathbf{V} = V_x\mathbf{i} + V_y\mathbf{j} + V_z\mathbf{k} \tag{16.20}$$

Utilizing Eq. (16.17), the absolute derivative of this vector becomes

$$\boxed{\frac{d\mathbf{V}}{dt} = \dot{V}_x\mathbf{i} + \dot{V}_y\mathbf{j} + \dot{V}_z\mathbf{k}} \tag{16.21}$$

Similarly, the description of **V** in the moving $(x'y'z')$ frame is

$$\mathbf{V} = V_{x'}\mathbf{i}' + V_{y'}\mathbf{j}' + V_{z'}\mathbf{k}' \tag{16.22}$$

The relative derivative of **V** can be obtained with the help of Eq. (16.18).

$$\left(\frac{d\mathbf{V}}{dt}\right)_{/\mathcal{B}} = \dot{V}_{x'}\mathbf{i}' + \dot{V}_{y'}\mathbf{j}' + \dot{V}_{z'}\mathbf{k}' \tag{16.23}$$

Note that $\dot{V}_x$, $\dot{V}_y$, $\dot{V}_z$, $\dot{V}_{x'}$, $\dot{V}_{y'}$, and $\dot{V}_{z'}$ are simply the derivatives of scalars. Since scalars do not depend on a coordinate system, the expression "relative derivative of a scalar" is meaningless.

Of primary interest is the expression for the *absolute* derivative of **V** when the vector is described in the moving reference frame. From Eq. (16.22) we have

$$\begin{aligned}\frac{d\mathbf{V}}{dt} &= \frac{d}{dt}(V_{x'}\mathbf{i}' + V_{y'}\mathbf{j}' + V_{z'}\mathbf{k}') \\ &= (\dot{V}_{x'}\mathbf{i}' + \dot{V}_{y'}\mathbf{j}' + \dot{V}_{z'}\mathbf{k}') + \left(V_{x'}\frac{d\mathbf{i}'}{dt} + V_{y'}\frac{d\mathbf{j}'}{dt} + V_{z'}\frac{d\mathbf{k}'}{dt}\right)\end{aligned}$$

Substituting for $d\mathbf{i}'/dt$, $d\mathbf{j}'/dt$, and $d\mathbf{k}'/dt$ from Eq. (16.19), we obtain

$$\frac{d\mathbf{V}}{dt} = (\dot{V}_{x'}\mathbf{i}' + \dot{V}_{y'}\mathbf{j}' + \dot{V}_{z'}\mathbf{k}') + V_{x'}(\boldsymbol{\omega} \times \mathbf{i}') + V_{y'}(\boldsymbol{\omega} \times \mathbf{j}') + V_{z'}(\boldsymbol{\omega} \times \mathbf{k}')$$

which can be written as

$$\frac{d\mathbf{V}}{dt} = (\dot{V}_{x'}\mathbf{i}' + \dot{V}_{y'}\mathbf{j}' + \dot{V}_{z'}\mathbf{k}') + \boldsymbol{\omega} \times \mathbf{V}$$

Identifying the term in parentheses as the relative derivative of **V** from Eq. (16.23), we obtain

$$\boxed{\frac{d\mathbf{V}}{dt} = \left(\frac{d\mathbf{V}}{dt}\right)_{/\mathcal{B}} + \boldsymbol{\omega} \times \mathbf{V}} \tag{16.24}$$

Observe that Eq. (16.24) gives the relationship between the absolute and relative derivatives of a vector.

The relationship between the absolute and relative second derivatives of **V** can be obtained by differentiating Eq. (16.24).

$$\frac{d^2\mathbf{V}}{dt^2} = \frac{d}{dt}\left(\frac{d\mathbf{V}}{dt}\right) = \frac{d}{dt}\left[\left(\frac{d\mathbf{V}}{dt}\right)_{/\mathcal{B}} + \boldsymbol{\omega} \times \mathbf{V}\right] \tag{16.25}$$

Applying Eq. (16.24) to the vector in the brackets and simplifying, we obtain

$$\boxed{\frac{d^2\mathbf{V}}{dt^2} = \left(\frac{d^2\mathbf{V}}{dt^2}\right)_{/\mathcal{B}} + 2\boldsymbol{\omega} \times \left(\frac{d\mathbf{V}}{dt}\right)_{/\mathcal{B}} + \dot{\boldsymbol{\omega}} \times \mathbf{V} + \boldsymbol{\omega} \times (\boldsymbol{\omega} \times \mathbf{V})} \tag{16.26}$$

e. Two special cases

The following two special cases are of special interest in the study of dynamics.

1. Nonrotating Reference Frame

If the $x'y'z'$-frame is not rotating—i.e., $\boldsymbol{\omega} = \mathbf{0}$—Eqs. (16.24) and (16.26) yield

$$\frac{d\mathbf{V}}{dt} = \left(\frac{d\mathbf{V}}{dt}\right)_{/\mathcal{B}} \tag{16.27}$$

and

$$\frac{d^2\mathbf{V}}{dt^2} = \left(\frac{d^2\mathbf{V}}{dt^2}\right)_{/\mathcal{B}} \tag{16.28}$$

Therefore, the absolute and relative derivatives are identical for a nonrotating reference frame, even if the frame is translating.

2. Vectors Embedded in the Moving Frame

If the vector $\mathbf{V}$ is embedded in the moving $x'y'z'$-frame (body $\mathcal{B}$), its components $V_{x'}$, $V_{y'}$, and $V_{z'}$ remain constant, which means that $(d\mathbf{V}/dt)_{/\mathcal{B}}$ and $(d^2\mathbf{V}/dt^2)_{/\mathcal{B}}$ are zero. In this case, the first and second derivative of $\mathbf{V}$ become

$$\frac{d\mathbf{V}}{dt} = \boldsymbol{\omega} \times \mathbf{V} \tag{16.29}$$

and

$$\frac{d^2\mathbf{V}}{dt^2} = \dot{\boldsymbol{\omega}} \times \mathbf{V} + \boldsymbol{\omega} \times (\boldsymbol{\omega} \times \mathbf{V}) \tag{16.30}$$

Note that the expressions in Eqs. (16.29) and (16.30) are similar to those for the first and second derivatives of the position vector $\mathbf{r}$ of a point in the body that is rotating about a fixed axis—see Eqs. (16.3) and (16.4). These results would be expected because the vector $\mathbf{r}$ is also embedded in the rotating body.

16.8 Motion Relative to a Rotating Reference Frame

This article discusses the relative motion between two points that do not belong to the same rigid body. As we will see, relative motion of this type can be described conveniently by using a rotating frame of reference. Although the present chapter has been devoted to plane motion, the following discussion also applies to three-dimensional motion, except as noted.

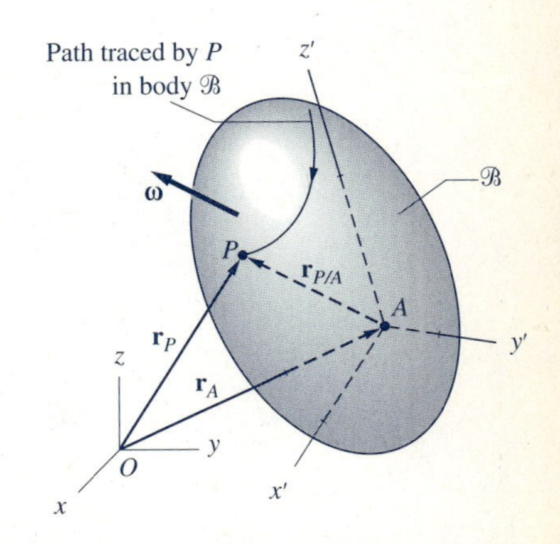

Fig. 16.14

Figure 16.14 shows a rigid body $\mathcal{B}$, the angular velocity of which is $\boldsymbol{\omega}$. Point P moves independently of $\mathcal{B}$ along a relative path that lies in $\mathcal{B}$. Point P' (not shown) is the point *embedded* in $\mathcal{B}$ that is coincident with P at the instant shown. The xyz-coordinate system is fixed, whereas the $x'y'z'$-coordinates are embedded in $\mathcal{B}$, point A being the origin.

From Fig. 16.14 we have $\mathbf{r}_P = \mathbf{r}_A + \mathbf{r}_{P/A}$, which on differentiation with respect to time yields

$$\mathbf{v}_P = \mathbf{v}_A + \mathbf{v}_{P/A} \tag{16.31}$$

where the relative velocity is

$$\mathbf{v}_{P/A} = \frac{d\mathbf{r}_{P/A}}{dt} \tag{16.32}$$

The relative velocity in Eq. (16.32) is measured relative to the fixed xyz-coordinate system. In many situations, however, the velocity of P relative to the moving body $\mathcal{B}$ is easier to describe. The relationship between these two velocities can be found by replacing $\mathbf{V}$ by $\mathbf{r}_{P/A}$ in Eq. (16.24) of the previous article, which gives

$$\frac{d\mathbf{r}_{P/A}}{dt} = \left(\frac{d\mathbf{r}_{P/A}}{dt}\right)_{/\mathcal{B}} + \boldsymbol{\omega} \times \mathbf{r}_{P/A} \tag{16.33}$$

where the subscript "$/\mathcal{B}$" denotes quantities that are measured relative to the body $\mathcal{B}$. Introducing the notation

$$\boxed{\mathbf{v}_{P/\mathcal{B}} = \left(\frac{d\mathbf{r}_{P/A}}{dt}\right)_{/\mathcal{B}}} \tag{16.34}$$

and substituting Eq. (16.33) into Eq. (16.31), the velocity of P becomes

$$\mathbf{v}_P = \mathbf{v}_A + \boldsymbol{\omega} \times \mathbf{r}_{P/A} + \mathbf{v}_{P/\mathcal{B}} \tag{16.35}$$

Since P' is embedded in the body $\mathcal{B}$, its velocity relative to A is given by

$$\boxed{\mathbf{v}_{P'/A} = \boldsymbol{\omega} \times \mathbf{r}_{P/A}} \tag{16.36}$$

Therefore, Eq. (16.35) can be written in the form

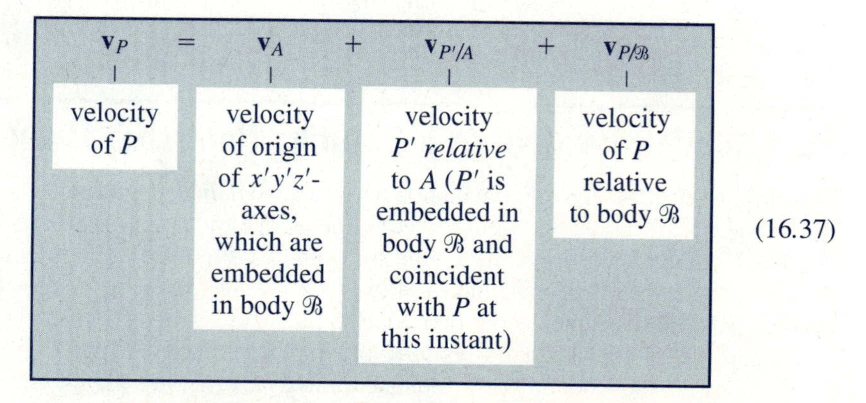

(16.37)

Figure 16.15 displays the terms that appear in Eq. (16.37) for the case of plane motion (the angular velocity $\boldsymbol{\omega}$ is assumed to be counterclockwise).

Observe that the motion is represented as a translation and a rotation of the body plus the velocity of P relative to the body. The velocity $\mathbf{v}_P$ is tangent to the absolute path of P, whereas $\mathbf{v}_{P/\mathcal{B}}$ is tangent to its relative path. The rotation term in Fig. 16.15 is shown as the vector cross product $\boldsymbol{\omega} \times \mathbf{r}_{P/A}$, but this could be replaced by the scalar representation in Fig. 16.8(d).

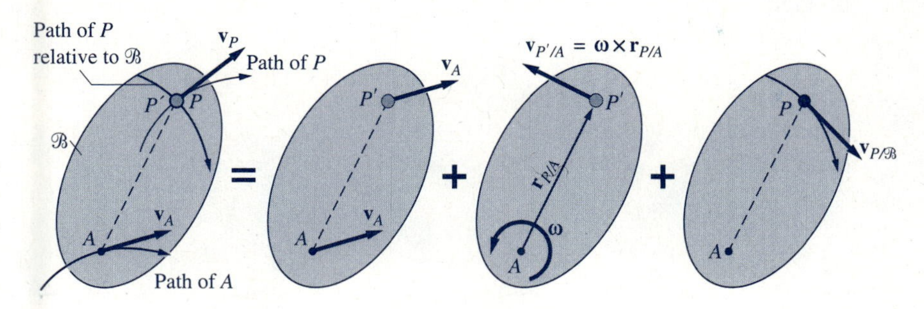

Fig. 16.15

The acceleration of P is found by differentiating Eq. (16.31).

$$\mathbf{a}_P = \mathbf{a}_A + \mathbf{a}_{P/A} \tag{16.38}$$

where the relative acceleration is

$$\mathbf{a}_{P/A} = \frac{d\mathbf{v}_{P/A}}{dt} = \frac{d^2\mathbf{r}_{P/A}}{dt^2} \tag{16.39}$$

Letting $\mathbf{r}_{P/A}$ replace $\mathbf{V}$ in Eq. (16.26) of the previous article, we see that the right-hand side of Eq. (16.39) may be written as

$$\mathbf{a}_{P/A} = \left(\frac{d^2\mathbf{r}_{P/A}}{dt^2}\right)_{/\mathcal{B}} + \dot{\boldsymbol{\omega}} \times \mathbf{r}_{P/A} + \boldsymbol{\omega} \times (\boldsymbol{\omega} \times \mathbf{r}_{P/A}) + 2\boldsymbol{\omega} \times \mathbf{v}_{P/\mathcal{B}} \tag{16.40}$$

The first term of Eq. (16.40) is the acceleration of P relative to body $\mathcal{B}$:

$$\mathbf{a}_{P/\mathcal{B}} = \left(\frac{d^2\mathbf{r}_{P/A}}{dt^2}\right)_{/\mathcal{B}} \tag{16.41}$$

The next two terms in Eq. (16.40) represent the acceleration of P' relative to A, namely

$$\mathbf{a}_{P'/A} = \dot{\boldsymbol{\omega}} \times \mathbf{r}_{P/A} + \boldsymbol{\omega} \times (\boldsymbol{\omega} \times \mathbf{r}_{P/A}) \tag{16.42}$$

The last term in Eq. (16.40), called the *Coriolis acceleration* (named after the French mathematician G. G. Coriolis), will be denoted by

$$\mathbf{a}_C = 2\boldsymbol{\omega} \times \mathbf{v}_{P/\mathcal{B}} \tag{16.43}$$

Observe that the Coriolis acceleration represents the interaction between the angular velocity of the body and the velocity of P relative to the body.

Substituting Eqs. (16.40)–(16.43) into Eq. (16.38), the acceleration of P becomes

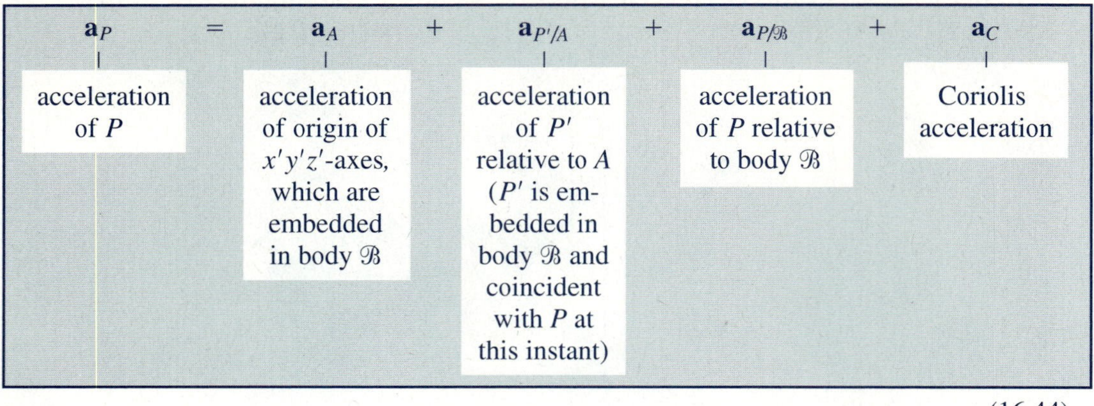

(16.44)

Figure 16.16 illustrates the terms that appear in this equation for the case of plane motion (the angular velocity $\boldsymbol{\omega}$ and the angular acceleration $\dot{\boldsymbol{\omega}}$ are assumed to be counterclockwise). Note that the motion is represented as a translation and rotation of the body plus the acceleration of P relative to the body plus the Coriolis acceleration. The rotation terms in Fig. 16.16 are shown using their vector representations, $\dot{\boldsymbol{\omega}} \times \mathbf{r}_{P/A}$ and $\boldsymbol{\omega} \times (\boldsymbol{\omega} \times \mathbf{r}_{P/A})$; they can be replaced by the scalar representations shown in Fig. 16.12(d). As seen in Fig. 16.16, the Coriolis acceleration $\mathbf{a}_C$ is perpendicular to both $\mathbf{v}_{P/\mathcal{B}}$ and $\boldsymbol{\omega}$. When the scalar representation is used, the magnitude of $\mathbf{a}_C$ is $2\omega v_{P/\mathcal{B}}$, and its direction can be determined by fixing the tail of $\mathbf{v}_{P/\mathcal{B}}$ and rotating this vector through 90° in the direction of ω.

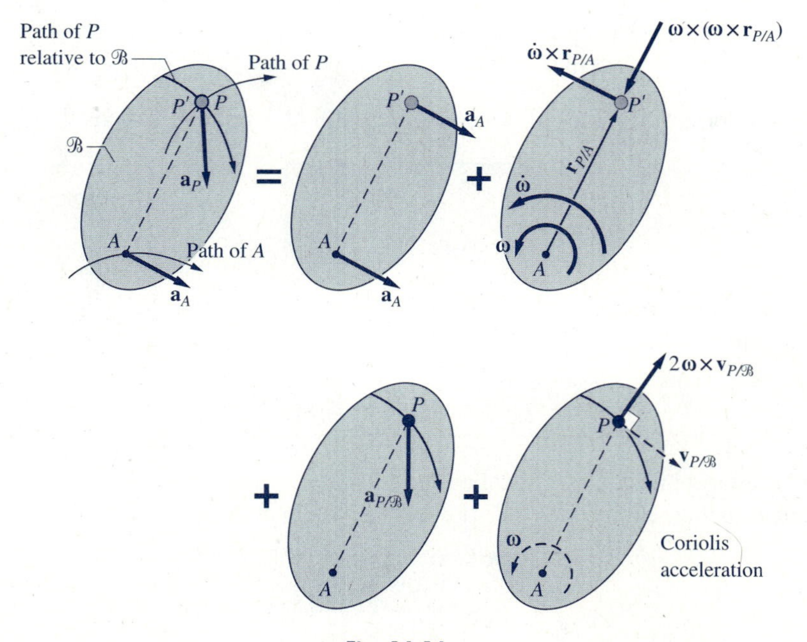

Fig. 16.16

Sample Problem 16.14

As shown in Fig. (a), a particle P slides from A toward B along a semicircular rod AB of radius 200 mm. The rod rotates about the pin at A, and the speed of P relative to the rod is constant at 120 mm/s. When the system is in the position shown, the angular velocity and angular acceleration of the rod are $\omega_{AB} = 0.8$ rad/s counterclockwise and $\alpha_{AB} = 0.5$ rad/s^2 clockwise. For this position, determine the velocity and acceleration vectors of P.

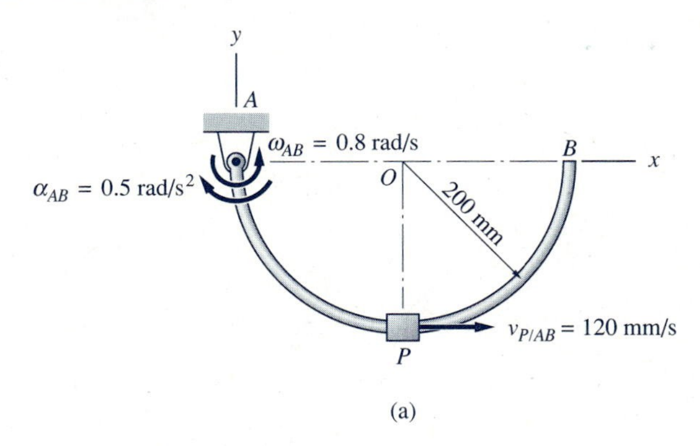

(a)

Solution

Preliminary Comments

This problem will be solved using semigraphical notation (Solution I) and vector notation (Solution II). In both solutions we employ point P', identified as the point on AB that coincides with P at the instant of concern. The relative position vectors required in the solutions are shown in Fig. (b).

Note that (1) the absolute path of P' is a circle that is centered at point A, and (2) the path of P relative to AB is a circle that is centered at point O.

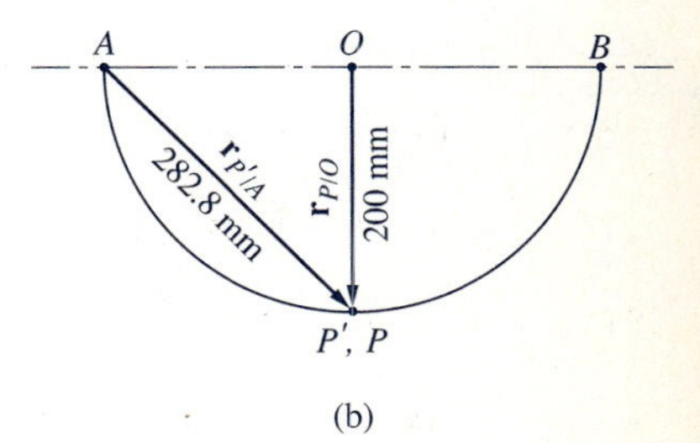

(b)

Solution I (using semigraphical notation)

Letting body $\mathcal{B}$ in Eq. (16.37) be the rod AB, the velocity of P becomes $\mathbf{v}_P = \mathbf{v}_A + \mathbf{v}_{P'/A} + \mathbf{v}_{P/AB}$. In semigraphical notation, this equation may be written as

$$\mathbf{v}_P = \mathbf{v}_A + \mathbf{v}_{P'/A} + \mathbf{v}_{P/AB} \qquad \text{(a)}$$

$\mathbf{v}_A = \mathbf{0}$; $\mathbf{v}_{P'/A}$: A, ω_{AB} = 0.8 rad/s, 282.8 mm, 45°, P', 282.8(0.8) = 226.2 mm/s; $\mathbf{v}_{P/AB}$: 120 mm/s →

from which we find that

$$(v_P)_x = 226.2 \sin 45° + 120 = 280 \text{ mm/s}$$

$$(v_P)_y = 226.2 \cos 45° = 160 \text{ mm/s}$$

or

$\mathbf{v}_P =$ 160, 322 mm/s, 280 *Answer*

Using Eq. (16.44), the acceleration of P is

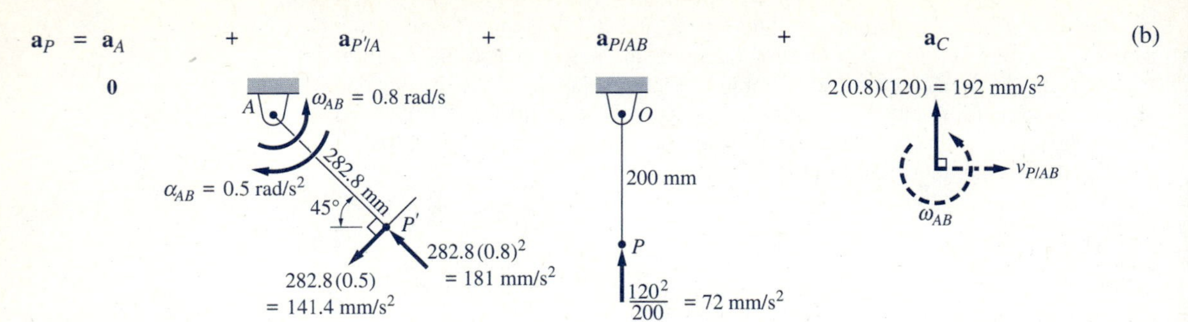

Note that in Eq. (b) the direction of the Coriolis acceleration $\mathbf{a}_C$ is found by fixing the tail of $\mathbf{v}_{P/AB}$ and then rotating this vector 90° in the direction of ω_{AB}. Furthermore, observe that $\mathbf{a}_{P/AB}$ contains only the normal component $v_{P/AB}^2/r_{P/O}$ because the magnitude of $\mathbf{v}_{P/AB}$ is constant. Evaluating the components of Eq. (b) gives

$$(a_P)_x = -181\cos 45^\circ - 141.4\sin 45^\circ = -228 \text{ mm/s}^2$$
$$(a_P)_y = 181\sin 45^\circ - 141.4\cos 45^\circ + 72 + 192 = 292 \text{ mm/s}^2$$

or

$\mathbf{a}_P$ = [370 mm/s²; components 228, 292] *Answer*

Solution II (using vector notation)

From Eq. (16.37) the velocity of P is

$$\mathbf{v}_P = \mathbf{v}_A + \mathbf{v}_{P'/A} + \mathbf{v}_{P/AB} \quad \text{(c)}$$

Since A is a fixed point, we have

$$\mathbf{v}_A = \mathbf{0} \quad \text{(d)}$$

Noting that P', being a point that is embedded in rod AB, travels along a circular path centered at A, we get

$$\mathbf{v}_{P'/A} = \boldsymbol{\omega}_{AB} \times \mathbf{r}_{P'/A} = (0.8\mathbf{k}) \times (200\mathbf{i} - 200\mathbf{j}) = 160\mathbf{j} + 160\mathbf{i} \text{ mm/s} \quad \text{(e)}$$

The velocity of P relative to bar AB is given as

$$\mathbf{v}_{P/AB} = 120\mathbf{i} \text{ mm/s} \quad \text{(f)}$$

Substituting Eqs. (d)–(f) into Eq. (c), the velocity of P becomes

$$\mathbf{v}_P = \mathbf{0} + (160\mathbf{j} + 160\mathbf{i}) + (120\mathbf{i})$$
$$= 280\mathbf{i} + 160\mathbf{j} \text{ mm/s} \quad \textit{Answer}$$

The acceleration of P is, according to Eq. (16.44),

$$\mathbf{a}_P = \mathbf{a}_A + \mathbf{a}_{P'/A} + \mathbf{a}_{P/AB} + \mathbf{a}_C \tag{g}$$

Since A is a fixed point, we have

$$\mathbf{a}_A = \mathbf{0} \tag{h}$$

Because the path of P' is a circle with its center at A, the acceleration of P' relative to A (which is, of course, also the absolute acceleration of P' considering that A is a fixed point) is

$$\begin{aligned}\mathbf{a}_{P'/A} &= \boldsymbol{\alpha}_{AB} \times \mathbf{r}_{P'/A} + \boldsymbol{\omega}_{AB} \times (\boldsymbol{\omega}_{AB} \times \mathbf{r}_{P'/A}) \\ &= (-0.5\mathbf{k}) \times (200\mathbf{i} - 200\mathbf{j}) + (0.8\mathbf{k}) \times (160\mathbf{i} + 160\mathbf{j}) \\ &= (-100\mathbf{j} - 100\mathbf{i}) + (128\mathbf{j} - 128\mathbf{i}) \\ &= -228\mathbf{i} + 28\mathbf{j} \text{ mm/s}^2 \end{aligned} \tag{i}$$

The acceleration of P relative to AB has only a normal component because $v_{P/AB}$ is constant. Since the normal component of the relative acceleration is directed toward the center of curvature of the relative path (i.e., toward point O), we find that

$$\mathbf{a}_{P/AB} = (\mathbf{a}_{P/AB})_n = \frac{v_{P/AB}^2}{r_{P/O}}\mathbf{j} = \frac{(120)^2}{200}\mathbf{j} = 72\mathbf{j} \text{ mm/s}^2 \tag{j}$$

The Coriolis acceleration from Eq. (16.43) is

$$\mathbf{a}_C = 2\boldsymbol{\omega}_{AB} \times \mathbf{v}_{P/AB} = 2(0.8\mathbf{k}) \times (120\mathbf{i}) = 192\mathbf{j} \text{ mm/s}^2 \tag{k}$$

Substituting Eqs. (h)–(k) into Eq. (g), we obtain

$$\begin{aligned}\mathbf{a}_P &= 0 + (-228\mathbf{i} + 28\mathbf{j}) + 72\mathbf{j} + 192\mathbf{j} \\ &= -228\mathbf{i} + 292\mathbf{j} \text{ mm/s}^2 \end{aligned} \qquad \textit{Answer}$$

Sample Problem 16.15

Crank AB of the quick-return mechanism shown in Fig. (a) rotates counterclockwise with a constant angular velocity $\omega_{AB} = 6$ rad/s. When the mechanism is in the position shown, calculate the velocity and acceleration of the slider B relative to arm DE and the angular velocity and acceleration of arm DE.

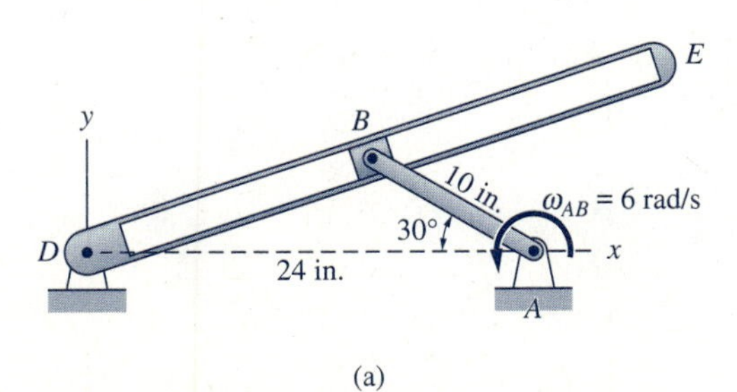

(a)

Solution

Introductory Comments

Note that the path of slider B relative to arm DE is the slot in arm DE. Since this relative path is a straight line, both $\mathbf{v}_{B/DE}$ and $\mathbf{a}_{B/DE}$ (the velocity and acceleration of B relative to arm DE) are directed along the slot.

We let B' be the point on DE that is coincident with B at the instant of concern. The velocity and acceleration of B from Eqs. (16.37) and (16.44), respectively, are $\mathbf{v}_B = \mathbf{v}_D + \mathbf{v}_{B'/D} + \mathbf{v}_{B/DE}$ and $\mathbf{a}_B = \mathbf{a}_D + \mathbf{a}_{B'/D} + \mathbf{a}_{B/DE} + \mathbf{a}_C$, where $\mathbf{a}_C$ is the Coriolis acceleration. Noting that D is a fixed point ($\mathbf{v}_D = \mathbf{0}$ and $\mathbf{a}_D = \mathbf{0}$), the velocity and acceleration equations become

$$\mathbf{v}_B = \mathbf{v}_{B'/D} + \mathbf{v}_{B/DE} \tag{a}$$

and

$$\mathbf{a}_B = \mathbf{a}_{B'/D} + \mathbf{a}_{B/DE} + \mathbf{a}_C \tag{b}$$

These equations will be analyzed using semigraphical notation (Solution I) and vector notation (Solution II).

Throughout the analyses it must be kept in mind that the paths of B and B' are circles that are centered at A and D, respectively. We assume that ω_{DE} and α_{DE} are counterclockwise and that $\mathbf{v}_{B/DE}$ and $\mathbf{a}_{B/DE}$ are both directed toward D, as indicated in Fig. (b). Figure (c) shows the relative position vectors required for the analysis. The distance $\overline{DB}$ and the angle between DE and the x-axis were determined by trigonometry.

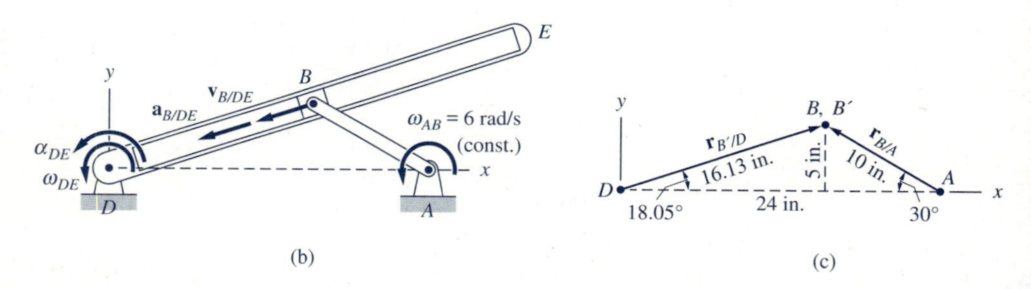

(b) (c)

The following vectors are involved in Eqs. (a) and (b).

$$\mathbf{r}_{B/A} = 10(-\cos 30°\mathbf{i} + \sin 30°\mathbf{j})$$
$$= -8.660\mathbf{i} + 5.00\mathbf{j} \text{ in.} \tag{c}$$

$$\mathbf{r}_{B'/D} = 16.13(\cos 18.05°\mathbf{i} + \sin 18.05°\mathbf{j})$$
$$= 15.34\mathbf{i} + 5.00\mathbf{j} \text{ in.} \tag{d}$$

$$\boldsymbol{\omega}_{AB} = 6\mathbf{k} \text{ rad/s} \qquad \boldsymbol{\alpha}_{AB} = \mathbf{0} \tag{e}$$

$$\boldsymbol{\omega}_{DE} = \omega_{DE}\mathbf{k} \text{ rad/s} \qquad \boldsymbol{\alpha}_{DE} = \alpha_{DE}\mathbf{k} \text{ rad/s}^2 \tag{f}$$

$$\mathbf{v}_{B/DE} = v_{B/DE} \swarrow 18.05°$$
$$= v_{B/DE}(-\cos 18.05°\mathbf{i} - \sin 18.05°\mathbf{j}) \text{ in./s} \tag{g}$$

$$\mathbf{a}_{B/DE} = a_{B/DE} \swarrow 18.05°$$
$$= a_{B/DE}(-\cos 18.05°\mathbf{i} - \sin 18.05°\mathbf{j}) \text{ in./s}^2 \tag{h}$$

Solution I (using semigraphical notation)

Velocity

In semigraphical notation, Eq. (a) becomes

$$\mathbf{v}_B \quad = \quad \mathbf{v}_{B'/D} \quad + \quad \mathbf{v}_{B/DE} \qquad \text{(i)}$$

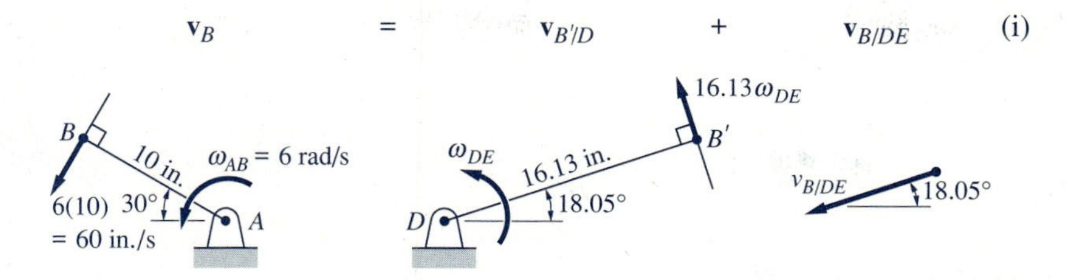

Inspection of Eq. (i) reveals that there are two unknowns, ω_{DE} and $v_{B/DE}$, which can be found by solving the two equivalent scalar equations.

Equating horizontal and vertical components of both sides of Eq. (i) gives

$$\xrightarrow{+} \quad -60 \sin 30^\circ = -16.13\omega_{DE} \sin 18.05^\circ - v_{B/DE} \cos 18.05^\circ \qquad \text{(j)}$$

$$+\uparrow \quad -60 \cos 30^\circ = 16.13\omega_{DE} \cos 18.05^\circ - v_{B/DE} \sin 18.05^\circ \qquad \text{(k)}$$

Solving Eqs. (j) and (k) yields $\omega_{DE} = -2.486$ rad/s and $v_{B/DE} = 44.63$ in./s, from which we find that

$$\boldsymbol{\omega}_{DE} = 2.486 \text{ rad/s} \circlearrowright \qquad \mathbf{v}_{B/DE} = 44.63 \text{ in./s} \;\; \swarrow 18.05^\circ \qquad \textit{Answer} \quad (1)$$

Acceleration

Written in semigraphical notation, Eq. (b) takes the form

$$\mathbf{a}_B \quad = \quad \mathbf{a}_{B'/D} \quad + \quad \mathbf{a}_{B/DE} \quad + \quad \mathbf{a}_C \qquad \text{(m)}$$

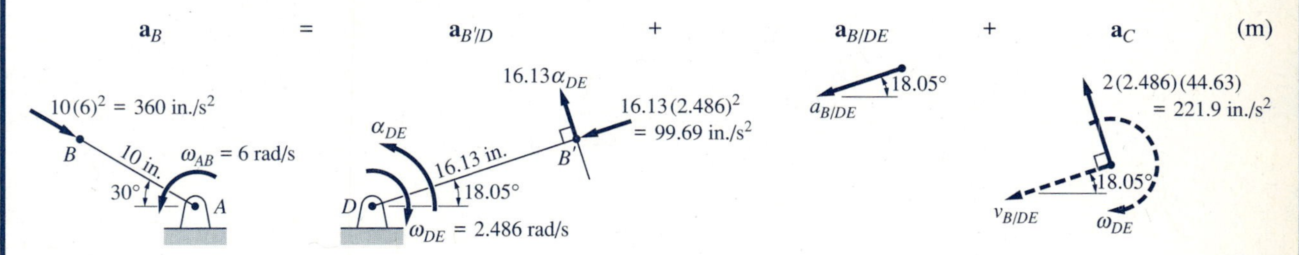

Note that in Eq. (m) the magnitude of the Coriolis acceleration is $a_C = 2\omega_{DE} v_{B/DE}$, and its direction is found by fixing the tail of $\mathbf{v}_{B/DE}$ and rotating this vector 90° in the direction of ω_{DE}. The unknowns in Eq. (m) are α_{DE} and $a_{B/DE}$, which can be solved by equating horizontal and vertical components.

$$\begin{aligned} \xrightarrow{+} \quad 360 \cos 30^\circ = & -16.13\alpha_{DE} \sin 18.05^\circ \\ & - 99.69 \cos 18.05^\circ - a_{B/DE} \cos 18.05^\circ \\ & - 221.9 \sin 18.05^\circ \end{aligned} \qquad \text{(n)}$$

$$+\uparrow \quad -360 \sin 30° = 16.13\alpha_{DE} \cos 18.05° - 99.69 \sin 18.05° - a_{B/DE} \sin 18.05° + 221.9 \cos 18.05° \qquad \text{(o)}$$

Solving Eqs. (n) and (o) simultaneously gives $\alpha_{DE} = -30.35 \text{ rad/s}^2$ and $a_{B/DE} = -340.3 \text{ in./s}^2$. Therefore the results are

$$\boldsymbol{\alpha}_{DE} = 30.35 \text{ rad/s}^2 \circlearrowright \quad \text{and} \quad \mathbf{a}_{B/DE} = 340 \text{ in./s}^2 \; \angle 18.05° \qquad \textit{Answer}$$

Solution II (using vector notation)

Velocity

Using vector algebra, the terms appearing in Eq. (a) are computed as described in the following.

Since B is a point on AB, we have

$$\begin{aligned} \mathbf{v}_B &= \boldsymbol{\omega}_{AB} \times \mathbf{r}_{B/A} \\ &= (6\mathbf{k}) \times (-8.660\mathbf{i} + 5.00\mathbf{j}) \\ &= -51.96\mathbf{j} - 30.00\mathbf{i} \text{ in./s} \end{aligned} \qquad \text{(p)}$$

Using the fact that B' is a point on DE, we have

$$\begin{aligned} \mathbf{v}_{B'/D} &= \boldsymbol{\omega}_{DE} \times \mathbf{r}_{B'/D} \\ &= (\omega_{DE}\mathbf{k}) \times (15.34\mathbf{i} + 5.00\mathbf{j}) \\ &= 15.34\omega_{DE}\mathbf{j} - 5.00\omega_{DE}\mathbf{i} \text{ in./s} \end{aligned} \qquad \text{(q)}$$

The relative velocity vector $\mathbf{v}_{B/DE}$ was found previously in Eq. (g):

$$\mathbf{v}_{B/DE} = v_{B/DE}(-\cos 18.05°\mathbf{i} - \sin 18.05°\mathbf{j}) \text{ in./s} \qquad \text{(r)}$$

Substituting Eqs. (p)–(r) into Eq. (a) and equating coefficients of **i** and **j**, respectively, we obtain

$$-30.00 = -5.00\omega_{DE} - v_{B/DE} \cos 18.05° \qquad \text{(s)}$$

$$-51.96 = 15.34\omega_{DE} - v_{B/DE} \sin 18.05° \qquad \text{(t)}$$

Solving Eqs. (s) and (t) simultaneously gives $\omega_{DE} = -2.486$ rad/s and $v_{B/DE} = 44.64$ in./s, from which we find that

$$\boldsymbol{\omega}_{DE} = -2.486\mathbf{k} \text{ rad/s} \qquad \textit{Answer}$$

and

$$\begin{aligned} \mathbf{v}_{B/DE} &= 44.64(-\cos 18.05°\mathbf{i} - \sin 18.05°\mathbf{j}) \\ &= -42.44\mathbf{i} - 13.83\mathbf{j} \text{ in./s} \end{aligned} \qquad \textit{Answer}$$

Acceleration

The terms in Eq. (b) will be computed next using vector notation.

Since B is a point on AB, the acceleration of B becomes

$$\begin{aligned} \mathbf{a}_B &= \boldsymbol{\omega}_{AB} \times (\boldsymbol{\omega}_{AB} \times \mathbf{r}_{B/A}) \\ &= (6\mathbf{k}) \times (-51.96\mathbf{j} - 30.00\mathbf{i}) \\ &= 311.8\mathbf{i} - 180\mathbf{j} \text{ in./s}^2 \end{aligned} \tag{u}$$

Because B' is a point on arm DE, we obtain

$$\begin{aligned} \mathbf{a}_{B'/D} &= \boldsymbol{\alpha}_{DE} \times \mathbf{r}_{B'/D} + \boldsymbol{\omega}_{DE} \times (\boldsymbol{\omega}_{DE} \times \mathbf{r}_{B'/D}) \\ &= (\alpha_{DE}\mathbf{k}) \times (15.34\mathbf{i} + 5.00\mathbf{j}) \\ &\quad + (-2.486\mathbf{k}) \times [(-2.486\mathbf{k}) \times (15.34\mathbf{i} + 5.00\mathbf{j})] \\ &= 15.34\alpha_{DE}\mathbf{j} - 5.00\alpha_{DE}\mathbf{i} \\ &\quad + (-2.486\mathbf{k}) \times (-38.14\mathbf{j} + 12.43\mathbf{i}) \\ &= 15.34\alpha_{DE}\mathbf{j} - 5.00\alpha_{DE}\mathbf{i} - 94.82\mathbf{i} - 30.90\mathbf{j} \text{ in./s}^2 \end{aligned} \tag{v}$$

The acceleration vector of B relative to arm DE was given in Eq. (h) to be

$$\mathbf{a}_{B/DE} = a_{B/DE}(-\cos 18.05^\circ\mathbf{i} - \sin 18.05^\circ\mathbf{j}) \text{ in./s}^2 \tag{w}$$

From Eq. (16.43), the Coriolis acceleration becomes

$$\begin{aligned} \mathbf{a}_C &= 2\boldsymbol{\omega}_{DE} \times \mathbf{v}_{B/DE} \\ &= 2(-2.486\mathbf{k}) \times (-42.44\mathbf{i} - 13.83\mathbf{j}) \\ &= 211.0\mathbf{j} - 68.76\mathbf{i} \text{ in./s}^2 \end{aligned} \tag{x}$$

Substituting Eqs. (u)–(x) into Eq. (b) and equating the coefficients of **i** and **j**, respectively, we obtain

$$311.8 = -5.00\alpha_{DE} - 94.82 - a_{B/DE}\cos 18.05^\circ - 68.76 \tag{y}$$

$$-180 = 15.34\alpha_{DE} - 30.90 - a_{B/DE}\sin 18.05^\circ + 211.0 \tag{z}$$

Solving Eqs. (y) and (z) simultaneously yields $\alpha_{DE} = -30.35$ rad/s^2 and $a_{B/DE} = -340.3$ in./s^2. Therefore, the results written in vector notation are

$$\boldsymbol{\alpha}_{DE} = -30.35\mathbf{k} \text{ rad/s}^2 \qquad \textit{Answer}$$

and

$$\begin{aligned} \mathbf{a}_{B/DE} &= 340.3(\cos 18.05^\circ\mathbf{i} + \sin 18.05^\circ\mathbf{j}) \\ &= 323.6\mathbf{i} + 105.4\mathbf{j} \text{ in./s}^2 \end{aligned} \qquad \textit{Answer}$$

PROBLEMS

16.80 Crank AD rotates with the constant clockwise angular velocity of 8 rad/s. For the position shown, determine the angular speed of rod BE and the velocity of slider D relative to BE.

16.81 In the position shown, the angular velocity of link AB is 2.5 rad/s clockwise. For this position, find the angular velocity of member EF and the velocity of bar BD relative to EF.

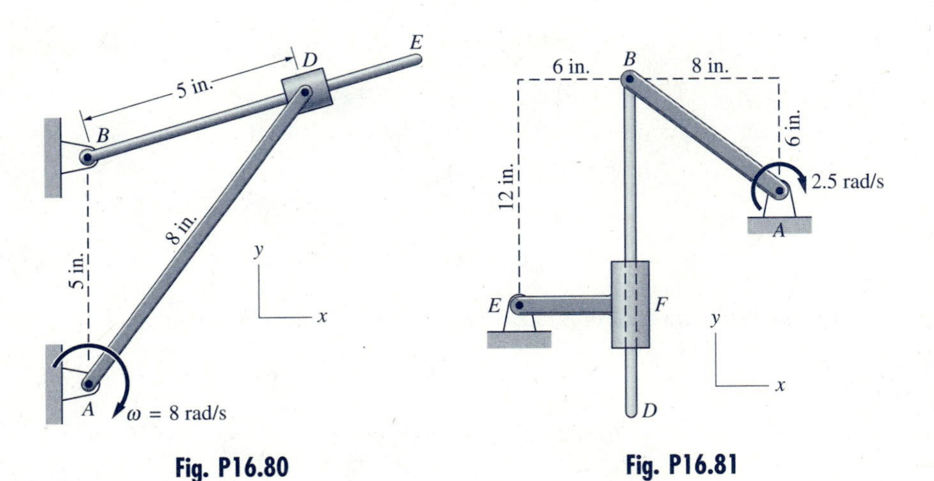

Fig. P16.80

Fig. P16.81

16.82 Crank DE of the mechanism rotates at 25 rad/s in the counterclockwise direction. For the position shown, determine the angular speed of bar AB and the velocity of slider A.

16.83 Both of the triangular frames $\mathcal{B}$ in (a) and (b) rotate about A with an angular velocity of 2 rad/s counterclockwise. At the same time, each slider P moves to the right relative to the frame with the relative speed of 0.2 m/s. Determine the acceleration of P for each case.

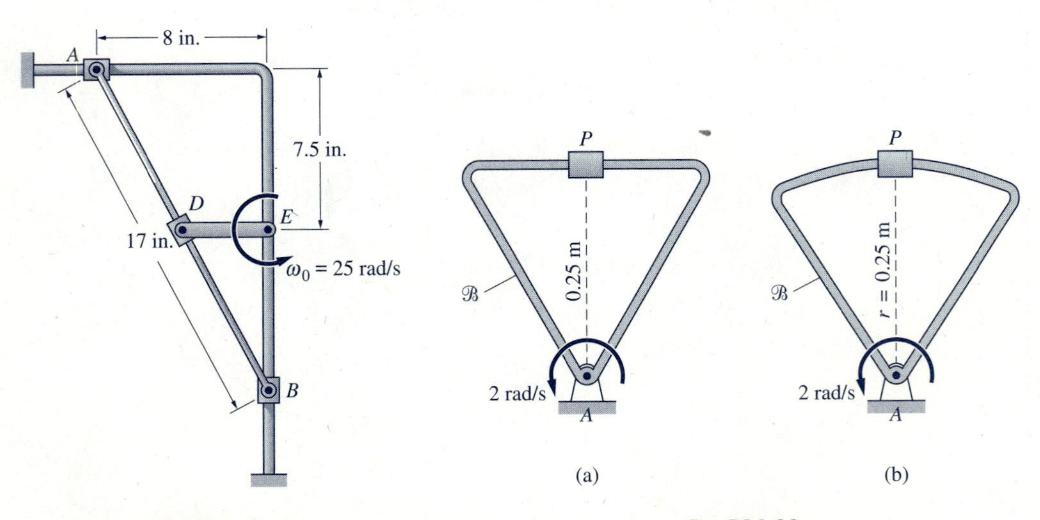

Fig. P16.82

Fig. P16.83

16.84 Rod OB rotates counterclockwise at the constant angular speed of 45 rev/min. At the same time, collar A is sliding toward B with the constant speed 2 ft/s relative to the rod. Calculate the acceleration of collar A when $R = 0.8$ ft and $\theta = 0$ by considering the rod OB to be a rotating reference frame. (This problem could also be solved using polar coordinates—see Prob. 13.19.)

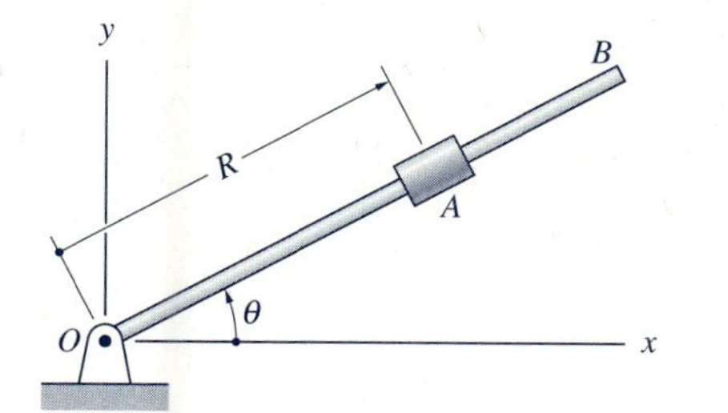

Fig. P16.84

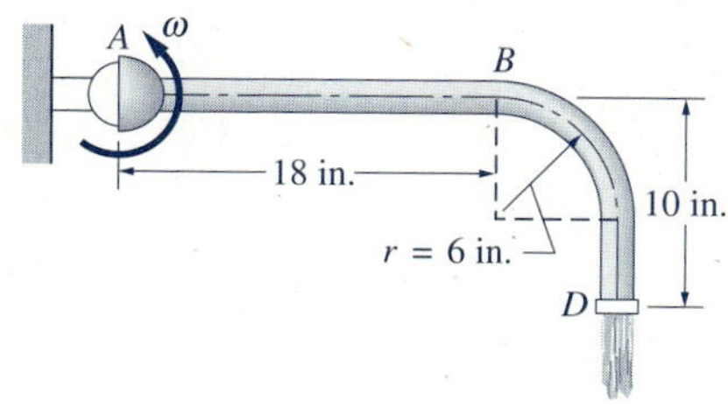

Fig. P16.85

16.85 Water entering the curved pipe at A is discharged at D. The pipe is rotating about A at the constant angular velocity $\omega = 10$ rad/s, and the water has a constant speed of 12 ft/s relative to the pipe. Determine the acceleration of the water (a) just after it enters the bend at B; and (b) just before it is discharged at D.

16.86 The figure shows a mechanism, called the Geneva stop, which converts the constant angular velocity of disk $\mathscr{A}$ into stop-and-go motion of the slotted disk $\mathscr{B}$. In the position shown, the pin P, which is attached to $\mathscr{A}$, is just entering a slot in disk $\mathscr{B}$. Compute the angular acceleration of $\mathscr{B}$ for this position. (Note that the angular velocity of $\mathscr{B}$ is zero at this instant.)

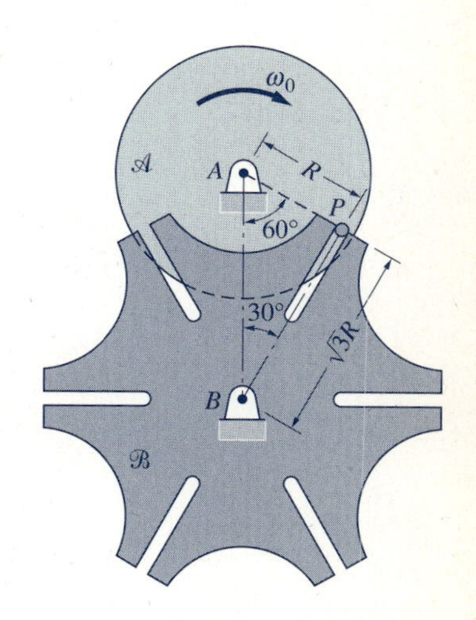

Fig. P16.86

16.87 Arm AB is rotating counterclockwise with the constant angular speed of 4 rad/s. At the same time, the disk is rotating clockwise with the angular speed 8 rad/s relative to AB. Determine the acceleration of point P on the rim of the disk by (a) considering AB as a rotating reference frame; and (b) using the relative acceleration method of Art. 16.6.

16.88 The disk $\mathscr{B}$ rotates about O at the constant angular speed ω. The particle P moves along the circular arc of radius ρ that is fixed in the disk. The speed $v_{P/\mathscr{B}}$ of P relative to the disk is constant. Show that the acceleration of P is always directed toward O if $v_{P/\mathscr{B}} = 2\omega\rho$.

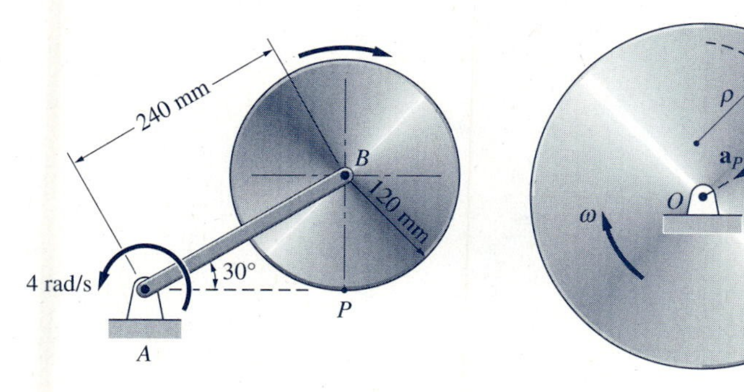

Fig. P16.87

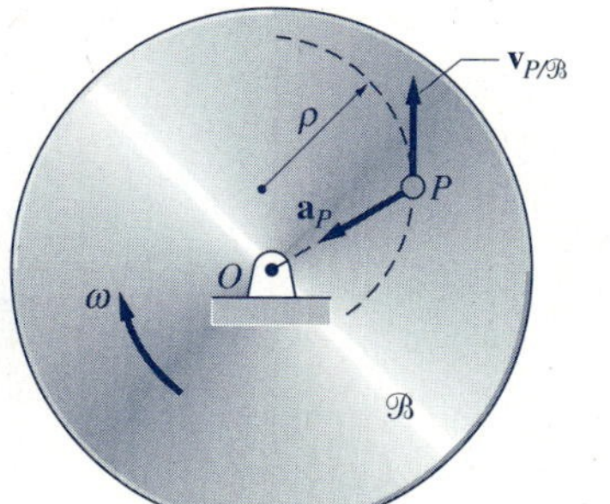

Fig. P16.88

16.89 Collar D of the mechanism is moving downward with the constant speed 15 in./s. In the position shown, determine the velocity and acceleration of rod DE relative to the collar B.

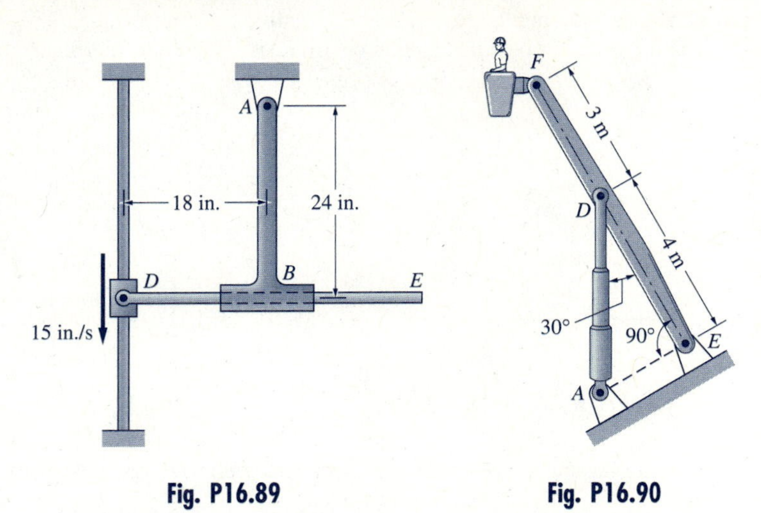

Fig. P16.89 **Fig. P16.90**

16.90 In the position shown, the hydraulic cylinder AD of the hoist is being extended at the constant rate of 0.25 m/s. For this position, calculate (a) the angular velocities of AD and the boom EF; and (b) the angular accelerations of AD and EF.

16.91 The crank AD rotates counterclockwise with the constant angular speed 12 rad/s. In the position shown, determine (a) the angular velocity of the slotted bar BE and the velocity of pin D relative to BE; and (b) the angular acceleration of BE and the acceleration of D relative to BE.

16.92 Rod AB of the mechanism rotates at the constant angular speed 8 rad/s clockwise. For the position shown, calculate (a) the angular velocity of rod BE; and (b) the angular acceleration of BE.

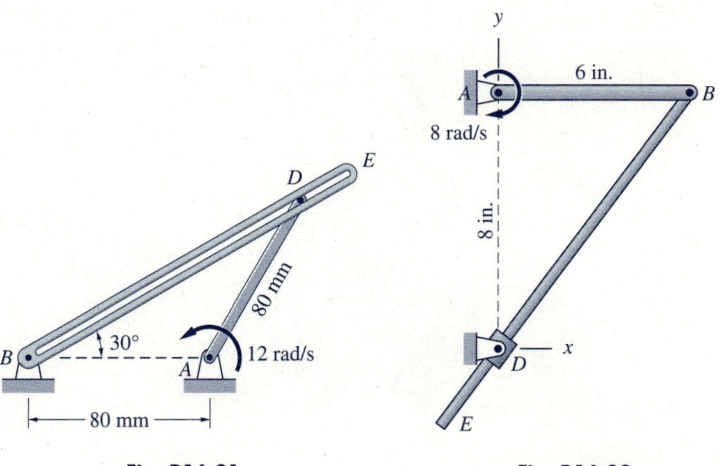

Fig. P16.91 **Fig. P16.92**

REVIEW PROBLEMS

16.93 In the position shown, velocities of corners A and B of the plate are $\mathbf{v}_A = 9.53\mathbf{j} - 4.0\mathbf{k}$ ft/s and $\mathbf{v}_B = u\mathbf{i} + 6.93\mathbf{k}$ ft/s, where u is an unknown. Knowing that the plate is rotating at a constant angular velocity about an axis that passes through O, determine (a) the angular velocity of the plate; and (b) the acceleration of corner A.

16.94 An electric motor reaches its operating speed in 5 rev after the power is turned on. Its angular acceleration during this period is $\alpha = \alpha_0 \cos(\theta/\theta_0)$, where $\alpha_0 = 5$ rad/s^2, $\theta_0 = 20$, and θ is the angular displacement of the rotor (in radians) measured from the rest position. Determine the operating speed of the motor.

16.95 Collar C slides on the horizontal guide rod with the constant velocity v_0. The rod CD is free to slide in sleeve B, which is rigidly attached to bar AB. Determine the angular velocity and angular acceleration of bar AB in terms of the length b and the angle θ.

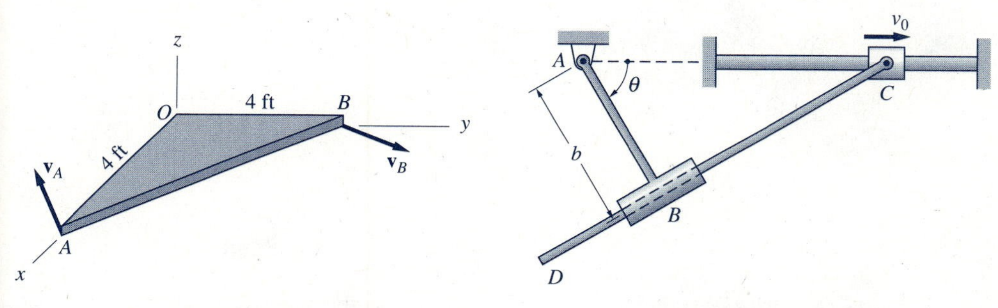

Fig. P16.93 **Fig. P16.95**

16.96 The bent rod ABC rotates about the axis AB. In the position shown, the angular velocity and acceleration of the rod are $\omega = 6$ rad/s and $\dot{\omega} = -25$ rad/s^2. Determine the velocity and acceleration of end C in this position.

16.97 The outer race of the ball bearing is rotating at 1500 rev/min; the inner race is stationary. The balls, which have a diameter of 20 mm, roll without slipping on both races. Determine the angular velocity of the balls and the speed of the cage (the cage maintains separation of the balls).

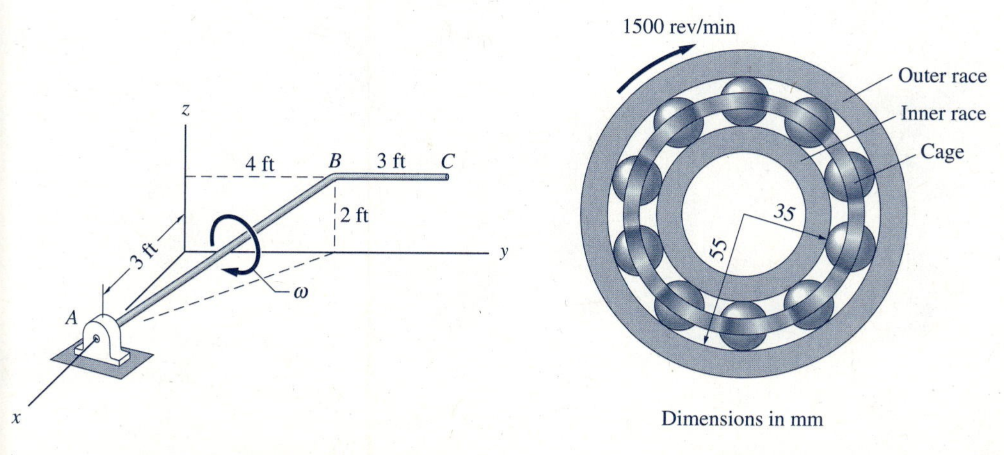

Fig. P16.96 **Fig. P16.97**

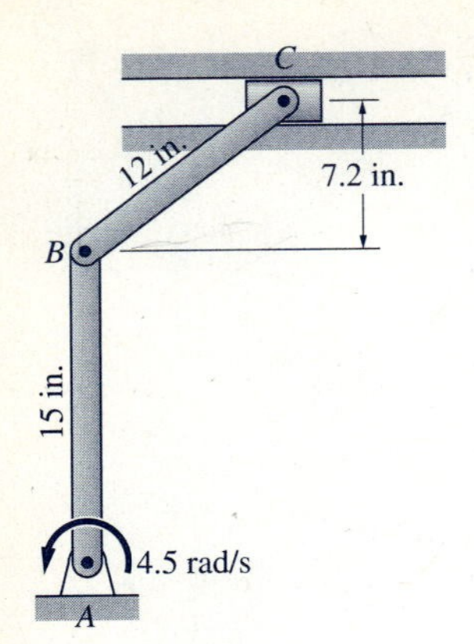

Fig. P16.98

16.98 Bar AB of the mechanism rotates with constant angular velocity of 4.5 rad/s. Determine the angular acceleration of bar BC and the acceleration of slider C in the position shown.

16.99 In the position shown, collar A is moving down with the velocity v_0. Assuming that the disk does not slip on the horizontal surface, determine the angular velocity of bar AB and the speed of the center of the disk in this position.

16.100 In the position shown, the bent rod AB is rotating about A with the angular velocity $\omega = 5$ rad/s and the angular acceleration $\dot{\omega} = 20$ rad/s^2. At the same time, collar D is sliding on the rod with the velocity $v = 2$ ft/s and acceleration $\dot{v} = -6$ ft/s^2 (measured relative to the rod). Determine the acceleration of collar D in this position.

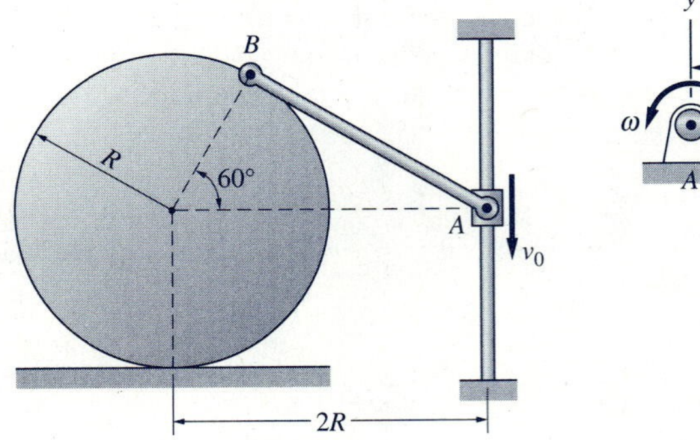

Fig. P16.99

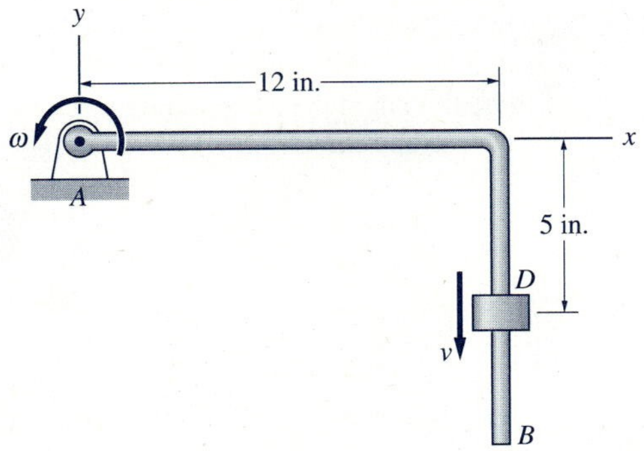

Fig. P16.100

16.101 The pin F, which is attached to the rod AF, engages a slot in bar BD of the parallelogram linkage $ABDE$. Bar AB of the linkage has a constant angular velocity of 15 rad/s counterclockwise. For the position shown, determine the angular acceleration of rod AF and the acceleration of pin F relative to bar BD.

16.102 Bar AB of the linkage rotates with the constant angular velocity of 10 rad/s. For the position shown, determine (a) the angular velocities of bars BC; and CD; and (b) the angular accelerations of BC and CD.

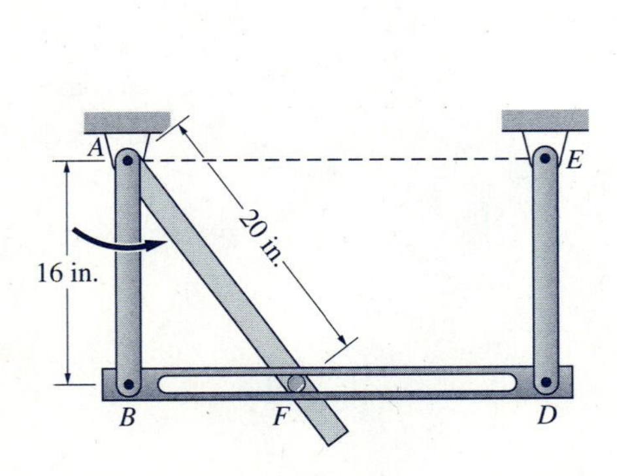

Fig. P16.101

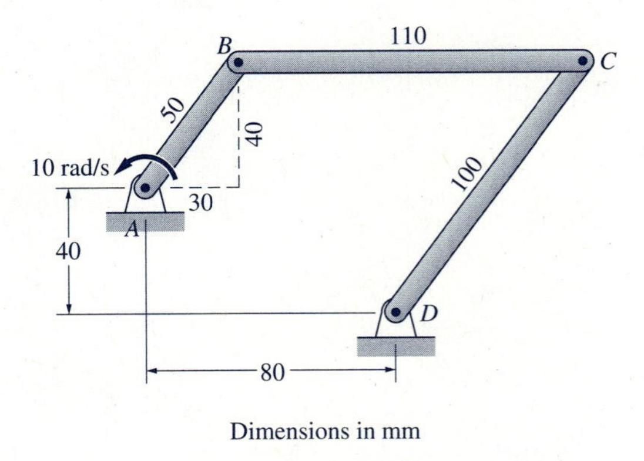

Fig. P16.102

16.103 The pin G in the center of the gear engages a slot in the arm AB while the arm rotates with the constant angular velocity $\omega = \dot{\theta}$. Determine the acceleration (magnitude and direction) of point D on the gear when $\theta = 60°$.

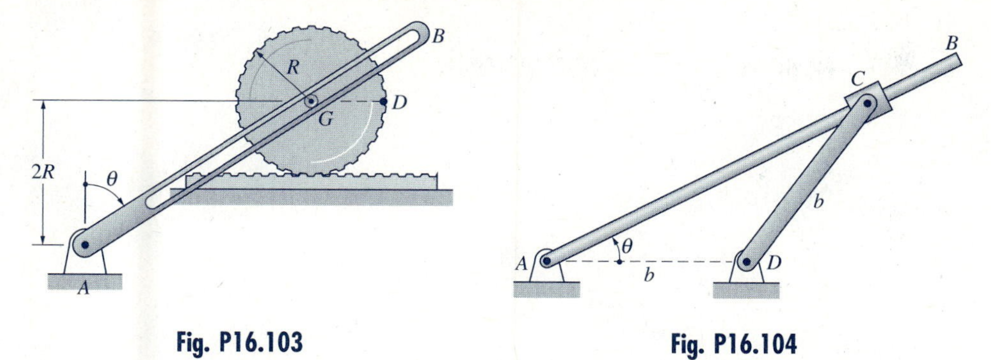

Fig. P16.103 **Fig. P16.104**

16.104 For the mechanism shown, determine the speed and the magnitude of the acceleration of collar C in terms of b, θ, $\dot{\theta}$, and $\ddot{\theta}$.

17 Planar Kinetics of Rigid Bodies: Force-Mass-Acceleration Method

17.1 Introduction

This chapter presents the force-mass-acceleration (FMA) method for the kinetic analysis of plane motion of rigid bodies. As mentioned previously, this method relates the forces that act on a body to its acceleration. The work-energy and impulse-momentum methods are discussed in the next chapter.

At the core of this chapter are the equations of motion that describe the plane motion of a rigid body. Derivation of these equations employs the mass-acceleration diagrams that were introduced in Chapter 12. Also included is an introduction to mass moment of inertia, a concept that arises in the moment equation of motion.

The chapter concludes with a discussion of analytical and numerical integration of the differential equations of motion.

17.2 Mass Moment of Inertia; Composite Bodies

In this article we introduce the mass moment of inertia of a body about an axis. A comprehensive discussion of mass moment of inertia, including its computation by integration, is contained in Appendix F.

a. Mass moment of inertia

Figure 17.1 shows a body of mass m that occupies the region $\mathcal{V}$; r is the perpendicular distance from the a-axis to the differential mass dm of the body. The mass moment of inertia of the body about the a-axis is defined as

$$I_a = \int_{\mathcal{V}} r^2 \, dm \qquad (17.1)$$

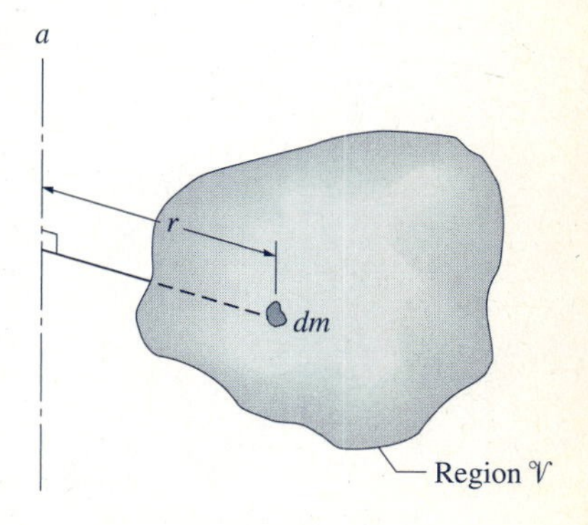

Fig. 17.1

It will be seen shortly that this integral is a measure of the ability of the body to resist a change in its angular motion about the a-axis, just as the mass of the body is a measure of its ability to resist a change in its translational motion.

From its definition we see that mass moment of inertia is a positive quantity with units equal to (mass) × (distance)^2. In SI units, I_a is measured in kg · m^2. In U.S. Customary units, I_a is measured in slug · ft^2, or equivalently, in lb · ft · s^2. (Slug · in.^{2} is never used because the units of length would be inconsistent—one slug is equivalent to 1 lb · s^2/ft.)

b. Radius of gyration

The radius of gyration k_a of the body about the a-axis is defined as

$$I_a = mk_a^2 \quad \text{or} \quad k_a = \sqrt{\frac{I_a}{m}} \tag{17.2}$$

Although the unit of radius of gyration is length (e.g., feet, meters), it is not a distance that can be measured physically. Instead, its value can be found only by computation using Eq. (17.2). The radius of gyration allows us to compare the rotational resistances of bodies that have the same mass.

c. Parallel-axis theorem

Consider the two parallel axes shown in Fig. 17.2. The location of the a-axis is arbitrary. We call the other axis, which passes through the mass center G of the body, the *central a-axis.**

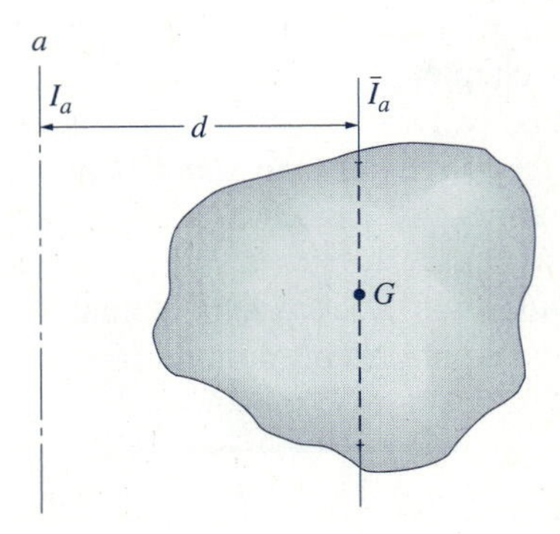

Fig. 17.2

Letting d be the distance between the two parallel axes, the parallel-axis theorem states that

$$I_a = \bar{I}_a + md^2 \tag{17.3}$$

where m is the mass of the body, I_a is the moment of inertia of the body about the a-axis, and $\bar{I}_a$ is its moment of inertia about the central a-axis. Note that if the moment of inertia about a central axis is known, the parallel-axis theorem can be used to calculate the moment of inertia about any parallel axis without resorting to integration. Table 17.1 lists the moments of inertia about central axes for a few homogeneous bodies.

* We will refer to any axis that passes through the mass center as a *central axis.* The *central a-axis* is the central axis that is parallel to the a-axis.

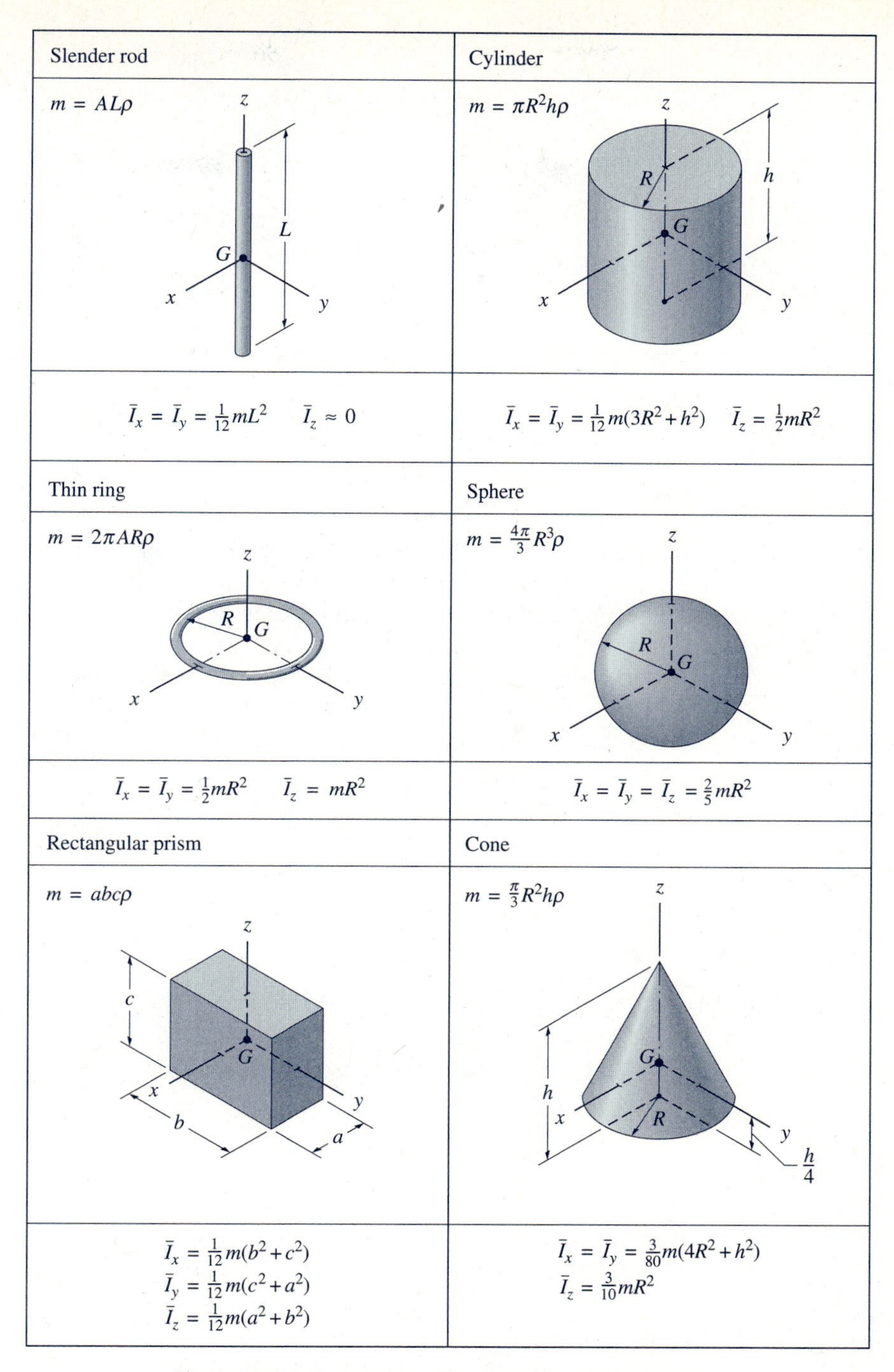

Slender rod	Cylinder
$m = AL\rho$	$m = \pi R^2 h\rho$
$\bar{I}_x = \bar{I}_y = \frac{1}{12}mL^2 \quad \bar{I}_z \approx 0$	$\bar{I}_x = \bar{I}_y = \frac{1}{12}m(3R^2 + h^2) \quad \bar{I}_z = \frac{1}{2}mR^2$
Thin ring	**Sphere**
$m = 2\pi AR\rho$	$m = \frac{4\pi}{3}R^3\rho$
$\bar{I}_x = \bar{I}_y = \frac{1}{2}mR^2 \quad \bar{I}_z = mR^2$	$\bar{I}_x = \bar{I}_y = \bar{I}_z = \frac{2}{5}mR^2$
Rectangular prism	**Cone**
$m = abc\rho$	$m = \frac{\pi}{3}R^2 h\rho$
$\bar{I}_x = \frac{1}{12}m(b^2 + c^2)$ $\bar{I}_y = \frac{1}{12}m(c^2 + a^2)$ $\bar{I}_z = \frac{1}{12}m(a^2 + b^2)$	$\bar{I}_x = \bar{I}_y = \frac{3}{80}m(4R^2 + h^2)$ $\bar{I}_z = \frac{3}{10}mR^2$

Table 17.1 Mass Moments of Inertia of Homogeneous Bodies

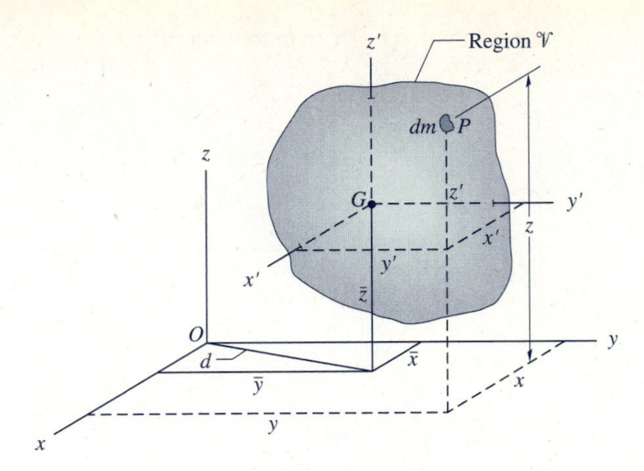

Fig. 17.3

To prove the parallel-axis theorem, consider the body of mass m that is shown in Fig. 17.3. The origin O of the xyz-axes is located arbitrarily, but the x', y', and z' axes are central axes that are parallel to the x, y, and z axes, respectively. The coordinates of the mass center G relative to the xyz-axes are denoted by $\bar{x}$, $\bar{y}$, and $\bar{z}$. Let dm be a differential mass element of the body that is located at P. Because the perpendicular distance from the z-axis to P is $r = (x^2 + y^2)^{1/2}$, the moment of inertia of the body about the z-axis is

$$I_z = \int_{\mathcal{V}} r^2\, dm = \int_{\mathcal{V}} (x^2 + y^2)\, dm \tag{a}$$

Substituting $x = x' + \bar{x}$ and $y = y' + \bar{y}$ yields

$$I_z = \int_{\mathcal{V}} [(x' + \bar{x})^2 + (y' + \bar{y})^2]\, dm \tag{b}$$

Expanding and rearranging terms, we obtain

$$I_z = \int_{\mathcal{V}} (x'^2 + y'^2)\, dm + \int_{\mathcal{V}} (\bar{x}^2 + \bar{y}^2)\, dm + 2\bar{x}\int_{\mathcal{V}} x'\, dm + 2\bar{y}\int_{\mathcal{V}} y'\, dm \tag{c}$$

Consider now each of the integrals that appear in this equation. The first integral is equal to $I_{z'}$, the moment of inertia about the z'-axis. Since z' is a central axis (passes through G), this term may be written as $\bar{I}_z$. Letting $d = (\bar{x}^2 + \bar{y}^2)^{1/2}$, the distance between the z- and z'-axes, the second integral in Eq. (c) equals md^2. The last two integrals in Eq. (c) vanish, since $\int_{\mathcal{V}} x'\, dm = 0$ and $\int_{\mathcal{V}} y'\, dm = 0$ when the y'- and x'-axes are central axes. Therefore, Eq. (c) becomes

$$I_z = \bar{I}_z + md^2 \tag{d}$$

Because the z-axis can be chosen arbitrarily, comparison of Eqs. (17.3) and (d) shows that the parallel-axis theorem has been proved.

d. Method of composite bodies

From Eq. (17.1) it can be seen that the computation of the mass moment of inertia requires that an integration be performed over the body. The various integration techniques are discussed in Appendix F. Here we will only consider the method of composite bodies, a method that follows directly from the property of definite integrals: The integral of a sum is equal to the sum of the integrals.* Using this property, it can be shown that if a body is divided into composite parts, the moment of inertia of the body about a given axis equals the sum of the moments of inertia of its parts about that axis. The following sample problems illustrate the application of this method.

* This property also formed the basis for the method of composite shapes discussed in Chapter 8.

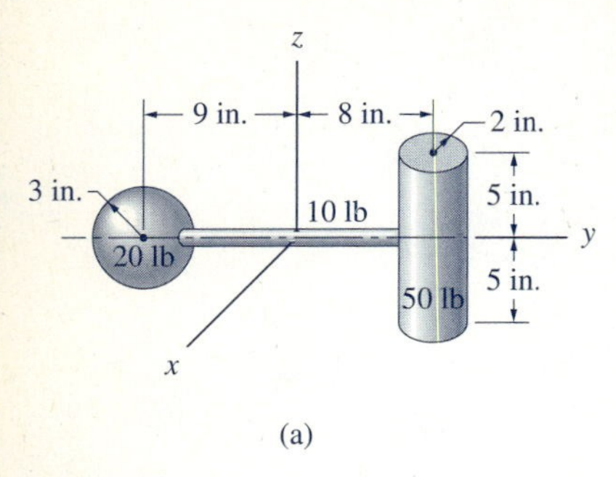

(a)

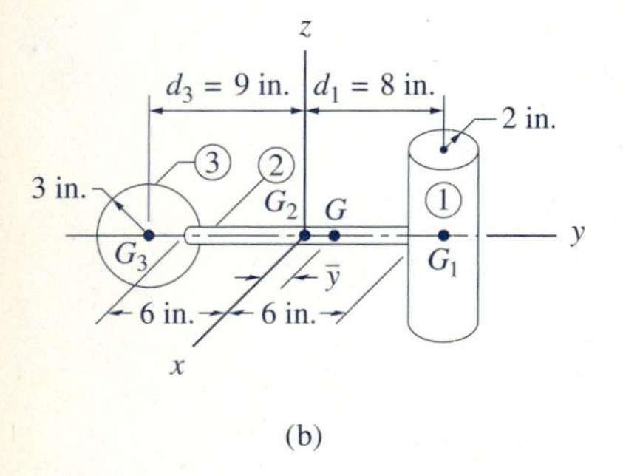

(b)

Sample Problem 17.1

The assembly in Fig. (a) is composed of three homogeneous bodies: the 50-lb cylinder, the 10-lb slender rod, and the 20-lb sphere. For this assembly, calculate (1) I_x, the mass moment of inertia about the x-axis; and (2) $\bar{I}_x$ and $\bar{k}_x$, the mass moment of inertia and radius of gyration about the central x-axis of the assembly.

Solution

The mass centers of the cylinder (G_1), the rod (G_2), and the sphere (G_3) are shown in Fig. (b). By symmetry, the mass center of the assembly (G) lies on the y-axis, with its coordinate $\bar{y}$ to be determined.

In the following computations, mass is calculated in slugs (or equivalently, $\text{lb} \cdot \text{s}^2/\text{ft}$) by dividing the weight in pounds by $g = 32.2 \text{ ft/s}^2$. To maintain consistent units, all distances are converted from inches into feet. The resulting units for moment of inertia are thus $\text{slug} \cdot \text{ft}^2$ (or equivalently, $\text{lb} \cdot \text{ft} \cdot \text{s}^2$).

Part 1

Cylinder

Using Table 17.1, the moment of inertia of the cylinder about its own central x-axis is

$$(\bar{I}_x)_1 = \frac{1}{12}m_1(3R^2 + h^2) = \frac{1}{12}\frac{50}{32.2}\left[3\left(\frac{2}{12}\right)^2 + \left(\frac{10}{12}\right)^2\right]$$
$$= 0.1006 \text{ slug} \cdot \text{ft}^2$$

Utilizing the parallel-axis theorem, the moment of inertia of the cylinder about the x-axis becomes

$$(I_x)_1 = (\bar{I}_x)_1 + m_1 d_1^2$$
$$= 0.1006 + \frac{50}{32.2}\left(\frac{8}{12}\right)^2 = 0.7908 \text{ slug} \cdot \text{ft}^2$$

Slender Rod

Because G_2 coincides with the origin of the xyz-axes, the moment of inertia of the rod about the x-axis is obtained directly from Table 17.1.

$$(I_x)_2 = (\bar{I}_x)_2 = \frac{1}{12}mL^2$$
$$= \frac{1}{12}\frac{10}{32.2}\left(\frac{12}{12}\right)^2 = 0.0259 \text{ slug} \cdot \text{ft}^2$$

Sphere

According to Table 17.1, the moment of inertia of the sphere about its own central x-axis is

$$(\bar{I}_x)_3 = \frac{2}{5}mR^2 = \frac{2}{5}\frac{20}{32.2}\left(\frac{3}{12}\right)^2 = 0.0155 \text{ slug} \cdot \text{ft}^2$$

Using the parallel-axis theorem, the moment of inertia about the x-axis is given by

$$(I_x)_3 = (\bar{I}_x)_3 + m_3 d_3^2$$
$$= 0.0155 + \frac{20}{32.2}\left(\frac{9}{12}\right)^2 = 0.3649 \text{ slug} \cdot \text{ft}^2$$

Assembly

The moment of inertia of the assembly about an axis equals the sum of the moments of inertia of its parts about that axis. Therefore, adding the values that were found above, we get

$$I_x = (I_x)_1 + (I_x)_2 + (I_x)_3 = 0.7908 + 0.0259 + 0.3649$$
$$= 1.1816 \text{ slug} \cdot \text{ft}^2 \qquad \textit{Answer}$$

Part 2

Referring to Fig. (b), the y-coordinate of G is

$$\bar{y} = \frac{\Sigma_i m_i y_i}{\Sigma_i m_i} = \frac{\Sigma_i W_i \bar{y}_i}{\Sigma_i W_i} = \frac{50(8) + 10(0) - 20(9)}{50 + 20 + 10}$$
$$= \frac{220}{80} = 2.750 \text{ in.}$$

Because $\bar{y}$ is the distance between the x-axis and the central x-axis of the assembly, the moment of inertia of the assembly about the latter axis is found from the parallel-axis theorem:

$$\bar{I}_x = I_x - m\bar{y}^2 = 1.1816 - \frac{80}{32.2}\left(\frac{2.750}{12}\right)^2$$
$$= 1.051 \text{ slug} \cdot \text{ft}^2 \qquad \textit{Answer}$$

The corresponding radius of gyration is

$$\bar{k}_x = \sqrt{\frac{\bar{I}_x}{m}} = \sqrt{\frac{1.051}{80/32.2}} = 0.650 \text{ ft} \qquad \textit{Answer}$$

Alternate Method for Computing $\bar{I}_x$

In the preceding solution, we first computed I_x for the assembly by summing the values of I_x for each part. Then $\bar{I}_x$ for the assembly was found by applying the parallel-axis theorem. An alternate method for computing $\bar{I}_x$ for the assembly is to first compute the moments of inertia for each part about the central x-axis of the assembly, and then sum these values. Using this approach, we obtain

$$\bar{I}_x = \left[(\bar{I}_x)_1 + m_1(d_1 - \bar{y})^2\right] + \left[(\bar{I}_x)_2 + m_2\bar{y}^2\right] + \left[(\bar{I}_x)_3 + m(d_3 + \bar{y})^2\right]$$

In this equation, note that $\bar{I}_x$ is the moment of inertia for the entire assembly about the central x-axis for the assembly (axis passing through G), whereas $(\bar{I}_x)_1$, $(\bar{I}_x)_2$, and $(\bar{I}_x)_3$ are the moments of inertia of the parts about their own central axes. The distances between G and the mass center of each part, namely $(d_1 - \bar{y})$, $\bar{y}$, and $(d_3 + \bar{y})$, are found from Fig. (b). Substituting the numerical values, we obtain

$$
\begin{aligned}
I_x &= \left[0.1006 + \frac{50}{32.2}\left(\frac{8 - 2.750}{12}\right)^2\right] \\
&\quad + \left[0.0259 + \frac{10}{32.2}\left(\frac{2.750}{12}\right)^2\right] + \left[0.0155 + \frac{20}{32.2}\left(\frac{9 + 2.750}{12}\right)^2\right] \\
&= 0.3978 + 0.0422 + 0.6110 = 1.051 \text{ slug} \cdot \text{ft}^2 \qquad \textit{Answer}
\end{aligned}
$$

which agrees with the result found previously.

Sample Problem 17.2

The 290-kg machine part in Fig. (a) is made by drilling an off-center, 160-mm hole through a homogeneous, 400-mm cylinder of length 350 mm. Determine (1) I_z (the mass moment of inertia of the part about the z-axis); and (2) $\bar{k}_z$ (the radius of gyration of the part about its central z-axis).

Dimensions in mm

(a)

Solution

The machine part in Fig. (a) can be considered to be the difference between the homogeneous cylinders A and B shown in Figs. (b) and (c), respectively. The mass density ρ of the machine part is

$$
\rho = \frac{m}{\pi(R_A^2 - R_B^2)h} = \frac{290}{\pi(0.20^2 - 0.08^2)(0.35)} = 7849 \text{ kg/m}^3
$$

Consequently, the masses of cylinders A and B are

$$
\begin{aligned}
m_A &= \rho\pi R_A^2 h = (7849)\pi(0.20)^2(0.35) = 345.2 \text{ kg} \\
m_B &= \rho\pi R_B^2 h = (7849)\pi(0.08)^2(0.35) = 55.2 \text{ kg}
\end{aligned}
$$

As a check on our computations, we note that $m_A - m_B = m$, as expected.

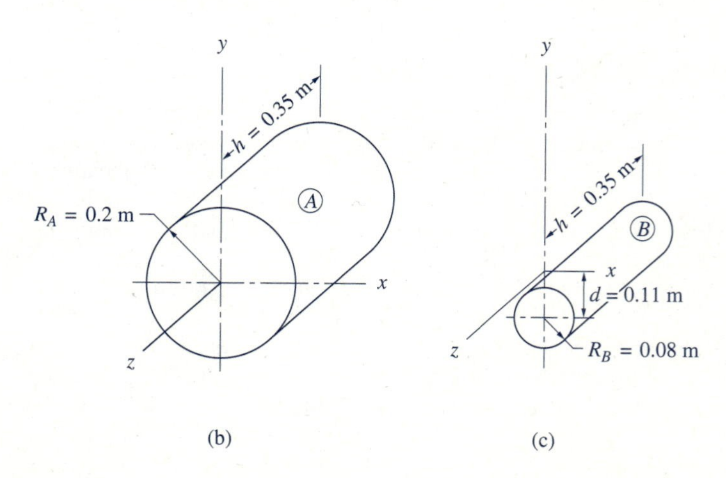

(b) (c)

Part 1

From Table 17.1, the moment of inertia of cylinder A about the z-axis, which coincides with its central z-axis, is

$$(I_z)_A = (\bar{I}_z)_A = \tfrac{1}{2}m_A R_A^2 = \tfrac{1}{2}(345.2)(0.20)^2 = 6.905\ \text{kg}\cdot\text{m}^2$$

The moment of inertia of cylinder B about its central z-axis is

$$(\bar{I}_z)_B = \tfrac{1}{2}m_B R_B^2 = \tfrac{1}{2}(55.2)(0.080)^2 = 0.177\ \text{kg}\cdot\text{m}^2$$

Because the distance between the z-axis and the central z-axis of B is $d = 0.11$ m, the moment of inertia of B about the z-axis is found from the parallel-axis theorem to be

$$(I_z)_B = (\bar{I}_z)_B + m_B d^2 = 0.177 + (55.2)(0.11)^2 = 0.845\ \text{kg}\cdot\text{m}^2$$

Therefore, the moment of inertia of the machine part about the z-axis is

$$I_z = (I_z)_A - (I_z)_B = 6.905 - 0.845 = 6.060\ \text{kg}\cdot\text{m}^2 \qquad \textit{Answer}$$

Part 2

By symmetry, the x- and z-coordinates of the mass center of the machine part are $\bar{x} = 0$ and $\bar{z} = -0.175$ m. The y-coordinate is found to be

$$\bar{y} = \frac{m_A \bar{y}_A - m_B \bar{y}_B}{m} = \frac{345.2(0) - 55.2(-0.11)}{290} = 0.020\,94\ \text{m}$$

The moment of inertia of the machine part about its central z-axis can now be found from the parallel-axis theorem:

$$\bar{I}_z = I_z - m\bar{y}^2 = 6.060 - 290(0.020\,94)^2 = 5.932\ \text{kg}\cdot\text{m}^2$$

The corresponding radius of gyration is

$$\bar{k}_z = \sqrt{\frac{\bar{I}_z}{m}} = \sqrt{\frac{5.932}{290}} = 0.1430\ \text{m} \qquad \textit{Answer}$$

PROBLEMS

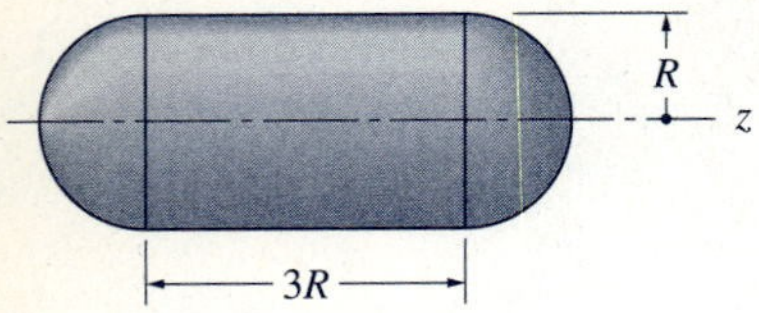

Fig. P17.1

17.1 The homogeneous body of total mass m consists of a cylinder with hemispherical ends. Calculate the moment of inertia of the body about the z-axis in terms of R and m.

17.2 Determine the moment of inertia of the truncated cone about the z-axis. The cone is made of wood that weighs 35 lb/ft^3.

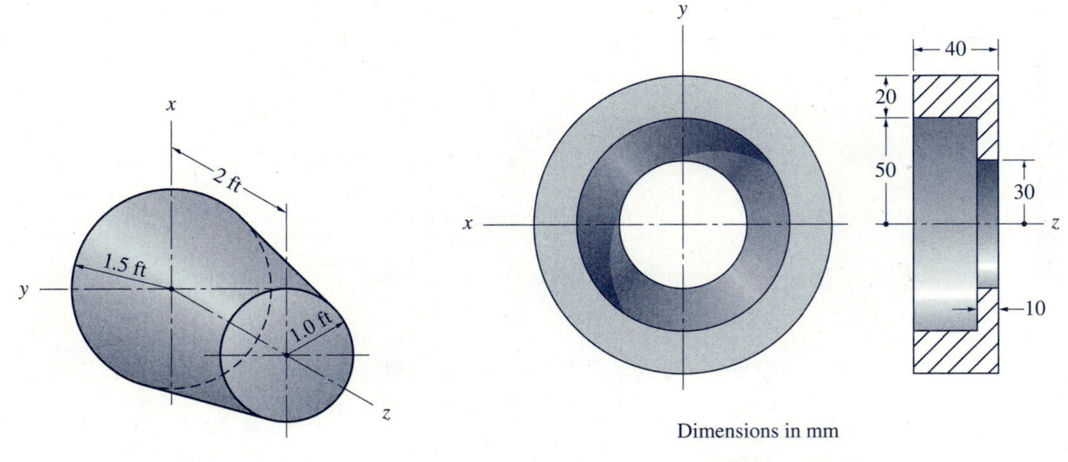

Fig. P17.2

Fig. P17.3

17.3 Compute the radius of gyration of the homogeneous wheel about the z-axis.

17.4 The inertial properties of a three-stage rocket are shown in the figure. Note that $\bar{k}_i$ is the radius of gyration for the ith stage about the axis that is parallel to the x-axis and passes through the center of gravity G_i of the stage. Find $\bar{z}$ and $\bar{I}_x$ of the rocket.

17.5 Two steel rods of different diameters are welded together as shown. Locate the mass center of the assembly and compute $\bar{I}_z$. For steel, $\rho = 7850$ kg/m^3.

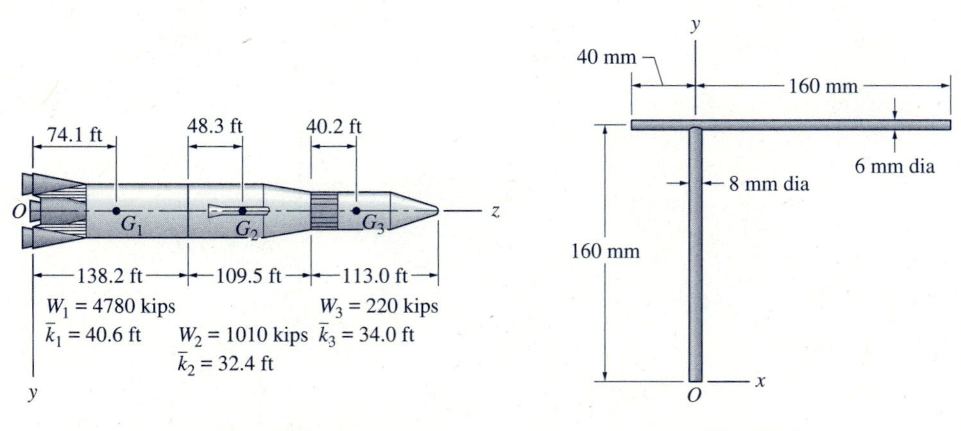

Fig. P17.4

Fig. P17.5

17.6 The weight density of the cast aluminum wheel is 165 lb/ft^3. Locate the mass center of the wheel, and calculate I_z and $\bar{I}_z$.

17.7 For the wheel in Prob. 17.6, compute the radii of gyration about (a) the x-axis; and (b) the y-axis.

17.8 The cage is formed from uniform slender wire. Determine I_z and k_z for the cage if its mass is 2.4 kg.

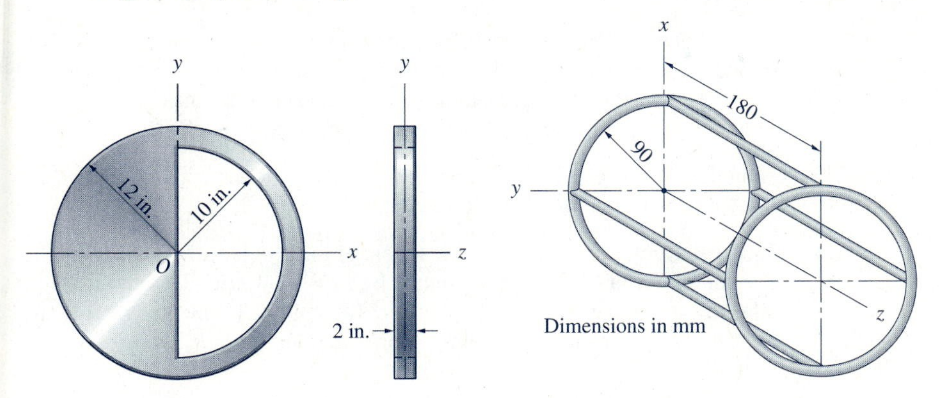

Fig. P17.6, P17.7

Fig. P17.8, P17.9

17.9 Calculate $\bar{I}_x$ for the cage in Prob. 17.8.

17.10 The solid body consists of a steel cylinder and a copper cone. The mass density of copper is 1.10 times the mass density of steel. Locate the mass center of the body and compute $\bar{k}_x$.

17.11 The machine part is made of steel with mass density $\rho = 7850$ kg/m^3. Compute $\bar{y}$, I_z, and $\bar{I}_z$.

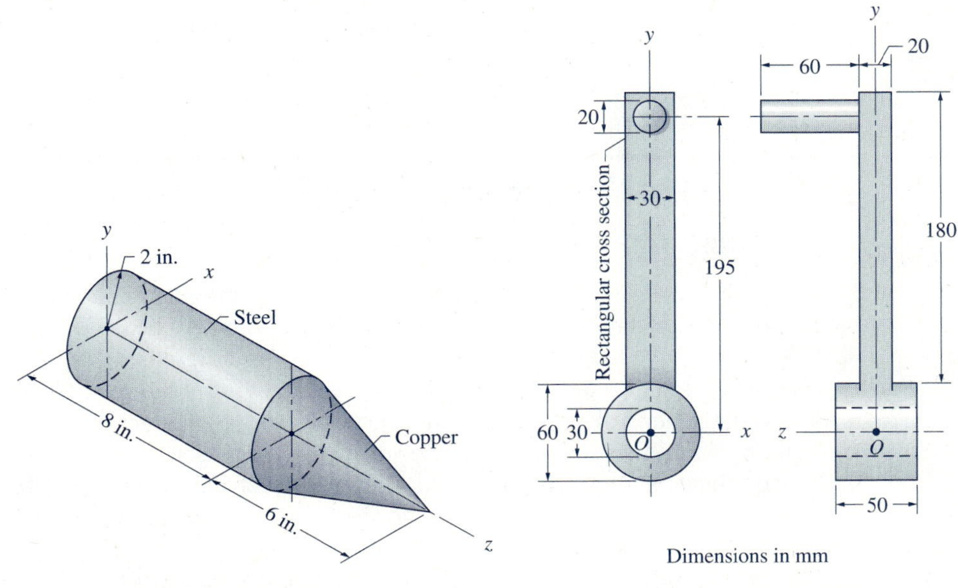

Fig. P17.10

Fig. P17.11

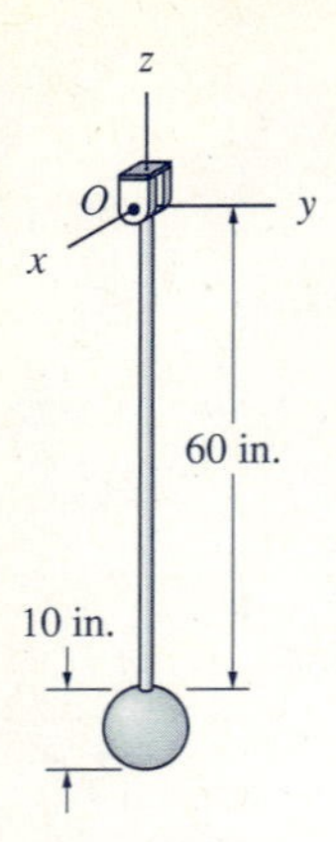

Fig. P17.13

17.12 Referring to Table 17.1, $\bar{I}_x$ for a cylinder can be approximated by $\bar{I}_x$ for a slender rod if the radius R of the cylinder is sufficiently small compared to its length h. Determine the largest ratio R/h for which the relative error for this approximation does not exceed 3%.

17.13 (a) Compute I_x for the pendulum, which consists of a 150-lb sphere attached to a 20-lb slender rod. (b) Determine the relative error in I_x if the mass of the rod is neglected and the sphere is approximated as a particle.

17.14 The moments of inertia of the 100-kg helicopter blade about the vertical axes passing through O and C are known from experiments to be 405.5 kg · m^2 and 147.5 kg · m^2, respectively. Determine the location of the mass center G and the moment of inertia about the vertical axis passing through G.

17.15 Using the properties of a sphere in Table 17.1, derive the expression for $\bar{I}_x$ of the homogeneous hemisphere of mass m.

***17.16** If the wall thickness t of the hollow sphere of mass m is sufficiently small, its moment of inertia can be approximated by $I_x = (2/3)mR^2$. Derive this result using the properties of a solid sphere in Table 17.1. (Hint: For $t \ll R$, the binomial series yields the following approximation: $R_o^n - R_i^n \approx nR^{n-1}t$).

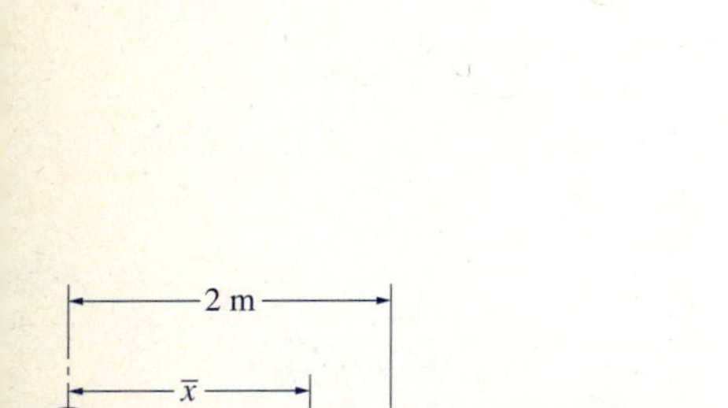

Fig. P17.14

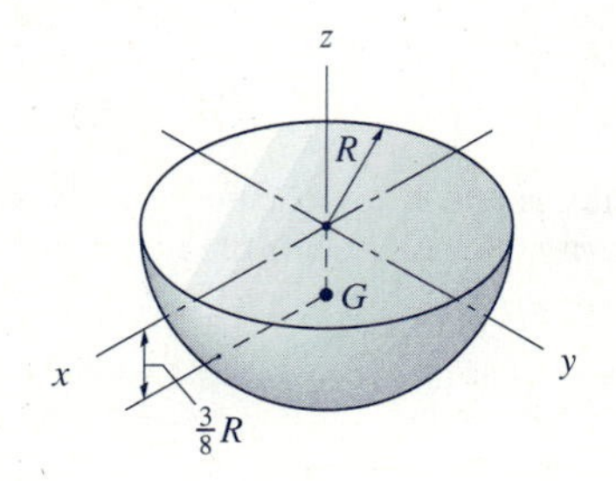

Fig. P17.15

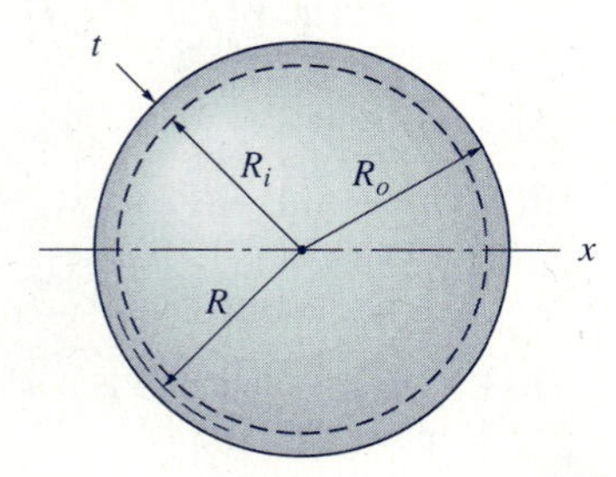

Fig. P17.16

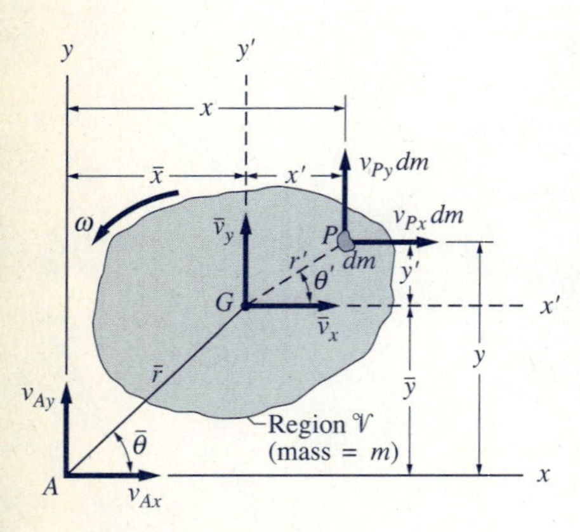

Fig. 17.4

17.3 Angular Momentum of a Rigid Body: Plane Motion

In this article we derive the expressions for the angular momentum of a rigid body that is undergoing plane motion. These expressions will be used in the next article to derive the moment equation of motion for a rigid body.

Figure 17.4 shows a rigid body of mass m occupying the region $\mathscr{V}$ that is undergoing general plane motion with counterclockwise angular velocity ω. An arbitrary point A, assumed to have the velocity $\mathbf{v}_A$, is chosen as the origin of the xy-coordinate system (A need not be fixed in the body*). The coordinates of the mass center G of the body are $\bar{x}$ and $\bar{y}$, or $\bar{r}$ and $\bar{\theta}$, and the components of its velocity vector are $\bar{v}_x$ and $\bar{v}_y$.

The rigid body is assumed to consist of an infinite number of infinitesimal particles. The mass of a typical particle P is dm, its position coordinates are

* In this chapter, the term *body* includes the *body extended*.

x and y, and $\mathbf{v}_P$ is its velocity vector. Therefore, the components of the linear momentum vector of the particle are $v_{Px}\,dm$ and $v_{Py}\,dm$, as shown in Fig. 17.4.

We know from Art. 14.7 that the angular momentum of a particle about A is defined as the moment of its linear momentum about A. Referring to Fig. 17.4, we obtain for the angular momentum of the particle P

$$\circlearrowleft\!+ \quad dh_A = v_{Py}x\,dm - v_{Px}y\,dm \tag{a}$$

The particle velocity can be related to the velocity of G by the relative velocity equation

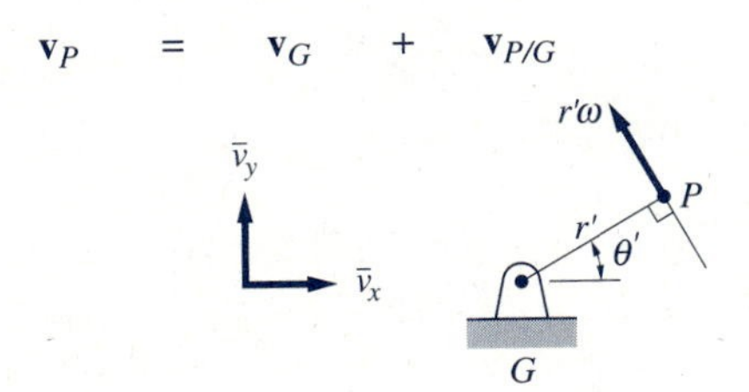

where the primed quantities refer to the coordinates of the particle relative to the $x'y'$-axes passing through G, as shown in Fig. 17.4. The components of $\mathbf{v}_P$ are

$$\left.\begin{aligned} v_{Px} &= \bar{v}_x - r'\omega\sin\theta' = \bar{v}_x - y'\omega \\ v_{Py} &= \bar{v}_y + r'\omega\cos\theta' = \bar{v}_y + x'\omega \end{aligned}\right\} \tag{b}$$

Substituting Eqs. (b), together with $x = \bar{x} + x'$ and $y = \bar{y} + y'$ (refer to Fig. 17.4), into Eq. (a) yields

$$dh_A = (\bar{v}_y + x'\omega)(\bar{x} + x')\,dm - (\bar{v}_x - y'\omega)(\bar{y} + y')\,dm \tag{c}$$

Carrying out all the multiplications in Eq. (c) and then integrating the result over the region $\mathcal{V}$, we obtain the angular momentum of the body about point A:

$$\begin{aligned} h_A = {}& \left(\bar{x}\bar{v}_y \int_{\mathcal{V}} dm + \bar{v}_y \int_{\mathcal{V}} x'\,dm + \bar{x}\omega \int_{\mathcal{V}} x'\,dm + \omega \int_{\mathcal{V}} x'^2\,dm\right) \\ & - \left(\bar{y}\bar{v}_x \int_{\mathcal{V}} dm + \bar{v}_x \int_{\mathcal{V}} y'\,dm - \bar{y}\omega \int_{\mathcal{V}} y'\,dm - \omega \int_{\mathcal{V}} y'^2\,dm\right) \end{aligned} \tag{d}$$

Noting that $\int_{\mathcal{V}} dm = m$, $\int_{\mathcal{V}} x'\,dm = 0$, $\int_{\mathcal{V}} y'\,dm = 0$, and $\int_{\mathcal{V}}(x'^2 + y'^2)\,dm = \bar{I}$, where $\bar{I}$ is the central mass moment of inertia of the body about the axis that is perpendicular to the plane of the motion, Eq. (d) simplifies to

$$\boxed{h_A = \bar{I}\omega + m\bar{x}\bar{v}_y - m\bar{y}\bar{v}_x \qquad \text{(A: arbitrary point)}} \tag{17.4}$$

Observe that in this equation point A is not necessarily fixed in the body.

Three convenient choices for the reference point A will be discussed next.

1. Point *A* is fixed in the body

If point A is fixed in the body, its velocity can be related to the velocity of G as follows.

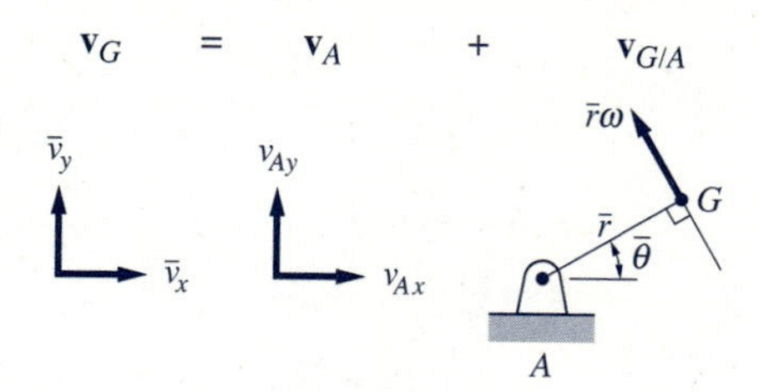

from which the components of $\mathbf{v}_G$ are

$$\left.\begin{aligned} \bar{v}_x &= v_{Ax} - \bar{r}\omega \sin\bar{\theta} = v_{Ax} - \bar{y}\omega \\ \bar{v}_y &= v_{Ay} + \bar{r}\omega \cos\bar{\theta} = v_{Ay} + \bar{x}\omega \end{aligned}\right\} \qquad \text{(e)}$$

Substituting Eqs. (e) into Eq. (17.4), we obtain

$$\begin{aligned} h_A &= \bar{I}\omega + m\bar{x}(v_{Ay} + \bar{x}\omega) - m\bar{y}(v_{Ax} - \bar{y}\omega) \\ &= [\bar{I} + m(\bar{x}^2 + \bar{y}^2)]\omega + m\bar{x}v_{Ay} - m\bar{y}v_{Ax} \end{aligned}$$

Recognizing that the coefficient of ω is equal to I_A by the parallel-axis theorem, the angular momentum of the body about a point A that is fixed in the body becomes

$$h_A = I_A\omega + m\bar{x}v_{Ay} - m\bar{y}v_{Ax} \qquad (A\text{: fixed in the body}) \qquad (17.5)$$

2. Point *A* is the instant center

If point A is the instant center (this includes the case where the body rotates about a fixed point A), then $\mathbf{v}_A = \mathbf{0}$. Consequently, Eq. (17.5) becomes

$$h_A = I_A\omega \qquad (A\text{: instant center}) \qquad (17.6)$$

3. Point *A* coincides with the mass center *G*

If the reference point A coincides with the mass center G (which is a point fixed in the body), the angular momentum of the body can be obtained by setting $\bar{x} = \bar{y} = 0$ and $I_A = \bar{I}$ in Eq. (17.5), which yields

$$h_G = \bar{I}\omega \qquad (17.7)$$

As will be seen in the following article, the expressions for the angular momentum presented in Eqs. (17.4)–(17.7) play a fundamental role in the development of the moment equation of motion for a rigid body.

17.4 Equations of Plane Motion

a. Introductory comments

In this article we derive the equations of motion for a rigid body that is undergoing plane motion. We begin by assuming that the body is made up of a large number of particles, with the internal forces between the particles occurring in equal and opposite, collinear pairs. By implication, the results described for systems of particles in Chapter 15 are thus also applicable to rigid bodies.* As we will see, there are three independent scalar equations that govern the plane motion of a rigid body. Two of the equations relate the motion of the mass center to the external forces (force equations of motion), whereas the third equation governs the angular motion of the body (moment equation of motion).

Figure 17.5 shows a rigid body of mass m in plane motion parallel to the xy-plane (the figure actually displays only the cross section of the body that contains its mass center G). The angular velocity and angular acceleration of the body at the instant shown are ω and α, respectively, both assumed to be counterclockwise. In order to keep the formulation two-dimensional, we assume that the body is symmetric with respect to the xy-plane.† The forces $\mathbf{F}_1$, $\mathbf{F}_2, \ldots$ are the projections of the external forces onto the xy-plane. Assuming that the xy-plane is horizontal, the weight of the body (which is an external force) is perpendicular to the xy-plane and does not appear in the figure. If the xy-plane were vertical, the weight of the body would have to be shown as an external force that acts at G.

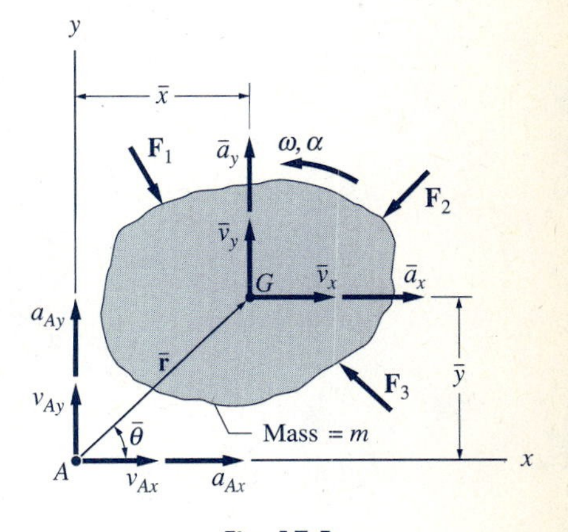

Fig. 17.5

As in Fig. 17.4, the origin A of the xy-coordinate system in Fig. 17.5 does not have to be fixed in the inertial frame, or fixed in the body. We do assume, however, that the *coordinate axes do not rotate*. In the derivations that follow, one must be careful not to confuse absolute velocities and accelerations (which refer to the inertial coordinate system not shown in Fig. 17.5) with velocities and accelerations measured relative to the xy reference frame. For example, $\bar{\mathbf{v}}$ denotes the absolute velocity of G, whereas $d\bar{\mathbf{r}}/dt$ is the velocity of G relative to A (i.e., $\mathbf{v}_{G/A}$), because $\bar{\mathbf{r}}$ appears in Fig. 17.5 as the position vector of G relative to A.

b. Force equation of motion

Since we have assumed a rigid body to be equivalent to a system of particles, we can use the same equations of motion for both. Therefore, Eq. (15.19) of Art. 15.4, namely, $\mathbf{R} = m\mathbf{a}_G$, is also applicable to a rigid body. Introducing the notation $\Sigma\mathbf{F} = \mathbf{R}$ for the resultant external force acting on the body, and $\bar{\mathbf{a}} = \mathbf{a}_G$ for the absolute acceleration of the mass center, the vector form of the force equations of motion becomes

$$\boxed{\Sigma\mathbf{F} = m\bar{\mathbf{a}}} \qquad (17.8)$$

* Although the particle model is not entirely accurate for a rigid body, it does yield the correct equations of motion. A more rigorous derivation based on the concept of stress is beyond the scope of this text. Historically, the equations of motion for a rigid body were not derived from particle mechanics, but were postulated outright. Presumably, the inspiration for the postulate came from the analysis of systems of particles.

† This assumption is overly restrictive. As will be explained in Chapter 19, the equations to be developed are valid as long as the central z-axis is a principal axis of inertia of the body.

c. Moment equation of motion

We first consider the moment equation of motion using an arbitrary reference point A; we then examine the special case where the reference point is fixed in the body.

1. Arbitrary Reference Point The moment equation of motion for a system of particles derived in Art. 15.7 was

$$(\Sigma \mathbf{M}_A)_{\text{ext}} = \dot{\mathbf{h}}_A + \mathbf{v}_A \times m\bar{\mathbf{v}} \qquad (A\text{: arbitrary point}) \qquad (15.34, \text{repeated})$$

where $(\Sigma \mathbf{M}_A)_{\text{ext}}$ is the resultant moment of the external forces about A, $\mathbf{h}_A$ is the angular momentum of the system about A, $\mathbf{v}_A$ is the velocity of point A, and $\bar{\mathbf{v}}$ represents the velocity of the mass center.

For plane motion in the xy-plane, all of the terms in Eq. (15.34) represent vectors that are perpendicular to the plane of the motion; i.e., the x- and y-components of the equation vanish. Substituting $\mathbf{v}_A = v_{Ax}\mathbf{i} + v_{Ay}\mathbf{j}$ and $\bar{\mathbf{v}} = \bar{v}_x\mathbf{i} + \bar{v}_y\mathbf{j}$ and expanding the cross product, the z-component of Eq. (15.34) becomes

$$\Sigma M_A = \dot{h}_A + mv_{Ax}\bar{v}_y - mv_{Ay}\bar{v}_x \qquad (A\text{: arbitrary point}) \qquad (17.9)$$

where, for the sake of brevity, we have dropped the subscript "ext" on the left side of the equation.

Equation (17.9) is valid both for systems of particles and for rigid bodies. We will now specialize the equation for rigid bodies only. We first introduce the angular momentum of a rigid body about an arbitrary point A, derived in the preceding article.

$$h_A = \bar{I}\omega + m\bar{x}\bar{v}_y - m\bar{y}\bar{v}_x \qquad (17.4, \text{repeated})$$

The time derivative of this expression is

$$\frac{dh_A}{dt} = \bar{I}\alpha + m\bar{x}\bar{a}_y - m\bar{y}\bar{a}_x + m\frac{d\bar{x}}{dt}\bar{v}_y - m\frac{d\bar{y}}{dt}\bar{v}_x \qquad \text{(a)}$$

Next, we observe (with the help of Fig. 17.5) that $d\bar{\mathbf{r}}/dt = \mathbf{v}_{G/A} = \bar{\mathbf{v}} - \mathbf{v}_A$, from which we get

$$v_{Ax} = \bar{v}_x - \frac{d\bar{x}}{dt} \quad \text{and} \quad v_{Ay} = \bar{v}_y - \frac{d\bar{y}}{dt} \qquad \text{(b)}$$

Substituting Eqs. (a) and (b) into Eq. (17.9) yields

$$\begin{aligned}\Sigma M_A = {} & \bar{I}\alpha + m\bar{x}\bar{a}_y - m\bar{y}\bar{a}_x + m\frac{d\bar{x}}{dt}\bar{v}_y - m\frac{d\bar{y}}{dt}\bar{v}_x \\ & + m\left(\bar{v}_x - \frac{d\bar{x}}{dt}\right)\bar{v}_y - m\left(\bar{v}_y - \frac{d\bar{y}}{dt}\right)\bar{v}_x\end{aligned}$$

which reduces to

$$\Sigma M_A = \bar{I}\alpha + m\bar{x}\bar{a}_y - m\bar{y}\bar{a}_x \qquad (A\text{: arbitrary point}) \tag{17.10}$$

This moment equation, which places no restrictions on the choice of A, is the basis of the FBD-MAD method of analysis discussed in the next article.

2. Reference Point Fixed in the Body Here we derive the moment equation for the special case where the reference point A is fixed in the body. The steps in the derivation are identical to those used in arriving at Eq. (17.10). We again start with Eq. (17.9), but this time we substitute the expression for h_A that is valid when A is fixed in the body. From the preceding article, this expression is

$$h_A = I_A\omega + m\bar{x}v_{Ay} - m\bar{y}v_{Ax} \qquad \text{(17.5, repeated)}$$

Recognizing that I_A is constant when A is fixed in the body, the time derivative of this equation is

$$\frac{dh_A}{dt} = I_A\alpha + m\bar{x}a_{Ay} - m\bar{y}a_{Ax} + m\frac{d\bar{x}}{dt}v_{Ay} - m\frac{d\bar{y}}{dt}v_{Ax} \tag{c}$$

which is to be substituted into Eq. (17.9).

From Eq. (b) we have

$$\bar{v}_x = v_{Ax} + \frac{d\bar{x}}{dt} \quad \text{and} \quad \bar{v}_y = v_{Ay} + \frac{d\bar{y}}{dt} \tag{d}$$

which is also to be substituted into Eq. (17.9).

Completing the substitutions, we get

$$\begin{aligned}\Sigma M_A =& I_A\alpha + m\bar{x}a_{Ay} - m\bar{y}a_{Ax} + m\frac{d\bar{x}}{dt}v_{Ay} - m\frac{d\bar{y}}{dt}v_{Ax}\\ &+ mv_{Ax}\left(v_{Ay} + \frac{d\bar{y}}{dt}\right) - mv_{Ay}\left(v_{Ax} + \frac{d\bar{x}}{dt}\right)\end{aligned}$$

which reduces to

$$\Sigma M_A = I_A\alpha + m\bar{x}a_{Ay} - m\bar{y}a_{Ax} \qquad (A\text{: fixed in the body}) \tag{17.11}$$

Equation (17.11) could be also be derived by substituting $\bar{\mathbf{a}} = \mathbf{a}_G = \mathbf{a}_A + \mathbf{a}_{G/A}$ into Eq. (17.10).

The following are the three cases for which Eq. (17.11) takes the simplified form

$$\Sigma M_A = I_A\alpha \tag{17.12}$$

Case 1: The reference point is the mass center G. Referring to Fig. 17.5, we see that $\bar{x} = \bar{y} = 0$ if the mass center G coincides with A. Consequently, Eq. (17.11) becomes

$$\Sigma M_G = \bar{I}\alpha \tag{17.13}$$

This is probably the most commonly used form of the moment equation.

Case 2: The acceleration of the reference point is zero. If $\mathbf{a}_A = \mathbf{0}$, it follows from Eq. (17.11) that $\Sigma M_A = I_A\alpha$. This special case is frequently used when the body is rotating about a fixed point A (i.e., A is fixed in the body as well as fixed in the inertial reference frame). However, note that the reference point need not be fixed; it only has to be the instant center of acceleration.

Case 3: The acceleration of A is directed along the line AG. Figure 17.6 shows the special case when $\mathbf{a}_A$ is directed along the line AG. From similar triangles we see that $a_{Ay}/a_{Ax} = \bar{y}/\bar{x}$, which gives $\bar{x}a_{Ay} = \bar{y}a_{Ax}$. After substituting this result into Eq. (17.11), the last two terms cancel, resulting in $\Sigma M_A = I_A\alpha$.

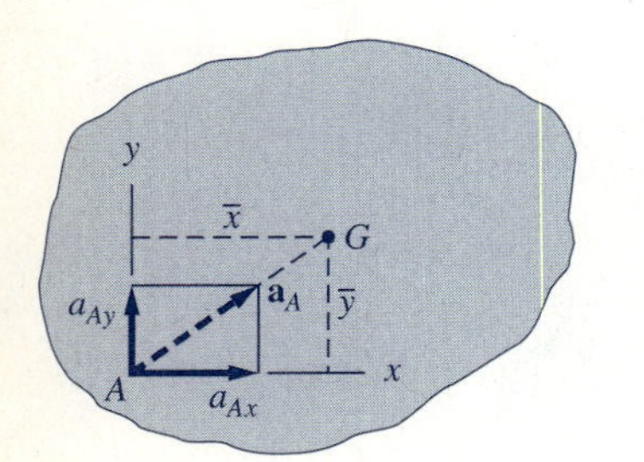

Fig. 17.6

Cases 1 and 2 describe the most commonly used reference points. The utility of Case 3 is mainly limited to the analysis of a balanced wheel that rolls without slipping. Since the acceleration of the contact point is toward the center of the wheel—i.e., toward the mass center—this reference point satisfies Case 3.

d. Summary

From the preceding discussion it can be seen that the plane motion of a rigid body is governed by two scalar force equations (e.g., $\Sigma F_x = m\bar{a}_x$ and $\Sigma F_y = m\bar{a}_y$) and a scalar moment equation (e.g., $\Sigma M_G = \bar{I}\alpha$).

However, the use of the moment equation can become fairly complicated unless the reference point A is chosen to be the mass center or a fixed point, in which case we have $\Sigma M_A = I_A\alpha$. Therefore, it is usually more convenient to derive all three equations of motion using the FBD-MAD approach described in the next article.

17.5 The FBD-MAD Technique

In this article, we describe a technique for obtaining the equations of motion for a body from its free-body and mass-acceleration diagrams. We begin with a discussion of general plane motion and then consider the special cases of translation and rotation about a fixed axis.

a. General plane motion

Figure 17.7(a) shows the free-body diagram (FBD) of a body that is undergoing general plane motion. The mass-acceleration diagram (MAD) of the body is shown in Fig. 17.7(b). The MAD consists of the vector $m\bar{\mathbf{a}}$ acting through the mass center G and the couple $\bar{I}\alpha$. (Since $\bar{I}\alpha$ is a couple, it can be shown anywhere in the plane of the motion. However, it is common practice to show $\bar{I}\alpha$ as if it acted at G.) We will refer to the vector $m\bar{\mathbf{a}}$ as the *inertia vector* of the body and call the couple $\bar{I}\alpha$ the *inertia couple*. The units of the inertia vector and inertia couple are the same as the units of a force and a moment.

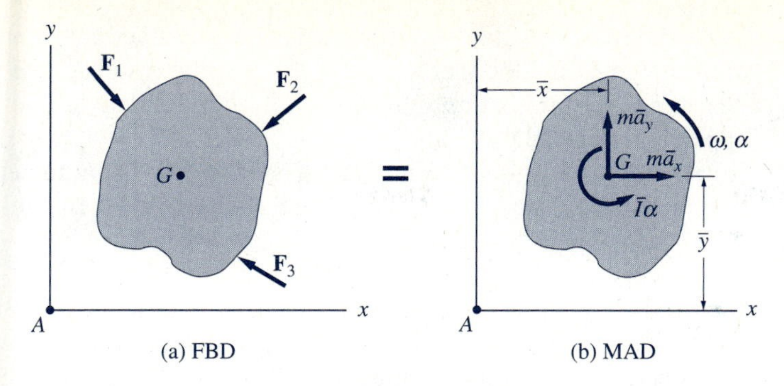

Fig. 17.7

The equal sign between the diagrams in Fig. 17.7 implies that the force systems in the FBD and in the MAD are equivalent. (We recall from statics that two force systems are equivalent if they reduce to the same resultant force and the same resultant couple at any point.)

We now prove that the conditions of equivalence of the FBD and the MAD represent the equations of motion for a rigid body. Referring to Fig. 17.7, we equate the resultant force $\Sigma\mathbf{F}$ on the FBD to the inertia vector on the MAD, which yields

$$\Sigma\mathbf{F} = m\bar{\mathbf{a}} \tag{a}$$

Let A be an arbitrary point in the plane of the motion. Equating the resultant moment ΣM_A about A on the FBD to the resultant moment about A on the MAD, we obtain

$$\Sigma M_A = \bar{I}\alpha + m\bar{x}\bar{a}_y - m\bar{y}\bar{a}_x \tag{b}$$

Since Eqs. (a) and (b) are identical to Eqs. (17.8) and (17.10), we conclude that the equivalence of the FBD and the MAD does indeed yield the equations of motion for the plane motion of a rigid body.

Utilizing the equivalence of the FBD and the MAD, the choice of the equations of motion is not limited to two force equations and one moment equation. Indeed, since the choice of A is arbitrary, there are an infinite number of equations of motion that can be obtained from the FBD and the MAD. However, as was the case in coplanar equilibrium in statics, only three of the equations are independent. The restrictions on the equations of motion that guarantee their independence are identical to those used for coplanar equilibrium—Eqs. (4.2)–(4.4) in Art. 4.4. For example, if three moment equations are used, the moment centers must not be collinear.

The steps that constitute the FBD-MAD technique for obtaining the equations of motion for a rigid body are as follows.

Step 1: Draw the free-body diagram of the body at the instant of interest.

Step 2: If the motion is subject to kinematic constraints, as is often the case, use relative acceleration analysis to find the kinematic relationship between $\bar{\mathbf{a}}$ and α. (In some instances, $\bar{\mathbf{a}}$ may also depend on the angular velocity ω, the determination of which may require relative velocity analysis.)

Step 3: Draw the MAD by showing the inertia vector $m\bar{\mathbf{a}}$ acting through the mass center G and the inertia couple $\bar{I}\alpha$. Use the results of the previous step to ensure that only kinematically independent accelerations appear on the MAD.

Step 4: Derive any three independent equations of motion using the equivalence of the FBD and the MAD. Solve the equations for the unknown forces and/or accelerations.

In general, if the total number of unknowns that appear in the FBD and MAD is three, all of them can be found from the equations of motion. However, in many instances, additional information must be derived from the history of the motion, e.g., by integrating the acceleration to determine the velocity at the instant of concern.

Note that three kinematic variables appear on the MAD: α and two components of $\bar{\mathbf{a}}$ (e.g., $\bar{a}_x$ and $\bar{a}_y$). If nothing is known about the acceleration of the body, then these three variables are unknown. If the motion is subject to kinematic constraints, then two or three of the kinematic variables can be related using relative acceleration equations (Step 2). In this case, the total number of unknowns on the MAD will be less than three.

One of the main advantages of the FBD-MAD technique is that the FBD displays the unknown forces, and the MAD displays the unknown accelerations. Consequently, one is less likely to attempt to derive and solve the equations of motion before the unknown variables have been correctly identified. Another obvious advantage of the FBD-MAD technique is that the moment equation about any point can easily be obtained without referring to Eq. (17.10).

As mentioned in Art. 12.4, the FBD-MAD method is a variation of d'Alembert's principle, where the inertia vector and the inertia couple are added to the FBD with their directions reversed. The equations of motion derived from this modified FBD are $\Sigma\mathbf{F} - m\bar{\mathbf{a}} = \mathbf{0}$ and $\Sigma M_A - \bar{I}\alpha + m\bar{x}\bar{a}_y - m\bar{y}\bar{a}_x = 0$. Therefore, the dynamic problem can be analyzed by equations that resemble equilibrium equations, with the resulting diagram said to represent a state of "dynamic equilibrium." As mentioned in Art. 12.4, we will not use d'Alembert's principle in this text because it offers no advantages over the FBD-MAD method.

b. Translation

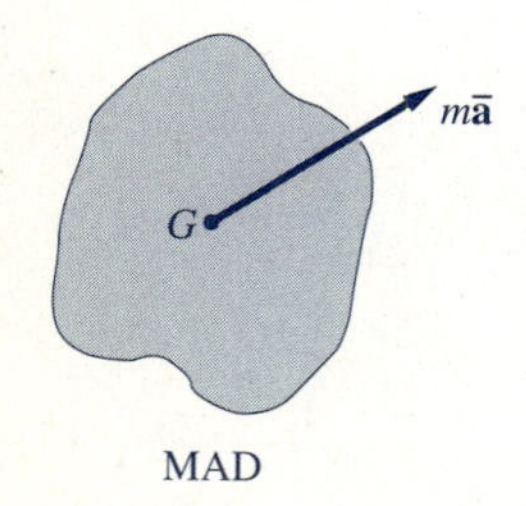

MAD

Fig. 17.8

Figure 17.8 shows the MAD for a rigid body that is translating. Since $\alpha = 0$, the MAD reduces simply to the inertia vector passing through G. It can be seen from this diagram that the resultant moment is zero about any point that lies on the same line as the inertia vector. For any other point, the resultant moment is equal to the moment of the inertia vector.

c. Rotation about a fixed axis

We distinguish between two types of rotation about a fixed axis—central and noncentral rotation.*

* The types of rotation are sometimes identified as *centroidal* and *noncentroidal* rotation. We avoid using the term centroid to eliminate the confusion that often exists between centroids, which are properties of geometric shapes, and mass centers, which are properties associated with mass.

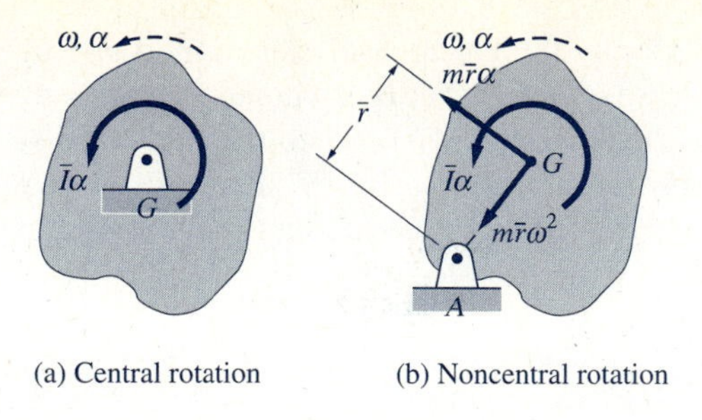

(a) Central rotation (b) Noncentral rotation

Fig. 17.9

In *central rotation,* the fixed axis passes through the mass center G. Since $\bar{\mathbf{a}} = \mathbf{0}$, the MAD reduces to the inertia couple, as shown in Fig. 17.9(a). In this case the inertia vector is zero and the resultant moment about every point is equal to $\bar{I}\alpha$.

In *noncentral rotation,* the axis of rotation passes through a fixed point A that is not the mass center. The MAD for this case is shown in Fig. 17.9(b), where the components of $\bar{\mathbf{a}}$ have been determined from the fact that the path of G is a circle centered at A. Since A is a fixed point, the special case of the moment equation, $\Sigma M_A = I_A\alpha$, could be used. However, even if this special case is not identified, the same moment equation is obtained by equating the resultant moments about A for the FBD and the MAD:

$$\overset{+}{\circlearrowleft} \quad \Sigma M_A = \bar{I}\alpha + (m\bar{r}\alpha)\bar{r} = (\bar{I} + m\bar{r}^2)\alpha = I_A\alpha$$

d. Systems of rigid bodies

When analyzing the motion of a system of connected rigid bodies, an FBD and an MAD could be drawn for each component body, since each component must satisfy a separate set of the three equations of motion. Therefore, a system containing N rigid bodies must satisfy $3N$ independent equations of motion.

Frequently it is convenient to use the FBD and MAD of the assembly, because then the internal forces do not appear in the FBD. For this reason, the FBD and the MAD for the entire system is often a good choice for beginning the analysis.

For systems of bodies, kinematic constraints require that special attention be paid to the MAD. Figure 17.10 shows the MAD for a system consisting of a slender bar AB that is pinned at B to a disk C (the MAD for the system is the sum of the MADs for each body). In this figure the properties (masses, accelerations, moments of inertia, and mass centers) of the bar and the disk are identified by the subscripts 1 and 2, respectively. Note that α_1 does not necessarily equal α_2 because of the pin at B. However, $\bar{\mathbf{a}}_1$, $\bar{\mathbf{a}}_2$, α_1, and α_2 are

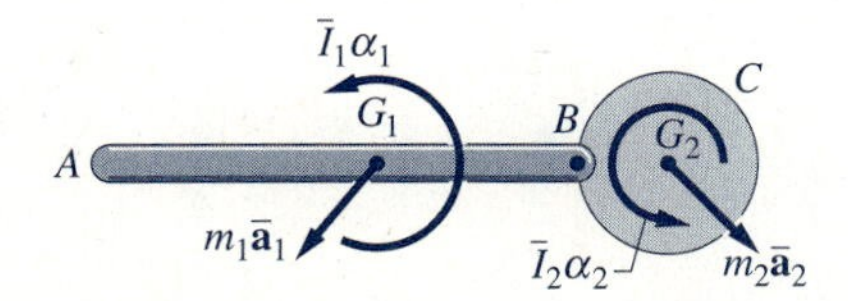

Fig. 17.10

related kinematically by the equation of constraint $\mathbf{a}_B = \bar{\mathbf{a}}_1 + \mathbf{a}_{B/G_1} = \bar{\mathbf{a}}_2 + \mathbf{a}_{B/G_2}$, which expresses the fact that B is a point on both bodies.

Figure 17.11(a) shows the MAD when the bar and disk are rigidly connected at B. In this case, $\alpha_1 = \alpha_2 = \alpha$. Furthermore, $\bar{\mathbf{a}}_1$ and $\bar{\mathbf{a}}_2$ are related by $\bar{\mathbf{a}}_2 = \bar{\mathbf{a}}_1 + \mathbf{a}_{G_2/G_1}$ because G_1 and G_2 are on the same rigid body.

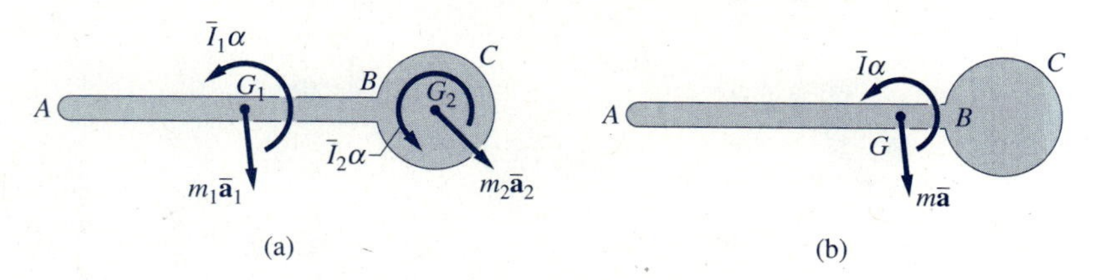

Fig. 17.11

Figure 17.11(b) shows an alternate form of the MAD for the rigid body in Fig. 17.11(a). Here, $m\bar{\mathbf{a}}$ and $\bar{I}\alpha$ are the inertia vector and inertia couple, respectively, for the entire body, and G is its mass center. Whether one employs the MAD in Fig. 17.11 parts (a) or (b) is a matter of personal preference, because the analysis using either diagram involves about the same amount of work.

Sample Problem 17.3

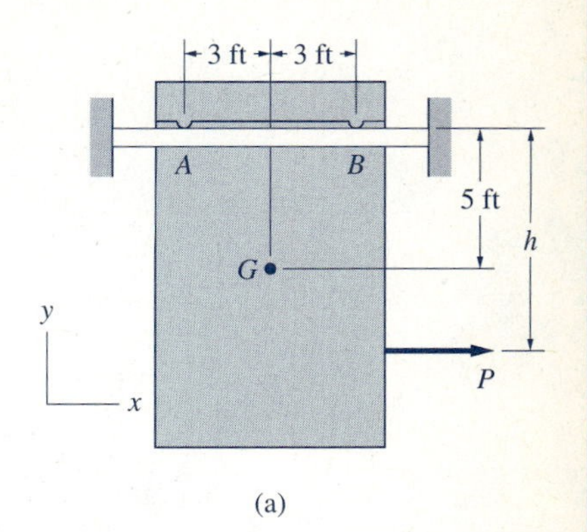

(a)

The mass center of the 400-lb sliding door in Fig. (a) is located at G. The door is supported on the horizontal rail by sliders at A and B. The coefficient of static as well as kinetic friction is 0.4 at A and 0.3 at B. The door was at rest before the horizontal force $P = 200$ lb was applied. (1) Find the maximum value of h for which the door will slide to the right without tipping and the corresponding acceleration of the door. (2) If $h = 5$ ft, find all forces acting on the door, and calculate its acceleration.

Solution

Part 1

The free-body and mass-acceleration diagrams for the door, shown in Fig. (b), are described in detail as follows.

Free-body diagram The FBD contains the following forces: the 400-lb weight acting at G, the applied force P, the normal force N_A, and the friction force $F_A = \mu_A N_A = 0.4N_A$. Since the door starts from rest, its velocity will be directed to the right (i.e., in the same direction as the acceleration), which means that F_A is directed to the left. There is no normal force and therefore no friction force at B, because the problem statement implies that the door is sliding to the right in a state of impending tipping about A.

Mass-acceleration diagram The MAD contains only the inertia vector of magnitude $m\bar{a}$ acting at G. There is no inertia couple because the door is translating ($\alpha = 0$).

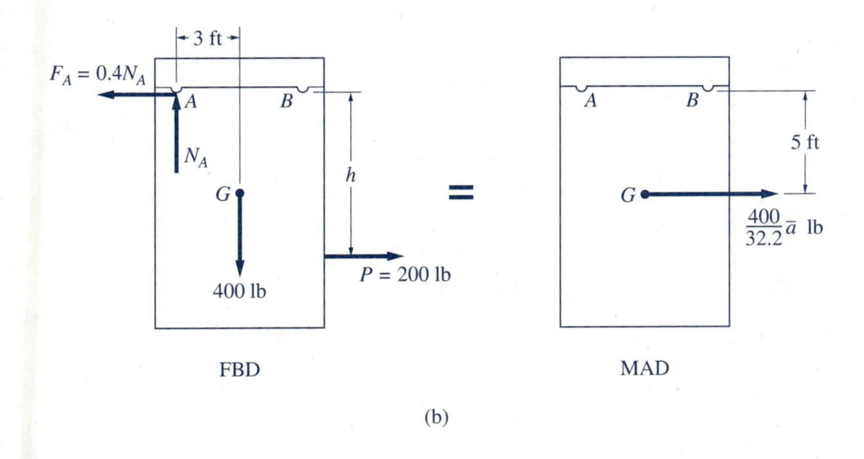

(b)

Inspection of Fig. (b) reveals that the unknowns are N_A and h on the FBD, and $\bar{a}$ on the MAD. These three unknowns can be computed by deriving and solving any three independent equations of motion. When employing the FBD-MAD technique, remember that (1) the resultant force on the FBD can be equated to the inertia vector $m\bar{\mathbf{a}}$ on the MAD, and (2) the resultant moment about any point on the FBD can be equated to the resultant moment about the same point on the MAD.

Equating the x- and y-components of the forces on the FBD to the corresponding components of the inertia vector, we obtain

$$\Sigma F_y = m\bar{a}_y \qquad +\uparrow \quad N_A - 400 = 0$$

$$N_A = 400 \text{ lb}$$

and

$$\Sigma F_x = m\bar{a}_x \qquad \xrightarrow{+} \qquad 200 - 0.4N_A = \frac{400}{32.2}\bar{a}$$

$$200 - 0.4(400) = \frac{400}{32.2}\bar{a}$$

which gives

$$\bar{a} = 3.22 \text{ ft/s}^2 \qquad \textit{Answer}$$

The third independent equation is a moment equation about any point. Choosing point A as the moment center, the resultant moment on the FBD is equated to the resultant moment on the MAD:

$$(\Sigma M_A)_{\text{FBD}} = (\Sigma M_A)_{\text{MAD}} \qquad \circlearrowleft + \qquad 200h - 400(3) = \left(\frac{400}{32.2}\bar{a}\right)5$$

With $\bar{a} = 3.22$ ft/s², we get

$$h = 7.00 \text{ ft} \qquad \textit{Answer}$$

Part 2

The FBD and MAD for $h = 5$ ft are shown in Fig. (c). The details of these diagrams are as follows.

Free-body diagram The FBD contains the 400-lb weight, the applied force $P = 200$ lb passing through G, the normal forces N_A and N_B, and the friction forces F_A and F_B. We assume that the door will be sliding to the right while maintaining contact at both A and B. The friction forces, determined by the kinetic coefficients of friction, are directed to the left, i.e., opposite to the motion.

Mass-acceleration diagram Since the door is sliding to the right without rotating, the MAD contains only the inertia vector acting through G.

The number of unknowns on the FBD and MAD is three: N_A, N_B, and $\bar{a}$, which can be found from any three independent equations of motion. One such set

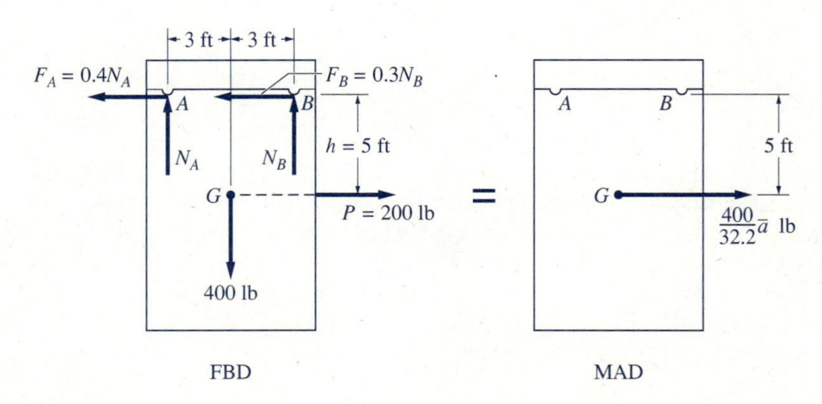

(c)

of equations is given below; their validity can be determined by referring to the FBD and MAD in Fig. (c).

$$\Sigma F_y = m\bar{a}_y \qquad +\uparrow \quad N_A + N_B - 400 = 0 \tag{a}$$

$$\Sigma M_G = 0 \qquad \circlearrowleft+ \quad N_B(3) + 0.3N_B(5) - N_A(3) + 0.4N_A(5) = 0 \tag{b}$$

$$\Sigma F_x = m\bar{a}_x \qquad \xrightarrow{+} \quad 200 - 0.4N_A - 0.3N_B = \frac{400}{32.2}\bar{a} \tag{c}$$

Solving Eqs. (a), (b), and (c) gives

$$N_A = 327.3 \text{ lb} \qquad N_B = 72.73 \text{ lb} \qquad \bar{a} = 3.80 \text{ ft/s}^2$$ *Answer*

The positive values confirm our assumption that the door will slide to the right without tipping.

Sample Problem 17.4

The homogeneous bar in Fig. (a) has a mass m and a length L. The bar, which is free to rotate in the vertical plane about a pin at 0, is released from rest in the position $\theta = 0$. Find the angular acceleration α when $\theta = 60°$.

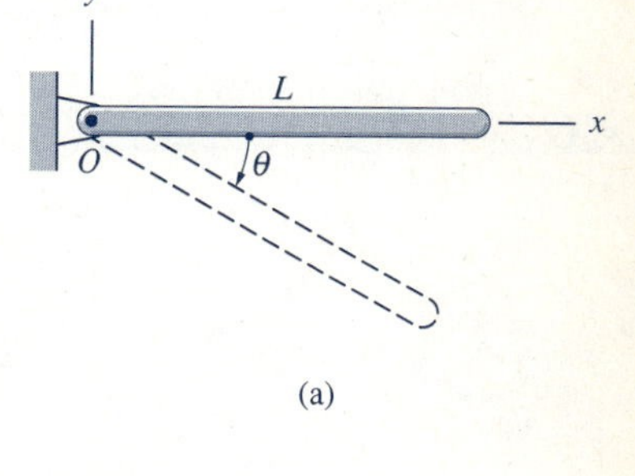

(a)

Solution

Figure (b) shows the FBD and the MAD of the bar when $\theta = 60°$. The FBD contains the weight W of the bar, acting at its mass center G (located at the midpoint of the bar) and the components of the pin reaction at O. In the MAD, the inertia couple $\bar{I}\alpha$ was drawn assuming that α is clockwise, and using $\bar{I} = mL^2/12$ from Table 17.1. The components of the inertia vector $m\mathbf{a}$ were found by noting that the path of G is a circle centered at O. Therefore, the normal and tangential components of $\bar{\mathbf{a}}$ are $\bar{a}_n = (L/2)\omega^2$ and $\bar{a}_t = (L/2)\alpha$.

FBD MAD

(b)

We note that there are a total of four unknowns in Fig. (b): O_x, O_y, α, and ω. Since there are only three independent equations of motion, we will not be able to determine all unknowns using only the FBD and the MAD. The reason for this is that ω depends on the history of motion: $\omega = \int \alpha \, dt + C$, where the constant of integration C is determined by the initial conditions. Therefore, the equations of motion at a specific position of the bar will not determine the angular velocity in that position. However, inspection of the FBD and MAD reveals that it is possible to determine the angular acceleration α, since it is the only unknown that appears in the moment equation when O is used as the moment center. Referring to the diagrams in Fig. (b), this moment equation is

$$(\Sigma M_O)_{\text{FBD}} = (\Sigma M_O)_{\text{MAD}}$$

$$\circlearrowleft+ \quad mg\frac{L}{2}\cos 60^\circ = \frac{mL^2}{12}\alpha + \left(m\frac{L}{2}\alpha\right)\frac{L}{2} = \frac{mL^2}{3}\alpha \tag{a}$$

from which we find that

$$\alpha = \frac{3g}{2L}\cos 60^\circ = 0.750\frac{g}{L} \qquad \textit{Answer} \tag{b}$$

Because the acceleration of O is zero, the above moment equation could have also been obtained from Eq. (17.6): $\Sigma M_O = I_O\alpha$, where

$$I_O = \bar{I} + md^2 = \frac{mL^2}{12} + m\left(\frac{L}{2}\right)^2 = \frac{mL^2}{3} \tag{c}$$

from the parallel-axis theorem. It is now seen that $\Sigma M_O = I_O\alpha$ will yield an equation that is identical to Eq. (a).

Sample Problem 17.5

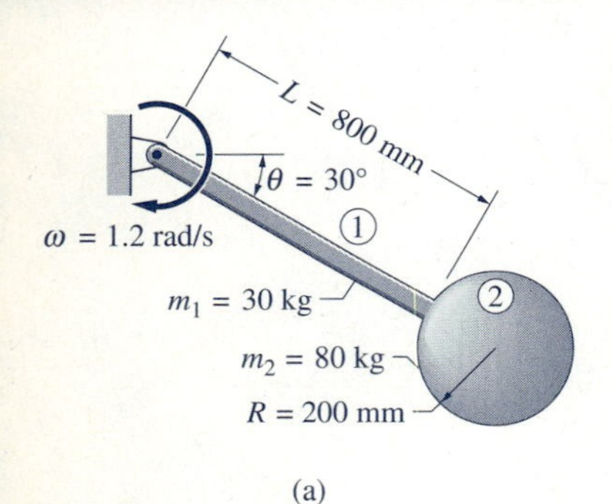

(a)

The body shown in Fig. (a) consists of the homogeneous slender bar ① that is rigidly connected to the homogeneous sphere ②. The body is rotating in the vertical plane about the pin at O. When the body is in the position where $\theta = 30^\circ$, its angular velocity is $\omega = 1.2$ rad/s clockwise. At this instant, determine the angular acceleration α and the magnitude of the pin reaction at O.

Solution

The FBD and MAD of the body in the position $\theta = 30^\circ$ are shown in Fig. (b). In these diagrams, the bar and the sphere are treated as separate entities, each with its

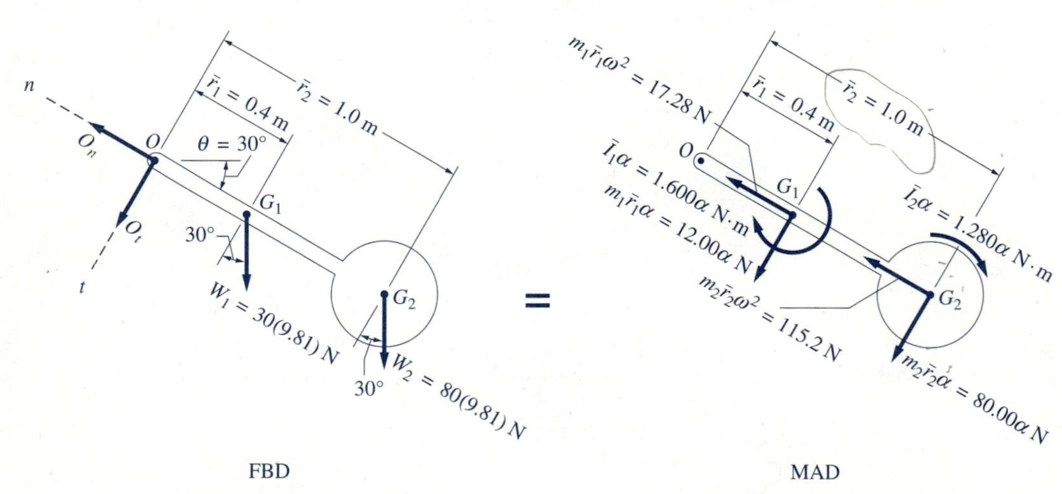

(b)

own inertia couple and inertia vector. [An equivalent form of the MAD would be obtained by showing the inertia couple and inertia vector for the assembly—refer to Fig. 17.11 parts (b) and (c).] Details of the diagrams are described in the following.

Free-body diagram The forces O_n and O_t are the components of the pin reaction relative to the n and t axes shown in the figure. The weights W_1 and W_2 of the bar and sphere, respectively, act at their mass centers G_1 and G_2. The distances $\bar{r}_1 = 0.4$ m and $\bar{r}_2 = 1.0$ m, measured from O to the mass centers, are deduced from the dimensions in Fig. (a).

Mass-acceleration diagram The MAD assumes that the angular acceleration α, measured in rad/s^2, is clockwise. Using the fact that the body rotates about the fixed point O, kinematic analysis enables us to express the accelerations of G_1 and G_2 in terms of α and ω of the body. The inertia terms that appear in the MAD have been computed in the following manner.

For the slender bar:

$$\bar{I}_1\alpha = \frac{m_1L^2}{12}\alpha = \frac{30(0.8)^2}{12}\alpha = 1.600\alpha \text{ N}\cdot\text{m}$$

$$m_1\bar{r}_1\omega^2 = 30(0.4)(1.2)^2 = 17.28 \text{ N}$$

$$m_1\bar{r}_1\alpha = 30(0.4)\alpha = 12.00\alpha \text{ N}$$

For the sphere:

$$\bar{I}_2\alpha = \frac{2}{5}m_2R^2\alpha = \frac{2}{5}(80)(0.2)^2\alpha = 1.280\alpha \text{ N}\cdot\text{m}$$

$$m_2\bar{r}_2\omega^2 = 80(1.0)(1.2)^2 = 115.2 \text{ N}$$

$$m_2\bar{r}_2\alpha = 80(1.0)\alpha = 80.00\alpha \text{ N}$$

In the MAD, the directions of the tangential components of the inertia vectors (those containing α) are consistent with the assumed clockwise direction of α. The normal components of the inertia vectors (those containing ω^2) are directed toward the center of rotation O, regardless of the direction of ω.

From Fig. (b) we see that there are two unknowns on the FBD (O_n and O_t) and one unknown (α) on the MAD. Therefore, all that remains is to derive and solve the three independent equations of motion for the unknowns.

Equating moments about O for the FBD and the MAD in Fig. (b), we obtain

$$(\Sigma M_O)_{\text{FBD}} = (\Sigma M_O)_{\text{MAD}}$$

$$\circlearrowright^{+} \quad 30(9.81)(0.4)\cos 30° + 80(9.81)(1.0)\cos 30°$$

$$= 1.600\alpha + (12.00\alpha)(0.4) + 1.280\alpha + (80.00\alpha)(1.0)$$

$$= 87.68\alpha$$

from which we find that

$$\alpha = 8.91 \text{ rad/s}^2 \qquad \textit{Answer}$$

Since the acceleration of point O is zero, this result could also have been derived using the special case $\Sigma M_O = I_O\alpha$.

Using $\alpha = 8.91 \text{ rad/s}^2$ and referring to Fig. (b), the force equations in the t and n direction give

$$\Sigma F_t = m\bar{a}_t$$

$$\overset{+}{\swarrow} \quad O_t + 30(9.81)\cos 30° + 80(9.81)\cos 30° = 12.00(8.91) + 80.00(8.91)$$

$$O_t = -114.8 \text{ N}$$

and

$$\Sigma F_n = m\bar{a}_n$$

$$\overset{+}{\nwarrow} \quad O_n - 30(9.81)\sin 30° - 80(9.81)\sin 30° = 17.28 + 115.2$$

$$O_n = 672.0 \text{ N}$$

Therefore, the magnitude of the pin reaction at O is

$$O = \sqrt{O_t^2 + O_n^2} = \sqrt{(-114.8)^2 + (672.0)^2} = 682 \text{ N} \qquad \textit{Answer}$$

Sample Problem 17.6

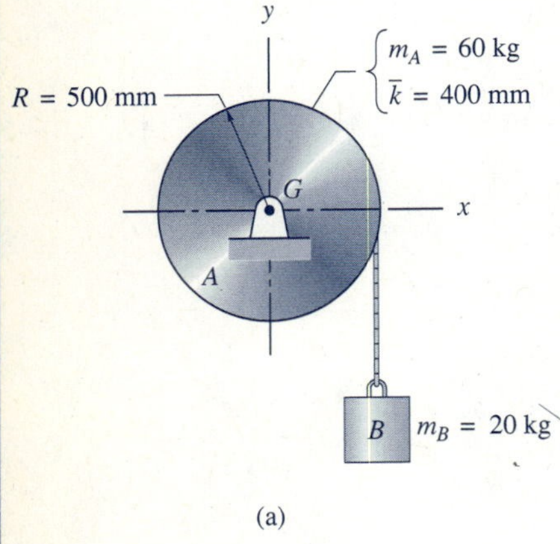

(a)

The cable connected to block B in Fig. (a) is wound tightly around disk A, which is free to rotate about the axle at its mass center G. The masses of A and B are 60 kg and 20 kg, respectively, and $\bar{k} = 400$ mm for the disk. Determine the angular acceleration of A and the tension in the cable.

Solution

We present two methods of solution. The first one applies the FBD-MAD technique to the entire system; the second one does not. In both solutions, the weights of the bodies are $W_A = 60(9.81) = 588.6$ N and $W_B = 20(9.81) = 196.2$ N. The angular acceleration α of the disk, measured in rad/s^2, is assumed to be clockwise. Therefore the inertia couple for A is $\bar{I}\alpha = m\bar{k}^2\alpha = 60(0.4)^2\alpha = 9.600\alpha$ N·m, also clockwise. Since the cable does not slip on the disk, the inertia vector for the block B becomes $m_B a_B = m_B R\alpha = 20(0.5)\alpha = 10\alpha$ N, directed downward.

Method I

The free-body and mass-acceleration diagrams for the system consisting of the disk and the block are shown in Fig. (b). The FBD contains the weights W_A and W_B, and

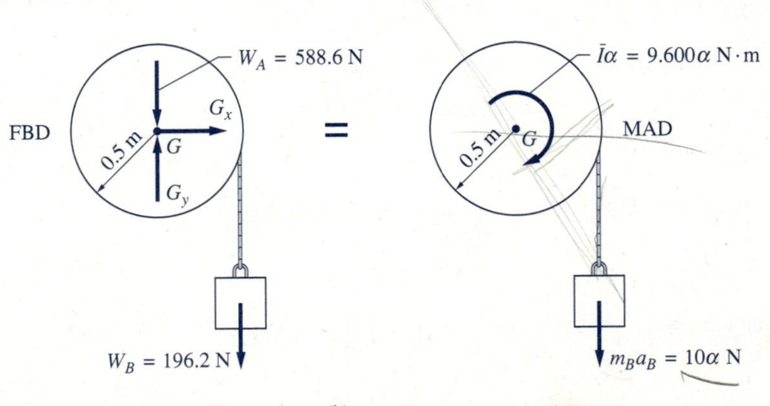

(b)

the components of the pin reaction at G. The tension in the cable does not appear on this FBD because it is an internal force. The MAD contains the inertia couple of the disk and the inertia vector of the block. There is no inertia vector for the disk because it is rotating about its mass center (i.e., the rotation is central).

The angular acceleration α can be found by equating the resultant moments about G on the FBD and the MAD.

$$(\Sigma M_G)_{\text{FBD}} = (\Sigma M_G)_{\text{MAD}} \qquad \circlearrowleft + \qquad 196.2(0.5) = 9.600\alpha + 10\alpha(0.5)$$

$$\alpha = 6.719 \text{ rad/s}^2 \qquad \textit{Answer}$$

To find the tension in the cable, we analyze the block separately (the disk could also be used). The FBD and MAD for the block are shown in Fig. (d), where T is the cable tension. Summing forces in the y-direction yields

$$\Sigma F_y = m\bar{a}_y \qquad +\downarrow \qquad 196.2 - T = 10\alpha = 10(6.719)$$

$$T = 129.0 \text{ N} \qquad \textit{Answer}$$

Method II

The FBD and MAD for the disk are shown in Fig. (c), and Fig. (d) contains the corresponding diagrams for the block. Note that T is the cable tension. Using Fig. (d) we get

$$\Sigma F_y = m\bar{a}_y \qquad +\downarrow \qquad 196.2 - T = 10\alpha \qquad \text{(a)}$$

Equating the resultant moments about G for the FBD and MAD in Fig. (c) yields

$$\Sigma M_G = \bar{I}\alpha \qquad \circlearrowleft + \qquad T(0.5) = 9.600\alpha \qquad \text{(b)}$$

Solving Eqs. (a) and (b), we obtain

$$\alpha = 6.719 \text{ rad/s}^2 \quad \text{and} \quad T = 129.0 \text{ N} \qquad \textit{Answer}$$

which, of course, agree with the values determined in the previous solution.

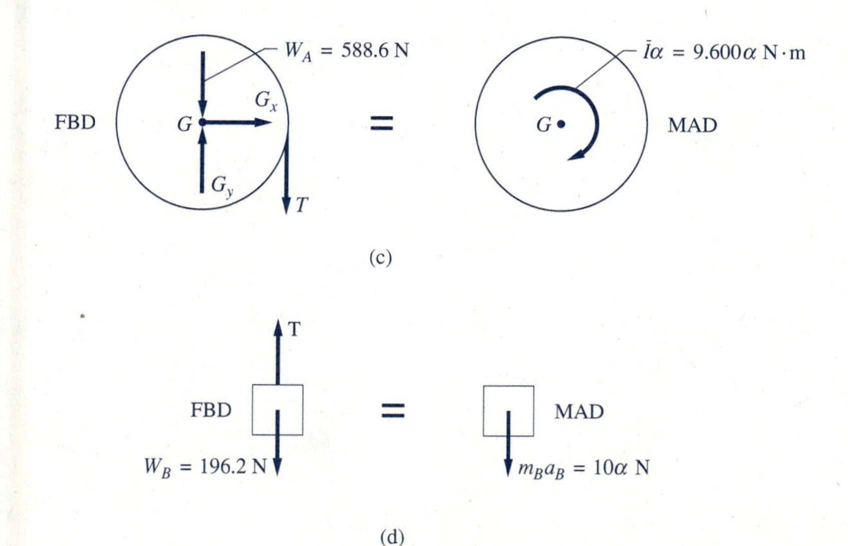

Sample Problem 17.7

The 40-kg unbalanced wheel in Fig. (a) is rolling without slipping under the action of a counterclockwise couple $C_0 = 20 \text{ N} \cdot \text{m}$. When the wheel is in the position shown, its angular velocity is $\omega = 2$ rad/s, clockwise. For this position, calculate the angular acceleration α and the forces exerted on the wheel at C by the rough horizontal plane. The radius of gyration of the wheel about its mass center G is $\bar{k} = 200$ mm.

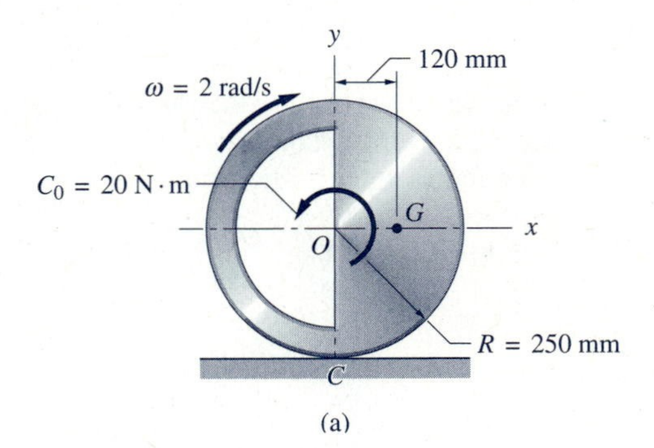

(a)

Solution

The free-body and mass-acceleration diagrams for the wheel, shown in Fig. (b), were constructed as follows.

Free-body diagram The FBD consists of the applied couple C_0, the weight $W = 40(9.81) = 392.4$ N, and the normal and friction forces that act at the contact point C, denoted by N_C and F_C, respectively. Observe that F_C has been assumed to be directed to the right.

We note that there are two unknown variables on the FBD: N_C and F_C. Since there are only three independent equations of motion, the number of unknown variables on the mass-acceleration diagram must be reduced to one by using kinematics.

Mass-acceleration diagram In the MAD of Fig. (b) the angular acceleration α, measured in rad/s^2, has been assumed to be clockwise. The corresponding inertia couple shown on this diagram is

$$\bar{I}\alpha = m\bar{k}^2\alpha = 40(0.200)^2\alpha = 1.600\alpha \text{ N} \cdot \text{m}$$

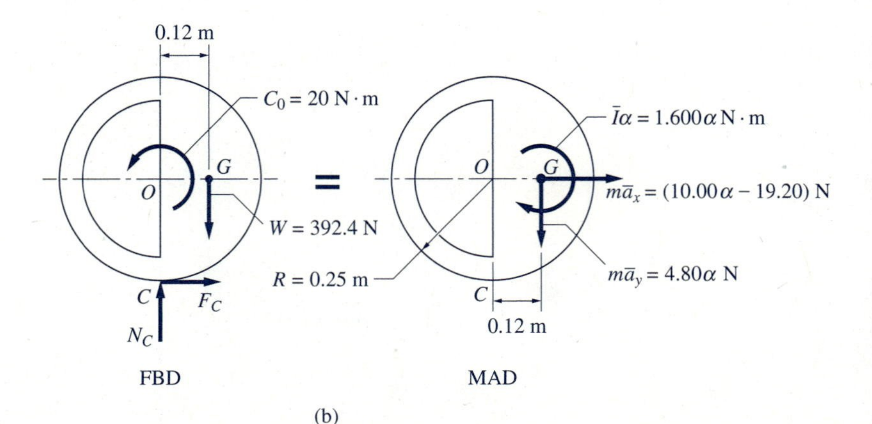

(b)

Because the wheel does not slip, the acceleration of its center is $a_O = R\alpha = 0.250\alpha$ m/s^2, directed to the right. Applying the relative acceleration equation between G and O, we obtain (the units of each term are m/s^2)

$$\bar{\mathbf{a}} = \mathbf{a}_G = \mathbf{a}_O + \mathbf{a}_{G/O}$$

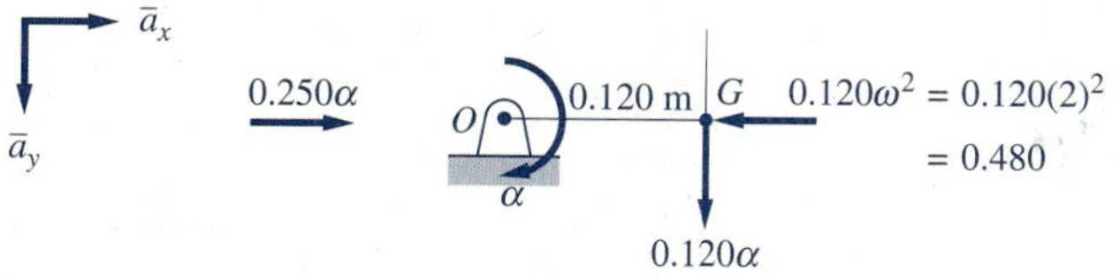

$\bar{a}_x$ $\bar{a}_y$ $\quad 0.250\alpha \quad$ O α $\quad 0.120$ m $\quad G \quad 0.120\omega^2 = 0.120(2)^2 = 0.480 \quad 0.120\alpha$

from which we find $\bar{a}_x = 0.250\alpha - 0.480$ m/s^2 and $\bar{a}_y = 0.120\alpha$ m/s^2. Multiplying these results by $m = 40$ kg, the components of the inertia vector become $m\bar{a}_x = (10.00\alpha - 19.20)$ N, directed to the right, and $m\bar{a}_y = 4.80\alpha$ N, directed downward.

The FBD and MAD in Fig. (b) now contain only three unknowns: N_C, F_C, and α, which can be found using any three independent equations of motion.

Since N_C and F_C act at C, it is convenient to use that point as a moment center, the corresponding moment equation being

$$(\Sigma M_C)_{\text{FBD}} = (\Sigma M_C)_{\text{MAD}}$$

$$\circlearrowright^{+} \quad -20 + 392.4(0.120) = 1.600\alpha + 0.250(10.00\alpha - 19.20) + 0.120(4.80\alpha)$$

The solution to this equation is

$$\alpha = 6.820 \text{ rad/s}^2 \qquad \textit{Answer}$$

Since α is positive, its direction is clockwise, as assumed.

The forces at C can now be found from force equations of motion:

$$\Sigma F_x = m\bar{a}_x \qquad \xrightarrow{+} \quad F_C = 10.00\alpha - 19.20 = 10.00(6.820) - 19.20$$

and

$$\Sigma F_y = m\bar{a}_y \qquad +\downarrow \quad 392.4 - N_C = 4.80\alpha = 4.80(6.820)$$

which yield

$$F_C = 49.0 \text{ N} \quad \text{and} \quad N_C = 360 \text{ N} \qquad \textit{Answer}$$

Since each force is positive, it is directed as shown in the FBD. It can be seen that the smallest coefficient of static friction for which the wheel will not slip in the position shown in Fig. (a) is $\mu_s = F_C/N_C = 49.0/360 = 0.136$.

Sample Problem 17.8

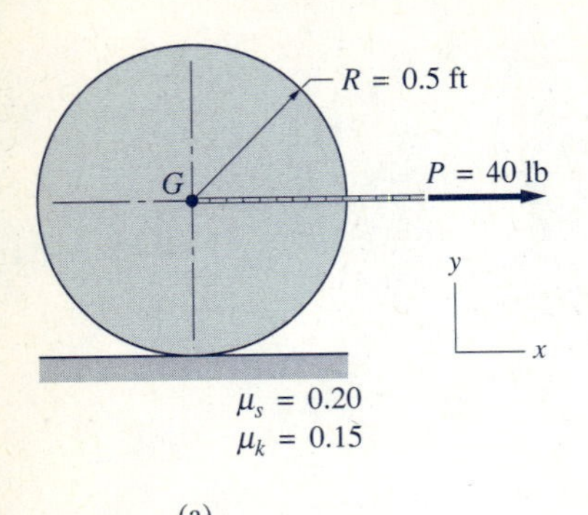

(a)

Figure (a) shows a 50-lb homogeneous disk of radius 0.5 ft. The disk is at rest before the horizontal force $P = 40$ lb is applied to its mass center G. The coefficients of static and kinetic friction for the surfaces in contact are 0.20 and 0.15, respectively. Determine the angular acceleration of the disk and the acceleration of G after the force is applied.

Solution

Two motions of the disk are possible: rolling without slipping, and rolling with slipping. We will solve the problem by assuming that the disk rolls without slipping. This assumption will then be checked by comparing the required friction force with its maximum static value.

The free-body diagram (FBD) and the mass-acceleration diagram (MAD) based on the no-slip assumption are shown in Fig. (b). The FBD contains the 50-lb weight, the 40-lb applied force, the normal force N, and the friction force F, assumed acting to the left. The MAD contains the inertia couple and inertia vector, where the angular acceleration α, measured in rad/s^2, has been assumed to be clockwise. The values of $\bar{I}\alpha$ and $m\bar{a}$ were computed as follows.

$$\bar{I}\alpha = \frac{mR^2}{2}\alpha = \frac{50(0.5)^2}{(32.2)2}\alpha = 0.1941\alpha \text{ lb}\cdot\text{ft}$$

$$m\bar{a} = mR\alpha = \frac{50}{32.2}(0.5)\alpha = 0.7764\,\alpha \text{ lb}$$

Note that $\bar{a} = R\alpha$ is a valid kinematic equation because the disk is assumed to be rolling without slipping. There are a total of three unknowns in the FBD and MAD: F, N, and α, which can be computed using any three independent equations of motion.

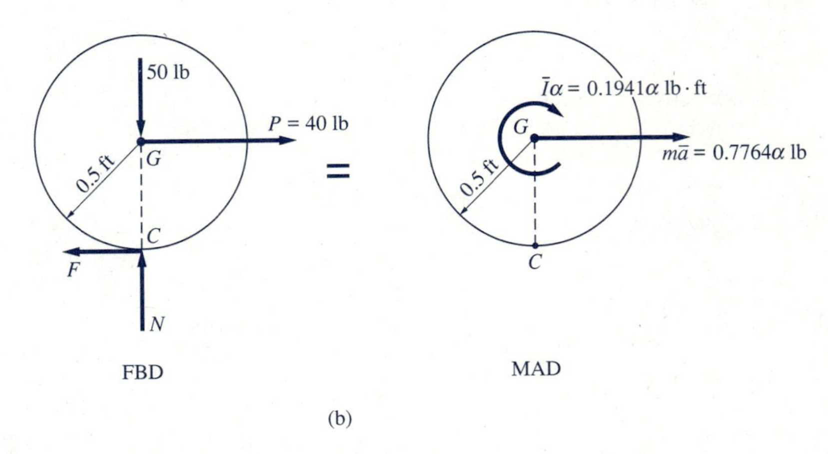

(b)

A convenient solution is to first equate the resultant moment about C on the FBD to the resultant moment about C on the MAD and then utilize the force equations of motion.

$$(\Sigma M_C)_{\text{FBD}} = (\Sigma M_C)_{\text{MAD}} \qquad \circlearrowright^{+} \quad 40(0.5) = 0.1941\alpha + 0.7764\alpha(0.5)$$

$$\alpha = 34.35 \text{ rad/s}^2$$

$$\Sigma F_x = m\bar{a}_x \qquad \xrightarrow{+} \qquad 40 - F = 0.7764\alpha = 0.7764(34.35)$$

$$F = 13.33 \text{ lb}$$

$$\Sigma F_y = m\bar{a}_y \qquad +\uparrow \qquad N - 50 = 0$$

$$N = 50 \text{ lb}$$

Since α, F, and N are all positive, their directions are as shown in Fig. (b).

Next we note that the maximum possible static friction force is $F_{max} = \mu_s N = 0.20(50) = 10.0$ lb. The friction force required for rolling without slipping is, according to our solution, $F = 13.33$ lb. Since $F > F_{max}$, we conclude that the disk does not roll without slipping, and we must reformulate the problem using the knowledge that the disk slips. (If F had been equal to F_{max}, the wheel would be rolling with impending slipping, and the preceding solution would be valid.)

The FBD and MAD for the case where the disk rolls and slips simultaneously are shown in Fig. (c). The friction force F in the FBD has been set equal to its kinetic value, $\mu_k N$. This force must be shown acting to the left in order to oppose slipping. The inertia couple $\bar{I}\alpha$ in the MAD is identical to that used in Fig. (b). However, the important difference here is that the magnitude of the inertia

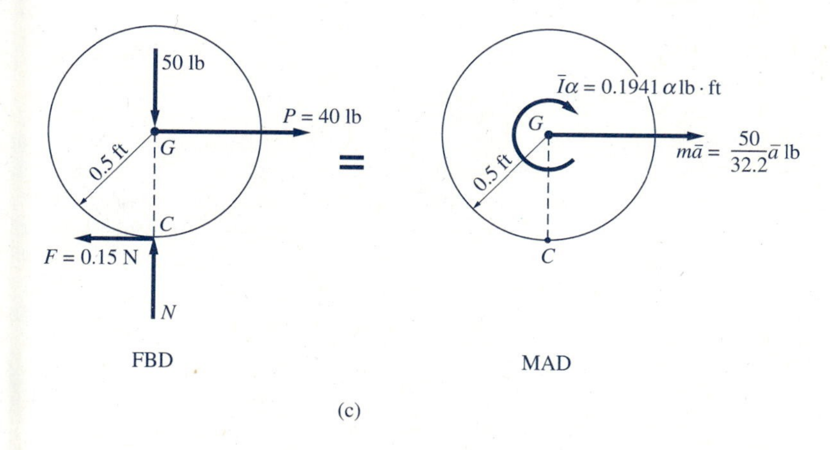

(c)

vector is now $m\bar{a} = (50/32.2)\bar{a}$ lb, where $\bar{a}$ is measured in ft/s^2. Since the disk is slipping, the kinematic constraint $\bar{a} = R\alpha$ does not apply. Once again we see that there are three unknowns on the FBD and the MAD, except that now the unknowns are N, α, and $\bar{a}$.

The three unknowns can be calculated as follows (of course, any other three independent equations could also be used):

$$\Sigma F_y = m\bar{a}_y \qquad +\uparrow \qquad N - 50 = 0 \qquad N = 50 \text{ lb} \tag{a}$$

$$\Sigma F_x = m\bar{a}_x \qquad \xrightarrow{+} \qquad 40 - 0.15N = \frac{50}{32.2}\bar{a} \tag{b}$$

$$\Sigma M_G = \bar{I}\alpha \qquad \circlearrowright + \qquad 0.5(0.15N) = 0.1941\alpha \tag{c}$$

Substituting $N = 50$ lb from Eq. (a) into Eqs. (b) and (c) yields

$$\bar{a} = 20.93 \text{ ft/s}^2 \quad \text{and} \quad \alpha = 19.32 \text{ rad/s}^2 \qquad \textit{Answer}$$

Sample Problem 17.9

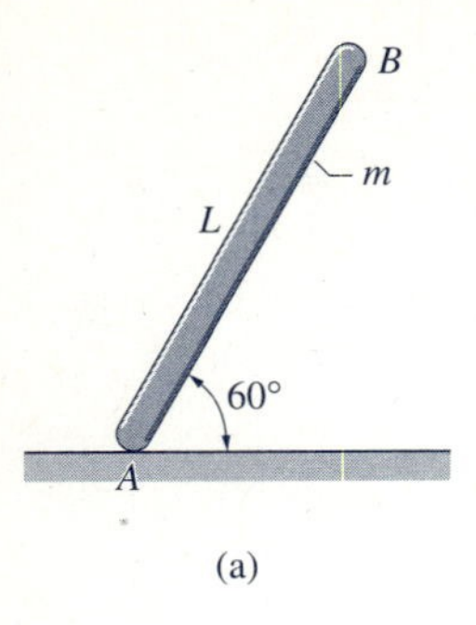

(a)

A homogeneous slender bar AB of mass m and length L is released from rest in the position shown in Fig. (a). For this position, determine the acceleration of end A, the reaction at A, and the angular acceleration of the bar. Assume that the horizontal plane is smooth.

Solution

FBD   MAD

(b)

The free-body diagram (FBD) and mass-acceleration diagram (MAD) of the bar at the instant of release are shown in Fig. (b). The FBD contains the weight of the bar, mg, and the vertical reaction N. The MAD contains the inertia couple, $\bar{I}\alpha$, and the inertia vector, $m\bar{\mathbf{a}}$. The latter consists of components ma_A and $m(L/2)\alpha$, which are obtained from kinematics.

Since A is known to move along the horizontal plane, it is a kinematically important point. The acceleration of the mass center G is related to $\mathbf{a}_A$ and α by the following relative acceleration equation.

$$\bar{\mathbf{a}} = \mathbf{a}_G = \mathbf{a}_A + \mathbf{a}_{G/A} \qquad \text{(a)}$$

In Eq. (a), the sense of $\mathbf{a}_A$ and α were assumed to be to the left and clockwise, respectively. The angular velocity ω is zero because the bar has just been released from rest in the position being considered. Multiplying the right-hand side of Eq. (a) by the mass m and placing the results at G gives the components of the inertia vector shown in the MAD of Fig. (b).

Inspection of Fig. (b) reveals that there are a total of three unknowns: N, a_A, and α. Therefore, the solution can be obtained by deriving and solving any three independent equations of motion.

Equating moments about A on the FBD and MAD in Fig. (b) yields

$$(\Sigma M_A)_{\text{FBD}} = (\Sigma M_A)_{\text{MAD}}$$

$$\circlearrowleft + \quad mg\left(\frac{L}{2}\cos 60°\right) = \frac{mL^2}{12}\alpha + m\frac{L}{2}\alpha\left(\frac{L}{2}\right) - ma_A\left(\frac{L}{2}\sin 60°\right)$$

which, on simplification, becomes

$$a_A = 0.7698L\alpha - 0.5774g \tag{b}$$

Referring again to Fig. (b), the force equation for the horizontal direction becomes

$$\Sigma F_x = m\bar{a}_x \qquad \xrightarrow{+} \quad 0 = -ma_A + m\frac{L}{2}\alpha \sin 60°$$

which reduces to

$$a_A = 0.4330L\alpha \tag{c}$$

Solving Eqs. (b) and (c) simultaneously yields

$$a_A = 0.742g \quad \text{and} \quad \alpha = 1.714\frac{g}{L} \qquad \textit{Answer}$$

Using the diagrams in Fig. (b), the force equation for the vertical direction is

$$\Sigma F_y = m\bar{a}_y \qquad +\downarrow \quad mg - N = m\frac{L}{2}\alpha \cos 60° \tag{d}$$

Substituting the expression for α found above, and solving for N, gives

$$N = 0.572mg \qquad \textit{Answer}$$

It must be emphasized that the values obtained for N, α, and a_A are valid only at the instant of release. Each of these variables will vary throughout the subsequent motion of the bar. However, it is interesting to note that there is never a horizontal force acting on the bar because the plane is smooth. Therefore, the path followed by the mass center G will be a vertical straight line.

PROBLEMS

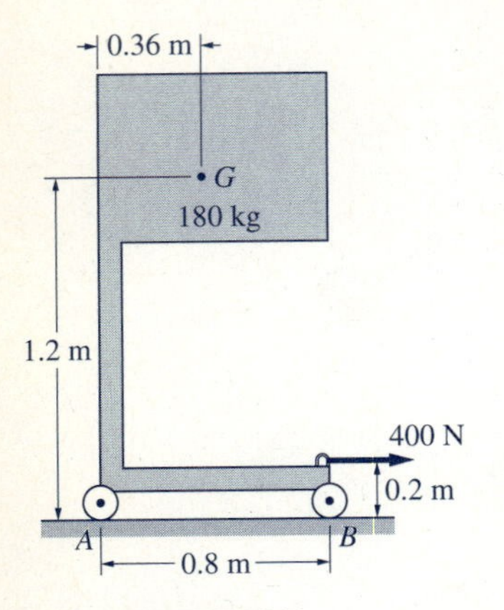

Fig. P17.17

17.17 The 400-N horizontal force is applied to the cabinet that is supported on frictionless casters at A and B. The mass of the cabinet is 180 kg, and G is its mass center. (a) Determine the acceleration of the cabinet assuming that it does not tip. (b) Verify that the cabinet does not tip by computing the reactions at A and B.

17.18 The combined mass center of the motorcycle and the cyclist is located at G. (a) Find the smallest acceleration for which the cyclist can perform a "wheelie," i.e., raise the front wheel off the ground. (b) What minimum coefficient of static friction between the tires and the road is required for the wheelie?

17.19 The mass centers of the 2400-lb car and 840-lb trailer are located at G_1 and G_2, respectively. The car and trailer are connected by a ball-and-socket hitch at C. Assuming that the car has rear-wheel drive, find the maximum possible acceleration of the car-trailer combination on a wet road where the coefficient of static friction is 0.35.

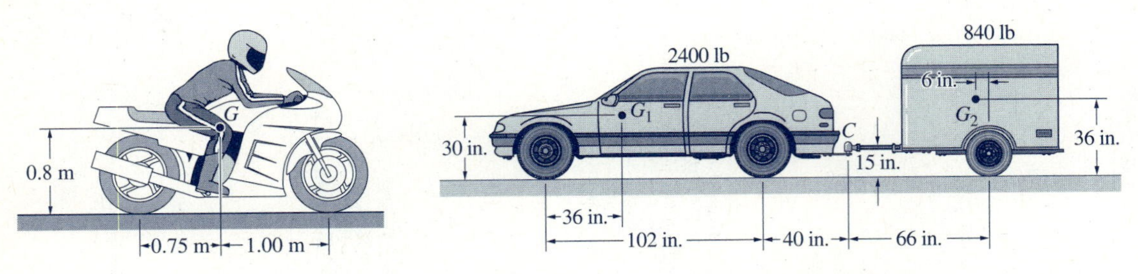

Fig. P17.18

Fig. P17.19, P17.20

17.20 Solve Prob. 17.19 assuming that the car has front-wheel drive.

17.21 For the car-trailer combination described in Prob. 17.19, determine (a) the maximum acceleration of the car for which the trailer does not pull up on the hitch; and (b) the corresponding horizontal force on the hitch.

17.22 The 40-lb homogeneous panel, pinned to a frictionless roller A and a light sliding collar B, is acted on by the 15-lb horizontal force. The coefficient of kinetic friction between B and the horizontal rod is 0.2. Determine the acceleration of the panel and the roller reaction at A, given that the velocity of the panel is 3.6 ft/s to the right.

17.23 Solve Prob. 17.22 if the velocity of the panel is 3.6 ft/s to the left.

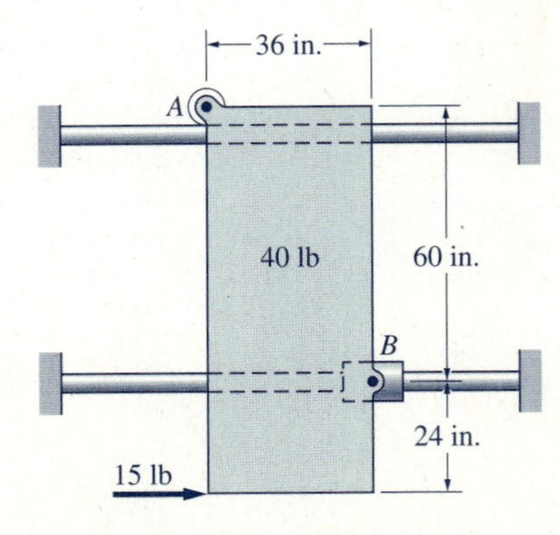

Fig. P17.22, P17.23

17.24 The homogeneous cylinder of mass m slides down the incline of slope angle β. The kinetic coefficient of friction between the cylinder and the incline is μ. Determine the expression for the smallest ratio d/h for which the cylinder will not tip.

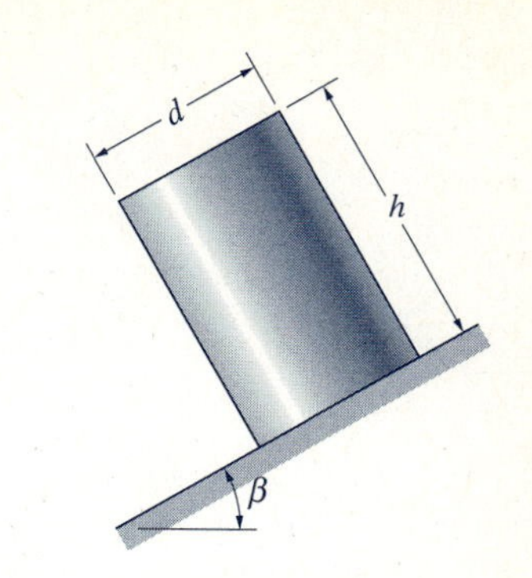

Fig. P17.24

17.25 The 20-kg pallet B carrying the 40-kg homogeneous box A slides freely down the inclined plane. The static and kinetic coefficients of friction between A and B are 0.4 and 0.35, respectively. (a) Show that the box will slide on the pallet, assuming that the dimensions of the box are such that it does not tip. (b) Determine the smallest ratio b/h for which the box will not tip.

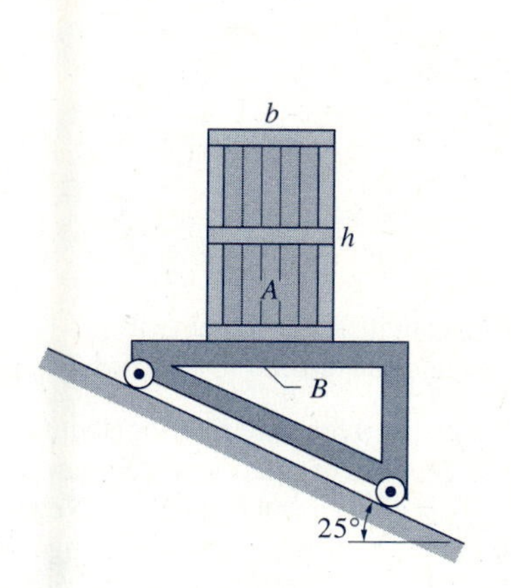

Fig. P17.25

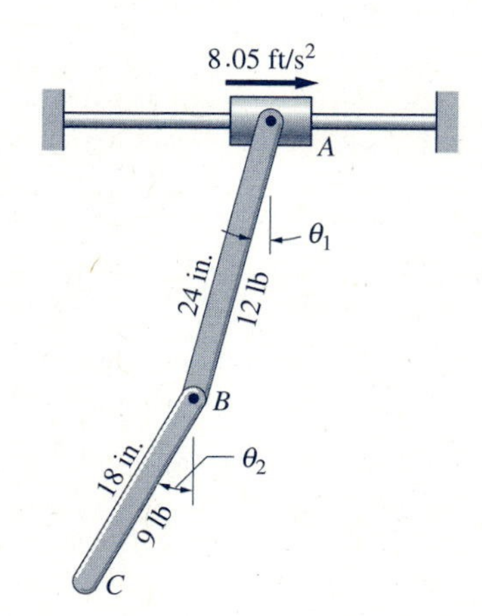

Fig. P17.26

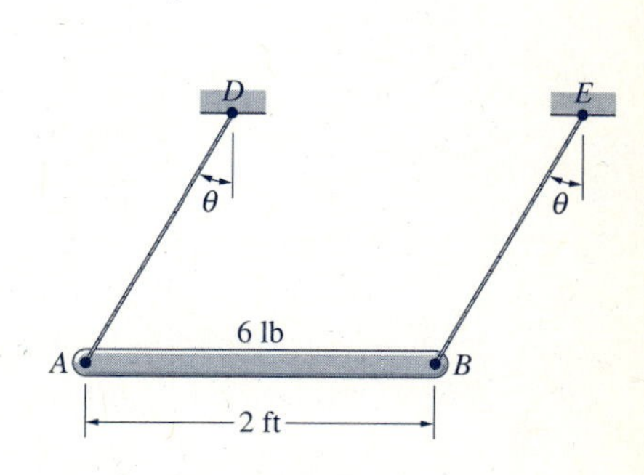

Fig. P17.27

17.26 The two homogeneous bars are connected by a pin at B. The upper bar is pinned to the sliding collar at A. The collar has a constant acceleration of 8.05 ft/s^2 to the right. Determine the angles θ_1 and θ_2, assuming that there is no oscillation (i.e., the angles are constant).

17.27 The 6-lb homogeneous bar AB is supported by the parallel strings attached at A and B. If the bar is released from rest when $\theta = 25°$, calculate the forces in the strings immediately after release.

17.28 The 125-kg refrigerator is being lowered to the ground by the platform C, which is controlled by the parallelogram linkage shown. Lips on the platform prevent the refrigerator from rolling on its wheels at A and B. Prove that the refrigerator will not tip when $\theta = 0$, $\omega = 1.2$ rad/s, and $\alpha = 1.6$ rad/s^2. (Hint: Assume that the refrigerator does not tip, and find the reactions at A and B.)

17.29 The angular velocity ω of the linkage that lowers the refrigerator in Prob. 17.28 is constant (i.e., $\alpha = 0$). Determine the maximum value of ω for which the refrigerator can be lowered without tipping, and the angle $\theta(0 < \theta < 90°)$ at which tipping impends.

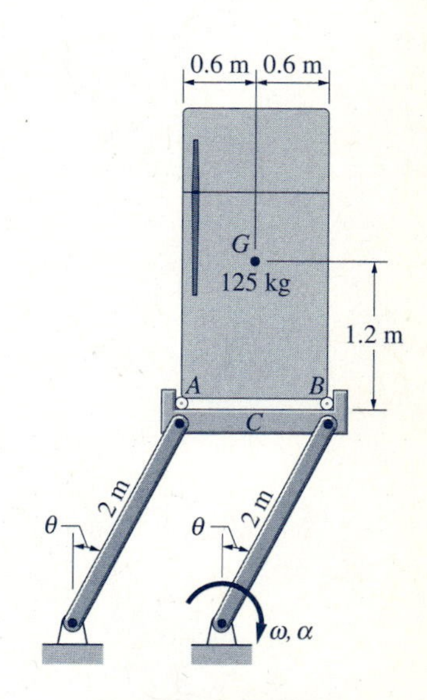

Fig. P17.28, P17.29

17.30 Determine the angular accelerations of the homogeneous pulleys shown in (a) and (b). The mass moment of inertia for each pulley about its mass center G is $0.36\text{kg}\cdot\text{m}^2$.

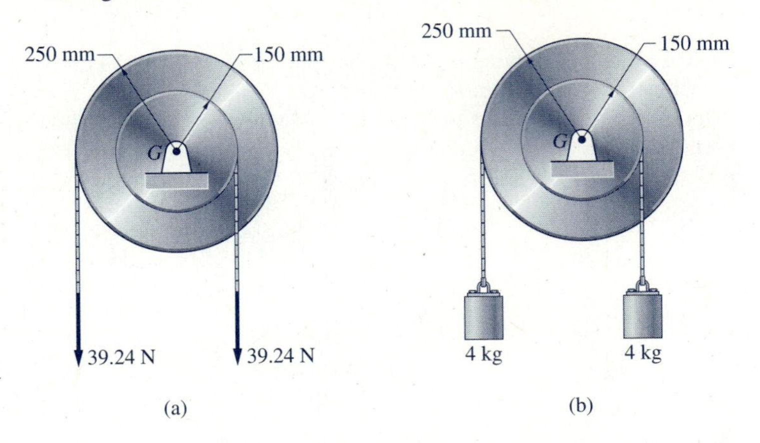

Fig. P17.30

17.31 The radius of gyration of the 36-lb pulley about its mass center G is 9 in. Compute the angular acceleration of the pulley and the tension in the cord AB.

17.32 Gears A and B, of masses 4 kg and 10 kg, respectively, are rotating about their mass centers. The radius of gyration about the axis of rotation is 100 mm for A and 300 mm for B. A constant couple $C_0 = 0.75\text{N}\cdot\text{m}$ acts on gear A. Neglecting friction, compute the angular acceleration of each gear and the tangential contact force between the gears at C.

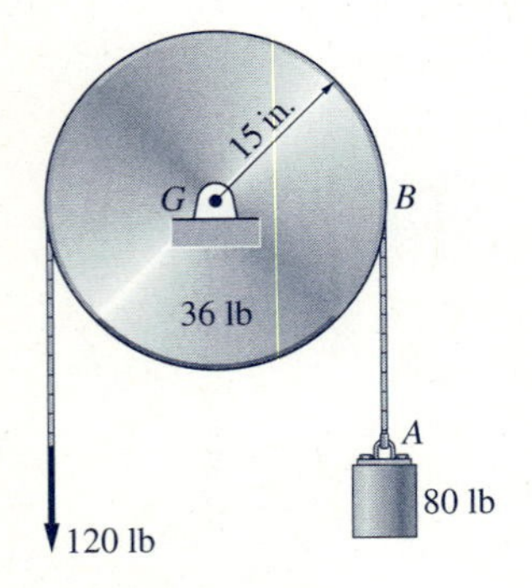

Fig. P17.31

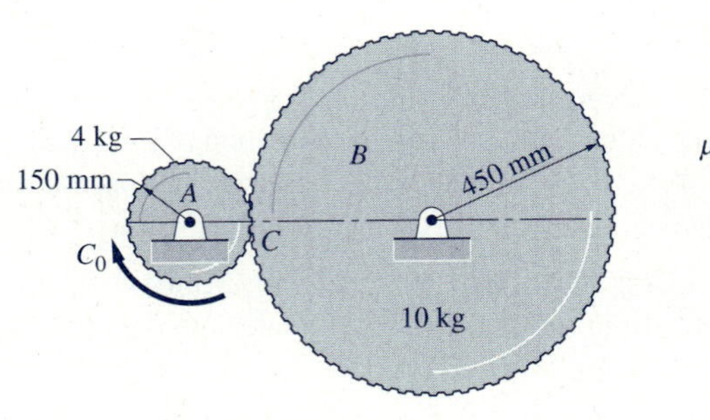

Fig. P17.32

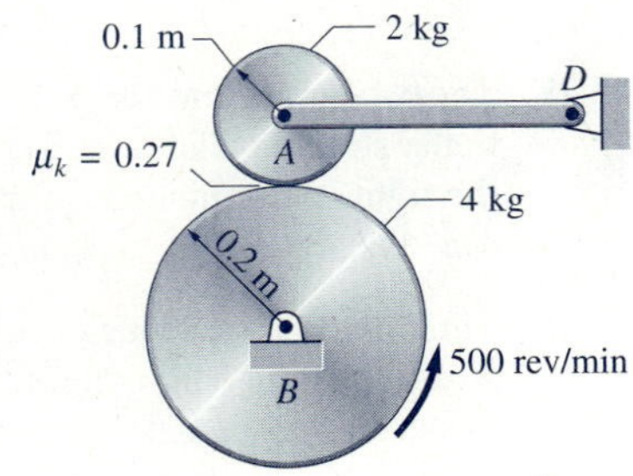

Fig. P17.33

17.33 Uniform disks A and B, having masses of 2 kg and 4 kg, respectively, can rotate about their mass centers. The kinetic coefficient of friction between the disks is 0.27. Disk B is spinning freely at 500 rev/min counterclockwise when it is placed in contact with the stationary disk A. Calculate the angular acceleration of each disk during the time that slipping occurs between the disks. Neglect the mass of bar AD.

17.34 The uniform disk of radius R and mass m is free to rotate about the pin at A. Determine the magnitude of the pin reaction at A immediately after the disk is released from rest when $\theta = 90°$.

17.35 The uniform disk of mass $m = 16$ kg and radius $R = 0.2$ m rotates freely about the pin at A. When $\theta = 40°$, the magnitude of the pin reaction is known to be 153 N. Compute the angular acceleration and the angular velocity of the disk in that position.

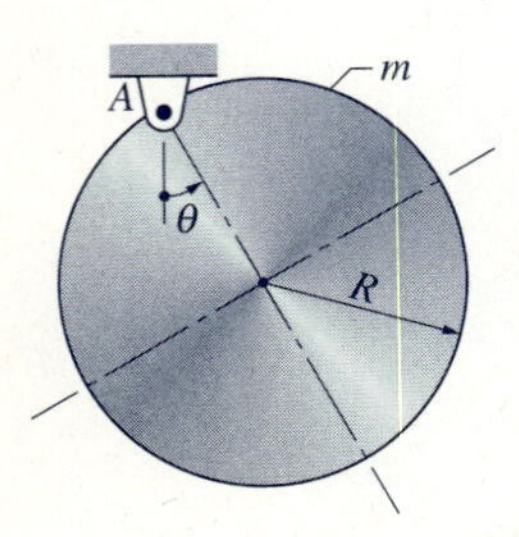

Fig. P17.34, P17.35

17.36 The solid steel cone of density 7850 kg/m^3 is released from rest when $\theta = 30°$. Assume that there is sufficient friction at A to prevent slipping. Determine the angular acceleration of the cone and the normal and friction forces at A immediately after release.

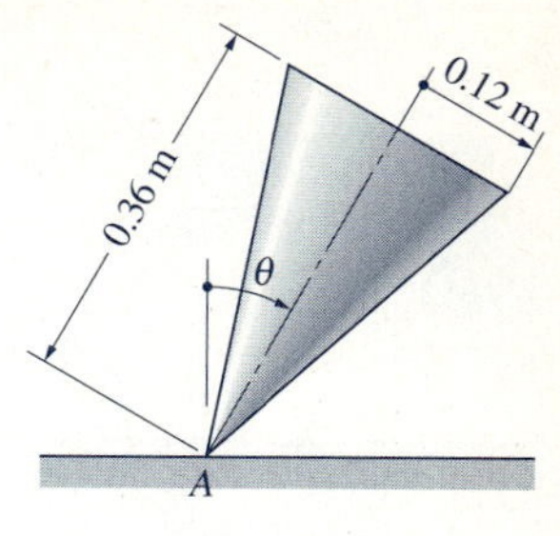

Fig. P17.36

17.37 The uniform 3-lb slender bar AB is mounted on a vertical shaft at C. A constant couple of 9 lb · in. is applied to the bar. Calculate the angular acceleration of the bar and the magnitude of the horizontal reaction at C at the instant when the angular velocity of the bar is 6 rad/s.

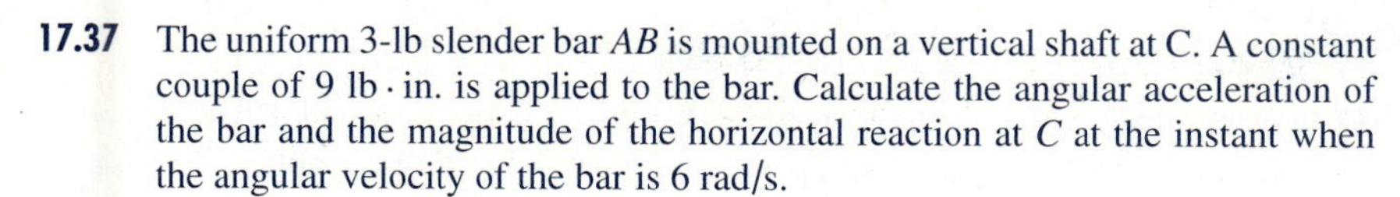

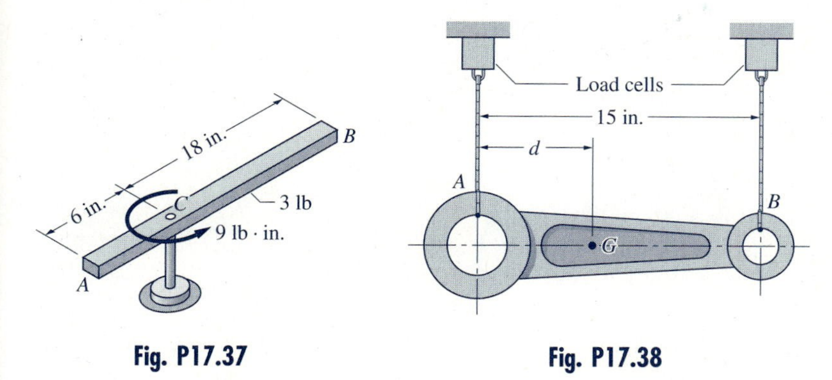

Fig. P17.37 Fig. P17.38

17.38 To determine the inertial properties of the connecting rod AB, it is suspended from two wires, one of which is subsequently cut. Load cells are used to measure the force in each wire. When the rod is hanging in the position shown, the forces in the wires are measured to be 6.80 lb at A and 5.20 lb at B. Immediately after the wire at B is cut, the force in the wire at A is reduced to 3.6 lb. Compute (a) the distance d that locates the mass center G; and (b) the radius of gyration about G.

17.39 The homogeneous 8-lb collar C is fastened to the uniform 4-lb rod AB. The mass moment of inertia of C about its center is $\bar{I}_z = 3.27 \times 10^{-3}$ slug · ft^2. The system is at rest in the position shown when the horizontal force P is applied through the mass center of the collar. Compute the distance d for which the pin reaction at A would not change immediately after P is applied.

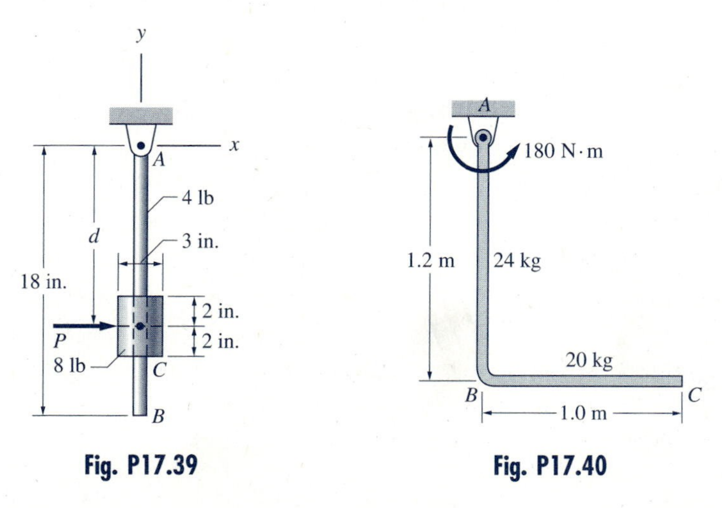

Fig. P17.39 Fig. P17.40

17.40 The constant 180-N· m counterclockwise couple acts on the L-shaped rod ABC, which rotates freely in the vertical plane about the pin at A. Determine the angular acceleration of the rod when it is in the position shown.

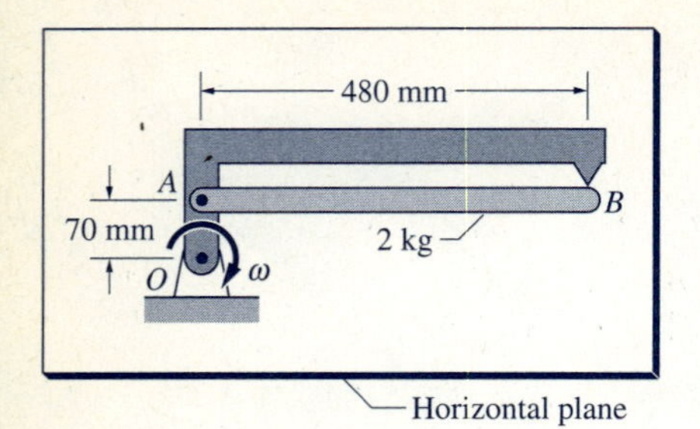

Fig. P17.41, P17.42

17.41 The uniform 2-kg rod AB is attached to the L-shaped frame by a pin at A, and it rests against the smooth protrusion at B. If the frame is rotating about a vertical axis at O with the constant angular velocity $\omega = 25$ rad/s, calculate all forces acting on AB.

17.42 Solve Prob. 17.41 given that $\omega = 0$ and $\alpha = 20$ rad/s^2 clockwise.

17.43 The axle of the 16-lb homogeneous disk is mounted on the end of bar AB, which rotates freely in the vertical plane about the pin at B. The cable BC maintains the disk in a fixed position relative to the bar. Find the cable tension immediately after the assembly is released from rest in the position shown. Neglect the mass of the bar AB.

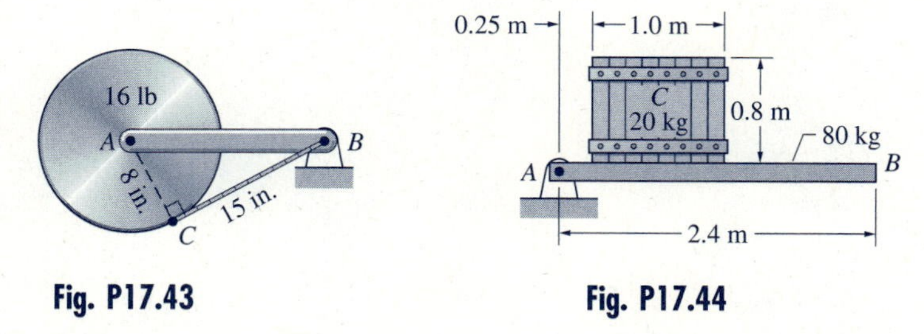

Fig. P17.43

Fig. P17.44

17.44 The 20-kg homogeneous box C rests on the 80-kg uniform slender bar AB that is free to rotate about the pin at A. (a) Assuming that the block does not slide relative to the bar, determine the initial angular acceleration of the system if it is released from rest in the position shown. (b) Determine the smallest coefficient of static friction that is consistent with the no-slip assumption in part (a).

17.45 The solid steel sphere, of density 7850 kg/m^3 and radius $R = 12.5$ mm, rolls down the rough, inclined plane without slipping. Determine the normal and friction forces that are exerted on the sphere by the inclined plane.

17.46 The homogeneous sphere of mass M and radius R is released from rest and moves down the rough, inclined plane. Calculate the acceleration of the center of the sphere if the coefficient of static friction is insufficient to prevent slipping. The coefficient of kinetic friction is 0.085.

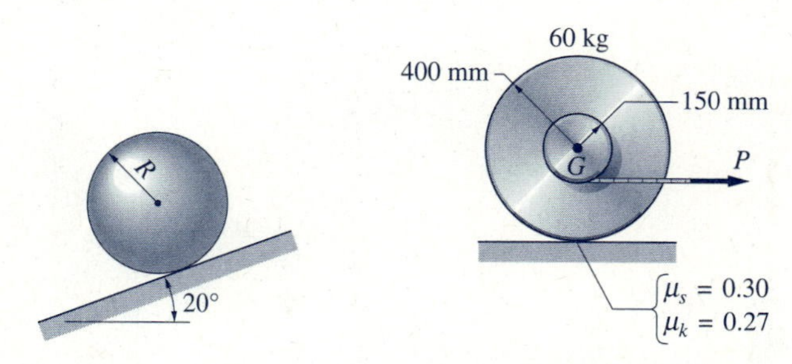

Fig. P17.45, P17.46

Fig. P17.47, P17.48

17.47 The mass moment of inertia of the 60-kg spool is $\bar{I} = 1.35$ kg · m^2. The static and kinetic coefficients of friction between the spool and the ground are 0.30 and 0.27, respectively. A cable wound around the hub of the spool is pulled with the constant horizontal force $P = 200$ N. Find the acceleration of the center of the spool.

17.48 Solve Prob. 17.47 if $P = 450$ N.

17.49 The rim of the wheel weighs 8 lb; the weights of the spokes and hub may be neglected. An 8-lb force, inclined at the angle β to the horizontal, is applied to the center of the wheel. The static and kinetic coefficients of friction between the wheel and the ground are 0.30 and 0.25, respectively. If $\beta = 0$, (a) show that the wheel slips on the ground; and (b) find the angular acceleration of the wheel and the acceleration of its center.

17.50 For the wheel described in Prob. 17.49, find the smallest angle β for which the wheel will roll without slipping, and determine the corresponding angular acceleration.

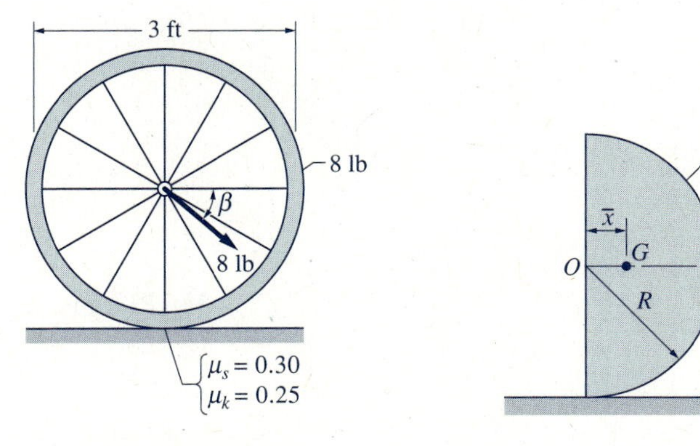

Fig. P17.49, P17.50 **Fig. P17.51**

17.51 The homogeneous semicylinder of mass m and radius R is released from rest in the position shown. Assuming no slipping, determine (a) the initial angular acceleration of the semicylinder; and (b) the smallest static coefficient of friction that is consistent with the no-slip condition. (Note: The mass center of the semicylinder is located at $\bar{x} = 4R/3\pi$; and $I_O = mR^2/2$).

17.52 The radius of gyration of the eccentric disk of mass M about its mass center G is 0.5 m. In the position shown, the angular acceleration of the disk is 4.0 rad/s^2. Assuming rolling without slipping, find the angular velocity of the disk for this position.

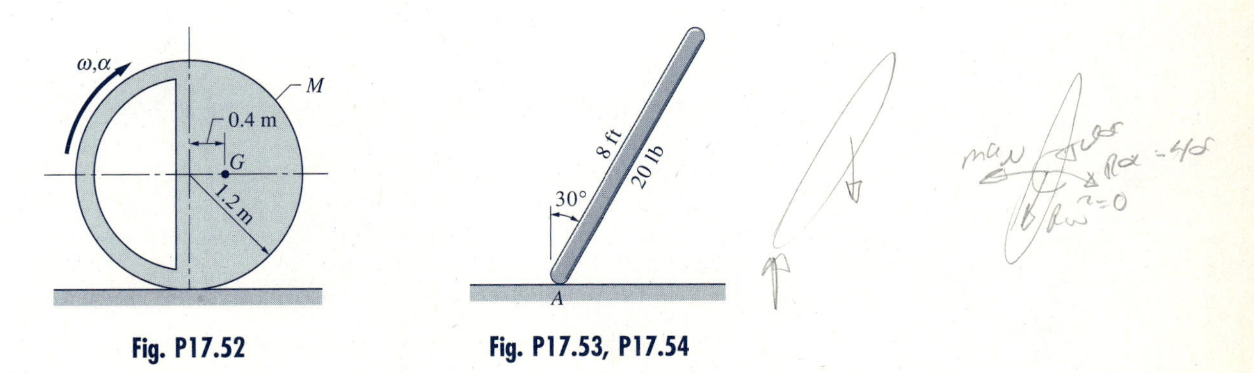

Fig. P17.52 **Fig. P17.53, P17.54**

17.53 If the uniform 20-lb bar is released from rest in the position shown, determine the initial angular acceleration of the bar. Neglect friction.

17.54 Solve Prob. 17.53 if the static and kinetic coefficients of friction between the bar and the horizontal surface are 0.45 and 0.40, respectively.

17.55 The radius of gyration of the 1620-lb spool about its mass center G is 1.75 ft. The cable that is wrapped tightly around the inner radius of the spool is attached to a rigid support as shown. If the spool is moving down the rough plane, determine its angular acceleration and the tension in the cable.

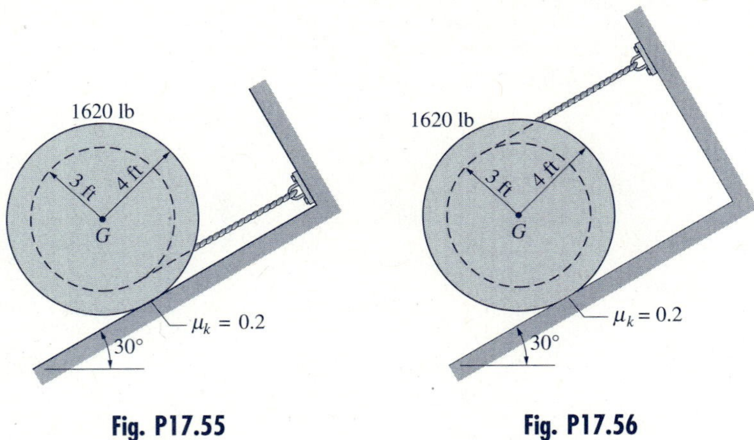

Fig. P17.55

Fig. P17.56

17.56 Repeat Prob. 17.55 if the cable unwinds from the top of the hub as shown.

17.57 The 3.6-kg homogeneous bar AB is pinned to the 2-kg slider at A. The system was at rest in the position $\theta = 0$ before the 12-N force was applied to the collar. Neglecting friction, compute the acceleration of the collar and the angular acceleration of the bar immediately after the 12-N force is applied.

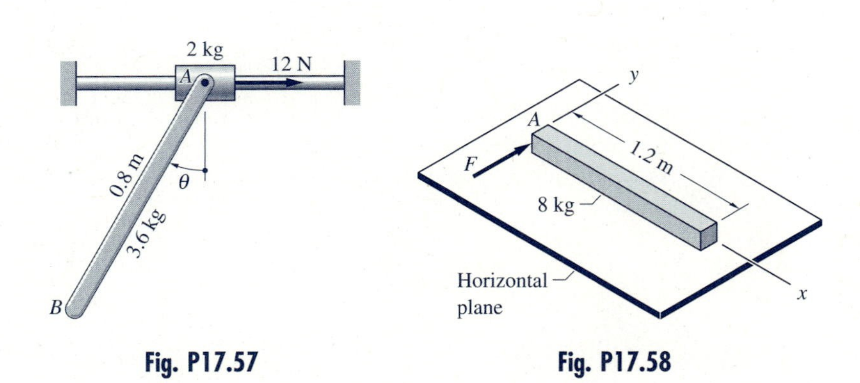

Fig. P17.57

Fig. P17.58

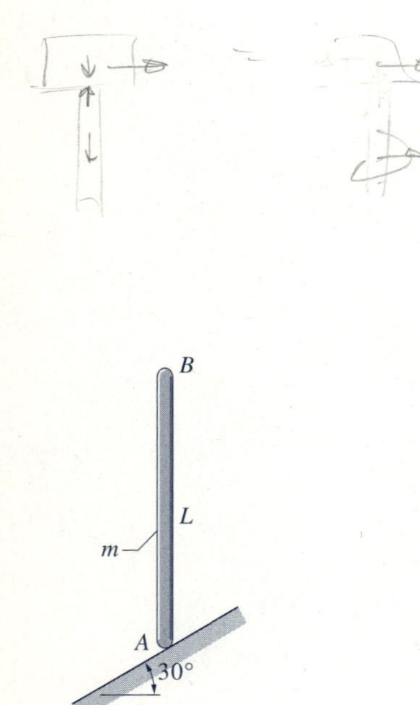

Fig. P17.59, P17.60

17.58 The 8-kg uniform slender bar was at rest on a smooth horizontal table before the application of the force $F = 16$ N. For the instant immediately after F was applied, determine (a) the acceleration of end A; and (b) the x-coordinate of the point on the bar that has zero acceleration.

17.59 The uniform bar AB of mass m and length L is released from rest in the position shown. If the inclined plane is smooth, calculate the initial acceleration of end A.

17.60 For the bar described in Prob. 17.59, determine the force at end A when end B is about to hit the inclined plane.

17.61 The homogeneous 400-lb box is released from rest in the position shown. Assuming that corner A slips on the rough inclined surface, calculate the initial acceleration of A.

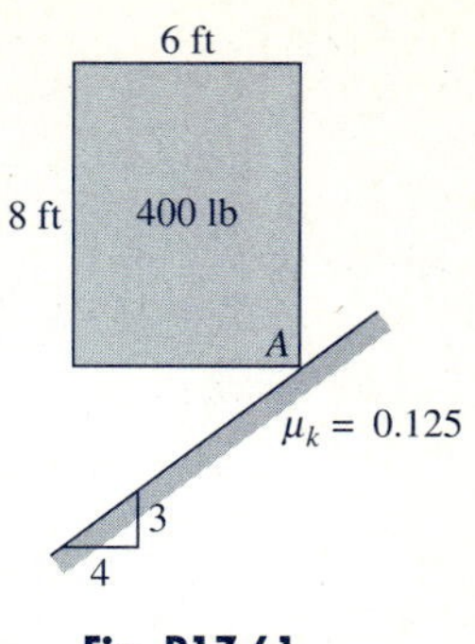

Fig. P17.61

17.62 The mass moment of inertia of the 128-kg disk about its mass center G is 20 $kg \cdot m^2$. The axle of the disk is supported by the lower half of a split bearing. The rope wrapped around the periphery of the disk is pulled horizontally with the speed v_0. Find the largest value of v_0 for which the axle of the disk will stay in the bearing. Neglect friction.

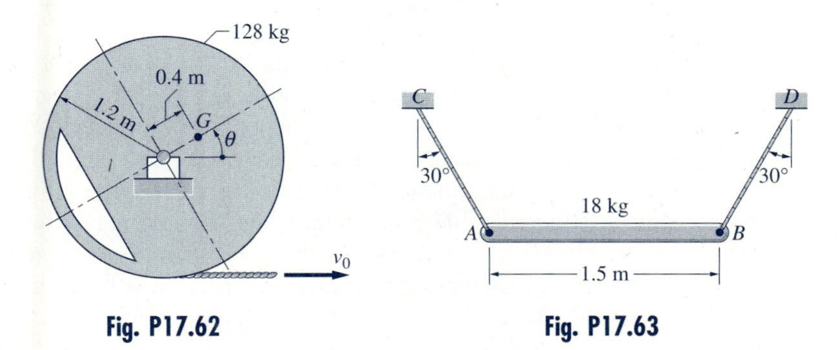

Fig. P17.62

Fig. P17.63

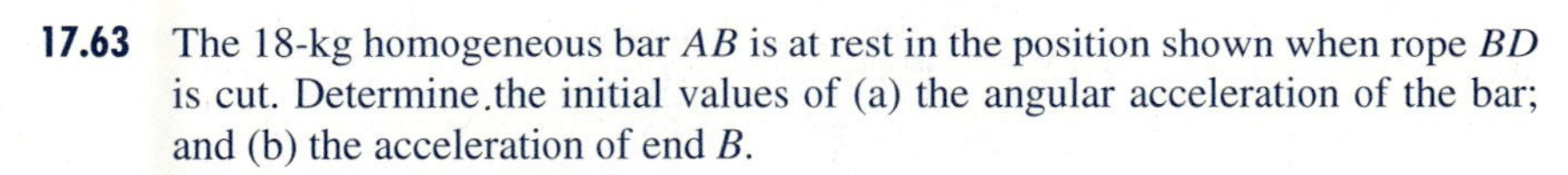

17.63 The 18-kg homogeneous bar AB is at rest in the position shown when rope BD is cut. Determine the initial values of (a) the angular acceleration of the bar; and (b) the acceleration of end B.

17.64 The uniform 20-lb bar AB is released from rest in the position shown. Neglecting friction, calculate the angular acceleration of the bar and the reactions at A and B immediately after the release.

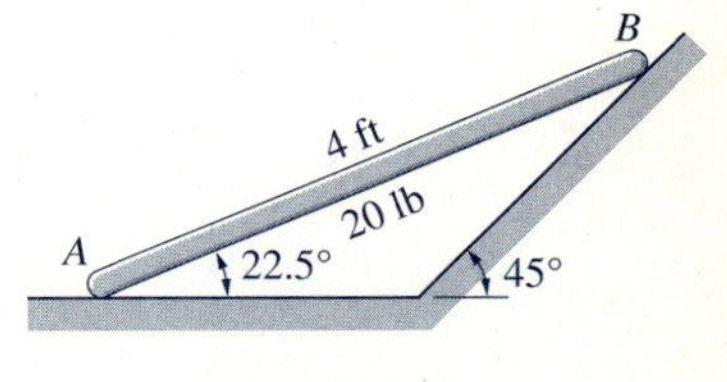

Fig. P17.64

17.65 The mechanism consists of two homogeneous bars of the weights shown, and the piston A of negligible weight. A varying horizontal force P acting on the piston maintains a constant clockwise angular velocity of 60 rad/s for bar BC. Neglecting friction, determine the magnitude and sense of the force P when the mechanism is in the position shown.

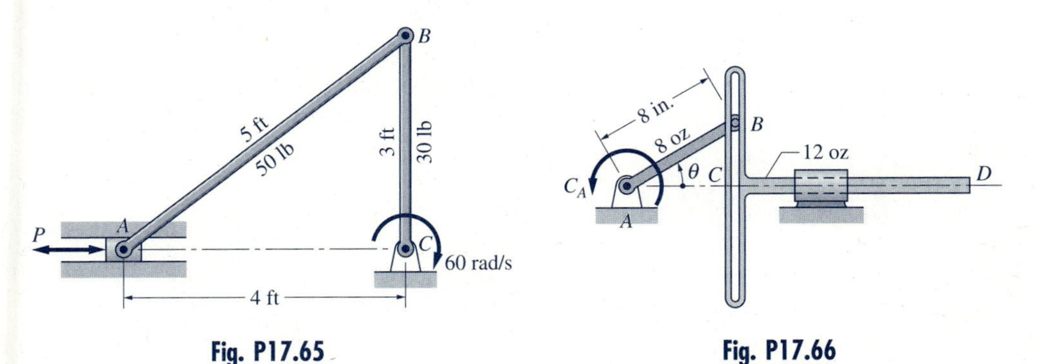

Fig. P17.65

Fig. P17.66

17.66 The pin B attached to the end of the uniform 8-oz crank AB slides in a vertical slot in the 12-oz slider CD. A constant counterclockwise angular velocity of 2000 rev/min is maintained by the couple C_A. Determine C_A as a function of the crank angle θ, and use this expression to show that the gravitational forces are negligible compared to the inertial forces. Neglect friction.

17.6 Differential Equations of Motion

The preceding article discussed the derivation of the equations of motion for a rigid body using the FBD-MAD method. The problems were restricted to the computation of the forces and accelerations at the instant when the body was in a *specified position.* In this article we consider the more practical problem of determining the motion of a rigid body as a function of time and/or position. As noted in previous chapters, the determination of motion involves two steps: the differential equations of motion must first be derived, and then they must be integrated (solved).

The differential equations of motion are obtained by applying the FBD-MAD method to an *arbitrary position* of the body. The forces appearing in the resulting equations of motion fall into two categories: applied loads and constraint forces. The applied loads are usually given as functions of time, position, or velocity. On the other hand, the constraint forces, such as pin reactions, are unknowns. Since the differential equations of motion are integrable when they contain only kinematic variables (the generalized coordinates and their derivatives) as the unknowns, the constraint forces must be eliminated from the equations of motion. The equations that remain after the elimination procedure are called the *differential equations of motion.*

As an illustration, consider the homogeneous bar of mass m and length L shown in Fig. 17.12(a). We are to determine the resulting motion when the bar is released from rest at $\theta = 0$. The FBD and MAD of the bar for an arbitrary value of θ are shown in Fig. 17.12(b). From these diagrams, we obtain the following three independent equations of motion:

$$\Sigma F_x = m\bar{a}_x \qquad \xrightarrow{+} \qquad O_x = -\frac{mL}{2}\omega^2 \cos\theta - \frac{mL}{2}\alpha \sin\theta \tag{a}$$

$$\Sigma F_y = m\bar{a}_y \qquad +\uparrow \qquad O_y - mg = \frac{mL}{2}\omega^2 \sin\theta - \frac{mL}{2}\alpha \cos\theta \tag{b}$$

$$\Sigma M_G = \bar{I}\alpha \qquad \circlearrowleft + \qquad O_x\frac{L}{2}\sin\theta + O_y\frac{L}{2}\cos\theta = \frac{mL^2}{12}\alpha \tag{c}$$

These equations of motion contain the components of the unknown pin reaction at O in addition to the kinematic variables θ, ω, and α. Substituting

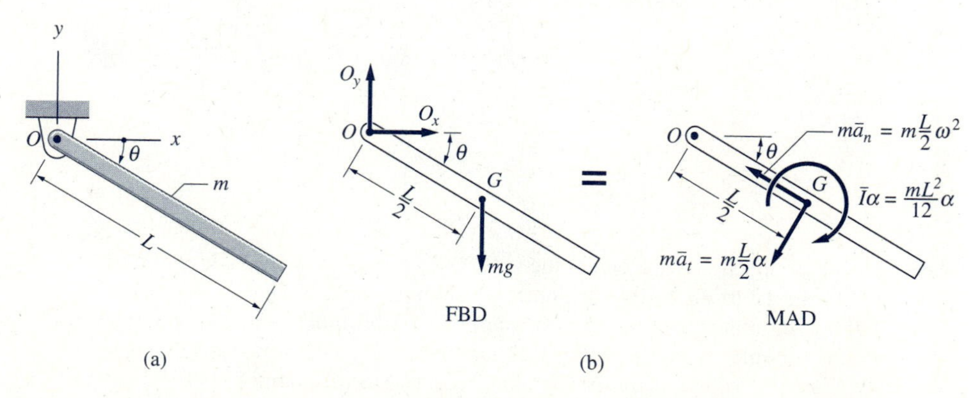

Fig. 17.12

the expressions for O_x and O_y obtainable from Eqs. (a) and (b) into Eq. (c) yields the differential equation of motion

$$\alpha = \ddot{\theta} = (3g/2L)\cos\theta \tag{d}$$

Equation (d) could also be obtained directly from the special case of the moment equation of motion: $\Sigma M_O = I_O\alpha$.

The following conclusions can now be drawn.

1. If a problem is well-posed—i.e., if it is solvable—the number of independent equations of motion equals the number of degrees of freedom (DOFs) plus the number of unknown forces. In the preceding illustration, the number of independent equations of motion was three; the body had one DOF (θ being the generalized coordinate), and there were two unknown components of the pin reaction at O.
2. Since an equation of motion is required to eliminate each of the unknown constraint forces, the number of differential equations of motion is equal to the number of DOFs. In the foregoing illustration, two equations, Eqs. (a) and (b), were used to eliminate O_x and O_y, resulting in the single differential equation of motion, Eq. (d).

The differential equations of motion can be solved analytically only in a few special cases, making numerical integration the primary method of solution. However, in this article we consider both analytical and numerical solutions. In either case, the solution of the differential equations of motion (note that they are second-order equations) requires the knowledge of two initial conditions for every DOF: the values of the generalized coordinates and the generalized velocities at some instant of time (usually at $t = 0$).

Sometimes it is possible to obtain a partial solution to the differential equations of motion in closed form, whereas the complete solution would require numerical integration. For example, to obtain the expression for ω versus θ for the bar shown in Fig. 17.12, we substitute α from Eq. (d) into $\alpha\, d\theta = \omega\, d\omega$, obtaining

$$\omega\, d\omega = (3g/2L)\cos\theta\, d\theta \tag{e}$$

Integrating both sides of this equation, and applying the initial condition $\omega = 0$ when $\theta = 0$, we obtain

$$\omega = \sqrt{(3g/2L)\sin\theta} \tag{f}$$

The relationships between the kinematic variables and time, however, cannot be determined this easily. To find θ versus time, for example, we substitute ω from Eq. (f) into $\omega = d\theta/dt$. Solving for dt, we find

$$dt = d\theta/\sqrt{(3g/2L)\sin\theta} \tag{g}$$

The right-hand side of Eq. (g) cannot be integrated in a closed form. Therefore, we see that even relatively simple problems may require numerical integration for the complete determination of the motion.

It should be mentioned that after the values of ω and α have been found as functions of θ, Eqs. (a) and (b) can be used to find O_x and O_y, also as functions of θ. Similarly, if θ, ω, and α have been found as functions of time, Eqs. (a) and (b) will determine O_x and O_y as functions of time.

Sample Problem 17.10

The 360-lb uniform plate shown in Fig. (a) rotates in the vertical plane about a smooth pin at A. The plate is released from rest when $\theta = 0$. (1) Show that the differential equation of motion for the plate is $\alpha = 0.9660(4\cos\theta - 3\sin\theta)$ rad/s^2. (2) Integrate the differential equation of motion analytically to obtain the angular velocity of the plate as a function of θ. (3) Find the maximum value of θ.

6 ft
360 lb
8 ft

(a)

Solution

Part 1

The mass of the plate is $m = 360/32.2 = 11.180$ slugs, and the moment of inertia about its mass center G is (see Table 17.1)

$$\bar{I} = \frac{1}{12}m(b^2 + c^2) = \frac{1}{12}(11.180)(8^2 + 6^2) = 93.17 \text{ slug} \cdot \text{ft}^2$$

Figure (b) shows the free-body diagram (FBD) and mass-acceleration diagram (MAD) for an arbitrary position of the plate. The FBD contains A_n and A_t, the unknown components of the pin reactions at A, and the 360-lb weight acting at G. The MAD consists of the inertia couple $\bar{I}\alpha$ and the components of the inertia vector $m\bar{\mathbf{a}}$ acting at G. Since the path of G is a circle of radius $r = 5$ ft centered at A, the normal component of $\bar{\mathbf{a}}$ is $\bar{a}_n = r\omega^2 = 5\omega^2$, and its tangential component is $\bar{a}_t = r\alpha = 5\alpha$. Observe that the angular acceleration α is assumed to be clockwise in the MAD. The units for ω and α are rad/s and rad/s^2, respectively.

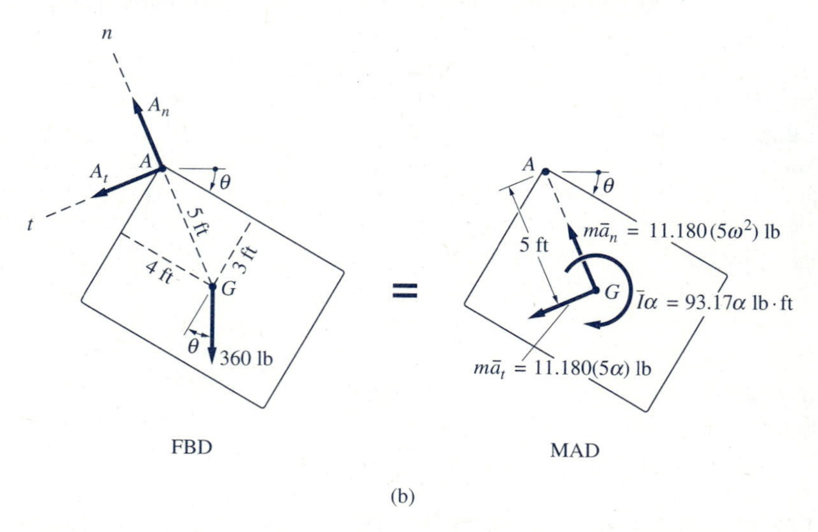

(b)

Since the angle θ completely determines the position of the plate, the plate has a single degree of freedom. Therefore, there is only one differential equation of motion. The most convenient method of deriving this equation is by equating moments about A in the FBD and MAD, thereby obtaining

$$(\Sigma M_A)_{\text{FBD}} = (\Sigma M_A)_{\text{MAD}}$$

$$\circlearrowright\!+ \quad (360\cos\theta)(4) - (360\sin\theta)(3) = 93.17\alpha + 11.180(5\alpha)5$$

which reduces to

$$\alpha = 0.9660(4\cos\theta - 3\sin\theta) \text{ rad/s}^2 \qquad \textit{Answer} \quad \text{(a)}$$

The identical result could be obtained by using the special case of the moment equation: $\Sigma M_A = I_A\alpha$.

Part 2

To find the angular velocity as a function of θ, we substitute α from Eq. (a) into $\omega\, d\omega = \alpha\, d\theta$, which yields

$$\omega\, d\omega = 0.9660(4\cos\theta - 3\sin\theta)\, d\theta$$

The result of integrating this equation analytically is

$$\frac{\omega^2}{2} = 0.9660(4\sin\theta + 3\cos\theta) + C$$

The constant of integration C is evaluated by applying the initial condition $\omega = 0$ when $\theta = 0$, which gives $C = -3(0.9660)$ rad^2/s^2. Therefore, the above equation becomes

$$\frac{\omega^2}{2} = 0.9660(4\sin\theta + 3\cos\theta - 3)$$

from which the angular velocity is found to be

$$\omega = 1.390\sqrt{4\sin\theta + 3\cos\theta - 3} \qquad \textit{Answer} \quad \text{(b)}$$

Part 3

The maximum value of θ occurs when $\omega = 0$. According to Eq. (b), this value is the nonzero root of the equation $4\sin\theta + 3\cos\theta - 3 = 0$. The numerical solution of this equation yields

$$\theta_{\text{max}} = 106.3^\circ \qquad \textit{Answer}$$

Only one equation of motion was used to determine the motion of the plate. The remaining two independent equations of motion—e.g., $\Sigma F_n = m\bar{a}_n$ and $\Sigma F_t = m\bar{a}_t$—could now be utilized to find the components of the pin reaction, A_n and A_t, in terms of θ, ω, and α. By substituting for α from Eq. (a) and ω from Eq. (b), we would then obtain A_n and A_t as functions of θ only.

Sample Problem 17.11

For the plate described in Sample Problem 17.10, (1) solve the differential equation of motion numerically from the time of release until θ reaches its maximum value for the first time; and (2) use the numerical solution to determine the maximum value of θ and the time when it first occurs.

Solution

Part 1

The differential equation of motion given in Sample Problem 17.10 is $\alpha = 0.9660(4\cos\theta - 3\sin\theta)$ rad/s^2. The initial conditions are $\theta = 0$ and $\omega = 0$ when $t = 0$ ($t = 0$ corresponds to the time of release). The period of integration

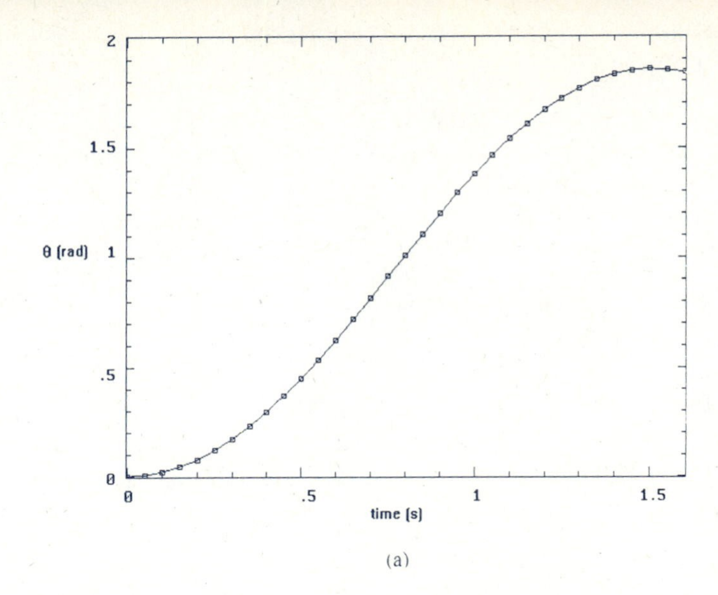

(a)

is from $t = 0$ until the time when θ reaches its maximum value for the first time. By trial and error, this time was found to be approximately 1.5 s. For the final integration, we used the fourth-order Runge-Kutta method with integration step size $\Delta t = 0.025$ s, completing 64 integration steps (only every other step was recorded). The resulting plot of θ versus time is shown in Fig. (a).

Part 2

To determine the time when θ reaches its maximum value, we must examine the numerical output of the solution in the region where ω changes sign.

t (s)	θ (rad)	ω (rad/s)
0.1500E + 01	0.1854E + 01	0.0398E + 00
0.1550E + 01	0.1852E + 01	−0.1532E − 01

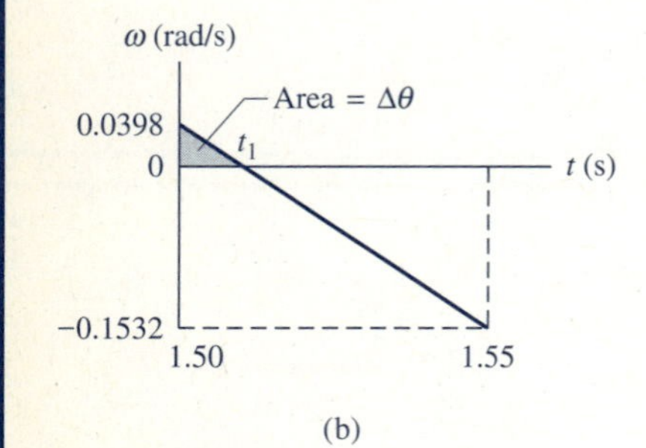

(b)

The corresponding straight-line plot of ω versus t is shown in Fig. (b). Letting t_1 be the time when $\omega = 0$, we obtain from similar triangles

$$\frac{t_1 - 1.50}{0.0398} = \frac{1.55 - 1.50}{0.0398 + 0.1532}$$

which gives

$$t_1 = 1.510 \text{ s} \qquad \textit{Answer}$$

The maximum value for θ can be approximated as the value of θ at $t = 1.50$ s plus $\Delta\theta$ (the change of θ between 1.50 s and 1.510 s). Since $\Delta\theta$ equals the shaded area in Fig. (b), we get

$$\Delta\theta = (1/2)(1.510 - 1.500)(0.0398) = 0.0002 \text{ rad}$$

from which we obtain

$$\theta_{max} = \theta|_{t=1.50\text{ s}} + \Delta\theta = 1.854 + 0.000$$
$$= 1.854 \text{ rad} = 106.2° \qquad \textit{Answer}$$

This answer agrees with that obtained by analytical means in Sample Problem 17.10.

Sample Problem 17.12

The uniform slender bar AB in Fig. (a) has a length L and weight W. The bar is released from rest when $\theta = \theta_0$ with end A in contact with the smooth horizontal plane. Note that the bar has two degrees of freedom, one possible choice of generalized coordinates being x_A (measured from a fixed point on the horizontal plane) and the angle θ. (1) Derive three independent equations of motion, and show that the angular motion of the bar is governed by the differential equation

$$\alpha = \frac{(2g/L)\sin\theta - \omega^2 \sin\theta\cos\theta}{(4/3) - \cos^2\theta} \qquad \text{(a)}$$

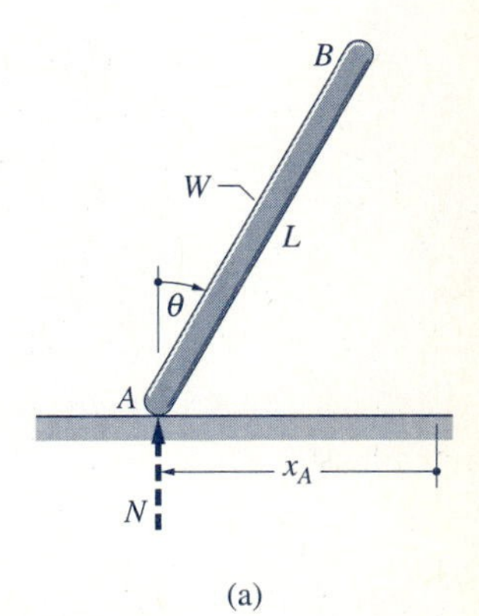

(a)

For Parts (2) and (3), use the following numerical data: $\theta_0 = 30°$, $W = 20$ lb, and $L = 8$ ft.

(2) Solve Eq. (a) numerically from time of release until the bar becomes horizontal. Plot θ versus time, and find the time of fall. (3) Plot the contact force N versus θ. Determine the maximum and minimum values of N and the corresponding values of θ.

Solution

Part 1

The free-body diagram (FBD) and mass-acceleration diagram (MAD) of the bar, drawn at an arbitrary position, are shown in Fig. (b). The FBD contains the weight of the bar acting at its mass center G and the normal force N exerted on the bar at A. The MAD contains the inertia couple $\bar{I}\alpha$ and the components of the inertia vector $m\bar{\mathbf{a}}$ acting at G. The components of $m\bar{\mathbf{a}}$ were found from kinematics using A as the

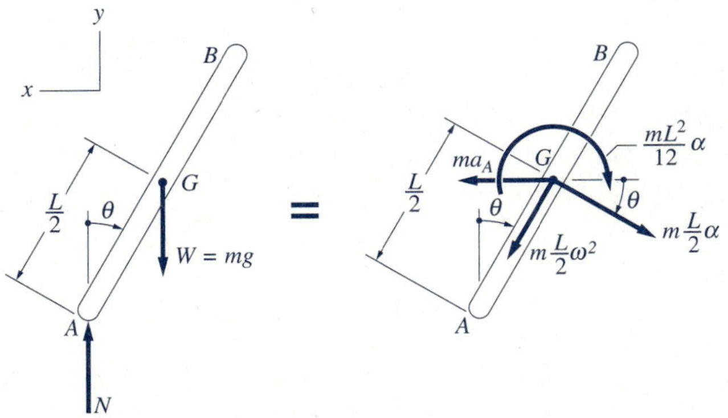

FBD MAD

(b)

reference point (note that the path of A is known to lie along the horizontal plane). The relative acceleration equation thus gives

$$\bar{\mathbf{a}} = \mathbf{a}_G = \mathbf{a}_A + \mathbf{a}_{G/A}$$

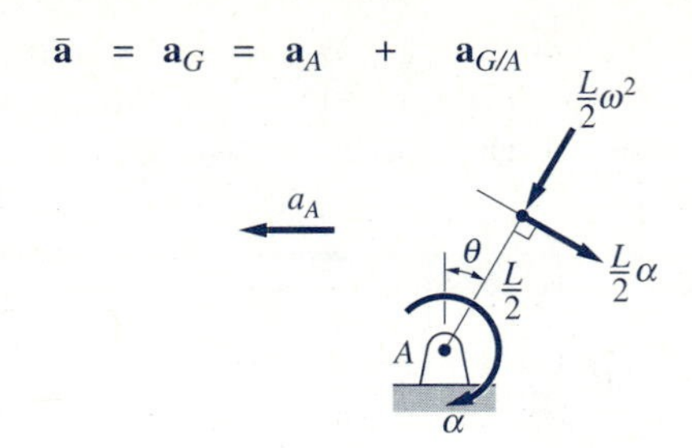

Observe that directions of α and $\mathbf{a}_A$ have been assumed to be clockwise and to the left, respectively. Multiplying each term on the right side of this equation by the mass m and placing the result at the mass center G gives the components of the inertia vector that are shown in Fig. (b).

Equating the resultant moment about A on the FBD to the resultant moment about the same point on the MAD, we obtain

$$(\Sigma M_A)_{\text{FBD}} = (\Sigma M_A)_{\text{MAD}}$$

$$\circlearrowright + \quad mg\frac{L}{2}\sin\theta = \frac{mL^2}{12}\alpha + m\frac{L}{2}\alpha\left(\frac{L}{2}\right) - ma_A\left(\frac{L}{2}\cos\theta\right)$$

which simplifies to

$$\alpha = \frac{3g}{2L}\sin\theta + \frac{3a_A}{2L}\cos\theta \tag{b}$$

Summing forces in the horizontal direction gives

$$\Sigma F_x = m\bar{a}_x \qquad \overset{+}{\leftarrow} \quad 0 = ma_A - m\frac{L}{2}\alpha\cos\theta + m\frac{L}{2}\omega^2\sin\theta$$

which reduces to

$$a_A = \frac{L}{2}\alpha\cos\theta - \frac{L}{2}\omega^2\sin\theta \tag{c}$$

Summing forces in the vertical direction yields

$$\Sigma F_y = m\bar{a}_y \qquad +\uparrow \quad N - mg = -m\frac{L}{2}\alpha\sin\theta - m\frac{L}{2}\omega^2\cos\theta \tag{d}$$

Equations (b)–(d) are the three independent equations of motion of the bar. Substituting Eq. (c) into Eq. (b) and simplifying yields the differential equation of motion given in Eq. (a).

The following additional comments should be made about the foregoing analysis. (1) If needed, the differential equation that governs the motion of end A could be obtained by substituting the expression for α from Eq. (a) into Eq. (c). (2) The third equation of motion, Eq. (d), is not required to derive the two differential equations of motion (it is needed only for the computation of N).

Part 2

When the given numerical values ($W = 20$ lb, $L = 8$ ft, and $g = 32.2$ ft/s^2) are substituted into Eq. (a), we obtain

$$\alpha = \frac{8.05 \sin\theta - \omega^2 \sin\theta \cos\theta}{(4/3) - \cos^2\theta} \text{rad/s}^2 \tag{e}$$

Equation (e) was integrated numerically using the fourth-order Runge-Kutta method and the initial conditions $\theta_0 = 30°$ (0.5236 rad) and $\omega = 0$ when $t = 0$. By trial and error it was found that an appropriate step size was $\Delta t = 0.01$ s, with 61 steps being required to cover the time of fall. The resulting plot of θ versus time is shown in Fig. (c).

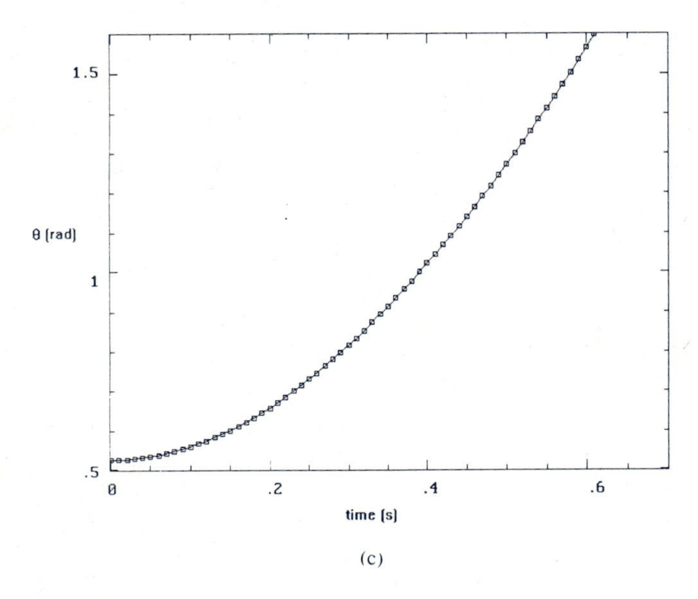

(c)

To determine the time of fall, we must find the time when $\theta = 90°$ (1.571 rad). A partial printout of the solution in the region of interest is given in the following.

t (s)	θ (rad)
0.600E + 00	0.1568E + 01
0.610E + 00	0.1600E + 01

Using the straight-line plot of these values shown in Fig. (d), we obtain from similar triangles

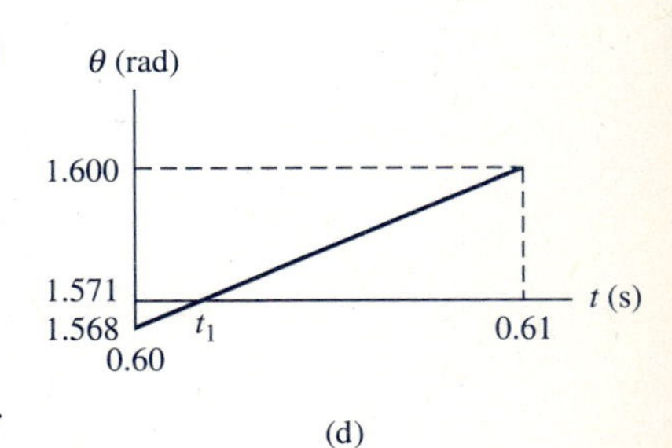

(d)

$$\frac{t_1 - 0.60}{1.571 - 1.568} = \frac{0.61 - 0.60}{1.600 - 1.568}$$

where t_1 is the time when $\theta = 1.571$ rad (90°). The solution is

$$t_1 = 0.601 \text{ s}$$ *Answer*

Part 3

Substituting the given numerical values into Eq. (d) and simplifying, the expression for the normal force N becomes

$$N = 20 - 2.4845(\alpha \sin\theta + \omega^2 \cos\theta) \text{ lb} \tag{f}$$

Figure (e) shows the plot of N versus θ that was obtained by substituting the values of θ, ω, and α produced by the numerical integration described in Part 2 into Eq. (f). Examination of the numerical output yields

$$N_{\text{max}} = 11.43 \text{ lb} \quad \text{when } \theta = 30° \ (0.5236 \text{ rad})$$
$$N_{\text{min}} = 3.86 \text{ lb} \quad \text{when } \theta = 69° \ (1.20 \text{ rad}), \text{ approximately}$$

Answer

Note from Fig. (e) that N remains positive. This means that end A stays in contact with the horizontal plane, as was assumed in the derivation of the equations of motion.

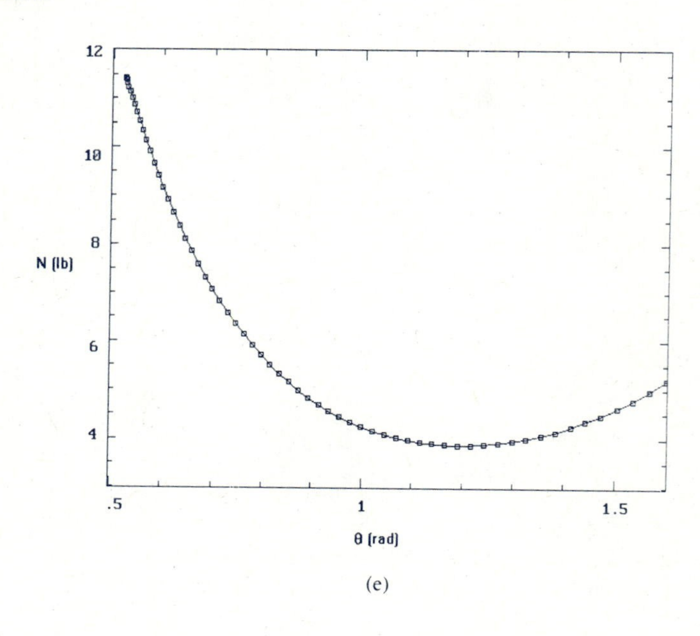

(e)

PROBLEMS

17.67 The uniform bar AB of mass m and length b, which is supported by two wires, is released from rest when $\theta = \theta_0$. Determine the expressions for $\ddot{\theta}$, $\dot{\theta}$, and the force in each wire as functions of θ.

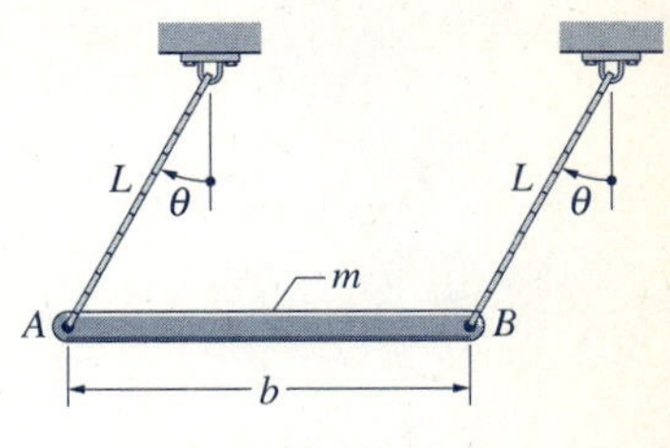

Fig. P17.67

17.68 The homogeneous box of mass m is at rest on a lubricated surface when the constant horizontal force P is applied. Find the smallest value of P for which the box will tip. Assume that the hydrodynamic drag force of the lubricant is (a) $F_D = c_D v$, where c_D is a positive, nonzero constant and v is the velocity of the box; and (b) $F_D = 0$.

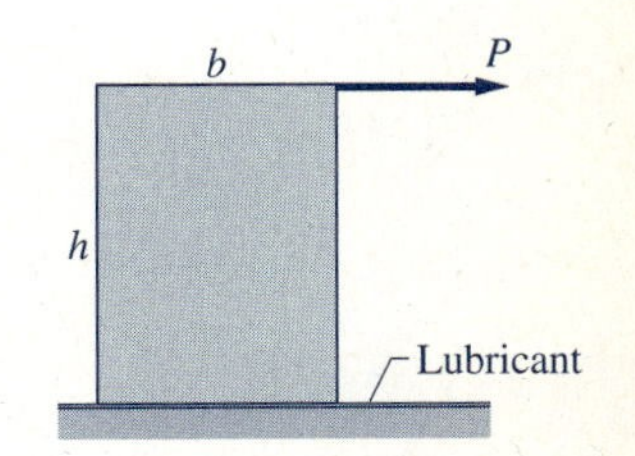

Fig. P17.68, P17.69

17.69 The mass of the homogeneous box is 50 kg, and its dimensions are $h = 1.2$ m, $b = 0.8$ m. The box is at rest on a lubricated surface when the constant horizontal force $P = 240$ N is applied at time $t = 0$. The hydrodynamic drag force of the lubricant during the ensuing motion is $F_D = 20v$ N, where v is the speed of the box in m/s. Determine (a) the time when the box tips over; and (b) the distance traveled before the tipping occurs.

17.70 The axle of the 2.4-kg homogeneous disk fits into a smooth, inclined slot. The disk is initially at rest when it is lowered onto the conveyor belt that is being driven at the constant speed of 6 m/s. The kinetic coefficient of friction between the disk and the belt is 0.2. Calculate (a) the angular acceleration of the disk during the time that it slips on the belt; and (b) the number of revolutions made by the disk before reaching its final angular speed.

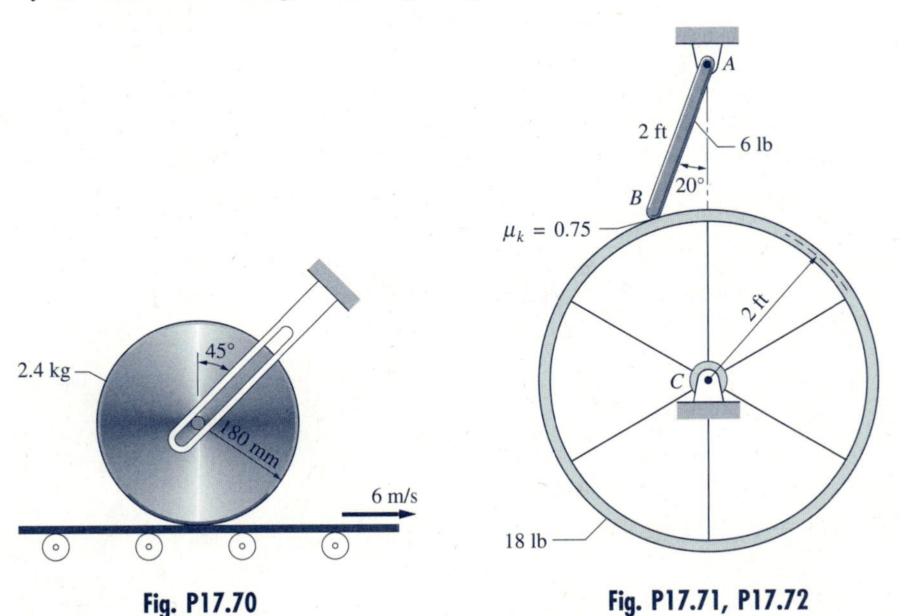

Fig. P17.70

Fig. P17.71, P17.72

17.71 The thin rim of the wheel weighs 18 lb; the weights of the spokes and hub may be neglected. Before the homogeneous 6-lb bar AB was lowered into the position shown, the wheel was rotating freely at 500 rev/min clockwise. If the kinetic coefficient of friction at B is 0.75, determine (a) the angular acceleration of the wheel; and (b) the time it takes the wheel to stop.

17.72 Repeat Prob. 17.71 if the wheel was initially rotating counterclockwise at 500 rev/min.

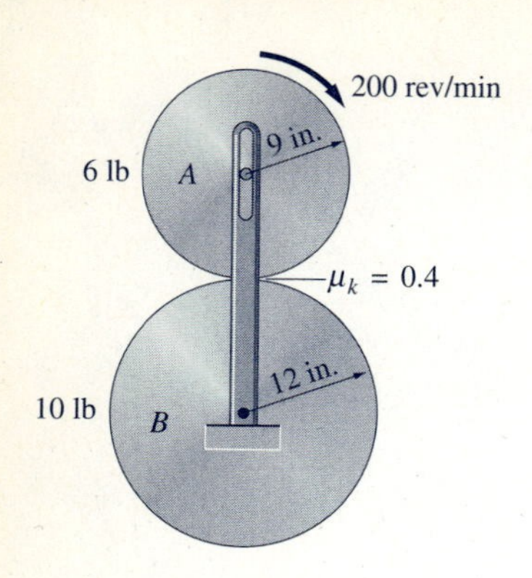

Fig. P17.73

17.73 The axle of the 6-lb uniform disk A can slide in the smooth vertical slot, whereas the axle of the 10-lb uniform disk B is fixed. The coefficient of kinetic friction between the disks is 0.4. Disk A is rotating clockwise at 200 rev/min when it is lowered onto the stationary disk B. Determine (a) the angular acceleration of each disk during the time that slipping occurs; and (b) the final angular speed of each disk.

17.74 The homogeneous bar AB of mass m and length L is released from rest in the position $\theta = \theta_0$. (a) Determine the angular acceleration and angular velocity of the bar as functions of θ. (b) Find the maximum value of the vertical component of the pin reaction at A and the value of θ at which it occurs.

17.75 The 24-lb uniform bar AB slides inside a smooth cylindrical surface. If the bar is released from rest in the position $\theta = 35°$, calculate (a) the angular acceleration and angular velocity as functions of θ; (b) the reaction N_A at A as a function of θ; and (c) the maximum and minimum values of N_A and the values of θ at which they occur.

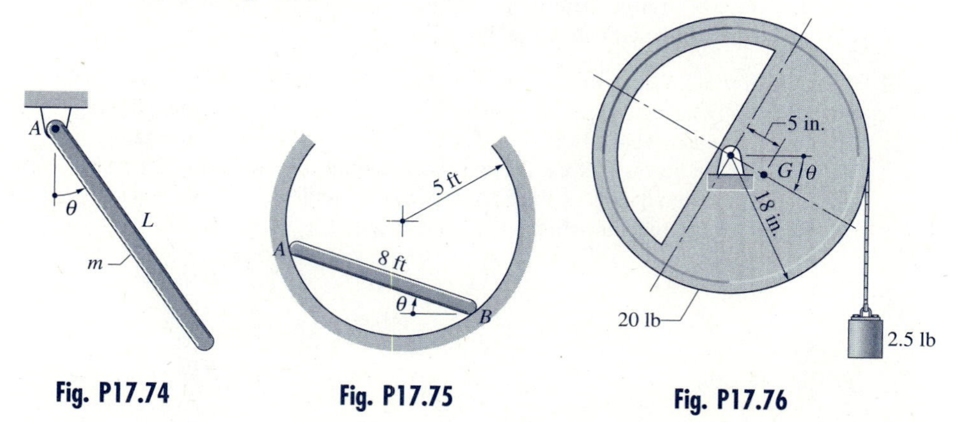

Fig. P17.74 Fig. P17.75 Fig. P17.76

17.76 The radius of gyration of the 20-lb pulley about its mass center G is 12 in. The 2.5-lb weight hangs from a cord that is wrapped around the pulley. (a) Derive the differential equation of motion for the pulley. (b) If the pulley is given an initial counterclockwise angular velocity of 10 rad/s when $\theta = 0$, find θ when the system comes to rest.

17.77 The L-shaped bar is released from rest when $\theta = 0$. (a) Show that the differential equation of motion is $\alpha = 8.017 \cos\theta + 2.088 \sin\theta$ rad/s^2. (b) Integrate the equation of motion analytically to obtain the expression for ω in terms of θ. (c) Determine the maximum value of ω and the corresponding value of θ. (d) Find the maximum value of θ.

17.78 (a) Solve the differential equation of motion in Prob. 17.77 from the time of release until θ reaches its maximum value, and plot θ versus time. (b) From the numerical solution, determine the maximum value of θ and the time when it first occurs.

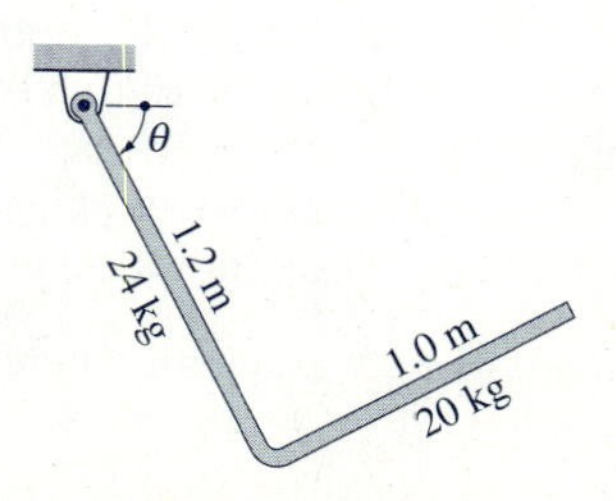

Fig. P17.77, P17.78

17.79 The 3-lb uniform rod BC is welded to the 10-lb drum A. The radius of gyration of the drum about its mass center C is 4 in. The constant couple M_0 lb · ft is applied when the system is at rest at $\theta = 0$. (a) Show that the differential equation of motion for the system is $\alpha = 9.580M_0 - 21.56\sin\theta$ rad/s^2. (b) Find the value of M_0 for which the bar reaches the position where both its angular velocity and angular acceleration are zero. Compute θ in that position.

17.80 Solve the differential equation of motion in Prob. 17.79 numerically from the time M_0 is applied until the rod BC returns to the original position for the first time. Plot ω versus θ and determine the maximum value of ω during this period. Using the plot, briefly describe the motion. Use (a) $M_0 = 1.5$ lb · ft; and (b) $M_0 = 2.0$ lb · ft.

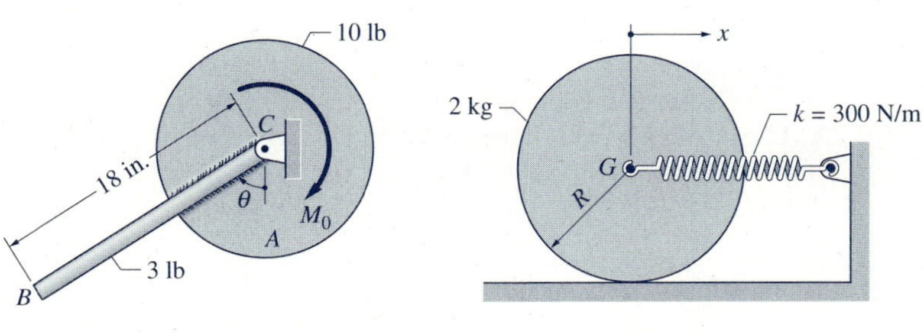

Fig. P17.79, P17.80

Fig. P17.81

17.81 The 2-kg uniform disk of radius R rolls without slipping on the rough, horizontal surface. The stiffness of the spring attached to the center G of the disk is 300 N/m. The disk is released from rest in the position shown with the spring stretched by 75 mm. (a) Determine the acceleration and velocity of G in terms of its coordinate x. (b) Compute the maximum velocity of G and the corresponding value of x. (c) Find the smallest static coefficient of friction that would prevent slipping.

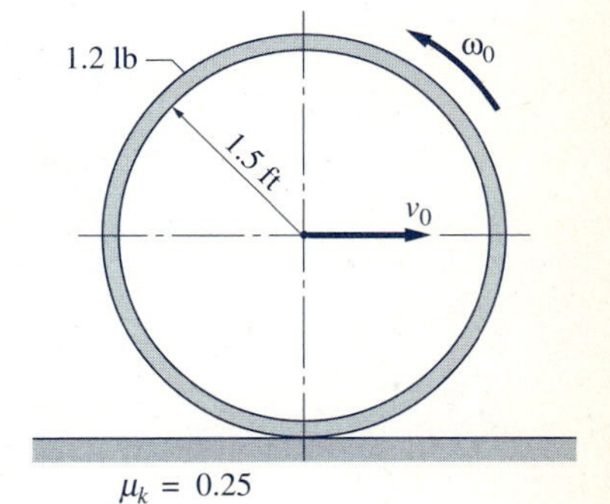

Fig. P17.82

17.82 The thin 1.2-lb hoop is launched on a horizontal surface with the velocity $v_0 = 10$ ft/s and the angular velocity $\omega_0 = 12$ rad/s, both directed as shown in the figure. The kinetic coefficient of friction between the hoop and the surface is 0.25. Calculate (a) the time elapsed before slipping stops; (b) the number of revolutions that the hoop makes during the slipping period; and (c) the final velocity of the center of the hoop.

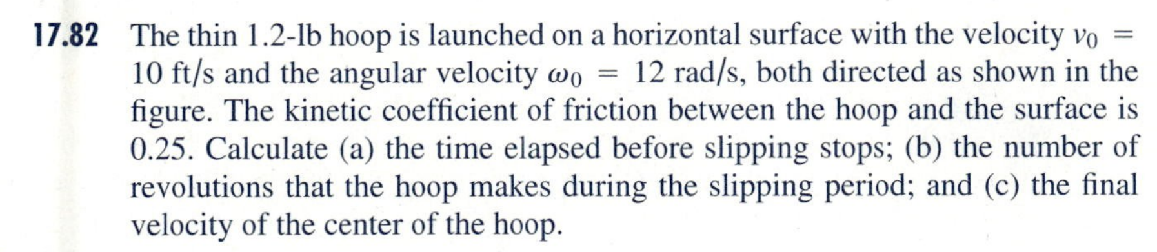

17.83 The eccentric wheel of mass M and radius R rolls without slipping. The radius of gyration of the wheel about its mass center is $\bar{k}$. (a) Show that the differential equation of motion for the wheel is

$$\alpha = \frac{(g + \omega^2 R)e\cos\theta}{\bar{k}^2 + R^2 + e^2 - 2Re\sin\theta}$$

(b) Knowing that $\omega = 2.69$ rad/s (clockwise) when $\theta = 0$, integrate the differential equation of motion numerically over the time period during which the wheel completes one revolution. From the solution determine the maximum and minimum angular velocity and the time required to complete one revolution. Use the data $R = 1.2$ m, $\bar{k} = 0.5$ m, and $e = 0.4$ m. (c) Plot the angular velocity versus time.

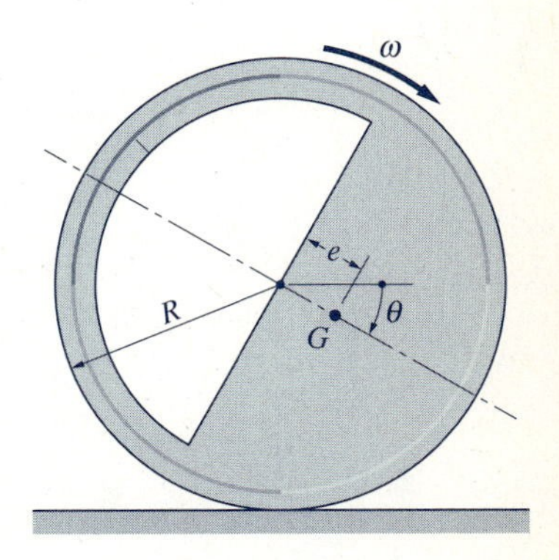

Fig. P17.83

17.84 The uniform ladder AB of length $L = 3$ m and mass $M = 25$ kg is released from rest when $\theta = 30°$. Friction between the ladder and the ground is negligible. (a) Derive the expressions for the angular velocity and angular acceleration of the bar, assuming that end A remains in contact with the vertical wall. (b) Determine the expression for the contact force at A as a function of θ. (c) At what value of θ will end A lose contact with the wall?

17.85 Friction between the uniform ladder of mass M and the ground is insufficient to prevent the ladder from sliding down. (a) Show that the differential equation of motion for the ladder is

$$\alpha = \frac{(g/L)(\sin\theta - 2\mu\cos\theta) + \mu\omega^2\cos^2\theta}{(2/3) - \mu\sin\theta\cos\theta}$$

where μ is the kinetic coefficient of friction at B. (b) Assuming that end A remains in contact with the vertical wall, solve the differential equation of motion numerically from the time when the ladder is released from rest at $\theta = 30°$ to the time when A hits the ground. Use $L = 3$ m, $M = 25$ kg and $\mu = 0.2$. (c) Use the numerical solution to determine if A leaves the wall, and if it does, find the corresponding value of θ.

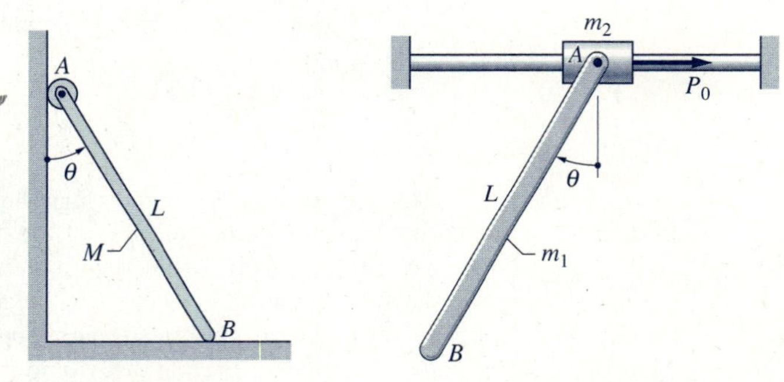

Fig. P17.84, P17.85 **Fig. P17.86, P17.87**

17.86 Bar AB of mass m_1 and length L is pinned to the sliding collar of mass m_2. The system is at rest with $\theta = 0$ when the constant horizontal force P_0 is applied to the collar. (a) Assuming that friction is negligible, show that the differential equation of motion for bar AB is

$$\alpha = \frac{2P_0\cos\theta - 2g(m_1 + m_2)\sin\theta - m_1 L\omega^2\sin\theta\cos\theta}{(4L/3)(m_1 + m_2) - m_1 L\cos^2\theta}$$

(b) Use numerical integration to find the maximum value of θ during the first two seconds of motion. Use the following data: $m_1 = 3.6$ kg, $m_2 = 2.0$ kg, $L = 0.8$ m, and $P_0 = 12$ N. (c) Plot θ versus time for the period of integration.

17.87 When the angular position θ of a body is small enough, its equations of motion can be simplified by the approximations $\sin\theta \approx \theta$ and $\cos\theta \approx 1$. Applying these approximations to the differential equation of motion in Prob. 17.86, re-solve parts (b) and (c). Compare the maximum value of θ with 24.6°, the value obtained if the small angle approximations are not used.

17.88 The steam engine consists of the balanced flywheel C pinned to the connecting rod AB. The rod, in turn, is pinned to the double-action piston A. The steam pressure is regulated so that the force P exerted on the piston varies with the angle θ as $P = P_0 \sin\theta$, where P_0 is a constant. The weight of the flywheel is W and its radius of gyration about its mass center is $\bar{k}$; the weights of the piston and the connecting rod may be neglected. (a) Show that the differential equation of motion for the flywheel is

$$\alpha = \frac{gP_0 R \sin^2\theta}{W\bar{k}^2}\left(1 + \frac{\cos\theta}{\sqrt{(L/R)^2 - \sin^2\theta}}\right)$$

(b) If the flywheel starts from rest when $\theta = 90°$, determine by numerical integration the time required for it to reach a speed of 10 rad/s. Use the following data: $W = 180$ lb, $\bar{k} = 0.6$ ft, $R = 0.75$ ft, $L = 1.5$ ft, $P_0 = 24$ lb. (c) Plot the angular speed versus time for the period of integration.

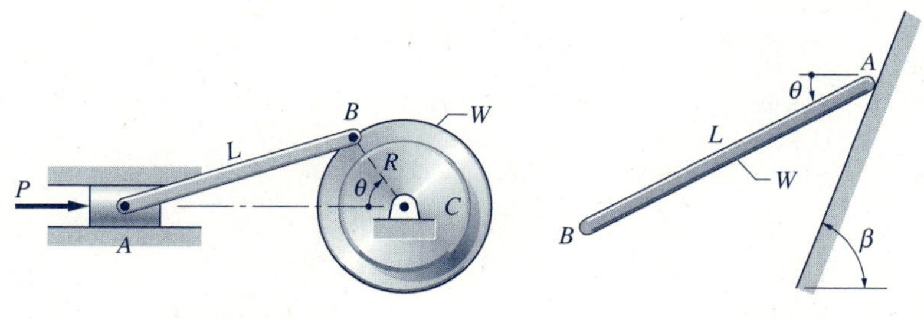

Fig. P17.88 **Fig. P17.89**

17.89 The uniform rod AB is released from rest when $\theta = 0$ with end A in contact with the smooth, inclined surface. (a) Assuming that end A maintains contact with the surface, show that the differential equation governing the angular motion of the rod is

$$\alpha = \frac{(2g/L)(\cos\beta - \sin\beta\sin\phi) - \omega^2\sin\phi\cos\phi}{(4/3) - \sin^2\phi}\ \text{rad/s}^2$$

where $\phi = \beta - \theta$. (b) Show that the expression for N, the contact force at A, is

$$N = W[\cos\beta - (L/2g)(\alpha\cos\phi + \omega^2\sin\phi)]$$

(c) Solve the differential equation of motion in part (a) numerically from the time of release until end B makes contact with the inclined surface. Use this solution and the expression for N to determine the minimum value of N. (d) Plot N versus θ. Use the data $L = 8$ ft, $W = 30$ lb, $\beta = 60°$.

17.90 The small collar C of mass m_2 slides with negligible friction on rod AB of mass m_1 and length L. The system is released from rest in the vertical plane when $\theta = \theta_0$ and $r = L$. (a) Show that the differential equations of motion of the system are

$$\ddot{r} = -g\sin\theta + r\dot{\theta}^2$$

$$\ddot{\theta} = -\frac{3}{2}\,\frac{g\cos\theta[L(m_1/m_2) + 2r] + 4r\dot{r}\dot{\theta}}{L^2(m_1/m_2) + 3r^2}$$

(b) Integrate the equations numerically from the time of release until the collar leaves the rod at B. From the solution, determine the minimum value of r and the value of θ when the collar is about to leave the rod. Use $\theta_0 = 60°$, $m_1/m_2 = 2$, $L = 1.5$ ft, $g = 32.2$ ft/s^2. (c) Plot r versus θ.

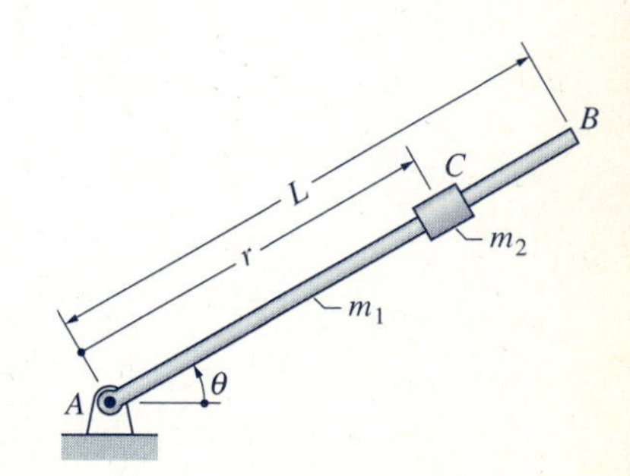

Fig. P17.90

17.91 The small collar D of mass m slides on the rod AB. A light spring of stiffness k joins the collar to end A of the rod. The spring is undeformed when the collar is in the position $r = r_0$. The rod and the frame ACB rotate freely about the vertical axis OC, their combined moment of inertia about that axis being I. A shaker (not shown) drives the base E in such a manner that the position of E is given by $y(t) = a \sin pt$, where a and p are constants. (a) Show that the differential equations of motion for the system are

$$\ddot{r} = -\frac{k}{m}(r - r_0) + r\dot{\theta}^2 + ap^2 \sin pt \sin\theta$$

$$\ddot{\theta} = \frac{r}{(I/m) + r^2}(ap^2 \sin pt \cos\theta - 2\dot{r}\dot{\theta})$$

(b) Solve the differential equations of motion numerically from $t = 0$ to 3 s. Assume that the motion begins at $t = 0$ with the system at rest in the position $\theta = 0$ and $r = r_0$. From the solution, find the time required for AB to rotate through 90°, and determine the direction of rotation (as viewed from above). Use the following data: $m = 0.125$ kg, $k = 3.125$ N/m, $r_0 = 50$ mm, $I = 312.5 \times 10^{-6}$ kg · m², $a = 10$ mm, and $p = 10$ rad/s. (c) Plot θ and $\dot{\theta}$ versus time for the period of integration.

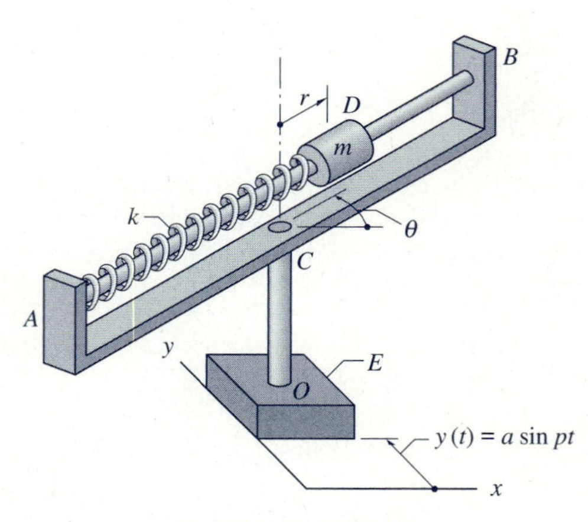

Fig. P17.91, P17.92

17.92 When the spring constant in Prob. 17.91 is increased to $k = 6.25$ N/m (all other data remaining unchanged), rod AB will begin rotating counterclockwise, as viewed from above. However, the direction of rotation will reverse within three seconds after the start of the motion. (a) Use numerical integration to determine the maximum counterclockwise angular displacement of rod AB. (b) Plot θ and $\dot{\theta}$ versus time for the period $t = 0$ to 3 s.

REVIEW PROBLEMS

17.93 The figure shown is made of uniform, thin wire; its mass is m. Determine (a) the x coordinates of the mass center; (b) the moment of inertia about the z-axis; and (c) the moment of inertia about the $\bar{z}$-axis (axis that passes through the mass center).

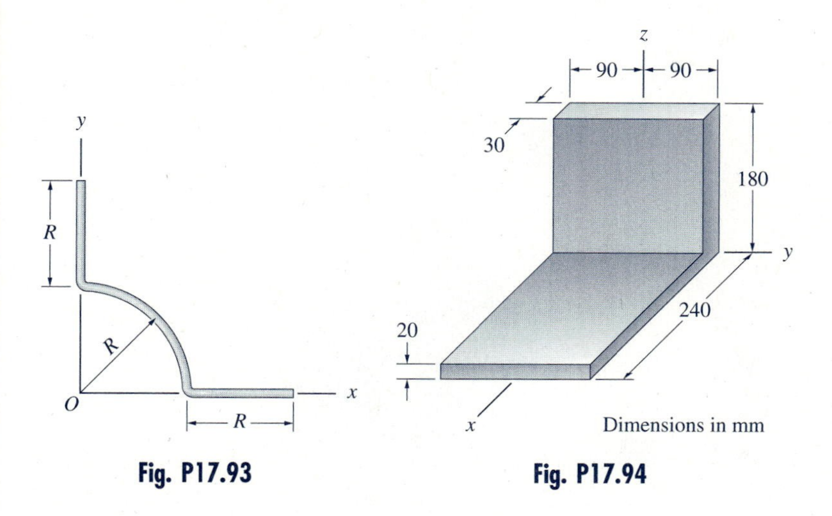

Fig. P17.93 **Fig. P17.94**

17.94 For the homogeneous block shown, determine (a) the x-coordinate of the mass center; (b) the radius of gyration about the z-axis; and (c) the radius of gyration about the $\bar{z}$-axis (axis that passes through the mass center).

17.95 The 50-kg block is pushed up the 30° incline with the constant force P. The coefficient of kinetic friction between the block and the incline is 0.25. Determine the largest P that can be applied without tipping the block.

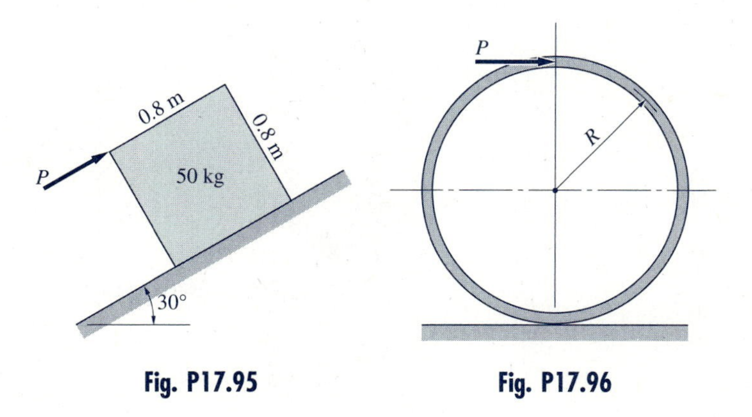

Fig. P17.95 **Fig. P17.96**

17.96 The slender hoop of radius R and mass m is pushed along the horizontal, rough surface by the force P acting at the top of the hoop. Show that the hoop will never slip regardless of the value of P, and determine its angular acceleration in terms of P, R, and m.

17.97 Disk A is spinning freely with the angular velocity of 20 rad/s when it is brought in contact with the stationary pulley B. The coefficient of kinetic friction between the disk and the pulley is 0.4. Determine the acceleration of block C during the period when there is slippage between the disk and the pulley.

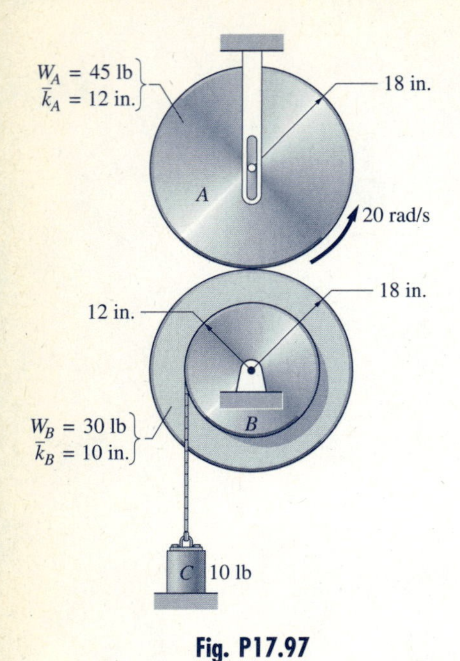

Fig. P17.97

17.98 The uniform bar AC has a mass of 8 kg. It is fastened to the rim of the rotating disk with a pin at A. A pin on the disk engages a slot in the bar at B. If the angular velocity and angular acceleration of the disk are $\omega = 4$ rad/s and $\dot{\omega} = 25$ rad/s^2, determine the pin reaction at B. Note that the disk lies in the horizontal plane.

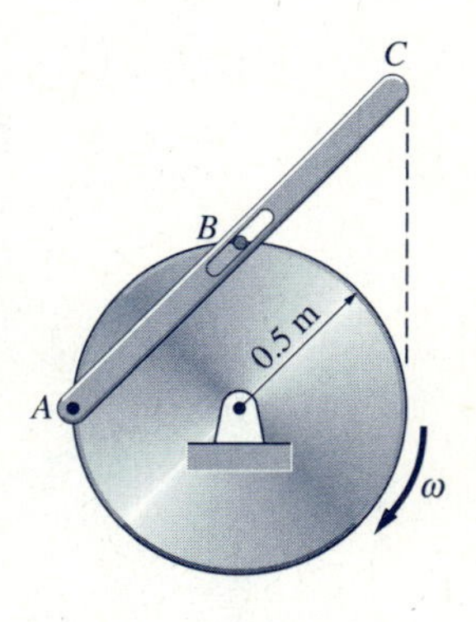

Fig. P17.98

17.99 The identical blocks A and B, each of mass m, are connected by a rod of length L and negligible mass. Block A is free to slide on the smooth, inclined surface. If the system is released from rest in the position shown, determine expressions for the acceleration of block A and the axial force in the rod immediately after the release.

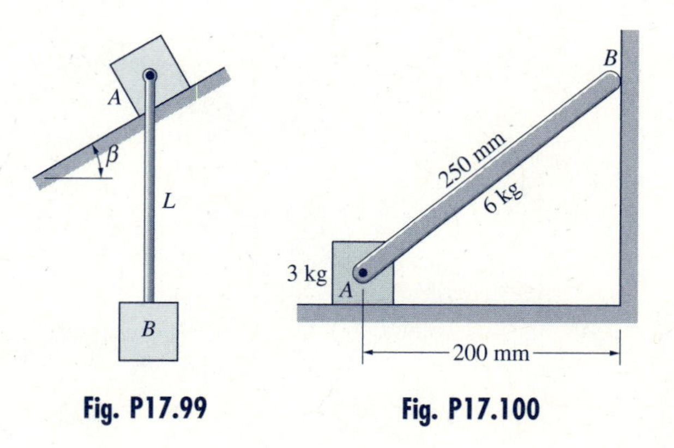

Fig. P17.99 **Fig. P17.100**

17.100 The 3-kg block and the 6-kg uniform rod are joined with a pin at A. Neglecting friction, determine the angular acceleration of rod AB immediately after the system is released from rest in the position shown.

17.101 The homogeneous, 10-lb disk rolls without slipping as it is pulled along the horizontal surface with the constant angular velocity ω_0. If a small 1-lb weight is attached to the rim of the disk, determine the maximum value of ω_0 for which the disk does not hop off the surface.

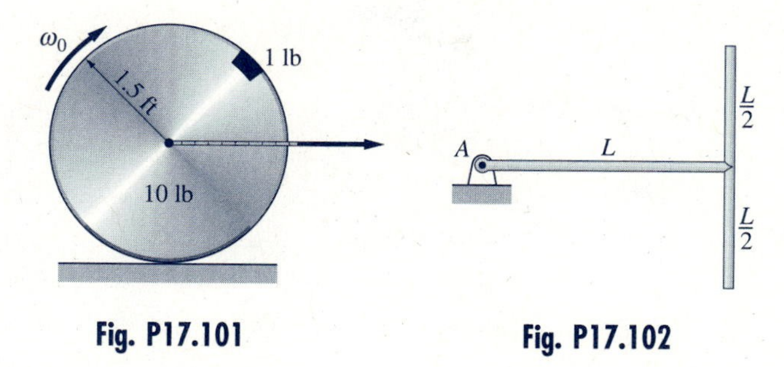

Fig. P17.101 **Fig. P17.102**

17.102 The T-bar consists of two identical rods, each of mass m and length L. Determine the pin reaction at A immediately after the bar is released from rest in the position shown.

17.103 The homogeneous, 12-lb disk is released from rest with the spring undeformed. Assuming no slipping, determine the maximum angular displacement of the disk.

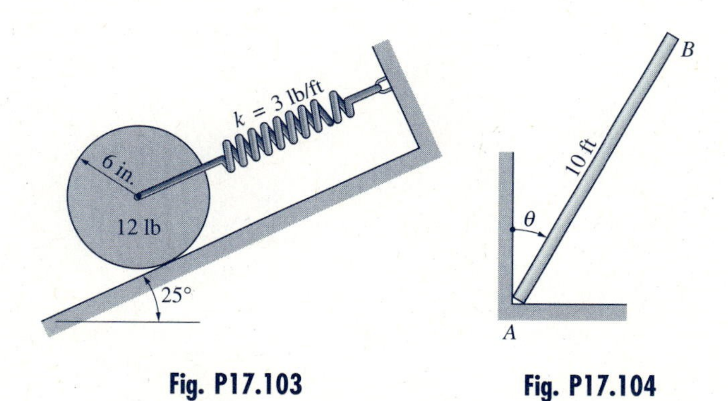

Fig. P17.103 **Fig. P17.104**

17.104 The uniform 120-lb log AB is released from rest at an unknown value of θ. Knowing that the vertical reaction at A is 60 lb when $\theta = 45°$, determine the angular speed of the log in this position.

18 Planar Kinetics of Rigid Bodies: Work-Energy and Impulse-Momentum Methods

18.1 Introduction

This chapter continues the kinetic analysis of plane motion of rigid bodies that was begun in the previous chapter.

Part A of the chapter extends the work-energy method for a system of particles (see Chapter 15) to rigid bodies. The work-energy method, when applied to rigid-body motion, relates the work done by the applied forces and *couples* to the change in the kinetic energy of the body. Therefore, our presentation of this method is preceded by discussions of the work done by a couple and the kinetic energy of a rigid body. As is the case for particle motion, the work-energy method is convenient for finding the change in the speed of the body as it moves between two spatial positions.

In Part B, the impulse-momentum method for systems of particles (see Chapter 15) is extended to rigid bodies. This method relates the linear and angular impulses of the applied forces and couples to the changes in the body's linear and *angular momenta*. The computation of angular momentum makes extensive use of momentum diagrams, which play a major role in this chapter.

As discussed in Chapter 15, the conservation of momentum (linear and/or angular) is one of the more useful concepts in dynamics. It is invaluable in rigid body impact, which is also included in Part B. As pointed out before, impact problems can be analyzed only by impulse-momentum techniques, since impact forces are generally unknown, and energy is not conserved during impact.

Part A: Work-Energy Method

18.2 Work and Power of a Couple

a. Work

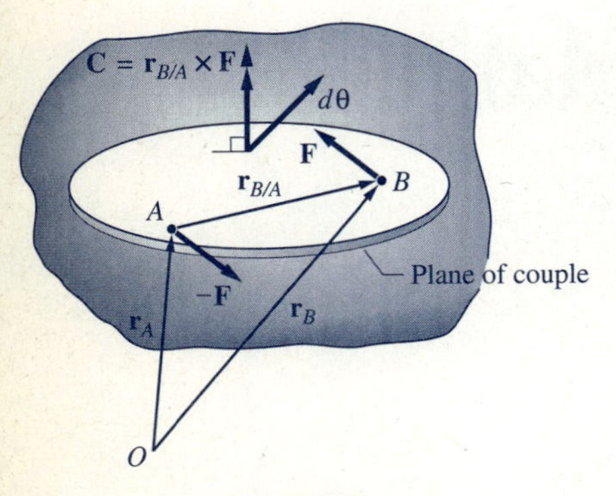

Fig. 18.1

When applying the work-energy method to rigid body motion, it is frequently necessary to calculate the work done by a couple. The work of a couple can be derived with the assistance of two tools already at our disposal: the definition of work of a force and rigid body kinematics.

Figure 18.1 shows a couple applied to a rigid body. The couple is represented by two parallel, but oppositely directed, forces of magnitude F, acting at points A and B. The corresponding couple-vector is

$$\mathbf{C} = \mathbf{r}_{B/A} \times \mathbf{F} \tag{a}$$

The work done by the couple is calculated by summing the work of the two forces that constitute the couple. Therefore, the incremental work of the couple during an infinitesimal displacement of the body is

$$dU = -\mathbf{F} \cdot d\mathbf{r}_A + \mathbf{F} \cdot d\mathbf{r}_B = \mathbf{F} \cdot (d\mathbf{r}_B - d\mathbf{r}_A) = \mathbf{F} \cdot d\mathbf{r}_{B/A} \tag{b}$$

where $d\mathbf{r}_A$ and $d\mathbf{r}_B$ are the incremental displacements of points A and B, respectively. According to rigid body kinematics, the relative velocity vector between A and B is

$$\mathbf{v}_{B/A} = \boldsymbol{\omega} \times \mathbf{r}_{B/A} \quad \text{or} \quad \frac{d\mathbf{r}_{B/A}}{dt} = \frac{d\boldsymbol{\theta}}{dt} \times \mathbf{r}_{B/A}$$

where $\boldsymbol{\omega}$ is the angular velocity vector and $d\boldsymbol{\theta}$ represents the incremental rotation of the body. Multiplying both sides of the last equation by dt, we obtain

$$d\mathbf{r}_{B/A} = d\boldsymbol{\theta} \times \mathbf{r}_{B/A}$$

which, on substitution into Eq. (b), yields

$$dU = \mathbf{F} \cdot (d\boldsymbol{\theta} \times \mathbf{r}_{B/A}) = (\mathbf{r}_{B/A} \times \mathbf{F}) \cdot d\boldsymbol{\theta}$$

Using Eq. (a), we finally get

$$\boxed{dU = \mathbf{C} \cdot d\boldsymbol{\theta}} \tag{18.1}$$

From Eq. (18.1) we see that the work of a couple depends on the rotation of the body and is independent of the translation. This conclusion was anticipated, since it is obvious that during rigid body translation the work of $\mathbf{F}$ is canceled by the work of $-\mathbf{F}$, because the displacements of A and B are equal.

For the special case of plane motion, **C** and $d\boldsymbol{\theta}$ are parallel, both being perpendicular to the plane of the motion. Consequently, $dU = C\,d\theta$, and the work during a finite displacement of the body is

$$U_{1\text{-}2} = \int_{\theta_1}^{\theta_2} C\,d\theta \tag{18.2}$$

where θ_1 and θ_2 are the initial and final angular positions of the body, measured from a convenient reference line. If the magnitude of the couple remains constant during the plane motion, its work becomes

$$U_{1\text{-}2} = C(\theta_2 - \theta_1) = C\,\Delta\theta \tag{18.3}$$

where $\Delta\theta = \theta_2 - \theta_1$ is the angular displacement of the body. It must be remembered that Eqs. (18.2) and (18.3) are valid only for plane motion.

In order to determine the correct sign for the work, the directions of C and the rotation must be compared. If C and the rotation are in the same direction, the work is positive; if C and the rotation have opposite directions, the work done is negative.

It should also be noted that the angular displacement must be measured in radians for Eqs. (18.1)–(18.3) to be valid.

b. Power

In Art. 14.5, the power P was defined to be the time rate at which work is done:

$$P = \frac{dU}{dt} \qquad \text{(14.30, repeated)}$$

It was also shown that the power of a force **F** may be expressed as

$$P = \mathbf{F}\cdot\mathbf{v} \qquad \text{(14.31, repeated)}$$

where **v** is the velocity vector of the point of application of **F**.

When a couple **C** acts on a rigid body, its power is, according to Eq. (18.1),

$$P = \frac{dU}{dt} = \frac{\mathbf{C}\cdot d\boldsymbol{\theta}}{dt} = \mathbf{C}\cdot\boldsymbol{\omega} \tag{18.4}$$

For plane motion, where **C** and $\boldsymbol{\omega}$ are parallel, the power of the couple becomes

$$P = C\omega \tag{18.5}$$

Observe that in Eq. (18.5) the power will be positive if C and ω are in the same direction, and negative if they are oppositely directed. As mentioned in Art. 14.5, the SI units for power are watts (1 W = 1 J/s = 1 N · m/s); the U.S. Customary units are lb · in./s, lb · ft/s, or horsepower (1 hp = 550 lb · ft/s).

We also recall that the efficiency η of a machine was defined in Art. 14.5 as

$$\eta = \frac{\text{output power}}{\text{input power}} \times 100\% \qquad \text{(14.32, repeated)}$$

18.3 Kinetic Energy of a Rigid Body in Plane Motion

This article develops several equations that may be used to calculate the kinetic energy of a rigid body that is undergoing plane motion.

Figure 18.2 shows a rigid body, occupying the region $\mathcal{V}$, that is moving parallel to the plane of the paper with the counterclockwise angular velocity ω. Point A is a reference point that is fixed in the body (or body extended). The components of the velocity vector of A relative to the xy-coordinate system are denoted by v_{Ax} and v_{Ay}. Point G is the mass center of the body; its coordinates are $\bar{x}$ and $\bar{y}$.

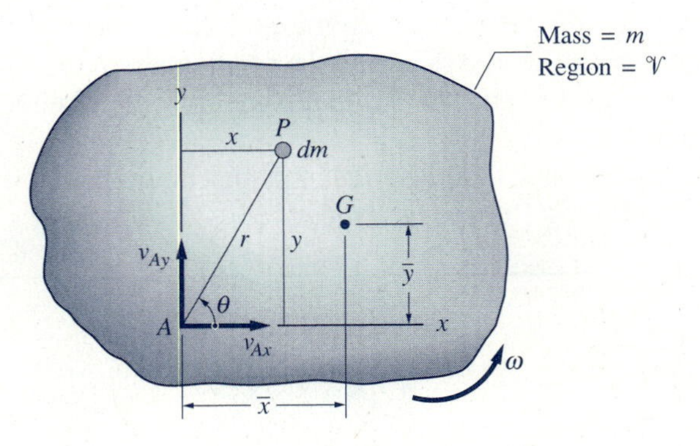

Fig. 18.2

We consider the body to be composed of an infinite number of infinitesimal particles. A typical particle P of mass dm is located by the rectangular coordinates x and y, or the polar coordinates r and θ.

According to particle mechanics, the kinetic energy of P is $dT = (1/2)v_P^2\,dm$, where v_P is the magnitude of its velocity vector. The kinetic energy of the body is obtained by summing the kinetic energies of all the particles in the region $\mathcal{V}$ occupied by the body, which yields

$$T = \int_{\mathcal{V}} \frac{1}{2} v_P^2\,dm \qquad \text{(a)}$$

Since all the particles P are in the same body, their velocities are not independent of each other, but are determined by the velocity of the reference point A and the angular velocity of the body as follows:

$$\mathbf{v}_P = \mathbf{v}_A + \mathbf{v}_{P/A} \qquad \text{(b)}$$

Therefore the components of $\mathbf{v}_P$ may be written as

$$\xrightarrow{+} \quad v_{Px} = v_{Ax} - r\omega \sin\theta = v_{Ax} - y\omega \qquad \text{(c)}$$

$$+\uparrow \quad v_{Py} = v_{Ay} + r\omega \cos\theta = v_{Ay} + x\omega \qquad \text{(d)}$$

Substituting Eqs. (c) and (d) into $v_P^2 = v_{Px}^2 + v_{Py}^2$ we obtain

$$\begin{aligned} v_P^2 &= (v_{Ax} - y\omega)^2 + (v_{Ay} + x\omega)^2 \\ &= (v_{Ax}^2 - 2yv_{Ax}\omega + y^2\omega^2) + (v_{Ay}^2 + 2xv_{Ay}\omega + x^2\omega^2) \\ &= (v_{Ax}^2 + v_{Ay}^2) + (x^2 + y^2)\omega^2 - 2yv_{Ax}\omega + 2xv_{Ay}\omega \end{aligned}$$

which yields

$$v_P^2 = v_A^2 + r^2\omega^2 - 2yv_{Ax}\omega + 2xv_{Ay}\omega \qquad \text{(e)}$$

Substituting Eq. (e) into Eq. (a) and taking the terms that are independent of x and y outside the integral signs, we obtain

$$T = \frac{1}{2}v_A^2 \int_{\mathcal{V}} dm + \frac{1}{2}\omega^2 \int_{\mathcal{V}} r^2\, dm - v_{Ax}\omega \int_{\mathcal{V}} y\, dm + v_{Ay}\omega \int_{\mathcal{V}} x\, dm \qquad \text{(f)}$$

After we make the substitutions,

$\int_{\mathcal{V}} dm = m$ (the mass of the body)

$\int_{\mathcal{V}} r^2\, dm = I_A$ (the moment of inertia of the body about the axis passing through A that is perpendicular to the plane of the motion)

$\int_{\mathcal{V}} y\, dm = m\bar{y}$ and $\int_{\mathcal{V}} x\, dm = m\bar{x}$

the kinetic energy of the body becomes

$$T = \frac{1}{2}mv_A^2 + \frac{1}{2}I_A\omega^2 - m\bar{y}v_{Ax}\omega + m\bar{x}v_{Ay}\omega \qquad \text{(18.6)}$$

(A: fixed in the body)

Note that Eq. (18.6) is applicable for any point A that lies in the body (or body extended). However, it is often convenient to compute the kinetic energy from the simpler formulas obtained by restricting the choice of A to one of the following special cases.

(1) Point A is the mass center G. For this case, $\bar{x} = \bar{y} = 0$, and Eq. (18.6) reduces to

$$T = \frac{1}{2}m\bar{v}^2 + \frac{1}{2}\bar{I}\omega^2 \qquad \text{(18.7a)}$$

where $\bar{v}$ is the speed of G, and $\bar{I}$ is the moment of inertia of the body about the central axis (the axis through G that is perpendicular to the plane of the motion). The first term on the right side of Eq. (18.7a) is called the *kinetic energy of translation,* whereas the second term is known as the *kinetic energy of rotation.* Note that the kinetic energy is independent of the *sense* of ω and the *direction* of $\bar{\mathbf{v}}$.

(2) Point A is the instant center of zero velocity. If the velocity of point A is zero, the kinetic energy in Eq. (18.6) becomes

$$T = \frac{1}{2} I_A \omega^2 \qquad (A\text{: instant center}) \tag{18.7b}$$

Equation (18.7b) is commonly used to calculate the kinetic energy of a body that is rotating about a fixed point A. However, note that point A need not be fixed for the equation to be valid—only its velocity need be zero at the instant of interest.

18.4 Work-Energy Principle and Conservation of Mechanical Energy

The work-energy principle for a system of particles, derived in Art. 15.5, was

$$(U_{1-2})_{\text{ext}} + (U_{1-2})_{\text{int}} = \Delta T \qquad \text{(15.24, repeated)}$$

where the subscripts 1 and 2 refer to the inital and final positions of the system, respectively, $(U_{1-2})_{\text{ext}}$ is the work done by the external forces (including the weights of the particles), $(U_{1-2})_{\text{int}}$ is the work done by the internal forces, and $\Delta T = T_2 - T_1$ is the change in the kinetic energy of the system.

Considering a rigid body to be made up of particles, Eq. (15.24) can be applied directly to the plane motion of a rigid body, or a system of connected rigid bodies, as described in the following.

(1) Single rigid body. The forces internal to a rigid body hold the body together; i.e., they are the constraint forces that impose the condition of rigidity. These internal forces occur in equal and opposite collinear pairs. Since the body is rigid, the distances between the particles do not change, which in turn implies that the distances between the points of application of the internal forces remain constant. Consequently, the internal forces do no work on a rigid body, so that the work-energy principle governing the motion of a single rigid body becomes

$$(U_{1-2})_{\text{ext}} = \Delta T \tag{18.8}$$

The work done by the external forces and couples can be computed using the methods described in Arts. 14.2 and 18.2. Equations (18.6)–(18.7) can be used to calculate the kinetic energy of the body in the initial and final positions.

(2) System of connected rigid bodies. As explained above, the forces internal to a rigid body are workless. However, the internal connections between rigid bodies may either be workless or capable of doing work. Inextensible strings and pins are examples of workless internal connectors, whereas springs and friction give rise to internal forces that can do work. (Workless internal forces and internal forces that can do work were discussed in Art. 15.5.) Therefore, the work-energy principle for the motion of a system of connected rigid bodies must be used in its general form,

$$(U_{1\text{-}2})_{\text{ext}} + (U_{1\text{-}2})_{\text{int}} = \Delta T \qquad (18.9)$$

In this equation, T represents the kinetic energy of the system, which equals the sum of the kinetic energies of the constituent bodies.

The principle of conservation of mechanical energy for a system of particles, stated in Art. 15.5, is also applicable to a rigid body (or a system of connected rigid bodies): If all forces, internal as well as external, are conservative (see Art. 14.4), the mechanical energy of the rigid body (or system of connected rigid bodies) is conserved. This principle may be written as

$$V_1 + T_1 = V_2 + T_2 \qquad (18.10)$$

where V_1 and V_2 are the initial and final potential energies, and T_1 and T_2 are the initial and final kinetic energies.

Sample Problem 18.1

The uniform 40-lb slender bar AC shown in Fig. (a) rotates in a vertical plane about the smooth pin at B. The ideal spring connected between end A and the fixed point D has a spring constant $k = 2$ lb/ft and an undeformed length $L_0 = 3$ ft. When the bar is at rest in the initial position $\theta = 0$, it is given a small angular displacement and released. Find the angular velocity of the bar when it reaches the horizontal position.

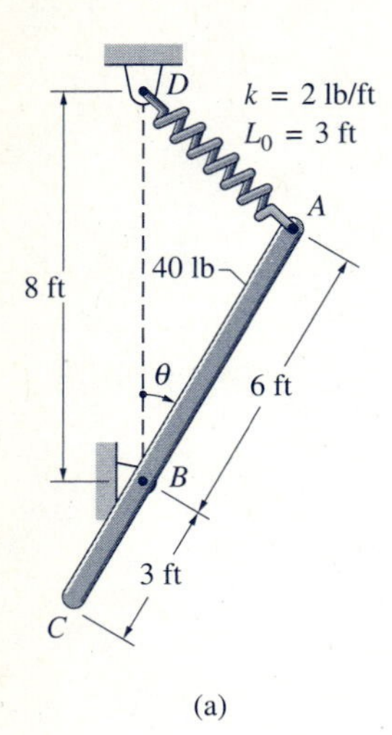

(a)

Solution

This problem is well suited for solution by the work-energy method, because it is concerned with the change in velocity that occurs during a change in position. Since the system is conservative, it may be analyzed by the work-energy or conservation of mechanical energy principles. We use the first of these methods; you may wish to try the second method as an exercise.

Figure (b) shows the initial and final positions of the bar, labeled ① and ②, respectively. Work is done on the bar by its weight and by the linear spring. Noting that the mass center G of the bar moves downward through a distance of 1.5 ft between positions ① and ②, the work of the weight is

$$U_{1\text{-}2} = 40(1.5) = 60 \text{ lb} \cdot \text{ft} \tag{a}$$

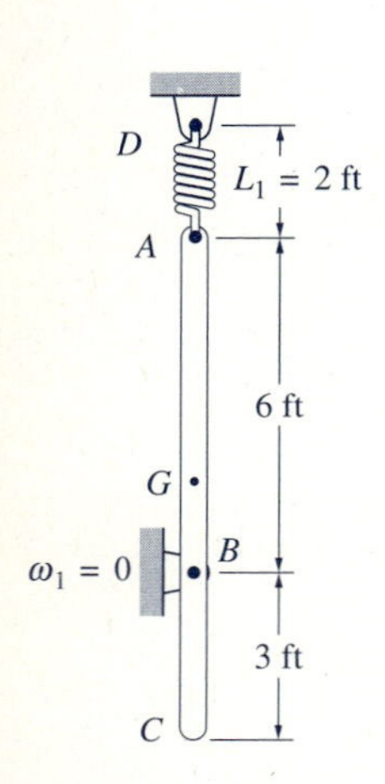

Position ①

From Fig. (b) we see that the length of the spring is $L_1 = 2$ ft in the initial position and $L_2 = \sqrt{6^2 + 8^2} = 10$ ft in the final position. Since the unstretched length of the spring is $L_0 = 3$ ft, the initial and final deformations of the spring are $\delta_1 = L_1 - L_0 = 2 - 3 = -1$ ft and $\delta_2 = L_2 - L_0 = 10 - 3 = 7$ ft. (The signs indicate that the spring is in compression in the initial position and in tension in the final position.) The work done by the spring on the bar, therefore, is

$$U_{1\text{-}2} = -\frac{1}{2}k(\delta_2^2 - \delta_1^2)$$

$$= -\frac{1}{2}(2)[7^2 - (-1)^2] = -48 \text{ lb} \cdot \text{ft} \tag{b}$$

Because point B is fixed, the kinetic energy of the bar can be calculated from $T = (1/2)I_B\omega^2$. Using the parallel-axis theorem we find that $I_B = \bar{I} + md^2 = (mL^2/12) + md^2 = [(40/32.2)(9^2)/12] + (40/32.2)(1.5)^2 = 11.180$ slug · ft^2. From Eqs. (a) and (b) and knowing that the bar is released from rest ($T_1 = 0$), the work-energy principle becomes

$$U_{1\text{-}2} = T_2 - T_1 = \frac{1}{2}I_B\omega_2^2 - 0$$

$$60 - 48 = \frac{1}{2}(11.180)\omega_2^2$$

which yields

$$\omega_2 = 1.465 \text{ rad/s} \qquad \textit{Answer}$$

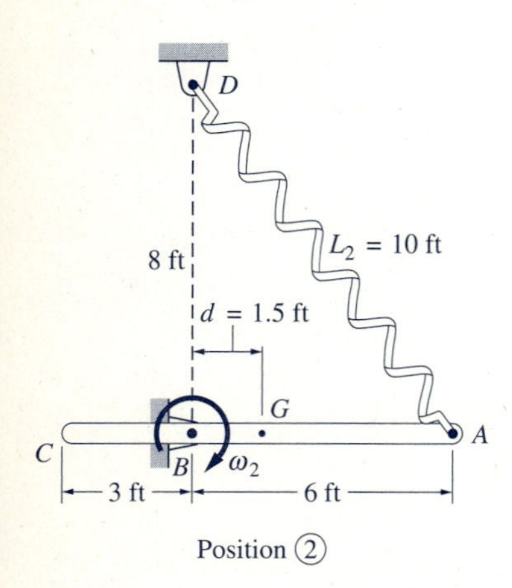

Position ②

(b)

Sample Problem 18.2

Figure (a) shows a homogeneous slender bar AB of mass m and length L. When the bar was in the position $\theta = 0$, it was displaced slightly and released from rest. Determine the angular velocity and angular acceleration of the bar as functions of the angle θ. Neglect friction and assume that end A does not lose contact with the vertical surface.

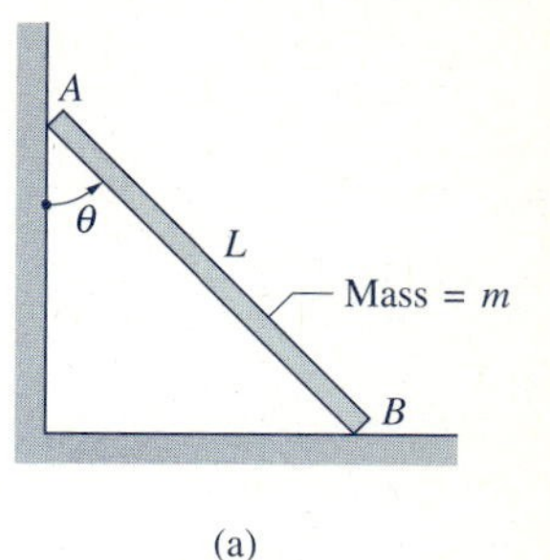

(a)

Solution

Our solution utilizes the fact that the mechanical energy is conserved (the weight of the bar is the only force that does work as the bar falls). An identical solution could be obtained just as easily from the work-energy principle.

Figure (b) shows the bar in the release ($\theta = 0$) position ① and in an arbitrary position ② defined by the angle θ. Choosing the horizontal plane as the datum for potential energy, the potential energies of the bar in the two positions are

$$V_1 = mg\frac{L}{2} \quad \text{and} \quad V_2 = mg\frac{L}{2}\cos\theta \tag{a}$$

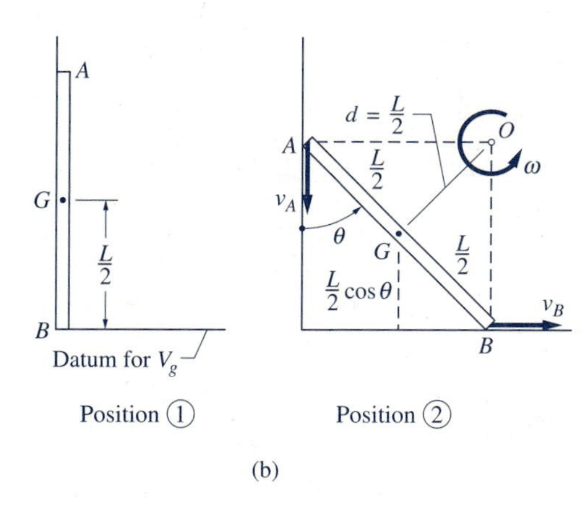

(b)

The initial kinetic energy is obviously $T_1 = 0$, since the bar is released from rest. The kinetic energy in position ② can be readily calculated by observing that the point O, shown in Fig. (b), is the instant center for velocities of the bar. (Point O is located where the perpendiculars to the velocity vectors of ends A and B intersect.) Utilizing the parallel-axis theorem, the moment of inertia of the bar about O is $I_O = \bar{I} + md^2 = (mL^2/12) + m(L/2)^2 = mL^2/3$. Consequently, the kinetic energies in the two positions are

$$T_1 = 0 \quad \text{and} \quad T_2 = \frac{1}{2}I_O\omega^2 = \frac{1}{2}\frac{mL^2}{3}\omega^2 \tag{b}$$

where ω is the angular velocity of the bar in position ②.

Because mechanical energy is conserved we have

$$V_1 + T_1 = V_2 + T_2$$

Substituting from Eqs. (a) and (b), we get

$$mg\frac{L}{2} + 0 = mg\frac{L}{2}\cos\theta + \frac{1}{2}\frac{mL^2}{3}\omega^2$$

from which the angular velocity in position ② is found to be

$$\omega = \left[\frac{3g}{L}(1 - \cos\theta)\right]^{1/2} \qquad \textit{Answer}$$

The angular acceleration α is obtained by taking the time derivative of ω, which yields

$$\alpha = \frac{d\omega}{dt} = \frac{1}{2}\left[\frac{3g}{L}(1 - \cos\theta)\right]^{-(1/2)} \frac{3g}{L}\dot{\theta}\sin\theta$$

Substituting the previously found expression for $\dot{\theta} = \omega$, the angular acceleration reduces to

$$\alpha = \frac{3g}{2L}\sin\theta \qquad \textit{Answer}$$

Sample Problem 18.3

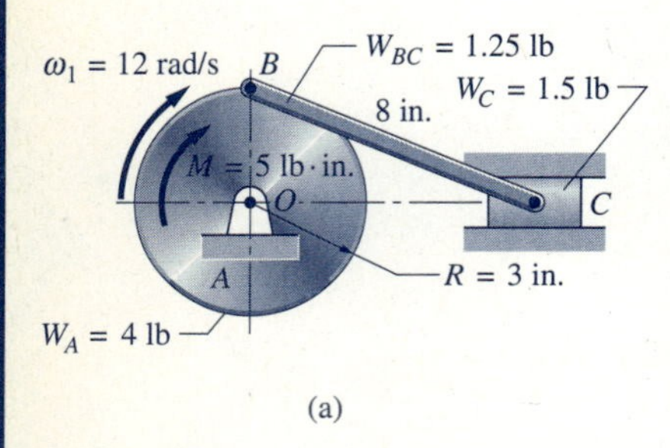

(a)

Figure (a) shows a slider-crank mechanism that is being driven by a constant clockwise couple $M = 5$ lb · in. All the components are homogeneous, with the weights and dimensions as indicated. When the mechanism is in the position shown in Fig. (a), the angular velocity of the crank is $\omega_1 = 12$ rad/s clockwise. Determine ω_2, the angular velocity of the crank after it has rotated 90°. Neglect friction and assume that motion is in a vertical plane.

Solution

The solution of this problem lends itself to a work-energy analysis because it is concerned with the change in velocity between two positions. The problem can be solved by using either the work-energy principle or by noting that mechanical energy is conserved (we employ the latter approach). Regardless of which method is used, it is convenient to analyze the entire mechanism, thereby eliminating the need to consider the work done at the connections (pins B and C). The initial and final positions of the mechanism, labeled ① and ②, respectively, are shown in Fig. (b).

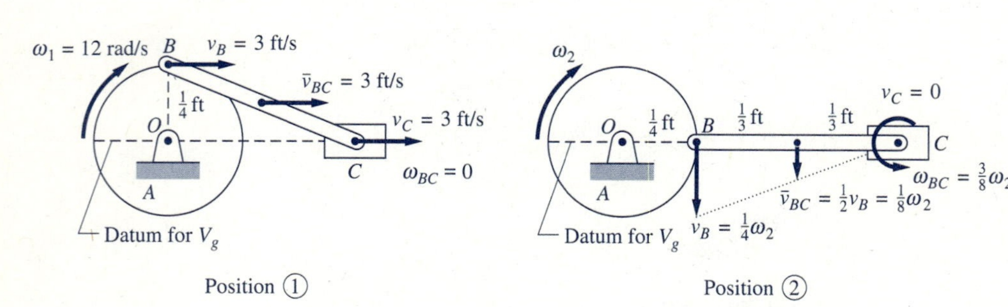

(b)

Kinematic Analysis

The steps used in the kinematic analysis that resulted in the velocities shown in Fig. (b) are described as follows.

Position ①

1. Since O is a fixed point, $v_B = R\omega_1 = (1/4)12 = 3$ ft/s.
2. Recognizing that both $\mathbf{v}_B$ and $\mathbf{v}_C$ are horizontal, we conclude that $\omega_{BC} = 0$; i.e., bar BC is translating in this position.
3. Since bar BC is translating, the velocities of all points on the bar are the same. In particular, the velocity of the mass center of BC is $\bar{v}_{BC} = 3$ ft/s.

Position ②

1. Since O is a fixed point, $v_B = R\omega_2 = (1/4)\omega_2$ ft/s (directed downward). It was assumed that ω_2 is directed clockwise.
2. Since $\mathbf{v}_B$ is vertical and the path of C is horizontal, we conclude that C is the instant center for bar BC; i.e., $v_C = 0$.
3. Because $v_C = 0$, we know that $\omega_{BC} = v_B/L_{BC} = (\omega_2/4)/(2/3) = 3\omega_2/8$ rad/s (counterclockwise) and that the velocity of the mass center of BC is $\bar{v}_{BC} = v_B/2 = \omega_2/8$ ft/s.

Potential Energy

The system possesses gravitational potential energy due to the weights of its components; in addition, there is the potential energy of the constant couple.

As shown in Fig. (b) we choose the horizontal plane passing through OC to be the datum for the gravitational potential energy V_g. In position ①, the mass centers of A and C lie in the datum plane, whereas the mass center of bar BC is 1/8 ft above this plane. Therefore we obtain $(V_g)_1 = W_{BC}h = 1.25(1/8) = 0.1563$ lb · ft. In position ②, there is no gravitational potential energy because the mass center of each part lies in the datum plane; i.e., $(V_g)_2 = 0$.

The constant couple M is conservative because its work $U_{1-2} = M\Delta\theta$ depends on only the magnitude M and the initial and final angular positions of crank A. The work done by the couple may be expressed by $U_{1-2} = -[(V_M)_2 - (V_M)_1]$, where V_M is the potential energy of the couple. Choosing position ① to be the datum, we have $(V_M)_2 = (V_M)_1 - U_{1-2} = 0 - (5/12)(\pi/2) = -0.6545$ lb · ft.

In summary, the initial and final potential energies are

$$V_1 = (V_g)_1 + (V_M)_1 = 0.1563 + 0 = 0.1563 \text{ lb} \cdot \text{ft} \tag{a}$$

and

$$V_2 = (V_g)_2 + (V_M)_2 = 0 + (-0.6545) = -0.6545 \text{ lb} \cdot \text{ft} \tag{b}$$

Kinetic Energy

The kinetic energy of the entire mechanism is the sum of the kinetic energies of the three parts:

$$T = \left(\frac{1}{2}\bar{I}\omega^2\right)_A + \left(\frac{1}{2}\bar{I}\omega^2 + \frac{1}{2}m\bar{v}^2\right)_{BC} + \left(\frac{1}{2}mv^2\right)_C \tag{c}$$

The central moments of inertia of the crank A and the arm BC are

$$\bar{I}_A = \left(\frac{mR^2}{2}\right)_A = \frac{1}{2}\left(\frac{4}{32.2}\right)\left(\frac{3}{12}\right)^2 = 3.882 \times 10^{-3} \text{ slug} \cdot \text{ft}^2$$

and

$$\bar{I}_{BC} = \left(\frac{mL^2}{12}\right)_{BC} = \frac{1}{12}\left(\frac{1.25}{32.2}\right)\left(\frac{8}{12}\right)^2 = 1.438 \times 10^{-3} \text{ slug} \cdot \text{ft}^2$$

Position ①
Substituting the values of $\bar{I}_A$, $\bar{I}_{BC}$, $\omega_A = \omega_1 = 12$ rad/s, $\omega_{BC} = 0$, and $\bar{v}_{BC} = v_C = 3$ ft/s into Eq. (c) we obtain

$$T_1 = \left[\frac{1}{2}(3.882 \times 10^{-3})(12)^2\right] + \left[0 + \frac{1}{2}\left(\frac{1.25}{32.2}\right)(3)^2\right] + \left[\frac{1}{2}\left(\frac{1.5}{32.2}\right)(3)^2\right]$$

which yields

$$T_1 = 0.2795 + 0.1747 + 0.2096 = 0.6638 \text{ lb} \cdot \text{ft} \quad \text{(d)}$$

Position ②
Substituting the values of $\bar{I}_A$, $\bar{I}_{BC}$, $\omega_A = \omega_2$ rad/s, $\omega_{BC} = 3\omega_2/8$ rad/s, $\bar{v}_{BC} = \omega_2/8$ ft/s, and $v_C = 0$ into Eq. (c) we find

$$T_2 = \left[\frac{1}{2}(3.882 \times 10^{-3})\omega_2^2\right] + \left[\frac{1}{2}(1.438 \times 10^{-3})\left(\frac{3}{8}\omega_2\right)^2 + \frac{1}{2}\left(\frac{1.25}{32.2}\right)\left(\frac{\omega_2}{8}\right)^2\right] + 0$$

which simplifies to

$$T_2 = (2.345 \times 10^{-3})\omega_2^2 \quad \text{(e)}$$

Note that the kinetic energy of bar BC in position ② could also be computed from $(1/2)I_C\omega^2$ since $v_C = 0$.

Conservation of Mechanical Energy

Utilizing the results of Eqs. (a), (b), (d), and (e), the principle of conservation of mechanical energy becomes

$$V_1 + T_1 = V_2 + T_2$$
$$0.1563 + 0.6638 = -0.6545 + (2.345 \times 10^{-3})\omega_2^2$$

from which the angular velocity in the final position is found to be

$$\omega_2 = 25.1 \text{ rad/s} \qquad \textit{Answer}$$

Sample Problem 18.4

The 10-kg block B shown in Fig. (a) is attached to a cable that runs around the pulley at C and is wrapped around the periphery of the unbalanced 40-kg disk A. The mass center of A is located at G, and its central moment of inertia is $12\ \text{kg}\cdot\text{m}^2$. When the disk is in the position shown in Fig. (a), its angular velocity is $\omega_1 = 2.5$ rad/s clockwise. Assuming that the disk rolls without slipping, calculate its angular velocity after it has rotated 180°.

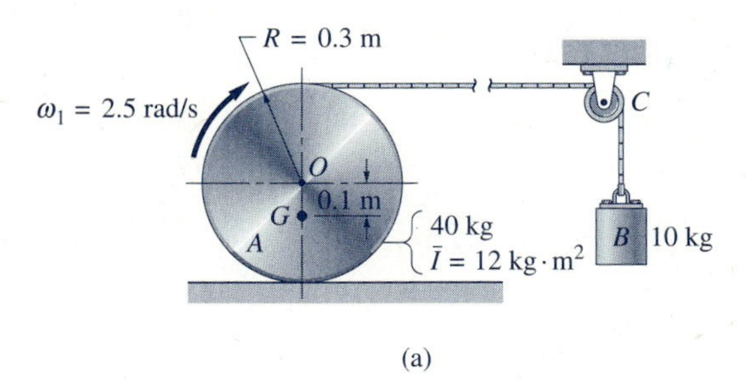

(a)

Solution

Because this problem is concerned with the change in velocity that occurs during a given change in position, the work-energy principle, or the principle of conservation of mechanical energy (note that the system is conservative), is the convenient method of analysis. In this problem we choose to employ the work-energy principle. It is again convenient to analyze the system as a whole, in which case the cable becomes a workless internal connection.

Kinematic Analysis

Figure (b) shows the system when it is in the initial and final positions, labeled ① and ②, respectively. Since the disk does not slip, the velocity of the point in contact with the ground is zero. Therefore, in position ① the velocity of G is $\bar{v} = \overline{DG}\omega_1 = 0.2(2.5) = 0.5$ m/s, and in position ② it is $\bar{v} = \overline{FG}\omega_2 = 0.4\omega_2$ m/s, where ω_2 is measured in rad/s. Since the velocity of B is equal to the velocity of the uppermost point on the disk, we have $v_B = \overline{DE}\omega_1 = 0.6(2.5) = 1.5$ m/s in position ① and $v_B = \overline{FH}\omega_2 = 0.6\omega_2$ m/s in position ②. Furthermore, the vertical distance moved by B between the two positions is $\Delta h = 2R\theta = 2(0.3)(\pi) = 1.885$ m. (For kinematic purposes, the motion of the disk can be thought of as a rigid-body translation equal to $R\theta$ to the right, followed by a clockwise rotation through the angle θ. Due to these two effects, the displacement of B is $2R\theta$.)

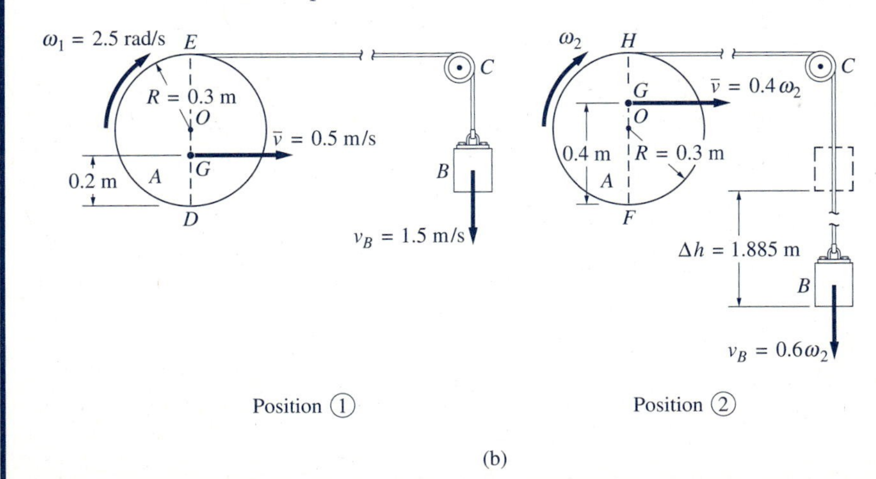

(b)

Work

Since the disk does not slip, the friction force at its point of contact with the ground does not do work. Consequently, the weights of A and B are the only forces that do work on the system (for this reason the system is conservative). Between the two positions shown in Fig. (b), the mass center G moves a distance of 0.2 m upward, and B moves downward the distance $\Delta h = 1.885$ m. Therefore, the total work done on the system by the weight of A and the weight of B is

$$U_{1\text{-}2} = -(40)(9.81)(0.2) + (10)(9.81)(1.885) = 106.4 \text{ J} \tag{a}$$

Kinetic Energy

The kinetic energy of the system, i.e., the sum of the kinetic energies of A and B, is

$$T = \left(\tfrac{1}{2}\bar{I}\omega^2 + \tfrac{1}{2}m\bar{v}^2\right)_A + \left(\tfrac{1}{2}mv^2\right)_B \tag{b}$$

(The kinetic energy of A could also be computed using $(1/2)I\omega^2$, where I is the moment of inertia of A about the point of contact, i.e., the instant center.)

The initial and final kinetic energies are evaluated by substituting the appropriate numerical values of $\bar{I}$, the masses, and the velocities into Eq. (b).

$$T_1 = \left[\tfrac{1}{2}(12)(2.5)^2 + \tfrac{1}{2}(40)(0.5)^2\right] + \left[\tfrac{1}{2}(10)(1.5)^2\right] = 53.75 \text{ J} \tag{c}$$

$$T_2 = \left[\tfrac{1}{2}(12)\omega_2^2 + \tfrac{1}{2}(40)(0.4\omega_2)^2\right] + \left[\tfrac{1}{2}(10)(0.6\omega_2)^2\right] = 11.00\omega_2^2 \text{ J} \tag{d}$$

Work-Energy Principle

Using Eqs. (a), (c), and (d), the work-energy principle becomes

$$U_{1\text{-}2} = T_2 - T_1$$
$$106.4 = 11.00\omega_2^2 - 53.75$$

from which the angular velocity of the disk in position ② is found to be

$$\omega_2 = 3.82 \text{ rad/s} \qquad \textit{Answer}$$

Sample Problem 18.5

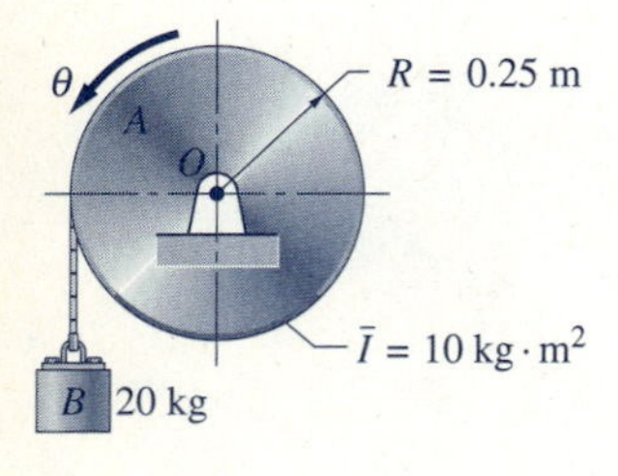

(a)

The 20-kg block B shown in Fig. (a) is suspended from a cable that is wrapped around the outside of the uniform cylinder A. The angular position of A is specified by the angle θ. If the system is released from rest at $\theta = 0$, determine the following as functions of θ: (1) the angular velocity of the cylinder; and (2) the power supplied to the system. Neglect friction.

Solution

Part 1

Figure (b) shows two positions of the system—the initial rest position, ①, and an arbitrary position, ②. Note that C is the point on the disk where the cable leaves the disk in the initial position. The angle θ locates the position of the line OC as the disk rotates.

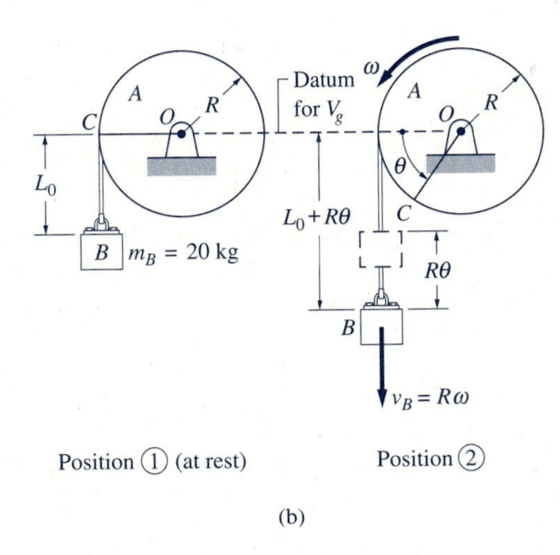

(b)

Since the system is conservative, we can use the conservation of energy principle to compute ω, the angular velocity of the cylinder in position ②. Of course, the work-energy method could also be used. We choose to apply the energy conservation principle to the entire system rather than to the individual parts. This approach eliminates the need to consider the work done by the cable (its work done on the cylinder is equal in magnitude but opposite in sign to its work done on the block).

As shown in Fig. (b), we choose the datum for the gravitational potential energy to be the horizontal plane that passes through the pin at O. The unknown distance of B below the datum in position ① is denoted by L_0; this distance increases to $L_0 + R\theta$ in position ②, where θ is measured in radians. Referring to Fig. (b) and assuming that L_0 is measured in meters, we see that the potential energies for the two positions of the system are

$$V_1 = -m_B g L_0 = -(20)(9.81)L_0 = -196.2L_0 \text{ J}$$

$$V_2 = -m_B g(L_0 + R\theta) = -(20)(9.81)(L_0 + 0.25\theta)$$

$$= -196.2L_0 - 49.05\theta \text{ J}$$

The kinetic energies of the system in the two positions are

$$T_1 = 0$$

$$T_2 = \left(\frac{1}{2}\bar{I}\omega^2\right)_A + \left(\frac{1}{2}mv^2\right)_B$$

$$= \frac{1}{2}(10)\omega^2 + \frac{1}{2}(20)(0.25\omega)^2$$

$$= 5.625\omega^2 \text{ J}$$

With the energies computed, the principle of conservation of mechanical energy becomes

$$V_1 + T_1 = V_2 + T_2$$

$$-196.2L_0 + 0 = (-196.2L_0 - 49.05\theta) + 5.625\omega^2$$

In the last equation, the terms containing L_0 cancel, as would be expected, leaving us with the result

$$\omega = 2.953\sqrt{\theta}\ \text{rad/s} \qquad \textit{Answer}$$

Part 2

Since the weight of block B is the only force that does work on the system, it is also the only force capable of providing power (recall that power is the time rate of doing work). Referring to Fig. (b) we see that B moves downward through the distance $R\theta$ between positions ① and ②. Therefore its work is

$$U(\theta) = U_{1\text{-}2} = m_B g(R\theta) = (20)(9.81)(0.25\theta) = 49.05\theta\ \text{J}$$

(The identical expression for the work would, of course, be obtained from $U_{1\text{-}2} = V_1 - V_2$.) Therefore the power supplied to the system is

$$P = \frac{dU}{dt} = 49.05\dot{\theta}$$

Substituting $\omega = \dot{\theta} = 2.953\sqrt{\theta}$ rad/s from the solution to Part 1, we get

$$P = 49.05(2.953\sqrt{\theta}) = 144.8\sqrt{\theta}\ \text{W} \qquad \textit{Answer}$$

PROBLEMS

18.1 Each homogeneous slender bar AB has a mass m and length L. The bar in Fig. (a) is guided by pins at G and B, which slide in slots, and the bar in Fig. (b) rotates about a pin at C. Calculate the kinetic energy of each bar in terms of its angular speed ω, m, and L.

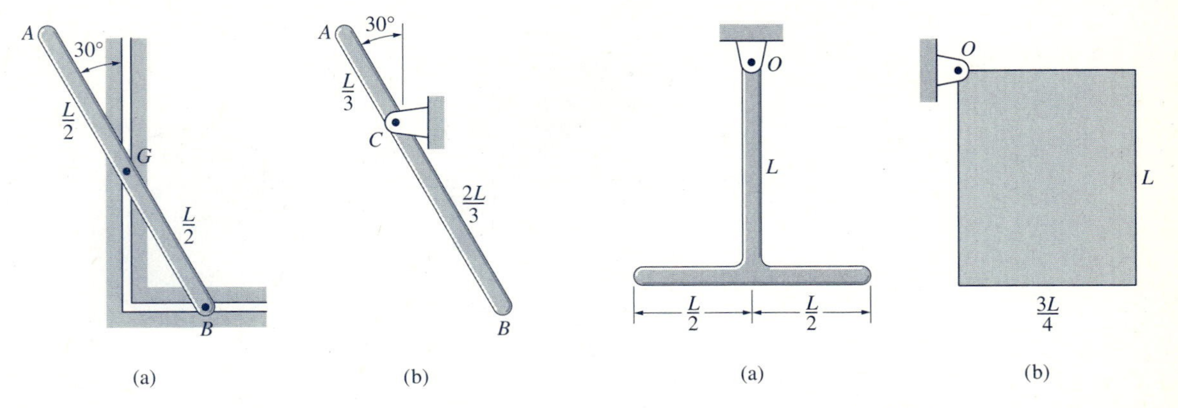

Fig. P18.1 **Fig. P18.2**

18.2 Each of the bodies has a mass m and rotates about a pin at O. The body in Fig. (a) is made from a uniform rod, and the body in Fig. (b) is a homogeneous rectangular plate. Determine the kinetic energy of each body in terms of its angular velocity ω, m, and L.

18.3 A couple (not shown) causes the eccentric disk of mass M to roll without slipping at the constant angular velocity ω_0. The radius of gyration of the disk about its mass center G is $\bar{k}$. Calculate (a) the kinetic energy of the disk as a function of θ (the angle between OG and the vertical); and (b) the maximum and minimum kinetic energy. Use the following data: $M = 40$ kg, $R = 240$ mm, $e = 50$ mm, $\bar{k} = 160$ mm, $\omega_0 = 10$ rad/s.

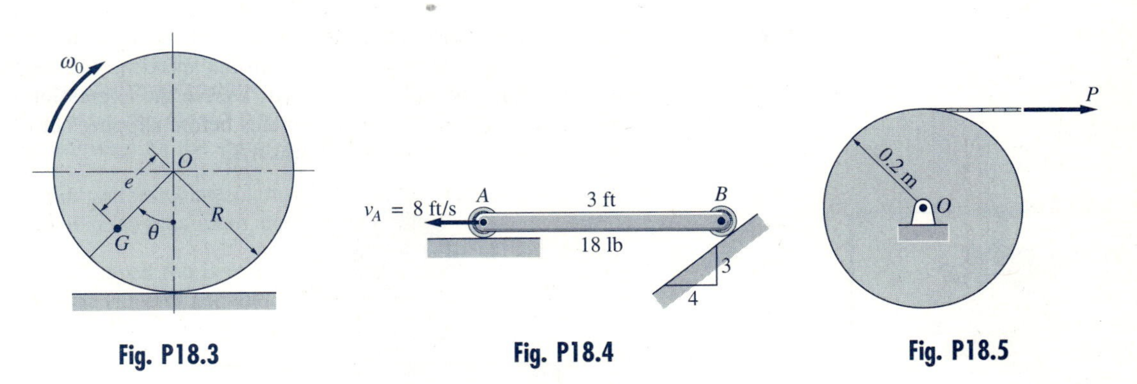

Fig. P18.3 **Fig. P18.4** **Fig. P18.5**

18.4 When the 18-lb uniform bar AB is in the position shown, the velocity of end A is 8 ft/s to the left. Determine the kinetic energy of the bar in this position.

18.5 The constant horizontal force P is applied to the rope that is wound tightly around the pulley. The mass of the pulley is 20 kg, and its central radius of gyration is $\bar{k} = 0.16$ m. What value of P will accelerate the pulley from rest to 300 rev/min in 4 revolutions?

18.14 The homogeneous disk of mass m and radius R is released from rest in the position shown. (a) Derive the expression for the angular velocity of the disk when CG is vertical. (b) Find the distance e that would maximize the angular velocity found in part (a); and (c) determine this maximum angular velocity.

18.6 The 120-lb bucket is suspended from a cable that is wrapped around the periphery of the 400-lb drum. The central radius of gyration of the drum is 0.8 ft. The system is released from rest in position ① and moves with negligible friction until the bucket reaches position ②, when a brake is applied to the drum that exerts a constant couple of 576 lb · ft. (a) Find the height h for which the bucket will land on the ground with zero velocity. (b) Determine the velocity of the bucket in position ②.

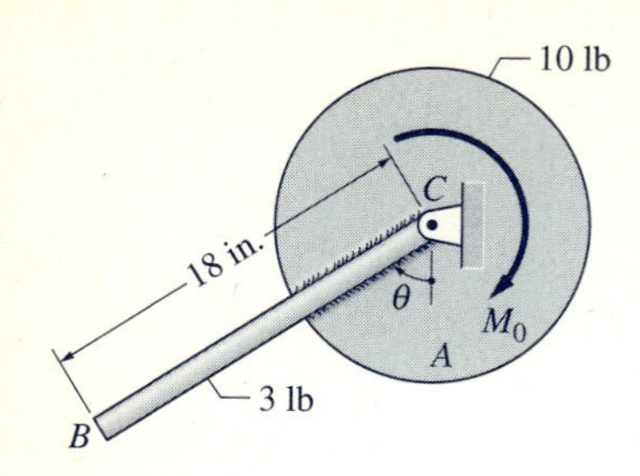

Fig. P18.15

18.15 The 3-lb uniform rod BC is welded to drum A, which weighs 10 lb and has a radius of gyration of 4 in. about C. The system is initially at rest in the position $\theta = 0$ when the constant clockwise couple $M_O = 1.5$ lb · ft is applied. (a) Derive the angular velocity (in rad/s) and angular acceleration (in rad/s^2) of the system as functions of θ. (b) Determine the maximum angular velocity and the value of θ at which it occurs.

18.16 A bicycle chain, 6 ft in length and weighing 0.5 lb/ft, hangs over a sprocket as shown. The sprocket weighs 6.4 lb, and its central radius of gyration is 7.42 in. If the sprocket is given a small clockwise angular displacement from the position shown and then released, calculate its angular velocity after the entire chain has left the sprocket.

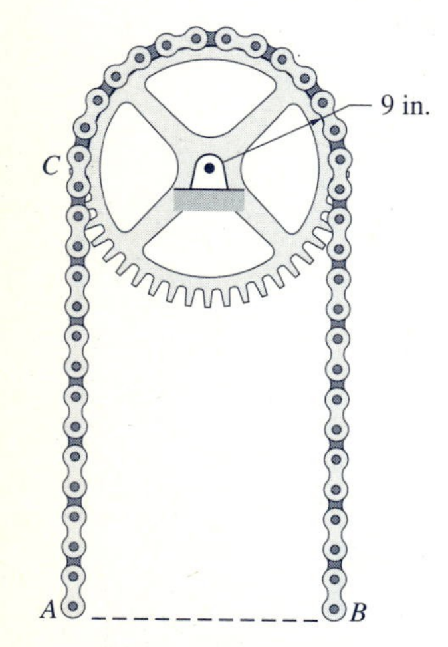

Fig. P18.16, P18.17

18.17 For the system described in Prob. 18.16, what is the angular velocity of the sprocket when end A of the chain has reached point C?

18.18 A cylinder of radius R_C and mass m_C, and a sphere of radius R_S and mass m_S are released from rest on a rough surface that is inclined at the angle β with the horizontal. Both bodies are homogeneous and roll without slipping. Determine $\bar{v}_C/\bar{v}_S$ (the ratio of the central velocities) after each body has moved the same distance d down the inclined plane.

18.19 The mass moment of inertia of the 60-kg spool about its mass center G is 1.35 kg · m^2. The spool is at rest on a rough surface when the constant force P is applied to the cable that is wound around its hub. Determine P if the spool is to have an angular speed of 8 rad/s after turning through one revolution. Assume rolling without slipping.

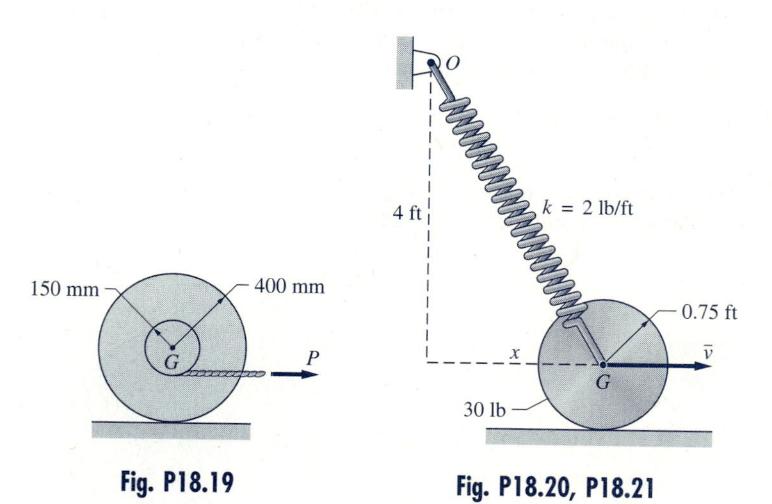

Fig. P18.19

Fig. P18.20, P18.21

18.20 The spring attached to the homogeneous 30-lb disk has an undeformed length of 1.0 ft, and its stiffness is 2 lb/ft. When $x = 0$, the velocity of the mass center G is $\bar{v} = 5$ ft/s to the right. Assuming rolling without slipping, find the velocity of G when $x = 3$ ft.

18.21 For the system described in Prob. 18.20, derive the acceleration of point G in ft/s^2 as a function of x, where x is measured in feet. (Hint: Express the total mechanical energy of the system as a function of x, and differentiate it with respect to time.)

18.22 The torsional pendulum in Fig. (a) consists of the 1.5-kg bar attached to the slender elastic shaft AB that acts as a torsional spring. Figure (b) shows the relationship between the resisting torque (couple) M and the rotation θ of the bar. If the pendulum oscillates with the amplitude $\theta_{max} = 60°$, determine the maximum angular velocity of the bar by numerical integration.

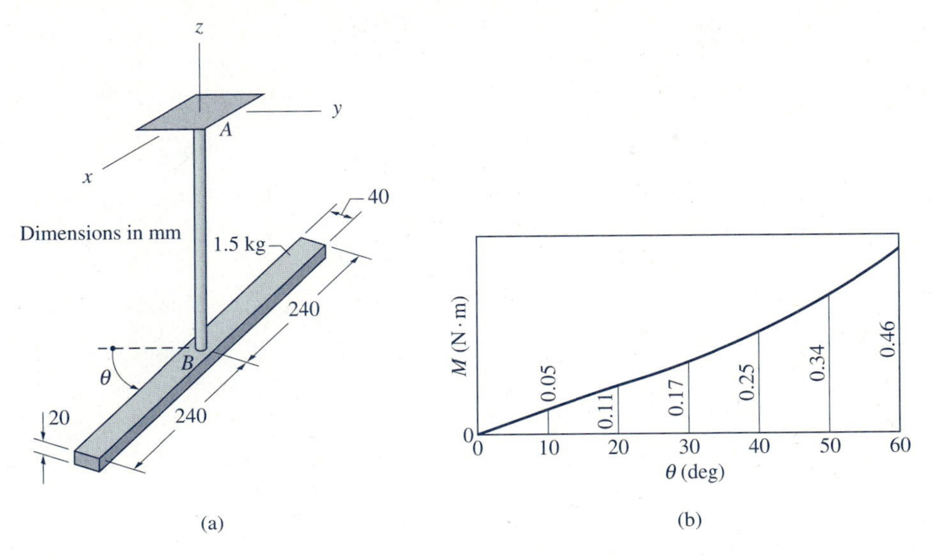

Fig. P18.22

18.23 The eccentric wheel of mass $M = 300$ kg and radius $R = 1.2$ m rolls without slipping. The radius of gyration of the wheel about its mass center G is $\bar{k} = 0.5$ m, and its eccentricity is $e = 0.4$ m. Knowing that the angular velocity of the wheel is $\omega = 2.69$ rad/s (clockwise) when $\theta = 0$, calculate (a) the maximum angular velocity; and (b) the minimum angular velocity.

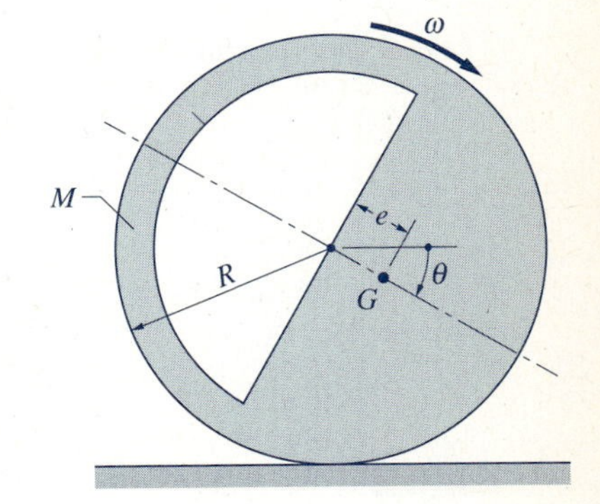

Fig. P18.23

18.24 The 24-lb uniform bar AB slides inside a smooth cylindrical surface. If the bar is released from rest in the position $\theta = 35°$, determine its angular velocity (in rad/s) and angular acceleration (in rad/s^2) as functions of θ.

18.25 End A of the uniform rod AB of mass m and length L is placed in the smooth corner between the ground and the vertical wall. The rod is released from rest at $\theta = 60°$. (a) Assuming that end A does not leave the ground, derive the angular velocity and angular acceleration of the rod as functions of θ. (b) Derive the expression for the vertical reaction at A in terms of θ, and show that A remains in contact with the ground.

18.26 The uniform 60-lb bar AB is released from rest in the position shown. Neglecting friction, determine the velocity of end A when end B reaches the corner C.

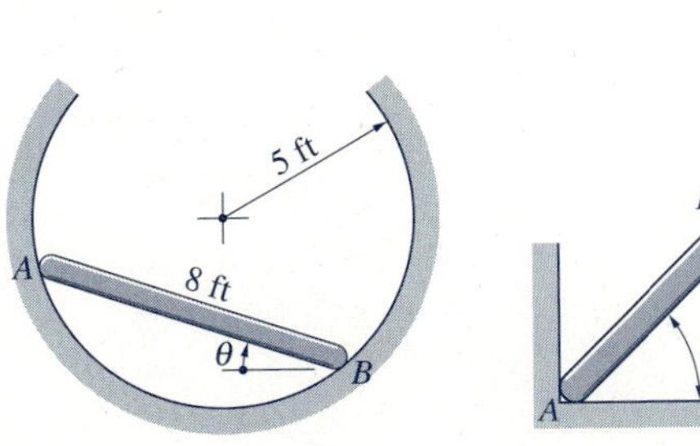

Fig. P18.24

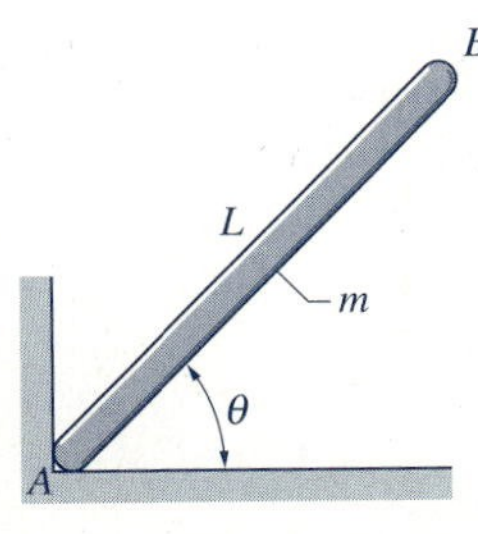

Fig. P18.25

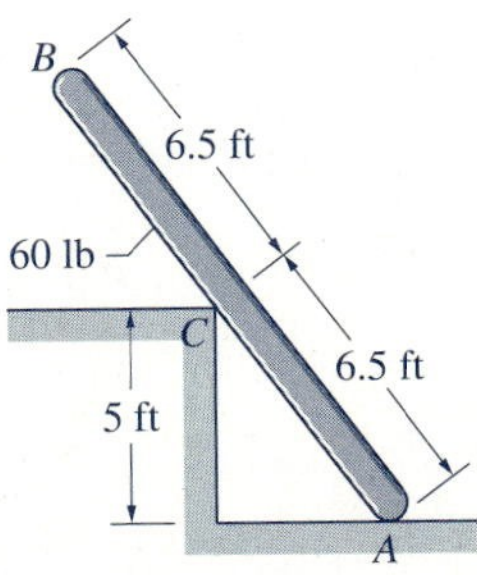

Fig. P18.26

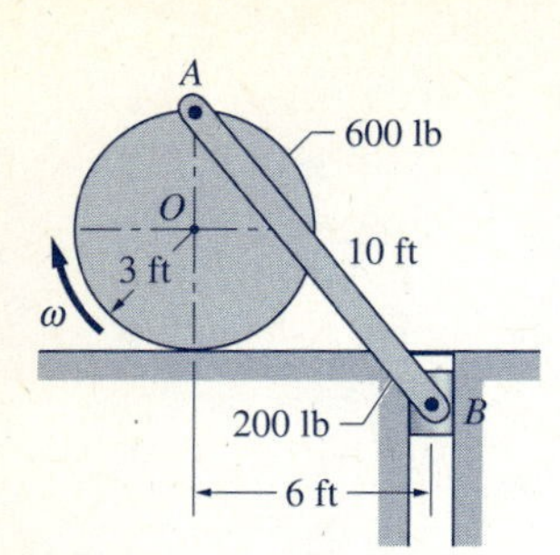

Fig. P18.27, P18.28

18.27 The 200-lb uniform bar AB is attached to the 600-lb homogeneous disk by a pin at A. End B of the bar is pinned to a slider of negligible weight. When the disk is in the position shown, its angular velocity is $\omega = 4$ rad/s. Assuming that the disk rolls without slipping, calculate the kinetic energy of the system in this position.

18.28 The 200-lb uniform bar AB is attached to the 600-lb homogeneous disk by a pin at A. The weight of the slider attached at B is negligible. The disk rolls without slipping along the horizontal plane. In the position shown, the kinetic energy of the system is 1800 lb · ft. After the disk has rolled 180° clockwise, calculate (a) the kinetic energy of the system; and (b) the angular velocity of the disk.

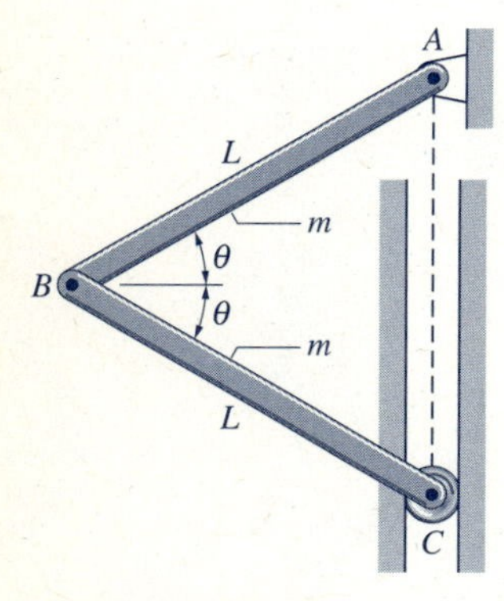

Fig. P18.29, P18.30

18.29 The linkage shown consists of two identical bars, each of length L and mass m. If the linkage is released from rest at $\theta = 0$, find the angular velocity of each link when $\theta = 90°$. Neglect friction.

18.30 Referring to Prob. 18.29, find the angular velocity of each link when $\theta = 30°$.

18.31 The mechanism consisting of two uniform rods is released from rest in the position shown. Neglecting friction, find the velocity of roller C when AB reaches the vertical position.

18.32 The mechanism consisting of two uniform rods moves in the horizontal plane. The spring connected between the massless slider A and the pin C has a stiffness of 200 N/m, and its free length is 150 mm. If the mechanism is released from rest in the position shown, determine the angular speed of rod BC when slider A is closest to C.

18.33 The parallelogram mechanism moves in the horizontal plane, driven by the constant counterclockwise couple M_0. The spring connected between pins B and D has a stiffness of 4 lb/ft and a free length of 6 in. Knowing that the angular speed of bar AB is 6 rad/s when $\theta = 0$, and 2 rad/s when $\theta = 90°$ (both counterclockwise), determine M_0.

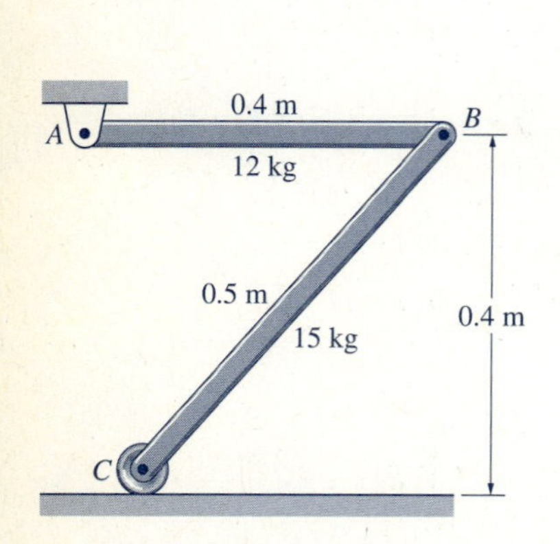

Fig. P18.31

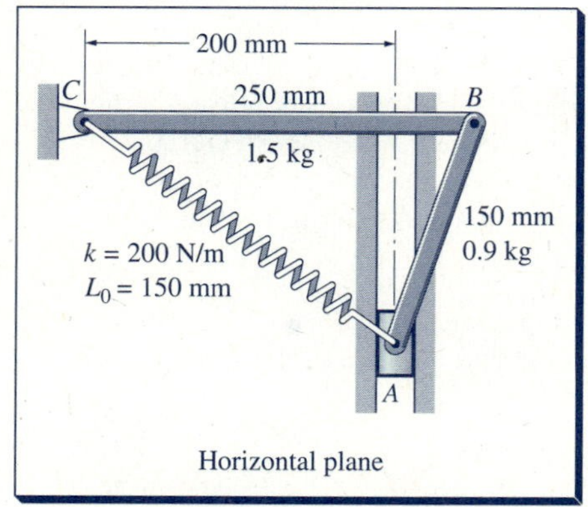

Fig. P18.32

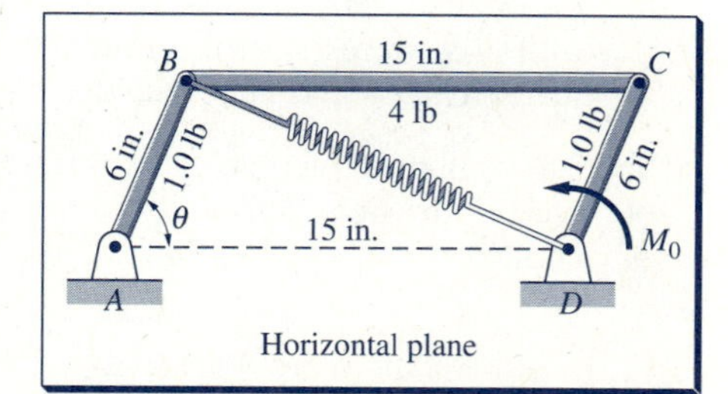

Fig. P18.33

18.34 The linkage consists of two uniform bars AB and BC of the masses and lengths shown; the mass of the slider at C is negligible. A constant counterclockwise couple M_0 acts on bar AB. Knowing that the angular velocity of AB is 8 rad/s counterclockwise when $\theta = 0$ and when $\theta = 90°$, determine the value of M_0.

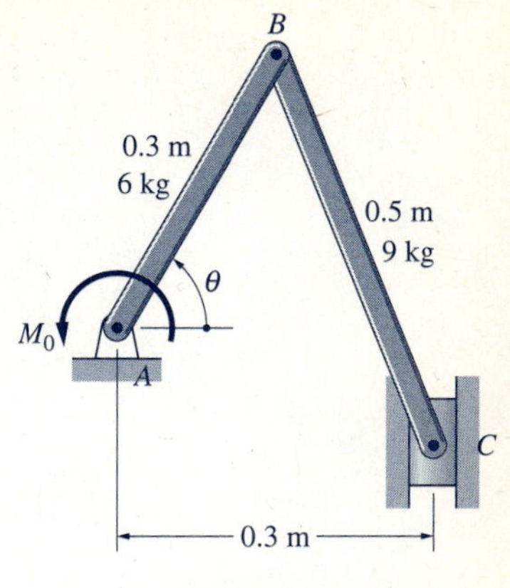

Fig. P18.34

18.35 The three bars that make up the linkage $ABCD$ are homogeneous and of equal weight. If the linkage is released from rest in the position shown, calculate the angular velocity of bar AB when it has reached the horizontal position.

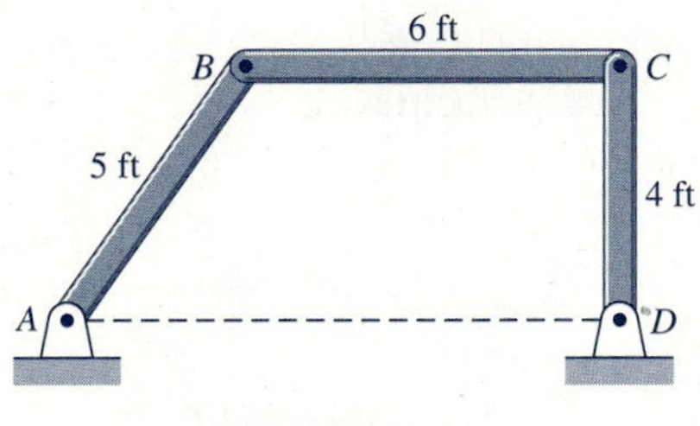

Fig. P18.35

18.36 An electric motor of 80% efficiency consumes 10 kW of power. What is the torque carried by the output shaft when the speed of the motor is (a) 2000 rev/min; and (b) 4000 rev/min.

18.37 A 25-lb flywheel with a central radius of gyration of 9 in. is driven with constant power and efficiency. If the angular acceleration of the flywheel is 2 rad/s^2 at 600 rev/min, calculate the angular acceleration at 800 rev/min.

18.38 The machine B is belt-driven by the electric motor A that has an efficiency of 85%. When the motor is running at 25 rad/s, the tension in the upper part of the belt is 20 lb greater than in the lower part. Determine the power consumption of the motor in horsepower (hp).

18.39 When the system described in Prob. 18.38 is running at 40 rad/s, the power consumption of the motor is 2.4 hp. Calculate the torque in (a) the output shaft of motor A; and (b) the input shaft of machine B.

18.40 The figure shows a plot of the power output P of a reciprocating engine versus the shaft speed ω. (a) Plot the torque (couple) developed by the engine versus ω. (b) From the plot in part (a), estimate the maximum torque and the corresponding angular speed.

18.41 The torque (couple) M developed by a turbine as a function of its angular speed ω is shown in the figure. (a) Plot the horsepower of the turbine versus ω. (b) Use the plot in part (a) to estimate the maximum power and the corresponding angular velocity.

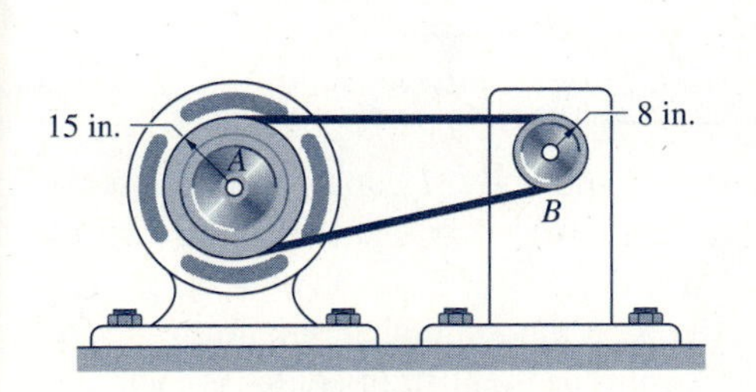

Fig. P18.38, P18.39

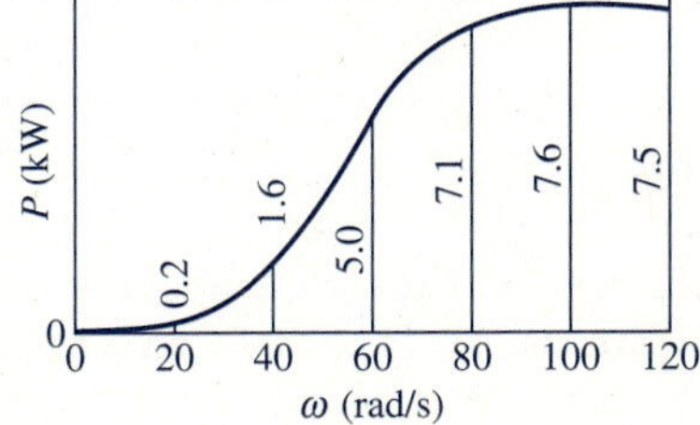

Fig. P18.40

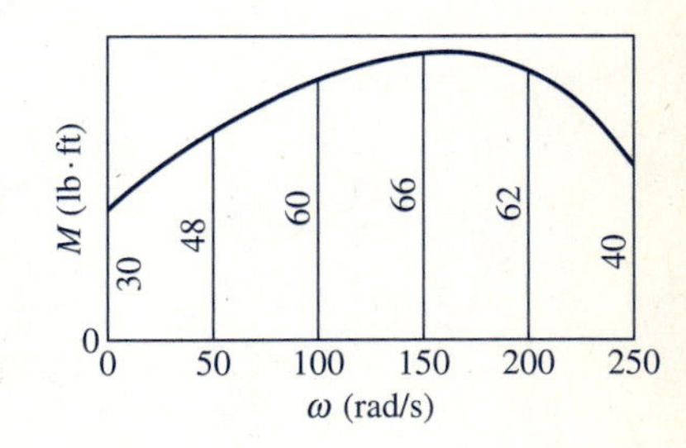

Fig. P18.41

Part B: Impulse-Momentum Method

18.5 Momentum of a Rigid Body in Plane Motion

a. Linear and angular momentum

Figure 18.3 shows a rigid body of mass m that is moving in the xy-plane with the angular velocity ω. The velocities of its mass center G and the arbitrary reference point A are denoted by $\bar{v}$ and $\mathbf{v}_A$, respectively.

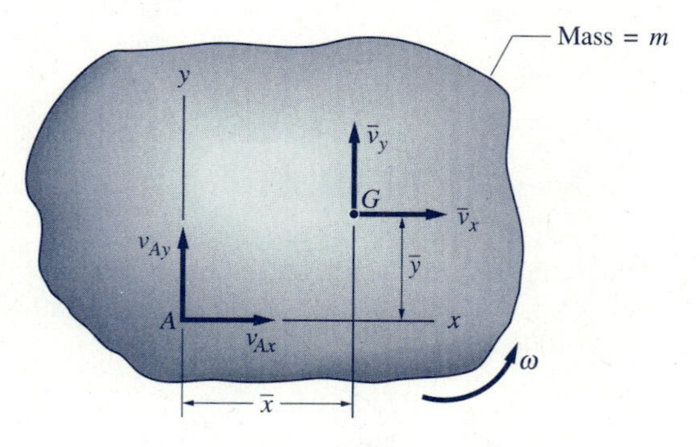

Fig. 18.3

The linear momentum of a system of particles has been discussed in Art. 15.6. From that discussion, and considering the rigid body to be a system of particles, we conclude that the linear momentum $\mathbf{p}$ of the body is

$$\mathbf{p} = m\bar{\mathbf{v}} \tag{18.11}$$

The angular momentum of a rigid body in plane motion was derived in Art. 17.3. The following repeats the different expressions that were obtained for motion in the xy-plane.

$$h_A = \bar{I}\omega + m\bar{x}\bar{v}_y - m\bar{y}\bar{v}_x \qquad (A\text{: arbitrary point}) \tag{18.12}$$

$$h_A = I_A\omega + m\bar{x}v_{Ay} - m\bar{y}v_{Ax} \qquad (A\text{: fixed in the body}) \tag{18.13}$$

$$h_A = I_A\omega \qquad (A\text{: instant center for velocities}) \tag{18.14}$$

$$h_G = \bar{I}\omega \qquad (G\text{: mass center}) \tag{18.15}$$

In Eqs. (18.12)–(18.15), m is the mass of the body; I_A and $\bar{I}$ are the moments of inertia of the body about axes parallel to the z-axis passing through A and G, respectively; and $\bar{x}$ and $\bar{y}$ are the coordinates of G, as shown in Fig. 18.3.

From Eqs. (18.12)–(18.15) we see that the angular momentum of a body is easiest to evaluate about the mass center or the instant center for velocities. However, as shown in the next section, the angular momentum about any point can be calculated easily if you employ the concept of a momentum diagram.

b. Momentum diagrams

The momentum diagram for a rigid body in plane motion is a sketch of the body showing the linear momentum vector $m\bar{\mathbf{v}}$ passing through the mass center G and the central angular momentum $\bar{I}\omega$ (a couple representing the angular momentum about G), as seen in Fig. 18.4.*

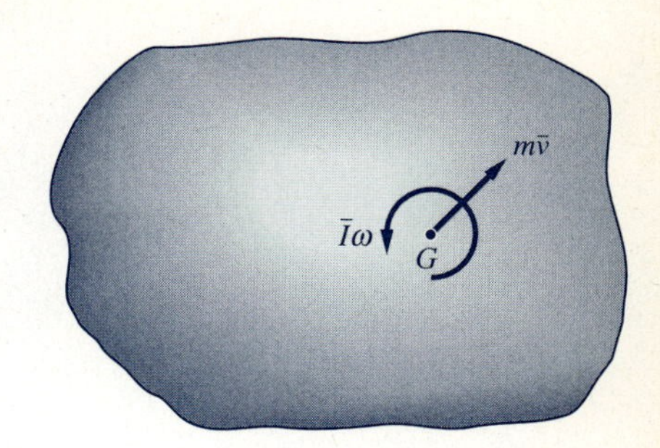

Fig. 18.4

In order to compute the angular momentum of the body about an arbitrary point A, we simply calculate the resultant moment of the momenta shown in Fig. 18.5 about A:

$$(+)\quad h_A = \bar{I}\omega + m\bar{v}_y\bar{x} - m\bar{v}_x\bar{y}$$

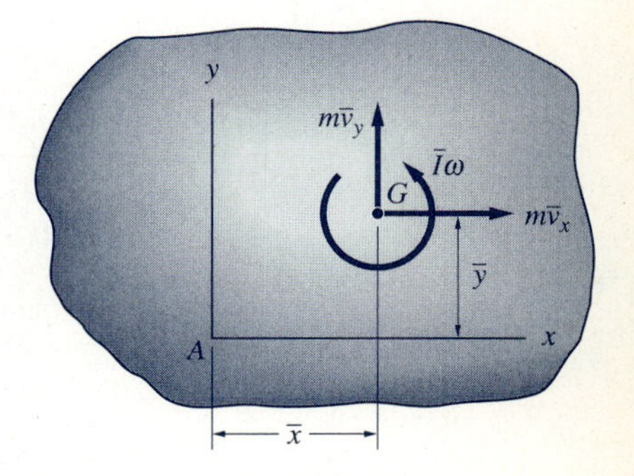

Fig. 18.5

which is identical to Eq. (18.12). We have thus demonstrated that the momentum diagram contains all of the information necessary for computing the angular momentum about any point.

It is recommended that a momentum diagram be used whenever one wishes to compute the momenta of a rigid body. Momentum diagrams not only provide a convenient pictorial representation of the linear and angular momenta of the body, but they also eliminate the need to memorize Eq. (18.12).

To illustrate the use of momentum diagrams, consider the homogeneous slender bar of mass m and length L shown in Fig. 18.6(a). The bar is rotating about the fixed point A with the counterclockwise angular velocity ω at the instant shown. The momentum diagram for the bar, shown in Fig. 18.6(b), consists of the linear momentum $m\bar{v} = m(L/2)\omega$ and the couple $\bar{I}\omega = (mL^2/12)\omega$. The resultant angular momentum (moment of the momentum) about A is

$$(+)\quad h_A = \frac{mL^2}{12}\omega + \left(m\frac{L}{2}\omega\right)\frac{L}{2} = \frac{mL^2}{3}\omega$$

Since $I_A = mL^2/3$, we see that $h_A = I_A\omega$, as would be expected, since the velocity of A is zero.

Referring again to Fig. 18.6(b), the angular momentum about B is

$$(+)\quad h_B = \frac{mL^2}{12}\omega - \left(m\frac{L}{2}\omega\right)\frac{L}{2} = -\frac{mL^2}{6}\omega$$

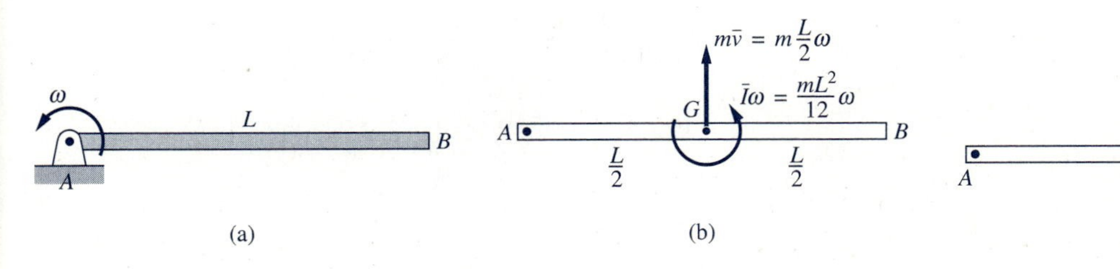

Fig. 18.6

* Since $\bar{I}\omega$ is a couple, it is free to act anywhere in the plane of the motion. However, it is common practice to show $\bar{I}\omega$ as if it acted at G.

Note that the angular momentum about B is negative, i.e., clockwise (recall that the angular velocity of the bar is counterclockwise). Furthermore, observe that h_B is not equal to $I_B\omega = (mL^2/3)\omega$. Because the velocity of B is not zero, Eq. (18.14) is clearly not applicable.

An equivalent form of the momentum diagram is shown in Fig. 18.6(c), where the momenta in Fig. (b) have been reduced to an equivalent single momentum vector $m\bar{\mathbf{v}}$ acting at point P. The distance e that locates P is found from the condition $h_p = 0$; i.e., $m\bar{v}e - \bar{I}\omega = 0$. Substituting $\bar{v} = (L/2)\omega$ and $\bar{I} = mL^2/12$, we obtain $e = L/6$. If a body rotates about a fixed axis, a point about which the angular momentum of the body is zero, such as P in Fig. 18.6(c), is called a *center of percussion.*

18.6 Impulse-Momentum Principles

a. Impulse-momentum relations

Assuming that a rigid body consists of a large number of particles, our previous discussions of impulse and momentum for particle systems can be applied to rigid-body motion. The equations developed in this article are valid for three-dimensional motion, but we restrict the applications to plane motion, with one notable exception.

Repeating the results obtained for particle systems in Art. 14.6, the linear impulse-momentum equation for rigid-body motion is

$$\mathbf{L}_{1-2} = \mathbf{p}_2 - \mathbf{p}_1 = \Delta\mathbf{p} \tag{18.16}$$

where $\mathbf{L}_{1-2}$ represents the linear impulse of external forces acting on the body during the time interval from time t_1 to time t_2, and $\Delta\mathbf{p}$ is the change in the linear momentum of the body during the same time interval. The linear momentum of the body is

$$\mathbf{p} = m\bar{\mathbf{v}} \tag{18.17}$$

where m is the mass of the body and $\bar{\mathbf{v}}$ is the velocity of its mass center.

Applying the angular impulse-momentum equation for systems of particles (Art. 14.7) to rigid-body motion, we have

$$(\mathbf{A}_A)_{1-2} = (\mathbf{h}_A)_2 - (\mathbf{h}_A)_1 = \Delta\mathbf{h}_A \quad (A\text{: fixed point* or mass center}) \tag{18.18}$$

where $(\mathbf{A}_A)_{1-2}$ is the angular impulse of the external forces about point A for the time interval t_1 to t_2, and $\Delta\mathbf{h}_A$ is the change in the angular momentum of the rigid body about A during the same period. Note That Eqs. (18.12)–(18.18) are also valid for three-dimensional motion of a rigid body.

* Point A is fixed in space; it is not necessarily a point in the body.

The preceding results can also be applied to systems of rigid bodies provided that (1) the impulses refer to only forces that are external to the system, (2) the momenta are interpreted as the sum of the momenta of the bodies that constitute the system, and (3) the mass center referred to in Eq. (18.18) is the mass center of the system.

As is the case with systems of particles, the linear momentum for a rigid body (or system of rigid bodies) is conserved when the linear impulse of the external forces is zero. Similarly, the angular momentum of a rigid body (or system of rigid bodies) about a point is conserved when the angular impulse of the external forces about that point is zero. The reference point must either be fixed in space or be the mass center, since Eq. (18.18) is valid only if A is restricted to these points. Of course, the best method for determining whether the linear or angular impulse about a point vanishes is to examine the free-body diagram of the body or system of bodies.

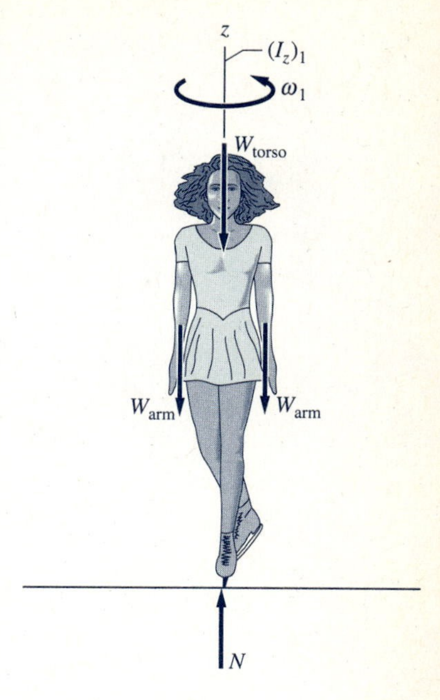

Initial position

b. Special case of three-dimensional motion

Here we examine a special case of three-dimensional motion that can be solved by planar analysis. Figure 18.7 shows two positions of a skater who is spinning freely about the vertical z-axis that passes through the skater's center of gravity. In the initial position, the skater is holding her arms against her body, and, in the final position, she has raised her arms to the horizontal position. The spin velocities of the skater about the z-axis in the two positions are ω_1 and ω_2, respectively, and the corresponding moments of inertia about the z-axis are denoted by $(I_z)_1$ and $(I_z)_2$. The reaction N shown in the drawings is the force exerted on the skater by the smooth skating surface. (The magnitude of N is the same in both positions because the mass center of the skater is stationary in each position.)

Observe that the skater is undergoing plane motion in both the initial and final positions. However, the motion of each arm is not planar as it moves from its initial to its final position. Therefore, we have here a special case where three-dimensional motion is necessary to move from one planar motion into another planar motion.

From Fig. 18.7 we see that the angular impulse is zero about the z-axis, since all the forces are always parallel to that axis. Because the z-axis is fixed in space, it follows that angular momentum about the z-axis is conserved throughout the motion. Therefore, it is permissible to equate the angular momenta about the z-axis in the initial and final positions, yielding

$$(I_z)_1\omega_1 = (I_z)_2\omega_2$$

In this example we have been able to apply the conservation of angular momentum equation in a two-dimensional setting (the motion in both the initial and final positions is planar), although the motion between these two positions is three-dimensional.

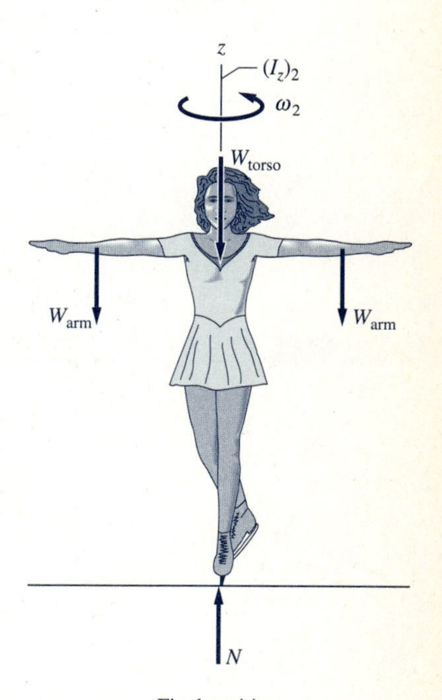

Final position

Fig. 18.7

Sample Problem 18.6

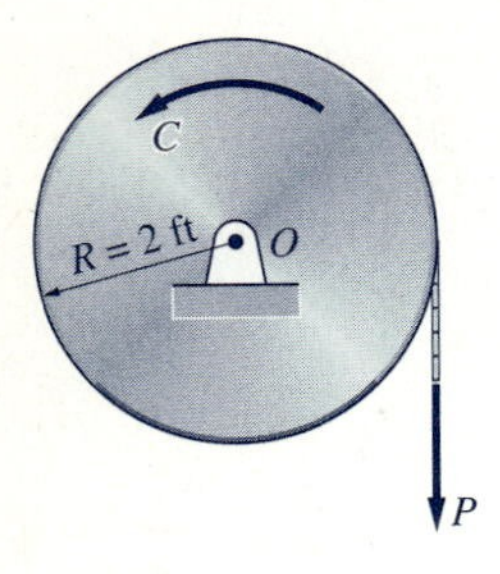

The homogeneous disk weighing 300 lb is rotating about the smooth pin at O. The disk is acted on by the constant downward force P (applied to a rope wound around the disk) and the counterclockwise couple $C = 6t^3 - 9t^2 + 4t$ lb · ft, where t is the time in seconds. Find P, knowing that the resultant angular impulse acting on the disk about O is zero during the time interval from $t = 4$ s to $t = 6$ s.

Solution

The resultant angular impulse about O equals the sum of the angular impulses about O for all of the forces that act on the disk. Noting that the pin reaction and the weight of the disk have no angular impulse about O (they pass through O), we need to calculate only the angular impulses of the force P and the couple C.

Since the force P is constant in magnitude and direction, its angular impulse about O for the time interval $\Delta t = t_2 - t_1 = 6 - 4 = 2$ s is

$$\circlearrowleft + \quad (A_O)_{1-2} = -RP\Delta t = -(2)(P)(2) = -4P \text{ lb} \cdot \text{ft} \cdot \text{s} \qquad \text{(a)}$$

where P is measured in pounds.

The angular impulse of C is found by integrating the expression for C over the given time interval.

$$\begin{aligned} \circlearrowleft + \quad (A_O)_{1-2} &= \int_{t_1}^{t_2} C\,dt = \int_4^6 (6t^3 - 9t^2 + 4t)\,dt \\ &= 1.5t^4 - 3t^3 + 2t^2\Big]_4^6 = 1368 - 224 \\ &= 1144 \text{ lb} \cdot \text{ft} \cdot \text{s} \qquad \text{(b)} \end{aligned}$$

Since the resultant angular impulse about O is known to be zero, Eqs. (a) and (b) yield $1144 - 4P = 0$, which gives

$$P = 286 \text{ lb} \qquad \textit{Answer}$$

Sample Problem 18.7

When the unbalanced 40-kg wheel is in the position shown in Fig. (a), its angular velocity is 8 rad/s clockwise. Assuming that the wheel is rolling without slipping, calculate its angular momentum about the following points: the mass center G, the geometric center O, and the point of contact C.

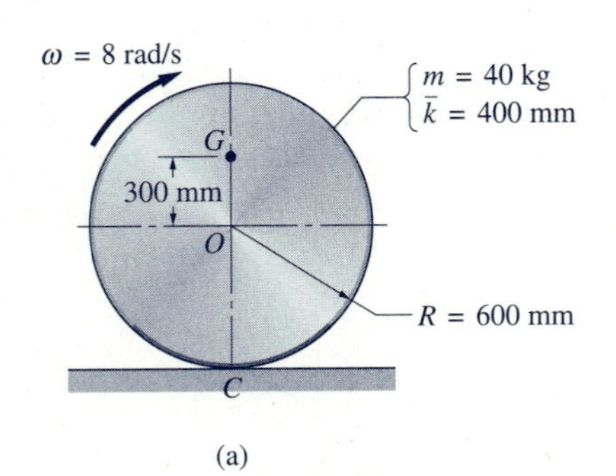

(a)

Solution

The momentum diagram of the wheel, shown in Fig. (b), consists of the central angular momentum $\bar{I}\omega$ and the linear momentum vector $m\bar{\mathbf{v}}$ passing through G. The numerical values shown in the figure have been calculated as described here.

Because the wheel rolls without slipping, its instant center for velocities is the contact point C. Therefore, $\bar{v} = \overline{CG}\omega = 0.9(8) = 7.200$ m/s, directed to the right (the direction is consistent with the clockwise direction of ω). Hence the magnitude of the linear momentum vector is $m\bar{v} = 40(7.200) = 288.0$ N · s. Using $\bar{I} = m\bar{k}^2 = 40(0.400)^2 = 6.400$ kg · m^2, the magnitude of the central angular momentum is $\bar{I}\omega = 6.400(8) = 51.20$ N · m · s; its clockwise direction is determined by the direction of ω.

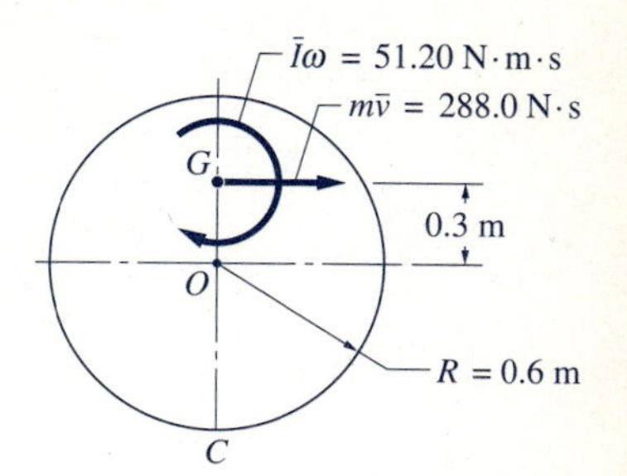

(b) Momentum diagram

The angular momentum of the wheel about any point can now be computed by summing the moments of the momenta about that point. Therefore, referring to Fig. (b), we obtain

$$\circlearrowright + \quad h_G = \bar{I}\omega = 51.20 \text{ N} \cdot \text{m} \cdot \text{s} \qquad \textit{Answer}$$

$$\circlearrowright + \quad h_O = 51.20 + (288.0)(0.3) = 137.6 \text{ N} \cdot \text{m} \cdot \text{s} \qquad \textit{Answer}$$

$$\circlearrowright + \quad h_C = 51.20 + (288.0)(0.9) = 310.4 \text{ N} \cdot \text{m} \cdot \text{s} \qquad \textit{Answer}$$

Since $v_C = 0$, we could have computed the last of these values using $h_C = I_C\omega$—see Eq. (18.14). With $I_C = \bar{I} + md^2 = 6.400 + 40(0.9)^2 = 38.80$ N · m^2, we find that $h_C = I_C\omega = 38.80(8) = 310.4$ N · m · s, which is identical to the result found. The use of the special equation $I_C\omega$ for the angular momentum about the instant center is, of course, a matter of personal preference. (You may find it instructive to show that $h_O \neq I_O\omega$ for this problem.)

Sample Problem 18.8

Figure (a) shows two uniform gears A and B that are supported by pins at D and E, respectively. The gears are stationary when the constant counterclockwise couple C_0 is applied at time $t = 0$. Determine the value of C_0 for which the angular velocity of gear A will be 50 rad/s when $t = 5$ s, and determine the corresponding tangential contact force between the gears.

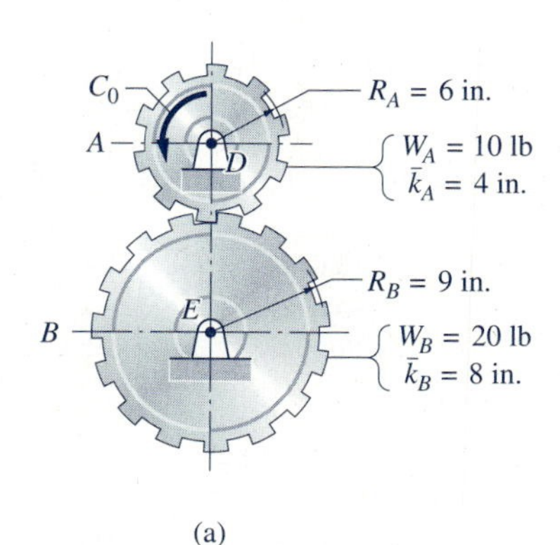

(a)

Solution

Figure (b) shows the FBDs, and the momentum diagrams at $t = 0$ and $t = 5$ s for each gear. The FBDs contain the weights of the gears, the applied couple C_0, the tangential contact force F, and the support reactions. (Since C_0 is constant, F is also constant.) Note that each momentum diagram contains only the central angular momentum $\bar{I}\omega$, because each gear rotates about its mass center; i.e., $m\bar{\mathbf{v}} = \mathbf{0}$.

The force-mass-acceleration (FMA) method, described in Chapter 17, or the impulse-momentum method may be used with equal facility in solving this problem. The FMA method would be rather convenient here, because the forces, and therefore the accelerations, are constant. The impulse-momentum method, which we will employ, is equally convenient, since we are required to calculate the change in velocity that occurs during a given time interval.

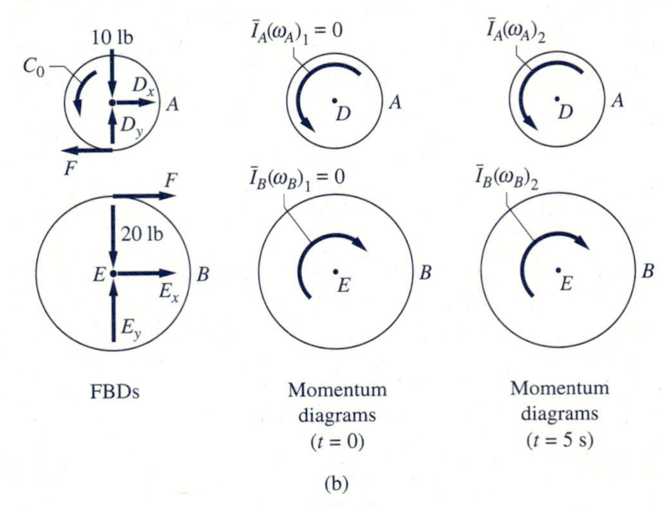

(b)

The central moments of inertia of the gears are $\bar{I}_A = m_A\bar{k}_A^2 = (10/32.2) \times (4/12)^2 = 0.0345$ slug · ft² and $\bar{I}_B = m_B\bar{k}_B^2 = (20/32.2)(8/12)^2 = 0.2761$ slug · ft². From Fig. (b) the angular impulse-momentum equation for the gear A about point D is

$$(A_D)_{1-2} = \Delta h_D$$

$$\circlearrowleft + \quad C_0(\Delta t) - FR_A\Delta t = \bar{I}_A[(\omega_A)_2 - (\omega_A)_1]$$

$$C_0(5) - F\left(\frac{6}{12}\right)(5) = 0.0345(50 - 0) \tag{a}$$

Note that the angular impulses were easy to compute, because C_0 and F are constant in both magnitude and direction.

Observing that the point of contact on each gear has the same velocity, we conclude that $6\omega_A = 9\omega_B$, which gives $(\omega_B)_2 = (2/3)(\omega_A)_2 = (2/3)(50) = 33.33$ rad/s. Therefore, from Fig. (b), the angular impulse-momentum equation for gear B about point E is

$$(A_E)_{1-2} = \Delta h_E$$

$$\circlearrowright + \quad FR_B\Delta t = \bar{I}_B[(\omega_B)_2 - (\omega_B)_1]$$

$$F\left(\frac{9}{12}\right)(5) = 0.2761(33.33 - 0) \tag{b}$$

From Eq. (b) we find that the contact force is

$$F = 2.454 \text{ lb} \qquad \textit{Answer}$$

Substituting this value into Eq. (a), we obtain

$$C_0 = 1.572 \text{ lb} \cdot \text{ft} \qquad \textit{Answer}$$

Sample Problem 18.9

Figure (a) shows a rod B of mass m_B that is placed inside a tube A of mass m_A. Both bodies are slender, homogeneous, and of length L. Initially, B is held inside A ($a = 0$) by a lightweight cap (not shown) that covers the end of A while the assembly is spinning freely with the angular velocity ω_1 about the z-axis. The cap then falls off, allowing the rod to slide out of the tube. Determine the angular velocity ω_2 of the assembly when the rod is fully out of the tube ($a = L$). Neglect friction.

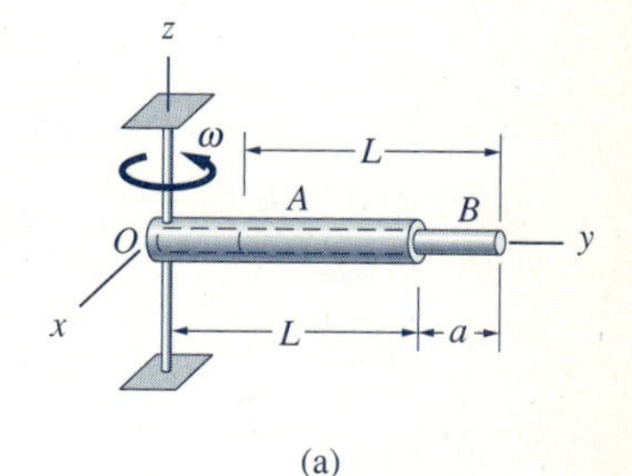

(a)

Solution

Free-Body Diagram (FBD)

Figure (b) shows the FBD of the assembly when the rod B extends an arbitrary distance a beyond the end of the tube A. This FBD shows only the forces that act in the xy-plane. The complete FBD would also include the weights of the bodies, a reaction O_z, and a reactive couple C_x acting at O. However, since the weights, O_z, and C_x have no effect on the motion in the xy-plane, they have been omitted. It should be noted that the contact forces between the tube and rod do not appear on the FBD because they are internal to the system. This is the primary reason why we choose to analyze the motion of the entire assembly, instead of considering each body separately.

FBD (arbitrary position)

Initial Momentum Diagram

Figure (b) also shows the initial momentum diagram, with rod B being entirely inside tube A, before the cap falls off. Note that this diagram includes the angular and linear momentum vectors of A and B. The angular velocities of both bodies are ω_1, so that the velocities of their mass centers are both equal to $(L/2)\omega_1$.

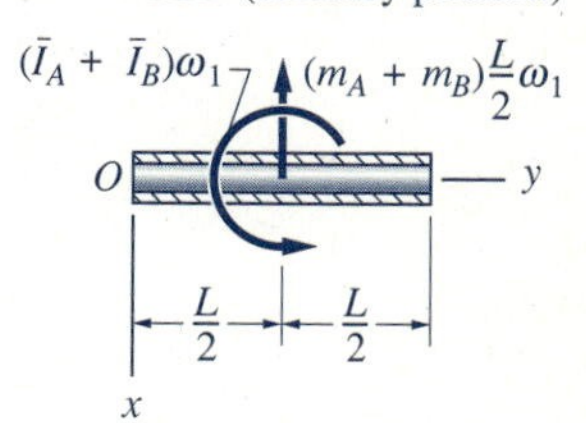

Initial momentum diagram

Final Momentum Diagram

In the final momentum diagram, rod B is about to leave the tube A with the relative velocity $v_{B/A}$. The angular velocities of the two bodies are both equal to ω_2, giving rise to the angular momenta shown. The linear momentum vector of A follows from the fact that the velocity of its mass center is $(L/2)\omega_2$. The two components of the linear momentum vector of B correspond to the polar components of the velocity of its mass center, namely $v_R = v_{B/A}$ and $v_\theta = (3L/2)\omega_2$.

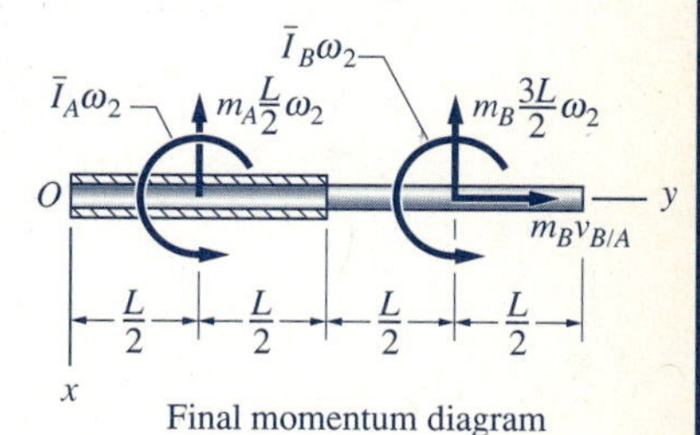

Final momentum diagram

(b)

Impulse-Momentum Analysis

The remainder of the analysis consists of writing down and solving the impulse-momentum equations using the three diagrams shown in Fig. (b). Since we are interested only in ω_2, the most convenient solution uses the angular impulse-momentum equation with O as the reference point, thereby eliminating the unknown reactions O_x and O_y. Because the FBD in Fig. (b) is valid throughout the duration of the motion, we conclude that $(A_O)_{1-2} = 0$ (O_x and O_y do not contribute to the angular impulse about O). Noting that O is a fixed point, we have $(A_O)_{1-2} = \Delta h_O$, which in our case gives $\Delta h_O = 0$. Therefore, angular momentum of the system about O is conserved.

Equating the moments of the momenta about O in the two momentum diagrams shown in Fig. (b), we get

$$(h_O)_1 = (h_O)_2$$

$$\circlearrowleft + \quad (\bar{I}_A + \bar{I}_B)\omega_1 + (m_A + m_B)\frac{L}{2}\omega_1\left(\frac{L}{2}\right)$$

$$= \left[\bar{I}_A\omega_2 + \left(m_A\frac{L}{2}\omega_2\right)\left(\frac{L}{2}\right)\right] + \left[\bar{I}_B\omega_2 + \left(m_B\frac{3L}{2}\omega_2\right)\left(\frac{3L}{2}\right)\right]$$

Substituting $\bar{I} = mL^2/12$ for each bar and solving for ω_2, we obtain

$$\omega_2 = \frac{m_A + m_B}{m_A + 7m_B}\omega_1 \qquad \textit{Answer}$$

It should be mentioned that this solution would be valid even if there were friction between the rod and tube—provided, of course, that the coefficient of friction were small enough to permit relative motion between them.

Sample Problem 18.10

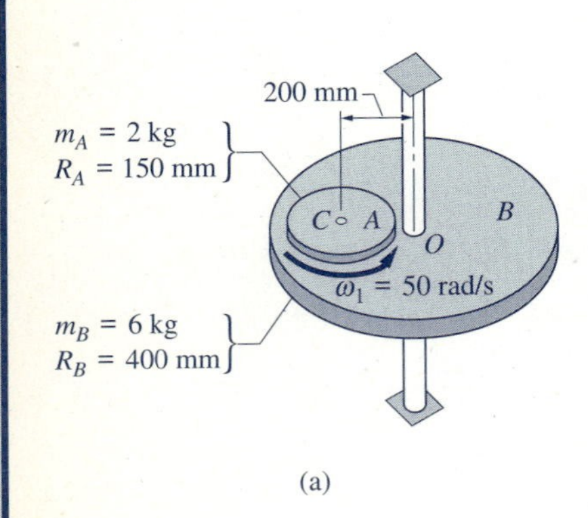

(a)

Figure (a) shows the 6-kg disk B that rotates in a horizontal plane supported by smooth bearings at O. The 2-kg disk A is mounted eccentrically onto B by means of a smooth axle at C. Both disks are homogeneous. Disk A is rotating with the angular velocity $\omega_1 = 50$ rad/s, and B is at rest when an internal brake is applied to the axle at C, causing A to slow down and B to start rotating. Determine the angular velocity ω_2 of the disks after the relative motion between them has stopped.

Solution

Figure (b) contains the FBD of the system, which shows only the forces that act in the plane of the motion—namely, O_x and O_y—two components of the bearing reaction at O. The forces at bearing C and the braking couple do not appear on the FBD because they are internal to the system. The weights of the disks, being perpendicular to the plane of the motion, also do not appear.

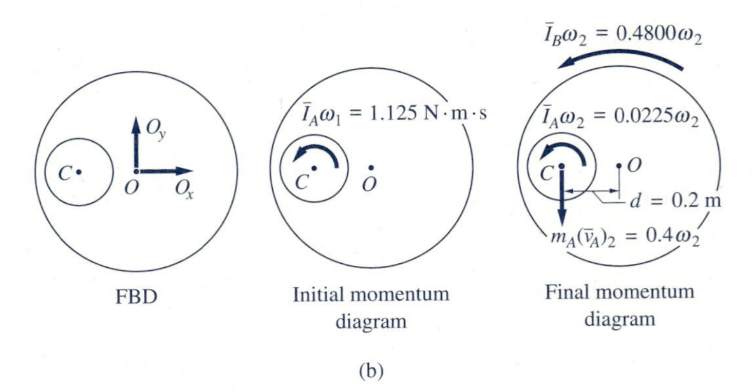

(b)

Figure (b) also shows the initial momentum diagram (before the brake is applied) and the final momentum diagram (after the relative motion has stopped). The central moments of inertia of the disks are $\bar{I}_A = m_A R_A^2/2 = 2(0.150)^2/2 = 0.0225$ kg · m^2 and $\bar{I}_B = m_B R_B^2/2 = 6(0.400)^2/2 = 0.4800$ kg · m^2. Using these results, the momenta in Fig. (b) are

$$\bar{I}_A\omega_1 = 0.0225(50) = 1.125 \text{ N} \cdot \text{m} \cdot \text{s}$$

$$\bar{I}_A\omega_2 = 0.0225\omega_2 \text{ N} \cdot \text{m} \cdot \text{s}$$

$$\bar{I}_B\omega_2 = 0.4800\omega_2 \text{ N} \cdot \text{m} \cdot \text{s}$$

$$m_A(\bar{v}_A)_2 = m_A d\omega_2 = 2(0.2)\omega_2 = 0.4\,\omega_2 \text{ N} \cdot \text{s}$$

where ω_2 is measured in rad/s.

From the FBD in Fig. (b) we see that the angular impulse about O is zero throughout the motion since the only external forces, O_x and O_y, pass through O. Because O is a fixed point, $(A_O)_{1-2} = \Delta h_O$ is applicable in this case, yielding $\Delta h_O = 0$ (the angular momentum about O is conserved), or

$$(h_O)_1 = (h_O)_2$$

Referring to Fig. (b) to compute the moments of the momenta about O, we get

$$\circlearrowleft + \quad 1.125 = [0.0225\omega_2 + 0.4\omega_2(0.2)] + 0.4800\omega_2$$

which yields

$$\omega_2 = 1.931 \text{ rad/s} \qquad \textit{Answer}$$

Sample Problem 18.11

The assembly shown in Fig. (a) consists of an arm AOC, to which are pinned two homogeneous slender rods AB and CD. The assembly rotates about the z-axis in a smooth bearing at O. An internal mechanism (not shown in the figure) can position and lock the two rods at any angle θ. The moment of inertia of the arm AOC about the z-axis is 0.8 slug · ft^2, and rods AB and CD weigh 3 lb each. Initially the assembly is rotating freely about the z-axis with the angular velocity $\omega_1 = 10$ rad/s with $\theta = 0$. Calculate the angular velocity of the assembly when the rods have been raised to the position $\theta = 90°$.

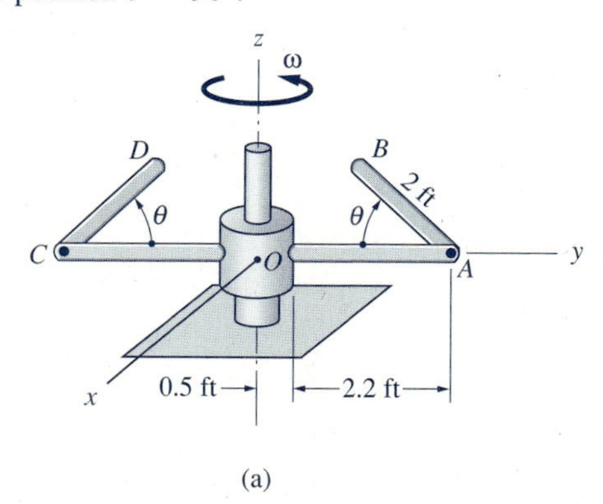

(a)

Solution

Figure (b) contains the FBD of the assembly, showing only forces that act in the xy-plane—namely, O_x and O_y, two components of the bearing reaction at O. Note that this FBD does not change during the motion of the assembly. Also shown in Fig. (b) are the initial ($\theta = 0$) and final ($\theta = 90°$) momentum diagrams of the assembly. The numerical values shown in the momentum diagrams have been computed as described in the following.

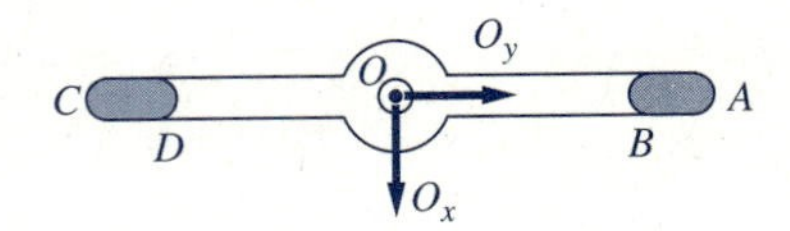

FBD

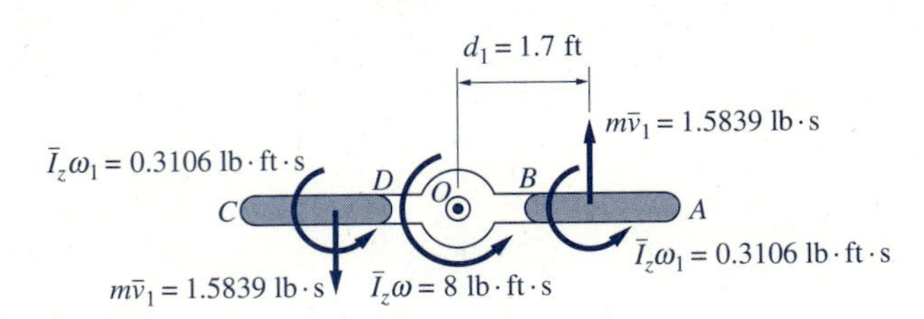

Initial momentum diagram

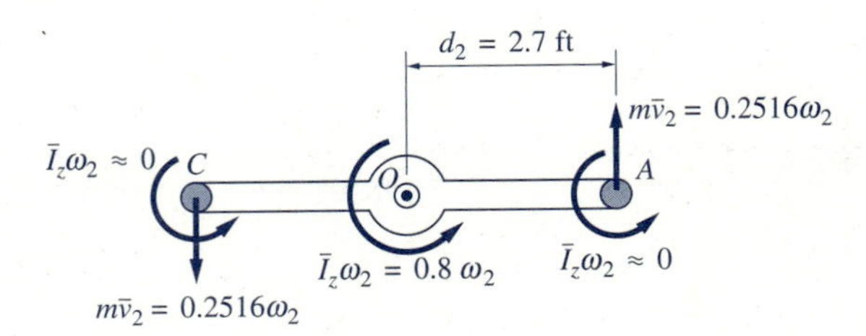

Final momentum diagram

(b)

Initial Momentum Diagram

Arm AOC: $\bar{I}_z\omega_1 = 0.8(10) = 8$ lb · ft · s

Rod AB or CD (both rods lie in the xy-plane):

$$\bar{I}_z = \frac{mL^2}{12} = \frac{1}{12}\left(\frac{3}{32.2}\right)(2)^2 = 0.031\,06 \text{ slug} \cdot \text{ft}^2$$

$$\bar{I}_z\omega_1 = (0.031\,06)(10) = 0.3106 \text{ lb} \cdot \text{ft} \cdot \text{s}$$

$$d_1 = 2.2 + 0.5 - 1.0 = 1.7 \text{ ft}$$

$$m\bar{v}_1 = md_1\omega_1 = \left(\frac{3}{32.2}\right)(1.7)(10) = 1.5839 \text{ lb} \cdot \text{s}$$

Final Momentum Diagram

Arm AOC: $\bar{I}_z\omega_2 = 0.8\omega_2$ lb · ft · s

Rod AB or BC (both rods are perpendicular to the xy-plane):

$$\bar{I}_z \approx 0$$

$$d_2 = 0.5 + 2.2 = 2.7 \text{ ft}$$

$$m\bar{v}_2 = md_2\omega_2 = \left(\frac{3}{32.2}\right)(2.7)\omega_2 = 0.2516\omega_2 \text{ lb} \cdot \text{s}$$

Impulse-Momentum Analysis

The motion of the assembly between the initial and final positions is the special case that is discussed in Art. 18.6: Both the initial and final motions are two-dimensional, although the motion between these two positions is three-dimensional. From the FBD in Fig. (b) we see that there is no angular impulse about the z-axis throughout the motion, which means that angular momentum of the system about that axis is conserved. Referring to the momentum diagrams in Fig. (b), we obtain

$$(h_z)_1 = (h_z)_2$$

$$\circlearrowleft + \quad 8 + 2[0.3106 + 1.7(1.5839)] = 0.8\omega_2 + 2(0.2516\omega_2)(2.7)$$

which gives

$$\omega_2 = 6.49 \text{ rad/s} \qquad \textit{Answer}$$

PROBLEMS

18.42 The magnitude and the direction of the force P_0 are constant, but its point of application C moves along the x-axis with the constant speed v_0. Calculate the following for the time during which C travels from A to B: (a) the angular impulse of the force about A; and (b) the linear impulse of the force.

18.43 The force of constant magnitude P_0 has a fixed point of application B. The line of action of the force rotates with the constant clockwise angular speed ω_0. For the period during which the line of action rotates from $\theta = 0$ to $\theta = 90°$, determine (a) the angular impulse of the force about A; and (b) the linear impulse of the force.

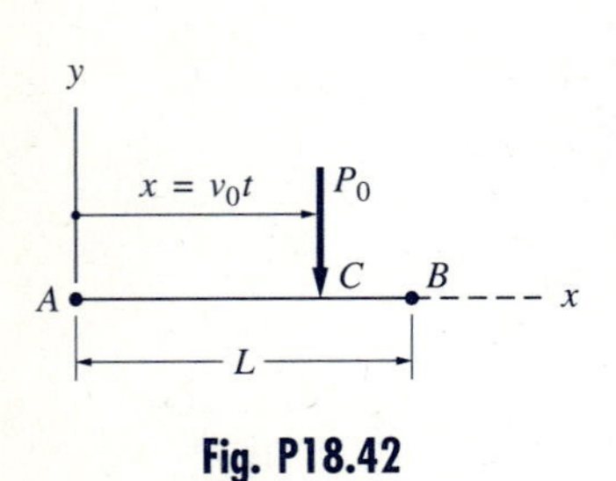

Fig. P18.42

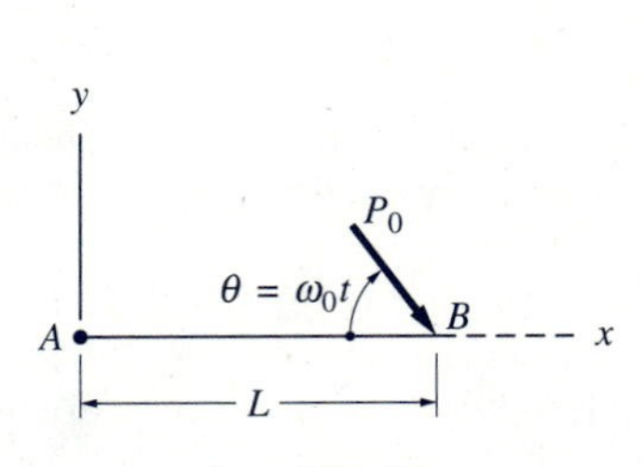

Fig. P18.43

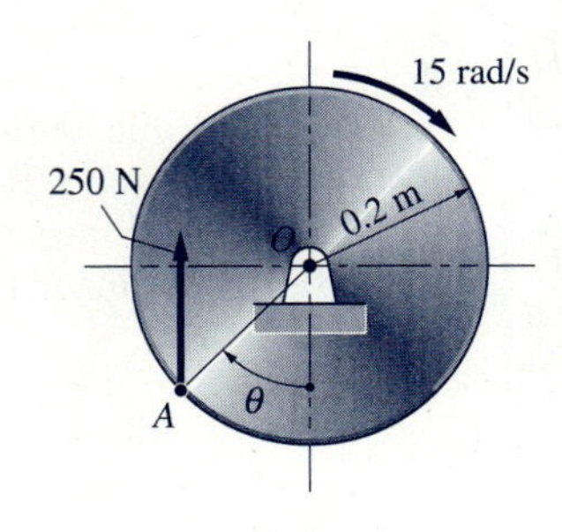

Fig. P18.44

18.44 The disk of radius 0.2 m rotates about O with a constant angular velocity of 15 rad/s. The constant vertical force of magnitude 250 N acts at point A that is fixed on the rim of the disk. Determine the angular impulse of the force about O for the period during which the disk rotates from the position $\theta = 0$ to $\theta = 90°$.

18.45 The T-shaped body is made of two identical uniform rods, each of mass $m/2$. If the body rotates about O with the angular velocity ω, calculate its angular momentum about (a) the mass center of the body; (b) point O; and (c) point A.

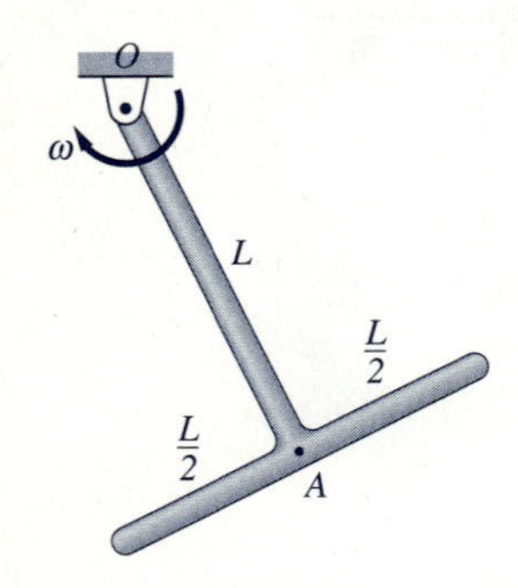

Fig. P18.45

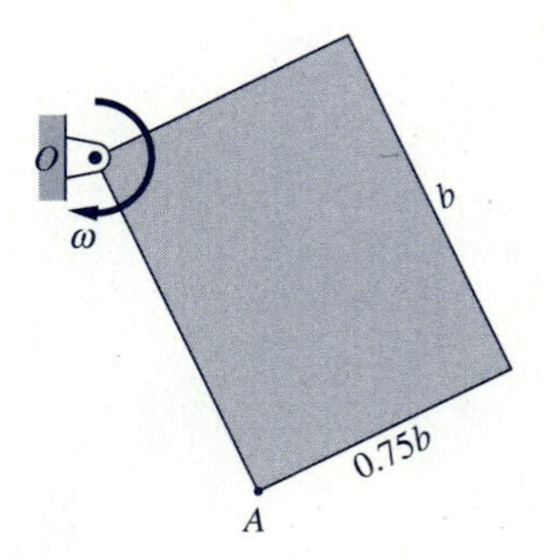

Fig. P18.46

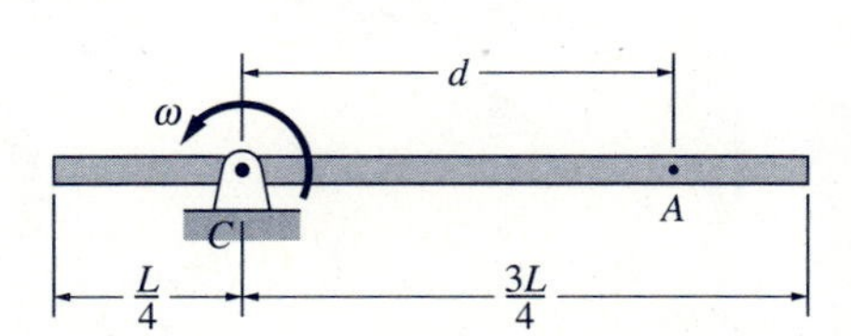

Fig. P18.47

18.46 The uniform plate of mass m rotates about its corner O with the angular speed ω. Determine the angular momentum of the plate about (a) the mass center; (b) the corner O; and (c) the corner A.

18.47 The uniform bar of mass m and length L rotates about the pin at C with the angular velocity ω. Find the distance d that locates the center of percussion A of the bar (the point about which the angular momentum is zero).

18.48 The uniform disk of mass m and radius R rotates about the pin at C, located a distance d from its center. Determine d so that the center of percussion will be at point A on the rim of the disk (the point about which the angular momentum is zero).

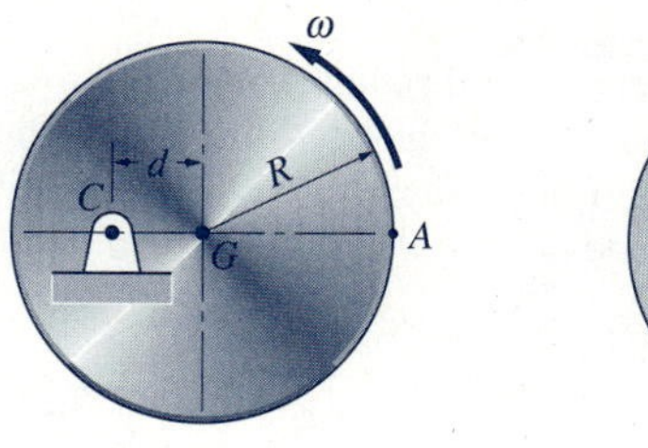

Fig. P18.48

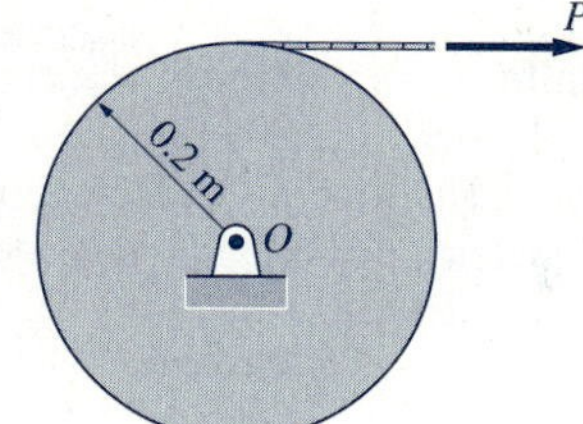

Fig. P18.49

18.49 The 20-kg drum has a central radius of gyration of 0.16 m. The end of a rope wrapped about the drum is pulled with the constant force P. Determine P that would cause the drum to accelerate from rest to 300 rev/min in 2 s.

18.50 Torque M acting on the input shaft of a generator varies with time t as $M(t) = M_0 \exp(-t/t_0)$, where $M_0 = 5.2$ N · m and $t_0 = 4.8$ s. If the generator is at rest at $t = 0$, calculate (a) its terminal angular speed; and (b) the maximum power input and the time when it occurs. Assume that the generator rotates without resistance, and use $\bar{I} = 0.84$ kg · m^2 for its rotor.

18.51 The uniform disk of mass m and radius R is initially at rest on a smooth horizontal surface. The constant force of magnitude P is applied at time $t = 0$ to the cord wrapped around the disk. For an arbitrary time t, determine the expression for the velocity of B (the point on the rim where the cord leaves the disk).

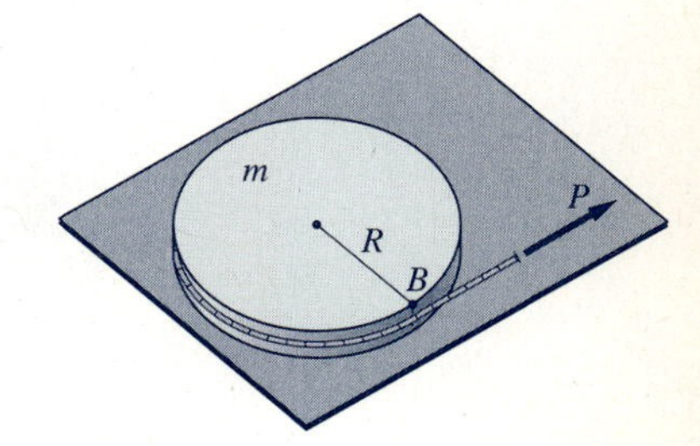

Fig. P18.51

18.52 A cylinder and a sphere are released simultaneously from rest on an inclined surface. Calculate the ratio of their central velocities, $\bar{v}_{cyl}/\bar{v}_{sph}$, at an arbitrary time after the release. Assume that both bodies are homogeneous and roll without slipping.

18.53 The 36-lb homogeneous disk A is attached to the 12-lb uniform rod BC by a bearing at B, the axis of which is vertical. The rod rotates freely about the vertical axis at C. The system is at rest when the constant couple $M_0 =$ 4 lb · ft is applied to the rod. Determine the angular velocity of rod BC after 2 s, assuming that the bearing at B is (a) free to turn; and (b) locked.

18.54 The homogeneous disk of mass m and radius R is at rest when it is lowered onto the conveyor belt that is being driven at the constant speed v_0. The coefficient of kinetic friction between the disk and the belt is μ, and the mass of arm BC is negligible. Derive the expression for the time that it takes for the disk to reach its final angular velocity.

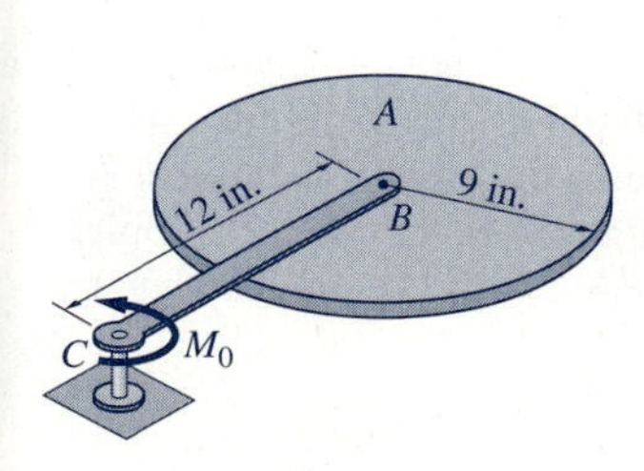

Fig. P18.53

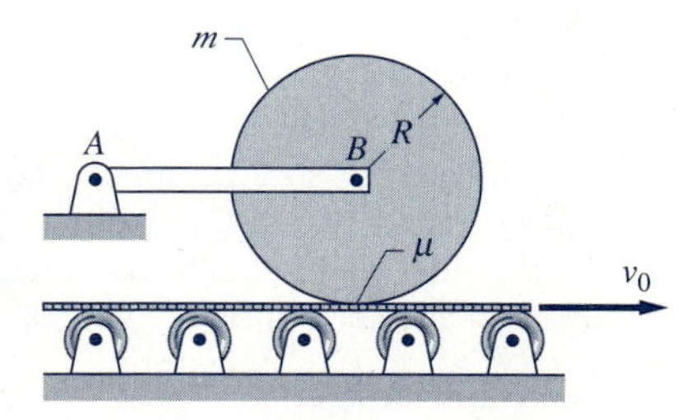

Fig. P18.54

18.55 The 2-kg upper disk is initially at rest when it is brought into contact with the 4-kg lower disk that is spinning freely at 500 rev/min. Find the time during which there is slipping between the disks, and the final angular speed of each disk. Assume that both disks are homogeneous, and neglect the weight of arm *AD*.

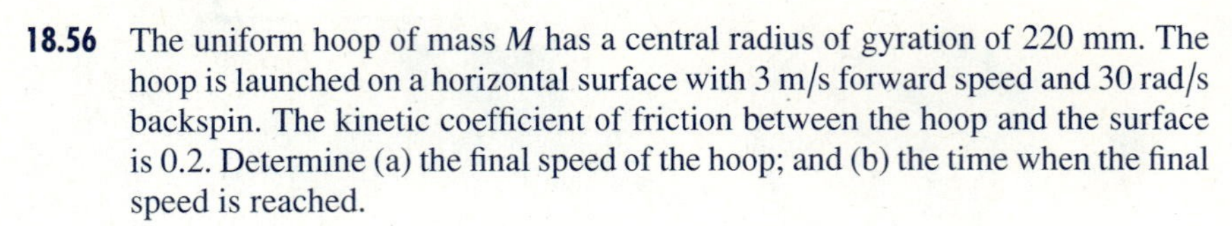
18.56 The uniform hoop of mass *M* has a central radius of gyration of 220 mm. The hoop is launched on a horizontal surface with 3 m/s forward speed and 30 rad/s backspin. The kinetic coefficient of friction between the hoop and the surface is 0.2. Determine (a) the final speed of the hoop; and (b) the time when the final speed is reached.

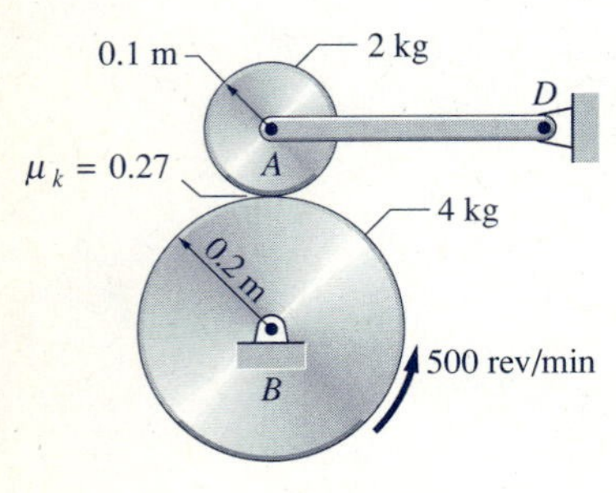

Fig. P18.55

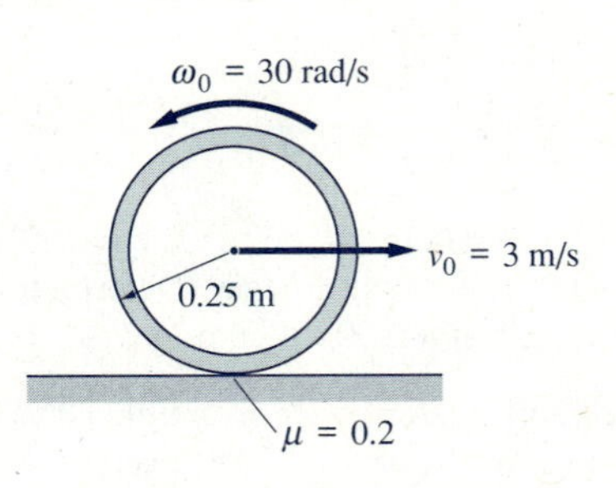

Fig. P18.56

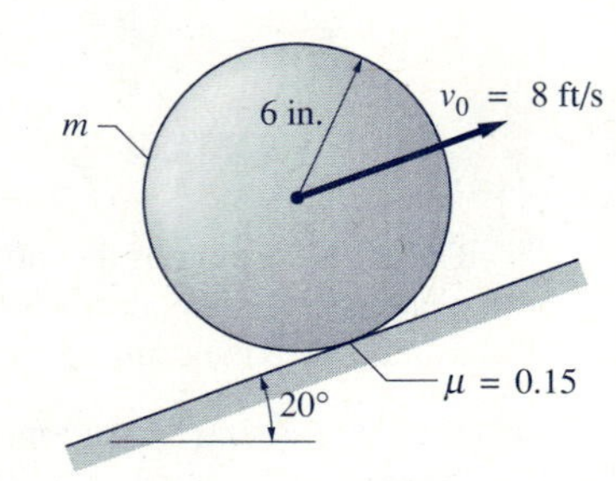

Fig. P18.57

18.57 The homogeneous, solid ball of mass *m* is launched on the inclined plane at time $t = 0$ with the forward speed of 8 ft/s and no spin. The kinetic coefficient of friction between the ball and the plane is 0.15. (a) Calculate the time when the ball stops slipping on the plane and the angular velocity of the ball at that time. (b) Find the time when the ball comes to a stop.

18.58 The mass moment of inertia of the 60-kg spool about its mass center *G* is 1.35 $kg \cdot m^2$. The spool is at rest on the rough surface when the constant force $P = 200$ N is applied to the cable that is wound around its hub. Calculate the angular velocity of the spool 2 s later, assuming that the static as well as the kinetic coefficient of friction between the spool and the surface is (a) $\mu = 0.2$; and (b) $\mu = 0.1$.

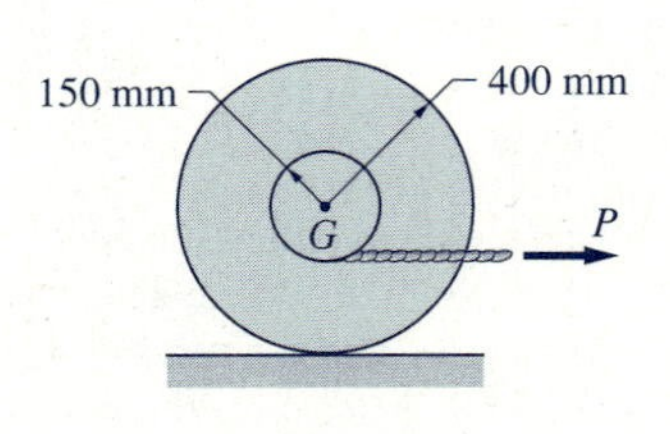

Fig. P18.58

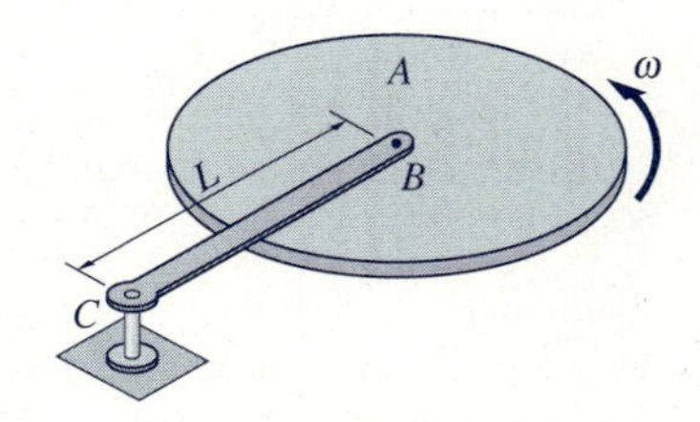

Fig. P18.59

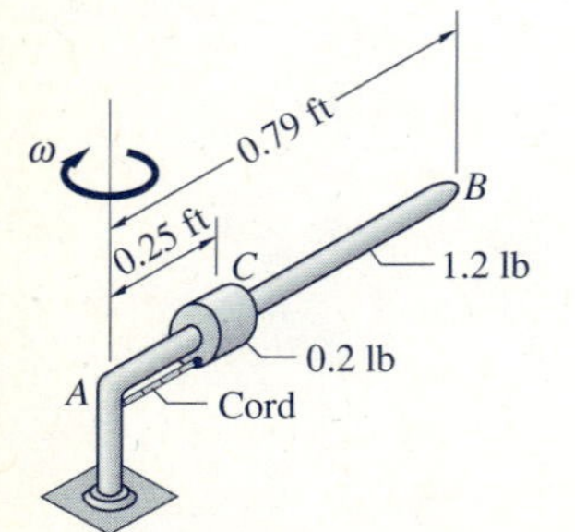

Fig. P18.60

18.59 The uniform disk *A* has a mass *m*, and its radius of gyration about the vertical axis at *B* is $\bar{k}$. The weight of arm *BC* is negligible. The disk is spinning freely with the angular speed ω on the stationary arm *BC* when a brake that is built into the hub of the disk (not shown in the figure) is activated. Determine the final angular speed of arm *BC* in terms of ω, $\bar{k}$, and *L*. Assume that the arm is free to rotate about the vertical axis at *C*.

18.60 The 1.2-lb uniform rod *AB* and the 0.2-lb small slider *C* rotate freely about the vertical axis at *A*. The angular velocity of the system is 5 rad/s when the cord holding *C* breaks. Determine the angular velocity of *AB* when (a) *C* is just about to leave the rod; and (b) just after *C* has left the rod.

18.61 The pipe AB of inner diameter 10 mm is rotating without friction about the vertical axis at A. Water flows through the pipe with the constant speed 1.8 m/s relative to the pipe. Find the couple that must be applied to the pipe in order to maintain its angular speed at 6 rad/s.

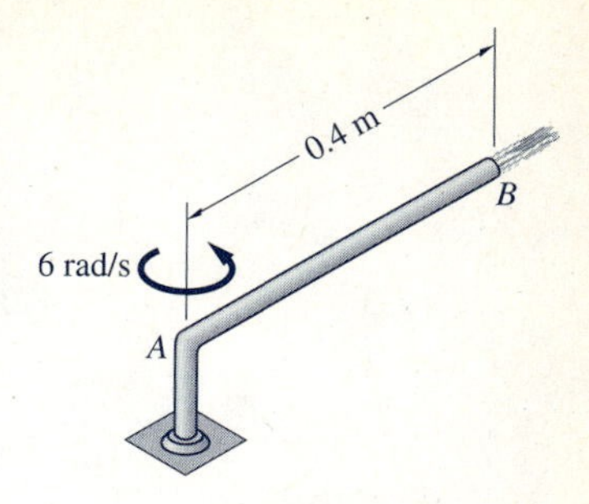

Fig. P18.61

18.62 Each of the uniform pulleys weighs 18 lb and has a central radius of gyration of 0.9 ft. Pulley A rotates about a fixed axis, whereas pulley B is suspended from a belt that runs around both pulleys. A constant couple C_0 acts on pulley A. Recognizing that the force in the belt cannot be compressive, determine (a) the shortest possible time in which the pulleys can be accelerated from rest to 1000 rev/min; and (b) the corresponding value of C_0.

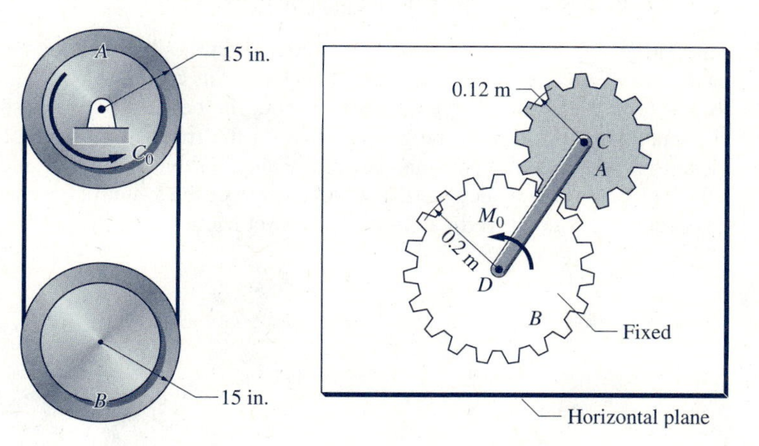

Fig. P18.62

Fig. P18.63

18.63 The 4-kg gear A has a central radius of gyration of 90 mm. A constant couple M_0 acts on the arm CD of negligible weight, causing the arm to accelerate from 200 rev/min to 320 rev/min in 2.5 s. Note that gear B is fixed and that the system lies in the horizontal plane. Calculate (a) the tangential contact force between the gears; and (b) the magnitude of M_0.

18.64 The cart consists of the 12-kg body A and four wheels attached to axles at B and C. The two wheels connected to axle B are homogeneous disks of mass 3 kg each, and the masses of the other two wheels are negligible. If the constant force $P = 15$ N is applied when the cart is at rest, find its speed 2 s later. Assume that the wheels do not slip.

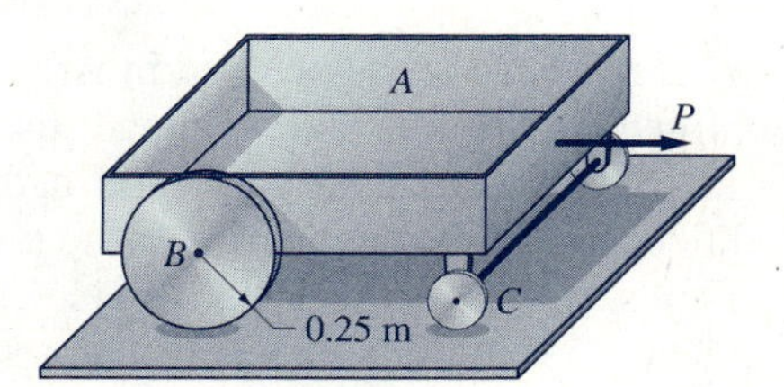

Fig. P18.64

18.65 The combined weight of the rider and the bicycle without its wheels is 172 lb. Each wheel weighs 5 lb, which is due primarily to the weights of the rim and the tire. If the bicycle starts from rest, determine its speed after 10 s, assuming that the chain provides a constant couple of 15 lb · ft on the rear wheel and that the wheels do not slip on the ground.

Fig. P18.65

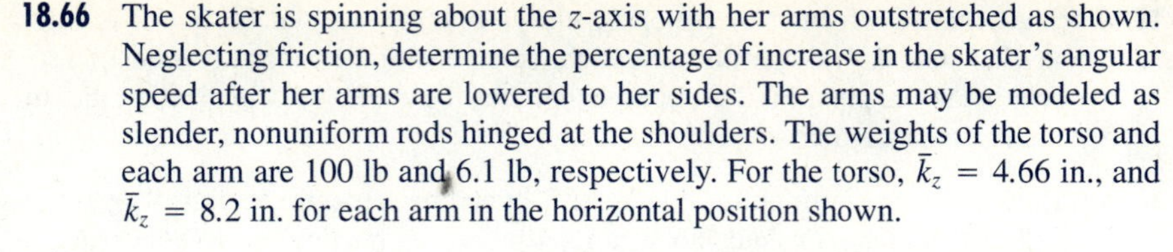

18.66 The skater is spinning about the z-axis with her arms outstretched as shown. Neglecting friction, determine the percentage of increase in the skater's angular speed after her arms are lowered to her sides. The arms may be modeled as slender, nonuniform rods hinged at the shoulders. The weights of the torso and each arm are 100 lb and 6.1 lb, respectively. For the torso, $\bar{k}_z = 4.66$ in., and $\bar{k}_z = 8.2$ in. for each arm in the horizontal position shown.

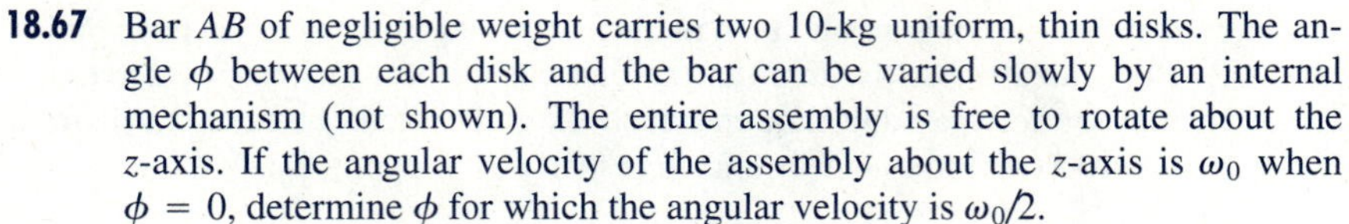

18.67 Bar AB of negligible weight carries two 10-kg uniform, thin disks. The angle ϕ between each disk and the bar can be varied slowly by an internal mechanism (not shown). The entire assembly is free to rotate about the z-axis. If the angular velocity of the assembly about the z-axis is ω_0 when $\phi = 0$, determine ϕ for which the angular velocity is $\omega_0/2$.

18.68 The 250-lb thin rectangular plate A is mounted on the U-shaped frame B by bearings at C and D. The frame weighs 160 lb, and its radius of gyration about the z-axis is 1.25 ft. An internal mechanism in the bearing at C (not shown) can change the position of the plate relative to the frame. When the plate is locked in the horizontal position shown, the angular velocity of the assembly is $\omega = 120$ rev/min. Calculate the angular speed of the system after the plate has been rotated to the vertical position. Neglect friction.

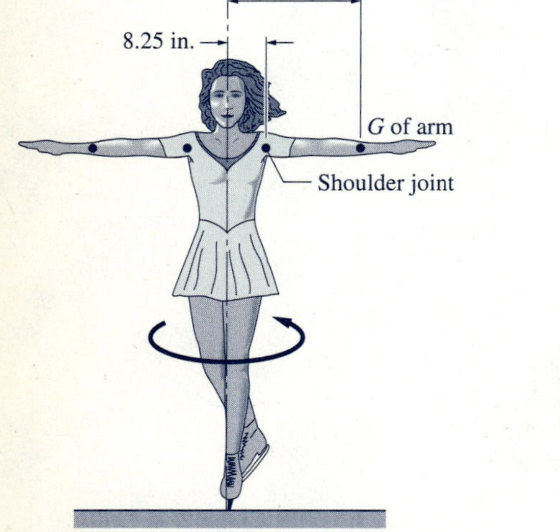

Fig. P18.66

Fig. P18.67

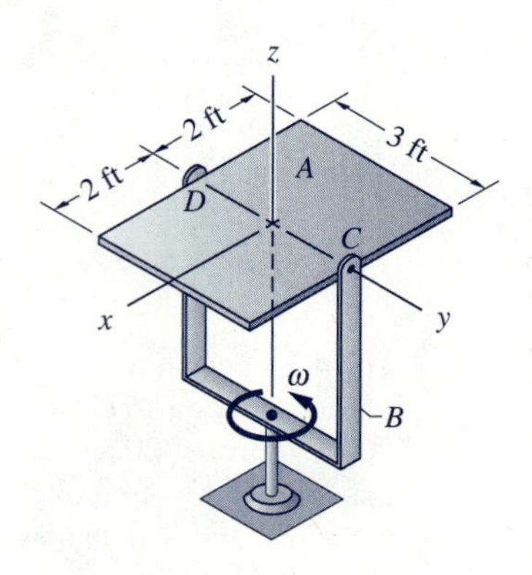

Fig. P18.68

18.7 Rigid-Body Impact

The impact of systems of particles was discussed in Arts. 15.8–15.10, where we introduced a simplified analysis of elastic impact that utilized the coefficient of restitution, an experimental constant. Rigid-body impact is a more complex problem that depends on the geometries of the impacting bodies and their surface characteristics, as well as their relative velocities. Any attempt to extend the concept of a constant coefficient of restitution to rigid-body impact greatly oversimplifies the real problem and can render the results meaningless. Therefore, we will not consider problems of rigid-body impact that require the use of an experimental constant analogous to the coefficient of restitution.

A useful simplification arises in the analysis of rigid-body impact when the motion is assumed to be impulsive, meaning that the duration of the impact is negligible—see Art. 15.9. As we have stated before, the expression "angular impulse equals change in angular momentum" is, in general, valid only about the mass center or a fixed point. However, for impulsive motion, "angular impulse equals change in angular momentum" is valid about *all* points. The reason for this simplification is that by assuming the time of impact to be infinitesimal, we are neglecting all displacements during the impact. Consequently, all points are, in effect, fixed during the impact. A formal proof of this assertion appears as follows.

We begin by recalling the moment equation of motion for a system of particles, which also applies to a rigid body:

$$(\Sigma \mathbf{M}_A)_{\text{ext}} = \dot{\mathbf{h}}_A + \mathbf{v}_A \times m\bar{\mathbf{v}} \qquad (A\text{: arbitrary point}) \qquad (15.34, \text{repeated})$$

Substituting $\dot{\mathbf{h}}_A = d\mathbf{h}_A/dt$ and $\bar{\mathbf{v}} = d\bar{\mathbf{r}}/dt$, where $\bar{\mathbf{r}}$ is the position vector of the mass center relative to an inertial reference frame, multiplying both sides by dt, and integrating over the time of impact, we obtain

$$\int_{t_1}^{t_2} (\Sigma \mathbf{M}_A)_{\text{ext}}\, dt = \int_{(\mathbf{h}_A)_1}^{(\mathbf{h}_A)_2} d\mathbf{h}_A + m\mathbf{v}_A \times \int_{\bar{\mathbf{r}}_1}^{\bar{\mathbf{r}}_2} d\bar{\mathbf{r}} \qquad \text{(a)}$$

The subscripts 1 and 2 refer to the values of the variables immediately before and immediately after the impact, respectively. Equation (a) may be rewritten as

$$(\mathbf{A}_A)_{1\text{–}2} = \Delta\mathbf{h}_A + m\mathbf{v}_A \times \int_{\bar{\mathbf{r}}_1}^{\bar{\mathbf{r}}_2} d\bar{\mathbf{r}} \qquad \text{(b)}$$

where $(\mathbf{A}_A)_{1\text{–}2}$ is the angular impulse of the external forces. Since the assumption of impulsive motion implies that the body occupies the same spatial position before, during, and after the impact, $\bar{\mathbf{r}}$ may be considered to be constant; i.e., $d\bar{\mathbf{r}} = \mathbf{0}$. Therefore the integral in Eq. (b) vanishes, and we are left with

$$(\mathbf{A}_A)_{1\text{–}2} = \Delta\mathbf{h}_A \qquad (A\text{: arbitrary point for impulsive motion}) \qquad (18.16)$$

The general steps in the analysis of rigid-body impact problems parallel those given for particle impact in Art. 15.9.

Step 1: Draw the FBD of the impacting bodies and/or of the system of impacting bodies. Identify the impulsive forces—use a special symbol, such as a caret (^) to label each impulsive force. (It is advisable to redraw the FBD, showing only the impulsive forces.)

Step 2: Draw the momentum diagrams for the bodies at the instant immediately before impact.

Step 3: Draw the momentum diagrams for the bodies at the instant immediately after impact.

Step 4: Using the diagrams drawn in Steps 1–3, derive and solve the appropriate impulse-momentum equations for the individual bodies and/or the system of bodies.

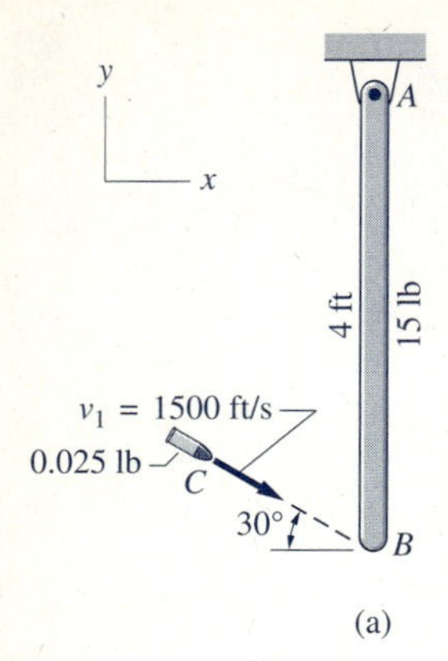

(a)

Sample Problem 18.12

Figure (a) shows a 0.025-lb bullet C that is fired at end B of the 15-lb homogeneous slender bar AB. The bar is initially at rest, and the initial velocity of the bullet is $v_1 = 1500$ ft/s, directed as shown. Assuming that the bullet becomes embedded in the bar, calculate (1) the angular velocity ω_2 of the bar immediately after the impact; (2) the impulse exerted on the bar at A during the impact; and (3) the percentage loss of energy as a result of the impact. Neglect the duration of the impact.

Solution

Introductory Comments

Because the time of impact is negligible, the motion is impulsive, with the bar and bullet occupying essentially the same positions before, during, and after the impact.

The FBDs for the bullet and bar during the impact are shown in Fig. (b), with only the impulsive forces (denoted with carets) shown. Because the weights of C and AB are finite forces, they will not enter into the impact analysis and consequently are omitted from the FBD. Observe that the FBD contains $\hat{B}_x$ and $\hat{B}_y$, the components of the impulsive contact force at B, and $\hat{A}_x$ and $\hat{A}_y$, the components of the impulsive pin reaction at A.

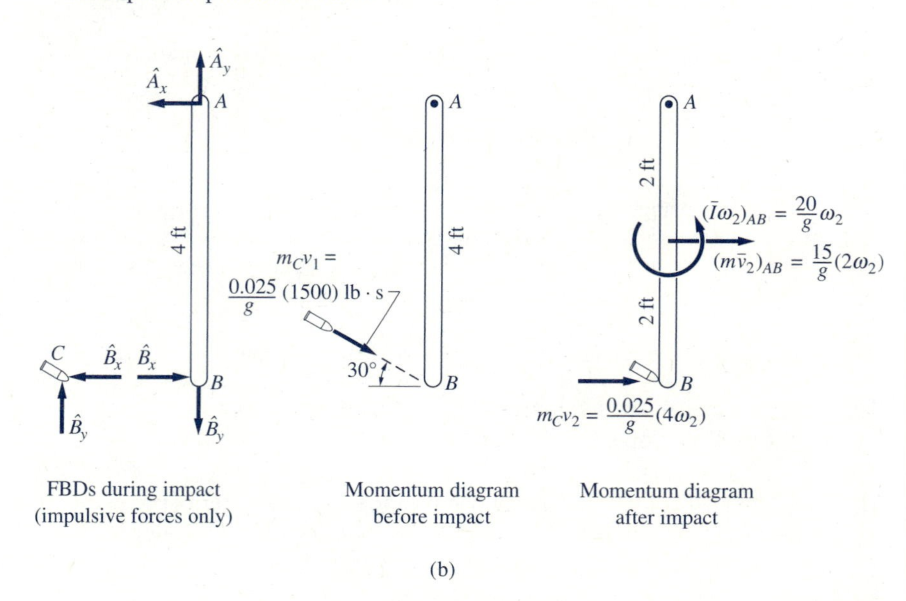

(b)

Figure (b) also contains the momentum diagrams immediately before and after the impact. The momentum diagram before impact contains only the initial linear momentum of the bullet C.

The momentum diagram after the impact contains both the final momentum of the bullet and the linear and angular momentum of the bar AB (the angular velocity ω_2 is assumed to be measured in rad/s). Note that the kinematic relationships used in the diagram, $v_2 = 4\omega_2$ and $(\bar{v}_2)_{AB} = 2\omega_2$, follow from the fact that A is a fixed point. Also, the central moment of inertia of AB is $\bar{I}_{AB} = mL^2/12 = (1/12)(15/g)(4)^2 = 20/g$ slug · ft², which gives $(\bar{I}\omega_2)_{AB} = (20/g)\omega_2$ lb · ft · s, as indicated in the diagram.

Inspection of Fig. (b) reveals that there are a total of five unknowns: the impulses of $\hat{A}_x$, $\hat{A}_y$, $\hat{B}_x$, and $\hat{B}_y$, and the angular velocity ω_2. There are also a total of five independent impulse-momentum equations: two for the bullet C and three for the bar AB. Therefore all five unknowns can be determined from the five independent equations. However, since we are required to find only three of the unknowns, ω_2 and the impulses of $\hat{A}_x$ and $\hat{A}_y$, it will not be necessary to use all the equations.

Part 1

The most efficient means of computing ω_2 is to consider that the system consists of both the bullet and the rod, as opposed to considering each of them separately. For the system, $\hat{B}_x$ and $\hat{B}_y$ are internal forces; consequently, $\hat{A}_x$ and $\hat{A}_y$ are the only external forces. The angular impulse acting on the system about A is, therefore, zero, which leads us to conclude that angular momentum about A is conserved. Referring to the momentum diagrams in Fig. (b), we obtain for the system

$$(h_A)_1 = (h_A)_2$$

$$\circlearrowleft + \quad \frac{0.025}{g}(1500\cos 30^\circ)(4) = \left[\frac{20}{g}\omega_2 + \frac{15}{g}(2\omega_2)(2)\right] + \left[\frac{0.025}{g}(4\omega_2)(4)\right]$$

Solving for ω_2 yields

$$\omega_2 = 1.616 \text{ rad/s} \qquad \textit{Answer}$$

Part 2

As mentioned previously, $\hat{A}_x$ and $\hat{A}_y$ are the only external forces that act on the system during the impact, because $\hat{B}_x$ and $\hat{B}_y$ are internal forces. Referring to Fig. (b), the x-component of the linear impulse-momentum equation for the system is

$$(L_x)_{1-2} = (p_x)_2 - (p_x)_1$$

$$\xrightarrow{+} \quad -\int \hat{A}_x\, dt = \left[\frac{15}{g}(2\omega_2) + \frac{0.025}{g}(4\omega_2)\right] - \left[\frac{0.025}{g}(1500\cos 30^\circ)\right]$$

Substituting $\omega_2 = 1.616$ rad/s from the solution to Part 1 gives

$$\int \hat{A}_x\, dt = -0.502 \text{ lb}\cdot\text{s} \qquad \text{(a)}$$

The negative sign means, of course, that the direction of the impulse of $\hat{A}_x$ is opposite to the direction of $\hat{A}_x$ assumed in the FBD.

The y-component of the linear impulse-momentum equation for the system is

$$(L_y)_{1-2} = (p_y)_2 - (p_y)_1$$

$$+\uparrow \quad \int \hat{A}_y\, dt = 0 - \left[-\frac{0.025}{g}(1500\sin 30^\circ)\right]$$

or

$$\int \hat{A}_y\, dt = 0.582 \text{ lb}\cdot\text{s} \qquad \text{(b)}$$

From the results given in Eqs. (a) and (b), the resultant impulse acting on the bar at A is

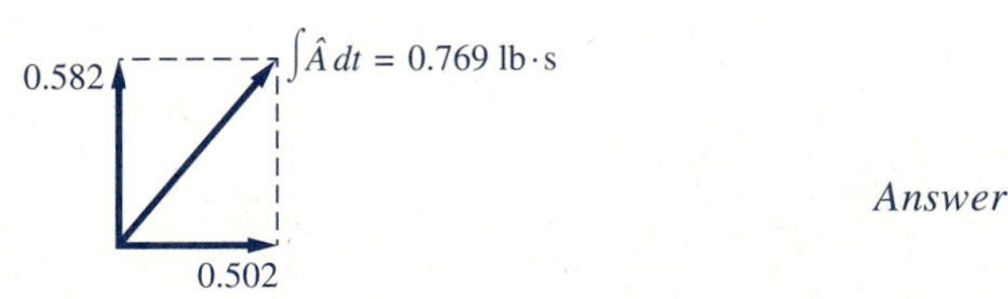

Answer

Part 3

The kinetic energy of the system before impact is

$$T_1 = \frac{1}{2}m_C v_1^2 = \frac{1}{2}\left(\frac{0.025}{32.2}\right)(1500)^2 = 873.45 \text{ lb}\cdot\text{ft}$$

After impact, the kinetic energy is

$$T_2 = \frac{1}{2}m_C v_2^2 + \left[\frac{1}{2}\bar{I}\omega_2^2 + \frac{1}{2}m\bar{v}_2^2\right]_{AB}$$

which, on substitution of the numerical values, becomes

$$\begin{aligned} T_2 &= \frac{1}{2}\left(\frac{0.025}{32.2}\right)(4\times 1.616)^2 + \left[\frac{1}{2}\left(\frac{20}{32.2}\right)(1.616)^2 + \frac{1}{2}\left(\frac{15}{32.2}\right)(2\times 1.616)^2\right] \\ &= 3.26 \text{ lb}\cdot\text{ft} \end{aligned}$$

Therefore, the percentage loss of energy during the impact is

$$\frac{T_1 - T_2}{T_1}\times 100\% = \frac{873.45 - 3.26}{873.45}\times 100\% = 99.6\% \qquad \textit{Answer}$$

Alternate Method of Computing $\int \hat{A}_x\, dt$

The impulse of reaction $\hat{A}_x$ could also be computed from an angular impulse-momentum equation. Referring to Fig. (b) and considering either the entire system or only bar AB, we have

$$(A_B)_{1-2} = (h_B)_2 - (h_B)_1$$

$$\circlearrowleft + \quad 4\int \hat{A}_x\, dt = \left[\frac{20}{g}\omega_2 - \frac{15}{g}(2\omega_2)(2)\right] - 0$$

which yields

$$\int \hat{A}_x\, dt = -\frac{10}{g}\omega_2 = -\frac{10}{32.2}(1.616) = -0.502 \text{ lb}\cdot\text{s} \qquad \textit{Answer}$$

which agrees with the result given in Eq. (a). Note that it is legitimate to use B as the reference point, although it is neither a fixed point nor the mass center. As proven in Art. 18.7, there are no restrictions on the location of the reference point for impulsive motion (all points are, in effect, "fixed" during the infinitesimal period of impact).

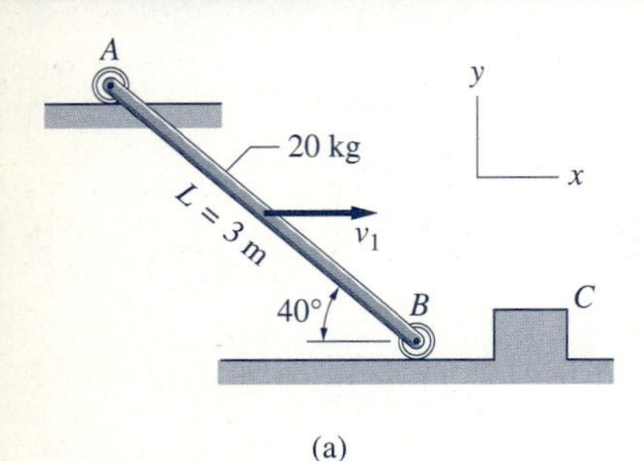

(a)

Sample Problem 18.13

The 20-kg uniform slender bar in Fig. (a) is moving to the right on frictionless rollers at A and B with the velocity v_1 when the roller at B strikes the small obstruction C without rebounding. Compute the minimum value of v_1 for which the bar will reach the vertical position after the impact.

Solution

Figure (b) shows the FBD of the bar during the impact, and the momentum diagrams of the bar before and immediately after the impact, labeled ① and ②, respectively. It also shows the final, vertical position ③, which is a rest position. We will neglect the time of impact, which means that the motion is assumed to be impulsive. Therefore, the bar occupies the same spatial position throughout the duration of the impact.

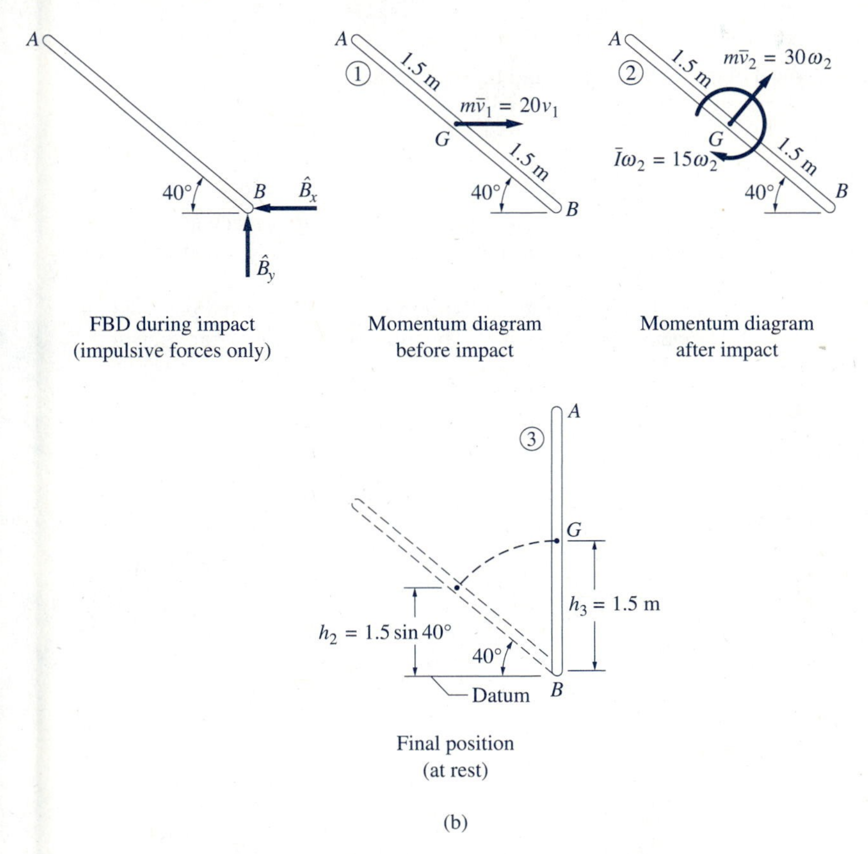

(b)

We must use the impulse-momentum method to analyze the impact that occurs between positions ① and ②. To analyze the motion between positions ② and ③, we employ the principle of conservation of mechanical energy.

The FBD in Fig. (b) shows the reactions $\hat{B}_x$ and $\hat{B}_y$, which are the only impulsive forces that act on the bar during impact. The weight of the bar is omitted from the FBD, since it is a finite force whose impulse may be neglected because of the very small duration of impact. The roller reaction at A does not appear in the FBD for the same reason (the reaction is not impulsive because the roller is about to lift off the plane during the impact).

As shown in Fig. (b), the momentum diagram for position ① consists only of the linear momentum vector, the magnitude of which is $mv_1 = 20v_1$ N · s, where v_1 is measured in m/s. The momentum diagram for position ② contains both the linear and angular momenta. We have $\bar{I} = mL^2/12 = 20(3)^2/12 = 15$ N · m^2, so that the central angular momentum is $\bar{I}\omega_2 = 15\omega_2$ N · m · s, where ω_2 is measured in rad/s. Since the bar is rotating about end B after the impact, $\bar{v}_2 = 1.5\omega_2$ m/s; consequently, the magnitude of the linear momentum vector is $m\bar{v}_2 = 20(1.5)\omega_2 = 30\omega_2$ N · s.

From the FBD in Fig (b) we see that the angular impulse about B is zero, since both $\hat{B}_x$ and $\hat{B}_y$ pass through B. Because angular impulse equals the change in angular momentum about every point during the period of impact, we conclude that the angular momentum about B is conserved between positions ① and ②. Referring to the momentum diagrams in Fig. (b) we thus obtain

$$(h_B)_1 = (h_B)_2$$

$$\circlearrowleft + \quad 20v_1(1.5 \sin 40°) = 15\omega_2 + 30\omega_2(1.5)$$

which gives the following relationship between v_1 and ω_2:

$$v_1 = 3.111\omega_2 \tag{a}$$

Since all the forces acting on the bar after the impact are conservative, the value of ω_2 can be computed by applying the conservation of mechanical energy principle between position ② and position ③ (recall that the bar is at rest in the vertical position).

As shown in Fig. (b), we choose the datum for the gravitational potential energy to be the horizontal plane through B. Therefore, the potential energies in positions ② and ③ are

$$V_2 = Wh_2 = 20(9.81)(1.5 \sin 40°) = 189.2 \text{ J} \tag{b}$$

$$V_3 = Wh_3 = 20(9.81)(1.5) = 294.3 \text{ J} \tag{c}$$

The kinetic energies for the two positions are

$$\begin{aligned} T_2 &= \frac{1}{2}\bar{I}\omega_2^2 + \frac{1}{2}m\bar{v}_2^2 \\ &= \frac{1}{2}(15)\omega_2^2 + \frac{1}{2}(20)(1.5\omega_2)^2 \\ &= 30.0\omega_2^2 \text{ J} \end{aligned} \tag{d}$$

and

$$T_3 = 0 \tag{e}$$

Utilizing Eqs. (b)–(e), the conservation of mechanical energy yields

$$V_2 + T_2 = V_3 + T_3$$

$$189.2 + 30.0\omega_2^2 = 294.3 + 0$$

from which

$$\omega_2 = 1.872 \text{ rad/s}$$

Substituting this value into Eq. (a), we find that the smallest initial velocity for which the bar will reach the vertical position is

$$v_1 = 3.111\omega_2 = 3.111(1.872) = 5.82 \text{ m/s} \qquad \textit{Answer}$$

PROBLEMS

18.69 The impact tester consists of the 20-kg striker B that is attached to the 16-kg uniform slender rod AC. The tester is released from an inclined position and breaks the test specimen D. Find the distance h for which the pin reaction at A will have no horizontal component during the impact with the specimen.

18.70 The impact tester described in Prob. 18.69 is released at $\theta = 55°$ with the striker set at $h = 2.5$ m. After breaking the specimen, the tester travels to $\theta = -38°$ before stopping. For the impact, determine the magnitude of the impulse exerted by (a) the striker; and (b) the pin at A.

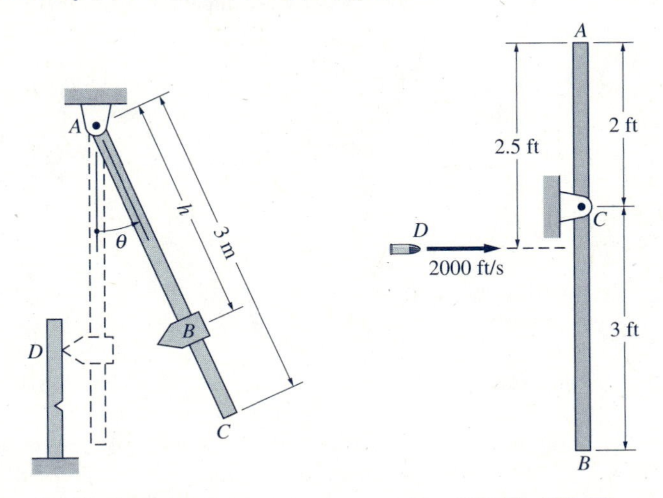

Fig. P18.69, P18.70 **Fig. P18.71, P18.72**

18.71 The 30-lb uniform rod AB is stationary when it is hit by the 0.12-lb bullet D that is traveling horizontally at 2000 ft/s. Assuming that the bullet becomes embedded in the rod, calculate (a) the angular velocity of AB immediately after the impact; and (b) the maximum angular displacement of AB following the impact.

18.72 Solve Prob. 18.71 assuming that the bullet passes through the rod and exits with a horizontal velocity of 1200 ft/s.

18.73 The uniform rod AB of mass m is falling vertically with no angular velocity. Just before point C hits the obstruction, the speed of the rod is 7 m/s. For the instant after impact, determine (a) the angular velocity of AB; and (b) the velocity of C. Assume that no energy is lost during the impact.

18.74 Assuming that point C in Prob. 18.73 does not rebound, determine (a) the angular velocity of the rod AB immediately after the impact; and (b) the percentage of kinetic energy lost during the impact.

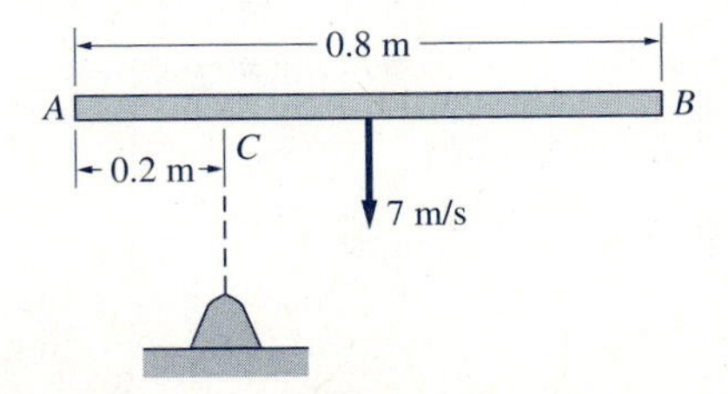

Fig. P18.73, P18.74

18.75 The uniform rod AB of mass m and length L is released from rest with end A positioned in a smooth corner. The rod strikes the smooth obstruction at C with the clockwise angular velocity ω_1 and does not rebound. (a) Determine the largest distance d for which the angular velocity of the rod will continue to be clockwise after the impact. (b) Find the value of d that maximizes the clockwise angular speed of the rod after the impact.

18.76 The homogeneous rod AB of mass $3m$ is free to rotate about the pin at C. The rod is stationary when the ball D of mass m hits end A of the rod with the vertical velocity v_1. Knowing that the velocity of D immediately after the impact is zero, determine (a) the velocity of A immediately after the impact; and (b) the percentage of kinetic energy lost during the impact.

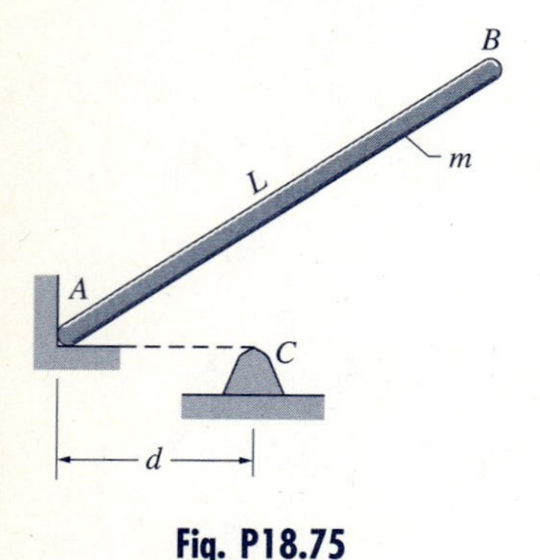

Fig. P18.75

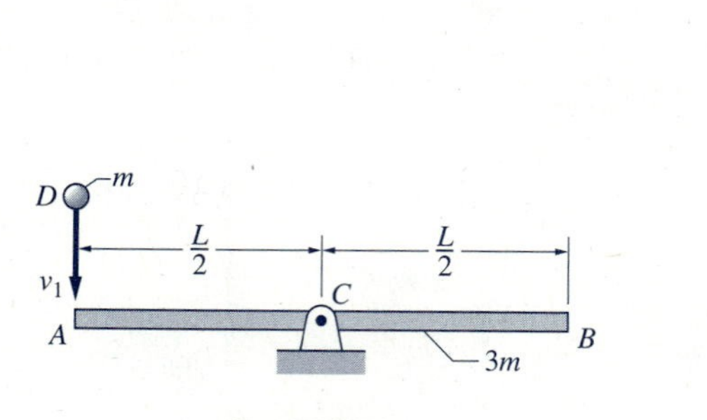

Fig. P18.76

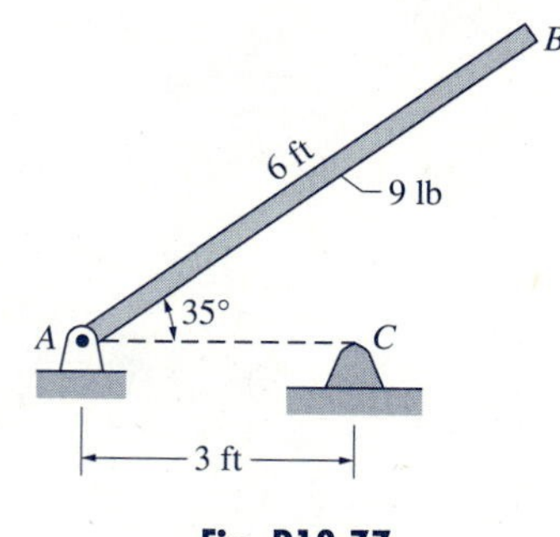

Fig. P18.77

18.77 The 9-lb uniform rod AB rotates freely about the pin at A. The rod is released from rest in the position shown and collides with the obstruction at C. Calculate the resulting impulses acting on the rod at A and C, assuming that (a) the rod does not rebound; and (b) no energy is lost during the impact.

18.78 The uniform bar AB of mass m is sliding on the horizontal surface with the velocity $v_1 = 4.2$ m/s when end B hits the 30° incline. Determine the velocity of end A immediately after the impact. Neglect friction and assume that the rod never loses contact with the surfaces.

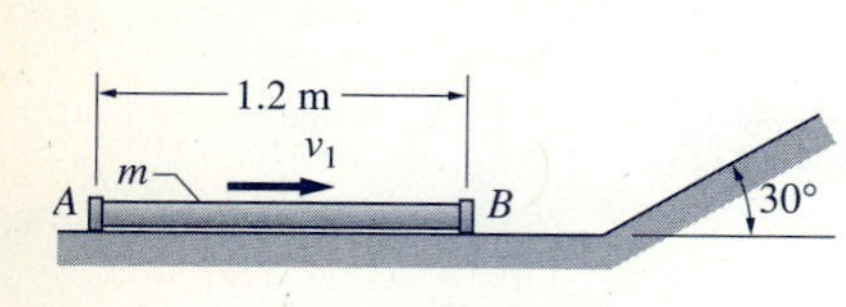

Fig. P18.78

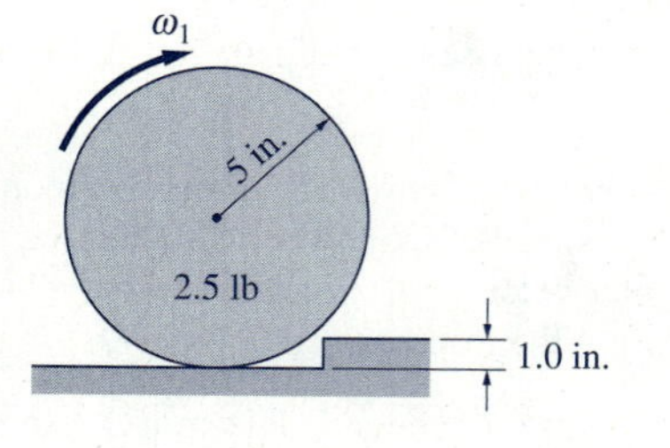

Fig. P18.79

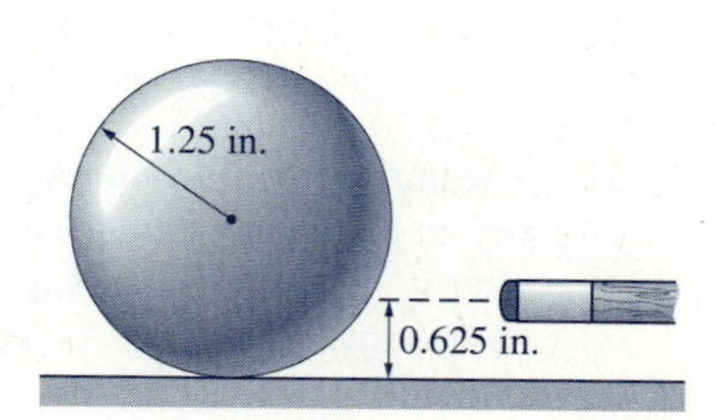

Fig. P18.80

18.79 The 2.5-lb uniform disk rolls without slipping with the angular velocity ω_1 prior to hitting the 1-in. curb. Assuming no rebound and no slipping between the disk and the curb, find the minimum value of ω_1 for which the disk will mount the curb.

18.80 The billiard ball of mass m is at rest on a table when it is struck horizontally by the cue. Immediately after the impact, the central velocity of the ball is $v = 3$ ft/s. Find the angular velocity of the ball (a) immediately after the impact; and (b) after the ball stops slipping on the table.

18.81 The uniform 5-kg box is hit near its top by the 2.0-g bullet traveling at the speed v_1. Determine the minimum value of v_1 that will topple the box about the corner at C. Assume that the bullet becomes embedded and that the box does not bounce out of the corner.

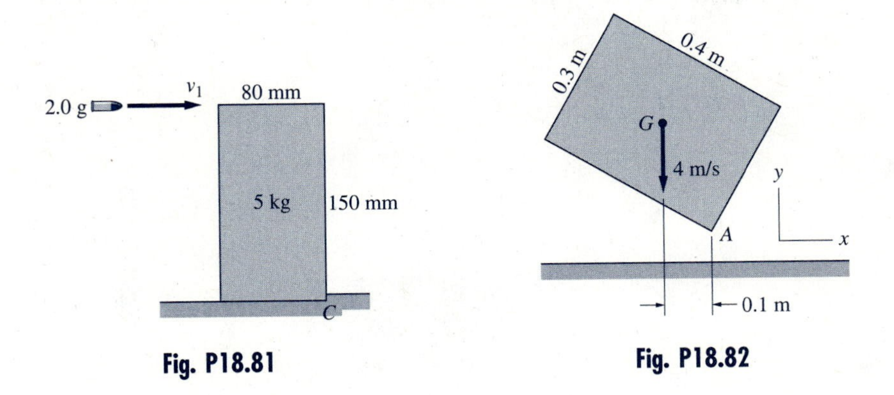

Fig. P18.81

Fig. P18.82

18.82 The uniform crate of mass m is falling in the position shown without rotating. The velocity of the crate is 4 m/s just before corner A hits the smooth floor. For the instant after the impact, determine (a) the central and angular velocities of the crate; and (b) the velocity vector of corner A. Assume that half of the initial kinetic energy of the crate is lost during the impact.

18.83 The uniform 80-lb bar AB is initially at rest in the horizontal position, supported by the pin at B and the spring of stiffness $k = 25$ lb/in. at A. The small (relative to the length of the bar) 40-lb sandbag C is tossed onto the bar with the initial velocity shown. Assuming that the sandbag does not rebound, calculate (a) the angular velocity of AB immediately after the impact; and (b) the maximum angular displacement of AB following the impact, assuming that it is a small angle.

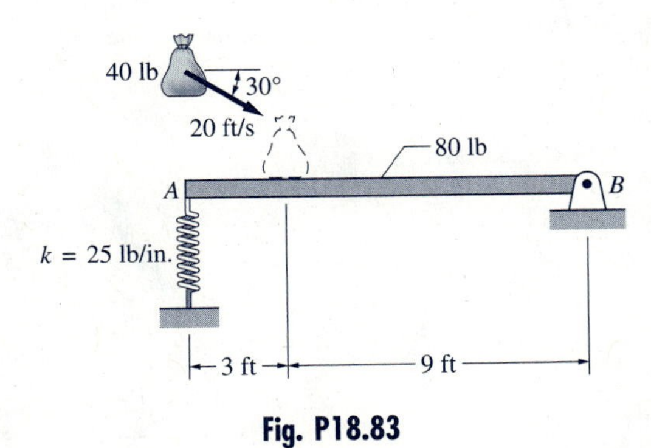

Fig. P18.83

REVIEW PROBLEMS

18.84 The uniform disk of mass m and radius R is attached to a spring of stiffness k. If the disk is released from rest with the spring undeformed, determine the expression for the maximum angular velocity of the disk. Assume that the disk rolls without slipping on the inclined surface.

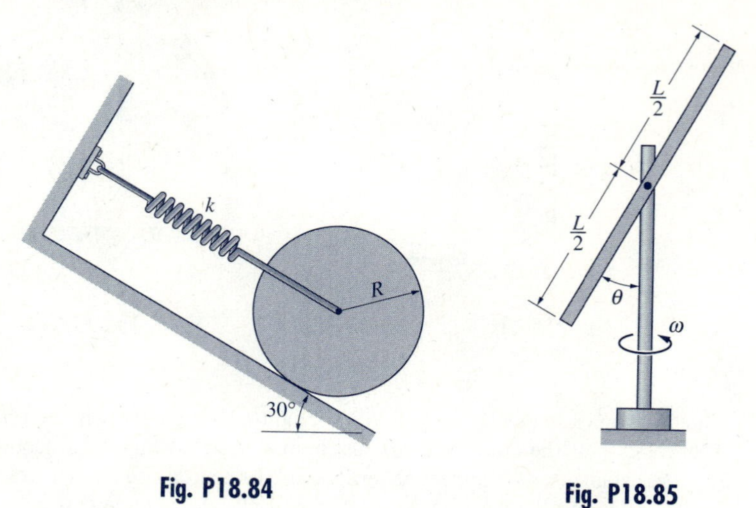

Fig. P18.84 **Fig. P18.85**

18.85 The uniform bar of mass m and length L is attached to a vertical shaft of negligible mass. The angle θ between the bar and the shaft can be changed by an internal mechanism (not shown). If the assembly is rotating freely with the angular velocity ω when $\theta = 90°$, determine the angular velocity of the assembly when θ is changed to 45°.

18.86 The 20-lb uniform bar AB is suspended from a roller at B, which travels on a horizontal rail. The bar is at rest in the vertical position when it is hit at end A by the 0.05-lb bullet moving horizontally at 3000 ft/s. Assuming that the bullet becomes embedded in the bar, determine the maximum angular displacement of the bar after the impact. Neglect the masses of the bullet and the roller in comparison to the mass of the rod.

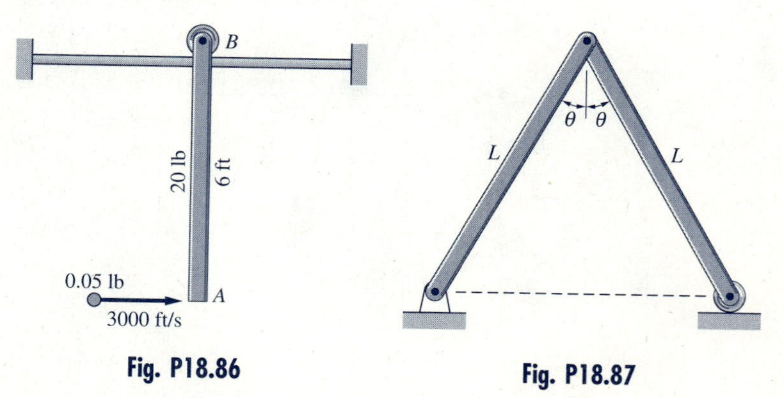

Fig. P18.86 **Fig. P18.87**

18.87 The mechanism consists of two uniform bars, each of mass m and length L. Derive the expression for the kinetic energy of the mechanism in terms of θ and $\dot{\theta}$.

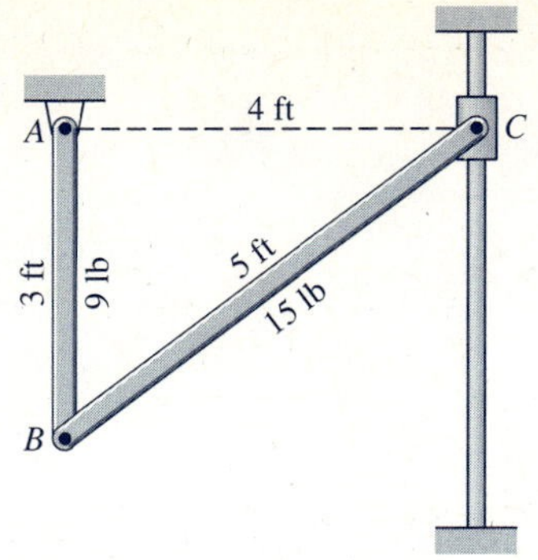

Fig. P18.88

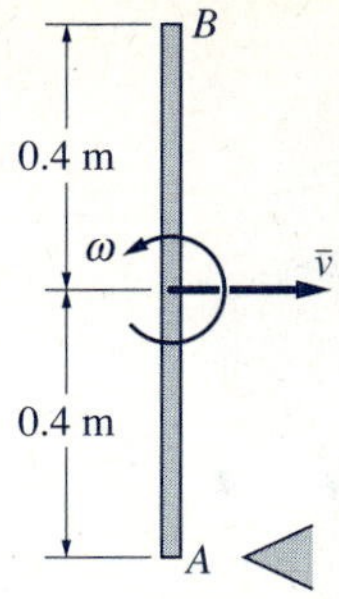

Fig. P18.89, P18.90

18.88 The mechanism is released from rest in the position shown. Determine the velocity of slider C when bar BC has become horizontal.

18.89 The 15-kg uniform bar AB is in the vertical position shown and moving with velocity $\bar{v} = 16$ m/s and angular velocity $\omega = 120$ rad/s when end A strikes a rigid obstruction. Determine the velocity of end B just after the impact, assuming that the bar does not rebound.

18.90 Solve Problem 18.89 assuming that no kinetic energy is lost during the impact.

18.91 The central radius of gyration of the spool is $\bar{k}$, and its inner radius is R. If the spool is released from rest, determine the velocity of the spool as a function of its displacement x. Neglect the friction between the spool and the vertical wall.

18.92 Disk B is rotating at 60 rad/s when the identical, stationary disk A is lowered into contact with it. If the coefficient of kinetic friction between the disks is 0.25, determine the time when the slipping between the disks stops and the final angular velocities of the disks.

18.93 The 20-lb uniform disk is launched along the horizontal surface with the velocity 15 ft/s and the backspin 45 rad/s. The coefficient of kinetic friction between the disk and the surface is 0.45. Determine (a) the final speed of the disk; (b) the time required to reach the final speed; and (c) the percentage of energy lost due to slipping.

18.94 The homogeneous half-cylinder is released from rest in the position shown. Assuming no slipping, determine the maximum angular velocity of the half-cylinder.

18.95 Solve Problem 18.94 assuming negligible friction between the half-cylinder and the horizontal surface.

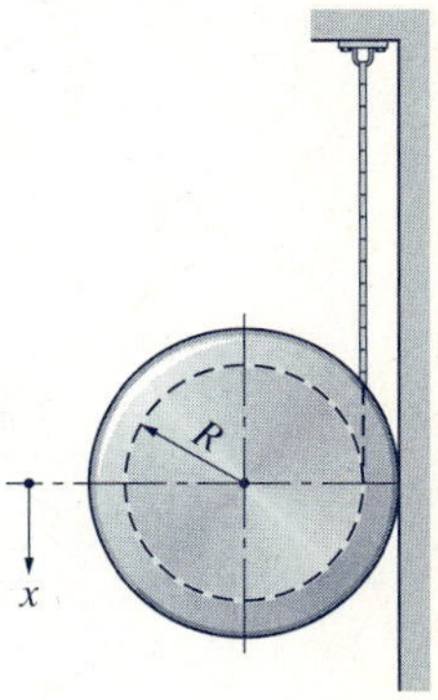

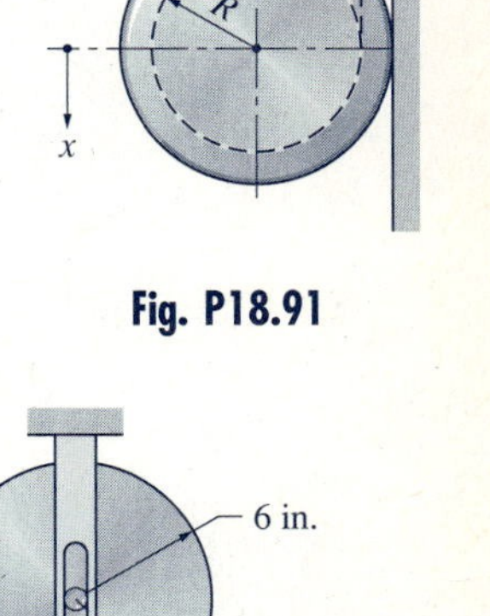

Fig. P18.91

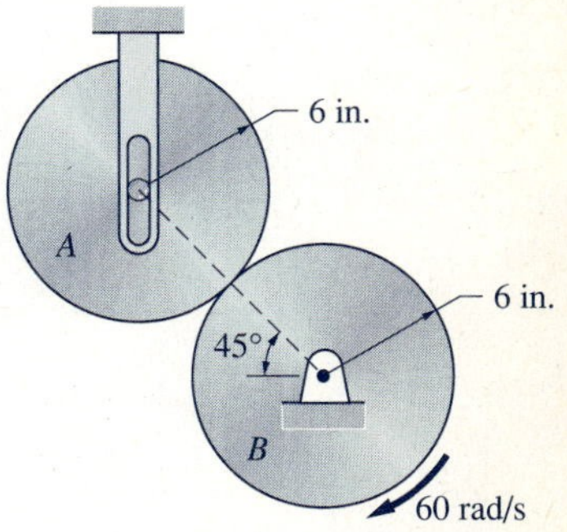

Fig. P18.92

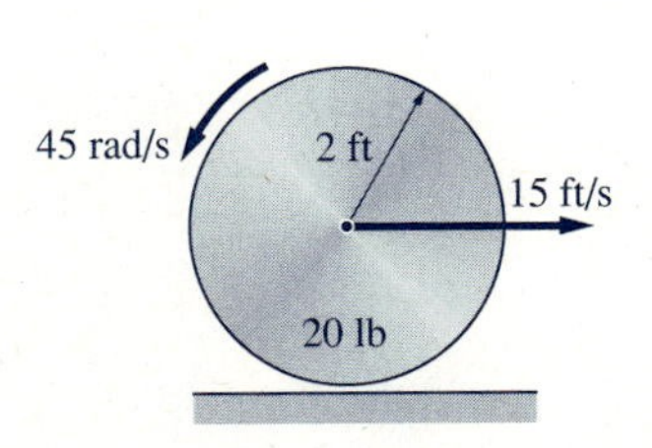

Fig. P18.93

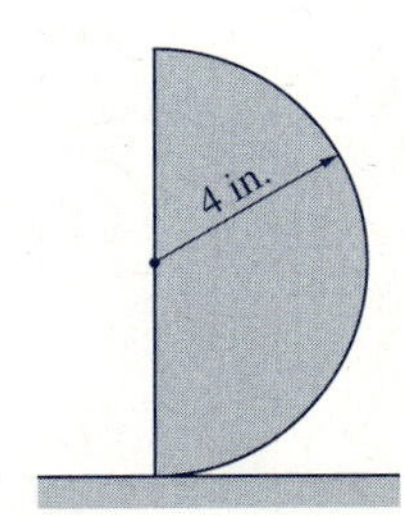

Fig. P18.94, P18.95

19 Rigid-Body Motion in Three Dimensions

*19.1 Introduction

As seen in previous chapters, there are many practical engineering problems that lend themselves to plane motion analysis. However, there are important applications (e.g., the motion of satellites and gyroscopes) for which three-dimensional analysis must be used. This chapter is intended to serve as an introduction to spatial motion; in-depth treatment of this topic is the domain of textbooks on advanced dynamics.

Article 19.2 presents the kinematic analysis of a body that is not restricted to planar motion. The fundamental kinematic equations derived in Chapter 16 are applicable without changes; the only new concept presented here is the instant axis of rotation. The remainder of the chapter presents the three procedures of kinetic analysis: impulse-momentum, work-energy, and force-mass-acceleration methods. The important special case of an axisymmetric body, including the motion of a gyroscope, is also discussed.

Most of the fundamental concepts, equations, and methods of analysis that are required for the analysis of spatial motion have been presented in previous chapters. We will refer to these topics throughout this chapter.

*19.2 Kinematics

Up to now, our discussion of rigid bodies has been restricted to plane motion. However, most of the kinematic equations in Chapter 16 were derived without assuming that the motion was two-dimensional. Consequently, these equations may also be applied to three-dimensional kinematics without modification. In this article, we review the equations of rigid-body kinematics that were developed previously and illustrate their application to three-dimensional motion. The only new concept introduced here is the instant axis of rotation.

a. Relative velocity and acceleration

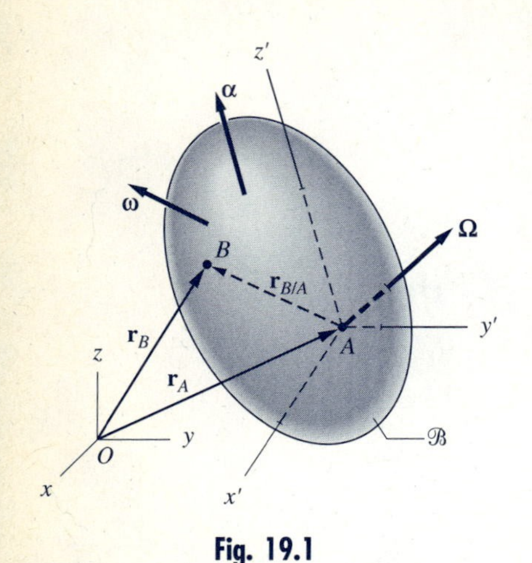

Fig. 19.1

Figure 19.1 shows a rigid body $\mathcal{B}$ that is moving in three dimensions. The angular velocity and angular acceleration of the body are $\boldsymbol{\omega}$ and $\boldsymbol{\alpha} = d\boldsymbol{\omega}/dt$, respectively. The position vectors of points A and B of the body (or body extended) are related to the relative position vector $\mathbf{r}_{B/A}$ by

$$\mathbf{r}_B = \mathbf{r}_A + \mathbf{r}_{B/A}$$

In Art 16.1 we have shown that successive time derivatives of the above relationship yield the following relative velocity and acceleration equations.

$$\boxed{\mathbf{v}_B = \mathbf{v}_A + \mathbf{v}_{B/A} = \mathbf{v}_A + \boldsymbol{\omega}\times\mathbf{r}_{B/A}} \tag{19.1}$$

and

$$\boxed{\mathbf{a}_B = \mathbf{a}_A + \mathbf{a}_{B/A} = \mathbf{a}_A + \boldsymbol{\omega}\times(\boldsymbol{\omega}\times\mathbf{r}_{B/A}) + \boldsymbol{\alpha}\times\mathbf{r}_{B/A}} \tag{19.2}$$

The terms appearing in Eqs. (19.1) and (19.2) were depicted in Fig. 16.1. It is important to note that these equations are valid only if A and B are points that are fixed in the same body (or body extended).

b. Vector differentiation in a rotating reference frame

As discussed in Art 16.7, it is sometimes convenient to describe the relative motion terms in Eqs. (19.1) and (19.2) within a reference frame that translates and rotates, rather than employing a frame that is fixed in space. Figure 19.1 shows two reference frames: the xyz-system, which is fixed in space, and the $x'y'z'$-system, the origin of which is attached to point A. If the $x'y'z'$-axes are embedded in the body (that is, the origin of these axes moves with A and the coordinate system rotates with the body), they are referred to as a *body frame,* whereas the fixed xyz-axes are called a *space frame.* Note that the angular velocity of the body frame is equal to the angular velocity $\boldsymbol{\omega}$ of the body. If the components of a vector $\mathbf{V}$ are described relative to the body frame, then its absolute derivative can be computed using the following identity derived in Art. 16.7.

$$\frac{d\mathbf{V}}{dt} = \left(\frac{d\mathbf{V}}{dt}\right)_{/\mathcal{B}} + \boldsymbol{\omega}\times\mathbf{V} \tag{19.3}$$

where the notation "$/\mathcal{B}$" is used to indicate that the derivative is to be evaluated relative to the body frame, i.e., as seen by an observer who translates and rotates with the body $\mathcal{B}$.

If the $x'y'z'$-coordinate system rotates with the angular velocity $\boldsymbol{\Omega}$ that is not necessarily equal to the angular velocity $\boldsymbol{\omega}$ of the body, then Eq. (19.3) must be modified as follows.

$$\frac{d\mathbf{V}}{dt} = \left(\frac{d\mathbf{V}}{dt}\right)_{/x'y'z'} + \boldsymbol{\Omega}\times\mathbf{V} \tag{19.4}$$

By using the notation "$/x'y'z'$" we draw attention to the fact that the relative derivative is now referred to a coordinate system that is not necessarily a body frame.

c. Instant axis of rotation

Let the velocity of a point C in a rigid body (or body extended) be zero at a particular instant (in plane motion, this point is called the instant center for velocities). As shown in Fig. 19.2, the line passing through C that is parallel to the angular velocity $\boldsymbol{\omega}$ of the body is called the *instant axis of rotation,* or simply the *instant axis.** With $\mathbf{v}_C = \mathbf{0}$, the velocity of any point P in the body is $\mathbf{v}_P = \mathbf{v}_{P/C} = \boldsymbol{\omega} \times \mathbf{r}_{P/C}$, as shown in Fig. 19.2. It follows that the velocities of all points behave as if the body were rotating about the instant axis, with the velocities of points on this axis being zero. A point that has zero velocity always exists in plane motion and for a body rotating about a fixed point. However, it can be shown that such a point need not exist for all motions.†

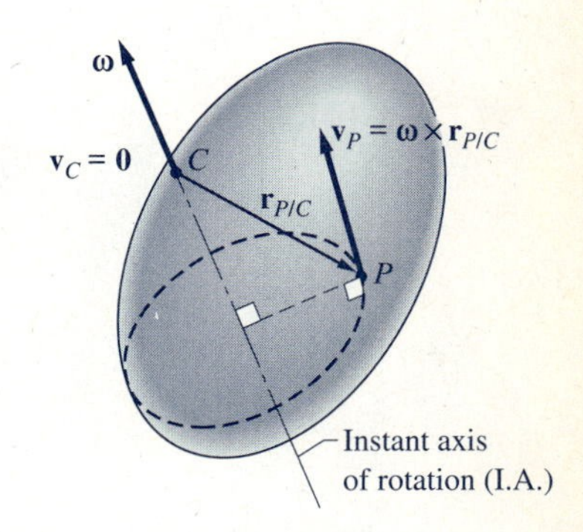

Fig. 19.2

When analyzing velocities, it is advantageous to utilize the instant axis of rotation when its location can be determined by inspection. For example, Fig. 19.3(a) shows a wheel of radius R that is rolling without slipping along a circular path on a horizontal plane. The horizontal axle OA of length L is attached to a collar that rotates on the fixed vertical shaft. As shown in Fig. 19.3(a), the angular velocity of the wheel is given by $\boldsymbol{\omega} = \boldsymbol{\omega}_1 + \boldsymbol{\omega}_2$, where $\boldsymbol{\omega}_1$ is the *spin velocity* of the wheel (its angular velocity relative to the axle) and $\boldsymbol{\omega}_2$ is the angular velocity of the axle.

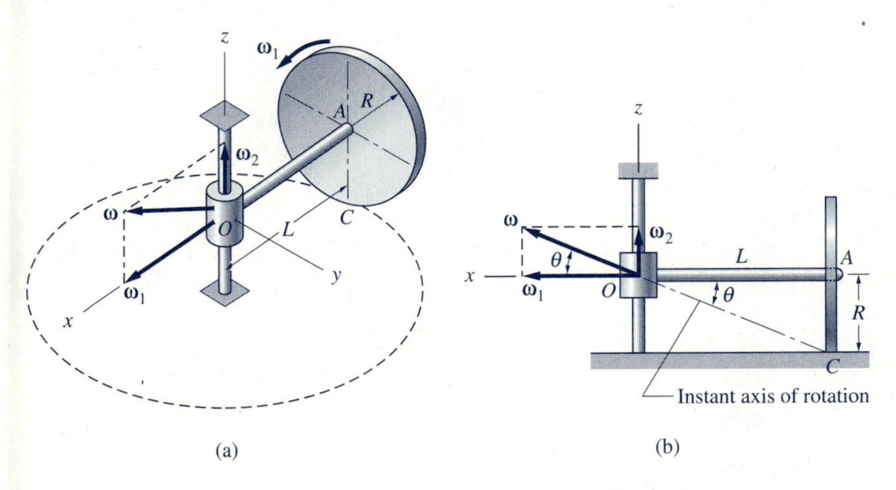

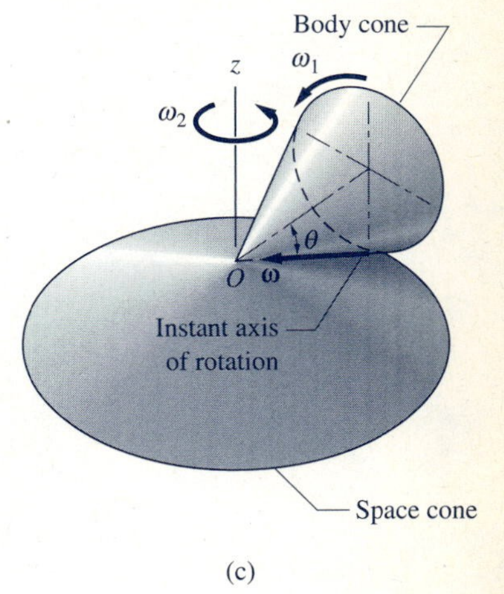

Fig. 19.3

Figure 19.3(b) shows another view of the assembly in which the instant axis of the wheel has been identified. This axis is known to pass through points C and O, because both are points on the wheel and have zero velocity (the body must be "extended" to include O). Referring to Fig. 19.3(b), we see that the

* In two-dimensional motion, $\boldsymbol{\omega}$ is perpendicular to the plane of the motion. Therefore, the instant axis of rotation is also perpendicular to this plane, and it passes through the instant center for velocities.

† See, for example, *Advanced Engineering Dynamics,* J. H. Ginsberg, Harper & Row, 1988, p. 115.

angle θ can be determined if L and R are known, which means that $\boldsymbol{\omega}$ and $\boldsymbol{\omega}_2$ can be found if the spin velocity $\boldsymbol{\omega}_1$ is known. (An alternate method for computing the angular velocities is to write the relative velocity equation using points O and C—see Sample Problem 19.3.)

As the wheel moves, the instant axis of rotation traces out a three-dimensional surface in space. Since the wheel undergoes rotation about the fixed point O, this surface is a cone, called the *space cone,* with its apex at O. The trace of the instant axes in the "extended" wheel is also a conical surface, known as the *body cone.* Both cones are shown in Fig. 19.3(c). Inspecting this figure, we see that the kinematics of the wheel and the body cone are identical if the latter is made to roll without slipping on the space cone with the spin velocity $\boldsymbol{\omega}_1$.

Sample Problem 19.1

Figure (a) shows an arm OA of length L that is rotating about a fixed vertical shaft with the angular velocity and angular acceleration $\boldsymbol{\omega}_1$ and $\dot{\boldsymbol{\omega}}_1$, respectively. At the same time, the thin disk of radius R is spinning freely relative to the arm OA with angular velocity and angular acceleration $\boldsymbol{\omega}_2$ and $\dot{\boldsymbol{\omega}}_2$, respectively. Calculate the angular velocity $\boldsymbol{\omega}$ and angular acceleration $\boldsymbol{\alpha}$ of the disk.

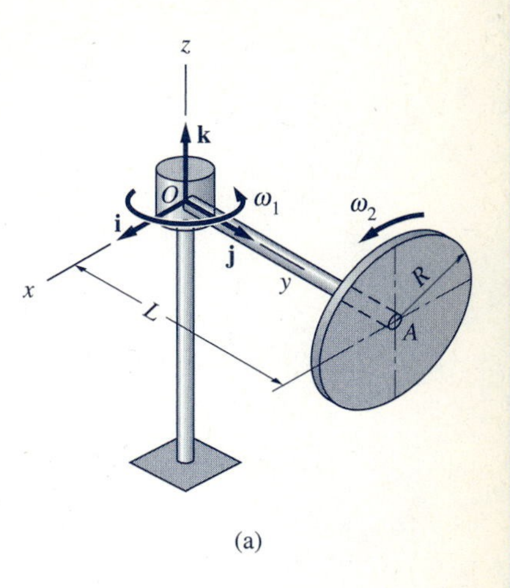

(a)

Solution

We will solve this problem using two different reference frames. Because the solutions will require time derivatives of vectors, it is very important to clearly specify the coordinate system that is being utilized.

Method I: Reference Frame Embedded in Arm *OA*

We assume that the origin of the xyz reference frame shown in Fig. (a) is attached to point O and that the coordinate axes are embedded in arm OA. The angular motion of this frame is thus identical to that of OA. The base vectors for the rotating frame are denoted by $\mathbf{i}$, $\mathbf{j}$, and $\mathbf{k}$.

From Fig. (a) we see that the angular velocity of the disk is

$$\boldsymbol{\omega} = \boldsymbol{\omega}_1 + \boldsymbol{\omega}_2 \tag{a}$$

from which we obtain

$$\boldsymbol{\omega} = \omega_1\mathbf{k} + \omega_2\mathbf{j} \qquad \text{(valid for all time)} \qquad \textit{Answer} \tag{b}$$

Equation (b) is true for all time, because the xyz reference frame rotates with arm OA, so that the y- and z-axes are always directed along OA and the vertical shaft, respectively. Indicating which expressions are valid for all time will help us avoid the common error of attempting to differentiate an expression that is true only at a particular instant.

The angular acceleration of the disk can be found by differentiating the expression for $\boldsymbol{\omega}$ given in Eq. (b). Noting that the absolute derivatives of the base vectors are not necessarily zero, we get

$$\begin{aligned}\boldsymbol{\alpha} = \dot{\boldsymbol{\omega}} &= \dot{\omega}_1\mathbf{k} + \omega_1\dot{\mathbf{k}} + \dot{\omega}_2\mathbf{j} + \omega_2\dot{\mathbf{j}} \\ &= \dot{\omega}_1\mathbf{k} + \omega_1(\boldsymbol{\Omega} \times \mathbf{k}) + \dot{\omega}_2\mathbf{j} + \omega_2(\boldsymbol{\Omega} \times \mathbf{j})\end{aligned}$$

where $\boldsymbol{\Omega}$ is the angular velocity of the xyz reference frame. The derivatives of the rotating base vectors, $\dot{\mathbf{j}} = \boldsymbol{\Omega} \times \mathbf{j}$ and $\dot{\mathbf{k}} = \boldsymbol{\Omega} \times \mathbf{k}$, have been obtained from Eq. (16.19). Since the frame rotates with OA, we have

$$\boldsymbol{\Omega} = \boldsymbol{\omega}_1 = \omega_1\mathbf{k}$$

yielding

$$\boldsymbol{\alpha} = \dot{\omega}_1\mathbf{k} + \omega_1(\omega_1\mathbf{k} \times \mathbf{k}) + \dot{\omega}_2\mathbf{j} + \omega_2(\omega_1\mathbf{k} \times \mathbf{j})$$

which simplifies to

$$\boldsymbol{\alpha} = \dot{\omega}_1\mathbf{k} + \dot{\omega}_2\mathbf{j} - \omega_1\omega_2\mathbf{i} \qquad \textit{Answer} \tag{c}$$

It should be noted that Eq. (c) is also valid for all time.

An alternate method for evaluating the time derivative of the angular velocity is to use Eq. (19.4).

$$\boldsymbol{\alpha} = \dot{\boldsymbol{\omega}} = (\dot{\boldsymbol{\omega}})_{/xyz} + \boldsymbol{\Omega} \times \boldsymbol{\omega} \tag{d}$$

where $(\dot{\boldsymbol{\omega}})_{/xyz}$ is the angular acceleration of the disk relative to the xyz-frame.

From Eq. (b), we observe that $(\dot{\boldsymbol{\omega}})_{/xyz} = \dot{\omega}_1\mathbf{k} + \dot{\omega}_2\mathbf{j}$ (because the base vectors $\mathbf{k}$ and $\mathbf{j}$ are fixed in the xyz-frame, their derivatives relative to that frame vanish). Furthermore, since $\boldsymbol{\Omega} = \boldsymbol{\omega}_1$ we obtain $\boldsymbol{\Omega} \times \boldsymbol{\omega} = \omega_1\mathbf{k} \times (\omega_1\mathbf{k} + \omega_2\mathbf{j}) = -\omega_1\omega_2\mathbf{i}$. Substituting this result into Eq. (d), we find that

$$\boldsymbol{\alpha} = -\omega_1\omega_2\mathbf{i} + \dot{\omega}_2\mathbf{j} + \dot{\omega}_1\mathbf{k} \qquad \textit{Answer}$$

which agrees with the result obtained in Eq. (c). The primary advantage in using Eq. (d) is that the absolute derivatives of the base vectors need not be formally evaluated.

Method II: Reference Frame Embedded in the Disk

The origin of the xyz-frame shown in Fig. (b) is attached to point A, and the coordinate axes are embedded in the disk. In addition to the base vectors $\mathbf{i}$, $\mathbf{j}$, and $\mathbf{k}$, we introduce the unit vector $\boldsymbol{\lambda}$, which is fixed in the vertical shaft. Therefore, the angular velocity of the disk, $\boldsymbol{\omega} = \boldsymbol{\omega}_1 + \boldsymbol{\omega}_2$, can be written as

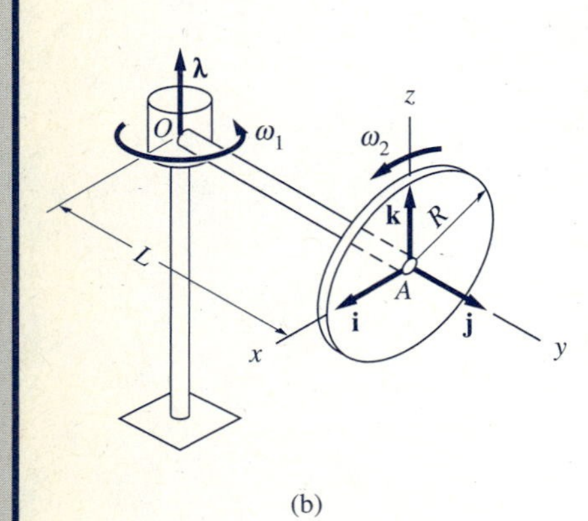

(b)

$$\boldsymbol{\omega} = \omega_1\boldsymbol{\lambda} + \omega_2\mathbf{j} \quad \text{(valid for all time)} \qquad \textit{Answer} \tag{e}$$

Because $\boldsymbol{\lambda} = \mathbf{k}$ in the position shown in Fig. (b), the angular velocity of the disk in this position is

$$\boldsymbol{\omega} = \omega_1\mathbf{k} + \omega_2\mathbf{j} \quad \text{(for the position shown)} \tag{f}$$

The angular acceleration of the disk can be found by differentiating the expression for $\boldsymbol{\omega}$ given in Eq. (e). Note that Eq. (f) cannot be differentiated, because it is valid only for a particular instant of time. Taking the absolute derivative of Eq. (e), we get

$$\begin{aligned}\boldsymbol{\alpha} = \dot{\boldsymbol{\omega}} &= \dot{\omega}_1\boldsymbol{\lambda} + \omega_1\dot{\boldsymbol{\lambda}} + \dot{\omega}_2\mathbf{j} + \omega_2\dot{\mathbf{j}} \\ &= \dot{\omega}_1\boldsymbol{\lambda} + \dot{\omega}_2\mathbf{j} + \omega_2(\boldsymbol{\Omega} \times \mathbf{j})\end{aligned} \tag{g}$$

where $\boldsymbol{\Omega} = \boldsymbol{\omega}$ is the angular velocity of the xyz-frame. In the above equation we have used $\dot{\boldsymbol{\lambda}} = \mathbf{0}$ since $\boldsymbol{\lambda}$ is fixed in space. Using $\boldsymbol{\lambda} = \mathbf{k}$ for the position shown in Fig. (b), Eq. (g) becomes

$$\boldsymbol{\alpha} = \dot{\omega}_1\mathbf{k} + \dot{\omega}_2\mathbf{j} + \omega_2\left[(\omega_1\mathbf{k} + \omega_2\mathbf{j}) \times \mathbf{j}\right]$$

which reduces to

$$\boldsymbol{\alpha} = \dot{\omega}_1\mathbf{k} + \dot{\omega}_2\mathbf{j} - \omega_1\omega_2\mathbf{i} \qquad \textit{Answer} \tag{h}$$

This expression is valid only for the position shown in Fig. (b), because the xyz-coordinate axes are rotating with the wheel.

Summary

This sample problem was solved using the two reference frames shown in Figs. (a) and (b), the former rotating with arm OA and the latter rotating with the disk.

Using either reference frame, the angular velocity and angular acceleration of the disk in the position shown were found to be

$$\boldsymbol{\omega} = \omega_1\mathbf{k} + \omega_2\mathbf{j}$$

and

$$\boldsymbol{\alpha} = \dot{\boldsymbol{\omega}} = \dot{\omega}_1\mathbf{k} + \dot{\omega}_2\mathbf{j} - \omega_1\omega_2\mathbf{i}$$

If the reference frame is embedded in arm OA, these results are also valid for all time. However, this is not the case when the frame is embedded in the disk.

Sample Problem 19.2

The figure repeats the mechanism that was described in Sample Problem 19.1. Calculate the velocity and acceleration of point P on the disk in the position shown.

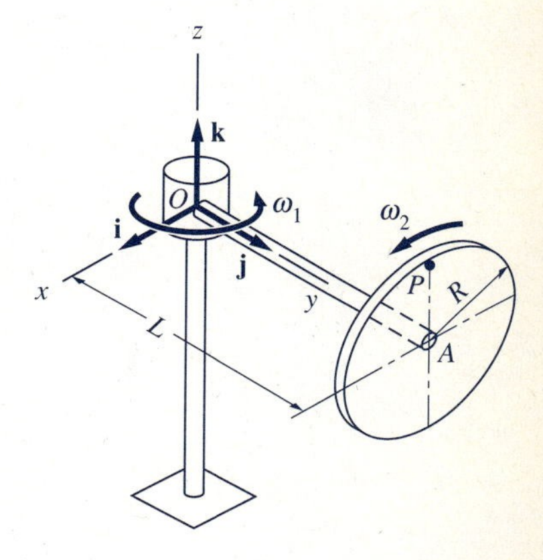

Solution

We assume that the xyz reference frame shown in the figure is attached to point O and embedded in the arm OA. (A reference frame that rotates with the disk would be equally convenient.) From the solution of Sample Problem 19.1, we know that the angular velocity and acceleration of the disk are

$$\boldsymbol{\omega} = \omega_1\mathbf{k} + \omega_2\mathbf{j} \tag{a}$$

and

$$\boldsymbol{\alpha} = \dot{\boldsymbol{\omega}} = \dot{\omega}_1\mathbf{k} + \dot{\omega}_2\mathbf{j} - \omega_1\omega_2\mathbf{i} \tag{b}$$

Since both expressions are valid for all time, we may use them for the position shown in the figure.

The motion of point P can be analyzed by relating it to the motion of point A. Referring to the figure, the position vector from the fixed point O to point A is $\mathbf{r}_{A/O} = L\mathbf{j}$. Since the angular velocity and acceleration of arm OA are $\boldsymbol{\omega}_1$ and $\dot{\boldsymbol{\omega}}_1$, the velocity and acceleration of point A are

$$\mathbf{v}_A = \boldsymbol{\omega}_1 \times \mathbf{r}_{A/O} = \omega_1\mathbf{k} \times L\mathbf{j} = -L\omega_1\mathbf{i} \tag{c}$$

and

$$\begin{aligned}\mathbf{a}_A &= \dot{\boldsymbol{\omega}}_1 \times \mathbf{r}_{A/O} + \boldsymbol{\omega}_1 \times (\boldsymbol{\omega}_1 \times \mathbf{r}_{A/O}) \\ &= \dot{\omega}_1\mathbf{k} \times L\mathbf{j} + \omega_1\mathbf{k} \times (-L\omega_1\mathbf{i}) \\ &= -L\dot{\omega}_1\mathbf{i} - L\omega_1^2\mathbf{j}\end{aligned} \tag{d}$$

The position vector of P relative to A is seen from the figure to be $\mathbf{r}_{P/A} = R\mathbf{k}$. Since both P and A belong to the disk, their relative velocity and acceleration vectors become

$$\mathbf{v}_{P/A} = \boldsymbol{\omega} \times \mathbf{r}_{P/A} = (\omega_1\mathbf{k} + \omega_2\mathbf{j}) \times R\mathbf{k} = R\omega_2\mathbf{i} \tag{e}$$

and

$$\begin{aligned}\mathbf{a}_{P/A} &= \boldsymbol{\alpha} \times \mathbf{r}_{P/A} + \boldsymbol{\omega} \times (\boldsymbol{\omega} \times \mathbf{r}_{P/A}) \\ &= [(\dot{\omega}_1\mathbf{k} + \dot{\omega}_2\mathbf{j} - \omega_1\omega_2\mathbf{i}) \times R\mathbf{k}] \\ &\quad + [(\omega_1\mathbf{k} + \omega_2\mathbf{j}) \times R\omega_2\mathbf{i}] \\ &= (R\dot{\omega}_2\mathbf{i} + R\omega_1\omega_2\mathbf{j}) + \left(R\omega_1\omega_2\mathbf{j} - R\omega_2^2\mathbf{k}\right) \qquad \text{(f)}\end{aligned}$$

The velocity of P is found by substituting Eqs. (c) and (e) into the relative velocity equation $\mathbf{v}_P = \mathbf{v}_A + \mathbf{v}_{P/A}$:

$$\mathbf{v}_P = (-L\omega_1 + R\omega_2)\mathbf{i} \qquad \textit{Answer} \quad \text{(g)}$$

Similarly, substituting Eqs. (d) and (f) into the relative acceleration equation, $\mathbf{a}_P = \mathbf{a}_A + \mathbf{a}_{P/A}$, the acceleration vector of P becomes (after some rearrangement of terms)

$$\mathbf{a}_P = (-L\dot{\omega}_1 + R\dot{\omega}_2)\mathbf{i} + \left(-L\omega_1^2 + 2R\omega_1\omega_2\right)\mathbf{j} - R\omega_2^2\mathbf{k} \qquad \textit{Answer} \quad \text{(h)}$$

The velocity and acceleration of point P could also be calculated by recognizing that the disk is rotating about the fixed point O (note that the length of a line connecting point O to any point on the disk remains constant). In this case, it is convenient to consider the disk to be extended so as to include point O. Utilizing this concept, the velocity and acceleration of point P could be computed using $\mathbf{v}_P = \boldsymbol{\omega} \times \mathbf{r}_{P/O}$ and $\mathbf{a}_P = \boldsymbol{\alpha} \times \mathbf{r}_{P/O} + \boldsymbol{\omega} \times (\boldsymbol{\omega} \times \mathbf{r}_{P/O})$. In this case $\boldsymbol{\omega}$ and $\boldsymbol{\alpha}$ are as given in Eqs. (a) and (b), respectively, and the position vector of P relative to O is seen from the figure to be $\mathbf{r}_{P/O} = L\mathbf{j} + R\mathbf{k}$. It may be verified that the results are identical to those given in Eqs. (g) and (h).

Sample Problem 19.3

The gear A shown in Fig. (a) is rolling around the fixed gear B as it spins freely about arm C. If the arm is rotating about the vertical at the constant rate $\omega_C = 15$ rad/s, calculate (1) the angular velocity of gear A; and (2) the angular acceleration of gear A.

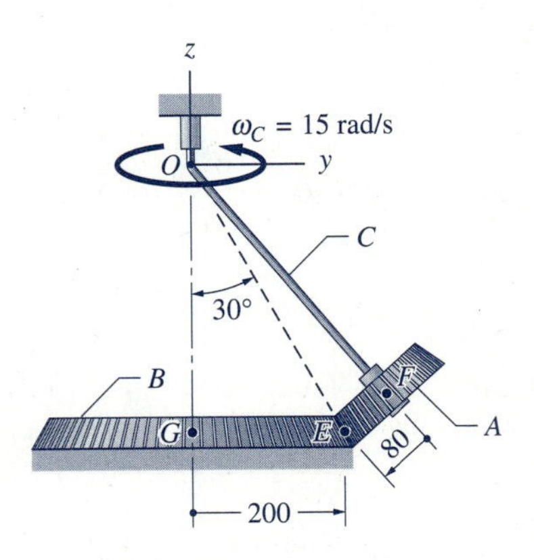

Dimensions in mm

(a)

Solution

Let the reference frame shown in Fig. (a) be attached to arm C. The geometry of the system is shown in Fig. (b), where the distances and angles were obtained directly from Fig. (a) or were computed using trigonometry.

(b)

Part 1

From Fig. (a) we see that gear A is rotating about the fixed point O (note that a line from O to any point on gear A does not change its length). Since the point of contact E also has no velocity, we conclude that the instant axis of rotation of gear A is the line OE, as indicated in Fig. (b) (we assume that the gear is extended to include O).

The angular velocity of gear A is given by

$$\boldsymbol{\omega}_A = \boldsymbol{\omega}_C + \boldsymbol{\omega}_{A/C} \tag{a}$$

where $\boldsymbol{\omega}_C$ is the angular velocity of arm C, and $\boldsymbol{\omega}_{A/C}$ is the spin velocity of gear A, i.e., the angular velocity of gear A relative to the arm C.

Referring to Fig. (a), we see that the angular velocity of arm C is

$$\boldsymbol{\omega}_C = \omega_C \mathbf{k} = 15\mathbf{k} \text{ rad/s} \quad \text{(valid for all time)} \tag{b}$$

Because the direction of the spin velocity of gear A coincides with the unit vector $\boldsymbol{\lambda}_{FO}$ shown in Fig. (b), it may be written as

$$\boldsymbol{\omega}_{A/C} = \omega_{A/C}\boldsymbol{\lambda}_{FO} \quad \text{(valid for all time)} \tag{c}$$

Substituting Eqs. (b) and (c) into Eq. (a), the angular velocity of gear A becomes

$$\boldsymbol{\omega}_A = 15\mathbf{k} + \omega_{A/C}\boldsymbol{\lambda}_{FO} \quad \text{(valid for all time)} \tag{d}$$

The vector diagram representing Eq. (d) is shown in Fig. (c). Observe that $\boldsymbol{\omega}_A$ lies along the instant axis of rotation, and $\boldsymbol{\omega}_{A/C}$ is directed along the line FO, the axis of arm C. The angles α, β, and γ in Fig. (c) were computed from the angles shown in Fig. (b).

$\alpha = 60 - 48.46 = 11.54°$
$\beta = 30°$
$\gamma = 90 + 48.46 = 138.46°$

We see that Fig. (c) contains only two unknown variables, ω_A and $\omega_{A/C}$, which can be computed by trigonometry. Applying the law of sines yields

$$\frac{15}{\sin 11.54°} = \frac{\omega_{A/C}}{\sin 30°} = \frac{\omega_A}{\sin 138.46°}$$

from which we obtain

$$\omega_{A/C} = 37.49 \text{ rad/s} \quad \text{and} \quad \omega_A = 49.72 \text{ rad/s} \tag{e}$$

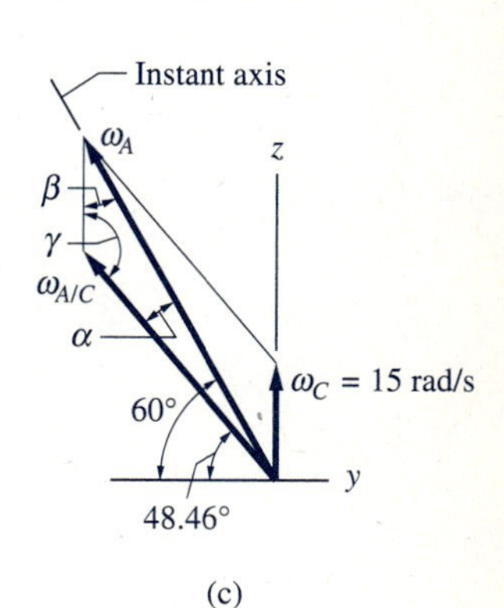

(c)

Using the fact that the angular velocity vector of gear A coincides with the unit vector $\boldsymbol{\lambda}_{EO}$ in Fig. (b), it may be written as

$$\boldsymbol{\omega}_A = 49.72\boldsymbol{\lambda}_{EO} = 49.72(-\sin 30°\mathbf{j} + \cos 30°\mathbf{k})$$

or

$$\boldsymbol{\omega}_A = -24.9\mathbf{j} + 43.1\mathbf{k} \text{ rad/s} \qquad \textit{Answer} \tag{f}$$

Note that Eq. (f) is valid for all time.

An alternate method of calculating the angular and spin velocities of A is to solve the vector equation $\mathbf{v}_E = \boldsymbol{\omega}_A \times \mathbf{r}_{E/O} = \mathbf{0}$, where $\boldsymbol{\omega}_A$ is given in Eq. (d) and $\mathbf{r}_{E/O} = 200\mathbf{j} - 346.4\mathbf{k}$ mm.

Part 2

The angular acceleration of gear A can be computed by differentiating its angular velocity.

$$\boldsymbol{\alpha}_A = \dot{\boldsymbol{\omega}}_A = (\dot{\boldsymbol{\omega}}_A)_{/xyz} + \boldsymbol{\Omega} \times \boldsymbol{\omega}_A \tag{g}$$

where $(\dot{\boldsymbol{\omega}}_A)_{/xyz}$ is the angular acceleration of A relative to the xyz reference frame and $\boldsymbol{\Omega}$ is the angular velocity of the frame.

Using Eq. (f), $\boldsymbol{\omega}_A = -24.9\mathbf{j} + 43.1\mathbf{k}$ rad/s (which is valid for all time and can thus be differentiated), we obtain $(\dot{\boldsymbol{\omega}}_A)_{/xyz} = \mathbf{0}$, since the xyz-frame was assumed to be rotating with arm C. Furthermore, we have $\boldsymbol{\Omega} = \boldsymbol{\omega}_C$, and Eq. (g) becomes

$$\boldsymbol{\alpha}_A = \dot{\boldsymbol{\omega}}_A = \boldsymbol{\omega}_C \times \boldsymbol{\omega}_A = 15\mathbf{k} \times (-24.9\mathbf{j} + 43.1\mathbf{k})$$

which yields

$$\boldsymbol{\alpha}_A = 374\mathbf{i} \text{ rad/s}^2 \qquad \textit{Answer}$$

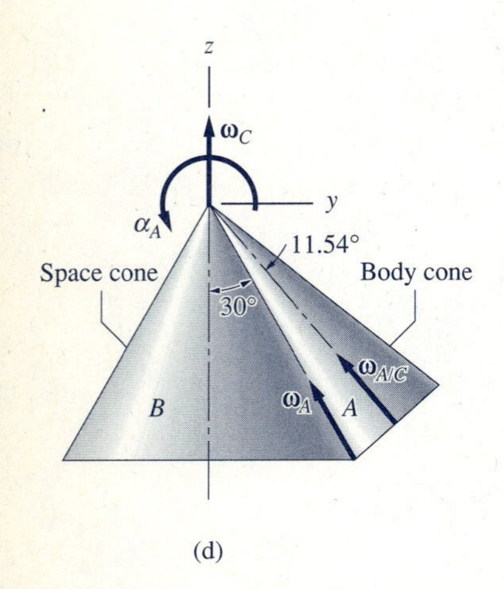

(d)

As shown in Fig. (d), the motion of gear A can be modeled as the body cone (representing gear A) rolling without slipping on the outside of the space cone (representing the fixed gear B). The figure also shows the angular velocity $\boldsymbol{\omega}_A$, its components $\boldsymbol{\omega}_C$ and $\boldsymbol{\omega}_{A/C}$, and the angular acceleration $\boldsymbol{\alpha}_A$. Note that $\boldsymbol{\alpha}_A$ is always perpendicular to $\boldsymbol{\omega}_A$.

Sample Problem 19.4

The mechanism shown in Fig. (a) consists of the crank PQ, which rotates about axis OP, and the control rod B connected to the crank and the sliding collar C. (1) If the connections at Q and C are ball-and-socket joints, compute the velocity of collar

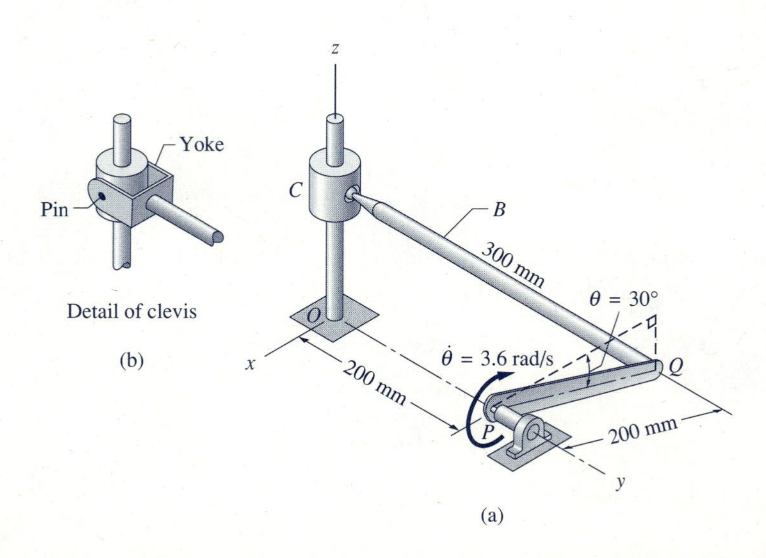

(a)

C and the angular velocity of rod B, given that $\theta = 30°$ and $\dot{\theta} = 3.6$ rad/s. (2) Re-solve Part 1 assuming that the connection at C is a clevis, the details of which are shown in Fig. (b).

Solution

Part 1

The relative velocity equation between points C and Q (note that both are points on rod B) is

$$\mathbf{v}_C = \mathbf{v}_Q + \mathbf{v}_{C/Q} = \boldsymbol{\omega}_{PQ} \times \mathbf{r}_{Q/P} + \boldsymbol{\omega}_B \times \mathbf{r}_{C/Q} \tag{a}$$

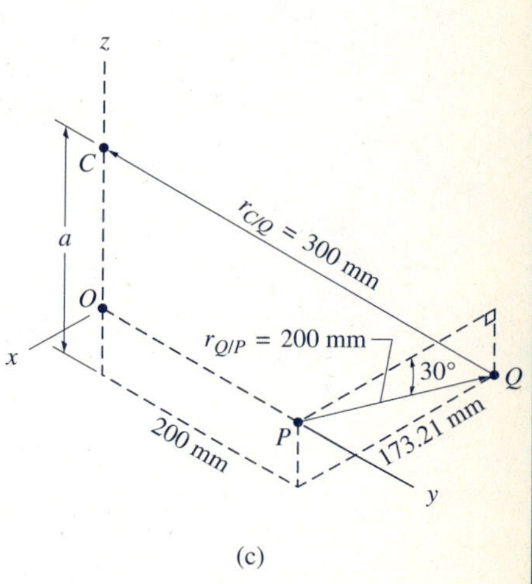

(c)

where $\boldsymbol{\omega}_{PQ}$ and $\boldsymbol{\omega}_B$ are the angular velocities of the crank and control rod, respectively, and the position vectors $\mathbf{r}_{Q/P}$ and $\mathbf{r}_{C/Q}$ are defined in Fig. (c).

Assuming that the velocity of the collar is upward, we have

$$\mathbf{v}_C = v_C\mathbf{k} \tag{b}$$

With $\mathbf{r}_{Q/P} = -173.21\mathbf{i} - 100.0\mathbf{k}$ mm and $\boldsymbol{\omega}_{PQ} = -3.6\mathbf{j}$ rad/s, the velocity of Q is

$$\begin{aligned}\mathbf{v}_Q = \boldsymbol{\omega}_{PQ} \times \mathbf{r}_{Q/P} &= -3.6\mathbf{j} \times (-173.21\mathbf{i} - 100.0\mathbf{k}) \\ &= 360.0\mathbf{i} - 623.6\mathbf{k} \text{ mm/s}\end{aligned} \tag{c}$$

Before we can find $\mathbf{r}_{C/Q}$, it is necessary to compute the distance a shown in Fig. (c). From the geometry of that figure, we see that $a = [(300)^2 - (173.21)^2 - (200)^2]^{1/2} = 141.42$ mm, which gives

$$\mathbf{r}_{C/Q} = 173.21\mathbf{i} - 200.0\mathbf{j} + 141.42\mathbf{k} \text{ mm} \tag{d}$$

Substituting Eqs. (b)–(d) into Eq. (a), and using the determinant form for the second cross product, we obtain

$$v_C\mathbf{k} = (360.0\mathbf{i} - 623.6\mathbf{k}) + \begin{vmatrix} \mathbf{i} & \mathbf{j} & \mathbf{k} \\ \omega_x & \omega_y & \omega_z \\ 173.21 & -200.0 & 141.42 \end{vmatrix}$$

where $\boldsymbol{\omega}_B = \omega_x\mathbf{i} + \omega_y\mathbf{j} + \omega_z\mathbf{k}$ is the angular velocity of rod B. Expanding the determinant, and equating like components, yields the following three scalar equations:

$$\left.\begin{aligned} 0 &= 360.0 \qquad\qquad\qquad + 141.42\,\omega_y + 200.0\omega_z \\ 0 &= \qquad -141.42\omega_x \qquad\qquad\qquad + 173.21\,\omega_z \\ v_C &= -623.6 - 200.0\omega_x - 173.21\,\omega_y \end{aligned}\right\} \tag{e}$$

These equations contain four unknown variables: v_C, ω_x, ω_y, and ω_z. Therefore, a solution cannot be obtained without additional information. The physical reason for our inability to find all four variables from the given information is the presence of ball-and-socket joints at C and Q. These joints allow rod B to spin about its axis with arbitrary angular velocity. Therefore, the component of $\boldsymbol{\omega}_B$ along CQ is not defined. However, note that this component does not affect the velocity of slider C.

We will complete our solution by assuming that the spin velocity of rod B about its axis is zero. This means the component of $\boldsymbol{\omega}_B$ along CQ vanishes; i.e., $\boldsymbol{\omega}_B \cdot \mathbf{r}_{C/Q} = 0$, or

$$(\omega_x\mathbf{i} + \omega_y\mathbf{j} + \omega_z\mathbf{k}) \cdot (173.21\mathbf{i} - 200.0\mathbf{j} + 141.42\mathbf{k}) = 0$$

$$173.21\omega_x - 200.0\omega_y + 141.42\,\omega_z = 0 \tag{f}$$

Equations (e) and (f) represent four scalar equations in four unknowns, the solution of which gives $v_C = -182.6$ mm/s, $\omega_x = -0.980$ rad/s, $\omega_y = -1.414$ rad/s, and $\omega_z = -0.800$ rad/s. Expressed in vector form, the results are

$$\left.\begin{aligned} \mathbf{v}_C &= -182.6\mathbf{k} \text{ mm/s} \\ \boldsymbol{\omega}_B &= -0.980\mathbf{i} - 1.414\mathbf{j} - 0.800\mathbf{k} \text{ rad/s} \end{aligned}\right\} \qquad \textit{Answer}$$

Part 2

Inspection of Fig. (b) reveals the following important characteristics of the clevis at C.

1. The pin is parallel to the xy-plane and perpendicular to rod B, i.e., to $\mathbf{r}_{C/Q}$.

A unit vector in the direction of the pin is therefore of the form $\boldsymbol{\lambda} = \lambda_x\mathbf{i} + \lambda_y\mathbf{j}$. Substituting for $\mathbf{r}_{C/Q}$ from Eq. (d) into $\boldsymbol{\lambda} \cdot \mathbf{r}_{C/Q} = 0$ gives $(\lambda_x\mathbf{i} + \lambda_y\mathbf{j}) \cdot (173.21\mathbf{i} - 200.0\mathbf{j} + 141.42\mathbf{k}) = 0$, from which we obtain $173.21\lambda_x - 200.0\lambda_y = 0$. Combining this equation with $\lambda_x^2 + \lambda_y^2 = 1$ yields

$$\boldsymbol{\lambda} = \pm 0.7559\mathbf{i} \pm 0.6547\mathbf{j} \tag{g}$$

2. The rotation of rod B relative to slider C can occur only about the pin of the clevis.

This constraint means that $\boldsymbol{\omega}_{B/C}$, the angular velocity of rod B relative to slider C, must be parallel to the pin. Using $\boldsymbol{\omega}_{B/C} = \omega_{B/C}\boldsymbol{\lambda}$ and choosing the positive signs in Eq. (g), we obtain

$$\boldsymbol{\omega}_{B/C} = 0.7559\omega_{B/C}\mathbf{i} + 0.6547\omega_{B/C}\mathbf{j} \tag{h}$$

The angular velocity of B may now be written as $\boldsymbol{\omega}_B = \boldsymbol{\omega}_C + \boldsymbol{\omega}_{B/C}$ where $\boldsymbol{\omega}_C = \omega_C\mathbf{k}$ is the angular velocity of the slider C and $\boldsymbol{\omega}_{B/C}$ is given in Eq. (h). Therefore,

$$\boldsymbol{\omega}_B = 0.7559\,\omega_{B/C}\mathbf{i} + 0.6547\,\omega_{B/C}\mathbf{j} + \omega_C\mathbf{k} \tag{i}$$

The relative velocity equation between points C and Q on rod B is

$$\mathbf{v}_C = \mathbf{v}_Q + \mathbf{v}_{C/Q} = \mathbf{v}_Q + \boldsymbol{\omega}_B \times \mathbf{r}_{C/Q} \tag{j}$$

The velocities $\mathbf{v}_C$ and $\mathbf{v}_Q$ are the same as found previously in Eqs. (b) and (c), respectively. Using Eq. (d) for $\mathbf{r}_{C/Q}$ and Eq. (i), the relative velocity term in Eq. (j) becomes

$$\boldsymbol{\omega}_B \times \mathbf{r}_{C/Q} = \begin{vmatrix} \mathbf{i} & \mathbf{j} & \mathbf{k} \\ 0.7559\,\omega_{B/C} & 0.6547\,\omega_{B/C} & \omega_C \\ 173.21 & -200.0 & 141.42 \end{vmatrix}$$

which on expansion becomes

$$\boldsymbol{\omega}_B \times \mathbf{r}_{C/Q} = \mathbf{i}(92.59\omega_{B/C} + 200.0\omega_C) + \mathbf{j}(-106.90\omega_{B/C} + 173.21\omega_C) + \mathbf{k}(-264.6\omega_{B/C}) \quad \text{(k)}$$

Substituting Eqs. (b), (c), and (k) into Eq. (j), and equating like components, yields the following scalar equations.

$$\begin{aligned} 0 &= 360.0 + 92.59\omega_{B/C} + 200.0\omega_C \\ 0 &= -106.90\omega_{B/C} + 173.21\omega_C \\ v_C &= -623.6 - 264.6\omega_{B/C} \end{aligned} \quad \text{(l)}$$

The solution of Eqs. (l) gives $v_C = 182.5$ mm/s, $\omega_{B/C} = -1.667$ rad/s, and $\omega_C = -1.029$ rad/s. The angular velocity of rod B is found by substituting these values into Eq. (i). The final results are

$$\begin{aligned} \mathbf{v}_C &= -182.5\mathbf{k} \text{ mm/s} \\ \boldsymbol{\omega}_B &= -1.260\mathbf{i} - 1.091\mathbf{j} - 1.029\mathbf{k} \text{ rad/s} \end{aligned} \quad \textit{Answer}$$

Comparing the answers for Parts 1 and 2, we see that the velocity of C is the same (except for minor round-off error). However, the angular velocity of rod B is different for the two connections, as would be expected.

PROBLEMS

19.1 Points A and B belonging to a rigid body are separated by the distance d. Point C, which lies on the straight line connecting A and B, is located at the distance βd from A, where β is a constant. Show that the velocities of the points are related by the equation $\mathbf{v}_C = (1 - \beta)\mathbf{v}_A + \beta \mathbf{v}_B$.

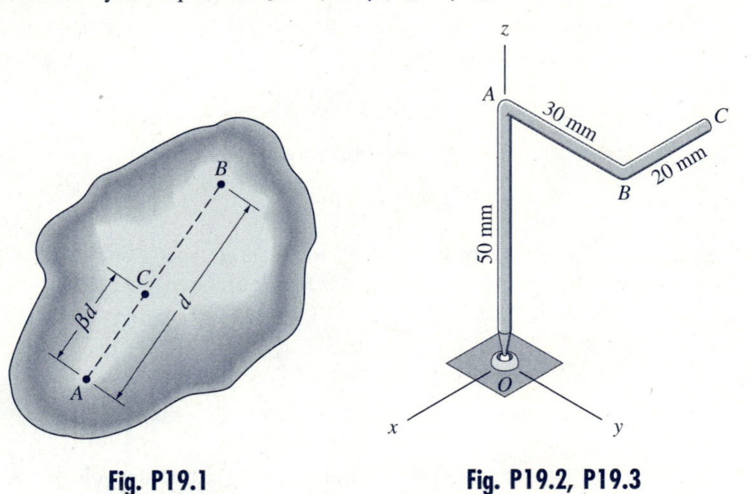

Fig. P19.1 **Fig. P19.2, P19.3**

19.2 Bar $OABC$ rotates about a ball-and-socket joint at O. In the position shown, the angular velocity vector of the rod is perpendicular to the line OC, and the velocity of point C is $\mathbf{v}_C = 3\mathbf{i} - 2\mathbf{j} + v_z\mathbf{k}$ m/s. For this position, determine v_z and the angular velocity vector.

19.3 Bar $OABC$ is rotating about the ball-and-socket joint at O. In the position shown, the velocities of points A and C are $\mathbf{v}_A = -2\mathbf{i} + a\mathbf{j}$ m/s and $\mathbf{v}_C = 3\mathbf{i} - 3\mathbf{j} + c\mathbf{k}$ m/s. For this position, calculate the constants a and c and the angular velocity vector of the bar.

19.4 The cone rolls on the xy-plane without slipping. The spin velocity $\omega_1 = 2.4$ rad/s is constant. For the position shown, determine (a) the angular velocity of the cone; (b) the angular acceleration of the cone; and (c) the velocity and acceleration of point P.

19.5 Cone A rolls without slipping inside the stationary cone B with the constant spin velocity $\omega_1 = 3.6$ rad/s. Calculate (a) the angular velocity of cone A; and (b) the angular acceleration of cone A.

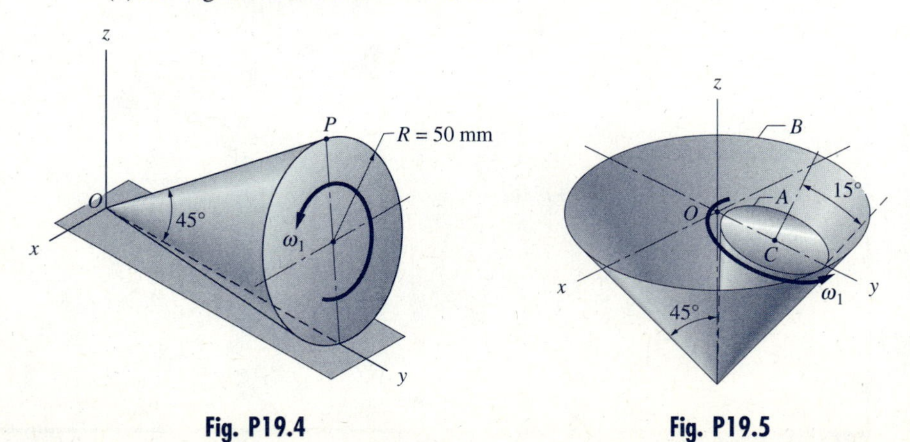

Fig. P19.4 **Fig. P19.5**

19.6 Disk A is free to spin about the bent axle B, which is rotating with the constant angular velocity ω_0 about the z-axis. Assuming that the disk rolls without slipping on the horizontal surface, determine the following for the position shown: (a) the angular velocity of the disk; (b) the angular acceleration of the disk; and (c) the velocity and acceleration of point P on the disk.

19.7 Disk A of the gyroscope spins about its axis, which is inclined at 23.6° relative to the xy-plane, at the constant angular speed $\omega_1 = 50$ rad/s. At the same time, frame B rotates about the z-axis with the variable angular velocity ω_2. At the instant when $\omega_2 = 20$ rad/s and $\dot{\omega}_2 = 600$ rad/s^2, calculate (a) the angular velocity of disk A; and (b) the angular acceleration of disk A.

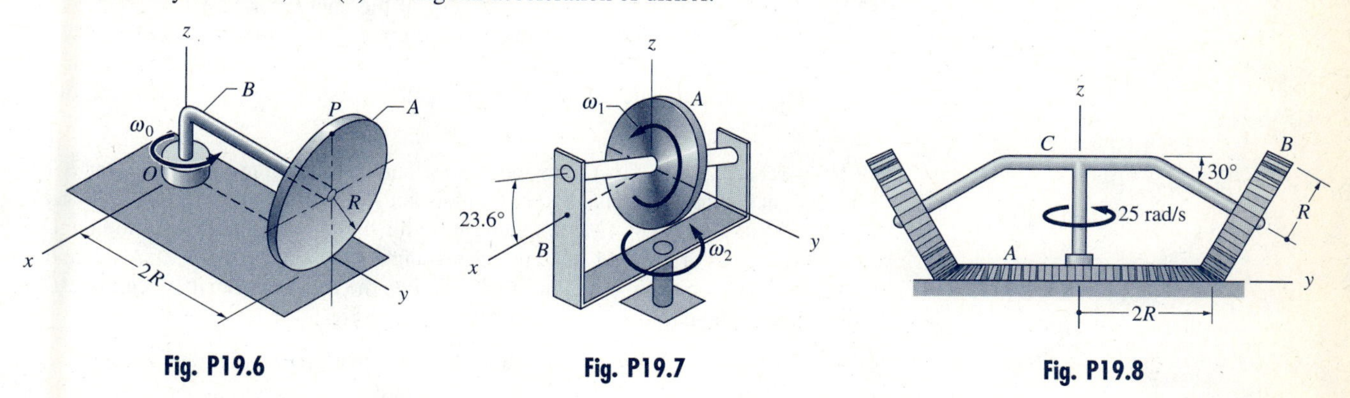

Fig. P19.6 **Fig. P19.7** **Fig. P19.8**

19.8 Two bevel gears attached to the arm C roll on the fixed gear A. The arm rotates about the z-axis at the constant angular velocity of 25 rad/s. Compute (a) the angular velocity of gear B; and (b) the angular acceleration of gear B.

19.9 Gears A and B spin freely on the bent shaft D, whereas gear C is fixed. The shaft D rotates about the y-axis with the constant angular velocity ω_0. For the position shown, calculate the angular velocity of (a) gear A; and (b) gear B.

19.10 Arm B of the cutting tool rotates about the x-axis with the constant angular velocity $\omega_1 = 15$ rad/s. When the arm is in the horizontal position shown, the cutting wheel A is spinning at the angular velocity $\omega_2 = 60$ rad/s, which is changing at the rate $\dot{\omega}_2 = -120$ rad/s^2. When the cutting wheel is in this position, determine (a) its angular velocity $\boldsymbol{\omega}_A$; (b) its angular acceleration $\boldsymbol{\alpha}_A$; and (c) the angle between $\boldsymbol{\omega}_A$ and $\boldsymbol{\alpha}_A$.

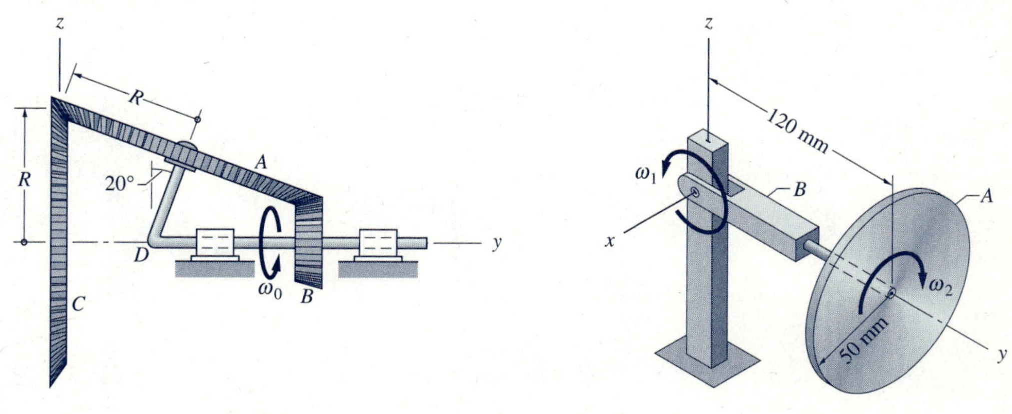

Fig. P19.9 **Fig. P19.10**

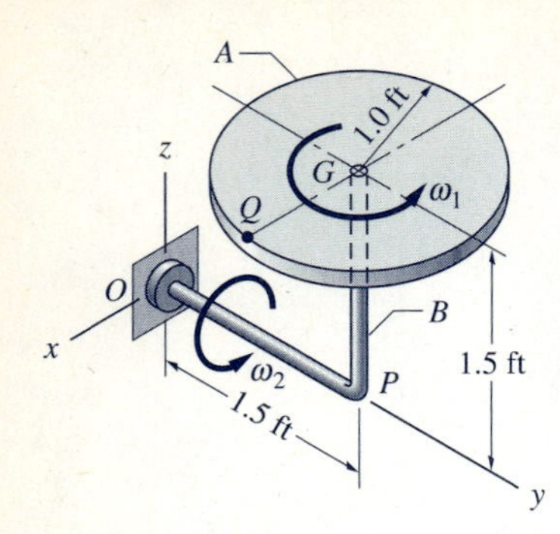

Fig. P19.11

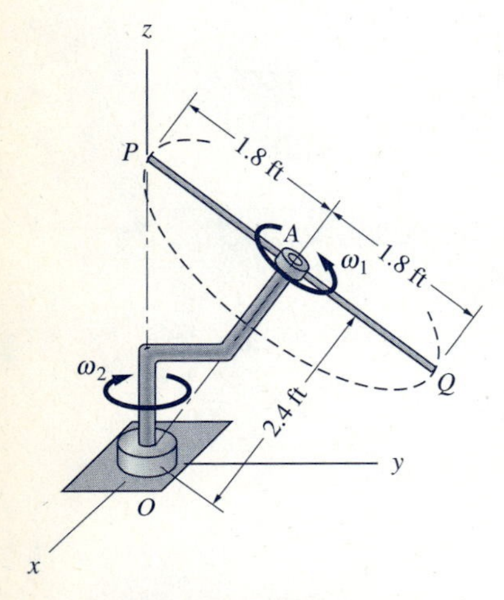

Fig. P19.12, P19.13

19.11 Disk A spins with respect to arm B with the angular speed ω_1 as the arm rotates about the y-axis with the angular speed ω_2, neither speed being constant. When the assembly is in the position shown, $\omega_1 = 3$ rad/s, $\omega_2 = 4$ rad/s, $\dot{\omega}_1 = -16$ rad/s^2, and $\dot{\omega}_2 = 25$ rad/s^2. For this position, determine (a) the velocity of point Q on the disk; and (b) the acceleration of point Q.

19.12 Bar PQ spins about the axis OA with the constant angular velocity $\omega_1 = 20$ rad/s. At the same time, OA rotates about the z-axis with the constant angular velocity $\omega_2 = 12$ rad/s. For the position shown, determine (a) the velocity of end P; and (b) the acceleration of end P.

19.13 The bar PQ spins about the arm OA at the constant rate $\omega_1 = 20$ rad/s. At the same time, OA rotates about the z-axis at the constant angular velocity ω_2. When bar PQ is in the position shown, the velocity of end Q is zero. For this position, calculate (a) the angular velocity of bar PQ; and (b) the acceleration of end Q.

19.14 Cranks AB and CD rotate about axes that are parallel to the y-axis. Bar BD is attached to the cranks by ball-and-socket joints. If the angular speed of AB is constant at $\omega_0 = 12$ rad/s, determine for the position shown: (a) the angular velocities of CD and BD; and (b) the angular acceleration of CD. Assume that bar BD is not spinning about its axes; i.e., the angular velocity vector of BD is perpendicular to BD.

19.15 The rigid body consists of three thin rods that are welded together at C. The motion of the body is constrained by a ball-and-socket joint at O and the fixed guide-slots at A and B. If the slope angle of the slot at A is $\beta = 90°$, determine the unit vector that defines the direction of the instant axis of rotation when the body is in the position shown.

19.16 When the body described in Prob. 19.15 is in the position shown, the speed of point C on the body is 25 in./s. Determine the angular velocity vector of the body if the inclination of the slot in A is $\beta = 30°$.

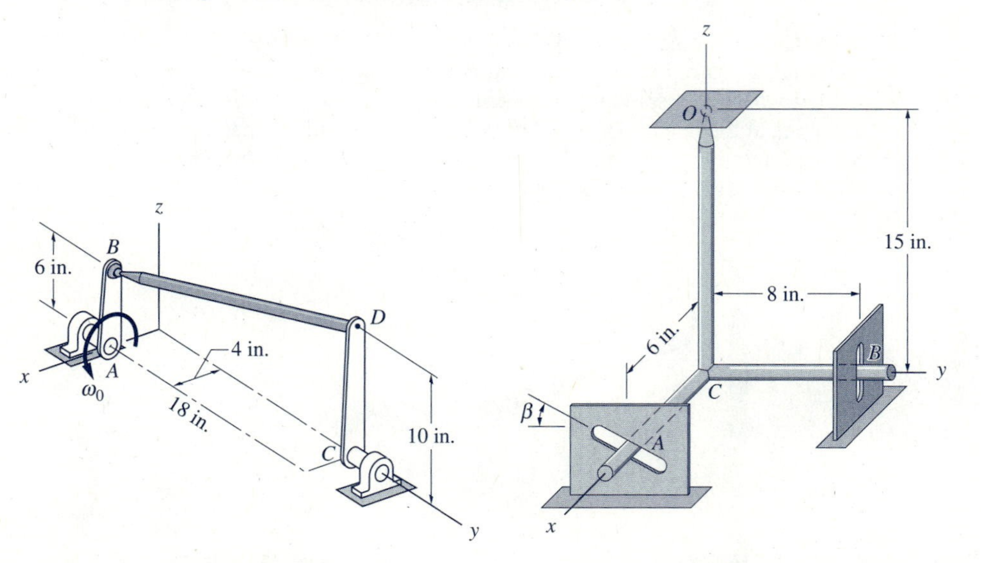

Fig. P19.14

Fig. P19.15, P19.16

19.17 Rod C is connected to collar A by a clevis and to collar B by a ball-and-socket joint. In the position shown, collar A is moving to the right with the speed v_A = 15 in./s. Find the speed of collar B for this position. (Note: It is not necessary to compute the angular velocity of rod C.)

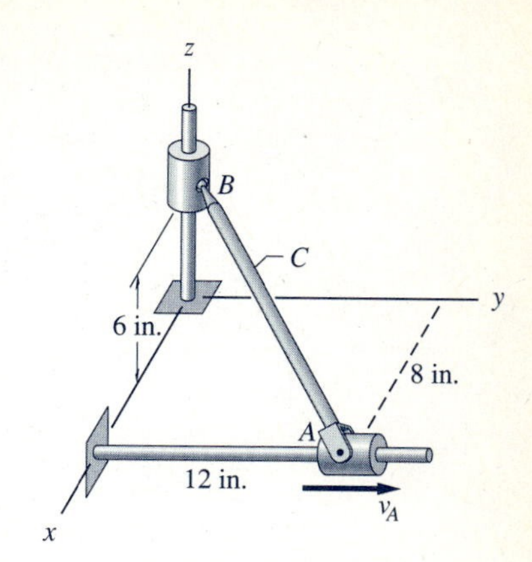

Fig. P19.17–P19.19

19.18 For the mechanism described in Prob. 19.17, determine the angular velocities of collar A and rod C in the position shown.

19.19 Assume that the connections for the rod described in Prob. 19.17 are interchanged (i.e., the ball-and-socket is at A and the clevis is at B). For the position shown, determine the angular velocity of rod C and the velocity of collar B.

19.20 Rod C of length $3R$ is attached to the rim of the disk with a ball-and-socket joint. The other end of C is joined to the collar B by a clevis. If the disk rotates with the constant angular speed ω_0, directed as shown, calculate the velocity of collar B and the angular velocity of rod C in the position shown.

19.21 Consider the mechanism described in Prob. 19.20 at the instant when the ball-and-socket is located at A'. For this position, calculate (a) the angular velocity of bar C and the velocity of collar B; and (b) the acceleration of collar B.

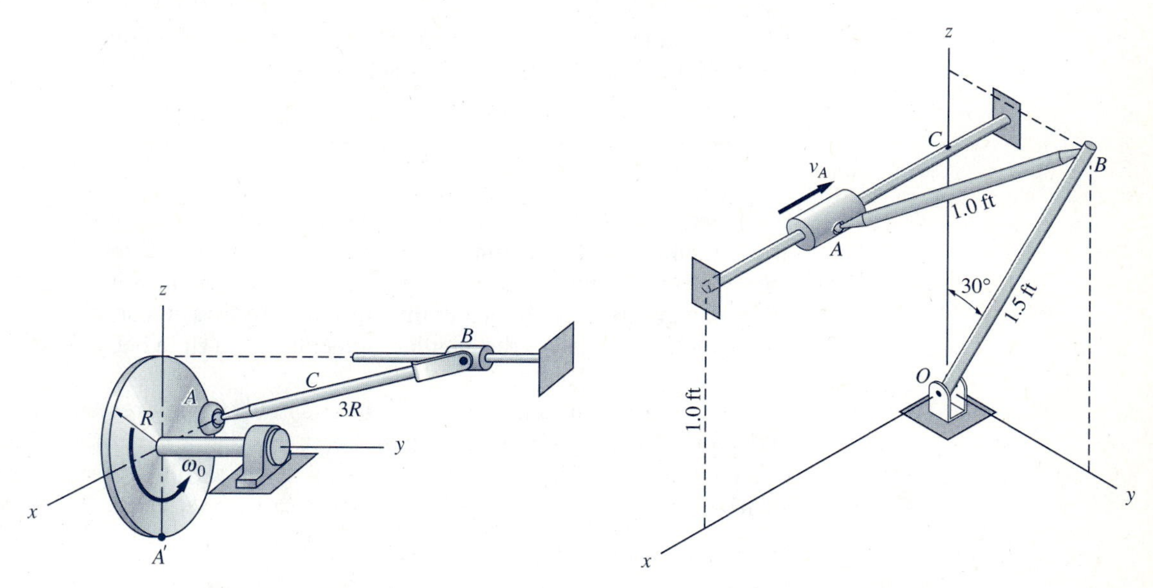

Fig. P19.20, P19.21

Fig. P19.22, P19.23

19.22 Rod OB rotates in the yz-plane about a pin at O as collar A moves with constant velocity v_A = 6 in./s, directed as shown. Ball-and-socket joints connect rod AB to the collar A and to the rod OB. Assuming that the rod AB does not spin about its axis, determine the angular velocity vectors of rods OB and AB in the position shown.

19.23 Consider the mechanism described in Prob. 19.22 at the instant when the collar A has reached point C. For this position, calculate (a) the angular velocity of bar AB (assume that AB does not spin about its axis); and (b) the acceleration of joint B and the angular acceleration of rod OB. (Hint: Joint B has no velocity in this position.)

*19.3 Impulse-Momentum Method

Chapter 18 extended the impulse-momentum method from a system of particles to a rigid body undergoing plane motion. Here we extend the method to general, three-dimensional motion of a rigid body. We assume, as we did in Chapter 18, that a rigid body consists of an infinite number of infinitesimal particles, with distances between the particles remaining constant during motion.

a. Linear momentum

The expression for linear momentum of a system of particles in Art. 15.6 is also applicable to a rigid body. Therefore, we have

$$\mathbf{p} = m\bar{\mathbf{v}} \tag{19.5}$$

where m is the mass of the body and $\bar{\mathbf{v}}$ is the velocity of its mass center. (This equation is identical to Eq. (18.11), which was used for plane motion.)

b. Angular momentum

From Art. 14.7, the angular momentum of a particle about a point (also known as the moment of momentum) is defined as the moment of its linear momentum. The angular momentum of a rigid body about a point is obtained by summing (integrating) the angular momenta of the constituent particles. Using this approach, we can derive the angular momentum of a rigid body about any reference point.

Figure 19.4 shows a rigid body of mass m occupying region $\mathcal{V}$, which is undergoing three-dimensional motion. Point O is the origin of the xyz inertial reference frame. Other points shown in the figure are the mass center G and an arbitrary reference point A (A is not necessarily fixed in the body).

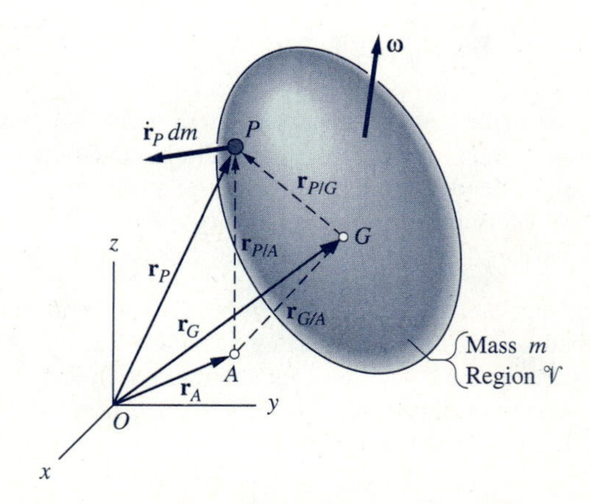

Fig. 19.4

The linear momentum of a typical particle P of mass dm is $dm\,\dot{\mathbf{r}}_P$, where $\mathbf{r}_P$ is the position vector of P. Therefore, the angular momentum of P about A is $d\mathbf{h}_A = \mathbf{r}_{P/A} \times dm\,\dot{\mathbf{r}}_P$, where $\mathbf{r}_{P/A}$ is the position vector of P relative to A. Integrating over the region $\mathcal{V}$ gives the angular momentum of the body about A:*

$$\mathbf{h}_A = \int_{\mathcal{V}} \mathbf{r}_{P/A} \times \dot{\mathbf{r}}_P\, dm \qquad (A\text{: arbitrary point}) \tag{19.6}$$

If point A is fixed in the rigid body, the velocity of P is related to the velocity of A by $\dot{\mathbf{r}}_P = \dot{\mathbf{r}}_A + \boldsymbol{\omega} \times \mathbf{r}_{P/A}$, which yields

$$\mathbf{h}_A = \int_{\mathcal{V}} \mathbf{r}_{P/A} \times \dot{\mathbf{r}}_A\, dm + \int_{\mathcal{V}} \mathbf{r}_{P/A} \times (\boldsymbol{\omega} \times \mathbf{r}_{P/A})\, dm \qquad (A\text{: fixed in the body}) \tag{19.7}$$

Since the mass center G is fixed in the rigid body, the angular momentum about G can be obtained by replacing $\mathbf{r}_{P/A}$ in Eq. (19.7) by $\mathbf{r}_{P/G}$, the position vector of P relative to G. Making this substitution and recognizing that $\int_{\mathcal{V}} \mathbf{r}_{P/G}\, dm = 0$ (the first moment of the mass about G is zero), the angular momentum about G becomes

$$\mathbf{h}_G = \int_{\mathcal{V}} \mathbf{r}_{P/G} \times (\boldsymbol{\omega} \times \mathbf{r}_{P/G})\, dm \qquad (G\text{: mass center}) \tag{19.8}$$

Consider the special case where the origin O of the inertial reference frame is fixed in the body (or body extended); i.e., the body rotates about the fixed point O. To obtain the angular momentum about O, we replace $\mathbf{r}_{P/A}$ in Eq. (19.7) by $\mathbf{r}_P$ and $\dot{\mathbf{r}}_A$ by $\dot{\mathbf{r}}_O = \mathbf{0}$, the result being

$$\mathbf{h}_O = \int_{\mathcal{V}} \mathbf{r}_P \times (\boldsymbol{\omega} \times \mathbf{r}_P)\, dm \qquad (O\text{: fixed in the body and in space}) \tag{19.9}$$

Equation (19.9) is valid even if point O is not fixed in space, as long as its velocity is zero at the instant of concern.

c. Momentum diagrams

There is a relatively simple relationship between the angular momenta about an arbitrary point and about the mass center. From Fig. 19.4, we see that $\mathbf{r}_{P/A} = \mathbf{r}_{P/G} + \mathbf{r}_{G/A}$ and $\mathbf{r}_P = \mathbf{r}_G + \mathbf{r}_{P/G}$. Therefore, Eq. (19.6) can be written as

$$\mathbf{h}_A = \int_{\mathcal{V}} [(\mathbf{r}_{P/G} + \mathbf{r}_{G/A}) \times (\dot{\mathbf{r}}_G + \dot{\mathbf{r}}_{P/G})]\, dm$$

* Since the condition of rigidity has not yet been introduced, Eq. (19.6) is valid even if the body is not rigid.

which on expansion becomes

$$\mathbf{h}_A = \left(\int_{\mathcal{V}} \mathbf{r}_{P/G}\, dm \times \dot{\mathbf{r}}_G\right) + \left(\int_{\mathcal{V}} \mathbf{r}_{P/G} \times \dot{\mathbf{r}}_{P/G}\, dm\right)$$
$$+ \left(\mathbf{r}_{G/A} \times \dot{\mathbf{r}}_G \int_{\mathcal{V}} dm\right) + \left(\mathbf{r}_{G/A} \times \int_{\mathcal{V}} \dot{\mathbf{r}}_{P/G}\, dm\right)$$

The first and fourth terms on the right-hand side vanish in view of the definition of the mass center: $\int_{\mathcal{V}} \mathbf{r}_{P/G}\, dm = \mathbf{0}$. On substituting $\dot{\mathbf{r}}_{P/G} = \boldsymbol{\omega} \times \mathbf{r}_{P/G}$, the second term equals $\mathbf{h}_G$—see Eq. (19.8). The third term is equal to $\mathbf{r}_{G/A} \times m\bar{\mathbf{v}}$, where $\bar{\mathbf{v}} = \dot{\mathbf{r}}_G$, the velocity of the mass center. Consequently, we obtain

$$\mathbf{h}_A = \mathbf{h}_G + \mathbf{r}_{G/A} \times m\bar{\mathbf{v}} \qquad (A\text{: arbitrary point}) \tag{19.10}$$

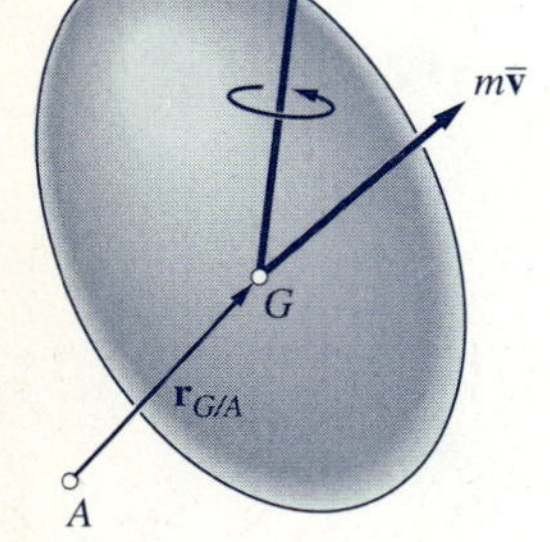

Fig. 19.5

The pictorial representation of Eq. (19.10), called the *momentum diagram,* is shown in Fig. 19.5. This diagram is the three-dimensional version of Fig. 18.4, which we used for plane motion. The momentum diagram consists of the couple $\mathbf{h}_G$, representing the angular momentum about the mass center, and the linear momentum vector $m\bar{\mathbf{v}}$ attached to the mass center. It can be seen by inspection that Eq. (19.10) is equivalent to summing moments of the momenta in Fig. 19.5 about an arbitrary point A.

d. Inertial properties

We introduce here the inertial properties of a rigid body described in three dimensions (this topic is discussed more fully in Appendix F). As we will see later, these properties arise in the computation of the angular momentum.

1. Moments and Products of Inertia and the Inertia Tensor The *mass moment of inertia* of a body about the a-axis was defined in Eq. (17.1): $I_a = \int_{\mathcal{V}} r^2\, dm$, where $\mathcal{V}$ is the region occupied by the body and r is the perpendicular distance from the a-axis to the differential mass dm. It can be seen in Fig. 19.6 that the perpendicular distances from the x-, y-, and z-axes to dm are

$$(y^2 + z^2)^{1/2} \qquad (z^2 + x^2)^{1/2} \qquad \text{and} \qquad (x^2 + y^2)^{1/2}$$

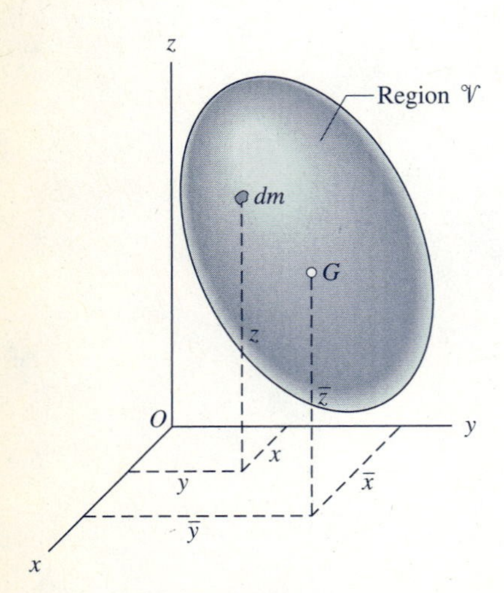

Fig. 19.6

respectively. Therefore, the mass moments of inertia of the body about the three coordinate axes are

$$\begin{aligned} I_x &= \int_{\mathcal{V}} (y^2 + z^2)\, dm \\ I_y &= \int_{\mathcal{V}} (z^2 + x^2)\, dm \\ I_z &= \int_{\mathcal{V}} (x^2 + y^2)\, dm \end{aligned} \tag{19.11}$$

The *products of inertia* of the body with respect to the rectangular axes shown in Fig. 19.6 are defined as

$$I_{xy} = I_{yx} = \int_{\mathcal{V}} xy\, dm$$
$$I_{yz} = I_{zy} = \int_{\mathcal{V}} yz\, dm \qquad (19.12)$$
$$I_{zx} = I_{xz} = \int_{\mathcal{V}} zx\, dm$$

The units of the inertial properties defined in Eqs. (19.11) and (19.12) are mass $\times$ (length)^2, e.g., kg $\cdot$ m^2 or slug $\cdot$ ft^2 (lb $\cdot$ ft $\cdot$ s^2). Although moments of inertia are always positive, products of inertia may be positive, negative, or zero.

The following matrix of inertial properties is called the *inertia tensor* of the body *at point O* (the origin of the coordinate axes).

$$\mathbf{I} = \begin{bmatrix} I_x & -I_{xy} & -I_{xz} \\ -I_{yx} & I_y & -I_{yz} \\ -I_{zx} & -I_{zy} & I_z \end{bmatrix} \qquad (19.13)$$

Note that the off-diagonal terms are the negatives of the products of inertia. (The minus signs must be included for **I** to satisfy certain conventions of tensor algebra, a topic that is beyond the scope of this text.)

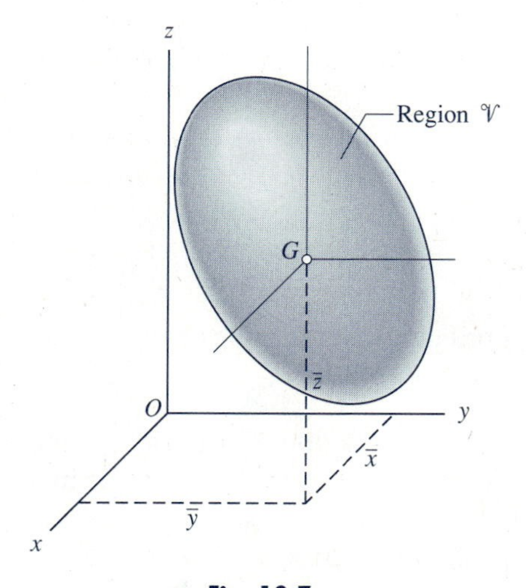

Fig. 19.7

2. Parallel-Axis Theorem for Moments of Inertia The *parallel-axis theorem* for the moment of inertia was derived in Chapter 17 [see Eq. (17.3)]: $I_a = \bar{I}_a + md^2$, where I_a is the moment of inertia about the a-axis, $\bar{I}_a$ is the moment of inertia about the central a-axis (the axis that passes through the mass center of the body and is parallel to the a-axis), m is the mass of the body, and d is the distance between the two axes. Referring to Fig. 19.7, we see that the distances between the x-, y-, and z-axes and the corresponding central axes are $(\bar{y}^2 + \bar{z}^2)^{1/2}$, $(\bar{z}^2 + \bar{x}^2)^{1/2}$, and $(\bar{x}^2 + \bar{y}^2)^{1/2}$, respectively. Letting $\bar{I}_x$, $\bar{I}_y$, and $\bar{I}_z$

denote the moments of inertia about the central axes, the parallel-axis theorem for moments of inertia becomes

$$\begin{aligned} I_x &= \bar{I}_x + m(\bar{y}^2 + \bar{z}^2) \\ I_y &= \bar{I}_y + m(\bar{z}^2 + \bar{x}^2) \\ I_z &= \bar{I}_z + m(\bar{x}^2 + \bar{y}^2) \end{aligned} \tag{19.14}$$

3. Parallel-Plane Theorem for Products of Inertia Referring again to the body in Fig. 19.7, it can be shown that the products of inertia with respect to the two sets of axes are related by

$$\begin{aligned} I_{xy} &= \bar{I}_{xy} + m\bar{x}\bar{y} \\ I_{yz} &= \bar{I}_{yz} + m\bar{y}\bar{z} \\ I_{zx} &= \bar{I}_{zx} + m\bar{z}\bar{x} \end{aligned} \tag{19.15}$$

where $\bar{I}_{xy}$, $\bar{I}_{yz}$, and $\bar{I}_{zx}$ denote the products of inertia about the central axes. This theorem, which is known as the *parallel-plane theorem,* is proved in Appendix F.

4. Principal Moments of Inertia It can be shown that it is always possible to find three perpendicular axes at any given point O such that the products of inertia with respect to these axes vanish. These axes are called the *principal axes at point O,* and the corresponding moments of inertia, denoted by I_1, I_2, and I_3, are known as the *principal moments of inertia* of the body *at point O*. Referred to the principal axes, the inertia tensor thus assumes the form of a diagonal matrix:

$$\mathbf{I} = \begin{bmatrix} I_1 & 0 & 0 \\ 0 & I_2 & 0 \\ 0 & 0 & I_3 \end{bmatrix} \tag{19.16}$$

The determination of principal axes and the computation of principal moments of inertia are explained in Appendix F.

5. Body with a Plane of Symmetry Consider a homogeneous body that is symmetric about a plane, say the xy-plane. Let the mass center of the body, which lies in the plane of symmetry, be the origin of the xyz-frame. Symmetry implies that for every differential mass dm with coordinates (x, y, z), there exists another dm with coordinates $(x, y, -z)$. It follows that $\bar{I}_{yz} = \int_{\mathcal{V}} yz\, dm = 0$ and $\bar{I}_{zx} = \int_{\mathcal{V}} zx\, dm = 0$, since the integral over the region of $\mathcal{V}$ where $z < 0$ cancels the integral over the region where $z > 0$. Because $\bar{I}_{yz} = \bar{I}_{xz} = 0$ regardless of the orientation of the xy-axes, the z-axis must be a principal axis of inertia. In this special case, the parallel-plane theorem for product of inertia, Eqs. (19.15), becomes

$$I_{xy} = \bar{I}_{xy} + m\bar{x}\bar{y} \qquad I_{yz} = m\bar{y}\bar{z} \qquad I_{zx} = m\bar{z}\bar{x} \tag{19.17}$$

In general, an axis that is perpendicular to a plane of symmetry and passes through the mass center G of the body is a principal axis of the body at point G.

e. Rectangular components of angular momentum

Equations (19.8) and (19.9) have the form

$$\mathbf{h} = \int_{\mathcal{V}} \mathbf{r} \times (\boldsymbol{\omega} \times \mathbf{r})\, dm \tag{19.18}$$

where

$\mathbf{r} = \mathbf{r}_P$ if the reference point O is fixed in the body and in space

or

$\mathbf{r} = \mathbf{r}_{P/G}$ if the reference point is the mass center G

This notation will enable us to treat both cases simultaneously in the following discussion.

The rectangular representations of the vectors appearing in Eq. (19.18) are

$$\mathbf{h} = h_x\mathbf{i} + h_y\mathbf{j} + h_z\mathbf{k}$$
$$\boldsymbol{\omega} = \omega_x\mathbf{i} + \omega_y\mathbf{j} + \omega_z\mathbf{k}$$
$$\mathbf{r} = x\mathbf{i} + y\mathbf{j} + z\mathbf{k}$$

In the last equation, x, y, and z are the rectangular components of either $\mathbf{r}_P$ or $\mathbf{r}_{P/G}$, as appropriate. The cross product of $\boldsymbol{\omega}$ and $\mathbf{r}$ now becomes

$$\boldsymbol{\omega} \times \mathbf{r} = \mathbf{i}(\omega_y z - \omega_z y) + \mathbf{j}(\omega_z x - \omega_x z) + \mathbf{k}(\omega_x y - \omega_y x)$$

Consequently, the integrand in Eq. (19.18) is

$$\mathbf{r} \times (\boldsymbol{\omega} \times \mathbf{r}) = \begin{vmatrix} \mathbf{i} & \mathbf{j} & \mathbf{k} \\ x & y & z \\ \omega_y z - \omega_z y & \omega_z x - \omega_x z & \omega_x y - \omega_y x \end{vmatrix} \tag{a}$$

Expansion of this determinant gives

$$\begin{aligned} \mathbf{r} \times (\boldsymbol{\omega} \times \mathbf{r}) &= \mathbf{i}\left[(y^2 + z^2)\,\omega_x - xy\,\omega_y - xz\,\omega_z\right] \\ &\quad + \mathbf{j}\left[-xy\,\omega_x + (z^2 + x^2)\,\omega_y - yz\,\omega_z\right] \\ &\quad + \mathbf{k}\left[-zx\,\omega_x - zy\,\omega_y + (x^2 + y^2)\,\omega_z\right] \end{aligned} \tag{b}$$

Substituting Eq. (b) into Eq. (19.18), we obtain

$$\begin{aligned} h_x &= \omega_x \int_{\mathcal{V}} (y^2 + z^2)\, dm - \omega_y \int_{\mathcal{V}} xy\, dm - \omega_z \int_{\mathcal{V}} xz\, dm \\ h_y &= -\omega_x \int_{\mathcal{V}} xy\, dm + \omega_y \int_{\mathcal{V}} (z^2 + x^2)\, dm - \omega_z \int_{\mathcal{V}} yz\, dm \\ h_z &= -\omega_x \int_{\mathcal{V}} xz\, dm - \omega_y \int_{\mathcal{V}} yz\, dm + \omega_z \int_{\mathcal{V}} (x^2 + y^2)\, dm \end{aligned} \tag{c}$$

Recognizing that the integrals in Eq. (c) are the components of the inertia tensor—see Eqs. (19.11) and (19.12)—we obtain

$$\begin{aligned} h_x &= I_x\omega_x - I_{xy}\omega_y - I_{xz}\omega_z \\ h_y &= -I_{yx}\omega_x + I_y\omega_y - I_{yz}\omega_z \\ h_z &= -I_{zx}\omega_x - I_{zy}\omega_y + I_z\omega_z \end{aligned} \tag{19.19}$$

Recall that Eqs. (19.19) are valid only if the reference point—i.e., the origin of the coordinate system—is either a fixed point (fixed in the body as well as space) or the mass center of the body. If the xyz-axes are the principal axes of inertia at the reference point, the products of inertia vanish, and we are left with

$$\begin{aligned} h_x &= I_x\omega_x \\ h_y &= I_y\omega_y \\ h_z &= I_z\omega_z \end{aligned} \tag{19.20}$$

In general, the angular momentum vector **h** is not in the same direction as the angular velocity vector **ω**. For example, if the xyz-axes are principal axes of inertia, we have

$$\begin{aligned} \mathbf{h} &= I_x\omega_x\mathbf{i} + I_y\omega_y\mathbf{j} + I_z\omega_z\mathbf{k} \\ \boldsymbol{\omega} &= \omega_x\mathbf{i} + \omega_y\mathbf{j} + \omega_z\mathbf{k} \end{aligned} \tag{19.21}$$

From Eqs. (19.21) we see that the directions of **h** and **ω** coincide only for the following special cases.

1. The principal moments of inertia are equal (e.g., a homogeneous sphere with mass center being the origin of the coordinate system).
   Observe that with $I_x = I_y = I_z = I$, the first of Eqs. (19.21) gives $\mathbf{h} = I\boldsymbol{\omega}$.
2. The direction of **ω** is parallel to one of the principal axes of inertia.
   If **ω** is parallel to a principal axis, say the z-axis, then Eqs. (19.21) yield $\mathbf{h} = I_z\omega_z\mathbf{k}$ and $\boldsymbol{\omega} = \omega_z\mathbf{k}$. This case applies to plane motion of a rigid body. It also demonstrates that plane motion can exist only if the coordinate axis perpendicular to the plane of motion is a principal axis of inertia.

The rectangular components of the angular momentum about a fixed point, or the mass center, can be computed from Eqs. (19.19). If the angular momentum about some other point A is required, it can be obtained by first calculating $\mathbf{h}_G$ from Eqs. (19.19) and then applying Eq. (19.10): $\mathbf{h}_A = \mathbf{h}_G + \mathbf{r}_{G/A} \times m\bar{\mathbf{v}}$. Note again that Eqs. (19.19) are not directly applicable for an arbitrary reference point.

f. Impulse-momentum principles

The impulse-momentum principles for plane motion, discussed in Art. 18.6, are also applicable to three-dimensional motion. Here we restate these principles, without repeating the derivations given in Art. 18.6.

The linear impulse-momentum equation for a rigid body is

$$\boxed{\mathbf{L}_{1-2} = \mathbf{p}_2 - \mathbf{p}_1 = \Delta\mathbf{p}} \tag{19.22}$$

where $\mathbf{L}_{1-2}$ is the linear impulse of the external forces acting on the body during the time interval t_1 to t_2 and $\Delta\mathbf{p}$ is the change in the linear momentum of the body during this time interval. As defined in Eq. (19.5), the linear momentum of a body of mass m is $\mathbf{p} = m\bar{\mathbf{v}}$, where $\bar{\mathbf{v}}$ is the velocity of its mass center.

The angular impulse-momentum equation for a rigid body is

$$\boxed{(\mathbf{A}_A)_{1-2} = (\mathbf{h}_A)_2 - (\mathbf{h}_A)_1 = \Delta\mathbf{h}_A \qquad (A\text{: fixed in space or mass center})} \tag{19.23}$$

where $(\mathbf{A}_A)_{1-2}$ is the angular impulse of the external forces about point A during the time interval t_1 to t_2, and $\Delta\mathbf{h}_A$ is the change in angular momentum about A during this time period. The angular momentum can be computed from Eqs. (19.19) or (19.20).

As mentioned in Art. 18.6, the impulse-momentum equations can also be applied to systems of rigid bodies provided that (1) the impulses refer only to forces that are external to the system; (2) the momenta are interpreted as the momenta of the system (obtained by summing the momenta of the bodies that constitute the system); and (3) the mass center referred to in Eq. (19.23) is the mass center of the system.

According to Eq. (19.22), linear momentum is conserved if the linear impulse is zero. Similarly, we conclude from Eq. (19.23) that the angular momentum about a fixed point or the mass center is conserved if the angular impulse about that point is zero.

If the motion is impulsive (infinite forces acting over infinitesimal time intervals), which is the approximation used in impact problems, Eq. (19.23) is valid for any reference point (a proof of this statement is given in Art. 18.7). Since the duration of the impact is assumed to be infinitesimal, only the impulsive forces need be taken into account, the impulses of finite forces being negligible.

*19.4 Work-Energy Method

In this article, we discuss the application of the work-energy principle developed in Chapter 18 to three-dimensional motion. As in the preceding article, we begin with the equations for particle motion and derive the corresponding equations for a rigid body by assuming that the body is made up of an infinite number of infinitesimal particles.

a. Kinetic energy

Figure 19.8 shows a rigid body of mass m that occupies region $\mathcal{V}$. The particle P of the body has mass dm, and its position vector is $\mathbf{r}_P$, measured from the origin O of the xyz inertial reference frame. The position vector of the mass center G of the body is $\mathbf{r}_G$.

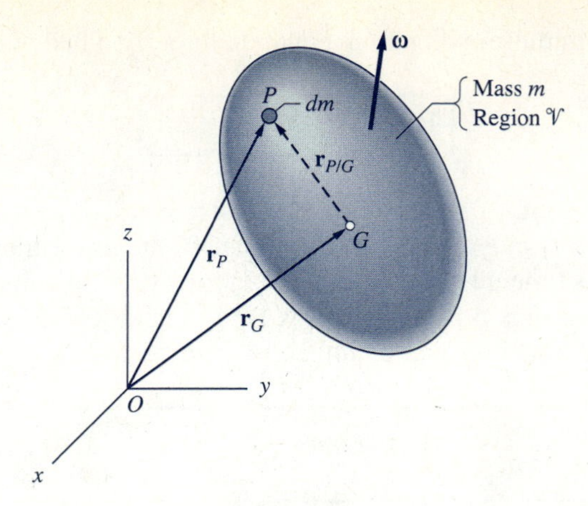

Fig. 19.8

The kinetic energy of the particle P is $dT = (1/2)\, dm\, v_P^2$, where v_P is the magnitude of its velocity vector. Integrating this expression over region $\mathcal{V}$ yields the kinetic energy of the body.

$$T = \frac{1}{2}\int_{\mathcal{V}} v_P^2\, dm \tag{19.24}$$

A convenient form of this equation is obtained if we introduce $\bar{\mathbf{v}}$, the velocity of the mass center. From Fig. 19.8 we see that $\mathbf{r}_P = \mathbf{r}_G + \mathbf{r}_{P/G}$. The time derivative of this equation yields $\dot{\mathbf{r}}_P = \dot{\mathbf{r}}_G + \dot{\mathbf{r}}_{P/G}$, or $\mathbf{v}_P = \bar{\mathbf{v}} + \dot{\mathbf{r}}_{P/G}$, where $\bar{\mathbf{v}} = \dot{\mathbf{r}}_G$. Therefore,

$$v_P^2 = (\bar{\mathbf{v}} + \dot{\mathbf{r}}_{P/G})\cdot(\bar{\mathbf{v}} + \dot{\mathbf{r}}_{P/G}) \tag{a}$$

Because points P and G both belong to the rigid body, we can write $\dot{\mathbf{r}}_{P/G} = \boldsymbol{\omega}\times\mathbf{r}_{P/G}$, which on substitution into Eq. (a) and expansion of the dot product yields

$$v_P^2 = \bar{v}^2 + 2\bar{\mathbf{v}}\cdot(\boldsymbol{\omega}\times\mathbf{r}_{P/G}) + (\boldsymbol{\omega}\times\mathbf{r}_{P/G})\cdot(\boldsymbol{\omega}\times\mathbf{r}_{P/G}) \tag{b}$$

Utilizing the vector identity $\mathbf{A}\cdot(\mathbf{B}\times\mathbf{C}) = \mathbf{B}\cdot(\mathbf{C}\times\mathbf{A})$ to rewrite the last term of Eq. (b), we get

$$v_P^2 = \bar{v}^2 + 2\bar{\mathbf{v}}\cdot(\boldsymbol{\omega}\times\mathbf{r}_{P/G}) + \boldsymbol{\omega}\cdot[\mathbf{r}_{P/G}\times(\boldsymbol{\omega}\times\mathbf{r}_{P/G})]$$

Substitution of this equation into Eq. (19.24) then gives

$$T = \frac{1}{2}\bar{v}^2\int_{\mathcal{V}} dm + \bar{\mathbf{v}}\cdot\left(\boldsymbol{\omega}\times\int_{\mathcal{V}}\mathbf{r}_{P/G}\, dm\right) + \frac{1}{2}\boldsymbol{\omega}\cdot\left[\int_{\mathcal{V}}\mathbf{r}_{P/G}\times(\boldsymbol{\omega}\times\mathbf{r}_{P/G})\, dm\right] \tag{c}$$

The second term on the right-hand side of Eq. (c) is zero, because $\int_{\mathcal{V}}\mathbf{r}_{P/G}\, dm = \mathbf{0}$ (G is the mass center). Referring to Eq. (19.8), we see that the last integral in Eq. (c) equals $\mathbf{h}_G$, the angular momentum of the body about its mass center. The kinetic energy of a rigid body may, therefore, be expressed as

$$T = \frac{1}{2}m\bar{v}^2 + \frac{1}{2}\boldsymbol{\omega}\cdot\mathbf{h}_G \tag{19.25}$$

It is sometimes convenient to write the kinetic energy as

$$T = \frac{1}{2}m\bar{v}^2 + T_{\text{rot}} \tag{19.26}$$

where

$$T_{\text{rot}} = \frac{1}{2}\boldsymbol{\omega}\cdot\mathbf{h}_G \tag{19.27}$$

is called the *rotational kinetic energy.* The term $\frac{1}{2}m\bar{v}^2$ is referred to as the *kinetic energy of translation.*

Using rectangular representation for $\boldsymbol{\omega}$ and $\mathbf{h}_G$, and substituting for the components of $\mathbf{h}_G$ from Eqs. (19.19), the kinetic energy of rotation becomes*

$$\begin{aligned} T_{\text{rot}} &= \frac{1}{2}\boldsymbol{\omega}\cdot\mathbf{h}_G \\ &= \frac{1}{2}(\omega_x\mathbf{i} + \omega_y\mathbf{j} + \omega_z\mathbf{k})\cdot\big[(\bar{I}_x\omega_x - \bar{I}_{xy}\omega_y - \bar{I}_{xz}\omega_z)\mathbf{i} \\ &\qquad + (-\bar{I}_{yx}\omega_x + \bar{I}_y\omega_y - \bar{I}_{yz}\omega_z)\mathbf{j} \\ &\qquad + (-\bar{I}_{zx}\omega_x - \bar{I}_{zy}\omega_y + \bar{I}_z\omega_z)\mathbf{k}\big] \end{aligned}$$

After evaluating the dot products, the rotational kinetic energy reduces to

$$\begin{aligned} T_{\text{rot}} = \frac{1}{2}(&\bar{I}_x\omega_x^2 + \bar{I}_y\omega_y^2 + \bar{I}_z\omega_z^2 - 2\bar{I}_{xy}\omega_x\omega_y \\ &- 2\bar{I}_{yz}\omega_y\omega_z - 2\bar{I}_{zx}\omega_z\omega_x) \end{aligned} \tag{19.28}$$

If the direction of one of the rectangular axes through G, say the z-axis, is taken to be parallel to the angular velocity vector (i.e., $\omega_x = \omega_y = 0$ and $\omega_z = \omega$), Eq. (19.28) reduces to $\frac{1}{2}\bar{I}_z\omega^2$, and the kinetic energy is

$$T = \frac{1}{2}m\bar{v}^2 + \frac{1}{2}\bar{I}_z\omega^2 \quad (\text{valid if } \boldsymbol{\omega} = \omega\mathbf{k}) \tag{19.29}$$

This equation is equivalent to Eq. (18.7a), which was used in the analysis of plane motion. It is interesting to note that Eq. (19.29) is valid even if the z-axis is not a principal axis of inertia.

* Bars are placed over the I's to remind us that the inertial properties are to be computed at the mass center.

Equation (19.25) is the general expression for the kinetic energy of a rigid body; i.e., there are no restrictions on its applicability. We now consider the special case where the origin of the coordinate system has zero velocity (either a fixed point or a point on the instant axis of rotation).

If the origin O of the inertial coordinate system (see Fig. 19.8) has no velocity, the velocity of the mass center of the body is

$$\bar{\mathbf{v}} = \boldsymbol{\omega} \times \mathbf{r}_G \tag{d}$$

Equation (19.10) then becomes (with O replacing A)

$$\mathbf{h}_G = \mathbf{h}_O - \mathbf{r}_G \times m\bar{\mathbf{v}} = \mathbf{h}_O - \mathbf{r}_G \times m(\boldsymbol{\omega} \times \mathbf{r}_G) \tag{e}$$

Substituting Eqs. (d) and (e) into Eq. (19.25), we get

$$T = \frac{1}{2}m(\boldsymbol{\omega} \times \mathbf{r}_G) \cdot (\boldsymbol{\omega} \times \mathbf{r}_G) + \frac{1}{2}\boldsymbol{\omega} \cdot \mathbf{h}_O - \frac{1}{2}\boldsymbol{\omega} \cdot [\mathbf{r}_G \times m(\boldsymbol{\omega} \times \mathbf{r}_G)]$$

Using the identity $\mathbf{A} \cdot (\mathbf{B} \times \mathbf{C}) = \mathbf{B} \cdot (\mathbf{C} \times \mathbf{A})$, the first term in this expression can be written as

$$\frac{1}{2}m\boldsymbol{\omega} \cdot [\mathbf{r}_G \times (\boldsymbol{\omega} \times \mathbf{r}_G)]$$

which cancels the last term, leaving us with

$$T = \frac{1}{2}\boldsymbol{\omega} \cdot \mathbf{h}_O \qquad (O\text{: point in the body that has zero velocity}) \tag{19.30}$$

Although Eqs. (19.27) and (19.30) appear to be similar, the former gives only the rotational term of the kinetic energy, whereas the latter gives the total kinetic energy.

The kinetic energy in Eq. (19.30) can be expanded in terms of the components of $\boldsymbol{\omega}$ and $\mathbf{h}_O$ by repeating the steps that gave us Eq. (19.28). The result is

$$T = \frac{1}{2}(I_x\omega_x^2 + I_y\omega_y^2 + I_z\omega_z^2 - 2I_{xy}\omega_x\omega_y - 2I_{yz}\omega_y\omega_z - 2I_{zx}\omega_z\omega_x) \tag{19.31}$$

(origin of the xyz-axes is a point in the body that has zero velocity)

In this equation the inertial properties of the body are to be computed about axes passing through the point that has zero velocity, not about the mass center G.

If the body rotates about a fixed axis, say the z-axis, then $\omega_x = \omega_y = 0$ and $\omega_z = \omega$. Consequently, Eq. (19.31) becomes

$$T = \frac{1}{2} I_z \omega^2 \qquad (z\text{-axis is the instant axis of rotation}) \tag{19.32}$$

We used this equation in the analysis of plane motion—see Eq. (18.7b). Again we observe that it is valid even if the z-axis is not a principal axis of inertia.

b. Work-energy principle and the conservation of mechanical energy

The principles of work-energy and conservation of mechanical energy for plane motion, presented in Art. 18.4, are also applicable to three-dimensional motion. Therefore, the following discussion is a review of the fundamental equations covered in Art. 18.4.

Letting the subscripts 1 and 2 refer to the initial and final positions of a rigid body, the work-energy principle is

$$(U_{1\text{-}2})_{\text{ext}} = \Delta T \tag{19.33}$$

where $(U_{1\text{-}2})_{\text{ext}}$ is the work done by external forces and couples and ΔT is the change in kinetic energy. The work can be computed using the methods described in Arts. 14.2 and 18.2, and Eqs. (19.25)–(19.32) can be used to calculate the kinetic energy.

For a system of connected rigid bodies, the work-energy principle is

$$(U_{1\text{-}2})_{\text{ext}} + (U_{1\text{-}2})_{\text{int}} = \Delta T \tag{19.34}$$

where $(U_{1\text{-}2})_{\text{ext}}$ and $(U_{1\text{-}2})_{\text{int}}$ represent the work done on the system by external and internal forces, respectively.

If all the forces that act on a rigid body or a system of connected rigid bodies are conservative, then the mechanical energy is conserved. The principle of conservation of mechanical energy

$$V_1 + T_1 = V_2 + T_2 \tag{19.35}$$

can then be used in place of the work-energy principle. In Eq. (19.35) V_1 and V_2 are the initial and final potential energies, and T_1 and T_2 are the initial and final kinetic energies.

Sample Problem 19.5

The uniform slender rods denoted by ①, ②, and ③ are welded together to form the rigid body shown in Fig. (a). The body, which is supported by bearings at A and B, is being driven at the constant angular velocity $\omega = 30$ rad/s. When the body is in the position shown, calculate the following: (1) the angular momentum about point C; (2) the angular momentum about point A; and (3) the kinetic energy. The mass per unit length of each rod is $\rho = 600$ g/m.

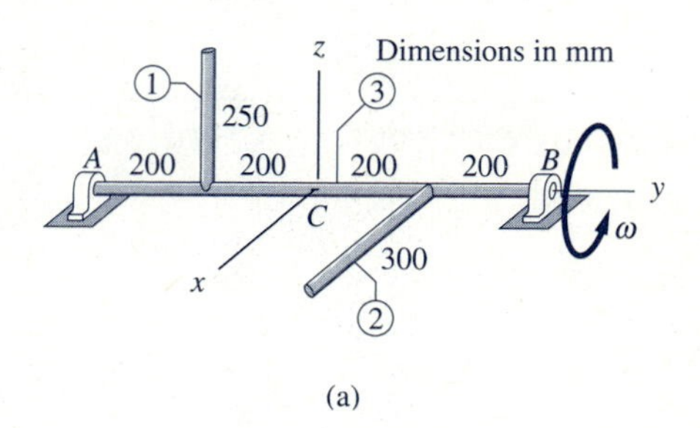

(a)

Solution

Preliminaries

Since the only nonzero component of the angular velocity vector is $\omega_y = \omega$, we see from Eq. (19.19) that the components of the angular momentum vector about either C or A reduce to the form

$$h_x = -I_{xy}\omega \qquad h_y = I_y\omega \qquad h_z = -I_{zy}\omega \tag{a}$$

When applying Eq. (a) it is convenient to move the origin of the coordinate system to the point about which the angular momentum is to be computed, as indicated in Figs. (a) and (c).

Because the body rotates about the fixed y-axis, its kinetic energy, from Eq. (19.31), is

$$T = \frac{1}{2}I_y\omega^2 \tag{b}$$

The computations of the moments and products of inertia required in Eqs. (a) and (b) are shown in the table. Note that the results for the body are obtained by summing the properties of rods ① and ② only, because the relevant properties of rod ③ are zero.

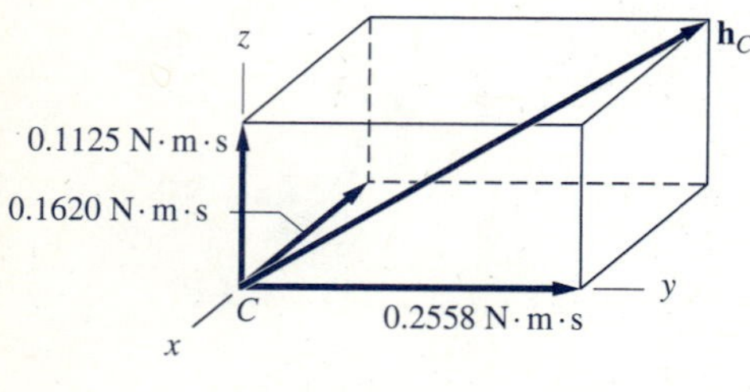

(b)

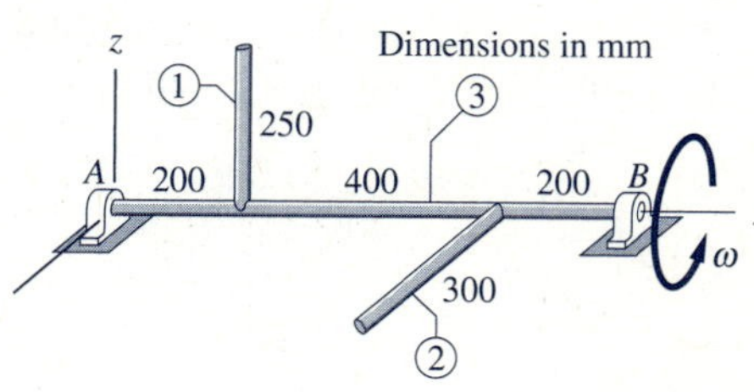

(c)

Computation of Inertial Properties

	Rod ①	Rod ②	Totals
$m = \rho L$	$(0.6)(0.25) = 0.15$ kg	$(0.6)(0.3) = 0.18$ kg	
	Part (1): x-, y-, z-axes at point C		
$\bar{x}$	0	0.15 m	
$\bar{y}$	-0.2 m	0.2 m	
$\bar{z}$	0.125 m	0	
$I_y = \frac{1}{3}mL^2$	$\frac{1}{3}(0.15)(0.25)^2 = 0.003\,125$ kg · m^2	$\frac{1}{3}(0.18)(0.3)^2 = 0.005\,40$ kg · m^2	$0.008\,525$ kg · m^2
$I_{xy} = m\bar{x}\bar{y}$	$(0.15)(0)(-0.2) = 0$	$(0.18)(0.15)(0.2) = 0.005\,40$ kg · m^2	$0.005\,40$ kg · m^2
$I_{zy} = m\bar{y}\bar{z}$	$(0.15)(-0.2)(0.125) = -0.003\,75$ kg · m^2	$(0.18)(0.2)(0) = 0$	$-0.003\,75$ kg · m^2
	Part (2): x-, y-, z-axes at point A		
$\bar{x}$	0	0.15 m	
$\bar{y}$	0.2 m	0.6 m	
$\bar{z}$	0.125 m	0	
$I_y = \frac{1}{3}mL^2$	$0.003\,125$ kg · m^2	$0.005\,40$ kg · m^2	$0.008\,525$ kg · m^2
$I_{xy} = m\bar{x}\bar{y}$	$(0.15)(0)(0.2) = 0$	$(0.18)(0.15)(0.6) = 0.0162$ kg · m^2	0.0162 kg · m^2
$I_{zy} = m\bar{y}\bar{z}$	$(0.15)(0.2)(0.125) = 0.003\,75$ kg · m^2	$(0.18)(0.6)(0) = 0$	$0.003\,75$ kg · m^2

Part 1

Substituting the moments and products of inertia for point C and $\omega = 30$ rad/s into Eq. (a), the components of the angular momentum vector about C become

$$h_x = -I_{xy}\omega = -(54.00 \times 10^{-4})(30) = -16.20 \times 10^{-2}\ \text{N} \cdot \text{m} \cdot \text{s}$$

$$h_y = I_y\omega = (85.25 \times 10^{-4})(30) = 25.58 \times 10^{-2}\ \text{N} \cdot \text{m} \cdot \text{s}$$

$$h_z = -I_{zy}\omega = -(-37.5 \times 10^{-4})(30) = 11.25 \times 10^{-2}\ \text{N} \cdot \text{m} \cdot \text{s}$$

Therefore, the magnitude of the angular momentum about C is

$$h_C = (10^{-2})\sqrt{(-16.20)^2 + (25.58)^2 + (11.25)^2}$$

$$= 32.3 \times 10^{-2}\ \text{N} \cdot \text{m} \cdot \text{s} \qquad \textit{Answer}$$

The vector $\mathbf{h}_C$ and its components are shown in Fig. (b).

Part 2

The results of substituting the moments and products of inertia at point A and ω = 30 rad/s into Eq. (a) are

$$h_x = -I_{xy}\omega = -(162 \times 10^{-4})(30) = -48.60 \times 10^{-2}\ \text{N} \cdot \text{m} \cdot \text{s}$$
$$h_y = I_y\omega = (85.25 \times 10^{-4})(30) = 25.58 \times 10^{-2}\ \text{N} \cdot \text{m} \cdot \text{s}$$
$$h_z = -I_{zy}\omega = -(37.50 \times 10^{-4})(30) = -11.25 \times 10^{-2}\ \text{N} \cdot \text{m} \cdot \text{s}$$

Therefore, the magnitude of the angular momentum about A is

$$h_A = (10^{-2})\sqrt{(-48.60)^2 + (25.58)^2 + (-11.25)^2}$$
$$= 56.1 \times 10^{-2}\ \text{N} \cdot \text{m} \cdot \text{s} \qquad \textit{Answer}$$

The vector $\mathbf{h}_A$ and its components are shown in Fig. (d).

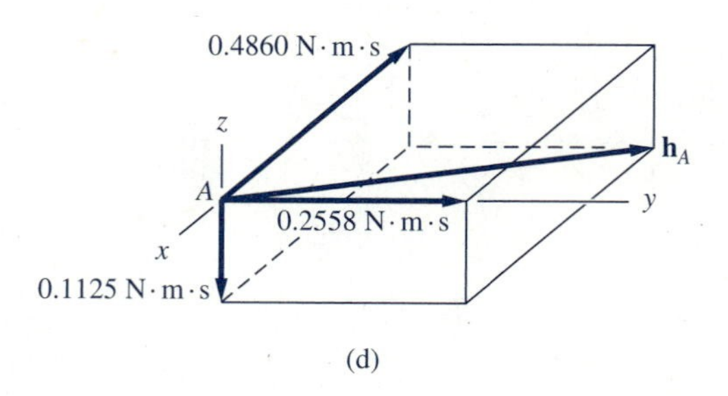

(d)

Part 3

The kinetic energy of the body is found by substituting I_y (either point C or A may be used, of course) and ω = 30 rad/s into Eq. (b), with the result being

$$T = \frac{1}{2}(85.25 \times 10^{-4})(30)^2 = 3.84\ \text{J} \qquad \textit{Answer}$$

Sample Problem 19.6

The uniform slender rod AB shown in Fig. (a) has the weight $W = 16$ lb and the length $L = 2.4$ ft. The weights of the sliding collars A and B, to which the rod is connected with ball-and-socket joints, may be neglected. When collar A is in the position $\theta = 0$, its velocity is 6 ft/s in the positive x-direction. Neglecting friction, determine the velocity vector of A when it has moved to the position $\theta = 60°$.

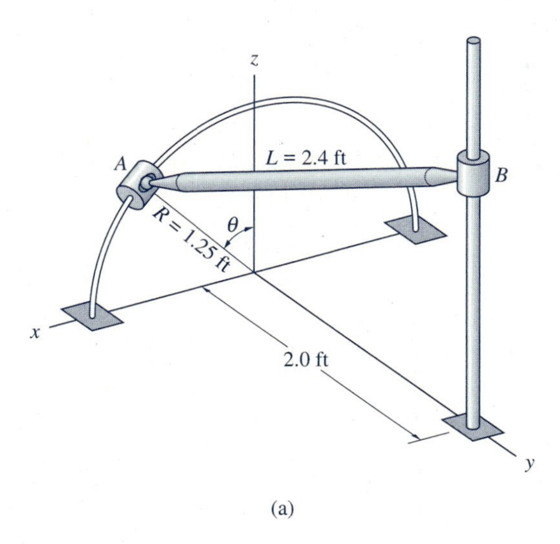

(a)

Solution

This problem is suited to the work-energy method of analysis, because it is concerned with the change in speed between two positions. Since the weight of the rod, which is a conservative force, is the only force that does work on the system, mechanical energy is conserved. Therefore, we can apply the principle of conservation of mechanical energy: $T_1 + V_1 = T_2 + V_2$, where T_1 and T_2 are the initial and final kinetic energies, and V_1 and V_2 are the initial and final gravitational potential energies. Figure (b) shows the reference plane that has been chosen for the potential energy V.

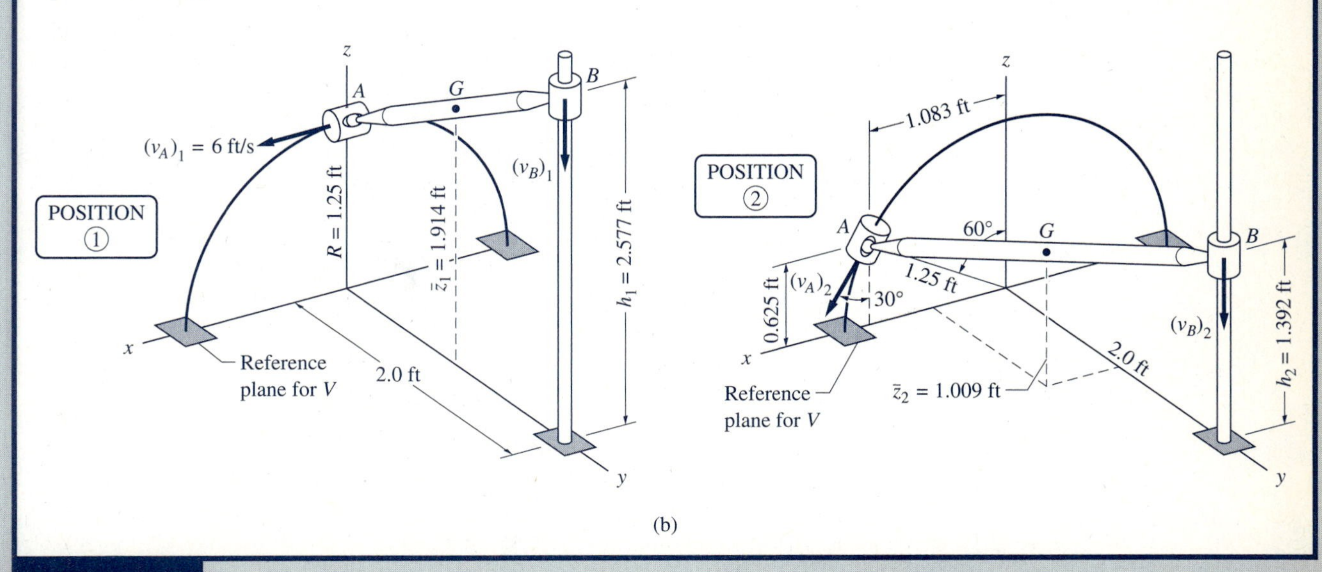

(b)

Position ①

Potential Energy

Referring to position ① in Fig. (b), the height h_1 of collar B above the reference plane must satisfy the geometric relation $(h_1 - 1.25)^2 + (2.0)^2 = (2.4)^2$, which gives $h_1 = 2.577$ ft. Therefore, the height of the mass center G is $\bar{z}_1 = (1.25 + 2.577)/2 = 1.914$ ft, and the initial potential energy becomes

$$V_1 = W\bar{z}_1 = 16(1.914) = 30.62 \text{ lb} \cdot \text{ft} \tag{a}$$

Kinematic Analysis

Kinematic analysis is required to find the angular velocity of the rod knowing that $(v_A)_1 = 6$ ft/s. The relative velocity equation for the kinematically important points A and B is

$$(\mathbf{v}_B)_1 = (\mathbf{v}_A)_1 + \boldsymbol{\omega}_1 \times (\mathbf{r}_{B/A})_1 \tag{b}$$

where $(\mathbf{v}_B)_1$ and $\boldsymbol{\omega}_1$ are the velocity of collar B and the angular velocity of the rod, respectively.

From Fig. (b), we see that

$$(\mathbf{r}_{B/A})_1 = 2\mathbf{j} + (2.577 - 1.25)\mathbf{k} = 2\mathbf{j} + 1.327\mathbf{k} \text{ ft}$$

Substituting $(\mathbf{v}_A)_1 = 6\mathbf{i}$ ft/s, $(\mathbf{v}_B)_1 = -(v_B)_1\mathbf{k}$, and writing $\boldsymbol{\omega}_1 = (\omega_x)_1\mathbf{i} + (\omega_y)_1\mathbf{j} + (\omega_z)_1\mathbf{k}$, Eq. (b) becomes

$$-(v_B)_1\mathbf{k} = 6\mathbf{i} + \begin{vmatrix} \mathbf{i} & \mathbf{j} & \mathbf{k} \\ (\omega_x)_1 & (\omega_y)_1 & (\omega_z)_1 \\ 0 & 2 & 1.327 \end{vmatrix}$$

Expanding the determinant, and equating like components, gives the following three scalar equations.

$$\begin{aligned} 0 &= 6 + 1.327(\omega_y)_1 - 2(\omega_z)_1 \\ 0 &= -1.327(\omega_x)_1 \\ -(v_B)_1 &= 2(\omega_x)_1 \end{aligned} \tag{c}$$

Equations (c) contain four unknowns: $(\omega_x)_1$, $(\omega_y)_1$, $(\omega_z)_1$, and $(v_B)_1$. Although all the unknowns cannot be found here, we can see that $(\omega_x)_1 = 0$ and $(v_B)_1 = 0$. The fact that B is the instant center of zero velocity of the rod is also apparent by inspection of Fig. (b).

To find $(\omega_y)_1$ and $(\omega_z)_1$, an additional kinematic equation is needed. This equation can be obtained by assuming any convenient value, say zero, for the spin velocity of the rod. Note that the moment of inertia of the rod about its axis (i.e., the line AB) is negligible. Therefore, the rate at which the rod is spinning, or rotating, about that axis will not enter into the calculation of the kinetic energy. Assuming the spin velocity of the rod to be zero, $\boldsymbol{\omega}_1$ is perpendicular to the rod—i.e., $\boldsymbol{\omega}_1 \cdot (\mathbf{r}_{B/A})_1 = 0$, or $[(\omega_y)_1\mathbf{j} + (\omega_z)_1\mathbf{k}] \cdot (2\mathbf{j} + 1.327\mathbf{k}) = 2(\omega_y)_1 + 1.327(\omega_z)_1 = 0$—from which we find that

$$(\omega_z)_1 = -1.507(\omega_y)_1 \tag{d}$$

Solving Eqs. (c) and (d) yields

$$(v_B)_1 = 0 \qquad (\omega_x)_1 = 0$$
$$(\omega_y)_1 = -1.382 \text{ rad/s} \qquad (\omega_z)_1 = 2.083 \text{ rad/s}$$

from which the magnitude of the initial angular velocity is

$$\omega_1 = \sqrt{(0)^2 + (1.382)^2 + (2.083)^2} = 2.500 \text{ rad/s} \tag{e}$$

Kinetic Energy

Since the velocity of B was found to be zero, and the angular velocity vector $\boldsymbol{\omega}_1$ is perpendicular to the rod, we conclude that the instant axis of rotation of the rod passes through B and is perpendicular to AB. Therefore, the kinetic energy in position ① can be obtained from Eq. (19.32).

$$T_1 = \tfrac{1}{2} I \omega_1^2 \tag{f}$$

where $I = mL^2/3$ is the moment of inertia of the rod about the instant axis of rotation. Substituting the numerical values in Eq. (f), we obtain

$$T_1 = \frac{1}{2}\frac{16(2.4)^2}{(32.2)3}(2.500)^2 = 2.981 \text{ lb} \cdot \text{ft} \tag{g}$$

Position ②

Potential Energy

When the system is in position ② shown in Fig. (b), the coordinates of collar A are $x = 1.25 \sin 60° = 1.083$ ft, $y = 0$, and $z = 1.25 \cos 60° = 0.625$ ft. Using these values, the height h_2 of B above the reference plane is obtained from the geometric relationship $(1.083)^2 + (2)^2 + (h_2 - 0.625)^2 = (2.4)^2$, which gives $h_2 = 1.392$ ft. The height of the mass center G above the reference plane is, therefore, $\bar{z}_2 = (0.625 + 1.392)/2 = 1.009$ ft. It follows that the gravitational potential energy of the rod is

$$V_2 = W\bar{z}_2 = 16(1.009) = 16.14 \text{ lb} \cdot \text{ft} \tag{h}$$

Kinematic Analysis

The purpose of the kinematic analysis is to relate $\boldsymbol{\omega}_2$ (the angular velocity of the rod) and $\bar{\mathbf{v}}_2$ (the velocity of its mass center) to $(\mathbf{v}_A)_2$ (the velocity of collar A). We utilize the relative velocity equation between A and B:

$$(\mathbf{v}_B)_2 = (\mathbf{v}_A)_2 + \boldsymbol{\omega}_2 \times (\mathbf{r}_{B/A})_2 \tag{i}$$

where $(\mathbf{v}_B)_2$ is the velocity of B. The position vector of A relative to B in position ② is, according to Fig. (b), $(\mathbf{r}_{B/A})_2 = -1.083\mathbf{i} + 2\mathbf{j} + (1.392 - 0.625)\mathbf{k} = -1.083\mathbf{i} + 2\mathbf{j} + 0.767\mathbf{k}$ ft. Again, referring to Fig. (b), we see that $(\mathbf{v}_A)_2 = (v_A)_2 \sin 30°\mathbf{i} - (v_A)_2 \cos 30°\mathbf{k}$. Expressing the angular velocity vector as $\boldsymbol{\omega}_2 = (\omega_x)_2\mathbf{i} + (\omega_y)_2\mathbf{j} + (\omega_z)_2\mathbf{k}$, Eq. (i) becomes

$$-(v_B)_2\mathbf{k} = (v_A)_2 \sin 30°\mathbf{i} - (v_A)_2 \cos 30°\mathbf{k} + \begin{vmatrix} \mathbf{i} & \mathbf{j} & \mathbf{k} \\ (\omega_x)_2 & (\omega_y)_2 & (\omega_z)_2 \\ -1.083 & 2 & 0.767 \end{vmatrix} \tag{j}$$

Expanding the determinant in Eq. (j), and equating like components, yields the following equations:

$$
\begin{aligned}
0 &= (v_A)_2 \sin 30° + 0.767(\omega_y)_2 - 2(\omega_z)_2 \\
0 &= -0.767(\omega_x)_2 - 1.083(\omega_z)_2 \\
-(v_B)_2 &= -(v_A)_2 \cos 30° + 2(\omega_x)_2 + 1.083(\omega_y)_2
\end{aligned} \tag{k}
$$

A fourth equation is obtained by setting the spin velocity of the rod to zero, as was done in position ①. This gives $\boldsymbol{\omega}_2 \cdot (\mathbf{r}_{B/A})_2 = 0$, or $[(\omega_x)_2\mathbf{i} + (\omega_y)_2\mathbf{j} + (\omega_z)_2\mathbf{k}] \cdot (-1.083\mathbf{i} + 2\mathbf{j} + 0.767\mathbf{k}) = 0$, which becomes

$$
-1.083(\omega_x)_2 + 2(\omega_y)_2 + 0.767(\omega_z)_2 = 0 \tag{l}
$$

Equations (k) and (l) can be solved in terms of $(v_A)_2$, with the results being

$$
\left.
\begin{aligned}
(\omega_x)_2 &= -0.2451(v_A)_2 \\
(\omega_y)_2 &= -0.1993(v_A)_2 \\
(\omega_z)_2 &= 0.1736(v_A)_2
\end{aligned}
\right\} \omega_2 = 0.3605(v_A)_2 \tag{m}
$$

and

$$
(v_B)_2 = 1.572(v_A)_2 \tag{n}
$$

The relationship between $\bar{\mathbf{v}}_2$ and $(\mathbf{v}_A)_2$ could be found from the relative velocity equation $\bar{\mathbf{v}} = (\mathbf{v}_A)_2 + \boldsymbol{\omega}_2 \times (\mathbf{r}_{G/A})_2$ and Eq. (m). However, it is simpler to use the fact that G is the midpoint of rod AB, so that its velocity is given by

$$
\begin{aligned}
\bar{\mathbf{v}}_2 &= \tfrac{1}{2}[(\mathbf{v}_A)_2 + (\mathbf{v}_B)_2] \\
&= \tfrac{1}{2}[(v_A)_2(\sin 30°\mathbf{i} - \cos 30°\mathbf{k}) - 1.572(v_A)_2\mathbf{k}] \\
&= (v_A)_2(0.250\mathbf{i} - 1.219\mathbf{j})
\end{aligned}
$$

which yields

$$
\bar{v}_2^2 = 1.549(v_A)_2^2 \tag{o}
$$

Kinetic Energy

The kinetic energy of the rod can be obtained from Eq. (19.29):

$$
T = \tfrac{1}{2}m\bar{v}^2 + \tfrac{1}{2}\bar{I}\omega_2^2
$$

where $\bar{I} = mL^2/12$ is the moment of inertia of the rod about an axis perpendicular to the rod (parallel to $\boldsymbol{\omega}_2$) at G. Substituting the numerical values, we get

$$
\begin{aligned}
T_2 &= \frac{1}{2}\left(\frac{16}{32.2}\right)(1.549)(v_A)_2^2 \\
&\quad + \frac{1}{2}\left(\frac{16}{32.2}\,\frac{(2.4)^2}{12}\right)[0.3605(v_A)_2]^2 \\
&= (0.3848 + 0.0155)(v_A)_2^2 = 0.4003(v_A)_2^2
\end{aligned} \tag{p}
$$

Conservation of Mechanical Energy

Substituting Eqs. (a), (g), (h), and (p) into the conservation principle,

$$T_1 + V_1 = T_2 + V_2$$

we obtain

$$2.981 + 30.62 = 0.4003(v_A)_2^2 + 16.14$$

from which the velocity of A in position ② is found to be

$$(v_A)_2 = 6.60 \text{ ft/s} \qquad \textit{Answer}$$

Sample Problem 19.7

Figure (a) shows a 2.0-kg homogeneous square plate that hangs from a ball-and-socket joint at O. The plate is at rest when it is struck by a hammer at A. The force $\hat{\mathbf{P}}$ applied by the hammer is impulsive, its impulse being $\int \hat{\mathbf{P}}\,dt = 1.20\mathbf{i}$ N · s. Determine the angular velocity and the kinetic energy of the plate immediately after the impact. Observe that the x-axis is perpendicular to the plate, and the y- and z-axes lie along the edges of the plate.

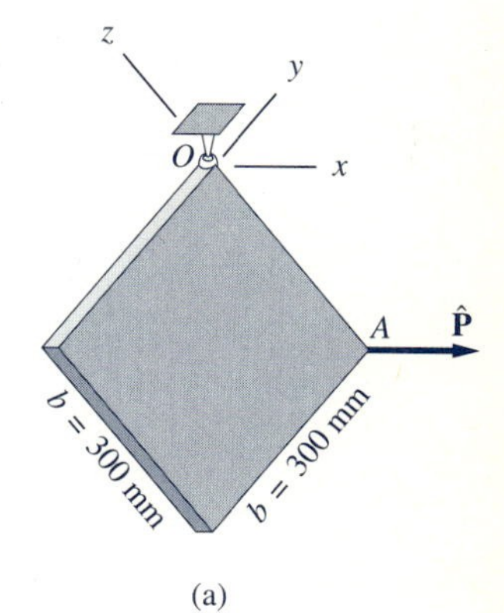

(a)

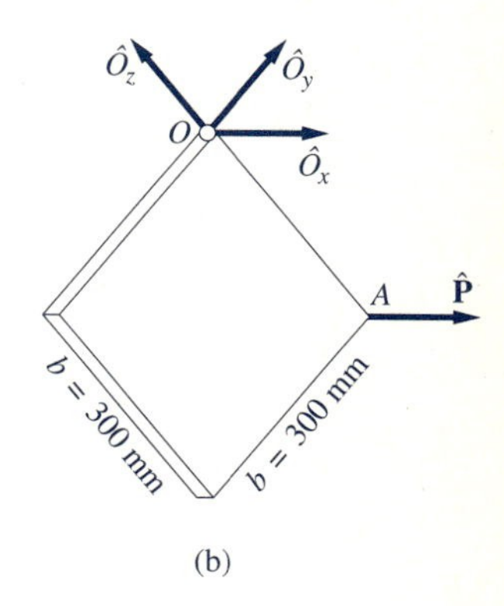

(b)

Solution

This problem has to be analyzed by the impulse-momentum method, because the applied force is impulsive.

Figure (b) shows the free-body diagram (FBD) of the plate during the impact. Since the duration of the impact is negligible, the plate occupies essentially the same spatial position before, during, and after the impact. The FBD includes the applied force $\hat{\mathbf{P}}$ and the components of the impulsive reaction at O. The weight of the plate was omitted, since it is not an impulsive force.

We recall that for impulsive motion the angular impulse about any point equals the change in angular momentum of the body about the same point. In our case, point O is a convenient choice for the reference point, because the impulsive reaction passes through that point. Letting the subscripts 1 and 2 refer to the instants immediately before and immediately after the impact, respectively, the angular impulse-momentum principle becomes

$$(\mathbf{A}_O)_{1\text{-}2} = \Delta\mathbf{h}_O = (\mathbf{h}_O)_2 - (\mathbf{h}_O)_1 = (\mathbf{h}_O)_2 \qquad \text{(a)}$$

Note that $(\mathbf{h}_O)_1 = \mathbf{0}$ because the plate is at rest before the impact.

From the FBD we see that the angular impulse (i.e., the moment of the linear impulse) about O is

$$(\mathbf{A}_O)_{1\text{-}2} = \mathbf{r}_{A/O} \times \int \hat{\mathbf{P}}\,dt = -0.3\mathbf{k} \times 1.2\mathbf{i} = -0.36\mathbf{j} \text{ N} \cdot \text{m} \cdot \text{s} \qquad \text{(b)}$$

To determine the angular momentum of the plate about point O, we must first compute the inertial properties of the plate at that point. The central moments of inertia can be obtained from Table 17.1 (we must set $a = 0$ since the plate is thin): $\bar{I}_x = (1/12)m(2b^2) = mb^2/6$ and $\bar{I}_y = \bar{I}_z = mb^2/12$. Also, due to symmetry we

assembly about (a) point O, and (b) point B.

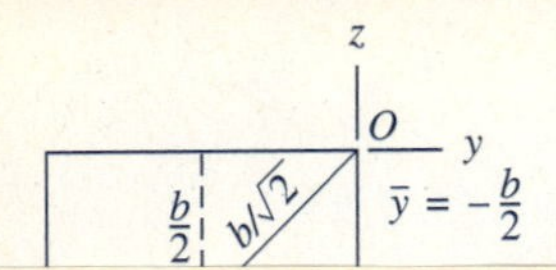

have $\bar{I}_{xy} = \bar{I}_{yz} = \bar{I}_{zx} = 0$. The moments and products of inertia about O can now be calculated from the parallel-axis theorem and Fig. (c):

$$I_x = \bar{I}_x + m\bar{r}^2 = \frac{mb^2}{6} + m\left(\frac{b}{\sqrt{2}}\right)^2 = \frac{2mb^2}{3}$$

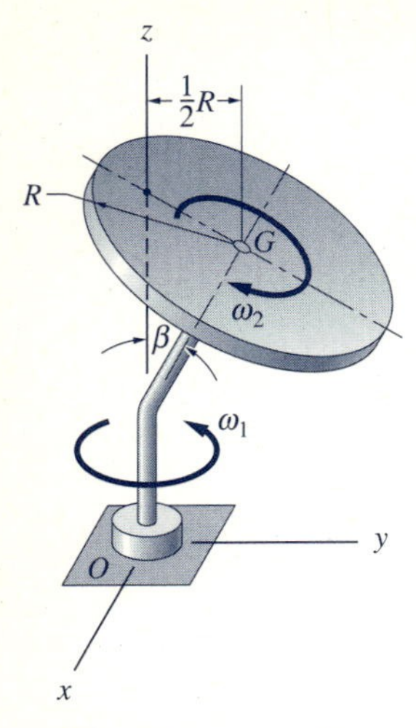

Fig. P19.28, P19.29

19.28 The thin uniform disk of mass m and radius R spins about the bent shaft OG with the angular speed ω_2. At the same time, the shaft rotates about the z-axis with the angular speed ω_1. If the angle between the bent portion of the shaft and the z-axis is $\beta = 35°$, find the ratio ω_2/ω_1 for which the angular momentum of the disk about its mass center G is parallel to the z-axis.

19.29 Determine the kinetic energy of the disk described in Prob. 19.28 if $\omega_1 = \omega_2$ and $\beta = 35°$.

19.30 The uniform 12-lb thin disk spins about the axle OG as it rolls on the horizontal plane without slipping. End O of the axle is welded to sliding collar that rotates about the fixed vertical rod with the constant angular velocity of 6 rad/s. Calculate the angular momentum of the disk about point O.

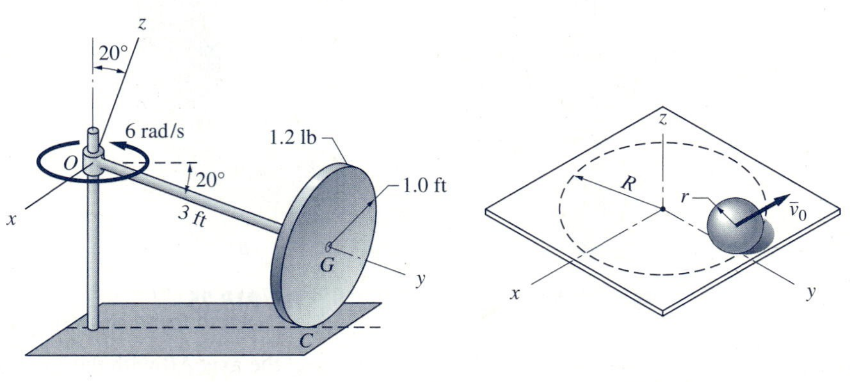

Fig. P19.30

Fig. P19.31

19.31 The homogeneous sphere of radius r and mass m rolls on a flat surface without slipping at the constant speed $\bar{v}_0$. The path of the center of the sphere is a circle of radius R. Determine the angle between the instant axis of rotation of the sphere and the surface so that the kinetic energy of the sphere is minimized, and find the corresponding kinetic energy.

19.32 The 10-kg uniform slender rod AB is connected to sliding collars at A and B by ball-and-socket joints. In the position shown, the velocity of collar A is $v_A = 1.8$ m/s. For this position, calculate (a) the angular momentum of the rod about its mass center; and (b) the kinetic energy of the rod.

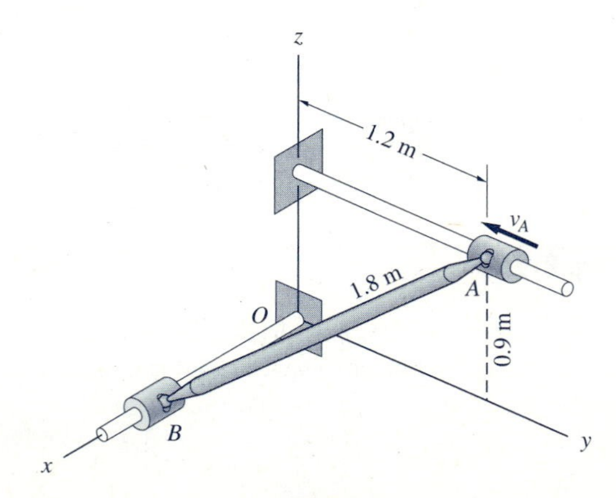

Fig. P19.32

19.33 The slender bar AB of length L and mass m is suspended from two strings, each of length L. If the bar is released from rest when $\theta = 90°$, find its maximum angular velocity.

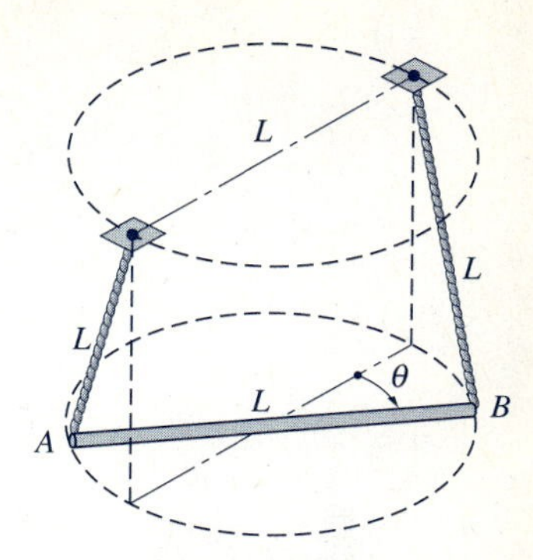

Fig. P19.33

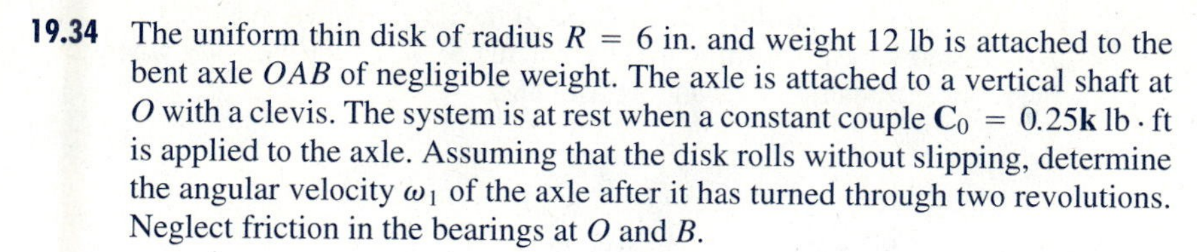

19.34 The uniform thin disk of radius $R = 6$ in. and weight 12 lb is attached to the bent axle OAB of negligible weight. The axle is attached to a vertical shaft at O with a clevis. The system is at rest when a constant couple $\mathbf{C}_0 = 0.25\mathbf{k}$ lb · ft is applied to the axle. Assuming that the disk rolls without slipping, determine the angular velocity ω_1 of the axle after it has turned through two revolutions. Neglect friction in the bearings at O and B.

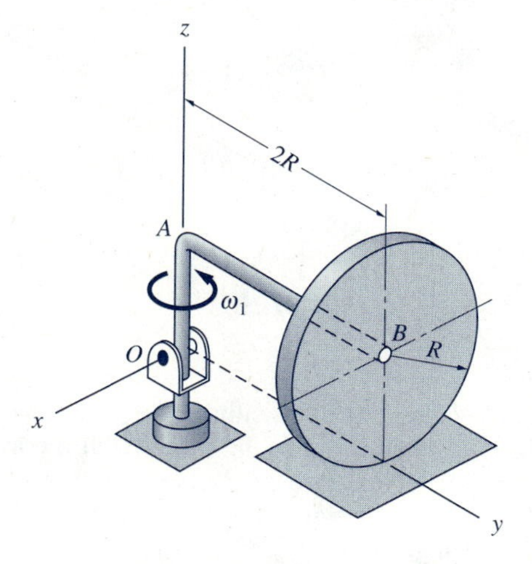

Fig. P19.34

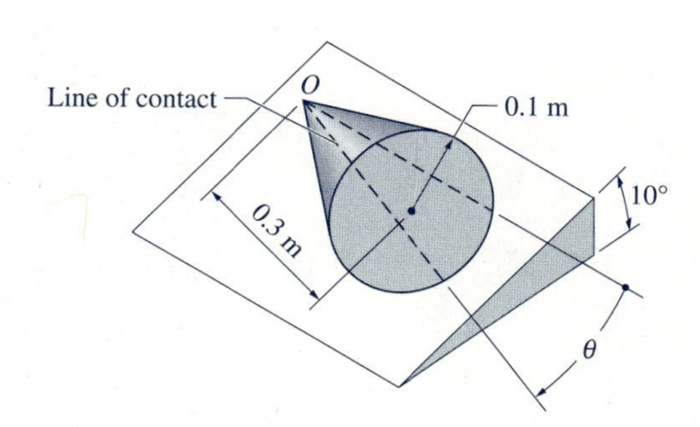

Fig. P19.35

19.35 The homogeneous cone of mass m and height 0.3 m is released from rest on the 10° inclined plane in the position $\theta = 0$. Assuming rolling without slipping, calculate $d\theta/dt$ when $\theta = 90°$.

19.36 The mechanism consists of two homogeneous slender bars AB and BC of masses 1.8 kg and 1.2 kg, respectively, and the 2-kg slider C. The connections at B and C are ball-and-socket joints. If the mechanism is released from rest at $\theta = 0$, determine the angular velocity of bar AB when $\theta = 90°$. Neglect friction.

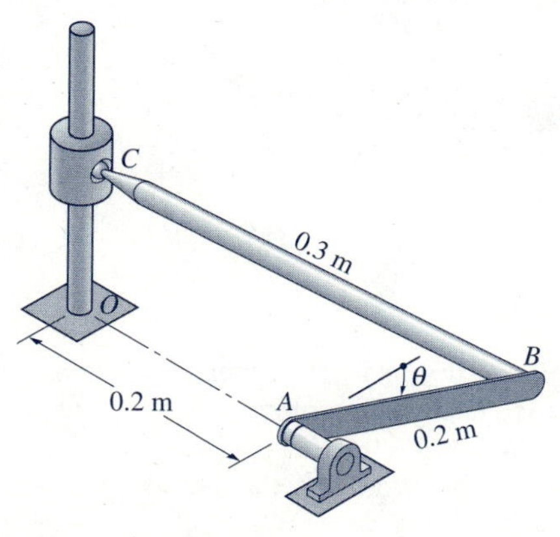

Fig. P19.36

19.37 The thin rim of the flywheel C weighs 5 lb, with the weights of the spokes and hub being negligible. The flywheel is joined to the 5-lb slider B with the 3-lb connecting rod AB. The joints at A and B are ball-and-sockets. A constant force $P = 60$ lb acts on the slider B as shown. If the flywheel has an angular velocity $\omega = 20$ rad/s in the position shown, calculate its angular velocity when joint A reaches the position A'. Neglect friction.

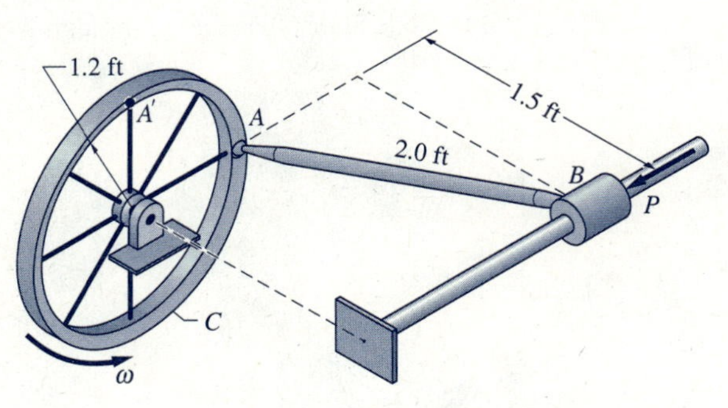

Fig. P19.37

19.38 The radii of gyration of the 360-kg satellite are $\bar{k}_x = \bar{k}_y = 0.72$ m, $\bar{k}_z = 0.54$ m. The satellite is spinning about the z-axis with the angular velocity $\omega_1 = 1.2$ rad/s when it receives an impulse of $-200\mathbf{i}$ N · s from a jet of gas emitted from a nozzle at A. Assuming that the duration of the impulse is very short, calculate the new angular velocity of the satellite.

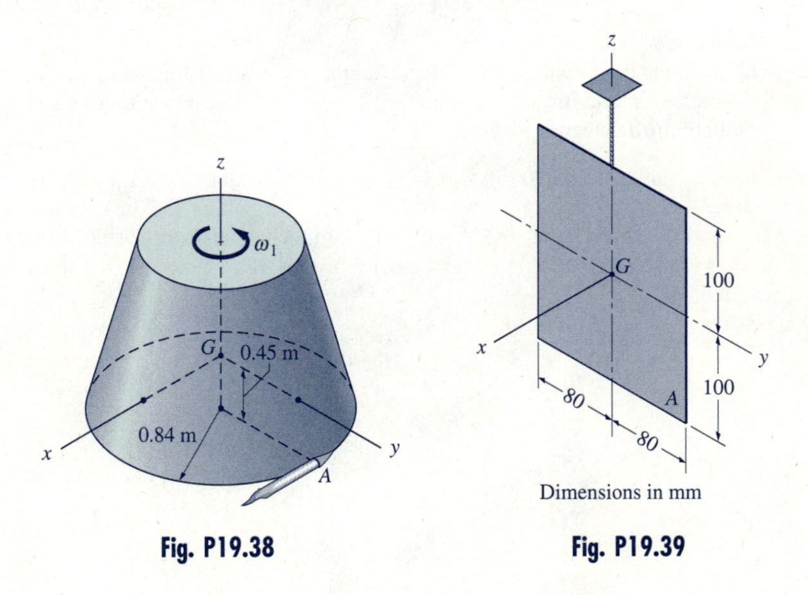

Fig. P19.38 **Fig. P19.39**

19.39 The 1.2-kg uniform thin plate is suspended from a string and is initially at rest in the position shown. Determine the following immediately after corner A receives an impulse of $-0.096\mathbf{i}$ N · s: (a) the velocity of the mass center G; (b) the angular velocity of the plate; and (c) the kinetic energy of the plate.

19.40 A 0.4-ft rod of negligible mass is welded to the 3.2-lb thin uniform plate and suspended from a ball-and-socket joint at O. The assembly is rotating about the z-axis with the angular speed of 6 rad/s when corner A of the plate hits a rigid obstruction. Assuming that A does not rebound, determine the angular velocity of the assembly immediately after the impact.

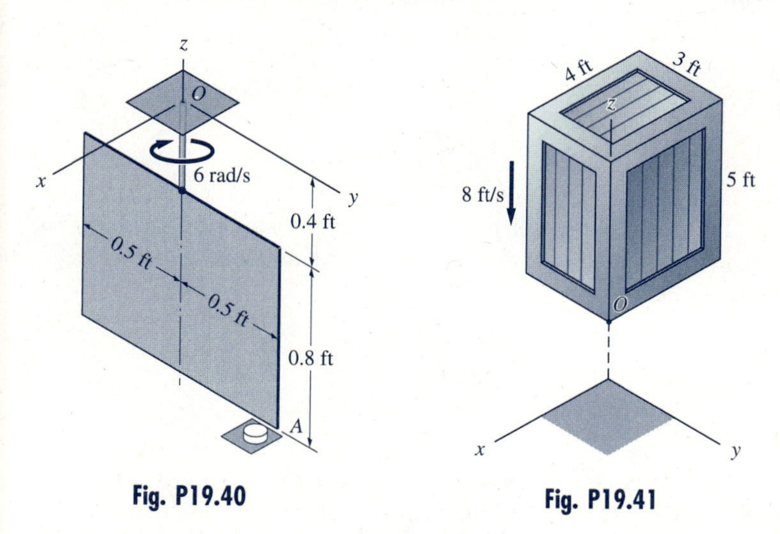

Fig. P19.40 **Fig. P19.41**

19.41 The uniform box of mass m is falling with a speed of 8 ft/s and no angular velocity when corner O hits a rigid obstruction. Assuming plastic impact (i.e., no rebound), determine (a) the angular velocity of the box immediately after the impact; and (b) the percentage of kinetic energy lost during the impact.

19.42 The uniform bent wire of total mass m is suspended from a ball-and-socket joint at O. The wire is stationary when an impulse is applied at corner A in the x-direction. Find the unit vector in the direction of the instant axis of rotation immediately after the impulse is received.

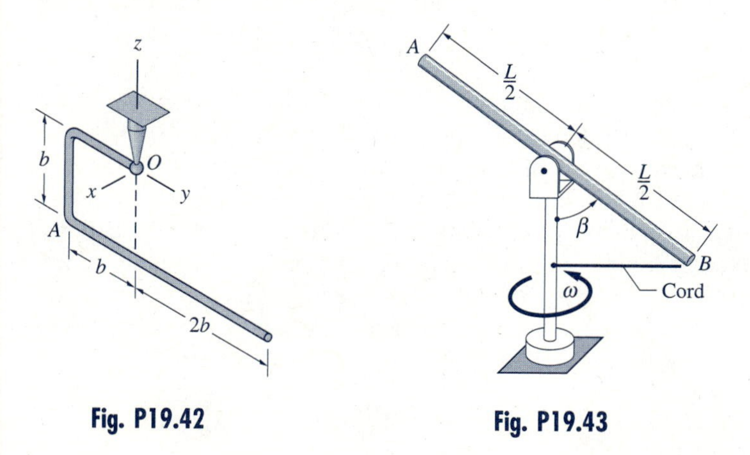

Fig. P19.42 **Fig. P19.43**

19.43 The slender rod AB of mass m and length L is attached to a vertical shaft with a clevis. The shaft is rotating freely at the angular speed ω when the cord, which maintained the angle $\beta = \beta_1$ between the rod and the shaft, breaks. Determine the expressions for ω and $d\beta/dt$ when the rod reaches the position $\beta = 90°$.

*19.5 Force-Mass-Acceleration Method

a. Equations of motion

The basic equations that govern the three-dimensional motion of a rigid body follow from our discussions of particle systems (Chapter 15) and plane motion of rigid bodies (Chapter 18). The force equation of motion is

$$\Sigma \mathbf{F} = m\bar{\mathbf{a}} \tag{19.36}$$

where $\Sigma \mathbf{F}$ is the resultant of the external forces acting on the body, m is the mass of the body, and $\bar{\mathbf{a}}$ is the acceleration of the mass center. The moment equation of motion is

$$\Sigma \mathbf{M}_A = \frac{d\mathbf{h}_A}{dt} \tag{19.37}$$

where $\Sigma \mathbf{M}_A$ is the resultant moment of the external forces acting on the body and $\mathbf{h}_A$ is the angular momentum of the body about point A. The latter must be a point in the body that is either fixed in space or is the mass center of the body. (This restriction on the choice of A must be kept in mind; for the sake of brevity, we will not mention it continually.) The other valid choices for point A that are described in Art. 15.7 (point with zero velocity, etc.) are of little practical importance.

When using Eq. (19.37), any convenient reference frame can be used to describe the angular momentum vector $\mathbf{h}_A$. For example, if $\mathbf{\Omega}$ is the angular velocity of such a reference frame, the moment equation of motion can be written as

$$\Sigma \mathbf{M}_A = \frac{d\mathbf{h}_A}{dt} = \left(\frac{d\mathbf{h}_A}{dt}\right)_{/xyz} + \mathbf{\Omega} \times \mathbf{h}_A \tag{19.38}$$

where the notation "$/xyz$" indicates that the derivative of $\mathbf{h}_A$ is to be evaluated relative to the xyz reference frame.

The choice of reference frame should be such that $(d\mathbf{h}_A/dt)_{/xyz}$ can be easily evaluated. In particular, we wish to avoid having to evaluate the time derivatives of moments and products of inertia. These complications can be eliminated by letting the xyz-axes be a body frame, i.e., a reference frame that is embedded in the body (has the same angular velocity as the body). If $\boldsymbol{\omega}$ is the angular velocity of the body, then for a body frame we have $\mathbf{\Omega} = \boldsymbol{\omega}$. Consequently, the moment equation, Eq. (19.38), becomes

$$\Sigma \mathbf{M}_A = \frac{d\mathbf{h}_A}{dt} = \left(\frac{d\mathbf{h}_A}{dt}\right)_{/xyz} + \boldsymbol{\omega} \times \mathbf{h}_A \tag{19.39}$$

According to Eq. (19.19), the components of the angular momentum about a fixed point, or about the mass center, are

$$\begin{aligned} h_x &= I_x\omega_x - I_{xy}\omega_y - I_{xz}\omega_z \\ h_y &= -I_{yx}\omega_x + I_y\omega_y - I_{yz}\omega_z \\ h_z &= -I_{zx}\omega_x - I_{zy}\omega_y + I_z\omega_z \end{aligned} \tag{19.40}$$

If the xyz-axes constitute a body frame, then the moments and products of inertia do not vary with time, and Eq. (19.39) becomes

$$\begin{aligned}\Sigma\mathbf{M}_A &= (I_x\dot{\omega}_x - I_{xy}\dot{\omega}_y - I_{xz}\dot{\omega}_z)\mathbf{i} \\ &\quad + (-I_{yx}\dot{\omega}_x + I_y\dot{\omega}_y - I_{yz}\dot{\omega}_z)\mathbf{j} \\ &\quad + (-I_{zx}\dot{\omega}_x - I_{zy}\dot{\omega}_y + I_z\dot{\omega}_z)\mathbf{k} \\ &\quad + \begin{vmatrix} \mathbf{i} & \mathbf{j} & \mathbf{k} \\ \omega_x & \omega_y & \omega_z \\ h_x & h_y & h_z \end{vmatrix}\end{aligned}$$

Substituting for the components of the angular momentum from Eq. (19.40) and expanding the determinant (also recalling that $I_{xy} = I_{yx}$, $I_{yz} = I_{zy}$, and $I_{xz} = I_{zx}$), the scalar components of the general moment equation are

$$\begin{aligned}\Sigma M_x &= I_x\dot{\omega}_x + \omega_y\omega_z(I_z - I_y) + I_{xy}(\omega_z\omega_x - \dot{\omega}_y) \\ &\quad - I_{xz}(\dot{\omega}_z + \omega_x\omega_y) - I_{yz}(\omega_y^2 - \omega_z^2) \\ \Sigma M_y &= I_y\dot{\omega}_y + \omega_z\omega_x(I_x - I_z) + I_{yz}(\omega_x\omega_y - \dot{\omega}_z) \\ &\quad - I_{xy}(\dot{\omega}_x + \omega_z\omega_y) - I_{xz}(\omega_z^2 - \omega_x^2) \\ \Sigma M_z &= I_z\dot{\omega}_z + \omega_x\omega_y(I_y - I_x) + I_{xz}(\omega_y\omega_z - \dot{\omega}_x) \\ &\quad - I_{yz}(\dot{\omega}_y + \omega_x\omega_z) - I_{xy}(\omega_x^2 - \omega_y^2)\end{aligned} \tag{19.41}$$

Equations (19.41) are first-order, nonlinear differential equations, which are very difficult to solve analytically, except for a few special cases.

b. Euler's equations

The moment equations simplify somewhat if we choose the xyz-axes to be principal axes of inertia at the reference point. Then, the products of inertia vanish, and Eqs. (19.41) reduce to

$$\begin{aligned}\Sigma M_x &= I_x\dot{\omega}_x + \omega_y\omega_z(I_z - I_y) \\ \Sigma M_y &= I_y\dot{\omega}_y + \omega_z\omega_x(I_x - I_z) \\ \Sigma M_z &= I_z\dot{\omega}_z + \omega_x\omega_y(I_y - I_x)\end{aligned} \tag{19.42}$$

These equations, known as *Euler's equations,* are among the more useful equations in rigid-body dynamics. When using Eqs. (19.42), it must be remembered that the xyz-axes are a body frame and coincide with the principal axes of inertia at the reference point (fixed point or mass center).

If the angular velocity of the body and its time derivative are known, the resultant external moment applied to the body can be calculated in a straightforward manner from Eqs. (19.42). However, if the resultant moment is known, the equations must be integrated in order to find the angular velocities, a task that must be performed numerically in most problems. An exception is the special case where the angular velocity $\boldsymbol{\omega}$ is constant, when Eqs. (19.42) take the form of algebraic rather than differential equations.

c. Modified Euler's equations

An important problem in dynamics is the motion of an axisymmetric rigid body, such as a spinning top or gyroscope. Consider a body that possesses an axis of rotational symmetry, and let one of the coordinate axes be embedded in the body so that it will always coincide with the axis of symmetry. We allow the coordinate axes to rotate with an angular velocity that is different from that of the body. Since only one of the coordinate axes is embedded in the body, the axes are not a body frame. However, due to the symmetry of the body, each of the coordinate axes will always be a principal axis of inertia.

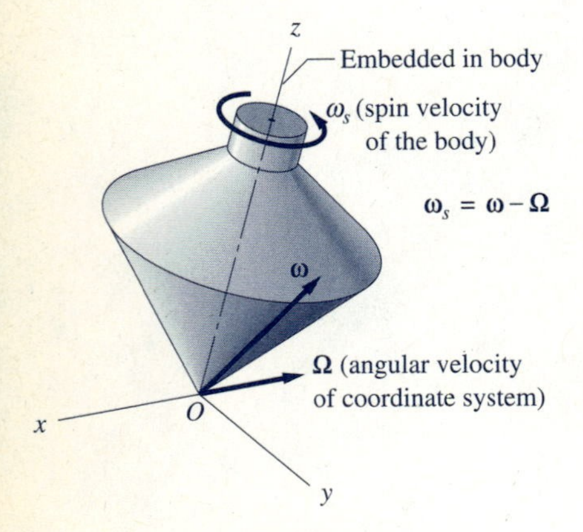

Fig. 19.9

As an illustration consider the axisymmetric top in Fig. 19.9, where the z-axis was chosen to be the axis of symmetry. It is seen that, no matter how the coordinate system rotates, the xyz-axes will always be principal axes of inertia at point O if the z-axis remains embedded in the body.

As before, we let $\mathbf{\Omega} = \Omega_x\mathbf{i}+\Omega_y\mathbf{j}+\Omega_z\mathbf{k}$ be the angular velocity of the xyz reference frame, and let $\boldsymbol{\omega} = \omega_x\mathbf{i} + \omega_y\mathbf{j} + \omega_z\mathbf{k}$ be the angular velocity of the body. These angular velocities will differ by the angular velocity of the body relative to the xyz-frame, called the *spin velocity.* The spin velocity vector will be parallel to the axis that is embedded in the body, i.e., the axis of symmetry. For example, if z is the axis of symmetry of the body, as in Fig. 19.9, the spin velocity is $\boldsymbol{\omega}_s = \omega_s\mathbf{k}$. Consequently, $\omega_x = \Omega_x, \omega_y = \Omega_y$, and $\omega_z = \Omega_z+\omega_s$.

The moment equation of motion given in Eq. (19.38) is

$$\Sigma\mathbf{M}_A = \left(\frac{d\mathbf{h}_A}{dt}\right)_{/xyz} + \mathbf{\Omega}\times\mathbf{h}_A \qquad \text{(19.38, repeated)}$$

Since the xyz-axes are principal axes, the components of the angular momentum of the body about the reference point A (fixed point or mass center) are

$$h_x = I_x\omega_x \qquad h_y = I_y\omega_y \qquad h_z = I_z\omega_z \qquad \text{(a)}$$

Then, the cross product in Eq. (19.38) becomes

$$\mathbf{\Omega}\times\mathbf{h}_A = \begin{vmatrix} \mathbf{i} & \mathbf{j} & \mathbf{k} \\ \Omega_x & \Omega_y & \Omega_z \\ I_x\omega_x & I_y\omega_y & I_z\omega_z \end{vmatrix} \qquad \text{(b)}$$

Using Eqs. (a) and (b), and the fact that the moments of inertia are independent of time, the scalar components of Eq. (19.38) are

$$\begin{aligned} \Sigma M_x &= I_x\dot{\omega}_x + I_z\Omega_y\omega_z - I_y\Omega_z\omega_y \\ \Sigma M_y &= I_y\dot{\omega}_y - I_z\Omega_x\omega_z + I_x\Omega_z\omega_x \\ \Sigma M_z &= I_z\dot{\omega}_z + I_y\Omega_x\omega_y - I_x\Omega_y\omega_x \end{aligned} \qquad \text{(19.43)}$$

These equations are called the *modified Euler's equations.* Observe that the moments ΣM_x, ΣM_y, and ΣM_z are referred to xyz-axes, which do not represent a body frame.

d. A note on angular acceleration

The terms $\dot{\omega}_x$, $\dot{\omega}_y$, and $\dot{\omega}_z$ that appear in the various moment equations are not necessarily the components of the angular acceleration vector of the body. In order to clarify this statement, let us consider first the case where the xyz-axes are a body frame, and then examine the situation where the body spins relative to the xyz-frame.

1. The xyz-axes form a body frame: $\mathbf{\Omega} = \boldsymbol{\omega}$ (Euler's equations).

Recall that the absolute derivative of a vector $\mathbf{V}$ is $(d\mathbf{V}/dt) = (d\mathbf{V}/dt)_{/xyz} + \boldsymbol{\omega} \times \mathbf{V}$, where $\boldsymbol{\omega}$ is the angular velocity of the xyz-frame. Therefore, the absolute derivative of the angular velocity vector $\boldsymbol{\omega}$ (i.e., the angular acceleration) of the body becomes

$$\frac{d\boldsymbol{\omega}}{dt} = \left(\frac{d\boldsymbol{\omega}}{dt}\right)_{/xyz} + \boldsymbol{\omega} \times \boldsymbol{\omega} = \left(\frac{d\boldsymbol{\omega}}{dt}\right)_{/xyz}$$

This result means that the absolute derivative of the angular velocity vector is identical to its derivative relative to the body frame. It follows that if $\boldsymbol{\omega} = \omega_x\mathbf{i} + \omega_y\mathbf{j} + \omega_z\mathbf{k}$, then

$$\dot{\boldsymbol{\omega}} = \dot{\omega}_x\mathbf{i} + \dot{\omega}_y\mathbf{j} + \dot{\omega}_z\mathbf{k}$$

which gives

$$(\dot{\boldsymbol{\omega}})_x = \dot{\omega}_x \qquad (\dot{\boldsymbol{\omega}})_y = \dot{\omega}_y \qquad (\dot{\boldsymbol{\omega}})_x = \dot{\omega}_z \tag{19.44}$$

Therefore, $\dot{\omega}_x$, $\dot{\omega}_y$, and $\dot{\omega}_z$ are the components of the angular acceleration $\dot{\boldsymbol{\omega}}$ of the body, a conclusion that can be very useful when applying Euler's equations.

2. The body spins relative to the xyz-axes: $\mathbf{\Omega} \neq \boldsymbol{\omega}$ (modified Euler's equations).

In this case, the angular acceleration vector of the body becomes $d\boldsymbol{\omega}/dt = (d\boldsymbol{\omega}/dt)_{/xyz} + \mathbf{\Omega} \times \boldsymbol{\omega}$, which may be written in the form

$$\dot{\boldsymbol{\omega}} = \dot{\omega}_x\mathbf{i} + \dot{\omega}_y\mathbf{j} + \dot{\omega}_z\mathbf{k} + \begin{vmatrix} \mathbf{i} & \mathbf{j} & \mathbf{k} \\ \Omega_x & \Omega_y & \Omega_z \\ \omega_x & \omega_y & \omega_z \end{vmatrix}$$

Expanding the above determinant, and equating like components, yields

$$\begin{aligned} (\dot{\boldsymbol{\omega}})_x &= \dot{\omega}_x + (\Omega_y\omega_z - \Omega_z\omega_y) \\ (\dot{\boldsymbol{\omega}})_y &= \dot{\omega}_y - (\Omega_x\omega_z - \Omega_z\omega_x) \\ (\dot{\boldsymbol{\omega}})_z &= \dot{\omega}_z + (\Omega_x\omega_y - \Omega_y\omega_x) \end{aligned} \tag{19.45}$$

It can be seen here that the components of the angular acceleration vector $\dot{\boldsymbol{\omega}}$ are not equal to the time derivatives of ω_x, ω_y, and ω_z.

e. Plane motion

Let the xyz-axes be a body frame with the z-axis remaining perpendicular to a fixed plane. In this case, $\boldsymbol{\omega} = \omega_z \mathbf{k}$ and $\omega_x = \omega_y = 0$, and the general moment equations, Eqs. (19.41), become

$$\begin{aligned} \Sigma M_x &= -I_{xz}\dot{\omega}_z + I_{yz}\omega_z^2 \\ \Sigma M_y &= -I_{yz}\dot{\omega}_z - I_{xz}\omega_z^2 \\ \Sigma M_z &= I_z\dot{\omega}_z \end{aligned} \tag{19.46}$$

If, in addition, the z-axis is a principal axis of inertia of the body, these equations further simplify to

$$\Sigma M_x = 0 \qquad \Sigma M_y = 0 \qquad \Sigma M_z = I_z\dot{\omega}_z \tag{19.47}$$

These equations are identical to the moment equations for plane motion discussed in Chapter 17.

f. Rotation about a fixed axis

Equations (19.46) and (19.47) are, of course, also valid for the special case where a body is rotating about an axis that is fixed in space. It is instructive to consider the case where a body is mounted on a shaft that is supported by a bearing at each end. Let the shaft coincide with the z-axis, assumed to be fixed in space, and the body rotate with *constant* angular speed ω_z. If the xyz body axes are not principal axes of inertia, Eqs. (19.46) yield (with $\dot{\omega}_z = 0$)

$$\Sigma M_x = I_{yz}\omega_z^2 \qquad \Sigma M_y = -I_{xz}\omega_z^2 \qquad \Sigma M_z = 0 \tag{19.48}$$

In this case, we see that the reactions at the two bearings must provide the moments ΣM_x and ΣM_y. These bearing reactions, which rotate with the xyz reference frame (i.e., with the body), are called *dynamic bearing reactions,* and the body is said to be *dynamically unbalanced.* Because the dynamic bearing reactions are proportional to the square of the angular speed, they can reach large magnitudes and cause severe vibrations of the body. To dynamically balance the body, the products of inertia in Eq. (19.48) must be made to vanish. This can be accomplished by redistributing the mass of the body or by adding additional mass to the body at appropriate locations. A common example of the latter method is the addition of small weights to the rim of an automobile wheel when it is being "spin balanced."

Sample Problem 19.8

Figure (a) shows a body that is formed by welding together three uniform rods labeled ①, ②, and ③, each with the mass shown. (This body also appeared in Sample Problem 19.5.) Support for the body is provided by smooth bearings at A and B, with only the bearing at A being capable of providing an axial thrust. A small motor at A (not shown) is driving the body at the constant angular velocity ω_0 rad/s. For the position shown in Fig. (a), determine the bearing reactions and the output torque (couple) T of the motor. Also find the dynamic bearing reactions, i.e., the bearing reactions caused by the rotation of the body.

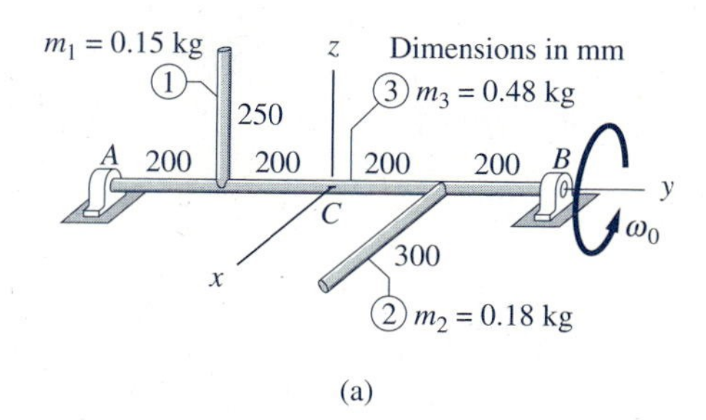

(a)

Solution

The free-body diagram of the body, shown in Fig. (b), displays (1) the bearing reactions at A and B; (2) the weight of each rod: $W_1 = 0.15(9.81) = 1.472$ N, $W_2 = 0.18(9.81) = 1.766$ N, $W_3 = 0.48(9.81) = 4.709$ N; and (3) the output torque T of the motor required to drive the body at constant angular velocity. The origin of the xyz reference frame, which is assumed to rotate with the body, is attached to C, the midpoint of rod ③.

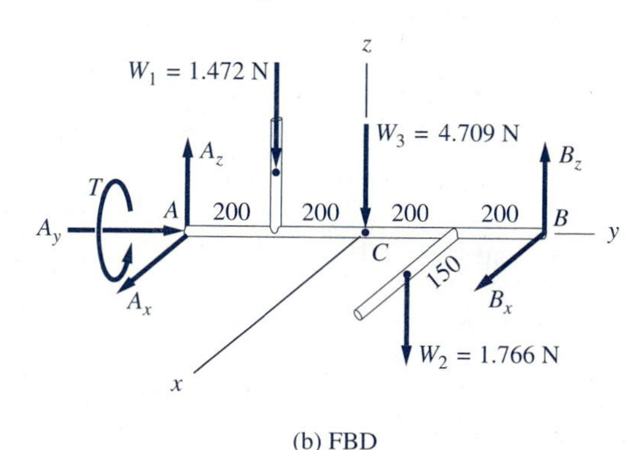

(b) FBD

Since the rotation occurs about a fixed axis, we can use Eqs. (19.48). However, we must first convert Eqs. (19.48) from rotation about the z-axis, which was assumed in their derivation, to rotation about the y-axis. This conversion can be accomplished with the aid of the following diagram.

x — Axis of rotation — z — y

z — Axis of rotation — y — x

From this diagram, we see that Eqs. (19.48) must be changed by replacing x by z, y by x, and z by y. The results are

$$\Sigma M_x = -I_{yz}\omega_y^2 \qquad \Sigma M_y = 0 \qquad \Sigma M_z = I_{xy}\omega_y^2$$

The products of inertia about point C were computed in Sample Problem 19.5: $I_{xy} = 0.005\,40\ \text{kg}\cdot\text{m}^2$ and $I_{yz} = -0.003\,75\ \text{kg}\cdot\text{m}^2$. Consequently, referring to the FBD in Fig. (b) and using $\omega_y = \omega_0$, the moment equations of motion become

$$\Sigma M_x = -I_{yz}\omega_0^2 \qquad -A_z(0.400) + B_z(0.400) + 1.472(0.200) - 1.766(0.200) = -(-0.003\,75\omega_0^2) \tag{a}$$

$$\Sigma M_y = 0 \qquad T + 1.766(0.150) = 0 \tag{b}$$

$$\Sigma M_z = I_{xy}\omega_0^2 \qquad A_x(0.400) - B_x(0.400) = 0.005\,40\omega_0^2 \tag{c}$$

From Eq. (b), the output torque of the motor for the position shown is

$$T = -0.2649\ \text{N}\cdot\text{m} \qquad \textit{Answer}$$

Note that the magnitude of this torque varies because the position of the body changes.

We next apply the force equation of motion $\Sigma\mathbf{F} = m\bar{\mathbf{a}}$ to the body, where m is the mass of the body and $\bar{\mathbf{a}}$ is the acceleration of its mass center. The inertia vector for the body is the sum of the inertia vectors of the individual rods. Since the mass center of rod ① moves with constant speed on a circle centered on the y-axis, the acceleration of its mass center consists of the normal acceleration $\bar{a}_n = r\omega_0^2$; i.e., $\bar{\mathbf{a}}_1 = -0.125\omega_0^2\mathbf{k}$. Similarly, for rod ② we get $\bar{\mathbf{a}}_2 = -0.150\omega_0^2\mathbf{i}$. The mass center of rod ③ is stationary. Therefore, the inertia vector for the body becomes

$$\begin{aligned} m\bar{\mathbf{a}} &= m_1\bar{\mathbf{a}}_1 + m_2\bar{\mathbf{a}}_2 \\ &= 0.15(-0.125\omega_0^2)\mathbf{k} + 0.18(-0.150\omega_0^2)\mathbf{i} \\ &= (-0.018\,75\omega_0^2)\mathbf{k} - (0.027\,00\omega_0^2)\mathbf{i} \end{aligned} \tag{d}$$

Using the above results and the FBD in Fig. (b), the force equations of motion become

$$\Sigma F_x = m\bar{a}_x \qquad A_x + B_x = -0.027\,00\omega_0^2 \tag{e}$$

$$\Sigma F_y = m\bar{a}_y \qquad A_y = 0 \qquad \textit{Answer} \tag{f}$$

$$\Sigma F_z = m\bar{a}_z \qquad A_z + B_z - 1.472 - 1.766 - 4.709 = -0.018\,75\omega_0^2 \tag{g}$$

Solving Eqs. (a) and (g), and Eqs. (c) and (e), the x- and z-components of the bearing reactions in Fig. (b) are found to be

$$\begin{aligned} A_x &= -0.006\,75\omega_0^2\ \text{N} \\ B_x &= -0.020\,25\omega_0^2\ \text{N} \\ A_z &= -0.014\,06\omega_0^2 + 3.900\ \text{N} \\ B_z &= -0.004\,688\omega_0^2 + 4.047\ \text{N} \end{aligned} \qquad \textit{Answer}$$

These components of the bearing reactions apply only when the body is in the position shown. However, it can be seen that the terms with ω_0^2 are valid for all positions of the body. These terms, called the dynamic bearing reactions, are caused by the rotation of the body. Figure (c) shows the dynamic bearing reactions together with the inertia vectors for bars ① and ②. Each vector shown in Fig. (c) rotates with the xyz-axes, i.e., with the body.

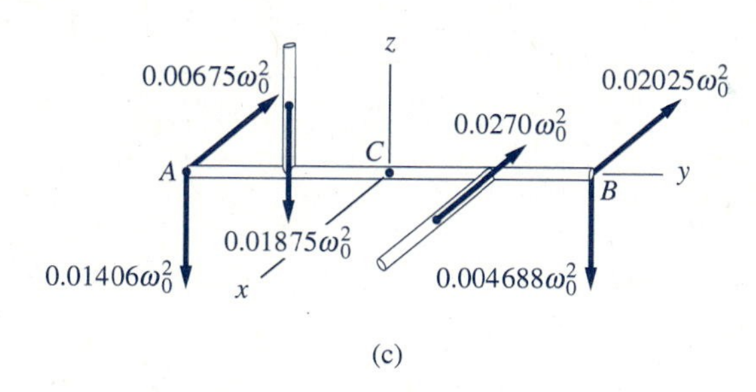

(c)

The so-called *static bearing reactions* (the terms that are independent of ω_0) support the weight of the body and do not rotate with the coordinate system. However, their magnitudes vary as the position of the body changes. The total reaction at a bearing is found by adding the static and dynamic bearing reactions.

Sample Problem 19.9

The platform A is rotating about the fixed vertical axis with an angular velocity ω_2 and angular acceleration $\dot{\omega}_2$. At the same time, an internal motor (not shown) spins the uniform disk B about its axle, which is rigidly mounted to the platform, at the angular velocity ω_1 and angular acceleration $\dot{\omega}_1$. Determine the components of the couple that is applied to the disk B by its axle.

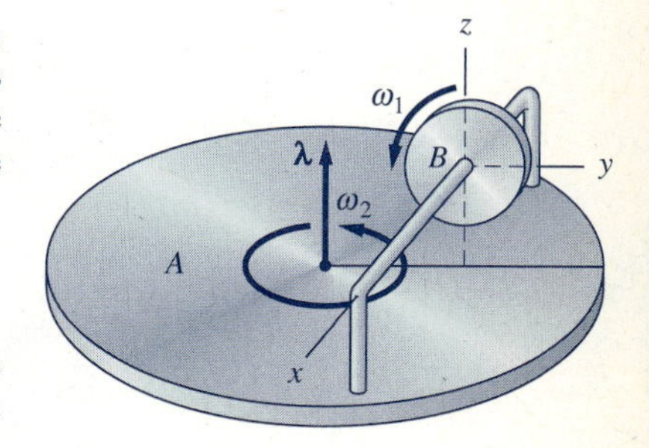

Solution

We solve this problem using two different reference frames. First, the xyz-axes are embedded in disk B (Euler's equations); then the xyz-axes are embedded in the axle, i.e., in the platform extended (modified Euler's equations).

Method I: *xyz* Rotating with the Disk

The xyz-axes shown in the figure are principal axes at the mass center of disk B. If these axes are assumed to rotate with the disk, they constitute a body frame, which means that Euler's equations can be used.

It is convenient to write the angular velocity $\boldsymbol{\omega}$ of disk B as

$$\boldsymbol{\omega} = \omega_1 \mathbf{i} + \omega_2 \boldsymbol{\lambda} \quad \text{(valid for all time)} \tag{a}$$

where $\boldsymbol{\lambda}$ is a unit vector directed perpendicular to the platform. Because Eq. (a) is valid for all time, it can be differentiated, resulting in the angular acceleration of disk B

$$\dot{\boldsymbol{\omega}} = \dot{\omega}_1 \mathbf{i} + \omega_1 \dot{\mathbf{i}} + \dot{\omega}_2 \boldsymbol{\lambda} + \omega_2 \dot{\boldsymbol{\lambda}} \tag{b}$$

We note that $\dot{\boldsymbol{\lambda}} = \mathbf{0}$. Furthermore, at the instant shown, we have $\boldsymbol{\lambda} = \mathbf{k}$, and $\dot{\mathbf{i}} = \boldsymbol{\omega} \times \mathbf{i} = (\omega_1 \mathbf{i} + \omega_2 \mathbf{k}) \times \mathbf{i} = \omega_2 \mathbf{j}$. Therefore, Eq. (b) becomes

$$\dot{\boldsymbol{\omega}} = \dot{\omega}_1 \mathbf{i} + \dot{\omega}_2 \mathbf{k} + \omega_1 \omega_2 \mathbf{j} \quad \text{(at this instant)} \tag{c}$$

Because the xyz-axes constitute a body frame, we have, according to Eqs. (19.44), $\dot{\omega}_x = (\dot{\boldsymbol{\omega}})_x$, $\dot{\omega}_y = (\dot{\boldsymbol{\omega}})_y$, and $\dot{\omega}_z = (\dot{\boldsymbol{\omega}})_z$. It follows from Eqs. (a) and (c) that the components of $\boldsymbol{\omega}$ and $\dot{\boldsymbol{\omega}}$ that are to be substituted into Euler's equations are

$$\begin{aligned} \omega_x &= \omega_1 & \dot{\omega}_x &= \dot{\omega}_1 \\ \omega_y &= 0 & \dot{\omega}_y &= \omega_1\omega_2 \\ \omega_z &= \omega_2 & \dot{\omega}_z &= \dot{\omega}_2 \end{aligned} \tag{d}$$

Substituting Eqs. (d) into Eqs. (19.42) and recognizing that $I_y = I_z$ for disk B, we obtain the moments acting on the disk about its mass center (the origin of the coordinate system).

$$\begin{aligned} \Sigma M_x &= I_x\dot{\omega}_x + \omega_y\omega_z(I_z - I_y) = I_x\dot{\omega}_1 \\ \Sigma M_y &= I_y\dot{\omega}_y + \omega_z\omega_x(I_x - I_z) = I_y\omega_1\omega_2 + \omega_1\omega_2(I_x - I_z) = I_x\omega_1\omega_2 \qquad \textit{Answer} \\ \Sigma M_z &= I_z\dot{\omega}_z + \omega_x\omega_y(I_y - I_x) = I_z\dot{\omega}_2 \end{aligned}$$

Since the only support for disk B is provided by its axle, these moments represent the components of the couple that is exerted on the disk by the axle. The moment about the x-axis must be applied by the motor, whereas the moments about the y- and z-axes are provided by bearings that support the disk. Since the coordinate axes rotate with the disk, the above expressions for the moments are valid only in the position shown in the figure.

Method II: *xyz* Rotating with the Platform

Because disk B is spinning about its axis of symmetry (the x-axis), modified Euler's equations are applicable. Assuming that the xyz-axes in the figure rotate with platform A, the angular velocity of the reference frame is

$$\boldsymbol{\Omega} = \omega_2\mathbf{k} \quad \text{(valid for all time)} \tag{e}$$

and the angular velocity of disk B becomes

$$\boldsymbol{\omega} = \omega_1\mathbf{i} + \omega_2\mathbf{k} \quad \text{(valid for all time)} \tag{f}$$

The components of $\boldsymbol{\omega}$, namely $\omega_x = \omega_1$, $\omega_y = 0$, and $\omega_z = \omega_2$, are differentiable because Eq. (f) is valid for all time. Therefore, the components of $\boldsymbol{\omega}$, $\dot{\boldsymbol{\omega}}$, and $\boldsymbol{\Omega}$ that are to be substituted into the modified Euler's equations—Eqs. (19.43)—are

$$\begin{aligned} \omega_x &= \omega_1 & \dot{\omega}_x &= \dot{\omega}_1 & \Omega_x &= 0 \\ \omega_y &= 0 & \dot{\omega}_y &= 0 & \Omega_y &= 0 \\ \omega_z &= \omega_2 & \dot{\omega}_z &= \dot{\omega}_2 & \Omega_z &= \omega_2 \end{aligned} \tag{g}$$

Completing the substitutions, Eqs. (19.43) yield

$$\begin{aligned} \Sigma M_x &= I_x\dot{\omega}_x + I_z\Omega_y\omega_z - I_y\Omega_z\omega_y = I_x\dot{\omega}_1 \\ \Sigma M_y &= I_y\dot{\omega}_y - I_z\Omega_x\omega_z + I_x\Omega_z\omega_x = I_x\omega_1\omega_2 \qquad \textit{Answer} \\ \Sigma M_z &= I_z\dot{\omega}_z + I_y\Omega_x\omega_y - I_x\Omega_y\omega_x = I_z\dot{\omega}_2 \end{aligned}$$

These moments are, of course, identical to those obtained previously by Method I. The difference is that the foregoing expressions are valid for all time, not just for the position shown in the figure.

PROBLEMS

19.44 The thin uniform disk of mass m and radius R is mounted at O on the end of a vertical shaft. The plane of the disk is inclined at the angle β to the horizontal. Determine the dynamic reactions acting on the disk at O when it is rotating about the z-axis with the constant angular velocity ω_0.

19.45 Repeat Prob. 19.44 assuming that a small mass $m_A = m/16$ is attached to the rim of the disk at A.

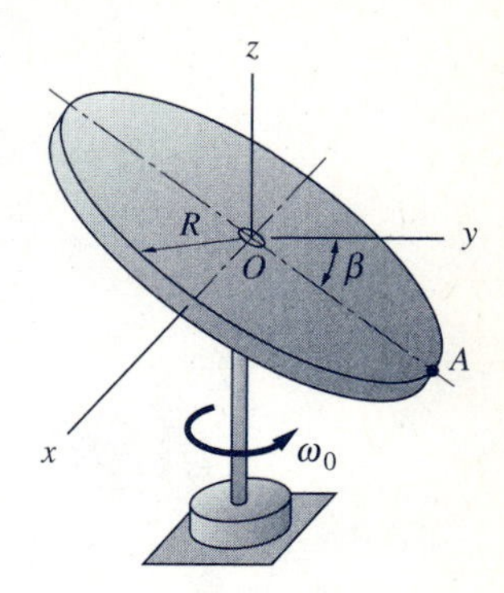

Fig. P19.44, P19.45

19.46 The small mass m is attached to the rim of a disk of mass M and radius R. The disk is being made to rotate about the axis AB with the constant angular speed ω_0. Calculate the dynamic pin reactions acting on the disk at A and B.

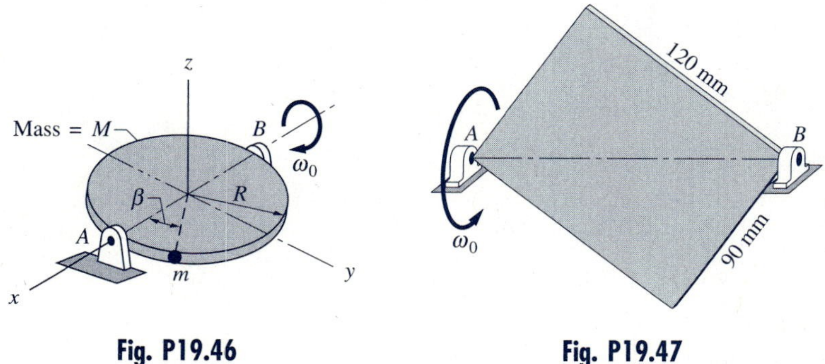

Fig. P19.46 Fig. P19.47

19.47 The 0.5-kg thin homogeneous plate is driven about the axis AB with the constant angular velocity $\omega_0 = 25$ rad/s. Determine the dynamic bearing reactions acting on the plate at A and B.

19.48 The crank is made of a uniform slender rod of total mass 8 kg. If the crank rotates about the y-axis with the constant angular speed $\omega = 40$ rad/s, find the dynamic bearing reactions acting on the crank at A and B.

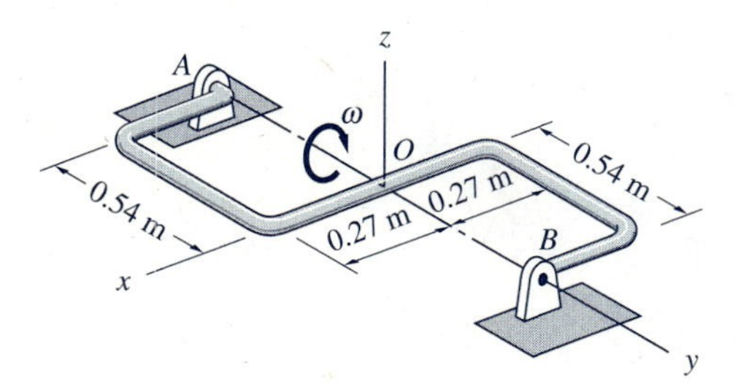

Fig. P19.48, P19.49

19.49 The 8-kg crank is made from a uniform slender rod. A constant couple $\mathbf{C}_0 = -240\mathbf{j}$ N · m is applied to the crank when it is at rest. Determine the angular acceleration of the crank and the bearing reactions at A and B acting on the crank immediately after the couple has been applied.

19.50 The thin uniform triangular plate of mass m is being driven about the y-axis with the constant angular velocity $\boldsymbol{\omega} = \omega_0\mathbf{j}$. Find the distance c for which the magnitudes of the dynamic bearing reactions at A and B are equal.

19.51 The 12-lb thin uniform triangular plate is initially at rest in the position shown. The plate is given a small angular displacement and then released. When the plate has become horizontal, calculate (a) its angular velocity; and (b) the magnitudes of the bearing reactions at A and B. Use $b = c = 1.5$ ft.

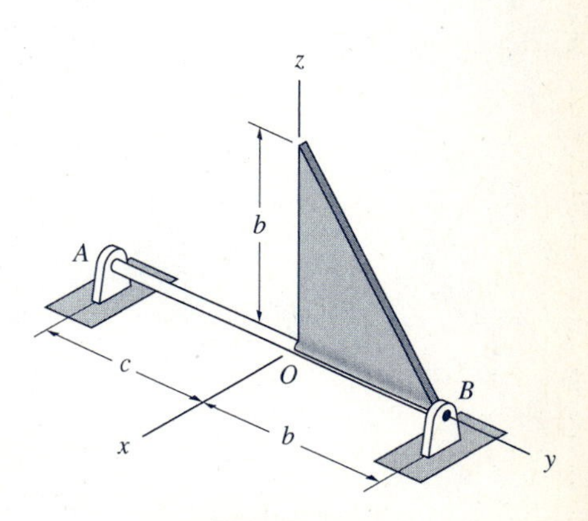

Fig. P19.50, P19.51

19.52 The uniform slender rod OAB of total mass $2m$ rotates about the z-axis with the constant angular velocity ω_0. Determine the dynamic reactions acting on the rod at O.

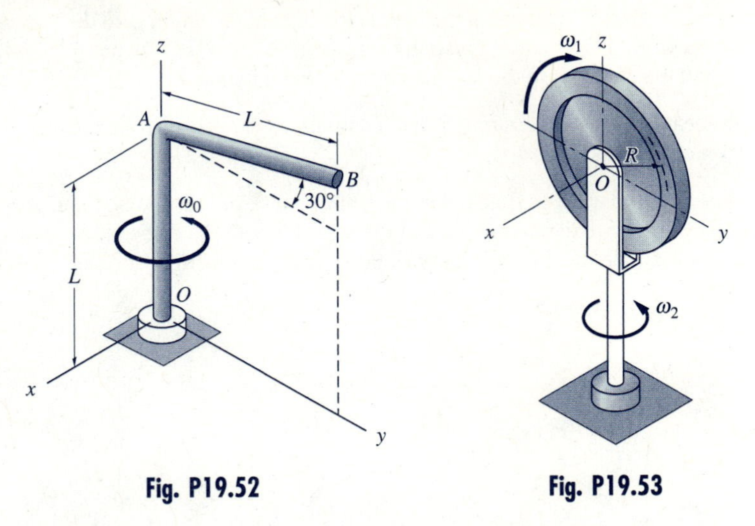

Fig. P19.52

Fig. P19.53

19.53 The 0.25-kg mass of the wheel is concentrated primarily in its thin, uniform rim of mean radius $R = 60$ mm. The wheel spins about the axle at O at the constant angular speed $\omega_1 = 200$ rad/s. At the same time, the mounting fork is rotating about the z-axis at the constant angular speed $\omega_2 = 60$ rad/s. Calculate the couple exerted on the wheel by the axle.

19.54 The 18-lb uniform disk spins about the axle AG with the constant angular velocity of 20 rad/s. The axle is supported by a ball-and-socket joint at A, and it rotates about the vertical axis with the constant angular velocity ω_1. Find the value of ω_1 for which the axle will remain horizontal during the motion. Neglect the weight of the axle.

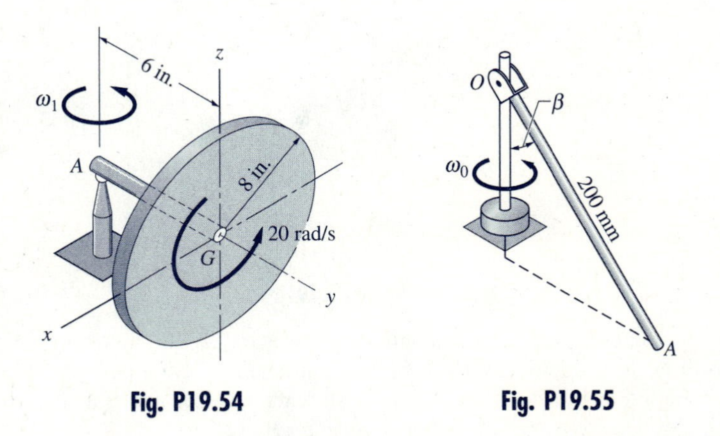

Fig. P19.54

Fig. P19.55

19.55 The uniform slender rod OA of mass m is attached to the vertical shaft with a clevis. Determine the constant angular velocity ω_0 of the shaft for which the rod will maintain a constant angle $\beta = 30°$ with the vertical.

19.56 The position of the 1.2-kg uniform slender rod OA is controlled by two small electric motors. A motor at B rotates the vertical shaft at the constant angular speed of 1.8 rad/s, and a motor at O increases the angle β between OA and the vertical at the constant rate $d\beta/dt = 1.5$ rad/s. Compute the output torque of each motor when the rod is in the position $\beta = 30°$. Neglect the masses of the motors.

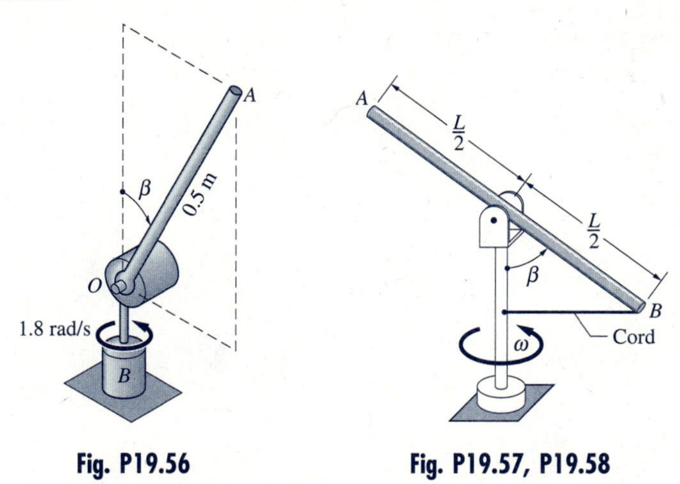

Fig. P19.56 **Fig. P19.57, P19.58**

19.57 The homogeneous slender rod AB of mass m and length L is attached to the vertical shaft with a clevis. The shaft is rotating at the constant angular speed ω. The angle β between the rod and the shaft is kept constant by the cord that connects end B to the shaft. Determine the expression for the tension in the cord.

19.58 The slender uniform rod AB of mass m and length L is attached to the vertical shaft with a clevis. The shaft is rotating freely with the angular speed ω when the cord that connects end B to the shaft breaks. Determine the expressions for $d\omega/dt$ and $d^2\beta/dt^2$ immediately after the break occurs.

19.59 The uniform 16-kg slender rod PQ spins about the axis BD at the constant angular speed $\dot{\theta} = 12$ rad/s. At the same time, the bracket AB is rotating about the vertical axis AD at the constant rate of 4 rad/s. The angular speeds are maintained by small electric motors (not shown) at A and D. Calculate the maximum torque that each motor must develop and the corresponding value of the angle θ. Neglect the masses of the bracket and the motors.

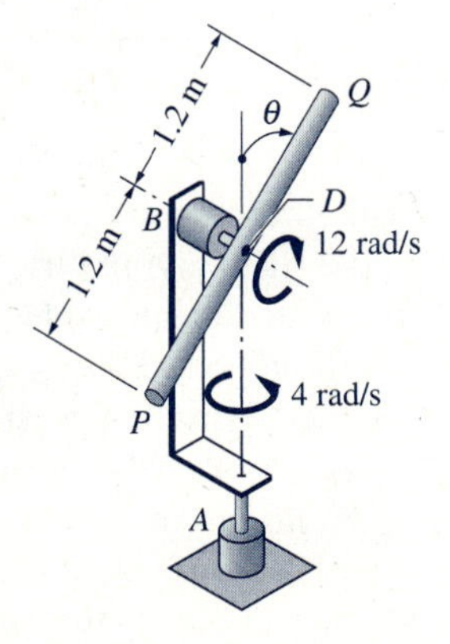

Fig. P19.59

19.60 The uniform disk of radius $R = 6$ in. and weight $W = 12$ lb is attached to the bent axle OAB of negligible weight. The axle is joined to a vertical shaft with a clevis at O. Assuming that the axle rotates freely about the z-axis with the constant angular velocity $\omega_1 = 6$ rad/s, and that the disk rolls without slipping, determine the vertical force exerted on the wheel by the horizontal surface.

19.61 The 40-lb uniform disk can spin freely about the axis BG. The mounting arm ABG, which has negligible weight, is free to rotate about axis AB. In the position shown, the spin velocity of the disk is $\omega_1 = 3$ rad/s and the angular velocity of the arm is $\omega_2 = 4$ rad/s. For this position, determine $d\omega_1/dt$, $d\omega_2/dt$, and the bearing reaction acting on the disk at G. (Hint: Draw separate free-body diagrams for the arm and the disk.)

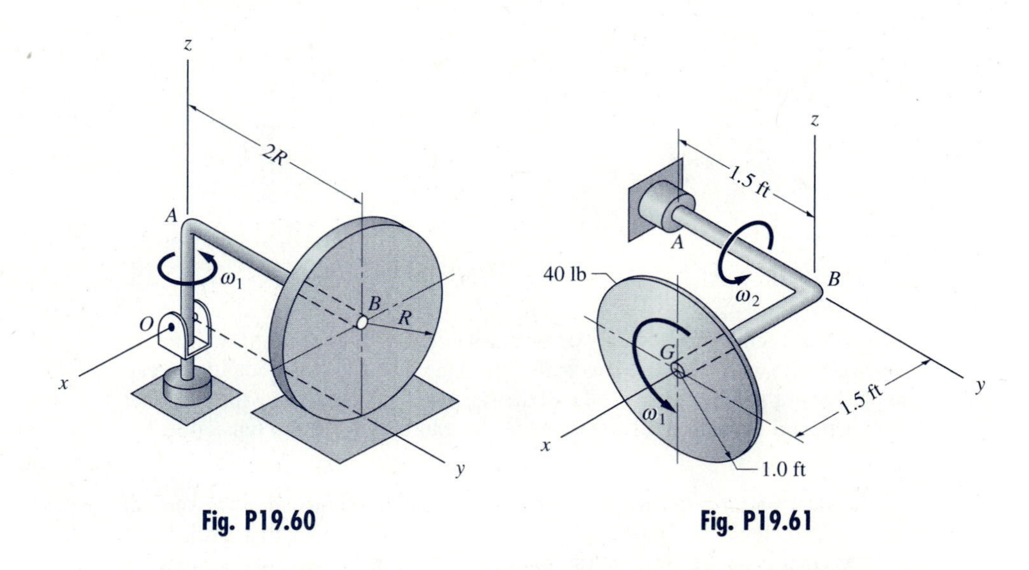

Fig. P19.60

Fig. P19.61

*19.6 Motion of an Axisymmetric Body

In this article we discuss the motion of axisymmetric bodies, which includes several important applications, such as gyroscopes, satellites, and projectiles. The equations of motion employed here are the modified Euler equations, but these equations will be reformulated by introducing new kinematic variables known as Euler's angles.

a. Euler's angles and angular velocity

In the modified Euler equations, the rotation of a rigid body was considered to consist of the rotation of an xyz reference frame plus the rotation of the body relative to that frame. If the body is axisymmetric, it is convenient to let the axis of symmetry be the z-axis, which is embedded in the body.

Figure 19.10(a) shows the rotations of the xyz reference frame that we will adopt in this article. The XYZ-coordinates are fixed in space, with the Z-axis (its direction can be chosen arbitrarily) being known as the *invariable line.* The x-axis, referred to as the *nodal line,* always remains in the XY-plane. The angle ϕ between the X- and x-axes is called the *precession angle,* and the angle θ between the Z- and z-axes is called the *nutation angle.* The angular velocity components $\dot{\boldsymbol{\phi}}$ and $\dot{\boldsymbol{\theta}}$ are called the rates of precession and nutation, respectively.

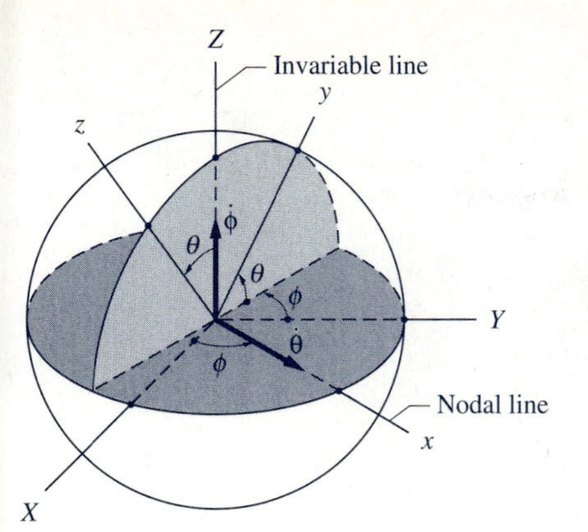

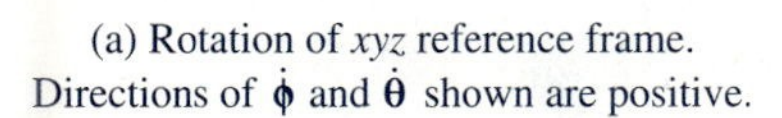
(a) Rotation of xyz reference frame. Directions of $\dot{\boldsymbol{\phi}}$ and $\dot{\boldsymbol{\theta}}$ shown are positive.

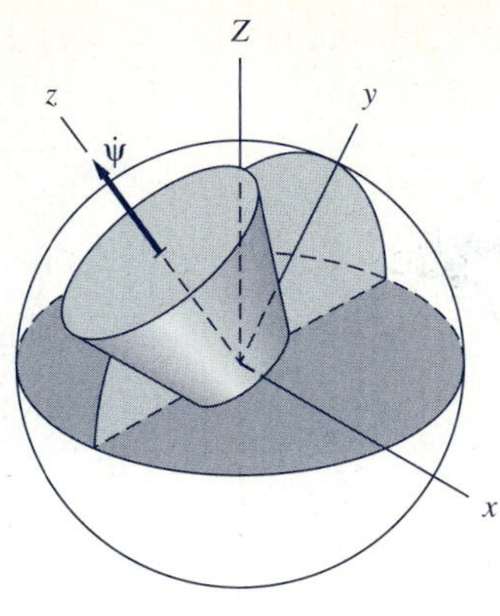

(b) Spinning of the body relative to xyz reference frame. Direction of $\dot{\boldsymbol{\psi}}$ shown is positive.

Fig. 19.10

Since the z-axis is embedded in the body, the rotation of the body relative to the xyz-frame is confined to a rotation, or spin, about the z-axis. The rate of this rotation, called the rate of spin and denoted by $\dot{\boldsymbol{\psi}}$, is shown in Fig. 19.10(b). The *spin angle* ψ (not shown) can be measured from any convenient reference. The angles ϕ, θ, and ψ, which are called *Euler's angles,* are useful kinematic variables for describing the motion of an axisymmetric body.

Before proceeding, we must reiterate that the xyz-axes shown in Fig. 19.10 are not a body frame, since the body is allowed to spin relative to these axes about its axis of symmetry (the z-axis). From Fig. 19.10(a) we see that the angular velocity of the xyz reference frame is

$$\boldsymbol{\Omega} = \dot{\boldsymbol{\phi}} + \dot{\boldsymbol{\theta}} \tag{19.49}$$

Since the body spins at the rate $\dot{\psi}$ relative to the xyz-frame, its angular velocity is $\boldsymbol{\omega} = \boldsymbol{\Omega} + \dot{\boldsymbol{\psi}}$, or

$$\boldsymbol{\omega} = \dot{\boldsymbol{\phi}} + \dot{\boldsymbol{\theta}} + \dot{\boldsymbol{\psi}} \tag{19.50}$$

Utilizing Fig. 19.10 parts (a) and (b), we deduce that the components of $\boldsymbol{\Omega}$ and $\boldsymbol{\omega}$ relative to the xyz-axes are

$$\begin{aligned} \Omega_x &= \dot{\theta} & \omega_x &= \dot{\theta} \\ \Omega_y &= \dot{\phi}\sin\theta & \omega_y &= \dot{\phi}\sin\theta \\ \Omega_z &= \dot{\phi}\cos\theta & \omega_z &= \dot{\phi}\cos\theta + \dot{\psi} \end{aligned} \tag{19.51}$$

b. Moment equations of motion

Because the z-axis is assumed to be an axis of symmetry for the body, we have $I_{xy} = I_{yz} = I_{xz} = 0$ and $I_x = I_y$. From now on, we will use the following notation:

$$I_x = I_y = I \tag{19.52}$$

Substituting Eqs. (19.51) and (19.52) into the modified Euler equations, Eqs. (19.43), we obtain

$$\begin{aligned}\Sigma M_x &= I\ddot{\theta} + (I_z - I)\dot{\phi}^2 \sin\theta\cos\theta + I_z\dot{\phi}\dot{\psi}\sin\theta \\ \Sigma M_y &= I\ddot{\phi}\sin\theta + 2I\dot{\theta}\dot{\phi}\cos\theta - I_z\dot{\theta}(\dot{\psi} + \dot{\phi}\cos\theta) \\ \Sigma M_z &= I_z(\ddot{\psi} + \ddot{\phi}\cos\theta - \dot{\phi}\dot{\theta}\sin\theta) \\ &= I_z\frac{d}{dt}(\dot{\psi} + \dot{\phi}\cos\theta)\end{aligned} \tag{19.53}$$

Sometimes it is convenient to use the following equations, obtained by substituting $\omega_z = \dot{\phi}\cos\theta + \dot{\psi}$ from Eqs. (19.51) into Eqs. (19.53).

$$\begin{aligned}\Sigma M_x &= I\ddot{\theta} + I_z\omega_z\dot{\phi}\sin\theta - I\dot{\phi}^2\sin\theta\cos\theta \\ \Sigma M_y &= I\ddot{\phi}\sin\theta + 2I\dot{\theta}\dot{\phi}\cos\theta - I_z\dot{\theta}\omega_z \\ \Sigma M_z &= I_z\dot{\omega}_z\end{aligned} \tag{19.54}$$

When using Eqs. (19.53) or (19.54) it must be recalled from the previous article that the modified Euler's equations are valid only if the origin of the xyz-axes is located at the mass center of the body, or at a fixed point (fixed in the body and in space).

c. Steady precession

The special motion that arises when $\dot{\psi}$, $\dot{\phi}$, and θ are constants is known as *steady precession.* In this case, Eqs. (19.53) simplify considerably.

$$\begin{gathered}\Sigma M_x = (I_z - I)\dot{\phi}^2\sin\theta\cos\theta + I_z\dot{\phi}\dot{\psi}\sin\theta \\ \Sigma M_y = 0 \qquad \Sigma M_z = 0\end{gathered} \tag{19.55}$$

whereas the equivalent equations, Eqs. (19.54), reduce to

$$\begin{gathered}\Sigma M_x = I_z\omega_z\dot{\phi}\sin\theta - I\dot{\phi}^2\sin\theta\cos\theta \\ \Sigma M_y = 0 \qquad \Sigma M_z = 0\end{gathered} \tag{19.56}$$

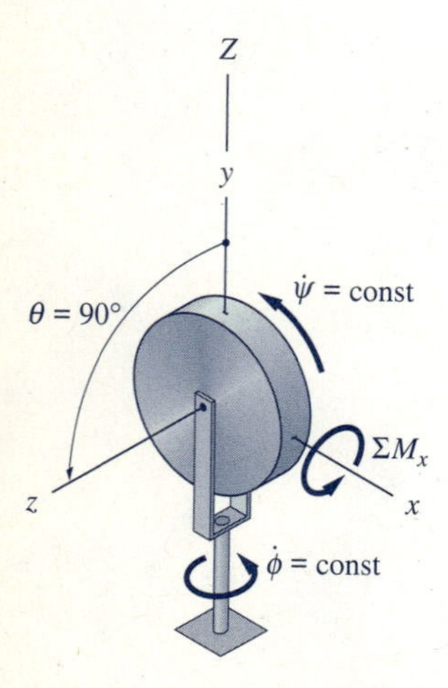

Fig. 19.11

From either Eqs. (19.55) or (19.56) we see that for a body to undergo steady precession, it must be acted upon by forces that provide a constant moment about the x-axis (the nodal line) with no moments acting about the other two axes. Therefore, the direction of the moment vector must be perpendicular to both the precession axis (Z) and the spin axis (z).

An interesting special case of steady precession occurs when the precession axis (Z) is perpendicular to the spin axis (z), as depicted in Fig. 19.11. Setting $\theta = 90°$ in the first of Eqs. (19.55), we find that the moment required to maintain the steady precession is

$$\Sigma M_x = I_z\dot{\phi}\dot{\psi} \tag{19.57}$$

d. Torque-free motion

If the resultant moment of the external forces about the mass center is zero, the body is said to undergo torque-free motion. *Torque-free motion* is thus characterized by $\Sigma \mathbf{M}_G = d\mathbf{h}_G/dt = \mathbf{0}$, from which we conclude that the angular momentum of the body about its mass center G remains constant in both magnitude and direction. Examples of torque-free motion are projectiles (with air resistance neglected) and space vehicles in unpowered flight.

Figure 19.12 shows an axisymmetric projectile in free flight. For mathematical convenience, it is customary to choose the fixed Z-axis to be in the direction of $\mathbf{h}_G$. The xyz-coordinate system is attached to G, with z being the axis of symmetry of the body. In accordance with Fig. 19.10, the x-axis is perpendicular to the plane formed by the Z- and z-axes. Although the z-axis is embedded in the body, it is important to recall that the xyz-axes do not constitute a body frame, because the body can rotate (spin) about the z-axis relative to the xyz-frame.

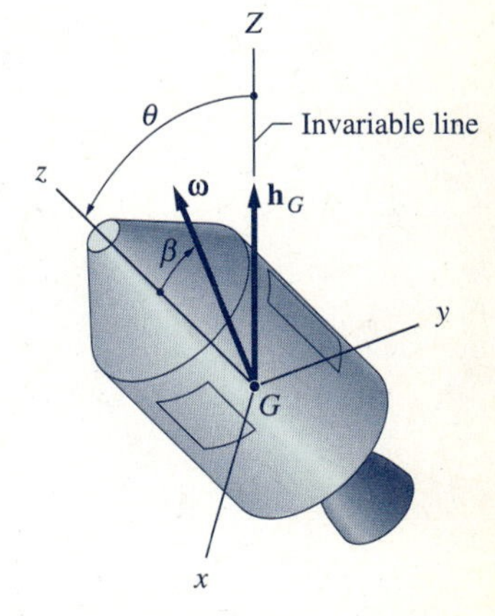

Fig. 19.12

As shown in Fig. 19.12, we let β be the angle between the angular velocity vector $\boldsymbol{\omega}$ of the body and the z-axis. The angle θ between the Z- and z-axes is the Euler angle that was defined in Fig. 19.10. The components of $\mathbf{h}_G$ relative to the xyz-axes thus are

$$h_x = 0 \qquad h_y = h_G \sin\theta \qquad h_z = h_G \cos\theta \tag{19.58}$$

Due to symmetry, the xyz-axes are principal axes of inertia of the projectile at G. Using the notation $\bar{I}_x = \bar{I}_y = \bar{I}$, the components of $\mathbf{h}_G$ take the form [see Eq. (19.20)]

$$h_x = \bar{I}\omega_x \qquad h_y = \bar{I}\omega_y \qquad h_z = \bar{I}_z\omega_z \tag{19.59}$$

Comparing Eqs. (19.58) and (19.59) we conclude that

$$\omega_x = 0 \qquad \omega_y = \frac{h_G \sin\theta}{\bar{I}} \qquad \omega_z = \frac{h_G \cos\theta}{\bar{I}_z} \tag{19.60}$$

Since $\omega_x = \dot{\theta}$ according to Eq. (19.51), we conclude that θ remains constant, which indicates that the motion is a steady precession about the Z-axis.

A useful relationship between the angles θ and β can be derived by noting that (see Fig. 19.12) $\tan\beta = \omega_y/\omega_z$. Utilizing Eqs. (19.60) we obtain

$$\tan\beta = \frac{\omega_y}{\omega_z} = \frac{h_G \sin\theta/\bar{I}}{h_G \cos\theta/\bar{I}_z} = \frac{\bar{I}_z}{\bar{I}}\tan\theta$$

or

$$\boxed{\tan\theta = \frac{1}{\lambda}\tan\beta \quad \text{where } \lambda = \frac{\bar{I}_z}{\bar{I}}} \tag{19.61}$$

It is customary to distinguish between cases of steady precession, depending on whether λ in Eq. (19.61) is greater or less than one.

Case 1: Direct (regular) precession: $\lambda < 1$.

For $\lambda < 1$, it follows from Eq. (19.61) that $\bar{I} > \bar{I}_z$, which is the case for an elongated body such as the rocket shown in Fig. 19.13. From Eq. (19.61) we also see that $\theta > \beta$, indicating that the angular velocity vector $\boldsymbol{\omega}$ lies within the angle formed by the positive Z- and z-axes, as shown in Fig. 19.13(a). Note that the projection of $\dot{\boldsymbol{\psi}}$ onto $\dot{\boldsymbol{\phi}}$ is in the same direction as $\dot{\boldsymbol{\phi}}$, which is a characteristic of direct precession. The system of vectors shown in Fig. 19.13(a) precesses at a constant rate $\dot{\phi}$ about the Z-axis, while the angular velocity $\boldsymbol{\omega}$ sweeps out a cone in space, the axis of the cone being Z.

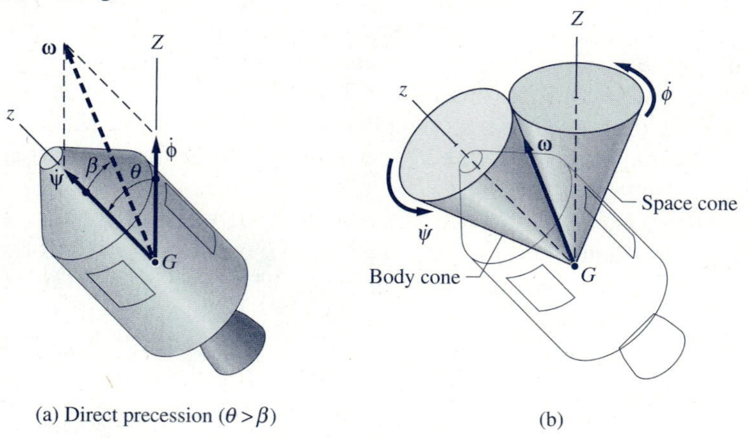

(a) Direct precession ($\theta > \beta$)

(b)

Fig. 19.13

This motion may be represented by the geometric model shown in Fig. 19.13(b), which consists of two right circular cones. The body cone is fixed in the body, with its axis being the axis of symmetry of the body (z-axis). The space cone is stationary, with Z being its axis. The angular velocity $\boldsymbol{\omega}$ lies along the line of contact between the two cones. Because the instant axis of rotation is the locus of points that have zero velocity, we see that the body cone rolls without slipping on the outside of the space cone. Therefore, the z-axis and angular velocity $\boldsymbol{\omega}$ precess about the Z-axis at the rate $\dot{\phi}$, and the body cone spins about the z-axis at the rate $\dot{\psi}$. By direct comparison of parts (a) and (b) of Fig. 19.13, we conclude that the motions of the body cone and the physical body (i.e., the rocket) are identical.

Case 2: Retrograde precession: $\lambda > 1$.

If $\lambda > 1$, we see from Eq. (19.61) that $\bar{I} < \bar{I}_z$, which would be the case for a flattened body such as the orbiting space vehicle in Fig. 19.14. Since $\lambda > 1$ implies that $\theta < \beta$, the angular velocity vector $\boldsymbol{\omega}$ lies outside the angle formed by the positive Z- and z-axes, as shown in Fig. 19.14(a). We see also that the direction of the projection of $\dot{\boldsymbol{\psi}}$ onto $\dot{\boldsymbol{\phi}}$ is opposite to the direction of $\dot{\boldsymbol{\phi}}$. The cone model for retrograde precession in Fig. 19.14(b) shows that the space cone is on the inside of the body cone.

For a body for which all three principal moments of inertia are equal to $\bar{I}$—i.e., for $\lambda = 1$ (as is the case for a uniform sphere)—it can be seen from Eqs. (19.20) that $\mathbf{h}_G = \bar{I}\boldsymbol{\omega}$. Therefore, once launched with an initial rotation about a given axis, the body simply continues to rotate about that axis with constant angular velocity.

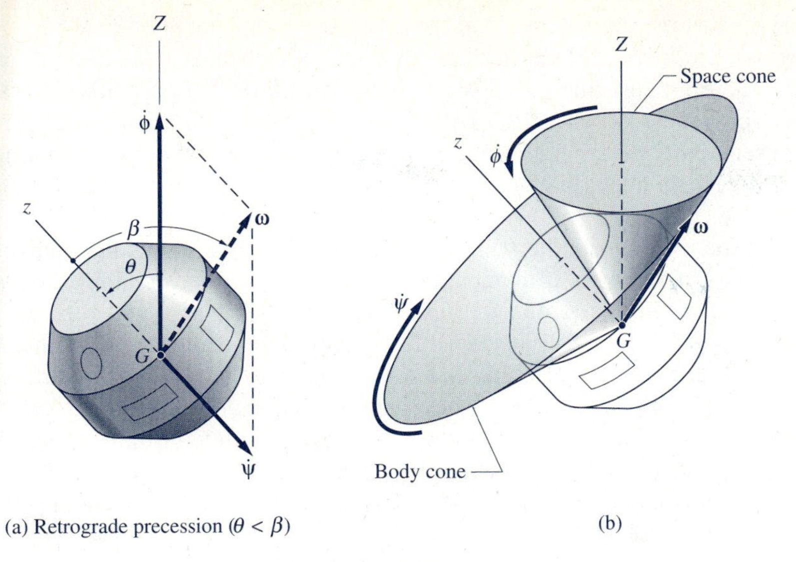

(a) Retrograde precession ($\theta < \beta$) (b)

Fig. 19.14

e. Gyroscopes

A gyroscope consists of an axisymmetric rotor, or disk, that is mounted in such a way that the rotor is free to spin about its axis of symmetry. Figure 19.15 shows a gyroscope that is mounted in a so-called Cardan's suspension, a widely used design for inertial guidance systems, gyrostabilizers, etc. The elements of this suspension are as follows: the rotor of mass m, which spins about the axis AB, which is fixed in the inner gimbal, or ring; the inner gimbal, which is free to rotate about the axis CD relative to the outer gimbal; and the outer gimbal, which can rotate about the Z-axis. Figure 19.15 also shows the manner in which the orientations of the gimbals correspond to the Eulerian angles ϕ

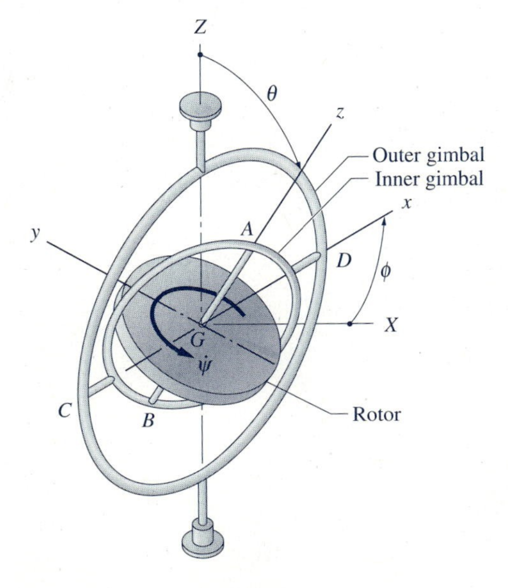

Fig. 19.15

and θ, and to the rate of spin $\dot{\psi}$ of the rotor. Assuming that the Z-axis has a fixed orientation in space, the rotor has three degrees of rotational freedom, and it can, therefore, assume all possible angular positions. Of particular interest is the fact that the three axes of rotation intersect at the mass center G of the rotor. Consequently, the motion of the rotor is torque-free, assuming that all of the bearings have negligible friction and that no external forces are applied to the gimbals.

If the rotor in an otherwise stationary gyroscope is set spinning about the z-axis, its initial angular momentum $\mathbf{h}_G$ will also be directed along the z-axis. Since the motion is torque-free, the direction of that axis will remain fixed ($\Sigma \mathbf{M}_G = d\mathbf{h}_G/dt = \mathbf{0}$). The capability of a gyroscope rotor to maintain a fixed direction serves as the principle of operation in many navigational instruments.

A force applied to one of the gimbals can result in a moment about the mass center of the rotor. In this case, the rotor will undergo steady precession provided that suitable initial conditions were present. The direction of the precession can be deduced from $\Sigma \mathbf{M}_G = d\mathbf{h}_G/dt$, where $\Sigma \mathbf{M}_G$ is the moment of the applied force about the mass center G.

Sample Problem 19.10

Figure (a) shows an axisymmetric body (z-axis is the axis of symmetry) that is undergoing torque-free motion. The fixed Z-axis is chosen to coincide with $\mathbf{h}_G$, the angular momentum about the mass center G. The angular velocity vector $\boldsymbol{\omega}$ is inclined at the angle β from the z-axis.

(1) Letting $\lambda = \bar{I}_z/\bar{I}$, derive the relationships

$$\dot{\psi} = \omega(1 - \lambda)\cos\beta \tag{a}$$

and

$$\dot{\phi} = \omega\cos\beta\sqrt{\lambda^2 + \tan^2\beta} \tag{b}$$

where $\dot{\psi}$ and $\dot{\phi}$ are the spin and precession rates, respectively. Use Eqs. (a) and (b) to show that $\dot{\phi}$, $\dot{\psi}$, and θ are related by

$$(1 - \lambda)\dot{\phi}\cos\theta = \lambda\dot{\psi} \tag{c}$$

where θ is the Euler angle shown in Fig. (a).

(2) Sketch the body and space cones, and calculate the rates of spin and precession for a satellite given that $\lambda = 1.8$, $\omega = 1.2$ rad/s, and $\beta = 25°$. Repeat the procedure for a rocket for which $\lambda = 0.2$, $\omega = 0.8$ rad/s, and $\beta = 15°$.

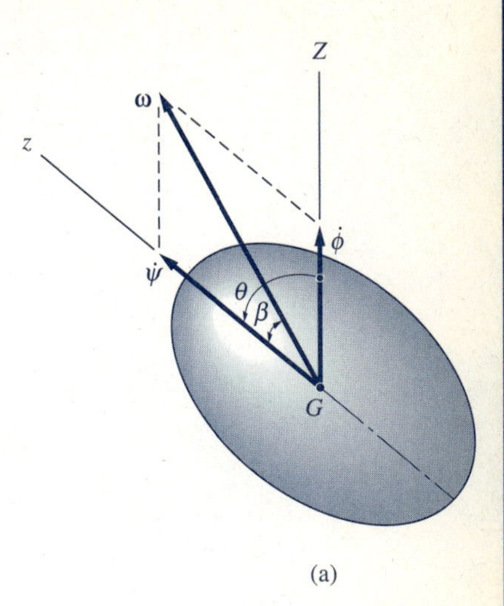

(a)

Solution

Part 1

Applying the law of sines to the velocity diagram shown in Fig. (b), we obtain

$$\frac{\dot{\psi}}{\sin(\theta - \beta)} = \frac{\omega}{\sin(\pi - \theta)}$$

which yields

$$\dot{\psi} = \omega\frac{\sin\theta\cos\beta - \cos\theta\sin\beta}{\sin\theta} = \omega\left(\cos\beta - \frac{\sin\beta}{\tan\theta}\right)$$

Substituting $\tan\theta = (1/\lambda)\tan\beta$ [see Eq. (19.61)], we obtain

$$\dot{\psi} = \omega\left(\cos\beta - \frac{\sin\beta}{(1/\lambda)\tan\beta}\right) = \omega(1 - \lambda)\cos\beta$$

which agrees with Eq. (a).

Application of the law of sines to Fig. (b) also yields

$$\frac{\dot{\phi}}{\sin\beta} = \frac{\omega}{\sin(\pi - \theta)}$$

or

$$\dot{\phi} = \omega\frac{\sin\beta}{\sin\theta} \tag{d}$$

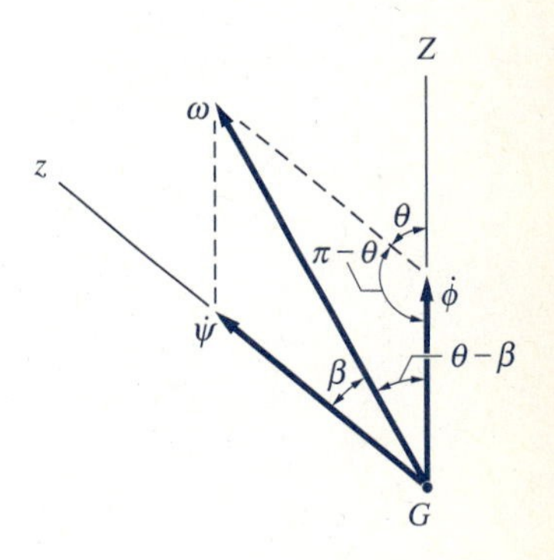

(b)

Utilizing, as before, the relationship $\tan\theta = (1/\lambda)\tan\beta$, $\sin\theta$ may be written as

$$\sin\theta = \frac{\tan\theta}{\sqrt{1+\tan^2\theta}} = \frac{(1/\lambda)\tan\beta}{\sqrt{1+[(1/\lambda)\tan\beta]^2}} = \frac{\tan\beta}{\sqrt{\lambda^2+\tan^2\beta}} \tag{e}$$

Substituting Eq. (e) into Eq. (d) gives

$$\dot{\phi} = \omega\sin\beta\,\frac{\sqrt{\lambda^2+\tan^2\beta}}{\tan\beta} = \omega\cos\beta\,\sqrt{\lambda^2+\tan^2\beta}$$

which agrees with Eq. (b).

Dividing Eq. (a) by Eq. (b) we obtain

$$\frac{\dot{\psi}}{\dot{\phi}} = \frac{\omega(1-\lambda)\cos\beta}{\omega\cos\beta\,\sqrt{\lambda^2+\tan^2\beta}} = \frac{1-\lambda}{\sqrt{\lambda^2+\lambda^2\tan^2\theta}}$$

$$= \frac{1-\lambda}{\lambda\sec\theta} = \frac{(1-\lambda)\cos\theta}{\lambda}$$

or

$$\lambda\dot{\psi} = (1-\lambda)\dot{\phi}\cos\theta$$

which is identical to Eq. (c).

Part 2

The Euler angle θ between the fixed Z-axis and the axis of symmetry is found by using Eq. (19.61). Equations (a) and (b) can be used to calculate the spin rate $\dot{\psi}$ and the precession rate $\dot{\phi}$. When the given parameters for the satellite and the rocket are substituted into these equations, the results are as follows. For the satellite:

$$\theta = \tan^{-1}[(1/\lambda)\tan\beta] = \tan^{-1}\left[(1/1.8)\tan 25^\circ\right] = 14.52^\circ$$

$$\dot{\psi} = \omega(1-\lambda)\cos\beta = 1.2(1-1.8)\cos 25^\circ = -0.870\text{ rad/s}$$

$$\dot{\phi} = \omega\cos\beta\,\sqrt{\lambda^2+\tan^2\beta}$$

$$= 1.2\cos 25^\circ\,\sqrt{(1.8)^2+\tan^2 25^\circ} = 2.022\text{ rad/s}$$

For the rocket:

$$\theta = \tan^{-1}[(1/\lambda)\tan\beta] = \tan^{-1}\left[(1/0.2)\tan 15^\circ\right] = 53.26^\circ$$

$$\dot{\psi} = \omega(1-\lambda)\cos\beta = 0.8(1-0.2)\cos 15^\circ = 0.618\text{ rad/s}$$

$$\dot{\phi} = \omega\cos\beta\,\sqrt{\lambda^2+\tan^2\beta}$$

$$= 0.8\cos 15^\circ\,\sqrt{(0.2)^2+\tan^2 15^\circ} = 0.258\text{ rad/s}$$

The space and body cones for the satellite and rocket are shown in Figs. (c) and (d), respectively. Observe that the satellite undergoes retrograde precession ($\dot{\psi}$ is negative), whereas the precession of the rocket is direct ($\dot{\psi}$ is positive).

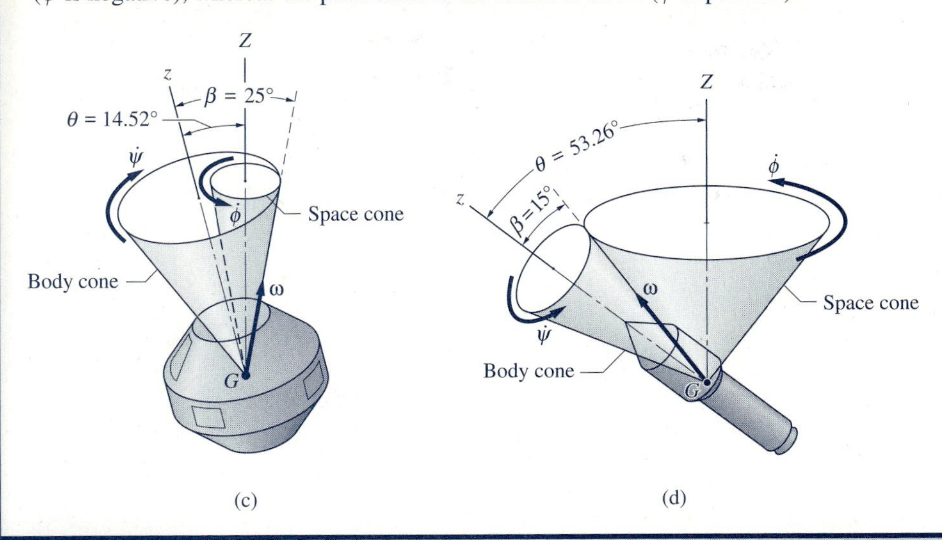

(c)

(d)

Sample Problem 19.11

Figure (a) shows a uniform sphere of radius R and mass m that is welded to the rod AB of length L (the mass of AB may be neglected). The clevis at B connects the rod to the vertical shaft BC. The assembly is initially rotating about the vertical at the angular velocity ω, with the sphere resting against the shaft. Assuming that ω is gradually increased, determine the critical angular velocity ω_{cr} at which contact between the sphere and rod is lost. Neglect friction.

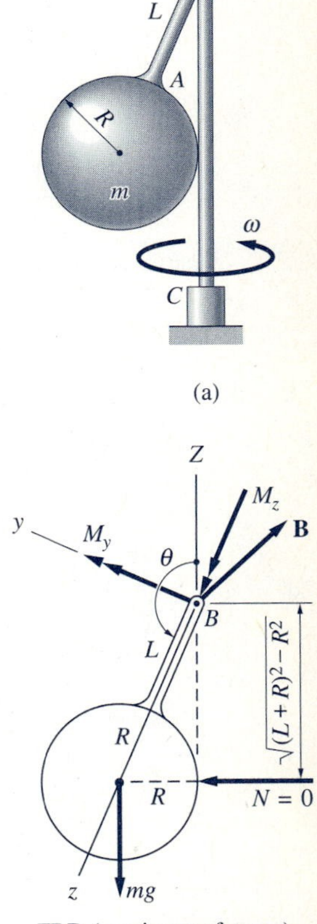

(a)

FBD (x-axis out of paper)

(b)

Solution

The free-body diagram of the sphere and the rod, drawn at the instant when the assembly is rotating at $\omega = \omega_{cr}$, is shown in Fig. (b). The xyz-axes are assumed to be attached to rod AB with the origin at B (the x-axis is out of the paper). In addition to the weight mg of the sphere, the FBD also contains the reactions provided by the clevis at B: the pin force **B** and the two moment components M_y and M_z ($M_x = 0$ because the pin of the clevis is smooth). If the angular velocity were less than the critical angular velocity, the FBD would also contain the normal force N that is exerted on the sphere by the vertical shaft. However, when $\omega = \omega_{cr}$, then $N = 0$.

It is convenient to choose the Z-axis to coincide with the vertical shaft, as shown in Fig. (b). The Euler angle θ, which was defined as the angle between the Z- and z-axes, is also shown in the figure. When the sphere is about to lose contact with the vertical shaft, its motion consists of a rotation about the Z-axis at the rate ω_{cr}. The spin rate is zero, since the sphere cannot rotate relative to the rod AB. Therefore, the motion of the sphere can be described as a steady precession with no spin; i.e., $\dot{\phi} = \omega_{cr}$, $\dot{\psi} = \dot{\theta} = 0$. As a result, the steady precession equation, Eq. (19.55), becomes

$$\Sigma M_x = (I_z - I)\omega_{cr}^2 \sin\theta \cos\theta \qquad \text{(a)}$$

The inertial properties of the sphere about point B are

$$I_z = \frac{2}{5}mR^2 \quad \text{and} \quad I = I_y = \frac{2}{5}mR^2 + m(L+R)^2$$

from which we obtain

$$I_z - I = -m(L+R)^2 \tag{b}$$

Referring to Fig. (b), we find that the moment of the external forces (the weight) about the x-axis is

$$\Sigma M_x = mgR \tag{c}$$

From the same figure we also deduce that

$$\sin\theta = \sin(\pi - \theta) = \frac{R}{L+R} \tag{d}$$

and

$$\cos\theta = -\cos(\pi - \theta) = -\frac{\sqrt{(L+R)^2 - R^2}}{L+R} \tag{e}$$

Substituting Eqs. (b)–(e) into Eq. (a) and solving for the critical angular velocity, we obtain

$$\omega_{\text{cr}} = \frac{\sqrt{g}}{\sqrt[4]{(L+R)^2 - R^2}} \qquad \textit{Answer}$$

Sample Problem 19.12

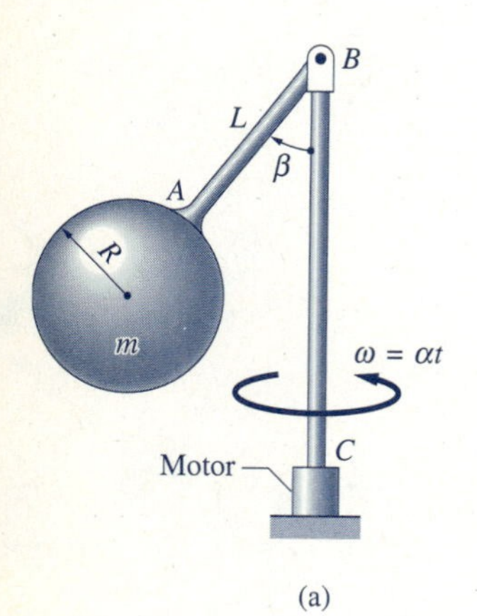

(a)

Figure (a) shows the same assembly that was described in Sample Problem 19.11. The assembly is initially stationary with the sphere resting against the shaft BC. A motor at C is then activated that drives the shaft with the constant angular acceleration α (the resulting angular velocity of the shaft is $\omega = \alpha t$). Letting t_0 be the time when the sphere loses contact with the shaft, (1) derive the equation of motion for the sphere in terms of the angle β for the period $t \geq t_0$ and state the initial conditions; (2) solve the equations numerically for the time interval $t = t_0$ to $t = t_0 + 2$ s; and (3) use the numerical solution to find the maximum value of β for the time interval given in Part 2. Use $m = 7$ kg, $L = R = 60$ mm, and $\alpha = 5.5$ rad/s^2.

Solution

Part 1

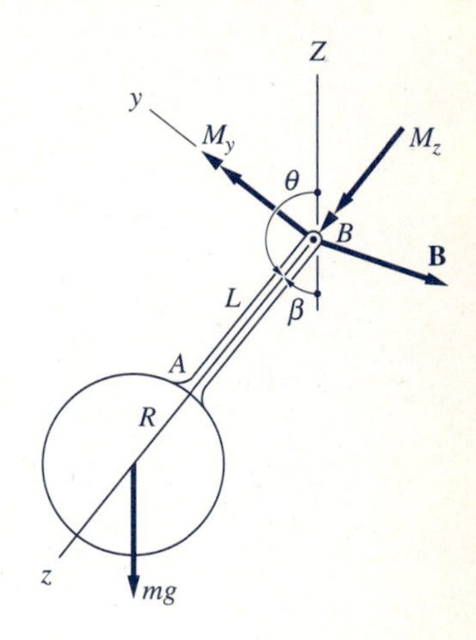

(b) FBD

The free-body diagram (FBD) of the rigid unit containing the sphere and the rod AB is shown in Fig. (b). The xyz-axes are assumed to be attached to rod AB with the origin at B (the x-axis is out of the paper). This FBD displays the weight of the sphere and the reactions at B: the pin force **B** and the moment components M_y and M_z (the smooth pin of the clevis does not provide a moment component about the x-axis). The fixed Z-axis is assumed to be directed along the vertical shaft as shown in Fig. (b), with θ being the Euler angle between the Z- and z-axes. Since the sphere and rod AB rotate as a rigid unit, the spin rate $\dot{\psi}$ is zero. Furthermore, comparing Figs. (a) and (b) with Fig. 19.10, the precession rate is found to be $\dot{\phi} = \omega = \alpha t$.

The modified Euler equation that governs β is the first of Eqs. (19.53) (the remaining two equations could be used to find M_y and M_z).

$$\Sigma M_x = I\ddot{\theta} + (I_z - I)\dot{\phi}^2 \sin\theta \cos\theta + I_z\dot{\phi}\dot{\psi}\sin\theta \qquad \text{(a)}$$

Using $L = R$ and $\theta = \pi - \beta$, the various terms in Eq. (a) become

$$\Sigma M_x = 2mgR\sin\beta \quad \text{(from the FBD)}$$

$$I_z = \frac{2}{5}mR^2$$

$$I = I_y = \frac{2}{5}mR^2 + m(R+R)^2 = \frac{22}{5}mR^2$$

$$\ddot{\theta} = -\ddot{\beta} \qquad \dot{\phi} = \alpha t \qquad \dot{\psi} = 0$$

Substituting these expressions into Eq. (a), we get

$$2mgR\sin\beta = \frac{22}{5}mR^2(-\ddot{\beta}) + (-4mR^2)(\alpha t)^2 \sin\beta(-\cos\beta) + 0$$

which, on canceling the mass m and rearranging terms, reduces to

$$\ddot{\beta} = \frac{10}{11}\alpha^2 t^2 \sin\beta\cos\beta - \frac{10g}{22R}\sin\beta \qquad \text{(b)}$$

When the numerical values $\alpha = 5.5\ \text{rad/s}^2$, $g = 9.81\ \text{m/s}^2$, and $R = 0.060$ m are substituted into Eq. (b), the equation of motion becomes

$$\ddot{\beta} = 27.50t^2 \sin\beta\cos\beta - 74.32\sin\beta \quad (\text{for } t \geq t_0) \qquad \textit{Answer} \quad \text{(c)}$$

From the solution to Sample Problem 19.11, we know that ω_{cr}, the critical angular velocity at which contact between the sphere and vertical shaft is lost, is

$$\omega_{cr} = \frac{\sqrt{g}}{\sqrt[4]{(L+R)^2 - R^2}} = \frac{\sqrt{9.81}}{\sqrt[4]{(0.120)^2 - (0.060)^2}} = 9.716\ \text{rad/s}$$

Consequently, the time at which contact is lost is $t_0 = \omega_{cr}/\alpha = 9.716/5.5 = 1.7665$ s. The initial value of β (the value when the sphere touches the vertical shaft) is $\beta_0 = \tan^{-1}[R/(L + R)] = \tan^{-1}(R/2R) = \tan^{-1}(1/2) = 30° = 0.5236$ rad. Therefore, the initial conditions are

$$t_0 = 1.7665 \text{ s} \qquad \beta_0 = 0.5236 \text{ rad} \qquad \dot{\beta}_0 = 0 \qquad \textit{Answer} \quad \text{(d)}$$

Part 2

For the numerical solution, we used the fourth-order Runge-Kutta method with integration step size 0.02 s (100 steps), recording the results of every step. The resulting plot of β versus time is shown in Fig. (c).

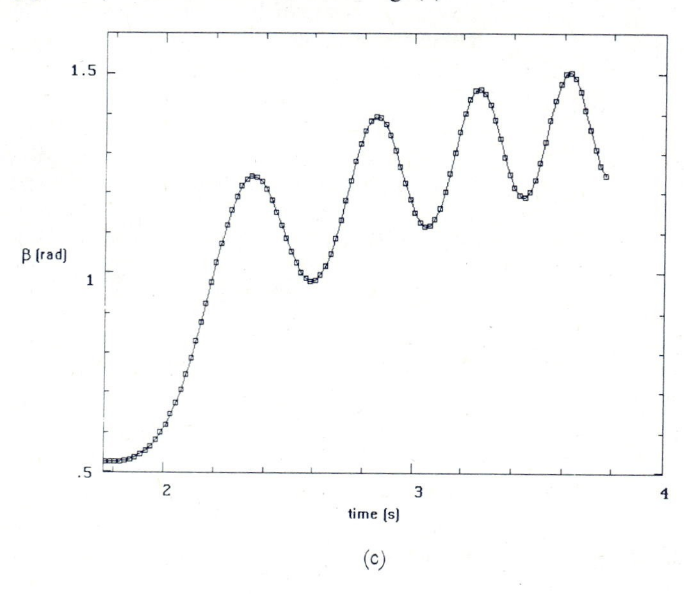

(c)

Part 3

From Fig. (c) we see that the maximum value of β occurs in the period of interest at approximately $t = 3.6$ s. Since β reaches a maximum value when $\dot{\beta}$ changes from positive to negative, a more accurate value of this time is obtained by examining the numerical output shown in the following.

t (s)	β (rad)	$\dot{\beta}$ (rad/s)
0.3607E+01	0.1500E+01	0.7196E+00
0.3627E+01	0.1504E+01	−0.2842E+00

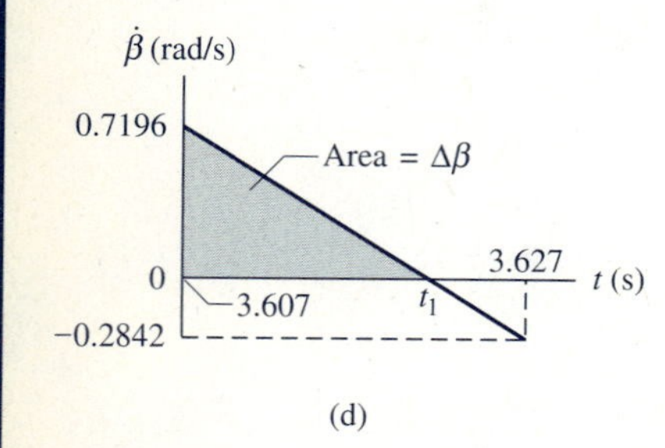

(d)

The corresponding straight-line plot of $\dot{\beta}$ versus t is shown in Fig. (d). Letting t_1 be the time when $\dot{\beta} = 0$, we obtain from similar triangles

$$\frac{t_1 - 3.607}{0.7196} = \frac{3.627 - 3.607}{0.7196 + 0.2842}$$

from which we find that $t_1 = 3.621$ s. The maximum value of β is $\beta_{max} = \beta(3.607 \text{ s}) + \Delta\beta$, where $\Delta\beta$ is the change in β between $t = 3.607$ s and 3.621 s. Since $\Delta\beta$ equals the shaded area in Fig. (d), we obtain $\Delta\beta = (1/2)(3.621 - 3.607)(0.7196) = 0.005$ rad. Consequently,

$$\beta_{max} = 1.500 + 0.005 = 1.505 \text{ rad} = 86.2° \qquad \textit{Answer}$$

PROBLEMS

19.62 The homogeneous cylinder of mass m, radius R, and length R spins with angular velocity ω_1 relative to its axle, which is inclined at the angle θ from the vertical. The axle rotates in the bearing at O at an angular speed ω_2. Determine the ratio ω_1/ω_2 for which no moment is exerted on the cylinder by the axle. Note that the mass center G of the cylinder is directly above O, and assume that $\theta \neq 0$ and $\omega_2 \neq 0$.

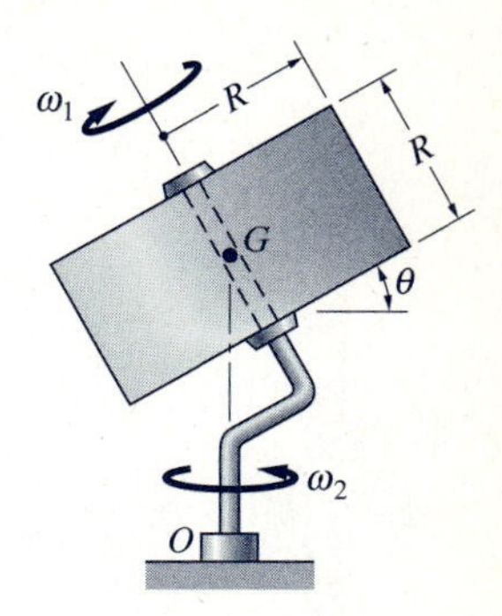

Fig. P19.62

19.63 The mass of the thin uniform disk is m and its radius is R. The light rod AB of length L is rigidly attached to the disk at A and connected by a clevis to the vertical shaft BC. The entire assembly rotates about the vertical at the constant angular velocity ω. Show that contact between the disk and BC will not be lost for any value of ω if $L \leq R/2$.

19.64 The weight of the thin uniform disk is 12 lb and its radius is $R = 4$ in. The light rod AB of length $L = 8$ in. is rigidly attached to the disk at A and connected by a clevis to the vertical shaft BC. The entire assembly rotates about the vertical with the constant angular velocity $\omega = 5$ rad/s. Calculate the normal force that acts between the disk and the shaft at C.

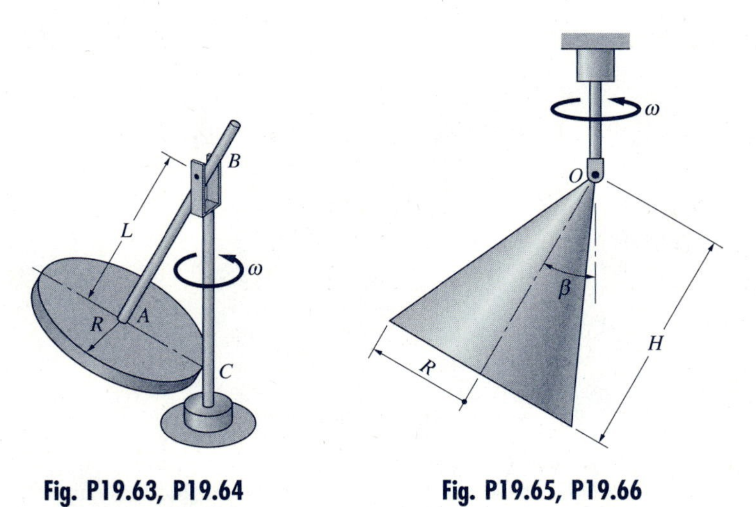

Fig. P19.63, P19.64 **Fig. P19.65, P19.66**

19.65 The homogeneous cone of mass m is suspended from a clevis at O. The vertical shaft attached to the clevis rotates with the constant angular velocity ω. Determine the expression for the smallest value of ω for which the cone may assume a position where the angle β is constant, other than $\beta = 0$.

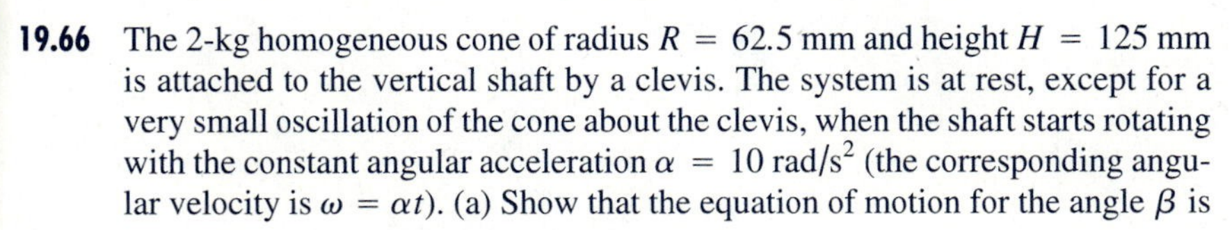

19.66 The 2-kg homogeneous cone of radius $R = 62.5$ mm and height $H = 125$ mm is attached to the vertical shaft by a clevis. The system is at rest, except for a very small oscillation of the cone about the clevis, when the shaft starts rotating with the constant angular acceleration $\alpha = 10$ rad/s^2 (the corresponding angular velocity is $\omega = \alpha t$). (a) Show that the equation of motion for the angle β is

$$\ddot{\beta} = 88.24t^2 \sin\beta \cos\beta - 92.33 \sin\beta \text{ rad/s}^2$$

and state the initial conditions (do not neglect the small initial oscillation of the cone). (b) Integrate the equation of motion from $t = 0$ to $t = 3$ s. Use the results to determine the time when the cone reaches the position $\beta = \pi/2$. (c) Plot β versus t for the period of integration. (d) What would happen to the numerical solution if the small initial motion of the cone were neglected?

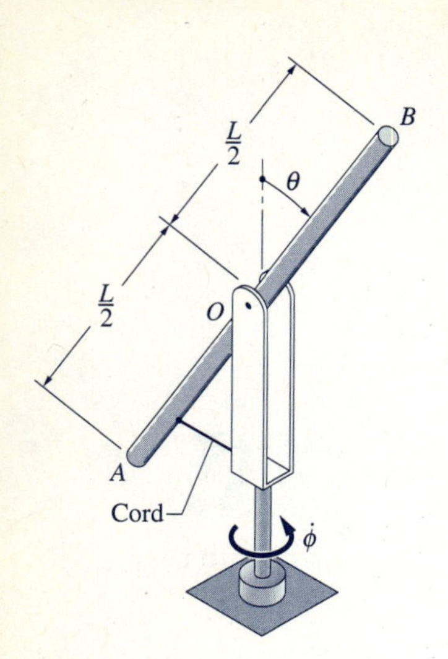

Fig. P19.67

19.67 The slender rod AB of mass m and length L is pinned at O to the fork that is attached to the vertical shaft. The masses of the fork and shaft may be neglected. The cord keeps rod AB at the angle $\theta = 30°$ with the vertical. The shaft is rotating freely with the angular velocity $\dot{\phi} = 200$ rad/s when the cord suddenly breaks. (a) Show that the equations that govern the motion of the rod after the break are

$$\ddot{\theta} = \dot{\phi}^2 \sin\theta \cos\theta \quad \text{and} \quad \ddot{\phi} = -2\dot{\theta}\dot{\phi}\cot\theta$$

and state the initial conditions. (b) Integrate the equations of motion numerically for a time period during which the rod completes at least two full oscillations about the pin. From the numerical solution, find the period of oscillation of the rod, and the range of values of $\dot{\theta}$ and $\dot{\phi}$. (c) Plot θ and $\dot{\phi}$ versus time over the period of integration.

19.68 The uniform cylindrical rotor weighing 576 lb is spinning at 10 000 rev/min. If the rotor is located at 42° latitude with its axis aligned with the meridian, calculate the gyroscopic couple that acts on the rotor as a result of the rotation of the earth.

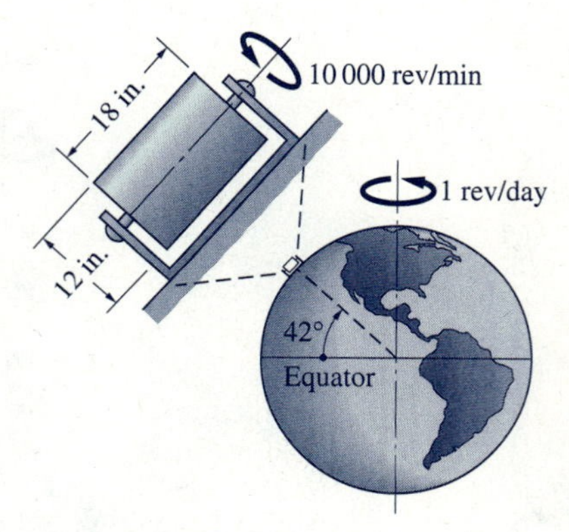

Fig. P19.68

19.69 The top consists of a 6-lb thin uniform disk that is free to spin about a rod of negligible weight. If the spin velocity of the disk is 200 rad/s and the rod is horizontal ($\theta = 90°$), find the rate of steady precession about the vertical.

19.70 The top described in Prob. 19.69 is spinning at 200 rad/s, and the rod is precessing steadily about the vertical at the rate of 3.5 rad/s. Calculate the angle θ between the rod and the vertical.

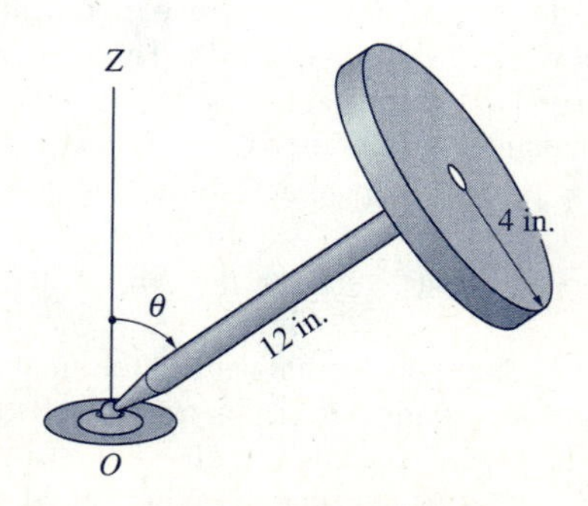

Fig. P19.69, P19.70

19.71 Determine the spin velocity of the uniform cone of mass M that is precessing about the Z-axis at the rate of 2 cycles per second.

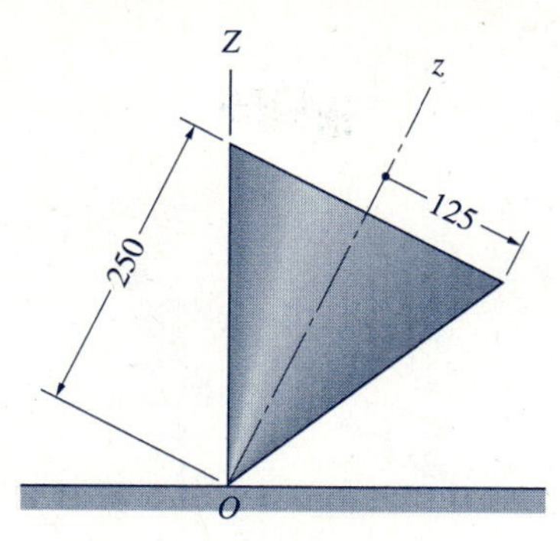

Fig. P19.71

19.72 The moments of inertia of the 0.5-kg top about axes passing through O are $I = 20 \times 10^{-4}$ kg · m^2 and $I_z = 5 \times 10^{-4}$ kg · m^2. If the top is spinning at $\dot{\psi} =$ 120 rad/s, determine the smallest steady precessional velocity $\dot{\phi}$ that would maintain the angle $\theta = 30°$ between the axis of the top and the vertical.

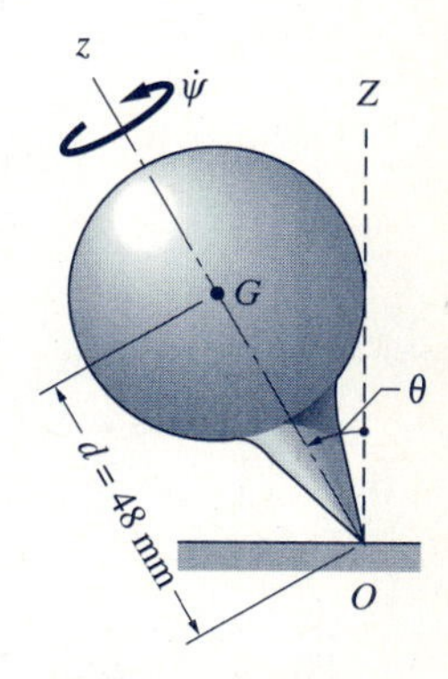

Fig. P19.72–P19.75

19.73 The top described in Prob. 19.72 is released at $\theta = 30°$ with the initial velocities $\dot{\theta} = 0$, $\dot{\phi} = 0$, and $\dot{\psi} = 120$ rad/s. (a) Derive the equations of motion in terms of the three Euler angles. (b) Integrate the equations of motion numerically from $t = 0$ to $t = 0.6$ s. Use the results to find the range of θ. (c) Plot θ and $\dot{\psi}$ versus t for the period of integration.

19.74 It can be shown that the steady precession rate of the top described in Prob. 19.72 is $\dot{\phi} = 4.330$ rad/s when $\theta = 30°$ and $\dot{\psi} = 120$ rad/s. Assume that air resistance causes a small frictional couple $M_z = -\mu I_z \omega_z$ about the z-axis, where $\mu = 0.5$ s^{-1}. (a) State the equations of motion in terms of the three Euler angles. (b) Integrate the equations numerically from $t = 0$ to $t = 1.0$ s, using the steady precession values as the initial conditions. From the results, determine the initial and final values of ω_z. (c) Show that the analytical solution for ω_z is $\omega_z = (\omega_z)_0 e^{-\mu t}$, where $(\omega_z)_0$ is its initial value. Verify that the values of ω_z found in part (b) agree with this result. (d) Plot θ and $\dot{\phi}$ versus t.

19.75 The moments of inertia of the top about axes passing through O are I_z and I. The weight of the top is W, and d is the distance between its mass center G and point O. Show that the top can have a steady precession at a given angle $\theta \neq 0$ only when $\omega_z \geq (\omega_z)_{cr}$, where $(\omega_z)_{cr} = (2/I_z)\sqrt{IWd \cos\theta}$. [Note: The top is said to be *spin-stabilized* if $\omega_z \geq (\omega_z)_{cr}$.]

19.76 Because the axis of the projectile is not aligned with its velocity vector, the resultant aerodynamic force P passes through the point C that is located a distance d from the mass center G. Determine the expression for the smallest value of ω_z for which the projectile will be spin-stabilized. (Hint: Refer to the result stated in Prob. 19.75.)

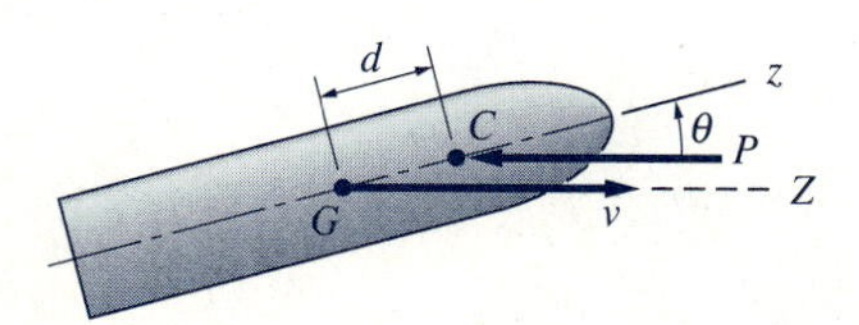

Fig. P19.76

19.77 The gyroscope consists of a thin disk of radius R and mass m that is supported by two gimbals of negligible mass. A constant vertical force F is applied to the inner gimbal. Assuming steady precession with $\theta = 90°$, derive an expression for $\dot{\phi}$ in terms of $\dot{\psi}$.

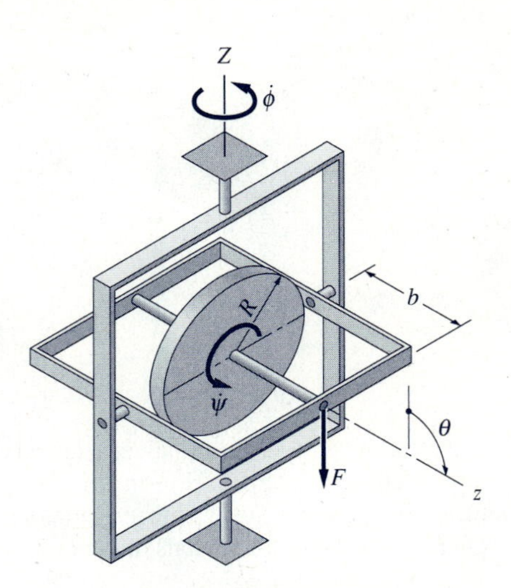

Fig. P19.77, P19.78

19.78 The thin homogeneous disk of the gyroscope weighs 6 lb and its radius is $R = 4.5$ in. The weights of the two gimbals may be neglected. Before the constant vertical force $F = 18$ lb was applied to the inner gimbal with $b = 4$ in., the disk was spinning at $\dot{\psi} = 500$ rev/min with its axis horizontal ($\theta = 90°$) and no precession ($\dot{\phi} = 0$). (a) Derive the equations of motion in terms of Euler angles, and state the initial conditions. (b) Integrate the equations of motion from $t = 0$ (the time when F was applied) to $t = 0.12$ s. From the result determine the range of values of θ, $\dot{\phi}$, and $\dot{\psi}$. (c) Plot $\dot{\phi}$ versus θ and $\dot{\psi}$ versus t.

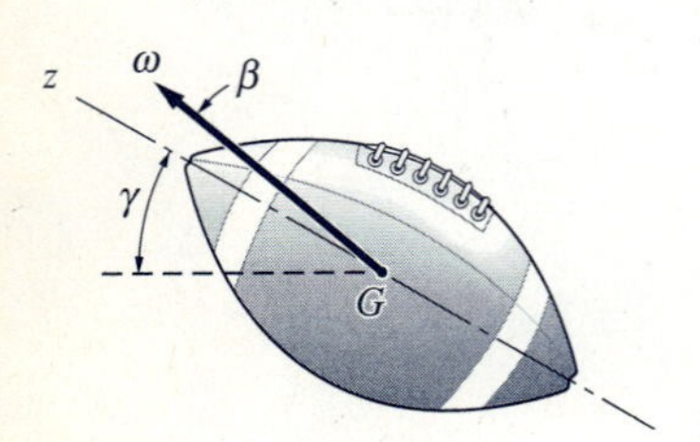

Fig. P19.79

19.79 The football is thrown with the angular velocity $\omega = 12$ rad/s, directed at the angle $\beta = 5.2°$ with the z-axis. The z-axis makes an initial angle of $\gamma = 15°$ with the horizontal. (a) Assuming that $\bar{I}_z = \bar{I}/4$ and neglecting air resistance, find the angle θ that locates the precession axis, and calculate the precession and spin rates. (b) Employing a sketch of the body and space cones, find the range of γ during the flight.

19.80 The axisymmetric satellite is rotating at the angular velocity $\omega = 0.6$ rad/s, directed at $\beta = 30°$ from the z-axis. Given that $\bar{I}_z = 2\bar{I}$, (a) determine the angle θ that locates the axis of precession, and find the rates of spin and precession; and (b) sketch the space and body cones.

19.81 The axisymmetric satellite is rotating with the angular velocity vector $\boldsymbol{\omega}$, where the angle β between $\boldsymbol{\omega}$ and the z-axis is very small. Show that the precession and spin rates of the satellite are $\dot{\phi} = \omega\bar{I}_z/\bar{I}$ and $\dot{\psi} = \omega(1 - \bar{I}_z/\bar{I})$, respectively.

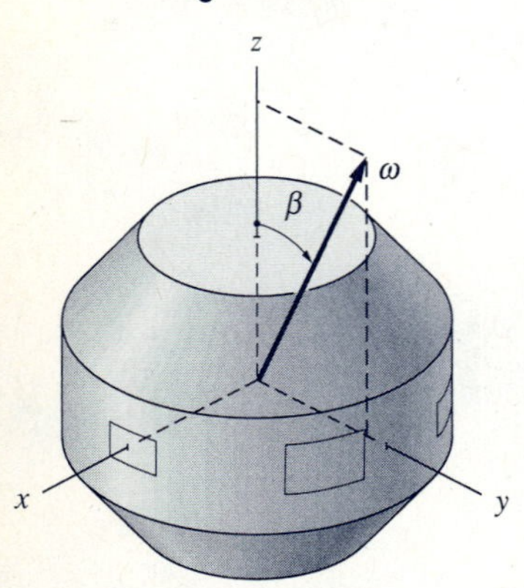

Fig. P19.80, P19.81

19.82 During its free flight, the rocket is observed to precess steadily about the horizontal Z-axis at the rate of 1 cycle every 3 minutes. Knowing that the moments of inertia about axes passing through the mass center G are related by $\bar{I} = 8\bar{I}_z$, calculate the magnitude and direction of the angular velocity vector of the rocket.

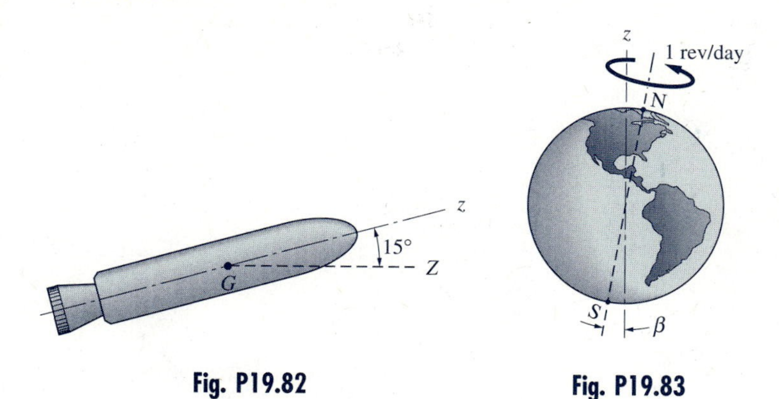

Fig. P19.82

Fig. P19.83

19.83 Since the earth is slightly flattened at the poles, its moment of inertia $\bar{I}_z$ about the z-axis (the axis of symmetry) is slightly larger than its moment of inertia $\bar{I}$ about an equatorial diameter. In addition, the polar axis of the earth, about which the earth rotates at the rate of one revolution per day, forms a small angle β with the z-axis. Knowing that the poles precess about the z-axis at the approximate rate of one complete cycle every 430 days, estimate the ratio $\bar{I}_z/\bar{I}$. (Hint: When using the formulas given in Sample Problem 19.10, be certain to correctly identify the variable that represents the rate of precession of the poles about the z-axis.)

20 Vibrations

20.1 Introduction

Vibration refers to the oscillation of a body or a mechanical system about its equilibrium position. Some vibrations are desirable, such as the oscillation of the pendulum that controls the movement of a clock, or the vibration of a string on a musical instrument. The majority of vibrations, however, are deemed to be objectionable or harmful, ranging from the annoying (vibration-induced noise) to the catastrophic (structural failure of aircraft). Excessive vibration of machines or structures can cause loosening of joints and connections, premature wear, and metal fatigue (breakage due to cyclic loading).

The study of vibrations is so extensive that entire textbooks are devoted to the subject. It is our intention here to introduce the fundamentals of vibrations that should be understood by all engineers and that will serve as the basis for further study. We consider only the simplest case: vibration of one-degree-of-freedom systems, i.e., problems where the motion can be described in terms of a single position coordinate.

The two basic components of all vibratory systems are the mass and the restoring force. The restoring force, often provided by an elastic mechanism, such as a spring, tends to return the mass to its equilibrium position. When the mass is displaced from its equilibrium position and released, it overshoots the equilibrium position, comes to a momentary stop on the other side of the equilibrium position, and then reverses direction. This oscillation between two stationary positions is a simple example of vibratory motion.

In general terms, vibrations are categorized as forced or free, and damped or undamped. When the vibration of a system is maintained by an external force, the vibration is said to be *forced.* If no external forces are driving the system, the motion is referred to as *free* vibration. *Damped* vibrations refer to a system where energy is being removed by friction or a viscous damper (resistance caused by the viscous drag of a fluid). If damping is absent, the motion is called *undamped.*

In free vibrations that are undamped, no energy is supplied to or dissipated from the system; consequently, the motion will continue forever, at least in theory. In reality, there is always some damping present, however small, that will eventually stop the vibration. In forced vibration, the oscillation can continue even if there is damping present, because the applied force provides energy to the system that can compensate for any energy that is removed by the damping.

In this chapter, we first consider the vibrations of particles in the following order: undamped free, undamped forced, damped free, and damped forced. In each case, the analysis begins by deriving the equation of motion, based on the free-body diagram of the particle when it is displaced from its equilibrium position. If the restoring force is linear, the resulting equation of motion will also be linear, and its solution can be found by analytical means. Nonlinear restoring forces result in nonlinear differential equations that can usually be solved only by numerical methods. The chapter concludes with rigid body vibrations and the application of energy methods (including Rayleigh's method) to vibration analysis.

20.2 Undamped Free Vibrations of Particles

Figure 20.1(a) shows a weight of mass m that is suspended from an ideal spring of stiffness k. If we consider only the vertical movement, there is one degree of freedom, represented by the position coordinate x, measured downward from the equilibrium position of the weight. When the weight is at the equilibrium position $x = 0$, the elongation of the spring, called the *static extension,* is $\Delta = mg/k$.

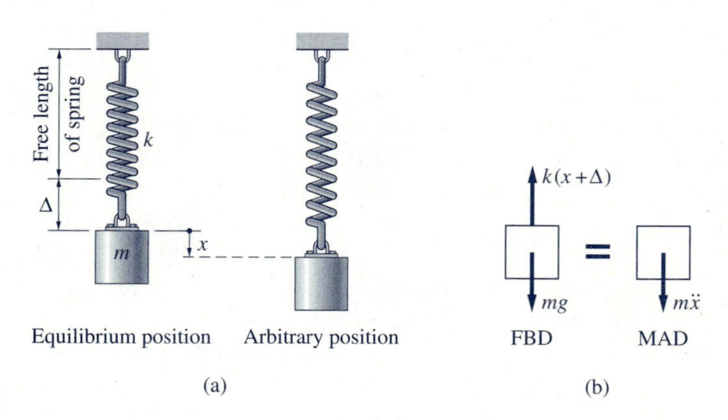

Fig. 20.1

Figure 20.1(b) shows the free-body diagram (FBD) and mass-acceleration diagram (MAD) of the weight in an arbitrary, nonequilibrium position. In addition to the weight mg, the FBD contains the force $k(x + \Delta)$ exerted by the spring. Observe that the FBD is valid for both positive and negative values of x. The MAD contains $m\ddot{x}$, the inertia vector of the weight.

Referring to Fig. 20.1(b), we see that the equation of motion of the weight is

$$\Sigma F_x = ma_x \qquad +\downarrow \quad mg - k(x + \Delta) = m\ddot{x}$$

Since $mg - k\Delta = 0$, this equation reduces to

$$m\ddot{x} + kx = 0 \tag{20.1}$$

which can be written as

$$\boxed{\ddot{x} + p^2 x = 0} \tag{20.2}$$

where

$$\boxed{p = \sqrt{\frac{k}{m}}} \tag{20.3}$$

The solution of the second-order linear differential equation in Eq. (20.2) is (you can verify this by direct substitution)

$$x = A\cos pt + B\sin pt \tag{20.4}$$

where the constants A and B can be determined from the initial conditions.

It is sometimes convenient to rewrite the solution by substituting $A = E\sin\alpha$ and $B = E\cos\alpha$. Equation (20.4) then becomes $x = E\sin\alpha\cos pt + E\cos\alpha\sin pt$. Using the trigonometric identity $\sin(a + b) = \sin a\cos b + \cos a\sin b$, we obtain the alternative form of the solution:

$$x = E\sin(pt + \alpha) \tag{20.5}$$

where A and B are related to E and α by

$$E = \sqrt{A^2 + B^2} \quad \text{and} \quad \tan\alpha = A/B \tag{20.6}$$

Let x_0 and v_0 be the initial position and velocity of the weight, respectively. Substituting $x = x_0$, $\dot{x} = v_0$, and $t = 0$ into Eqs. (20.4) and (20.5) yields

$$\begin{aligned} A &= x_0 \qquad B = v_0/p \\ E &= \sqrt{x_0^2 + (v_0/p)^2} \qquad \tan\alpha = x_0 p/v_0 \end{aligned} \tag{20.7}$$

Therefore, the motion is described by

$$\boxed{x = x_0\cos pt + (v_0/p)\sin pt} \tag{20.8}$$

or, equivalently,

$$\boxed{x = \sqrt{x_0^2 + (v_0/p)^2}\sin[pt + \tan^{-1}(x_0 p/v_0)]} \tag{20.9}$$

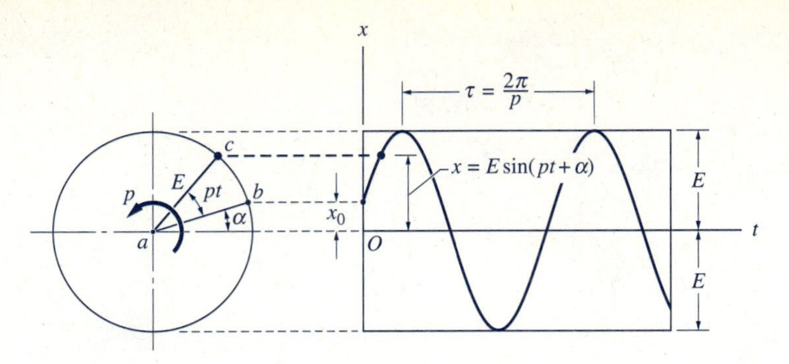

Fig 20.2

A convenient graphical representation of Eq. (20.5) is shown in Fig. 20.2. Consider the motion of point c along the circle of radius E, the radial line ac having the constant angular velocity p. If the point starts at b at time $t = 0$, its vertical position at an arbitrary time t is $x = \overline{ac}\sin(pt + \alpha) = E\sin(pt + \alpha)$, which is identical to Eq. (20.5). Note that $\overline{ab}\sin\alpha = E\sin\alpha = x_0$ represents the position at $t = 0$.

The plot of x versus t in Fig. 20.2 shows that the weight oscillates, or vibrates, about its equilibrium position $x = 0$. Since the motion repeats itself over equal intervals of time, it is called *periodic motion.* Furthermore, motion that is described in terms of the circular functions, sine and cosine, is known as *harmonic motion.* (All harmonic motion is periodic, but not all periodic motion is harmonic.) The parameter p is referred to as the (natural) *circular frequency, E* is called the *amplitude,* and α is known as the *phase angle.* As shown in Fig. 20.2, τ denotes the *period* of the motion, i.e., the time taken by one complete cycle of the motion. Therefore, $p\tau = 2\pi$, which gives

$$\boxed{\tau = \frac{2\pi}{p}} \tag{20.10}$$

The *frequency* of the motion is the number of cycles completed per unit time:

$$\boxed{f = \frac{1}{\tau} = \frac{p}{2\pi}} \tag{20.11}$$

As we noted previously, the motion of a particle undergoing undamped free vibration continues indefinitely from a theoretical viewpoint. In reality, there is always some resistance to the motion (called damping) caused by friction, air resistance, etc., that will eventually stop the motion.

The differential equation that describes the motion of the spring-mass system, Eq. (20.2), is linear because the restoring force kx is a linear function of the displacement x. Vibrations that are described by linear differential equations are called *linear vibrations.*

As an example of a nonlinear vibration consider the simple pendulum shown in Fig. 20.3(a), which consists of a particle of mass m attached to the end of a string of length L and negligible mass. The angular displacement of the pendulum from the vertical is measured by the angle θ. The free-body diagram in Fig. 20.3(b) shows that the forces acting on the mass are the tension

T and its weight mg. The normal and tangential (n-t) components of the inertia vector are shown in the mass-acceleration diagram in Fig. 20.3(b). Note that the restoring force, i.e., the force tending to return the pendulum to its equilibrium position, is $mg \sin\theta$, which is a nonlinear function of the angular displacement θ. Summing forces in the tangential direction, we obtain

$$\Sigma F_t = ma_t \qquad \overset{+}{\nearrow} \qquad -mg\sin\theta = ma_t = mL\ddot{\theta}$$

which becomes

$$\ddot{\theta} + \frac{g}{L}\sin\theta = 0 \tag{20.12}$$

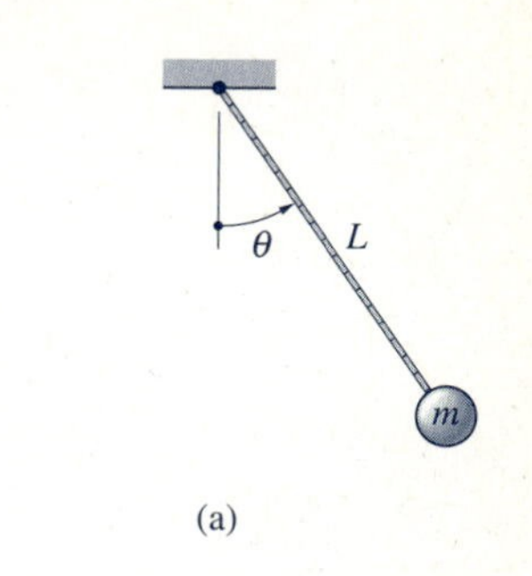

(a)

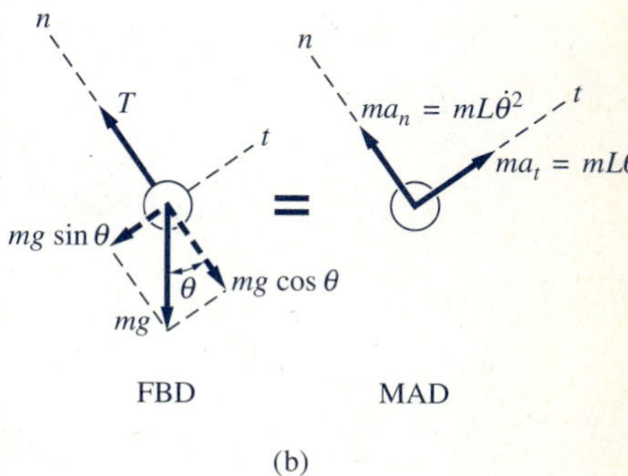

(b)

Fig. 20.3

The solution of this nonlinear differential equation must be obtained numerically. Although the motion of the pendulum is periodic, it is not harmonic; i.e., the solution of Eq. (20.12) cannot be expressed in terms of sine and cosine functions. The motion of the pendulum can be approximated by a harmonic solution only if the amplitude of the vibration is assumed to be small. Using $\sin\theta \approx \theta$, an approximation that is sufficiently accurate for $\theta < 6°$ for most applications, Eq. (20.12) becomes

$$\ddot{\theta} + \frac{g}{L}\theta = 0 \tag{20.13}$$

which has the same form as Eq. (20.2). Therefore, the motion of the simple pendulum is harmonic for small oscillations, the circular frequency being $p = \sqrt{g/L}$.

It turns out that many vibration problems are nonlinear if the amplitude is large, but simplify to a linear form if the amplitude is assumed to be sufficiently small. But be forewarned: Not all vibration problems can be linearized in this fashion—it must be demonstrated that a linear equation of motion is a valid approximation for small amplitudes.

Sample Problem 20.1

Three identical springs, each of stiffness k, support a block of mass m, as shown in Fig. (a). Deformation of bar AB may be neglected. (1) Find the equivalent spring stiffness k_0, i.e., the stiffness of a single spring as shown in Fig. (c), that can replace the original springs without changing the displacement characteristics of the block. (2) If the block weighs 0.5 lb and $k = 60$ lb/ft, find the circular frequency, the frequency, and the period of free vibration.

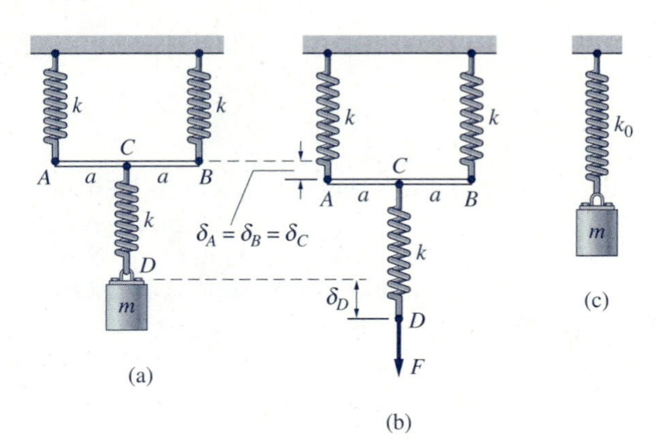

Solution

Part 1

Figure (b) shows a method for calculating the equivalent spring stiffness. First, a static vertical force F is applied at point D, and the vertical movement δ_D is computed. The equivalent spring stiffness k_0 is then calculated using $F = k_0\delta_D$. Following this procedure, the mass in Fig. (c) will have the same displacement characteristics as the mass in Fig. (a).

From equilibrium analysis of Fig. (b) we see that the spring forces are equal to F in the lower spring and $F/2$ in each of the upper springs (C is the midpoint of bar AB). The elongations of the upper springs are, therefore, identical, and the vertical displacements δ_A, δ_B, and δ_C are each equal to $(F/2)/k$ (note that bar AB remains horizontal). Next we observe that the vertical movement of D equals the vertical movement of C plus the elongation of the lower spring; i.e., $\delta_D = \delta_C + (F/k)$. The equivalent spring stiffness thus becomes

$$k_0 = \frac{F}{\delta_D} = \frac{F}{\delta_C + (F/k)} = \frac{F}{(F/2k) + (F/k)}$$

$$= \frac{1}{(1/2k) + (1/k)} = \frac{2k}{3} \qquad \textit{Answer}$$

Part 2

Using the given data and the equivalent system shown in Fig. (c), the circular frequency, frequency, and period of free vibration are

$$p = \sqrt{\frac{k_0}{m}} = \sqrt{\frac{2k/3}{m}} = \sqrt{\frac{2(60)/3}{0.5/32.2}} = 50.75 \text{ rad/s} \qquad \textit{Answer}$$

$$f = \frac{p}{2\pi} = \frac{50.75}{2\pi} = 8.077 \text{ Hz} \qquad \textit{Answer}$$

$$\tau = \frac{1}{f} = \frac{1}{8.077} = 0.1238 \text{ s} \qquad \textit{Answer}$$

Sample Problem 20.2

Figure (a) shows two elastic cords, each of length L, that are connected to a small ball of mass m that slides on a smooth horizontal surface. The cords are stretched to an initial tension T between rigid supports. The ball is given a small displacement $x = x_0$ perpendicular to the cords, and then released from rest at $t = 0$. (1) Derive the equation of motion for the ball, and show that it undergoes simple harmonic motion. (2) Using $mg = 3$ lb, $T = 50$ lb, $L = 6$ ft, and $x_0 = 3$ in., calculate the frequency and period of the vibration and the maximum velocity and maximum acceleration of the ball, and plot x versus time.

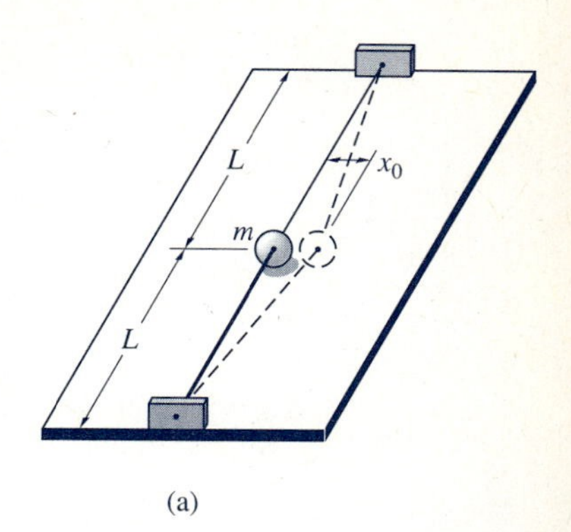

(a)

Solution

Part 1

The free-body diagram (FBD) and mass-acceleration diagram (MAD) of the ball when it is displaced an arbitrary distance x are shown in Fig. (b) (only forces that act in the plane of the motion are shown). Since the change in the tension can be neglected if x is small enough, the forces in the strings are equal to T. Letting θ be the angle shown on the FBD, we get

$$\Sigma F_x = ma_x \qquad \xrightarrow{+} \qquad -2T \sin\theta = m\ddot{x}$$

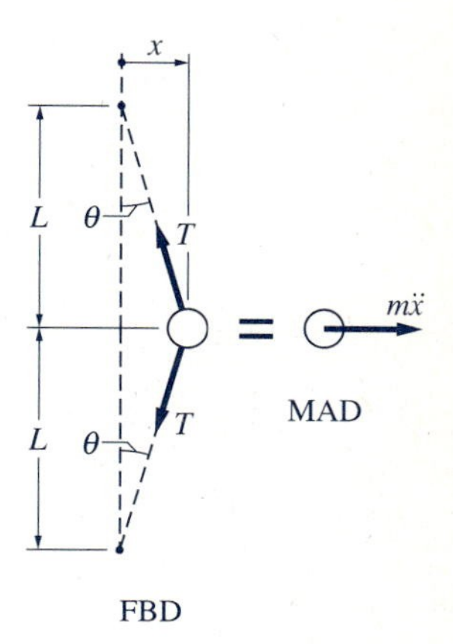

(b)

Using the small angle approximation $\sin\theta \approx x/L$, the equation of motion may be written as

$$\ddot{x} + \frac{2T}{mL}x = 0 \qquad \textit{Answer}$$

Comparing this equation with Eq. (20.2), $\ddot{x} + p^2x = 0$, we see that the motion of the ball is simple harmonic vibration with circular frequency $p = \sqrt{2T/mL}$.

Part 2

For the given data, the circular frequency, frequency, and period of the vibration are

$$p = \sqrt{\frac{2T}{mL}} = \sqrt{\frac{2(50)}{(3/32.2)(6)}} = 13.37 \text{ rad/s} \qquad \textit{Answer}$$

$$f = \frac{p}{2\pi} = \frac{13.37}{2\pi} = 2.128 \text{ Hz} \qquad \textit{Answer}$$

$$\tau = \frac{1}{f} = \frac{1}{2.128} = 0.470 \text{ s} \qquad \textit{Answer}$$

The solution of the equation of motion is given in Eq. (20.8): $x = x_0 \cos pt + (v_0/p) \sin pt$. Substituting $x_0 = 3$ in., $v_0 = 0$ (the ball is released from rest), and $p = 13.37$ rad/s, the displacement of the ball as a function of time is

$$x(t) = 3\cos pt = 3\cos 13.37t \text{ in.}$$

Successive differentiation with respect to time gives the following velocity and acceleration.

$$\dot{x}(t) = -3p\sin pt = -40.1\sin 13.37t \text{ in./s}$$
$$\ddot{x}(t) = -3p^2\cos pt = -536\cos 13.37t \text{ in./s}^2$$

Therefore, the maximum velocity and acceleration are

$$\dot{x}_{max} = 40.1 \text{ in./s} \qquad \ddot{x}_{max} = 536 \text{ in./s}^2 \qquad \textit{Answer}$$

The plot of x versus time is given in Fig. (c).

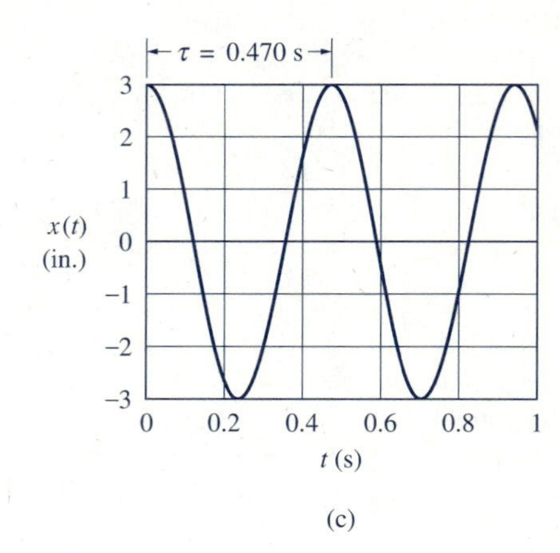

(c)

Sample Problem 20.3

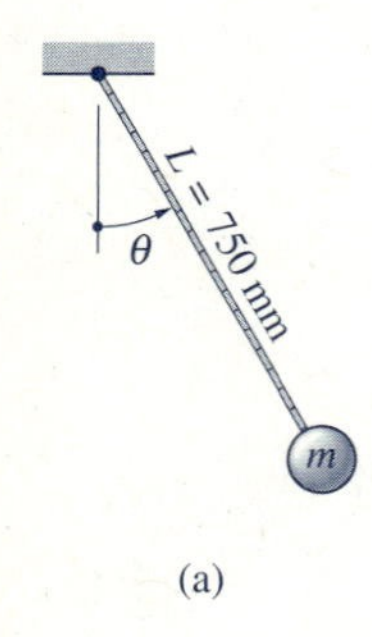

(a)

The simple pendulum in Fig. (a) consists of a small mass m that is attached to the end of a string. The pendulum is released from rest when $\theta = 30°$. Using numerical integration of the equation of motion, calculate (1) the period of oscillation; and (2) the maximum angular velocity. Compare the maximum angular velocity to the exact value that can be computed by the work-energy method.

Solution

Part 1

Because the pendulum is a conservative system, it oscillates between $\theta = \pm 30°$. The motion is periodic, but it is not harmonic, because θ is not a small angle, i.e., not less than 6°. We calculate the period of the pendulum by noting that the time taken by the pendulum to travel from the initial position ($\theta = 30°$) to the vertical position ($\theta = 0$) equals one-fourth of the period.

From Eq. (20.12), the differential equation of motion is $\ddot{\theta} = -(g/L)\sin\theta = -(9.81/0.750)\sin\theta = -13.08\sin\theta$. The initial conditions are $\theta = 30°$ ($\pi/6$ rad) and $\dot{\theta} = 0$ when $t = 0$ (the time of release). The integration interval extends from $t = 0$ until the time when $\theta = 0$ for the first time. By trial and error, this time was found to be approximately 0.5 s. For the final integration, we used the fourth-order Runge-Kutta method with integration step size $\Delta t = 0.01$ s, completing 50 integration steps. To determine an accurate value of the time when $\theta = 0$, we must examine the numerical output of the solution in the region when θ changes sign.

t (s)	θ (rad)	$\dot{\theta}$ (rad/s)	$\ddot{\theta}$ (rad/s^2)
0.44	0.003 53	−1.872	−0.0462
0.45	−0.015 19	−1.871	0.1986

The corresponding straight-line plot of θ versus time is shown in Fig. (b). Letting t_1 be the time when $\theta = 0$, we obtain from similar triangles

$$\frac{t_1 - 0.44}{0.003\,53} = \frac{0.45 - 0.44}{0.003\,53 + 0.015\,19}$$

which yields $t_1 = 0.4419$ s. Since this time equals one-fourth of the total period τ, we get

$$\tau = 4t_1 = 4(0.4419) = 1.768 \text{ s} \qquad \textit{Answer}$$

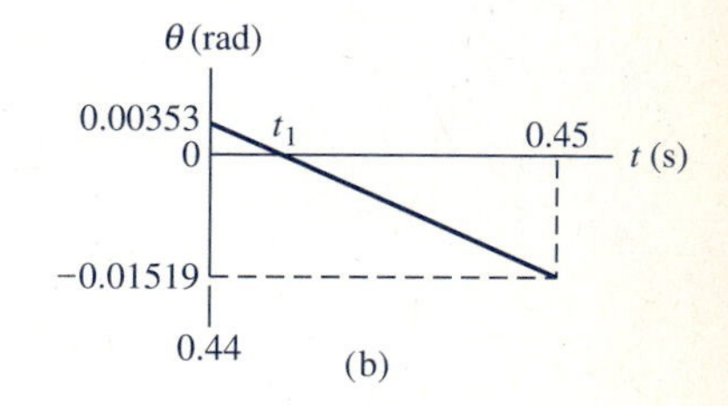

(b)

By comparison, if the motion were assumed to be linear ($\sin\theta$ approximated by θ), the circular frequency from Eq. (20.13) would be $p = \sqrt{g/L} = \sqrt{9.81/0.750} = 3.617$ rad/s, and the period would be $\tau = 2\pi/p = 2\pi/3.617 = 1.737$ s.

Part 2

Figure (c) shows the straight-line plot of $\ddot{\theta}$ versus time for the numerical output given in Part 1. The maximum angular velocity $\dot{\theta}_{max}$ can be approximated as the value of $\dot{\theta}$ at $t = 0.44$ s plus $\Delta\dot{\theta}$, the change in $\dot{\theta}$ between 0.44 s and 0.4419 s. Since $\dot{\theta} = \int \ddot{\theta}\, dt$, $\Delta\dot{\theta}$ equals the shaded area in Fig. (c), which yields

$$\Delta\dot{\theta} = -\frac{1}{2}(0.4419 - 0.44)(0.0462) \approx 0$$

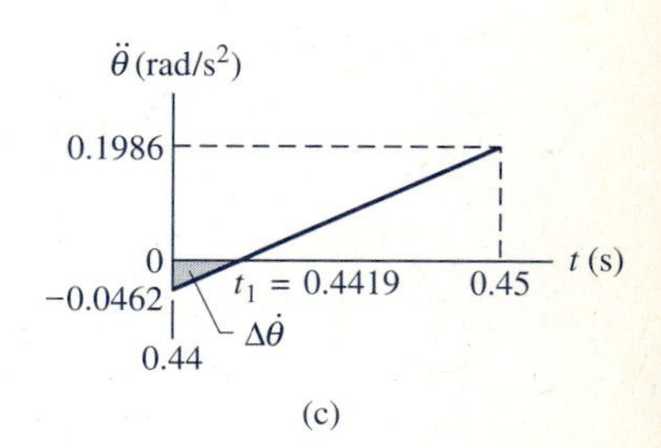

(c)

Therefore, the maximum angular velocity is approximately equal to the angular velocity at $t = 0.44$ s, i.e.,

$$\dot{\theta}_{max} = \dot{\theta}|_{t=0.44\text{ s}} = -1.872 \text{ rad/s} \qquad \textit{Answer}$$

The negative sign indicates that the angular velocity is clockwise at the time when the pendulum passes through $\theta = 0$ for the first time, as would be expected.

To calculate the exact value for $\dot{\theta}_{max}$ using the work-energy method, we let the subscripts 1 and 2 denote the positions $\theta = \pi/6$ rad and $\theta = 0$, respectively. The corresponding kinetic energies are $T_1 = 0$ (pendulum is stationary in position 1) and $T_2 = (1/2)m(L\dot{\theta}_{max})^2$, where $L\dot{\theta}_{max}$ is the velocity of the pendulum in position 2. The work done by the weight of the mass as it moves from position 1 to position 2 is $U_{1\text{-}2} = mg(L - L\cos\theta) = mgL[1 - \cos(\pi/6)]$, since the vertical distance between the two positions equals $L - L\cos\pi/6$. Applying the work-energy principle, we obtain

$$U_{1\text{-}2} = T_2 - T_1$$

$$mgL\left(1 - \cos\frac{\pi}{6}\right) = \frac{1}{2}m(L\dot{\theta}_{max})^2 - 0$$

from which the value of $\dot{\theta}_{max}$ is found to be

$$\dot{\theta}_{max} = \sqrt{\frac{2g}{L}\left(1 - \cos\frac{\pi}{6}\right)} = \sqrt{\frac{2(9.81)}{0.750}\left(1 - \cos\frac{\pi}{6}\right)}$$

$$= 1.872 \text{ rad/s} \quad \text{(clockwise)}$$

We see that the value for $\dot{\theta}_{max}$ obtained by numerical integration agrees with the above result within the four significant digits used, thereby verifying the accuracy of the numerical integration.

It is instructive to note that although the angular velocity at any position can be obtained analytically, the time of travel between two positions must be computed numerically.

PROBLEMS

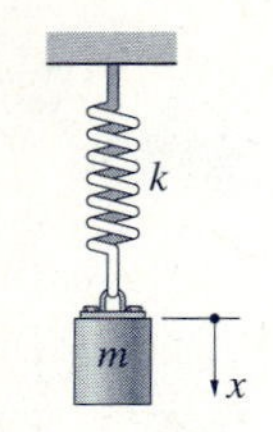

Fig. P20.1–P20.3

20.1 The mass $m = 20$ kg is suspended from an ideal spring of stiffness $k = 500$ N/m. If the mass is set into motion at $t = 0$ with the initial conditions $x_0 =$ 45.5 mm and $v_0 = -104$ mm/s, calculate (a) the amplitude of the motion; and (b) the time when the mass stops for the first time. Assume that x is measured from the equilibrium position of the mass.

20.2 Repeat Prob. 20.1 assuming that the direction of the initial velocity is reversed.

20.3 The mass m is suspended from an ideal spring of stiffness k and is set into motion with the initial conditions $x_0 = 9$ mm and $v_0 = 1.2$ m/s. Assume that x is measured from the equilibrium position of the mass. If the amplitude of the vibration is 12 mm, determine (a) the frequency; and (b) x as a function of time.

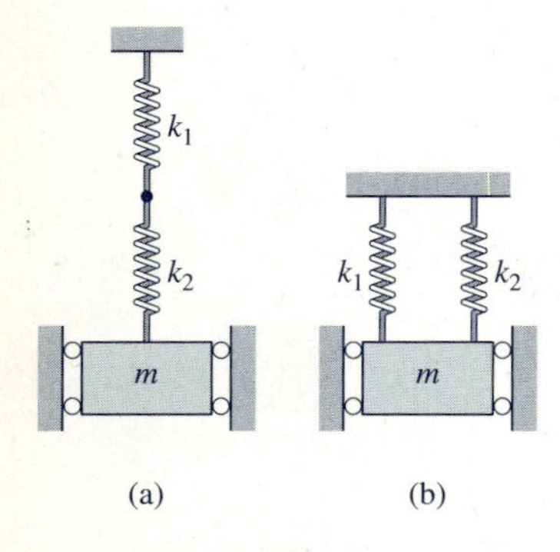

Fig. P20.4

20.4 The mass m is suspended from two springs of stiffnesses k_1 and k_2. Determine the expression for the circular frequency of the mass if the springs are arranged as shown in (a) and in (b).

20.5 A mass suspended from an ideal spring is undergoing simple harmonic motion. At a certain instant, its displacement (measured from the equilibrium position), velocity, and acceleration are $x = 3$ in., $v = 18$ in./s, and $a = -17.7$ in./s^2. Find the period and amplitude of the motion.

20.6 When only mass B is attached to the ideal spring, the frequency of the system is 3.90 Hz. When mass C is added, the frequency decreases to 2.55 Hz. Determine the ratio m_B/m_C of the two masses.

20.7 The ideal spring of stiffness $k = 120$ lb/ft is connected to mass B through a hole in mass C. The weights of B and C are 0.4 lb and 0.8 lb, respectively. Find the smallest vibrational amplitude for which C will lose contact with B. Would the result change if the weights of B and C were interchanged?

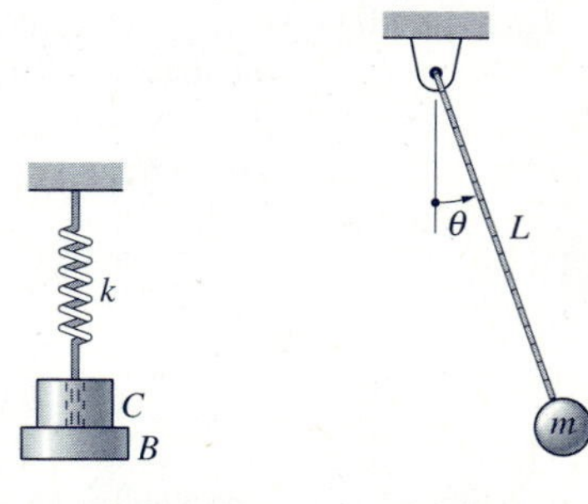

Fig. P20.6, P20.7 **Fig. P20.8**

20.8 The simple pendulum is released from rest at $\theta = \theta_0$. Determine the expressions for the maximum values of $\dot{\theta}$ and $\ddot{\theta}$ (a) assuming that θ is small (simple harmonic motion); and (b) without making any simplifying assumptions. (c) Compare the expressions found in (a) and (b) for $\theta_0 = 5°$, 10°, and 15°.

20.9 The particle of mass m slides without friction in a trough. The profile of the trough is described by $y = s^2/b$, where s is measured along the surface and b is a constant. If the particle is released from rest at a position above the bottom of the trough, show that the resulting motion is simple harmonic, and determine the corresponding circular frequency of the vibration.

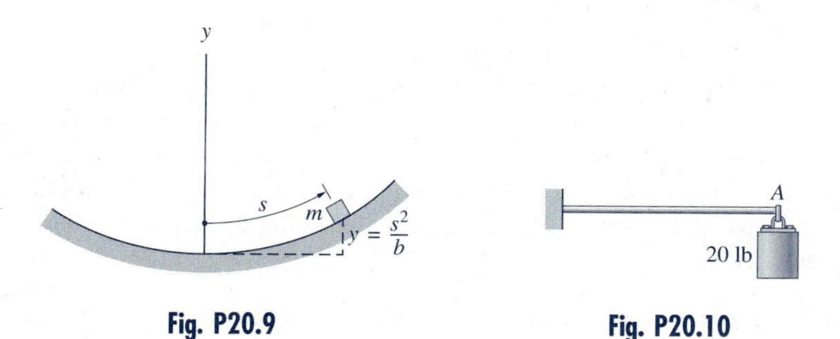

Fig. P20.9 **Fig. P20.10**

20.10 When the 20-lb weight is attached to end A of the elastic beam, its static deflection is 0.25 in. Neglecting the weight of the beam, calculate the frequency of the system assuming simple harmonic motion.

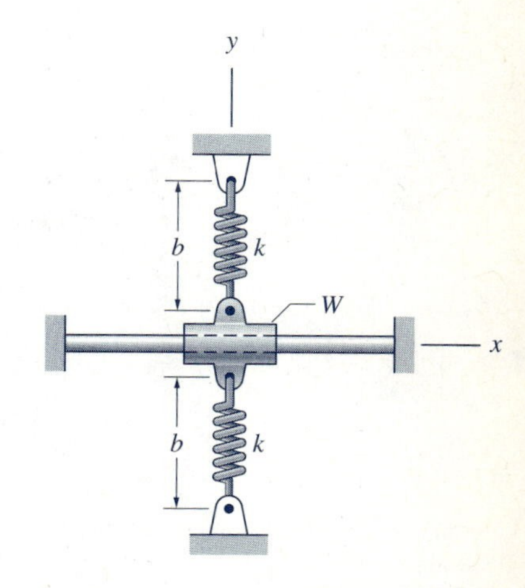

Fig. P20.11, P20.12

20.11 Two identical springs of free length L_0 and stiffness k are attached to the collar of weight W. Friction between the collar and the horizontal rod may be neglected. Assuming small amplitudes, (a) derive the equation of motion; and (b) find the frequency given that $W = 0.2$ lb, $L_0 = 4$ in., $b = 6$ in., and $k = 0.4$ lb/in.

20.12 For the collar described in Prob. 20.11, (a) show that the equation of motion for large amplitudes is

$$\ddot{x} = -\frac{2gk}{W}\left(1 - \frac{L_0}{\sqrt{b^2 + x^2}}\right)x$$

(b) Using the data given in Prob. 20.11, determine the period if the amplitude is 6 in.

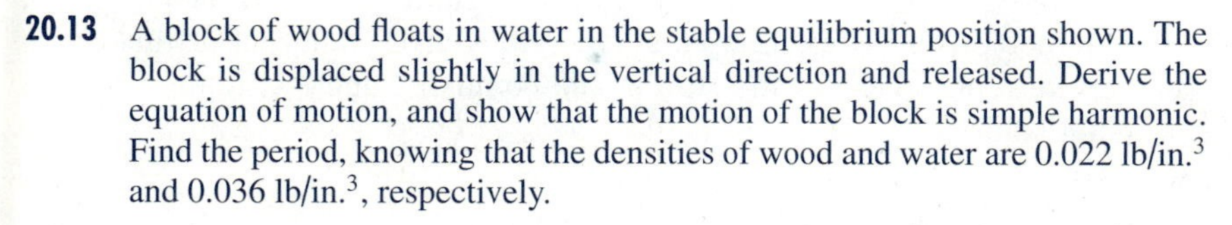

20.13 A block of wood floats in water in the stable equilibrium position shown. The block is displaced slightly in the vertical direction and released. Derive the equation of motion, and show that the motion of the block is simple harmonic. Find the period, knowing that the densities of wood and water are 0.022 lb/in.^{3} and 0.036 lb/in.^{3}, respectively.

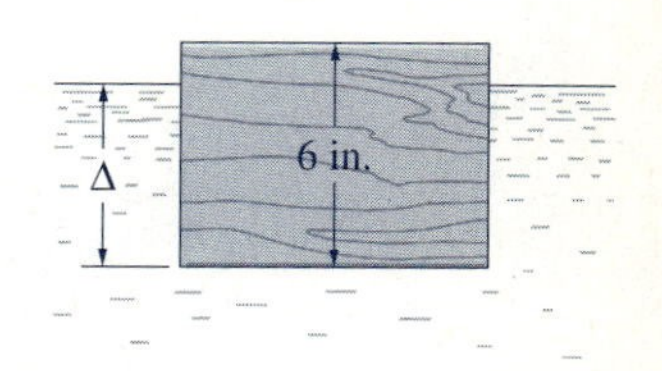

Fig. P20.13

20.14 The homogeneous conical float slides freely along the vertical guide rod. The float, which is initially in static equilibrium in the position $x = 0.4$ m, is displaced to the position $x = 0.5$ m and released. (a) Show that the equation of the ensuing motion is

$$\ddot{x} = 9.81(1 - 15.625x^3) \text{ m/s}^2$$

(b) Verify that the float moves between the limits $x = 0.280$ m and $x = 0.500$ m.

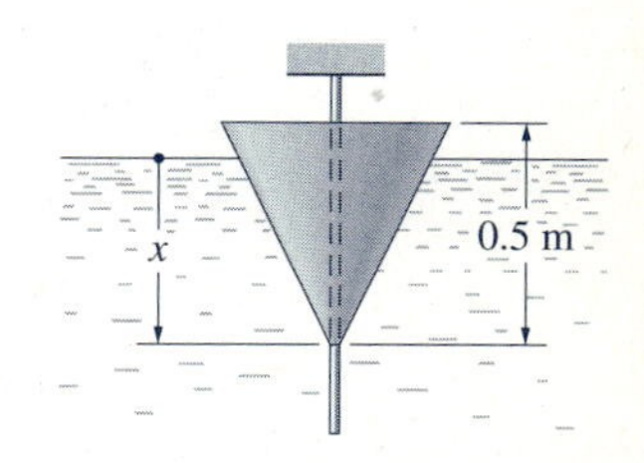

Fig. P20.14, P20.15

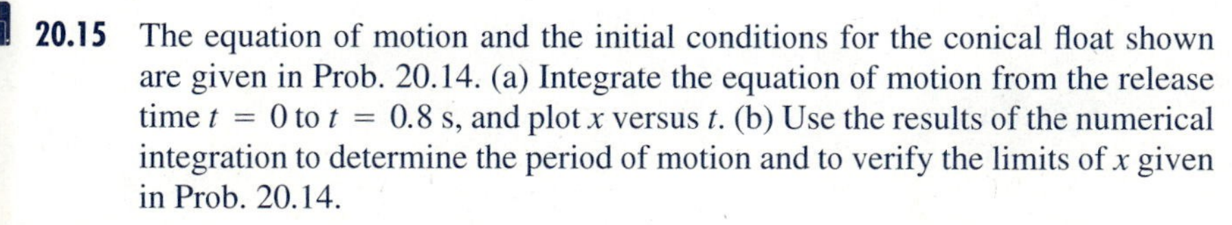

20.15 The equation of motion and the initial conditions for the conical float shown are given in Prob. 20.14. (a) Integrate the equation of motion from the release time $t = 0$ to $t = 0.8$ s, and plot x versus t. (b) Use the results of the numerical integration to determine the period of motion and to verify the limits of x given in Prob. 20.14.

20.3 Undamped Forced Vibrations of Particles

In free vibrations, the oscillations are initiated by a momentary disturbance that gives rise to an initial displacement, initial velocity, or both. No external forces are required to maintain the motion. In forced vibration, a sustained external source is responsible for maintaining the vibration. A common example of forced vibration is the automobile "rattle," caused either by the reciprocation of the engine or irregularities of the road surface. Here we consider forced vibrations that arise from either a harmonic (i.e., sinusoidally varying) forcing function, or a harmonic support displacement. Apart from being important by itself, harmonic input is also useful in the analysis of more general forced motion, because any forcing function can be decomposed into harmonic components, i.e., expressed as a Fourier series.

a. Harmonic forcing function

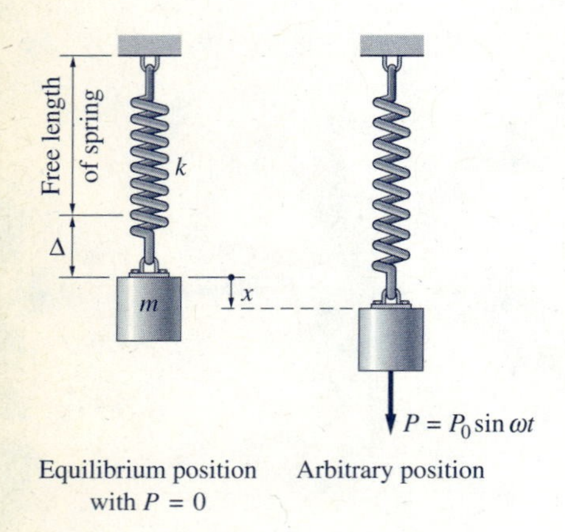

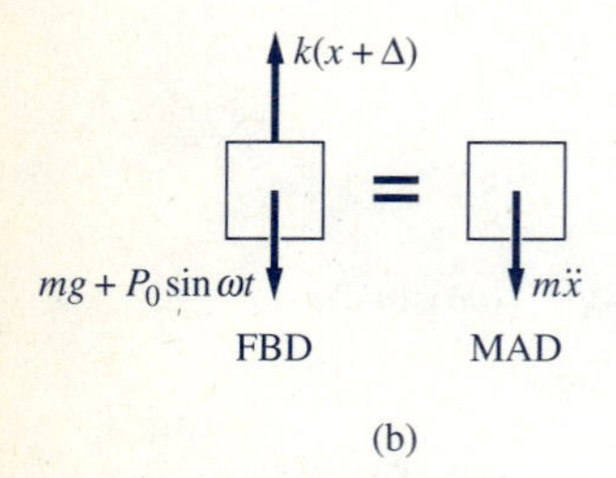

Fig. 20.4

Figure 20.4(a) shows a spring-mass system that is subjected to a time-dependent force $P = P_0 \sin \omega t$, where P_0 is the magnitude of the force and ω is its circular frequency. The force P is referred to as the *harmonic forcing function,* and ω is called the *forcing frequency.* The static deflection of the spring is Δ, and x is the displacement of the mass from its equilibrium position. From the free-body and mass-acceleration diagrams shown in Fig. 20.4(b), we obtain for the equation of motion

$$\Sigma F_x = ma_x \qquad +\downarrow \quad mg + P_0 \sin \omega t - k(x + \Delta) = m\ddot{x}$$

Using the equilibrium equation $mg = k\Delta$, the equation of motion simplifies to

$$\boxed{m\ddot{x} + kx = P_0 \sin \omega t} \tag{20.14}$$

Equation (20.14) is a nonhomogeneous (right-hand side is not zero), second-order linear differential equation. The general solution of this equation can be represented as the sum of the complementary solution x_c and a particular solution x_p; i.e.,

$$x = x_c + x_p \tag{20.15}$$

The complementary solution of Eq. (20.14) is a solution of the homogeneous equation (obtained by setting the right-hand side equal to zero), and the particular solution is any solution of the complete equation.

We note that the homogeneous equation is identical to Eq. (20.1). Therefore, for the complementary solution of Eq. (20.14) we can choose either Eq. (20.4) or (20.5), whichever is more convenient. Using the latter, we have

$$x_c = E \sin(pt + \alpha) \tag{20.16}$$

where E and α are constants to be determined by the initial conditions and $p = \sqrt{k/m}$ is the natural circular frequency.

As a particular solution of Eq. (20.14), we try $x_p = X \sin \omega t$. Substituting this expression into Eq. (20.14) yields

$$m(-X\omega^2 \sin \omega t) + k(X \sin \omega t) = P_0 \sin \omega t$$

which can be solved for X. After some algebraic manipulation, we get

$$X = \frac{P_0/k}{1 - (\omega/p)^2} \quad \text{for } \omega \neq p \tag{20.17}$$

The numerator P_0/k is called the *zero-frequency deflection* of the spring-mass system—deflection caused by a constant force of magnitude P_0 (not to be confused with $\Delta = mg/k$, the static deflection). The term ω/p, the ratio of the forcing frequency to the undamped natural frequency, is called the *frequency ratio.* Note that Eq. (20.17) is not valid if $\omega = p$, because the denominator will be zero.

Summing the complementary and particular solutions, we obtain the complete solution of Eq. (20.14).

$$x = E \sin(pt + \alpha) + \frac{P_0/k}{1 - (\omega/p)^2} \sin \omega t \tag{20.18}$$

We see that the complete solution is the sum of two sine curves of different frequencies. The complementary solution (frequency p) is called the *transient vibration,* since in practice it will die out due to the unavoidable presence of damping. The particular solution (frequency ω) is referred to as the *steady-state vibration,* because it continues even if damping is present. The value of X in Eq. (20.17) is positive if $(\omega/p) < 1$, in which case the forcing function and the steady-state vibration are said to be *in phase.* If $(\omega/p) > 1$, then X is negative, and the forcing function and the steady-state vibration are 180° *out of phase.*

Note that the initial conditions of the motion determine the constants E and α that appear in the transient vibration. However, the steady-state vibration is independent of the manner in which the motion was initiated.

The steady-state vibration is characterized by the *magnification factor,* which is defined as

$$\text{Magnification Factor} = \frac{\text{Amplitude of steady-state vibration}}{\text{Zero-frequency deflection}} \tag{20.19}$$

Using Eq. (20.17), the magnification factor becomes

$$\text{Magnification Factor} = \frac{|X|}{P_0/k} = \left| \frac{1}{1 - (\omega/p)^2} \right| \tag{20.20}$$

(The absolute value of X is used because the amplitude is by convention a positive number.)

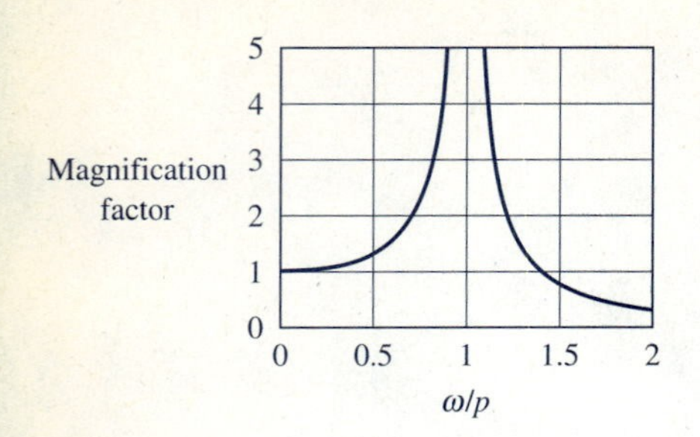

Fig. 20.5

The magnification factor, plotted in Fig. 20.5, is convenient for studying the effect that the forcing frequency has on the amplitude of the steady-state vibration. We see that the magnification factor (and thus the amplitude of the steady-state vibration) approaches infinity as $\omega \rightarrow p$, i.e., as $\omega/p \rightarrow 1$, a condition that is known as *resonance*. When the forcing frequency approaches the resonant frequency, the amplitude will increase and eventually exceed the linear range of the response (e.g., the restoring force of a spring ceases to be proportional to the elongation). In that case, the linear analysis will not apply.

b. Harmonic support displacement

Here we consider the case where a harmonic displacement is imposed on the support of a spring-mass system. Problems of this type are encountered, for example, in the analysis of vibration measuring devices, such as seismographs and accelerometers. They also serve as approximate models of vehicle suspension systems—the movement of the support represents the up-and-down motion of a wheel traveling along an uneven highway.

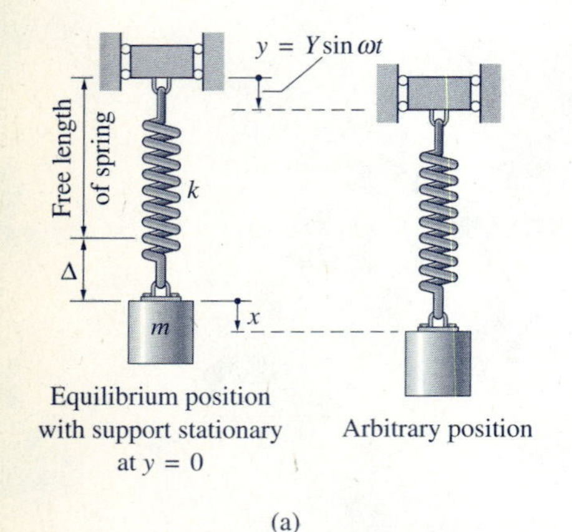

(a)

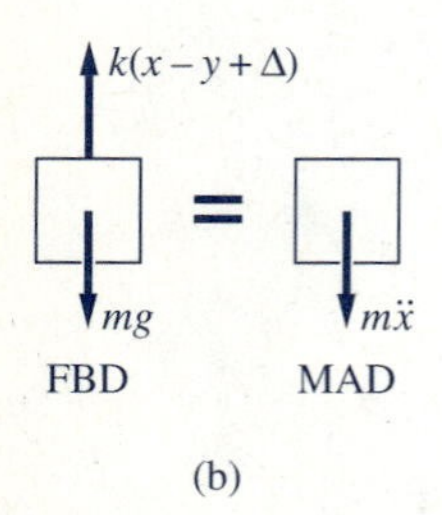

(b)

Fig. 20.6

Figure 20.6(a) shows a spring-mass system where the support undergoes the prescribed harmonic displacement $y = Y \sin \omega t$. In the free-body and mass-acceleration diagrams in Fig. 20.6(b), x is the displacement of the mass from its equilibrium position. Note that the deformation of the spring depends on the static deflection Δ and the relative distance $(x - y)$ between the mass and the support. The equation of motion of the mass is

$$\Sigma F_x = ma_x \qquad +\downarrow \quad mg - k(x - y + \Delta) = m\ddot{x}$$

which, utilizing the equilibrium equation $mg = k\Delta$, becomes

$$m\ddot{x} + kx = ky \tag{a}$$

Substituting $y = Y \sin \omega t$ into Eq. (a) would give the equation of motion in terms of the coordinate x. However, a more useful form of the equation of motion is obtained by introducing $z = x - y$, which represents the position of the mass *relative* to the support (the relative position z is usually easier to measure than the absolute position x). Substituting $x = z + y$, Eq. (a) becomes

$$m(\ddot{z} + \ddot{y}) + k(z + y) = ky$$

which reduces to

$$m\ddot{z} + kz = -m\ddot{y} \tag{b}$$

Substituting $y = Y \sin \omega t$ into Eq. (b), the equation of motion in terms of the relative position z becomes

$$\boxed{m\ddot{z} + kz = mY\omega^2 \sin \omega t} \tag{20.21}$$

Comparing Eqs. (20.21) and (20.14), we see that all of our results for a harmonic forcing function are also applicable to harmonic support displacement if we replace P_0 by $mY\omega^2$. Making this substitution in Eq. (20.18), we obtain for the solution of Eq. (20.21)

$$\boxed{z = E\sin(pt + \alpha) + Z\sin\omega t} \tag{20.22}$$

where $p = \sqrt{k/m}$. According to Eq. (20.17), the amplitude of the steady-state vibration is

$$\boxed{Z = \frac{mY\omega^2/k}{1 - (\omega/p)^2} = Y\frac{(\omega/p)^2}{1 - (\omega/p)^2}} \tag{20.23}$$

The constants E and α that appear in the transient solution can, as before, be determined from the initial conditions.

Sample Problem 20.4

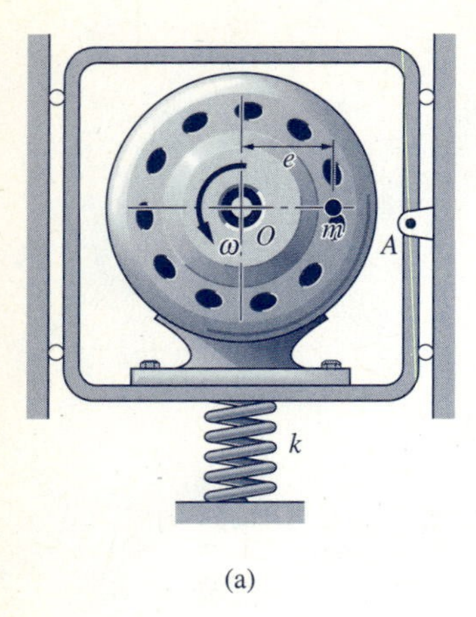

(a)

The electric motor and its frame shown in Fig. (a) have a combined mass M. The unbalance of the rotor is equivalent to a mass m (included in M), located at the distance e from the center O of the shaft. The frame is supported by vertical guides, a spring of stiffness k, and a pin at A. (The pin at A was inserted when the motor was at rest in the static equilibrium position.) The motor is running at the constant angular speed ω when the pin is withdrawn at the instant when the mass m is in the position shown. (1) Derive the equation of motion of the assembly. (2) Plot the transient, steady-state, and total displacement versus time, using the data $M = 40$ kg, $m = 1.2$ kg, $e = 150$ mm, $k = 900$ N/m, $\omega = 8$ rad/s.

Solution

The rotating unbalance causes the position of the mass center of the assembly to vary with time, thereby giving rise to vibrations in the vertical direction (the vertical guides prevent horizontal motion of the frame).

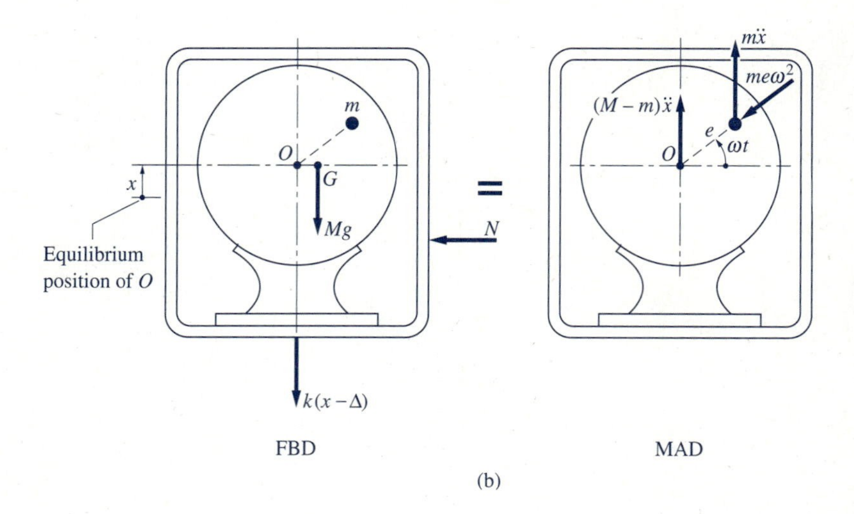

(b)

Part 1

The free-body diagram (FBD) of the assembly is shown in Fig. (b). Note that x was chosen to be the displacement of O from its equilibrium position. The FBD contains the total weight Mg of the assembly, acting at its mass center G, the spring force $k(x - \Delta)$, where Δ is the static deflection of the spring, and the resultant horizontal force N exerted by the vertical guides. The mass-acceleration diagram in Fig. (b) displays the inertia vectors $(M - m)\ddot{x}$ and $m\ddot{x}$ associated with the vertical motion of the assembly, and the radial inertia vector $me\omega^2$ caused by the rotation of the mass m about O. Equating the resultant vertical forces on the FBD and the MAD, we get

$$+\uparrow \quad -Mg - k(x - \Delta) = (M - m)\ddot{x} + m\ddot{x} - me\omega^2 \sin\omega t \qquad \text{(a)}$$

After eliminating Δ by utilizing the equilibrium equation $Mg - k\Delta = 0$, the equation of motion becomes

$$M\ddot{x} + kx = me\omega^2 \sin\omega t \qquad \textit{Answer} \quad \text{(b)}$$

Comparing Eq. (b) with Eq. (20.14), we see that the equations are identical if we replace the magnitude P_0 of the forcing function by $me\omega^2$ (the term $me\omega^2$ is sometimes referred to as the centrifugal force due to the unbalanced mass).

Part 2

With $P_0 = me\omega^2$, the solution of the equation of motion, Eq. (20.18), becomes

$$x = E\sin(pt + \alpha) + \frac{me\omega^2/k}{1 - (\omega/p)^2}\sin\omega t \tag{c}$$

where the first term represents the transient vibration and the second term is the steady-state vibration.

Using the given numerical values, we obtain

$$p = \sqrt{k/M} = \sqrt{900/40} = 4.743 \text{ rad/s}$$

$$\frac{me\omega^2/k}{1 - (\omega/p)^2} = \frac{1.2(0.15)(8)^2/900}{1 - (8/4.743)^2} = -0.006\,938 \text{ m} = -6.938 \text{ mm}$$

Substituting these values into Eq. (c) yields

$$x(t) = E\sin(4.743t + \alpha) - 6.938\sin 8t \text{ mm} \tag{d}$$

Taking the time derivative, we obtain for the velocity

$$\dot{x}(t) = 4.743\,E\cos(4.743t + \alpha) - 55.50\cos 8t \text{ mm/s} \tag{e}$$

Substituting the initial conditions ($x = 0$ and $\dot{x} = 0$ when $t = 0$) into Eqs. (d) and (e) gives $E\sin\alpha = 0$ and $4.743E\cos\alpha - 55.50 = 0$, which yields $\alpha = 0$ and $E = 11.701$ mm. Therefore the description of the motion is

$$x = 11.70\sin 4.74t - 6.94\sin 8t \text{ mm} \qquad \textit{Answer} \tag{f}$$

Figure (c) shows plots of the transient vibration (with the period $\tau_t = 2\pi/p = 2\pi/4.743 = 1.325$ s), the steady-state vibration (period $\tau_s = 2\pi/\omega = 2\pi/8 = 0.785$ s), and the superposition of the two. Because the coefficient of $\sin 8t$ in Eq. (f) is negative, the steady-state vibration is 180° out of phase with the rotation of the unbalanced mass.

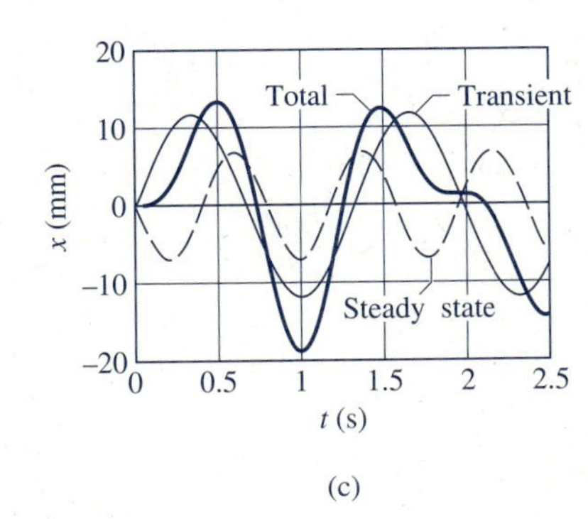

(c)

PROBLEMS

Fig. P20.16, P20.17

20.16 The spring-mass system starts from rest in the equilibrium position $x = 0$ when $t = 0$. The harmonic force applied to the mass is $P(t) = P_0 \sin \omega t$, where $P_0 =$ 100 N and $\omega = 25$ rad/s. (a) Derive the expression for $x(t)$. (b) Plot x versus t from $t = 0$ to $t = 0.5$ s. (c) Find the maximum displacement of the steady-state solution.

20.17 The spring-mounted mass is driven by the force $P = P_0 \sin \omega t$, where $P_0 =$ 100 N. Calculate the two values of ω for which the amplitude of the steady-state vibration is 50 mm.

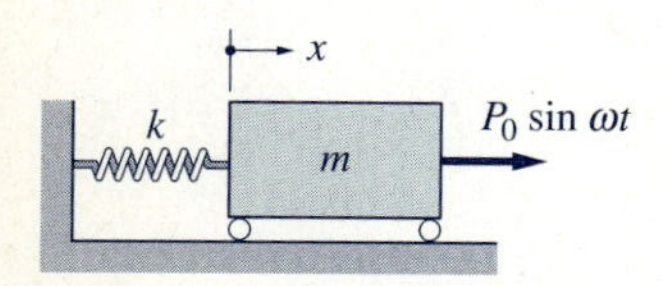

Fig. P20.18, P20.19

20.18 The system shown consists of a linear spring and a mass m. The system starts from rest in the equilibrium position $x = 0$ when $t = 0$. The forcing frequency ω equals the natural circular frequency p of the system (resonance). Verify that the following expression for $x(t)$ satisfies the equation of motion and the initial conditions.

$$x(t) = \frac{P_0}{2m\omega^2}(\sin \omega t - \omega t \cos \omega t)$$

(Observe that the amplitude increases linearly with time.)

20.19 The system shown consists of the mass $m = 6$ kg and a nonlinear spring. The force-extension relationship of the spring is $F = kx[1 + (x/b)^2]$, where x is measured in meters, $k = 1200$ N/m, and $b = 0.25$ m. The amplitude and frequency of the driving force are $P_0 = 60$ N and $\omega = 14.142$ rad/s, respectively, the latter being equal to the natural circular frequency of the system for small amplitudes. (Note that if $x \ll b$, then $F \approx kx$, and $p = \sqrt{k/m} = \sqrt{1200/6} = 14.142$ rad/s.) The system starts from rest in the equilibrium position $x = 0$ when $t = 0$. (a) Show that the equation of motion is

$$\ddot{x} = -200x(1 + 16x^2) + 10 \sin 14.142t \text{ m/s}^2$$

(b) Integrate the equation of motion numerically from $t = 0$ to $t = 2$ s, and plot x versus t. (c) Plot the expression for $x(t)$ given in Prob. 20.18 (valid for small amplitudes only) on the same graph that was drawn in part (b).

20.20 The block of weight $W = 2.4$ lb is connected to the shaker table by a spring of stiffness $k = 12$ lb/in. When the spring is unstretched, the distance between the block and a stop attached to the table is $b = 1.0$ in. If the table is being driven at $y = 0.25 \sin \omega t$ in., determine the range of ω for which the block will not hit the stop. Consider only the steady-state vibration.

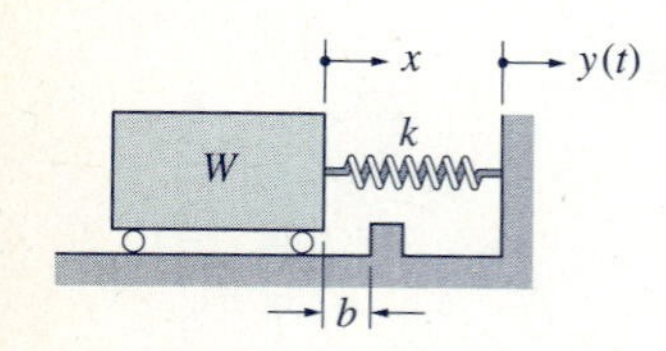

Fig. P20.20, P20.21

20.21 The system described in Prob. 20.20 is at rest in the equilibrium ($x = 0$) position when the displacement $y = 0.25 \sin 60t$ in. is imposed on the shaker table. (a) Derive the expression for the displacement of the weight relative to the table as a function of time (ignore the presence of the stop that is attached to the table). (b) Plot the expression derived in part (a) from $t = 0$ to $t = 2$ s, and determine whether the weight will hit the stop.

20.22 An electric motor and its base, with a combined mass of $M = 12$ kg, are supported by four identical springs, each of stiffness $k = 480$ kN/m. The unbalance of the rotor is equivalent to a mass $m = 0.005$ kg located at a distance $e = 90$ mm from its axis. (a) Calculate the angular speed of the motor that would cause resonance. (b) Compute the maximum steady-state displacement of the motor when its angular speed is 99 percent of the speed at resonance.

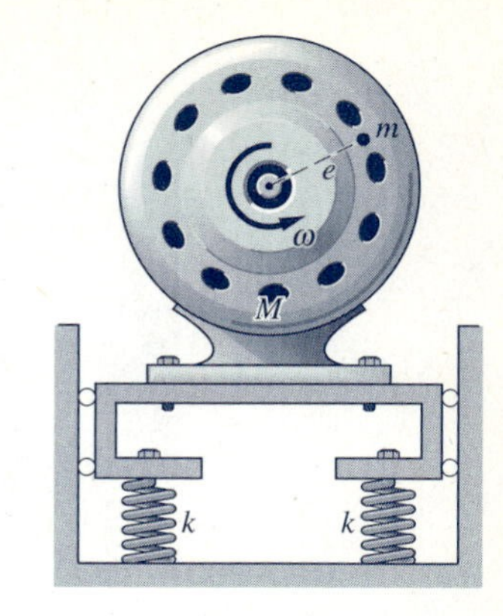

Fig. P20.22, P20.23

20.23 The electric motor and its base of combined mass M are mounted on four identical springs, each of stiffness k. The slight unbalance of the rotor is equivalent to the mass m at a distance e from its axis. When the motor is running at 1000 rev/min, its steady-state amplitude is measured to be 0.8 mm. When the speed is increased to 1500 rev/min, the amplitude changes to 2.4 mm. Determine the speed of the motor at resonance.

20.24 The pendulum of length L and mass m is suspended from a sliding collar. If the horizontal displacement $y(t) = Y \sin \omega t$ is imposed on the collar, show that for sufficiently small θ, the steady-state amplitude of the pendulum is

$$\theta_{\max} = \frac{\omega^2 L}{|g - \omega^2 L|} \frac{Y}{L}$$

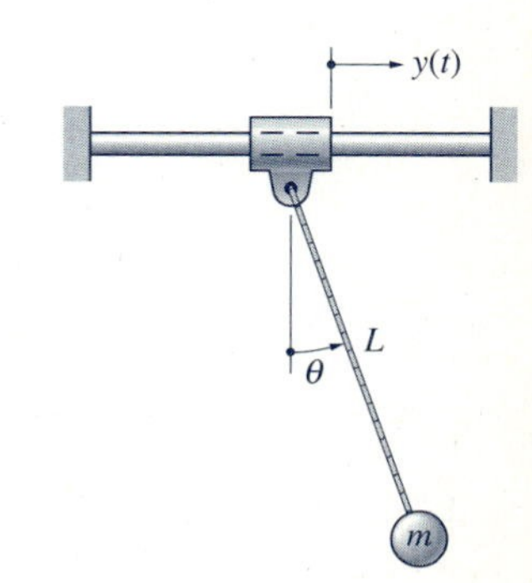

Fig. P20.24–P20.26

20.25 The length of the pendulum in Prob. 20.24 is $L = 1.09$ m. (a) Using the expression given in Prob. 20.24, find Y as a function of ω if $\theta_{\max} = 5°$. (b) Plot Y versus ω for part (a), and mark the domains in the ω-Y plane for which $\theta_{\max}$ is less than 5°.

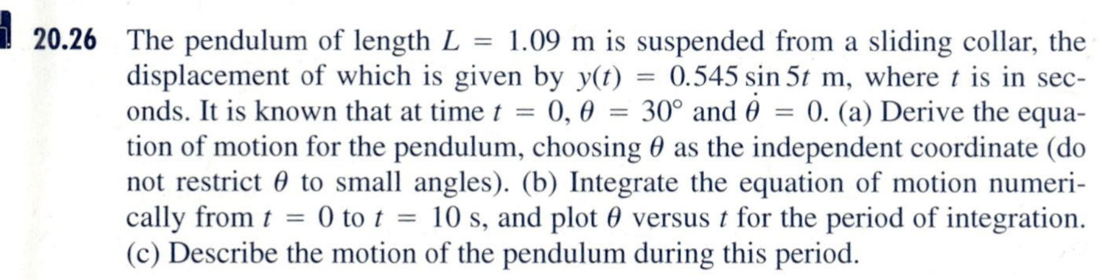

20.26 The pendulum of length $L = 1.09$ m is suspended from a sliding collar, the displacement of which is given by $y(t) = 0.545 \sin 5t$ m, where t is in seconds. It is known that at time $t = 0$, $\theta = 30°$ and $\dot{\theta} = 0$. (a) Derive the equation of motion for the pendulum, choosing θ as the independent coordinate (do not restrict θ to small angles). (b) Integrate the equation of motion numerically from $t = 0$ to $t = 10$ s, and plot θ versus t for the period of integration. (c) Describe the motion of the pendulum during this period.

20.27 A slightly unbalanced motor of weight W is attached to the middle of a light elastic beam. The unbalance is equivalent to a weight $W/400$ located at a distance $e = 8$ in. from the axis of the motor. When the motor is rotating at $\omega = 1280$ rev/min, which is known to be less than the speed at resonance, its steady-state amplitude is $x_{\max} = 0.04$ in. Find the angular speed of the motor at which resonance will occur.

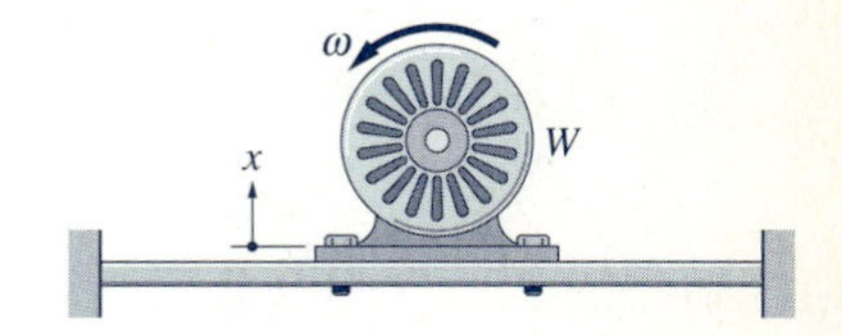

Fig. P20.27, P20.28

20.28 Solve Prob. 20.27 if the given ω is known to be greater than the speed at resonance.

20.29 The trailer is towed with the constant velocity v over a "washboard" road that may be approximated by a sine curve. The weight of the trailer and its load is 800 lb, with the weights of the wheels being negligible. The trailer is mounted on two leaf springs, each of stiffness 240 lb/in. Considering only the steady-state motion, determine the range of v for which the wheels would lose contact with the ground.

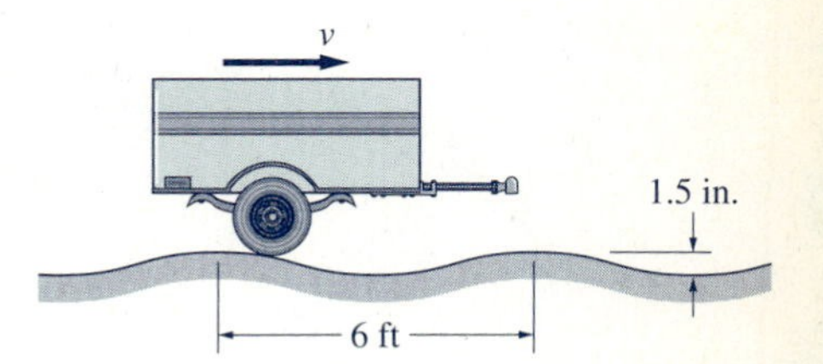

Fig. P20.29, P20.30

20.30 If the velocity of the trailer described in Prob. 20.29 is $v = 12$ mi/h, calculate the maximum force in each spring.

20.4 Damped Free Vibrations of Particles

When energy is dissipated from a vibrating system, the motion is said to be *damped.* The common forms of damping are viscous, Coulomb, and solid damping. Viscous damping describes resistance to motion that is proportional to the first power of the velocity. (Incidental damping, such as air resistance, is often assumed to be viscous. However, a damping force proportional to the square of the velocity would be a more accurate description.) Coulomb damping arises from dry friction between sliding surfaces. Solid damping is caused by internal friction within the body itself. In this text, we consider only viscous damping.

In viscous damping, the damping force F_d is proportional to the velocity; i.e., $F_d = -cv$, where v is the velocity and c is a constant of proportionality, called the *coefficient of viscous damping.* The negative sign indicates that the damping force always opposes the velocity.

A common example of a viscous damper (also known as a *dashpot*) is the automobile shock absorber. When an automobile hits a bump, the shock absorbers carry most of the impact loading, thus preventing the springs from "bottoming out." The shock absorbers are also responsible for damping out the ensuing up-and-down oscillations of the vehicle. An automobile shock absorber consists of a piston that is encased in oil and mounted between the wheel and the frame of the automobile. As the piston moves, oil is forced to flow through a hole from one side of the piston to the other. The amount of damping (due to the viscous resistance of the oil) depends largely on the size of the orifice being used.

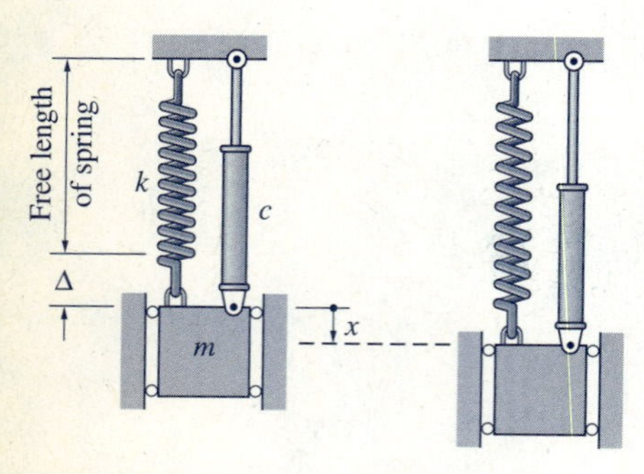

(a)

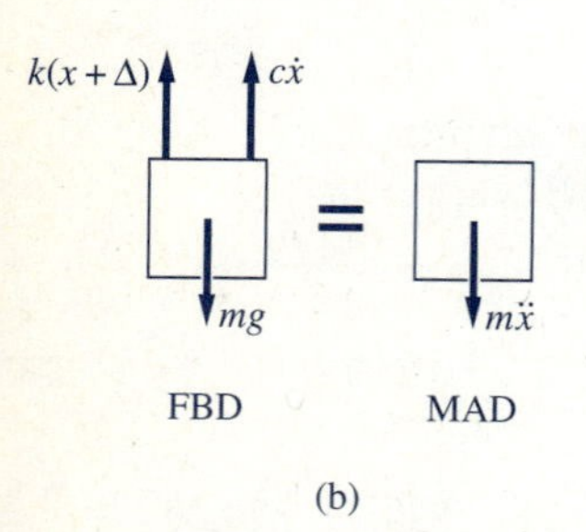

(b)

Fig 20.7

Figure 20.7(a) shows a spring-mass system. We have added a dashpot with damping coefficient c. Choosing x to be the downward movement of the mass, measured from its equilibrium position, results in the FBD shown in Fig. 20.7(b), where Δ is the static deflection of the spring. The mass-acceleration diagram in Fig. 20.7(b) consists of the inertia vector $m\ddot{x}$. The equation of motion is

$$\Sigma F_x = ma_x \qquad +\downarrow \quad mg - k(x + \Delta) - c\dot{x} = m\ddot{x}$$

Utilizing the equilibrium equation $mg - k\Delta = 0$, we obtain

$$\boxed{m\ddot{x} + c\dot{x} + kx = 0} \tag{20.24}$$

A linear differential equation with constant coefficients, such as Eq. (20.24), admits a solution of the form

$$x = Ae^{\lambda t}$$

where A and λ are constants. Substituting this into Eq. (20.24) and dividing each term by $Ae^{\lambda t}$ results in the *characteristic equation*

$$m\lambda^2 + c\lambda + k = 0 \tag{20.25}$$

the roots of which are

$$\left.\begin{matrix}\lambda_1\\ \lambda_2\end{matrix}\right\} = -\frac{c}{2m} \pm \sqrt{\left(\frac{c}{2m}\right)^2 - \frac{k}{m}} \tag{20.26}$$

The *critical damping coefficient* c_{cr} is defined as the value of c for which the radical in Eq. (20.26) vanishes. Therefore, we find that

$$c_{cr} = 2mp \tag{20.27}$$

where $p = \sqrt{k/m}$, the undamped circular frequency of the system. It is convenient to introduce the damping factor ζ, defined as the ratio of the actual damping to the critical damping; i.e.,

$$\zeta = \frac{c}{c_{cr}} = \frac{c}{2mp} \tag{20.28}$$

Now, Eq. (20.26) can be written as

$$\left.\begin{matrix}\lambda_1\\ \lambda_2\end{matrix}\right\} = p\left(-\zeta \pm \sqrt{\zeta^2 - 1}\right) \tag{20.29}$$

The general solution of Eq. (20.24) is any linear combination of the two solutions corresponding to λ_1 and λ_2:

$$x = A_1 e^{\lambda_1 t} + A_2 e^{\lambda_2 t}$$

where A_1 and A_2 are arbitrary constants. After substitution for the λ's from Eq. (20.29), the solution becomes

$$x = A_1 e^{(-\zeta + \sqrt{\zeta^2 - 1})pt} + A_2 e^{(-\zeta - \sqrt{\zeta^2 - 1})pt} \tag{20.30}$$

There are three categories of damping, determined by the value of the damping factor ζ.*

1. Overdamping: $\zeta > 1$ The roots λ_1 and λ_2 in Eq. (20.29) are real and distinct. Consequently, the motion is nonoscillatory and decaying with time, as shown in Fig. 20.8. Motion of this type is called *aperiodic,* or *dead-beat,* motion.

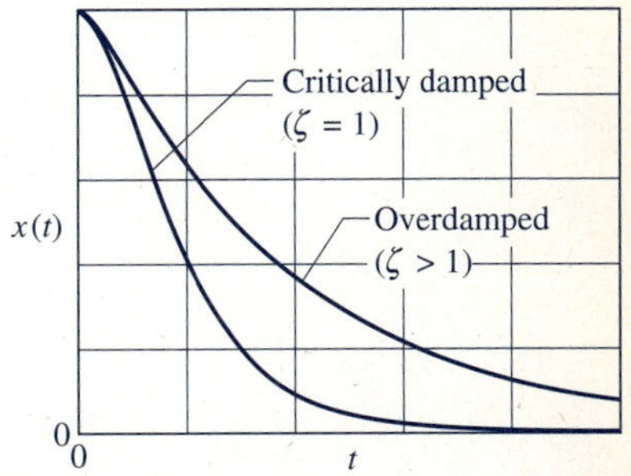

2. Critical damping: $\zeta = 1$ The roots λ_1 and λ_2 in Eq. (20.29) are both equal to $-p$, and the solution can be shown to take the form

$$x = (A_1 + A_2 t)e^{-pt} \tag{20.31}$$

where A_1 and A_2 are arbitrary constants. The motion is again aperiodic, as shown in Fig. 20.8.

Critically damped ($\zeta = 1$)
Overdamped ($\zeta > 1$)
$x(t)$
0
0
t

Both curves drawn for the initial condition $\dot{x}_0 = 0$

Fig 20.8

* The three forms of solutions are stated here without proof. For a complete discussion, see a textbook on differential equations.

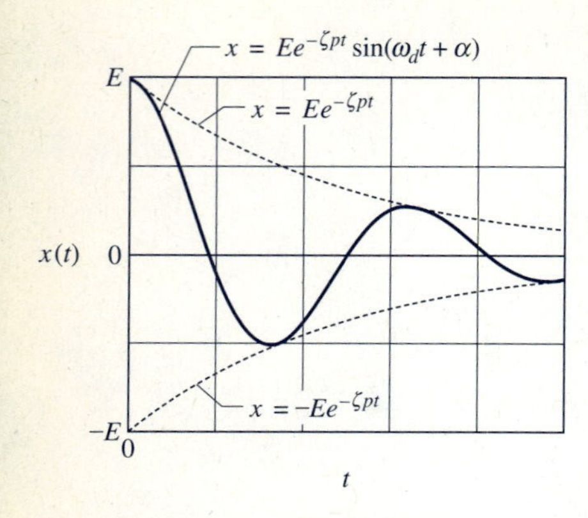

Fig 20.9

3. Underdamping: $\zeta < 1$ The roots λ_1 and λ_2 in Eq. (20.29) are complex conjugates. It can be shown that Eq. (20.30) can be written in the form

$$x = Ee^{-\zeta pt}\sin(\omega_d t + \alpha) \tag{20.32}$$

where E and α are arbitrary constants, and

$$\omega_d = p\sqrt{1-\zeta^2} \tag{20.33}$$

The motion represented by Eq. (20.32) is oscillatory with decreasing amplitude, as shown in Fig. 20.9. Observe that the plot is tangent to the curves $x = \pm Ee^{-\zeta pt}$. Although the motion does not repeat itself, ω_d is called the *damped circular frequency,* and the corresponding *damped period* is given by

$$\tau_d = \frac{2\pi}{\omega_d} \tag{20.34}$$

According to Eq. (20.33), the damped circular frequency ω_d is smaller than the circular frequency p. Consequently, the damped period τ_d is larger than the period of the undamped free vibration.

Sample Problem 20.5

Figure (a) shows a block of mass m that is attached to two ideal springs and a viscous damper. (1) Derive the equation of motion for the block, assuming that x is measured from the position where the springs are undeformed. (2) If $x = 0$ and $\dot{x} = 4$ m/s when $t = 0$, plot $x(t)$ for $\zeta = 2.5$, 1.0, and 0.25. Use $m = 0.2$ kg, $k_1 = 20$ N/m, and $k_2 = 30$ N/m.

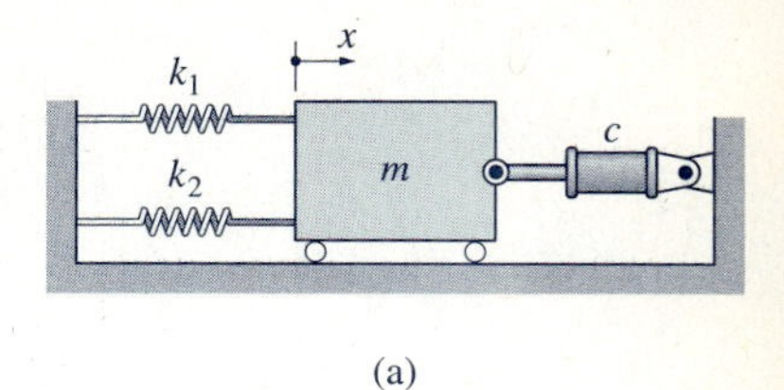

(a)

Solution

Part 1

The free-body (FBD) and mass-acceleration (MAD) diagrams of the block in an arbitrary position are shown in Fig. (b). The forces acting on the mass are its weight mg, the resultant normal reaction N, the two spring forces, and the force exerted by the damper. The equation of motion, obtained by summing forces in the x- direction, is

$$\Sigma F_x = ma_x \qquad \xrightarrow{+} \qquad -k_1 x - k_2 x - c\dot{x} = m\ddot{x}$$

or

$$m\ddot{x} + c\dot{x} + (k_1 + k_2)x = 0 \qquad \textit{Answer} \quad \text{(a)}$$

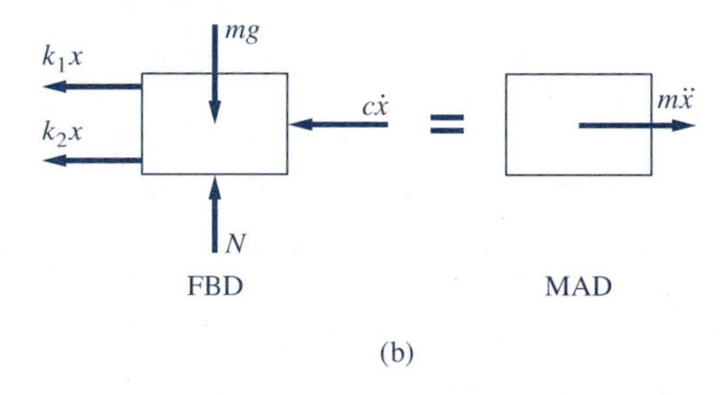

(b)

Part 2

By comparing Eq. (a) with Eq. (20.24), we deduce that the undamped circular frequency is

$$p = \sqrt{\frac{k_1 + k_2}{m}} = \sqrt{\frac{20 + 30}{0.2}} = 15.81 \text{ rad/s} \qquad \text{(b)}$$

The Case of $\zeta = 2.5$

Since $\zeta > 1$, the motion is overdamped, and the displacement is described by Eq. (20.30).

$$x = A_1 e^{\left(-\zeta + \sqrt{\zeta^2 - 1}\right)pt} + A_2 e^{\left(-\zeta - \sqrt{\zeta^2 - 1}\right)pt}$$

where A_1 and A_2 are constants to be determined by the initial conditions. For the given data

$$\left(-\zeta \pm \sqrt{\zeta^2 - 1}\right)p =$$

$$\left(-2.5 \pm \sqrt{(2.5)^2 - 1}\right)(15.81) = \begin{cases} -3.300 \text{ rad/s} \\ -75.75 \text{ rad/s} \end{cases}$$

Therefore, the displacement and the velocity become

$$x(t) = A_1 e^{-3.300t} + A_2 e^{-75.75t} \tag{c}$$

and

$$\dot{x}(t) = -3.300A_1 e^{-3.300t} - 75.75A_2 e^{-75.75t} \tag{d}$$

Applying the initial conditions $x(0) = 0$ and $\dot{x}(0) = 4000$ mm/s to Eqs. (c) and (d) yields

$$x(0) = A_1 + A_2 = 0$$

$$\dot{x}(0) = -3.300A_1 - 75.75A_2 = 4000$$

Solving these equations, we get $A_1 = -A_2 = 55.23$ mm, which on substitution into Eq. (c) gives the displacement

$$x(t) = 55.23\left(e^{-3.300t} - e^{-75.75t}\right) \text{ mm} \qquad \textit{Answer}$$

The Case of $\zeta = 1.0$

The motion is critically damped, the displacement being given by Eq. (20.31):

$$x = (A_1 + A_2 t)e^{-pt}$$

Substituting $p = 15.81$ rad/s from Eq. (b), the displacement and the velocity become

$$x(t) = (A_1 + A_2 t)e^{-15.81t} \tag{e}$$

and

$$\dot{x}(t) = -15.81(A_1 + A_2 t)e^{-15.81t} + A_2 e^{-15.81t} \tag{f}$$

Substituting the initial conditions into Eqs. (e) and (f) gives $x(0) = A_1 = 0$ and $\dot{x}(0) = -15.81A_1 + A_2 = A_2 = 4000$ mm/s. Therefore, the displacement in Eq. (e) becomes

$$x(t) = 4000te^{-15.81t} \text{ mm} \qquad \textit{Answer}$$

The Case of $\zeta = 0.25$

Since the motion is underdamped ($\zeta < 1$), the displacement is given by Eq. (20.32):

$$x = Ee^{-\zeta pt}\sin(\omega_d t + \alpha)$$

where E and α are constants to be determined from the initial conditions. Using the given data, we get $\zeta p = 0.25(15.81) = 3.953$ rad/s, and $\omega_d = p\sqrt{1 - \zeta^2} = 15.81\sqrt{1 - (0.25)^2} = 15.31$ rad/s. Therefore, $x(t)$ and $\dot{x}(t)$ become

$$x(t) = Ee^{-3.953t}\sin(15.31t + \alpha) \tag{g}$$

and

$$\begin{aligned}\dot{x}(t) = {} & Ee^{-3.953t}\cos(15.31t + \alpha)15.31 \\ & - 3.953Ee^{-3.953t}\sin(15.31t + \alpha)\end{aligned} \tag{h}$$

Applying the initial conditions to Eqs. (g) and (h) gives

$$x(0) = E \sin \alpha = 0$$

$$\dot{x}(0) = 15.31E \cos \alpha - 3.953E \sin \alpha = 4000$$

Solving for E and α, we obtain $\alpha = 0$ and $E = 4000/15.31 = 261.3$ mm. Substituting these values into Eq. (g) yields

$$x(t) = 261.3e^{-3.953t} \sin 15.31t \text{ mm} \qquad \textit{Answer}$$

The displacements $x(t)$ for the three cases are plotted in Fig. (c). As expected, the motion for the underdamped case ($\zeta = 0.25$) oscillates, but in the other two cases the motion is aperiodic.

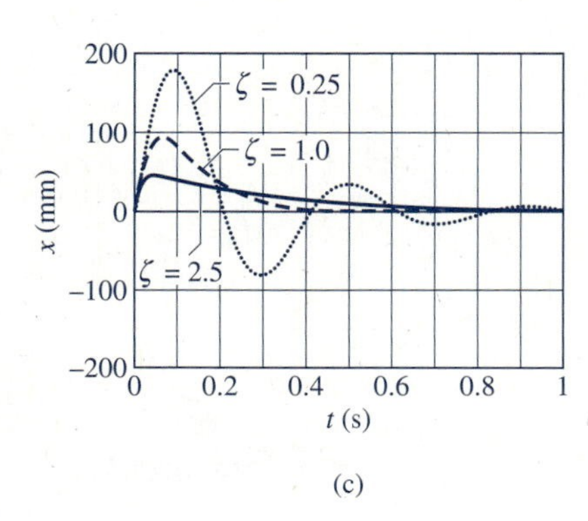

(c)

PROBLEMS

20.31 As shown in the figure, we let the peak displacements of an underdamped system (measured from equilibrium) be x_n, $n = 1, 2, \ldots$. Show that the ratio of two successive peaks, x_{n+1}/x_n, is constant, and that the natural logarithm of this ratio, called the *logarithmic decrement*, is

$$\ln(x_{n+1}/x_n) = -\frac{2\pi\zeta}{\sqrt{1-\zeta^2}}$$

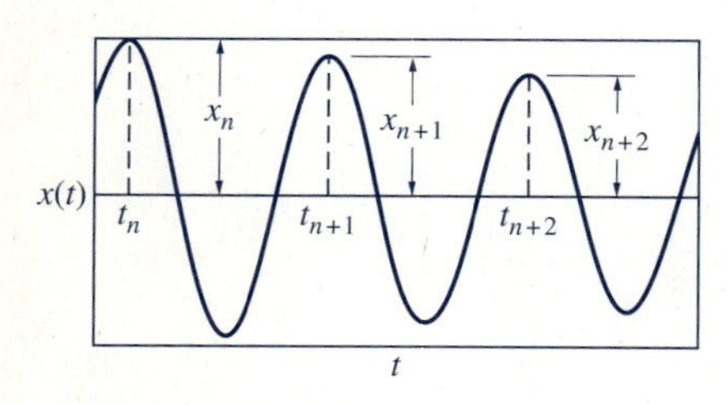

Fig. P20.31

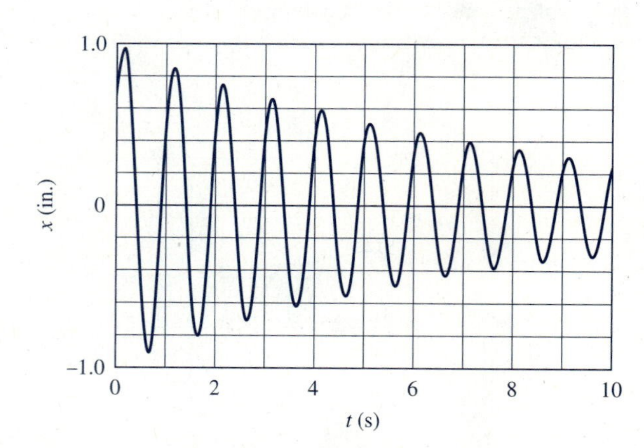

Fig. P20.32

20.32 (a) Using the expression for the logarithmic decrement given in Prob. 20.31, show that

$$\ln(x_{n+k}/x_n) = -\frac{2k\pi\zeta}{\sqrt{1-\zeta^2}}$$

where k is a positive integer greater than 1. (b) Utilizing the result of part (a), estimate the damping factor ζ for a system that has the displacement-time curve shown in the figure.

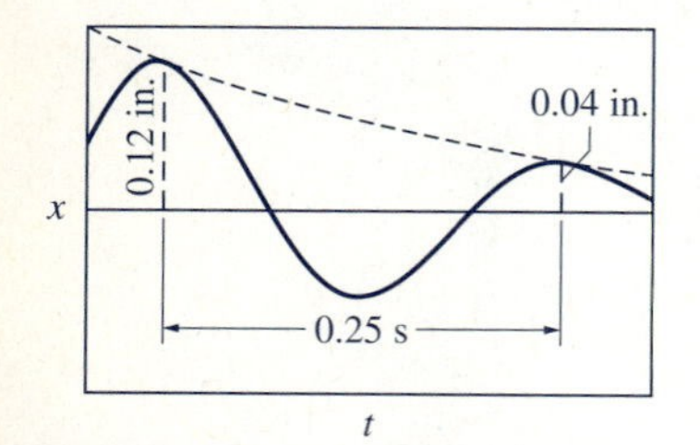

Fig. P20.33

20.33 The figure shows the displacement-time response of a damped oscillating system (x is measured from the equilibrium position). Knowing that the weight of the oscillator is 0.5 lb, find the spring stiffness k and the damping coefficient c. (Hint: Use the logarithmic decrement given in Prob. 20.31.)

20.34 (a) Use the logarithmic decrement given in Prob. 20.31 to show that the relationship between ΔE, the percentage of energy lost per cycle, and the damping factor ζ of an underdamped oscillator is

$$\Delta E = \left(1 - e^{-4\pi\zeta/\sqrt{1-\zeta^2}}\right) \times 100\%$$

(b) Using the results of part (a), compute the damping factor that would cause 10% energy loss per cycle.

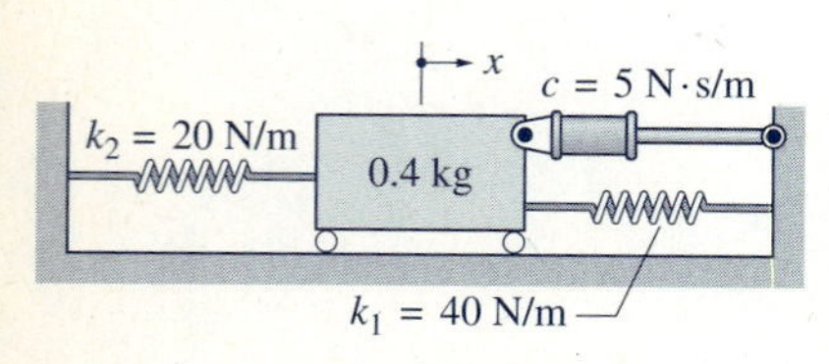

Fig. P20.35

20.35 The system is released from rest at time $t = 0$ with the initial displacement $x_0 = 50$ mm. Both springs are unstretched when $x = 0$. Determine the expression for $x(t)$.

20.36 The system is set into motion at time $t = 0$ with $x_0 = 0$ and $\dot{x}_0 = 2$ m/s, where x is measured from the unstretched position of the spring. (a) Determine the expression for $x(t)$. (b) Compute the maximum displacement of the mass.

20.37 Solve Prob. 20.36 if c_2 is changed to 3 N · s/m.

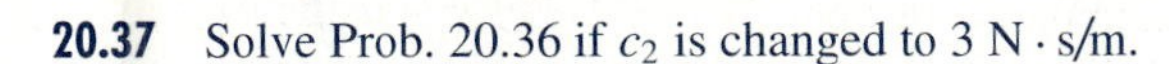

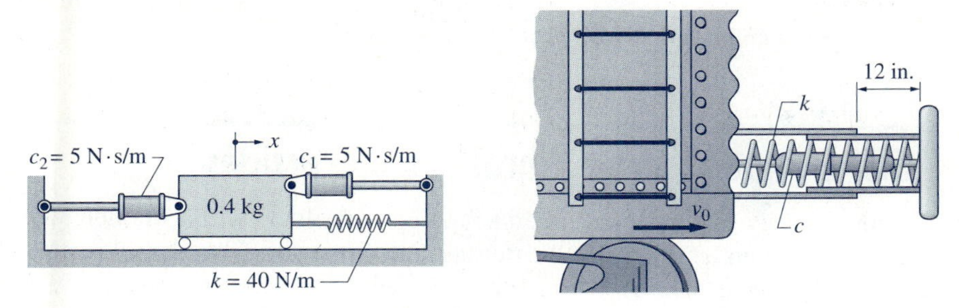

Fig. P20.36, P20.37 **Fig. P20.38**

20.38 Each of the two bumpers mounted on the end of a 180 000-lb railroad car has a spring stiffness of $k = 12 \times 10^3$ lb/ft and a coefficient of viscous damping equal to 45×10^3 lb · s/ft. Recognizing that a bumper will "bottom out" when its deformation exceeds 12 in., find the largest velocity v_0 with which the car can safely hit a rigid wall.

20.39 A critically damped oscillator is released from rest with the initial displacement x_0 (measured from the equilibrium position). (a) Derive the expressions for the displacement and the velocity of the oscillator in terms of x_0, p, and t. (b) Determine the expression for the maximum speed of the oscillator in terms of x_0 and p.

20.40 Repeat Prob. 20.39 if the oscillator is overdamped, the damping factor being $\zeta = 2$.

20.41 The mass of the pendulum is $m = 0.5$ kg and its length is $L = 1.5$ m. At a certain time, the amplitude of the pendulum was measured to be 5°; one hour later, it was 3°. Assuming that the damping responsible for the decay of amplitude is viscous, calculate the damping coefficient.

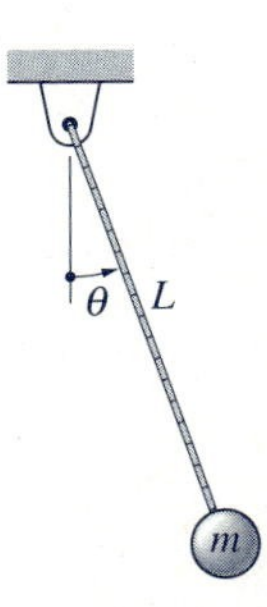

Fig. P20.41

20.42 Viscous damping is not an accurate representation of the resistance experienced by a body that is moving through a low viscosity fluid, such as air or water. Experiments indicate that the damping force is actually proportional to the square of the velocity: $F_d = cv^2$. For a one-inch diameter sphere moving in air, the approximate value of the damping constant is $c = 2.6 \times 10^{-6}$ lb · s²/ft². (a) Use this information to derive the equation of motion for the pendulum shown (the bob is made of steel). Neglect the damping effect of the string. (b) Integrate the equation of motion numerically over two periods of vibration, assuming that the pendulum is released from rest when $\theta = 30°$. (c) Use the results of the integration to calculate the percent loss of amplitude over the first two periods. (Note: To verify that the loss of amplitude is real and not due to numerical errors in the integration procedure, it is recommended that you repeat parts (b) and (c) with zero damping—there should be no loss of amplitude.)

20.43 Repeat Prob. 20.42 if the bob of the pendulum is replaced with a one-inch diameter styrofoam ball that weighs 4.5×10^{-4} lb.

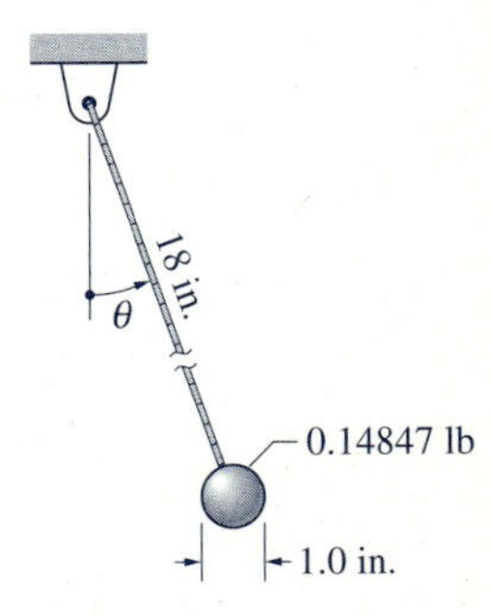

Fig. P20.42, P20.43

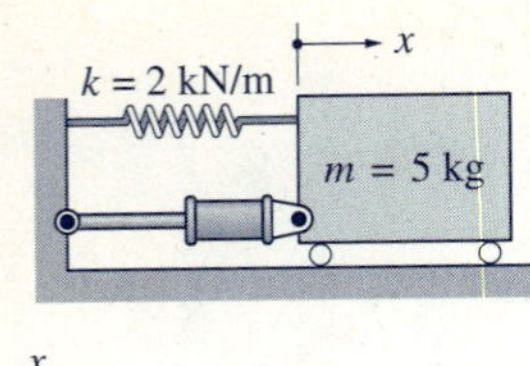

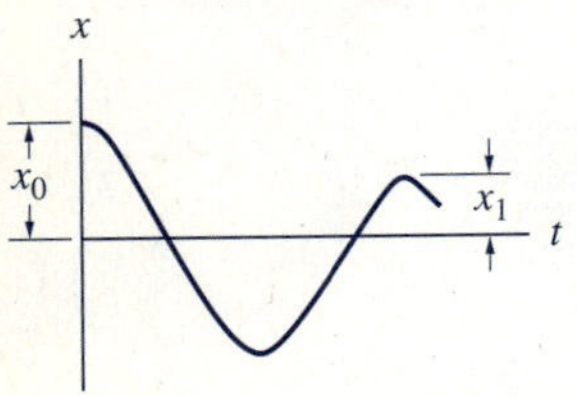

Fig. P20.44

20.44 The system shown is released from rest with the initial displacement $x_0 = 0.1$ m, measured from the equilibrium position. (a) Assuming that the damping force exerted by the dashpot is $F_d = c_2\dot{x}^2$, where $c_2 = 50\ \text{N} \cdot \text{s}^2/\text{m}^2$, determine the ratio x_1/x_0. (b) If the damping force were $F_d = c_1\dot{x}$, determine the value of c_1 that would give the same ratio x_1/x_0 as in part (a). (c) Plot x versus t over the period $t = 0$ to 1.0 s for the two cases of damping. What is the main difference in the amplitude decay between the two types of damping?

20.5 Damped Forced Vibrations of Particles

Because energy is dissipated when vibrations are damped, the motion will eventually stop unless there is an external source of energy. Here we will consider the case where the vibration is maintained either by a harmonic force or a harmonic support displacement.

a. Harmonic forcing function

Figure 20.10(a) shows a damped spring-mass system that is subjected to the harmonic force $P = P_0 \sin \omega t$. Letting x be the displacement of the mass from its equilibrium position, we obtain the following equation of motion from the free-body and mass-acceleration diagrams shown in Fig. 20.10(b):

$$\Sigma F_x = ma_x \qquad +\downarrow \quad mg + P_0 \sin \omega t - k(x + \Delta) - c\dot{x} = m\ddot{x}$$

Since $mg = k\Delta$ (the equilibrium equation), this reduces to

$$\boxed{m\ddot{x} + c\dot{x} + kx = P_0 \sin \omega t} \tag{20.35}$$

Equation (20.35) is a nonhomogeneous, second-order linear differential equation [we previously encountered such equations in the analysis of undamped forced vibrations—see Eq. (20.14)]. Because the homogeneous equation (right-hand side equal to zero) of Eq. (20.35) is identical to Eq. (20.24), we

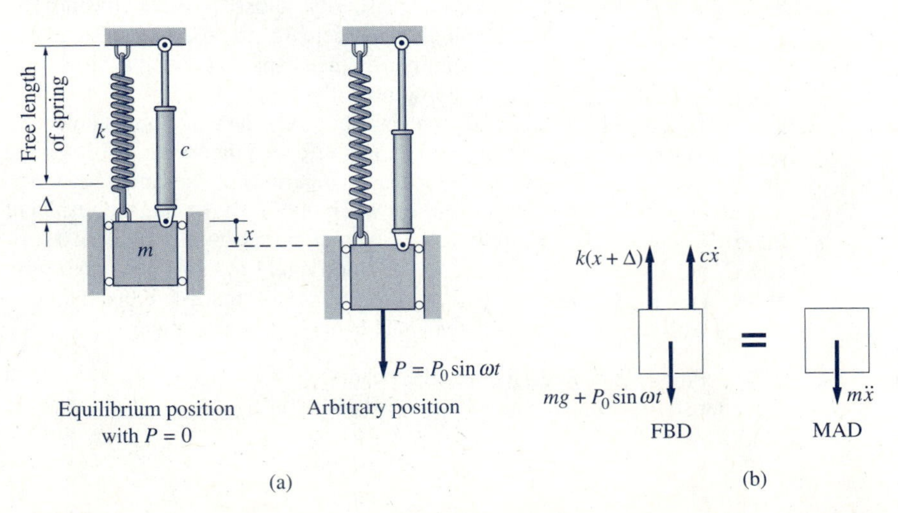

Fig. 20.10

conclude that the complementary solution is given by Eq. (20.30), (20.31), or (20.32), depending on the damping factor. The complementary solution, also called the transient vibration, is generally not interesting from a practical viewpoint, because it decays with time. From here on, we restrict our attention to the particular solution, which constitutes the steady-state vibration.

By direct substitution, it can be shown that a particular solution of Eq. (20.35) is

$$x = X \sin(\omega t - \phi) \tag{20.36}$$

where the amplitude X is given by

$$X = \frac{P_0/k}{\sqrt{[1 - (\omega/p)^2]^2 + (2\zeta\omega/p)^2}} \tag{20.37}$$

and the phase angle ϕ (the angle by which x lags P) is

$$\phi = \tan^{-1}\left[\frac{2\zeta\omega/p}{1 - (\omega/p)^2}\right] \tag{20.38}$$

As noted previously, P_0/k represents the zero frequency deflection, ω/p is the frequency ratio, and $\zeta = c/2mp$ is the damping factor, where $p = \sqrt{k/m}$.

As we noted in Art. 20.3, the initial conditions on the motion will determine the constants of integration that appear in the transient vibration, whereas the steady-state vibration is independent of how the motion was started.

The magnification factor (the ratio of the amplitude of the steady-state vibration divided by the zero-frequency deflection) is found from Eq. (20.37) to be

$$\text{Magnification Factor} = \frac{1}{\sqrt{[1 - (\omega/p)^2]^2 + (2\zeta\omega/p)^2}} \tag{20.39}$$

The magnification factor is plotted versus the frequency ratio in Fig. 20.11. As expected, the magnification factor is largest for frequencies near resonance if the damping coefficient is sufficiently small (the magnification factor becomes infinite at resonance in the absence of damping).

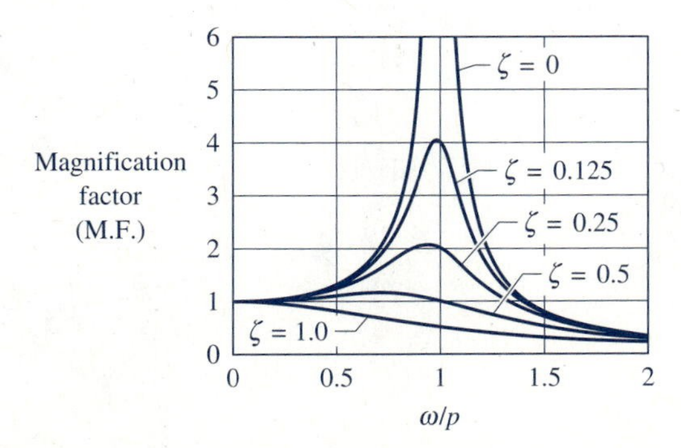

Fig 20.11

Additional information about the plots in Fig. 20.11 can be obtained by differentiating Eq. (20.39) with respect to (ω/p) and setting the result equal to zero. This procedure yields the following information: (1) all curves in Fig. 20.11 are tangent to the horizontal at $\omega/p = 0$ and as $\omega/p \to \infty$; (2) if $\zeta > 0.707$, the maximum magnification factor is 1.0, occurring at $\omega/p = 0$; (3) if $\zeta < 0.707$, the magnification factor attains its maximum value at $(\omega/p) = \sqrt{1 - 2\zeta^2}$, not at resonance.

b. Harmonic support displacement

Figure 20.12(a) shows a viscously damped spring-mass system with the support undergoing the prescribed harmonic displacement $y = Y \sin \omega t$. The free-body and mass-acceleration diagrams are shown in Fig. 20.12(b). Note that x is the displacement of the mass from its equilibrium position (with support stationary at $y = 0$) and Δ is the static deflection of the spring. We see that the relative displacement $x - y$ and the relative velocity $\dot{x} - \dot{y}$ determine the spring and damping forces, respectively. Applying Newton's second law to the FBD and MAD, we obtain

$$\Sigma F_x = ma_x \qquad +\downarrow \quad mg - c(\dot{x} - \dot{y}) - k(x - y + \Delta) = m\ddot{x} \tag{a}$$

Noting that $mg = k\Delta$ (the equilibrium equation) and introducing the relative position coordinate $z = x - y$, Eq. (a) simplifies to

$$m\ddot{z} + c\dot{z} + kz = -m\ddot{y} \tag{b}$$

Substituting $y = Y \sin \omega t$ into Eq. (b), the equation of motion becomes

$$\boxed{m\ddot{z} + c\dot{z} + kz = mY\omega^2 \sin \omega t} \tag{20.40}$$

Comparing Eqs. (20.35) and (20.40), we see that our previous analysis with a harmonic force is also applicable to harmonic support displacement, provided

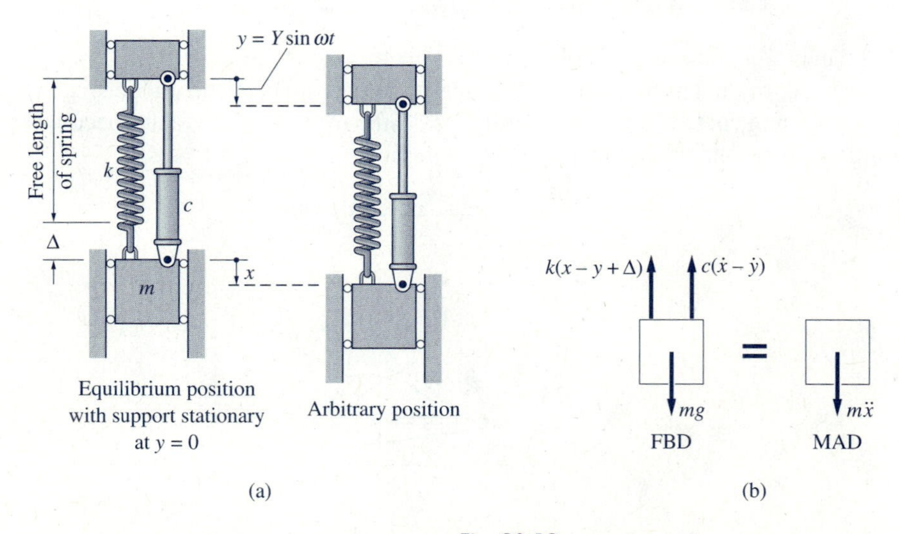

Fig. 20.12

that P_0 is replaced by $mY\omega^2$ and x by $z = x - y$. Making these substitutions, Eqs. (20.36)–(20.38) describing the steady-state vibration become

$$z = Z\sin(\omega t - \phi) \tag{20.41}$$

where the amplitude Z and phase angle ϕ (the angle by which z lags y) are given by

$$\begin{aligned} Z &= \frac{mY\omega^2/k}{\sqrt{[1-(\omega/p)^2]^2 + (2\zeta\omega/p)^2}} \\ &= Y\frac{(\omega/p)^2}{\sqrt{[1-(\omega/p)^2]^2 + (2\zeta\omega/p)^2)}} \end{aligned} \tag{20.42}$$

and

$$\phi = \tan^{-1}\left[\frac{2\zeta\omega/p}{1-(\omega/p)^2}\right] \tag{20.43}$$

where $p = \sqrt{k/m}$ and $\zeta = c/2mp$.

Sample Problem 20.6

The block of weight W is connected in a rigid frame between a linear spring and a viscous damper. The frame is subjected to the time-dependent vertical displacement $y(t) = Y \sin \omega t$. The displacement x of the block is measured from its static equilibrium position (with support stationary at $y = 0$). Determine the steady-state solution for (1) the relative displacement $z = x - y$; and (2) the absolute displacement x. Use $Y = 1.5$ in., $\omega = 400$ rad/s, $W = 6$ lb, $k = 1500$ lb/in., and $c = 40$ lb · s/ft.

$y = Y \sin \omega t$

Solution

Part 1

Because the given system is equivalent to that shown in Fig. 20.12, Eqs. (20.41)–(20.43) can be used to determine the relative displacement z. The parameters appearing in these equations are

$$p = \sqrt{\frac{k}{m}} = \sqrt{\frac{1500(12)}{6/32.2}} = 310.8 \text{ rad/s}$$

$$\zeta = \frac{c}{2mp} = \frac{40}{2(6/32.2)(310.8)} = 0.3453 \quad \text{(underdamped)}$$

$$\frac{\omega}{p} = \frac{400}{310.8} = 1.287$$

The relative amplitude of the steady-state solution is now obtained from Eq. (20.42)

$$\begin{aligned} Z &= Y \frac{(\omega/p)^2}{\sqrt{[1-(\omega/p)^2]^2 + (2\zeta\omega/p)^2}} \\ &= 1.5 \frac{(1.287)^2}{\sqrt{[1-(1.287)^2]^2 + [2(0.3453)(1.287)]^2}} \\ &= 2.249 \text{ in.} \end{aligned}$$

and the phase angle can be computed from Eq. (20.43)

$$\begin{aligned} \phi &= \tan^{-1}\left[\frac{2\zeta\omega/p}{1-(\omega/p)^2}\right] = \tan^{-1}\left[\frac{2(0.3453)(1.287)}{1-(1.287)^2}\right] \\ &= -0.9347 \text{ rad} \end{aligned}$$

The relative displacement thus becomes

$$z(t) = 2.249 \sin(400t + 0.9347) \text{ in.}$$

Answer

Note that $z(t)$ leads $y(t)$ by 0.9347 rad (53.6°).

Part 2

Using the trigonometric identity $\sin(a + b) = \sin a \cos b + \cos a \sin b$, the expression for $z(t)$ in Part (1) can be written as

$$\begin{aligned} z(t) &= 2.249(\cos 0.9347 \sin 400t + \sin 0.9347 \cos 400t) \\ &= 1.336 \sin 400t + 1.809 \cos 400t \text{ in.} \end{aligned}$$

Therefore, the expression for the absolute displacement of the mass becomes

$$\begin{aligned} x &= z + y \\ &= (1.336 \sin 400t + 1.809 \cos 400t) + 1.5 \sin 400t \\ &= 2.836 \sin 400t + 1.809 \cos 400t \text{ in.} \end{aligned}$$

From Eqs. (20.4)–(20.6), we know that $x = A \cos pt + B \sin pt$ can be written as $E = \sin(pt + \alpha)$, where $E = \sqrt{A^2 + B^2}$ and $\tan \alpha = A/B$. Thus the above expression for x can be written as

$$x = \sqrt{(2.836)^2 + (1.809)^2} \sin\left[400t + \tan^{-1}\left(\frac{1.809}{2.836}\right)\right]$$

or

$$x = 3.364 \sin(400t + 0.5678) \text{ in.} \qquad \textit{Answer}$$

We see that the amplitude of the absolute displacement of the block is 3.364 in. and $x(t)$ leads $y(t)$ by 0.5678 rad (32.5°).

PROBLEMS

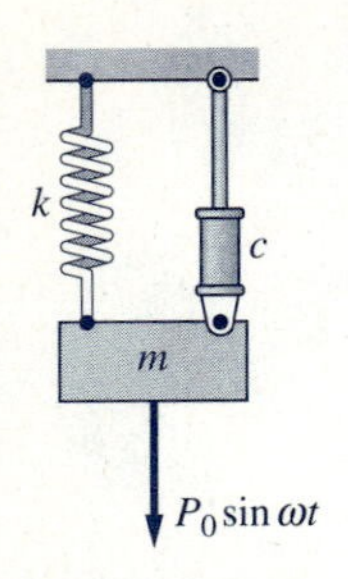

Fig. P20.45–P20.49

20.45 For the damped system shown, prove that the maximum steady-state amplitude for a given damping factor ζ occurs at the frequency ratio $\omega/p = \sqrt{1 - 2\zeta^2}$ for $\zeta^2 \leq 1/2$. Also show that the corresponding maximum amplitude is

$$x_{\max} = \frac{P_0/k}{2\zeta\sqrt{1-\zeta^2}}$$

20.46 For the system shown, $m = 0.2$ kg and $k = 2880$ N/m. When the system is driven by the harmonic force of amplitude P_0, it is observed that the amplitude of the steady-state vibration is the same at $\omega = 96.0$ rad/s and $\omega = 126.4$ rad/s. Calculate the damping coefficient c.

20.47 For the system shown, $m = 0.5$ slugs, $k = 10$ lb/in., and $c = 18.6$ lb · s/ft. (a) Determine the magnification factor if the circular frequency ω of the applied force equals the resonant frequency p of the system. (b) Find the maximum possible magnification factor and the corresponding value of ω.

20.48 Repeat Prob. 20.47 if $c = 6$ lb · s/ft.

20.49 The mass $m = 0.5$ slugs is suspended from a nonlinear spring and a viscous damper with the damping coefficient $c = 18.6$ lb · s/ft. The force deformation relationship of the spring is $F = kx[1 + (x/b)^2]$ lb, where $k = 10$ lb/in., $b = 2$ in., and x is measured in inches from the undeformed position of the spring. The amplitude of the applied force is $P_0 = 10$ lb and its circular frequency is $\omega = 12$ rad/s. (a) Show that the equation of motion of the mass is

$$\ddot{x} = -240x(1 + 36x^2) - 37.2\dot{x} + 32.2 + 20 \sin 12t \text{ ft/s}^2$$

where x is the displacement in feet measured from the undeformed position of the spring (not the equilibrium position) and t is the time in seconds. (b) Use numerical integration to determine the maximum and minimum values of x for the steady-state motion. (Hint: Start with the system at rest in the equilibrium position and integrate until the transient term has been damped out.) Is the motion symmetric about the equilibrium position?)

20.50 Determine the steady-state response $x(t)$ of the block if $Y = 25$ mm and $\omega = 30$ rad/s. Does $x(t)$ lead or lag $y(t)$?

20.51 Determine the expression for the steady-state displacement $x(t)$ of the block if $P_0 = 0.1$ lb and $\omega = 60$ rad/s. Does $x(t)$ lead or lag the applied force?

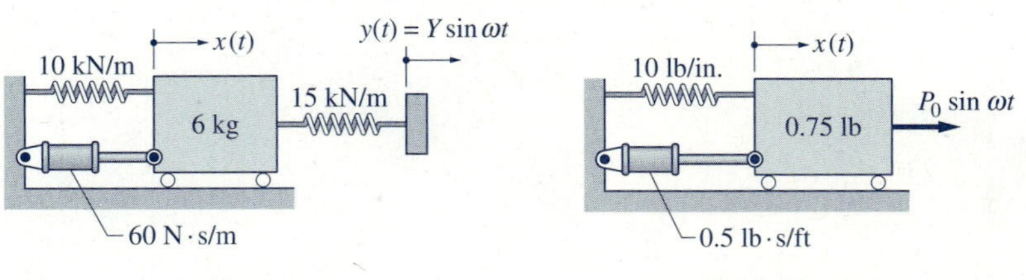

Fig. P20.50 **Fig. P20.51**

20.52 Find the expression for the steady-state response $x(t)$ of the block if $Y = 10$ mm and $\omega = 600$ rad/s. Does $x(t)$ lead or lag the imposed displacement $y(t)$?

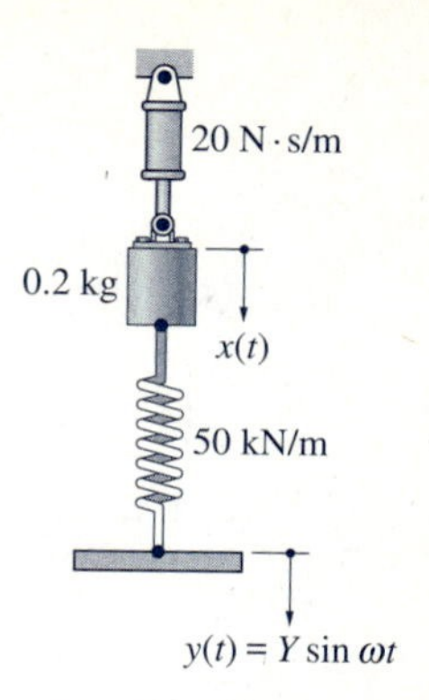

Fig. P20.52

20.53 Block A is connected to the shaker table B with a spring and a viscous damper. When the horizontal displacement $y(t) = Y \sin \omega t$ is imposed on the table, the resulting steady-state displacement of the block relative to the table is $z(t) = Z \sin(\omega t - \phi)$. Show that if the damping factor $\zeta \geq 1/\sqrt{2}$, then Z never exceeds Y regardless of the value of ω.

20.54 For the system described in Prob. 20.53, determine the largest possible value of Z and the corresponding frequency ratio ω/p if $\zeta = 1/2$.

20.55 For the system shown, $W = 12$ lb, $k = 160$ lb/ft, and $c = 2.5$ lb · s/ft. The horizontal displacement imposed on the shaker table B is $y(t) = 1.0 \sin 18t$ in., where the time t is measured in seconds. Determine the steady-state amplitude of the block.

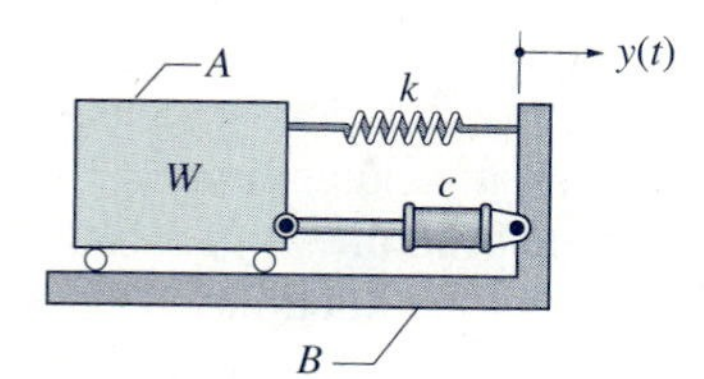

Fig. P20.53–P20.55

20.56 An electric motor and its base have a combined mass of $M = 12$ kg. Each of the four springs attached to the base has a stiffness $k = 480$ kN/m and a viscous damping coefficient c. The unbalance of the motor is equivalent to a mass $m = 0.005$ kg located at the distance $e = 90$ mm from the center of the shaft. When the motor is running at 400 rad/s, its steady-state amplitude is 1.8 mm. Determine (a) the damping coefficient of each spring; and (b) the phase angle between the displacement of the motor and ωt.

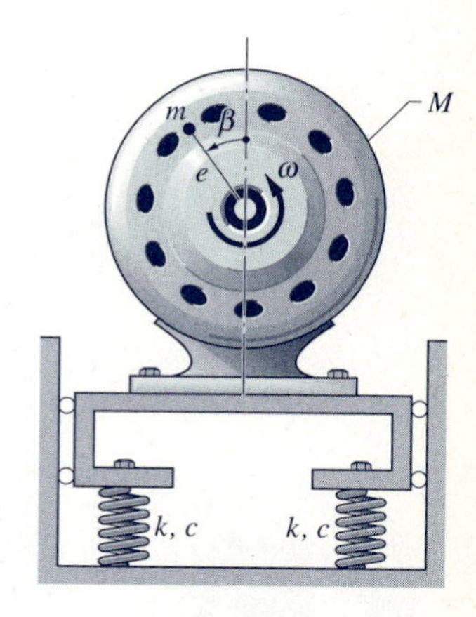

Fig. P20.56, P20.57

***20.57** The slightly unbalanced electric motor and its base have a total mass of $M = 18$ kg. The unbalance is equivalent to a small mass m located at the distance e from the axis of the motor. Each of the four springs attached to the base has a stiffness k and a viscous damping coefficient c. When the motor is running at 950 rev/min, it is observed that the maximum vertical displacement of the motor occurs when m is located at $\beta = 12.5°$. When the speed is 1400 rev/min, this angle changes to $-63.8°$. Determine the spring parameters k and c.

20.58 An iron cylinder is suspended from a spring and placed in a container filled with fluid. An electromagnet at the base of the container applies the force $P(t) = P_0 \sin \omega t$ to the cylinder, where $P_0 = 1.0$ kN and $\omega = 500$ rad/s. The damping force acting on the cylinder due to the fluid is $F_d = c\dot{x}^2$, where $c = 250$ N · s²/m². (a) Derive the equation of motion of the cylinder. (b) Estimate the steady-state amplitude of the cylinder by numerical integration of the equation of motion. [Hint: Start with the initial condition $x(0) = \dot{x}(0) = 0$, and integrate until the transient motion has been damped out.]

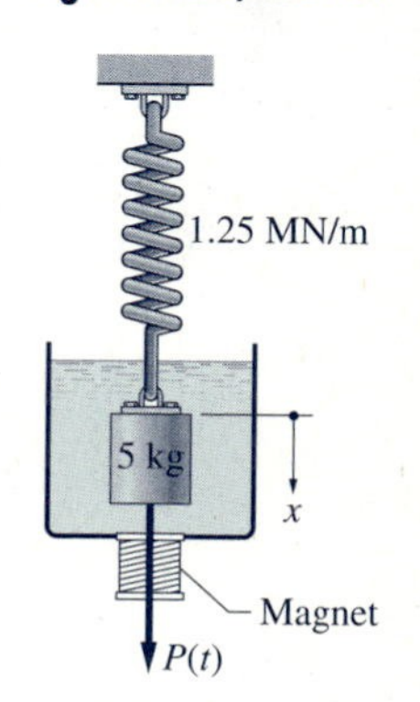

Fig. P20.58

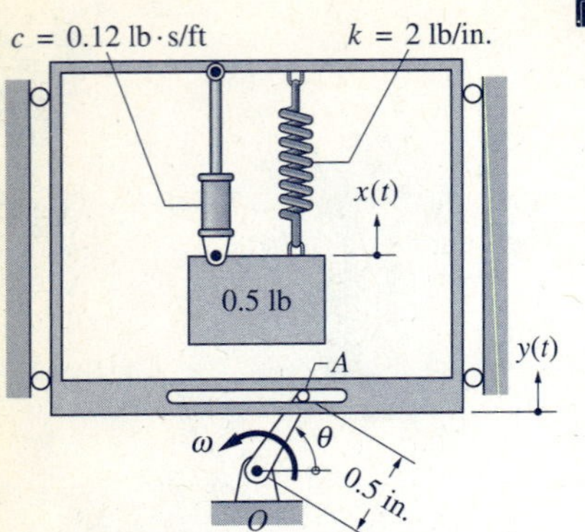

Fig. P20.59

20.59 The 0.5-lb weight is suspended from a rigid frame as shown. Pin A at the end of the rotating arm OA engages a slot in the frame, causing the frame to oscillate in the vertical direction. The arm is accelerated uniformly from rest at $t = 0$ and $\theta = 0$ at the rate $\dot{\omega} = 100$ rad/s^2. (a) Show that the equation of motion of the weight in terms of the relative displacement $z = x - y$ is

$$\ddot{z} = -1546z - 7.73\dot{z} - 4.17(\cos 50t^2 - 100t^2 \sin 50t^2) \text{ ft/s}^2$$

where z and t are measured in feet and seconds, respectively. (b) Use numerical integration to obtain the plot of z versus t from $t = 0$ to $t = 1.0$ s. (c) Use the numerical results of part (b) to determine the largest value of z.

20.6 Rigid-Body Vibrations

The analysis of rigid-body vibrations is fundamentally no different from the analysis of vibrating particles. We first derive the equation of motion utilizing the free-body and mass-acceleration diagrams of the body, and then seek a solution of this equation. If the system is linear, the equation of motion will be a second-order linear differential equation with constant coefficients. For a harmonically driven system with a single degree of freedom, this equation will have the following general form.

$$M\ddot{q} + C\dot{q} + Kq = F_0 \sin \omega t \tag{20.44}$$

where q is a variable that defines the position of the body (either a linear or an angular position coordinate). Because Eq. (20.44) has the same form as the equation of particle motion, $m\ddot{x} + c\dot{x} + kx = P_0 \sin \omega t$, we can write down its solution by analogy. For example, the solution of forced, steady-state motion is

$$q = Q \sin(\omega t - \phi) \tag{20.45a}$$

where

$$Q = \frac{F_0/K}{\sqrt{[1 - (\omega/p)^2]^2 + (2\zeta\omega/p)^2}} \tag{20.45b}$$

and

$$\phi = \tan^{-1}\left[\frac{2\zeta\omega/p}{1 - (\omega/p)^2}\right] \tag{20.45c}$$

The parameters appearing in the above equations for Q and ϕ are

$$\left.\begin{aligned} \text{Undamped circular frequency} \quad p &= \sqrt{\frac{K}{M}} \\ \text{Critical damping constant} \quad C_{cr} &= 2Mp \\ \text{Damping factor} \quad \zeta &= \frac{C}{C_{cr}} = \frac{C}{2Mp} \end{aligned}\right\} \tag{20.45d}$$

The expression for the transient vibration can be written in the same manner (its form would depend on whether the motion is over, under, or critically damped).

As an example, consider the nonhomogeneous disk of radius R and mass m shown in Fig. 20.13(a). The mass center G of the disk is located at the distance e from the pin O, and its moment of inertia about O is I_O. A linear spring and a viscous damper are attached to the periphery of the disk at A and B, respectively. A clockwise couple $M_0 \sin \omega t$ also acts on the disk. We assume that the spring tension is adjusted so that the line OG is horizontal when the disk occupies its static equilibrium position. Therefore, the static deflection Δ of the spring at equilibrium satisfies the equation $mge = kR\Delta$.

To derive the equation for rotational motion of the disk, we draw the FBD shown in Fig. 20.13(b). The angular position of the disk is indicated by the angle θ, with clockwise rotation assumed to be positive. Throughout the analysis we assume θ to be small, so that we can use $\sin \theta \approx \theta$ and $\cos \theta \approx 1$. With these approximations, the displacement of point A is $R\theta$ (downward) and the velocity of point B is $R\dot{\theta}$ (upward). This gives the spring and damping forces shown in the FBD. Without the need to draw the mass-acceleration diagram of the disk, we can sum moments about O to obtain the equation of motion

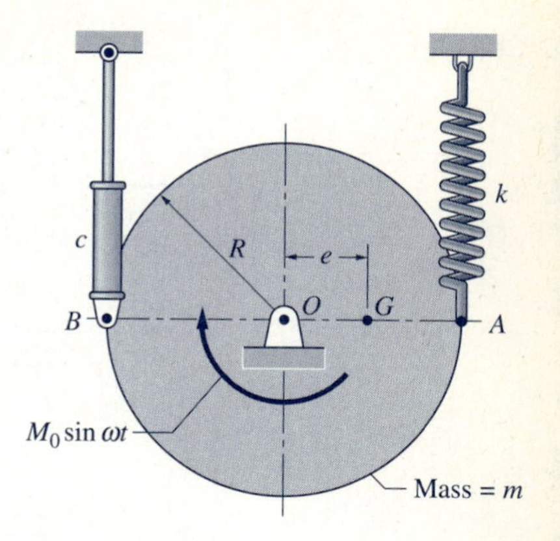

(a) Equilibrium position

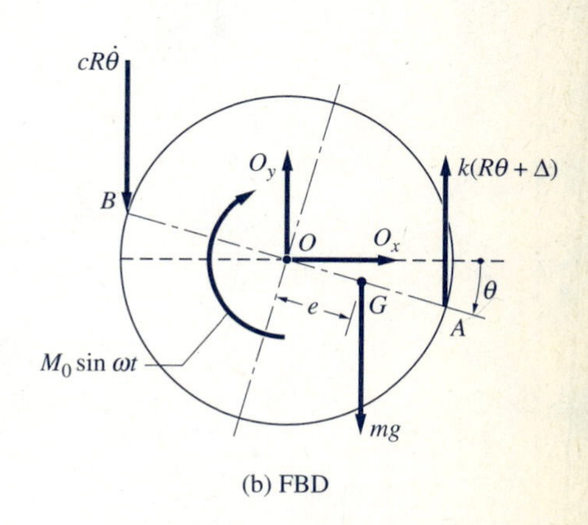

(b) FBD

Fig. 20.13

$$\Sigma M_O = I_O \ddot{\theta} \qquad \circlearrowright^{+} \quad M_0 \sin \omega t - k(R\theta + \Delta)R - (cR\dot{\theta})R + mge = I_O \ddot{\theta}$$

Using the equilibrium equation $mge = kR\Delta$, and rearranging the terms, yields

$$I_O \ddot{\theta} + cR^2 \dot{\theta} + kR^2 \theta = M_0 \sin \omega t \tag{a}$$

Comparing Eq. (a) with Eq. (20.44), we find that they are of the same form, the coefficients of Eq. (20.44) being

$$M = I_O \qquad C = cR^2 \qquad K = kR^2 \qquad F_0 = M_0 \tag{b}$$

We conclude, therefore, that the steady-state vibration of the disk is harmonic for small oscillations, and the motion can be determined directly from Eqs. (20.45). For example, the undamped circular frequency is

$$p = \sqrt{\frac{K}{M}} = \sqrt{\frac{kR^2}{I_O}}$$

The critical damping coefficient can be obtained from $C_{cr} = 2Mp$, which on substitution from Eq. (b) becomes

$$(cR^2)_{cr} = 2I_O p$$

yielding

$$c_{cr} = \frac{2I_O p}{R^2}$$

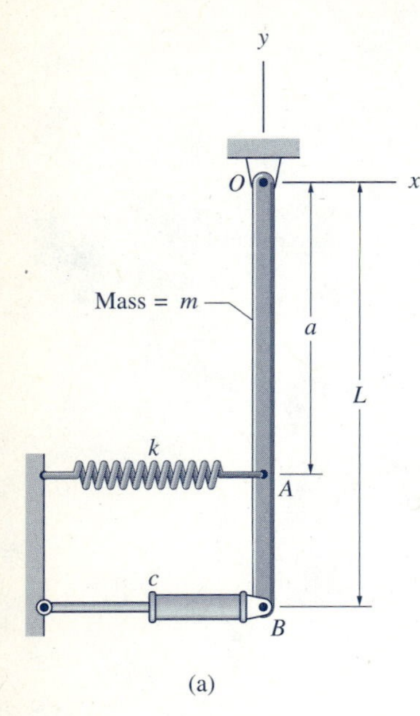

(a)

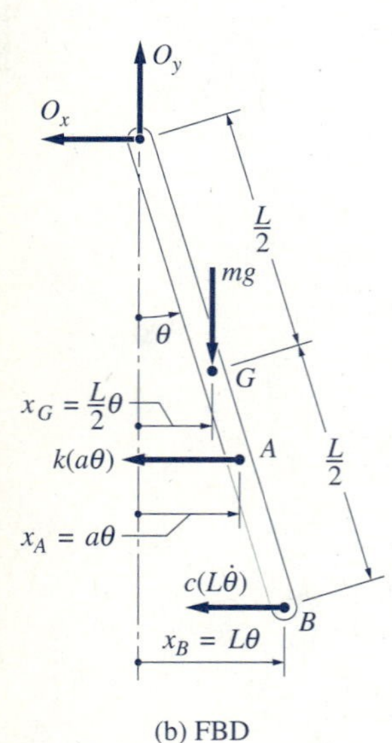

(b) FBD

Sample Problem 20.7

The homogeneous slender bar of mass m and length L in Fig. (a) is supported by a pin at O. The bar is also connected to an ideal spring and viscous damper at points A and B, respectively. The bar is in equilibrium in the position shown with the spring undeformed. (1) Derive the equation of motion for small angular displacements of the bar. (2) Determine whether the bar is overdamped or underdamped, given that $m = 12$ kg, $L = 800$ mm, $a = 400$ mm, $k = 80$ N/m, and $c = 20$N · s/m.

Solution

Part 1

Figure (b) shows the free-body diagram (FBD) of the bar when it is displaced a small angle θ from the vertical. The horizontal displacements x_G (G is the mass center), x_A, and x_B shown in the figure were obtained from the small angle approximation $\sin\theta \approx \theta$. The forces acting on the bar are its weight mg, the spring force $k(a\theta)$ (recall that the spring is undeformed when $x_A = 0$), and the damping force $c(L\dot{\theta})$ due to the dashpot.

We derive the equation of motion by summing moments about point O, thereby eliminating the pin reaction. Since O is a fixed point, a valid equation of motion is $\Sigma M_O = I_O\ddot{\theta}$. From the FBD in Fig. (b) and the approximation $\cos\theta \approx 1$, we obtain

$$\Sigma M_O = I_O\ddot{\theta} \qquad \circlearrowleft + \qquad -mg\frac{L}{2}\theta - (ka\theta)a - (cL\dot{\theta})L = I_O\ddot{\theta}$$

which, on rearranging the terms, becomes

$$I_O\ddot{\theta} + cL^2\dot{\theta} + \left(ka^2 + \frac{mgL}{2}\right)\theta = 0 \qquad \textit{Answer} \quad \text{(a)}$$

Part 2

Comparing Eq. (a) with Eq. (20.44), we conclude that

$$M = I_O \qquad C = cL^2 \qquad K = ka^2 + \frac{mgL}{2}$$

Therefore, the undamped circular frequency is [see Eq. (20.45d)]

$$p = \sqrt{\frac{K}{M}} = \sqrt{\frac{ka^2 + (mgL/2)}{I_O}}$$

With $I_O = mL^2/3 = 12(0.8)^2/3 = 2.560$ kg · m², this becomes

$$p = \sqrt{\frac{80(0.4)^2 + [12(9.81)(0.8)/2]}{2.560}} = 4.837 \text{ rad/s}$$

From Eq. (20.45d) we also obtain $C_{cr} = 2Mp$, which in our case becomes $c_{cr}L^2 = 2I_Op$, yielding

$$c_{cr} = \frac{2I_Op}{L^2} = \frac{2(2.560)(4.837)}{(0.8)^2} = 38.70 \text{ N} \cdot \text{s/m}$$

Since the given damping constant $c = 20$ N · s/m is less than c_{cr}, we conclude that the bar is *underdamped*. The damping factor is $\zeta = c/c_{cr} = 20/38.70 = 0.5168$, and the damped circular frequency is

$$\omega_d = p\sqrt{1-\zeta^2} = 4.837\sqrt{1-(0.5168)^2} = 4.14 \text{ rad/s}$$

PROBLEMS

20.60 Determine the natural frequency of the uniform circular sector in terms of R, α, and g. Assume oscillations of small amplitude about the pin at O.

20.61 The rectangular homogeneous plate is suspended from two identical wires. Calculate the period of vibration if the plate is released from rest in the position shown, where θ is a small angle.

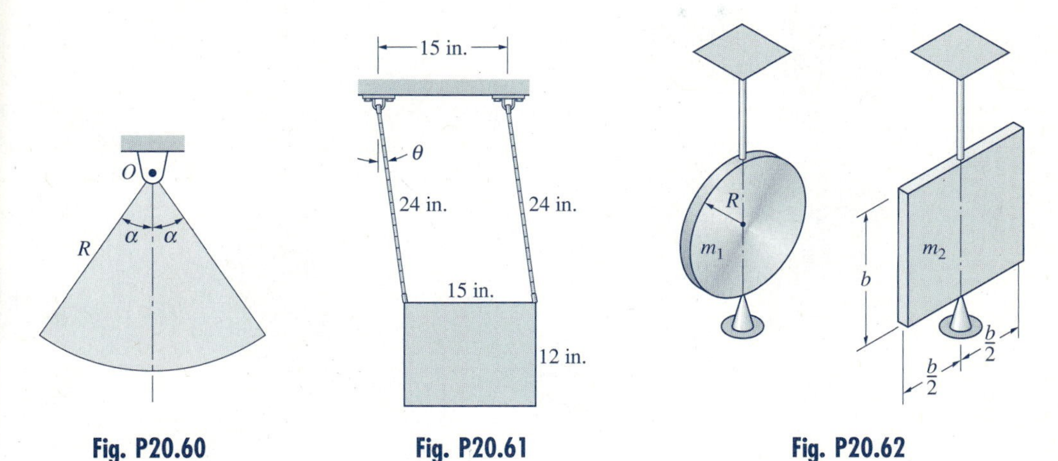

Fig. P20.60 Fig. P20.61 Fig. P20.62

20.62 Each of the two thin homogeneous plates is attached to an identical elastic rod that acts as a linear torsion spring (i.e., the restoring torque is proportional to the angular displacement). If the frequency of torsional oscillation of the circular plate of mass m_1 is f_1, find the frequency f_2 of the square plate of mass m_2.

20.63 The thin ring of radius R and the bent wire of length $2b$ are suspended from rough pegs at A and B, respectively. Determine the ratio b/R for which the two bodies would have the same period of oscillation for small amplitudes.

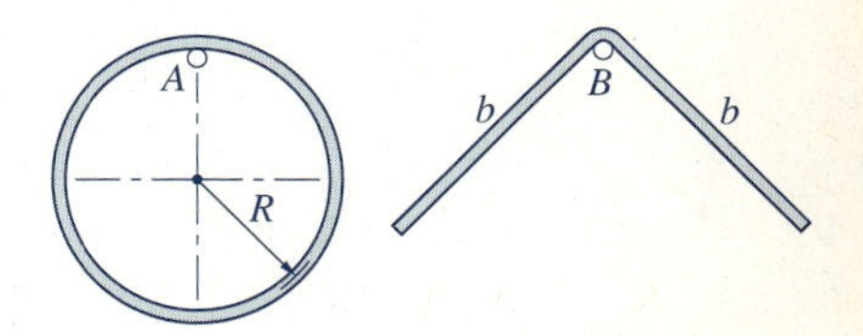

Fig. P20.63

20.64 A rigid body that is suspended from a pin at A is displaced slightly from its equilibrium position and released. (a) Show that the circular frequency of the resulting vibration is

$$p = \sqrt{\frac{gy}{\bar{k}^2 + y^2}}$$

where y is the distance from A to the mass center G and $\bar{k}$ is the radius of gyration of the body about G. (b) Determine the largest possible circular frequency and the corresponding value of y.

20.65 When the pin is at A, the period for small oscillations of the pendulum is 1.066 s. When the pin is at B, the period changes to 0.984 s. Using the equation given in Prob. 20.64, calculate (a) the distance d between A and the mass center G of the pendulum; and (b) the radius of gyration of the pendulum about G.

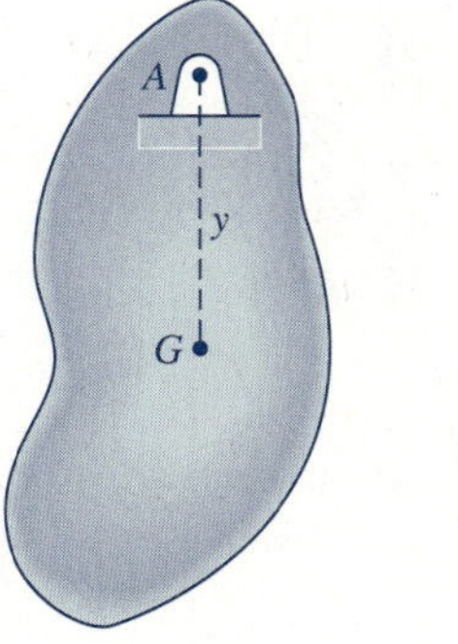

Fig. P20.64

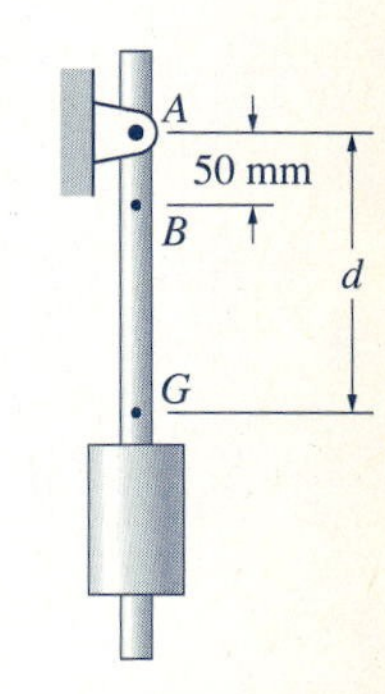

Fig. P20.65

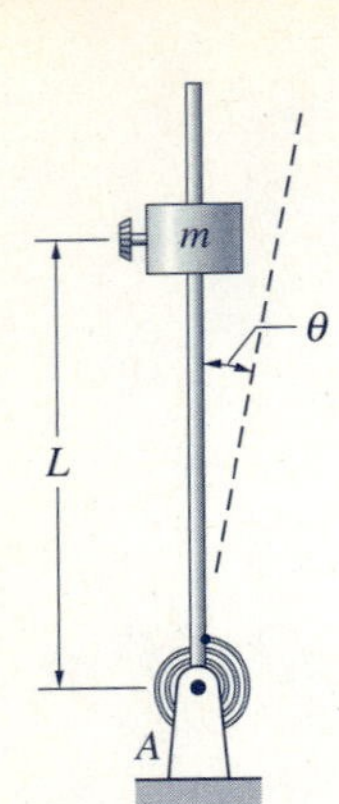

Fig. P20.66, P20.67

20.66 The metronome consists of the mass $m = 0.12$ kg attached to a light rod that is pivoted at A. The torsional spring at A applies the restoring couple $M = -k\theta$ to the rod, where $k = 0.25$ N · m/rad and θ is the angular displacement of the rod as shown. The period of the metronome can be adjusted by changing the length L between the mass and pivot A. (a) Determine the period for small amplitudes if $L = 180$ mm. (b) What is the largest L for which the metronome will oscillate?

20.67 (a) Show that the equation of motion for the metronome of length $L = 180$ mm in Prob. 20.66 is $\ddot{\theta} = -(64.3\theta - 54.5 \sin \theta)$ rad/s^2. (b) Use numerical integration to find the period of the metronome if the amplitude is $\theta_{\max} = \pi/4$ rad.

20.68 The uniform bar of length 0.75 m rests in a smooth circular trough of radius 0.5 m. If the bar is displaced slightly from the equilibrium position and released, calculate the natural frequency of the ensuing oscillation.

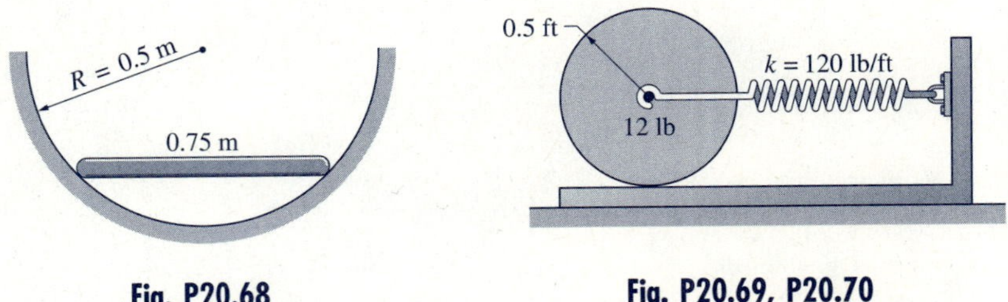

Fig. P20.68

Fig. P20.69, P20.70

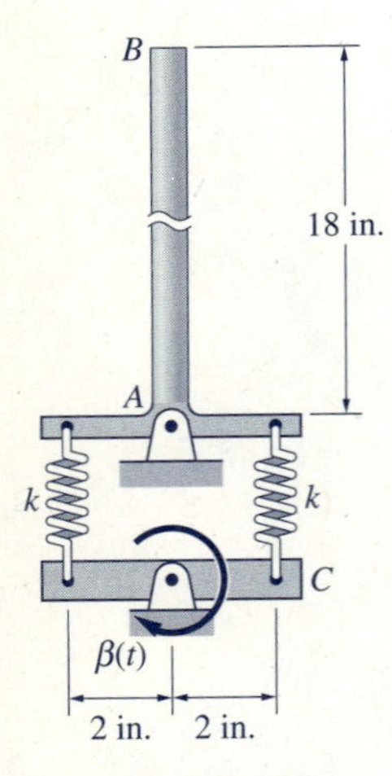

Fig. P20.71

20.69 When the equilibrium of the 12-lb uniform disk is disturbed, it rolls back and forth on the rigid base without slipping. If the maximum displacement of the center of the disk from its equilibrium position is 1.0 in., find (a) the circular frequency of the oscillation; and (b) the minimum coefficient of friction between the disk and the base.

20.70 The 12-lb uniform disk rolls back and forth on the base without slipping. If the base undergoes the horizontal displacement $y(t) = 3 \sin 6t$ ft, determine the maximum steady-state displacement of the center of the disk relative to the base.

20.71 When the 1.8-lb uniform rod AB is in the vertical position, the two springs of stiffness $k = 100$ lb/ft are unstretched. The weight of the horizontal bar attached to the rod at A is negligible. If the support C undergoes the harmonic angular displacement $\beta(t) = \beta_0 \sin \omega t$, where $\beta_0 = 2°$ and $\omega = 7.5$ rad/s, find the steady-state angular amplitude of bar AB.

20.72 Determine the damping coefficient c for which the uniform 24-kg bar would be critically damped for small oscillations.

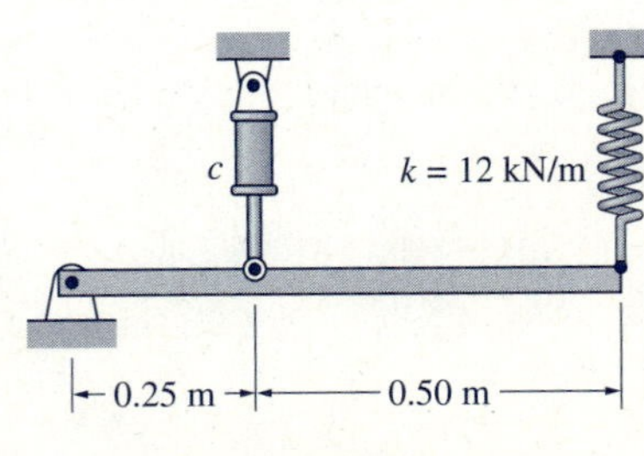

Fig. P20.72

20.73 The mass C is attached to the light arm ABC, which is pinned at A and supported by a spring at B. The spring tension is adjusted so that the system is in equilibrium in the position shown. If the base undergoes the vertical displacement $y(t) = Y \sin \omega t$, where $Y = 1.0$ mm and $\omega = 30$ rad/s, determine the maximum steady-state displacement of point B relative to the base.

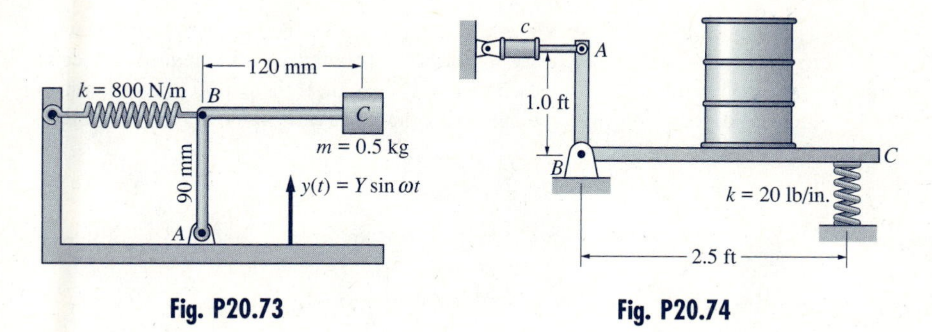

Fig. P20.73 **Fig. P20.74**

20.74 The platform, supported by a pin at B and a spring at C, is in equilibrium in the position shown. When the viscous damper at A is disconnected, the frequency of the system for small amplitudes is 2.52 Hz. Determine the damping coefficient c that would critically damp the system.

20.75 The 12-kg uniform bar AOB is pinned to the carriage at O. The carriage is undergoing the harmonic displacement $y(t) = Y \sin \omega t$, where $Y = 10$ mm and $\omega = 2$ rad/s. Calculate the steady-state angular amplitude of the bar for small amplitudes.

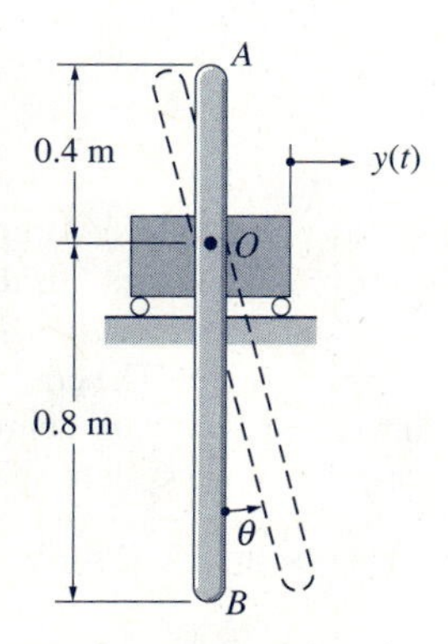

Fig. P20.75, P20.76

20.76 The 12-kg uniform bar AOB is pinned to the carriage at O. The bar is at rest in the vertical position at $t = 0$ when the harmonic displacement $y(t) = Y \sin pt$ is imposed on the carriage, where p is the resonant frequency for small amplitudes and $Y = 10$ mm. (a) Show that the equation of motion for the bar is

$$\ddot{\theta} = -12.263 \sin \theta + 0.153\,31 \cos \theta \sin 3.502t \text{ rad/s}^2$$

(b) Integrate the equation of motion numerically from $t = 0$ to $t = 50$ s, and plot θ versus t. (c) By inspection of the plot, estimate the maximum value of θ.

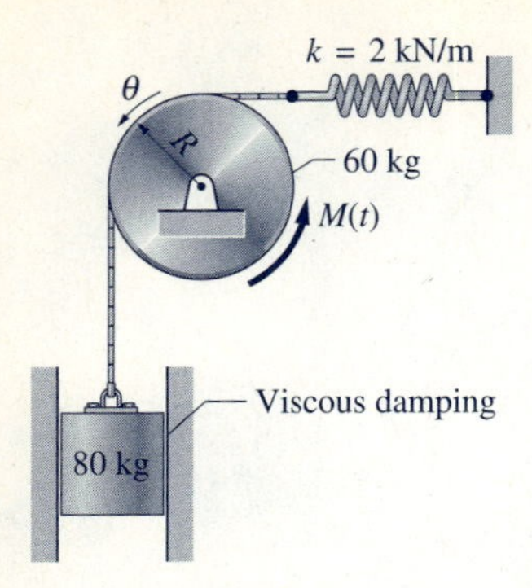

Fig. P20.77

20.77 The radius of the 60-kg uniform disk is $R = 500$ mm. The space between the 80-kg mass and the vertical slot is lubricated, providing viscous damping with a damping factor of $\zeta = 0.15$. If the couple $M(t) = M_0 \sin \omega t$ is acting on the disk, where $M_0 = 40$ N · m and $\omega = 5$ rad/s, determine the steady-state angular displacement $\theta(t)$ of the disk. Assume that the cable remains taut and does not slip on the disk.

20.78 The 6-lb uniform disk oscillates about the pin at A. The pin is lubricated with heavy grease, which gives rise to rotational resistance equivalent to a viscous damping couple $M_d = -c\dot{\theta}$, where c is a constant. If the period of oscillation is measured to be 1.468 s, calculate (a) the damping ratio; and (b) the constant c.

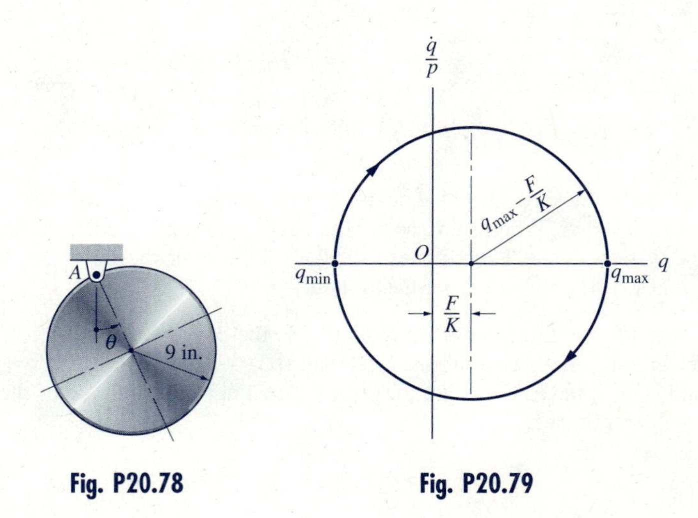

Fig. P20.78

Fig. P20.79

20.79 The equation of motion for an undamped system with constant external forces, from Eq. (20.44), is $M\ddot{q} + Kq = F$, where F is the constant generalized force. Prove that the plot of q versus $\dot{q}/p$ is the circle shown, where p is the natural circular frequency of the system. (Note: The circle is called the *phase plane* plot of the system. Plots of this type can be very helpful in describing the behavior of complex, single-degree-of-freedom systems. See, for example, Prob. 20.80.)

***20.80** The equation of motion for a system that is subjected to Coulomb damping (dry friction) of constant magnitude is

$$M\ddot{q} + Kq = \begin{cases} -F & \text{if } \dot{q} > 0 \\ F & \text{if } \dot{q} < 0 \end{cases}$$

where F is the generalized constant friction force. Assume that the system is released from rest at $q = q_0$, and let $F = 0.15Kq_0$. (a) Draw the phase plane plot for the system. (Hint: See Prob. 20.79.) (b) From the phase plane plot, determine the reduction of amplitude in each cycle of oscillation, and the point where the system comes to rest.

*20.7 Energy Method and Rayleigh's Principle

Here we examine two methods that are convenient for analyzing undamped, free vibrations with a single degree of freedom: the energy method and Rayleigh's principle. Both techniques are based on the principle of conservation of mechanical energy. The energy method is a means of obtaining the equation of motion, whereas Rayleigh's principle allows us to calculate the natural frequency without having to derive the equation of motion.

a. Energy method

If the forces that do work on a mechanical system are conservative (see Art. 14.4), such as spring forces and weights, the total mechanical energy—i.e., the sum of the kinetic energy T and the potential energy V—will remain constant: $T + V = \text{constant}$. Differentiation with respect to time yields

$$\frac{d}{dt}(T + V) = 0 \tag{20.45}$$

If the kinetic and potential energies are expressed as functions of an appropriate position coordinate (linear or angular), performing the differentiation indicated in Eq. (20.45) will yield the equation of motion. Denoting this position coordinate by q, the potential energy of a linear system has the form*

$$V(q) = V_0 + F_0 q + \frac{1}{2}Kq^2 \tag{20.46}$$

where V_0, F_0, and K are constants. The units of V_0 are the units of energy, whereas the units of F_0 and K depend on whether q is a linear or angular coordinate.

Henceforth we assume that $q = 0$ corresponds to the equilibrium position of the system. Then we see from Eq. (20.46) that V_0 represents the potential energy of the system in the equilibrium position. Furthermore, since $dV/dq = 0$ at equilibrium—see Eq. (10.17)—we conclude that $F_0 = 0$ (the static equilibrium equation for the system), and the potential energy becomes

$$V(q) = V_0 + \frac{1}{2}Kq^2 \tag{20.47}$$

The kinetic energy of a body in an arbitrary position has the form

$$T = \frac{1}{2}M\dot{q}^2 \tag{20.48}$$

where the constant M corresponds to mass if q is a linear coordinate or moment of inertia if q is an angular coordinate.

* The potential energy of a linear system is always a quadratic form in the generalized coordinate.

Substituting Eqs. (20.47) and (20.48) into Eq. (20.45) gives

$$\frac{d}{dt}\left[(V_0 + \frac{1}{2}Kq^2) + \frac{1}{2}M\dot{q}^2\right] = 0$$

which becomes

$$\dot{q}(M\ddot{q} + Kq) = 0$$

Discarding the trivial solution $\dot{q} = 0$, we obtain the equation of motion

$$\boxed{M\ddot{q} + Kq = 0} \tag{20.49}$$

The natural circular frequency of the system can be found from Eq. (20.49) to be $p = \sqrt{K/M}$.

In summary, the procedure for deriving the equation of motion using the energy method is as follows.

1. Choose a coordinate q that describes the position of the rigid body, noting that q must equal zero in the equilibrium position.
2. Write the expression for the potential energy in an arbitrary position. When making small-displacement approximations, remember that quadratic as well as linear terms in q must be kept. Compare the potential energy with $V = V_0 + F_0 q + (1/2)Kq^2$, and identify the constant K.
3. Write the expression for the kinetic energy in an arbitrary position. By comparing it with $T = (1/2)M\dot{q}^2$, identify the constant M.
4. Write the equation of motion: $M\ddot{q} + Kq = 0$.

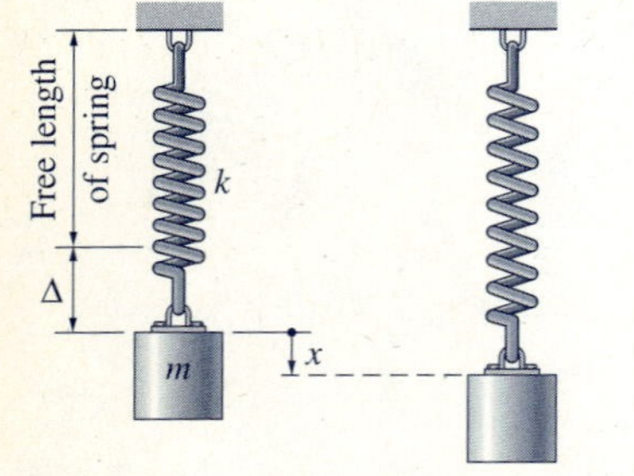

Fig. 20.14

As a simple illustration of the energy method, reconsider the undamped spring-mass system shown in Fig. 20.14. Note that the position coordinate x is measured from its equilibrium position. Letting $x = 0$ be the datum position, the potential energy in an arbitrary position is $V = -mgx + (1/2)k(x + \Delta)^2 = (1/2)k\Delta^2 - (mg - k\Delta)x + (1/2)kx^2$, where Δ is the static deflection of the spring. The kinetic energy is $T = (1/2)k\dot{x}^2$. Identifying x with q and comparing the expressions for V and T with Eqs. (20.46) and(20.48), we find that $F_0 = -mg + k\Delta$, $K = k$, $V_0 = \frac{1}{2}k\Delta^2$, and $M = m$. The equation of motion, Eq. (20.49), thus becomes $m\ddot{x} + kx = 0$, as expected. Note that $F_0 = 0$ yields $mg = k\Delta$, which is the static equilibrium equation of the system.

b. Rayleigh's principle

The energy method is sometimes tedious due to the geometric difficulties encountered when deriving the energy of the system in an arbitrary position. In particular, the kinematic analysis required for computation of kinetic energy may become quite complex (the same difficulties would be encountered in the conventional analysis using the free-body diagram and Newton's second law). Therefore, we present here another method of calculating the natural frequency, called *Rayleigh's principle.* It requires the computation of the energy at specific positions of the system, which is usually easier than using an arbitrary position.

Rayleigh's principle is essentially a restatement of the principle of conservation of mechanical energy. We assume that the system is linear, has a single degree of freedom, and $q = 0$ at the equilibrium position. Since we consider only conservative systems (no damping and no forcing function), $T + V =$ constant, which leads to the following conclusions concerning the maximum potential and kinetic energies.

1. The maximum kinetic energy T_{max} occurs at the position where the potential energy V is minimum. But according to Eq. (20.47), $V_{min} = V_0$ at $q = 0$ (the equilibrium position). Therefore, we conclude that T_{max} occurs at the equilibrium position, its value being $T_{max} = (1/2)M\dot{q}^2_{max}$, where $\dot{q}_{max}$ is the maximum velocity.
2. By inspection of Eq. (20.47) it is evident that the maximum potential energy V_{max} occurs when $q = q_{max}$; consequently, $\dot{q} = 0$. Therefore, $T = T_{min} = 0$ in this position.

Conservation of mechanical energy requires that $T_{max} + V_{min} = T_{min} + V_{max}$. Since we have found that $T_{min} = 0$ and $V_{min} = V_0$, the energy balance becomes

$$\boxed{T_{max} = V_{max} - V_0} \tag{20.50}$$

which is *Rayleigh's principle.* If the datum of the potential energy is chosen such that $V_0 = 0$, Rayleigh's principle assumes the more conventional form

$$T_{max} = V_{max} \tag{20.51}$$

If the motion is simple harmonic, we have $q = A\sin(pt + \alpha)$. We now deduce that $q_{max} = A$ and $\dot{q}_{max} = Ap$; i.e.,

$$\boxed{\dot{q}_{max} = pq_{max}} \tag{20.52}$$

This result can be substituted in $T_{max} = \frac{1}{2}m\dot{q}^2_{max}$, enabling us to solve Eq. (20.50) or (20.51) for the natural circular frequency p.* Sample Problem 20.9 illustrates the application of the method.

* The proof of this statement follows from the fact that $T_{max} = (1/2)M\dot{q}^2_{max} = 1/2M(pq_{max})^2$ and $V_{max} = V_0 + (1/2)Kq^2_{max}$. Substituting into Eq. (20.50) yields $(1/2)Mpq^2_{max} = [V_0 + (1/2)Kq^2_{max}] - V_0$, from which $p = \sqrt{K/M}$, which we recognize as the natural circular frequency.

Sample Problem 20.8

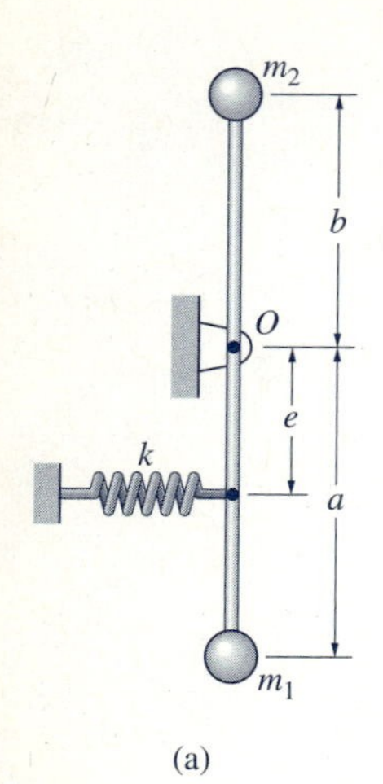

(a)

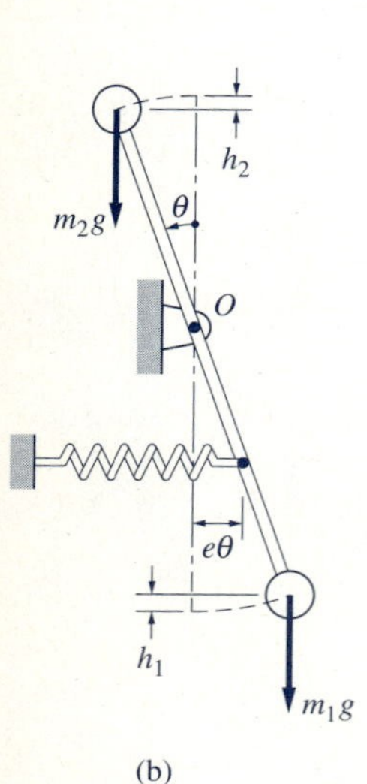

(b)

The rigid body in Fig. (a) consists of the small masses m_1 and m_2, which are connected to the ends of a light bar that can rotate about a pin at O. The ideal spring is undeformed when the bar is vertical. Determine the circular frequency of small vibrations using the energy method.

Solution

Figure (b) shows the body when it has rotated through a small angle θ. Since O is a fixed point, the kinetic energy of the bar (as discussed in Art. 18.3) may be written as

$$T = \frac{1}{2}I_O\dot{\theta}^2 \tag{a}$$

where $I_O = m_1a^2 + m_2b^2$ is the moment of inertia of the body about O.

When computing the gravitational potential energy V_g, we note from Fig. (b) that the lower ball has moved upward through the distance $h_1 = a(1 - \cos\theta)$. Replacing $\cos\theta$ by the series $\cos\theta = 1 - (\theta^2/2) + (\theta^4/24) - \ldots$ and keeping the first two terms (recall that all terms up to the quadratic must be kept), we have $h_1 = a\theta^2/2$, valid for small angles. A similar argument can be used to show that $h_2 = b\theta^2/2$. Letting the vertical equilibrium position in Fig. (a) be the datum position for gravitational potential energy, we obtain

$$V_g = m_1gh_1 - m_2gh_2 = (m_1ga - m_2gb)\frac{\theta^2}{2}$$

From Fig. (b) we see that the elongation of the spring for small angles is $e\theta$, which means that the elastic potential energy is $V_e = (1/2)k(e\theta)^2$. The total potential energy is

$$V = V_g + V_e = \left(m_1ga - m_2gb + ke^2\right)\frac{\theta^2}{2} \tag{b}$$

The condition for conservation of energy, $d(T + V)/dt = 0$, now becomes

$$\frac{d}{dt}\left\{\frac{1}{2}\left[I_O\dot{\theta}^2 + \left(m_1ga - m_2gb + ke^2\right)\theta^2\right]\right\} = 0$$

$$\dot{\theta}\left[I_O\ddot{\theta} + \left(m_1ga - m_2gb + ke^2\right)\theta\right] = 0$$

After discarding the trivial solution $\dot{\theta} = 0$, we obtain the following equation of motion.

$$I_O\ddot{\theta} + \left(m_1ga - m_2gb + ke^2\right)\theta = 0$$

Comparing this equation with the equation for undamped particle vibration, Eq. (20.49), we see that $M = I_O$ and $K = m_1ga - m_2gb + ke^2$. Therefore, the circular frequency of the bar is $p = \sqrt{K/M}$, or

$$p = \sqrt{\frac{m_1ga - m_2gb + ke^2}{I_O}} \qquad \textit{Answer}$$

If $m_1ga + ke^2 < m_2gb$, the value obtained for p is imaginary. For this condition, the body will not vibrate about the position in Fig. (a) because it is an unstable equilibrium position. Note that the moment that tends to return the body to its equilibrium position is $m_1g(a\theta) + ke(e\theta)$, and the moment that tends to move the body away from the equilibrium position is $m_2g(b\theta)$. For the equilibrium position to be stable, the first of these moments must be larger than the second.

Sample Problem 20.9

Figure (a) shows a homogeneous semicylinder that rocks back and forth without slipping on the horizontal plane. The angular position of the semicylinder is defined by the angle θ. Determine the natural circular frequency of the oscillations for small amplitudes, assuming the motion to be simple harmonic. The mass center G is located at the distance $4R/3\pi = 0.4224R$ from point O.

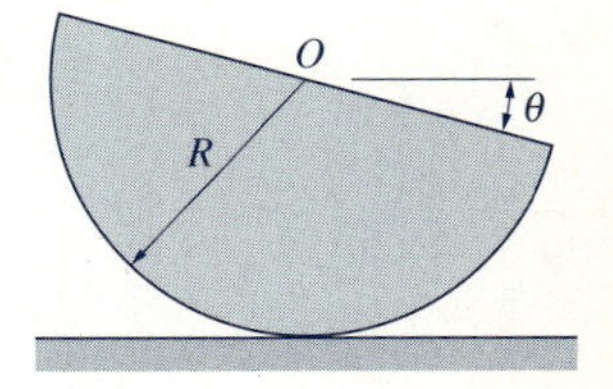

(a)

Solution

The natural circular frequency of the rocking motion can be obtained by using either the energy method or Rayleigh's principle. The energy method requires that the kinetic energy be obtained for an arbitrary position of the body, a fairly complicated task for this problem. It is easier to employ Rayleigh's principle, for which only the position of maximum kinetic energy needs to be considered.

According to Eq. (20.50), Rayleigh's principle is $T_{max} = V_{max} - V_0$, where T_{max} and V_{max} are the maximum kinetic and potential energies and V_0 is the potential energy in the equilibrium position.

Figure (b) shows the body as it passes through the equilibrium position. The potential energy of the body is gravitational. If we choose the horizontal surface as the datum, the potential energy is

$$V_0 = mgh_1 = 0.5756mgR \tag{a}$$

As discussed in Art. 20.7, the maximum kinetic energy and the minimum potential energy occur in the equilibrium position. This implies that the angular velocity is also maximum; i.e., $\dot{\theta} = \dot{\theta}_{max}$ in the equilibrium position. Noting that the contact point C in Fig. (b) is the instant center for velocity, the kinetic energy is $T_{max} = (1/2)I_C\dot{\theta}^2_{max}$. Using $I_O = mR^2/2$ and the parallel-axis theorem, it can be shown that $I_C = 0.6512mR^2$. Therefore, the kinetic energy becomes

$$\begin{aligned} T_{max} &= \frac{1}{2}I_C\dot{\theta}^2_{max} = \frac{1}{2}(0.6512mR^2)\dot{\theta}^2_{max} \\ &= 0.3256mR^2\dot{\theta}^2_{max} \end{aligned} \tag{b}$$

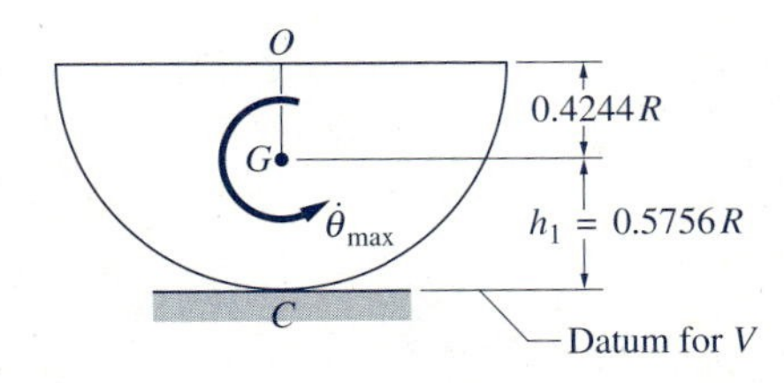

(b) Passing through equilibrium position ($V = V_0$, $T = T_{max}$)

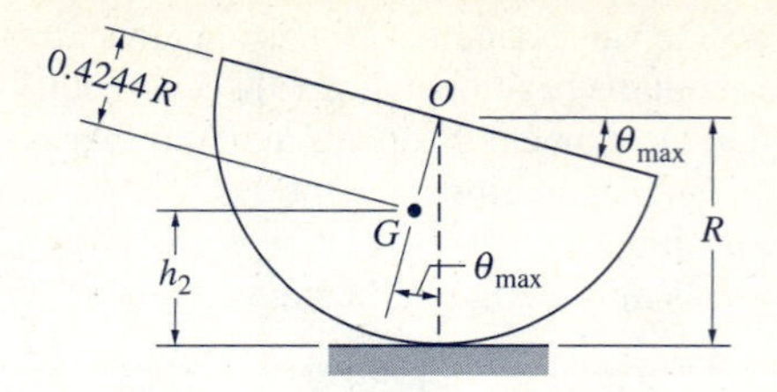

(c) Position of maximum displacement ($V = V_{max}$, $T = 0$)

Figure (c) shows the semicylinder when θ has reached its maximum value θ_{max}. This position corresponds to zero kinetic energy ($\dot{\theta} = 0$) and maximum potential energy. The potential energy is $V_{max} = mgh_2 = mg(R - 0.4244R\cos\theta_{max})$.

Using $\cos\theta_{max} \approx 1 - (1/2)\theta_{max}^2$, valid for small amplitudes, we obtain

$$V_{max} = mgR(0.5756 + 0.2122\theta_{max}^2) \tag{c}$$

Rayleigh's principle, $T_{max} = V_{max} - V_0$, now becomes

$$0.3256mR^2\dot{\theta}_{max}^2 = mgR(0.5756 + 0.2122\theta_{max}^2) - 0.5756mgR$$

which reduces to

$$0.3256R\dot{\theta}_{max}^2 = 0.2122g\theta_{max}^2 \tag{d}$$

If the motion is assumed to be simple harmonic,* then $\dot{\theta}_{max} = p\theta_{max}$—see Eq. (20.52), and Eq. (d)—becomes

$$0.3256Rp^2\theta_{max}^2 = 0.2122g\theta_{max}^2$$

yielding

$$p = 0.807\sqrt{g/R} \qquad \textit{Answer}$$

* The motion can be shown to be simple harmonic for small θ, but it is not easy—see, e.g., *Engineering Mechanics,* W. W. Hagerty and H. J. Plass, Jr., D. Van Nostrand Company, Inc., 1967, p. 416.

PROBLEMS

20.81 Determine the circular frequency of motion for the particle in Prob 20.9 by the energy method.

20.82 Solve Prob. 20.61 by the energy method.

20.83 Solve Prob. 20.66 by the energy method.

20.84 The homogenous semicircular hoop rocks back and forth on the horizontal surface without slipping. Determine the natural circular frequency of oscillations for small amplitudes.

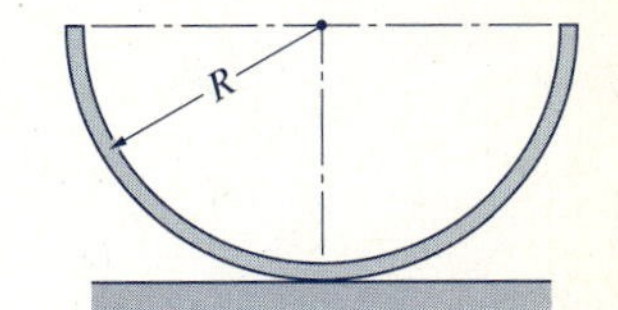

Fig. P20.84

20.85 Pulley B can be approximated as a uniform disk, and the mass of pulley C can be neglected. Determine the expression for the natural frequency of the system.

20.86 The hoop to which the pendulum is rigidly attached rolls without slipping on the horizontal surface. Neglecting the masses of the hoop and the rod, calculate the period for small oscillations about the equilibrium position shown.

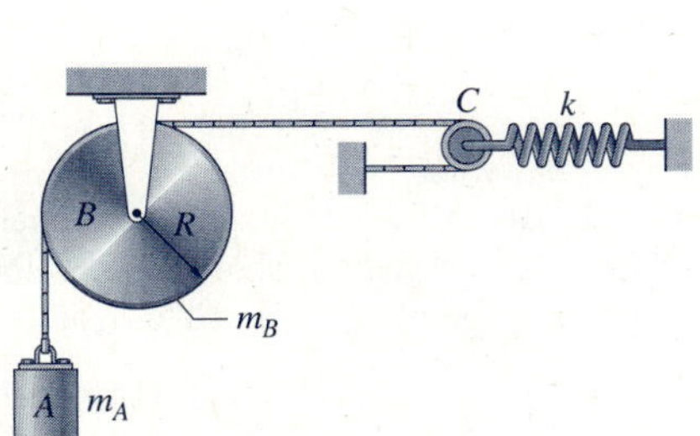

Fig. P20.85

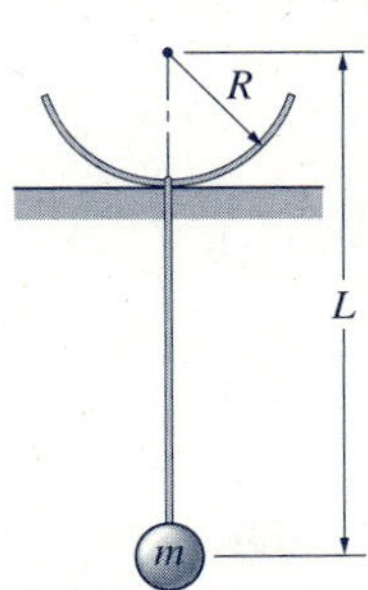

Fig. P20.86

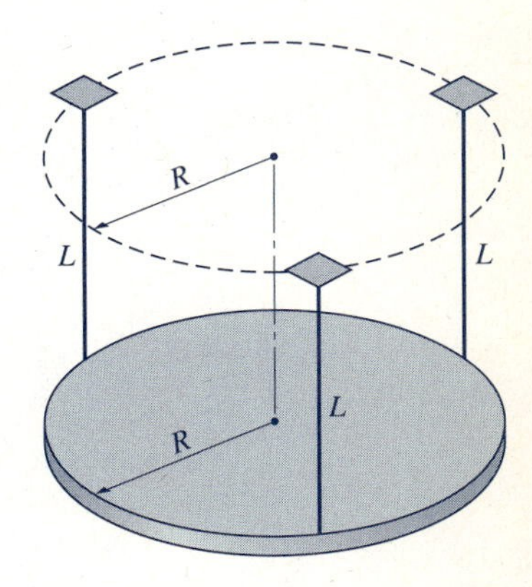

Fig. P20.87

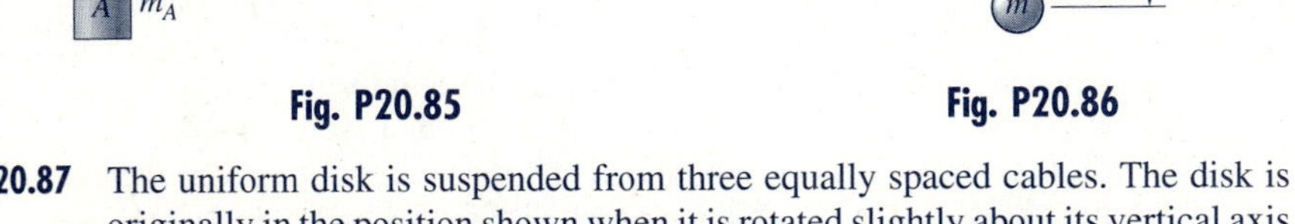

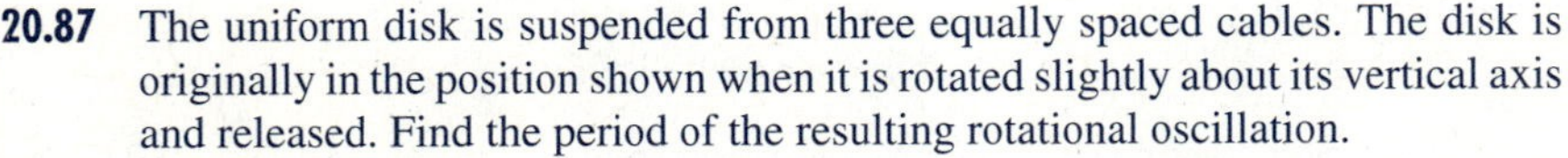

20.87 The uniform disk is suspended from three equally spaced cables. The disk is originally in the position shown when it is rotated slightly about its vertical axis and released. Find the period of the resulting rotational oscillation.

20.88 The pendulum consists of a uniform disk attached to a rod of negligible weight. Determine the two values of the distance L for which the period of the pendulum is 1.0 s for small amplitudes.

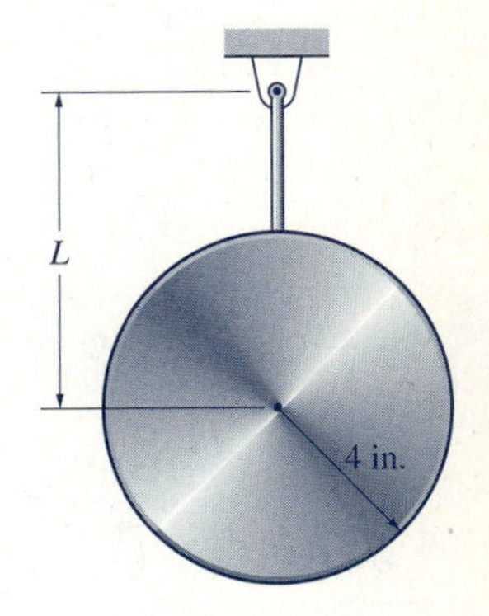

Fig. P20.88

20.89 The uniform slender bar AB is in equilibrium in the position shown. Calculate the period of small oscillations.

20.90 If the uniform slender bar is in equilibrium in the position shown, calculate the period of vibration for small amplitudes.

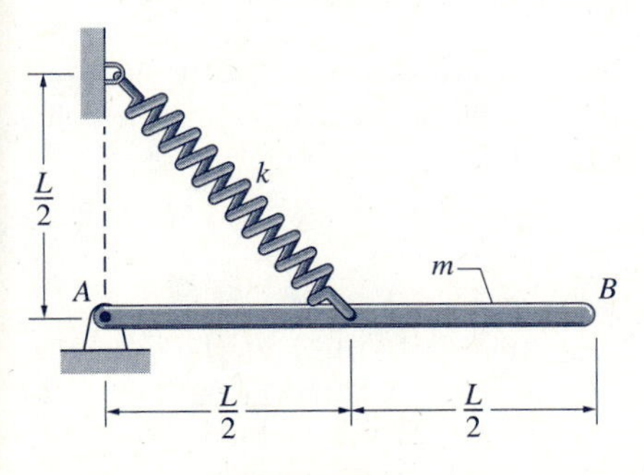

Fig. P20.89

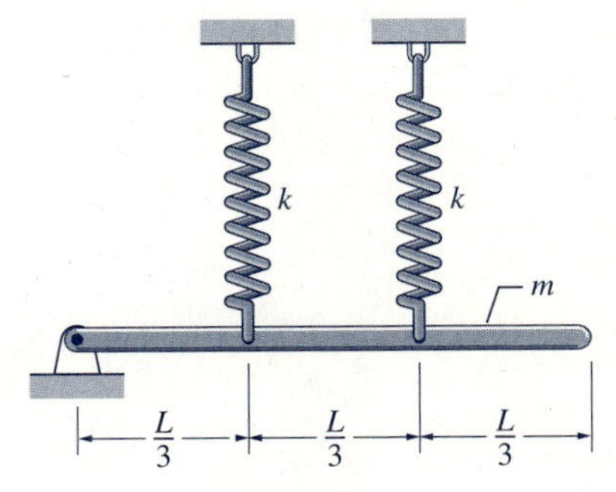

Fig. P20.90

20.91 The small mass m is attached to the end of the L-shaped rod of negligible mass. The spring tension is adjusted so that the system is in equilibrium in the position shown. Determine the expression for the natural frequency of the system assuming small amplitudes. Assume that the plane of the figure is vertical.

20.92 Solve Prob. 20.91 assuming that the plane of the figure is horizontal.

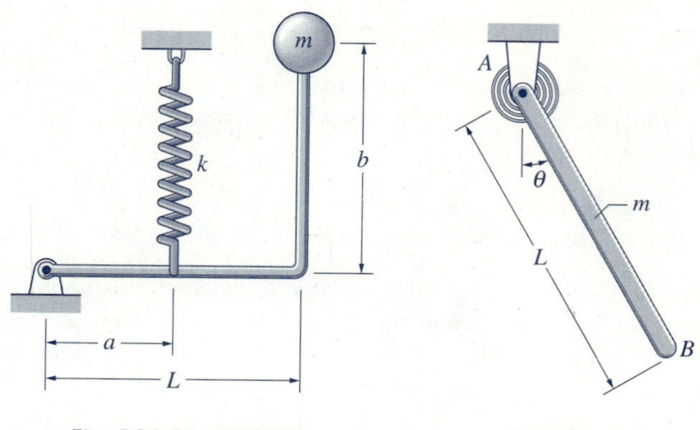

Fig. P20.91, P20.92 **Fig. P20.93**

20.93 The torsional spring at end A of the uniform slender bar is adjusted so that the bar is in equilibrium in the position $\theta = \theta_0$, where θ_0 is not necessarily small. If the rotational stiffness of the spring is k (torque/rad), determine the expression for the natural frequency of the bar for small oscillations about the equilibrium position. Assume that the plane of the figure is (a) vertical; (b) horizontal.

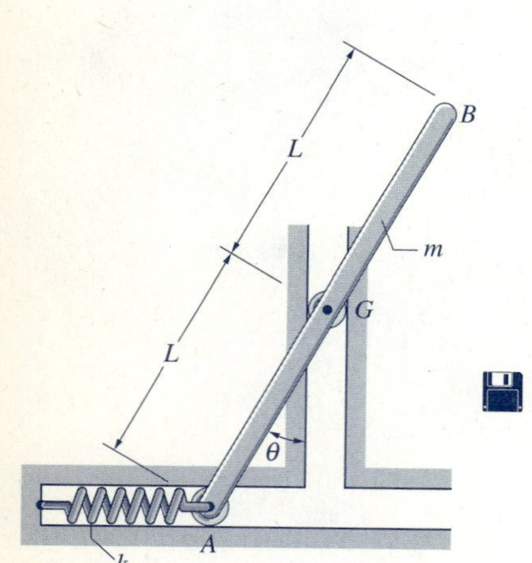

Fig. P20.94, P20.95

20.94 The rollers at A and G are pinned to the uniform bar AB and move freely in fixed slots as shown. The spring at A is initially unstretched, and the bar is at rest when $\theta = 0$. (a) Derive the equation of motion for the bar, assuming that the angle θ remains small. (b) Compute the period of oscillation if $m = 15$ kg, $L = 1.2$ m, and $k = 490.5$ N/m. (c) Using the values of L and k given in part (b), find the largest value of m for which the angle θ will remain small.

20.95 (a) Show that the equation of motion for the bar described in Prob. 20.94 is

$$\ddot{\theta} = -\left[\left(\dot{\theta}^2 + \frac{k}{m}\right)\cos\theta - \frac{2g}{L}\right]\frac{3\sin\theta}{1 + 3\sin^2\theta}$$

(b) Use numerical integration with $m = 15$ kg, $L = 1.2$ m, and $k = 490.5$ N/m to calculate the period of oscillation if the amplitude is (i) small; (ii) 30°.

20.96 The slender uniform rod is balanced on a rough cylindrical surface as shown. If the rod is tipped slightly and released, determine the frequency of the resulting oscillation.

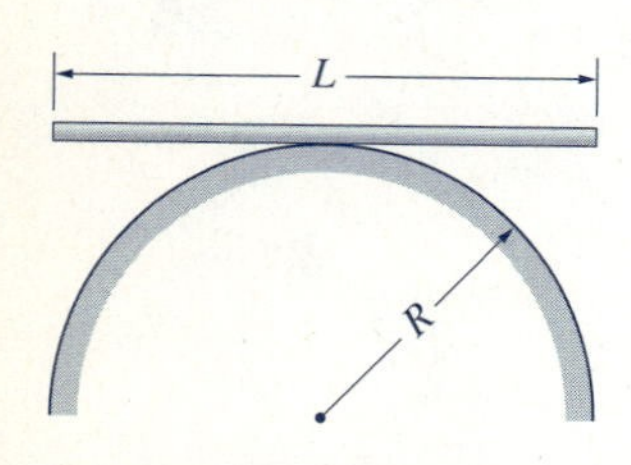

Fig. P20.96, P20.97

20.97 The slender rod described in Prob. 20.96 is released from rest with an initial angle of tilt $\theta = \theta_0$ where θ is measured from the horizontal. (a) Show that the resulting motion is described by

$$\ddot{\theta} = -\frac{12R\theta(g\cos\theta + R\dot{\theta}^2)}{L^2 + 12R^2\theta^2}$$

Assume that the rod does not slide on the cylindrical surface. (b) Integrate the equation of motion for at least one period using the following data: $L = 9$ ft, $R = 3$ ft, $\theta_0 = 15°$. (c) Linearize the equation of motion and repeat part (b). (d) Using the same axes, plot θ versus time for the solutions obtained in parts (b) and (c).

D Proof of the Relative Velocity Equation for Rigid-Body Motion

Here we prove Eq. (16.1), $\mathbf{v}_{B/A} = \boldsymbol{\omega} \times \mathbf{r}_{B/A}$, for rigid-body motion—where $\boldsymbol{\omega}$ is called the *angular velocity* of the body. In a three-dimensional setting, this result is neither intuitively obvious, nor is its proof trivial.

Figure D.1 shows four points A, B, C, and D belonging to the same rigid body. These points may be chosen arbitrarily, except for the following restrictions: (1) the reference point A must not lie on the line BC or on the line CD, and (2) the four points must not lie in the same plane.

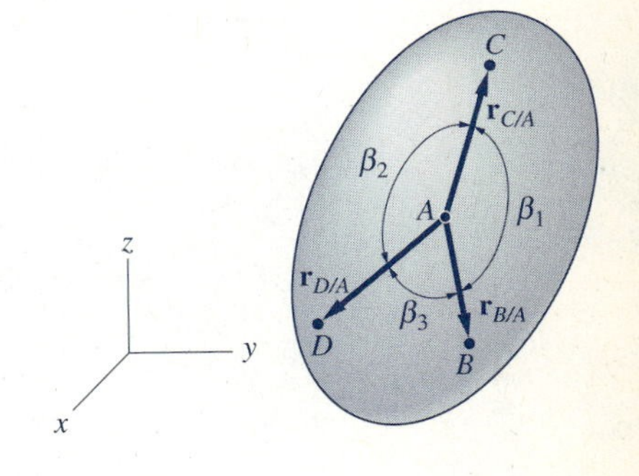

Fig. D.1

The rigidity of the body imposes the following constraints on the motion: (1) the lengths of the relative position vectors shown in Fig. D.1 remain constant, and (2) the angles β_1, β_2, and β_3 between the relative position vectors remain constant. It will be demonstrated that these conditions can be satisfied only if the relative velocity between any two points on the body has the form of Eq. (16.1).

Consider first the requirement that the magnitude of vector $\mathbf{r}_{B/A}$ is constant: $\mathbf{r}_{B/A} \cdot \mathbf{r}_{B/A} = |\mathbf{r}_{B/A}|^2 =$ constant. Taking the time derivative of both sides of this equation, we get $(\mathbf{r}_{B/A} \cdot \dot{\mathbf{r}}_{B/A}) + (\dot{\mathbf{r}}_{B/A} \cdot \mathbf{r}_{B/A}) = 0$, or $2\mathbf{r}_{B/A} \cdot \mathbf{v}_{B/A} = 0$. A similar argument can be applied to $\mathbf{r}_{C/A}$ and $\mathbf{r}_{D/A}$. Consequently, we obtain

$$\begin{aligned} \mathbf{r}_{B/A} \cdot \mathbf{v}_{B/A} &= 0 \\ \mathbf{r}_{C/A} \cdot \mathbf{v}_{C/A} &= 0 \\ \mathbf{r}_{D/A} \cdot \mathbf{v}_{D/A} &= 0 \end{aligned} \tag{D.1}$$

The first of these equations is satisfied if either $\mathbf{v}_{B/A} = \mathbf{0}$ or if $\mathbf{r}_{B/A}$ is perpendicular to $\mathbf{v}_{B/A}$. In either case, we conclude that the relative velocity vector must be of the form $\mathbf{v}_{B/A} = \boldsymbol{\omega}_1 \times \mathbf{r}_{B/A}$, where $\boldsymbol{\omega}_1$ is a vector. Observe that if $\boldsymbol{\omega}_1 = \mathbf{0}$ or if $\boldsymbol{\omega}_1$ is parallel to $\mathbf{r}_{B/A}$, we obtain $\mathbf{v}_{B/A} = \mathbf{0}$. Otherwise $\mathbf{v}_{B/A}$ is perpendicular to $\mathbf{r}_{B/A}$ (this follows from the properties of the cross product). Applying similar arguments to the remaining two cases in Eqs. (D.1), we obtain

$$\begin{aligned} \mathbf{v}_{B/A} &= \boldsymbol{\omega}_1 \times \mathbf{r}_{B/A} \\ \mathbf{v}_{C/A} &= \boldsymbol{\omega}_2 \times \mathbf{r}_{C/A} \\ \mathbf{v}_{D/A} &= \boldsymbol{\omega}_3 \times \mathbf{r}_{D/A} \end{aligned} \tag{D.2}$$

So far, there are no restrictions on the vectors $\boldsymbol{\omega}_1$, $\boldsymbol{\omega}_2$, and $\boldsymbol{\omega}_3$. Equations (D.2) are necessary and sufficient for the relative position vectors to remain constant during the motion of the body.

Consider next the requirement that the angle β_1, between $\mathbf{r}_{B/A}$ and $\mathbf{r}_{C/A}$, remains constant. Utilizing the properties of the dot product, we have $\mathbf{r}_{B/A} \cdot \mathbf{r}_{C/A} = |\mathbf{r}_{B/A}||\mathbf{r}_{C/A}| \cos \beta_1$. Taking the time derivative of this equation and imposing the condition that $|\mathbf{r}_{B/A}|$, $|\mathbf{r}_{C/A}|$, and β_1 are constants, we obtain

$$(\mathbf{r}_{B/A} \cdot \dot{\mathbf{r}}_{C/A}) + (\dot{\mathbf{r}}_{B/A} \cdot \mathbf{r}_{C/A}) = 0$$

or

$$(\mathbf{r}_{B/A} \cdot \mathbf{v}_{C/A}) + (\mathbf{r}_{C/A} \cdot \mathbf{v}_{B/A}) = 0$$

Similar arguments can also be applied to β_2 and β_3, the results being

$$\begin{aligned}(\mathbf{r}_{B/A} \cdot \mathbf{v}_{C/A}) + (\mathbf{r}_{C/A} \cdot \mathbf{v}_{B/A}) &= 0\\ (\mathbf{r}_{C/A} \cdot \mathbf{v}_{D/A}) + (\mathbf{r}_{D/A} \cdot \mathbf{v}_{C/A}) &= 0\\ (\mathbf{r}_{D/A} \cdot \mathbf{v}_{B/A}) + (\mathbf{r}_{B/A} \cdot \mathbf{v}_{D/A}) &= 0\end{aligned} \tag{D.3}$$

Using Eqs. (D.2) to eliminate the relative velocities from Eqs. (D.3), we obtain

$$(\mathbf{r}_{B/A} \cdot \boldsymbol{\omega}_2 \times \mathbf{r}_{C/A}) + (\mathbf{r}_{C/A} \cdot \boldsymbol{\omega}_1 \times \mathbf{r}_{B/A}) = 0$$
$$\text{(etc.)}$$

Using the properties of the scalar triple product, these equations simplify to

$$\begin{aligned}(\boldsymbol{\omega}_1 - \boldsymbol{\omega}_2) \cdot (\mathbf{r}_{B/A} \times \mathbf{r}_{C/A}) &= 0\\ (\boldsymbol{\omega}_2 - \boldsymbol{\omega}_3) \cdot (\mathbf{r}_{C/A} \times \mathbf{r}_{D/A}) &= 0\\ (\boldsymbol{\omega}_3 - \boldsymbol{\omega}_1) \cdot (\mathbf{r}_{D/A} \times \mathbf{r}_{B/A}) &= 0\end{aligned} \tag{D.4}$$

These three scalar equations contain nine unknowns: the three components of $\boldsymbol{\omega}_1$, $\boldsymbol{\omega}_2$, and $\boldsymbol{\omega}_3$.

The general solution to Eqs. (D.4) is (this may be verified by substitution)

$$\begin{aligned}\boldsymbol{\omega}_1 &= \boldsymbol{\omega} + k_1\mathbf{r}_{B/A}\\ \boldsymbol{\omega}_2 &= \boldsymbol{\omega} + k_2\mathbf{r}_{C/A}\\ \boldsymbol{\omega}_3 &= \boldsymbol{\omega} + k_3\mathbf{r}_{D/A}\end{aligned} \tag{D.5}$$

where $\boldsymbol{\omega}$ is called the angular velocity of the body and the k's are undetermined constants.

When Eqs. (D.5) are substituted into Eqs. (D.2), the k's vanish (since $\mathbf{r}_{B/A} \times \mathbf{r}_{B/A} = \mathbf{0}$, etc.), and the final results are

$$\begin{aligned}\mathbf{v}_{B/A} &= \boldsymbol{\omega} \times \mathbf{r}_{B/A}\\ \mathbf{v}_{C/A} &= \boldsymbol{\omega} \times \mathbf{r}_{C/A}\\ \mathbf{v}_{D/A} &= \boldsymbol{\omega} \times \mathbf{r}_{D/A}\end{aligned} \tag{D.6}$$

Equations (D.6) not only prove Eq. (16.1), but also show that $\boldsymbol{\omega}$ (the angular velocity) is a property of the body that does not depend on the chosen points (recall that the choice of the points was arbitrary).

E Numerical Differentiation

E.1 Introduction

Numerical differentiation enables us to compute the derivatives of a given function $y = f(x)$ without using differential calculus. As is the case with numerical integration (Appendix A), numerical differentiation is a procedure that only estimates the value of the derivative. The need for numerical differentiation often arises in the following cases:

1. The function $f(x)$ is known only at discrete values of x, as is often the case for experimental data.
2. The differentiation of $f(x)$ is very tedious algebraically, which is not unusual if the function must be differentiated more than once.

However, the utility of numerical differentiation is by no means limited to these two cases, since the technique is also the basis of several branches of numerical analysis, such as the finite difference method of solving boundary value problems.

The most common approximations for the derivatives of $f(x)$ are the *finite difference formulas.* The main distinction among the various forms of these formulas is the accuracy of the approximation. Unfortunately, the more accurate expressions are also the more complex ones. We confine our attention to a particular set of formulas that offer a good compromise between accuracy and complexity, known as *formulas with error of order* $(\Delta x)^2$.

E.2 Central Difference Formulas with Error of Order $(\Delta x)^2$

Consider the case for which the values of $y = f(x)$ are known at the n equally spaced values of x: x_i ($i = 1, 2, \ldots, n$), as shown in Fig. E.1(a). The values of the function at these points are denoted by f_i; i.e., $f_i = f(x_i)$, and Δx is the

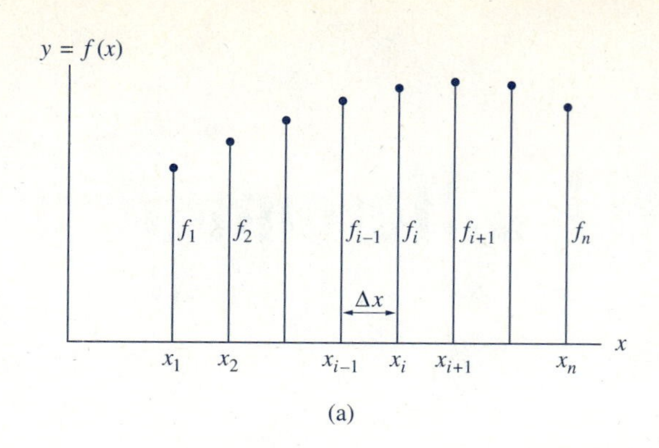

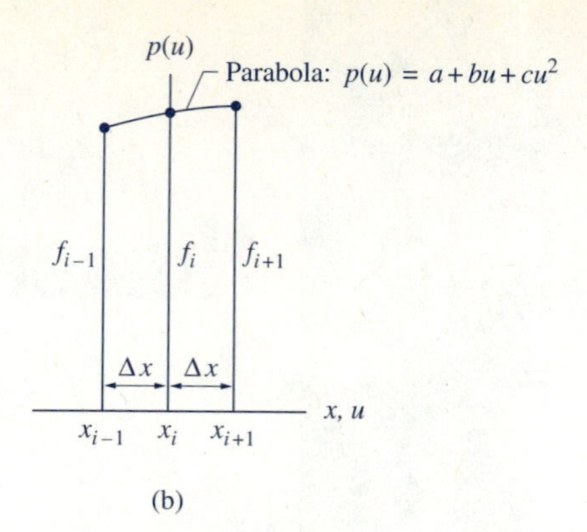

Fig. E.1

spacing between the points. To estimate f_i', the derivative of $f(x)$ at $x = x_i$, we first find a parabola that passes through the point (x_i, f_i) and its neighboring points (x_{i-1}, f_{i-1}) and (x_{i+1}, f_{i+1}), as shown in Fig. E.1(b) (recall that three points uniquely define a parabola). The slope of the parabola at $x = x_i$ can then be used as an approximation to f_i'.

We can simplify the computations by introducing a new coordinate u with its origin at x_i, as shown in Fig. E.1(b). It can be shown that the equation of the parabola is $p(u) = a + bu + cu^2$, where $a = f_i$, $b = (f_{i+1} - f_{i-1})/(2\Delta x)$, and $2c = (f_{i-1} - 2f_i + f_{i+1})/(\Delta x)^2$.* Therefore, the first derivative of $f(x)$ at $x = x_i$ is $f_i' \approx (dp/du)_{u=0} = b$, or

$$f_i' \approx \frac{f_{i+1} - f_{i-1}}{2\Delta x} \tag{E.1}$$

The parabola can also be used to estimate the second derivative of $f(x)$ at x_i: $f_i'' \approx (d^2p/du^2)_{u=0} = 2c$, which leads to

$$f_i'' \approx \frac{f_{i-1} - 2f_i + f_{i+1}}{(\Delta x)^2} \tag{E.2}$$

A more detailed analysis, based on a Taylor series expansion of $f(x)$ in the neighborhood of x_i, shows that the error involved in Eq. (E.1) or (E.2) is proportional to $(\Delta x)^2$. This error is called *truncation error,* because its source can be traced to a truncation of the Taylor series. It follows that the accuracy of the approximations improves with decreasing Δx.

However, if Δx is made very small, the values of f_{i-1}, f_i, and f_{i+1} become almost equal. Consequently, the numerator of Eq. (E.1) represents the difference of two nearly equal numbers, which can lead to the loss of several significant digits during its computation. The same situation exists when the numerator of Eq. (E.2) is evaluated. Since this error is due to the

* We note that the equation for the approximating parabola can also be used to derive Simpson's rule of numerical integration by showing that

$$\int_{-\Delta x}^{\Delta x} p(u)\,du = (f_{i-1} + 4f_i + f_{i+1})\Delta x/3$$

rounding off that occurs when noninteger numbers are stored, it is called *round-off error.* The round-off error can be reduced by increasing the number of significant figures used in the computations. (Truncation and round-off errors are also discussed in Art. 12.8.)

E.3 Forward and Backward Difference Formulas with Error of Order $(\Delta x)^2$

The central difference formula cannot be used to evaluate the derivatives at the endpoints x_1 and x_n because the values of $f(x)$ are unknown outside the interval $x_1 \leq x \leq x_n$. This difficulty is overcome by using the so-called forward and backward difference formulas. To have the same truncation error as the central difference expressions, these formulas use cubic polynomials that pass through four neighboring points. It can be shown that the *forward difference formulas* are*

$$f_1' \approx \frac{-3f_1 + 4f_2 - f_3}{2\Delta x} \tag{E.3}$$

$$f_1'' \approx \frac{2f_1 - 5f_2 + 4f_3 - f_4}{(\Delta x)^2} \tag{E.4}$$

and the *backward difference formulas* are

$$f_n' \approx \frac{3f_n - 4f_{n-1} + f_{n-2}}{2\Delta x} \tag{E.5}$$

$$f_n'' \approx \frac{2f_n - 5f_{n-1} + 4f_{n-2} - f_{n-3}}{(\Delta x)^2} \tag{E.6}$$

* See, for example, James, M. L., Smith, G. M., and Wolford, J. C., *Applied Numerical Methods for Digital Computation,* Harper and Row, New York, 3rd ed., 1985.

Sample Problem E.1

Estimate the first and second derivatives of the function $f(x) = \sin x$ at $x = 0$, $\pi/4$, and $\pi/2$. Use the values of the function that are tabulated as follows.

x	$f(x) = \sin x$
0	0
$\pi/16$	0.1951
$\pi/8$	0.3827
$3\pi/16$	0.5556
$\pi/4$	0.7071
$5\pi/16$	0.8315
$3\pi/8$	0.9239
$7\pi/16$	0.9808
$\pi/2$	1.0000

Solution

The uniform spacing of the x-values is $\Delta x = \pi/16$, with $x = 0$ and $x = \pi/2$ being the endpoints of the range for x.

Since $x = 0$ is the left endpoint, we have no choice but to use the forward difference formulas, Eqs. (E.3) and (E.4), which yield

$$f'(0) \approx \frac{-3(0) + 4(0.1951) - 0.3827}{2(\pi/16)} = \underset{(1.0000)}{1.0127} \qquad \textit{Answer}$$

$$f''(0) \approx \frac{2(0) - 5(0.1951) + 4(0.3827) - 0.5556}{(\pi/16)^2} = \underset{(0.0000)}{-0.0078} \qquad \textit{Answer}$$

The numbers in parentheses are the "exact" values computed from $f'(x) = \cos x$ and $f''(x) = -\sin x$.

Because $x = \pi/4$ is not an endpoint, we can use the forward, central, or backward difference formulas. Since all three methods have the same basic accuracy, we choose the central difference formulas, Eqs. (E.1) and (E.2), due to the fact that they contain the fewest terms. The results are

$$f'(\pi/4) \approx \frac{0.8315 - 0.5556}{2(\pi/16)} = \underset{(0.7071)}{0.7026} \qquad \textit{Answer}$$

$$f''(\pi 4) \approx \frac{0.5556 - 2(0.7071) + 0.8315}{(\pi/16)^2} = \underset{(-0.7071)}{-0.7029} \qquad \textit{Answer}$$

For $x = \pi/2$, the right endpoint, we must use the backward difference formulas, Eqs. (E.5) and (E.6), with the results yielding

$$f'(\pi/2) \approx \frac{3(1.0000) - 4(0.9808) + 0.9239}{2(\pi/16)} = \underset{(0.0000)}{0.0018} \qquad \textit{Answer}$$

$$f''(\pi/2) \approx \frac{2(1.0000) - 5(0.9808) + 4(0.9239) - 0.8315}{(\pi/16)^2} = \underset{(-1.0000)}{-1.0349} \qquad \textit{Answer}$$

F Mass Moments and Products of Inertia

F.1 Introduction

The concept of mass moment of inertia was first introduced in Art. 17.2. In that article, we considered only those inertial properties that were required for a discussion of the plane motion of a rigid body—specifically, the definitions of mass moment of inertia and the radius of gyration, the parallel-axis theorem, and the method of composite bodies. We used these concepts extensively throughout the analysis of plane motion in Chapters 17 and 18. The inertial properties of a rigid body described in three dimensions were introduced in Art. 19.3, including the definition of mass product of inertia, the parallel-axis theorem for products of inertia, and the principal moments of inertia. The techniques for determining moment of inertia by integration, which were not discussed, are included in this appendix. This appendix concludes with a discussion of the inertia tensor, including the principal moments of inertia and the principal directions.

F.2 Review of Mass Moment of Inertia

The mass moment of inertia of a rigid body of mass m about an axis, such as the axis a-a in Fig. F.1, was defined as

$$I_a = \int_{\mathcal{V}} r^2 \, dm \qquad (17.1, \text{repeated})$$

where r is the distance from the axis to the mass element dm and the integral is taken over the region $\mathcal{V}$ occupied by the body. Methods for computing the integral in Eq. (17.1) are discussed in Art. F.4.

The radius of gyration of the body with respect to the axis a-a was defined as

$$k_a = \sqrt{I_a/m} \qquad (17.2, \text{repeated})$$

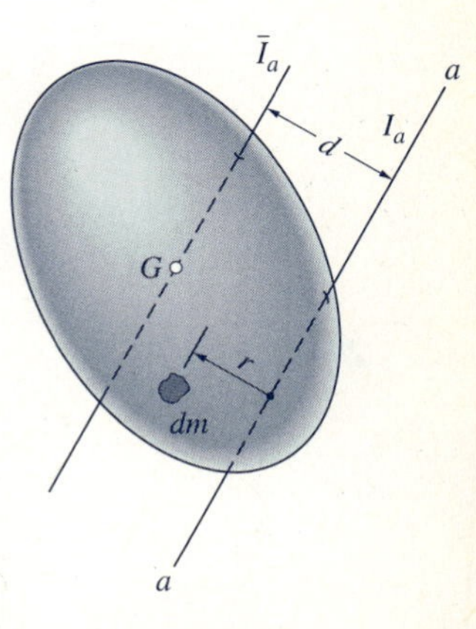

Fig. F.1

The parallel-axis theorem (proved in Art. 17.2) states that

$$I_a = \bar{I}_a + md^2 \qquad \text{(17.3, repeated)}$$

As indicated in Fig. F.1, $\bar{I}_a$ is the moment of inertia of the body about the axis that is parallel to the axis *a-a* and passes through the mass center G of the body (we refer to this axis as a *central axis*), and d is the distance between the two axes. In Art. 17.2, the use of this theorem was restricted to the computation of moments of inertia of composite bodies. As we see in Art. F.4, the parallel-axis theorem is also very useful when computing the moment of inertia of a body by integration.

F.3 Moments of Inertia of Thin Plates

Here we present a convenient method of calculating the mass moments of inertia of homogeneous thin plates from the second moments of their surface areas (properties of areas are discussed in Chapter 9). Not only are thin plates important in their own right, but their inertial properties are also useful in the calculation of the inertial properties of solids by integration.

Figure F.2 shows a thin plate of thickness t. The surface of the plate is the plane region $\mathscr{A}$ of area A, and the solid region enclosing the plate is denoted by $\mathscr{V}$. Letting ρ be the mass density (mass per unit volume), the mass of the plate is

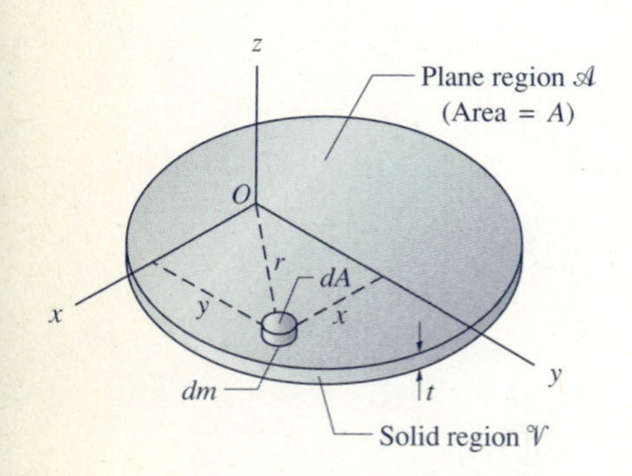

Fig. F.2

$$m = \rho t A \qquad \text{(F.1)}$$

Using the definition in Eq. (17.1) and referring to Fig. F.2, the moments of inertia of the differential mass element dm about the coordinate axes are

$$dI_x = y^2\,dm \qquad dI_y = x^2\,dm \qquad dI_z = r^2\,dm \qquad \text{(a)}$$

where r is measured from the origin O of the coordinate system. Substituting $dm = \rho t\,dA$ and integrating over the region $\mathscr{A}$ yield

$$I_x = \rho t \int_{\mathscr{A}} y^2\,dA \qquad I_y = \rho t \int_{\mathscr{A}} x^2\,dA \qquad I_z = \rho t \int_{\mathscr{A}} r^2\,dA \qquad \text{(b)}$$

Referring to Eqs. (9.2) and (9.3), we see that the integrals in Eqs. (b) represent the moments of inertia of the surface area of the plate*: $\int_{\mathscr{A}} y^2\,dA = (I_x)_{\text{area}}$, $\int_{\mathscr{A}} x^2\,dA = (I_y)_{\text{area}}$, and $\int_{\mathscr{A}} r^2\,dA = (J_O)_{\text{area}}$. Therefore, mass moments of inertia of the plate are related to the area moments of inertia by

$$I_x = \rho t (I_x)_{\text{area}} \qquad I_y = \rho t (I_y)_{\text{area}}$$
$$I_z = \rho t (J_O)_{\text{area}} \qquad \text{(F.2)}$$

* In this appendix, we use the subscript "area" to distinguish the moment of inertia of the plate area from the mass moment of inertia of the plate.

Substituting $\rho t = m/A$ from Eq. (F.1), we obtain the following alternative forms of Eqs. (F.2).

$$I_x = \frac{m}{A}(I_x)_{\text{area}} \qquad I_y = \frac{m}{A}(I_y)_{\text{area}}$$
$$I_z = \frac{m}{A}(J_O)_{\text{area}} \tag{F.3}$$

Since $r^2 = x^2 + y^2$, Eqs. (b) yield the identity

$$I_z = I_x + I_y \tag{F.4}$$

It is important to note that Eq. (F.4) is valid only for thin plates. In general, it is not true for bodies of arbitrary shape.

As an illustration, consider the homogeneous thin disk shown in Fig. F.3(a). The surface area of the disk is the circle of radius R in Fig. F.3(b). From Table 9.1, the moments of inertia of a circular area are $(I_x)_{\text{area}} = (I_y)_{\text{area}} = \pi R^4/4$. Therefore, the mass moments of inertia of the plate can be computed as follows.

$$I_x = I_y = \frac{m}{A}(I_x)_{\text{area}} = \frac{m}{\pi R^2}\frac{\pi R^4}{4} = \frac{mR^2}{4}$$

and

$$I_z = I_x + I_y = 2\left(\frac{mR^2}{4}\right) = \frac{mR^2}{2}$$

The above result for I_z could also have been found using $I_z = (m/A)(J_O)_{\text{area}}$, where $(J_O)_{\text{area}} = (I_x)_{\text{area}} + (I_y)_{\text{area}}$.

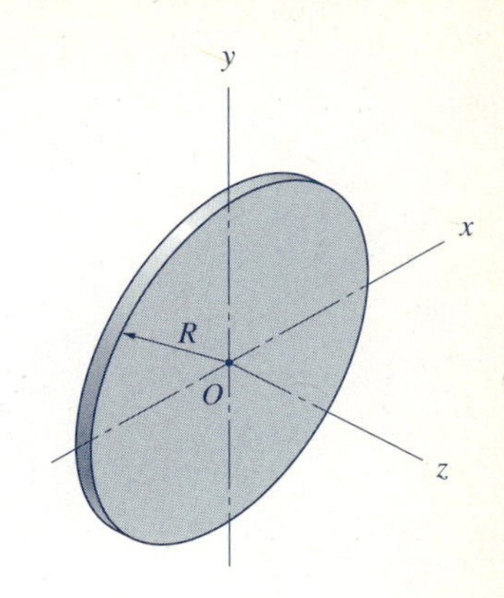

(a) Homogeneous thin circular disk

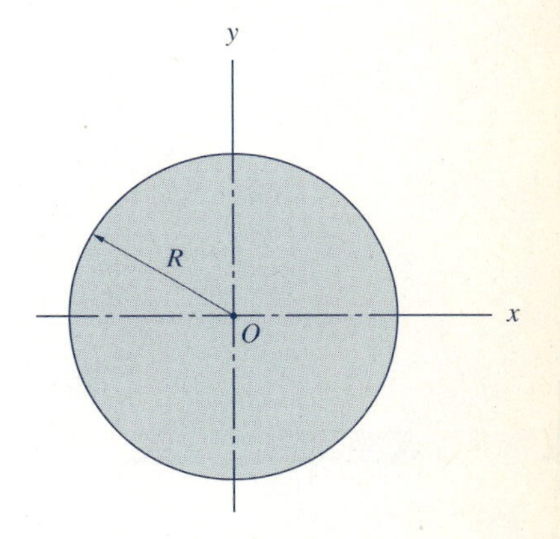

(b) Surface area of disk

Fig. F.3

F.4 Mass Moment of Inertia by Integration

According to Eq. (17.1), the mass moment of inertia of a body that occupies a region $\mathcal{V}$ is obtained by evaluating an integral of the form $\int_{\mathcal{V}} r^2\, dm$, which, in general, represents a triple integral. Using rectangular coordinates, for example, we have $dm = \rho\, dV = \rho\, dx\, dy\, dz$, where ρ is the mass density at the point whose coordinates are (x, y, z). Techniques of evaluating multiple integrals are presented in introductory calculus texts (you will find that most of those texts use moment of inertia as a practical application of integration performed over spatial regions).

Here we consider only bodies whose symmetry permits us to evaluate their moments of inertia with a single integration. As will be seen in Sample Problem F.3, the single integration technique is based on the inertial properties of thin plates that have been discussed in the preceding article.

Sample Problem F.1

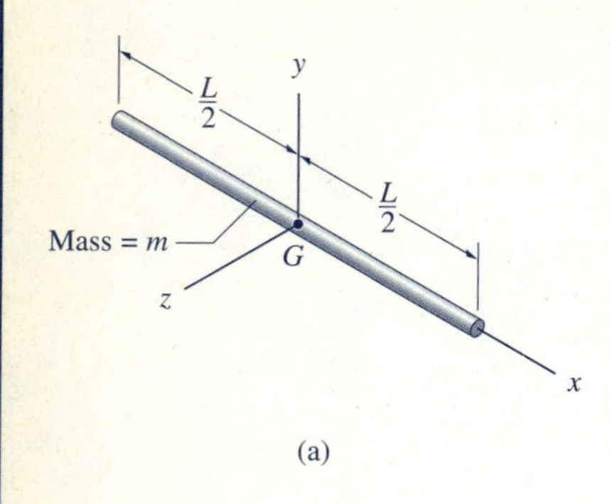

(a)

Figure (a) shows a homogeneous slender rod of mass m and length L. Determine the moments of inertia of the rod about the x-, y-, and z-axes that pass through its mass center G.

Solution

The term *slender* implies that the cross-sectional dimensions of the rod are negligible compared with its length. Therefore, the mass of the rod may be considered as being distributed along the x-axis, which means that its moment of inertia about the axis is negligible; i.e.,

$$I_x \approx 0 \qquad \textit{Answer}$$

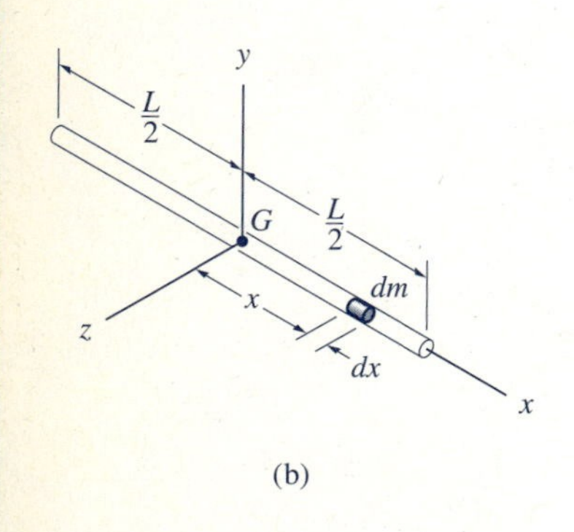

(b)

The differential mass dm chosen for integration is shown in Fig. (b). Since x is the perpendicular distance from both the y-axis and the z-axis to dm, the moments of inertia about these two axes are identical. Letting ρ be the (constant) mass of the rod *per unit length,* we have $dm = \rho\, dx$. The definition of moment of inertia gives

$$I_y = I_z = \int x^2\, dm = \rho \int_{-L/2}^{L/2} x^2\, dm = \frac{\rho}{3}\left(\frac{L^3}{8} + \frac{L^3}{8}\right) = \frac{\rho L^3}{12}$$

Since the mass of the rod is $m = \rho L$, this result may be written as

$$I_y = I_z = \frac{mL^2}{12} \qquad \textit{Answer}$$

Sample Problem F.2

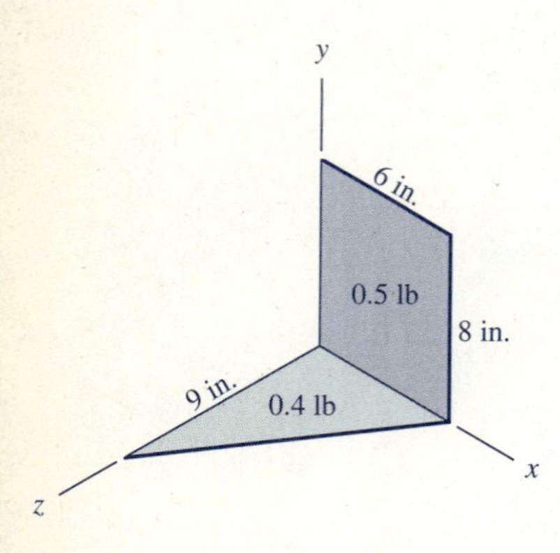

As shown in the figure, an assembly is formed by joining a 0.5-lb rectangular plate to a 0.4-lb triangular plate. Assuming that both plates are thin and homogeneous, calculate the moment of inertia of the assembly about each of the three coordinate axes without integrating.

Solution

We use Eqs. (F.3) to compute the mass moment of inertia of each plate from the properties of areas listed in Table 9.1. Summing the results for the two plates then gives the moments of inertia for the assembly.

Rectangular Plate

$$m = \frac{0.5}{32.2} = 15.53 \times 10^{-3} \text{ slugs}$$

$$A = 6 \times 8 = 48.0 \text{ in.}^2 = 333.3 \times 10^{-3} \text{ ft}^2$$

$$(I_x)_{\text{area}} = \frac{6 \times 8^3}{3} = 1024 \text{ in.}^4 = 49.38 \times 10^{-3} \text{ ft}^4$$

$$(I_y)_{\text{area}} = \frac{8 \times 6^3}{3} = 576.0 \text{ in.}^4 = 27.78 \times 10^{-3} \text{ ft}^4$$

Since the plate lies in the xy-plane,

$$
\begin{aligned}
(J_O)_{\text{area}} &= (I_x)_{\text{area}} + (I_y)_{\text{area}} \\
&= (49.38 + 27.78)(10^{-3}) = 77.16 \times 10^{-3} \text{ ft}^4
\end{aligned}
$$

Substituting the above results into Eqs. (F.3) gives

$$
I_x = \frac{m}{A}(I_x)_{\text{area}} = \frac{15.53}{333.3}(49.38 \times 10^{-3}) = 2.301 \times 10^{-3} \text{ slug} \cdot \text{ft}^2
$$

$$
I_y = \frac{m}{A}(I_y)_{\text{area}} = \frac{15.53}{333.3}(27.78 \times 10^{-3}) = 1.294 \times 10^{-3} \text{ slug} \cdot \text{ft}^2
$$

$$
I_z = \frac{m}{A}(J_O)_{\text{area}} = \frac{15.53}{333.3}(77.16 \times 10^{-3}) = 3.595 \times 10^{-3} \text{ slug} \cdot \text{ft}^2
$$

Triangular Plate

$$
m = \frac{0.4}{32.2} = 12.42 \times 10^{-3} \text{ slugs}
$$

$$
A = \frac{6 \times 9}{2} = 27.0 \text{ in.}^2 = 187.5 \times 10^{-3} \text{ ft}^2
$$

$$
(I_x)_{\text{area}} = \frac{6 \times 9^3}{12} = 364.5 \text{ in.}^4 = 17.58 \times 10^{-3} \text{ ft}^4
$$

$$
(I_z)_{\text{area}} = \frac{9 \times 6^3}{12} = 162.0 \text{ in.}^4 = 7.813 \times 10^{-3} \text{ ft}^4
$$

Noting that the plate lies in the xz-plane, we have

$$
\begin{aligned}
(J_O)_{\text{area}} &= (I_x)_{\text{area}} + (I_z)_{\text{area}} \\
&= (17.58 + 7.813)(10^{-3}) = 25.39 \times 10^{-3} \text{ ft}^4
\end{aligned}
$$

Substituting the above results into Eqs. (F.3) (modified to take into account the fact that the area lies in the xz-plane) gives

$$
I_x = \frac{m}{A}(I_x)_{\text{area}} = \frac{12.42}{187.5}(17.58 \times 10^{-3}) = 1.165 \times 10^{-3} \text{ slug} \cdot \text{ft}^2
$$

$$
I_y = \frac{m}{A}(J_O)_{\text{area}} = \frac{12.42}{187.5}(25.39 \times 10^{-3}) = 1.682 \times 10^{-3} \text{ slug} \cdot \text{ft}^2
$$

$$
I_z = \frac{m}{A}(I_z)_{\text{area}} = \frac{12.42}{187.5}(7.813 \times 10^{-3}) = 0.518 \times 10^{-3} \text{ slug} \cdot \text{ft}^2
$$

Assembly

Summing the results obtained for the two plates gives the following values for the inertial properties of the assembly.

$$
\begin{aligned}
I_x &= (2.301 + 1.165)(10^{-3}) = 3.466 \times 10^{-3} \text{ slug} \cdot \text{ft}^2 \\
I_y &= (1.294 + 1.682)(10^{-3}) = 2.976 \times 10^{-3} \text{ slug} \cdot \text{ft}^2 \qquad \textit{Answer} \\
I_z &= (3.595 + 0.518)(10^{-3}) = 4.113 \times 10^{-3} \text{ slug} \cdot \text{ft}^2
\end{aligned}
$$

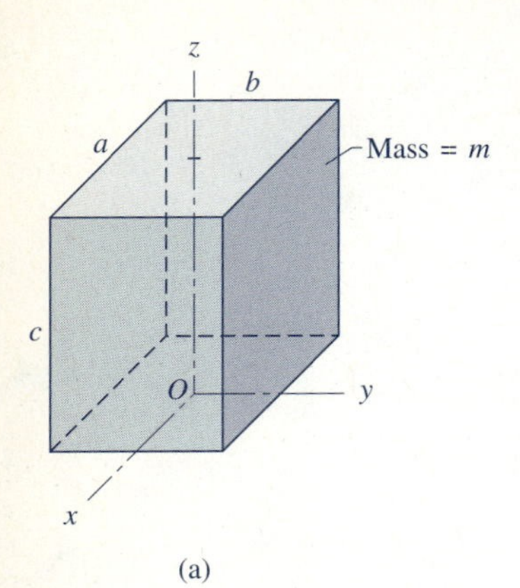

(a)

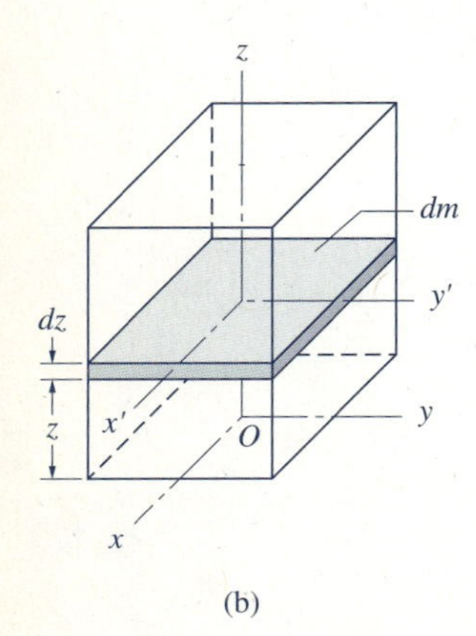

(b)

Sample Problem F.3

Figure (a) shows a homogeneous block of mass m. Using integration, calculate its mass moments of inertia about each of the coordinate axes shown. The origin O is located at the center of the bottom face.

Solution

We select for the differential element the plate of mass dm that is shown in Fig. (b). Since the thickness dz of this element is infinitesimal, all parts of the plate are a distance z from the xy-plane. The x'- and y'-axes shown in Fig. (b) are central axes of the element.

Applying the parallel-axis theorem to the element, we get

$$dI_x = dI_{x'} + z^2\, dm \tag{a}$$

where dI_x and $dI_{x'}$ are the mass moments of inertia of dm about the x- and x'-axes, respectively. According to Eqs. (F.3), we have

$$dI_{x'} = \rho\, dz\, (I_{x'})_{\text{area}} = \rho\, dz\, \frac{ab^3}{12} \tag{b}$$

where ρ is the (constant) mass density of the block; the expression for $(I_{x'})_{\text{area}}$ has been taken from Table 9.1.

Substituting Eq. (b) and $dm = \rho ab\, dz$ into Eq. (a), we obtain

$$dI_x = \rho\, dz \frac{ab^3}{12} + z^2(\rho ab\, dz) = \rho ab\left(\frac{b^2}{12} + z^2\right)dz \tag{c}$$

Integrating Eq. (c) from $z = 0$ to c yields

$$I_x = \rho ab \int_0^c \left(\frac{b^2}{12} + z^2\right) dz = \rho abc\left(\frac{b^2}{12} + \frac{c^2}{3}\right)$$

Recognizing that the mass of the block is $m = \rho abc$, the mass moment of inertia about the x-axis may be written as

$$I_x = m\left(\frac{b^2}{12} + \frac{c^2}{3}\right) \qquad \textit{Answer} \tag{d}$$

The computation of I_y is identical to that of I_x, except that the dimensions a and b are interchanged. Therefore, we can deduce from Eq. (d) that

$$I_y = m\left(\frac{a^2}{12} + \frac{c^2}{3}\right) \qquad \textit{Answer} \tag{e}$$

From Eqs. (F.2), the mass moment of inertia of the element dm about the z-axis is

$$dI_z = \rho\, dz\, (J_O)_{\text{area}} \tag{f}$$

where $(J_O)_{\text{area}}$ is the polar moment of inertia of the area of the element. Substituting

$$(J_O)_{\text{area}} = (I_{x'})_{\text{area}} + (I_{y'})_{\text{area}} = \frac{ab^3}{12} + \frac{a^3b}{12} \tag{g}$$

into Eq. (f), we obtain

$$dI_z = \rho\frac{ab}{12}\left(a^2 + b^2\right) dz$$

Integrating between $z = 0$ and c, we find that

$$I_z = \frac{\rho ab}{12}\left(a^2 + b^2\right)\int_0^c dz = \frac{\rho abc}{12}\left(a^2 + b^2\right)$$

which, on substituting $m = \rho abc$, may be written as

$$I_z = \frac{m}{12}\left(a^2 + b^2\right) \qquad \textit{Answer}$$

PROBLEMS

F.1 The thin plate weighs 4 oz. Calculate its moment of inertia about each coordinate axis.

F.2 The thin plate of mass m has the shape of a circular segment. Determine its moment of inertia about the z-axis.

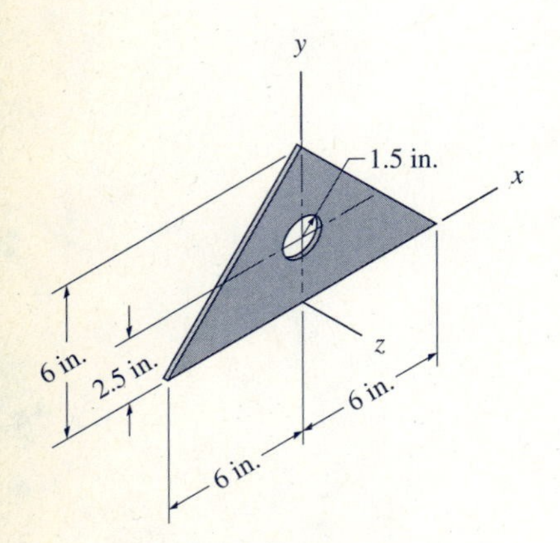

Fig. PF.1

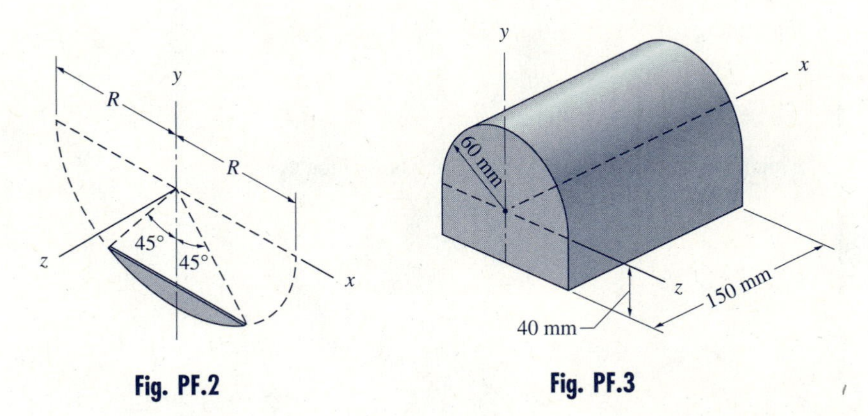

Fig. PF.2

Fig. PF.3

F.3 Calculate the moment of inertia of the iron casting about the x-axis. The mass density of cast iron is 7200 kg/m^3.

F.4 The bracket of mass m has a uniform thickness. Calculate its moment of inertia about each coordinate axis.

F.5 The part shown is formed by slitting and bending a thin plate. If the total mass of the plate is 7.5 kg, determine its moment of inertia about the three coordinate axes.

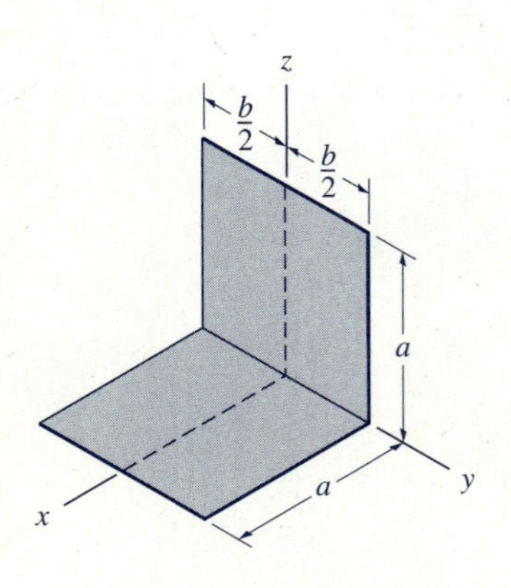

Fig. PF.4

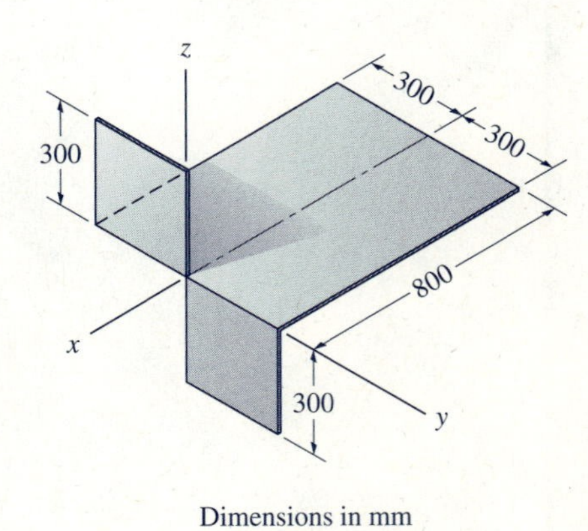

Fig. PF.5

Fig. PF.6

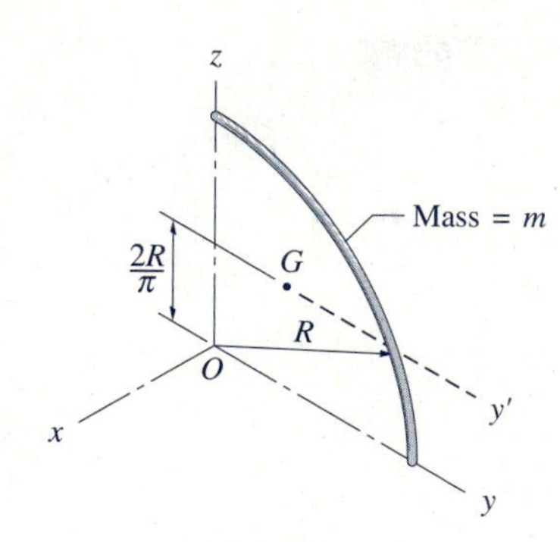

Fig. PF.7

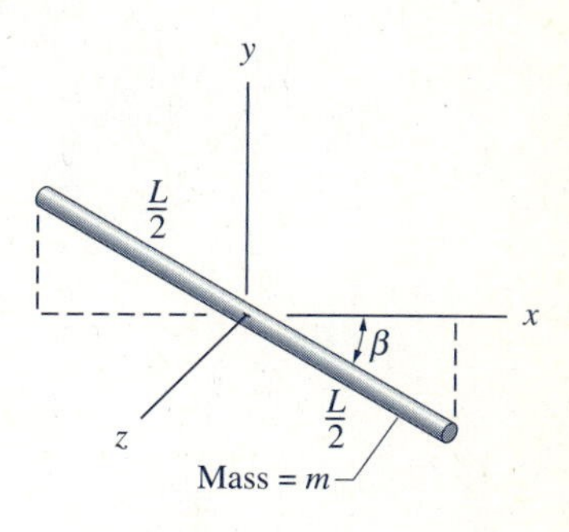

Fig. PF.8

F.6 The rocket casing consists of a 120-kg cylindrical shell and four triangular fins, each of mass 15 kg. Assuming all components to be thin and of uniform thickness, determine the moment of inertia of the casing about the z-axis.

F.7 Without integrating, find the moment of inertia of the wire about the y'-axis, which passes through the mass center G.

F.8 (a) Find the moment of inertia of the slender rod about the x-axis by integration. (b) Using the results of part (a), determine the moments of inertia about the other two coordinate axes.

F.9 The slender rod of mass m lies in the xy-plane. Using integration, determine its moment of inertia about the x-axis.

F.10 (a) Determine the moment of inertia for the homogeneous cylinder about the z-axis by integration. (b) Obtain the moment of inertia about the y-axis using Eq. (F.3).

F.11 Using integration, find the moment of inertia of the paraboloid of revolution about the x-axis.

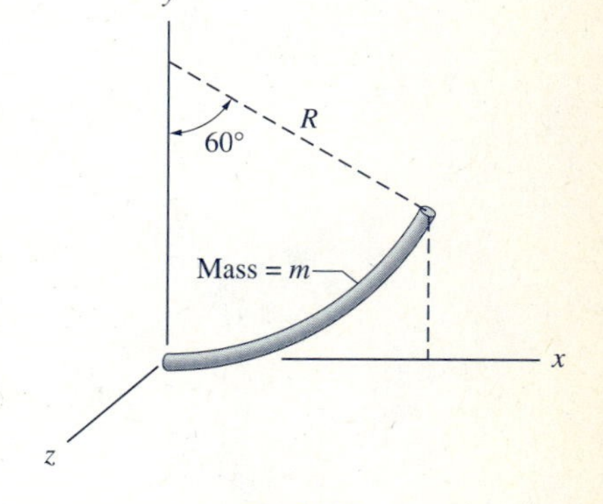

Fig. PF.9

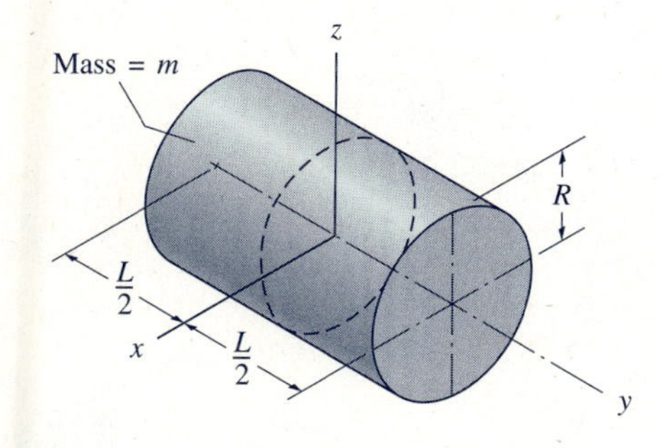

Fig. PF.10

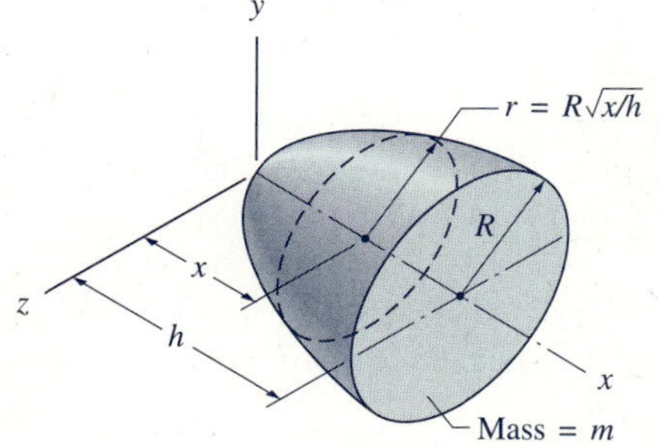

Fig. PF.11

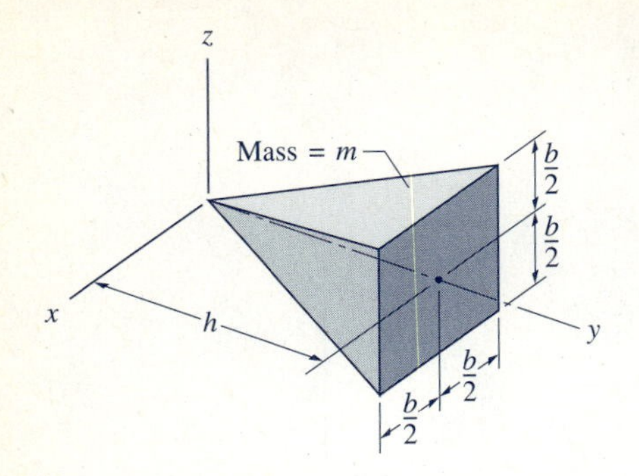

Fig. PF.12

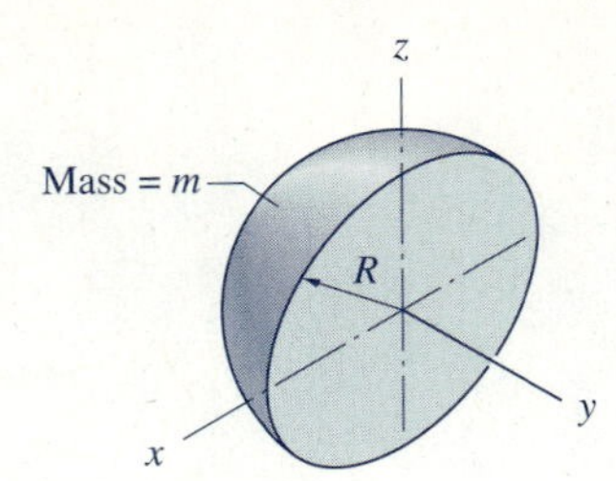

Fig. PF.13

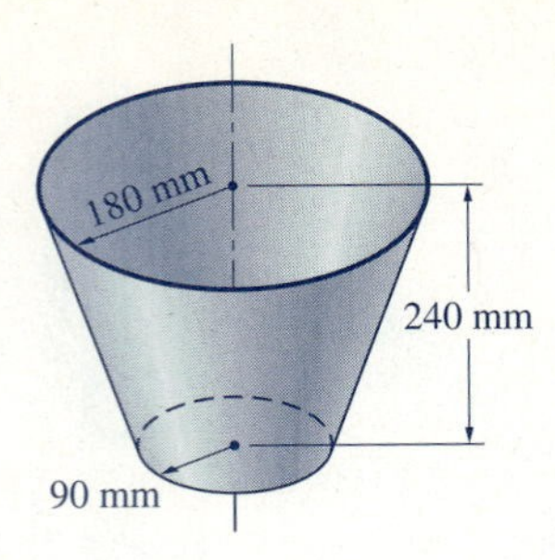

Fig. PF.14

F.12 Use integration to determine the moment of inertia of the rectangular pyramid about (a) the z-axis; and (b) the y-axis.

F.13 (a) Determine the moment of inertia of the homogeneous hemisphere of mass m about the y-axis using integration. (b) Using the result found in part (a), find the moment of inertia of a sphere of mass M about a diameter.

F.14 The mass of the truncated conical shell of constant wall thickness is 1.5 kg. Use integration to find the moment of inertia about the axis of the shell.

F.15 Determine the moment of inertia of the thin cylindrical panel about the z-axis using integration.

F.16 The cover of the football may be approximated by a thin homogeneous shell of mass m. Determine its moment of inertia about the z-axis.

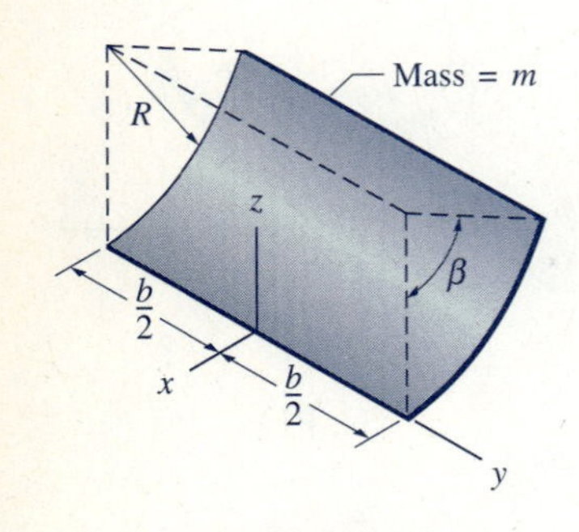

Fig. PF.15

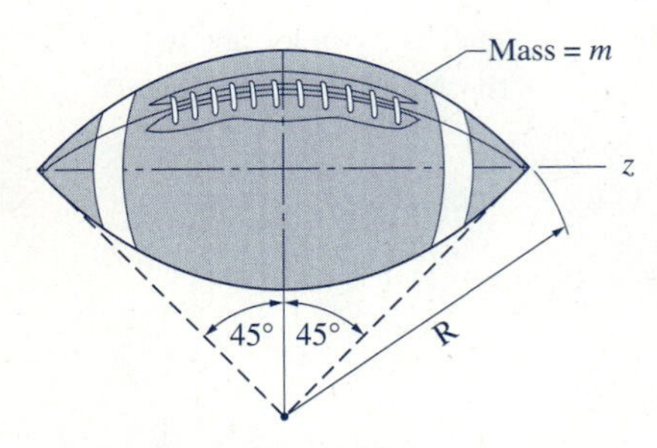

Fig. PF.16

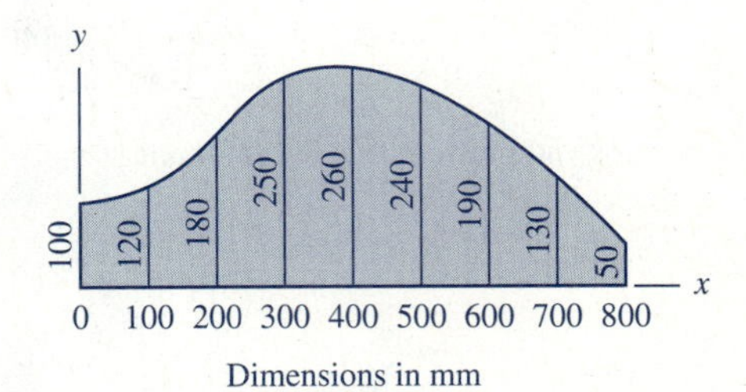

Fig. PF.17, PF.18

F.17 A thin steel plate of mass 80 kg is cut from the pattern shown. Calculate the moment of inertia of the plate about the y-axis.

F.18 An axisymmetric cavity is formed in a sand mold by rotating the pattern shown about the x-axis. A casting is then made by filling the cavity with aluminum of density 2650 kg/m^3. Calculate the moment of inertia of the casting about the x-axis.

F.5 Mass Products of Inertia; Parallel-axis Theorems

Figure F.4 shows a body of mass m that occupies a region $\mathcal{V}$. The mass products of inertia of this body relative to the coordinate axes shown are defined as

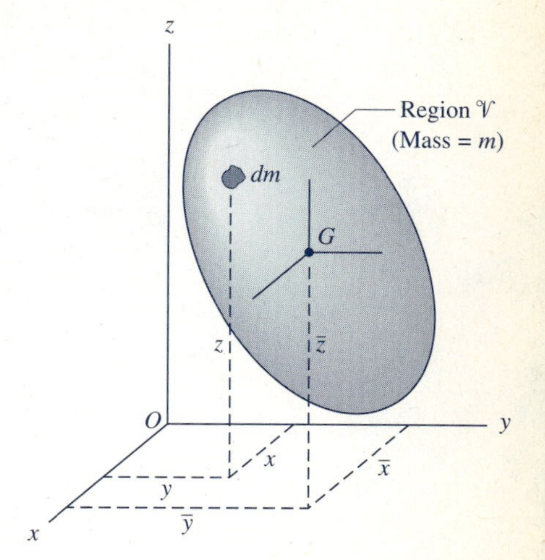

Fig. F.4

$$
\begin{aligned}
I_{xy} &= I_{yx} = \int_{\mathcal{V}} xy\, dm \\
I_{yz} &= I_{zy} = \int_{\mathcal{V}} yz\, dm \\
I_{zx} &= I_{xz} = \int_{\mathcal{V}} zx\, dm
\end{aligned}
\tag{F.5}
$$

The units for product of inertia are the same as those for moment of inertia, i.e., mass × (length)^2 (slug · ft^2 or kg · m^2). Whereas the moment of inertia of a body is always positive, its product of inertia can be positive, negative, or zero.

Products of inertia satisfy the following parallel-axis theorems, which are similar to the parallel-axis theorems for moment of inertia.

$$
\begin{aligned}
I_{xy} &= \bar{I}_{xy} + m\bar{x}\bar{y} \\
I_{yz} &= \bar{I}_{yz} + m\bar{y}\bar{z} \\
I_{zx} &= \bar{I}_{zx} + m\bar{z}\bar{x}
\end{aligned}
\tag{F.6}
$$

In the first of Eqs. (F.6), $\bar{I}_{xy}$ is the product of inertia with respect to central axes that are parallel to the x- and y-axes, respectively; and $\bar{x}$ and $\bar{y}$ are the coordinates of the mass center G—see Fig. F.4. The terms in the other two equations are defined in an analogous manner.

To prove the parallel-axis theorem, we consider Fig. F.5, which shows the same body viewed along the positive z-axis. The $x'y'$-coordinate system has its origin at G, and its axes are parallel to the xy-axes. From Fig. F.5, we see that $x = \bar{x} + x'$ and $y = \bar{y} + y'$, which, when substituted into the definition $I_{xy} = \int_{\mathcal{V}} xy\, dm$, gives

$$I_{xy} = \int_{\mathcal{V}} (\bar{x} + x')(\bar{y} + y')\, dm$$

Carrying out the multiplication, we obtain

$$I_{xy} = \bar{x}\bar{y}\int_{\mathcal{V}} dm + \bar{y}\int_{\mathcal{V}} x'\, dm + \bar{x}\int_{\mathcal{V}} y'\, dm + \int_{\mathcal{V}} x'y'\, dm$$

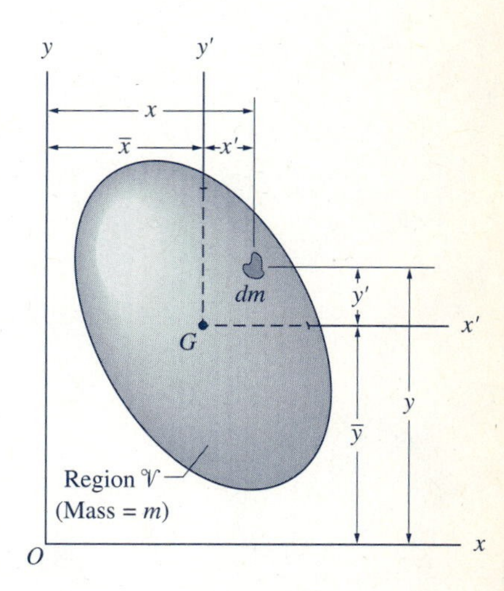

Fig. F.5

Note that since the x'- and y'-axes pass through G, we have $\int_{\mathcal{V}} x'y'\, dm = \bar{I}_{xy}$, $\int_{\mathcal{V}} x' dm = 0$, and $\int_{\mathcal{V}} y' dm = 0$. Consequently, the above equation reduces to $I_{xy} = \bar{I}_{xy} + m\bar{x}\bar{y}$. This completes the proof of the parallel-axis theorem.

The method of composite bodies for mass products of inertia is equivalent to the same method for moments of inertia—the product of inertia of a composite body equals the sum of the products of inertia of its parts. The proof of this statement follows directly from the definition of product of inertia: The integral of a sum equals the sum of the integrals.

F.6 Products of Inertia by Integration; Thin Plates

The evaluation of the integrals that define the mass products of inertia of a body, e.g., $\int_{\mathcal{V}} xy\,dm$, generally require triple integration. As done in Art. F.4 for mass moments, we restrict our attention here to bodies whose symmetry permits their products of inertia to be evaluated with only a single integration utilizing the properties of thin plates.

The mass products of inertia of the homogeneous thin plate depicted in Fig. F.2 are

$$I_{xy} = \int_{\mathcal{V}} xy\,dm \qquad I_{yz} = \int_{\mathcal{V}} yz\,dm \qquad I_{zx} = \int_{\mathcal{V}} zx\,dm \tag{a}$$

Using $dm = \rho t\,dA$, Eqs. (a) become

$$I_{xy} = \rho t\int_{\mathcal{A}} xy\,dA \qquad I_{yz} = \rho t\int_{\mathcal{A}} yz\,dA \qquad I_{zx} = \rho t\int_{\mathcal{A}} zx\,dA \tag{b}$$

From Eq. (9.11), we recall that the integrals in Eqs. (b) are the products of inertia of the plane region $\mathcal{A}$ with respect to the coordinate axes. Again using the label "area" to refer to area properties, Eqs. (b) become

$$I_{xy} = \rho t(I_{xy})_{\text{area}} \qquad I_{yz} = \rho t(I_{yz})_{\text{area}} \qquad I_{zx} = \rho t(I_{zx})_{\text{area}} \tag{F.7}$$

Substituting $\rho t = m/A$ yields the alternative forms of Eqs. (F.7):

$$I_{xy} = \frac{m}{A}(I_{xy})_{\text{area}} \qquad I_{yz} = \frac{m}{A}(I_{yz})_{\text{area}} \qquad I_{zx} = \frac{m}{A}(I_{zx})_{\text{area}} \tag{F.8}$$

Before attempting to use Eqs. (F.7) or (F.8), you should review the discussion of products of inertia of areas in Art. 9.3. The products of inertia for commonly encountered plane areas are given in Tables 9.1 and 9.2.

F.7 Inertia Tensor; Moment of Inertia About an Arbitrary Axis

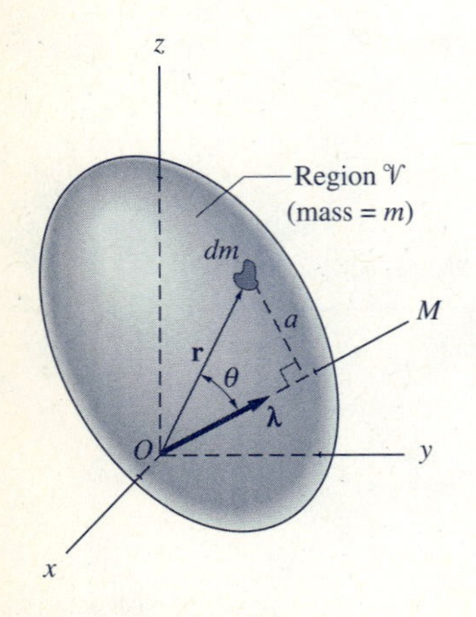

Fig. F.6

As defined in Art. 19.3, the inertia tensor of a body at point O (the origin of the coordinate axes) is the matrix

$$\mathbf{I} = \begin{bmatrix} I_x & -I_{xy} & -I_{xz} \\ -I_{yx} & I_y & -I_{yz} \\ -I_{zx} & -I_{zy} & I_z \end{bmatrix} \tag{19.13, repeated}$$

In this article, we show that the inertia tensor at point O completely determines the moment of inertia about any axis that passes through O.

Figure F.6 shows a rigid body of mass m that occupies the region $\mathcal{V}$. Point O, a point on the body or body extended, is chosen as the origin of the xyz-

coordinate system. The axis OM is an arbitrary axis that passes through O. The angle between the position vector $\mathbf{r}$ of the differential mass dm and the axis OM is denoted by θ. Furthermore, we let $\boldsymbol{\lambda}$ be a unit vector in the direction of OM. Note that the magnitude of the cross product of $\mathbf{r}$ and $\boldsymbol{\lambda}$ is $|\mathbf{r}\times\boldsymbol{\lambda}| = r\sin\theta = a$, the perpendicular distance between OM and dm, as shown in Fig. F.6.

The moment of inertia of the body with respect to the axis OM is

$$I_{OM} = \int_{\mathcal{V}} a^2\, dm = \int_{\mathcal{V}} (\mathbf{r}\times\boldsymbol{\lambda})\cdot(\mathbf{r}\times\boldsymbol{\lambda})\, dm \tag{a}$$

where we have used the fact that the dot product of a vector with itself equals the square of the magnitude of the vector; i.e., $(\mathbf{r}\times\boldsymbol{\lambda})\cdot(\mathbf{r}\times\boldsymbol{\lambda}) = |\mathbf{r}\times\boldsymbol{\lambda}|^2 = a^2$. We let x, y, and z be the coordinates of the location of dm and l, m, and n be the components of $\boldsymbol{\lambda}$. The cross product of $\mathbf{r} = x\mathbf{i} + y\mathbf{j} + z\mathbf{k}$ and $\boldsymbol{\lambda} = l\mathbf{i} + m\mathbf{j} + n\mathbf{k}$ then becomes

$$\mathbf{r}\times\boldsymbol{\lambda} = (yn - zm)\mathbf{i} + (lz - xn)\mathbf{j} + (xm - yl)\mathbf{k}$$

and the expression for a^2 is

$$a^2 = (\mathbf{r}\times\boldsymbol{\lambda})\cdot(\mathbf{r}\times\boldsymbol{\lambda}) = (yn - zm)^2 + (zl - xn)^2 + (xm - yl)^2$$

Expanding the squares and collecting terms, we get

$$a^2 = (y^2 + z^2)l^2 + (x^2 + z^2)m^2 + (x^2 + y^2)n^2 - 2xylm - 2xzln - 2yzmn \tag{b}$$

Substituting Eq. (b) into Eq. (a), and identifying the expressions for the moments and products of inertia, the moment of inertia about the axis OM becomes

$$\boxed{I_{OM} = I_x l^2 + I_y m^2 + I_z n^2 - 2I_{xy}lm - 2I_{yz}mn - 2I_{zx}nl} \tag{F.9}$$

We have now arrived at the following important conclusion: If the components of the inertia tensor (I_x, I_{xy}, I_{xz}, etc.) are known at a point, the moment of inertia about any axis through the point can be computed from Eq. (F.9).

F.8 Principal Moments and Principal Axes of Inertia

In general, the components of the inertia tensor (I_x, I_{xy}, etc.) vary with the location of the reference point O and with the orientation of the xyz-axes. Throughout this article, we assume that the reference point does not change, and we study the effect of changing the orientation of the coordinate axes.

It can be shown* that there exists at least one orientation of the xyz-axes for which the inertia tensor has the following *diagonal form.*

$$\mathbf{I} = \begin{bmatrix} I_1 & 0 & 0 \\ 0 & I_2 & 0 \\ 0 & 0 & I_3 \end{bmatrix} \tag{F.10}$$

* In this article, we state some important properties of inertia tensors. Proofs may be found in most advanced dynamics texts—see, for example, *Advanced Engineering Dynamics,* Ginsberg, J. H., Harper & Row, 1988.

where I_1, I_2, and I_3 are called the *principal moments of inertia* at point O and the corresponding coordinate axes are called the *principal axes* (or *principal directions*) *of inertia* at point O. Note that the products of inertia are zero with respect to the principal axes.

The principal moments of inertia and the direction cosines of each principal axis (denoted l, m, and n) can be found by solving the following four equations for I, l, m, and n.

$$\begin{aligned} (I_x - I)l \quad -I_{xy}m \quad -I_{xz}n &= 0 \\ -I_{xy}l \quad +(I_y - I)m \quad -I_{yz}n &= 0 \\ -I_{zx}l \quad -I_{zy}m \quad -(I_z - I)n &= 0 \\ l^2 + m^2 + n^2 &= 1 \end{aligned} \tag{F.11}$$

The first, second, and third equations can be shown to represent the conditions for zero products of inertia with respect to the principal axes, and the fourth equation must be satisfied for l, m, and n to be direction cosines.

Note that the first three equations of Eqs. (F.11) are linear and homogeneous (right side equal to zero) in the unknowns l, m, and n. Therefore, the three values of I (representing I_1, I_2, and I_3) can be obtained by solving the equation that results from setting the determinant of the coefficients of l, m, and n equal to zero; i.e.,

$$\begin{vmatrix} I_x - I & -I_{xy} & -I_{xz} \\ -I_{yx} & I_y - I & -I_{yz} \\ -I_{zx} & -I_{zy} & I_z - I \end{vmatrix} = 0 \tag{F.12}$$

Once the principal moments of inertia have been found, the direction cosines of the principal axes can be obtained from Eqs. (F.11).

In linear algebra, the computation of the principal moments of inertia and the principal axes is an example of a *matrix eigenvalue problem,* in which the principal moments are the *eigenvalues* of the inertia tensor, and the unit vectors in the direction of the principal axes are its *eigenvectors.* Equation (F.12) is referred to as the *characteristic equation* of the eigenvalue problem. Since matrix eigenvalue problems occur in many branches of the physical sciences, their properties have been thoroughly studied, and several numerical methods have been developed for their solution (e.g., the Jacobi method).

Using the known characteristics of eigenvalue problems, it can be shown that the solutions of Eqs. (F.11) possess the following properties.

1. The eigenvalues, i.e., I_1, I_2, and I_3, are real and positive.
2. Assuming that the eigenvalues are ordered so that $I_3 > I_2 > I_1$, then I_3 and I_1 are the maximum and minimum moments of inertia, respectively, at point O. In other words, I_1 and I_3 are the extrema (extreme values) of I_{OM} in Eq. (F.9) with respect to changes in the direction of OM.
3. If the eigenvalues are distinct, i.e., if Eq. (F.12) has no double roots, then the eigenvectors (principal axes) are mutually perpendicular.

Sample Problem F.4

Using integration, compute the products of inertia of the homogeneous body in Fig. (a) with respect to the axes shown. Express the results in terms of the mass m of the body.

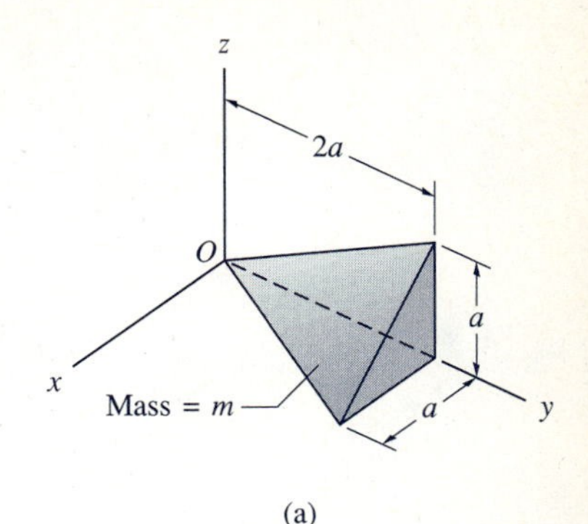

(a)

Solution

We choose the integration element shown in Fig. (b). This element can be considered to be a thin triangular plate of thickness dy, with the surface area A, as shown in Fig. (c). Recognizing the similar triangles in Fig. (b), we see that $(x/y) = (a/2a)$ and that $(z/y) = (a/2a)$, which gives

$$x = z = \frac{y}{2} \tag{a}$$

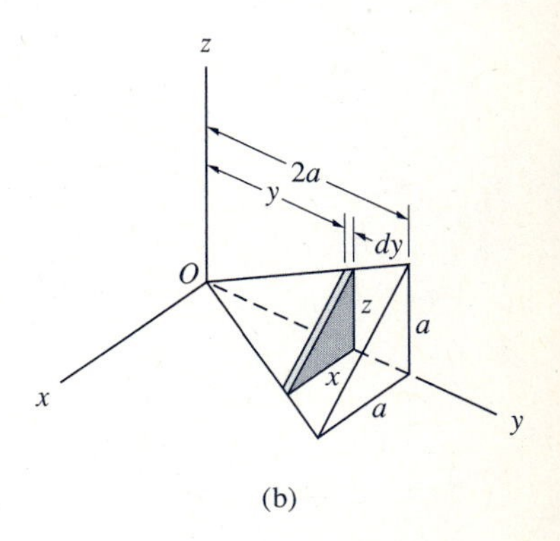

(b)

Next we use Eqs. (F.7) to relate the products of inertia of the mass of the element to the properties of its area. Therefore, we turn our attention to finding the products of inertia of the area shown in Fig. (c).

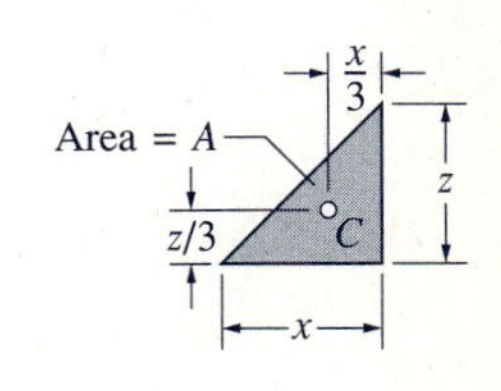

(c)

Products of Inertia of Area A

Since the plane of area A is parallel to the xz-plane, $(\bar{I}_{xy})_{\text{area}} = (\bar{I}_{yz})_{\text{area}} = 0$. And from Table 9.1 we find that $(\bar{I}_{zx})_{\text{area}} = x^2z^2/24$. Using the relationships in Eq. (a), the products of inertia of A become

$$(\bar{I}_{xy})_{\text{area}} = 0 \qquad (\bar{I}_{yz})_{\text{area}} = 0$$

$$(\bar{I}_{zx})_{\text{area}} = \frac{x^2z^2}{24} = \frac{1}{24}\left(\frac{y}{2}\right)^2\left(\frac{y}{2}\right)^2 = \frac{y^4}{384} \tag{b}$$

In terms of the coordinate y, the area A is

$$A = \frac{xz}{2} = \frac{1}{2}\left(\frac{y}{2}\right)\left(\frac{y}{2}\right) = \frac{y^2}{8} \tag{c}$$

and the coordinates of its centroid C are

$$\bar{x} = \frac{x}{3} = \frac{y}{6} \qquad \bar{y} = y \qquad \bar{z} = \frac{z}{3} = \frac{y}{6} \tag{d}$$

We now use the parallel-axis theorem and Eqs. (b)–(d) to compute the products of inertia of A with respect to the coordinate axes.

$$(I_{xy})_{\text{area}} = (\bar{I}_{xy})_{\text{area}} + A\bar{x}\bar{y} = 0 + \frac{y^2}{8}\left(\frac{y}{6}\right)(y) = \frac{y^4}{48} \tag{e}$$

$$(I_{yz})_{\text{area}} = (\bar{I}_{yz})_{\text{area}} + A\bar{y}\bar{z} = 0 + \frac{y^2}{8}(y)\left(\frac{y}{6}\right) = \frac{y^4}{48} \tag{f}$$

$$(I_{zx})_{\text{area}} = (\bar{I}_{zx})_{\text{area}} + A\bar{z}\bar{x} = \frac{y^4}{384} + \frac{y^2}{8}\left(\frac{y}{6}\right)\left(\frac{y}{6}\right) = \frac{7y^4}{1152} \tag{g}$$

Products of Inertia of Mass

We let ρ be the mass density of the body, and we reconsider the differential mass element—the thin plate of thickness dy shown in Fig. (b). Substituting Eqs. (e)–(g) into Eqs. (F.7), the inertial properties of the mass element become

$$dI_{xy} = \rho\, dy\, (I_{xy})_{\text{area}} = \rho\, dy\, \frac{y^4}{48} \tag{h}$$

$$dI_{yz} = \rho\, dy\, (I_{yz})_{\text{area}} = \rho\, dy\, \frac{y^4}{48} \tag{i}$$

$$dI_{zx} = \rho\, dy\, (I_{zx})_{\text{area}} = \rho\, dy\, \frac{7y^4}{1152} \tag{j}$$

Integrating with respect to y between the limits 0 and $2a$ yields

$$I_{xy} = I_{yx} = \frac{\rho}{48}\int_0^{2a} y^4\, dy = \frac{\rho}{48}\frac{(2a)^5}{5} = \frac{2\rho a^5}{15} \tag{k}$$

$$I_{zx} = \frac{7\rho}{1152}\int_0^{2a} y^4\, dy = \frac{7\rho}{1152}\frac{(2a)^5}{5} = \frac{7\rho a^5}{180} \tag{l}$$

We note that $dm = \rho\, dV = \rho A\, dy = \rho(y^2/8)\, dy$, which, when integrated between the limits $y = 0$ and $2a$, gives $m = \rho a^3/3$. Therefore, Eqs. (k) and (l) become

$$I_{xy} = I_{yz} = \frac{2\rho a^5}{15} \cdot \frac{m}{\rho a^3/3} = \frac{2}{5}ma^2 \qquad \textit{Answer}$$

$$I_{zx} = \frac{7\rho a^5}{180} \cdot \frac{m}{\rho a^3/3} = \frac{7}{60}ma^2 \qquad \textit{Answer}$$

Sample Problem F.5

The assembly shown is formed by joining two pieces of $\frac{1}{16}$-in. steel plate. Determine its products of inertia with respect to the axes shown. The weight density of steel is $\gamma = 490$ lb/ft^3.

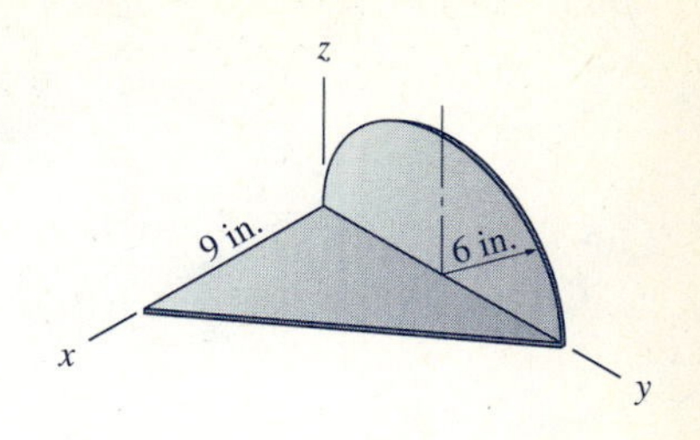

Solution

In this problem, we are justified in using the thin plate approximation of Eqs. (F.2) or (F.3), since the thickness of each component plate is much smaller than its in-plane dimensions.

Triangular Plate

Because the triangular area lies in the xy-plane, $(I_{yz})_{\text{area}} = (I_{zx})_{\text{area}} = 0$. From Table 9.1 we obtain

$$(I_{xy})_{\text{area}} = \frac{(12)^2(9)^2}{24} = 486.0 \text{ in.}^4$$

Using Eqs. (F.2), the mass products of inertia of the triangular plate become

$$\begin{aligned} I_{xy} &= \rho t (I_{xy})_{\text{area}} = \frac{490}{32.2}\left[\frac{1}{16(12)}\right]\frac{486.0}{(12)^4} = 18.58 \times 10^{-4} \text{ slug} \cdot \text{ft}^2 \\ I_{yz} &= I_{zx} = 0 \end{aligned} \qquad \text{(a)}$$

In the calculation of I_{xy}, note that 32.2 ft/s^2 is required to convert the weight density to mass density and that all the dimensions have been changed from inches to feet.

Semicircular Plate

Noting that the semicircular area lies in the yz-plane, we have $(I_{xy})_{\text{area}} = (I_{zx})_{\text{area}} = 0$. We use the parallel-axis theorem to compute $(I_{yz})_{\text{area}}$. Referring to Table 9.1, the centroid of the semicircular area is located at $\bar{x} = 0$, $\bar{y} = 6$ in., $\bar{z} = 4(6)/3\pi$ in. Also observe that $(\bar{I}_{yz})_{\text{area}} = 0$, because the centroidal z-axis is an axis of symmetry. The parallel-axis theorem thus yields

$$\begin{aligned} (I_{yz})_{\text{area}} &= (\bar{I}_{yz})_{\text{area}} + A\bar{y}\bar{z} \\ &= 0 + \frac{\pi(6)^2}{2}(6)\frac{4(6)}{3\pi} = 864.0 \text{ in.}^4 \end{aligned}$$

The mass products of inertia of the semicircular plate can now be computed from Eqs. (F.2)—remembering to convert inches to feet and weight density to mass density:

$$\begin{aligned} I_{xy} &= I_{zx} = 0 \\ I_{yz} &= \rho t (I_{yz})_{\text{area}} = \frac{490}{32.2}\left[\frac{1}{16(12)}\right]\frac{864.0}{(12)^4} = 33.02 \times 10^{-4} \text{ slug} \cdot \text{ft}^2 \end{aligned} \qquad \text{(b)}$$

Assembly

The mass products of inertia of the assembly are found by adding the results for the triangular and semicircular plates in Eqs. (a) and (b), which gives

$$\begin{aligned} I_{xy} &= 18.58 \times 10^{-4} \text{ slug} \cdot \text{ft}^2 \\ I_{yz} &= 33.02 \times 10^{-4} \text{ slug} \cdot \text{ft}^2 \\ I_{zx} &= 0 \end{aligned} \qquad \textit{Answer}$$

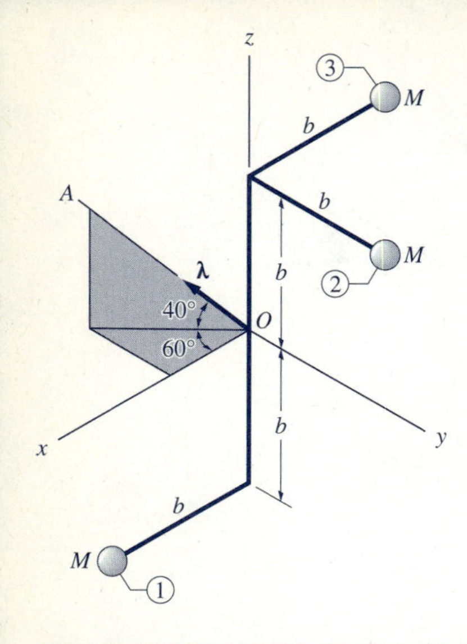

Sample Problem F.6

The assembly consists of three small balls, each of mass M, that are attached to slender rods of negligible mass. Calculate the moment of inertia of the assembly about the axis OA.

Solution

The moment of inertia about the axis OA can be found from Eq. (F.9). In order to use this equation, we must first calculate the inertia tensor at point O and the direction cosines of the axis OA.

The following table shows the computation of the inertia tensor at point O of the assembly, obtained by summing the inertia tensors of the individual balls. Since each ball is small, its moments and products of inertia with respect to axes passing through its mass center may be neglected. Therefore, the parallel-axis theorems simplify to $I_x = \bar{I}_x + M(\bar{y}^2 + \bar{z}^2) = M(y^2 + z^2)$, etc., and $I_{xy} = \bar{I}_{xy} + M\bar{x}\bar{y} = Mxy$, etc.

	Ball ①	Ball ②	Ball ③	Totals
Coordinates	$x = b,\ y = 0,\ z = -b$	$x = 0,\ y = b,\ z = b$	$x = -b,\ y = 0,\ z = b$	
$I_x = M(y^2 + z^2)$	$M[0 + (-b)^2] = Mb^2$	$M(b^2 + b^2) = 2Mb^2$	$M(0 + b^2) = Mb^2$	$4Mb^2$
$I_y = M(z^2 + x^2)$	$M[(-b)^2 + b^2] = 2Mb^2$	$M(b^2 + 0) = Mb^2$	$M[b^2 + (-b)^2] = 2Mb^2$	$5Mb^2$
$I_z = M(x^2 + y^2)$	$M(b^2 + 0) = Mb^2$	$M(0 + b^2) = Mb^2$	$M[(-b)^2 + 0] = Mb^2$	$3Mb^2$
$I_{xy} = Mxy$	$M(b)(0) = 0$	$M(0)(b) = 0$	$M(-b)(0) = 0$	0
$I_{yz} = Myz$	$M(0)(-b) = 0$	$M(b)(b) = Mb^2$	$M(0)(b) = 0$	Mb^2
$I_{zx} = Mzx$	$M(-b)(b) = -Mb^2$	$M(b)(0) = 0$	$M(b)(-b) = -Mb^2$	$-2Mb^2$

The unit vector $\boldsymbol{\lambda}$ that is directed along the axis OA is

$$\begin{aligned}\boldsymbol{\lambda} &= (\cos 40^\circ \cos 60^\circ)\mathbf{i} - (\cos 40^\circ \sin 60^\circ)\mathbf{j} + \sin 40^\circ \mathbf{k} \\ &= 0.3830\mathbf{i} - 0.6634\mathbf{j} + 0.6428\mathbf{k}\end{aligned}$$

Therefore, the direction cosines of OA are $l = 0.3830$, $m = -0.6634$, and $n = 0.6428$.

Substituting the inertia properties of the assembly (computed in the table) and the direction cosines into Eq. (F.9), we get

$$\begin{aligned}I_{OA} &= I_x l^2 + I_y m^2 + I_z n^2 - 2I_{xy}lm - 2I_{yz}mn - 2I_{zx}nl \\ &= Mb^2[4(0.3830)^2 + 5(-0.6634)^2 + 3(0.6428)^2 \\ &\qquad - 2(0)(0.3830)(-0.6634) - 2(1)(-0.6634)(0.6428) \\ &\qquad - 2(-2)(0.6428)(0.3830)] \\ &= 5.86Mb^2\end{aligned}$$

Answer

Sample Problem F.7

The assembly in Fig. (a) consists of two identical, uniform, thin plates, each of mass M and thickness t. For point O, determine (1) the inertia tensor with respect to the axes shown; and (2) the principal moments of inertia and the principal axes.

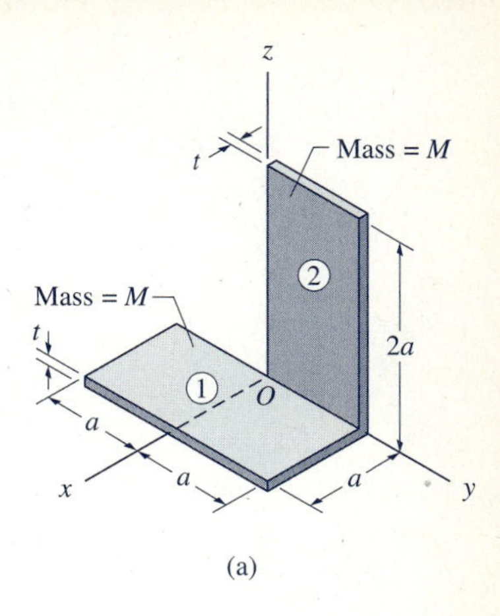

(a)

Solution

Part 1

The inertia tensor at point O for the assembly is found by summing the moments and products of inertia about the xyz-axes of the two plates. We utilize the thin plate approximations described in Arts. F.3 and F.6.

Plate ①

Referring to Fig. (a) and Table 9.1, the moments and products of inertia for the area of Plate ① are

$$(I_x)_{\text{area}} = \frac{a(2a)^3}{12} = \frac{2a^4}{3}$$

$$(I_y)_{\text{area}} = \frac{2a(a)^3}{3} = \frac{2a^4}{3}$$

$$(J_O)_{\text{area}} = (I_x)_{\text{area}} + (I_y)_{\text{area}} = \frac{2a^4}{3} + \frac{2a^4}{3} = \frac{4a^4}{3}$$

$$(I_{xy})_{\text{area}} = 0 \qquad (x \text{ is an axis of symmetry})$$

$$(I_{yz})_{\text{area}} = (I_{zx})_{\text{area}} = 0 \qquad (\text{area lies in } xy\text{-plane})$$

Using the thin plate equations [Eqs. (F.2) and (F.7)] and $M = \rho A t = \rho 2a^2 t$, where $A = 2a^2$ is the plate area and ρ the mass density, the mass properties become

$$I_x = \rho t (I_x)_{\text{area}} = \rho t \frac{2a^4}{3} = \frac{1}{3}Ma^2$$

$$I_y = \rho t (I_y)_{\text{area}} = \rho t \frac{2a^4}{3} = \frac{1}{3}Ma^2$$

$$I_z = \rho t (J_O)_{\text{area}} = \rho t \frac{4a^4}{3} = \frac{2}{3}Ma^2$$

$$I_{xy} = I_{yz} = I_{zx} = 0$$

Plate ②

The area moments and products of inertia for Plate ② are

$$(I_y)_{\text{area}} = \frac{a(2a)^3}{3} = \frac{8a^4}{3}$$

$$(I_z)_{\text{area}} = \frac{2a(a)^3}{3} = \frac{2a^4}{3}$$

$$(J_O)_{\text{area}} = (I_y)_{\text{area}} + (I_z)_{\text{area}} = \frac{8a^4}{3} + \frac{2a^4}{3} = \frac{10a^4}{3}$$

$$(I_{xy})_{\text{area}} = (I_{zx})_{\text{area}} = 0 \qquad (\text{the area lies in the } yz\text{-plane})$$

$$I_{yz} = \frac{a^2(2a)^2}{4} = a^4$$

Substituting these results and $M = \rho At = \rho 2a^2 t$ into Eqs. (F.2) and (F.7), the mass properties of Plate ② become

$$I_x = \rho t(J_O)_{\text{area}} = \rho t\frac{10a^4}{3} = \frac{5}{3}Ma^2$$

$$I_y = \rho t(I_y)_{\text{area}} = \rho t\frac{8a^4}{3} = \frac{4}{3}Ma^2$$

$$I_z = \rho t(I_z)_{\text{area}} = \rho t\frac{2a^4}{3} = \frac{1}{3}Ma^2$$

$$I_{xy} = I_{zx} = 0$$

$$I_{yz} = \rho t(I_{yz})_{\text{area}} = \rho t a^4 = \frac{1}{2}Ma^2$$

Assembly

The mass properties of the assembly are found by summing the properties of the two plates.

$$I_x = Ma^2\left(\frac{1}{3} + \frac{5}{3}\right) = 2Ma^2$$

$$I_y = Ma^2\left(\frac{1}{3} + \frac{4}{3}\right) = \frac{5}{3}Ma^2$$

$$I_z = Ma^2\left(\frac{2}{3} + \frac{1}{3}\right) = Ma^2$$

$$I_{xy} = 0 + 0 = 0$$

$$I_{yz} = 0 + \frac{1}{2}Ma^2 = \frac{1}{2}Ma^2$$

$$I_{zx} = 0 + 0 = 0$$

Substituting these values into Eq. (19.13), the inertia tensor at point O for the assembly becomes

$$\mathbf{I} = \begin{bmatrix} 2 & 0 & 0 \\ 0 & \frac{5}{3} & -\frac{1}{2} \\ 0 & -\frac{1}{2} & 1 \end{bmatrix} Ma^2 \qquad \textit{Answer}$$

Part 2

Substituting the inertial properties of the assembly obtained in Part 1 into Eqs. (F.11) yields the following equations that must be solved for the principal moments of inertia and the direction cosines of the principal axes.

$$(2Ma^2 - I)l + 0m + 0n = 0 \qquad \text{(a)}$$

$$0l + \left(\frac{5}{3}Ma^2 - I\right)m - \frac{Ma^2}{2}n = 0 \qquad \text{(b)}$$

$$0l - \frac{Ma^2}{2}m + (Ma^2 - I)n = 0 \qquad \text{(c)}$$

$$l^2 + m^2 + n^2 = 1 \qquad \text{(d)}$$

The characteristic equation is obtained by setting the determinant of the coefficients in Eqs. (a)–(c) equal to zero:

$$\begin{vmatrix} 2Ma^2 - I & 0 & 0 \\ 0 & \frac{5}{3}Ma^2 - I & -\frac{Ma^2}{2} \\ 0 & -\frac{Ma^2}{2} & Ma^2 - I \end{vmatrix} = 0 \tag{e}$$

Expanding the determinant and simplifying, we get

$$(2Ma^2 - I)\left(I^2 - \frac{8}{3}Ma^2 I + \frac{17}{12}M^2a^4\right) = 0 \tag{f}$$

Solving Eq. (f) and labeling the roots as I_1, I_2, and I_3 we obtain

$$I_1 = 0.7324Ma^2 \qquad I_2 = 1.9343Ma^2 \qquad I_3 = 2.0000Ma^2 \qquad \textit{Answer}$$

as the principal moments of inertia (eigenvalues).

The direction cosines of a principal axis can be found by substituting the corresponding eigenvalue into Eqs. (a)–(d) and then solving these equations for the direction cosines l, m, and n. Substituting $I = I_3 = 2.0Ma^2$, Eqs. (a)–(c) become

$$0 = 0$$

$$-\frac{1}{3}m - \frac{1}{2}n = 0$$

$$-\frac{1}{2}m - n = 0$$

Since the second and third equations are linearly independent, their solution is $m = n = 0$. Combining these results with Eq. (d), $l^2 + m^2 + n^2 = 1$, we conclude that $l = \pm 1$. Hence the unit vector in the direction of the axis associated with I_3, i.e., the third eigenvector, is

$$\boldsymbol{\lambda}_3 = \pm\mathbf{i} \qquad \textit{Answer}$$

The plus-or-minus sign in this answer reflects the fact that an eigenvector defines only the direction of the principal axis; the sense of the unit vector along that axis is arbitrary. In other words, if $\boldsymbol{\lambda}$ is an eigenvector, then $-\boldsymbol{\lambda}$ is also an eigenvector.

With $I = I_2 = 1.9343Ma^2$, Eqs. (a)–(c) are

$$0.0657l = 0$$

$$-0.2676m - 0.5000n = 0$$

$$-0.5000m - 0.9343n = 0$$

From the first equation we get $l = 0$. The second and third equations are not independent, since both yield $n = -0.5352m$. Substituting these results into Eq. (d), we obtain

$$0^2 + m^2 + (-0.5352m)^2 = 1$$

which yields $m = \pm 0.8817$. Consequently, $n = -0.5352(\pm 0.8817) = \mp 0.4719$, from which the second eigenvector is

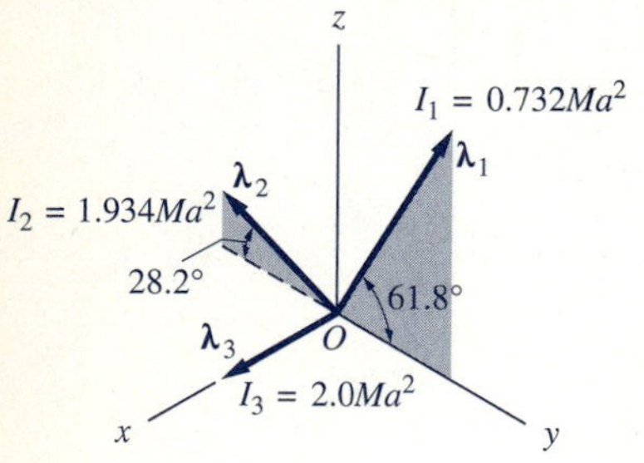

$$\boldsymbol{\lambda}_2 = \pm 0.882\mathbf{j} \mp 0.472\mathbf{k} \qquad \textit{Answer}$$

The direction of the first eigenvalue, corresponding to $I = I_1 = 0.7324Ma^2$, can be obtained in an analogous manner, with the result being

$$\boldsymbol{\lambda}_1 = \pm 0.472\mathbf{j} \pm 0.882\mathbf{k} \qquad \textit{Answer}$$

The three principal axes are shown in Fig. (b). Note that these axes are mutually orthogonal, as they should be, since the principal moments of inertia are distinct. Furthermore, as shown in Fig. (b), the signs of the eigenvectors are usually chosen so that $\boldsymbol{\lambda}_1$, $\boldsymbol{\lambda}_2$, and $\boldsymbol{\lambda}_3$ form a right-handed triad.

(b)

PROBLEMS

F.19 Use integration to determine the products of inertia of the homogeneous slender rod of mass m and length L about the axes shown.

F.20 Calculate I_{xy} for the slender rod in Prob. F.9 by integration.

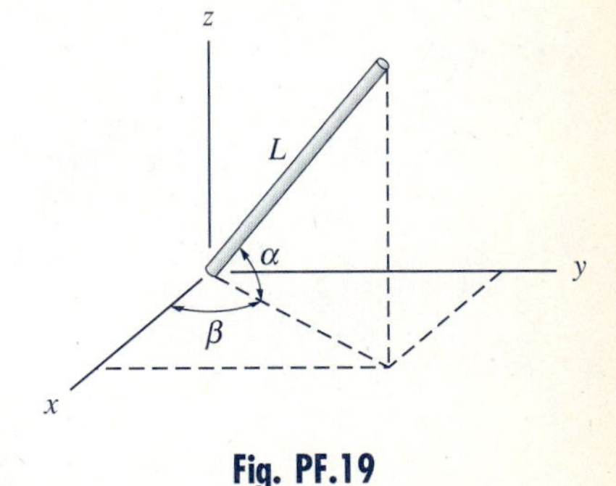

Fig. PF.19

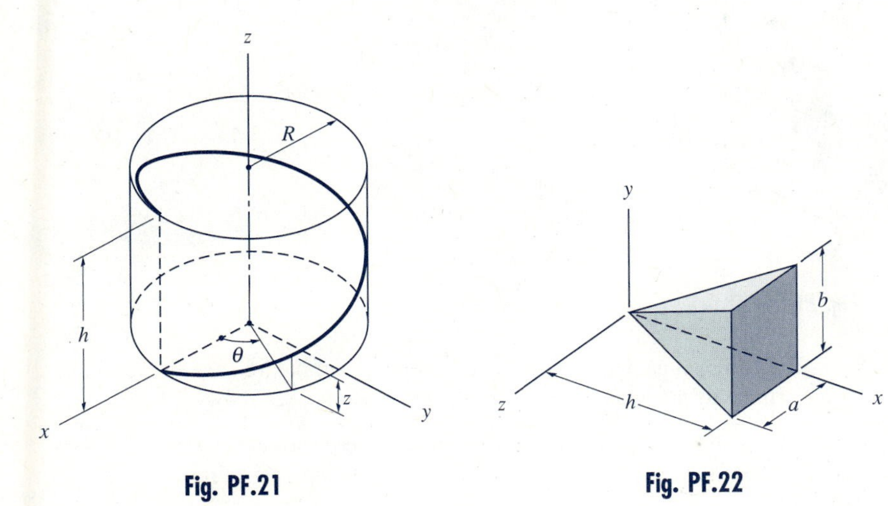

Fig. PF.21

Fig. PF.22

F.21 The uniform slender wire of mass m is bent into the shape of a helix. Using integration, find the products of inertia about the axes shown. (Note that the equation of the helix is $z = h\theta/(2\pi)$, where θ is measured in radians.)

F.22 Use integration to determine the products of inertia of the homogeneous solid about the axes shown.

F.23 By integration, find the products of inertia of the homogeneous prism about the axes shown.

F.24 Determine the products of inertia of the homogeneous solid of mass m about the axes shown. Use integration.

F.25 The mass of the thin homogeneous plate is m. Determine its inertia tensor at point O with respect to the axes shown.

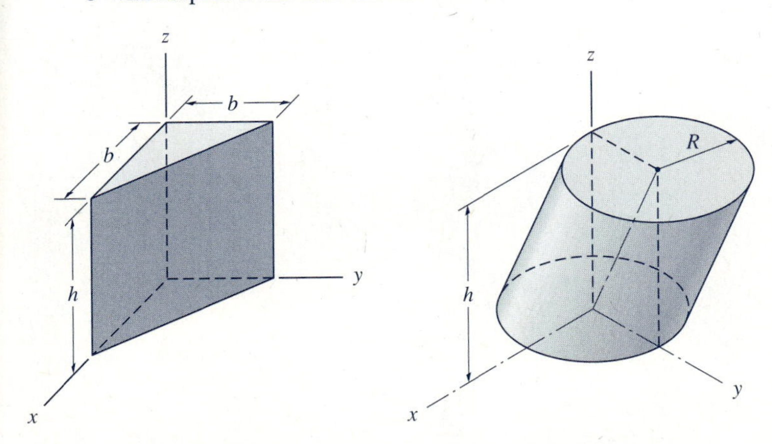

Fig. PF.23

Fig. PF.24

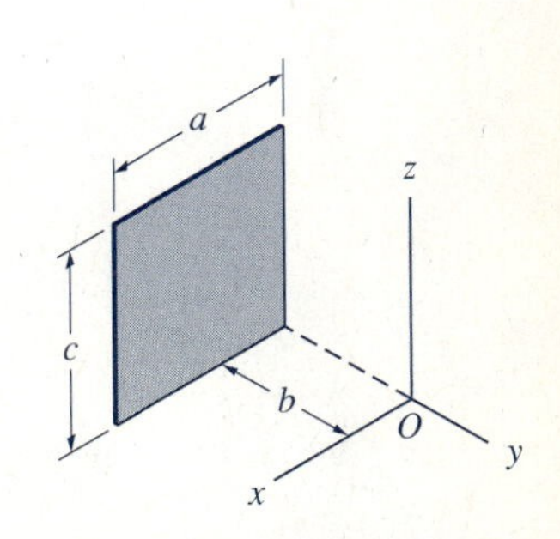

Fig. PF.25

F.26 The steel bracket has a uniform thickness of $\frac{1}{16}$ in. Calculate its products of inertia about the axes shown. (For steel, $\gamma = 490$ lb/ft^3.)

F.27 Determine the moment of inertia of the 18-lb homogeneous disk about axis AB.

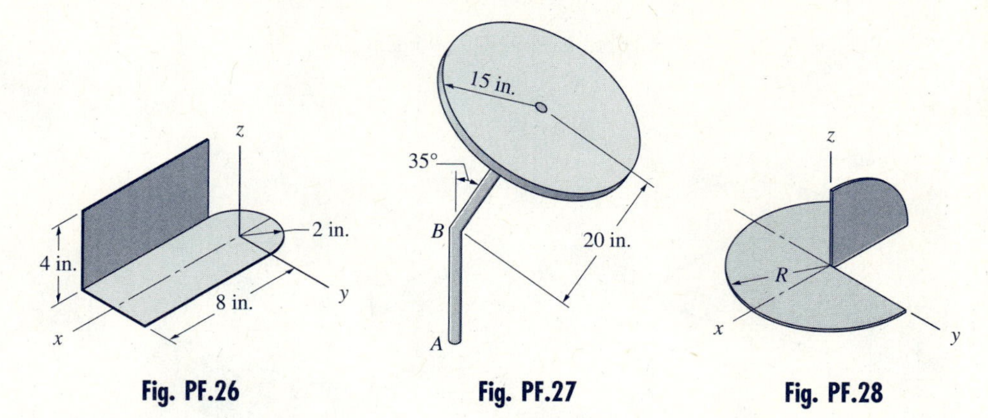

Fig. PF.26 **Fig. PF.27** **Fig. PF.28**

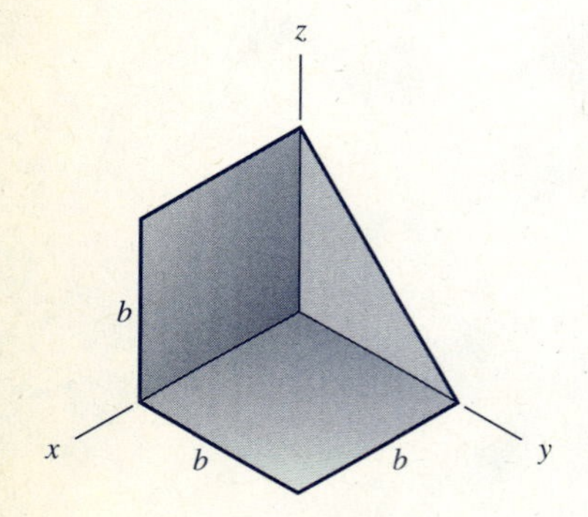

Fig. PF.29

F.28 The part shown is made by slitting and bending a thin circular plate of mass m. Determine its products of inertia with respect to the axes shown.

F.29 The body of mass m is fabricated from thin sheet metal of constant thickness. Calculate its products of inertia with respect to the axes shown.

F.30 The assembly is made by welding together three pieces of a uniform slender rod. If the mass of the assembly is m, determine its moment of inertia about the axis AB.

F.31 Calculate the moment of inertia of the homogeneous cone about the axis OA.

F.32 The mass of the uniform cube is m. (a) Determine the moment of inertia about the axis OA. (b) Show that OA is a principal axis of inertia at O. (Hint: Substitute your answer from part (a) into the characteristic equation for point O.)

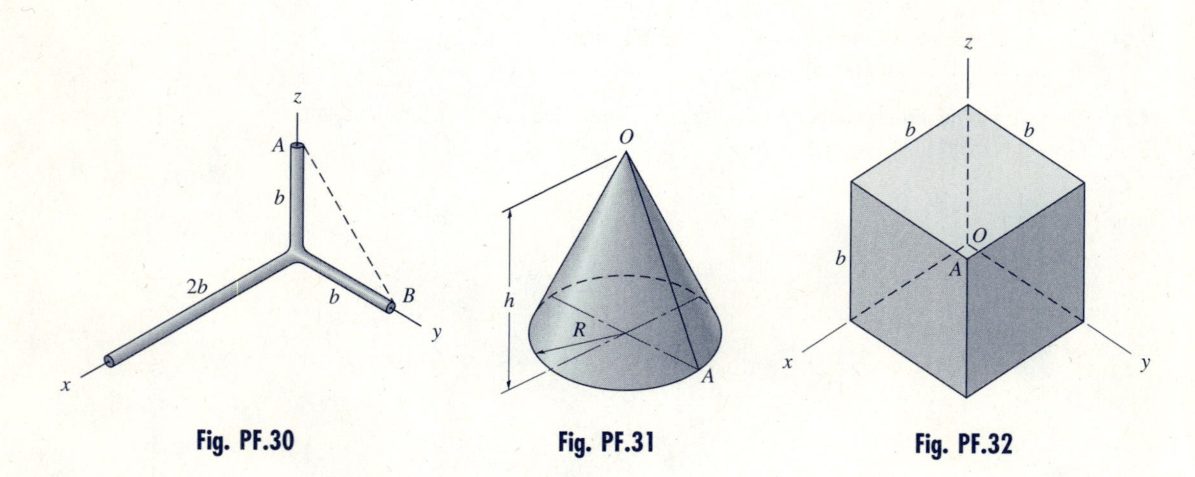

Fig. PF.30 **Fig. PF.31** **Fig. PF.32**

F.33 Determine the ratio h/R for the homogeneous cylinder so that the moments of inertia about all axes passing through its mass center are equal.

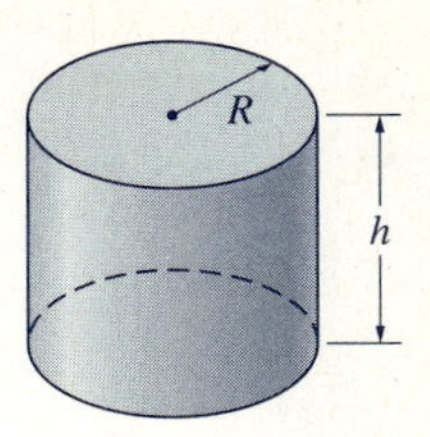

Fig. PF.33

F.34 Find the moment of inertia for the homogeneous thin plate about the side AB from the properties of its area.

F.35 The dimensions of the 0.36-kg thin uniform plate are $b = 240$ mm and $h = 150$ mm. For the mass center of the plate, (a) compute the principal moments of inertia; and (b) locate the principal axis corresponding to the smallest moment of inertia.

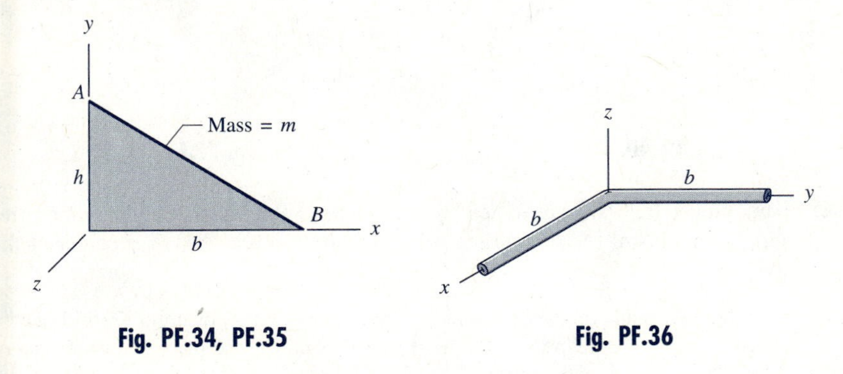

Fig. PF.34, PF.35

Fig. PF.36

F.36 The mass of the slender bent bar is m. Determine the principal axes and the principal moments of inertia at the mass center of the bar. (Hint: The principal axes can be located by inspection from symmetry.)

F.37 The small masses are joined by a light rod. For point O, determine (a) the inertia tensor with respect to the axes shown; and (b) the principal moments of inertia and principal axes at O. Show the principal axes on a sketch of the body.

F.38 The two small masses are joined by a rigid rod of negligible mass. Determine the principal moments of inertia of the system at its mass center. (Hint: It is unnecessary to calculate the inertia tensor because the principal axes can be found by inspection.)

F.39 The mass of the uniform bent rod is m. For point O, determine (a) the inertia tensor with respect to the axes shown; (b) the principal moments of inertia; and (c) the principal axis associated with the smallest moment of inertia.

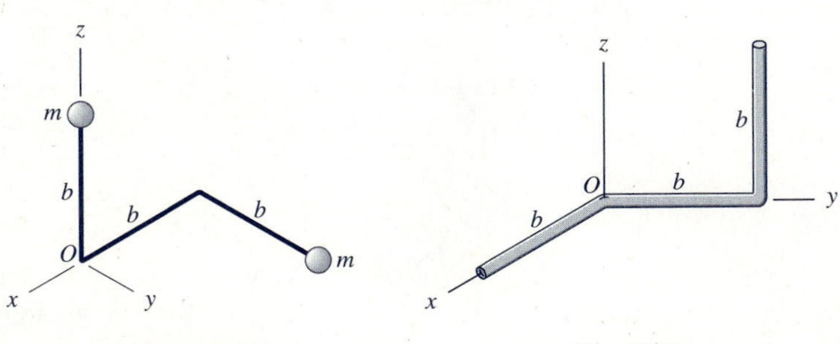

Fig. PF.37, PF.38

Fig. PF.39

F.40 The three small balls are joined by rods of negligible mass. For point O, determine (a) the inertia tensor with respect to the axes shown; and (b) the principal moments of inertia. Each rod is parallel to a coordinate axis.

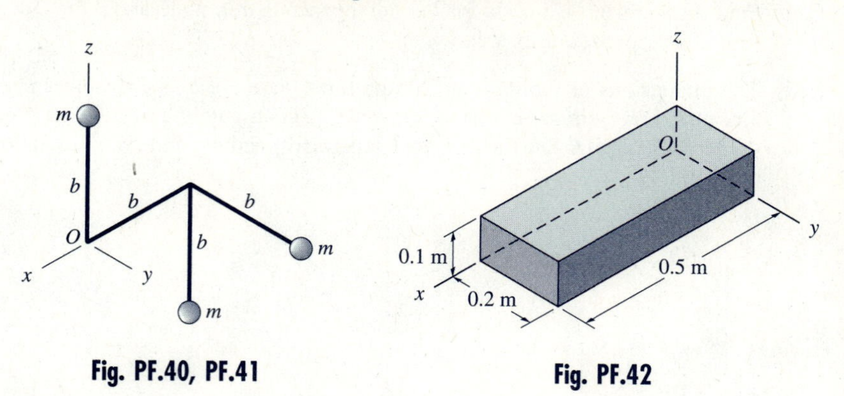

Fig. PF.40, PF.41

Fig. PF.42

F.41 The three small balls are joined by rods of negligible mass. Calculate the principal moments of inertia and the principal directions at the mass center of the system.

F.42 The block is made of steel for which $\rho = 7850\ \text{kg/m}^3$. For point O, find (a) the inertia tensor with respect to the axes shown; and (b) the principal moments of inertia and the principal axes.

Answers to Even-Numbered Problems

Chapter 12

12.2 (a) 17 m/s; (b) 10.0 s, 80.0 m; (c) 2.0 m/s^2

12.4 (b) 2.67 in.

12.6 (a) $v = -\frac{1}{2}R\omega \sin \omega t$, $a = -\frac{1}{2}R\omega^2 \cos \omega t$; (b) v_{max} would double and a_{max} would quadruple; (c) $2R$

12.8 (a) $a = -g(r_0/r)^2$; (b) $v_0 = \sqrt{2gr_0}$; (c) 36 700 ft/s

12.10 (a) $v = v_0\sqrt{1 + (2hx/b^2)^2}$; (b) $a = 2hv_0^2/b^2 \downarrow$; (c) $v_{max} = 82.5$ km/h, $a_{max} = 0.0772$ m/s^2

12.12 (a) $v = R\omega^2 t$; (b) 57.5°

12.14 $v = 15.15$ m/s, $a = 189.5$ m/s^2

12.16 (a) $\mathbf{v} = (4 + 6t)\mathbf{i} + (3 - 8t)\mathbf{j} - 6\mathbf{k}$ ft/s, $\mathbf{a} = 6\mathbf{i} - 8\mathbf{j}$ ft/s^2

12.18 (a) $v = -7.5$ ft/s, $a = 2.0$ ft/s^2; (b) $v = 534$ ft/s, $a = 105.0$ ft/s^2

12.20 (a) $v_A = b\dot{\theta}\sec^2\theta$; (b) $a_A = b\sec^2\theta(\ddot{\theta} + 2\dot{\theta}^2\tan\theta)$

12.22 (a) $v = -R\omega\sin\theta[1 + \cos\theta(9 - \sin^2\theta)^{-1/2}]$; (b) $v_{max} = 1.05R\omega$, $a_{max} = 1.3R\omega^2$

12.24 (b) -662 m/s^2

12.26 $\theta = \tan^{-1}(a/g)$

12.28 191.6 lb

12.30 (a) $x = v_0^2/(2\mu g)$; (b) 200 ft

12.32 (a) 45.08 ft/s; (b) 721 ft

12.34 (a) 1.0 m; (b) 600 m/s

12.36 $0.693m/k$

12.38 6.55 s

12.40 0.358 s

12.42 988 kN/m

12.44 $v = 1.082\sqrt{Vq/m}$

12.46 81.2 ft/s

12.48 (a) $y = \frac{1}{2}b[1 - (x/b)^2]$; (b) $F = -m(g + 8\pi^2 b/t_0^2)$

12.50 (a) $N_1 = 1.862mb\omega^2$; (b) $N_2 = mb\omega^2$

12.52 73.2°, 90.3 ft

12.54 37.6 ft/s, 29.1°

12.56 7.17 ft

12.58 (a) $t = \dfrac{m}{c}\ln\left(1 + \dfrac{c}{mg}v_0\sin\alpha\right)$; (b) $v = \dfrac{v_0\cos\alpha}{1 + (c/mg)v_0\sin\alpha}$

12.60 (a) $v = 404$ m/s, $h = 4040$ m; (b) 61.2 s; (c) 12 350 m

12.62 (a) 67.7 ft; (b) 45.2 ft/s ↗ 27.7°

12.64 $v_0 = 961$ ft/s, $d = 7690$ ft, $h = 1030$ ft

12.66 (a) 88.0 ft/s; (b) 1173 ft

12.68 (a) 20 m/s; (b) 138.7 m

12.70 (a) $v = 0.0$, $x = 0.500a_0t_0^2$; (b) $v = 0.250a_0t_0$, $x = 0.583a_0t_0^2$

12.72 (a) 7.15 ft/s; (b) 3.20 ft

12.74 (a) 532 ft/s; (b) 8690 ft; (c) 29.5 s

12.76 (b) 136.0 m

12.78 (b) 0.505 s

12.80 (b) 0.258 s

12.82 (a) 61.2 s; (b) 12 350 m

12.84 6.57 s

12.86 62.0 in./s

12.88 (a) $x_{max} = 1.555$ m, $v_{max} = 2.00$ m/s

12.90 (a) $a = -480x - 3.924\,\text{sgn}(v)$; (b) 0.287 s, 67.3 mm

12.92 (a) $a_x = 0.5x(x^2 + y^2)^{-3/2}$, $a_y = 0.5y(x^2 + y^2)^{-3/2}$; $x = 0.3$ m, $y = 0.4$ m, $v_x = 0$, $v_y = -2$ m/s at $t = 0$; (b) $x = 0.340$ m, $v = 1.751$ m/s

12.94 (No answer)

12.96 (a) See Prob. 12.92; (b) $x = 0.360$ m, $v = 1.796$ m/s

12.98 (a) $a_x = -40x[1 - 0.5(x^2 + y^2)^{-1/2}]$, $a_y = -9.81 - 40y[1 - 0.5(x^2 + y^2)^{-1/2}]$; $x = 0.5$ m, $y = -0.5$ m, $v_x = 0$, $v_y = 0$ at $t = 0$; (b) -0.40 m $\leq x \leq 0.50$ m, -1.02 m $\leq y \leq -0.46$ m

12.100 (a) $a_x = -0.05v_x\sqrt{v_x^2 + v_y^2} + 1.6v_y$, $a_y = -0.05v_y\sqrt{v_x^2 + v_y^2} - 1.6v_x - 32.2$; (b) $t = 1.051$ s, $x = 19.38$ ft

12.102 $a = -15/2 + x/16$ ft/s^2

12.104 23.2 m/s

12.106 16 540 m

12.108 (a) 0.050 lb · s/ft; (b) 18.61 s

12.110 (a) $F_{max} = 2mg$; (b) $v_{max} = g\sqrt{m/k}$

12.112 $0.0457g$

Chapter 13

13.2 (a) $\mathbf{v}_A = 4\mathbf{j}$ m/s, $\mathbf{v}_B = 8\mathbf{j}$ m/s; (b) $\mathbf{a}_A = -32\mathbf{i} + 12\mathbf{j}$ m/s^2, $\mathbf{a}_B = -64\mathbf{i} + 24\mathbf{j}$ m/s^2

13.4 (No answer)

13.6 (a) 0.2 m; (b) 22.4 m/s^2

13.8 72.5 m

13.10 $v = 16.13$ ft/s, $a_t = 9.95$ ft/s^2

13.12 (a) $dy/dx = 0.75$; (b) $\rho = 0.1563$ m

13.14 (a) $a_A = v_0^2/b$; (b) $a_B = 0.420v_0^2/b$

13.16 $a_{max} = 2.01$ ft/s^2, $s = 500$ ft

13.18 303 ft

13.20 $\mathbf{v} = R_0\omega e^{\omega t}(\mathbf{e}_R + \mathbf{e}_\theta)$, $\mathbf{a} = 2R_0\omega^2 e^{\omega t}\mathbf{e}_\theta$

13.22 (a) $v = 1.795b\omega$ 9.04°; (b) $a = 1.903b\omega^2$ 72.75°; (c) $1.710b$

13.24 $\dot{\theta} = (v_0/b)(1 + \theta^2)^{-1/2}$

13.26 $v = 3.85b\omega$ 45°, $a = 17.38b\omega^2$ 19.6°

13.28 $a = 3v_0^2/b$ 45°

13.30 $h = 5000$ ft, $v = 476$ ft/s

13.32 $v = \dot{R}\csc\theta$, $a = -(\dot{R}^2/b)\cot^3\theta$

13.34 49.6 in./s

13.36 $v = 8.58$ ft/s, $a = 44.2$ ft/s^2

13.38 (a) $v = \dfrac{h\omega}{2\pi\cos\beta}\sqrt{1 + \theta^2\sin^2\beta}$; (b) $a_R = -\dfrac{h\omega^2}{2\pi}\theta\tan\beta$, $a_\theta = \dfrac{h\omega^2}{\pi}\tan\beta$, $a_z = 0$

13.40 80.2 ft/s

13.42 $v = \sqrt{Rg\tan\beta}$

13.44 (a) $\ddot{\theta} = -4.905\sin\theta$ rad/s^2; (b) $v = 6.26\sqrt{\cos\theta - 0.866}$ m/s

13.46 (a) 135.1 N; (b) 6.68 m/s

13.48 (a) 416 lb; (b) 312 lb

13.50 11.18 m/s

13.52 $\theta = \mu/2$

13.54 (a) 39.0 lb; (b) 468 lb · ft

13.56 $0.0451 \text{ m} \le R \le 0.1276$ m

13.58 (a) $F_R = -400$ N, $F_\theta = 480$ N; (b) $F_R = -720$ N, $F_\theta = 0$

13.60 70 N/m

13.62 (a) $N_1 = mg\tan\theta + 2mb\omega^2\sec^2\theta\tan^2\theta$; (b) $N_2 = mg\sec\theta + 2mb\omega^2\sec^3\theta\tan\theta$

13.64 (a) $F_R = -4.50$ N, $F_\theta = 0$, $F_z = 5.89$ N; (b) 0.433

13.66 $F_R = -1033$ lb, $F_\theta = 675$ lb, $F_z = 15$ lb

13.68 1.854 s

13.70 (b) 120.2°

13.72 (a) $(\pi^4/144)R\sin^2\pi t - 32.2\sin[(\pi/12)\cos\pi t]$; (c) $t = 3.41$ s, $\dot{R} = 4.78$ ft/s

13.74 (c) 2.07 ft

13.76 (c) $R_{max} = 1.282$ m, $\theta = 130.9°$

13.78 0.621

13.80 42.7 N

13.82 98.6 ft/s

13.84 $N = mg(3\cos\theta - 2)$

13.86 0.537

13.88 6.10 rad/s

Chapter 14

14.2 (a) $U = F_0b/2$; (b) $U = F_0b/2$

14.4 (a) $U = -0.414kR^2$; (b) $U = WR$

14.6 $U = -0.1186\mu kb^2$

14.8 (a) $x = v_0^2/(2\mu g)$; (b) 200 ft

14.10 10.91 m

14.12 150.1 mm

14.14 78.5 N

14.16 (a) 15.73 ft/s; (b) 12.81 ft

14.18 (a) 0.966 m; (b) 1.795 m

14.20 (a) $x_{max} = (2mg/k)(\sin\theta - \mu\cos\theta)$; (b) $v_{max} = g\sqrt{m/k}(\sin\theta - \mu\cos\theta)$, $x = (mg/k)(\sin\theta - \mu\cos\theta)$

14.22 114.2 ft/s

14.24 90.3 lb

14.26 988 kN/m

14.28 (b) $v = 1.082\sqrt{Vq/m}$

14.30 4.91 ft/s

14.32 2.02×10^3 lb

14.34 $v = 1.794[10 - (x^2 + 36)^{1/2}]^{1/2}$ ft/s

14.36 $a = -8.05x(x^2 + 0.5625)^{-1/2}$ ft/s^2

14.38 $a = 9.81 - 400x$ m/s^2

14.40 $a = -1100[x - 0.15x(0.0225 + x^2)^{-1/2}]$ m/s^2

14.42 8.99 ft/s

14.44 4570 km

14.46 (a) 8.86 m/s; (b) 9.90 m/s

14.48 2620 N (at point O)

14.50 6.82 rad/s

14.52 $P_{max} = 47.7$ W, $t = 0.25$ s

14.54 (a) 0.816 m/s; (b) 9.81 m/s^2

14.56 42.7 hp

14.58 51.4 hp

14.60 (a) $P = -500x\sqrt{0.25 - x^2}\,\text{sgn}(v)$ lb · ft/s; (b) $P_{max} = 62.5$ lb · ft/s, $x = \pm 4.24$ in.

14.62 0.514 s

14.64 $18\mathbf{i} - 31.2\mathbf{j}$ m/s

14.66 5.35 s

14.68 (a) 21.9 ft/s; (b) 0.230 lb · s ↑

14.70 (a) $v_{max} = 5.00$ m/s, $t = 2.00$ s; (b) 0

14.72 (a) 3.00 s; (b) 13.42 ft/s

14.74 1.631 s

14.76 (a) 45.5 m/s; (b) 100.0 m/s

14.78 707 m/s

14.80 3.92 m/s

14.82 (a) $\dot{\theta} = 20 + 8.54[1 - \cos(\pi t/4)]$ rad/s; (b) 37.1 rad/s

14.84 $\dot{\theta} = 32.0$ rad/s, $\ddot{\theta} = 16.0$ rad/s^2

14.86 $-0.497\mathbf{k}$ lb · ft

14.88 6980 m/s

14.90 $h_x = 212$ N · m · s, $h_y = 170$ N · m · s, $h_z = 108$ N · m · s

14.92 3580 km

14.94 $R_{min} = 1.119$ m, $R_{max} = 1.645$ m

14.96 (a) -36.5 rad/s; (b) 9.00 N (T)

14.98 2860 lb · ft

14.100 (a) $\ddot{\theta} = 0$; (b) $\ddot{\phi} = (\dot{\theta}^2 \cos\phi - g/L)\sin\phi$

14.102 $R_{min} = 225 \times 10^3$ mi, $R_{max} = 251 \times 10^3$ mi

14.104 (b) 68.3 years; (c) 4.87×10^9 km

14.106 6.47 s

14.108 (a) 6710 ft/s; (b) The spacecraft would crash.

14.110 (b) 2.23 km/s

14.112 46.5°

14.114 (a) $\ddot{R} = R\dot{\theta}^2 - KM_e/R^2$, $\ddot{\theta} = -2\dot{R}\dot{\theta}/R$; $R = 23.46 \times 10^6$ ft, $\dot{R} = 2.237 \times 10^3$ ft/s, $\theta = 0$, $\dot{\theta} = 1.0899 \times 10^{-3}$ rad/s at $t = 0$; (c) 7020 s; (d) $H_{min} = 316$ mi, $H_{max} = 1606$ mi

14.116 2.5

14.118 6.50 m/s

14.120 6.02 m

14.122 $v = 64.4t(1 + 0.010\,83t)/(1 - 0.0433t)$ ft/s

14.124 1.930 m/s

14.126 (a) $C = 1288 \sin 2\phi$ lb · ft; (b) 4830 lb · ft

Chapter 15

15.2 (a) 22.6°; (b) 12 min

15.4 8.97 mi/h

15.6 $v_A = 226$ mi/h, $v_B = 304$ mi/h

15.8 (No answer)

15.10 (a) 9.0 in./s →; (b) 2.0 in./s ↑

15.12 32.0 in./s ↓

15.14 $v = 0.1033$ ft/s ↑, $a = 0.0441$ ft/s^2 ↓

15.16 $v = 2.12$ m/s, $a = 0.265$ m/s^2

15.18 52.1°

15.20 $v_{A/B} = R\omega_0 \cos\theta$ ↑, $a_{A/B} = R\omega_0^2 \sin\theta$ ↓

15.22 $\mathbf{v}_{B/A} = 1.6\mathbf{i}$ m/s, $\mathbf{a}_{B/A} = -8.0\mathbf{i} - 3.2\mathbf{j}$ m/s^2

15.24 (a) $\dot{R} = 0.447v_0$, $\ddot{R} = -0.537v_0^2/b$; (b) $\dot{R} = -v_0$, $\ddot{R} = 0$

15.26 $v_A = 11.43$ ft/s ↗ 64.5°, $v_B = 9.83$ ft/s ←

15.28 (a) 5.30 m/s^2; (b) 6.36 kN

15.30 $a = 2.92$ m/s^2, $P = 0.566$ N

15.32 32.0 lb

15.34 $a_A = 3.057$ m/s^2, $a_B = 1.057$ m/s^2, $a_C = 2.057$ m/s^2

15.36 $x_A = 4.0$ ft, $x_B = 8.0$ ft

15.38 $a_A = 0$, $a_B = 24.15$ ft/s^2

15.40 $a_A = 2.45$ m/s^2 ↓, $a_B = 4.25$ m/s^2 →

15.42 0.522 m/s^2 ↑

15.44 $a_B = 0.325g$ →, $a_{A/B} = 0.717g$ 30° ↙

15.46 (a) $a_1 = 92.74 - 154.56(x_1 - x_2)$, $a_2 = 154.56(x_1 - x_2)$; $x_1 = x_2 = v_1 = v_2 = 0$ at $t = 0$; (b) $v_1 = 7.23$ in./s, $v_2 = 2.05$ in./s, $P = 0.717$ lb

15.48 (a) $a_A = -\frac{1}{3}(x_B - x_A)^{-2}$, $a_B = \frac{1}{2}(x_B - x_A)^{-2}$; $x_A = 0$, $x_B = 0.5$ m, $v_A = 0$, $v_B = -2$ m/s at $t = 0$; (b) $d_{min} = 0.228$ m, $v_A = v_B = 0.800$ m/s ←

15.50 (a) $a_A = \frac{10}{3}(v_B - v_A)$, $a_B = -2000(v_B - v_A)$; $x_A = x_B = v_A = 0$, $v_B = 600$ m/s at $t = 0$; (b) $v_B = 83.1$ m/s, $x_A = 0.568$ mm

15.52 (a) $a_A = -1000x_A - 2.943\,\mathrm{sgn}(v_A - v_B)$, $a_B = 1.4715\,\mathrm{sgn}(v_A - v_B)$; (c) $(v_A)_{max} = 0.539$ m/s at $t = 0.0497$ s, $(v_B)_{max} \approx 0.132$ m/s at $t \approx 0.09$ s

15.54 (c) $v = 5.87$ ft/s, $\theta = 118.4°$

15.56 Block A hits first, $v_A = 4.91$ ft/s

15.58 $v_A = 1.692$ ft/s ←, $v_B = 2.538$ ft/s →

15.60 1.833 ft/s

15.62 (a) 1.232 m/s; (b) 1.445 m

15.64 (a) 40.2 m; (b) 3.08 m/s

15.66 8.79 ft/s

15.68 1.926 ft

15.70 6340 lb

15.72 (a) 2.72 m/s ←; (b) 1.087 m/s →

15.74 (a) (26.2 m, 29.2 m); (b) $49.7\mathbf{i} + 111.85\mathbf{j}$ N · s; (c) 3240 N · m · s CCW; (d) 1768 N · m · s CCW

15.76 (a) 10.80 s; (b) 103.1 rev

15.78 $\dot{\theta} = (2v_0/L)(m_A - m_B)/(m_A + m_B)$ CCW

15.80 $\mathbf{v}_A = \frac{1}{4}v_0(-\mathbf{i} + 3\mathbf{j})$, $\mathbf{v}_B = \frac{3}{4}v_0(\mathbf{i} + \mathbf{j})$

15.82 $R = 0$ and any $\dot{\theta}$, $\dot{\theta} = \sqrt{k/m}$ and any R

15.84 (c) 0.283 m $\leq R \leq$ 0.4 m, 30.0 rad/s $\leq \dot{\theta} \leq$ 40.0 rad/s

15.86 1008 m/s

15.88 (a) 0.203 m; (b) 99.8%

15.90 (a) 1.732 ft/s; (b) 81.25%

15.92 0.865 ft

15.94 120 lb

15.96 (a) $u = v\cos\alpha$; (b) $\frac{1}{2}mv^2 \sin^2\alpha$

15.98 (a) 0.419 ft/s; (b) $-97.60\mathbf{i} - 4660\mathbf{j}$ lb · s

15.100 1/3

15.102 (No answer)

15.104 (No answer)

15.106 $v_A = 1.80$ ft/s →, $v_B = 4.64$ ft/s →, $v_C = 6.56$ ft/s →

15.108 (No answer)

15.110 $v_A = 8.91$ ft/s ↑, $v_B = 25.87$ ft/s ↗ 39.4°

15.112 (a) 0.5; (b) $v_A = 4.10$ m/s →, $v_B = 5.64$ m/s ↑

15.114 5.69 rad/s

15.116 (a) 8230 ft/s; (b) 5350 ft/s

15.118 (a) 127.2×10^3 ft; (b) 560.1×10^3 ft

15.120 (a) 7070 N; (b) 5940 N

15.122 10.51 lb ↓ 75.0°

15.124 29.3×10^3 lb ←

15.126 46.6 kN

15.128 1045 lb $\rightarrow$

15.130 155.7 N

15.132 (a) $\sqrt{2P/\rho}$; (b) Energy is lost in impacts between links at the bend.

15.134 (a) $P = 0.0994(18v - v^2)$ lb · ft/s; (b) $P_{\max} = 8.05$ lb · ft/s, $v = 9.0$ ft/s

15.136 (a) $\mathbf{r}_{B/A} = -4.713\mathbf{i} + (8.60t - 120)\mathbf{j}$ m; (b) 57.8 m

15.138 70.5 mi/h 73.5°

15.140 $a_A = 0.273g \downarrow$, $a_B = 0.455g \uparrow$, $a_C = 0.636g \downarrow$

15.142 (a) 0.796; (b) 5.16 m/s

15.144 $\dot{R} = 1.422b\omega_0$

15.146 0.641 m/s

Chapter 16

16.2 48.0 s

16.4 (a) 180 rad/s; (b) 229 rev

16.6 (a) 5.0 rad/s^2; (b) 14.32 rev

16.8 $v_C = 3.60$ m/s, $a_C = 37.7$ m/s^2

16.10 $v_B = 28.0$ in./s, $a_B = 448$ in./s^2

16.12 $\mathbf{v}_B = -0.604\mathbf{i} - 0.725\mathbf{k}$ m/s, $\mathbf{a}_B = -3.55\mathbf{i} + 2.47\mathbf{j} + 0.19\mathbf{k}$ m/s^2

16.14 $\boldsymbol{\omega} = -2.500\mathbf{i} + 0.625\mathbf{j}$ rad/s

16.16 3.49 in./s^2

16.18 $v_D = 82.7$ in./s $\rightarrow$, $a_D = 2070$ in./s^2 $\rightarrow$

16.20 $\dot{\theta}_2 = -1.239$ rad/s, $\ddot{\theta} = 1.946$ rad/s^2

16.22 (a) $\dot{\theta} = (v_0 \sin^2\theta)/(R\cos\theta)$; (b) $\ddot{\theta} = (1 + \cos^2\theta)\tan^3\theta(v_0/R)^2$

16.24 $a_A = 2R\alpha_0(\sin\theta + 2\theta\cos\theta) \downarrow$

16.26 1.5

16.28 (a) $v_A = R\omega_0 \sin\theta\left(1 + \cos\theta/\sqrt{4 - \sin^2\theta}\right)$; (b) $|v_A|_{\max} = 1.12R\omega_0$, $\theta = 68°$ and 292°

16.30 2.22 rad/s

16.32 $v_B = L\sqrt{\dot{\theta}_1^2 + \dot{\theta}_2^2 + 2\dot{\theta}_1\dot{\theta}_2\cos(\theta_2 - \theta_1)}$

16.34 $v_B = 96.0$ in./s $\rightarrow$, $v_D = 48.0$ in./s $\uparrow$

16.36 $\omega_B = \omega_0(r_A + r_B)/r_B$ CCW

16.38 $\omega_B = 7.50$ rad/s CCW, $\omega_{AB} = 3.33$ rad/s CW

16.40 (a) $\omega = 0.600\sin^2\theta$ rad/s; (b) $\alpha = 0.720\sin^3\theta\cos\theta$ rad/s^2

16.42 $\omega_{BD} = \omega_{DE} = 1.143$ rad/s CCW

16.44 32.5 rad/s

16.46 4.57 rad/s CCW

16.48 (a) 5.31°; (b) 3.17 m/s

16.50 (0.0583 m, −0.050 m)

16.52 (a) (4.0 ft, 3.0 ft); (b) (0, 36.125 in.)

16.54 (No answer)

16.56 $\omega = 57.7$ rad/s CCW, $v_B = 1.732$ m/s, $v_C = 1.732$ m/s 30°

16.58 0.553 m/s

16.60 1/4

16.62 $\omega_{BE} = 4.57$ rad/s CCW, $v_{BE/D} = 94.2$ in./s

16.64 1.0 m/s^2 30°

16.66 (a) $\alpha_{AB} = \alpha_{BC} = 5.0$ rad/s^2 CCW; (b) 0.515 m/s^2

16.68 5.10 m/s^2

16.70 $a_G = L\alpha/2$ θ

16.72 $a_G = 11.94$ m/s^2 15.1°

16.74 $\alpha_{AB} = 48.0$ rad/s^2 CW, $\alpha_{BD} = 144.0$ rad/s^2 CCW

16.76 $\alpha_{BD} = 2.45$ rad/s^2 CCW, $\alpha_{DE} = 6.12$ rad/s^2 CW

16.78 (b) $\alpha_{BD} = 3.53$ rad/s^2 CCW, $a_D = 0.177$ m/s^2 $\rightarrow$

16.80 $\omega_{BE} = 10.24$ rad/s CW, $v_{D/BE} = 38.4$ in./s

16.82 $\omega_{AB} = 9.94$ rad/s CCW, $v_A = 149.1$ in./s $\leftarrow$

16.84 25.9 ft/s^2 46.7°

16.86 $\alpha = \omega_0^2/\sqrt{3}$ CCW

16.88 (No answer)

16.90 (a) $\omega_{AD} = 0.093\,75$ rad/s CW, $\omega_{EF} = 0.125$ rad/s CW; (b) $\alpha_{AD} = 0.001\,69$ rad/s^2 CW, $\alpha_{EF} = 0.006\,76$ rad/s^2 CW

16.92 (a) 2.88 rad/s CW; (b) 8.60 rad/s^2 CCW

16.94 14.14 rad/s

16.96 $\mathbf{v}_C = -6.69\mathbf{i} - 10.03\mathbf{k}$ ft/s, $\mathbf{a}_C = -16.8\mathbf{i} - 48.4\mathbf{j} + 71.6\mathbf{k}$ ft/s^2

16.98 $\alpha_{BC} = 31.6$ rad/s^2 CCW, $a_C = 228$ in./s^2 $\leftarrow$

16.100 $40\mathbf{i} + 437\mathbf{j}$ in./s^2

16.102 (a) $\omega_{BC} = 0$, $\omega_{CD} = 5.0$ rad/s CCW; (b) $\alpha_{BC} = 28.41$ rad/s^2 CCW, $\alpha_{CD} = 18.75$ rad/s^2 CCW

16.104 $v_C = 2b\dot{\theta}$, $a_C = 2b\dot{\theta}^2\sqrt{4 + (\ddot{\theta}/\dot{\theta}^2)^2}$

Chapter 17

17.2 9.01 slug · ft^2

17.4 $\bar{z} = 100.8$ ft, $\bar{I}_z = 861 \times 10^6$ slug · ft^2

17.6 $\bar{x} = -0.1881$ ft, $I_z = 1.018$ slug · ft^2, $\bar{I}_z = 0.956$ slug · ft^2

17.8 $I_z = 0.019\,44$ kg · m^2, $k_z = 0.0900$ m

17.10 $\bar{z} = 5.19$ in., $\bar{k}_x = 3.24$ in.

17.12 0.1015

17.14 $\bar{x} = 1.645$ m, $\bar{I} = 134.9$ kg · m^2

17.16 (No answer)

17.18 (a) 9.20 m/s^2; (b) 0.938

17.20 4.91 ft/s^2

17.22 $a = 11.63$ ft/s^2 $\rightarrow$, $N_A = 37.2$ lb $\uparrow$

17.24 $d/h = \mu$

17.26 $\theta_1 = \theta_2 = 14.04°$

17.28 (No answer)

17.30 (a) 10.90 rad/s^2 CCW; (b) 5.61 rad/s^2 CCW

17.32 $\alpha_A = 5.36$ rad/s^2, $\alpha_B = 1.79$ rad/s^2, $F_C = 3.57$ N

17.34 $\frac{1}{3}mg$

17.36 $\alpha = 16.57$ rad/s^2 CW, $N_A = 513$ N $\uparrow$, $F_A = 165.1$ N $\rightarrow$

17.38 (a) 6.50 in.; (b) $\bar{k} = 4.26$ in.

17.40 1.743 rad/s^2 CCW

17.42 $A = 4.25$ N 48.8°, $B = 6.40$ N $\downarrow$

17.44 (a) 6.38 rad/s^2 CW; (b) 0.507

17.46 $\bar{a} = 2.57$ m/s^2

17.48 $\bar{a} = 4.85$ m/s^2 $\rightarrow$

17.50 $\beta = 28.1°$, $\alpha = 9.47$ rad/s^2 CW

17.52 2.69 rad/s

17.54 3.02 rad/s^2

17.56 $\alpha = 0.768$ rad/s^2 CW, $T = 414$ lb

17.58 (a) 8.0 m/s^2; (b) 0.80 m

17.60 $0.217mg$

17.62 5.94 m/s

17.64 $\alpha = 2.42$ rad/s^2, $N_A = 10.54$ lb, $N_B = 9.46$ lb

17.66 $C_A = \underbrace{454 \sin\theta \cos\theta}_{\text{inertia effect}} + \underbrace{0.1667 \cos\theta}_{\text{due to weight}}$

17.68 (a) $\frac{1}{2}mgb/h$; (b) mgb/h

17.70 (a) 18.17 rad/s^2 CCW; (b) 4.87 rev

17.72 (a) 0.565 rad/s^2 CW; (b) 92.6 s

17.74 (a) $\alpha = \frac{3}{2}(g/L)\sin\theta$ CW, $\omega = \pm\sqrt{3(g/L)(\cos\theta - \cos\theta_0)}$; (b) $A_{max} = \frac{1}{2}mg(5 - 3\cos\theta_0)\uparrow$, $\theta = 0$

17.76 (a) $\alpha = 4.150 + 9.222\cos\theta$ rad/s^2; (b) $-711°$

17.78 (b) $\theta_{max} = 209.2°$, $t = 1.375$ s

17.80 (a) $\omega_{max} = 3.16$ rad/s, system oscillates between $\theta = 0$ and 98°; (b) $\omega_{max} = 15.52$ rad/s, system rotates CW with increasing ω.

17.82 (a) 1.739 s; (b) 2.03 rev CCW; (c) 4.0 ft/s $\leftarrow$

17.84 (a) $\omega = 3.132\sqrt{0.8660 - \cos\theta}$ rad/s, $\alpha = 4.905\sin\theta$ rad/s^2; (b) $N_A = 183.94(3\cos\theta - 1.7320)\sin\theta$ N; (c) 54.7°

17.86 (b) 24.6°

17.88 (b) 2.27 s

17.90 (b) $r_{min} = 0.595$ ft, $\theta = -92.1°$

17.92 (a) 70.9°

17.94 (a) 67.5 mm; (b) 111.6 mm; (c) 88.8 mm

17.96 $\alpha = P/(mR)$

17.98 163.1 N

17.100 27.4 rad/s^2

17.102 $3.59mg\uparrow$

17.104 1.067 rad/s

Chapter 18

18.2 (a) $\frac{17}{48}mL^2\omega^2$; (b) $\frac{25}{96}mL^2\omega^2$

18.4 21.2 lb · ft

18.6 (a) 20.0 ft; (b) 39.5 ft/s

18.8 $v_0^2/(4\mu gR)$ rad

18.10 6.28 rad/s

18.12 6.01 rad/s

18.14 (a) $2\sqrt{ge/(R^2 + 2e^2)}$; (b) $R/\sqrt{2}$; (c) $1.189\sqrt{g/R}$

18.16 11.09 rad/s

18.18 0.966

18.20 3.87 ft/s

18.22 3.70 rad/s

18.24 $\omega = 3.67\sqrt{\cos\theta - 0.819}$ rad/s, $\alpha = -6.74\sin\theta$ rad/s^2

18.26 13.36 ft/s

18.28 (a) 3140 lb · ft; (b) 5.0 rad/s

18.30 $\omega_{AB} = \omega_{BC} = 0.961\sqrt{g/L}$

18.32 4.20 rad/s

18.34 17.39 N · m

18.36 (a) 38.2 N · m; (b) 19.1 N · m

18.38 1.337 hp

18.40 (b) $C_{max} \approx 92$ N · m, $\omega \approx 73$ rad/s

18.42 (a) $A_A = \frac{1}{2}(P_0L^2/v_0)$ CW; (b) $\mathbf{L} = -(P_0L/v_0)\mathbf{j}$

18.44 3.33 N · m · s CW

18.46 (a) $h_G = 0.1302mb^2\omega$ CW; (b) $h_O = 0.5208mb^2\omega$ CW; (c) $h_A = 0.0208mb^2\omega$ CW

18.48 $d = R/2$

18.50 (a) 29.7 rad/s; (b) $P_{max} = 38.6$ W, $t = 3.33$ s

18.52 0.933

18.54 $t = v_0/(2\mu g)$

18.56 (a) 1.583 m/s; (b) 2.34 s

18.58 (a) 9.13 rad/s CW; (b) 9.56 rad/s CCW

18.60 (a) 3.50 rad/s; (b) 3.50 rad/s

18.62 (a) 2.11 s; (b) 45.0 lb · ft

18.64 1.429 m/s

18.66 155.8%

18.68 211 rev/min

18.70 (a) 38.6 N · s; (b) 2.68 N · s $\rightarrow$

18.72 (a) 0.686 rad/s CCW; (b) 15.0°

18.74 (a) 15.0 rad/s CW; (b) 57.1%

18.76 (a) $v_A = v_1$; (b) 0%

18.78 3.78 m/s

18.80 (a) 36.0 rad/s CW; (b) 10.29 rad/s CCW

18.82 (a) $\bar{v} = 0.0805$ m/s $\uparrow$, $\omega = 19.59$ rad/s CCW; (b) $4.49\mathbf{i} + 2.04\mathbf{j}$ m/s

18.84 $\omega_{max} = (g/R)\sqrt{m/(6k)}$

18.86 82.7°

18.88 15.71 ft/s $\downarrow$

18.90 32.0 ft/s $\rightarrow$

18.92 $t = 0.824$ s, $\omega = 30.0$ rad/s for both disks

18.94 11.22 rad/s

Chapter 19

19.2 $v_z = 2.40$ m/s, $\boldsymbol{\omega} = 45.3\mathbf{i} + 52.1\mathbf{j} - 13.2\mathbf{k}$ rad/s

19.4 (a) $2.22\mathbf{j}$ rad/s; (b) $2.04\mathbf{i}$ rad/s^2; (c) $\mathbf{v}_P = 0.1568\mathbf{i}$ m/s, $\mathbf{a}_P = -0.144\mathbf{j} - 0.204\mathbf{k}$ m/s^2

19.6 (a) $(-2\mathbf{j} + \mathbf{k})\omega_0$; (b) $2\omega_0^2\mathbf{i}$; (c) $\mathbf{v}_P = -4\omega_0R\mathbf{i}$, $\mathbf{a}_P = (-6\mathbf{j} - 4\mathbf{k})\omega_0^2R$

19.8 (a) $\boldsymbol{\omega}_B = -43.3\mathbf{j} + 50.0\mathbf{k}$ rad/s; (b) $\boldsymbol{\alpha}_B = 1082\mathbf{i}$ rad/s^2

19.10 (a) $15\mathbf{i} - 60\mathbf{j}$ rad/s; (b) $120\mathbf{j} - 900\mathbf{k}$ rad/s^2; (c) 97.4°

19.12 (a) $36.0\mathbf{i}$ ft/s; (b) $-288\mathbf{j} - 432\mathbf{k}$ ft/s^2

19.14 $\boldsymbol{\omega}_{CD} = 7.20\mathbf{j}$ rad/s, $\boldsymbol{\omega}_{BD} = \mathbf{0}$; (b) $34.6\mathbf{j}$ rad/s^2

19.16 $1.370\mathbf{i} + 0.949\mathbf{j} - 1.780\mathbf{k}$ rad/s

19.18 $\boldsymbol{\omega}_A = -2.4\mathbf{j}$ rad/s, $\boldsymbol{\omega}_C = 0.90\mathbf{i} - 2.40\mathbf{j} + 1.20\mathbf{k}$ rad/s

19.20 $\mathbf{v}_B = 0.378R\omega_0\mathbf{j}$, $\boldsymbol{\omega}_C = (-0.1889\mathbf{i} + 0.50\mathbf{j} + 1.889\mathbf{k})\omega_0$

19.22 $\boldsymbol{\omega}_{OB} = -0.393\mathbf{i}$ rad/s, $\boldsymbol{\omega}_{AB} = -(0.374\mathbf{i} + 0.0245\mathbf{j} + 0.6765\mathbf{k})$ rad/s

19.24 (a) $(0.1083\mathbf{j} + 0.4375\mathbf{k})mR^2\omega_0$; (b) $0.219mR^2\omega_0^2$

19.26 (a) $\mathbf{h}_O = \left(\frac{1}{2}\mathbf{i} + \frac{2}{3}\mathbf{j}\right)mb^2\omega_0$; (b) $\mathbf{h}_B = \left(\frac{1}{2}\mathbf{i} + \frac{2}{3}\mathbf{j}\right)mb^2\omega_0$

19.28 0.410

19.30 $-3.15\mathbf{j} + 19.43\mathbf{k}$ lb · ft · s

19.32 (a) $1.35\mathbf{i} - 1.63\mathbf{j} + 3.66\mathbf{k}$ N · m · s; (b) 13.26 N · m

19.34 3.28 rad/s

19.36 15.64 rad/s

19.38 $0.48\mathbf{j} + 2.80\mathbf{k}$ rad/s

19.40 $1.02\mathbf{j} - 2.45\mathbf{k}$ rad/s

19.42 $0.776\mathbf{j} - 0.631\mathbf{k}$

19.44 $\mathbf{R} = \mathbf{0}, \mathbf{C} = -\frac{1}{4}mR^2\omega_0^2 \sin\beta \cos\beta\mathbf{i}$

19.46 $\mathbf{R}_A = -\frac{1}{2}mR\omega_0^2 \sin\beta(1 + \cos\beta)\mathbf{j}$,
$\mathbf{R}_B = -\frac{1}{2}mR\omega_0^2 \sin\beta(1 - \cos\beta)\mathbf{j}$

19.48 $\mathbf{R}_A = -648\mathbf{i}$ N, $\mathbf{R}_B = 648\mathbf{i}$ N

19.50 $0.75b$

19.52 $\mathbf{R} = -0.433mL\omega_0^2\mathbf{j}, \mathbf{C} = 0.577mL^2\omega_0^2\mathbf{i}$

19.54 3.62 rad/s

19.56 $C_O = 1.612$ N · m, $C_B = 0.117$ N · m

19.58 $\dot{\omega} = 0, \ddot{\beta} = \omega^2 \sin\beta\cos\beta$

19.60 22.1 lb

19.62 $\frac{1}{3}\cos\theta$

19.64 2.72 lb

19.66 (b) 1.916 s; (d) Numerical solution would yield $\beta = 0$ for all t.

19.68 $C = 0.1139$ lb · ft CCW

19.70 55.8°

19.72 4.33 rad/s

19.74 (a) $\ddot{\theta} = (0.75\dot{\phi}^2\cos\theta - 0.25\dot{\phi}\dot{\psi} + 117.70)\sin\theta$,
$\ddot{\phi} = (-1.75\dot{\phi}\cos\theta + 0.25\dot{\psi})\dot{\theta}\csc\theta$,
$\ddot{\psi} = (1.75\dot{\phi}\cos\theta - 0.25\dot{\psi})\dot{\theta}\cot\theta + \dot{\theta}\dot{\phi}\sin\theta - 0.5(\dot{\phi}\cos\theta + \dot{\psi})$;
(b) $\omega_z = 123.75$ rad/s at $t = 0$, $\omega_z = 75.06$ rad/s at $t = 1.0$ s

19.76 $(2/I_z)\sqrt{IPd\cos\theta}$

19.78 (a) $\ddot{\theta} = (-\dot{\phi}^2\cos\theta - 2\dot{\phi}\dot{\psi} + 915.9)\sin\theta$, $\ddot{\phi} = 2\dot{\theta}\dot{\psi}\csc\theta$,
$\ddot{\psi} = (-2\dot{\psi}\cot\theta + \dot{\phi}\sin\theta)\dot{\theta}$;
$\theta = \pi/2$ rad, $\phi = \psi = \dot{\theta} = \dot{\phi} = 0$, $\dot{\psi} = 52.36$ rad/s at $t = 0$;
(b) 1.571 rad $\leq \theta \leq$ 1.734 rad, $0 \leq \dot{\phi} \leq 17.49$ rad/s, 52.36 rad/s $\leq \dot{\psi} \leq$ 55.20 rad/s

19.80 (a) $\theta = 16.1°$, $\dot{\psi} = -0.520$ rad/s, $\dot{\phi} = 1.082$ rad/s

19.82 $\omega = 0.267$ rad/s ↗ 16.92°

Chapter 20

20.2 (a) 0.050 m; (b) 0.0858 s

20.4 (a) $p = \sqrt{k_1k_2/[m(k_1 + k_2)]}$; (b) $p = \sqrt{(k_1 + k_2)/m}$

20.6 0.747

20.8 (a) $\dot{\theta}_{max} = p\theta_0, \ddot{\theta}_{max} = p^2\theta_0$; (b) $\dot{\theta}_{max} = p\sqrt{2(1 - \cos\theta_0)}$,
$\ddot{\theta}_{max} = p^2\sin\theta_0$

20.10 6.26 Hz

20.12 (b) 4.38 Hz

20.14 (No answer)

20.16 (a) $x = 0.026\,67(-0.5\sin 50t + \sin 25t)$ m; (c) 34.6 mm

20.18 (No answer)

20.20 $\omega < 39.3$ rad/s or $\omega > 50.8$ rad/s

20.22 (a) 400 rad/s; (b) 1.847 mm

20.24 (No answer)

20.26 (a) $\ddot{\theta} = -9.0\sin\theta + 12.5\sin 5t\cos\theta$

20.28 905 rev/min

20.30 967 lb

20.32 (b) $\xi \approx 0.020$

20.34 (b) 0.008 38

20.36 (a) $x = 0.1333(e^{-5t} - e^{-20t})$ m; (b) 0.0630 m

20.38 16.99 ft/s

20.40 (a) $x = x_0(1.0774e^{-0.286pt} - 0.077\,37e^{-3.732pt})$,
$\dot{x} = -2.887px_0(e^{-0.268pt} - e^{-3.732pt})$;
(b) $|\dot{x}|_{max} = 0.219px_0$

20.42 (a) $\ddot{\theta} = -(8.458 \times 10^{-4})\dot{\theta}|\dot{\theta}| - 21.47\sin\theta$; (c) 0.25%

20.44 (a) 0.2671; (b) 41.12 N · s/m

20.46 12.01 N · s/m

20.48 (a) 1.291; (b) $(\text{M.F.})_{max} = 1.400$, $\omega = 12.96$ rad/s

20.50 $x = 0.019\,05\sin(30t - 0.0916)$ m, x lags y

20.52 $x = 0.019\,95\sin(600t + 0.499)$ m, x leads y

20.54 $Z_{max} = 1.155Y, \omega/p = \sqrt{2}$

20.56 (a) 25.0 N · s/m; (b) 90°

20.58 (a) $\ddot{x} = 200\sin 500t - 50\dot{x}|\dot{x}| - (250 \times 10^3)x$ m/s²;
(b) 0.0044 m

20.60 $f = (1/\pi)\sqrt{(g\sin\alpha)/(3R\alpha)}$

20.62 $f_2 = (f_1R/b)\sqrt{3m_1/m_2}$

20.64 (b) $p_{max} = \sqrt{g/(2\bar{k})}, y = \bar{k}$

20.66 (a) 2.01 s; (b) 0.212 m

20.68 0.725 Hz

20.70 0.403 ft

20.72 5.580 kN · s/m

20.74 189.5 lb · s/ft

20.76 (c) 41°

20.78 (a) 0.600; (b) 1.009 lb · ft · s

20.80 (b) $\Delta q = 0.6q_0, q = -0.1q_0$

20.82 1.566 s

20.84 $p = 0.936\sqrt{g/R}$ rad/s

20.86 $\tau = 2\pi(1 - R/L)\sqrt{L/g}$

20.88 8.89 in., 0.90 in.

20.90 $\tau = 2\pi\sqrt{(3m)/(5k)}$

20.92 $f = \dfrac{1}{2\pi}\sqrt{(k/m)[a^2/(L^2 + b^2)]}$

20.94 (a) $\ddot{\theta} = -3(k/m - 2g/L)\theta$; (b) 0.897 s; (c) 30.0 kg

20.96 $f = \sqrt{3gR/(\pi L)}$

Appendix F

F.2 $I_z = 0.792mR^2$

F.4 $I_y = ma^2/3, I_x = I_z = m(b^2 + 2a^2)/12$

F.6 53.5 kg · m²

F.8 (a) $I_x = (mL^2\sin^2\beta)/12$; (b) $I_y = (mL^2\cos^2\beta)/12$,
$I_z = mL^2/12$

F.10 (a) $I_z = m(L^2 + 3R^2)/12$; (b) $I_y = mR^2/2$

F.12 (a) $I_z = m(b^2 + 12h^2)/20$; (b) $I_y = mb^2/10$

F.14 $0.304 \text{ kg} \cdot \text{m}^2$

F.16 $I_z = 0.0584mR^2$

F.18 $6.22 \text{ kg} \cdot \text{m}^2$

F.20 $I_{xy} = 3mR^2/(8\pi)$

F.22 $I_{xy} = 3mbh/10, I_{zx} = 3mah/10, I_{yz} = 3mab/20$

F.24 $I_{xy} = I_{zx} = 0, I_{yz} = mRh/3$

F.26 $I_{xy} = -0.979 \times 10^{-3} \text{ slug} \cdot \text{ft}^2$
$I_{yz} = -0.489 \times 10^{-3} \text{ slug} \cdot \text{ft}^2$,
$I_{zx} = 0.979 \times 10^{-3} \text{ slug} \cdot \text{ft}^2$

F.28 $I_{xy} = mR^2/(8\pi), I_{yz} = 0, I_{zx} = -mR^2/(8\pi)$

F.30 $I_{AB} = 3mb^2/4$

F.32 (a) $I_{OA} = mb^2/6$

F.34 $I_{AB} = (m/6)[b^2h^2/(b^2 + h^2)]$

F.36 $\boldsymbol{\lambda}_1 = (\mathbf{i} + \mathbf{j})/\sqrt{2}$
$\boldsymbol{\lambda}_2 = (-\mathbf{i} + \mathbf{j})/\sqrt{2}, \boldsymbol{\lambda}_3 = \mathbf{k}$;
$\bar{I}_1 = mb^2/6, \bar{I}_2 = mb^2/24, \bar{I}_3 = 5mb^2/24$

F.38 $\bar{I}_1 = 0, \bar{I}_2 = \bar{I}_3 = 3mb^2/2$

F.40 (a) $\mathbf{I} = mb^2 \begin{bmatrix} 3 & 1 & -1 \\ 1 & 4 & 0 \\ -1 & 0 & 3 \end{bmatrix}$;

(b) $I_1 = 1.75mb^2$,
$I_2 = 4.80mb^2$,
$I_3 = 3.45mb^2$

F.42 (a) $\mathbf{I} = \begin{bmatrix} 1.308 & -1.963 & -0.981 \\ -1.963 & 6.803 & -0.393 \\ -0.981 & -0.393 & 7.588 \end{bmatrix} \text{kg} \cdot \text{m}^2$;

(b) $I_1 = 0.522 \text{ kg} \cdot \text{m}^2$,
$I_2 = 7.416 \text{ kg} \cdot \text{m}^2$,
$I_3 = 7.762 \text{ kg} \cdot \text{m}^2$;
$\boldsymbol{\lambda}_1 = 0.941\mathbf{i} + 0.303\mathbf{j} + 0.148\mathbf{k}$,
$\boldsymbol{\lambda}_2 = -0.330\mathbf{i} + 0.919\mathbf{j} + 0.216\mathbf{k}$,
$\boldsymbol{\lambda}_3 = -0.070\mathbf{i} - 0.252\mathbf{j} + 0.965\mathbf{k}$

Index

A

B

C

D

E

F

G

H

I

K

Q

R